COMPREHENSIVE HANDBOOK ON HYDROSILYLATION

Related Pergamon Titles of Interest

BOOKS

BARTON (ed.)
Comprehensive Organic Chemistry

CARRUTHERS
Cycloaddition Reactions in Organic Synthesis

CROMPTON
Analysis of Polymers

DAVIES
Organotransition Metal Chemistry: Applications to Organic Synthesis

TROST (ed.)
Comprehensive Organic Synthesis

WILKINSON (ed.)
Comprehensive Organometallic Chemistry

JOURNALS

European Polymer Journal

Polyhedron

Progress in Polymer Science

Tetrahedron

Tetrahedron Letters

COMPREHENSIVE HANDBOOK ON HYDROSILYLATION

Edited by
BOGDAN MARCINIEC

Contributing authors
BOGDAN MARCINIEC, JACEK GULIŃSKI,
WŁODZIMIERZ URBANIAK and
ZYGMUNT W. KORNETKA

Faculty of Chemistry, Adam Mickiewicz University, Poznań, Poland

PERGAMON PRESS

OXFORD · NEW YORK · SEOUL · TOKYO

U.K.	Pergamon Press Ltd, Headington Hill Hall, Oxford OX3 0BW, England
U.S.A.	Pergamon Press, Inc., 660 White Plains Road, Tarrytown, New York 10591–5153, U.S.A.
KOREA	Pergamon Press Korea, KPO Box 315, Seoul 110–603, Korea
JAPAN	Pergamon Press Japan, Tsunashima Building Annex, 3–20–12 Yushima, Bunkyo-ku, Tokyo 113, Japan

First edition 1992

Transferred to digital printing 2006

Library of Congress Cataloging-in-Publication Data

Comprehensive handbook on hydrosilylation / edited by Bogdan Marciniec.
p. cm.
Includes bibliographical references.
1. Hydrosilylation. I. Marciniec, Bogdan
QD281.H84C66 1991 547'.08--dc20 91–31306

British Library Cataloguing in Publication Data

Comprehensive handbook on hydrosilylation
I. Marciniec, Bogdan
547.08

ISBN 0–80–040272–0

Printed and bound by Antony Rowe Ltd, Eastbourne

Dedicated to

DR JOHN L. SPEIER

on the occasion of his 70th birthday

Preface

CONSIDERABLE interest has been aroused by organosilicon compounds, some of which are manufactured on a large scale. Hydrosilylation is one of the most fundamental and elegant methods for the laboratory and industrial synthesis of organosilicon compounds, and other related silicon compounds which are subjected directly to organic syntheses. Despite this broad interest, manifested by a growing number of research papers and excellent reviews, no comprehensive book on hydrosilylation incorporating various aspects of the subject has yet been published.

This present book has been written with the intention of providing researchers and industrial chemists with a complete survey of the literature on hydrosilylation processes up to the beginning of 1990. The aim of this volume, which contains experimental and theoretical material gleaned from the published literature, is to present to the reader the state of the art concerning hydrosilylation reactions and the directions in which synthetic and mechanistic studies, as well as practical applications of these processes, are proceeding.

This monograph is an extension of the Polish version of *Hydrosilylation*, written in 1976–78 and published in 1981. However, as the amount of information on hydrosilylation reactions published since that date, as well as that available to us, has increased considerably, we have decided to completely revise and extend the text. The present monograph, written in 1988–90, consists of two basic parts. Part I, comprising six chapters, is purely descriptive, presenting the catalytic, mechanistic, structural and synthetic aspects of hydrosilylation. This section also considers the application of hydrosilylation reactions and their products in organic, organosilicon and polymer chemistry.

Part II, presented in a tabular form, sets out information concerning the reaction conditions – substrates, catalysts, detailed reaction conditions, products and their yields and/or selectivities – and references chosen for selected trisubstituted silanes from more than 2100 papers and patents published in the past 25 years (1965–90).

The preparation of this book would have been impossible without the kind and great efforts of many collaborators and students of the Faculty of Chemistry, Adam Mickiewicz University, who worked there in the period 1987–90. I am especially grateful to Mrs Teresa Nowicka and to Mrs Elżbieta Śliwińska for help in collecting and tabulating the table contents (Part II). A special note of thanks is due to Mrs Teresa Nowicka for the donation of her time when the book was being prepared and written, particularly in typing the manuscript; thanks are also due to Mrs Hanna Ewertowska for providing software to set up the tables (Part II) and to prepare this monograph for publication. I am also grateful to all those who devoted so much effort and time towards preparing the manuscript.

Poznań, October 1990

BOGDAN MARCINIEC
Professor of Chemistry

Contents

Part I General Survey

Part II A Compilation of Information on Hydrosilylation Studies over the Period from 1965 to 1990

Part I

GENERAL SURVEY

CHAPTER 1

Introduction

HYDROSILYLATION (American – Hydrosilation) is a term that describes addition reactions of organic and inorganic silicon hydrides to multiple bonds such as carbon–carbon, carbon–oxygen, carbon–nitrogen, nitrogen–nitrogen and nitrogen–oxygen and which occur according to the following scheme:

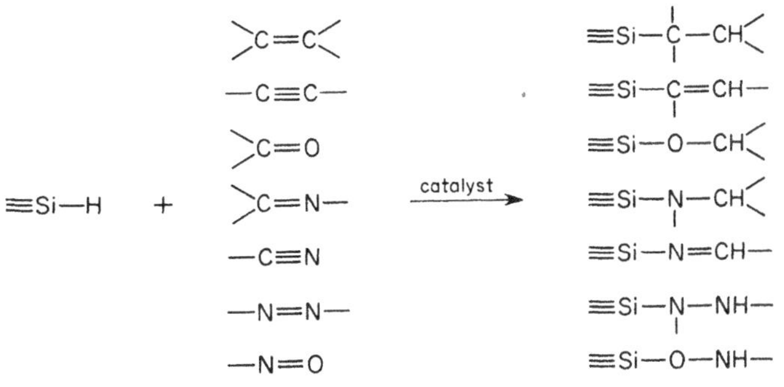

$$
\equiv Si-H \;+\;
\begin{matrix}
>C{=}C< \\
-C{\equiv}C- \\
>C{=}O \\
>C{=}N- \\
-C{\equiv}N \\
-N{=}N- \\
-N{=}O
\end{matrix}
\;\xrightarrow{\;catalyst\;}\;
\begin{matrix}
\equiv Si-\overset{|}{\underset{|}{C}}-CH< \\
\equiv Si-\overset{|}{C}{=}CH- \\
\equiv Si-O-CH< \\
\equiv Si-\underset{|}{N}-CH< \\
\equiv Si-N{=}CH- \\
\equiv Si-\underset{|}{N}-NH- \\
\equiv Si-O-NH-
\end{matrix}
\qquad (1)
$$

The first instance of such a reaction, i.e. that between trichlorosilane and 1-octene, was reported in 1947 [1129]:

$$
Cl_3SiH + CH_2{=}CHC_6H_{13} \xrightarrow{(CH_3COO)_2} Cl_3SiCH_2CH_2C_6H_{13} \qquad (2)
$$

Since then the reaction has developed into one of the most fundamental methods for synthesizing organosilicon compounds and organic silyl derivatives, and is now of considerable industrial significance.

Hydrosilylation reactions can proceed according to a free-radical mechanism (as indicated mainly in earlier reports) but when catalysed by transition metal salts and complexes, nucleophilic–electrophilic and other catalysts the mechanism is predominantly polar. A discovery of hexachloroplatinic acid as a very efficient catalyst by John L. Speier from Dow Corning Co. in 1957 became a starting point for wide and common application of reaction as one of the most effective and elegant methods for synthesizing the organosilicon and related compounds. Such processes are exothermic, although little detailed information on the reaction enthalpy has been published. The heat of hydrosilylation of a series of unsaturated organic and organosilicon compounds by methyldiphenylsilane, methyldipentylsilane and hydrosiloxanes has been shown to range from 22 to 40 kcal/mole [1166]. These results confirm an earlier figure determined by Speier *et al.* for the hydrosilylation of ethylene by trichlorosilane catalysed by hexachloroplatinic acid ($-\Delta H$ = 38 kcal/mole) [1133].

Scientific literature provides a number of surveys of hydrosilylation reactions or of particular aspects of such processes. These include general reviews, chapters devoted to hydrosilylation, and articles reviewing specific problems related to the process. These works, which amount to 44 references in total, are listed in the Bibliography under the heading "Reviews". The reader should note, however, that those references which attempt to provide a comprehensive coverage of the literature were published some time ago. Thus, Lukevics and Voronkov [Rev.20] compiled a bibliography on hydrosilylation up to 1963 and tabulated 2283 examples, including free-radical-initiated hydrosilylation reactions (696 examples), catalysed by chloroplatinic acid (729 examples), and by metals and their salts or by organic bases (858 examples). In 1968 a comprehensive review by Eaborn and Bott appeared under the title "Addition of silicon hydrides to alkenes and alkynes" as a separate chapter in Part I entitled *Synthesis and Reactions of the Silicon–Carbon Bond* [Rev.13]. Synthetic aspects of hydrosilylation concerning the preparation of carbofunctional organosilanes were reviewed by Pomerantseva *et al.* in 1971 [Rev.37]. In 1977 Lukevics *et al.* published a further comprehensive review which covered data on hydrosilylation published between 1964 and 1975 (1809 references, including papers, patents, symposium communications and Ph.D. theses) [Rev. 22]. This work complemented the review article by Lukevics, which appeared in *Uspekhi Khimii* in 1977 [Rev.21]. Finally, as stated above, another book entitled *Hydrosilylation*, which provides a comprehensive review of the literature covering the years 1965–77, was published in Polish in 1981 [Rev.26] (1250 references).

The problem of hydrosilylation has also been considered within the general framework of organosilicon compounds by Eaborn [Rev.12], Bazant *et al.* [Rev.4], Petrov *et al.* [Rev.36], and Voronkov [Rev.40], and in connection with the subject of silicon hydrides by Hajos [Rev.16] and in the book edited by Wiberg and Amberger [Rev.42]. In addition to the above publications, other works have also been published in which certain aspects of hydrosilylation are discussed. Thus, early catalytic, synthetic and mechanistic views on hydrosilylation processes have been analysed in review articles by Benkeser [Rev.5], Frainnet [Rev.14], Calas [Rev.8], Chalk [Rev.9], Cundy and Kingston [Rev.11], Meals [Rev.27] and Pomerantseva *et al.* [Rev.37]. More extensive reviews, mainly on catalytic aspects of hydrosilylation, were produced by Harrod and Chalk [Rev.17] in 1977, by Belyakova *et al.* [Rev.6] in 1982, by Voronkov and Pukhnarevich [Rev.40, Rev.41] in 1982 and 1983, and above all, by Speier [Rev.38] in 1979. Key intermediates of some transition metal complex catalysed hydrosilylation of the C=C bond were presented in 1988 [Rev.23]. The addition reactions of silyl radicals to various unsaturated compounds were reviewed in 1987 by Alberti and Pedulli [Rev.1]. The crosslinking via hydrosilylation was reviewed by Gorshkov and Dontsov in 1983 [Rev.15], and by Arkles and Crosby in 1990 [Rev.3]. Reviews largely covering the synthetic aspects of hydrosilylation of some unsaturated organic compounds have been published by Ojima [Rev.31] and by Nagai [Rev.28], Itoh *et al.* [Rev.18], and Yur'ev and Salimgareeva [Rev.43]. Andrianov *et al.* have summarized the addition and polyaddition of hydrosiloxanes to carbon–carbon multiple bonds in organic and organosilicon compounds [Rev.2]. Asymmetric hydrosilylation was extensively discussed in numerous reviews by Ojima *et al.* [Rev.30, Revs.33–5], by Kagan [Rev.19] in 1975, by Brunner [Rev.5] in 1983, and by Nogradi [Rev.29] in 1987. The current directions in which the study and application of hydrosilylation processes are aimed, based on literature published between 1978 and 1982, were reviewed in 1985 [Rev.24]. The reactions of hydrosilylation of vinylsilanes were summarized in 1989 [Rev.25]. Very recently, a chapter entitled "The hydrosilylation reaction", written by Ojima – regarding synthetic aspects of the hydrosilylation of alkenes, alkynes, conjugated dienes, carbonyl compounds and carbon–nitrogen multiple bonds – has appeared in the book *The Chemistry of Organic Silicon Compounds*. It constitutes a part of a monograph entitled "The chemistry of functional groups" [Rev.32].

The aim of the present monograph is to establish, on the basis of literature data published particularly during the last 25 years (1965–1990), the basic fundamental relationships between the type and structure of the substrate (unsaturated compounds and silicon hydrides) on the one hand, and the reactivity of the substrate, its reaction pathways, and the yields of various products, on the other (Chapters 3 and 4). These structural aspects together, with a comprehensive treatment of various catalytic problems (Chapter 2) which form the basis of hydrosilylation reactions, are discussed essentially from two points of view: elucidation of the reaction mechanism and evaluation of optimum conditions necessary for synthesis of organosilicon compounds. Most reports, especially patents, provide information only on the latter aspect of the problem, and indicate either a new pathway (catalyst, detailed reaction conditions) for the synthesis of a well-known organosilicon compound, or a method for preparing a hitherto unknown compound by hydrosilylation usually in the presence of a common catalyst such as H_2PtCl_6. Hydrosilylation of organosilicon monomers and polymers as a way of effecting the addition of Si–H to carbon–carbon bonds has been extensively developed and is treated separately in Chapter 5. Some applications of the reaction products of such additions and of the process itself in organic and polymer chemistry are discussed in Chapter 6 in order to focus attention on hydrosilylation products known as silane coupling agents, on the reduction of organic compounds, and on the modification of polymers by hydrosilylation. The circular diagram (p. 6) illustrates the various aspects of hydrosilylation processes discussed.

The reactivity of the Si–H bond in organic and inorganic silicon hydrides when added to various multiple bonds depends on the physicochemical properties of these hydrides and on the nature of the silicon atom [Rev.4, Rev.12]. A smaller electronegativity of the silicon atom in comparison to the carbon atom and the ability of the d-orbitals of the silicon atom to participate in bonding (d_{π}–p_{π} interaction and sp^3d or sp^3d^2 hybridization) are two characteristic features which differentiate silicon from carbon and make it similar to the rest of the elements in

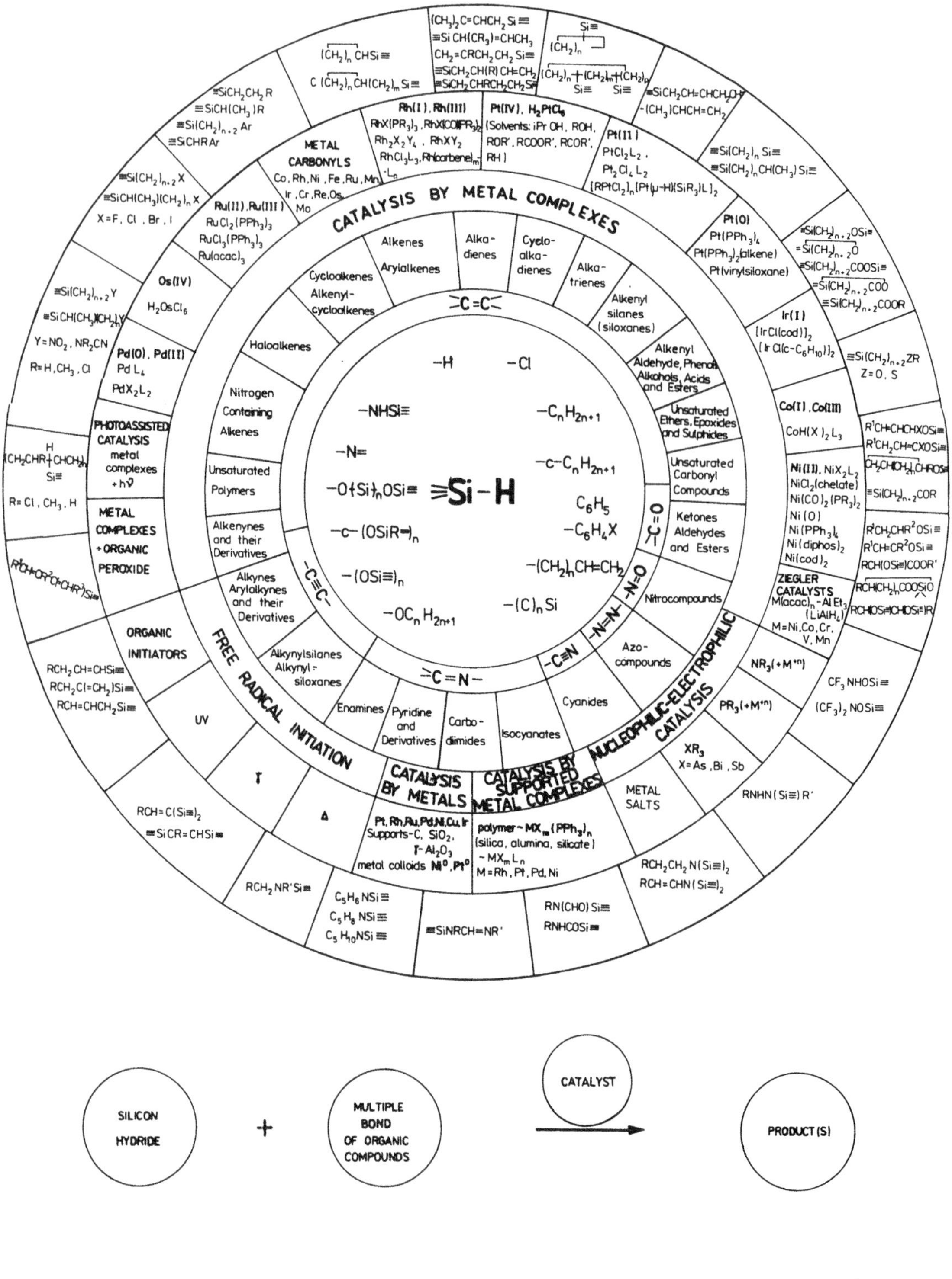
≡Si–H
–H
–Cl
–NHSi≡
–N=
–O(Si)nOSi≡
–C–(OSiR=)n
–(OSi≡)n
–OCnH2n+1
–CnH2n+1
–C–CnH2n+1
C6H5
–C6H4X
–(CH2)nCH=CH2
–(C)nSi
>C=C<
–C≡C–
>C=N–
–C≡N
–N=N–
–N=O
>C=O
Alkenes
Arylalkenes
Alka-dienes
Cyclo-alka-dienes
Alka-trienes
Alkenyl silanes (siloxanes)
Alkenyl Aldehyde, Phenols, Alcohols, Acids and Esters
Unsaturated Ethers, Epoxides and Sulphides
Unsaturated Carbonyl Compounds
Ketones Aldehydes and Esters
Nitrocompounds
Azo-compounds
Cyanides
Isocyanates
Carbo-diimides
Pyridine and Derivatives
Enamines
Alkynylsilanes Alkynyl-siloxanes
Alkynes Arylalkynes and their Derivatives
Alkenynes and their Derivatives
Unsaturated Polymers
Nitrogen Containing Alkenes
Haloalkenes
Cycloalkenes Alkenyl-cycloalkenes
CATALYSIS BY METAL COMPLEXES
FREE RADICAL INITIATION
CATALYSIS BY METALS
CATALYSIS BY SUPPORTED METAL COMPLEXES
NUCLEOPHILIC-ELECTROPHILIC CATALYSIS
METAL CARBONYLS
Co, Rh, Ni, Fe, Ru, Mn, Ir, Cr, Re, Os, Mo
Rh(I), Rh(III)
RhX(PR3)3, RhX(CO)(PR3)2
Rh2X2Y4, RhXY2
RhCl3L3, Rh(carbene)m-Ln
Pt(IV), H2PtCl6
(Solvents: iPrOH, ROH, ROR', RCOOR', RCOR', RH)
Pt(II)
PtCl2L2, Pt2Cl4L2
[RPtCl2]n [Pt(μ-H)(SiR3)L]2
Pt(0)
Pt(PPh3)4
Pt(PPh3)2(alkene)
Pt(vinylsiloxane)
Ir(I)
[IrCl(cod)]2
[IrCl(c-C6H10)]2
Co(I), Co(III)
CoH(X)2L3
Ni(II), NiX2L2
NiCl2(chelate)
Ni(CO)2(PR3)2
Ni(0)
Ni(PPh3)4
Ni(diphos)2
Ni(cod)2
ZIEGLER CATALYSTS
M(acac)n – AlEt3 (LiAlH4)
M=Ni, Co, Cr, V, Mn
Ru(II), Ru(III)
RuCl2(PPh3)3
RuCl3(PPh3)3
Ru(acac)3
Os(IV)
H2OsCl6
Pd(0), Pd(II)
PdL4
PdX2L2
PHOTOASSISTED CATALYSIS metal complexes + hν
METAL COMPLEXES + ORGANIC PEROXIDE
ORGANIC INITIATORS
UV
γ
Δ
Pt, Rh, Ru, Pd, Ni, Cu, Ir
Supports-C, SiO2, γ-Al2O3
metal colloids Ni0, Pt0
polymer ~ MXm(PPh3)n
(silica, alumina, silicate)
~ MXmLn
M = Rh, Pt, Pd, Ni
METAL SALTS
XR3
X = As, Bi, Sb
PR3(+M+n)
NR3(+M+n)
(CH2)n CHSi≡
C(CH2)nCH(CH2)mSi≡
(CH3)2C=CHCH2Si≡
≡SiCH(CR3)=CHCH3
CH2=CRCH2CH2Si≡
≡SiCH2CH(R)CH=CH2
≡SiCH2CHRCH2CH2Si≡
Si≡
(CH2)n
(CH2)n(CH2)m(CH2)p
Si≡ Si≡
≡SiCH2CH=CHCH2OH
–(CH3)CHCH=CH2
≡Si(CH2)nSi≡
≡Si(CH2)nCH(CH3)Si≡
≡Si(CH2)n+2OSi≡
≡Si(CH2)n+2O
≡Si(CH2)n+2COOSi≡
≡Si(CH2)n+2COO
≡Si(CH2)n+2COOR
≡Si(CH2)n+2ZR
Z = O, S
R1CH=CHCHXOSi≡
R1CH2CH=CXOSi≡
CH2CH(CH2)nCHROSi≡
≡Si(CH2)n+2COR
R1CH2CHR2OSi≡
R1CH=CR2OSi≡
RCH(OSi≡)COOR'
RCH(CH2)nCOOSiO
RCH(OSi≡)CH(OSi≡)R
CF3NHOSi≡
(CF3)2NOSi≡
RNHN(Si≡)R'
RCH2CH2N(Si≡)2
RCH=CHN(Si≡)2
RN(CHO)Si≡
RNHCOSi≡
≡SiNRCH=NR'
C5H6NSi≡
C5H8NSi≡
C5H10NSi≡
RCH2NR'Si≡
RCH=C(Si≡)2
≡SiCR=CHSi≡
RCH2CH=CHSi≡
RCH2C(=CH2)Si≡
RCH=CHCH2Si≡
R1CH=CR2CH=CHR3Si≡
H
(CH2CHR–CHCH2)n
Si≡
R = Cl, CH3, H
≡Si(CH2)n+2Y
≡SiCH(CH3)(CH2)nY
Y = NO2, NR2CN
R = H, CH3, Cl
≡Si(CH2)n+2X
≡SiCH(CH3)(CH2)nX
X = F, Cl, Br, I
≡SiCH2CH2R
≡SiCH(CH3)R
≡Si(CH2)n+2Ar
≡SiCHRAr
SILICON HYDRIDE
+
MULTIPLE BOND OF ORGANIC COMPOUNDS
CATALYST
PRODUCT(S)

Group IVb of the Periodic Table, particularly to germanium and tin. The electronegativity of silicon is not only lower than that of carbon but also lower than that of hydrogen, leading to a reversal in the polarity of the covalent $Si^{\delta+}$–$H^{\delta-}$ bond in comparison to the $C^{\delta-}$–$H^{\delta+}$ bond. The Si–H bond is about 3% ionic (C–H 4%), but the absolute value and even the direction of polarity depend on the nature of the substituents on the silicon atom.

This brief description of the Si–H bond demonstrates that the majority of the reactions of silicon hydrides occur by nucleophilic attack on the silicon atom and/or electrophilic attack on the hydride hydrogen. However, in the presence of electron-withdrawing substituents on the silicon, the partly ionic nature of the Si–H bond can be reduced. Furthermore, since the energy of this bond is relatively low (76 kcal/mole) in comparison to that of the C–H bond (99 kcal/mole), this facilitates the homolytic cleavage of the Si–H bond and leads to a radical mechanism for the substitution and addition reactions.

A given hydrosilylation reaction, e.g. the hydrosilylation of 1,3-butadiene in the presence of Ni, Pd or Rh complexes, may be commented upon several times: e.g. from the standpoint of the choice of catalyst, the nature of the unsaturated bond or the structure of the silicon hydride (Chapters 2–4). In such cases the details of the mechanism or synthesis are mentioned in only one of the above listed chapters, depending on the catalyst or substrates involved.

CHAPTER 2

Catalytic Aspects of Hydrosilylation

As mentioned previously, the addition of the Si–H bond to unsaturated compounds proceeds either in the presence of free-radical initiators (homolytic addition) or with various other catalysts which almost always accomplish the process through a heterolytic mechanism. These catalysts may be nucleophilic, such as tertiary amines; Lewis acids, such as metal salts; supported metals; metals reduced *in situ*; and Group VIII metal complexes; transition metal carbonyls and Ziegler catalysts which form a very important class of catalyst for this reaction. The Group VIII metal complexes can be employed in homogeneous systems or attached to inorganic and polymer substrates. In addition to the above, a new class of catalysts, i.e. photo- and peroxide-initiated transition metal complexes, is also discussed in this chapter.

In the following discussion the mechanistic pathways by which a hydrosilylation reaction may proceed in the presence of such catalysts are considered in detail, since to a large extent the choice of catalyst determines the basic course of the reaction. Because of the somewhat different nature of homolytic processes initiated by silyl radicals, the description of the radical mechanism and stereochemistry of Si–H addition to various multiple bonds has also been included in this chapter.

2.1 Free-radical Addition

In early reports it was suggested that the addition of hydrosilanes to multiple bonds could be thought of as a free-radical chain process because of the relatively low energy of the Si–H bond in comparison to that of the C–H bond [Rev.4]. For this reason many synthetic and mechanistic studies have been devoted to the hydrosilylation of carbon–carbon (C=C, C≡C), carbon–heteroatom (C=O, C=N), and heteroatom–heteroatom (N=O) bonds initiated by free radicals generated in the reaction mixture.

2.1.1 Methods of Silyl Radical Generation

Chain-propagating silyl radicals generated by various common chemical and physical methods may be employed as direct initiators of radical addition reactions. Such chemical methods are based on the thermal cleavage of the Si–H bond in silicon hydrides

$$\equiv\text{Si–H} \overset{\Delta H}{\rightarrow} \equiv\text{Si}^{\bullet} + \text{H}^{\bullet} \qquad (3)$$

and/or the homolytic decomposition of these hydrides under milder conditions in the presence of free-radical initiators.

As mentioned previously, the low energy of the Si–H bond, as well as other properties of silicon in its compounds, contributes towards the high sensitivity of silicon hydrides (and in particular the Si–H bond) as far as homolytic cleavage is concerned. Thus silicon tetrahydride (monosilane), SiH_4, decomposes spontaneously at temperatures within the range 380–400°C, and the higher Si_nH_{2n+2} homologues show considerably less stability. However, the alkylsilanes, e.g. $C_2H_5SiH_3$, $(C_2H_5)_2SiH_2$, and $(C_2H_5)_3SiH$, are more stable than monosilane. Heating unsubstituted silanes and halogenosilanes above 100°C gives rise to the direct formation of silyl radicals, and thus in some cases thermal initiation of a free-radical process leading to addition to a C=C bond is possible. However, in order to achieve a lower initiation temperature, compounds with a relatively low dissociation energy (20–40 kcal/mole) are most often employed. These can dissociate sufficiently to produce organic free radicals even at 40°C.

Compounds which commonly initiate hydrosilylation reactions are these which also initiate polymerization and other radical processes in organic chemistry, and include organic peroxides such as acetyl [335], tert-butyl [338, 347, 1215] and benzoyl [119, 168, 1297], and also azo-compounds such as azo-bis-isobutyronitrile [1132]. The two elementary stages leading to the initiation of silyl free radicals in the presence of benzoyl peroxide (4) and azo-bis-isobutyronitrile (5) as initiators are given below: decomposition of the compound to yield organic free radicals (4.1, 5.1) and a subsequent reaction in which these radicals abstract hydrogen from the hydrosilane (4.2, 5.2).

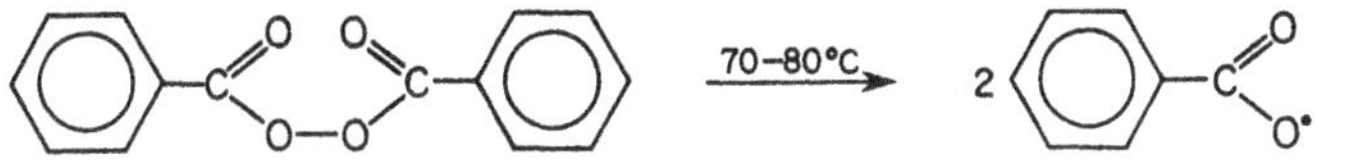

(4.1)

(4.2)

(4.3)

$$\text{CH}_3\text{–}\underset{\text{CN}}{\overset{\text{CH}_3}{\text{C}}}\text{–N=N–}\underset{\text{CN}}{\overset{\text{CH}_3}{\text{C}}}\text{–CH}_3 \xrightarrow{40\text{–}80^{\circ}\text{C}} 2\ \text{CH}_3\text{–}\underset{\text{CN}}{\overset{\text{CH}_3}{\text{C}^{\bullet}}} + \text{N}_2 \qquad (5.1)$$

$$\equiv Si-H + CH_3-\underset{CN}{\overset{CH_3}{|}}{C^{\bullet}} \longrightarrow \equiv Si^{\bullet} + CH_3-\underset{CN}{\overset{CH_3}{|}}{CH} \quad (5.2)$$

Other than halogen radicals, those with an unpaired electron localized on the oxygen atom are the best acceptors of hydrogen (oxygen forms a stronger bond with hydrogen than does carbon to the extent of about 84 kJ/mole [521]), and are, therefore, the most efficient initiators of hydrosilylation processes. However, such radicals have a tendency to isomerization; the unpaired electron is transferred from oxygen to carbon during such intramolecular rearrangements. This happens, for example, in the oxy-tert-butyl radical which cleaves to form a methyl radical and acetone, and in the oxybenzoyl radical which, after decomposition, produces a phenyl radical and carbon dioxide (4.3). Carbon radicals generated in this way are also active in hydrosilylation, since the energy with which they are bound to hydrogen is greater than that of the Si–H bond. Silyl free radicals can also be formed from silicon hydrides when the latter absorb UV radiation. However, since some silicon hydrides (mainly halogenosilanes) are not capable of absorbing UV light above a wavelength of 185 nm, photodissociation of such compounds can occur only in the presence of sensitizers which are capable of transmitting the absorbed energy to the silanes. Thus, the production of silyl radicals in the presence of mercury vapour as a photosensitizer proceeds as follows:

$$Hg \xrightarrow{h\nu} (Hg)^* \quad (6.1)$$

$$(Hg)^* + \equiv Si-H \rightarrow Hg + (\equiv Si-H)^* \quad (6.2)$$

$$(\equiv Si-H)^* \rightarrow \equiv Si^{\bullet} + H^{\bullet} \quad (6.3)$$

The photolytic addition of trichlorosilane to olefins is presumably initiated as a result of very weak photoabsorption by olefins in the UV region of the spectrum, since $HSiCl_3$ has been found to be almost transparent in this region [407]. The absorption of γ-radiation is also a highly effective method of producing silyl free radicals. A cobalt bomb containing ^{60}Co is often used as a source of γ-radiation; this emits two gamma quants for every decay:

$$\equiv Si-H \xrightarrow{\gamma(^{60}Co)} \equiv Si^{\bullet} + H^{\bullet} \quad (7)$$

Combination of two methods of initiator generation, i.e. photolysis in the presence of UV light together with organic peroxides, enables such reactions to proceed under mild conditions and in a more effective manner. The basis of this procedure is that initiation proceeds by photochemical means rather than by the thermal decomposition of the organic initiators.

$$(CH_3)_3C-O-O-C(CH_3)_3 \xrightarrow{h\nu} 2\,(CH_3)_3C-O^{\bullet} \quad (8.1)$$

$$CH_3COCH_3 \xrightarrow{h\nu} 2\ CH_3^{\bullet} + CO \qquad (8.2)$$

The tert-butoxy and methyl radicals thus initiated generate silyl radicals during the next elementary stage of the process.

Other chemical initiators have also been used in homolytic hydrosilylation, e.g. ozone, peroxides and organometallic compounds (see Rev.13, Rev.20, 382).

Kinetic data and product studies reported recently for SiH_4 pyrolysis in the presence of olefins and acetylene showed that silyl radicals and H atoms are not involved in the reactions of the silane with acetylene and olefins. A formation of hydrosilylation products occurs, according to these results, due to mechanism involving silene species as initiators of the reaction [348a]:

$$SiH_4 + (M) \rightleftharpoons SiH_2 + H_2 + (M) \qquad (9.1)$$

$$SiH_2 + C_3H_6 \rightleftharpoons \overset{*}{Pr}SiH \qquad (9.2)$$

$$PrSiH + SiH_4 \rightleftharpoons [PrSiH_2SiH_3] \rightarrow PrSiH_3 + SiH_2 \qquad (9.3)$$

$$PrSiH + C_3H_6 \rightleftharpoons Pr_2Si: \qquad (9.4)$$

$$Pr_2Si: + SiH_4 \rightleftharpoons [Pr_2SiHSiH_3] \rightarrow Pr_2SiH_2 + SiH_2 \qquad (9.5)$$

$$SiH_2 + C_3H_6 \rightarrow \text{products} \qquad (9.6)$$

* The propyl groups can be either n-propyl or isopropyl.

Data on the reactivity of substituted silanes with olefins and acetylenes should explain a real role of silene intermediate in hydrosilylation reactions.

2.1.2 *Hydrosilylation of the C=C Bond: Mechanism, Stereochemistry and Examples*

The free-radical addition of hydrosilanes resembles the addition of hydrogen bromide to alkenes and always occurs according to the anti-Markownikov rule. When applied to hydrosilylations this rule is called Farmer's rule [352] and can be formulated as follows: during the addition of a silane molecule (Si–H bond) to an unsymmetrical carbon–carbon double bond, the silyl radical is joined to the carbon atom of the double bond which has the greater number of hydrogen atoms attached to it. In deriving the mechanism of the free-radical addition it is assumed that the unpaired electron remains in the position of the greatest possible electron delocalization. Thus, it is of considerable importance to establish which of the following organosilicon free radicals is the more stable:

$$\equiv Si^{\bullet} + CH_2{=}CH{-}R \rightarrow {}^{\bullet}CH_2{-}\underset{\underset{\equiv Si}{|}}{CH}{-}R \qquad (10.1)$$

$$\equiv Si^{\bullet} + CH_2{=}CH{-}R \rightarrow \equiv Si{-}CH_2{-}\dot{C}H{-}R \qquad (10.2)$$

According to the dissociation energies of the C–H bond the radical with the unpaired electron at the secondary carbon atom is more stable than that with the electron at the primary carbon atom. The stability of a radical increases in the following order: primary < secondary < tertiary. Consequently, the silyl group is attached to the olefin in the terminal position (10.2).

Later stages of kinetic chain propagation may involve addition accompanied by telomerization (or polymerization). This occurs when the alkylsilyl radical reacts further with one or more olefin molecules:

$$\equiv SiCH_2{-}\dot{C}H{-}R \xrightarrow{\equiv Si-H} \equiv SiCH_2CH_2R + \equiv Si^{\bullet} \qquad (11.1)$$

$$\equiv SiCH_2{-}\dot{C}H{-}R \xrightarrow{CH_2=CHR} \equiv SiCH_2CH(R)CH_2\dot{C}HR \qquad (11.2)$$

$$\equiv SiCH_2CH(R)CH_2\dot{C}HR + (n-1)CH_2{=}CHR \rightarrow \equiv Si(CH_2CHR)_nCH_2\dot{C}HR \qquad (12)$$

The propagation process may continue until the alkylsilyl radical participates in chain transfer with a hydrosilane:

$$\equiv Si(CH_2CHR)_nCH_2\dot{C}HR \xrightarrow{\equiv Si-H} \equiv Si\ (CH_2CHR)_nCH_2CH_2R + \equiv Si^{\cdot} \qquad (13)$$

Dohmaru and Nagata have reported [325] a series of kinetic studies on the photolytic reactions of trichlorosilane with olefins, in which addition to acetone was usually employed as a reference reaction. Such studies have revealed it to an appreciable extent in the gaseous phase. The enthalpy of reaction (ΔH) for the addition to ethylene has been estimated independently as about 16 or 18 kcal/mole [325, 521]. In the absence of acetone the ratio of the yields of the respective adduct R_0 was derived on the basis of the following mechanism [320]:

$$^{\bullet}SiCl_3 + CR_1R_2{=}CH_2 \underset{k_{-2}}{\overset{k_2}{\rightleftharpoons}} {}^{\bullet}CR_1R_2CH_2SiCl_3 \qquad (14.1)$$

$$^{\bullet}CR_1R_2CH_2SiCl_3 + HSiCl_3 \xrightarrow{k_3} HCR_1R_2CH_2SiCl_3 + {}^{\bullet}SiCl_3 \qquad (14.2)$$

For the $HSiCl_3$/ethylene/acetone reaction the next two steps in the reaction

$$^{\bullet}SiCl_3 + (CH_3)_2CO \xrightarrow{k_4} (CH_3)_2\dot{C}OSiCl_3 \qquad (14.3)$$

$$(CH_3)_2\dot{C}OSiCl_3 + HSiCl_3 \xrightarrow{k_5} (CH_3)_2CHOSiCl_3 + {}^{\bullet}SiCl_3 \qquad (14.4)$$

are assumed to occur concurrently [319, 322, 325], since the observed reaction products are almost exclusively alkyltrichlorosilane (the adduct of trichlorosilane to the given 1-alkene) and isopropoxytrichlorosilane (the adduct of trichlorosilane to acetone). The mechanism for the reaction in the absence of acetone leads to the rate equation:

$$R_o = \frac{k_2[\text{olefin}][^{\bullet}\text{SiCl}_3]}{1 + k_{-2}/k_3[\text{HSiCl}_3]} \qquad (15)$$

Several series of kinetic measurements have allowed the Arrhenius parameters k_2/k_4 and k_{-2}/k_3 for various 1-alkenes [322, 324] and 2-alkenes [319] to be measured over the temperature range of 378–525K as well as the relative rate constants for radical addition to olefins [319, 322]. In the case of ethylene, E_4–E_2 (see eqs 14) was found to be equal to 1.4 kcal/mole (recalculated as 0.8 kcal/mole) [324], while the values for the next three homologues of ethylene were within the range 1.2–1.4 kcal/mole, and that for isobutylene was 2.4 kcal/mole [322, 324]. For addition to 2-olefins, the E_4–E_2 values were predominantly equal to zero, while for cyclopentene it was 2.5 kcal/mole [319]. The activation energy, E_2, for the addition of trichlorosilyl radicals to olefins may be readily evaluated from the given data, assuming that the absolute value of E_4 is 3 kcal/mole [319]. It should be noted that in the liquid phase k_2/k_4 for competitive addition of methyldichlorosilane to 1-hexene and to acetone strongly depends on acetone concentration and its temperature dependence does not give a straight line in Arrhenius coordinates. It can be regarded as kinetic evidence for formation of a complex by methyldichlorosilyl radical and acetone [687].

Choo and Gaspar have estimated the activation energy for the addition of the trimethylsilyl radical to ethylene in solution as 2.5 ± 0.2 kcal/mole [261]. However, this research did not detect the presence of any triethylsilyl radicals in reactions with ethylene initiated by di-tert-butyl peroxide and UV photolysis. This suggests that step (10.2) in this hydrosilylation [624, 625] occurs very readily. With trialkylsilyl radicals the two chain propagating processes, i.e. (9.2) and (10.1), are much slower than the corresponding processes involving trichlorosilyl radicals, presumably because the former are more stable. In the presence of a tert-butyl peroxide initiator, the relative reactivity of trichlorosilane addition in comparison with the addition of triethylsilane is 2.3–3.1 at 140°C [347].

Hydrosilylation involving a radical mechanism may be promoted by any agent which produces free radicals in the system. On the other hand, under such conditions, particularly at elevated temperatures, other by-products in addition to telomers and polymers may be formed. Thus, alkenes and alkadienes react with trichlorosilane to yield alkenylchlorosilanes as the main products or as by-products of a dehydrocondensation reaction [Rev.36]. In reactions of substituted olefins, e.g. allyl chloride, other by-products may be isolated besides the adduct $Cl_3SiCH_2CH_2CH_2Cl$; these occur as a result of the dehydrohalogenation of the adduct ($Cl_3SiCH_2CH{=}CH_3$) or its substitution ($Cl_3SiCH_2CH_2CH_3$, $Cl_3SiCH_2CH_2CH_2SiCl_3$). Whilst the generation of all these by-products often leads to a decrease in the adduct yield, they can often be utilized as means of effecting telomerization or even polymerization. In general, telomerization occurs most readily during thermally initiated addition processes particularly in the presence of excess olefin, although polymeric products may also be formed when the reaction mixture is exposed to either γ-radiation [688] or UV light, or heated with peroxides [Rev.13, Rev.20, Rev.36]. The products (telomers or polymers) possess terminal tri(organo, alkoxy, halogeno)silyl groups in all cases.

To protect the reaction from undesirable telomerization an excess of silane should be added, since this increases the probability of the competitive reaction of alkylsilyl radicals with olefins (11.2). For instance, if the yield of monomeric adducts can be substantially increased so that the trialkylsilane/alkene ratio is more than 6/1 [347], the telomerization reaction (11.2) may be practically eliminated. When the reaction occurs in the liquid phase the formation of monomeric adducts can be accelerated by using the silane as the solvent.

The radical hydrosilylation of a C=C bond is stereospecific. This can be explained by assuming the formation of a ring radical in which the silicon atom is attached to both carbon atoms. This bridge radical adds on the hydrogen from the Si–H bond of a second silane molecule (Scheme 16) in the *trans* position to give a *cis* adduct:

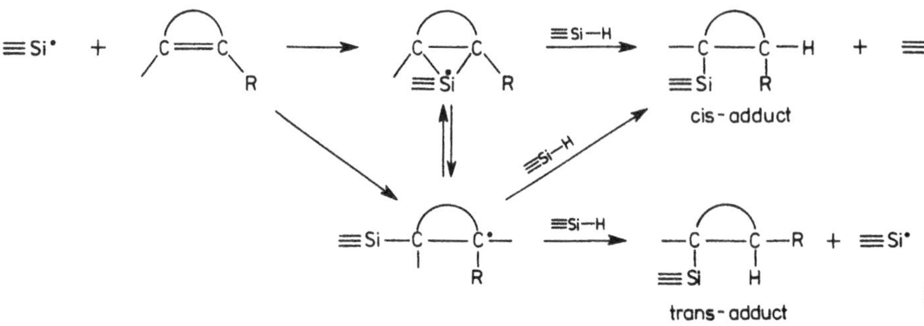

(16)

However, under certain conditions the bridge radical isomerizes into a non-stereospecific open radical, and as a result a mixture of *cis* and *trans* products may be obtained. The stereospecificity of this reaction is exemplified by the hydrosilylation of 1-methylcyclohexene-1 by trichlorosilane, which may be initiated by free-radical methods. When the reaction mixture is heated with peroxides or exposed to ultraviolet light, the *cis*/*trans* ratio of adducts is 6:1; thermal addition, however, gives a *cis*/*trans* ratio of 2.5:1. This latter effect appears to be due to the formation of a linear free radical, although the formation of methylcyclohexyltrichlorosilane can be accounted for by a partial ionic mechanism involving a carbanionic intermediate [1035]:

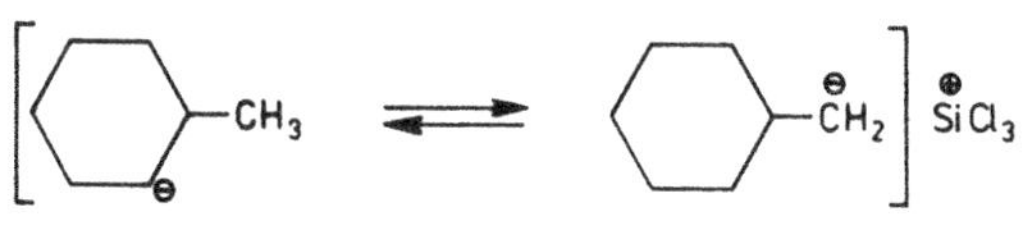

(17)

Photochemically initiated $^{\bullet}SiCl_3$ radicals can induce *cis–trans* isomerization of 2-alkenes in the gaseous phase via a radical mechanism with the simultaneous formation of *sec*-$C_4H_9SiCl_3$ [318]:

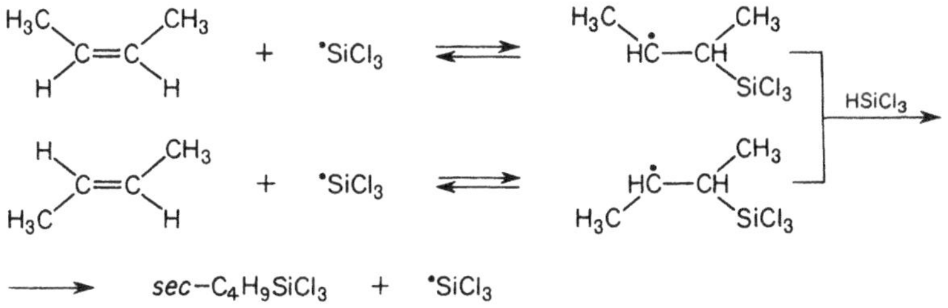

(18)

The radical adduct formed in this way is capable of rotationing on the C–C bond. The equilibrium constant, K, for the *cis–trans* isomerization has been shown to be 3.2. The addition of $^{\bullet}SiCl_3$ radicals to *cis*-2-pentene also occurs via radical isomerization to yield 2- and 3-trichlorosilylpentanes, with a ratio of 0.47 in the gaseous phase and about 1.2 when the reaction is carried out in cyclohexane [319, 321].

Thermal initiation of addition and telomerization reactions requires prolonged heating of the reaction mixtures at elevated temperatures and often at elevated pressures. Addition and simultaneous telomerization reactions occur when inactive aliphatic olefins (ethylene, propylene) [826, 1085] and their halogen derivatives (e.g. $CF_2{=}CF_2$) are hydrosilylated by inorganic silanes, principally $HSiCl_3$ and SiH_4, and less often by CH_3SiHCl_2 [388, 826, 1082, 1292]:

$$SiH_4 + RCH{=}CH_2 \xrightarrow{120^{\circ}C,\ 4\ h} R(CH_2)_2SiH_3 \qquad (19)$$

$$2\ RSiH_3 \xrightarrow{370^{\circ}C,\ 4\ h} R_2SiH_2 + SiH_4 \qquad (20)$$

$$HSiCl_3 + n\ CH_2{=}CH_2 \xrightarrow[280^{\circ}C]{200\ atm} Cl_3Si(CH_2CH_2)_nH \qquad (21)$$

$$CH_3HSiCl_2 + n\ CH_2{=}CH_2 \xrightarrow[260^{\circ}C]{560\ atm} CH_3SiCl_2(CH_2CH_2)_nH \qquad (22)$$

Recent investigations on SiH_4 pyrolysis in the presence of propylene are in accord with propylsilane formation via propylsilene formed by silylene addition to propylene (see Scheme 9) [348a]. Halogenohydrosilanes also undergo thermal addition to higher alkenes, alkadienes, and unsaturated organosilicon compounds [38, 1085, 1233], e.g.

$$HSiCl_3 + CH_3\underset{\displaystyle CH_3}{\underset{|}{C}}{=}CHCH_2CH_3 \xrightarrow[48\%]{} CH_3\overset{\displaystyle CH_3}{\overset{|}{C}H}\underset{\displaystyle SiCl_3}{\underset{|}{C}H}CH_2CH_3 + CH_3\overset{\displaystyle CH_3}{\overset{|}{\underset{\displaystyle SiCl_3}{\underset{|}{C}}}}CH_2CH_2CH_3 \qquad (22.1)$$

Syntheses of organosilicon compounds involving homolytic hydrosilylation in the presence of organic initiators mostly occur when trichlorosilane is the substrate

$$HSiCl_3 + CH_2{=}CH(CH_2)_nCH_3 \xrightarrow{(C_6H_5COO)_2 \text{ or } (CH_3COO)_2} Cl_3SiCH_2CH_2(CH_2)_nCH_3 \quad (23)$$

with $n=1$ [1026], $n=4$ [192], $n=6$ [903] or $n=7$ [1212]. Only highly reactive olefins such as styrene or acrylonitrile can undergo hydrosilylation or telomerization under such initiation conditions [1134]:

$$HSiCl_3 + CH_2{=}CHC_6H_5 \xrightarrow{(C_6H_5COO)_2} Cl_3Si[CH(C_6H_5)CH_2]_nH \quad (24)$$

$n = 59$.

An interesting reaction from the synthetic viewpoint is the addition of trichlorosilane and silicon hydrides with substituents which are more electron-donating substituents, e.g. trialkoxysilanes and phenylsilanes, to unsaturated organic acids and esters, to alkenes with other functional groups [541, 1133, 1223], and to unsaturated organosilicon compounds [193, 541]:

$$C_6H_5SiH_3 + CH_2{=}CHCH_2OCOCH_3 \xrightarrow{CH_3COOCOC_6H_5} C_6H_5SiH_2(CH_2)_3OCOCH_3 \quad (25)$$

$$(C_2H_5O)_3SiH + CH_2{=}CHCH_2P(C_2H_5)_2 \xrightarrow{[(CH_3)_3CO]_2} (C_2H_5O)_3Si(CH_2)_3P(C_2H_5)_2 \quad (26)$$

$$X_3SiH + CH_2{=}CHSiR_3 \xrightarrow{(CH_3COO)_2,\ (C_6H_5COO)_2} X_3SiCH_2CH_2SiR_3 \quad (27)$$

$R = CH_3, C_2H_5$

Similar reactions occur with polyalkenes, e.g. polybutadiene [882].

A few works in recent years have been undertaken on hydrosilylation reactions initiated by organic peroxides. Such reactions are mostly carried out in order to compare hetero- and homolytic mechanisms and the stereochemistry of these processes [110, 261, 786]. Recent examinations by the ESR method on triethyl- and triethoxysilane interaction with 3,5-di-tert-butyl-*o*-benzoquinone show oxidation of silicon hydride to cation-radical $[Si{-}H]^{\cdot +}$ which seems to be an active intermediate in the hydrosilylation of olefins [573].

Silyl free radicals initiated photochemically add to carbon–carbon double bonds under much milder conditions. Of particular interest there are substituted olefins of the general formula $CH_2{=}CHX$, where the functional group X is $(CH_2)_9Cl$ [334], $(CH_2)_nF$ [407, 408], $(CH_2)_nOR$ [198, 199], $(CH_2)_8COCl$, $(CH_2)_nOCOR$ [194], CH_2Br [1027], $(CH_2)_nCF_3$ [S-017], pinene [375], and perfluoroethylene and its derivatives [69, 442, 473, 479, 480, G-013, G-032, S-017]. The examples of such reactions are given below:

$$\underset{}{CF_3CH{=}CHF} + HSiCl_3 \xrightarrow[64^\circ C,\ 15\ days]{UV} \underset{46\%}{CF_3CH(SiCl_3)CH_2F} + \underset{16\%}{CF_3CH_2CHFSiCl_3} \quad (28)\ [479]$$

$$CF_3CH{=}CH_2 + SiH_4 \xrightarrow[96\ h]{UV} \underset{26\%}{CF_3(CH_2)_2SiH_3} + \underset{12\%}{[CF_3(CH_2)_2]_2SiH_2} + \underset{10\%}{[CF_3(CH_2)_2]_3SiH} + \underset{3\%}{[CF_3(CH_2)_2]_4Si} \quad (29)$$

[475]

When the substrates involved in these reactions are irradiated with UV light, telomers may also be obtained [474]:

$$(CH_3)_3SiH + CF_2{=}CF_2 \xrightarrow{h\nu} (CH_3)_3Si(CF_2CF_2)_nH \quad (30)$$

$n = 1,2.$

$$CH_2{=}CH(CH_2)_3\,SiCl_2H \rightarrow \overline{(CH_2)_5SiCl_2} + \text{polymers} \quad (31)$$

Photochemical initiation and intramolecular hydrosilylation have also been employed for the addition of silanes to large molecules [1293] such as polyfluorocycloalkenes [S-014].

The relative merits of initiation by UV irradiation and peroxide decomposition have been compared in the hydrosilylation of 1-methylcyclohexene by trichlorosilane. When the reaction mixture was heated with acetyl peroxide for 9 h a 100% yield of adducts was obtained. Exposure to UV light for 44 h, however, afforded only a 49% yield [1035]. The simultaneous use of both methods for silyl free-radical initiation enabled adducts to be obtained under milder conditions [126, 196, 261, 338]. Thus, photochemical initiation to tert-butylperoxide radicals enabled the reaction temperature to decrease from 110–130°C to 40°C. Further variation of the reaction temperature within the range 0–40°C had no effect upon the rate of the reaction of trichlorosilane with 1-octene, the yield of n-octyltrichlorosilane being 78% during a 14 h reaction at 40°C, 79% during 16 h at 29°C and 76% during 9 h at 0°C [338]. Photosensitization of trichlorosilane with bis(trimethylsilyl)-mercury has increased the yield of n-octyltrichlorosilane to 100% just after 2 h [126], as compared to 78% after 14 h in the absence of the sensitizer.

Numerous examples also exist of hydrosilylation initiated by a wide range of doses of γ-radiation over long reaction periods to give different yields of adducts (30–60%) [189, 344, 345, 640, 1030, 1210, 1360], e.g.

$$HSiCl_3 + CH_3COOCH{=}CH_2 \xrightarrow{\gamma(^{60}Co)} Cl_3Si(CH_2)_2OCOCH_3 + \quad (32)$$

$$CH_3CH_2O(CH_2)_2SiCl_3 + Cl_3SiCH_2CH(OCOCH_3)(CH_2)_2OCOCH_3$$

$$CH_3Cl_2SiH + CH_2{=}CHCl \xrightarrow{\gamma} ClCH_2CH_2SiCl_2CH_3 \quad (33)$$

This initiation technique is more effective than UV absorption and, unlike the situation with organic initiators, it is possible to regulate the rate by thermal methods, the yield and the selectivity of the addition:

$$CH_3Cl_2SiH + CH_3CH{=}CH_2 \xrightarrow[100\%]{80°C} CH_3CH_2CH_2SiCl_2CH_3 \quad (34)$$

[1360]

$$SiH_4 + CH_2{=}CH_2 \xrightarrow{\gamma(^{60}Co)} C_2H_5SiH_3 + (C_2H_5)_2SiH_2 + Si_2H_6 \quad (35)$$

[1025]

As for other methods, the hydrosilylations of halogen derivatives of ethylene and higher alkenes, styrene and α-methylstyrene, when initiated by γ-radiation, all proceed to yield olefin telomerization products [345, 685, 686, 688].

Studies of the elementary reactions involved in the telomerization of ethylene [686] and hexene [685], and in the co-telomerization both of olefins [688] during hydrosilylation (initiated by γ-radiation) of ethyldichlorosilane have enabled the relative rate constants of the process under study to be determined, and the reactivity of radicals involved in the chain propagation and transfer reaction to be characterized.

2.1.3 Hydrosilylation of the C≡C Bond: Mechanism, Stereochemistry and Examples

The addition of hydrosilanes to alkynes and their derivatives occurs in two stages. Initially, alkenylsilanes are produced when the reagent concentration is suitable:

$$\equiv Si^{\bullet} + HC{\equiv}CR \longrightarrow R{-}\dot{C}{=}CH{-}Si{\equiv} \quad (36)$$

$$R{-}\dot{C}{=}CH{-}Si{\equiv} \xrightarrow{+\equiv Si{-}H} \equiv SiCH{=}CHR + \equiv Si^{\bullet} \quad (37.1)$$

$$R{-}\dot{C}{=}CH{-}Si{\equiv} \xrightarrow{+HC{\equiv}CR} \equiv SiCH{=}CR{-}CH{=}\dot{C}R \quad (37.2)$$

These may then undergo further addition reactions:

$$\equiv SiCH{=}CHR + \equiv Si^{\bullet} \rightarrow \equiv Si\dot{C}HCHRSi{\equiv} \quad (38.1)$$

$$\equiv Si\dot{C}HCHRSi{\equiv} + \equiv SiH \rightarrow \equiv SiCH_2CHRSi{\equiv} + \equiv Si^{\bullet} \quad (38.2)$$

or telomerization or chain transfers:

$$\equiv SiCH{=}CRCH{=}\dot{C}R + (n{-}1)CH{\equiv}CR \rightarrow \equiv Si(CH{=}CR)_n(CH{=}\dot{C}R) \quad (39.1)$$

$$\equiv Si(CH{=}CR)_n(CH{=}\dot{C}R) + \equiv SiH \rightarrow \equiv Si(CH{=}CR)_n(CH{=}CHR) + \equiv Si^{\bullet} \quad (39.2)$$

$$\equiv Si(CH{=}CR)_n(CH{=}CHR) + \equiv Si^{\bullet} \rightarrow \equiv Si(CH{=}CR)_n\dot{C}HCHRSi{\equiv} \quad (39.3)$$

$$\equiv Si(CH{=}CR)_n\dot{C}HCHRSi{\equiv} + \equiv SiH \rightarrow \equiv Si(CH{=}CR)_nCH_2CHRSi{\equiv} + \equiv Si^{\bullet} \quad (39.4)$$

The mechanism of such silane addition reactions has been considered by Benkeser *et al.* They have also established the stereochemistry of the processes in which $HSiCl_3$ was added to five acetylene derivatives: pentyne-1, hexyne-1, heptyne-1, 3-methylbutyne-1 and 3,3-dimethylbutyne-1. Trichlorosilane addition initiated by benzoyl peroxide gave *cis* products predominantly, except for 3,3-dimethyl-1-butyne where the products were almost entirely *trans* after 20 h reaction, together with a di-adduct – the product of alkenylsilane hydrosilylation. The stereochemical rule regarding *trans* addition of silyl radicals to alkynes has been confirmed in this case, as well as when other free-radical initiators (mechanisms) have been used [Rev.5, Rev.42].

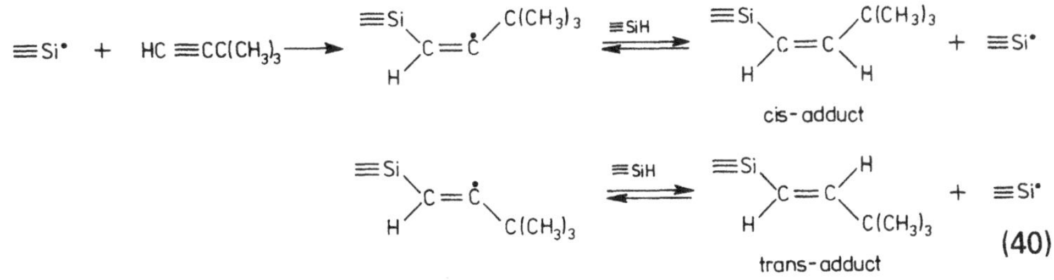

(40)

It has been shown that only *cis* products are formed initially, although some of them isomerize to *trans*, and the rest react with trichlorosilane to yield a di-adduct [Rev.5]. Similar results were obtained in the hydrosilylation of the above mentioned alkynes by dihydrosilanes (see [110], for instance).

Some interesting results have been reported regarding the stereochemistry of the photochemically initiated addition of trichlorosilane and trimethylsilane to the fluoro derivatives of acetylene. Thus, hydrosilylation of 1,3,3,3-tetrafluoropropyne-1 leads predominantly to the *cis* adduct giving a 4:1 *cis/trans* ratio with $SiCl_3$ substituent and a 2.5:1 ratio with $Si(CH_3)_3$, while the addition of these hydrosilanes to 3,3,3-trifluoro-1-propene yields *trans* adducts in the most part (a 3.7:1 ratio with $SiCl_3$ and a 2.2:1 ratio with $Si(CH_3)_3$) [481]. The second stage in hydrosilylation addition to alkynes, i.e. addition to the C=C bond as discussed above, mainly yields *cis* adducts [115]. Far less interest has been shown in the radical hydrosilylation of alkynes than in the corresponding reaction with a C=C bond. The addition of trichlorosilane and other hydrosilanes to acetylene [192, 1041] and to its organic (RC≡CH) and organosilicon ($R_3SiC{\equiv}CR$) derivatives has been studied comprehensively when initiated by peroxides. The products obtained have usually been found to consist of a mixture of *cis* and *trans* unsaturated organosilicon isomers [192, 1041], e.g.

$$Cl_3SiH + CH{\equiv}CH \xrightarrow{[(CH_3)_3CO]_2} Cl_3SiCH{=}CH_2 \quad (41)$$

$$Cl_3SiH + CH_3(CH_2)_3C{\equiv}CH \xrightarrow{(C_6H_5COO)_2} Cl_3SiCH{=}CH(CH_2)_3CH_3 \quad (42)$$

$$R^1Cl_2SiH + R^2_3SiC{\equiv}CR^3 \xrightarrow{(C_6H_5COO)_2} R^2_3SiC(SiCl_2R^1){=}CHR^3 \quad (43)$$

$$HSiCl_3 + ClC{\equiv}CCl \xrightarrow{[CH_3COO]_2} Cl_3SiC(Cl){=}CHCl \qquad (44)$$

Thermal hydrosilylation of the C≡C bond has recently been applied to the synthesis of some complicated organosilicon compounds, e.g. halogen-substituted derivatives of silaphenalene [256]:

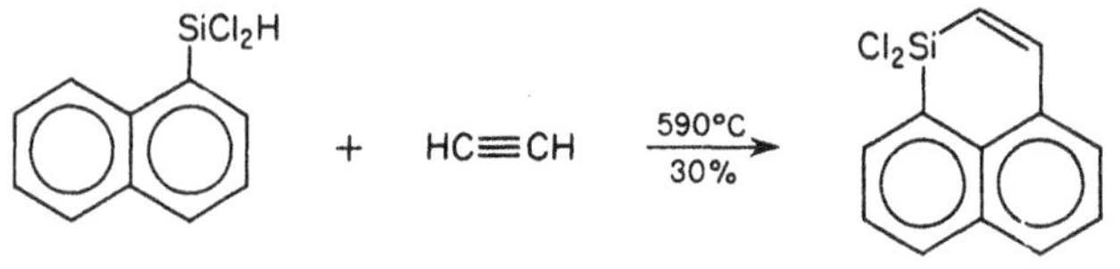

(45)

Much more than in the case of the C=C bond, radical addition of Si–H to the C≡C bond is accompanied by numerous side-reactions, e.g. telomerization, polymerization, substitution and isomerization of unsaturated compounds, which all occur competitively through the agency of various unsaturated free radicals generated in the reaction mixture.

2.1.4 Addition of Silicon Hydrides to Other Multiple Bonds

Radical hydrosilylation of a carbonyl group proceeds as follows:

$$\equiv Si^{\bullet} + R{-}\underset{\underset{O}{\|}}{C}{-}R' \rightarrow R{-}\underset{\underset{OSi\equiv}{|}}{\dot{C}}{-}R' \qquad (46.1)$$

$$R{-}\underset{\underset{OSi\equiv}{|}}{\dot{C}}{-}R' + \equiv SiH \rightarrow R{-}\underset{\underset{OSi\equiv}{|}}{CH}{-}R' + \equiv Si^{\bullet} \qquad (46.2)$$

Regardless of how they are initiated, silyl free radicals are always added to the oxygen of the carbonyl group. This arises because the Si–O bond is stronger than the only alternative available, i.e. the Si–C bond.

Silyl ethers can be obtained by hydrosilylation of ketones, the reactions being most often initiated photochemically [77, 326, 502, 1309] but also thermally [304, 526, 889] and via γ-radiation [1211]. Thermal hydrosilylation of fluoro-substituted ketones by dihydrosilanes occurs readily [296, 304, 889]. Thus, hexafluorocyclobutanone when heated with dipropylsilane under these conditions give a 48% yield of adducts [720].

$$2\ \begin{array}{l} F_2C{-}CF_2 \\ \ \ |\qquad | \\ F_2C{-}C{=}O \end{array} + (C_3H_7)_2SiH_2 \xrightarrow{100°C,\ 16\ h} \left[\begin{array}{l} F_2C{-}CF_2 \\ \ \ |\qquad | \\ F_2C{-}CHO \end{array}\right]_2 Si\ (C_3H_7)_2 \qquad (47)$$

The addition of diphenylsilane, triphenylsilane and tribenzylsilane to the C=O bond of quinones gives yields of 30–80% over the temperature range of 270–

340°C with various reagents [84, 889]. As described previously, photolysis of acetone for 0.5 h combined with a temperature rise from 78 to 253°C as a means of a free-radical initiation in the presence of trichlorosilane improves the yield of the trichloro-*iso*-propoxysilane adduct from 3% to 90%. The radical mechanism for this reaction, proposed on the basis of the GLC analysis of products, may be written as follows [326]:

$$CH_3COCH_3 \rightarrow 2\ CH_3^{\bullet} + CO \qquad (48.1)$$

$$CH_3^{\bullet} + HSiCl_3 \rightarrow CH_4 + {}^{\bullet}SiCl_3 \qquad (48.2)$$

$${}^{\bullet}SiCl_3 + CH_3COCH_3 \rightarrow (CH_3)_2\dot{C}OSiCl_3 \qquad (48.3)$$

$$(CH_3)_2\dot{C}OSiCl_3 + HSiCl_3 \rightarrow (CH_3)_2CHOSiCl_3 + {}^{\bullet}SiCl_3 \qquad (48.4)$$

$$2\ (CH_3)_2\dot{C}OSiCl_3 \rightarrow \text{products} \qquad (48.5)$$

The exposure of the reaction mixture to UV light provides an effective method of hydrosilylation of not only aliphatic ketones [197, 200, U-129] but also aromatic ketones [84, 369], esters and aldehydes. Employing different reaction times (2.5–185 h), the photochemical addition of mono- and di-substituted silanes to benzoyltrimethylsilane leads to the efficient formation of adducts [84]:

$$(CH_3)_3Si{-}\underset{\underset{O}{\|}}{C}{-}C_6H_5 + RR'SiH_2 \xrightarrow{h\nu} (CH_3)_3Si{-}\underset{\underset{H_5C_6}{|}}{CH}{-}O\overset{\overset{R}{|}}{\underset{\underset{H}{|}}{Si}}{-}R' \qquad (49)$$

($R = H, C_6H_5$, $R' = C_6H_5, C_6H_5CH_2$) with yield 52–98% (UV exposure time 2.5 h). However, di- and triethylsilane do not form addition products under such conditions.

Reported examples of hydrosilane addition to N=N and N=O multiple bonds [587, 676] provide substantial possible extensions of this reaction:

$$X_3SiH + CF_3NO \xrightarrow{UV} \underset{(17\text{–}80\%)}{CF_3NHOSiX_3} \qquad (50)$$

X = Cl, F [735], CH₃ [388].

$$(C_6H_5)_3SiH + R{-}N{=}N{-}R \xrightarrow{UV\ +\ (PhCOO)_2} \underset{(21\text{–}77\%)}{(C_6H_5)_3Si{-}\overset{\overset{R}{|}}{N}{-}NHR} \qquad (51)$$

The following propagating chains have been proposed in the addition of triorganosilanes ($R^1 = C_6H_5, C_2H_5$) to hydrazine derivatives ($R^2 = COOC_2H_5$, $COOCH_3$) [857, 858]:

$$R^1_3Si^{\bullet} + R^2{-}N{=}N{-}R^2 \rightarrow R^1_3Si{-}\overset{R^2}{\overset{|}{N}}{-}\overset{R^2}{\overset{|}{N}}{}^{\bullet} \qquad (52.1)$$

$$R^1_3Si{-}\overset{R^2}{\overset{|}{N}}{-}\overset{R^2}{\overset{|}{N}}{}^{\bullet} + R^1_3SiH \rightarrow R^1_3Si{-}\overset{R^2}{\overset{|}{N}}{-}\overset{R^2}{\overset{|}{N}}{-}H + R^1_3Si^{\bullet} \qquad (52.2)$$

In summary, the radical addition of an Si–H bond to multiple bonds has only limited synthetic and technological applications, mainly because of side-reactions but also since the reaction may proceed in an uncontrollable manner in the presence of initiators.

2.2 Nucleophilic–Electrophilic Catalysis

As already mentioned, most substitution reactions involving the Si–H bond occur via an ionic mechanism either involving the attack of a nucleophile on the silicon atom or electrophilic interaction with hydridic hydrogens. Such heterolytic cleavage of the Si–H bond stems from the fact that silicon is more electropositive than carbon and that the d-orbitals of silicon are capable of participating in bond formation. However, strong electronegative substituents at silicon, e.g. chlorine, can lead to a decrease in the electron density at the central atom, and may reverse the polarization of the Si–H bond [116] $Cl_3\overset{\delta-}{-}Si\overset{\delta+}{-}H$, particularly in the absence of any nucleophilic reagent. The different properties of the Si–H bond, depending on the substituents at silicon, may account for the two different mechanistic pathways for nucleophilic (basic) catalysis by organic bases (nucleophiles), e.g. tertiary amines, phosphines and arsines, for the addition of hydrosilanes to C=C and C≡C bonds. Nucleophilic hydrosilylation is the predominant mechanism during the addition of trichlorosilane, less than for dichloro(methyl, phenyl, hydro)silane [U-004, U-007, U-015], to vinyl derivatives, e.g. acrylonitrile [U-007], styrene [U-004] and vinyltrichlorosilane [U-004] and, to a lesser extent, to allyl derivatives and other organic compounds with C=C and C≡C bonds [U-007]. The following bases have been used as catalysts:

Amines: Ph_3N [U-015], $(C_4H_9)_3N$ [841, U-004, U-011], $(C_3H_7)_3N$ [905, U-004, U-007, U-015], $(C_5H_{11})_3N$ [U-004], pyramidone [F-013], antipyrine [F-013], $(CH_2{=}CHCH_2)_3N$ [S-009], $(CH_2{=}CHCH_2)_2N(CH_2)_3Si(OC_2H_5)_3$ [S-009], C_5H_5N [841, 842, 843], $\overline{(CH_2)_3N}C_2H_5$ [U-004], $\overline{(CH_2)_5N}C_2H_5$ [U-004], $\overline{(CH_2)_4N}C_2H_5$ [U-004], $(CH_3)_2NCH_2CH_2N(CH_3)_2$ [959].
Phosphines: Ph_3P [905, U-007, U-015], $(C_2H_5)_3P$ [U-015], $(C_4H_9)_3P$ [905, U-015], Ph_2PCl [905], $PhPCl_2$ [905].
Arsines: Ph_3As [905, U-007, U-015].
Bismuthines: Ph_3Bi [905].
Stibines: Ph_3Sb [905].
Others: $[(CH_3)_2N]_3P{=}O$ [J-008].

Some of these organic bases are efficient catalysts for the hydrosilylation of acetylene derivatives [Rev.5, 118]. If the organic substrate is a base, e.g. vinylpyridine, no additional catalyst is necessary to enable the reaction to take place [532]. This donor type of catalysis usually proceeds in the absence of solvents, although solvents such as benzene and acetonitrile are sometimes used [141, 843, 944]. Strong inorganic bases react spontaneously with hydrosilanes, particularly under solvolytic conditions, and for this reason cannot be used as hydrosilylation catalysts.

The mechanism for the hydrosilylation of acrylonitrile originally presented (Scheme 53) suggests that a tertiary amine functioning as a base produces trichlorosilyl anion (or rather a trichlorosilyl anion–protonated amine cation ion pair) by removing a proton from trichlorosilane [841, 842, 843].

$$Cl_3SiH + B \rightarrow Cl_3Si^- + HB^+ \quad (53.1)$$

$$Cl_3Si^- + \overset{+}{C}H_2{=}CH{-}C{\equiv}\bar{N} \rightarrow Cl_3SiCH_2\bar{C}HCN \quad (53.2)$$

$$Cl_3SiCH_2\bar{C}HCN + BH^+ \rightarrow Cl_3SiCH_2CH_2CN + B \quad (53.3)$$

Such abstraction of hydrogen from trichlorosilane and the formation of a trichlorosilane anion has been demonstrated by a spectroscopic study of the trichlorosilane–tertiary amine system, *viz.* by the disappearance of the ν_{SiH} band in the IR spectrum [734] and of the ^1H NMR signal corresponding to the silyl hydrogen [121], as well as by the simultaneous appearance of signals corresponding to a nitrogen-bonded hydrogen typical of an ammonium salt. Thus, the addition of trichlorosilane to vinyl derivatives can be explained by acid–base catalysis, although it is not possible to rule out the coordination of tertiary amines or other nucleophilic reagents with the silicon atom to form pentacoordinate or hexacoordinate species which can act as intermediates [115]:

$$R_3N + HSiX_3 \rightleftharpoons H{-}\overset{X\ \ X}{\underset{X}{\overset{\ominus}{Si}}}{-}\overset{\oplus}{N}R_3 \rightleftharpoons R_3\overset{\oplus}{N}H\overset{\ominus}{Si}X_3 \quad (54)$$

A formation of such active intermediates was confirmed recently by regiospecific addition of trichlorosilane to enamines without catalyst. An explanation of the mechanism is based on nitrogen–silicon complexation followed by intermolecular hydride transfer [1114]:

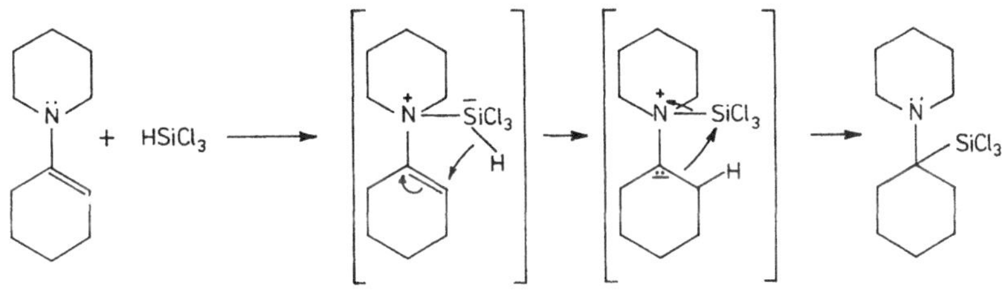

(55)

It seems probable that sp^3d intermediates formation as depicted in equilibrium (54) predominate in hydrosilanes which contain fewer electronegative substituents at silicon, where the hydrogen is initially more hydride-like and which are therefore capable of reacting with a relatively weak base. Hence, it seems reasonable to propose a competitive reaction between the hydride intermediates and acrylonitrile to give an α-adduct:

$$
\overset{\oplus}{R_3N}-\overset{|}{Si}-H + CH_2{=}CHCN \longrightarrow \left[\begin{array}{c} \overset{\oplus}{CH_2}{\cdots}CH{-}CN \\ \vdots \\ H{\cdots}Si{\equiv} \\ | \\ NR_3 \end{array} \right]^{\ast} \longrightarrow \underset{\underset{\alpha\text{-adduct}}{Si{\equiv}}}{CH_3{-}\overset{|}{CH}{-}CN} + NR_3
$$

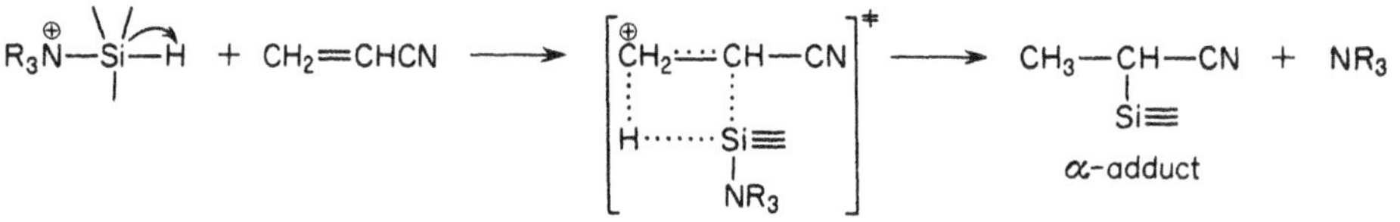

(56)

A spectroscopic study of the methyldichlorosilane–triethylamine adduct has shown that a fairly broad ν_{SiH} band is retained in the spectrum, although a ν_{N-H} band also appears [734]. It has been suggested that the last step in equilibrium (54) involves direct intermolecular hydride transfer from silicon to nitrogen [115].

An increase in Si–H reactivity towards carbonyl groups was reported by Corriu *et al.* [150]. The silicon hydrides of well-defined trigonal bipyramidal geometry with Si–H bonds occupying equatorial positions react with aldehydes and ketones without any catalyst:

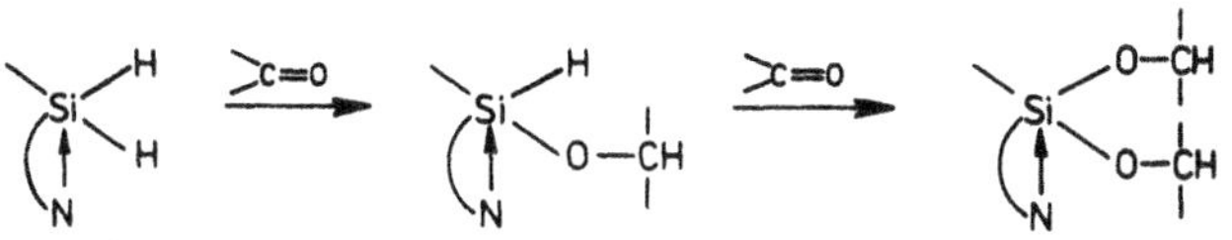

(57)

Corresponding monohydrogenosilane also reacts with aldehydes, although tetra-coordinated hydrogenosilanes do not react with carbonyl groups even in the presence of a catalyst.

The study of the hydrosilylation of acetylene derivatives by trichlorosilane has shed more light on the mechanism of this reaction. Pike has observed that in the tributylamine-catalysed addition of trichlorosilane to 1-hexyne and phenylacetylene, which occurs when the reaction is undertaken in a bomb at 148°C, *trans* adduct formation is accompanied by di-adducts and *cis* adducts [904]. Benkeser and Muench [114] identify the products of the latter reaction as the di-adducts, α,β-bis(trichlorosilyl)ethyl benzene (I), and three mono-adducts, i.e. α-trichlorosilylstyrene (II) and *cis*- and *trans*-β-trichlorosilylstyrene (III) and (IV), respectively [Rev.5, 118]. This reaction was carried out under much milder conditions than those employed by Pike because an excess of amine was employed:

$$PhC{\equiv}CH \ + \ HSiCl_3 \ \xrightarrow{(C_4H_9)_3N} \ \underset{Cl_3Si}{\overset{Ph}{>}}CH{-}CH_2SiCl_3 \ + \ \underset{Cl_3Si}{\overset{Ph}{>}}C{=}CH_2$$

(I)   (II)

$$+ \ \underset{H}{\overset{Ph}{>}}C{=}C{\overset{SiCl_3}{\underset{H}{<}}} \ + \ \underset{H}{\overset{Ph}{>}}C{=}C{\overset{H}{\underset{SiCl_3}{<}}}$$

(III)   (IV)

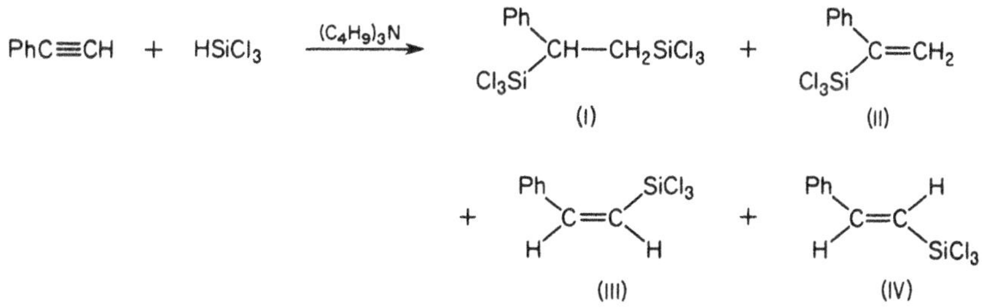

(58)

The first adduct to be detected (after 3–5 h) was *cis*-β-trichlorosilylstyrene (III), followed by the α, β-di-adduct (I). After longer reaction times the *trans* isomer (IV) appeared and some traces of α-trichlorosilylstyrene (II) were also found.

In addition, the easy isomerization of the *cis* (II) isomer to the *trans* (IV) isomer was observed experimentally in the absence of trichlorosilane. It was therefore suggested that the di-adduct (I) is formed mostly from the *cis* adduct (III) rather than from the *trans* adduct (IV). The detection of the above products during the addition of trichlorosilane to phenylacetylene, together with the observed facts, indicates that at least two essential pathways involving catalysis by amines are plausible, both involving a trichlorosilyl anion–protonated amine adduct and/or a pentacoordinate silicon complex as reactive intermediates (see Scheme 54). These both react with phenylacetylene followed by attack of the protonate amine (or ion pair) on the α-carbon of the substrate [Rev.5]:

$$R_3\overset{\oplus}{N}H\overset{\ominus}{S}iCl_3 \ + \ PhC{\equiv}CH \ \underset{-\overset{\oplus}{HNR_3}}{\rightleftarrows} \ \underset{H}{\overset{Ph}{>}}\overset{\bullet}{C}{=}C{\overset{SiCl_3}{\underset{}{<}}} \ \underset{+NR_3}{\overset{+\overset{\oplus}{HNR_3}}{\rightleftarrows}}$$

$$\rightleftarrows \ \underset{H}{\overset{Ph}{>}}C{=}C{\overset{SiCl_3}{\underset{H}{<}}} \ \xrightarrow{\text{isomerization}} \ \underset{H}{\overset{Ph}{>}}C{=}C{\overset{H}{\underset{SiCl_3}{<}}}$$

cis–adduct (III)   *trans*–adduct (IV)

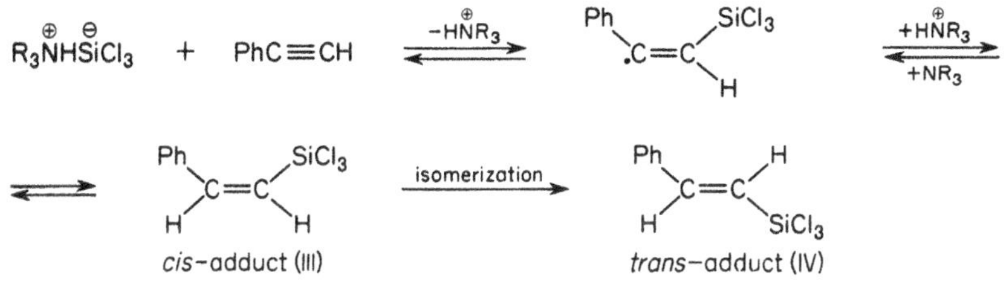

(59)

In view of the stereochemistry thus established, it seems reasonable to reject the simple four-centred process originally proposed [904] since this would involve predominantly *trans* isomer formation.

The second type of mechanism appears to account for α-adduct formation. This assumes hydride attack of a pentacoordinate silane–amine intermediate on the β-carbon of phenylacetylene, which occurs via a possible four-centred transition state to yield finally α-trichlorosilylstyrene:

$$\underset{Cl}{\overset{Cl \ \ Cl}{H{-}\overset{|}{S}i{-}\overset{\oplus}{N}R_3}} \ + \ \overset{\ominus}{P}hC{\equiv}\overset{\oplus}{C}H \ \longrightarrow \ \begin{bmatrix} Ph{-}C{\cdots}CH \\ \ \ \vdots \ \ \ \vdots \\ Cl_3Si{\cdots}H \\ \overset{|}{N}R_3 \end{bmatrix}^{\ddagger} \ \xrightarrow{-NR_3}$$

$$\longrightarrow \ \underset{Cl_3Si}{\overset{Ph}{>}}C{=}CH_2$$

α–adduct (II)

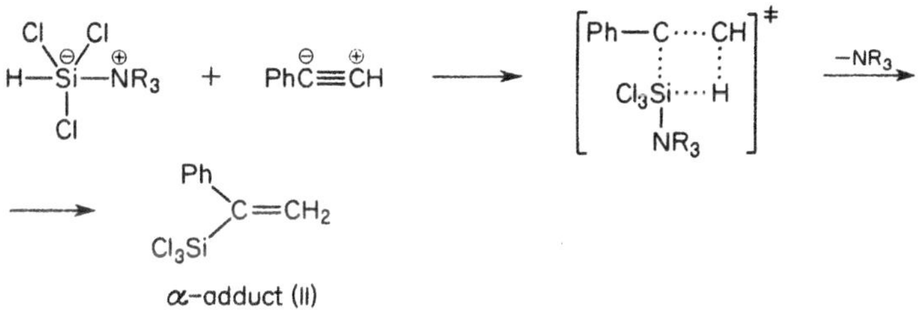

(60)

Both products of the first hydrosilylation step react with trichlorosilane to give the di-adduct:

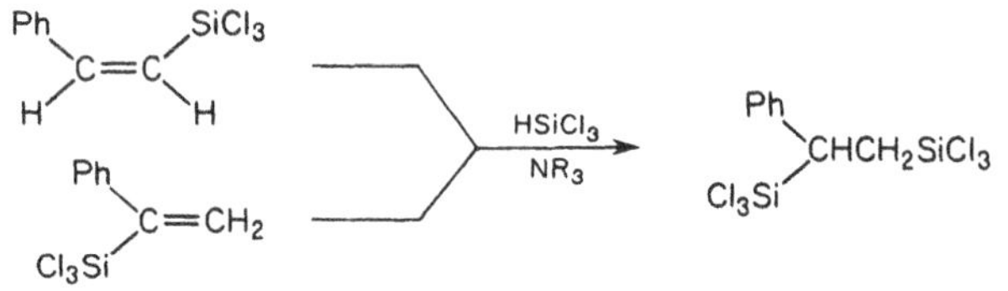

(61)

The following relative order of mono-adduct reactivity with trichlorosilane has been established experimentally: II $\gg$ III > IV > I. Thus, it is believed that the *trans* adduct is present in the final reaction products as a result of the isomerization of the mono-adducts. The above concept of two polar mechanisms is supported by the subsidiary effect on the carbon of acetylene. The following experimental reaction rate order

$$m\text{-}CF_3C_6H_4C{\equiv}CH > PhC{\equiv}CH > p\text{-}CH_3OC_6H_4C{\equiv}CH \quad (62)$$

indicates that the electron-withdrawing group of an aryl substituent at carbon leads to an increase in α- and *cis*-β-adduct formation with the final α,β-di-adduct which is the predominant product [Rev.5]. The positive effect of polar solvents observed in the case of weak bases may be explained in terms of the stabilization of the polar transition state caused by solvation [904].

From the mechanistic viewpoint it is interesting to note the numerous reported examples of base (nucleophile) catalysed addition, simultaneously promoted by electrophilic metal ions [D-036]. A characteristic feature of these reactions is their occurrence in the absence of metal salts, as well as the fact that the mechanism remains basically unchanged when electrophilic reagents are added. In principle, electrophilic catalysis by metal salts is no different from the catalytic activity of tertiary amine hydrochlorosilanes, not to mention base catalysis by the amines themselves.

A typical example of nucleophilic–electrophilic catalysis is provided by the addition of hydrochlorosilanes to alkenes, which occurs in the presence of Cu(II) and/or Cu(I) salts, Cu_2O and tertiary amines [141, 959, 1165]. Such ions exhibit only a limited ability to form Π-bonds with olefins and acetylenes and hence such electrophilic catalysis must differ fundamentally from the coordinative catalysis by transition metal complexes discussed in Section 2.4. The electrophilic properties of such metal ions result from their ability to form complexes with substrates and provide the driving force for the reaction by withdrawing electrons from the reaction centre (superacid catalysis). This ability is characterized by the following stability sequence given by Irvin and Wiliamson:

$$Zn^{2+} > Cu^{2+} > Ni^{2+} > Co^{2+} > Fe^{2+} > Mn^{2+} \quad (63)$$

The above sequence readily accounts for the catalytic activity of zinc, copper and nickel salts as electrophiles in polar hydrosilylation processes. However, nickel (as well as other transition metal ions containing unfilled d-orbitals) is capable of playing a double role as a catalyst since it has a distinct ability to form Π-bonds with Π-acceptor ligands (back-donation), a fundamental requirement in homogeneous catalysis by transition metal complexes.

In the presence of relatively strong bases, e.g. tertiary amines, Cu(I) and Cu(II) salts mainly catalyse the α-addition of trichlorosilane to acrylonitrile yielding β-adduct, e.g. Cu(I) + NEt_3 gives a 91% yield [1103] and Cu_2O + $Me_2NCH_2CH_2NMe_2$ even 95–100% of β-adduct [959]. The latter heterogeneous reaction is promoted by ultrasounds. Such behaviour appears to result from stabilization of the trichlorosilyl anion by cuprous and cupric ions so that the reaction pathway could occur according to mechanism (54). It is quite likely that in the presence of relatively weak bases (e.g. pyridine, chelate phosphines [183, 560, 1165, 1325]) the role of the metal is to facilitate electrophilic interactions on the hydridic hydrogen of the pentacovalent intermediate (see Scheme 56) leading to the principal formation of an α-adduct. Blustein has reported a three-component catalyst system containing a mixture of a copper salt, a tertiary amine and a diamine, which is mainly active in the hydrosilylation of acrylonitrile, not only by trichlorosilane but also by other hydrosilanes with less electron-withdrawing substituents such as phenyl and methyl [104, 135a, 1103]. In order that a Cu(I) salt can catalyse a reaction efficiently, the concentration of Cu(I) ions should be lower than that of the tertiary amine. This rule holds in both two- and three-component catalyst systems [Rev.33].

In connection with nucleophilic–electrophilic interactions, it is possible to discuss other metal salts [18, 381, 412, 413, 648, 649, 655, 656] (mainly $ZnCl_2$, but also $SnCl_2$ [412, 413], $PdCl_2$ [811], $NiCl_2$ [413], $InCl_3$ [201, 999] and $GaCl_3$ [999], $AlCl_3$ [848], $M^+ [AlR_4]^-$ + MCl where M = Li, Na, K [U-174]) and non-metal compounds $BF_3 \cdot Et_2O$ [592] as electrophilic catalysts which are chiefly effective in the hydrosilylation of carbon–oxygen and carbon–nitrogen bonds but also in C=C double bond, e.g. [848, U-174]. Such reactions may be carried out in the presence or absence of solvents.

According to the above mechanism, the unsaturated substrate acts as a nucleophile capable of coordinating to a hydrosilane molecule through oxygen or nitrogen, thus forming a pentacovalent reactive intermediate of the type depicted in Scheme (56).

Calas *et al.* have extended the study of the hydrosilylation of compounds containing C=O, C=N and C≡N bonds, in particular with triethylsilane, by carrying out the reaction in the presence of $ZnCl_2$ which is thought to act mainly as an electrophilic catalyst [Rev.8, 196]. Cuprous Cu(I) salts catalyse the addition of $C_6H_5Cl_2SiH$ and CH_3Cl_2SiH to vinyl pyridines in a similar manner [U-003]. However, these olefins also act as basic catalysts.

Alkali metal fluorides such as CsF, KF or other salts, e.g. HCOOK, KCNS [152–154, 286, 1350], have been shown to be novel catalysts, which are mainly active in the hydrosilylation of the C=O bond. Fluoride ion-catalysed and/or induced addition of silicon hydride to carbonyl compounds (aldehydes, ketones, esters) is of current interest from both synthetic [151–154, 286, 393, 418, 1351] and mechanistic [286, 393, 395, 396, 1307, 1350] viewpoints. The reaction is carried out without solvent [151–153, 286] or in aprotic [393] solvents. The efficiency of the salts [262, 307, 396, 418] increases with decreasing cation–anion interaction as follows:

$$CsF > KF > LiF \qquad (64)$$

The mechanism of catalysis with these salts is still uncertain, although polar interactions of such ionic species with hydrosilanes are essentially of the donor–acceptor type.

Corriu *et al.* assume that anionic activation of the Si–H bond by fluoride ions occurs according to the following scheme [286]:

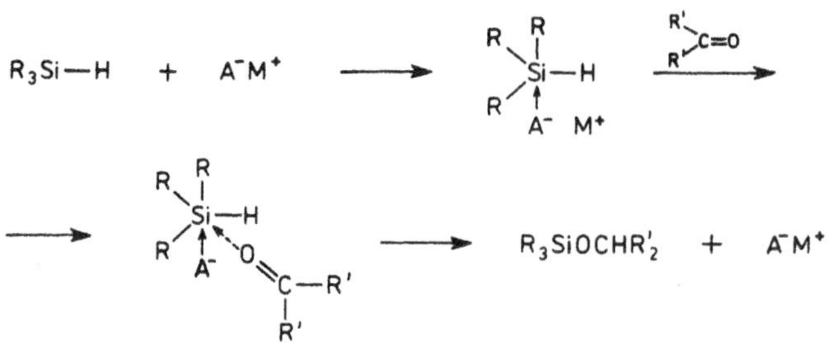

(65)

However, the metal cation (M^+) also seems capable of supporting the process through its interaction as an electrophile with hydridic hydrogen. For this reason the above mechanism has been decided to support such a function, making it similar to that of nucleophilic–electrophilic catalysis.

The pentavalent complex is regarded as an excellent hydride reducing agent which can also act as a single electron transfer reagent to give minor amounts of radical-derived reaction products [1307]. Fujita and Hiyama found that hydrosilylation of ketones and aldehydes can occur in the presence of catalytic amounts of tetrabutylammonium fluoride in HMPA as solvent, concluding that the solvent-coordinated hexavalent silicate is an active hydride species formed by equation (66) [395]:

$$H{-}SiR_3 + F^- \rightarrow [H{-}SiR_3F]^- \qquad (66.1)$$

$$[H{-}SiR_3F]^- + HMPA \rightarrow [H{-}SiR_3F(HMPA)]^- \qquad (66.2)$$

Very recently Lukevics *et al.* have reported effective experiments on the hydrosilylation of aromatic and heteroaromatic aldehydes and ketones proceeding under extremely mild conditions (at room temperature and in low-polarity solvent, CH_2Cl_2) induced by CsF as the source of F^- and 18–crown–6 as the phase-transfer agent [418].

In conclusion, it is worth saying that, in the past, few synthetic applications of base-catalysed hydrosilylation have been known, since it is mostly limited to vinyl derivatives, nitriles and trichlorosilane. The addition processes frequently require elevated temperatures (above 150°C) and pressures. Although the use of electrophilic metal ions and a multicomponent system allows the reactions to proceed under milder conditions, and allows an improvement in both the rate and yield, the selectivity of such processes is sometimes reduced. However, recent reveals of the increase in the Si–H bond reactivity associated with extension of the co-ordination at silicon (in the pentacoordinated and/or hexacoordinated state) creates also new perspectives in nucleophile (base)-catalysed hydrosilylation of multiple bonds, e.g. catalysis by F^- ions, by unsaturated amines or by pentacoordinated silicon hydride derivatives of amine.

2.3 Metal Catalysts

Another group of catalysts active in the addition of the Si–H bond to unsaturated compounds are metals supported on inorganic materials or carbon. The application of a supported platinum catalyst by Wagner and Strother has considerably extended the possibilities of organosilicon synthesis through hydro-silylation. Initially only a platinum catalyst supported on carbon, silicates and silica appeared to be effective in the reactions of trichlorosilane with ethylene, acetylene, butadiene, allyl chloride and vinylidene fluoride [Rev.36]. However, it was soon established that other metals could also be used to catalyse hydrosilylation reactions, *viz.* Rh [896], Ru [896], Pd [268, 688, 859, 1095, 1096], Ni [688, 898] and Ir [435]. These metals are usually supported on active carbon, γ-Al_2O_3, SiO_2 or $CaCO_3$. γ-Al_2O_3 supported platinum has been employed effectively in the hydrosilylation of acetylene, allyl derivatives [D-002, U-007], acrylonitrile [U-007] and ethylacrylate [U-007] by various tri-substituted silanes. Moreover, it has been shown that α-Al_2O_3-supported platinum (5%) is inactive in the addition of trichlorosilane to acetylene, whereas in the presence of platinum supported on γ-Al_2O_3 a 79% yield of the adduct is obtained.

The two-stage hydrosilylation of acetylene by trichlorosilane at 150°C in the gaseous phase using a flow method has provided a test reaction for the catalytic activity of various metals supported on γ-Al_2O_3 and active carbon. The yields of the double hydrosilylation adduct, bis(trichlorosilyl)ethane, obtained in the presence of the following metals, may thus be placed in the following order of catalyst efficiency [435]:

$$\gamma\text{-}Al_2O_3 \text{ support:} \quad Pd > Pt > Ru > Rh > Ir \tag{67}$$

$$\text{carbon support:} \quad Rh > Ir > Pd > Ru > Pt \tag{68}$$

In contrast, the activity of the same catalysts in the reaction of trimethylsilane with pyridine or picoline is given by the following sequence [268]:

$$Pd/C > Rh/C > PdCl_2 \gg \text{Raney Ni, Pt/C, Ru/C} \tag{69}$$

Platinum supported on carbon (usually in 5% concentration) is the most common and most efficient metal catalyst for polyaddition [196] and hydrosilylation of C=C and C≡C bonds [Rev.8, 54, 93, 302, 436, 1030, 1097, 1136, 1352], and of C≡N [859] and C=O bond [Rev.36].

Studies of the stereospecifity of hydrosilylation, as well as experimental evidence that these reactions are not retarded by free-radical inhibitors, demonstrate that when catalysed by supported metals (and by platinum in particular) the mechanism of the reaction is generally heterolytic [111, 1128, 1133], although Petrov *et al.* [899] have stated that under such conditions a radical mechanism for the addition cannot be excluded. The stereochemistry of the addition to 1-acetylenes in the presence of various types of catalysts has been studied by Benkeser *et al.* [Rev.5, 111, 124]. As already mentioned, they found a peroxide catalyst mainly induced *trans* addition to give *cis* adducts, whereas carbon-supported platinum and soluble chloroplatinic acid induced *cis* addition giving *trans* products. This demonstrates that platinum-catalysed additions occur by a different (non-radical) mechanism than that observed for peroxide-initiated processes.

The addition of the optically active silane, α-naphthylphenylmethylsilane, R_3Si^*H, to 1-octene occurs with a high degree of stereospecificity in the presence of a 5% Pt–C catalyst, and resembles the reaction homogeneously catalysed by chloroplatinic acid and $[(C_2H_4)PtCl_2]_2$. All three catalysts permit retention of the configuration at the asymmetric silicon centre. However, these experiments are incapable of supplying the answer to the question of whether Pt–C functions as a heterogeneous catalyst, or whether traces of Pt(II) or Pt(IV) species on this surface are the real catalysts [1128].

The side-reactions, e.g. disproportionation and dehydrogenation of silane substrates, are a negative feature of metal-catalysed hydrosilylation. The selectivity of hydrosilylation with a Pt–C catalyst depends to a substantial extent on the hydrosilane and unsaturated compound employed. Such a catalyst appears to be very selective towards the addition of methyldiethylsilane to divinylglycolic ethers (the yield was 83%); in the presence of a platinum complex two adducts were obtained with the predominating mono-adduct [1097], whereas in free-radical initiators (UV, peroxide) no adducts were observed at all. Whilst reactions of halogen-substituted alkenes with hydrosilanes on Pt–C catalysts are accompanied by polymerization of such simple olefins as $CF_2{=}CF_2$, $CF_2{=}CFCl$, $CF_2{=}CH_2$ [900], far fewer polymeric products are observed in the hydrosilylation of active olefins such as styrene, acrylonitrile or butadiene than in the free-radical-initiated process [934].

Recent studies by Boudjouk *et al.* on the effect of sonic waves on the yield and rates of reaction have also included the platinum-catalysed hydrosilylation of C=C and C≡C bonds by various tri-substituted silanes under very mild conditions [467]. Although carbon-supported palladium may be utilized to catalyse the hydrosilylation of the N≡C bond in isocyanates [859], under ultra-high-vacuum conditions palladium surfaces effected the hydrosilylation of acetylene adsorbed thereon [406].

Metals, which are usually colloidal and formed *in situ* by the reduction of metal salts (mainly nickel, but also including Pt, Pd and others) provide similar catalysts to supported metals. Indeed, the most important catalyst for hydrosilylation processes is a pyrophoric nickel which is much more reactive than the Raney nickel catalysts employed for hydrogenation, and can be generated by refluxing tri-substituted silanes (principally triethylsilane) with anhydrous $NiCl_2$. This reduced nickel has been widely utilized by several research groups, mainly Calas and Frainnet [146–148, 196, 223, 372, 373, 374, 376, 370], Lapkin [645–647, 652, 657] and Kharitonov and Glushkova [411, 414–416], to catalyse the hydrosilylation of aldehydes and ketones, leading to the corresponding alkoxysilanes or enoxysilanes [372]:

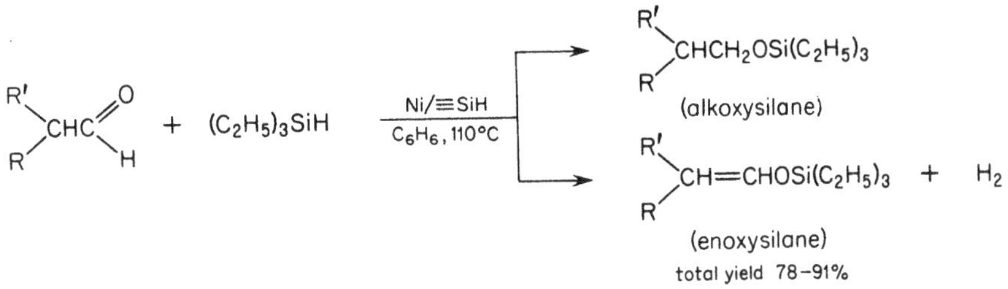

$$
\begin{array}{c}
\overset{R'}{\underset{R}{\diagup}}CHC\overset{O}{\underset{H}{\diagdown}} + (C_2H_5)_3SiH \xrightarrow[C_6H_6, 110°C]{Ni/\equiv SiH}
\begin{cases}
\overset{R'}{\underset{R}{\diagup}}CHCH_2OSi(C_2H_5)_3 \quad \text{(alkoxysilane)} \\[2mm]
\overset{R'}{\underset{R}{\diagup}}CH{=}CHOSi(C_2H_5)_3 + H_2 \quad \text{(enoxysilane)}
\end{cases}
\end{array}
$$

(70)

total yield 78–91%

Other compounds such as disilanes have also been used as reducing agents [370], but sulphur derivatives, e.g. PhSH or Et_2S, act as inhibitors [196, 223, 370]. This catalytic system is also active in additions to the C=N bond:

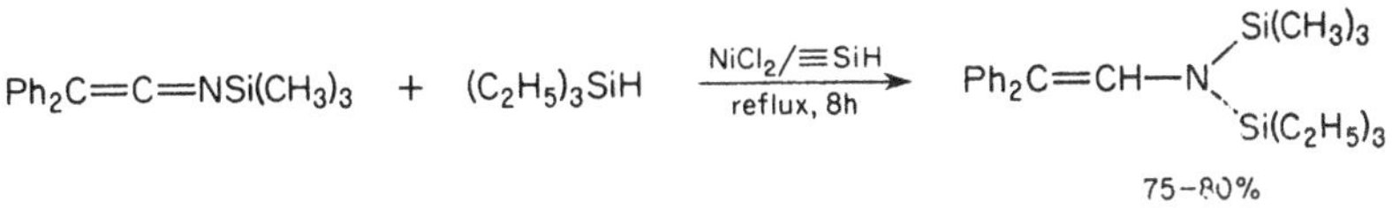

$$
Ph_2C{=}C{=}NSi(CH_3)_3 + (C_2H_5)_3SiH \xrightarrow[\text{reflux, 8h}]{NiCl_2/\equiv SiH} Ph_2C{=}CH{-}N\underset{Si(C_2H_5)_3}{\overset{Si(CH_3)_3}{\diagdown}}
$$

(71)

75–80%

The reduced nickel prepared from nickel iodide by lithium powder in the presence of ultrasound dispersed in an inert reaction medium (THF) can be commonly used as an effective catalyst for hydrosilylation of C=C bond (1-hexene, acrylonitrile, styrene, methylmethacrylate, vinyl acetate, vinylbutylether) [E-049].

Early concepts regarding the extremely efficient catalysis of hydrosilylation by chloroplatinic acid involved the reduction of the complex to colloidal platinum, which was believed to be the real catalyst. But only recent reports by Lewis and Lewis showed, for the first time, strong evidence that the mechanism for hydrosilylation mediated by highly active catalysts involves formation of colloidal platinum as the key step [667]. The reaction of $Pt(cod)Cl_2$ with triethoxysilane demonstrates that colloid formation is the key step in hydrosilylation. Colloids are generated from the reactions of $Pt(cod)_2$ and $Pt(cod)Cl_2$ with hydrosilanes and analysed by light scattering, TEM and ESCA. The proposed mechanism (Scheme 72) is based on the formation of colloidal platinum as the active catalyst during the induction period, which predominantly precedes the exothermic hydrosilylation reaction:

$$
Pt^{+2a}X_{2a}L_b \xrightarrow{HSiR_3} H_2L + 2aXSiR_3 + [Pt^0]_x
$$

(72)

where $a = 0, 1$; $b = 0–4$; X = halogen, pseudohalogen; L = "reducible" ligand (olefin).

According to the authors' conclusions it is quite likely that the mechanism based on colloid formation is the most important one where formation of Pt(O), free of ligands, is readily achieved [667]. Thus colloid formation can account for the high activity of a recently reported hydrosilylation catalyst based on complexes of Pt(O) with olefins such as divinyltetramethyldisiloxane (Karstedt catalyst) [U-064] and cyclooctadiene [668].

It is still believed that the hydrosilylation by such catalysts as metal–phosphine complexes, e.g. $(Ph_3P)_4Pt$, occur homogeneously. However, in light of the above results, all processes where reduction of ligand and metal are feasible should be re-examined and colloid catalyst intermediates should be considered. Since 1986 the activity of various platinum group colloids as hydrosilylation catalysts has been investigated regardless of which starting species is used as the catalyst [667–669, 876].

Lewis and Uriarte have very recently described the reactivity of platinum group metal colloids prepared according to equations [668]:

$$(CH_3)_2C_2H_5OSiH(\text{excess}) + MCl_x + DVi_4 \rightarrow H_2 + \text{Si products} + \text{M colloid} \quad (73)$$

where M = Pt, Rh, Ru, Ir, Os, Pd. DVi_4 serves (except in the Pt case) as a colloid stabilizer. The general trend in relative activities for catalysts of platinum group metal for hydrosilylation, based on various studies, is as follows:

$$\text{Pt} > \text{Rh} > \text{Ru}{=}\text{Ir} > \text{Os, Pd} \quad (74)$$

The proposed mechanism of the hydrosilylation catalysed by metal colloid presumably involves the formation of the metal colloid R_3SiH intermediate followed by nucleophilic attack of the olefin. This also explains a co-catalysis by dioxygen which, coordinating to the intermediate, makes it more electrophilic to attack by the nucleophile. Oxygen can also prevent irreversible colloid agglomeration [669]. However, the thorough studies on the mechanism of homogeneous catalysis by transition metal complexes described in the next section have demolished this particular approach.

2.4 Homogeneous Catalysis by Transition Metal Complexes

2.4.1 Background to Homogeneous Catalysis by Transition Metal Complexes

A characteristic feature of transition metals is their ability to form complexes with a variety of neutral compounds, including unsaturated organic compounds in which donor–acceptor interactions take place between the metal and the ligand. According to the Dewar–Chatt description based on molecular orbital theory, such bonding consists of two independent components: (1) overlap of the Π-electron density of the olefin or σ-electron density arising from lone-pair donation by the ligand; (2) back-bonding resulting from the flow of high electron density from filled metal d_x^2 or other d_π–p_π hybrid orbitals (particularly those of low oxidation states) into antibonding, low-lying empty Π-orbitals of the ligand atom, e.g. carbon. These qualitative concepts relating to metal–ligand bonding are illustrated in Scheme (75.2) (metal–carbon oxide bonding) and Scheme (75.1) (metal–olefin bonding) [287, 288], respectively.

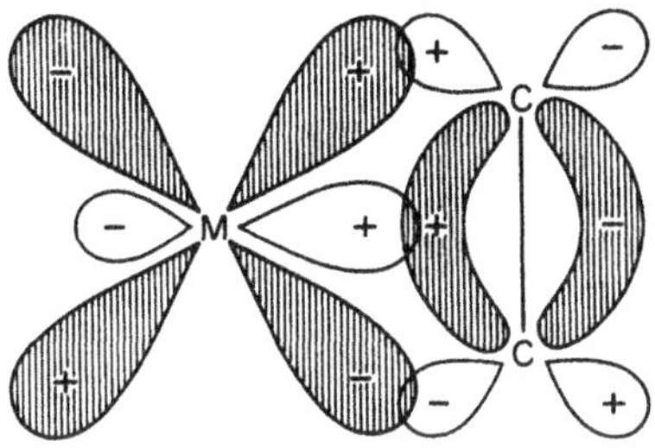

(75.1)

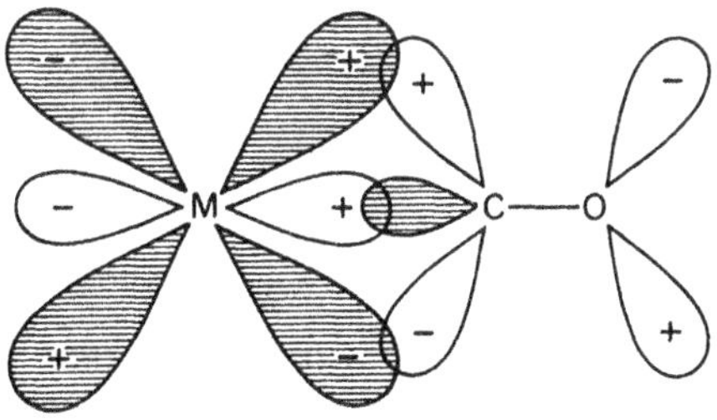

(75.2)

More complicated systems have also been considered, i.e. those involving polyolefinic ligands with unconjugated double bonds, and those in which allylic and other delocalized Π-systems such as carbocyclic groups (the commonest being the cyclopentadienyl ligand) and carbenes are bound to the metal. The subject of transition organometallic compounds, including the problem of the bonding involved, has been covered comprehensively by Cotton and Wilkinson [287, 288]. The great efficacy of such organometallic compounds as catalysts appears to be linked to the range of available metal orbitals (s, p_x, p_y, p_z, d_{z_2}, $d_{z_2-y_2}$, d_{x_2}, d_{y_2}, d_{xy}) which facilitate orbital symmetry restrictions, and also to the relative lability of the bond between the organic species and the metal (M-alkyl 150–200 kJ/mole). These are labile organometallic compounds that play an important role as active intermediates in many of the reactions of organic and other compounds which are catalysed by metal complexes. From the catalytic point of view, transition metals have the important possibility of being able to form relatively stable bonds, basically of the σ-type, with very reactive substrates, such as hydrides, which may be generated and stabilized by transition metal compounds under mild conditions. The coordination of such reagents to the metal atom markedly influences a proportion of the ligand bonding, thus changing their reactivity and enabling them to react while in the coordination sphere of the metal. As free substrates they have reacted only very slowly or not at all.

Coordination of the reagent to the metal may proceed by a variety of methods;

1. Coordinative addition of the substrate to the coordinatively unsaturated metal:

$$L_nM + X \rightleftharpoons L_nM{-}X \qquad (76)$$

However, when the reaction occurs in a solvent, the latter occupies the unfilled coordination sites on the metal, which means that solvent molecules have to be displaced by the incoming reactant molecule.

2. Ligand exchange, particularly when the metal is coordinatively saturated

$$L_nM + X \rightleftharpoons L_{n-1}M{-}X + L \tag{77}$$

Such exchange occurs in five- or six-coordinate metal complexes where coordination sites may be made available thermally or photochemically only by dissociation of one or more ligands.

3. Oxidative addition, i.e.

$$L_nM^n + XY \rightleftharpoons L_nM^{n+2}(X)(Y) \tag{78}$$

This reaction involves oxidation of the metal at the same time as its coordination number is increased. Generally, both the oxidation and coordination numbers are increased by two. The reverse reaction is termed a reductive elimination. This reaction occurs mainly with the d^8 and d^{10} electron configurations of metal complexes having four-coordinate square geometry, i.e. Fe(O), Ru(O), Os(O), Co(I), Ir(I), Rh(I), Ni(O), Pd(O), Pt(O), Pd(II), Pt(II) as well as with the d^9 Co(O) system. In some cases two metal atoms may react with an HX molecule and increase the oxidation state of each metal by one, e.g.

$$Co_2(CO)_8 + HX \rightleftharpoons HCo^I(CO)_4 + XCo^I(CO)_4 \tag{79}$$

4. An insertion reaction. This involves the migration of atoms or groups of atoms from the metal to a ligand within the coordination sphere of the metal

$$\underset{\substack{|\\ Z}}{L_nM}{-}XY \rightleftharpoons L_nM{-}XY{-}Z \tag{80}$$

In addition, other reactions of coordinated ligands are also likely to occur; thus a hydrogen atom may initially transfer intramolecularly from the ligand to the metal atom but is subsequently lost, or nucleophilic attack of bases, e.g. amines, on the coordinated ligands (directly or by prior coordination to the metal atom) may occur and be followed by insertion.

The principles of homogeneous catalysis by transition metal complexes have been expounded in greater detail elsewhere [494, 755, 814], but in summary it may be said that all the reactions of both organic and other substrates catalysed by transition metal complexes in solution proceed in accordance with the foregoing elementary processes which lead to the formation of metal–substrate intermediates activating the coordinated species, and thus enabling the reactions to occur.

2.4.2 Mechanism, Stereochemistry and Side-reactions

In 1957 J. L. Speier from Dow Corning Co. discovered that chloroplatinic acid was an effective homogeneous catalyst for the hydrosilylation process. As a result it ultimately replaced all previous catalytic systems, thus considerably extending the synthetic possibilities of these reactions [1133]. However, it was only after the publication in 1965 of a general mechanism for the catalysis of hydrosilylation by transition metal complexes proposed by Chalk and Harrod [226] that any increase in interest in the catalysis of hydrosilylation was shown. The proposed mechanism, which was originally derived from studies of chloroplatinic acid as a catalyst, provided a qualitative rational generalization for other transition metal complexes. The general scheme for the addition of silicon hydrides to the C=C bond outlined below highlights the following stages [Rev.9, Rev.11, Rev.17, Rev.31]:

1. Activation of the initial complex during the induction period, e.g. reduction of the Pt(IV) complex (chloroplatinic acid) to form a d^8 Pt(II) active catalyst.
2. Coordination of the unsaturated olefin compound to the metal centre

$$- \overset{|}{\underset{|}{M}} - \; + \; R'CH{=}CH_2 \rightleftharpoons \left[\begin{array}{l} R'CH\ | \\ \;\; \| \; {-}M{-} \\ CH_2\ | \end{array} \right] \tag{81.1}$$

$$\text{(I)} \qquad\qquad\qquad\qquad \text{(II)}$$

One ligand of the square-planar complex (I), e.g. a solvent molecule, is only weakly attached and therefore can readily be replaced by the olefin. Such olefin coordination occurs as result of bonding involved a α-type electron donation from the olefin to the dsp^2 hybridized orbitals of the metal, and a back-donation of electron density from the metal orbitals to the antibonding II-orbitals of the olefin.

3. Oxidative addition of the hydrosilane to the metal centre

$$\begin{array}{l} R'CH\ | \\ \;\; \| \; {-}M{-} \\ CH_2\ | \end{array} \; + \; HSiR_3 \; \underset{}{\overset{(2)}{\rightleftharpoons}} \; \begin{array}{l} \quad\;\; SiR_3 \\ R'CH\ |\,{\diagup}H \\ \;\; \| \; {-}M{-} \\ H_2C\ {\diagup}\,| \end{array} \tag{81.2}$$

$$\text{(II)} \qquad\qquad\qquad\qquad\qquad \text{(III)}$$

Such reactions are readily undergone by d^8 and d^{10} complexes. The reversibility of this stage has been well established, e.g. by H/D exchange reactions on silicon during chloroplatinic acid catalysis [992]. The stability of the complex (III) depends on the nature of the metal ligands. An increase in the σ-basicity and/or a decrease in the II-acidity of some or all of the ligands reduces the promotion energy of the metal atom and facilitates oxidative addition. Hence, the third row of transition metals usually has lower promotion energies than the second row of metals. The values of the latter are in turn lower than the respective figures for the first row of transition metals.

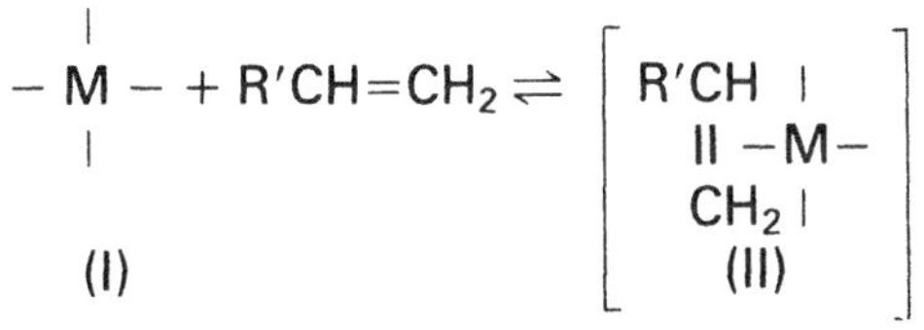

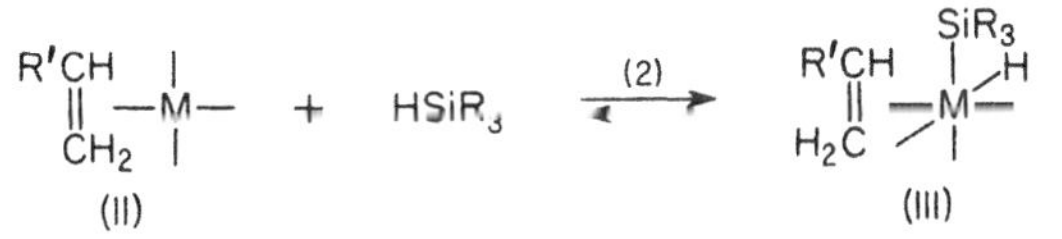

4. *cis*-Insertion of the ligand (Π–σ rearrangement)

$$
\begin{array}{c}
\underset{\text{(III)}}{\overset{\displaystyle \underset{H_2C}{\overset{R'CH}{\Vert}}\!\!\!\diagup\overset{SiR_3}{\underset{|}{M}}\diagdown\!\!\overset{H}{\diagup}}{}} \; + \quad \xrightleftharpoons{\;(3)\;} \quad \underset{\text{(IV)}}{R'CH_2CH_2\!\!-\!\!\overset{\overset{\displaystyle SiR_3}{|}}{M}\!\!\diagdown}
\end{array}
\qquad (81.3)
$$

It has been suggested that this Π–σ rearrangement be the rate-determining step of the hydrosilylation reaction. It is also assumed to be reversible, so that the observed olefin isomerization could be regarded as a side-reaction.

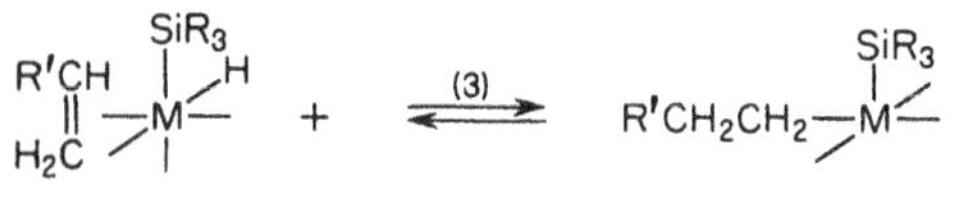

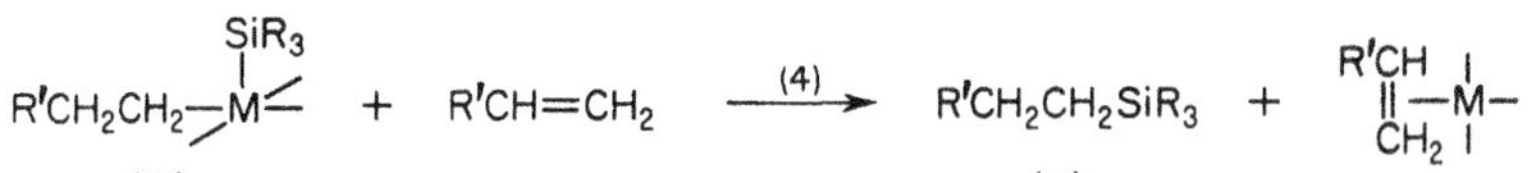

$$
\underset{\text{(IV)}}{R'CH_2CH_2\!\!-\!\!\overset{\overset{\displaystyle SiR_3}{|}}{M}\!\!\diagdown} \; + \; R'CH\!\!=\!\!CH_2 \quad \xrightarrow{\;(4)\;} \quad \underset{\text{(V)}}{R'CH_2CH_2SiR_3} \; + \; \underset{\text{(II)}}{\overset{\displaystyle \underset{CH_2}{\overset{R'CH}{\Vert}}\!\!\!-\!\!\overset{|}{\underset{|}{M}}\!\!-}{}}
\qquad (81.4)
$$

The complex (IV) reacts with excess of olefin, regenerating the real catalyst (II) and forming the organosilicon compound (V).

The stereochemistry of the addition to acetylenes has been studied for various types of homogeneous catalyst [63, 111, 124, 315, 992]. In practice, *cis* addition seems to be a common feature of hydrosilylation catalysed by transition metal complexes, although the mechanistic scheme given above does not require *cis* addition. Benkeser *et al.* [111, 124] have found that the addition of trichlorosilane to 1-alkynes in the presence of heterogeneous (Pt–C) and homogeneous (chloroplatinic acid) catalysts induces a *cis* approach and results in the *trans* product. The observed difference between peroxide and platinum catalysts may be an expression of the non-radical nature of hydrosilylation brought about by both types of platinum catalyst [Rev.27]. Selin and West [1036] have reported stereospecific *cis* addition of trichlorosilane to the 1-methylcyclohexene-d$_3$ ring in the presence of chloroplatinic acid. Recently, Asatiani and Zurabishvili have found that hydrosilylation of ferrocenylacetylene by triorganosilanes, in the presence of chloroplatinic acid, occurs stereospecifically forming a *trans* adduct [63].

Although Haszeldine *et al.* [351] found *cis*- and *trans*-n-butyltriethylsilylethenes products of the reaction of 1-hexene with triethylsilane catalysed by RhCl(PPh$_3$)$_3$, the occurrence of *cis* stereoselectivity is easily rationalized. Mechanistic schemes have been proposed to account for those cases where *cis* addition leads to the *trans* isomer as the initial product of hydrosilylation, followed by *trans* → *cis* isomerization. A more detailed explanation of the observed stereochemistry will be presented below when catalysis by rhodium complexes is considered.

Several series of experiments using optically active trisubstituted silanes have indicated that, when catalysed by platinum complexes, the hydrosilylation process occurs overall with retention of configuration at the asymmetric silicon centre [1124–1126, 1128]. The above data, and other experimental data, are consistent with the mechanistic scheme advanced by Chalk and Harrod, and enable further implications to be suggested [1128], such as:

1. Oxidative addition of silicon hydrides (Scheme 81.2), with the corresponding reductive elimination occurring with retention of configuration at silicon.
2. Step (3) (Scheme 81.3), where rearrangement of the Π-complex (III) to the σ-alkyl complex (IV) involves *cis* addition of the metal–hydrogen bond to the carbon–carbon double bond.
3. The formation of the hydrosilylated product (E) which proceeds with retention of configuration at both silicon and carbon.

The general mechanism for the addition of silicon hydrides to alkenes catalysed by Pt complexes, originally described by Chalk and Harrod, has been widely accepted and adapted with various modifications to provide a rational explanation of the hydrosilylation of other multiple bonds (C≡C, C=O, C≡N, etc.) which is homogeneously catalysed by various metal complexes. The latter include Π-allyl metal complexes and carbonyl complexes which generally form σ-bonds with olefins.

Although most catalytic cycles containing the active intermediates are still based on Chalk's and Harrod's scheme, some new postulations have appeared in the literature, e.g. olefin insertion into the M–Si bond as a crucial step explaining the formation of unsaturated (dehydrogenated) products [964, 1028]. Present study on the catalysis of hydrosilylation by transition metal complexes, e.g. platinum, rhodium, cobalt, ruthenium and palladium, aims at the elucidation of a composition and structure key intermediates, which can define more accurately individual steps and detailed mechanisms of the catalytic processes [Rev.23]. Mechanisms involving such intermediates will be presented in subsequent chapters considering particular transition metal complexes as catalysts.

The high catalytic reactivity of transition metal complexes in reactions in which unsaturated compounds or silicon hydrides take part, gives rise to different selectivities in hydrosilylation processes. It is pertinent to note here that the side-reactions which accompany metal complex-catalysed hydrosilylations involve compounds with one or more multiple bonds. These reactions include isomerization, oligo- and polymerization, hydrogenation, metathesis and/or reactions of silicon hydrides such as redistribution or dehydrogenation, as well as reactions in which both substrates take part, e.g. telomerization and dehydrogenative silylation. Some reactions mentioned above are useful in the search for novel and efficient synthetic methods, and for this reason require separate consideration.

Although many transition metal complexes catalysed olefin isomerization, early studies showed that the presence of silane markedly affects such reactions which occur concurrently with hydrosilylation. Double-bond migration, which is a characteristic feature of most coordination catalytic reactions, may be illustrated schematically as in Scheme (82) as a side-reaction occurring during hydrosilylation [Rev.11, Rev.17].

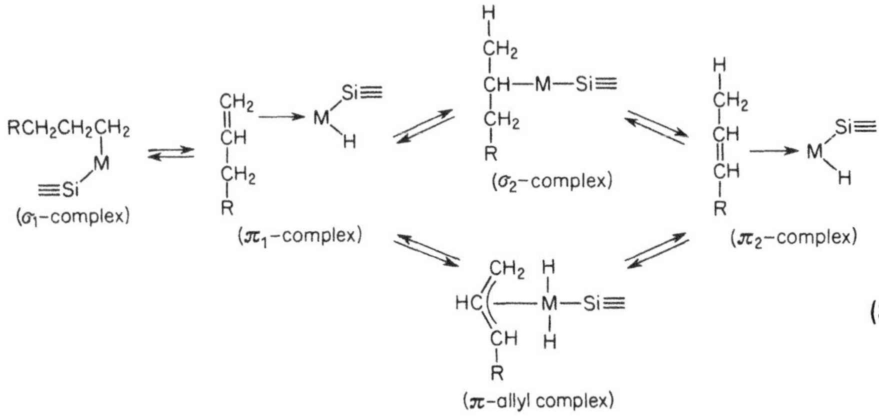

(82)

According to this scheme, olefin isomerization occurs via a Π–σ reversible rearrangement and/or through the formation of a Π-allyl complex. In this regard the numerous experimental data obtained from hydrosilylation studies could provide some support. Thus, it has been shown that olefins with an internal double bond yield addition products with a terminal silyl group. Similarly, the effect of a substituent at silicon on the isomerization process has been studied. Finally, it has been observed that the olefin recovered from the reaction system isomerizes to yield the isomer composition which is sustainable thermodynamically. What is interesting is that in most cases the process of isomerization is considerably faster than that of hydrosilylation, particularly when trialkyl and trichloro-substituted silanes are involved, and that during the addition of a trialkoxysilane to a 1-alkene an almost olefin-like isomerization is observed [Rev.17]. From these results it appears that β-hydride transfer in the σ_1-metal alkyl intermediate (Scheme 82) occurs at a slower rate than the reductive elimination which yields the hydrosilylation product.

The redistribution reactions at the silicon atom which are catalysed by transition metal complexes have been reviewed by Curtis and Epstein [301], while those occurring under hydrosilylation conditions have been discussed by Speier [Rev.38]. It is apparent from the numerous examples listed in the two reviews that the by-products formed during hydrosilylation arise from the following kinds of redistribution reactions:

(a) H/Cl exchange [89, 628]

$$2\,CH_3SiHCl_2 \xrightleftharpoons{NiCl_2(PPh_3)_2} CH_3SiCl_3 + CH_3SiH_2Cl \quad (83)$$

(in the presence of Speier catalyst $(CH_3)_2SiHCl$ and gaseous products are also observed [89])

(b) Vinyl/H exchange [Rev.38]

$$R_3SiH + R_3'SiCH{=}CH_2 \xrightleftharpoons{H_2PtCl_6} R_3SiCH{=}CH_2 + R_3'SiH \quad (84)$$

(c) $(CH_3)_3SiO/CH_3$ exchange [Rev.38]

$$2[(CH_3)_3SiO]_2SiHCH_3 \rightleftharpoons (CH_3)_3SiOSiH(CH_3)_2 + [(CH_3)_3SiO]_3SiH \quad (85)$$

(d) CH_3/H exchange [309]

$$2 =NSiH(CH_3)_2 \rightarrow =NSiH_2(CH_3) + =NSi(CH_3)_3 \quad (86)$$

(e) H/OR exchange [Rev.24]

$$2\,(RO)_3SiH \xrightleftharpoons{\text{catalyst}} (RO)_2SiH_2 + (RO)_4Si \quad (87)$$

(f) RO/R'O exchange [745]

$$(RO)_3SiH + (R'O)_3SiCH{=}CH_2 \xrightleftharpoons{RuCl_2(PPh_3)_2} (RO)_n(R'O)_{3-n}SiH + (R'O)_n(RO)_{3-n}SiCH{=}CH_2 \quad (88)$$

(g) H/D exchange [992]

$$3.5\ Cl_3SiD + CH_2{=}C(CH_3)_2 \rightarrow Cl_3SiC_4H_{6.5}D_{2.5} + 1.5\ Cl_3SiH + Cl_3SiD \quad (89)$$

(h) SiR_3/H [174]

$$2\,R_3SiH \xrightarrow{\text{Pt,Rh,Ir,Pd complex}} R_3SiSiR_3 + H_2 \quad (90)$$

Some of these redistributions may be initiated either by homolysis or catalytic activation, or even by conventional acidic or basic catalysts, but a number proceed more readily in the presence of olefins. These reactions are fast and yield a large number of products which also contain adducts formed during the preliminary redistribution of the silane substrate(s) prior to their hydrosilylation. Redistributions of silicon compounds catalysed by transition metals occur when at least one silicon–hydrogen bond is present in the molecule, since the Si–H bond is the most labile of those undergoing oxidative addition to yield a silyl metal hydride. Metals in the stable +1 or +3 oxidation state can redistribute ligands at silicon via a series of oxidative additions and reductive eliminations, thereby generating new M(I) species.

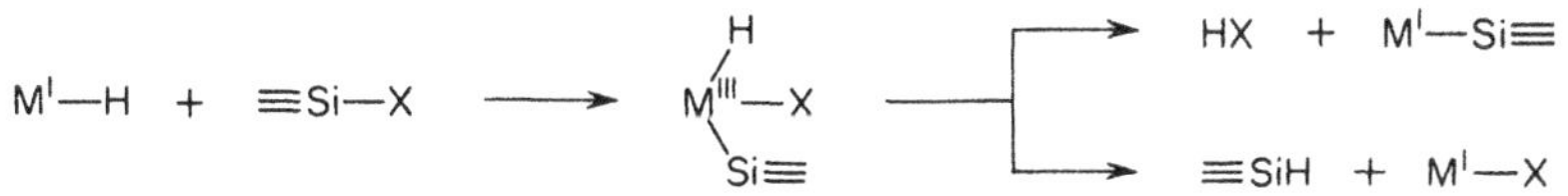

(91)

Interactions of a silyl hydride with zero-valent metal species give a M^{II}(silyl)-hydride complex. In this case, in order to exchange ligands on the Si atom, a further oxidative addition is required [301]:

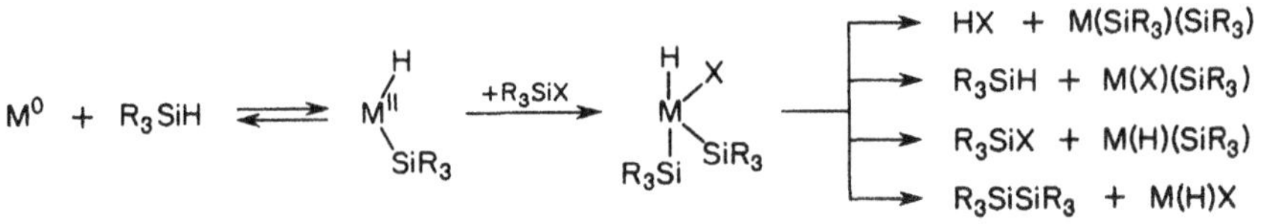

$$M^0 + R_3SiH \rightleftharpoons M^{II} \overset{H}{\underset{SiR_3}{<}} \xrightarrow{+R_3SiX} \overset{H}{\underset{R_3Si}{\overset{|}{M}}}\overset{X}{\underset{SiR_3}{<}} \longrightarrow \begin{cases} HX + M(SiR_3)(SiR_3) \\ R_3SiH + M(X)(SiR_3) \\ R_3SiX + M(H)(SiR_3) \\ R_3SiSiR_3 + M(H)X \end{cases} \quad (92)$$

The oxidative addition of silicon hydride to a divalent metal species leads to the complex in which the metal exists in the formal +4 oxidation state:

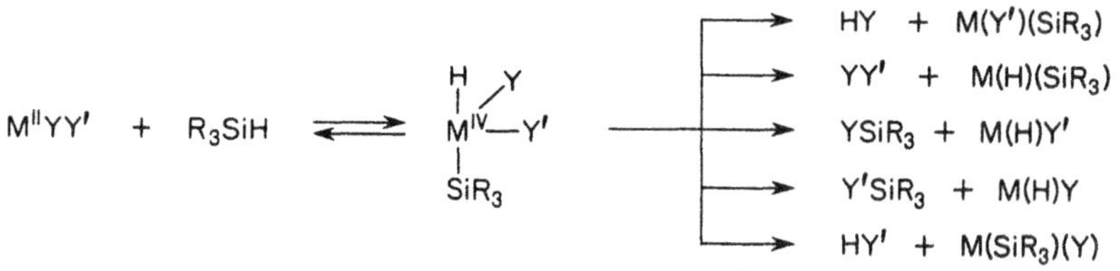

$$M^{II}YY' + R_3SiH \rightleftharpoons \overset{H}{\underset{SiR_3}{\overset{|}{M^{IV}}}}{-}\overset{Y}{\underset{}{Y'}} \longrightarrow \begin{cases} HY + M(Y')(SiR_3) \\ YY' + M(H)(SiR_3) \\ YSiR_3 + M(H)Y' \\ Y'SiR_3 + M(H)Y \\ HY' + M(SiR_3)(Y) \end{cases} \quad (93)$$

As was shown in equation (93), depending on the nature of the YY' groups the complex formed can undergo reductive elimination via five distinct pathways.

All the metal species listed above (equations 83–93) can react further by oxidative addition or reductive elimination to yield finally the spectrum of by-products that often accompanies hydrosilylation. Another side-reaction observed in these systems has been termed dehydrogenative (double) silylation, and occurs according to the general scheme;

$$X_3SiH + 2\ RCH{=}CH_2 \rightarrow X_3SiCH{=}CHR + RCH_2CH_3 \quad (94)$$

This reaction, which allows the direct production of unsaturated silyl compounds, has recently been the subject of separate studies. The synthetic aspects of this reaction are also being examined at present. In most cases the group R in equation (94) depicts an electron-withdrawing substituent in such olefins, e.g. styrene or a substituted styrene [274, 859, 1033, 1295], a vinyl tri-substituted silane [399, 733, 740, 745, 844, 912], a siloxane [Rev.38], or trifluoropropene [858]. Such olefins react with hydrosilane in the presence of ruthenium, rhodium, iron, osmium and platinum complexes. However, dehydrogenative silylation of alkenes also leads to the formation of alkenylsilanes in the presence of iron carbonyls [383, 833] and rhodium complexes [771, 1150]. A comprehensive examination of the hydrosilylation of vinyltrialkoxysilanes by trialkoxysilanes, particularly in the presence of ruthenium complex catalysts, has shown that an unsaturated product is not formed during the catalytic cycle which induces the double dehydrogenative hydrosilylation of the vinylsilane [723, 736]. The detection of ethylene among the reaction products suggests that the competitive metathesis of vinyltrialkoxysilanes occurs according to the general equation [737]:

$$2\ (RO)_3SiCH{=}CH_2 \xrightarrow{Ru^{II}, Ru^{III}} (RO)_3SiCH{=}CHSi(OR)_3 + CH_2{=}CH_2 \quad (95)$$

$$R = CH_3,\ C_2H_5,\ C_3H_7,\ iso{-}C_3H_7.$$

Hydrosilylation in the presence of transition metal complex catalysts can also be accompanied by other side-reactions such as the reduction of the organic compounds [Rev. 28, 100].

$$CF_3COOCH_2CH{=}CH_2 + CH_3SiHCl_2 \xrightarrow{H_2PtCl_6} CF_3COOCH_2CH_2CH_3 + CF_3COOSiCH_3Cl_2 + CH_2 = CHCH_3 \quad (96)$$

This is particularly observed in the reduction of allyl acetate and halides to propene (where H/Cl exchange also occurs) [Rev.13, 301]

$$CH_2{=}CHCH_2Cl + HSiCl_3 \xrightarrow{H_2PtCl_6} Cl(CH_2)_3SiCl_3 + C_3H_7SiCl_3 + SiCl_4 \quad (97)$$

as does condensation with the evolution of hydrogen.

$$RC{\equiv}CH + (C_2H_5)_3SiH \xrightarrow{H_2PtCl_6 + LiI\ (I_2, LiI_3, R_3SiI)} RC{\equiv}CSi(C_2H_5)_3 + H_2 \quad [957, 1259, 1242] \quad (98)$$

$$CH_2{=}CHR + R_3'SiH \xrightarrow{Rh_2Cl_2(CO)_4} RCH{=}CHSiR_3' + H_2 \quad [U\text{-}173] \quad (99)$$

$$CH_2{=}CHCH_2NH_2 + (C_2H_5O)_3SiH \xrightarrow{H_2PtCl_6} CH_2{=}CHCH_2NHSi(OC_2H_5)_3 + H_2 \quad [93] \quad (100)$$

2.4.3 Chloroplatinic Acid and Other Platinum Complexes

Although a wide range of catalysts has now been studied for hydrosilylation reactions, most research and industrial syntheses are carried out in the presence of platinum complexes.

Hexachloroplatinic acid $H_2PtCl_6 \cdot 6H_2O$, is most commonly used to provide the initial precursor for platinum catalysts. The acid is usually dissolved in an organic solvent, and can be used as the hydrate or in the partially dehydrated form. A solution of chloroplatinic acid hexahydrate dissolved in isopropyl alcohol (1–10%) and referred to as Speier's catalyst [1133], is the most widely used platinum catalyst. In addition to isopropanol, other alcohols have also been used in the preparation of active catalytic species from chloroplatinic acid; thus methanol [D-034, D-054, Dd-001, N-003, U-039], ethanol [D-016, D-068, J-016], n-propanol [S-019], n-butanol [D-001, J-026], isobutanol [D-001], tert-butanol [S-021, U-006], pentanol [D-001, S-021], *iso*-octanol [J-005], n-octanol [D-001, F-002, U-014], 2-ethylhexanol [J-005, U-096], benzyl alcohol [D-042] and higher aliphatic alcohols [F-002]. Solutions of chloroplatinic acid in n-octyl alcohol are often referred to as a Lamoreaux catalyst [F-002, U-014]. Effective and relatively stable catalysts may be obtained as the products of the reaction of hexachloroplatinic acid with ketones [D-041, G-022, S-011] which possess no multiple carbon–carbon bonds, e.g. cyclohexanone [F-006, U-001, U-072], acetophenone, acetone and methylethylketone [D-041, U-072], and with aldehydes [F-002], mono- and dialkylethers of ethylene glycol [B-001, D-053, F-006, S-011, U-039, U-085], unsaturated ethers [S-012], esters such as methylbenzoate [D-023, G-003], dimethyl phthalate [778, 1167, 1169, F-001, U-002, U-026, U-031] and butyrolactone [E-037] and butylcarbitol acetate [U-035], and with organic acids [S-011]. Recently such catalysts have been obtained from reactions with THF [55, D-087, E-037], hydrocarbons such as benzene [F-021], toluene [G-021] and xylene, and from petroleum ether [G-021].

A general characteristic of the H_2PtCl_6–solvent catalytic system is the induction period which occurs in its presence and which lasts for some time, being followed by a very fast exothermic hydrosilylation reaction. Although this system has often been used for catalytic hydrosilylation in the laboratory, and under industrial conditions, little study has been undertaken of the composition of the active catalytic species formed during the initial induction period. Early work on this subject by Voronkov *et al.* [944, 1254] and Lahaye and Lagarde [639] suggested that a classic Speier catalyst containing chloroplatinic acid was formed initially, and was thus reduced by the alcohol to chloroplatinous acid with the formation of acetone and hydrochloric acid according to the following equation:

$$H_2PtCl_6{\cdot}6H_2O + iso\text{-}C_3H_7OH \rightarrow H_2PtCl_4 + (CH_3)_2CO + 2HCl + 6H_2O \qquad (101)$$

Very elegant and detailed experiments carried out in Benkeser's laboratory [112] have shown that the Speier catalyst contains the complex $H(C_3H_6)PtCl_3$; the stoichiometry of the reaction can thus be expressed as follows:

$$H_2PtCl_6{\cdot}6H_2O + 2\ iso\text{-}C_3H_7OH \rightarrow H(C_3H_6)PtCl_3 + 3HCl + 7H_2O + (CH_3)_2CO \qquad (102)$$

The spectroscopic methods (IR, UV–VIS, ESCA) employed in this study confirmed the suggestion that dimerization to $[(C_3H_6)PtCl_2]_2$ (an orange solid) occurs, and that this dimer dissociates in isopropanol–hydrochloric acid solutions to the monomeric form. The chloride ion liberated as a result of the transformation shown in equation (102) stabilizes the anionic platinum(II) complex, shifting the equilibrium depicted in equation (103) to give the $[PtCl_3(C_3H_6)]^-$ complex which is stable under these conditions and active in hydrosilylation.

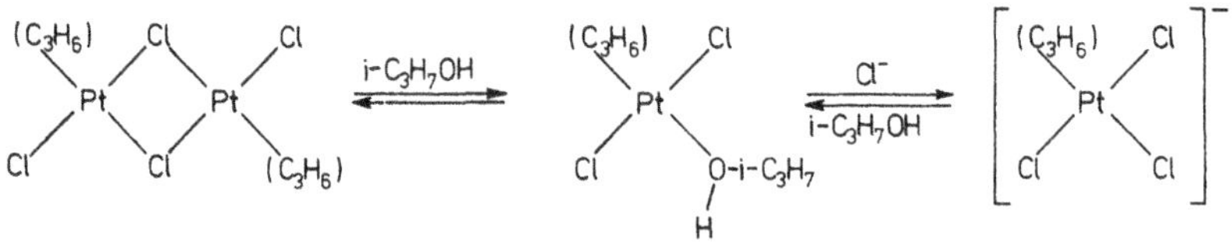

(103)

The detailed mechanism of active catalyst formation given above explains why, in contrast to an alcoholic solution of the dimer $(C_3H_6)_3Pt_2Cl_4$ and the Zeise type dimer $(C_2H_4)_2Pt_2Cl_4$, which give the same result as the Speier catalyst in terms of rate, yield and hydrosilylation products, the complex $[PtCl_3(C_3H_6)]^-$ is stable for up to 2.5 years and shows no tendency to generate a black precipitate [112].

^{195}Pt NMR method was used to follow the reaction of H_2PtCl_6 (as well as $PtCl_4$) with isopropanol, confirming a formation of $[(C_3H_6)PtCl_3]^-$ anion but simultaneously revealing the following equilibrium present in the solution [892]:

$$[(C_3H_6)PtCl_3]^- + iso\text{-}C_3H_7OH \rightleftharpoons Cl^- + [(C_3H_6)PtCl_2(iso\text{-}C_3H_7OH)] \qquad (104)$$

which is dependent on concentration of Cl^- (and HCl) in the reaction mixture.

The interaction of tert-butylacetylene with chloroplatinic acid in propyl alcohol [Rev. 40] leads to an analogous Pt(II) dimer containing alkoxycarbene ligands,

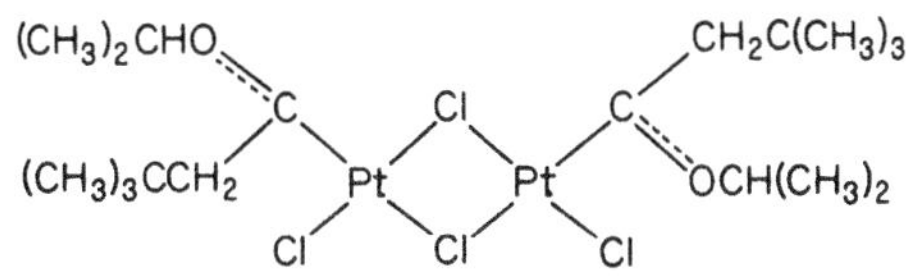

(105)

while the replacement of tert-butylacetylene by trimethylsilylacetylene leads to the formation of a monomeric complex with two carbene ligands attached to the platinum centre as shown below [948, 1143].

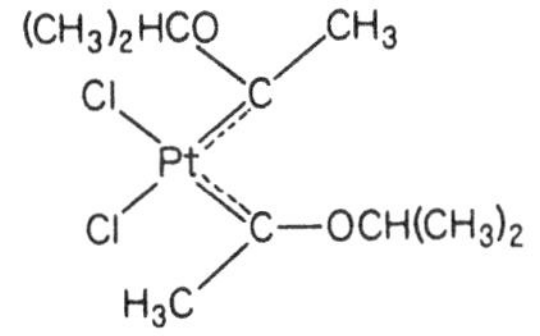

(106)

During the preparation of chloroplatinic acid in THF (which has recently become a common catalyst [55, 641, 1204, D-087, E-037] for hydrosilylation), reduction of Pt(IV) to Pt(II) and Pt(O) species is observed which generate several organic products during the oxidation of THF [1254, U-097].

Under hydrosilylation conditions chloroplatinic acid is reduced by silicon hydride. Thus, in the absence of an olefin, Pt(O) is precipitated from hexane after a period of several days, whilst in its presence no detectable precipitation of Pt(O) is observed at room temperature. However, if the temperature is increased to about 80–100°C, this causes the solution to darken from yellow to brown with the formation of colloidal platinum [Rev.38]. The formation of platinum black as a result of the reduction of numerous platinum species has also been observed. Activation of chloroplatinic acid with oxygen to enhance, e.g. hydrosilylation of unactivated alkyl-, dialkyl- and trialkylsilanes with 1-alkenes to synthesize tetraalkylsilanes, has recently been found [876] but the effect can be accounted for by metal colloid formed [667, 668, 669].

Although chloroplatinic acid prepared in a given solvent is commonly used as a catalyst for hydrosilylation, the presence of a co-catalyst is often necessary to enhance the catalytic activity and the rate of adduct formation, and to vary the regioselectivity. On the other hand, there is a need to find an inhibitor that will protect the platinum catalyst from premature ageing at room temperature [1287]. A more detailed discussion of the ageing or curing of organosilicon polymers by hydrosilylation is presented in Section 5.3, where numerous catalytic systems which have been widely used in such processes are also described. The activation of chloroplatinic acid by orthotitanates during the hydrosilylation of vinylsilanes is a good example of promotion [924]. This effect can be attributed to the nucleophilic interaction between the titanate oxygen and silicon, which increases the hydridic character of the Si–H bond (structure (I) in equation 107.1):

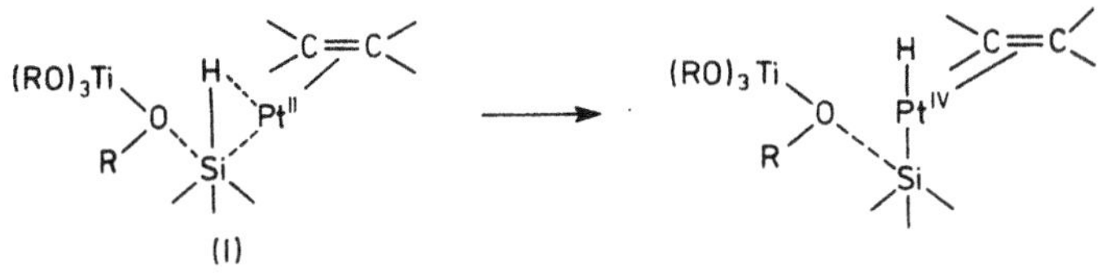

(107.1)

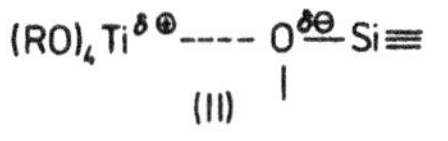

(107.2)

According to the authors, the absence of a promoting role of orthotitanate during the hydrosilylation of vinylsiloxanes is due to the competitive nucleophilic interaction of the siloxane oxygen at the titanium centre and a decrease in the d_{Π}–p_{Π} back-donation of the Si–O bond in the siloxanes [925] (structure (II) in equation 107.2). Unsaturated organic and organosilicon compounds, e.g. vinylcyclohexene [F-036], vinylsilanes [Pl-023] and vinylsiloxanes [D-102, E-027, U-028], can also activate the H_2PtCl_6 precursor. Chlorides of metals such as Al, Ge, Ga, In, Sn, Ce, Nb, Te and Co appear to be effective co-catalysts of the chloroplatinic acid precursor during the hydrosilylation of olefins, and also in the hydrosilylation of acetylene and its mono-derivatives [1234, 1262, 1264, D-079, D-085]. Triphenylphosphine and other phosphines and phosphites are also well known as co-catalysts of chloroplatinic acid, and have been used with good effect in the regioselective hydrosilylation of allylbenzene by tri(chloro,methyl)silanes [1218] and of styrene by trichlorosilane [213] and of vinylcyclohexene by trichlorosilane [C-042, C-039]. Tertiary and other amines [F-015, U-071], e.g. butylamine or triethylenediamine, are capable of increasing the activity of H_2PtCl_6, particularly in the presence of mono- and di-functional silicon hydrides as reducing agents [C-024]. Phosphonium and ammonium salts of platinum acids with the general formula $(REPh_3)_2PtCl_n$, where E = P, N and n = 3, 4, 6 [170, 171, 517a] as well as complexes of chloroplatinic acid with aminocyclophosphasenes having NH group in the substituent have also been used as regioselective catalysts of the hydrosilylation process. Stable complexes of chloroplatinic acid and sodium chloroplatinate with 18-crown-6 ethers and 1,7,10,16-tetra-oxo-4,13-diazocyclooctadiene, slightly soluble or insoluble in common organic solvents were recently found to be active catalysts of hydrosilylation and can be recovered and reused several times [670]. Examples for promotion of hydrosilylation reactions by hydrosilane or chlorosilane present in reaction mixture have recently been reported [672, U-168].

All recent experiments with chloroplatinic acid as a precursor catalyst have suggested that, in most cases, the Pt(II) intermediate generated *in situ* is responsible for hydrosilylation catalysis. For this reason it is expected that Pt(II) complexes with a square planar symmetry have been applied for several years as initial catalysts in hydrosilylations. In addition to the classic Zeise's salt $[Pt(C_2H_4)Cl_3]^-$ [966], most of these catalysts are either monomeric complexes of the formula $PtCl_2L_2$ or dimeric species with the general formula $Pt_2Cl_4L_2$. A comprehensive study [965, 972] by Reikhsfel'd *et al.* on dinuclear platinum–styrene complexes $[(XC_6H_4CH{=}CH_2)PtCl_2]_2$ (where X = H, CH_3O, CH_3, Cl, NO_2 and others) employed as initial catalysts in the hydrosilylation of styrene has led to the conclusion that the dimeric catalyst may be cleaved by either an olefin or a silicon hydride. In fact, the catalytic cycle involves reactions which occur via monomeric and dimeric complexes, as in the following equilibrium:

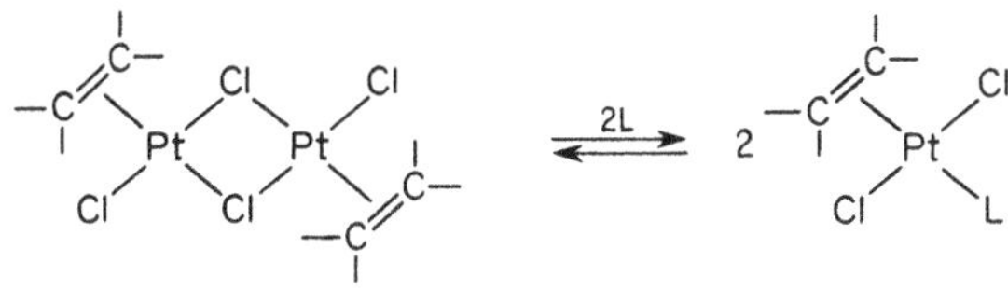

(108)

At ambient temperature (above 20°C), the equilibrium is shifted to the right, producing the less active monomeric species; thus *cis*-$PtCl_2(PPh_3)_2$ was found to be more efficient than the *trans* isomer in the reaction under study [976]. Kinetic measurements have revealed that the reaction is first-order with respect to catalyst, but zero-order with respect to reagents [969, 970]. This could be constitued as providing further evidence in confirmation of the Π–σ rearrangement suggested in Chalk and Harrod's mechanism (Scheme 81.3) being the rate-determining step in the reaction. The rate and induction period of the reaction is influenced not only by temperature but also by the kind of solvent employed, the concentration ratio of the reagents, and the order in which these reagents are introduced into the reaction mixture [68, 968, 1232].

As far as the nature of the ligands L in the complexes $PtCl_2L_2$ and $Pt_2Cl_4L_2$ is concerned, Π-acceptor ligands capable of forming σ and Π bonds with metal orbitals may be used with such compounds as ethylene [1282, D-006, D-026, D-065], propene [F-003, U-005], styrene and its derivatives [D-006, D-008], cyclohexenes [A-002, D-001, D-056, F-018, N-001], nitriles [10, 220, 221, 1105, D-045], thioethers [1226, D-026, D-030, D-065, D-066, F-011, G-025, G-033, U-086], tertiary phosphines R_3P [171, 1298, 1299, 1301, U-013, U-035, U-090, U-116], tertiary amines [D-093, D-094], 1-vinylazols [612], vinylsilanes, vinylsiloxanes, polyvinylsiloxanes [D-026, D-086, D-102, D-105, G-028, U-028, U-064, U-103], and other functional siloxanes [D-064], alkynes and their derivatives [36], unsaturated ethers [97], ketones [1361] and diketones [J-011, U-055, U-111]. Complexes with miscellaneous ligands, L, have also been used, e.g. (amine)-$Pt(alkene)Cl_2$ [D-006, D-105], $(amine)_2Pt(alkene)Cl_2$ [D-093, D-094] and (phosphine)$Pt(alkene)Cl_2$ [1300, 1301]. The precursor *cis*-$PtCl_2(styrene)_2$ complex has recently been successfully employed in the hydrosilylation of various terminal unfunctional olefins and acetylene. In the hydrosilylation of styrene the procursor is reduced to $Pt(PhCH=CH_2)_2$ with concomitant formation of 2 equivalents of R_3SiCl and 1 equivalent of $PhCH_2CH_3$ [10, 220, 221].

Diketones and other chelating ligands form active and relatively stable platinum complexes of the general formula $[RPtCl_2]_n$ where R is an unsaturated compound, e.g. cyclooctadienyl, norbornodienyl [J-34], mesityl oxide [D-023], butanone, forone [D-023] or acac [G-011]. $Pt(acac)_2$, which catalyses the addition of silicon hydride to the C=C bond of several unsaturated compounds [D-011, U-055], has also been used as crosslinking catalyst [U-055]. Platinum(II) complexes of the general formula $[L^*PtCl_2]_2$ or cis-$C_2H_4L^*PtCl_2$ have been found to be effective catalysts for the asymmetric hydrosilylation of prochiral olefins [1301, 1303] and ketones [298, 490, 584, 1299]. In these cases the following chiral phosphine as well as phosphine–amine ligands were used: (R)-benzylmethyl-phenylphosphine (BMPP), (R)-menthylphenyl-n-propylphosphine (MPPP) amino-phosphinoferrocene [298], methyldiphenylphosphine (MDPP) [490, 1301, 1303] and (S)-amphos [584]. Asymmetric hydrosilylation by methyldichlorosilane has been successful only for 1,1-disubstituted olefins, (i.e. α-methylstyrene), 2,3-dimethyl-1-butene and 2-methyl-1-butene. The same complex $[L^*PtCl_2]_2$ also catalyses the intromolecular hydrosilylation of 4-pentenyldimethylsilane to yield an optically active silacyclopentane in addition to silacyclohexane [1303].

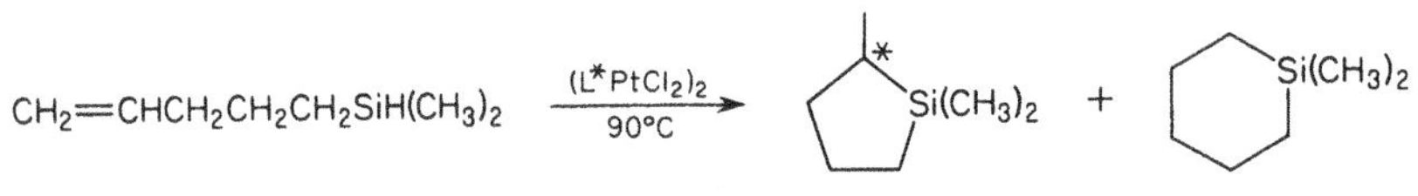

$$CH_2=CHCH_2CH_2CH_2SiH(CH_3)_2 \xrightarrow[90°C]{(L^*PtCl_2)_2} \overset{*}{\bigcirc}\!\!\!Si(CH_3)_2 \quad + \quad \bigcirc\!\!\!Si(CH_3)_2 \quad (109)$$

The optical yields of products obtained in the hydrosilylation of olefins [Rev.7] and of alkyl(phenyl)ketones [298, 490, 584] in the presence of platinum(II) catalyst are relatively low in comparison with the yields from reactions catalysed by rhodium complexes.

A new class of the hydride-bridged platinum complexes involving silyl or germyl ligands of the following structure has recently been discovered:

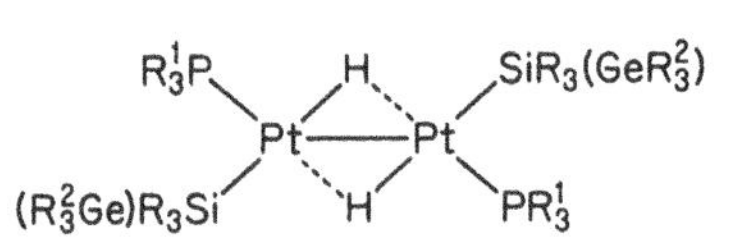

$$\begin{array}{c} R_3^1P \diagdown \quad \overset{H}{\cdots} \quad \diagup SiR_3(GeR_3^2) \\ Pt\text{——}Pt \\ (R_3^2Ge)R_3Si \diagup \quad {}^H \quad \diagdown PR_3^1 \end{array} \qquad (110)$$

$$R_3=(C_2H_5)_3, (c-C_6H_{11})_3, [C_6H_5CH_2(CH_3)_2]; R^1, R^2 = CH_3$$

Such complexes have been used by Stone *et al.* for catalysing addition reactions to olefins [429], acetylenes [430, 1205, 1206], α,β-unsaturated aldehydes and ketones [79].

Detailed dynamic examinations by Tsipis [1205, 1206] have thrown further light on the catalytic mechanism of the hydrosilylation of terminal and internal alkynes $RC≡CH$, $RC≡CR$ and $RC≡CR'$ by trisubstituted silanes. This reaction, when catalysed by the above Pt catalyst (Scheme 110), proceeds stereospecifically via *cis*-anti Markovnikov addition of the silicon hydride to give *trans*-$RCH=CHSiX_3$ during the hydrosilylation of terminal alkynes (minor amounts of internal adducts are also formed), and to give *cis*-RC(SiX_3)=CHR during the hydrosilylation of symmetrically substituted internal alkynes. Both reactions occur under very mild conditions (20–65°C). Although the concentrations of silane and alkyne do not control the regioselectivity of the addition, they markedly influence the rate. These facts may be explained in terms of the following equilibrium which has been suggested as being involved in the catalytic cycle proposed for the hydrosilylation of alkynes [430, 1205] (Scheme 111).

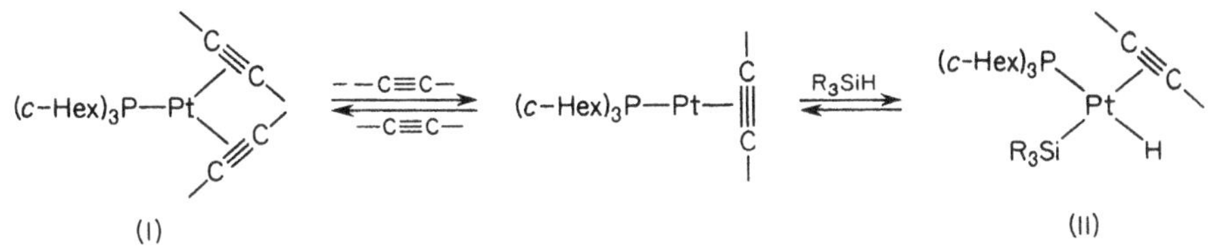

(111)

In this scheme it is quite clear that an increase in the alkyne concentration will increase the concentration of (I), and thus lower the concentration of (II) as well as the reaction rate. However, any increase in the silane concentration will increase that of complex (II) which is the catalytically active intermediate, and hence will accelerate the hydrosilylation process. Experimental data confirm these considerations. To explain the regioselectivity, Tsipis has applied a qualitative molecular orbital treatment to the transition state formed during the insertion reaction which corresponds to *cis* addition of the hydrosilane suggests that a four-centred six-electron (4c–6e) transition state is presumably formed.

The electron density distribution in the frontier orbitals of the catalyst depends on the nature of the ligands attached to the central atom. In the presence of the strong basic tricyclohexylphosphine ligand, an increase in the electron density distribution of the frontier orbitals, brought about by the silane, can be readily accounted for. The electron density distribution in the frontier orbitals of the alkyne (Π-HOMO and Π*-LUMO) depends on the substituents attached to the sp carbon atoms. The difference between the ^{13}C chemical shifts for the sp carbon atoms providing a quantitative measure of the differences between the relative magnitudes of the p coefficients of the Π-HOMO and Π*-LUMO. The largest differences (15–20 ppm) are observed for mono-substituted alkynes, RC≡CH; for asymmetrically di-substituted acetylenes, RC≡CR, the differences are 3–5 ppm; while for symmetrically disubstituted acetylenes, RC≡CR, there is no difference in the ^{13}C chemical shifts.

Scheme 112 illustrates the type of interaction which occurs between the frontier orbital of the catalyst, but the Π*-LUMO of the alkyne interacts with the bonding of (a) terminal alkynes and (b) symmetrically di-substituted acetylenes. The frontier orbital Π-HOMO of the alkyne interacts with an unoccupied d^2–p^2 hybrid orbital of the catalyst, but the Π-*LUMO of the alkyne interacts with the bonding σ-MO ($d^2p_\sigma^2 - s_\sigma$) of the catalyst. For simplicity, only two of four frontier orbitals Π-HOMO and Π*-LUMO are shown in Scheme (112).

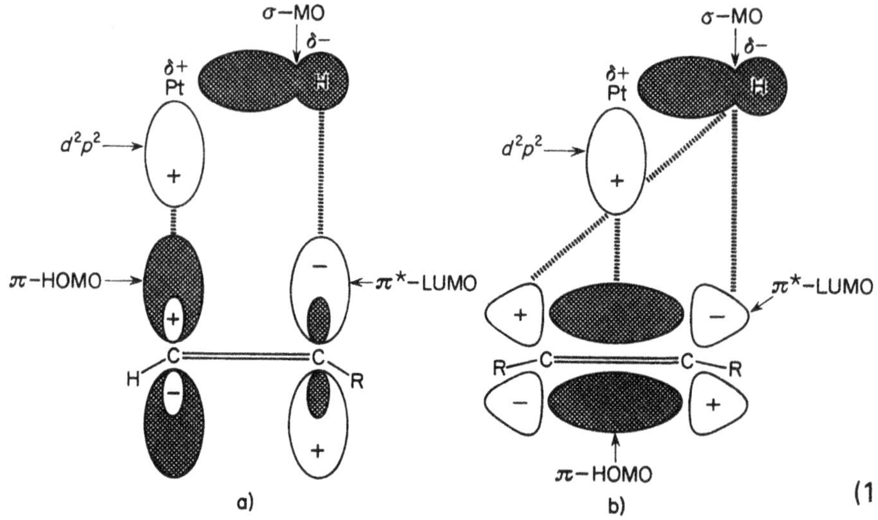

(112)

According to this mechanism, for terminal alkynes (case (a)) nucleophilic attack by the hydride can take place on the C (2) atom since the largest p coefficient of the Π*-LUMO is at C (1)). For di-substituted alkynes (case (b)), the differences between the relative values of the sp coefficients in the Π-HOMO and the Π*-LUMO are smaller. In the extreme case of symmetrical RC≡CR alkynes, no differences are observed between these values, and hence migration of the hydride to either the C (1) or C (2) atoms is equally probable. It should be noted that although the above model theoretically accounts for the experimental facts, no steric effects are taken into consideration [1205].

In fact, the catalysis of hydrosilylation by $[Pt(\mu-H)SiR_3(PR_3)]_2$ occurs via Pt(O)→Pt(II) oxidative addition as shown in Scheme 110. Several types of Pt(O) complexes, in particular phosphine complexes such as $Pt(PPh_3)_4$ [50, 96, 977, 1176, 1290, 1349] and $Pt(O_2)(PPh_3)_2$ [1349], are commonly used as hydrosilylation catalysts. All these phosphine–platinum complex catalysts can be promoted by dioxygen [1349]. $Pt(PPh_3)_2(CH_2{=}CH_2)$ appeared to be a versatile catalyst in hydrosilylation of α-alkenes (with 5–22 C atoms) as well as functionalized alkenes and conjugated dienes [1300, 748].

In the presence of platinum phosphine complexes, hydrosilylation is accompanied by several side-reactions such as H/Cl exchange and hydrosilane dehydrogenation. The latter proceeds according to the following equation [Rev.11, 1174]:

$$Pt(PPh_3)_4 + CH_3Cl_2SiH \xrightarrow[-PPh_3]{} CH_3Cl_2SiPt(H)(PPh_3)_2 \xrightarrow{CH_3Cl_2SiH} (CH_3Cl_2Si)_2Pt(PPh_3)_2 + H_2 \qquad (113)$$

Catalyst oxygenation also proceeds very readily. Current investigations on catalysis of hydrosilylation often involve initial platinum(O) complexes with olefin and acetylene ligands without phosphines, e.g. Pt–vinylsiloxane [U-028, E-027],

$Pt[C_6H_5C{\equiv}CCMe(C_6H_5)OH]_2$ [E-014, E-024]. A very attractive catalyst appeared to be the Karstedt catalyst $Pt_2\{[(CH_2{=}CH)Me_2Si]_2O\}_3$ [U-064] patented in 1973, but widely used only recently. Its crystal structure was determined [230a] but according to Lewis's considerations these types of Pt(O) catalysts act as colloid platinum formed *in situ* in the presence of hydrosilanes.

A recent study of the thermal decomposition of an iodotrimethylplatinum(IV) tetramer in the presence of substrates commonly present in hydrosilylation processes has allowed the following logical scheme for catalyst generation to be drawn up:

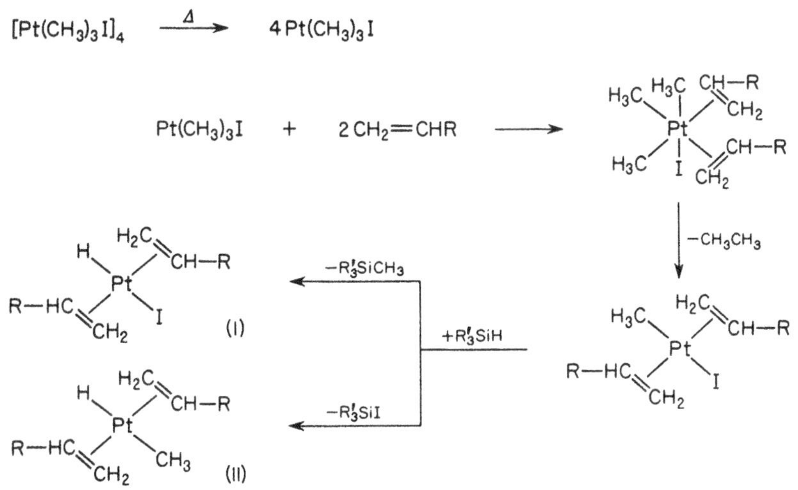

(114)

Both products are postulated as initial complexes in the catalytic cycle involved in hydrosilylation [937].

In summary, $d^8Pt(II)$ and $d^{10}Pt(O)$ complexes have been widely studied for more than 30 years as catalysts for hydrosilylation. However, the composition, structure and oxidation state of the metal in the active catalytic species have only recently become subjects of extensive experiments and considerations. Catalytic cycles containing active intermediates based on Chalk and Harrod's general scheme are still fashionable.

2.4.4 Rhodium Complexes

Rhodium d^8 complexes have been well known as highly effective catalysts for the hydrosilylation of C=C, and C=O bonds in particular, for the past 25 years. Rhodium phosphine complexes are usually active and effective in the asymmetric hydrosilylation of olefins, ketones and aldehydes, allowing the virtual synthesis of optically active alkoxysilanes and organic compounds in high purity.

The high reactivity of *cis*-2-heptene (60%) in comparison to *trans*-2-heptene (5%) in the hydrosilylation catalysed by $(Ph_2PH)_3RhCl$, and other phosphine complexes [1166], illustrates the greater stereospecificity of *cis* addition. However, as already mentioned, studies of the addition of hydrosilanes to alkylacetylenes in the presence of Wilkinson's catalyst have revealed a different type of stereospecificity [870, 947, 1291]. In contrast to chloroplatinic acid, $RhCl(PPh_3)_3$ catalyses the *trans* addition of triethylsilane to the alkylacetylene (the *cis–trans* product ratio being 1:4) [870]. Ojima *et al.* have also shown that in the presence of this catalyst the hydrosilylation of alkylacetylenes by triorganosilanes leads predominantly to *cis* products [870]. Watanabe *et al.* have also observed the same relationships during the addition of a phenyldimethylsilane to phenylacetylene in the presence of $RhCl(CO)(PPh_3)_2$ as the catalyst [1291]. Originally it was suggested that the rhodium-catalysed reaction involves a stereoselective *trans* addition [870]. A later explanation of this "unusual" stereospecificity of rhodium-catalysed hydrosilylation was based on the assumption that a close similarity exists between such catalysed hydrosilylations and the homolytic addition of silicon hydrides initiated by peroxides or photochemical methods which also result in *trans* addition. As discussed previously, Haszeldine *et al.* have shown that during the hydrosilylation of alkylacetylene the predominant yield of *cis* adducts arises from *cis* addition of hydrosilane, followed by rhodium-catalysed *trans–cis* isomerization which leads to the final formation of the *cis*-isomer [870]. However, it has also been shown that the hydrosilylation of phenylacetylene is accompanied by an extensive reverse *cis–trans* isomerization. Furthermore, other reported data indicate that in the presence of rhodium catalysts a variety of *cis–trans* adduct ratios have been observed [610, 611, 947]. The study undertaken by Brady and Nile has shed more light on this problem [160]. They used the general conception of Hart and Schwartz, who assumed that the addition of hydrides to acetylenes always occurs in a *cis* manner, followed by isomerization of the vinylmetal compound to the *trans* adduct. This mechanism as adapted to hydrosilylations catalysed by rhodium complexes may be written as follows [160]:

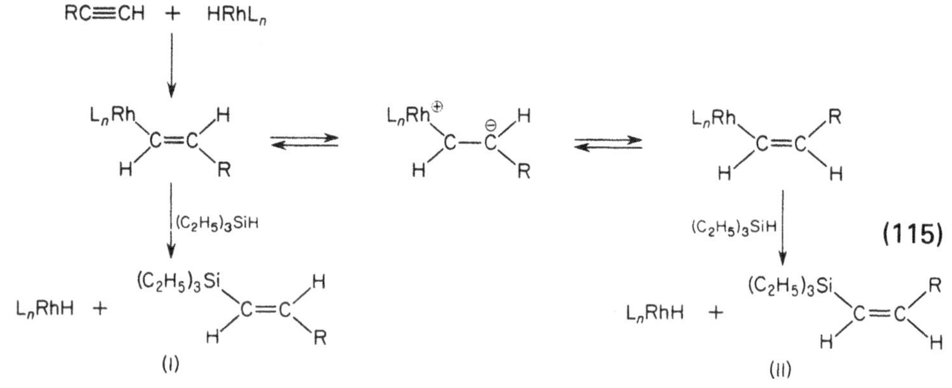

(115)

This suggests that the isomerization occurs via a dipolar intermediate which is stabilized by electron-releasing ligands on the rhodium centre. This proposition is supported experimentally by the observed reactivity order of the phosphine ligand, L:

$$P(m-C_6H_4CH_3)_3 \geqslant PPh_3 \approx P(o-C_6H_4CH_3)_3 \approx P(p-C_6H_4CH_3)_3 \gg P(OCH_3)_3 \approx P(OC_2H_5)_3 \approx P(OPh)_3 \qquad (116)$$

Hence, it is possible to generalize the observed stereochemical data as well as the supporting theory by stating that the presence of electron-releasing phosphine ligands leads to the generation of a greater amount of *cis*-1-triethylsilyl-1-pentene, while electron-withdrawing ligands cause more *trans*-1-triethylsilyl-1-pentene to be produced [160].

Two types of the rhodium phosphine catalysts are usually employed in hydrosilylation processes:

(a) $RhX(R_3P)_3$, where X is usually Cl and R is mainly Ph (Wilkinson's catalyst) or some other substituent [50, 204, 227, 232, 279–281, 339, 476, 630, 760, 868, 870, 873, 981, 983, 1166, 1180, 1207, 1290, 1321, C-004]; X = Br, I or H [981, 982, C-004].
(b) $RhX(CO)(R_3P)_2$, where X = H, R = Ph [982, 981, 1257]; X = Cl, R = Ph [232, 982, C-022]; X = Cl, $R_3 = CH_3[CH_2Si(CH_3)_3]_2$ [G-027].

In addition, dinuclear rhodium complexes containing Π-acceptor ligands not involving phosphines have been used:

(c) $Rh_2X_2Y_4$, where X = Cl, Y = C_2H_4 [336, 630, 757, 1167], X = Cl, Y = CO [S-045], X = Cl, Y = C_8H_{14} [277, 280], X = other olefins, Y = CO [278, 1257], X = Cl, Y = $P(OR)_3$ where R = methyl, *o*-tolyl [229] and mononuclear complexes with similar ligands:
(d) RhX_{2Y}, where X = acac, Y = CO [50, 1257, S-014, S-017, S-018], C_2H_4 [D-076].

Mention should also be made of the more unusual rhodium(I) catalysts, *viz.* Rh(acac)X where X = *o*-hydroquinone, *o*-hydroxyquinoline [S-017]; $Rh_2(C_4H_7N_2O_2)_4(PPh_3)_2$ [S-003]; $RhH(PPh_3)_4$ [605, 1290, J-019]; $Rh(CO)ClX_n$ where X = monoazadiene, $n = 2$ or diazadiene, $n = 1$, [166], and $RhCl(RNC)_n$ where R = alkyl or aryl and $n = 1–3$ [1].

Rhodium(III) complexes are often used as precursors for catalysts which are reduced to rhodium(I) species during the initiation process. These include $RhCl_3(PPh_3)_3$, $RhCl_3$ in C_2H_5OH, $(\Pi-C_3H_5)_2Rh(acac)$ [S-017], $RhCl_3[(C_2H_5)_2S]_3$ [D-076], $RhCl_3[RSCH_2Si(CH_3)_3]_3$ (where R = C_4H_9, C_2H_5 [D-072]), $RhCl_3$-$[(CH_3)_3SiCH_2PPh_2]_3$ [G-026] and $(RR^1C_2B_9H_9)RhH(PPh_3)_2$ [813]. All these rhodium catalysts are dissolved in non-polar solvents such as benzene or toluene. The use of more polar and, in particular, more nucleophilic solvents such as dioxane or THF usually leads to a marked decrease in the catalytic activity.

Although numerous rhodium catalysts have been used in the hydrosilylation of unsaturated hydrocarbons, e.g. 1-alkenes, styrene, or butadiene, it is impossible to select such complexes in a rational manner so that they will confirm correctly for a given process. Thus, in the addition of $(C_2H_5O)_3SiH$, $(C_2H_5)_3SiH$ and Cl_3SiH to 1-hexene, the catalytic activity decreases in the order:

$$RhX(PPh_3)_3\ [X = Cl>Br>I] > RhCl(CO)(PPh_3)_2 > RhH(CO)(PPh_3)_2 \qquad (117)$$

When the process occurs according to the anti-Markovnikov rule, in the hydrosilylation of styrene by trimethylsilane and triethoxysilane the activity order is as follows [981]:

$$RhCl(CO)(PPh_3)_2 > RhH(CO)(PPh_3)_2 \approx RhCl(PPh_3)_3 \gg RhI(PPh_3)_3 \quad (118)$$

In this case 1-silyl- and 2-silyl-1-phenylethanes are obtained as the reaction products, the predominant product again being formed according to the anti-Markovnikov rule [981]. In addition Haszeldine *et al.* have found the following activity sequence in the hydrosilylation of 1-hexene at room temperature [476]

$$Rh_2Cl_2X_4 > (R_3P)_2Rh(CO)Cl > (PPh_3)_3RhCl \quad (119)$$

with this order being reversed at 60°C.

In comparison with platinum, more research has been devoted to the chemistry of rhodium silyl hydride complexes as a consequence of the observed oxidative addition of Rh(I) to Rh(III) species. Few studies have been made of their stability and structure, and of the effects of various ligands on the catalytic activity [232, 476, 478]. Rhodium silyl hydrides are generated from mono- and dimeric complexes according to the equations:

$$Rh_2Cl_2X_4 + 2\,R_3SiH \longrightarrow 2\,R_3SiRh(H)ClX + 2\,X \quad (120)$$

$X = C_2H_4$.

$$RhCl(PPh_3)_3 + Ph_3SiH \xrightarrow[-PPh_3]{} Ph_3SiRhCl(H)(PPh_3)_2 \quad (121)$$

X-ray structural examinations of the latter adduct (Scheme 121) have indicated a trigonal bipyramid structure [232, 476]. Attempts to synthesize hexacoordinate complexes in the presence of excess ligand, e.g. CH_3PPh_2, C_2H_4 and CO, have usually led to the reductive elimination of hydrosilane [232, 476]. The presence of more electron-withdrawing substituents at silicon (Cl,OR) gives rise to rhodium silyl complexes of greater stability, e.g. $Cl_3SiRh(H)Cl(PPh_3)_2$

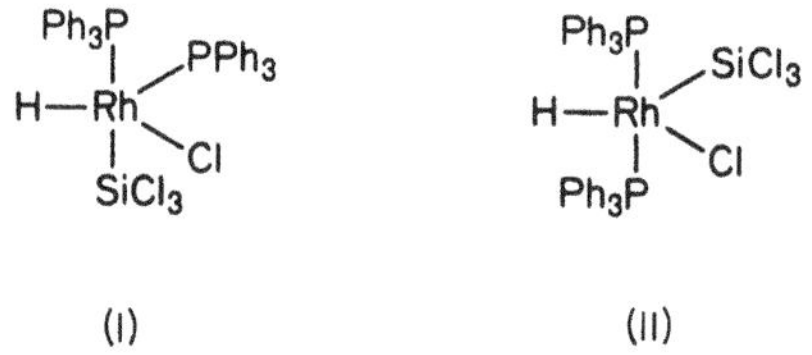

(I) (II)

(122)

However, the latter (structure (II) in Scheme 122) did not function as an active intermediate during the addition of trichlorosilane in the presence of Wilkinson's catalyst, since olefins are incapable of cleaving the Rh–H bond in this adduct [476]. The absence of activity may be associated with the inability of the catalyst to coordinate further ligands. It is suggested that this is due to the large steric requirements of the phosphine and chlorine ligands in the intermediate [Rev.17].

$RhH(PPh_3)_4$ has also been found to be an effective catalyst in numerous hydrosilylation reactions involving the C=C bond [605]. The oxidative addition of hydrosilane to this complex causes the isolation of yellow adducts of the general formula: $(R_3Si)RhH_2(PPh_3)_{4-n}$, where R_3 = $Ph(CH_3)_2$, $Ph(OCH_3)_2$, $(C_2H_5)_2H$ (n = 2) and $(CH_3O)_3$, $(C_2H_5O)_3$, (n = 3) [605].

As was already mentioned, the electronic and steric properties of ligands influence the stability of the adduct, as it can be demonstrated by the following sequences (Scheme 123) [478]

$$P(c{-}C_6H_{11})_3 < PhP(c{-}C_6H_{11})_2 < Ph_2P(c{-}C_6H_{11}) \leqslant Ph_2P(tol) \approx PPh_3;$$
$$Ph_2PCH_3 < Ph_2PC_2H_5 < Ph_2P(iso{-}C_3H_7) \approx Ph_3P;\ Ph_3As < Ph_3P \qquad (123)$$

The corresponding order of activity for rhodium phosphine catalysts containing such ligands during the hydrosilylation of 1-hexene can be expressed as follows:

$$P(c{-}C_6H_{11})_3 > P(CH_3)_2Ph > P(CH_3)Ph_2 \approx PPh_3 > P(OC_4H_9)_3 >$$
$$P(OPh)_3 \qquad (124)$$

From this order it may be concluded that the reactivity of adducts increases with an increase in the σ-donor (Π-acceptor) properties of their ligands [610, 981].

The addition of silicon hydrides to 1,3-butadiene in the presence of various rhodium(I) complexes as catalysts has been studied systematically by Rejhon and Hetflejs [982]. Rhodium catalysts such as $RhX(PPh_3)_3$ (X = Cl, Br, I), $RhY(CO)(PPh_3)_n$, n = 2, 3 (Y = Cl, H), $Rh(\Pi{-}C_3H_5)(PPh_3)_2$, $Rh(CH_3)(PPh_3)_3$ and $Rh_2Cl_2(PPh_3)_4$ do not significantly influence the type of product formed and only differ in their reactivity. The catalytic activity of $RhX(PPh_3)_3$ is greater than that of $RhX(CO)(PPh_3)_2$ during hydrosilylation by $(CH_3)_3SiH$ and Cl_3SiH, and diminishes in the sequence Cl > Br > I, i.e. the activity decreases. $RhX(CO)(PPh_3)_2$ is the most efficient catalyst for the addition of triethoxysilane to 1,3-butadiene. Regardless of the reaction conditions, i.e. temperature, catalyst concentration or molar ratios of the reagents four products were always obtained hydrosilylating butadiene with trimethylsilane, i.e. 4-trimethylsilyl-1-butene(I), 1-trimethylsilyl-*cis*- and *trans*-butenes (II, III), and unexpectedly 1,4-bis(trimethylsilyl)-2-butene (IV) (Scheme 125). As it can be seen, no adduct with an internal silyl group could be identified:

$$CH_2{=}CHCH{=}CH_2 + (CH_3)_3SiH \rightarrow \underset{(I)}{(CH_3)_3SiCH_2CH_2CH{=}CH_2} +$$
$$\underset{(II,\ III)}{(CH_3)_3SiCH_2CH{=}CHCH_3} + \underset{(IV)}{(CH_3)_3SiCH_2CH{=}CHCH_2Si(CH_3)_3} \qquad (125)$$

Product (IV) is derived from neither the hydrosilylation of the products (I)–(III) nor the dehydrogenation of bis(trimethylsilyl)-butane. Hydrosilylation of butadiene by $(C_2H_5O)_3SiH$ and Cl_3SiH leads selectively to the formation of 1-silyl-substituted *cis*-butene-2 [982]. However, a curiosity is the absence of the silyl-substituted derivatives of the octadienes amongst the products since these are generated in the analogous reactions catalysed by Ni and Pd complexes (see Scheme 168). The absence of C_8 adducts indicates that in the proposed mechanism the lifetime of the Π-allyl intermediate is too short to allow another butadiene molecule to be coordinated and hence to generate a C_8-intermediate with the following structure:

$$R_3Si\text{---}Rh$$
$$CH_3CH\text{==}CH\text{---}CH_2\text{---}CH_2$$

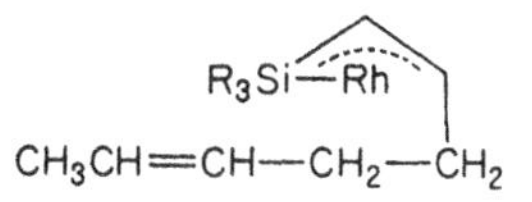

(126)

Rhodium phosphine complexes also effectively catalyse the isomerization of 1-alkenes. In addition to the hydrosilylation of 1-alkenes and dienes [852, 982], Wilkinson's catalyst and other phosphine and phosphinocarbonyl complexes of rhodium have been used to catalyse addition to styrene [630, 981, G-034], to olefins with functional groups [23, 43, 868, J-017], to the C=O bond of ketones, to unsaturated esters, and to the C≡N bond of Schiff bases and of nitriles [868, 873]. The hydrosilylation of dienes, unsaturated esters and acetylenes has recently been reviewed by Ojima and Kogure, who have probably made the greatest contribution to this field [Rev.31].

One of the most confusing problems in the catalysis of hydrosilylation by rhodium complexes is the influence of molecular oxygen as a co-catalyst. This effect was also reported earlier in rhodium-catalysed hydrogenation. Such a phenomenon is general, and also occurs in the presence of other transition metal complexes, particularly those with CO and phosphine ligands, e.g. of platinum and ruthenium. Careful removal of dioxygen from a reaction mixture containing Wilkinson's catalyst in which the P/Rh ratio was 3:1 led to the total deactivation of the catalyst. However, when the P/Rh ratio was equal to or less than 2:1 in the initial mixture, hydrosilylation occurred even under anaerobic conditions (under nitrogen or argon) [314]. The concerted mechanism of hydrosilylation processes catalysed by Wilkinson's catalyst involves predissociation of the phosphine from the complex. In this respect, molecular oxygen functions as a promoter since the dissociation of phosphine occurs more readily from $RhCl(O_2)(PPh_3)_3$, than from the original precursor. Apparently, the formation of the highly active, coordinatively unsaturated species, $RhCl(PPh_3)_2$, via preliminary oxygenation of phosphine to phosphine oxide, is the key to co-catalysis in metal phosphine complexes. As a ligand, phosphine oxide possesses very weak coordinating properties. The following scheme summarizes the process of co-catalysis by oxygen (Scheme 127) [350]:

$$M(PR_3)_{n+1} \xrightarrow{(O_2)} M(O_2)(PR_3)_n + PR_3 \xrightleftharpoons{\text{x}} M(PR_3)_n + P(O)R_3$$
$$\xrightarrow{\text{vacuum}} M(PR_3)_n + PR_3$$

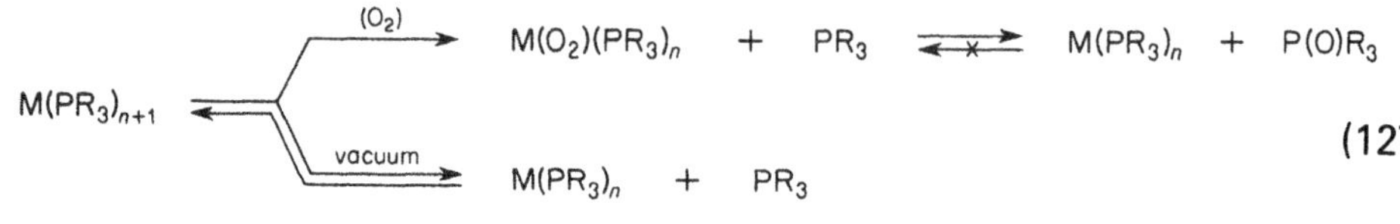

(127)

The intermediate $M(PR_3)_n$ readily undergoes oxidative addition of hydrosilane, which according to the general scheme is the initial step in the catalytic cycle for hydrosilylation [226]. This concept of co-activation by oxygen has been confirmed by the photoassisted catalysis, which is much enhanced by UV radiation, also observed during the oxygenation of phosphine to phosphine oxide [350]. Photocatalytic problems associated with hydrosilylation will be discussed later.

The detailed mechanism of the exemplified hydrosilylation of 1-hexene by triethoxysilane catalysed by RhCl(cod) (phosphine) based on kinetic studies (by visible spectrometry and GLC) of most of the individual steps of the reaction was reported and discussed [741].

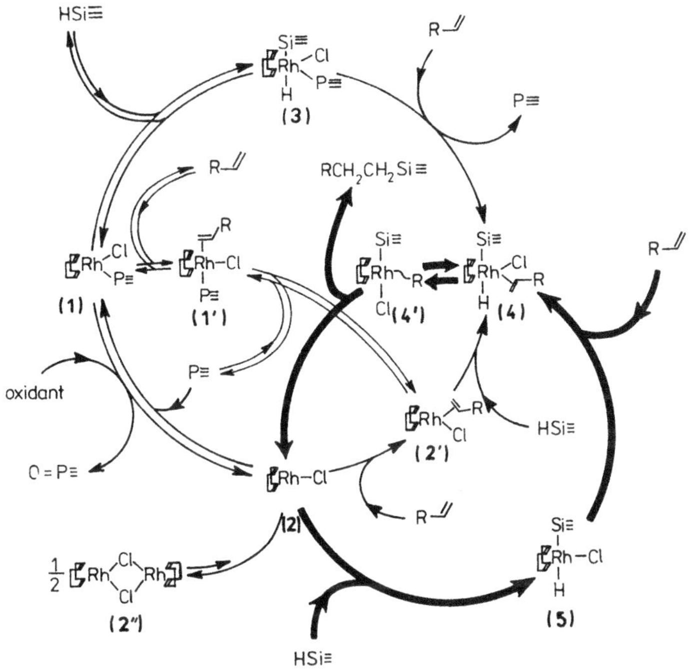

(128)

Individual rate constants have been determined from the reactions of the initial complex (1) as well as the isolated intermediates with various substrates and oxidant [331, 332, 331a]. Under oxidant-free conditions caused by careful removal of oxidants from the commercial 1-hexene the replacement of phosphine by 1-hexene ((1)→(1′)→(2′)) occurs as a two-step reversible reaction by associative mechanism. But if commercial 1-hexene is used and, when [oxidant] > [(1)] two distinct reaction steps are observed; oxidation of phosphine to phosphine oxide with generation of (2) and oxidation of rhodium dimer to Rh(III) species. The latter process is several times faster than the former one [331, 333, 331a]. It was concluded that the hydrosilylation consists of two distinct stages, the activation of the rhodium precursor by initiating the active intermediates followed by the usual catalytic cycle of hydrosilylation.

It was very interesting to discover that the proper catalytic cycle occurs at a rate several orders of magnitude greater than that of activation via intermediates without the phosphine ligand. These observations confirm the suggestion that for silica (or silicate) supported complex catalyst (see Section 2.5.4) with a ligand structure similar to that of siloxyphosphine used [disiloxyphosphines and trisiloxyphosphines) the anchored complex (via one molecule of phosphine) is released into the solution to catalyse the hydrosilylation homogeneously [741].

A catalytic overview of hydrosilylation processes must also involve novel complexes which are highly active and selective under mild conditions, as well as the mechanism by which these compounds catalyse hydrosilylation. Phosphine and phosphinocarbonyl complexes of Rh(I) and Rh(III) are the most commonly used catalysts in these contexts. Rhodium carbonyls will be discussed in Section 2.4.5. An important class of compounds appears to include all organometallic complexes with Π-bonding ligands, such as the rhodium–olefin complexes mentioned earlier which are analogous to the Zeise complexes. These include diene ligands such as octadiene, and ligands with delocalized electrons. Non-classical three-electron donors of this type have a tendency to form allyl metal compounds. The corresponding Π-allyl rhodium complexes have also been found to be effective during hydrosilylation; thus $Rh(\Pi-C_3H_5)[P(OCH_3)_3]_3$ has been studied in the addition of trialkylsilanes to 1-hexene and to the C=O bond [144].

Reactions catalysed by rhodium complexes containing one of the commonest carbocyclic ligands in transition metal chemistry, *viz.* the cyclopentadienyl ligand or its derivatives, have also been the subject of numerous investigations. Maitlis *et al.* have extended the study of the catalysis of hydrosilylation to dichloro-bis(pentamethylcyclopentadienyl)rhodium [722, 771, 772]. Hydrosilylation of 1-hexene (as well as of other 1-alkenes, e.g. styrene, 1-heptene and 1-nonene) occurs in the presence of this catalyst to yield the normal adduct, i.e. n-hexyl(trimethyl)silane, as well as two alkenylsilanes, E-hex-1-en-1-yl(triethyl)silane (major product) and E-hex-2-en-1-yl(triethyl)silane (minor product). Hexane is formed in proportion to the sum of both alkenylsilanes. The probable mechanism for this complex-catalysed hydrosilylation, involving pathways for alkylsilane formation, may be written as follows [771]:

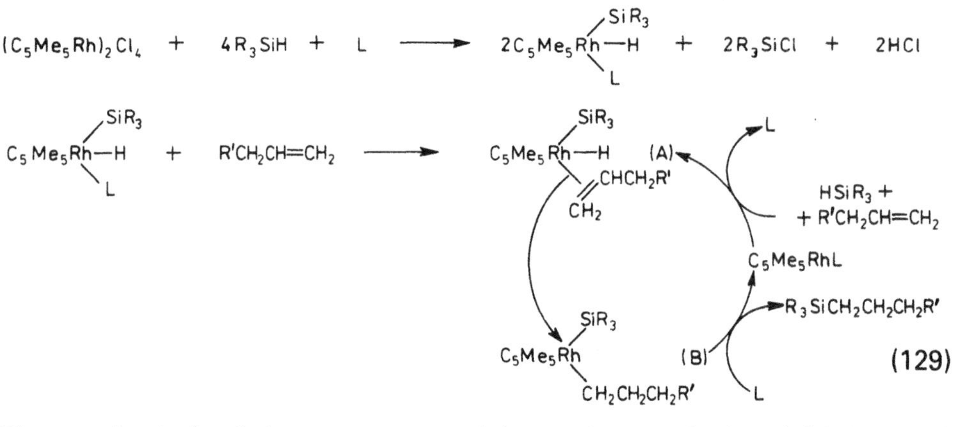

(129)

The reaction is clearly homogeneous, and since at least 90% of the initially added catalyst can be recovered it appears that the pentamethylcyclopentadienyl ligand probably remains attached to the metal throughout the various reaction cycles. On the other hand, the half-order reaction observed with respect to catalyst strongly implies that a mononuclear species is involved in the process (see Scheme

129). Since the formation of the vinylsilane is promoted by a high olefin/silane ratio and low temperature, related intermediates must be involved in the formation of alkenyl- and alkylsilanes. For this reason, in the presence of excess olefin the species (A) and (B) are suggested as being in equilibrium with the alkylsilylolefin rhodium complex (C) [771] (Scheme 130).

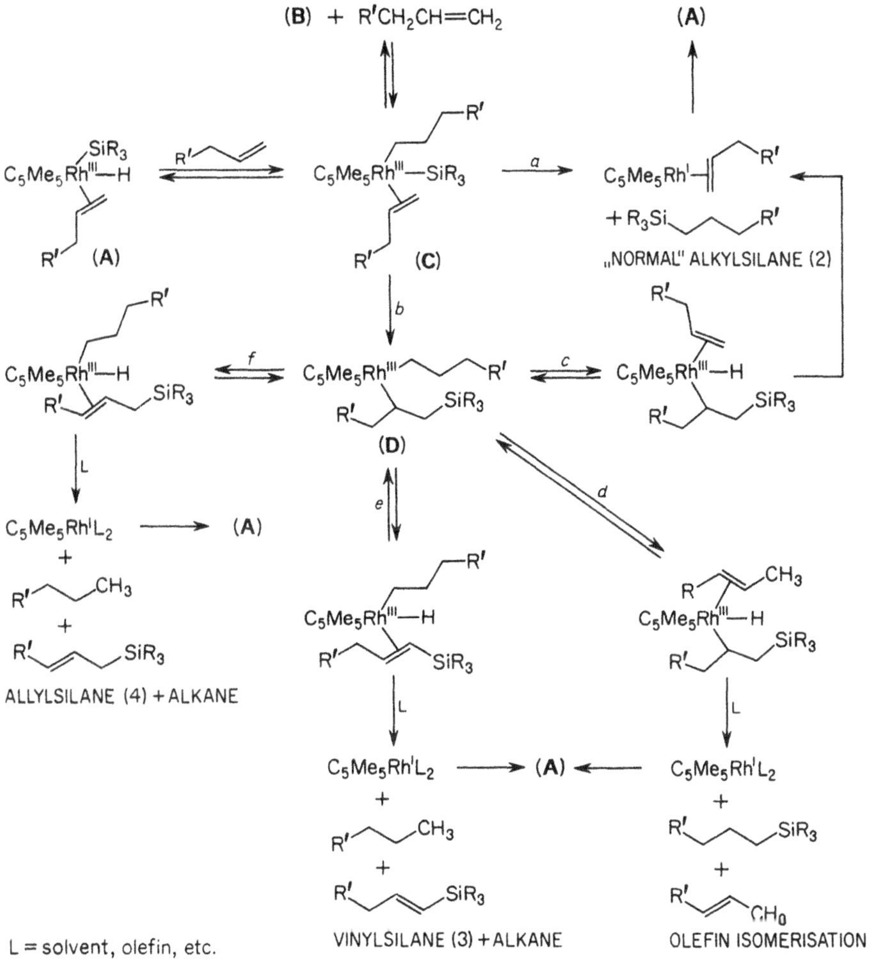

(130)

Complex (C) can react along pathway (a) to yield the alkylsilane and/or along pathway (b) to yield intermediate (D), where the coordinated olefin has been inserted into the Rh–Si bond. Intermediate (D) can subsequently react along several pathways involving β-hydrogen transfer from the alkyl ligand to the silyl–organic ligand (pathway c) to yield alkylsilane as a hydrosilylation product. Pathway (d) accounts for the formation of the internal hexene isomers, but the most interesting routes are shown by pathways (e) and (f), respectively. In both cases hydrogen migration proceeds from the silyl–organic ligand to hexene and vinylsilane (pathway e), and to hexene and alkylsilane (pathway f). However, experimentally vinylsilane is mainly formed, presumably due to preferential activation of hydrogen at the carbon atom α to the silicon in (D). The complex A (R = Et, Me, olefin = C_2H_4) was detected and characterized spectroscopically as intermediate in the thermal and photochemical reaction [986].

The above-mentioned hydrosilylation of 1-alkenes by triethoxysilane has also been studied in the presence of other catalysts not bearing $C_5(CH_3)_5$ ligands but yielding very similar products. In order to generalize this reaction model, the authors have suggested that the active intermediate for vinylsilane formation should require firmly bound ligands and a reasonably high electron density at the metal atom [771].

The catalytic activity of various rhodium-carbene complexes of general formula $L_nRh(:\overline{CN(R)CH_2CH_2N}R)_n$ has been noted in the hydrosilylation of olefins, acetylenes, dienes and ketones [541]. Hill and Nile have described the use of monocarbenerhodium(I) complexes [496], and their work has been supported by Lappert and Maskell who have also dealt with oligocarbenemetal complexes of rhodium(I) and rhodium(II) [660]. These two reports could well mark the beginning of the application of carbene–transition metal complexes as catalysts in hydrosilylation. The activity of the rhodium–carbene complexes is comparable to that of other rhodium(I) complexes in the hydrosilylation of unsaturated hydrocarbons and ketones. The product yield has been shown to be sensitive to the reaction conditions, the ligands involved and various kinds of silane or organic substrate [496].

The addition of triethylsilane to phenylacetylene or diphenylacetylene, catalysed by *cis*-$RhCl(cod)L^{Me}$ or *trans*-$RhCl(PPh3_3)_2L^{Me}$ ($L^{Me}=:\overline{CN(Me)(CH_2)_2N}Me$) proceeds stereoselectively via a *trans* addition.

Since the catalytic activity of the rhodium–carbene complexes is enhanced in air or by UV radiation, these authors have suggested that the hydrosilylation process proceeds according to a radical mechanism. In addition, the stereochemistry of the product from the hydrosilylation of stereochemistry of the product from the hydrosilylation of diphenylacetylene (a *trans*-silyl olefin) also requires a mechanism involving a radical intermediate [660].

During the past 15 years considerable interest has been shown in the asymmetric hydrosilylation of the C=C bond and predominantly of carbonyl-containing compounds particularly in respect of the preparation of both asymmetric alcohols and bifunctional asymmetric alkoxysilanes. Chiral rhodium phosphine catalysts predominate in the hydrosilylation of prochiral ketones. This subject has recently been comprehensively reviewed by numerous authors who have made major contributions to this field [Rev.7, Rev.19, Rev.34, Rev.35, 856] and it will be discussed only briefly here. The mechanism for the hydrosilylation of carbonyl compounds is also based on Chalk and Harrod's concept [Rev. 19] (See Scheme 131).

Accordingly, a suggested mechanism involving the introduction of asymmetry is shown in Scheme (131). This mechanism contains the following steps: (1) oxidative addition of the hydrosilane to the rhodium(I) complex; (2) coordination of the ketone with replacement of a solvent molecule (S) (such a sequence may be observed for monophosphine, but for diphosphines, as described, the order of these two transformations may be reversed); (3) insertion of the carbonyl substrate into the rhodium–silicon bond to produce the diastereomeric α-

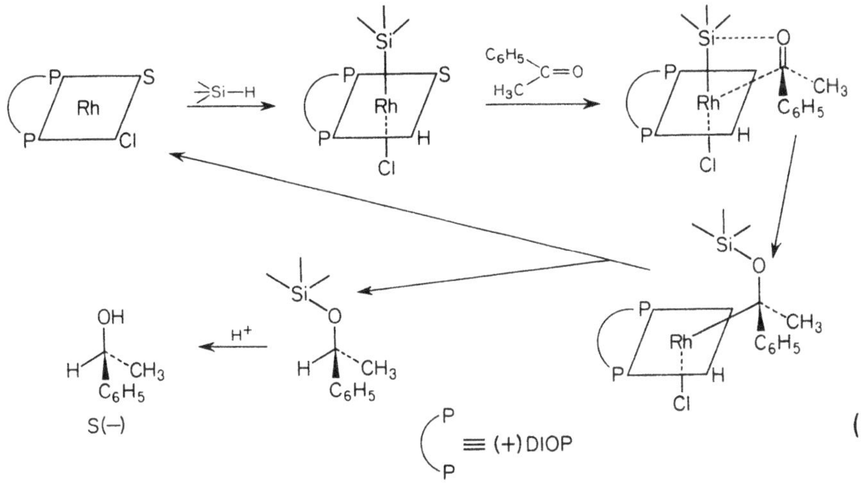

(131)

silyloxyalkylrhodium intermediate (I); (4) reductive elimination to yield an alkoxysilane as the primary product; (5) hydrolysis of the alkoxysilane affording the optically active alcohol. It is postulated that the configuration of intermediate (I) depends on the relative size of the ketone substituents, on the monodentate chiral phosphine and on the silyloxy groups; this is illustrated in Scheme (132) [865]. When the silyloxy group is bulkier than L or S, it should occupy the quasi-apical position which is the least hindered site.

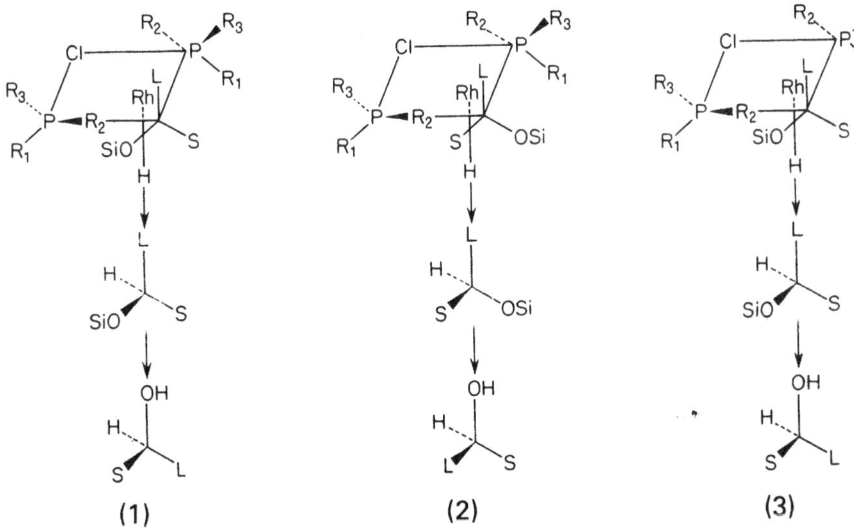

(1) (2) (3) (132)

The conformation which satisfies these requirements is denoted by C_1 in Scheme (132). When L > ≡SiO > S the most probable conformation is C_2, and when L > S > ≡SiO the most stable conformation is C_3. The alcohols derived from the alkoxysilanes C_1 and C_3 have the same configuration, whereas the alcohol derived from C_2 has the reverse configuration. These mechanistic considerations leading to the production of the preferred configuration have been assessed for alkylphenyl and dialkyl ketones [865].

Dihydrosilanes are mainly employed in the asymmetric reduction of ketones, giving rise to relatively high values for the enantiomeric excess more than 80%, although the optical yield still depends strongly on the nature of the dihydrosilane. Hydrosilylation of prochiral ketones by prochirally disubstituted silanes leads to asymmetry on the silicon atom as well as on the carbon atom, but with different optical yields. This may be depicted as follows:

$$R^1R^2SiH_2 + R^3R^4C{=}O \xrightarrow{Rh^*} R^1R^2Si^*(H)O\overset{*}{C}HR^3R^4 \qquad (133)$$

The reaction occurs in the presence of rhodium complexes with chiral phosphine or amine ligands. As a standard method the addition of diphenylsilane to acetophenone has been investigated. To obtain detailed insight into the origin of different enantioselectivity a large variety of chiral ligands and metals (mostly rhodium) have been studied. In most cases the direct dependence of the product configuration on the ligand configuration has been shown. The usual method used in asymmetric synthesis is the systematic variation of the catalytic system so as to optimize the activity and enantioselectivity. Thus the structure of the chiral ligands is usually modified. Recent work on the asymmetric catalysis of hydrosilylation has been aimed at obtaining novel efficient chiral ligands which are usually bidentate, (P,P– N,N–, and P,N–) with the centre of chirality as far removed as possible from the metal centre. Of the large number of optically active chelating ligands used in asymmetric catalysis, the following are the most common: 2,3-isopropylidenedioxy-1,4-bis(diphenylphosphino)butane [Rev. 19, 178, 280–282, 421, 537, 598, 845, 902] (DIOP) and its analogues derived from tartaric acid [105]; chiraphos [Rev. 7], chiral ferrocenylphosphines [488] (PPFA, MPFA, BPPFA), norphos [Rev. 7, 178], amphos [548], BPPM [Rev. 7, Rev. 31].

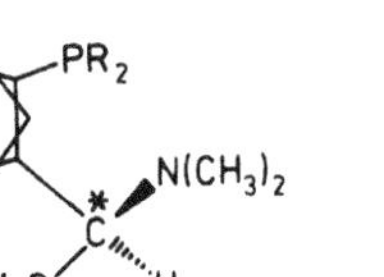
DIOP

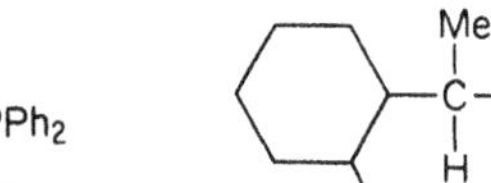
Chiraphos

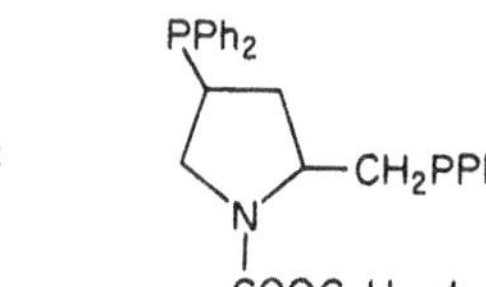
BPPFA (R, R′ = Ph)

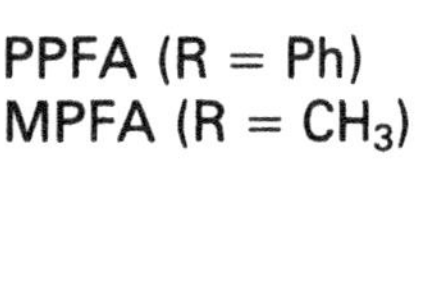
PPFA (R = Ph)
MPFA (R = CH₃)

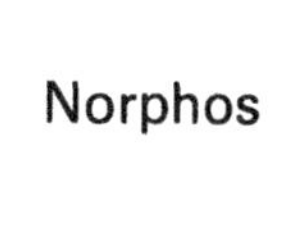
Norphos

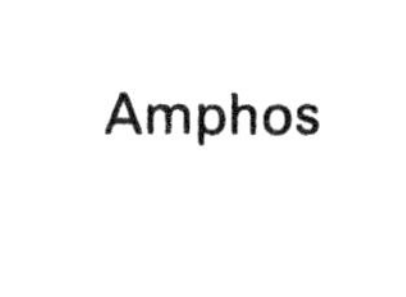
Amphos

BPPM

Chiral monophosphines used include menthyldiphenylphosphine (MDPP) [281, 282], neomenthyldiphenylphosphine (NMDPP) [281, 282] and benzylmethylphenylphosphine (BMPP) [486, 861, 865). In contrast to the situation in asymmetric hydrogenation, some of them, e.g. BMPP, are as effective as chiral diphosphines in asymmetric hydrosilylation.

The following diphosphines represent the major innovations obtained to date in the field of asymmetric hydrosilylation [527, 891]:

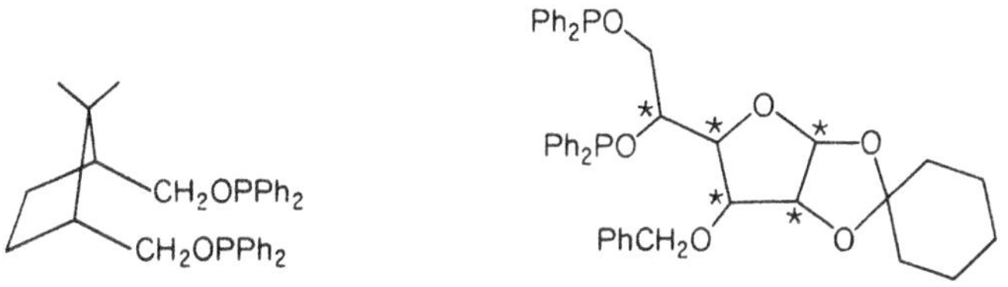

Camphinite Glucophinite (134)

Arguments have been presented in a recent review in favour of improving chirality transmission from optically active two-nitrogen ligands within a given catalyst. In particular, it is pointed out that the pyridine imine skeleton in ligands of the type

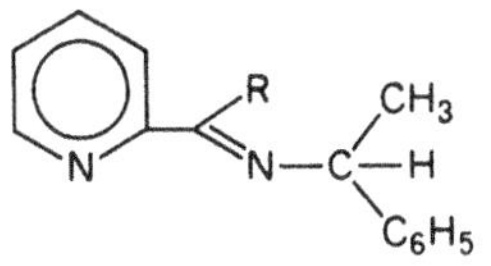

leads to a high optical purity in asymmetric hydrosilylation [Rev.7, 182]. In view of this it is surprising to find that the thiazolidine ligand gives the highest optical yield (97.6%) in the hydrosilylation of acetophenone with diphenylsilane [185, 179].

Very recently Brunner *et al.* used a lot of *N,N* chelate ligands with lateral chiral centres derived from optically active primary amine, aminoacids and aminoacid derivatives and other nitrogen-containing compounds, e.g. pyridinethioazolidines [179, 143, 177, 188], pyridineimidazolines [143, 177, 188], pyridineoxazolines [181, 829], bis(oxazalicyl)pyridynes [829], Schiff bases [143, 582] and bipyridines [143]. Aminophosphines, e.g. amphos, gave very encouraging optical yields (up to 72%) and thereore have also become the ligand under study [177, 182, 584, 762, 891, 1227]. Aminophosphites as ligands achieved optical activity up to 43% 551]. The first successful use of a tertiary arsine-containing complex for catalytic asymmetric hydrosilylation of $C_6H_5COCH_3$ and $CH_3COC(CH_3)_3$ has recently been reported (18–41% e.e.) [789].

The asymmetric hydrosilylation of prochiral ketones has also been carried out in the presence of chiral cationic rhodium complexes of the type $[RhL_2H_2S_2]^+$, where L is a relatively basic chiral phosphine and S is the solvent or another weakly bonded molecule [486]; L_2 may be a chelate diamine and S_2 a diene [188].

Most rhodium(I) complexes have been prepared *in situ* from RhCl(olefin) species, where the olefin is usually C_2H_4 [105] or dienes such as cod, 2,5-norbornadiene, hexadiene [865], and a chiral phosphorus or nitrogen ligand. Brunner commenced application of some transition metal complexes containing uncoordinated donor groups as co-catalysts in the enantioselective Rh-catalysed hydrosilylation of acetophenone reaching α-phenylethanol with an optical yield up to 25.9% e.e. [175].

2.4.5 Metal Carbonyls

Transition metal carbonyls, particularly octacarbonyl dicobalt(O) and $CoH(CO)_4$ [712, 721, D-032, U-019], but also carbonyls of nickel [215, 1217, D-032], rhodium [30, 31, 713, S-023, S-027], iron [383, 827], ruthenium [827, 858, F-016], manganese [U-019, 495, 635], iridium [30, 635, S-024, S-028], chromium [1, 718, 1293], and molybdenum [1, 564], have been found to be active catalysts for the hydrosilylation of olefins [718, 1169], dienes [272, 712, 721], unsaturated nitriles [227], esters and vinylsilanes [714], as well as the hydrosilylation of C=O and C≡N bonds.

Octacarbonyldicobalt(O) is a commonly used catalyst, usually at concentration about 1 mole% relative to the reagents, and is active under very mild conditions. Cobalt carbonyls function in a slightly different manner from platinum and rhodium complexes. The detailed mechanism of hydrosilylation catalysis by these complexes can be understood by considering their role in hydroformylation [Rev.17]. Thus, a general scheme for the suggested mechanism may be written as follows [Rev.6, Rev.17, 30, 31, 1169]:

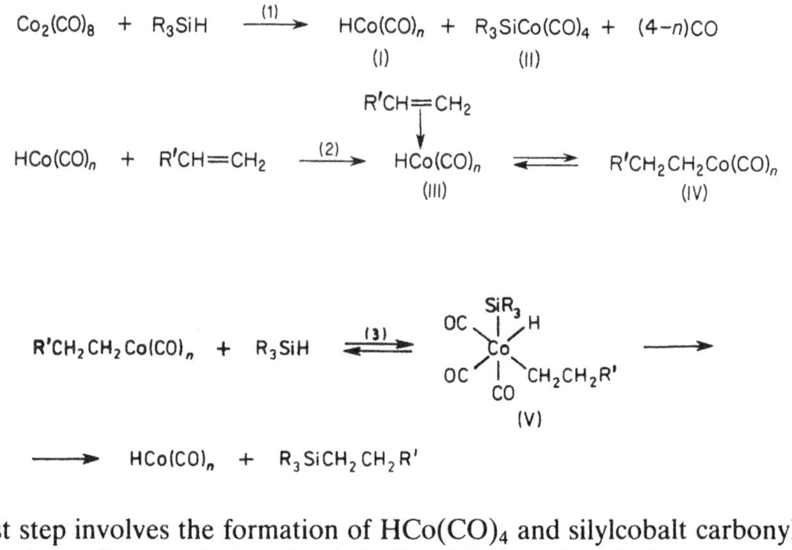

$$Co_2(CO)_8 + R_3SiH \xrightarrow{(1)} \underset{(I)}{HCo(CO)_n} + \underset{(II)}{R_3SiCo(CO)_4} + (4-n)CO$$

$$\underset{(III)}{HCo(CO)_n} + R'CH{=}CH_2 \xrightarrow{(2)} \overset{\displaystyle R'CH{=}CH_2}{\underset{(III)}{HCo(CO)_n}} \rightleftharpoons \underset{(IV)}{R'CH_2CH_2Co(CO)_n}$$

$$R'CH_2CH_2Co(CO)_n + R_3SiH \underset{(3)}{\rightleftharpoons} \underset{(V)}{\overset{SiR_3}{\underset{\underset{CO}{|}}{\overset{OC}{\underset{OC}{>}}Co\overset{H}{\underset{CH_2CH_2R'}{<}}}}} \longrightarrow$$

$$\longrightarrow HCo(CO)_n + R_3SiCH_2CH_2R' \tag{135}$$

The first step involves the formation of $HCo(CO)_4$ and silylcobalt carbonyl. The latter has been shown to be inactive in hydrosilylation, since it does not react with olefins [225]. In contrast, and by analogy to hydroformylation, the hydrocarbonyl complex of cobalt readily reacts with olefins to give an alkylcobalt carbonyl. This reaction seems to be the rate-determining step for the overall hydrosilylation process. The proposition that this path be reversible is supported by a concurrent olefin isomerization leading to the formation of 2- and 3-alkenes far more readily than with platinum or rhodium catalysts [Rev.38]. Compound (IV) is generally believed to be an effective intermediate when the reaction is carried out in an atmosphere of CO. Under these circumstances *n* is presumably 4, but in the absence of carbon monoxide *n* is generally believed to be equal to 3. The reaction of silylcobalt carbonyl with alkylcobalt carbonyl complexes, which has been proposed as an alternative path for the formation of hydrosilylation products, may be rejected since none of these products have been isolated in the reaction of $R_3SiCo(CO)_n$ with $HCo(CO)_n$ and olefins [Rev.17].

$$R'CH_2CH_2Co(CO)_n + R_3SiCo(CO)_4 \rightarrow R'CH_2CH_2SiR_3 + Co_2(CO)_8 \qquad (136)$$

The formation of inactive $R_3SiCo(CO)_4$ occurs when excess hydrosilane is present relative to the unsaturated compound. Under these circumstances the following side-reaction can occur:

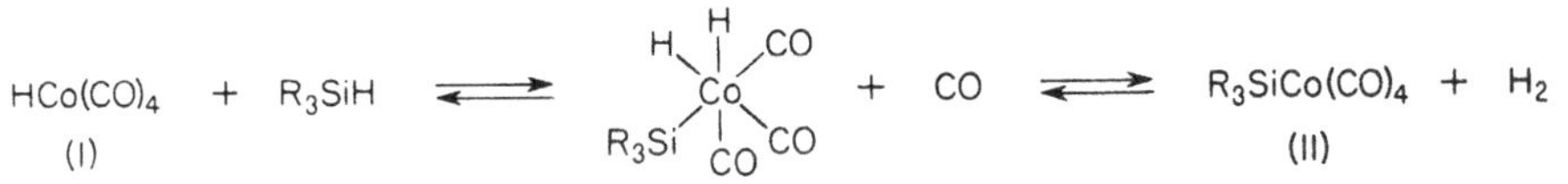

(137)

For this reason the sequence in which the reagents are added plays an important role in catalysis by metal carbonyls. Addition of the olefin markedly increases the effectiveness of the hydrosilylation process and hence prevents the above side-reaction.

The effect of numerous (σ-donor, Π-donor, Π-acceptor) co-catalysts, introduced as a ligand L, on the catalytic activity of $Co_2(CO)_8$ during the hydrosilylation of vinyltrimethylsilane (and also 1-hexene) by triethoxysilane is reflected in the following experimental order [714]:

$$(C_2H_5)_2O > I^- > Br^- > SbPh_3 > PhC{\equiv}CPh > THF > CO \approx C_6H_6 \approx PhCH{=}CHPh > AsPh_3 > C_5H_5N > (iso{-}C_3H_7)NH_2 > PPh_3 > (iso{-}C_4H_9O)_2HPO > SnCl_2 \approx (CH_3)_2SO > CS(NH_2)_2 \qquad (138)$$

The rate of the reaction may be assumed to increase with increasing stability of the intermediate (2) and/or of the transition state (2′) (equation 139).

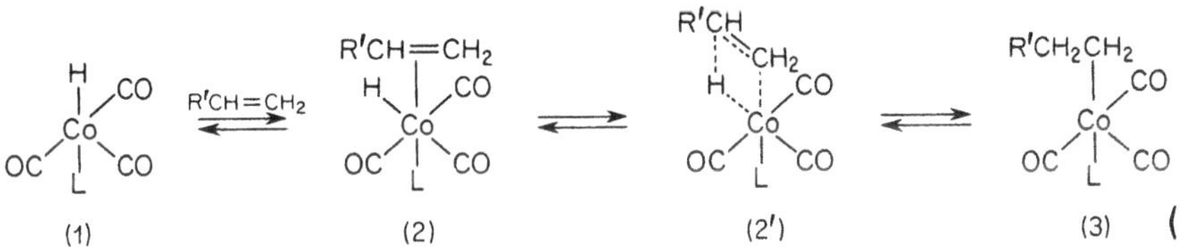

(139)

As a rule the activity decreases in accordance with a decrease in the Π-acceptor properties of the ligand. This applies to ligands having medium Π-acceptor properties, but not to THF, $(C_2H_5)_2O$, olefins or CO. Strong Π-acceptors, such as olefins and CO, are less active co-catalysts. Equally, the catalytic activity is also reduced when the σ-donor properties increase, e.g. strong σ-donors such as amines inhibit the reaction. In summary, ligands with slight σ-donor properties but virtually no acceptor abilities, e.g. $(C_2H_5)_2O$ and THF (when used as a solvent), and ligands with medium acceptor properties, such as I^-, Br^-, $SbPh_3$, appear to be strong promotors of cobalt carbonyl catalysis [714, 720]. As is well known $Co_2(CO)_8$ disproportionates in the presence of polar solvents, especially cyclic ethers such as THF [582], according to the equation:

$$3\,Co_2(CO)_8 \xrightarrow{THF} 2\,Co(THF)_6[Co(CO)_4]_2 + 8\,CO \quad (140)$$

In the presence of silanes, THF polymerizes according to the following mechanism [272]:

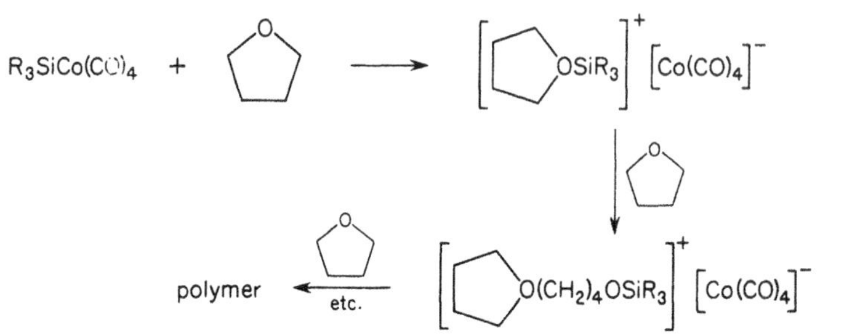

(141)

For these reasons the exact nature of the catalytic species in THF or in other polar solvents is somewhat obscure. At concentrations of 10^{-4} M, i.e. lower than that of $Co_2(CO)_8$, tetranuclear rhodium carbonyls $Rh_4(CO)_{12}$ exhibit similar or even better activity towards the hydrosilylation of vinyltrimethylsilane and of alkenes and their derivatives, cycloalkenes and styrene [31, 713, 715]. $Ni(CO)_4$ also appeared to be an effective catalyst for hydrosilylation of the C=C bond, for example in styrene. However, the carbonyl is decomposed to the metal during the reaction and the yield of the α-adduct is correspondingly reduced [213].

Iron pentacarbonyl was the first reported metal carbonyl catalyst for hydrosilylation [383, 827], and although this reaction occurs under ambient conditions (temperature below 100°C), it takes a somewhat unexpected course. In the presence of this catalyst the hydrosilylation of ethylene and its derivatives proceeds according to the following sequence:

$$X_3SiH + RCH{=}CH_2 \xrightarrow{Fe(CO)_5} \begin{cases} X_3SiCH_2CH_2R & (1) \\ X_3SiCH{=}CHR + H_2 & (2) \end{cases} \quad (142)$$

$X = Cl, OC_2H_5, C_2H_5$.

Since the alkene is present in excess, vinyl trisubstituted silanes are produced almost exclusively. However, a high ratio of hydrosilane to olefin favours the formation of alkyl trisubstituted silanes (hydrosilylation by triethylsilane). Generally, the presence of an electron-withdrawing substituent at silicon and substitution at ethylene results in the generation of greater amounts of substitution products such as vinylsilanes. Corriu *et al.* studied a reactivity of $Fe(CO)_4HSiR_3$ (R = Ph) which was recognized as an intermediate in photochemical hydrosilylation catalysed by iron pentacarbonyl. This complex undergoes carbonyl displacement with nucleophilic ligands (phosphines, phosphites) to give $Fe(CO)_3(H)(L)SiPh_3$. The versatile reactivity of the above complexes in various reactions suggests the mechanism for hydrosilylation of olefins can follow different pathways as a function of the experimental conditions and the nature of olefins. Nevertheless, in the thermal hydrosilylation a direct addition of the Fe–H group can occur by a radical or an ionic process [87].

Later reports on the catalysis of hydrosilylation by iron and rhodium carbonyls have confirmed the unusual direction taken by the reaction [827, 844, 858]. Hydrosilylation of vinyltrimethylsilane by tri(chloro,methyl)silanes at 70–80°C in an autoclave yields α- and β-adducts accompanied by up to 35% of the product of dehydrogenative silylation [844, 399]. In this case the Π-complex formed between vinyltrimethylsilane and iron carbonyl was used as the catalyst:

$$(CH_3)_3SiCH{=}CH_2 + R_3SiH \xrightarrow{\text{catalyst}} \begin{cases} (CH_3)_3SiCH_2CH_2SiR_3 \\ (CH_3)_3SiCH(CH_3)SiR_3 \\ (CH_3)_3SiCH{=}CHSiR_3 + C_2H_5Si(CH_3)_3 \end{cases} \quad (143)$$

$$\text{catalyst} = \begin{array}{l} (CH_3)_3SiCH{=}CH_2 \\ \qquad\qquad\quad | \\ \qquad\quad\;\; Fe(CO)_4 \end{array}$$

The order of silane activity observed was as follows:

$$Cl_3SiH > CH_3Cl_2SiH > (CH_3)_2ClSiH > (CH_3)_3SiH \quad (144)$$

$Ru_3(CO)_{12}$ appears to be a very active catalyst for the addition predominantly of trialkyl- and phenyldialkylsilanes (but also triethoxysilane) to styrene and its p-substituted derivatives, vinylnaphthalene [827, 1032], trifluoropropene and pentafluorostyrene [858] leading exclusively to the corresponding vinylsilanes. However, as was shown by Ojima *et al.* [858] and Seki *et al.* [1032, 1033], the alkylsilane/vinylsilane ratio in the products is highly dependent on the nature of the hydrosilane used. Hydrosilylation of trifluoropropene by triethoxysilane yields a β-adduct as the sole product, and this is also the predominant product if dichloromethylsilane is used, although the reactivity in this case is very low. The addition of trichlorosilane to this olefin does not proceed at all, even after heating at elevated temperature (150°C) for a prolonged period of time (70 h). Alkenes having a hydrogen atom at the allylic position (1-hexene, allylbenzene,

3-phenoxypropene-1, vinylcyclohexane, β-methylstyrene, α-methylstyrene, 2-hexene) formed with triallylsilanes mixtures of vinylsilanes and allylsilanes [1032].

Ojima *et al.* have also reported on the mechanism of the temperature-dependent reactions occurring between trisubstituted silanes and trifluoropropene or pentafluorostyrene in the presence of $Ru_3(CO)_{12}$ (and also $RhCl(PPh_3)_3$). For trifluoropropene the proposed mechanism involves the oxidative addition of the hydrosilane followed by a hydride shift from the oxidative adduct to the olefin, to give an α-R-ethylmetal intermediate. In contrast, pentafluorostyrene does not undergo a reductive elimination because of strong coordination with another molecule of $RCH{=}CH_2$ (Scheme 145) [858].

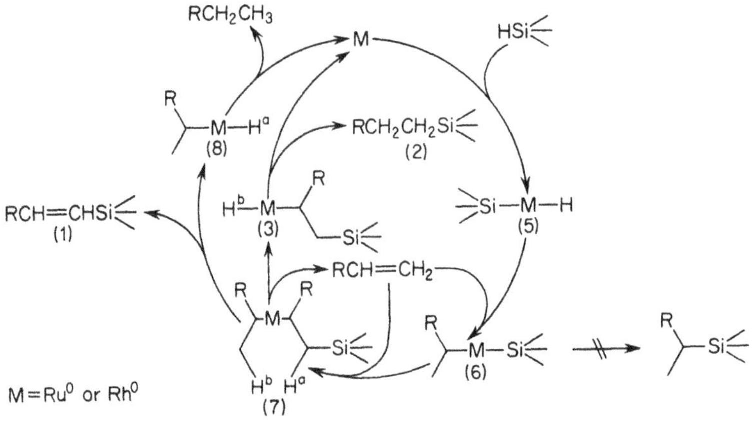

(145)

A second silicon shift from the metal to the β-carbon produces intermediate (7). Two distinct pathways lead from this intermediate. In the first, β-hydride (H^b) transfer to the metal releases the olefin to form an intermediate (3) which yields the hydrosilylation product (2) after reductive elimination. In the second pathway H^a is abstracted and the vinylsilane released forming the hydride intermediate (8). The latter undergoes reductive elimination, producing the RCH_2CH_3, and regenerating the catalyst $[Ru]^0$.

Those reactions catalysed by metal carbonyls which occur often under quite drastic thermal conditions, can be made to proceed at relatively low temperature (0–50°C) through the use of near-UV irradiation [1028, 1293]. Photocatalysis refers to the situation where the proper catalyst involved in the process is generated by irradiating the initial catalyst. The direct oxidative addition of silicon hydrides to $Fe(CO)_5$ occurs after UV radiation as follows:

$$Fe(CO)_5 + R_3SiH \rightarrow R_3Si(H)Fe(CO)_4 + CO \qquad (146.1)$$

with subsequent olefin substitution:

$$R_3Si(H)Fe(CO)_4 + \text{alkene} \rightarrow (R_3Si)Fe(H)(CO)_3(\text{alkene}) + CO \qquad (146.2)$$

It is suggested that the latter intermediate is the direct precursor of hydrosilylation or dehydrogenative hydrosilylation (silylation) products. Substitution of the olefin can occur prior to the oxidative addition of the silane. In addition, both reactions may be preceded by the photogeneration of a $Fe(CO)_4$ species:

$$Fe(CO)_5 \overset{h\nu}{\rightleftharpoons} Fe(CO)_4 + CO \qquad (147)$$

The trinuclear metal carbonyls $M_3(CO)_{12}$ (M = Fe, Ru, Os) have been reported as catalysing the hydrosilylation reactions of 1-pentene, yielding saturated and unsaturated products under photochemical conditions [72]. The trinuclear cluster anion $[HRu_3(CO)_{11}]^{-1}$ catalyses the reaction of ethylene or propylene with ethoxysilane, to yield several products, particularly vinyltriethoxysilane and allyltriethoxysilane, respectively [1144]. $Os_3(CO)_{12}$, $Ru_3(CO)_{12}$, $Re_2(CO)_{11}$ and $Ir_4(CO)_{12}$ appear to be effective photoactivated catalysts for the hydrosilylation of acetone, the latter two catalysts being active also in the photohydrosilylation of 2-heptanone. The structure of the catalyst generated through optical irradiation [1313] and the mechanism of photochemical assistance in hydrosilylation catalysis will be discussed in Section 2.6.

Very recent information has been reported on an enhanced activity of the silicon–hydrogen bond in organosilanes coordinated to the cobalt carbonyl complexes, e.g. ethynyldiorganosilanes [284, 812].

2.4.6 Iridium and Cobalt Complexes

In contrast to the corresponding rhodium(I) complexes, the iridium and cobalt members of the cobalt triad have seldom been employed as hydrosilylation catalysts.

Iridium(I) complexes with the d^8 metal configuration, e.g. $Ir(CO)Cl(PPh_3)_2$ (Vaska's complex) and $IrCO(H)(PPh_3)_3$, have been tested in the hydrosilylation of olefins [Rev.9, 226, 228, 470a, 471] and although in a similar manner to rhodium(I) complexes they can undergo oxidative addition with hydrosilane to yield adducts such as $IrHCl(SiR_3)(PPh_3)_2$. It was concluded that they were inactive towards hydrosilylation.

The promotion energy for iridium d^8 complexes is lower than that of rhodium; as a result the stability of the complex is higher; consequently addition of an olefin to an Ir–H bond cannot occur. In contrast, the adduct formed from Vaska's complex readily dissociates, eliminating the chlorosilane molecule according to Scheme (148).

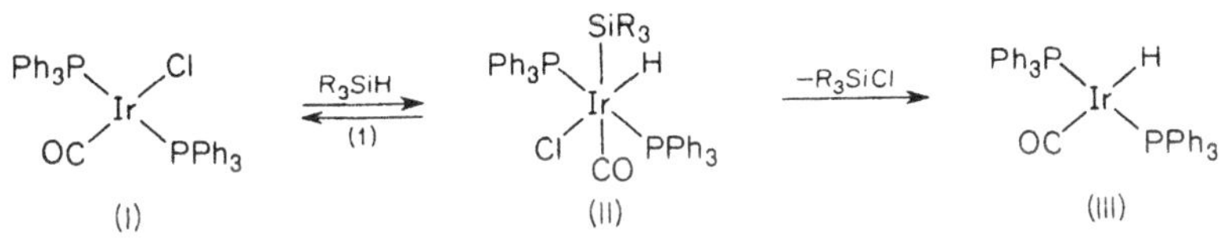

(148)

The equilibrium depends strongly on the electronegativity of the substituent attached to silicon; product (II) can be isolated only in the case of trichlorosilane [Rev.11, 226]. A high electron density at the iridium centre can apparently stabilize the Ir–Si bond. Olefin coordination to complex (I), which is a necessary step in the hydrosilylation step accomplished only by ligand exchange, lowers the electron density at the metal and decreases the formation constant for the complex of the type (II). In addition, the relatively stable Ir–H bond in complex (II) readily leads to the reductive elimination of chlorosilane. Hence, Vaska's complex catalyses H/Cl exchange quite effectively [Rev.11, 226].

In a recent and successful study of the catalysis of hydrosilylation by iridium complexes, non-phosphine ligands such as *cis*-cyclooctene (COE) were investigated. For this reason, phosphine ligand free complexes of iridium, e.g. $[IrCl(COE)_2]_2$, were principally used in the reaction as well as organic substrates such as 1-pentyne and 2,3-dimethylbuta-1,3-diene which are capable of forming stable complexes with iridium and thus preventing its reduction to the metal [59].

On the other hand, the addition of a stabilizing ligand such as phosphine is possible to substrates such as mono-olefins or ketones, which would not normally be expected to form stable complexes. However, in this case the P/Ir ratio always is less then 2:1 in order to prevent the formation of stable $IrHCl[Si(C_2H_5)_3](PPh_3)_2$ [59]. In the hydrosilylation of 1-alkenes (1-octene), even the addition of triphenylphosphine (1:1) does not make this catalyst effective (max. 30% yield of 1-octylsilane). $[Ir(COE)_2Cl]_2$ and $[Ir(COD)Cl]_2$ are also excellent catalysts for the hydrosilylation of α,β-unsaturated ketones, e.g. cyclohex-2-enone. This reaction occurs according to the mechanism advanced for catalysis by Wilkinson's complex [850]. The complexes of the composition [Ir(diene)X] appeared to be very effective catalyst for hydrosilylation of allyl chloride [U-172].

Very recent studies of the Fernandez group on application of iridium complexes in hydrosilylation have shown that systems derived from $[IrX(COD)_2]$ (X = OMe, Cl) + nL (L = triarylphosphine or triarylarsine) are active catalysts for the normal hydrosilylation of 1-hexene [880] and 1-hexyne [360] as well as for a dehydrogenative silylation. The latter process, well known for some rhodium, ruthenium, cobalt and iron carbonyl and other complexes, is enhanced by using triarylphosphines in the dehydrogenative silylation of 1-hexyne while triphenylarsine favours the dehydrogenative silylation of 1-hexene. The authors isolated the intermediate $IrH_2(SiEt_3)(COD)L$ (where L = PPh_3 or $AsPh_3$) which appeared to be an active catalyst for the dehydrogenative silylation of alkenes [361].

The $[Ir(COD)Cl]_2$ precursor in the presence of (S)-amphos as the chiral ligand catalyses asymmetric hydrosilylation of acetophenone with Ph_2SiH_2 showing high activity, but an optical yield of only 16.5% was obtained [584].

Moreover, a quaternary ammonium, hexachloroirrydate very effectively catalyses the addition of triethylsilane and phenyldimethylsilane to phenylacetylene, yielding adduct β- and α- in the ratio 5:1 and 2:1, respectively [517a].

In the mechanistic scheme for catalysis by $Co_2(CO)_8$ (Scheme 135) the alkylcobalt carbonyl, $RCH_2CH_2Co(CO)_n$, may be regarded as a d^8 Co(I) complex undergoing oxidative addition with the hydrosilane to yield an active Co(III) intermediate similar to the Rh(I) and Ir(I) species. Apparently, the critical property of the active M(I) species is its ability to interact reversibly with tertiary silanes. With the exception of cobalt carbonyls, little attention has been paid to other cobalt(I) or iridium(I) complexes as potential hydrosilylation catalysts, Complexes which are successful in this respect have the general formula $CoH(X)_2L_3$, where X = H and N, L = Ph_3P; complexes such as $CoH_2[Si(OC_2H_5)_3]L_3$ are also highly effective in the hydrosilylation of 1-hexene by triethoxysilane. Scheme (149), below, suggests a catalytic cycle starting from a 16-electron $CoHL_3$ species which can be formed from all the precursors given above [61].

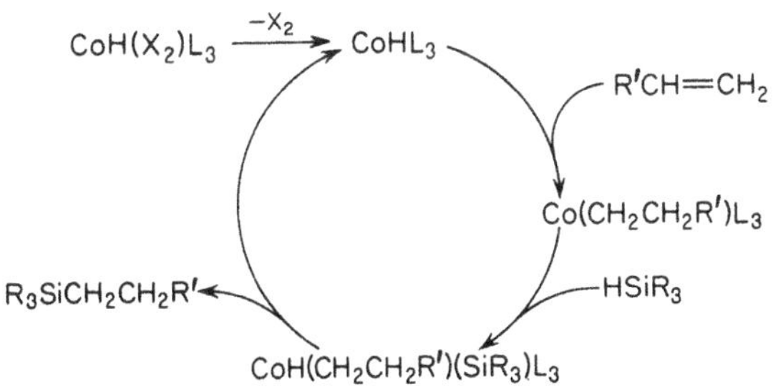

(149)

Small amounts (~10%) of by-products, such as hexane and hexenyltriehoxysilane, could result from the following pathways:

$$CoH(C_6H_{13})[Si(OC_2H_5)_3]L_3 \rightarrow Co\,[Si(OC_2H_5)_3]L_3 + C_6H_{14} \qquad (150)$$

$$Co[Si(OR)_3]L_3 + \text{1-hexene} \rightarrow CoHL_3 + C_6H_{11}SI(OR)_3 \qquad (151)$$

No catalysis has been observed in hydrosilylation reactions involving trichlorosilane, trifluorosilane and triethylsilane. The stability order of hydrosilanes with Co(I) complexes is as follows:

$$HSiX_3 > HSi(OR)_3 \gg HSiR_3 \ (X = F, Cl) \qquad (152)$$

2.4.7 Ruthenium and Osmium Complexes

Excluding the carbonyl M(O) catalysts discussed in Section 2.4.4, within the iron triad only ruthenium (I–III) complexes have been subject to recent investigation as catalysts for the hydrosilylation of carbon–carbon multiple bonds. The carbonyl clusters $Ru_3(CO)_{12}$ [827, 858], and $[HRu_3(CO)_{11}]^-$ [1144] have already been mentioned but phosphine complexes of ruthenium have now been employed in the addition of dichlorosilanc [1289, D-095] and trialkoxysilanes [733, 735, 736, PI-001, PI-015] to the C=C bond, as well as of trichlorosilane to acetylene [1290], and for the curing by polyhydrosiloxanes of polysiloxanes containing vinyl groups [D-032, F-012, F-016]. In addition, it was shown earlier that they are as efficient as the corresponding rhodium complexes for the hydrosilylation of ketones and aldehydes [279, 310, 339, 872]. Early studies of phosphine ruthenium precursors such as L_3RuHCl_2. L_3RuHCl, L_3RuCl_2 and $L_3Ru(CO)HCl$, where L = tertiary phosphine, as catalysts for the hydrosilylation of alkenes and of vinyl derivatives were not successful. The only reaction observed by Hetflejs *et al.* in the presence of dichlorotris-(triphenylphosphine)ruthenium(II) [1170] was the addition of trichlorosilane to acrylonitrile which proceeded with the high yield of 93%. Precise experiments have shown, however, that in this case the actual catalyst is the triphenylphosphine released on dissociation of the complex.

All the above experiments were carried out in the absence of air, and mostly in benzene. The elimination of solvent and the introduction of molecular oxygen appears to be necessary to make this catalyst effective in the addition of trialkoxysilanes to 1-alkenes, and to a lesser degree to some vinyl derivatives [733, 735, Pl-001]. These studies have shown that chloro-substituted silanes and alkylsilanes do not react with olefins under these particular reaction conditions. Hydrosilylation proceeds only with alkoxy-substituted silanes or, more generally, with oxygeneous substituents attached to silicon. The results of Kono *et al.* [605a] and Hetflejs *et al.* [1170] clearly show the effect of solvents on the composition and structure of complexes formed during the reaction of chlororuthenium(II) phosphine complexes with tri-substituted silanes. When the reaction proceeds in solvents such as benzene, toluene, nitrobenzene, CH_2Cl_2 or hexane, five- and, occasionally, six-coordinate phosphine complexes are formed (for example, a mixture of $RuH[Si(OC_2H_5)_3]_2(PPh_3)_3$ and $RuHCl(PPh_3)_3$ when triethoxysilane is employed as the reagent). But when the reaction is allowed to proceed in the absence of solvent, mainly four-coordinate complexes with the general formula $RuH(SiR_3)(PPh_3)_2$, where $R_3 = Cl_3$, Cl_2CH_3, $Ph(OCH_3)_2$, $(OC_2H_5)_3$, $(OCH_3)_3$, are obtained. Trialkoxysilyl-substituted complexes of ruthenium(II) are stable in air.

$$RuXCl(PPh_3)_3 + (RO)_3SiH \xrightarrow{7\ days,\ 20°C} RuH[Si(OR)_3](PPh_3)_2 \qquad (153)$$

X = H, Cl.

Haszeldine *et al.*, however, reported later that seven-coordinated complex $RuH_3[Si(OC_2H_5)_3](PPh_3)_3$ was obtained by the reaction of chlororuthenium(II) complex with triethoxysilane (in benzene and under nitrogen) [474a].

The electron-donating character of alkyl substituents reduces the possible formation of trialkylsilylruthenium complexes, and this explains the inactivity of Ru(II) complexes in the hydrosilylation of the C=C bond by trialkylsilanes. On the other hand, the inactivity of chloro-substituted silanes may be accounted for by their high stability, and by the migration of the chlorine atom from the silyl ligand to the ruthenium atom, followed by cleavage of the Si–Ru bond in the silylruthenium hydrides [605a]. Apparently, the electron-withdrawing character of alkoxy groups at silicon on the one hand, and d_π–p_π conjugation on the other, contribute to the predominance of trialkoxy-substituted silanes over trichloro- or alkyl-substituted silanes during the formation of ruthenium–silyl intermediates which are active in hydrosilylation [733]. Very similar substituent effects at silicon have been noted for the Co(I) complexes discussed previously [61] (Section 2.4.5). Schematic interactions of the silicon d-orbitals with metal d-orbitals and oxygen p-orbitals occur as follows [Rev.24]:

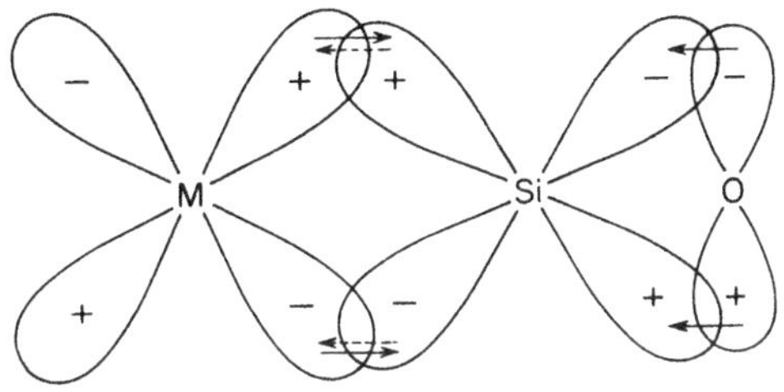

(154)

However, in contradiction to this concept, the stability of trialkoxysilylhydridobis(triphenylphosphine)ruthenium(II) complexes appears to be too great. Thus, it has been shown that these complexes do not add to olefins, nor are they efficient catalysts for hydrosilylation. Hence the only active ruthenium species is that formed by the reaction of ruthenium(II) complex with a trialkoxysilane in the absence of solvent but in an oxygen atmosphere. The product of this reaction (complex A) has been isolated and identified as a mixture of $Ru(O_2)[Si(OC_2H_5)_3](PPh_3)_n$ and $RuH[Si(OC_2H_5)_3]_2(PPh_3)_n$ where $n = 2$ or 3, and is assumed to result from the following equilibrium:

$$Ru(O_2)[Si(OC_2H_5)_3](PPh_3)_n \xrightleftharpoons{HSi(OC_2H_5)_3} RuH[Si(OC_2H_5)_3]_2(PPh_3)_n \quad (155)$$

The activity of complex (A) appears to be higher (74%) than that of its precursor (68%) in contrast to the inactivity of the corresponding ruthenium–silyl complex obtained and isolated under oxygen-free conditions [735]. The analytical and catalytic data obtained suggest the following scheme for olefin hydrosilylation catalysed by $RuCl_2(PPh_3)_3$ in the form of the ruthenium(II) precursor given in Scheme (156) [733].

It is suggested that the formation of active species (1) under hydrosilylation conditions occurs as a result of a series of complex reactions between trialkoxysilane and oxygen involving the precursor. The reactions include simultaneous dehalogenation and reduction, together with the elimination of HCl, Cl_2, H_2 and $ClSi(OR)_3$, as well as the oxygenation of phosphine and its final elimination. The addition of trialkoxysilane to the metal centre, together with the simultaneous elimination of molecular oxygen and/or oxygenation of phosphine to phosphine oxide, constitutes the second step (2) in the proposed catalytic cycle. This is followed by olefin coordination to this ruthenium(II) complex but succession can be inversed and the subsequent rearrangement of the Π-complex (3) to the σ-complex (4), in accordance with the general mechanism of Chalk and

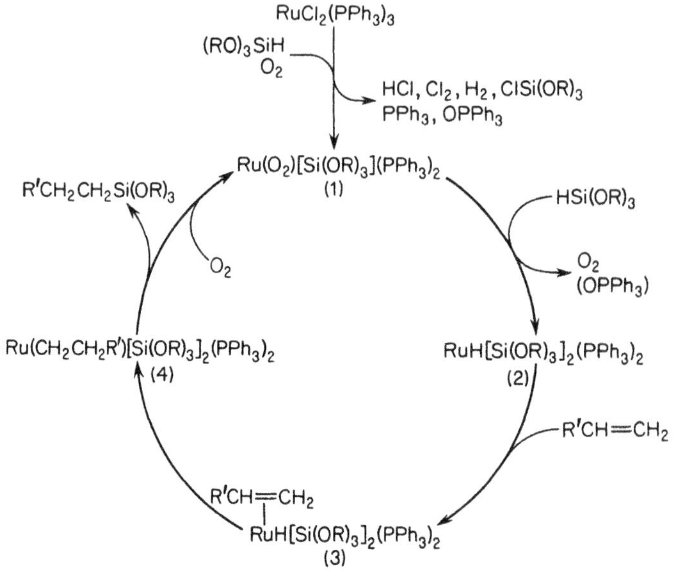

(156)

Harrod [226]. The intermediate (4) readily decomposes to give the β-adduct and, in the oxygen atmosphere, regenerates the active complex (1). The above mechanistic considerations indicate that ruthenium(III) complexes can be involved in the hydrosilylation of 1-alkenes, as demonstrated experimentally. Indeed, the reported data have shown that Ru(III) species exhibit a very high activity in the hydrosilylation of 1-hexene. Thus, in air, the following activities have been observed: $RuCl_3(PPh_3)_3$, 86%; $RuCl_3$, 77%; and $RuCl_3$ + PPh_3 (1:1 ratio), 96%. In contrast to Ru(III) precursors the initial Ru(II) complexes are also efficient in an oxygen-free atmosphere. Thus, under such circumstances, the following activities apply: $RuCl_3(PPh_3)_3$, 52%; $RuCl_3$, 60%; $RuCl_3$+PPh_3 (1:1 ratio), 63% [733, Pl-015]. Nevertheless, the dual role of oxygen in catalysis by an Ru(III) precursor appears to be similar to that which applies in Ru(II) systems. Oxygen stabilizes Ru(III) complexes by forming the $Ru(O_2)$ species (as occurs with Ru(II) precursors) and oxidizes phosphine to phosphine oxide, leading to the release of ligand molecules from the coordination site of the catalyst. However, since the presence of a tertiary phosphine is not necessary to promote the activity of $RuCl_3$, it plays only a minor role as far as the Ru(III) precursor is concerned. Further study is needed to clarify the real role of dioxygen and the solvent in the hydrosilylation of alkenes catalysed by ruthenium complexes.

A comprehensive examination of complexes which catalyse the hydrosilylation of vinyl-substituted silanes has shown a correspondingly high activity as is exhibited by Ru(II) and Ru(III) phosphine precursors, particularly when alkoxy-substituted hydrosilanes and vinylsilanes are present in the system. In contrast to 1-alkenes, the hydrosilylation of vinyl(alkoxy,alkyl)silanes can also occur in the presence of an excess of the unsaturated substrate, yielding an unsaturated product of formula ≡SiCH=CHSi≡ as a result of dehydrogenative silylation. However, more detailed analysis has revealed that this compound actually arises from the metathesis of vinyltrialkoxysilane [Rev.24, 733]. Further details regarding the competitive reactions occurring in the presence of Ru(II) and Ru(III) precursors will be discussed in Section 5.1.

Unlike trichlorosilane, dichlorosilane is very effective (68–99%) in adding to 1-alkenes (such as 1-hexene) at elevated temperatures (70–120°C) over prolonged time spans (15–25 h) in the presence of the phosphine and phosphine silyl complexes of Ru(II):

$$Cl_2SiH_2 + RCH{=}CH_2 \xrightarrow{Ru(II)} RCH_2CH_2SiHCl_2 \qquad (157)$$

However, the addition of the second olefin does not appear to be possible [1289]. On the other hand, the hydrosilylation of acetylene by trichlorosilane, in the presence of $RuCl_2(PPh_3)_3$ and in xylene as a solvent, proceeds exothermally at room temperature to give vinyltrichlorosilane selectively (85% yield after 2.5 h reaction) [1290].

A benzene solution of triethylsilane and $RuCl_2(PPh_3)_3$ generates the adduct $RuClH(PPh_3)_3 \cdot C_6H_6$ on refluxing. Replacement of benzene by acetophenone enables the corresponding hydrochloride to be isolated when the solvent is removed. It has been suggested that $RuClH(PPh_3)_3$ is an active intermediate in the hydrosilylation of carbonyl compounds, since the Ru–H bond of this species can add to the C=O bonds of aldehydes and ketones. The catalytic activity of $RuClH(PPh_3)_3$ appears to be somewhat lower than that of Wilkinson's complex [310, 339].

Although in catalysis by ruthenium, as well as other transition metals, phosphorus-containing ligands are of great importance, there are very recent reports on synthesis and catalysis of hydrosilylation by cyclooctadiene complexes of ruthenium with diazadiene (DAD) [1363] ligand and other unsaturated N-containing ligands [167]. These complexes catalyse hydrosilylation of isoprene but the reaction is strongly accompanied by olefin isomerization.

Few reports exist on the use of osmium complexes in the hydrosilylation process. However, in the presence of such complexes it has been reported that the reaction of hydrosilanes with 1-alkenes yields a dehydrogenation product rather than adducts (93%) [D 035]

$$CH_2{=}CHR + HSiX_3 \rightarrow X_3SiCH{=}CHR + H_2 \qquad (158)$$

but in the hydrosilylation of trimethylvinylsilane also the strong tendency of H_2OsCl_6 to form unsaturated adducts was observed. In this latter reaction, nine products and by-products were isolated according to the following stoichiometry:

$$[(CH_3)_3SiO]_2CH_3SiH + (CH_3)_3SiCH{=}CH_2 \xrightarrow{H_2OsCl_6}$$

$$\underset{(100\ parts)}{(CH_3)_3SiCH{=}CHSiCH_3[OSi(CH_3)_3]_2} +$$

$(CH_3)_3SiCH_2CH_2SiCH_3[OSi(CH_3)_3]_2$ +
(12.7 parts)

$[(CH_3)_3SiO]_2SiCH_3CH{=}CHSiCH_3[OSi(CH_3)_3]_2$+
(66.8 parts)

$(CH_3)_3SiCH{=}CHSi(CH_3)_3$+$[(CH_3)_3SiO]_2SiCH_3(C_2H_5)$ +
(32.4 parts) *(72.9 parts)*

$(CH_3)_3SiC_2H_5$ + $(CH_3)_3SiCH_2CH_2Si(CH_3)_3$ +
(113 parts) *(4 parts)*

two unidentified products (159)

This demonstrates that the side-reactions are accompanied by processes such as redistribution, e.g. $(CH_3)_3SiO/CH_3$ and vinyl/H exchange, and dehydrogenative hydrosilylation.

A recent report has confirmed that hexachloroplatinic acid can be replaced by H_2OsCl_6 in hydrosilylation catalysis. This catalyst was found to be slightly less active than H_2PtCl_6, but more regioselective in the addition of trichlorosilane, dichloroalkylsilane, triethylsilane and triethoxysilane to phenylacetylene (yields 69–86%, 110°C, 4 h) [Rev.38].

2.4.8 Nickel and Palladium Complexes

Numerous complexes of nickel(II) and nickel(O) with the d^8 configuration catalyse the addition of the Si–H bond to olefins. Among such catalysts may be listed monophosphine complexes of the $Ni(PR_3)_2X_2$ type, where X = Cl, I, NO_3, O and R = alkyl, aryl [588, 589, 1109, 1306, C-018, D-070, D-111, J-006], $Ni(PPh_3)_4$ [D-070, D-111], $Ni(CO)_2(PPh_3)_2$ [C-018] and bidentate complexes of $NiCl_2$(chelate), where the chelate ligand includes DMPC [627a], diamines [589, 589a] and other diphosphines [577, 588, 589], and also complexes of the type $Ni(acac)_2L$, where L = PPh_3 or $(Ph_2PCH_2)_2$ [206, 207, 1135, 1325, 1335]. Other nickel complexes which catalyse the hydrosilylation of olefins have also been reported, e.g. $Ni(cod)_2$ [205, 791, D-021], $Ni(cod)_2(PR_3)_2$ [U-152], Ni(α-ω-divinylpoly(dialkyl)siloxane)$(PR_3)_2$ [U-152], $[\Pi\text{-}C_5H_5(CO)Ni]_2$ [1171] and $Ni(diphos)_2$. The latter forms an ionic complex with trichlorosilane [588]:

$$(diphos)_2Ni + HSiCl_3 \rightarrow (diphos)_2NiH^+SiCl_3^- \qquad (160)$$

Nickel phosphine complexes catalyse olefin hydrosilylation in a specific manner. One characteristic feature of such processes is the interchange of hydrogen and chloride ions at the silicon atom according to the following equation [337, 588, 629]:

$$RCH{=}CH_2 + HSiX_2Cl \xrightarrow{Ni(II),Ni(O)} RCH_2CH_2SiX_2Cl + RCH_2CH_2SiX_2H + SiX_2Cl_2$$
X_2 = Cl_2, CH_3Cl or $(CH_3)_2$, (161)

whilst another is the formation of substantial amounts of internal adducts (II) in addition to terminal ones (I) arising from simple terminal olefins; for example [629a],

$$RCH{=}CH_2 + HSiCH_3Cl_2 \xrightarrow{Ni(DMPC)Cl_2} \underset{(II)}{RCH(CH_3)SiCH_3Cl_2} + \underset{(I)}{RCH_2CH_2SiCH_3Cl_2} \quad (162)$$

Mechanistic considerations given by Kumada *et al.* for this reaction are based on the general Chalk–Harrod mechanism and involve the isomerization of olefins via nickel–silyl complexes (see also Scheme 82) together with the establishment of an equilibrium between the four possible intermediates formed in the system, i.e. terminal σ-, terminal Π-, internal σ- and internal Π-complexes. The ratio of α/β adducts formed depends strongly upon the inductive effect of the substituents associated with the phosphorus ligand [588]. The mechanism of the H/Cl exchange is illustrated in Scheme (163) [588].

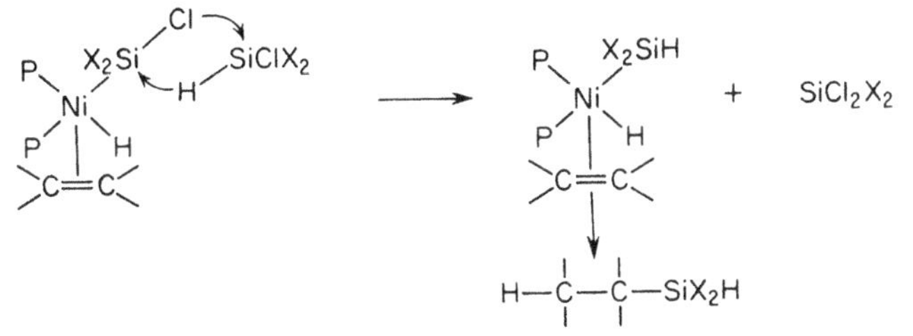

(163)

The interchange takes place between the chlorine attached to the silicon linked to the nickel(II) intermediate and the hydrogen of a second silane molecule. The Π-complexes of nickel with cyclooctadiene and cyclopentadiene–nickelocene have also been shown to be active in the hydrosilylation of 1-octene and styrene [791, 1171, J-012]. The hydrosilylation of styrene by trichlorosilane in the presence of $[Ni(cod)(CO)]_2$ as a catalyst leads to the selective formation of 1-trichlorosilyl-1-phenylethane (82%). As with $Ni(CO)_4$ and other carbonyl complex-catalysed reactions, the efficiency of these various catalysts decreases during the reaction as a result of decomposition leading to the precipitation of nickel [213]. Addition of phosphine stabilizes such systems, but generally, in the presence of phosphine and phosphine oxide complexes of nickel, the activity and selectivity of catalysts of this type are determined by the nature of the ligands on the metal atom; i.e. the method of coordination, the donor and steric characteristic and the presence of other coordinating groups in the molecule of the organophosphorus ligand [1109]. For example, $Ni(cpd)_2$ catalyses the addition of methyldichlorosilane to styrene (50°C) generating an α-adduct selectively (74% yield), while $Ni(cpd)_2$ + 2 PPh_3 leads to the formation of α- and β-adducts (in a 24/76 ratio) with a 77% yield. Finally, when an excess of triphenylphosphine (Ni/P = 1:5) is present, the β-adduct is also obtained selectively at 74% yield [791]. The mechanism of the reaction when catalysed by Π-complexes of the $Ni(cod)_2$ type, as well as by $Ni(acac)_2$ and other chelate complexes of nickel, seems to be similar to that proposed for electrophilic catalysis by simple nickel salts.

Studies of the reaction of triethoxysilane with 1,3-butadiene show different yields of the 1:1 adduct in the presence of particular nickel salts: thus NiF_2, 28%; $NiCl_2$, 61%; $NiBr_2$, 28%; NiI_2, traces; $Ni(acac)_2$, 92%; and nickel lactate, 69%. The addition of tertiary phosphine (Ni/P = 1:5) reduces the observed differences in the yield of the 1:1 adduct. Thus for PPh_3 the yield is 54–77% regardless of the salt employed, and for PBu_3 the yield is 1–5%. The only exception is the $Ni(lactate)$–PBu_3 system, in which the yield remains at 90% [205]. The above results provide evidence illustrating differences between the mechanistic pathways involved. In the absence of tertiary phosphine, nickel salts act as Lewis acids. However, as nickel–phosphine complexes they catalyse the reaction according to a Chalk and Harrod mechanism, and for this reason no anion effect is observed.

Phosphine complexes of nickel are used as catalysts in the hydrosilylation of olefins with functional groups, e.g. vinyl acetate [589, 1325], acrylonitrile [589, 1325] and methacrylate [589, 1325], as well as in the hydrosilylation of acetylene derivatives. In the addition of silicon hydride to a C≡C bond, *cis*-adducts are produced [589]:

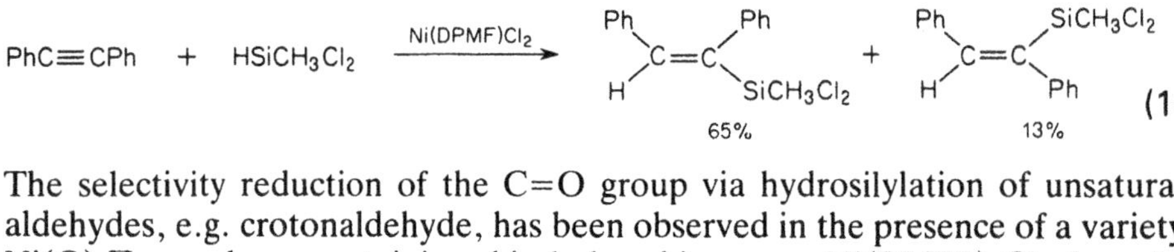

$$PhC \equiv CPh \ + \ HSiCH_3Cl_2 \ \xrightarrow{Ni(DPMF)Cl_2} \ \underset{65\%}{\overset{Ph}{\underset{H}{>}}C=C\overset{Ph}{\underset{SiCH_3Cl_2}{<}}} \ + \ \underset{13\%}{\overset{Ph}{\underset{H}{>}}C=C\overset{SiCH_3Cl_2}{\underset{Ph}{<}}} \qquad (164)$$

The selectivity reduction of the C=O group via hydrosilylation of unsaturated aldehydes, e.g. crotonaldehyde, has been observed in the presence of a variety of Ni(O) Π-complexes containing chiral phosphines, e.g. $Ni(BMPP)_2Cl_2$, have been used to promote asymmetric catalysis in the hydrosilylation catalysis in the presence of Ni(O) and Ni(II) complexes, Kumada *et al.* have suggested a mechanism involving the prior reduction of nickel(II) by silicon hydride to generate a phosphine olefin complex of nickel(O) in the presence of the olefin. This complex undergoes oxidative addition of hydrosilane in the second step of the process to produce a nickel(II) hydride(silyl) intermediate which then decomposes in accordance with the general scheme for such reactions. This commonly accepted concept of nickel-catalysed hydrosilylation is contradicted to some extent by data which indicate differences between the catalytic behaviour of Ni(O) and Ni(II) precursors, e.g. in the sequence of hydrosilane activity towards the same olefin [790].

Magnetic and spectral studies of the diamagnetic square complex $NiX_2(PBu_3)_2$ when used as a catalyst of the reaction between methyldichlorosilane and 1-heptene have not revealed the existence of a paramagnetic species during the course of hydrosilylation. The intermediate isolated was identified as $Ni[P(C_4H_9)_3]_2(SiCH_3Cl_2)_2$ tentatively but its catalytic activity was close to that of the precursor. On the basis of this result, authors such as Reikhsfel'd *et al.* have assumed the formation of silyl complexes of Ni(II), rather than reduction of Ni(II) to Ni(O), in the initial step of the hydrosilylation when catalysed by an Ni(II) precursor. It has been suggested that at elevated temperatures the formation of dimeric nickel complexes with the following structure occurs [790]:

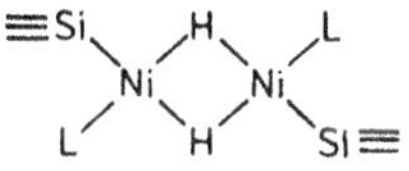

Further study is needed to clarify this situation.

Although their catalytic and chemical properties resemble those of the platinum complexes, palladium complexes are not generally regarded as good hydrosilylation catalysts because of the facility with which they are reduced to the metal by silicon hydrides. However, recent reports have suggested the following types of Pd(II) and Pd(O) complexes as being active in the addition of hydrosilanes, particularly with respect to addition to the C=C bond in alkenes (vinyl derivatives and alkadienes are most effective):

$Pd(PR_3)_4$, where R = Ph, alkyl [468, 1176, 1208, D-027, 738];
$PdX_2(PR_3)_2$, where X = Cl, Br; R = aryl, alkyl, aryloxyl, alkoxyl [D-060, 738, U-192];
$PdX_2(MPh_3)_2$, where X = Cl; M = As, Sb D-027];
$Pd(chelate)(PPh_3)_2$ [1176, J-004];
$PdCl_2(RCN)_2+PPh_3$, where R = aryl [123, 590, 644, 855, 858, 1221, D-060];
PdCl(cpd) [J-012];
$Pd_2(PPh_3)_nCl_2B_mCl_m$, where $n = 4, 6$; $m = 10, 12$ [1258];
Pd(II)salts + MR_3, where M = P, As, Sb [D-027];
Pd + MR_3 [D-027].

When used for the catalysis of hydrosilylation, palladium complexes must be prevented from being reduced to metal. This is achieved by coordinating the metal with suitable ligands, most commonly tertiary phosphines. For this reason no significant reduction is observed in the catalytic activity of $Pd(PPh_3)_4$ or of a mixture of palladium in the presence of excess triphenylphosphine: both systems lead to the formation of $[Pd(PPh_3)_2]_x$ [368, 468]. After completion of the reaction, the catalyst can be regenerated by a further addition of triphenylphosphine. Such reactions generally occur in solvents such as benzene, toluene, hexane, ethers or $CHCl_3$ but proceed equally well in their absence.

The hydrosilylation of 1-alkenes in the presence of $Pd(PPh_3)_4$ leads to the selective formation of products with terminal silyl groups. The activity of the catalyst increases as the electron-withdrawing character of the substituents at the phosphorus atom increase, i.e. in line with the following decrease in the σ-basicity of the ligands:

$$Ph_3P > (C_2H_5)_3P > (C_4H_9)_3P > (c{-}C_6H_{11})_3P \qquad (165)$$

In contrast to unsubstituted 1-alkenes, the hydrosilylation of their derivatives with electron-acceptor substituents, e.g. acrylonitrile, styrene [1208] or vinyltrichlorosilane [747, 749], in the presence of phosphine complexes of palladium, leads to the selective formation of α-adducts.

A general scheme advanced for the mechanism of palladium-catalysed hydrosilylation is also based on the Chalk and Harrod concept, and involves the initial reduction of divalent palladium followed by the oxidative addition of hydrosilane. This suggestion was confirmed at an early stage by the separate reaction of metallic palladium with phosphine in the presence of excess trichlorosilane giving yellow crystals of bis(triphenylphosphine)palladium [367, 368] at 120°C. The same product was also observed by Tsuji *et al.* in the reaction of $Pd(PPh_3)_4$ with trichlorosilane at 120°C [1208]. These results can be accommodated by a scheme which includes the formation of an active intermediate [738, 749].

Recent studies on characterization of such palladium–trichlorosilyl intermediates (see Scheme 166) suggest the idea of H/Cl exchange and silicon–hydrogen affinity at a metal centre what was confirmed by spectroscopic (IR, ^1H NMR) examinations.

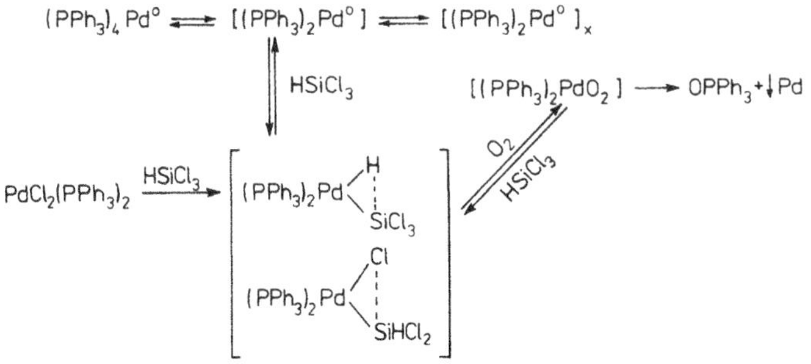

(166)

A comprehensive study of the hydrosilylation of vinyltri(chloro,methyl)silanes by tri(chloro,methyl)silanes catalysed by $Pd(PPd_3)_4$ and $PdCl_2(PPh_3)_2$ has shown that under such circumstances β-adducts are predominantly formed. The α-adduct can be obtained only if electron-withdrawing chloro-substituents are present at both silicon atoms. However, the replacement of succeeding chloro-substituents, particularly in hydrosilanes, markedly lowers the yield in such reactions, with the exception of the hydrosilylation of $CH_2{=}CHSi(CH_3)_{3-m}Cl_m$ (m = 0–3) by trichlorosilane. According to the general mechanistic scheme [226, 1176], in the presence of vinylsilane (olefin) the silyl hydride intermediate (1) (see Scheme 167.1), generated by previous or subsequent coordination with the vinylsilane, decomposes during the hydrosilylation cycle.

The electron-withdrawing character of chloro-substituents at silicon causes intermediate (1) to become stabilized. In contrast, the successive introduction of electron-donating methyl-substituents weakens the Pd–Si bond and leads to a considerable decrease in the reaction yield. The different regioselectivity observed in the hydrosilylation of vinyltrichlorosilane by trichlorosilane presumably arises from attack by a nucleophilic hydride on the β-carbon of the vinylsilane molecule coordinated to the metal centre. Interactions of H–Pd–Si and Cl–Pd–Si in the two (I and II) palladium silyl complexes (see Scheme 167.2) with a coordinating molecule of vinylsilane are proposed to involve a five covalent silicon bonded to a palladium centre (III) with hydridic hydrogen favouring hydride transfer from the silicon atom to the β-carbon of the vinylsilane molecule.

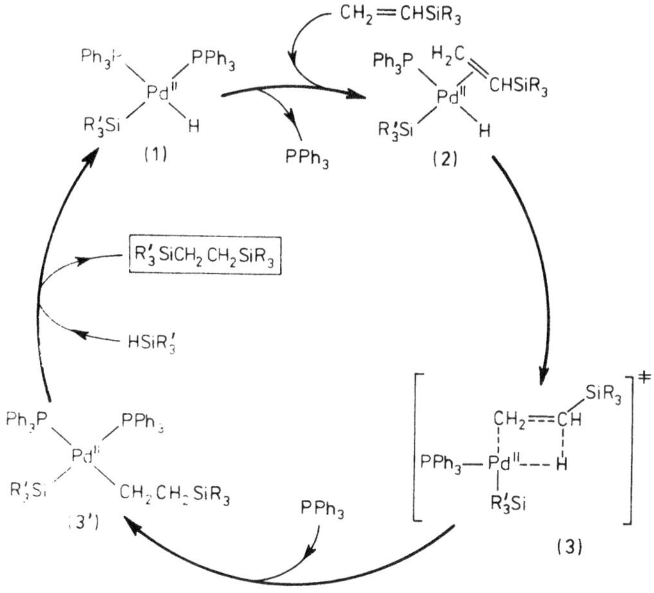

(167.1)

$R'_3 = Cl_2Me, ClMe_2$ $R_3 = Cl_2Me, ClMe_2, Me_3$

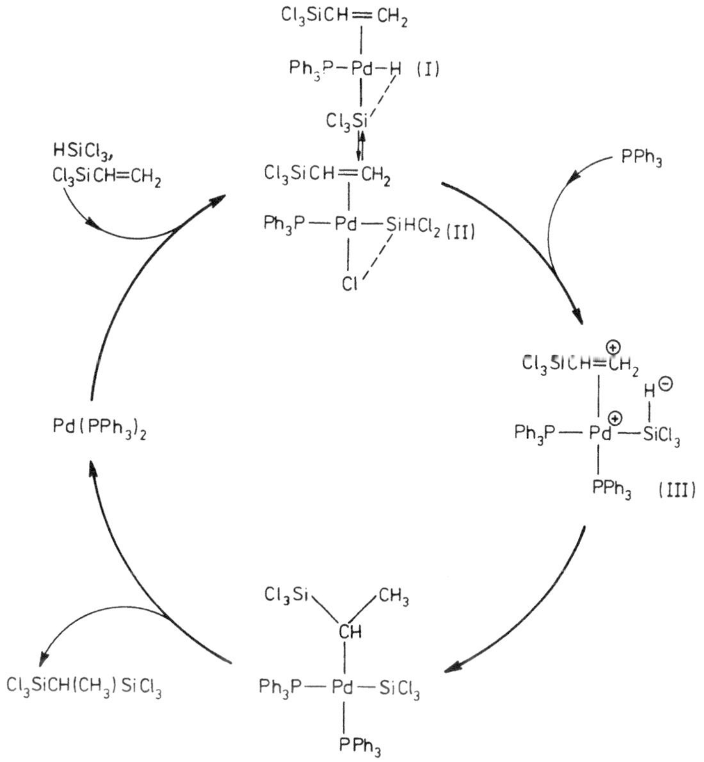

(167.2)

This leads to bonding to the β-silylethyl ligand to the palladium centre and finally to the generation of the α-adduct. Apparently, replacement of the chlorine by a methyl group in vinylsilanes, markedly decreasing the electron-withdrawing effect of the substituents at silicon reduces the possibility of palladium–silicon (III) intermediates and simultaneously the reaction path via such intermediates. In this case the reaction involves hydride attack on an α-carbon atom (the position of the vinyl-substituted silanes, that is more electron-deficient) which can occur via the hydrogen bonded to silicon in palladium intermediate (for trichlorosilane see Scheme 167.2) or via Pd–H intermediate (for the rest of the tri-substituted silanes see Scheme 167.1). Both routes lead to β-adduct formation via β-silylethylpalladium intermediate [738, 740, 749]. Different pathways are proposed for reactions involving substituents at the vinylsilane silicon which are less electronegative than those in vinyltrichlorosilane.

Complexes of palladium and nickel have been extensively studied as catalysts in the hydrosilylation of alkadienes [590, 1219, 1220, 1221], cycloalkadienes [590, 1217, 1304, C-015, D-027] and, particularly, of conjugated dienes, e.g. 1,3-butadiene [205, 206, 207, 1208, C-017, D-027] and isoprene [1222, 1335, C-017, C-018, D-027]. It has been proposed that both palladium- and nickel-catalysed hydrosilylations of 1,3-butadiene occur via an identical reaction mechanism involving the formation of Π-allyl intermediates [1176] giving silyl-substituted butenes and octadienes as the main products (Scheme 168):

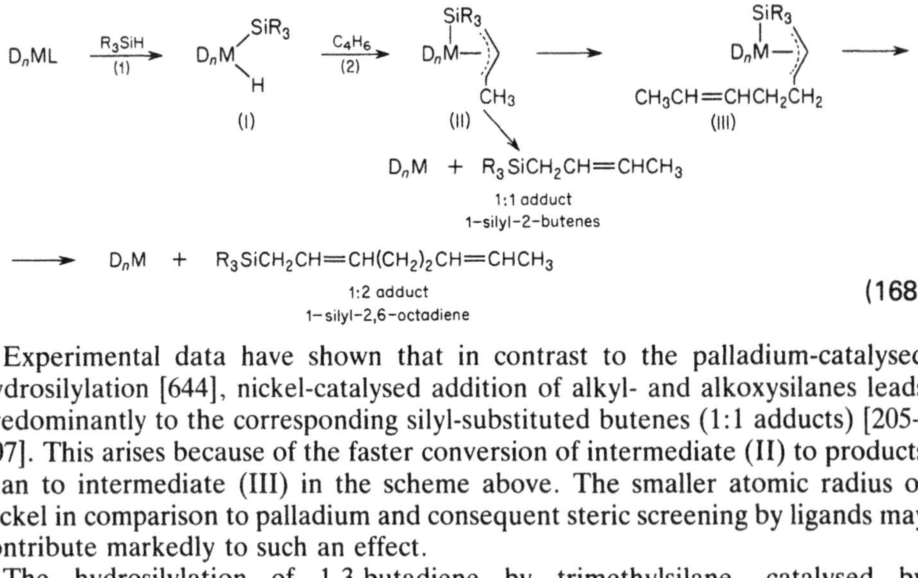

(168)

Experimental data have shown that in contrast to the palladium-catalysed hydrosilylation [644], nickel-catalysed addition of alkyl- and alkoxysilanes leads predominantly to the corresponding silyl-substituted butenes (1:1 adducts) [205–207]. This arises because of the faster conversion of intermediate (II) to products than to intermediate (III) in the scheme above. The smaller atomic radius of nickel in comparison to palladium and consequent steric screening by ligands may contribute markedly to such an effect.

The hydrosilylation of 1,3-butadiene by trimethylsilane, catalysed by $PdCl_2(PhCN)_2$ (100°C), has been studied in detail by Hetflejs *et al.*, and shown to follow a more complex course to yield 1-trimethylsilyl-2,6-octadiene and 4-trimethylsilyl-1-butene as the main products, accompanied by 1-trimethylsilylbutadiene, 1,4-bis(trimethylsilyl)-2-butene and 1-trimethylsilyl-2-butene as the minor products. The relative amounts of the two main products are strongly dependent on the reaction temperature, the solvent and the substrate ratio. If triphenylphosphine is introduced as co-catalyst, the yield of the 1:2

adduct increases. The mechanism proposed by Langova and Hetflejs [643] in an attempt to account for the observed products is as follows (Scheme 169)

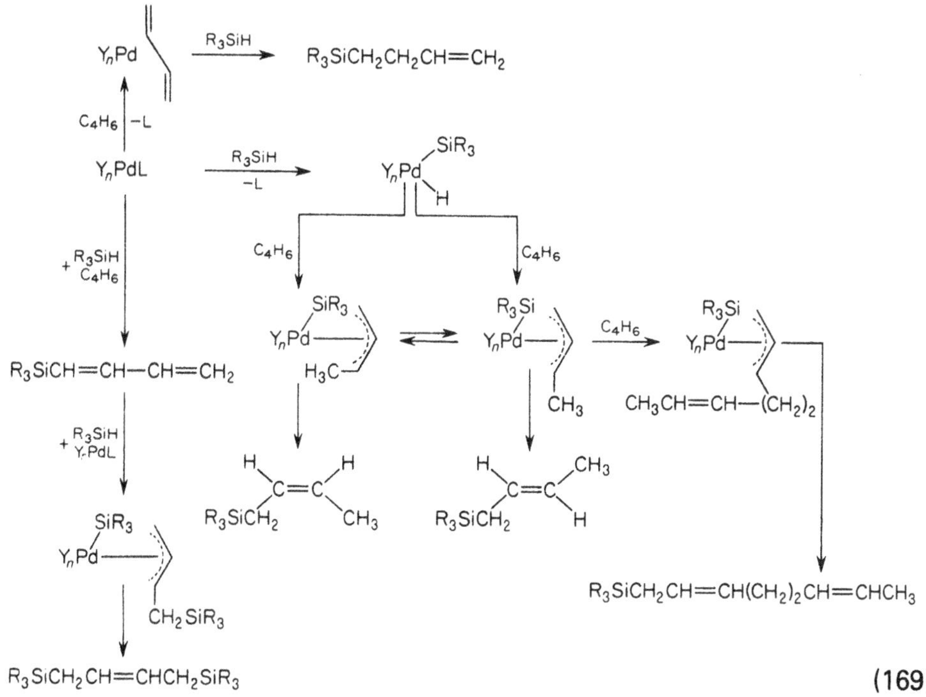

(169)

This scheme does not explain the formation of 4-trimethylsilyl-1-butene, one of the main adducts. This can, however, be produced by one of the following paths involving a Π-butadienyl and/or Π-allyl intermediates [643]:

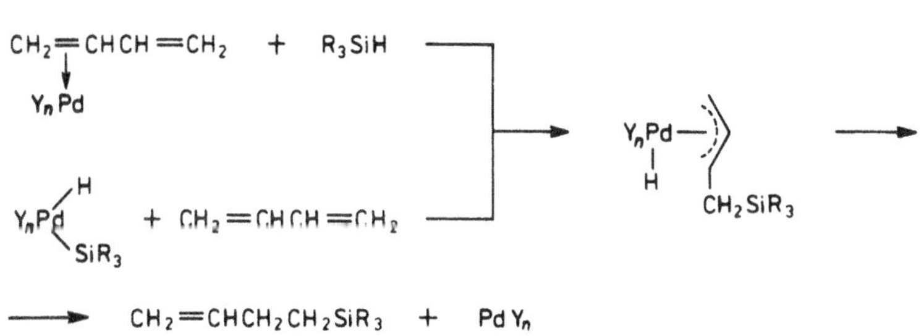

(170)

Numerous nickel and palladium complexes have also been employed as catalysts in the hydrosilylation of other dienes, e.g. isoprene [851, 855], phenylbutadiene [482], cyclopentadiene [1217, 1335], pentadiene [1335], 2,3-dimethyl-1,3-butadiene [1222], piperylene [1019] and trienes [1335]. Phosphine Pd complexes with chiral phosphines (MDPP, NMDPP) may also be used as catalysts for the asymmetric hydrosilylation of C=C bonds (styrene, norbornene, cyclopentadiene and dienes [485, 590, 1304]), affording a relatively high yield and optical purity. Less effective they appeared to be in asymmetric hydrosilylation of ketones [584].

2.4.9 *Ziegler Catalysts*

Ziegler or Ziegler–Natta catalysts, commonly applied in the polymerization of unsaturated compounds, are mainly heterogeneous and consist of aluminium alkyls or hydrides as co-catalysts with transition metal halides, alkoxides or organometallic complexes as the catalysts. In the presence of organometallic complexes alone, it is possible to carry out the reaction under homogeneous conditions. It is thought that the function of the aluminium alkyl is to facilitate the formation of an active transition metal species. The metal is alkylated when the chloride or other anion is replaced to give a coordinatively unsaturated organometallic intermediate. An olefin molecule can be added to the latter, yielding the Π-complex [281, 288].

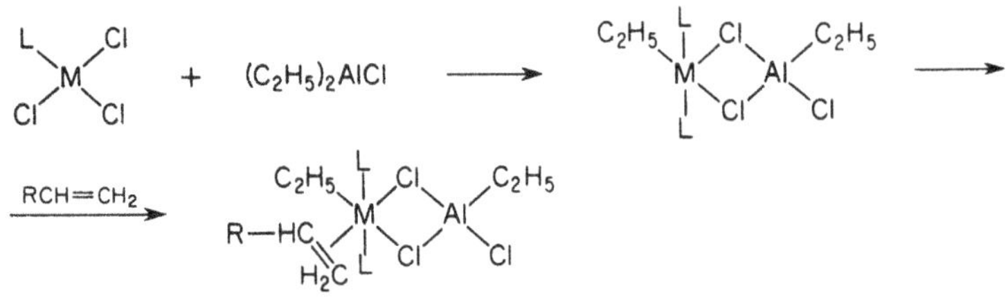

(171)

The Ziegler system ML_n–$Al(C_2H_5)_3$ (M = Ni, Co or Fe, $n = 2, 3$) has been mostly used in hydrosilylation. In this case the function of the aluminium alkyl now appears to be mainly that of a reducing agent, similar to its function during catalysis of polymerization by an Ni(II) compound which is reduced to a zero-valent complex during the process [493]. Thus, the definition of a Ziegler (or Ziegler–Natta) catalyst can be expanded to include binary systems involving transition metal salts (e.g. chloride, acac or complex) and a reducing co-catalyst. In exceptional cases silicon hydride can also act as a reducing agent when used as a substrate, the silyl–transition metal species formed *in situ* being responsible for the catalytic activity of the metal salts in hydrosilylation reactions.

The following Ziegler-type systems have been used as catalysts for hydrosilylation:

$Ni(acac)_2$–$Al(C_2H_5)_3$ [662, 663, 1017, 1327, 1328, 1330–1332, 1334, 1357],
$Ni(acac)_2$–$Al(OC_2H_5)(C_2H_5)_2$ [273],
$Co(acac)_2$–$Al(C_2H_5)_3$ [317, 644, 662, 1330, 1332],
$Ni(acac)_2$–$LiAlH_4$ [662],
$M(acac)_2$–$NaAl(OR)_2H_2$, M = Ni [207], M = Co [208],
CoX_2–$NaAlH_2X_2$ [663, 644, 1330, 1332, 1334],
$NiCl_2$–$Al(C_2H_5)_3$ [662],
$M(acac)_n$–$Al(C_2H_5)_3$ [663, 1324],
$V(acac)_2$–$Al(C_2H_5)_3$ [1326, 1327].

Organotitanium compounds, because of their remarkable activity in the polymerization of 1-olefins – the best commercial Ziegler–Natta catalyst – were discovered very recently as catalysts for hydrosilylation of ketones [815] as well as of olefins [472]. Tertiary systems involving a donor ligand (mainly PPh_3) have also been used quite commonly. This third component usually accelerates the reaction rate and has a marked effect upon the selectivity; for example, the three-component system $M(acac)_2$–[M = Ni, V]$Al(C_2H_5)_3$–PPh_3 is a highly active catalyst in the hydrosilylation of β-pinene [1327].

It is generally believed that in the presence of the two normal components of a Ziegler catalyst system a lower oxidation state of the transition metal is produced in the active intermediates generated [663]. Such complexes may be stabilized by suitable ligands such as dienes or acetylenes [524]. For this reason this type of unsaturated hydrocarbon has become the subject of many interesting studies.

Organometallic catalysts of the $Ni(acac)_2$–$Al(C_2H_5)_3$ type efficiently catalyse the addition of alkyl- and phenyl-silanes to conjugated dienes and trienes [273, 1330, 1334]. In the presence of this catalyst, hydrosilylation of 1,3-butadiene by trimethylsilane (substrate ratio 1:1, 60°C, 4 h) gives a 90% yield of 1-trimethylsilyl *cis*-2-butene (I). In the presence of excess butadiene (2.1:1), the di-adduct 1-trimethylsilyl-*cis*,*cis*-octa-2,6-diene is also formed (10%). The yield of the latter may be increased to 50% by the presence of triphenylphosphine in the system. The addition of dimethylsilane to butadiene also depends on the substrate, giving 1-dimethylsilyl-*cis*-2-butene as the product when the substrates are present, but with additional 1:2 adducts in the presence of excess butadiene and particularly in the presence of PPh_3.

Hydrosilylation studies of other conjugated dienes have led to the development of other highly efficient Ziegler systems of the type $M(acac)_2$–$Al(C_2H_5)_3$ (mainly Ni salts). Hydrosilylation of isoprene or penta-1,3-diene in the presence of such catalysts gives the 1,4-adducts, $CH_3CH{=}C(CH_3)CH_2SiX_3$ or $CH_3CH_2CH{=}CHCH_2SiX_3$, respectively as the major products, with the salts $NiCl_2$ and $Ni(acac)_2$ providing the best catalysts [543]. In the hydrosilylation of isoprene and trimethylsilane, $(CH_3)_2C{=}CHCH_2Si(CH_3)_3$ is a significant by-product of the reaction and for 1,3-pentadiene and triethoxysilane the corresponding by-product is $CH_3CH{=}CHCH[Si(OC_2H_5)_3]CH_3$ [543]. The formation of the latter may be suppressed by the addition of phosphines. With methyldichlorosilane and 1,3-diene, significant yields of the 1:2 diadduct are obtained [1018].

The addition of trimethylsilane to 3-methyl 1,4,6 heptatriene leads to the selective formation (in 90% yield) of 1-trimethylsilyl-5-methyl-*cis,cis*-hepta-2, 6-diene

$$CH_2{=}CHCH{=}CHCH(CH_3)CH{=}CH_2 \xrightarrow{(CH_3)_3SiH} \qquad (172)$$

$$\rightarrow (CH_3)_3SiCH_2CH{=}CHCH_2CH(CH_3)CH{=}CH_2$$

The $Ni(acac)_2$–$Al(C_2H_5)_3$ catalytic system is also very effective in the reaction of phenyl-substituted silanes with 1,3-dienes. Such reactions occur at low temperatures (−10 to +10°C), giving phenylsilyl-substituted adducts [1334]. Three products are observed in the hydrosilylation of cyclopentadiene by trichlorosilane: 3-trichlorosilylcyclopentene, 4-trichlorosilyl-cyclopentene and a 1:2 adduct [1217].

The three-component system $Ni(acac)_2 + AlR_3 + PPh_3$ (or another donor) has also been used as a catalyst for hydrosilylating olefins with functional groups [1014–1016] and of acetylene [663].

2.5   Catalysis by Supported Transition Metal Complexes

2.5.1   Background

During the past 20 years a great deal of work has been undertaken on the application of supported metal complexes as catalysts to many reactions in which homogeneous catalysts were previously employed [81, 914]. Much of the original interest focused on the potential commercial utility of this field, but studies have also focused on providing an understanding of the reaction mechanisms. Transition metal complexes may be supported heterogeneously on organic polymers such as polystyrene–divinylbenzene, polyvinyl chloride, polypropylene, polystyrene or a polyalcohol, on the surface of inorganic oxides such as silica or γ-Al_2O_3, and as is generally the case, on glass or molecular sieves. The metal complexes are attached to the supports by coordinating them with phosphine ($-PR_2$), amine ($-NR_2$) or other groups ($-SH$, $-CN$, etc.) linked to the organic or inorganic support. Such supported complexes constitute a new class of heterogeneous catalyst combining the advantages of the classic homogeneous catalysts such as high activity, selectivity, and the possibility of ligand selection, with those of heterogeneous systems, i.e. product separation on completion of the reaction.

Polymer-bound metal complexes have been mainly synthesized by the introduction of functional groups which act as ligands (e.g. phosphines) into organic polymers, followed by anchoring of the transition metal precursor onto such a "phosphinated" polymer according to the following scheme:

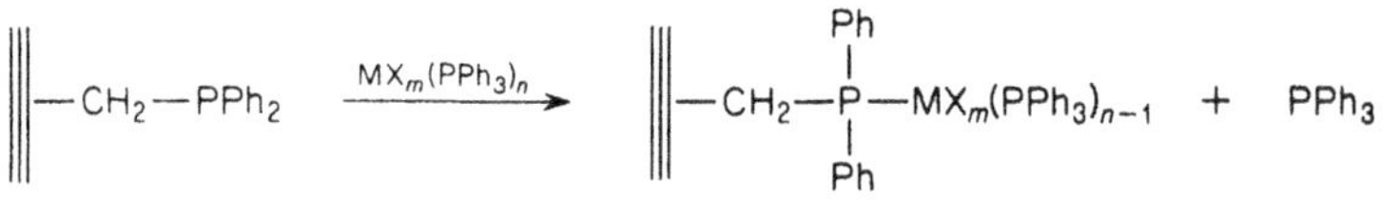

$$\left\Vert\!\!-CH_2-PPh_2 \xrightarrow{\;MX_m(PPh_3)_n\;} \left\Vert\!\!-CH_2-\!\!\!\underset{\underset{Ph}{|}}{\overset{\overset{Ph}{|}}{P}}\!\!\!-MX_m(PPh_3)_{n-1} \;+\; PPh_3 \tag{173}$$

where M denotes the metal (Pt, Pd, Rh, Ni, Ru). The application of polymer-supported catalysts has now been extended to the synthesis of complexes between transition metal derivatives and structurally ordered macromolecular ligands to give catalyst systems exhibiting high activity and stereoselectivity. The constitutional and configurational sequence of a macromolecular ligand in determining the properties of the corresponding transition metal complex is of particular importance in the design of catalytic systems, and controls their activity, selectivity and stereochemistry [219]. Polyamides were recently synthesized as a very suitable polymers for heterogenization which do not need any further functionalization and exhibit much higher thermal stability than conventional polystyrene supports [765].

Nevertheless, because of the lower thermal stability of numerous organic polymers and their swelling during reactions with solvents and reagents over extended time spans, organic supports have tended to be replaced by inorganic materials such as silica. Ligands may then be introduced by utilizing the following reactions of the silanol groups on silica [14]:

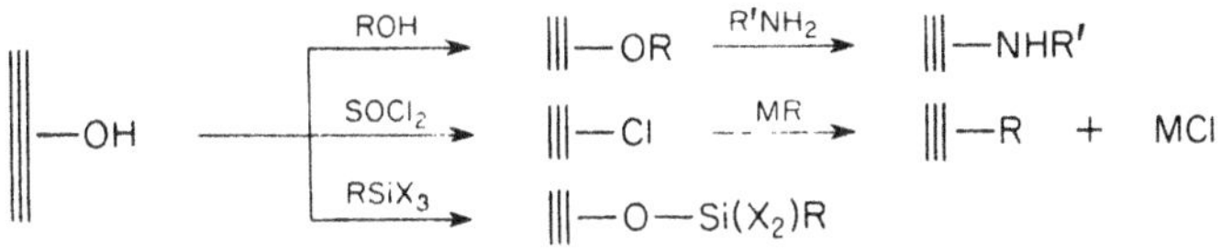

$$\text{|||}-\text{OH} \quad \xrightarrow{\begin{array}{c} \xrightarrow{ROH} \quad \text{|||}-OR \xrightarrow{R'NH_2} \text{|||}-NHR' \\ \xrightarrow{SOCl_2} \quad \text{|||}-Cl \xrightarrow{MR} \text{|||}-R \; + \; MCl \\ \xrightarrow{RSiX_3} \quad \text{|||}-O-Si(X_2)R \end{array}} \qquad (174)$$

where X is a hydrolyzable group (–Cl, –OR, –OCOR), R and R' are organic groups.

In the most common technique the metal complexes are anchored by means of siloxane chain (Scheme 174), where R can be, for example, $(CH_2)_2PPh_2$ [14, 204, 767], $(CH_2)_3CN$ [204], $(CH_2)_3Cl$ [746], $(CH_2)_2SR$ [D-083], $(CH_2)_3NR_2$ [746, F-025]. In this method the metal complexes may be attached in two ways:

1. By complex coordination on modified support:

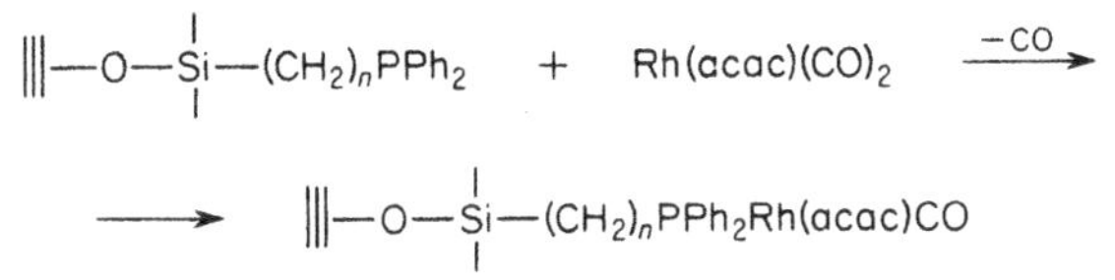

$$\text{|||}-O-\underset{|}{\overset{|}{Si}}-(CH_2)_nPPh_2 \; + \; Rh(acac)(CO)_2 \xrightarrow{-CO}$$

$$\longrightarrow \quad \text{|||}-O-\underset{|}{\overset{|}{Si}}-(CH_2)_nPPh_2Rh(acac)CO \qquad (175)$$

2. By modifying the support with an organosilicon–metal complex, prepared according to the equation:

$$(RO)_3Si(CH_2)_nPPh_2 + Rh(acac)(CO)_2 \xrightarrow{-CO} (RO)_3Si(CH_2)_nPPh_2Rh(acac)CO \qquad (176)$$

The organometallic species can also be attached directly to silica according to the following equation, e.g. allylmetals [230]:

$$SiO_2 + M(C_3H_5)_n \rightarrow SiO_2 \sim M(C_3H_5)_{n-1} + C_3H_6 \qquad (177)$$

Polyorganosiloxanes have recently been used as soluble and insoluble supports to provide heterogeneous catalysts from transition metal complexes, principally of rhodium [555, D-067, D-071]. Silsesquioxane, with the general formula $(RSiO_{1.5})_n$ (where R = H, alkyl, phenyl), is a suitable example of such a support [D-067]. There are also few reports on the use of supported chiral metal catalysts in asymmetric synthesis, mostly applying Rh as well as other (Pt, Pd) complexes.

The mechanism of catalysis by supported metal complexes is not yet clear. Thus, during the reaction the anchored complex may be reversibly detached from the support and can act as a homogeneous catalyst in solution [316] and/or it may act catalytically as part of a heterogeneous system [204]. Such immobilized complexes are commonly used as catalysts in hydroformylation, hydrogenation, oxidation, and in hydrosilylation reactions. Conventional precursors for supported hydrosilylation catalysts are frequently rhodium, platinum, palladium or nickel complexes.

2.5.2 Supported Platinum Complexes

Early studies of the preparation of heterogeneous metal complexes to provide supported hydrosilylation catalysts concentrated upon the immobilization of chloroplatinic acid by ion exchangers [974, 1232, 1344, 1345, 1348, D-071, S-016, S-025]. Such catalysts appear to be very effective in the addition of phenylmethylsilanes to 1-alkenes, 1-alkynes and ethyl acrylate, remaining active even after being used over 20 times [980]. The extensive work undertaken by Reikhsfel'd *et al.* was designed to evaluate the catalytic pathways of hydrosilylation by kinetic means and through studies of the substituent effect at silicon [974, 1344, 1345, 1348] during the reaction of triorganosilanes with 1-heptene using AW-17-8 anionite containing $PtCl_4^{-2}$ ions.

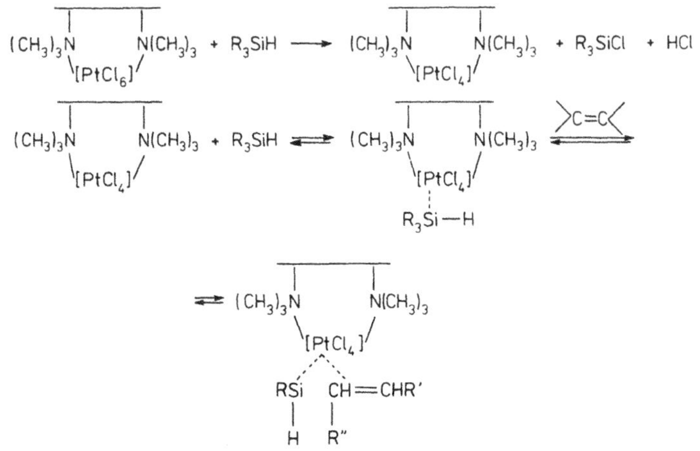

(178)

The proposed mechanism involves the preliminary reduction of Pt(IV) to Pt(II) followed by the formation of a silyl–platinum intermediate stabilized by olefin coordination. An increase in the catalyst concentration does not influence the induction period of the reaction [1348], thus suggesting that an adsorption–desorption process plays a decisive role in the catalysis [1344]. Cationic exchangers have also been tested as supports for platinum complexes [S-016].

Hexachloroplatinic acid is a common platinum precursor capable of support on silica and other inorganic materials [204, 505, 506, 746, 887, 1172, 1285, C-007, D-008, D-063] or on organic polymers [217, 293, 621, 764, D-056] and affording an efficient catalyst for hydrosilylation. Other complexes of platinum, e.g. Na_2PtCl_4 [343, 887, D-063, D-083, D-089], K_2PtCl_4 [243–245, 293] and $Pt(PPh_3)_4$ [767], have also been immobilized and used as hydrosilylation catalysts. The sorption of chloroplatinic acid from aqueous solution onto a macroreticular styrene–divinylbenzene copolymer substituted with cyanomethyl [621] or dimethylamine [764] groups, followed by successive activation of this precursor by heating it in a stream of acetylene, has led to the development of catalysts active towards the addition of trichlorosilane to acetylene which yields vinyltrichlorosilane. A kinetic study of reactions undertaken in the gaseous phase and in a flow system at 100°C has demonstrated the predominant role of the product in determining the reaction rate. The apparent activation energy, as evaluated from an Arrhenius plot over the temperature range 80–140°C, was found to be 4.4 kcal/mole [621]. Platinum (as well a palladium) derivatives of polymers functionalized with the ferrocene derivatives are effective catalysts for hydrosilylation of styrene and 1-hexene [293]. Generally, platinum catalysts prepared as outlined above can be used several times for various reactions involving hydrosilylation of C=C and C≡C bonds with virtually no change in catalytic activity [217, D-056].

Platinum complexes anchored to silica or other inorganic supports (porous glasses, zeolites, γ-Al_2O_3) by means of a siloxane chain and a coordinating group such as –SH [243, 1286], –SR [245, 504–506, D-082, U-105], $-NR_2$ [204, 505, 746, C-007], $-NH_2$ [986, 1285, D-089, U-146], $-PR_2$ [204, 244, 767, 938, 1203, 1295] or –CN [204, 679, C-013] are commonly used in hydrosilylation and in other reactions catalysed by metal complexes. Under these circumstances the catalytic activity is comparable to that with the homogeneous analogues, although an increase in selectivity with heterogeneous catalysts has been noted [204, C-007]. Polysiloxane-bound platinum complexes prepared by the reaction of chloroplatinic acid silica-supported with polysiloxane containing sulphur was very effective in the hydrosilylation of allyl chloride producing 3-chloropropyltrichlorosilane in 84% while in the presence of H_2PtCl_6 it gave only 60% yield [505]. Heterogenized Pt catalysts linked to inorganic supports via various functional groups have been very extensively studied for the hydrosilylation of 1-alkenes [204, 244, 293, 343, 434, 506, 765, 938, 939, 1285, 1295], styrene [293, 434, 1295, S-039], allyl derivatives [243, 244, 505, 679, 746, 1295 C-006, C-083], 1,3-dienes [C-007], and acetylene [504, 505, S-036, Dd-005].

A significant increase in selectivity has been reported for platinum complexes dissolved in polysiloxanes of the general formula $[OSi(CH_3)(CH_2)_2CN]_n$, followed by adsorption of the catalyst onto silica or active carbon. Catalysts prepared in this fashion can be reused after filtration. Carbon-adsorbed Pt complexes are more active than the silica-adsorbed analogues [Dd-003, Dd-005]. Platinum catalysts dissolved in polysiloxanes possessing amine [J-050] or vinyl [D-063] substituents exhibit a high efficiency for the activated cure of silicone rubber brought about by hydrosilylation processes. The 22-fold improvement in the activity of platinum–phosphine complexes supported on silica in comparison to the homogeneous complex presumably arises from the greater effect of triphenylphosphine dissociation from the complex under heterogeneous conditions [767].

Platinum(II) oxalate can be supported on silica via a trichlorophosphine anchoring ligand instead of PPh_3 groups which gives a more stable complex, and no leaching of the platinum was observed under catalytic conditions [938]. Ultraviolet irradiation of such a precursor in the presence of $(CH_3)_2SiHCl$ results in the loss of the oxalate ligand as CO_2 and the formation of a surface-attached PtL_2 species. Thus silica-bound Pt complex catalyses the hydrosilylation of olefins [938, 939, 1203] as well as other reactions.

Different types of supported catalyst have been prepared by anchored hexachloroplatinic acid to silsequioxanes. The latter can be obtained, for example, by hydrolysing 3-diphenylphosphinepropyltriethoxysilane [D-067], a process which leads to an insoluble support. The recently developed concept of heterogenizing metal complexes is based on the polycondensation of suitable trialkoxysilyl-substituted organo-sulphides, -phosphines and -amines bound as ligands to a given complex, such as a Pt complex [887, D-082]. Such a procedure leads to the synthesis of a new insoluble type of catalyst possessing an organopolysiloxane backbone. Such systems are discussed further below.

2.5.3 Supported Rhodium Complexes

As has been mentioned above, rhodium complexes immobilized on inorganic or organic supports constitute the majority of active "heterogenizing" hydrosilylation catalysts. As such they have developed into important catalyst systems and have been studied under hydrosilylation conditions to establish the mechanism of catalysis by supported metal complexes. The following precursors have been employed in the synthesis of such systems:

$RhCl(PPh_3)_3$ [216, 632, 739, 746, 750, 767, 1167, C-008, D-056];
$Rh_2Cl_2(CO)_4$ [316, 768, 769, C-022, Dd-002–004];
$RhCl_3 \cdot 2\,H_2O$ [204, 216, 217, 244, C-007, C-022, D-059, D-072, D-077];
$RhH(CO)(PPh_3)_2$ [767, C-007, Dd-002–004];
$RhCl(CO)(PPh_3)_2$ [C-022, Dd-003];
$Rh_2Cl_2(C_2H_4)_4$ [336, 769];
Rh_2Cl_2(c-octene)$_4$ [209, 211, 599, 941];
$RhX(C_2H_4)$(acac) [D-083, U-108].

Silica and various organic copolymers have been mainly used as supports. In contrast to silica, other inorganic supports such as γ-Al_2O_3, zeolites and porous glasses, do not influence the activity or selectivity [204, C-007, Dd-003]. Diphenylphosphine is a common functional group which has been linked to silica via an organosilicon chain, enabling the coordination of the rhodium precursor [204, 244, 743, 767, 941]. Polyligands with amine [204, 350, 743, C-007] and mercapto [96, 744] groups have also been tested in hydrosilylation. A separate group of catalysts exists in which rhodium complexes are sorbed onto silica or some other porous support, but such complexes are subject to removal from the support, by the solvent, as a result of repeated usage in a catalytic process [211, Dd-002].

The catalytic activity and reaction selectivity of all silica-supported rhodium catalysts studied under hydrosilylation conditions have been shown to be essentially similar to those pertaining in the corresponding homogeneous systems. The support has no influence on the mechanism of the reaction, although it enables more stable catalysts to be formed. An exception to this rule has been observed for a catalyst derived from $RhCl_3$ linked to phosphinated silica, which appears to be more active than the homogeneous Wilkinson's complex. Elemental analysis has indicated that the P–Rh ratio in this catalyst is equal to two, so that it appears that most of the rhodium atoms in the heterogeneous system are bound by only two phosphine groups, facilitating coordination of the reactants [767]. The catalytic activity also appears to depend on the structure of the functionalizing agent, which may be varied on the one hand, by changing the length of the phosphine–alkylene chain $-(CH_2)_nPPh_2$ (where n = 1–6) linked to the silica support or, on the other hand, by direct coordination to the rhodium precursor [769]. On the basis of ESCA and IR measurements, the following types of catalyst structure have been proposed:

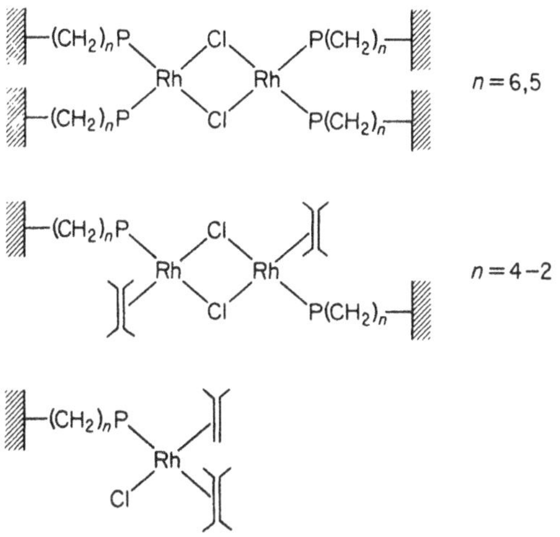

(179)

It has been calculated that for a silica with a surface area of 365 m^2/g and a phosphine concentration of 0.125 mmole PPh_2/g, the average intermolecular distance between two phosphine groups is about 20 A. This distance can be changed depending on the length of the hydrocarbon chain – $(CH_2)_nPPh_2$ incorporated in the structure of the complex adsorbed on the silica surface. When n = 5 or 6 replacement of the ethylene from the Cramer complex causes the formation of dimeric bridge structure. When n = 2, 3 or 4 the product is predominantly a dimer in which rhodium is linked to the silica by one phosphine group. When n = 1 the bridge structure is split and the system exhibits a monocoordinated, mononuclear structure with substantial activity towards hydrosilylation [769].

The amine functional group when linked by an organosilicon chain to a silica surface also displays a considerable efficiency in immobilizing rhodium complexes, and in a similar manner to phosphine derivatives influences the catalytic activity of the system [204, 746, C-007]. It has been shown that when Wilkinson's complex is immobilized on silica by means of a $-(CH_2)_3NCH_2CH_2OCH_2CH_2$ group, constant activity is guaranteed even after the catalyst has been used nine times.

An unexpected promotion of the catalytic activity of rhodium complexes anchored onto a silica surface previously modified by sulphur derivatives has been observed [D-072, D-083, 744]. Such behaviour has been noted for $Rh(C_2H_4)_2(acac)$ immobilized on silica pretreated with the disilazane sulphur derivative $HN[Si(CH_3)_2CH_2CH_2S(C_2H_5)]_2$. This catalyst appears to be highly active in the hydrosilylation of 1-decene by heptamethylhydrotrisiloxane, particularly after being used twice [D-083].

Organic polymers (particularly styrene–divinylbenzene copolymers) containing functional groups such as $-PPh_2$ [216, 217, 767, 1168, C-008], $-NR_2$, $-CN$, $-CH_2Cl$ [217, 316, D-056] and others have also been commonly used to immobilize rhodium complexes. The catalytic activity of the resulting catalyst appears to be very similar to that of the homogeneous analogue [216, 1168, C-008].

EXAFS (extended X-ray absorption fine structure) studies of the precursor of the catalytic site associated with a polymer-bound Wilkinson's catalyst have recently been made. The results suggest that a catalyst can aggregate to form binuclear clusters (Scheme 180) when attached to a polymer, particularly a 2% DVB crosslinked polymer [961, 962].

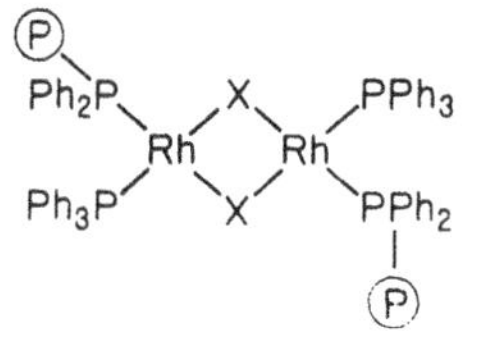

(180)

When the catalyst was supported on a 20% DVB crosslinked polymer it was found to be monomeric containing three phosphorus atoms and one X atom [962].

Polymer-supported rhodium catalysts, prepared by the reaction of tetracarbonyldichlorodirhodium with dimethylamino-methylated styrene–divinylbenzene copolymers with different degrees of crosslinking (10–60%), exhibit a selectivity comparable with that of the soluble rhodium complexes and an activity which increases as surface area and pore radius increase [316]. Interesting attempts have been made to characterize some rhodium precursors attached both polymeric and silica supports containing bidentate amine–phosphine ligands, i.e $Ph_2PC_6H_4CH_2N(CH_3)_2$ (PCN) and $Ph_2PC_6H_4N(CH_3)_2$ (PN). Spectroscopic (IR- and X-ray photoelectron) studies and measurements of the evolved carbon monoxide have shown that various rhodium carbonyl complexes are formed on the surface, depending on the degree of crosslinking and the ligand/Rh ratio. It has been concluded that the polymeric support acted either in bidentate fashion when the ligand sites were dispersed (low L/Rh ratio) or in monodentate manner through phosphorus, when the L/Rh ratio was high [768]. Polyamide-bound rhodium complex catalysts (of the same support as the platinum ones mentioned previously) were synthesized and tested in the hydrosilylation of 1-hexene [765].

Polyphenylenes (M_w = 1000–3000) have been used as macroligands associated with rhodium (and also Pd and Pt) complexes. Such ligands are bound to the metal through the Π-bonds of the benzene rings in the polymer. The resulting catalysts have been employed both as soluble compounds in polymers and/or solvents and as insoluble compounds. They exhibit very high activity and selectivity to yield a β-adduct in the hydrosilylation of acrylonitrile by trichlorosilane [D-077]. Bidentate polymeric metal complexes containing the chelate ligand bis-(18-hydroxy-5-quinolinyl)methane with the general structure (Scheme 181) where M is Rh, Pt, Pd or Ni, are effective catalysts in the hydrosilylation of 1-hexene, styrene and butadiene, but exhibit only a relatively low activity towards the addition of trichlorosilane to acrylonitrile and allyl chloride [1220].

Polyorganosiloxanes have also been used as matrices for rhodium catalysts in

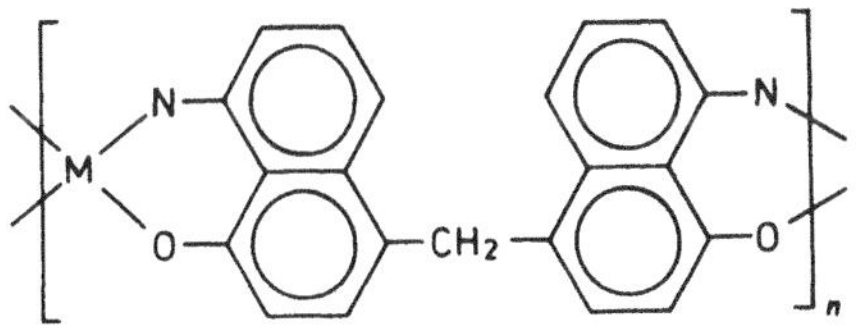

(181)

hydrosilylation reactions [750, D-098, D-113]. Kinetic data based on the reaction of 1-hexene with triethoxysilane have shown that modification of Wilkinson's complex by the use of a phosphinated polyalkylsesquisiloxane support and soluble silylalkylphosphine ligands has virtually no effect upon the mechanism of the reaction [750].

A few specially substituted silanes, for example bis(triethoxysilylpropyl)monosulphane [D-098], tris(triethoxysilylpropyl)amine [887, D-082] and tris(triethoxysilylpropyl)phosphine [D-113], are susceptible to condensation; their coordination to metals (Rh, Ir, Pt, Ru) results after the hydrolysis and polycondensation in the formation of solid catalysts containing organofunctional polysiloxanes. This sol-gel method was applied for the preparation of supported metal-complex catalyst illustrated in Scheme (182) [887, D-082].

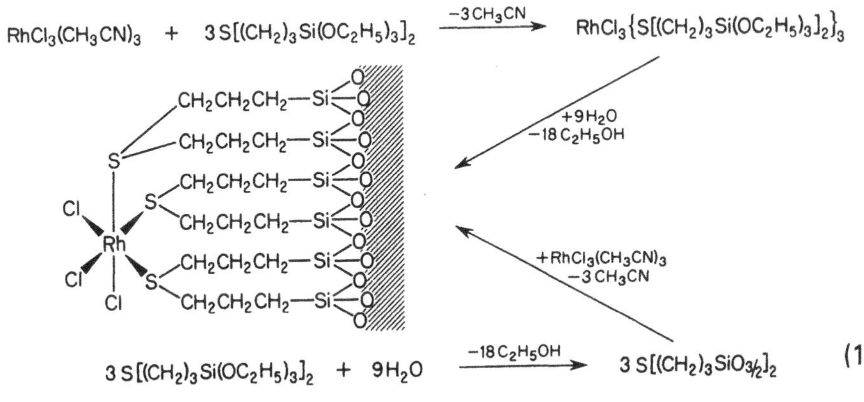

(182)

The supported catalysts thus obtained have been employed in the hydrosilylation of allyl chloride, for example. Their advantages include a higher temperature stability, non-swelling properties and resistance to organic reagents; furthermore they can be used in both aqueous and organic media [887, D-082, D-098, S-050]. Schubert *et al.* prepared such a type of a catalyst by polycondensation of $Rh(CO)Cl[PPh_2CH_2CH_2Si(OEt)_3]_2$ with tetraethoxysilane which exhibit similar catalytic activity in the hydrosilylation of 1-hexene to homogeneous catalyst attaching to the surface of SiO_2 [1029].

Quite novel supports for heterogenizing metal complexes, which have also been tested under hydrosilylation conditions, have been prepared by grafting organosilyl and organosiloxyanyl radicals onto mineral silicates, e.g. chrysotile asbestos [739]. An idealized formula for such an organosiloxanes–silicate support has been suggested. Thus Scheme (183) illustrates the arrangement which appears to be the most stable and effective in the hydrosilylation of 1-hexene by trichlorosilane in solution, and also in the reaction of acetylene with trichlorosilane and methyldichlorosilane in the gaseous phase [742–744].

Polyorganosiloxane chains, when grafted *in situ* in an acidic medium onto chrysotile structure, provide a support which retains certain elements of the original silicate structure, *viz.* its fibrous shape, which provides additional advantages in its use as a support for complex catalysts. Moreover, the processes are accompanied by a considerable increase in the surface area of the organosiloxy derivative of chrysotile [739].

The combination of the two latter concepts for synthesis of new polymeric inorganic supports, i.e. sol-gel process and grafting organosiloxanyl radicals aimed to immobilization of rhodium(I) complexes bound to the nitrogen of amine derivatives of polyorganosiloxanes grafted onto the silicate structure [743] (Scheme 183). These catalysts were successfully tested in hydrosilylation of 1-alkenes.

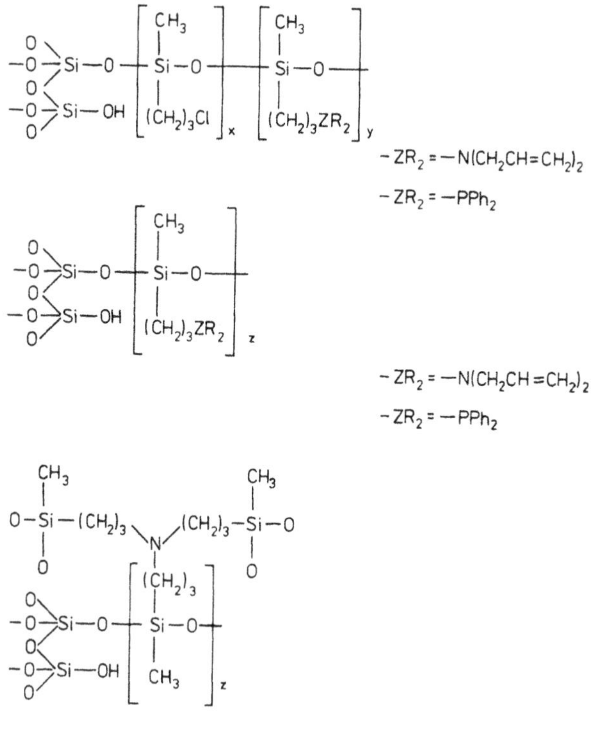

(183)

The asymmetric hydrosilylation of prochiral ketones in the presence of supported complexes has also been a subject of considerable investigation since 1973. Chiral rhodium(I) complexes, prepared *in situ*, have been bound coordinatively to inorganic [209, 211, 599] or organic polymers [336] (Scheme 184). It is noteworthy that although the immobilized chiral rhodium complexes sometimes exhibit a higher catalytic activity, their stereoselectivity is similar to that of the soluble chiral analogue. Several efficient rhodium(I) catalysts for the enantioselective hydrosilylation of ketones have been reported. These catalysts

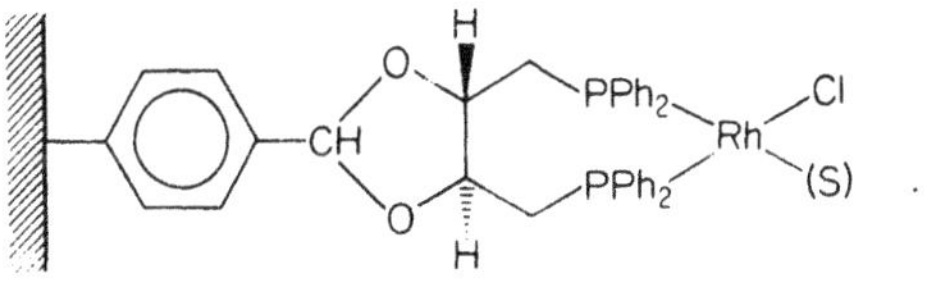

(184)

are highly active and exhibit an asymmetric efficiency. Such efficiency, and the activity of the catalyst, decreases during the course of the reaction, presumably because the metal complex becomes detached from the support [209, 211].

Immobilization of the complexes has no effect upon the stereoselectivity of the hydrosilylation of cyclohexene with, for example, $RhCl(PPh_3)_3$ and SiO_2 even after repeated use of the following catalyst [519].

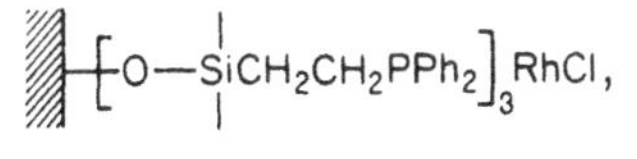

2.5.4 *Supported Complexes of Palladium, Nickel and Other Metals*

Supported complexes of palladium and nickel are less well known than platinum or rhodium as hydrosilylation catalysts. This is because of the ready abstraction of palladium from the support [204] as well as to the generally low selectivity of nickel complexes in hydrosilylation.

The majority of the methods employed for the immobilization of palladium and nickel complexes are the same as for the platinum and rhodium complexes described in preceding sections. The initial precursors for heterogenized palladium complexes are $PdCl_2$, $PdCl_2(PPh_3)_2$, and $PdCl_2(PhCN)_2$ [204, C-007, D-060], while for nickel catalysts they are $NiCl_2$ and $NiCl_2(PPh_3)_2$ [D-070, 1108]. The structure of the most readily heterogenizing catalysts of palladium and nickel was discussed when platinum and/or rhodium complexes, respectively, were considered above [204]. In order to stabilize silica-supported nickel complexes containing phosphinoxide ligands Reikhsfel'd *et al.* used organosilicon derivatives of phosphine oxides as their modifiers [1108].

Palladium and nickel complexes immobilized on various organic polymers appear to be highly efficient in the hydrosilylation of conjugated dienes by organosilicon hydrides of the general formula $RSiHCl_2$, $SiHCl_3$ and $(R'O)_3SiH$, where R = CH_3, C_6H_5 and R' = CH_3, C_2H_5, C_3H_7, $CH_3OC_2H_4$, $C_2H_5OC_2H_4$. Such reactions lead to the selective generation of unsaturated organosilicon compounds, e.g. butenylsilanes from 1,3-butadiene and trisubstituted silanes [D-060, D-070]. The stability of the heterogenizing palladium complexes depends markedly on the type of support used; for example, $PdCl_2$ attached to phosphinated SDVB [D-060] is much more stable than when supported on γ-Al_2O_3 [204]. The silica-supported alkyl–nickel proved to be a moderate catalyst for the hydrosilylation of dienes and 1-alkynes. The products and the proportion of isomers formed are essentially the same as those formed using Ziegler-Natta system, $Ni(acac)_2$–Et_3Al although the activity of the SiO_2–$Ni(C_3H_5)$ seems to be lower [230].

More complex supports were prepared by reaction of chloromethylated polystyrene with lithium derivatives of ferrocene and substituted ferrocenes to immobilize chloropalladium complex [293]. The catalyst was applied in stereoselective α-addition of trichlorosilane to styrene as well as to 1-hexene.

Cationic exchangers have also been used as suitable supports for Fe, Co and Ni cations which form complexes active in the hydrosilylation of 1-alkenes and styrene [S-033]. An octacarbonyldicobalt complex sorbed on γ-Al_2O_3 provides an example of an immobilized photocatalyst tested in the addition of hydrosilanes to 1-pentene [963].

2.6 Photo- and Peroxide-initiated Catalysis by Metal Complexes

Photogenerated catalysts, defined as the light-induced generation of a ground-state catalyst from a catalytically inactive precursor, have become a topic of current interest in catalysis by transition metal complexes [1020]. Such methods have also been applied in hydrosilylation. However, most of the literature reports on photo-assisted homogeneous catalysis, particularly, in hydrosilylation, deal with carbonyl and phosphine metal complexes and their substitution products as catalyst precursors [1293].

As far as carbonyl complexes are concerned the formation of the proper catalyst proper upon irradiation involves the loss of at least one CO ligand from the initial metal complex. The first photocatalyst discovered was $Cr(CO)_6$ which is active in the 1,4-hydrosilylation of 1,3-dienes to yield allylsilanes using a method of synthetic utility [1293]. Wrighton *et al.* have reported the high catalytic activity of $Fe(CO)_5$ upon UV irradiation under mild temperature conditions (0–50°C) in the hydrosilylation of alkenes [1028]. The truly photocatalytic nature of the reactions is demonstrated by the quantum yields which exceed unity and by the strong influence of temperature on the photocatalysis. The mechanism involves the generation of the active intermediate $H(R_3Si)Fe(CO)_3$(alkene) as shown in Scheme (185) (where $HSiR_3$ stands for $HSi(CH_3)_3$ and alkene for propylene) [1028].

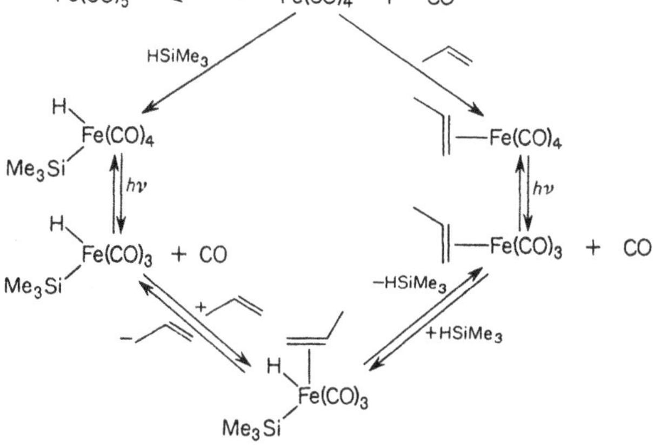

(185)

This scheme assumes the formation of an "$Fe(CO)_3$" species as the repeat unit, thus proving that the photodissociative role of irradiation is continuously needed to maintain the steady-state concentration of this active species. The removal of CO under such conditions allows inactive polynuclear iron carbonyl complexes to be formed. Irradiation of a mixture of 1,3-dienes, triorganosilanes and the $Fe(CO)_5$ precursor leads to the formation of isomeric butenyltriorganosilicon derivatives [366]. The proposed mechanism for the photogenerated catalysis involves the formation of an (η^3-anti-butenyl)tricarbonyliron–trialkoxysilyl species as the active intermediate, which also fits in with the above reaction scheme.

Contrary to commonly proposed Chalk and Harrod involving the key steps of insertion of olefin into M–H bond of an R_3Si–M–H complex followed by the reductive elimination of alkylsilane an alternative mechanism has recently been suggested for the photocatalysed hydrosilylation of alkenes using $Fe(CO)_5$ [1028], $M_3(CO)_{12}$ (M = Fe, Ru, Os) [72] or $(CO)_4CoSiR_3$ [964] (see Scheme 186) as catalysts [960, 1031].

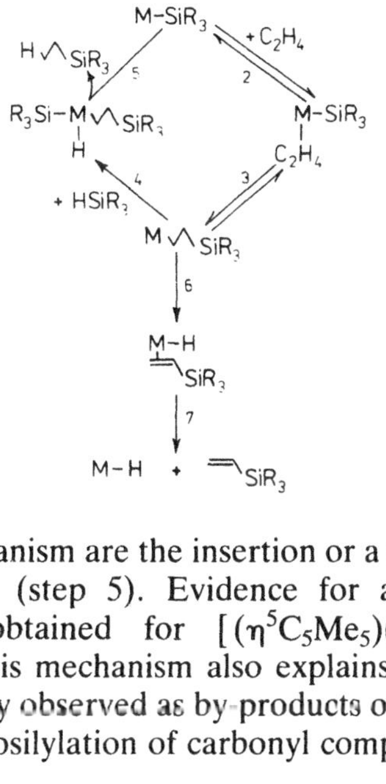

(186)

The key steps of this mechanism are the insertion or a reductive elimination of an alkyl and hydrido ligand (step 5). Evidence for all steps required by this mechanism has been obtained for $[(\eta^5C_5Me_5)(CO)_2FeSiR_3]$ [960] and $[(CO)_4CoSiR_3]$ [1031]. This mechanism also explains the formation of alkenylsilanes which are frequently observed as by-products of hydrosilylation reactions.

The photocatalytic hydrosilylation of carbonyl compounds occurs according to the following scheme:

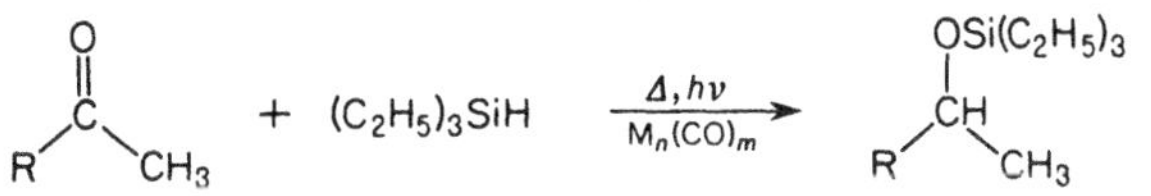

R=CH_3, C_5H_{11}; M=Re, Ir, Os, Ru, Fe... (187)

Ruthenium, osmium and iridium clusters are effective photocatalysts of the hydrosilylation of acetone, whereas $Fe_3(CO)_{12}$, $Co_2(CO)_8$ and $Co_4(CO)_{12}$ are known to be efficient photoactivated catalysts in the hydrosilylation of olefins. It was also demonstrated that $Ru(CO)_4PPh_3$ and $Ru(CO)_3(PPh_3)_2$ complexes can serve as catalysts for the photochemical hydrosilylation of benzaldehyde [1000].

The most active species in the hydrosilylation of acetone is one derived from $Re_2(Co)_{10}$, which exhibited some activity even after 20 h in the dark. The observed catalytic activity sequence $Re_2(CO)_{10} > Os_3(CO)_{12} > Ru_3(CO)_{12}$ partially accounts for the greater quantum yield in the $Re_2(CO)_{10}$ decomposition ($\phi = 0.6$) relative to that for the Os and Ru compounds ($\phi = 0.03$). In addition, it is suggested that a comparison of the relative activations of Ru and Os cluster precursors reflects the catalytic ability of the products of the primary photophysical reactions. Diradical complex formation takes place preferentially by metal–metal bond cleavage as fas as $Ru_3(CO)_{12}$ is concerned, (Scheme 188), while Os–CO dissociation is favoured in the case of $Os_3(CO)_{12}$, which forms a cluster containing a 16-electron centre (Scheme 189) [1313].

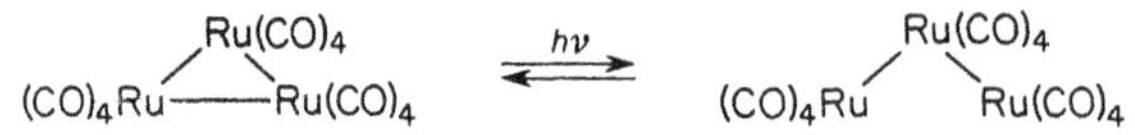

(188)

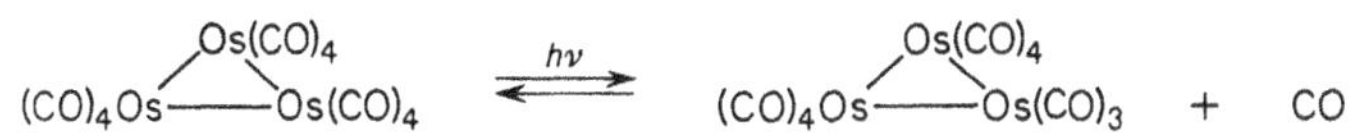

(189)

In order to inhibit cluster fragmentation during catalysis Pittman *et al.* have used a metal cluster bonded by stable non-fluxional bridging ligands as a heterometallic catalyst for the hydrosilylation of acetophenone by triethylsilane [915]. It was found that the quantum yields for 254 nm radiation were greater than for 255 nm radiation, as expected if the catalytic mechanism requires the loss of CO. HPLC examinations after the reaction (the cluster concentration did not change during the reaction) strongly suggested that the intact clusters are the active catalysts. Surprisingly, the initial cluster can be regenerated by reassociation of CO [915].

Low-valent metal complexes containing non-carbonyl ligands also exhibit d–d transitions in the near-UV, and give rise to similar photodissociative behaviour. When the labilized ligand is a phosphine, and the reaction is performed in air, a secondary photooxidation of the free ligand to phosphine oxide readily occurs in a manner characteristic of a thermal reaction.

We have already mentioned that for thermal processes, since phosphine oxides are not so strongly coordinated as their phosphine counterparts, UV radiation can lead to the generation of high concentrations of coordinatively unsaturated centres by ligand dissociation and free ligand modification to prevent back-reactions (compare Scheme 127 [350]). Photocatalytic experiments also confirm the previously suggested role of dioxygen as a Wilkinson complex co-catalyst. Faltynek has recently reported the effects of air and/or UV light on the catalytic activity of $RhCl(PPh_3)_3$ in the hydrosilylation of divinyltetramethyldisiloxane by sym-tetramethyldihydrosiloxane [350]. The reaction rate was considerably increased when the reaction mixture was photolysed in the presence of air or soluble oxidizing agents. Mechanistic considerations based on photocatalytic kinetic and spectrophotometric methods, and the preparation of intermediates, point to the suggestion that the generation of $RhCl(PPh_3)_2$ is the rate-determining step in the reaction. The relative sequence of rate constants $K_1 > K_2 > K_3 > K_4$

demonstrates that the direct photoconversion of the oxorhodium complex to $RhHCl(PPh_3)_2\ SiR_3$ is partly responsible for the observed increase in the rate, and that the oxidation of the ligand is more important than enhanced dissociation in these systems. It is suggested that these unsaturated intermediates are formed in four ways (Scheme 190).

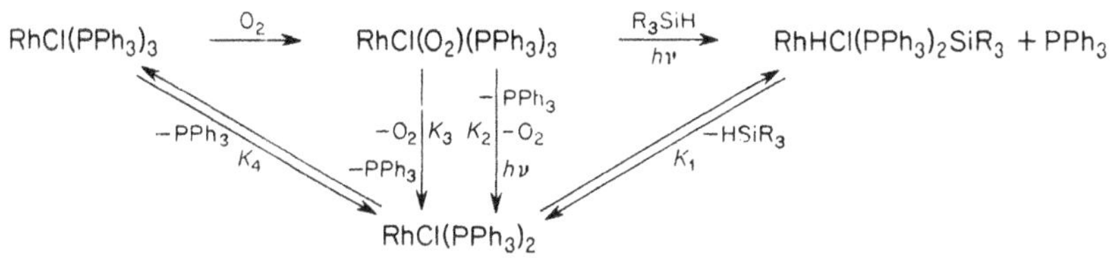

(190)

Photo-induced catalysis by the ruthenium complex $RuCl_2(PPh_3)_3$ in air leads to an almost three-fold increase in the rate of hydrosilylation of vinyltriethoxysilane with triethoxysilane, as well as to an increase in the rate of the metathesis accompanying the hydrosilylation. This can be partly explained not only by phosphine ligand photodissociation, but also by the generation of high concentrations of metal–carbenes as active metathetic species [736, 740]. The thermal hydrosilylation of a ketone or alkyne by triethylsilane catalysed by carbene–metal complexes is very slow at room temperature. However, on irradiating a reaction mixture containing diphenylacetylene, triethylsilane and a catalyst [*cis*-$RhCl(cod)\overline{C(CN)(CH_3)(CH_2)_2\ N}(CH_3)$] a 60% yield of the *trans* adduct can be obtained after 1 h at 20°C (compared with a 2% yield if the reaction mixture is not irradiated [660]).

Recently information on photochemically induced hydrosilylation by platinum compounds, e.g. chloroplatinic acid, was reported to increase the yield (above 90%) of cycloalkylsilane compounds and enable the process to occur at ambient temperature [E-038].

In most of the above-mentioned photocatalysed reactions, no hydrosilylation products were formed under control conditions either in the absence of UV radiation or in the absence of a catalyst. This proves that an active catalyst is photogenerated from the initial metal complex. The quite different role of UV light was demonstrated previously in the hydrosilylation occurring in the presence of silica-supported bis(trialkylphosphine)platinum oxalate precursor [938, 939, 1203]. UV irradiation yields SiO_2–L_2Pt or (in atmosphere of CO) [SiO_2]–$L_2Pt(CO)_2$ surface bound zero-valent platinum complexes that catalyse effectively the hydrosilylation of olefins.

Molecular oxygen has become a commonly used co-catalyst for inactive or weakly active transition metal complexes. In addition, other oxidizing agents, mainly peroxides, have recently been used to activate rhodium complexes in particular [202, 314, 350], but also other metal carbonyls [202], as catalysts for hydrosilylation. The catalytic activity of chlorobis(triphenylphosphine)carbonyl rhodium(I) in the hydrosilylation of the C=C and C=O bond can be greatly increased by the addition of about 50% molar excess of tert-butyl hydroperoxide. The Wilkinson's complex $RhCl(PPh_3)_3$ becomes moderately active at room temperature during the hydrosilylation of 1-octene with triethylsilane, when various organic peroxides are used as co-catalysts in the molar ratio ROOH/Rh = 8:1. Chromium triad carbonyls $M(CO)_6$, where M = Cr, Mo, W have been tested to examine the effect of various organic peroxides on the hydrosilylation of 2,3-dimethyl-1, 3-butadiene by triethyl-, triethoxy- and methyldiethoxysilanes. The results have shown that, provided a correct choice of organic peroxide is made, it is possible to enhance the activity of these hexacarbonyls [202].

The co-catalytic behaviour of organic peroxides, especially in association with carbonyl and/or phosphine complexes of transition metals, is similar to that of dioxygen, and has been attributed to the oxidation of carbon monoxide to carbon dioxide, and of phosphines to phosphine oxides. The oxidation products are known to possess much weaker coordinating ligands than those initially formed. The evidence for organic oxidant promotion of RhCl(cod)(phosphine) catalysed hydrosilylation of 1-hexene was previously demonstrated [331a, 741]. Its role was shown to oxidize phosphine to phosphine oxide yielding very readily a tricoordinated key intermediate.

The limited experimental work that has been conducted on this subject to date suggests that the application of various oxidizing agents as co-catalysts for metal complex precursors could lead to a new class of hydrosilylation catalysts which would be generally useful in synthesis and industrial applications.

CHAPTER 3

The Reactivity in Hydrosilylation of Particular Types of Organic Compounds with Multiple Bonds

THE descriptions of silicon hydride additions to multiple bonds given in the literature do not contain many examples of the direct structural effect of the unsaturated compound on the yield, selectivity and rate of the reaction studied.

For this reason the aim of this chapter is to consider the reactivity of the main classes of unsaturated organic compounds, especially those used in the synthesis of organosilicon compounds of general utility, with a view to characterizing the reactivity of particular multiple bonds, and the directions in which hydrosilylation and side-reactions involving such bonds proceed. Information on structural effects is also discussed. A number of recent comprehensive reviews of various classes of compounds, are available, especially those by Lukevics *et al.* (all types of organic compounds) [Rev.22], Ojima *et al.* (dienes, acetylenes and carbonyl compounds) [Rev.31, Rev.32], and Yur'ev and Salimgareeva (olefins) [Rev.43], and since Part II of this book lists examples of the hydrosilylation of these organic compounds (although ordered according to the silane), discussion in this chapter is limited to the fundamental types of unsaturated organic compounds which undergo hydrosilylation.

3.1 Hydrocarbons with an Isolated C=C Bond

3.1.1 Alkenes and Arylalkenes

The addition reaction of hydrosilanes to alkenes leads to alkyl derivatives of silicon. Thus 1-alkenes react with hydrosilanes in the presence of most catalysts (except for some Pt(II) and Pd(II) chiral phosphine complexes and some nickel complexes) to give 1-silylalkanes. Reactions of 2-alkenes, 3-alkenes, etc., with monohydrosilanes lead predominantly to alkylsilanes (silylalkanes) with a terminal silyl group; 2-silylalkanes can also be formed by such a reaction [1193]:

$$CH_3CH{=}CHCH_3 + C_6H_5SiHCl_2 \xrightarrow{H_2PtCl_6} \begin{cases} CH_3(CH_2)_3SiCl_2C_6H_5 \\ C_2H_5CH(CH_3)SiCl_2C_6H_5 \end{cases} \qquad (191)$$

Pentyltrichlorosilane may be obtained from the hydrosilylation of *cis*- and *trans*-2-pentenes with trichlorosilane in the presence of H_2PtCl_6 or $[(C_2H_4)_2RhCl]_2$ as the catalyst [1167]:

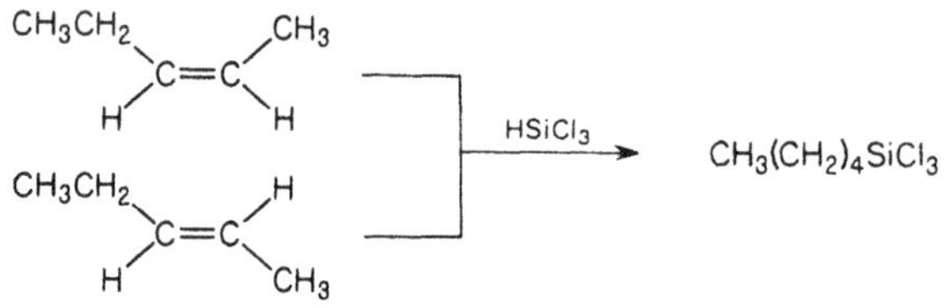

(192)

Similarly, triethoxysilane reacts with *trans*-2-hexene in the presence of supported [Dd-004] or soluble [Dd-003] rhodium complexes to yield n-hexyltriethoxysilane. Methylchlorosilanes with the general formula $(CH_3)_nSiHCl_{3-n}$ (n = 1–3) can also be added to 3-heptene to give n-heptyl-substituted silanes (H_2PtCl_6 catalyst) [995], whereas *cis*-2-octene is hydrosilylated by triethylsilane (using $Rh(acac)_3$ as the catalyst) to yield n-octyltriethylsilane [274].

All these examples illustrate the fact that, in the presence of transition metal complexes, hydrosilylation is accompanied by the isomerization of olefins, and that the formation of addition products with a terminal silyl group from olefins with an internal double bond can be readily accounted for by the migration of the C=C bond under the hydrosilylation conditions (see Scheme 82). The formation of adducts with an internal silyl group has also been reported [337, 588, 1301, S-033]. In the presence of chiral platinum complexes, e.g. *cis*-$(C_2H_4)[(+)(R)BzPhMeP]PtCl_2$ or $\{[(-)(R)C_3H_7P(C_6H_5)CH_3]PtCl_2\}_2$ asymmetric hydrosilylation of olefins occurs:

$$CH_3(R)C{=}CH_2 + CH_3SiHCl_2 \xrightarrow[40°C,\ 40\ h]{catalyst} CH_3SiCl_2CH_2{}^*CH(R)CH_3 \qquad (193)$$

R = C_2H_5, C_6H_5.

Small amounts of adducts with an internal silyl group (α-adducts) have been formed in the presence of nickel–phosphine complexes or nickel salts [589]. The yield increased markedly when the electron-accepting carboranyldiphosphine (DMPC) ligand was used [337].

$$RCH{=}CH_2 + CH_3SiHCl_2 \xrightarrow{Ni(DMPC)Cl_2} \underset{(\alpha\text{-}adduct)}{RCH(CH_3)SiCl_2CH_3} + \underset{(\beta\text{-}adduct)}{RCH_2CH_2SiCl_2CH_3} \qquad (194)$$

The α/β ratio obtained depends on the olefin chain and silicon hydride used. With methyldichlorosilane the ratio was found to range from 2:3 to 1:2 for n-alkenes [589], whereas with isobutene no α-adducts were formed at all [337].

The substitution of hydrogen by chlorine is a characteristic feature of catalysis by various nickel complexes, occurring as a side-reaction during the hydrosilylation of various alkenes, e.g. 1-alkenes [588, 589, 628, 655, 790–793, D-061, J-006], 2-alkenes [589], styrene [589, S-049], α-methylstyrene [1302, 1306], allylbenzene [S-048] and cycloalkenes [559, 589].

As a result of the addition of methyldichlorosilane to 1-hexene, two adducts and two products of the resulting substitution reaction [D-061] are obtained:

$$CH_3(CH_2)_3CH{=}CH_2 \xrightarrow[120°C,\ yield\ 73\%]{Ni[(CH_3)_3SiCH_2P(C_4H_9)_2]_2Br_2} \underset{(5\%)}{CH_3(CH_2)_3CH(CH_3)SiHClCH_3} + \qquad (195)$$

$$\underset{(20\%)}{CH_3(CH_2)_5SiHClCH_3} + \underset{(18\%)}{CH_3(CH_2)_3CH(CH_3)SiCl_2CH_3} + \underset{(65\%)}{CH_3(CH_2)_5SiCl_2CH_3}$$

where the percentages denote the selectivities. Generally, the yield of substitution products increases with decreasing reaction temperature and increasing reaction time. During the free-radical initiated addition of silicon hydride to alkenes, silyl radicals are added principally to the terminal carbon atom according to the mechanism given in Section 2.1, although the reaction is also accompanied by alkene telomerization (Schemes 10–12).

In addition to isomerization, H/Cl exchange in silanes and telomerization, other side-reactions can also occur, *viz.* dehydrocondensation, hydrogenation and olefin isomerization [265, 361, 880]. Hydrosilylation of 1-pentene by methyldiethylsilane in the presence of the iron pentacarbonyl complex leads to the following mixture of products [265]:

$$CH_3(CH_2)_2CH{=}CH_2 + CH_3(C_2H_5)_2SiH \rightarrow \begin{cases} CH_3(C_2H_5)_2Si(CH_2)_4CH_3 \\ CH_3(C_2H_5)_2SiCH{=}CH(CH_2)_2CH_3 \\ CH_3(CH_2)_3CH_3 \\ cis\text{- and } trans\text{-}CH_3CH{=}CHC_2H_5 \end{cases} \qquad (196)$$

The hydrosilylation of alkenes, e.g. 1-hexene with triethylsilane catalysed by iridium complexes, gives alkenylsilanes as a major product accompanied by the normal adduct and 1 alkane [361, 880]. Also in the presence of ruthenium with diazadienes, during the hydrosilylation of 1-alkenes, an isomerization of terminal alkenes to a *cis/trans* mixture of internal alkenes is a competing side-reaction [1363].

Reaction (196) confirms previous studies of competitive reactions, and demonstrates that formation of an unsaturated product is favoured by the presence of a high (5:1) molar excess of olefin in the reaction mixture [560] (see also Scheme 143).

Silicon dihydrides react with the C=C bond of alkenes to give products whose variety is also influenced by the olefin structure. Thus, the addition of dichlorosilane to 2-alkene leads to the formation of two adducts, with the reaction yield ranging from 51% to 92% for n = 1–3 [114] but decreasing as n increases further:

$$CH_3(CH_2)_nCH{=}CHCH_3 + H_2SiCl_2 \xrightarrow{H_2PtCl_6} \underset{\substack{|\\ SiHCl_2 \\ (64–69\%)}}{CH_3(CH_2)_nCH_2CHCH_3} + \quad (197)$$

$$\underset{\substack{|\\ SiHCl_2 \\ (31–36\%)}}{CH_3(CH_2)_nCHCH_2CH_3}$$

In general, the rate of reaction of 2- and 3-alkenes is much less than that of 1-alkenes in the presence of H_2PtCl_6 [114] and $[(C_2H_4)_2RhCl]_2$ [1167]. Symmetrical alkenes, e.g. 3-hexene, react with dichlorosilane to form an adduct containing an asymmetric carbon atom:

$$C_2H_5CH{=}CHCH_2H_5 + H_2SiCl_2 \xrightarrow[150°C, 8–24\ h]{H_2PtCl_6} \underset{(56–88\%)}{C_3H_7CH(C_2H_5)SiHCl_2} \quad (198)$$

The reactivity of the C=C bond can be exemplified by hydrosilylation with trichlorosilane (using H_2PtCl_6 as the catalyst), with the quantitative distribution of products being governed by the following sequences [215]:

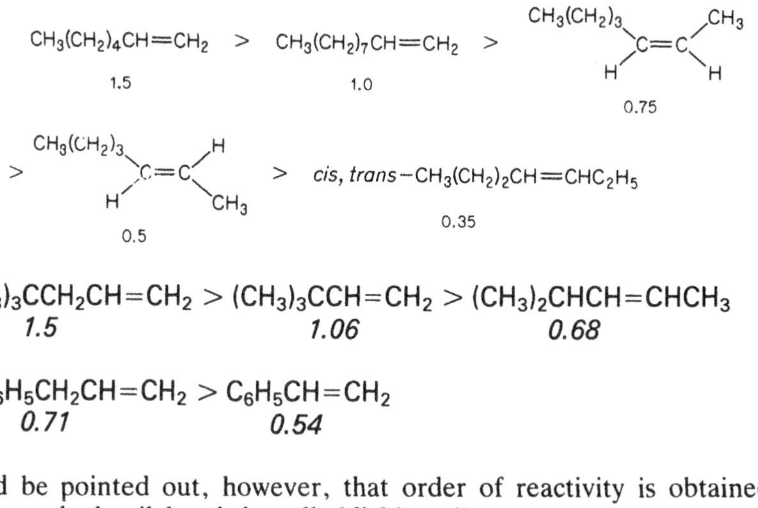

(199.1)

$$\underset{1.5}{(CH_3)_3CCH_2CH{=}CH_2} > \underset{1.06}{(CH_3)_3CCH{=}CH_2} > \underset{0.68}{(CH_3)_2CHCH{=}CHCH_3} \quad (199.2)$$

$$\underset{0.71}{C_6H_5CH_2CH{=}CH_2} > \underset{0.54}{C_6H_5CH{=}CH_2} \quad (199.3)$$

It should be pointed out, however, that order of reactivity is obtained when alkenes are hydrosilylated by alkyldichlorosilane, also in the presence of H_2PtCl_6 [427]:

$$CH_3(CH_2)_2CH{=}CH_2 > CH_3(CH_2)_3CH{=}CH_2 > CH_3(CH_2)_4CH{=}CH_2 \quad (200)$$

Tetraalkylsilanes carrying one short and three long alkyl groups can be synthesized via a two-step hydrosilylation procedure, which appears more suited for commercial production than the Grignard or alkyllithium route [877, U-162]; see the exemplary sequence of the reaction:

$$CH_3HSiCl_2 + CH_2{=}CHR \rightarrow CH_3SiCl_2CH_2CH_2R \quad (201.1)$$

$$2\ CH_3SiCl_2CH_2CH_2R + LiAlH_4 \rightarrow 2\ CH_3SiH_2CH_2CH_2R + AlCl_3 + LiCl \quad (201.2)$$

$$CH_3SiH_2CH_2CH_2R + 2\ CH_2{=}CHR \rightarrow CH_3Si(CH_2CH_2R)_3 \quad (201.3)$$

The final products were isolated by distillation in yields of ~75–86%.

The general scheme for the hydrosilylation of arylalkenes containing a terminal double bond, $C_6H_5(CH_2)_mCH{=}CH_2$ (m = 1, 2), or an internal double bond, $C_6H_5CH{=}CH(CH_2)_nCH_3$, may be written as follows:

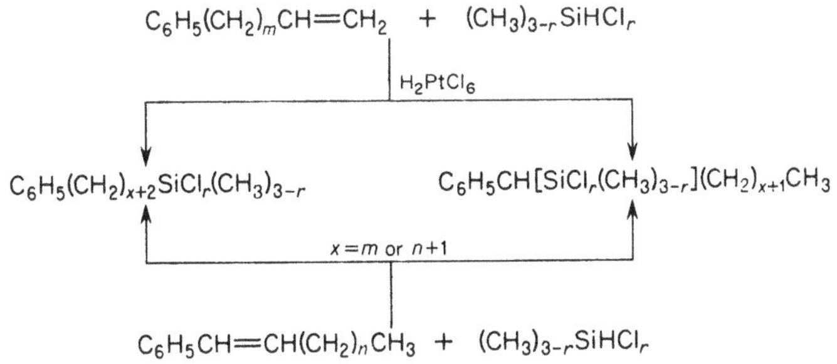

(202)

Following preliminary migration of the C=C bond, the reaction proceeds to the formation of α- and β-substituted adducts. As n increases, the yield of the ω-isomer increases, whereas an increase in the value of m leads to greater amounts of the α-isomer. An example of a typical reaction which occurs in the presence of H_2PtCl_6 is given below [215]:

$$C_6H_5CH_2CH{=}CH_2 + HSiCl_3 \xrightarrow{H_2PtCl_6} C_6H_5(CH_2)_3SiCl_3 + C_6H_5CH(SiCl_3)C_2H_5 \quad (203)$$

When catalysed by nickel complexes the reaction yields substitution products. The hydrosilylation of styrene by methyldichlorosilane gives three products [S-048] in a 29:41:30 ratio:

$$C_6H_5CH{=}CH_2 + CH_3SiHCl_2 \xrightarrow[120°C,\ 10\ h]{NiCl_2+(Me_3SiCH_2)_3PO} \begin{cases} C_6H_5CH_2CH_2SiCl_2CH_3 \\ C_6H_5CH(CH_3)SiCl_2CH_3 \\ C_6H_5CH_2CH_2SiHClCH_3 \end{cases} \quad (204)$$

The corresponding hydrosilylation of allylbenzene under these conditions leads to a 71% yield of the two γ-isomers, i.e. the normal adduct and the H/Cl exchange product, in a 77:23 ratio [S-048]:

$$C_6H_5CH_2CH{=}CH_2 + CH_3SiHCl_2 \rightarrow C_6H_5(CH_2)_3SiCl_2CH_3 + C_6H_5(CH_2)_3SiHClCH_3 \quad (205)$$

In contrast, hydrosilylation of the same reactant with triethylsilane in the presence of $Rh(acac)_3$ leads to the formation of the γ-adduct, β-adduct and an unsaturated dehydrocendensation product [274]:

$$C_6H_5CH_2CH{=}CH_2 + (C_2H_5)_3SiH \rightarrow \begin{cases} C_6H_5CH_2CH_2CH_2Si(C_2H_5)_3 & (46\%) \\ C_6H_5CH_2CH(CH_3)Si(C_2H_5)_3 & (20\%) \\ C_6H_5CH_2CH{=}CHSi(C_2H_5)_3 & (12\%) \end{cases} \quad (206)$$

This distribution of products is also obtained in the hydrosilylation of styrene (and α-methylstyrene) in the presence of various rhodium, platinum and other transition metal complexes, e.g. $RhCl(PPh_3)_3$ (I, 25–34%; II, 2–47%; III, 2–43%) [878], $RhHCO(PPh_3)_3$ (I, 4–27%; II, 0–51%; III, 1–44%) [878] and $Rh(acac)_3$ (I, 24%; II, 75%) [950], $PtCl_2(PhCH{=}CH_2)_2$ (I, 73–75%; II, 1–2%; III, 11–13%; IV, 11–13%) [19, 220, 221], H_2PtCl_6 + $RCONR_1R_2$ (I, 60%; II, 40%) [U-178], ion-exchanger supported complexes AW-17-[$PtCl_6$] (I, 98.3%; II, 1.7% [828]), AW-17-Rh_2Cl_9 (I, 61.2%; II, 38.8% [828]).

$$C_6H_5CH{=}CH_2 + (C_2H_5)_3SiH \rightarrow \begin{cases} C_6H_5CH_2CH_2Si(C_2H_5)_3 & (I) \\ C_6H_5CH(CH_3)Si(C_2H_5)_3 & (II) \\ C_6H_5CH{=}CHSi(C_2H_5)_3 & (III) \\ C_6H_5CH_2CH_3 & (IV) \end{cases} \quad (207)$$

Hydrosilylating styrene with tri(isopropyl)silane in the presence of $RhCl(PPh_3)_3$ (100°C, 6 h) leads to the selective and quantitative formation of $C_6H_5CH{=}CHSi[CH(CH_3)_2]_3$, whereas addition to α-methylstyrene ($[RhCl(C_2H_4)_2]_2$ catalyst, 60°C, 6 h) produces a 15% yield of $C_6H_5C({=}CH_2)CH_2Si[CH(CH_3)_2]_3$ [630] and the hydrosilylation of p-(t-butyl)styrene by $(CH_3)_2SiHCl$ ($PtCl_2(PhCN)_2$, 80°C) gives selectively p-(t-Bu)$C_6H_4CH_2CH_2SiCl(CH_3)_2$ (95%) [U-143]).

Platinum(II) complexes containing chiral phosphine ligands also catalyse the asymmetric addition of trichlorosilane to styrene [590]. When *trans*-$[(+)(R)(PhCH_2)CH_3PPh]_2NiY_2$ was used as the catalyst, the hydrosilylation of α-methylstyrene with methyldichlorosilane gave a 31% yield of the β-adduct with an 18% enantiomeric excess of the R isomer [1306].

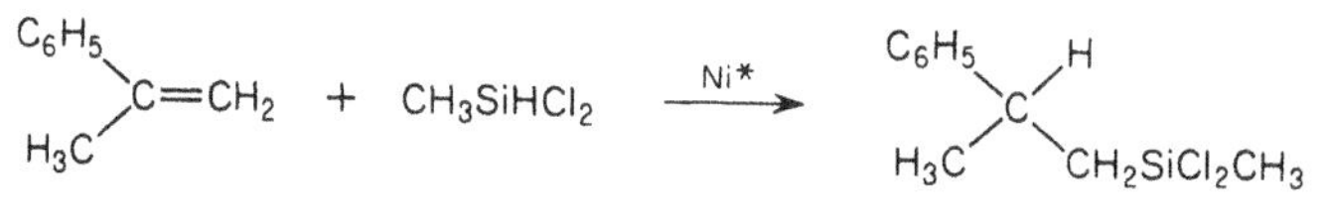

(208)

In the hydrosilylation of olefins which are susceptible to polymerization, this latter reaction may become dominant. The hydrosilylation of styrene with trichlorosilane in the presence of diacyl peroxide as the catalyst leads only to high boiling products.

3.1.2 Cycloalkenes and Alkenylcycloalkenes

Whereas the addition of trichlorosilane to 1-methylcyclohexene proceeds readily in the presence of H_2PtCl_6, its addition to 4-methylcyclohexene is relatively slow. This difference arises from the different induction periods observed in the two reactions. In both cases trichloro(cyclohexylmethylene)silane is formed as the major product, and this is accompanied by traces of 1-(trichlorosilyl)-2-methylcyclohexane and 1-(trichlorosilyl)-3-methylcyclohexane [958, E-004].

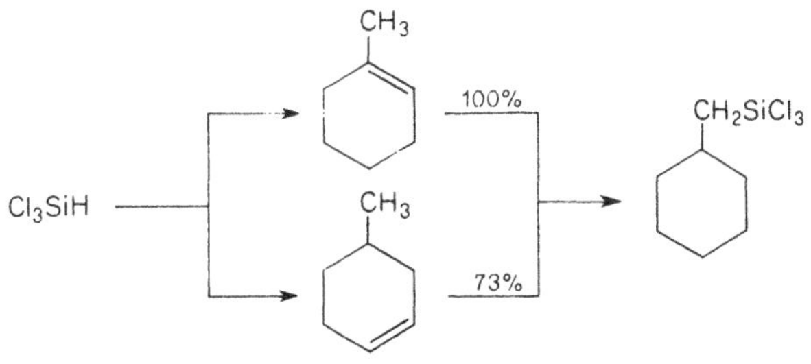

(209)

If the hydrogen of the methyl group is substituted by deuterium, this tends to suppress the rearrangement and promote competitive addition to the ring [1035].

The rearrangement reaction proceeds via intermoleculr hydrogen transfer and depends on the nature of the silane. Thus, methyldichlorosilane reacts more slowly with 1-methylcyclohexene than trichlorosilane (56% after 48 h) [996], the products being (cyclohexylmethylene)dichloromethylsilane (84%) and (3-methylcyclohexyl)dichloromethylsilane (16%). The reaction of 1-propylcyclohexene with trichlorosilane in the presence of H_2PtCl_6 gives 3-cyclohexylpropyltrichlorosilane:

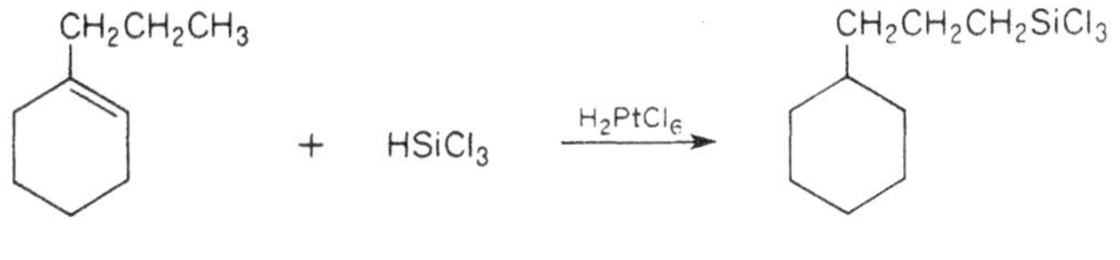

(210)

but isomerization of the olefin also occurs quite readily [119].

The hydrosilylation of 1- and 4-tert-butylcyclohexene and 1-methyl-1-isopropenylcyclohexene with trichlorosilane catalysed by $[(C_2H_4)PtCl_2]_2$ proceeds according to a mechanism that requires the migration of a methyl group (Scheme 211).

Radical initiation of the reaction leads to the formation of a β-adduct [106].

The hydrosilylation of β-pinene (2-methylene-6,6-dimethylbicyclo[3.2.1]heptane) has been the subject of extensive study [Rev.20, 484, 556, 557, 1181, 1283, 1327, 1335]. Depending on the catalytic system and silicon hydride employed, the reaction usually leads to one of the products shown below, or to a mixture of these products (Scheme 212).

Phosphine complexes of Ni(II) [557] and three-component catalytic systems containing Ni or V, PPh_3 and $AlEt_3$ [1327] are highly effective in this particular reaction, which involves the skeletal rearrangement of the cyclobutane ring according to the following suggested mechanism (Scheme 213):

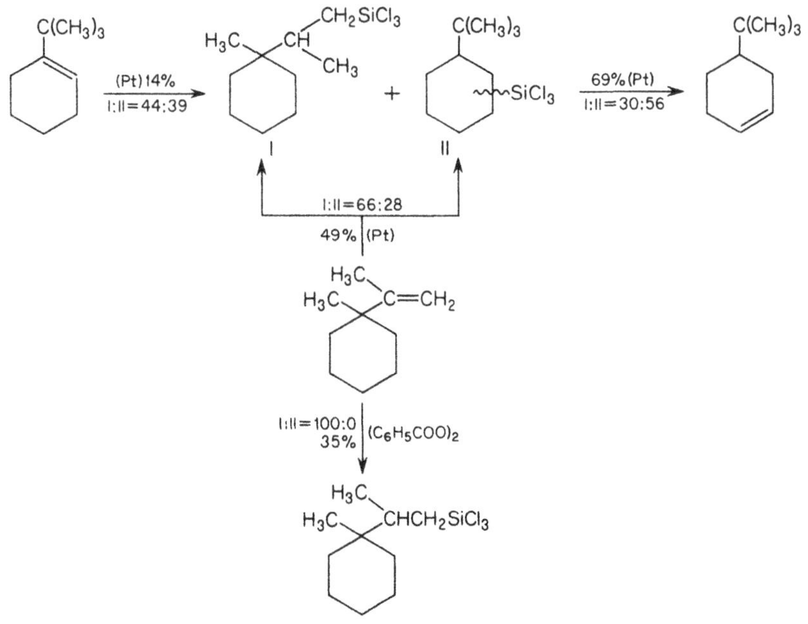

(211)

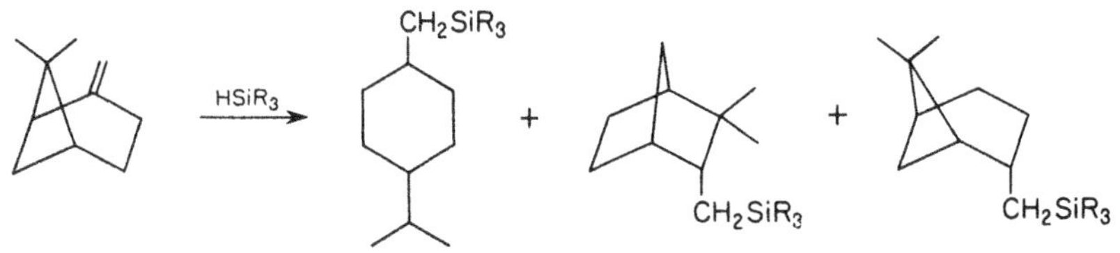

(212)

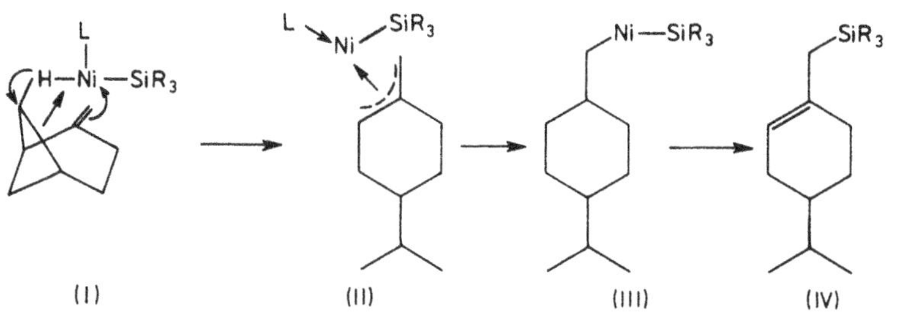

(213)

The reaction scheme assumes the formation of a Π-allyl intermediate (II) which, as a result of cleavage of the cyclobutane ring, leads to a σ-allyllnickel intermediate (III) which subsequently rearranges to give the adduct IV [1327]. The addition of methyldichlorosilane and/or dimethylchlorosilane to β-pinene in the presence of H_2PtCl_6 induces formation of the adduct V (Scheme 214) [1283]. When the same silicon hydrides are added to (−)-β-pinene, the corresponding optically active organosilicon compounds are obtained.

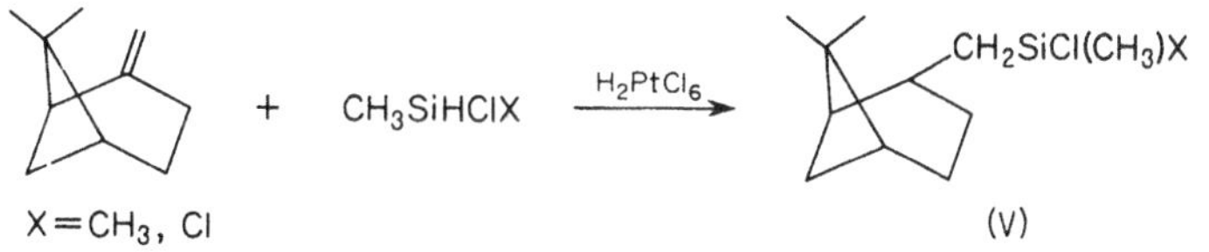

(214)

Since (−)-β-pinene has the 1S,5S-configuration, the silanes have the 1S,2S,5S-configuration [1283].

The hydrosilylation of α-pinene with phenylsilane involves olefin isomerization followed by the addition of the silyl group to the ring:

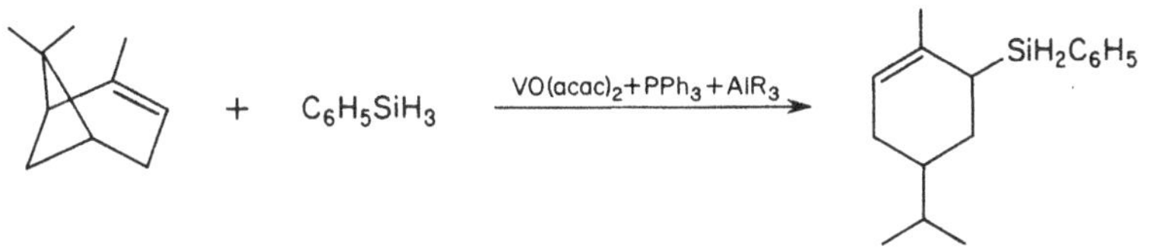

(215)

Addition of trichlorosilane to 5-*endo*-methylbicyclo-[2.2.1]-2-heptene-(*endo*-5-methylnorbornene) in the presence of $Ni(acac)_2$ + PPh_3 as the catalyst (100°C, 7 h) leads quantitatively to 2-*endo*-methyl-5-*exo*-(trichlorosilyl)bicyclo[2.2.1]heptane and 2-*endo*-methyl-6-*exo*-(trichlorosilyl)bicyclo[2.2.1]heptane [1329]:

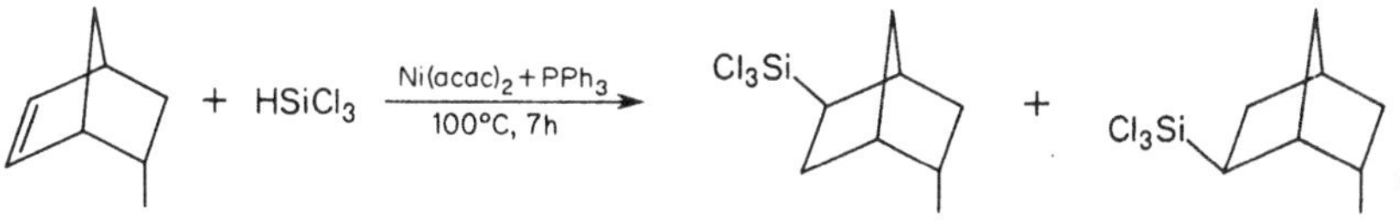

(216)

Hydrosilylation of 1-vinyladamantane in the presence of platinum catalyst yields selectively 2-(1-adamantyl)ethylsilane [708].

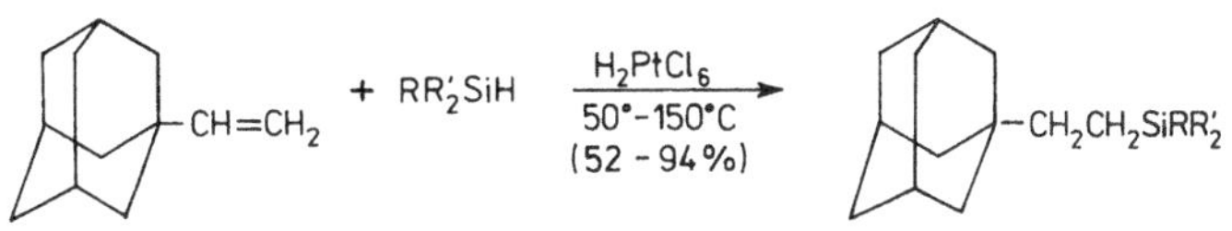

(217)

3.2 Hydrocarbons with Two or More C=C Bonds

Hydrosilylation of organic compounds containing two or more carbon–carbon double bonds, particularly dienes or polyenes, enables a comparison of the reactivity of such bonds with respect to their position in a given molecule. The following dienes have been the subject of a great number of studies: allene [289, 1208, S-014, S-034] and its derivatives [212, 1213], isoprene [167, 589, 662, 663, 1265, 1293, 1335], 1,3-butadiene [205–207, 210, 468, 982, 983, 1176, 1208, 1219–1221, C-003, C-007, C-010, C-017, C-018, D-027, D-060, D-070, J-001, U-080], 1,3-pentadiene [620, 1208, 1331, D-068, U-109] and higher homologues [306, 429, 482, 594, 662, 663, 881, 1266, 1293, 1324, 1332, 1334, D-067, U-106, U-118], and cyclic dienes [239, 589, 590, 628, 1217, 1297, 1302, 1304, 1333, 1335].

3.2.1 Dienes with Isolated Double Bonds

In the case of organic alkadienes and their derivatives which contain two equivalent bonds, the Si–H bond adds across the two C=C bonds, e.g. in the hydrosilylation of 1,5-hexadiene with methyldichlorosilane [306, 629, G-004] and 1,7-octadiene with triethylsilane [429], as well as of 1,7-octadiene [429, 881, D-068, U-109, U-118], 1,11-dodecadiene [881, D-068, U-109, U-118], 1,13-tetradecadiene [346, D-068, U-109, U-118] with trichlorosilane. The maximum yield of the mono-adduct can be obtained when the synthesis is carried out in the presence of a considerable excess of the alkadiene:

$$\underset{0.2\ moles}{HSiCl_3} + \underset{0.6\ moles}{CH_2{=}CH(CH_2)_4CH{=}CH_2} \xrightarrow{H_2PtCl_6} \underset{0.15\ moles}{CH_2{=}CH(CH_2)_6SiCl_3} + \quad (218)$$

$$\underset{0.1\ moles}{Cl_3Si(CH_2)_8SiCl_3}$$

As for 1,3-pentadiene, 1,4-pentadiene reacts with a silicon hydride in the presence of H_2PtCl_6 to form (in theory) five mono-adducts, i.e. $CH_2{=}CH(CH_2)_3SiR_3$, *cis*(Z)- and *trans*(E)-$CH_3CH{=}CHCH_2CH_2SiR_3$, *cis*(Z)- and *trans*(E)-$C_2H_5CH{=}CHCH_2SiR_3$. When the reaction is catalysed by the $Ni(acac)_2 + AlEt_3 + PPh_3$ system, isomerization of the diene is followed solely by hydrosilylation which yields 3-methyl-2-butenylsilane and 2-methyl-1,4-disilyl-2-butene [1173]. Some alkadienes undergo intermolecular hydrosilylation with dihydrosilanes to form linear [594] or cyclic products [722].

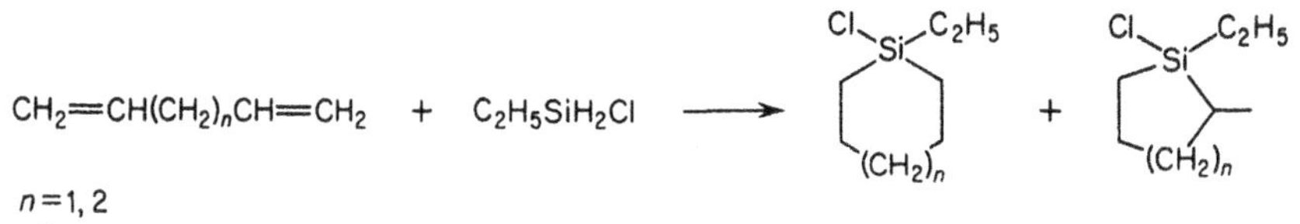

$n=1, 2$

(219)

The susceptibility of alkenylcycloalkenes, i.e. dienes containing C=C double bonds both in the ring and in the ring substituents, towards hydrosilylation is mainly associated with the substituent alkenyl group, e.g. 4-vinylcyclohexene couples with the silicon hydride in the presence of H_2PtCl_6 through the vinyl substituent while the ring double bond remains intact [123, 429, 556, 907, 1192, 1327, C-028, D-027, D-044, D-080, D-085, D-103, Dd-007].

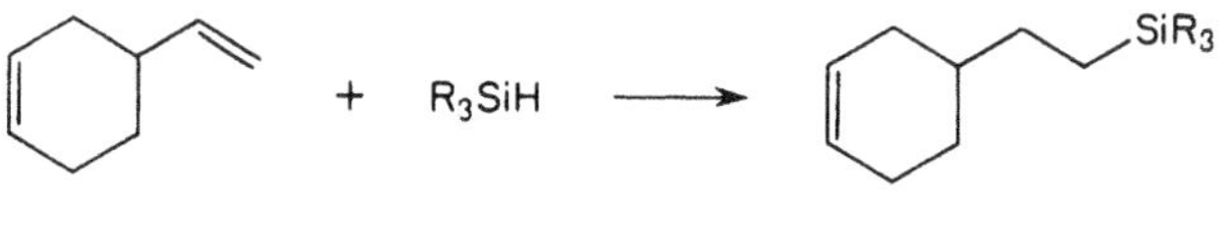

(220)

Ziegler catalysts based on nickel and vanadium compounds catalysed the hydrosilylation of the vinyl group to provide an adduct in 50–76% yield [1327]. The reaction of allyl and other alkenyl cycloalkenes occurs in the same way. Allylcyclopentadiene reacts with trichlorosilane and methyldichlorosilane in the presence of H_2PtCl_6 to give an adduct where a silyl group is contained in the substitutent chain, i.e. $C_6H_5(CH_2)_3SiCl_2X$ (X = Cl or CH_3) [725]. Limonene (1,8-methadiene) reacts with tri-substituted silanes according to the following equation:

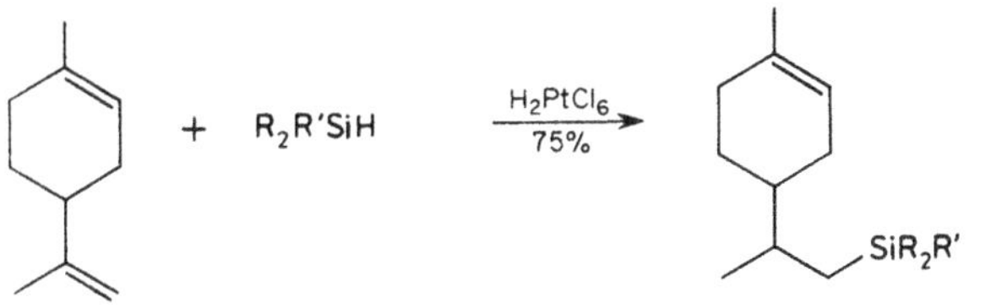

(221)

where $R_2R' = (C_2H_5O)_3$, $CH_3(C_2H_5O)_2$ [U-164].

These various results can be interpreted only by taking account of the relatively small reactivity of alkylcycloalkenes in the presence of chloroplatinic acid in comparison to that of alkenes. In addition, alkylcycloalkenes can be hydrosilylated only after preliminary migration of the C=C double bond, to finally yield adducts with a terminal silyl group in the aliphatic chain [106, 119, 996, 1036].

In the presence of excess of hydrosilane and over prolonged reaction times, limonene forms a disilyl adduct in the presence of chloroplatinic acid [D-080].

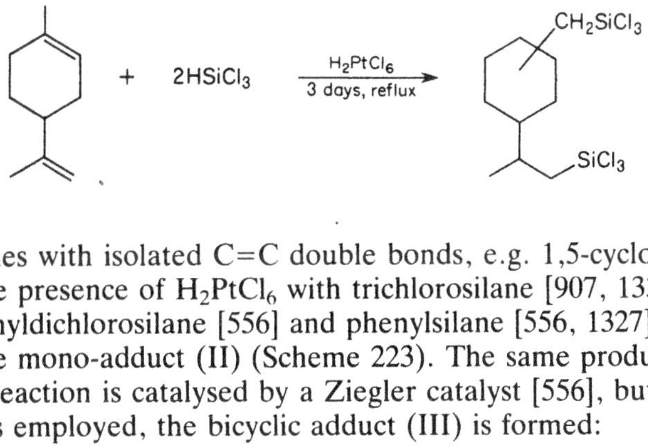

(222)

Cyclic dienes with isolated C=C double bonds, e.g. 1,5-cyclooctadiene (COD), react in the presence of H_2PtCl_6 with trichlorosilane [907, 1335], dimethylsilane [556], methyldichlorosilane [556] and phenylsilane [556, 1327] to give a 40–90% yield of the mono-adduct (II) (Scheme 223). The same product is also obtained when the reaction is catalysed by a Ziegler catalyst [556], but when free-radical initiation is employed, the bicyclic adduct (III) is formed:

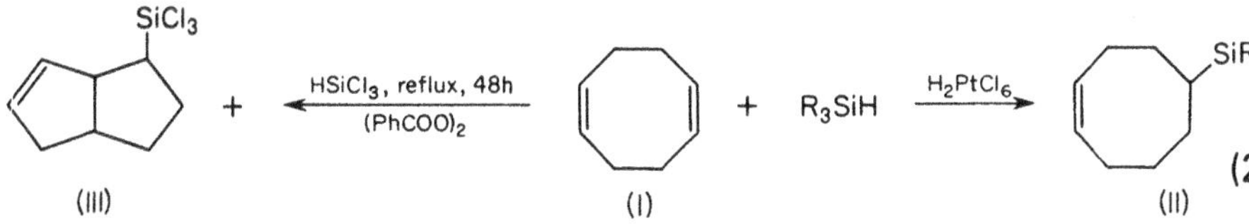

(223)

1,5-Dimethyl-1,5-cyclooctadiene reacts with phenylsilane to obtain mono-adduct [1326]:

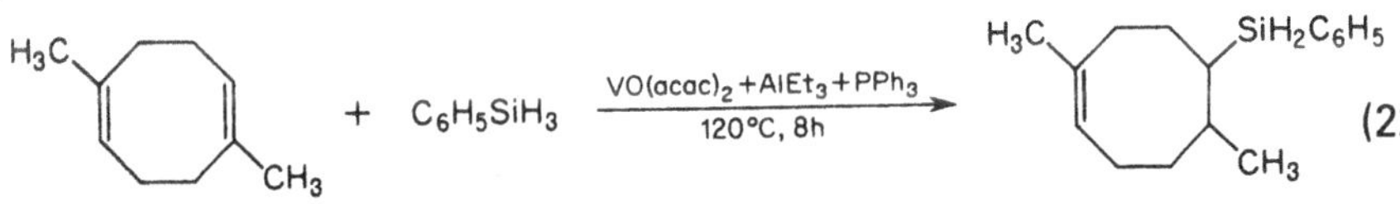

(224)

Norbornadiene (bicyclo[2.2.1]-2,5-heptadiene) can be hydrosilylated in the presence of Pt/C, H_2PtCl_6 and/or azoisobutyronitrile. The bicyclic dienes isomerize at the same time, with the result that the following products are obtained:

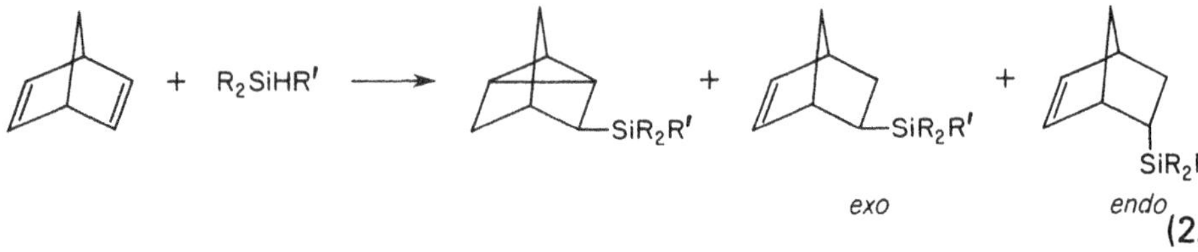

(225)

The addition reaction of tri(chloro, methyl, phenyl) silanes as well as triethoxysilane with 5-ethylidene-bicyclo [2.2.1]-2-heptene leads in the presence of platinum complex catalyst to the mixture of 1:1 adducts [E-019, J-069]:

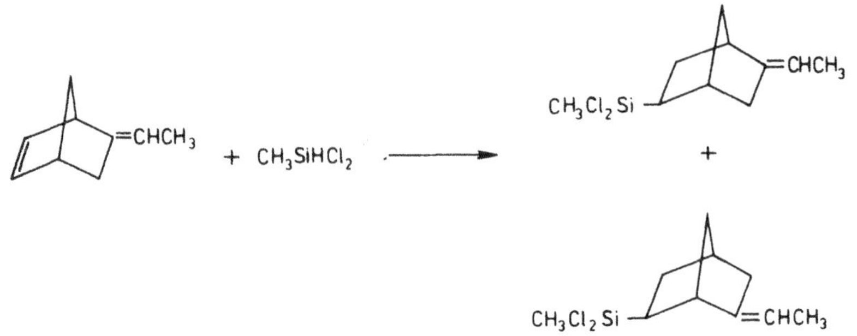

(226)

The hydrosilylation of tricyclo[5.2.1.0]dece-3.8-diene by tri(methyl,chloro)silanes and trialkoxysilanes in the presence of cobalt and rhodium carbonyls form cyclodecenylsilane according to the following equation [712, 721]:

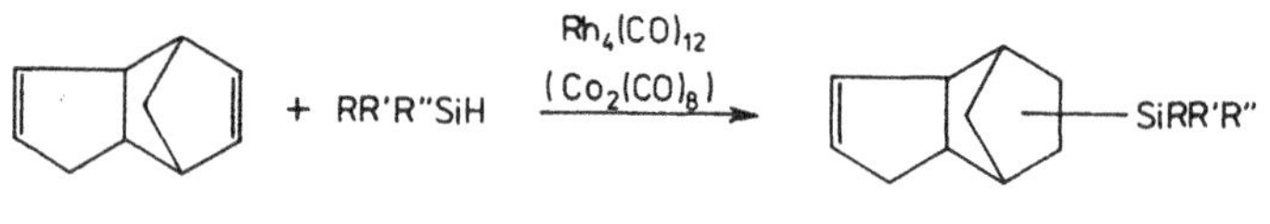

(227)

$R = R' = CH_3$, $R'' = Cl$; $R = R' = R'' = OC_2H_5$;
$R = CH_3$, $R' = R'' = OC_2H_5$; $R = CH_3$; $R' = R'' = OC_4H_9$;
$R = R' = Cl$, $R'' = CH_3$.

Hydrosilylation of 2-vinyl-5-norbornene by dimethylchlorosilane carried out at 70°C for 1 h gives 2-(5-norbornenyl)ethyl dimethylchlorosilane [U-193).

3.2.2 Conjugated Dienes

Hydrosilylation of conjugated dienes proceeds predominantly across positions 1 and 4, although 1,2-addition can also occur. Near-UV photolysis at 25°C in the presence of catalytic amounts of $Cr(CO)_6$, of degassed solutions containing a 1:1 molar ratio of a 1,3-diene (1,3-butadiene, isoprene, 1,3-pentadiene,2,3-dimethyl-1,3-butadiene) and a silane yields the product generated by hydrosilylation of the diene across the 1,4-positions. Hydrosilylation of 1,3-butadiene by trichlorosilane also leads to 1,4-addition, but whether the *cis*(Z) or *trans*(E) isomers of $Cl_3SiCH_2CH{=}CHCH_3$ is generated depends upon the catalyst used [205–207, 468, 982, 1176, 1208, 1220, 1221, C-003, C-007, C-017, C-018, D-027, D-060, D-070, J-001, S-029].

The addition of trichlorosilane to 1,3-butadiene can also proceed across the 1,2-positions, especially in the presence of $Pt(PPh_3)_4$ [162], but the use of $Pd(PPh_3)_4$ as a catalyst promotes 1,4-addition. Partial replacement of the triphenylphosphine ligand in the latter complex by maleic anhydride leads to a change in selectivity towards the formation of 1,2-addition products.

The hydrosilylation of 1,3-butadiene in the presence of palladium [1174, 1176, 1177, 1208, C-001, J-012] and nickel complexes (e.g. 1,3-cyclopentadiene–nickel) [C-001] occurs simultaneously with dimerization:

$$CH_2{=}CHCH{=}CH_2 + R_3SiH \rightarrow R_3SiCH_2CH{=}CH(CH_2)_2CH{=}CHCH_3 \quad (228)$$

The addition of trimethylsilane to 1,3-butadiene in a benzene solution of $PdCl_2$ gives a 94% yield of 1-(trimethylsilyl)-2,6-octadiene [C-001]. Takahashi *et al.* have proposed a mechanism for the formation of monomeric and dimeric adducts (obtained when the SiH/butadiene ratios are 1:1 and 1:2, respectively) in the presence of palladium and nickel catalysts (see Scheme 168) [1176].

When the reaction is catalysed by a Ziegler catalyst ($Ni(acac)_2 + AlR_3 + PPh_3$), 1,4-disilyl-2-butene is also formed. The reaction of butadiene with dihydrosilanes leads to the formation not only of the 1,4-adduct, but also of the cyclic and disilyl adducts, according to the following equation [123]:

$$CH_2{=}CHCH{=}CH_2 + H_2SiCl_2 \rightarrow \begin{cases} CH_3CH{=}CHCH_2SiHCl_2 & (I) \\ \overline{(CH_2)_4Si}Cl_2 & (II) \\ HSiCl_2(CH_2)_4SiHCl_2 & (III) \end{cases} \quad (229)$$

Overall, therefore, the hydrosilylation of 1,3-butadiene can lead to the following products:

$$CH_2{=}CHCH{=}CH_2 \xrightarrow{R_3SiH} \begin{cases} CH_3CH{=}CHCH_2SiR_3 \ (\textit{cis} \text{ and } \textit{trans}) \\ CH_2{=}CHCH_2CH_2SiR_3 \\ CH_3CH{=}CHCH_2CH_2CH{=}CHCH_2SiR_3 \\ R_3Si(CH_2)_4SiR_3 \\ R_3SiCH_2CH{=}CHCH_2SiR_3 \ (\textit{cis} \text{ and } \textit{trans}) \\ \overline{(CH_2)_4Si}R_2 \quad (\text{for } R_3 = R'_2H) \end{cases} \quad (230)$$

The hydrosilylation of isoprene (2-methyl-1,3-butadiene) by monohydrosilanes (using a 1:1 substrate ratio) gives the following products:

$$CH_2{=}C(CH_3)CH{=}CH_2 \xrightarrow{R_3SiH} \begin{cases} (CH_3)_2C{=}CHCH_2SiR_3 & (I) \\ R_3SiCH_2C(CH_3){=}CHCH_3 & (II) \\ CH_2{=}C(CH_3)CH_2CH_2SiR_3 & (III) \\ R_3SiCH_2CH(CH_3)CH{=}CH_2 & (IV) \end{cases} \quad (231)$$

The addition of silanes to isoprene [726, 1222] and to chloroprene [726] in the presence of H_2PtCl_6 occurs in the 1,4-position.

The 1,4-addition of triethylsilane to isoprene (130°C, 3 h) yields two adducts: (I) 20% and (II) 80% (see Scheme 231), whereas the hydrosilylation of 2,3-dimethylbutadiene leads to the formation of the 1,4-adduct (87%) and the 1,2-adduct, respectively [1222].

In the presence of ruthenium(II) complexes with unsaturated N-containing controlling ligands as co-catalyst, $cis[Ru(DAD)(COD)HCl]$ in the hydrosilylation of isoprene with triethoxysilane, adduct I is mainly yielded, accompanied by adducts IV and III [167].

Catalysis of the hydrosilylation of isoprene by other transition metal complexes gives predominantly 2-methyl-2-butenylsilane (product (II) in Scheme 231) [167, 589, 663, 1208, 1219, 1293, 1363]. Thus, when the reaction is conducted in the presence of a Pd complex prepared *in situ* [855], for example, it proceeds as follows:

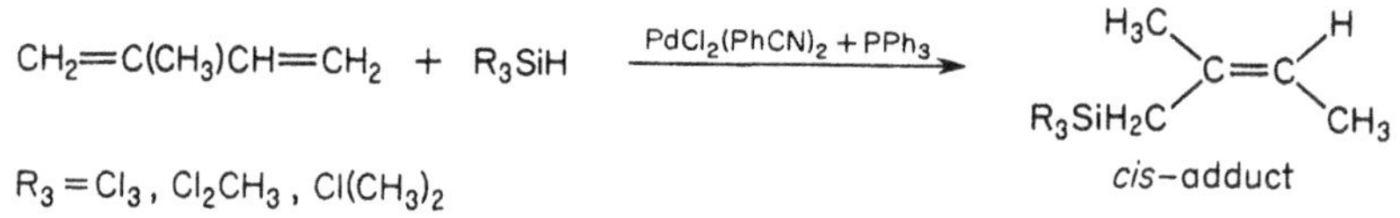

$$CH_2{=}C(CH_3)CH{=}CH_2 \; + \; R_3SiH \quad \xrightarrow{PdCl_2(PhCN)_2 \,+\, PPh_3}$$

$$R_3 = Cl_3,\; Cl_2CH_3,\; Cl(CH_3)_2 \qquad\qquad cis-\text{adduct} \qquad\qquad (232)$$

However, other trisubstituted silanes, e.g. triethoxysilane, tri(phenyl, methyl)silane and triethylsilane, do not add up to isoprene under these conditions. Ojima has advanced the following mechanism for this reaction in an attempt to account for the observed stereoselectivity (Scheme 233):

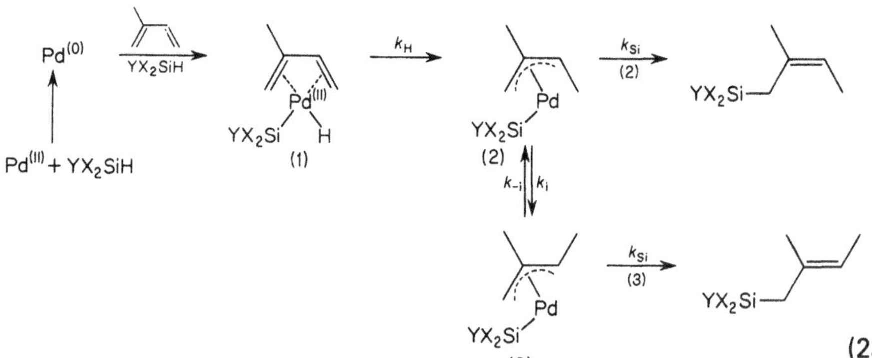

$$(233)$$

The reaction occurs via a Π-allyl intermediate of the type proposed for most diene reactions [493, 1177]. The *cis* adduct is obtained as a result of silyl group migration (for which the rate constant k_{Si} (2) may be written), and which occurs at a greater rate than the isomerization of the intermediate.

It is interesting to note that the addition of methyldichlorosilane to isoprene in the presence of nickel complexes, e.g. $Ni(DMPP)Cl_2$, proceeds without H/Cl exchange [589]. The nature of the resulting product (Scheme 232) may be attributed to a reaction similar to that occurring in the presence of Wilkinson's catalyst [852]. However, in the latter case, a *cis*-(II) by-product is also observed.

The mechanism of this reaction is similar to that proposed by Rejhon and Hetflejs for the hydrosilylation of butadiene [982]. Ojima and Kumagai have suggested an alternative mechanism (Scheme 234) which involves initiation via regioselective migration of the silyl group, as in the reactions of alkynes [852, 870] and α,β-unsaturated esters [868], to form the Π-allyl complexes (IV) and (V). Hydride transfer from the latter complex yields 3-methyl-2-butenylsilane (I) and *cis*-2-methyl-2-butenylsilane (II).

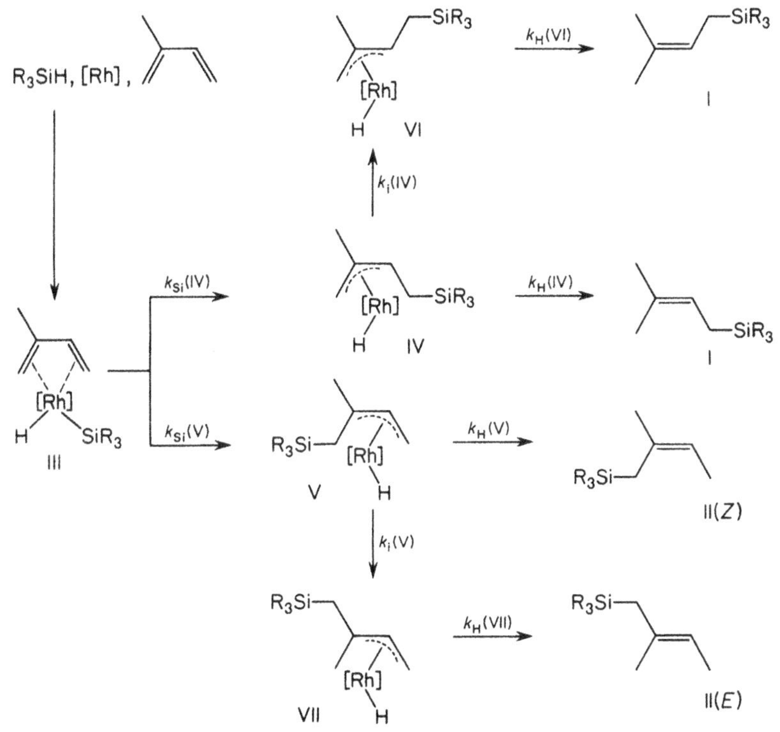

(234)

When this reaction is catalysed by a Ziegler catalyst ($Ni(acac)_2 + AlR_3 + PPh_3$), 1,4-disilyl-2-methyl-2-butene is also formed [1018]. In addition, a saturated product (1,4-disilyl-2-methylbutane) can also be obtained during the hydrosilylation of isoprene [122]. Hydrosilylation of 1,3-pentadiene always leads to an adduct containing terminal silyl groups, as for example in the presence of $Ni(acac)_2 + PPh_3$ [1335]. The reaction generally yields eight products of the following type:

$$\begin{array}{ll} cis\text{- and } trans\text{-}CH_3CH{=}CHCH_2CH_2Si{\equiv} & (I) \\ cis\text{- and } trans\text{-}(R \text{ and } S)CH_3CH{=}CHCH_2CH_2Si{\equiv} & (II) \\ cis\text{- and } trans\text{-}C_2H_5CH{=}CHCH_2Si{\equiv} & (III) \end{array} \qquad (235)$$

Asymmetric hydrosilylation of 1-arylbutadienes, e.g. 1-phenylbutadiene in the presence of a chiral ferrocenylphosphine–palladium catalyst gives optically active alkylsilane (Z)-1-aryl-1-silyl-2-butenes and their regioisomers [483].

The addition of a silyl hydride to cyclopentadiene can occur both across the 1,2- or 1,4-positions:

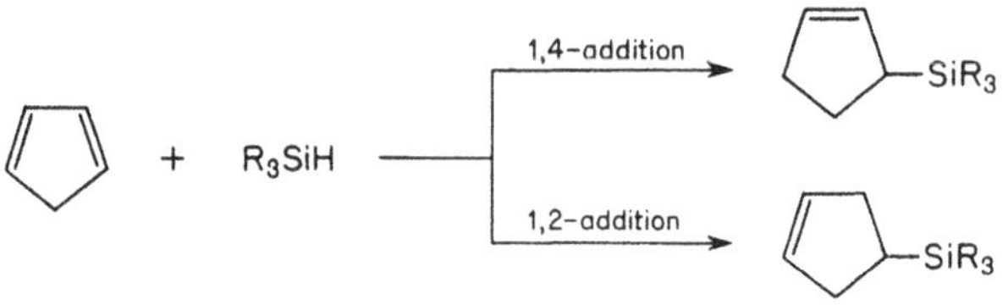

(236)

Thus, for example, when the addition of phenylsilane is catalysed by $Ni(acac)_2$ + $Al(C_2H_5)_3$ + PPh_3 the reactions occur in the 1,2-position, giving a 60% yield of 4-(phenylsilyl)cyclopentene [1333]. 1,4-Addition of cyclopentadiene is often accompanied by 1,2-addition when the following catalysts are employed: $Ni(CO)_4$ [1217, C-015], $Ni(CO)_4$ + PR_3 (R = C_4H_9, C_6H_5, C_2H_5O) [1217], $Pd(PPh_3)_4$ [1217, C-014], $PdCl_2(PhCN)_2$ [1217], $PdCl_2(PhCN)_2$ + PPh_3 [867] and H_2PtCl_6 (*iso*-PrOH) [784]. Selective 1,2-addition to pentadiene takes place in the presence of $Ni(CO)_2(PPh_3)_2$ [1217].

The hydrosilylation of 1,3- and 1,4-cyclohexadiene catalysed by $PdCl_2(PhCN)_2$ + PPh_3 gives similar products to those with cyclopentadiene; 3-silylcyclohexene is the major product in this case [590, 1302, 1304].

Saturated product 1,3-bis(triethoxysilyl)cyclohexane, can also be formed during the addition of $HSiCl_3$ to 1,3-cyclohexadiene in the presence of chloroplatinic acid followed by ethanolysis [239].

Asymmetric hydrosilylation of 1,3-cyclohexadiene (and of cyclopentadiene) with trichlorosilane catalysed by chiral phosphine complexes of Pt(II) occurs according to equation (237) [590]:

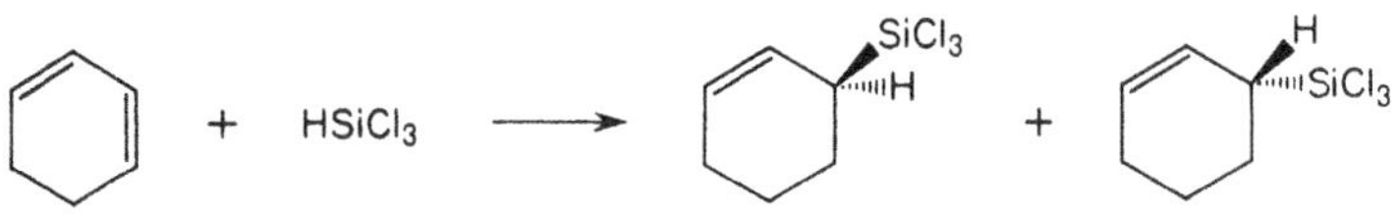

R(7%) *S*(64%) (237)

As discussed above, the major product has the S-configuration and the reaction mechanism assumes addition of a Pd–H bond to the cyclic diene yielding a Π-alkenyl–metal bond, followed by transfer of the silyl group from the metal to the Π-complex.

3.2.3 Allene

The addition of tri-substituted silanes to allene (propadiene) occurs predominantly across the 1,2-position in the presence of palladium complexes such as $PdCl_2(PPh_3)_2$ [D-027], $Pd(PPh_3)_4$ [1208] or $Pd(OCCH{=}CHCOO)(PPh_3)_2$ [289], and also in the presence of H_2PtCl_6 or $Rh(CO)_2(acac)$ [S-014]. The silyl product is also obtained but to a lesser extent:

$$CH_2{=}C{=}CH_2 + HSiCl_3 \xrightarrow{Pd(PPh_3)_4} CH_2{=}CHCH_2SiCl_3 + Cl_3Si(CH_2)_3SiCl_3 \quad (238)$$

When the reaction is catalysed by a rhodium complex, two other products can be formed [919]:

$$CH_2{=}C{=}CH_2 + R_3SiH \xrightarrow{Rh(CO)_2(acac)} \begin{cases} CH_2{=}CHCH_2SiR_3 & (I) \\ CH_3CH{=}CHSiR_3 & (II) \\ CH_2{=}C(CH_3)SiR_3 & (III) \end{cases} \qquad (239)$$

An internal adduct (III) is produced during hydrosilylation with trichlorosilane (120°C, 6 h, THF) in the presence of $Pd(\overline{COCH{=}CHCOO})(PPh_3)_2$ [289]. A 100% yield of the adduct $CH_2{=}CHCH_2SiH(C_6H_5)_2$ has been obtained when allene is allowed to react with diphenylsilane at room temperature [S-014].

When a rhodium catalyst is treated with allene the resulting IR spectrum does not contain a band characteristic of a C=C bond (1690 cm^{-1}) but indicates the formation of a new band at 2300 cm^{-1} corresponding to a C≡C bond. Apparently, an acetylenic structure is favoured in the metal–substrate complex [919].

Allene reacts with trichlorosilane, triethoxysilane [289], diphenylsilane [1074, S-014, S-034] and triethoxysilane [S-034]. When excess silane is present, addition to allene in the presence of $Rh(CO)_2(acac)$ leads to three isomeric disilylpropanes [380] (Schemes 240 and 241).

$$CH_2{=}C{=}CH_2 + (C_6H_5)_2SiH_2 \xrightarrow[\text{yield 23\%}]{100°C,\ 0.5\ h} \qquad (240)$$

$$\longrightarrow \begin{cases} (C_6H_5)_2SiH(CH_2)_3SiH(C_6H_5)_2 & (I) \\ (C_6H_5)_2SiHCH_2CH(CH_3)SiH(C_6H_5)_2 & (II) \\ (CH_3)_2C[SiH(C_6H_5)_2]_2 & (III) \end{cases}$$

I:II:III = 10.5:60:29 [S-034].

$$CH_2{=}C{=}CH_2 + (RO)_3SiH \xrightarrow[\text{yield 78.7\%}]{70°C,\ 0.5\ h} \begin{cases} (RO)_3Si(CH_2)_3Si(OR)_3 & (I) \\ (RO)_3SiHCH_2CH(CH_3)Si(OR)_3 & (II) \\ (CH_3)_2C[Si(OR)_3]_2 & (III) \end{cases}$$

R = C_2H_5; I:II:III = 28.8:54.0:17.2 [S-034]. (241)

3.2.4 Alkatrienes

Alkatrienes also undergo hydrosilylation reactions [468, 663, 725, 1332, 1335], their reactivity depending on the position of the C=C bond in the molecule. Thus, the reactivity sequence of the C=C bonds in myrcene during its reaction with trichlorosilane (with H_2PtCl_6 as the catalyst) is as follows [823]:

$$(CH_3)_2\underset{a}{C{=}C}HCH_2CH_2\underset{b}{C({=}CH_2)}\underset{c}{CH{=}CH_2}; \quad a > b > c \qquad (242)$$

When ethyldimethylsilane is added to the (*a*) and (*b*) C=C bonds of myrcene in the presence of $RhCl(PPh_3)_3$, a mixture consisting of 1,6,9-trimethyl-3-sila-2,6-dodecadiene [$(CH_3)_2C{=}CHCH_2CH_2C(CH_3){=}CHCH_2Si(CH_3)_2C_2H_5$] and

6-ethylidene-2-methyl-8-sila-2-decene [$(CH_3)_2C{=}CHCH_2CH_2C({=}CHCH_3)CH_2Si(CH_3)_2C_2H_5$] is obtained. Hydrosilylation of 5-methyl-1,3,6-heptatriene by trichlorosilane in the presence of $Ni(acac)_2PPh_3$ gives the 1,4-addition product [1335]:

$$CH_2{=}CHCH(CH_3)CH{=}CHCH{=}CH_2 + HSiCl_3 \xrightarrow{80°C,\ 6\ h} CH_2{=}CHCH(CH_3)CH_2CH{=}CHCH_2SiCl_3 \tag{243}$$

Adding dichlorosilane to this triene can lead to the intermolecular hydrosilylation product [1335].

1,3,7-Octatriene reacts with trichlorosilane (100°C, 3 h, 85% yield) in the presence of $Pd(PPh_3)_4$ to give the 1,4-addition product [D-027] containing an internal silyl group, i.e. 5-trichlorosilyl-1,6-octadiene. By analogy, cymene (3,7-dimethyl-1,3,6-octatriene) gives the 1,4-addition product with the silyl group attached to C4 [851]:

$$((CH_3)_2C{=}CHCH_2CH{=}C(CH_3)CH{=}CH_2 + CH_3SiHCl_2 \xrightarrow{\text{(yield 77\%) } PdCl_2(PhCN)_2\ +\ PPh_3} (CH_3)_2C{=}CHCH_2CH(SiCl_2CH_3)C(CH_3){=}CHCH_3 \tag{244}$$

However, addition of ethyldimethylsilane to this triene in the presence of Wilkinson's catalyst gives two 1,4-adducts with a silyl group at C1 and C4, respectively [851]:

$$(CH_3)_2C{=}CHCH_2CH{=}C(CH_3)CH{=}CH_2 + C_2H_5SiH(CH_3)_2 \xrightarrow{\text{(yield 82\%) } RhCl(PPh_3)_3} (CH_3)_2C{=}CHCH_2C(CH_3){=}CHCH_2Si(CH_3)_2C_2H_5 + (CH_3)_2C{=}CHCH_2CH[Si(CH_3)_2C_2H_5]C(CH_3){=}CHCH_3 \tag{245}$$

Hydrosilylation of 2,7-dimethyl-1,3,7-octatriene, $CH_2{=}C(CH_3)CH_2CH_2CH{=}CHC(CH_3){=}CH_2$, ($Ni(acac)_2$, 80–100°C, 3 h) proceeds via 1,4-addition to give a product with a terminal silyl group [1019].

3.3 Alkene and Cycloalkene Derivatives with Functional Groups

3.3.1 Haloalkenes and Haloarylalkenes

When the hydrosilylation of perfluoroethylene is initiated by γ-radiation, the accompanying formation of a dimeric product occurs:

$$CF_2{=}CF_2 + (CH_3)_3SiH \xrightarrow{\gamma(^{60}Co)} \underset{66\%}{CF_2HCF_2Si(CH_3)_3} + \underset{34\%}{CF_2HCF_2(CF_2)_2Si(CH_3)_3} \quad (246)$$

In contrast, the addition of trichlorosilane and/or methyldichlorosilane to 2-(trifluoromethyl)-3,3,3-trifluoropropene leads to the selective formation of the β-adduct [1360]. It is notable that no H/F exchange takes place during the hydrosilylation of unsaturated fluorine derivatives [1360]. Radical addition of silicon hydride to 1,1,3,3,3-pentafluoropropene ($CF_3CH{=}CF_2$) yields a greater amount of isopropylsilane (the α-adduct) than of the isomer with the terminal silyl group (the β-adduct); this arises because the –CH= carbon atom is more susceptible to attack by the silyl radical than the $CF_2{=}$ carbon atom. The relative reactivity of the unsaturated group can be expressed as:

$$-CH{=}CF_2 > -CF{=}CF_2 > -CH{=}CH_2 \quad (247)$$

although the ratio of various adducts formed also depends on the nature of the silane employed [1360].

Hydrosilylation of 3,3,3-trifluoropropene and homologues olefins $F(CF_2)_nCH{=}CH_2$ with dichloromethylsilane (and dimethylchlorosilane) catalysed by $PdCl_2(PhCN)_2$ + 2 PPh_3 and H_2PtCl_6 leads exclusively to the β-adduct [858, U-161], while in the presence of $RhCl(PPh_3)_3$ and $Ru_3(CO)_{12}$ quite significant amounts of the product arising from dehydrogenative hydrosilylation with triethoxysilane may be formed in addition to the β-adduct [1049].

Like styrene, vinylfluorobenzene can form the following three products:

$$C_6F_5CH{=}CH_2 \xrightarrow{R_3SiH} \begin{cases} C_6F_5CH_2CH_2SiR_3 & \text{(I)} \quad \beta\text{-adduct} \\ C_6F_5CH(CH_3)SiR_3 & \text{(II)} \quad \alpha\text{-adduct} \\ C_6F_5CH{=}CHSiR_3 & \text{(III)} \quad \text{unsaturated product} \end{cases} \quad (248)$$

Depending on the catalyst and silane employed various products may be obtained. Hence, when the reaction with, for example, dimethylchlorosilane is catalysed by H_2PtCl_6, the β-adduct (I) is formed [132, 149], whereas in the presence of $RhCl(PPh_3)_3$ different products result depending on the silane used, e.g. when R_3 is $Cl(CH_3)_2$ and reaction is conducted at 120°C for 4 h, (I) is obtained in 44% yield [132]; when R is C_2H_5, two products ((I) and (II)) are obtained in 10% and 8% yield, respectively [858]; and when R_3 is $(CH_3)_2C_2H_5$ and the reaction is conducted at 70°C for 18 h, (I) and (III) are obtained in 45% and 31% yield, respectively [858]. Catalysis of the reaction by $Ru_3(CO)_{12}$ leads to the unsaturated product in the greater yield, e.g. when R_3 is $(CH_3)_2C_2H_5$ and the reaction is conducted at 70°C for 18 h, (I) and (III) are obtained in 5% and 80% yields, respectively.

Fluorinated silanes as co-monomers for organosilicon copolymers can be synthesized by hydrosilylation of a mixture of homologous olefins $F(CF_2)_nCH{=}CH_2$ (n = 6, 8, 10, 12) by methyldichlorosilane or dimethylchlorosilane in the presence of chloroplatinic acid [U-155]. 1,1,2-Trifluoro-1,4-pentadiene adds a hydrosilane only across the unsubstituted C=C bond [1194]:

$$CF_2{=}CFCH_2CH{=}CH_2 + (CH_3)_2SiHCl \rightarrow \underset{85\%}{CF_2{=}CF(CH_2)_3SiCl(CH_3)_2} \quad (249)$$

Hydrosilylation of vinyl chloride with, for example, trichlorosilane [747, 1055, 1058, 1360, C-002, D-027, D-041, D-080, D-097, D-111, Pl-010, S-024], particularly in the presence of Pt/C [D-097], $NiCl_2(PPh_3)_2$ [750, U-122], $RhCl(PPh_3)_3$ [747], or $Pd(PPh_3)_4$ [1058] often leads to the formation of a product in which chlorine is absent from the alkyl chain ($C_2H_5SiCl_3$); this arises because of H/Cl exchange preceding hydrosilylation of ethylene.

When the reaction is initiated by means of a free-radical process, a telomeric product is often formed in addition:

$$CH_2{=}CHCl + HSiCl_3 \xrightarrow[100-180^\circ C]{\gamma(^{60}Co)} \underset{21\%}{CH_3CH_2SiCl_3} + \underset{7\%}{CH_3(CH_2)_3SiCl_3} + SiCl_4 \quad (250)$$

$CH_2ClCH_2SiMeCl_2$ was recently reported as the sole product of radiation-induced hydrosilylation of $CH_2{=}CHCl$ by CH_3SiHCl_2 [189]. However, when vinyl fluoride reacts with excess of trichlorosilane after UV initiation, 90–95% yield of product is obtained with no telomerization, the latter being prevented by the excess silane [270]. 2-Chloroethyltrichlorosilane can be synthesized (in 80–100% yield) by catalysis of the reaction by $Ni(PPh_3)_4$ [D-111], $Rh_4(CO)_{12}$ [449], $RhCl(PPh_3)_4$ [Pl-010] or H_2PtCl_6 [747, D-041]. Regardless of the nature of the silane, hydrosilylation of perfluorovinyl chloride initiated by γ-radiation always proceeds according to the following equation [1360]:

$$FClC{=}CF_2 + {\equiv}SiH \xrightarrow[20-30^\circ C]{} ClFCHCF_2Si{\equiv} + CFH_2CF_2Si{\equiv} + ClSi{\equiv} \quad (251)$$

showing that addition of a silyl radical occurs only at $CF_2{=}$ and is accompanied by H/Cl exchange at the silicon atom. The same phenomenon is observed during hydrosilylation of 1,2-dichloroethene in the presence of chloroplatinic acid [1055].

Tri-substituted silanes, especially trichlorosilane, react with allyl chloride by means of three competitive reactions, i.e. addition (I), condensation (II) and reduction (H/Cl exchange) (III) (see equation 252):

$$Cl_3SiCH_2CH_2CH_2SiCl_3$$
$$\uparrow HSiCl_3$$
$$CH_2{=}CHCH_2Cl + HSiCl_3 \rightarrow \begin{cases} Cl_3SiCH_2CH{=}CH_2 + HCl & (II) \\ Cl_3SiCH_2CH_2CH_2Cl & (I) \\ CH_3CH{=}CH_2 + SiCl_4 & (III) \end{cases}$$
$$\downarrow HSiCl_3$$
$$Cl_3SiCH_2CH_2CH_3 \quad (252)$$

Allyltrichlorosilane is efficiently synthesized (over 80% yield) by catalysis of the reaction in a basic medium using various metal salts, e.g. Cu_2Cl_2 [397]. When the hydrosilylation of allyl chloride is allowed to proceed in the presence of transition metal complexes, addition is accompanied by only H/Cl exchange. The ratio of the addition and reduction products decreases according to the following sequences:

$$F > Cl > Br > I \tag{253.1}$$

$$HSiCl_3 > CH_3SiHCl_2 > C_6H_5SiHCl_2 > (CH_3)_2SiHCl > CH_3(C_6H_5)SiHCl > (CH_3)_2C_6H_5SiH > (C_6H_5)_2SiHCl > CH_3(C_6H_5)SiHCH_2Cl > CH_3(C_6H_5)_2SiH \sim C_2H_5SiHCl_2 > (C_2H_5)_2SiHCl > (C_2H_5)_3SiH \text{ [98, 99]} \tag{253.2}$$

$$HSiCl_3 > (RO)_3SiH \text{ [613]} \tag{253.3}$$

3-Chloropropyltrichlorosilane and 3-chloropropyltrialkoxysilanes are well-known key intermediates for the commercial production of organofunctionalized silanes and polysiloxanes [313a] and their synthesis by hydrosilylation of allylchloride will be more comprehensively discussed in Section 6.1 due to their application as silane (and siloxane) coupling agents. For details see also especially Tables 1.3 and 9.3 (Part II).

3,3,3-Trichloropropene reacts with methyldichlorosilane (Speier's catalyst, 65°C, 48 h) to give not only the expected adduct (49% yield), but also isomerization products, i.e. $CCl_2{=}CHCH_2Cl$, and condensation products, i.e. $CCl_2{=}CHCH_2SiCl_2CH_3$ [D-037]. In the UV-initiated reaction of trichlorosilane with $CF_3CCl{=}CHF$ (20°C, 300 h), preliminary H/Cl exchange and fluorine migration are followed by hydrosilylation, resulting in $CF_3CCl{=}CHF$ (60%) and $CF_3CH_2CHFSiCl_3$ (29%) [479].

Migration of the C=C bond has often been observed during the hydrosilylation of monochloro-n-butenes and dichlorobutenes, regardless of the catalyst used [1113].

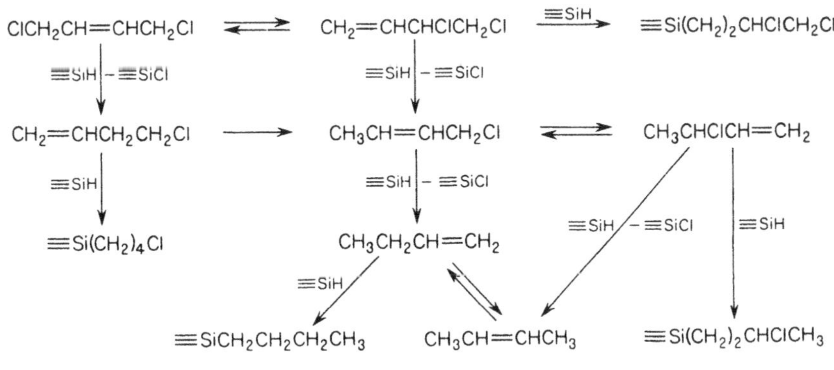

(254)

The isomerization may be followed by hydrosilylation and/or H/Cl exchange at the silicon atom. In the presence of H_2PtCl_6, methyldichlorosilane reacts with

3-chloro-1-butene or 4-chloro-1-butene to form 3-chlorobutylmethyldichlorosilane. Similar products are obtained from the hydrosilylation of the corresponding 3-and 4-chloro-2-butenes. Hydrosilylation of 4-chloro-1-butene is accompanied by C=C bond isomerization and chlorine migration, followed by H/Cl exchange. 3,4-Dichloro-1-butene reacts with silane to form predominantly 2,2,5,6-tetrachloro-2-silahexane as the adduct; a mixture of other products has also been observed at the same time, although 4-chlorobutyl derivatives are not present. The latter, e.g. 4-chlorobutylalkyldichlorosilane, may be formed as a result of the following reactions [730]:

$$RSiHCl_2 + ClCH_2CH{=}CHCH_2Cl \xrightarrow{H_2PtCl_6} ClCH_2CH(SiCl_2R)CH_2CH_2Cl \quad (255.1)$$

$$ClCH_2CH(SiCl_2R)CH_2CH_2Cl \xrightarrow{\beta\text{-elimination}} CH_2{=}CH(CH_2)_2Cl + RSiCl_3 \quad (255.2)$$

$$CH_2{=}CH(CH_2)_2Cl + RSiHCl_2 \xrightarrow{H_2PtCl_6} RSiCl_2(CH_2)_4Cl \quad (255.3)$$

$R = CH_3, C_2H_5$.

Hydrosilylation of 5,5,5-trichloropentene-1 occurs in the presence of chloroplatinic acid according to the equation:

$$CH_2{=}CH(CH_2)_2CCl_3 + HSi(C_2H_5)_3 \rightarrow (C_2H_5)_3Si(CH_2)_4CCl_3 \quad (256)$$

with 67% yield, and the addition is accompanied by by-products $(C_2H_5)_3SiCl$ (0.7%), $[(C_2H_5)_3Si]_2O$ (2.3%) and Z,E-isomers $CCl_3CH_2CH{=}CHCH_3$ (5%) [400]. Higher ω-chloroalkylsilanes can be prepared by hydrosilylation of corresponding alkenylchlorides by trisubstituted silanes, e.g. $Cl(CH_2)_{11}Si(OCH_3)_3$ [1294].

Reaction of trichlorosilane and methyldichlorosilane with 1-chloro-3,3,4,4-tetrafluorocyclobutene initiated by γ-radiation leads to the products arising both from addition and H/Cl exchange [1213]:

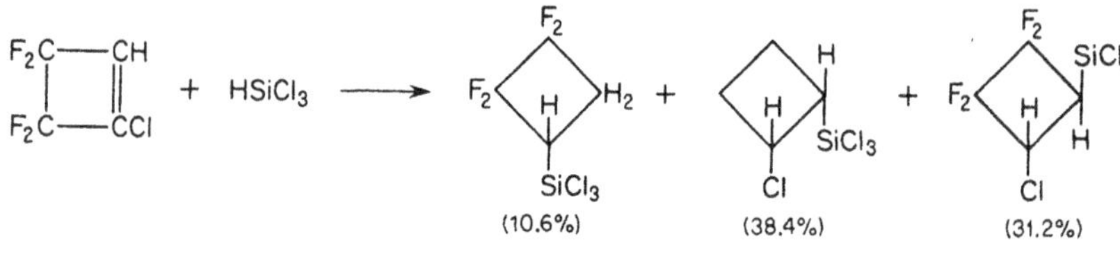

(257)

The replacement of Cl by H enables two addition products to be formed. Perfluorocyclobutene adds trialkylsilanes at 250°C without F/H exchange. On the other hand, 1,2-dichloro-3,3,4,4-tetrafluorobutene yields condensation and H/Cl substitution products rather than adducts [294]:

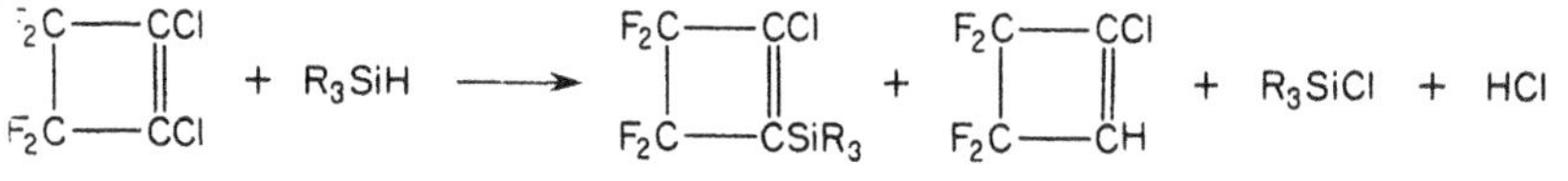

(258)

The addition of silicon hydride to cyclic and bicyclic halogen derivatives, e.g. 2-halogenomethylbicyclo[2.2.1]-5-heptene, in the presence of Speier's catalyst, leads to the following sequence of adduct yield [728]:

$$F > Cl > Br > I \qquad (259)$$

However, the structure of the olefin also plays a decisive role in the hydrosilylation of this class of compound. On reaction with silicon hydrides, bicyclic halogen derivatives give a higher yield of adducts than monocyclic derivatives [591]. 1,1-Dichloro-2-vinylcyclopropene may be hydrosilylated by trichlorosilane in the presence of Speier's catalyst without ring opening [1090].

3.3.2 Nitroalkenes and Nitroarylalkenes

The addition of trichlorosilane to 3-nitropropene in the presence of Speier's catalyst occurs at atmospheric or elevated pressure [839].

$$CH_2{=}CHCH_2NO_2 + HSiCl_3 \xrightarrow[\text{reflux, 5 h}]{H_2PtCl_6\ (iso\text{-}PrOH)} Cl_3Si(CH_2)_3NO_2 \qquad (260)$$

Changing the order of substrate addition has no effect upon the yield of the adduct, and the reactivity of silanes decreases in the sequence:

$$HSiCl_3 > CH_3SiHCl_2 > C_2H_5SiHCl_2 \qquad (261)$$

Organic compounds with more than one nitro group in the molecule can also be hydrosilylated [838]: thus 4,4-dinitro-1-butene and 4,4,4-trinitro-1-butene react with methyldichlorosilane in the presence of H_2PtCl_6 to form the expected adducts in yields of 94% and 62% respectively.

When diphenylmethylsilane is added to m-nitrostyrene in the presence of Speier's catalyst, formation of the β-adduct (46%) is accompanied by a small amount of aminostyrene (5%). Under the same conditions (1.5 h, 150°C), phenyldimethylsilane forms only the β-adduct in 58.5% yield [48].

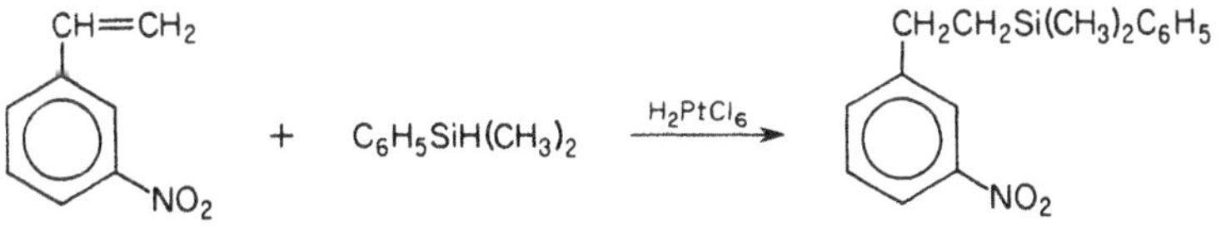

(262)

although the yield is reduced to 16% when reduction of the nitro group is accomplished with triphenylsilane [48]:

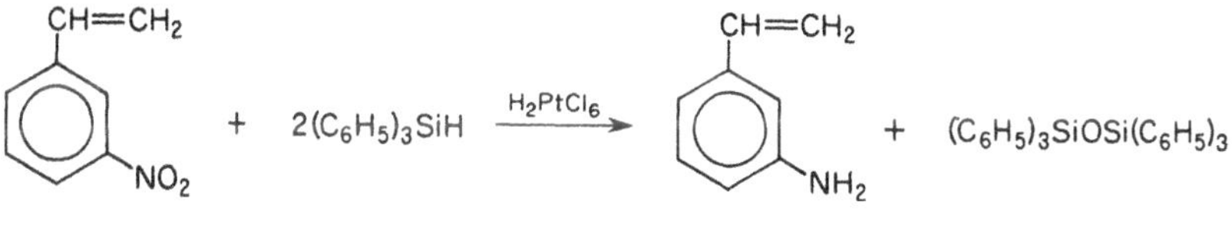

(263)

In the presence of $Rh_4(CO)_{12}$ as the catalyst, hydrosilylation of m-nitrostyrene with trichlorosilane and methyldichlorosilane occurs with surprising ease on heating for 1 h at 30–40°C, yielding 99% and 95% of the β-adducts, respectively [S-023].

The addition of tri-substituted silanes ($R^1R^2R^3SiH$, where R^1, R^2, R^3 = $(CH_3)_2Cl$, CH_3Cl_2, Cl_3, $C_6H_5(CH_3)Cl$ and $(C_2H_5O)_3$ to *o*- and *p*-nitrophenylalkenyl ethers [D-75], when catalysed by chloroplatinic acid, gives very high yields (85–91%) of the β-adduct. When the NO_2 group is not conjugated with the C=C bond undergoing hydrosilylation, the yields of β-adducts are generally very high. But in conjugated systems the NO_2 group is deactivated, presumably as a result of a decrease in electron density of the C=C bond.

3.3.3 Aminoalkenes

The widespread application of 3-aminopropylsubstituted silanes as coupling agents (see also Section 6.1) has led to an extensive study of the hydrosilylation of allylamine and its derivatives [91, 317, 748, 795, D-073, S-007, J-072]. Triethoxysilane can be added to allylamine in the presence of a variety of platinum catalysts in addition to H_2PtCl_6 [820, Dd-002, Pl-022], giving generally a 4:1 mixture of γ- and β-isomers (yield 60–80%):

$$CH_2{=}CHCH_2NH_2 + (C_2H_5O)_3SiH \longrightarrow \begin{cases} (C_2H_5O)_3Si(CH_2)_3NH_2 \quad \gamma\text{-isomer} \\ (C_2H_5O)_3SiCH(CH_3)CH_2NH_2 \quad \beta\text{-isomer} \end{cases} \qquad (264.1)$$

Such hydrosilylation is accompanied by dehydrocondensation [93, 820, Dd-002]:

$$RNH_2 + (C_2H_5O)_3SiH \xrightarrow{H_2PtCl_6} RNHSi(OC_2H_5)_3 + H_2 \qquad (264.2)$$

$$R = CH_2{=}CHCH_2{-},\ (C_2H_5O)_3Si(CH_2)_3{-}.$$

N-Triethoxysilyl-3-aminopropyltriethoxysilane undergoes further rearrangement during reaction and distillation [747a]:

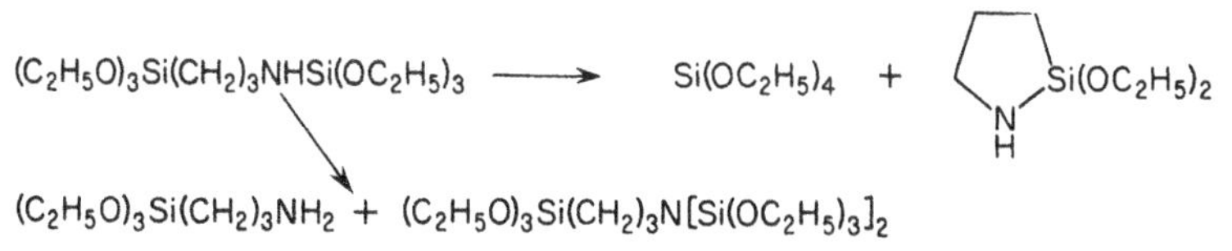

(264.3)

Hydrogen and propene have also been detected among the gaseous products [93].

The addition of triethoxysilane to allylamine has been reported as proceeding in the presence of other platinum catalysts, such as Pt/C (60% yield of the adduct), $PtCl_2 + C_2H_4 + C_5H_5N$ (63% yield of the adduct), and $Pt\{[(CH_3)_2C{=}CHCO(CH_3)]Cl_2\}_2$ (20% yield of the α-adduct) [D-049, G-023], $Pt(PPh_3)_2(C_2H_4)$ (≫50% yield of the γ-adduct) [748]. In the hydrosilylation of

N-(2-aminoethyl-)-3-aminopropene, a 1:3 mixture of the α- and β-adducts is also formed [D-049, G-023]:

$$CH_2{=}CHCH_2NHCH_2CH_2NH_2 + (CH_3O)_3SiH \begin{cases} \rightarrow (CH_3O)_3Si(CH_2)_3NHCH_2CH_2NH_2 \\ \rightarrow (CH_3O)_3SiCH(CH_3)CH_2NHCH_2CH_2NH_2 \end{cases} \quad (265)$$

3-Aminopropylalkoxysilanes can be synthesized by intra- and inter-hydrosilylation of silazane followed by alcoholysis (hydrolysis) of the Si–N bond [D-073].

Cobalt and rhodium carbonyls catalyse the addition of triethoxysilane to dimethylallylamine at room temperature, giving a high yield (86%) of *N,N*-dimethyl(3-triethoxysilylpropyl)amine.

When alkyl allylamines and their derivatives [11, 290, 690, 701, 402, 9919, D-038, D-058, D-101, D-114, J-013, J-079, S-006, S-007, U-123], and also dialkylallylamines [553], react with silicon hydrides in the presence of chloroplatinic acid, β-and γ-isomers are always obtained, with their ratio depending closely on the structure of the unsaturated amine and hydrosilane [255]; hence for $(C_2H_5O)_3SiC_3H_6NR^1R^2$ the following values of γ/β+γ were established:

R^1R^2	H C_2H_5	H C_4H_9	H $CH_2C_6H_5$	$(C_3H_7)_2$	C_2H_5 C_4H_9	$(CH_2C_6H_5)_2$	
$\frac{\gamma}{\beta+\gamma}$	0.32	0.53	0.66	0.93	0.94	1	(266.1)

However, for $(C_2H_5)_3SiC_3H_6NR^1R^2$, the corresponding ratios are:

R^1R^2	$(C_2H_5)_2$	$(CH_2CH{=}CH_2)_2$	$(C_3H_7)_2$	$(C_4H_9)_2$	
$\frac{\gamma}{\beta+\gamma}$	0.92	0.92	0.98	0.98	(266.2)

For silanes of the type $R^3R^4R^5SiC_3H_6NHC_2H_5$ the following sequence applies:

$R^3R^4R^5$	$(C_2H_5O)_3$	$CH_3(C_6H_5)_2$	$(C_2H_5)_3$	
$\frac{\gamma}{\beta+\gamma}$	0.32	0.8	1	(266.3)

The hydrosilylation of alkenylamines containing other functional groups, e.g. epoxide, cyanide [1151], ester [689], alkylsilanes [690], amine [D-106] and urethane [780], is also possible.

During the hydrosilylation of vinyl and other linear amines [Dd-006], heterocyclic amines containing vinyl groups [607, 699, 799, 819, 943] and vinyl imides [D-106, D-107] and their derivatives [1089], formation of the β-isomer alone or a mixture of α- and β-isomers has been observed. The hydrosilylation of the ammonium salts of allylamine derivatives or allylalkylcarboxylanes with trichloro- and alkoxy-silanes results in a low yield of product [613]. Similarly, organosilicon phosphinic salts can be obtained by hydrosilylation [265]:

$$CH_3SiHCl_2 + CH_2{=}CHCH_2P^+(C_6H_5)_3Br^- \xrightarrow[10\ h,\ CHCl_3]{H_2PtCl_6} CH_3SiCl_2(CH_2)_3P^+(C_6H_5)_3Br^- \quad (267)$$

3.3.4 Cyanoalkenes and Other Nitrogen-containing Alkenes

Studies of the hydrosilylation of various unsaturated bonds in compounds containing a cyanide functional group have also been performed [868, 869, 974, 988, 1151, J-018, G-017, C-043]. Of these acrylonitrile has received the most attention. A comprehensive examination of the hydrosilylation of acrylonitrile has been undertaken in the presence of base catalysts (Section 2.2). Catalysis of the reaction (with $SiHCl_3$) by an $RhCl_3$(THF)/polyolefin-supported complex, leads to the selective formation of the β-isomer (reflux, 3 h, 60% yield) [S-029]. β-Cyanoethyltrichlorosilane has also been obtained in the presence of a Pt/support system, benzoyl peroxide, tertiary amines [422, 840, 905, J-095] and various complexes of nickel [841] and rhodium [227, C-006].

β-Adduct was also recently synthesized by the reaction of acrylonitrile with R_3SiH (where R_3 = Cl_3, Cl_2CH_3, $Cl_2C_6H_5$) with high yield (70–100%), exclusively under mild conditions using a new type Blustein catalyst consisting of tetramethylethylenediamine and cuprous oxide. Ultrasonic waves significantly increased the rate of hydrosilylation [959].

In contrast, Wilkinson's catalyst, $RhCl(PPh_3)_3$, directs the addition of the hydrosilane (with the exception of trichlorosilane) towards the α-adduct [868], which presumably results from the addition of the hydride to the electron-deficient carbon atom, as proposed by Ojima *et al.* [868].

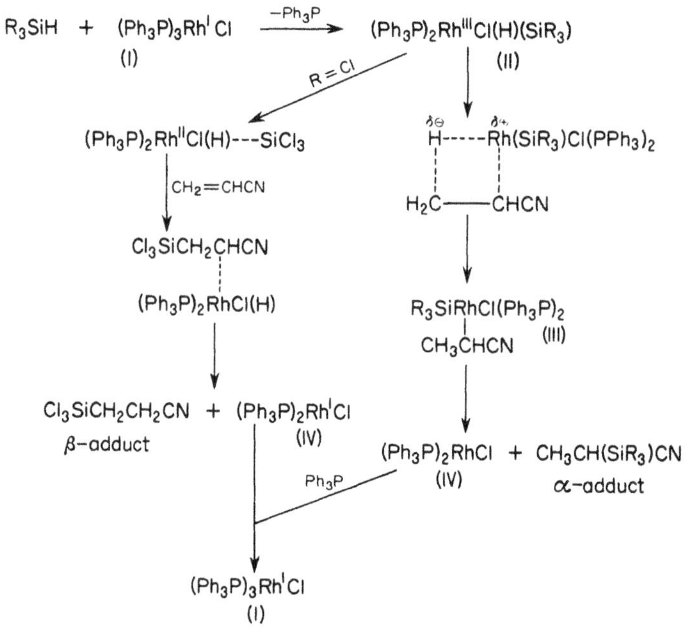

(268)

These authors have suggested that the reverse regioselectivity observed in the hydrosilylation with trichlorosilane results from a radical mechanism which occurs at the rhodium centre [868]. A dispersed nickel catalyses exclusive formation of α-adduct (96%) in the hydrosilylation of methyldichlorosilane conducted in the presence of ultrasound [E-049].

The α-adduct, $(CH_3)_2SiHCH(CN)CH_3$, is formed during the reaction of acrylonitrile with dimethylsilane in the presence of a Ziegler (Ni salt + AlR_3) catalyst [945] and with dialkyldimethyldisiloxane catalysed by chloroplatinic acid [988]. Hydrosilylation of 2-cyanobicyclo[2.2.2]heptene-6 by this siloxane gives bicyclic nitriles with the yield of 30–80% [990].

Simple unsaturated cyanides other than acrylonitrile, e.g. crotonitrile ($CH_3CH{=}CHCN$) [868, 869], methacrylonitrile, $CH_2{=}C(CH_3)CN$ [227], allyl cyanide ($CH_2{=}CHCH_2CN$) [821, 935, 971, 974, U-007, U-027, U-085] and β-phenylacrylonitrile ($C_6H_5CH{=}CHCN$) [869, 868], also undergo similar hydrosilylations.

Other organic nitrogen compounds are also capable of adding silanes to the C=C double bond in the alkenyl group, as illustrated by the following examples:

$$CH_2{=}CHCH_2OC_6H_4NO_2\text{-}p + CH_3SiHCl_2 \xrightarrow[90\text{-}100°C]{H_2PtCl_6} \underset{85\%}{CH_3SiCl_2(CH_2)_3OC_6H_4NO_2\text{-}p} \qquad (269.1)\ [D\text{-}075]$$

$$CH_2{=}CHCH_2NCO + HSiCl_3 \xrightarrow{H_2PtCl_6\ (EtOH)} \underset{73\%}{Cl_3Si(CH_2)_3NCO} \qquad (269.2)\ [D\text{-}024,\ U\text{-}039]$$

$$N_3COOCH_2CH{=}CH_2 + HSiCl_3 \xrightarrow[50°C,\ 16\ h]{H_2PtCl_6(i\text{-}PrOH)} \underset{76\%}{N_3COO(CH_2)_3SiCl_3} \qquad (269.3)\ [G\text{-}019]$$

$$N_3COOCH_2CH_2OCH_2CH{=}CH_2 + HSiCl_3 \xrightarrow[50°C,\ 20\ h]{H_2PtCl_6(i\text{-}PrOH)} N_3COO(CH_2)_2O(CH_2)_3SiCl_3 \qquad (269.4)\ [G\text{-}019]$$

$$(CF_3)_2NCH_2CH{=}CH_2 + CH_3SiHCl_2 \xrightarrow[80°C,\ 24\ h]{} \underset{85\%}{(CF_3)_2N(CH_2)_3SiCl_2CH_3} \qquad (269.5)\ [290,\ D\text{-}038]$$

$$CH_2{=}CHC_5H_4N + HSiCl_3 \xrightarrow[150°C,\ 5\ h]{H_2PtCl_6(i\text{-}PrOH)} \underset{80\text{-}90\%}{Cl_3SiCH_2CH_2C_5H_4N} \qquad (269.6)\ [819]$$

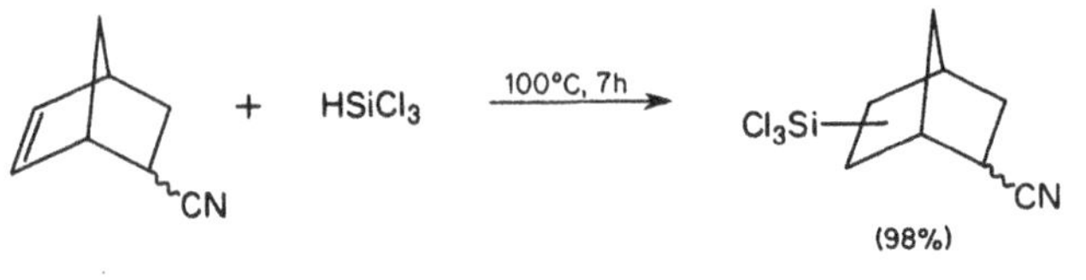

(269.7)
[1329]

3.3.5 Alkenyl Alcohols, Phenols, Aldehydes, Acids and Esters

The C=C bond of an alcohol is not hydrosilylated in the presence of chloroplatinic acid [337, S-004]. Instead, the molecule undergoes dehydrocondensation:

$$CH_2{=}CHCH_2OH + (C_2H_5)_3SiH \xrightarrow{H_2PtCl_6} CH_2{=}CHCH_2OSi(C_2H_5)_3 + H_2 \qquad (270)$$

In contrast to triethylsilane, triethylgermane $(C_2H_5)_3GeH$ does add across the C=C bond of an alkenyl alcohol [337]. In general, like other metal complexes, chloroplatinic acid catalyses not only the hydrosilylation but also the competitive reaction of the hydroxy groups of alcohols or carboxylic acids with Si–H bonds, to form alkenoxysilane and acyloxysilane compounds, respectively, with the evolution of hydrogen [Rev.12, Rev.36]. Trialkoxysilanes react with amino alcohols in the same manner even in the absence of a catalyst [766].

In order to obtain silyl alcohols as hydrosilylation products it is essential to block the hydroxy groups, particularly when chloro- and alkoxy-silanes have been added. In the synthesis of 3-(2-hydroxyethoxy-1-ethoxyethane), the preliminary product (1-(2-allyloxyethoxy)-1-ethoxyethane) may be obtained by reacting 2-allyloxyethanol and vinylethyl ether in the presence of HCl and propylene oxide at a temperature of 50°C [U-070]:

$$CH_2{=}CHCH_2OCH_2CH_2OH + CH_2{=}CHOC_2H_5 \xrightarrow[CH_3\overline{CHCH_2O}]{HCl} CH_2{=}CHCH_2OCH_2CH_2OCH(CH_3)OC_2H_5 \qquad (271)$$

In the presence of *p*-toluene sulphonic acid, the proper hydrosilylation reaction is followed by alcoholysis:

$$\begin{array}{c} CH_2{=}CHCH_2OCH_2CH_2OCH(CH_3)OC_2H_5 + (CH_3O)_3SiH \\ \downarrow H_2PtCl_6 \\ (CH_3O)_3Si(CH_2)_3OCH_2CH_2OCH(CH_3)OC_2H_5 \\ CH_3OH \downarrow p\text{-}CH_3C_6H_4SO_3H \\ (CH_3O)_3Si(CH_2)_3OCH_2CH_2OH + CH_3OCH(CH_3)OC_2H_5 \end{array} \qquad (272)$$

Vinyl ethers can also be used to block the hydroxy groups of unsaturated carboxylic acids [U-084]. However, silylation of hydroxy groups, e.g. with trimethylchlorosilane, is a commonly used as blocking method [G-013, G-015].

Contrary to alcohols silylated phenols can be prepared without blocking hydroxy group, e.g. [J-068]:

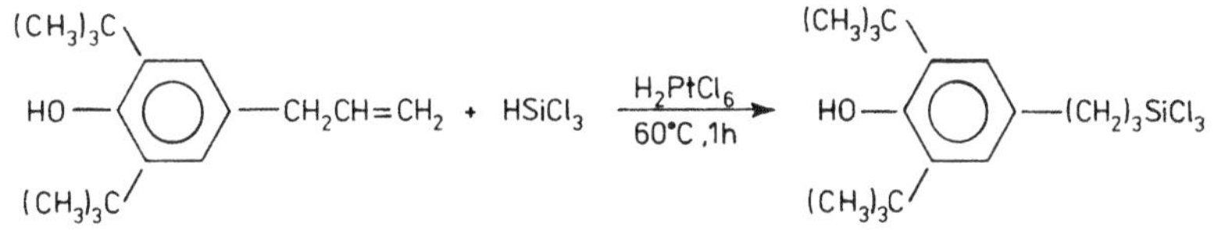

$$(CH_3)_3C$$

HO $-\langle\bigcirc\rangle-$ CH$_2$CH=CH$_2$ + HSiCl$_3$ $\xrightarrow[\text{60°C,1h}]{\text{H}_2\text{PtCl}_6}$ HO $-\langle\bigcirc\rangle-$ (CH$_2$)$_3$SiCl$_3$

$$(CH_3)_3C$$ $$(CH_3)_3C$$ (273)

Reaction of methacrylic and acrylic acids with tri-substituted silanes catalysed by H_2PtCl_6 and Pt/C results in dehydrocondensation, whereas the corresponding esters of these acids are readily hydrosilylated [456]. The addition of trimethyl- and tripropylsilane to acrolein acetals leads to the acetals of 3-trimethylsilyl- and 3-(tripropylsilyl)propionaldehyde [135]:

$$CH_2=CHCH(OR)_2 + R_3SiH \xrightarrow{H_2PCl_6} R_3SiCH_2CH_2CH(OR)_2 \qquad (274)$$

Di-substituted silanes, e.g. phenylmethylsilane, may be added to alkenols of the general formula $CH_2=CH(CH_2)_nOH$ (where $n = 1,2 \ldots , 9$) in the presence of Wilkinson's catalyst [132]. Reaction of phenylmethylsilane with undecylenol leads to the formation of 2-phenyl-2-methyl-1-oxo-2-silacyclododecane [132] as a result of a preliminary dehydrocondensation followed by hydrosilylation:

$$CH_2=CH(CH_2)_9OH + C_6H_5SiH_2CH_3 \xrightarrow{RhCl(PPh_3)_3} C_6H_5\overline{Si(CH_3)(CH_2)_{11}O} \qquad (275.1)$$

The reaction of a disubstituted silane with an alkene thiol proceeds similarly:

$$CH_2=CH(CH_2)_nSH + C_6H_5SiH_2CH_3 \xrightarrow{RhCl(PPh_3)_3} C_6H_5\overline{Si(CH_3)(CH_2)_{n+2}S} +$$

$$\overline{C_6H_5Si(CH_3)(CH_2)_{n+2}SSi(CH_3)(C_6H_5)(CH_2)_{n+2}S} \qquad (275.2)$$

Cyclic products can be obtained from reaction of 1,1,3,3-tetramethyldisiloxane with allyl alcohols or *ortho*-allylphenols in the presence of H_2PtCl_6 [S-004]. The 1-ω-alkenoxy-2-hydrodisiloxane adduct can be isolated before formation of the di adduct. Hydrosilylation of unsaturated carboxylic acids with di substituted silanes or dihydrosiloxanes also leads to cyclic compounds [357, S-004]:

$$CH_2=CH(CH_2)_nCOOH + O[SiH(CH_3)_2]_2 \xrightarrow[\text{80–137°C, 3 h}]{\text{H}_2\text{PtCl}_6 \ (iso-\text{PrOH})}$$

$$\overline{Si(CH_3)_2OSi(CH_3)_2(CH_2)_{n+2}COO} \qquad (275.3)$$

The yield depends on the value of n, being 73% when $n = 1$ and 60% when $n = 8$.

3.3.6 Unsaturated Ethers, Epoxides and Sulphides

The UV-initiated interaction of trichlorosilane with vinylethyl ether leads to the selective formation of a variety of adducts [198]. In most cases, however, the addition of silicon hydrides to aliphatic and aromatic vinyl ethers in the presence of chloroplatinic acid gives rise to two competitive reactions [1101], i.e. hydrosilylation and H/phenoxy exchange:

$$C_6H_5OCH{=}CH_2 + CH_3SiHCl_2 \xrightarrow{H_2PtCl_6\ (iso\text{-}PrOH)} \underset{76\%}{C_6H_5O(CH_2)_2SiCl_2CH_3} + \underset{16\%}{C_6H_5OSiCl_2CH_3} \qquad (276)$$

Methyl-*iso*-pentylphenoxyfluorosilane, which occurs as a by-product of the hydrosilylation of phenylvinyl ether with methyl-*iso*-pentylfluorosilane, has been detected in addition to the expected adduct (74%) [1120].

Trichlorosilane adds to 2-vinylfurane to give the β-adduct (60%):

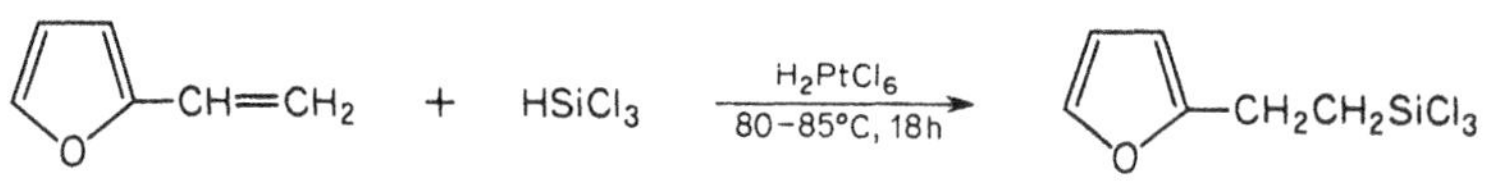

(277)

In the reaction with methyldichlorosilane and ethyldichlorosilane the product yield was 58–71%, whereas with ethoxy- and acetoxy-substituted silanes it was 50%, and only 30% with other trialkoxysilanes [693].

The hydrosilylation of allyl ethers $CH_2{=}CHCH_2OR'$ is commonly used for synthesis of silane coupling agents, e.g. 3-glycidoxypropyltrimethoxysilane (R′ = glycidyl), 3-metacryloxy-propyltrimethoxysilane (R′ = methacryl) [E-018, E-042] as well as for other γ-functional silanes, e.g. 3-acryloxytriphenylsilane (R′ = acryl)-3-allyloxypropyltriphenylsilane (R′ = allyl) [U-181], 7-[3-chlorodimethylsilyl)propoxy]-4-methylcoumarin [U-200]. The addition of dihydrodisiloxanes to allyl nitro-substituted alkyl and alkenoxy ethers leads to organosilicon nitriles with reactive silicon hydrogen bonds (mono-addition) [1160] or siloxane-containing dinitriles (di-addition) [1146].

The hydrosilylation of allylperfluoroethers with methyl(chloro)silanes is a frequently used method for synthesis finally (after hydrolysis and polycondensation) fluorine-containing silicone polymers [237, 681, J-104]. Polysiloxanes with oligo(oxyethylene) side-chains were prepared by treating hydro-substituted polysiloxanes with $CH_2{=}CHCH_2(OCH_2CH_2)_7OCH_3$ [1358] and bis(silylpropyl-substituted) ethylene polymers by hydrosilylation of $CH_2{=}CHCH_2(OCH_2CH_2)_n OCH_2CH{=}CH_2$, where $n = 2$–20 [J-114].

The hydrosilylation of allylaryl ethers is often accompanied by side-reactions [53] which can be expressed by the following series of equations (Scheme 278).

The Claisen rearrangement can be avoided by heating the reaction mixture at a lower temperature.

Hydrosilylation of 2-benzyloxypropene and 1-benzyloxypropene in the presence of chloroplatinic acid leads to both β- and γ-silylpropylbenzyl ethers, although the γ-isomer is always predominant [1023], indicating reversible isomerization between allyl and propenyl groups.

The addition of diethylsilane to divinylsulphides is accompanied by cyclization

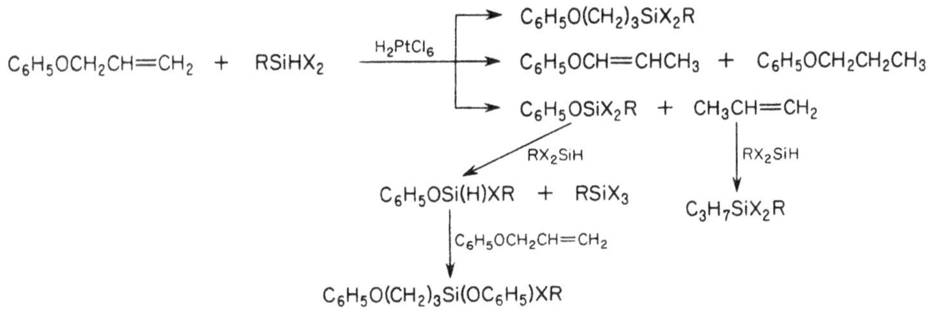

$R{=}X{=}OC_2H_5$; $R{=}CH_3$; $X{=}Cl$ (278)

and formation of the heterocyclic 2,4-dimethyl-3,3-diethyl-1-thio-3-silacyclobutane (III) (Scheme 279). In the presence of chloroplatinic acid, α- (II) and β- (I) adducts, as well as the product of the intermolecular hydrosilylation of the mono-adduct (III), are obtained [1239]:

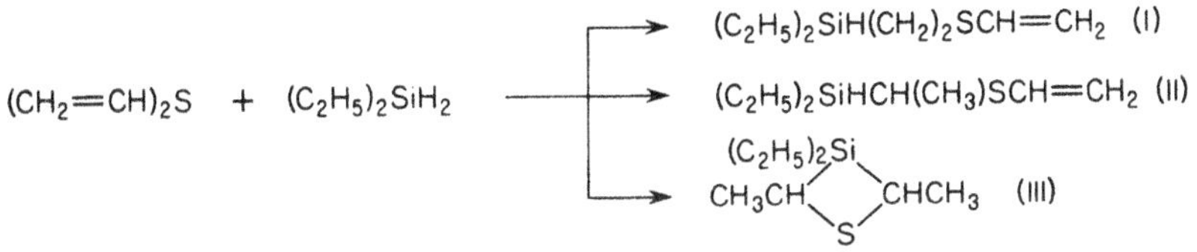

(279)

and the observed ratio I:II:III is equal to 4:4.5:1. The product (III) can be obtained directly by the intermolecular hydrosilylation of the α-adduct (II). In this case, a mixture of *cis* and *trans* adducts (1:1) is formed:

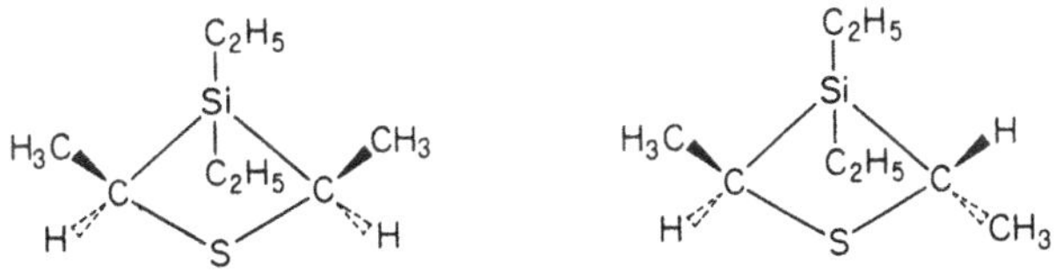

(280)

Although intermolecular hydrosilylation of the β-mono-adduct does occur, the yield of the resulting product thereof is extremely small.

Hydrosilylation of O-allyl ether oximes gives the corresponding 3-(triorganosilylo)propyl ether oximes [691, S-007, S-015]:

$$RR^1C{=}NOCH_2CH{=}CH_2 + R^2_3SiH \xrightarrow{H_2PtCl_6} RR^1C{=}NO(CH_2)_3SiR^2_3 \quad (281)$$

$R = H, CH_3$; $R^1 = CH_3$; $R^2 = C_2H_5, C_4H_9, OC_2H_5$.

The hydrosilylation of vinylallyl ethers, for example, by trimethoxysilane and triethoxysilane at temperatures of 20–50°C, occurs mostly at the C=C bond of the allyl group, and leads to the selective formation of tri-substituted 3-vinyloxypropyl silanes [D-047].

$$CH_2{=}CHCH_2OCH{=}CH_2 + (RO)_3SiH \xrightarrow{H_2PtCl_6} (RO)_3Si(CH_2)_3OCH{=}CH_2 \quad (282)$$

Heating the reaction mixture at a higher temperature (80°C) gives the expected product (in 65% yield when R = CH_3) accompanied by a disilyl adduct, i.e. 3,3,10,10-tetramethoxy-2,6,11-trioxa-3,10-disiladodecane, in 20% yield [D-047].

Alkenylalkylthioethers react with hydrosilanes in the same manner as ethers [53, 627, 693, 1101, 1120].

$$R'S(CH_2)_nCH{=}CH_2 + R_3SiH \rightarrow R'S(CH_2)_{n+2}SiR_3 \qquad (283)$$

The C=C bond of the allyl group in vinylallyl sulphide is mainly hydrosilylated [B-001]:

$$CH_2{=}CH(CH_2)_nSCH{=}CH_2 + (CH_3O)_3SiH \xrightarrow[\text{reflux, 16 h}]{H_2PtCl_6} (CH_3O)_3Si(CH_2)_{n+2}SCH{=}CH_2 \qquad (284)$$

In general, with a molecule containing a heteroatom and two or more carbon–carbon double bonds, hydrosilylation occurs at the bond which is as far as possible from the electronegative heteroatom (oxygen, sulphur) [B-001, D-047] or group [D-081].

3.4 Hydrocarbons with an Isolated C≡C Bond

Hydrosilylation of acetylene leads to the formation of the mono-adduct and/or 1,2-disilylethane (di-adduct). Apart from adducts, the reaction of monoorgano-substituted acetylenes with tri-substituted silanes yields a mixture of vicynal (I) (β-isomer) and geminal (II) (α-isomer) products, as well as the products of dehydrocondensation (III) and C=C migration (IV). This may be illustrated by the following set of equations:

$$R'CH_2C{\equiv}CH + R_3SiH \rightarrow \begin{cases} R'CH_2CH{=}CHSiR_3 \ (\beta\text{-isomer}) & (I) \\ R'CH_2C({=}CH_2)SiR_3 \ (\alpha\text{-isomer}) & (II) \\ R'CH_2C{\equiv}CSiR_3 & (III) \\ R'CH{=}CHCH_2SiR_3 & (IV) \end{cases} \qquad (285)$$

The addition of trichlorosilane to alkynes when catalysed by bases (tertiary amines and phosphines) mainly proceeds in the presence of basic solvents, e.g. nitriles [904]. In the presence of platinum complexes [117, 124, 134, 226, 390, 428, 430, 946, 947, 1205, 1346], and tertiary organic bases [904], tri-substituted silanes exhibit *cis* addition across the C=C bond to yield β-adducts with the *trans*(E) configuration. However, phenylacetylene reacts with silane in the presence of $Rh(C_4H_7N_2O_2)(PPh_3)_2$ to yield the *cis* β-isomer Z—$C_6H_5CH{=}CHSiCl_2C_2H_5$ [1048] as a result of *trans* addition. Such *trans* addition also takes place during the reaction of triethylsilane with octyne in the presence of $AlBr_3$ as an electrophile. This leads to a 90% yield of *cis*-3,3-diethyl-3-sila-4-undecane [1238].

Various transition metal complexes, soluble and supported, are well-known catalysts for hydrosilylation of acetylene by tri-substituted silanes in the liquid phase [504, 573, 941, 1286, 1290, D-056]. On the other hand, platinum, rhodium and ruthenium, heterogeneous supported complex catalysts, are extensively applied in this reaction in the gas phase [506, 616, 744, 772, F-007, Pl-016, Pl-018, Pl-019] where flow methods or a high-pressure autoclave are employed at elevated temperatures. The main products – vinyltri-substituted silanes – are mostly accompanied by products of double hydrosilylation of acetylene and redistribution of trichlorosilane followed by other by-products.

Wilkinson's complex $RhCl(PPh_3)_3$, immobilized on the silica especially via mercapto and phosphine ligands, appeared to be very effective and, under special conditions, a very selective catalyst for synthesis of vinyltrichlorosilane [744]. Moreover, the same complex anchored to phosphino-organosiloxane macromolecules grafted on to chrysotile asbestos was unusually effective and selective for the synthesis of vinylmethyldichlorosilane [742]. On the other hand, the addition of monosilane to acetylene catalysed by Pt/C at 100°C (in autoclave) gave a mixture of $CH_2{=}CHSiH_3$, $(CH_2{=}CH)_2SiH_2$ and $H_3SiCH_2CH_2SiH_3$ [J-079a].

Hydrosilylation of phenylacetylene in the presence of $(R_4N)_2PtCl_6$ and $(R_4N)_3IrCl_6$ yields predominantly β-adduct which is always accompanied by α-adducts [517a, 673]. Contrary to this, α-silyl styrene was a main product (82%) of the hydrosilylation of phenylacetylene by ferrocenyldiphenylsilane [885]. The hydrosilylation of monoalkyl-substituted acetylenes (see Scheme 285), where $R',R_3 = C_3H_7, (C_2H_5)_3$, catalysed by H_2PtCl_6 [1112], $RhCl(CO)(PPh_3)_2$ [947] and $RhH(CO)(PPh_3)_3$, results in the formation of C=C rearrangement products (IV). The addition of di-substituted silanes to monoargonoacetylenes catalysed by H_2PtCl_6 and Pt/C proceeds stereospecifically through *cis* addition to give *trans* adducts (34–56% yield) [110]:

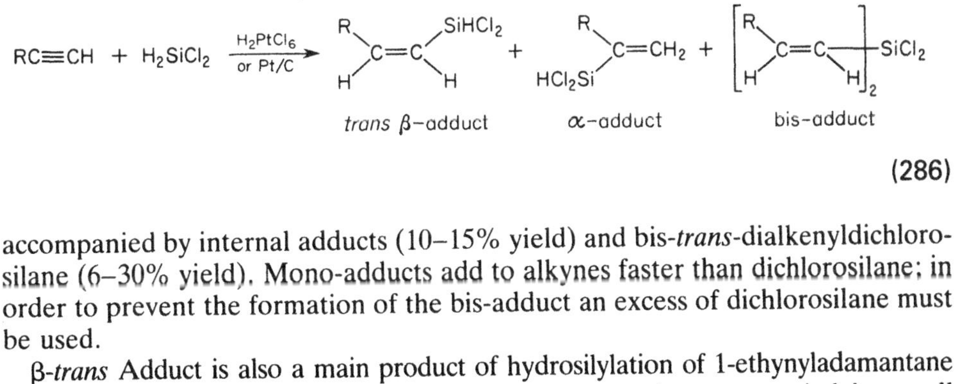

(286)

accompanied by internal adducts (10–15% yield) and bis-*trans*-dialkenyldichlorosilane (6–30% yield). Mono-adducts add to alkynes faster than dichlorosilane; in order to prevent the formation of the bis-adduct an excess of dichlorosilane must be used.

β-*trans* Adduct is also a main product of hydrosilylation of 1-ethynyladamantane by various tri-substituted silanes, but in some cases is accompanied by small amounts of α-adduct [708]. Trialkylsilanes undergo dehydrogenation as a main path of the reaction with alkyl- and aryl-acetylenes in the presence of chloroplatinic acid promoted by I_2, LiI, and R_3SiI [1242, 1259, 1267].

$$R_3SiH + HC{\equiv}CR' \rightarrow R_3SiC{\equiv}CR' + H_2 \qquad (287)$$

Radical initiators direct the reaction so, that *trans* addition yields the *cis* adducts [124] (see Section 2.1). Hence, the addition of trimethylsilane to various alkenes in the presence of benzoyl peroxide to give the isomers *cis* and *trans* in ratios extending from 3:1 to 4:1:

$$RC\equiv CH \ + \ (CH_3)_3SiH \ \xrightarrow{(C_6H_5COO)_2} \ \underset{(Z)}{\overset{R}{\underset{H}{>}}C=C\overset{Si(CH_3)_3}{\underset{H}{<}}} \ + \ \underset{(E)}{\overset{R}{\underset{H}{>}}C=C\overset{H}{\underset{Si(CH_3)_3}{<}}}$$

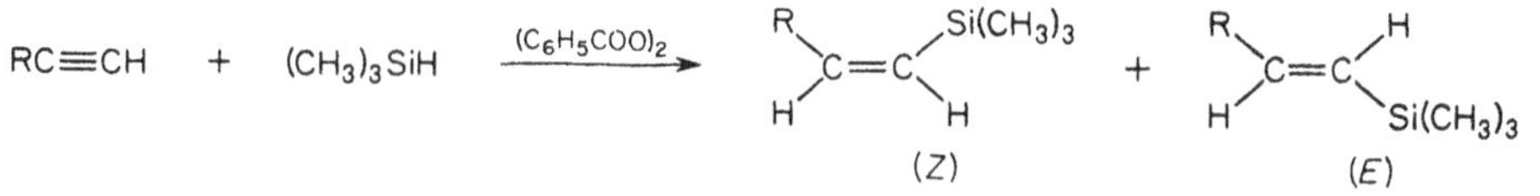

(288)

where R = $CH_3(CH_2)_2$, $CH_3(CH_2)_3$, $CH_3(CH_2)_4$ and $(CH_3)_2CH$.

However, the hydrosilylation of 3,3,3-trimethyl-1-pentyne leads to the selective formation of the *trans*(E) adduct [110]. As for free-radical hydrosilylation, when 1-alkynes (1-pentyne, 1-hexyne, 1-heptyne, 1-octyne) are hydrosilylated in the presence of rhodium complexes, *trans* addition gives largely *cis* adducts [870, 854] (although this does not happen when platinum catalysts are used), e.g.

$$CH_3(CH_2)_3C\equiv CH \ + \ (C_2H_5O)_3SiH \ \xrightarrow[100°C, 10h]{Rh(X)L(PPh_3)_3}$$

$$\longrightarrow \ \underset{\substack{(Z)\\ \text{main product}}}{\overset{H}{\underset{CH_3(CH_2)_3}{>}}C=C\overset{H}{\underset{Si(OC_2H_5)_3}{<}}} \ + \ \underset{\substack{(E)\\ \text{by–product}}}{\overset{H}{\underset{CH_3(CH_2)_3}{>}}C=C\overset{Si(OC_2H_5)_3}{\underset{H}{<}}}$$

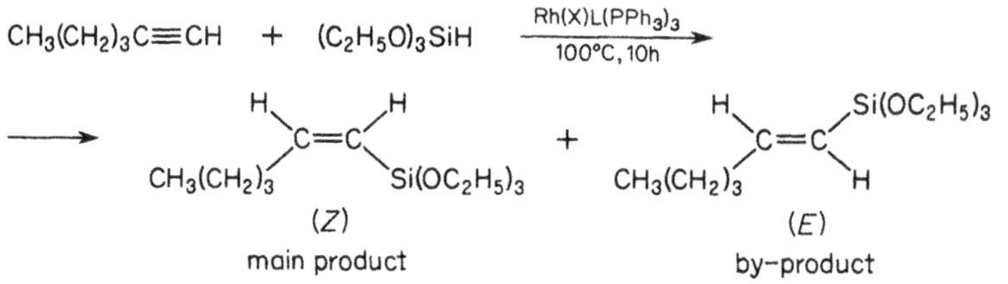

(289)

The above product ratio depends on the type of complex used, i.e. whether X = H, Cl or I, and whether L = CO.

The products of hydrosilylation of 1-hexyne by triethylsilane catalysed by rhodium phosphine complexes $\{RhCl[P(OR)_3]_2\}_2$ where R = methyl, tolyl also consist of *trans* and *cis*-1-triethylsilyl-1-hexenes [229]. In the presence of a Ziegler catalyst based on $Ni(acac)_2$+Lewis acid (Et_3Al, $LiAlH_4$, CH_3Li) [662, 663, G-030], hydrosilylation of mono-substituted alkynes is accompanied by the following dimerization:

$$2\,RC\equiv CH + X_3SiH \xrightarrow[Al(C_2H_5)_3]{Ni(acac)_2} CH_2=CRCR=CHSiX_3 + RCH=CHCR=CHSiX_3$$

(290)

Hexyne-1 undergoes reaction with triethylsilane in the presence of iridium catalysts to yield *trans* and *cis* silyl-1-hexenes accompanied by products of dehydrogenative (hydro)silylation and hydrogenation of 1-hexynes [360]. Amino-substituted silanes are added to mono-substituted acetylenes in the presence of platinum catalyst to form vinylaminosilanes [U-158, U-163] whereas 1-hydrosilatrane reacts effectively with monoorganoacetylenes only in the presence of Rh(acac)(CO)_2 catalyst [1260].

Dialkylacetylenes [274, 430, 991, 1205], arylalkylacetylenes and diarylacetylenes [458, 1184] add silicon hydrides to yield internal adducts. When 2-butyne is selectively hydrosilylated by triethylsilane, a high (92%) yield of E-2-silyl-2-butene [430] is obtained:

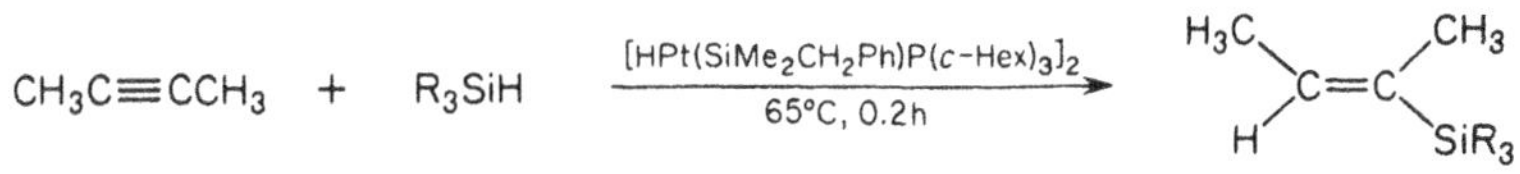

$$CH_3C{\equiv}CCH_3 \ + \ R_3SiH \xrightarrow[65°C, 0.2h]{[HPt(SiMe_2CH_2Ph)P(c-Hex)_3]_2} \begin{array}{c} H_3C \\ \diagdown \\ H \end{array} C{=}C \begin{array}{c} CH_3 \\ \diagup \\ \diagdown SiR_3 \end{array}$$

$$(E) \qquad (291)$$

where $R_3 = (C_2H_5)_3$, $H(C_6H_5)_2$ [430] or $Cl_2(CH_3)$ [430, 991]. However, the hydrosilylation of 2-pentyne with triethylsilane in the presence of the above catalyst gives two adducts, i.e. $C_2H_5CH{=}C(CH_3)Si(C_2H_5)_3$ (the β-adduct) in 63% yield and $C_2H_5C({=}CHCH_3)Si(C_2H_5)_3$ (the α-adduct) in 34% yield [1205].

As already mentioned, the addition of silicon hydrides to acetylene hydrocarbons may be accompanied by dehydrocondensation (Scheme 287) [878, 1269]:

$$C_6H_5C{\equiv}CH + (CH_3)_2NSiH(C_2H_5)_2 \xrightarrow[62-70\%]{H_2PtCl_6} C_6H_5C{\equiv}CSi(C_2H_5)_2N(CH_3)_2 + H_2$$

$$(292)$$

while the reaction with tolane (diphenylacetylene), which cannot form a condensation product, yields the adduct [458]:

$$C_6H_5C{\equiv}CC_6H_5 + (CH_3)_2NSiH(C_2H_5)_2 \xrightarrow{H_2PtCl_6}$$
$$C_6H_5CH{=}C(C_6H_5)Si(C_2H_5)_2N(CH_3)_2 \quad (293)$$
$$\textit{40–50\% yield}$$

The addition of alkylchloro-, alkylarylochloro- and trialkyl-silanes to diphenylacetylene, which occurs in the presence of Speier's catalyst, leads to *cis* and mono-adducts with an E-configuration [837]:

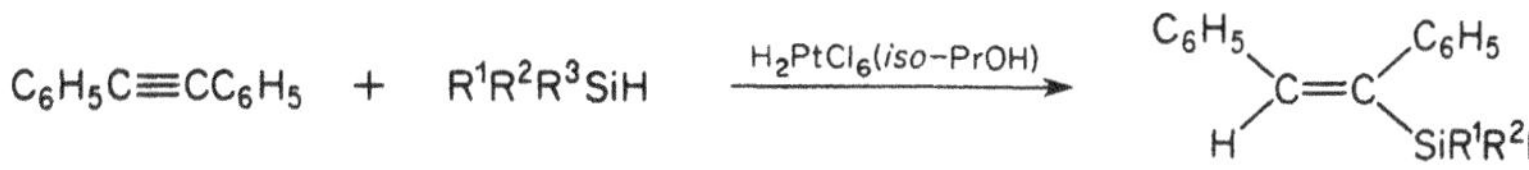

$$C_6H_5C{\equiv}CC_6H_5 \ + \ R^1R^2R^3SiH \xrightarrow{H_2PtCl_6(iso-PrOH)} \begin{array}{c} C_6H_5 \\ \diagdown \\ H \end{array} C{=}C \begin{array}{c} C_6H_5 \\ \diagup \\ \diagdown SiR^1R^2R^3 \end{array}$$

$$(E) \qquad (294)$$

The selective mono-addition of organosilanes to tolane is possible because the steric effect of the silyl and two phenyl groups in the product inhibit the second step in the process. The reactivity of the silane decreases with increasing size of the alkyl groups, but increases with the introduction of aromatic substituents [837]. Mainly mono-adducts are formed in the hydrosilylation of diphenylacetylene catalysed by a nickel complex [1184], whereas the replacement of the phenyl substituents by less bulky ones favours dehydrocondensation (Scheme 295), e.g. when $n = 2$ and $R,R^1 = C_6H_5$, the yield of (I) is 60% and that of (II) is 3%; when $R,R^1 = C_6H_5$, CH_3, the yield of (I) is 52% and that of (II) is 27%; and when $R,R^1 = C_4H_9$, the yield of (I) is 27% and that of (II) is 72%.

The reaction of phenylacetylene with a substituted disilane gives a heterocyclic compound of the formula (Scheme 296) [875].

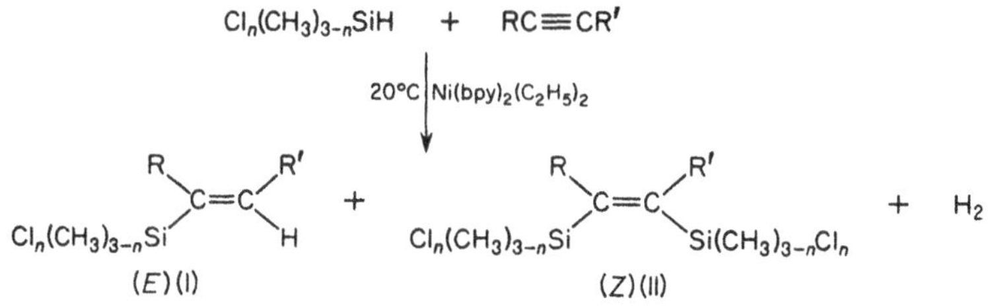

(295)

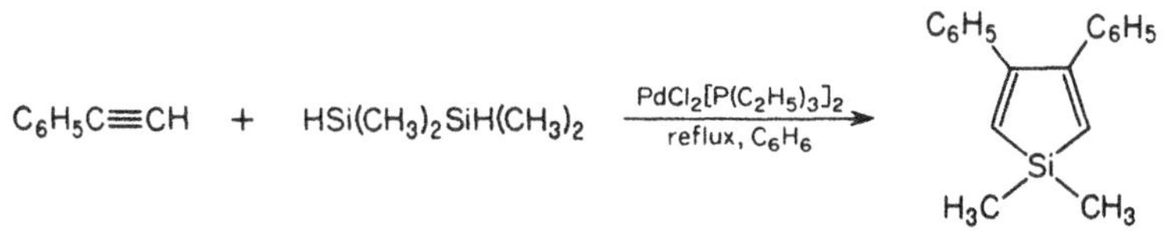

(296)

The addition of a substituted disiloxane to phenylacetylene and diphenylacetylene also leads to cyclic products [929]:

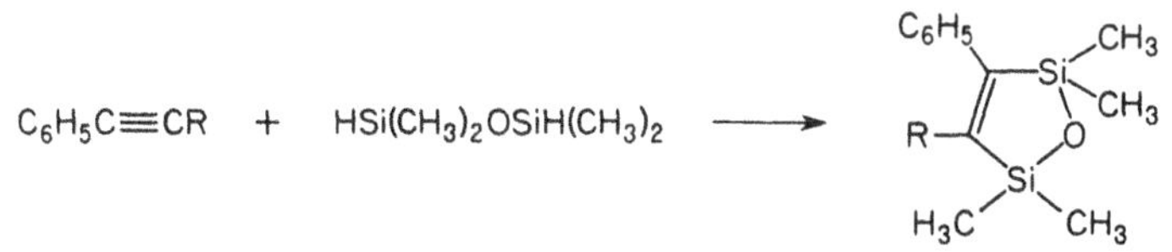

(297)

with 2,2,5,5-tetramethyl-3-phenyl-1-oxa-2,5-disilacyclopentene being formed when R = H and 2,2,5,5-tetramethyl-3,4-diphenyl-1-oxa-2,5-disilacyclopentene when R = C_6H_5.

3.5 Alkyne Derivatives with Functional Groups

Generally, the hydrosilylation of all mono-substituted acetylenes having the general formula $HC{\equiv}CY$ with tri-substituted silanes occurs in the presence of chloroplatinic acid, resulting in a mixture of functional *trans* (β-isomer) and geminal (α-isomer) products [73, 247, 250, 252, 522, 549, 601, 604, 611, 709, 773, 951, 1050, 1204, 1205, 1245, 1309]:

$$HC{\equiv}CY + R_3SiH \xrightarrow{H_2PtCl_6} \begin{cases} R_3SiCH{=}CHY \text{ (β-adduct)} \\ R_3SiC({=}CH_2)Y \text{ (α-adduct)} \end{cases} \qquad (298)$$

The β-isomer always has the (E) configuration as a result of *cis* addition [700, 701, 1309]. Surprisingly, the reaction of $CH_3OCH_2CH_2C(OH)CH_3C{\equiv}CH$ with triethylsilane catalysed by Speier's catalyst yields a mixture containing the α-isomer and the *cis* and *trans* β-isomers [445]. The two latter compounds are also formed when the reaction is irradiated by UV light [481].

In a similar manner to the hydrosilylation of monoalkylacetylene, the reaction addition to acetylenic compounds with functional groups may also be accompanied by dehydrocondensation products [1102]:

$$HC{\equiv}COC_6H_5 + (C_2H_5)_3SiH \xrightarrow{H_2PtCl_6(THF)} \begin{cases} (C_2H_5)_3SiCH{=}CHOC_6H_5 & (I) \\ (C_2H_5)_3SiC({=}CH_2)OC_6H_5 & (II) \\ (C_2H_5)_3SiC{\equiv}COC_6H_5 & (III) \end{cases} \quad (299)$$

Promotion of this catalyst by I_2 leads the reaction of propargyl chloride with triethylsilane predominantly to the dehydrogenation product [957].

Contrary to much of the evidence arising from the homogeneous or heterogeneous catalysis of hydrosilylation by transition metal complexes, the presence of tetrahydrofuran in the above-mentioned reaction brings about an increase in the yield of β-adduct, which can be illustrated as follows:

	Yield (%)			
	I	II	III	
No solvent	84.7	12.8	2.6	
THF	91.8	7.4	0.8	(300)

The interaction of alkynes as well as silanes with aprotic solvents may be a significant factor in determining the observed product ratios.

The hydrosilylation of mono-substituted acetylenes with the general formula $HC{\equiv}CR'$ with triethylsilane proceeds in the presence of H_2PtCl_6 to produce α- and β-adducts in the following ratios:

R′	α/β	
$COCH_3$	55/45	
C_6H_5	28/72	
OC_6H_5	13/87	
CH_2OH	52/48	
$CH(CH_3)OH$	45/55	
$C(CH_3)_2OH$	39/61	(301)

An increase in the yield of the β-adduct arises mainly as a result of the electron-donor properties of the R′ group, but it could also be associated with mesomeric effects. The addition of trialkylsilanes to alkynes with strong electron-donor substituents, e.g. $C(CH_3)_3$ [916], occurs stereospecifically according to Farmer's rule (to give the β-adduct).

The same general relationship is observed in the case of propargyl compounds. Thus, glycydoxy derivatives react with tri-substituted silanes to give exclusively the product arising from hydrosilylation of the C≡C bond with the terminal silyl group [543, 568, 806, 808, 997, 1162, 1250]:

$$HC{\equiv}CCH_2OCH_2\underset{\diagdown\,O\,\diagup}{CH{-}CH_2} + R_3SiH \xrightarrow{H_2PtCl_6} R_3SiCH{=}CHCH_2OCH_2\underset{\diagdown\,O\,\diagup}{CH{-}CH_2} \quad (302)$$

A β-adduct is also formed in the addition of triorganosilane to *N,N*-diethylpropargylamine in the presence of a platinum catalyst [368]. The platinum complex with the above amine acts as an active catalyst only at temperatures over the range 60–80°C [945]. The hydrosilylation of other propargyl-substituted amines, e.g. pyrrolidyne [700], piperidine [700, 701], azepine [701] and morpholine, produces γ-monoadducts with a *trans* configuration (Scheme 303).

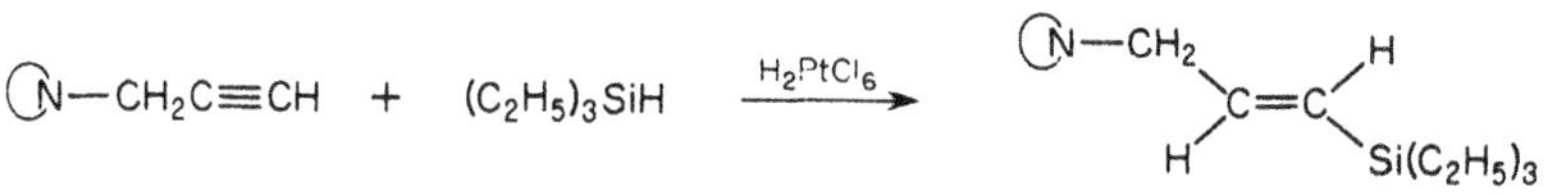

(303)

The addition of tetraorganodisiloxane dihydrides to propargyl glycidyl ether in the presence of chloroplatinic acid gave up to 4% epoxysilanes [1148]. Hydrosilylation of propargyl chloride by trichlorosilane in the presence of H_2PtCl_6 gives two isomers: 2-(trichlorosilyl)-3-chloropropene and *trans*-1-(trichlorosilyl)-3-chloroprene. However, the addition of trimethylsilane to this compound yields 2-(trimethylsilyl)-3-chloroprene (the α-isomer) [775] exclusively. Moreover, addition of triethylsilane to 1-chloro-2-propyne in the presence of H_2PtCl_6 leads to two adducts (α and β) [1245], but if the reaction is catalysed by $AlBr_3$ *cis*-triethyl-1-propenylsilane (0°C, 1 h, 25% yield) is formed [1238]. The addition of silicon hydride to dichloro derivatives of 2-butyne leads to a very smooth reaction [83, 752]:

$$ClCH_2C{\equiv}CCH_2Cl + RSiHR'_2 \xrightarrow[90\text{–}100^\circ C]{H_2PtCl_6} ClCH_2CH{=}C(SiR'_2R)CHCl \quad (304)$$

where R = R′ = Cl; R = CH_3, R′ = Cl; R = C_2H_5, R′ = Cl; R = R′ = C_2H_5.

Addition of silanes to acetylene alcohols can also lead to dehydrocondensation of the hydroxy group with the Si–H bond, but the rate of hydrosilylation is markedly higher than the rate of dehydrocondensation. Propargyl alcohol reacts with triethylsilane to yield four products [67, 247, 252, 295, 548, 803, 804, 951, 1049, 1063, 1072, 1081, 1205, 1245, 1263]:

$$CH{\equiv}CCH_2OH + (C_2H_5)_3SiH \xrightarrow{H_2PtCl_6} \begin{cases} (C_2H_5)_3SiCH{=}CHCH_2OSi(C_2H_5)_3 & (III) \\ (C_2H_5)_3SiCH{=}CHCH_2OH & (I) \\ CH_2{=}C(CH_2OH)Si(C_2H_5)_3 & (II) \\ CH_2{=}C[Si(C_2H_5)_3]CH_2OSi(C_2H_5)_3 & (IV) \end{cases}$$

(305)

The product ratio depends on the reaction conditions and polarity of the solvent [1263]. At lower temperatures, adduct (I) (the β-adduct) is formed, whereas at higher temperatures formation of the β-adduct is accompanied by the α-adduct (II), although both reaction are followed by dehydrocondensation.

The reaction of $CH{\equiv}CC(CH_3)_2OH$ with phenylmethylsilane catalysed by H_2PtCl_6 leads to the formation of the mono-β-adduct, $C_6H_5SiH(CH_3)CH{=}CHC(CH_3)_2OH$ (47%), the β-di-adduct, $C_6H_5Si(CH_3)[CH{=}CHC(CH_3)_2OH]_2$

(13%), and the dehydrocondensation product, $C_6H_5SiH(CH_3)OC(CH_3)_2C{\equiv}CH$ (11%).

Dehydrocondensation has also been reported as a side-reaction in the hydrosilylation of other alkynyl alcohols, e.g. $CH_3CH(OH)C{\equiv}CH$ [604], $CH{\equiv}CCH_2OCH_2CH_2OH$ [994] and $CH_2{=}CHC{\equiv}CCH_2CH_2OH$ [468], and of acetylene phenols, e.g. $4{-}HO{-}3{-}CH_3OC_6H_3C(CH_3)_2C{\equiv}CCH{=}CH_2$ [13].

The hydrosilylation of acetylene diols having a hydroxy groups in the 1,4-position [442, 444] or in the 1,7-position [185] leads to the following cyclic compounds:

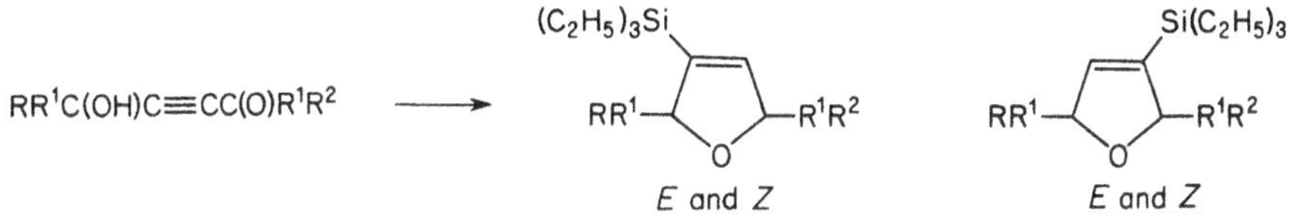

(306.1)

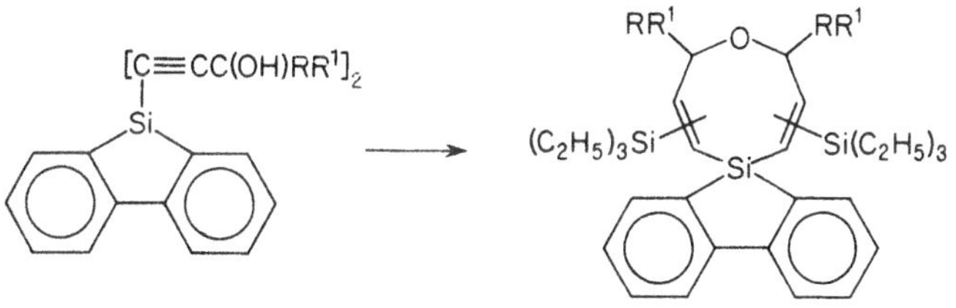

(306.2)

As with compounds of the type $R^1C{\equiv}CH$ (where $R^1 = C_4H_9$, *iso*-C_3H_7, C_6H_5 and C_6H_5O) catalysed by H_2PtCl_6 in THF, the rate of hydrosilylation of compounds with the general formula $YCH_2C{\equiv}CH$, where $Y = C_2H_5OCO$, C_6H_5O, Cl or $(C_2H_5)_2N$, has little dependence on the structure of the propargyl derivatives [945]. Propargyl peroxy compounds react with triphenylsilane to yield organosilicon alkenyl peroxides:

$$ROOC(CH_3)_2C{\equiv}CH + HSi(C_6H_5)_3 \xrightarrow[20-50°C]{H_2PtCl_6} ROOC(CH_3)_2C[Si(C_6H_5)_3]{=}CH_2 \tag{307}$$

where $R = C(CH_3)_3$, $C(CH_3)_2C_2H_5$, $C(CH_3)_2C_3H_7$, $C(CH_3)_2(CH_2)_4CH_3$. Derivatives of alkynes possessing two isolated carbon–carbon triple bonds of equal value add silicon hydrides at both sites [549, 626, 1070]:

$$\begin{aligned}&HC{\equiv}CCH_2(OCH_2CH_2)_nOCH_2C{\equiv}CH + (C_2H_5)_3SiH \rightarrow\\&\quad (C_2H_5)_3SiCH{=}CHCH_2(OCH_2CH_2)_nOCH_2CH{=}CHSi(C_2H_5)_3\end{aligned} \tag{308}$$

for $n = 1$ [549] or $n = 3$ [1070].

3.6 Hydrocarbons Containing C=C and C≡C Bonds and Their Derivatives

Generally, alkyne derivatives are more readily hydrosilylated than alkene derivatives. 1-Allyloxy-1-propargyloxycyclopentane reacts with triethylsilane to give the hydrosilylation mono-adduct in 71% yield [1080].

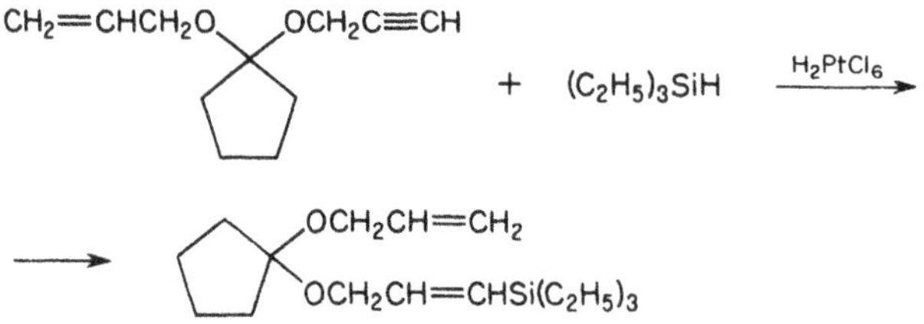

(309)

The reaction of trialkylsilanes with 1-propargyloxy-2-glycydoxy-3-allyloxypropane proceeds selectively at the C≡C bond to give an adduct with a terminal silyl group [1154]:

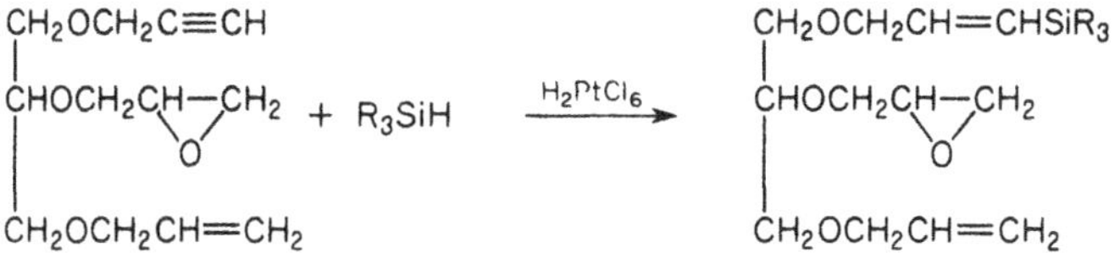

(310)

As with allylglycyl derivatives, the oxirane ring remains intact. In the reaction of 1-allyloxy-3-propargyloxy-propan-2-ol C≡C bond is also selectively hydrosilylated [1248]. In all cases of hydrosilylation of unsaturated compounds containing two or more C≡C and/or C=C bonds, high-boiling double hydrosilylation by-products are formed.

To evaluate the relative reactivity of ethylene and acetylene bonds in hydrosilylation, the reaction of silicon hydrides with 4-substituted vinylacetylene, $RC{\equiv}CCH{=}CH_2$, has been studied in the presence of platinum catalysts [453, 1001, 1007, 1159]. In such cases, when R = CH_2CH_2OH [453], $C(CH_3)_2OH$ [1007], $C(CH_3)_2OCH_2CH_2CN$ [1159], $C(CH_3)_2OCH_2\overline{CHCH_2O}$ [1159] and $CH(OH)\overline{CH(CH_2)_3O}$ [1007], addition of the silyl group occurs at C3.

1-(3,3-Dialkylpropyn-3-ol)cyclohexane reacts with triethylsilane (and with triethylgermane) in such a way that only the C≡C bond is hydrosilylated [447]. The respective diene is formed in each case:

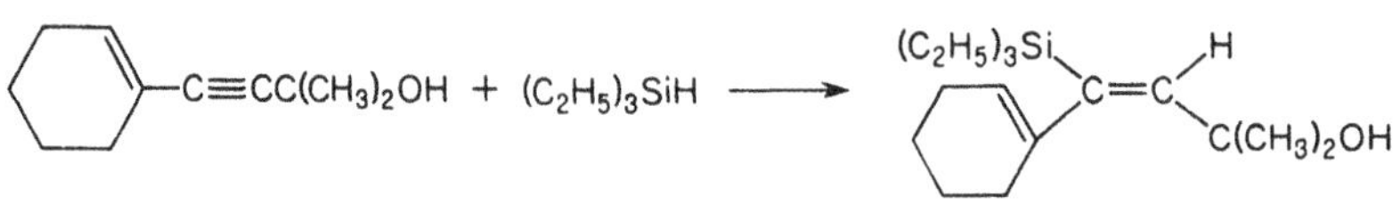

(311)

1-Alkene-3-ynes add hydrosilanes to the C≡C bond to give products with silyl groups attached at C3. Vinylacetylene [1111], 2-methyl-1-butene-3-yne [672],

1-alkoxyvinylacetylene ($HC{\equiv}CCH{=}CHR$, where R = OC_2H_5 or OC_4H_9 [1098]) and 1-(*N,N*-dialkylamine)vinylacetylene [1098] also add silicon hydrides to the C≡C bond to yield products with a terminal silyl group according to the following equation, e.g.:

$$CH{\equiv}CCH{=}CH_2 + R_3SiH \xrightarrow{H_2PtCl_6\ (iso\text{-}PrOH)} R_3SiCH{=}CHCH{=}CH_2 \tag{312}$$

$R_3 = (CH_3)_3,\ C_2H_5(CH_3)_2,\ (C_2H_5)_3$.

In competitive hydrosilylation of C=C and C≡C bonds in separate compounds by dimethyl(2-tienyl)silane only the reaction with acetylenes occurs while the olefin remains intact [707].

3.7 Saturated Carbonyl Compounds

Silicon hydrides are readily added to the C=O group of ketones, aldehydes, acids and esters in the presence of all the various kinds of catalyst mentioned above. The direction of the hydrosilylation is determined by the polarization of the C=O group, which always leads to a product in which hydrogen is added to the electrophilic carbon atom and the silyl group is added to the nucleophilic oxygen atom, leading finally to alkoxy- or aryloxy-substituted silanes:

$$\rangle C^{\delta+}{=}O^{\delta-} + H{-}Si{\equiv} \rightarrow {\equiv}Si{-}O{-}\overset{|}{\underset{|}{C}}{-}H \tag{313}$$

The radical addition of hydrosilanes to a C=O bond has been discussed in Section 2.1, but hydrosilylation of carbonyl compounds (especially ketones and aldehydes) can also be effected by reduced metals [200] (Section 2.2) and Lewis acids [200] (Section 2.3), as well as by chloroplatinic acid. However, the process occurs only under quite drastic conditions, and is accompanied by several side-reactions. The application of Wilkinson's catalyst and other rhodium, ruthenium and platinum complexes (Section 2.4) as well as fluoride ion (see Section 2.3) has increased markedly the possibility of hydrosilylating carbonyl compounds.

3.7.1 Ketones and Aldehydes

Since the product obtained by adding a ketone or an aldehyde (alkoxysilane or aryloxysilane) is sensitive towards hydrolysis (yielding a primary alcohol with aldehydes and a secondary alcohol with ketones), this reaction provides an efficient method of reduction. A study of the hydrosilylation of ketones with the general formula C_6H_5COR (where R = CH_3, C_2H_5, $CH(CH_3)_2$, $C(CH_3)_3$) has shown that the reaction yield is influenced by inductive and steric effects. This is illustrated by the example below:

$$C_6H_5COR + CH_3SiHCl_2 \xrightarrow[r.t.]{catalyst} C_6H_5CHROSiCl_2CH_3 \tag{314}$$

Catalyst/R	CH_3	C_2H_5	C_3H_7	$CH(CH_3)_2$	$C(CH_3)_3$
$[PtCl_2(MPPP)]_2$	71	83	74	57	24*
$[PtCl_2(BMPP)]_2$	81	81	77	55	33*

* 90°C, 10 days

The stereochemistry of hydrosilylation of 2- and 4-substituted alkylcyclohexanones with Ph_2SiH_2 is influenced by the position and size of alkyl groups, the catalyst concentration, the reaction temperature, and the types of ligand attached to Rh(I) complex [358].

Hydrosilylation of phenyl*iso*propyl ketone leads to the adduct accompanied by the dehydrogenation product, i.e. $C_6H_5C(OSi){=}C(CH_3)_2$ (10% yield).

The asymmetric reduction of ketones by hydrosilylation in the presence of a chiral catalyst, followed by hydrolysis, has been extensively studied during the past 10 years by several research groups. The results have been discussed in some excellent reviews and papers e.g. [Rev.7, Rev.19, Rev.32, Rev.34, Rev.35, 143, 175, 177–179, 181, 182, 533, 762, 789, 829, 876, 1227] which include a plausible mechanism for this asymmetric reaction. For example, the asymmetric hydrosilylation of alkylphenyl ketones leads to an optically active alkoxysilane, the reaction being influenced by the nature of the catalyst, the chiral phosphine ligands, and also by the initial ketone employed [Rev.7, Rev.19, Rev.32, Rev.34, Rev.35] (see also Section 2.4.4), e.g. a terpene ketone [Rev.35].

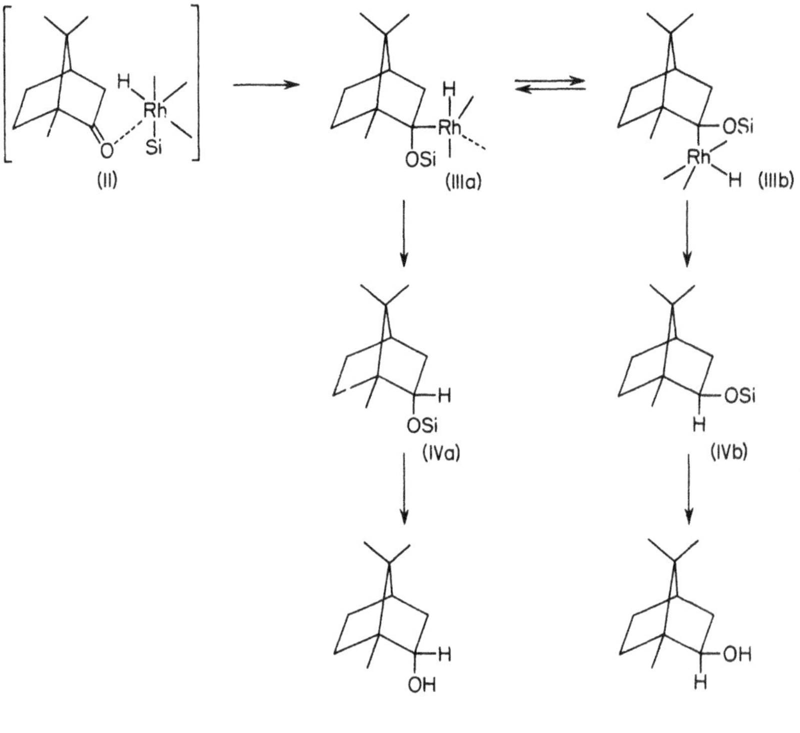

(315)

The hydrosilylation of camphor, followed by hydrolysis of the above products, leads (where R_3 = $(C_2H_5)_2H$, 92% yield) to a mixture of borneol (9%) and isoborneol (91%) [F-023]. Hydrosilylating the carbonyl group of an aldehyde also leads to an alkoxy- or aryloxy-silane:

$$R{-}C(=O)H + \equiv SiH \longrightarrow RCH_2OSi\equiv \qquad (316)$$

e.g. R = CH_3 [372], C_2H_5 [280], C_3H_7 [278, 280, 372], C_4H_9 [280, 372], XC_6H_4 [307, 339, 380, 411], C_6H_5 [8, 279, 280, 307, 339, 380, 418, 872, 1351], C_6H_{13} [280, 372], $C_6H_5CH_2$ [372], $C_6H_5CH(CH_3)$ [372] $(C_6H_5)_2CH$ [372] and c-C_6H_{11} [395].

Using saturated aldehydes and ketones it is also possible to obtain enol silyl ethers as a result of the dehydrocondensation that occurs in the presence of a number of catalysts, e.g. $Co_2(CO)_8$ [U-138] and complexes of Pd(II) [811], rhodium [853] and nickel [370, 372, 516]:

$$R^1CH_2COR^2 + R_3SiH \xrightarrow{PdCl_2+PhSH} R^1CH{=}C(OSiR_3)R^2 + H_2 \qquad (317.1)$$

$$CH_3COCH_2COCH_3 + (C_2H_5)_3SiH \xrightarrow[+p-CH_3C_6H_4SH]{RhCl(PPh_3)_3} (C_2H_5)_3SiOC(CH_3){=}CHCOCH_3 \qquad (317.2)$$

If there is no hydrogen attached to the α-carbon atom, the dehydrocondensation product cannot be obtained from the hydrosilylation of aldehydes [372]:

$$(CH_3)_3CCHO + (C_2H_5)_3SiH \xrightarrow[100-110°C,\ C_6H_6]{Ni+Et_2S} (CH_3)_3CCH_2OSi(C_2H_5)_3 \qquad (318)$$

Hydrosilylation of such carbonyl compounds sometimes produces a dimeric product [147, 374, 380],

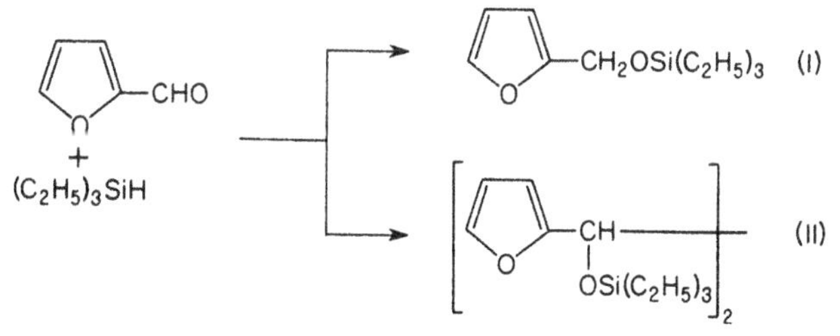

(319)

although CsF catalyses this reaction with $Ph(CH_3)_2SiH$ under extremely mild conditions yielding the mono-adduct of the typc I [418].

The ratio of the products, (I)/(II), depends on the co-catalyst and solvent used. This is a characteristic feature of electrophilic catalysis for hydrosilylation processes as well as of metal catalysts, e.g. nickel formed *in situ* [123, 656] (see also Sections 2.2 and 2.3). Such catalysts direct the reaction mainly towards competitive hydrosilylation and dehydrocondensation:

$$(C_2H_5)_3SiH \; + \; \underset{R^2}{\overset{R^1}{\diagdown}}CH-C\underset{H}{\overset{O}{\diagdown}} \longrightarrow \underset{R^2}{\overset{R^1}{\diagdown}}C=CHOSi(C_2H_5)_3 \; + \; H_2$$

$$\underset{R^2}{\overset{R^1}{\diagdown}}CHCH_2OSi(C_2H_5)_3$$

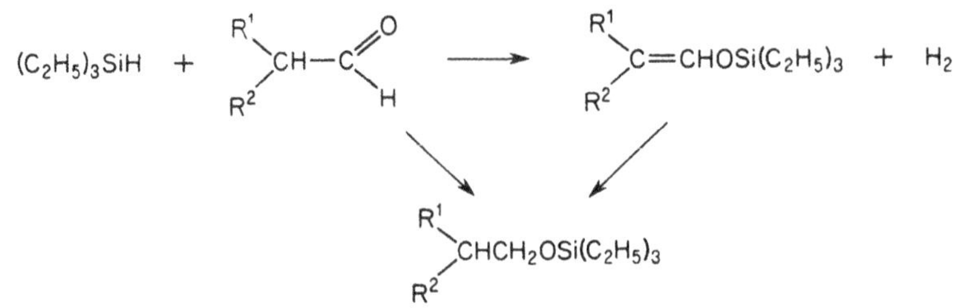

$$(320)$$

As a result of the hydrosilylation of 1,2,5,6-tetrahydrobenzoic aldehyde, $CH_2CH_2CH=CHCH_2CHCHO$, by triethylsilane in the presence of nickel (reflux, 12 h, C_6H_6), the following four products were formed: $(CH_2)_3CH=CHCHCH_2OSi(C_2H_5)_3$, $(CH_2)_5CHCH_2OSi(C_2H_5)_3$, $(CH_2)_5C=CHOSi(C_2H_5)_3$ and $C_6H_5CH_2OSi(C_2H_5)_3$ [372].

3.7.2 Other Carbonyl Compounds

In the presence of various catalysts, principally electrophilic acids or metal catalysts, alkyl esters of carboxylic acids react with tri-substituted silanes (e.g. triethoxysilane) [110] according to equation:

$$R'COOR + 2(C_2H_5O)_3SiH \xrightarrow{\text{catalyst}} R'CH_2OSi(OC_2H_5)_3 + ROSi(OC_2H_5)_3$$

$$(321)$$

In the presence of nickel, carboxylic acid esters react with silicon hydrides to give appropriate alkylsilylacetals [377]:

$$R^1COOR^2 + R_3SiH \xrightarrow[\text{reflux}]{\text{Ni}} R^1CH(OR^2)OSiR_3 \qquad (322)$$

but in the presence of Et_2S as a co-catalyst two other products are obtained [890]:

$$C_6H_5CH_2COOC_2H_5 + (C_2H_5)_3SiH \xrightarrow[\text{reflux, 72 h}]{\text{Ni}+Et_2S}$$

$$\begin{array}{ll} \longrightarrow C_6H_5CH=C[OSi(C_2H_5)_3]_2 & (32\%) \\ \longrightarrow C_6H_5CH_2COOSi(C_2H_5)_3 & (51\%) \end{array} \qquad (323)$$

When $ZnCl_2$ is employed as the catalyst the reaction leads to ethers rather than hydrosilylation products [648]:

$$R^1COOR^2 + R_3SiH \xrightarrow[-R^2 CHO]{ZnCl_2} (R^1CH_2)_2O + (R_3Si)_2O \qquad (324)$$

The yield (%) is given in parentheses for the sets of substituent groups listed below:

$R^1,R^2,R = C_3H_7, C_2H_5, C_2H_5$ (76); C_5H_{11}, CH_3, C_3H_7 (69); p–BrC_6H_4, C_2H_5, C_2H_5 (56); C_6H_5, C_2H_5, C_2H_5 (60); $C_{15}H_{31}, C_2H_5, C_2H_5$ (70).

Silyl esters [$(R^1COOCH_2O)_2Si(C_2H_5)_2$, where $R^1 = C_3H_7, C_6H_{13}$] also undergo this reaction [658].

However, di-substituted silanes react with carboxylic acid esters in the presence of $ZnCl_2$ to give a 46–58% yield of adducts [658]. In contrast, mono-substituted silanes, e.g. phenylsilane, react with dialkyl oxalate (at 90–100°C for 3–4 h) to give $C_6H_5Si[OCH(OR)COOR]_3$, the yields (%) in parentheses being obtained for various substituent groups R; R = $(CH_3)_3C(CH_2)_3$ (49); C_4H_9 (54); C_6H_{13} (43); C_7H_{15} (44) [659]. Compounds with two carbonyl groups, e.g. 2,3-butanedione, can react in such way that silicon hydride is added independently to both groups [873].

In the presence of a carbonyl group in the molecule, the addition of hydrosilane to a carbonyl group is catalysed by caesium fluoride [151], although the hydrosilylation of ketoesters also occurs in the presence of a $ZnCl_2$ catalyst [646, S-040]:

$$R^1CO(CH_2)_nCOOR + (C_2H_5)_3SiH \xrightarrow[100°C,\ 2-3\ h]{ZnCl_2} (C_2H_5)_3SiOCH(R^1)(CH_2)_nCOOR \quad (325)$$

where n = 0, 1, 2, 3 (yield 37–86%) and in the presence of nickel [649]:

$$ClC_6H_4COCOOC_2H_5 + (C_2H_5)_3SiH \xrightarrow[160-170°C,\ 6\ h,\ xylene]{Ni} (C_2H_5)_3SiOCH(C_6H_4Cl)COOC_2H_5 \quad (326)$$

The following reaction can also occur in the presence of nickel [645]:

$$2\ C_6H_5COCOOC_2H_5 + 2\ (C_2H_5)_3SiH \xrightarrow[reflux,\ 3-4\ h]{Ni} (C_2H_5)_3SiOC(C_6H_5)(COOC_2H_5)C(C_6H_5)(COOC_2H_5)OSi(C_2H_5)_3 \quad (327)$$

Ojima *et al.* [Rev.31] have found an effective way of reducing α-ketoesters (typically *n*-propyl pyruvate and ethyl benzoyl formate) asymmetrically by means of chiral rhodium complexes catalysed hydrosilylation as shown below:

$$R^1COCOOR^2 + R_2SiH_2 \xrightarrow{Rh^*} R^1\overset{*}{C}H(OSiHR_2)COOR^2 \quad (328)$$

$R^1 = CH_3, C_6H_5$; $R^2 = C_2H_5, C_3H_7$; Rh^* = $(BMPP)_2Rh(S)Cl$, $(DIOP)Rh(S)Cl$; $R = C_6H_5, C_{10}H_7$.

The optical yields of the lactates obtained are generally much higher than those obtained by other methods.

The reaction of the C=O bond in carbonyl compounds containing a hydroxy group in the presence of H_2PtCl_6 generally leads to dehydrocondensation products in preference to adducts [D-115]. Ketoacids, e.g. levulinic acid, undergo preliminary dehydrocondensation with trisubstituted silanes even in the absence of catalyst; this is followed by hydrosilylation of the keto group:

$$CH_3COCH_2CH_2COOH + R_3SiH \xrightarrow[6\ h\ (THF)]{70\%} CH_3COCH_2CH_2COOSiR_3 \xrightarrow[110°C,\ 4\ h]{Ni} R_3SiOCH(CH_3)CH_2CH_2COOSiR_3 \quad (329)$$

When $R_3 = (C_2H_5)_3$ the yield of hydrosilylation is 86%; when $R_3 = C_4H_9(C_2H_5)_2$ it is 90% [S-040].

The addition of disubstituted silanes to α-ketoacids yields five-membered cyclic compounds [653]:

$$RCOCOOH + (C_2H_5)_2SiH_2 \xrightarrow[reflux,\ 2-3\ h]{Ni} \overline{RCHCOOSi(C_2H_5)_2O} \quad (330)$$

In this case, when R = C_4H_9 the yield is 68%; when R = C_6H_{13} the yield is 73%; and when R = C_7H_{15} the yield is 64%.

Aromatic carboxylic acids react with trichlorosilane in the presence of triethylamine to give benzylsilane [108, 109]. Aliphatic and aromatic acid chlorides add trichlorosilane to give predominantly bissilylalkanes [U-049] according to the equation:

$$RCOCl + 2\,HSiCl_3 \xrightarrow[15-66°C,\ 4\ h]{(C_3H_7)_3N} RCH(SiCl_3)_2 \quad (331)$$

The following yields (%) being obtained for different R groups: CH_3 (55); C_2H_5 (57); $(CH_3)_2CH$ (41); $(CH_3)_3C$ (19); C_5H_{11} (72).

The addition of triethylsilane to p−$CH_3OC_6H_4COCl$ in the presence of $PdCl_2(PPh_3)_2$ (120°C, 5 h) leads to an alkoxysilane (27%) which does not contain chlorine, i.e. p−$CH_3OC_6H_4CH_2OSi(C_2H_5)_3$ [310]. Hydrosilylation of $C_6H_5COCH_2COOCN$ occurs with simultaneous dehydrocondensation as follows:

$$C_6H_5COCH_2COOCN + (C_2H_5)_3SiH \xrightarrow[60°C,\ 12\ h]{RhCl(PPh_3)_3} \begin{cases} (C_2H_5)_3SiOCH(C_6H_5)CH_2COOCN \\ (C_2H_5)_3SiOC(C_6H_5)=CHCOOCN \end{cases} \quad (332)$$

but the products of the addition of triethylsilane to C_6H_5COCN depend on the catalyst employed:

$$C_6H_5COCN + (C_2H_5)_3SiH \to \begin{cases} \xrightarrow{PdCl_2} C_6H_5CH(CN)OSi(C_2H_5)_3 \quad 45\% \\ \xrightarrow{RhCl(PPh_3)_3} [(C_2H_5)_3SiOC(C_6H_5)(CN)]_2 \quad 5\% \end{cases} \qquad (333)$$

Rh(I) complexes also catalyse homogeneously hydrosilylation of organotransition metal acyl complexes, e.g. $Cp(CO)_2FeCOR$ with dihydrosilanes to produce α-siloxyalkylderivatives [291]:

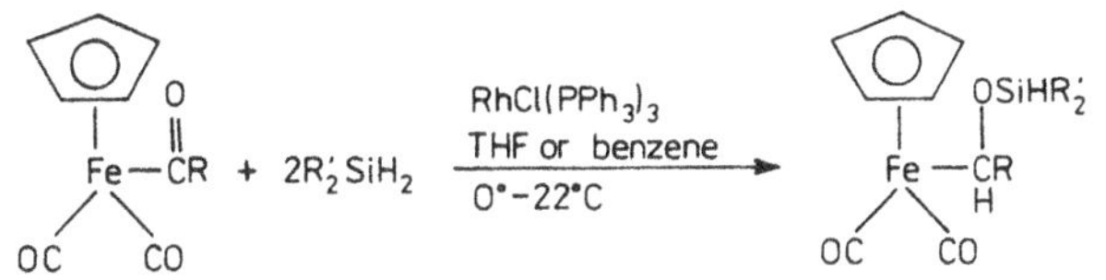

(334)

By analogy with hydroformylation, carbon monoxide reacts with alkenes and trialkylsilanes in the presence of transition metal complexes to yield silyl enol ethers [811]:

$$R^1CH{=}CH_2 + CO + R_3SiH \xrightarrow[140°C]{Co_2(CO)_8} R^1CH_2CH{=}CHOSiR_3 \qquad (335.1)$$

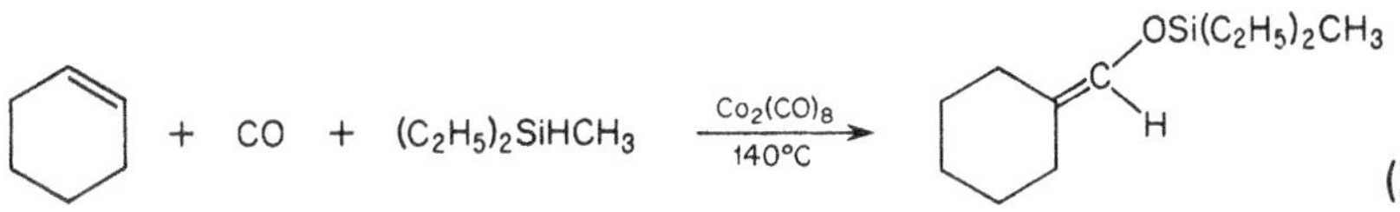

(335.2)

In addition, carbon dioxide reacts directly with triethylsilane in the presence of $[HRu_3(CO)_{11}]^-$ or $[HRu_3(CO)_{10}Si(C_2H_5)_3]^-$ clusters to yield the triethylsilyl ester of formic acid [1145]:

$$CO_2 + (C_2H_5)_3SiH \longrightarrow HC(=O)OSi(C_2H_5)_3 \qquad (336)$$

3.8 Unsaturated Carbonyl Compounds

Reaction of silicon hydrides with unsaturated carbonyl derivatives involves two bonds, i.e. C=O and C=C (as well as C≡C) whose effects can be in isolation or linked in conjugation or some other cumulative manner, and whose behaviour can lead to potential addition of an Si–H bond.

3.8.1 Aldehydes and Ketones

The hydrosilylation of aldehydes and ketones containing isolated as well as conjugated C=C bonds proceeds quite readily. In the presence of a cesium and potassium fluorides, such addition occurs at the C=C bond [151, 262, 286] as follows:

$$\mathrm{CH_2{=}CHCH_2CH_2COCH_3 + (C_2H_5O)_2SiHCH_3 \xrightarrow[60^{\circ}C]{KF,DMF}}$$
$$\mathrm{CH_2{=}CHCH_2CH_2CH(CH_3)OSi(OC_2H_5)_2CH_3} \quad \textit{85\% yield} \qquad (337)$$

Other unsaturated aldehydes and ketones, e.g. $(CH_3)_2C{=}CHCH_2CH_2COCH_3$ [262], $C_6H_5CH{=}CHCHO$ [151, 286] and $(CH_3)_2C{=}CHCH_2CH_2CH(CH_3)CH_2CHO$ [151, 262], also undergo this reaction.

In the presence of metallic nickel formed *in situ*, unsaturated aldehydes, such as that formed from undecylene, yield silyl ethers as a result of the migration of the C=C bond, whereas in the presence of Ni + $(C_2H_5)_2S$ as the catalyst silyl enol ether is produced in addition as follows [148]:

$$\mathrm{CH_2{=}CH(CH_2)_8CHO + (C_2H_5)_3SiH \xrightarrow[reflux]{Ni+(C_2H_5)_2S}}$$
$$\rightarrow \mathrm{CH_3CH{=}CH(CH_2)_8OSi(C_2H_5)_3}$$
$$\rightarrow \mathrm{CH_3CH{=}CH(CH_2)_6CH{=}CHOSi(C_2H_5)_3} \qquad (338)$$

A three-component system (phosphine palladium complex, diphenylsilane and $ZnCl_2$) is an efficient catalyst for conjugate reduction of broad range of α,β-unsaturated ketones and aldehydes [563]. Similar effects gave two-component catalyst (consisting of phenylsilane or diphenylsilane and $Mo(CO)_6$ in refluxing THF [564].

In contrast, an unsaturated ketone can react in the presence of H_2PtCl_6 to give two products, which arise from hydrosilylation of both the C=C and C=O bonds [602]:

$$\mathrm{CH_2{=}CHCH_2CH_2COCH_3 + (C_2H_5)_3SiH \xrightarrow[130\text{–}40^{\circ}C,\ 1.5\text{–}2\ h]{H_2PtCl_6}}$$
$$\rightarrow \mathrm{CH_2{=}CHCH_2CH_2CH(CH_3)OSi(C_2H_5)_3} \quad \textit{40\%} \qquad (339)$$
$$\rightarrow \mathrm{(C_2H_5)_3Si(CH_2)_4COCH_3} \quad \textit{34\%}$$

However, during the hydrosilylation of α,β-unsaturated aldehydes, e.g. crotonaldehyde [661] and cinnamaldehyde [873], in the presence of platinum complexes [347, 525, 661, 866, 873, 894, 895, 987], 1,4-addition occurs, leading to various *cis* and *trans* (1:1) adducts [785, 1243]. Similarly, crotonaldehyde [866] reacts with triethoxysilane in the presence of Wilkinson's catalyst to give *cis*- and *trans*-4,4-diethoxy-3,5-dioxa-4-sila-6-nonene:

$$\textit{cis}\text{-}\mathrm{CH_3CH{=}CHCHO + (C_2H_5O)_3SiH \xrightarrow{RhCl(PPh_3)_3} C_2H_5CH{=}CHOSi(OC_2H_5)_3}$$
$$\textit{cis} \text{ and } \textit{trans} \quad \textit{(84\% yield)} \qquad (340)$$

Not only silyl ethers such as $CH_3CH=CHOSi(C_2H_5)_2CH_3$ (35% yield) and $CH_3CH=CHOSiF(C_2H_5)_2$ are formed during the addition of diethylmethylsilane and fluorodiethylsilane to acrolein in the presence of Speier's catalyst, but also high-boiling compounds which result from telomerization, e.g. $CH_2=CH(CH_3)CHOSi(C_2H_5)_2CH_2CH(CH_3)CHO$ (15% yield) [1243], products arising from diene condensation of the adduct with acrolein to give $CH_3\overline{CHCH_2CH=CHOCHOSi(C_2H_5)_2R}$, others resulting from the addition of hydrosilane to the $C=C$ bond to give $RSi(C_2H_5)_2CH(CH_3)CH_2OSi(C_2H_5)_2R$ as well as the addition of silane to the cyclic dimer to give $\overline{CH_2CH_2CH=CHCH_2OCHCH_2OSi(C_2H_5)_2R}$. Separate addition of diethylmethylsilane and fluorodiethylsilane to the acrolein cyclodimer occurs exclusively at the carbonyl group [1243]:

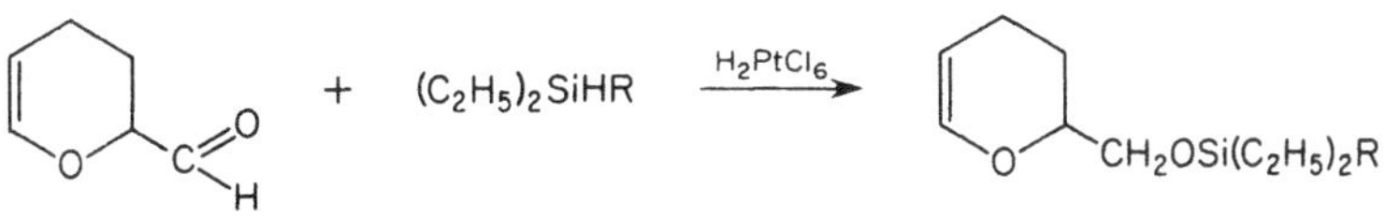

where $R=F, CH_3$

(341)

whereas hydrosilylation of α-ethoxyacrolein by fluorodiethylsilane leads ultimately to finally 1,4-addition products (*cis* and *trans*), as well as the 1,2-adduct and a cyclic product arising from an intramolecular reaction:

$$CH_2=C(OC_2H_5)CHO \quad + \quad (C_2H_5)_2SiH_2 \quad \longrightarrow \quad CH_2=C(OC_2H_5)CH_2OSiH(C_2H_5)_2$$

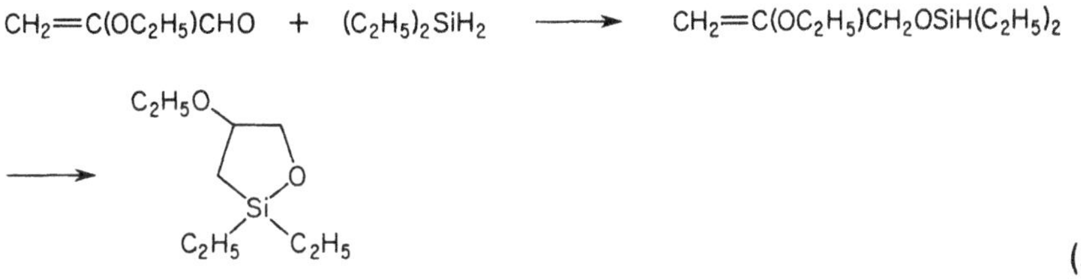

(342)

Diethylsilane appears to be generated as a result of the ready redistribution of fluorodiethylsilane.

With various nickel(O) complexes, e.g. $Ni(cod)_2$ and $Ni[P(OC_2H_5)_3]_4$ as catalysts, crotonaldehyde adds to triethoxysilane in the 1,4-position:

$$cis-CH_3CH=CHCHO + (RO)_3SiH \xrightarrow{Ni(O)} cis-CH_3CH=CHCH_2OSi(OR)_3$$

(343)

However, on reaction with diethylmethylsilane, two unsaturated adducts are formed; *viz.* *cis*-$C_2H_5CH=CHOSi(C_2H_5)_2CH_3$ and *trans*-$CH_3CH=CHCH_2OSi(C_2H_5)_2CH_3$ [661].

The hydrosilylation of α,β-unsaturated ketones when catalysed by Rh(I) complexes results in addition principally in the 1,4-position [873], e.g. triethylsilane adds to benzylidene acetophenone ($C_6H_5CH=CHOC_6H_5$) to give a product in 20% yield after reaction in benzene for 15 h [620] but this may also be accompanied by 1,2-addition [595, 863, F-023]. In contrast, reaction of dimethylcyclohexenone with mono-, di- and trisubstituted silanes leads to isomerization of the C=C bond as follows:

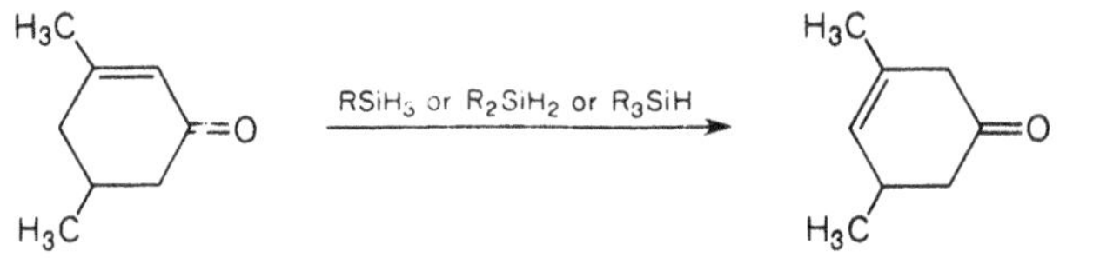

(344)

Indeed, under such conditions only the reaction of a simple cyclohexen-2-one results in adducts. The direction in which addition proceeds depends on the nature of the solvent employed, the temperature, and on the structure and concentration of the silane [562]. The addition of phenyl-1-naphthylsilane to carvone occurs at the carbonyl group [595]:

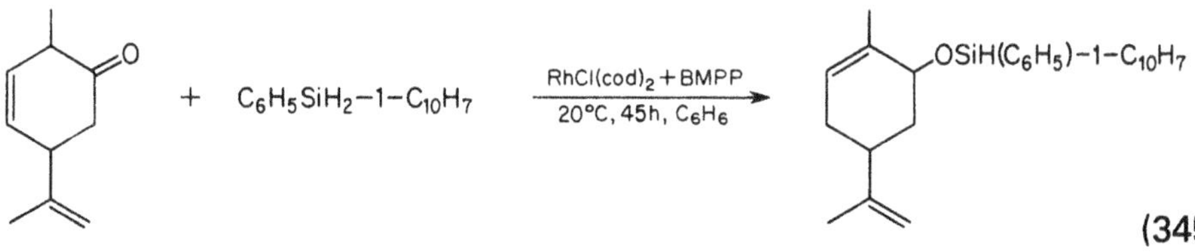

(345)

whereas hydrosilylation of (1-methyl-4-isopropylidene-3-cyclo-hexanone) (I) and β-ionone (II) (see Scheme 346) occurs at the 1,4- and 1,2-positions in an approximate ratio of 1:1 or 3:1 [850] when catalysed by Wilkinson's catalyst. The lack of conjugation between the C=C bond of the cyclohexenyl ring and other unsaturated bonds in α-ionone (III) (Scheme 346) results in only the 1,2-adduct being formed (in 96% yield) [863, F-023]:

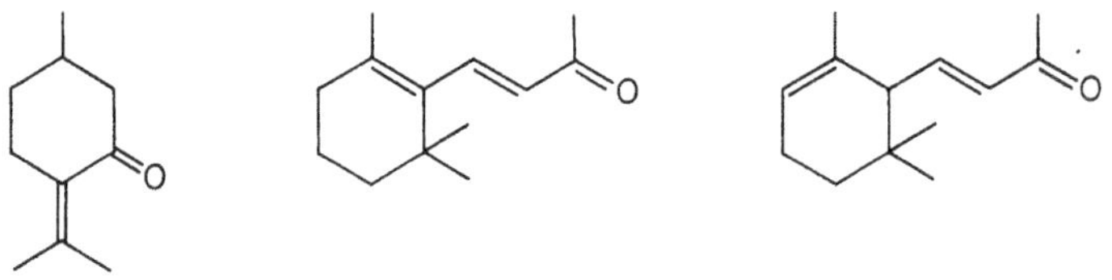

Pulegone (I) β-Ionone (II) α-Ionone (III) (346)

α,β-Unsaturated carbonyl compounds such as $CH_3CH{=}CHCOC_6H_5$, $CH_3CH{=}CHCHO$, $C_6H_5CH{=}CHCHO$ and $(CH_3)_2C{=}CHCOCH_3$ give 1,4-addition products with triorganosilanes and products resulting from selective 1,2-addition with mono- and di-organosilanes [Rev.31, 857, F-023].

Indeed, the application of such discrimination could provide an efficient method for the regioselective 1,2- or 1,4-reduction of α,β-unsaturated carbonyl substrates, with the C=C bond remaining unhydrosilylated. A detailed discussion of this problem has already been presented [Rev.31].

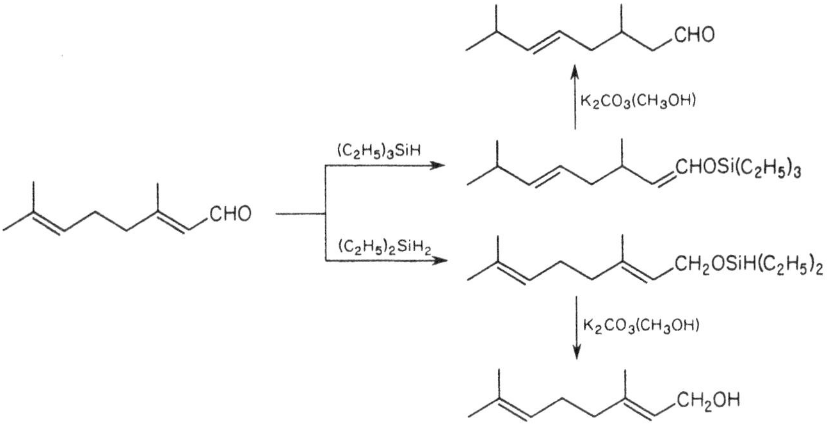

(347)

The asymmetric hydrosilylation of α,β-unsaturated carbonyl compounds is catalysed mainly by chiral (cationic) rhodium complexes. The selectivity of the 1,2- and 1,4-addition of hydrosilanes to this class of compounds depends closely on the type of hydrosilane used [Rev.31]. The optical yields resulting from such processes are similar to those pertaining in the asymmetric hydrosilylation of simple ketones previously described (Section 3.8.1). The general equation for this reaction may be written as follows:

$$>C{=}\overset{|}{C}{-}\underset{\underset{O}{\|}}{C}{-}R + H_2SiR'_2 \xrightarrow{Rh^*} >C{=}\overset{|}{C}{-}\underset{\underset{OSiHR'_2}{|}}{CH}{-}R \qquad (348)$$

$Rh^* = [(R){-}BMPP]_2Rh(S)Cl$ or $(+){-}DIOPRh(S)Cl$.

It should be noted that addition is mainly favoured for (R)-ketones [Rev.31].

In contrast to the addition of silicon hydrides to the alkenyl derivatives of carbonyl compounds, the corresponding addition to propargyl aldehydes in the presence of H_2PtCl_6 occurs first at the C≡C bond:

$$(C_2H_5)_3SiCH_2CH{=}CHOSi(C_2H_5)_3 \qquad (III)$$
$$\uparrow$$
$$\rightarrow (C_2H_5)_3SiCH{=}CHCHO \qquad (I)$$
$$\beta\text{-adduct } 26\%$$
$$CH{\equiv}CCHO + (C_2H_5)_3SiH \xrightarrow[54-57^\circ C,\ 2\,h]{H_2PtCl_6} \qquad (349)$$
$$\rightarrow CH_2{=}C(CHO)Si(C_2H_5)_3 \qquad (II)$$
$$\alpha\text{-adduct } 28\%$$
$$\downarrow$$
$$(C_2H_5)_3SiC(CH_3){=}CHOSi(C_2H_5)_3 \qquad (IV)$$

although in the presence of excess triethylsilane, this is followed by C=O hydrosilylation (yielding products (III) and (IV)) [601].

Preferential hydrosilylation of the C≡C bond occurs during the reaction of acetylacetylene with silicon hydrides in the presence of H_2PtCl_6 [753]. However, the reaction of ethyldimethylsilane with pyruvic nitrile (containing C=O and C≡N bonds) when catalysed by Wilkinson's complex (80°C, 2 h) gives a high yield of the selective product resulting from C=O hydrosilylation, i.e. 2-cyano-4, 4-dimethyl-3-oxa-4-silahexane (90% yield) [873]. The reaction of hydrosilanes and hydrosiloxanes with ketones proceeds via a 1,2-addition across the C=O group (Scheme 350) [376, 1054]; thus, trimethylsilane and tripropylsilane react additively in the presence of Ni or $NiCl_2$ [376]:

$$(C_6H_5)_2C{=}C{=}O + R_3SiH \xrightarrow[\text{reflux, 5 h}]{\text{Ni}} (C_6H_5)_2C{=}CHOSiR_3 \qquad (350)$$

In this reaction, when R = C_2H_5 a 90% yield is obtained, but when R = C_3H_7 the yield is reduced to 70% [376].

Ketones with different substituents, i.e. $R^1R^2C{=}C{=}O$, give a mixture of E- and Z-vinylsilyl ethers with an overall 80% yield:

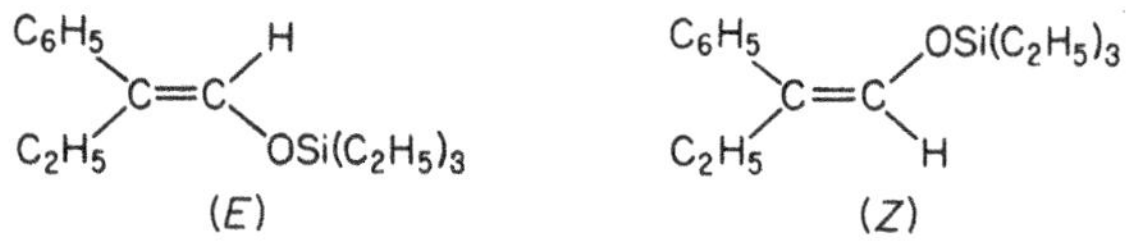

(351)

3.8.2 Unsaturated Esters

In the presence of H_2PtCl_6 and Pt/C, hydrosilylation of unsaturated esters containing isolated C=C and C=O bonds involves addition of the SiH linkage across both multiple bonds to give two separate adducts:

$$CH_2{=}CH(CH_2)_nCOOCH_3 + R_3SiH \rightarrow R_3SiCH_2CH_2(CH_2)_nCOOCH_3 \qquad (352.1)$$

$$CH_3COO(CH_2)_nCH{=}CH_2 + R_3SiH \rightarrow CH_3COO(CH_2)_nCH_2CH_2SiR_3 \qquad (352.2)$$

With ω-olefins the silyl group is added to the terminal carbon, but with unsaturated acid esters such addition occurs with internal C=C bonds [1011]:

$$CH_3(CH_2)_nCH{=}CH(CH_2)_mCOOCH_3 + R_3SiH \rightarrow \begin{cases} CH_3(CH_2)_{n+1}CH(SiR_3)(CH_2)_mCOOCH_3 \\ CH_3(CH_2)_nCH(SiR_3)(CH_2)_{m+1}COOCH_3 \end{cases} \qquad (353)$$

Initiation of the addition of trichlorosilane to vinyl acetate by γ-radiation leads to the simultaneous reduction of the carbonyl group to methylene one [1210]. When the reaction is carried out with trichlorosilane [C-006] in the presence of Lamoreaux catalyst (H_2PtCl_6 in $C_8H_{17}OH$ with reflux for 4 h), or with triethoxy- and triethylsilane [S-029] in the presence of a polyolefin-supported Rh complex (with reflux for 1.5 or 4 h), the final product is 2-silylethyl acetate (80–90% yield).

Allyl esters of saturated acids have been extensively studied because of the potential application as a silane coupling agent of the product arising from hydrosilylation of their C=C bonds. Hydrosilylation of this class of compound is usually accompanied by several side-reactions, such as dehydrogenation and redistribution of silanes, isomerization and the decomposition of unsaturated carbonyl compounds, etc. The addition of silicon hydrides to allyl chloroformate [786], allyl carbonate [1051, A-003, U-023], allyl formate [1052, A-004, D-023, G-018] and allyl acetate [1121, J-099] occurs selectively at the C=C bond according to Farmer's rule, giving the appropriate β-adducts (I) although propene and acetoxysilane (II) are also formed as by-products.

$$R^1COOCH_2CH{=}CH_2 + R_nSiHCl_{3-n} \xrightarrow[\text{or } H_2PtCl_6]{Pt/Al_2O_3} \quad (354)$$

$$\rightarrow R^1COOCH_2CH_2CH_2SiR_nCl_{3-n} \quad (I)$$

$$\rightarrow CH_3CH{=}CH_2 + R^1COOSiR_nCl_{3-n} \quad (II)$$

where R = CH_3, C_2H_5, C_2H_5O or siloxanyl. Through the use of this reaction it is possible to obtain a 20–35% yield of 3-(trichlorosilyl)propyl chloroformate. In the presence of tert-butyl peroxide this yield can be increased to 52%, but in this case CO_2, propene and silane redistribution products are obtained. In the presence of a 100% excess of trichlorosilane (25°C, 45 h), the yield of 3-(trichlorosilyl)propyl chloroformate is increased to 40% [G-030]. During the hydrosilylation of allylchloroacetate in the presence of homogeneous and heterogeneous platinum catalysts a 65–85% yield of the β-adduct is produced [Pl-002, Pl-003, Pl-007, Pl-008] and no substitution of the chlorine atom by the hydrogen of the silane takes place [613].

The hydrosilylation of allyl acetals and their α-chloro, phenoxy, alkoxy and allyloxy derivatives by tri(chloro,alkoxy)silanes, which takes place in the presence of homogeneous and heterogeneous platinum catalysts, leads to the formation of silyl propyl acetates and several by-products. With these various derivatives, the yield of 3-(trichlorosilyl)propyl acetate, $YCH_2COOCH_2CH_2CH_2SiCl_3$, decreases in the following sequence [613]:

Y	H	= C_6H_5O	> Cl	> CH_3O	> C_2H_5O	> n-C_3H_7O	(355)
yield (%)	83.8	83.8	78.8	69.0	51.0	47.0	

The hydrosilylation of allyl diazoacetate can be initiated by azo-bis-isobutyronitrile [G-016]:

$$N_2CHCOOCH_2CH{=}CH_2 + (CH_3O)_3SiH \xrightarrow[60°C,\ 16\ h]{C_6H_6} (CH_3O)_3SiCH_2CH_2CH_2OCOCHN_2 \quad (356)$$

while the reaction of allylazidoformate with trichlorosilane catalysed by Speier's catalyst (50°C, 16 h) leads to the formation of 3-(trichlorosilyl)propyl azidoformate (76% yield) [G-019].

Chloroplatinic acid also catalyses the addition of trichlorosilane to allyl esters containing a nitro group [D-052, F-022]:

$$m{-}O_2NC_6H_4COOCH_2CH{=}CH_2 + HSiCl_3 \xrightarrow[C_6H_5CH_3,\ 60{-}90°C,\ 2h]{H_2PtCl_6} m{-}O_2NC_6H_4COO(CH_2)_3SiCl_3 \quad (357)$$

The hydrosilylation of allyl esters of perfluoro acids with alkylchlorosilanes in the presence of chloroplatinic acid leads to yields of up to 70% [103, S-005]. The addition of methyldichlorosilane to allyltrifluoroacetate results in the formation of methyl ((3-trifluoroacetoxy)propyl)dichlorosilane but the reaction is accompanied by the formation of several by-products [100]:

$$\begin{aligned} &CF_3COOCH_2CH{=}CH_2 + CH_3SiHCl_2 \xrightarrow{H_2PtCl_6} \underset{52.2\%}{CF_3COO(CH_2)_3SiCl_2CH_3} + \\ &\underset{3.3\%}{CF_3COOCH_2CH_2CH_3} + \underset{13\%}{CF_3COOSiH(Cl)CH_3} + \underset{9.4\%}{CH_3SiCl_3} + \\ &\underset{12.2\%}{C_3H_7SiCl_2CH_3} + \underset{7.3\%}{CF_3COO(CH_2)_3SiCl(OCOCF_3)CH_3} \end{aligned} \quad (358)$$

Further side-reactions occur during the hydrosilylation of allyl perfluoroacetate by trichloro-, tri(alkyl-,alkoxy)- and trialkoxysilanes (the yield of β-adduct being 19–45%). In the case of triethoxysilane the following products were identified: $CF_3COO(CH_2)_3Si(OC_2H_5)_3$, $CF_3COOSi(OC_2H_5)_3$, $CF_3COOC_2H_5$, $CF_3COOCH_2CH_2CH_3$, $(CF_3CH_2O)_2Si(OC_2H_5)_2$ and $Si(OC_2H_5)_4$ [931].

The hydrosilylation of α,β-unsaturated esters by tri-substituted silanes generally gives the following four products: the α- and β-adducts resulting from 1,2-addition to the C=C bond and two adducts arising from 1,2- and 1,4-addition across the carbonyl group.

$$R^1CH{=}CHCOOR^2 + R_3SiH \rightarrow \begin{cases} R^1CH_2CH(SiR_3)COOR^2 \\ R^1CH(SiR_3)CH_2COOR^2 \\ R^1CH{=}CHCH(OR^2)OSiR_3 \\ R^1CH_2CH{=}C(OR^2)OSiR_3 \end{cases} \quad (359)$$

Only the hydrosilylation of methyl crotonate and its homologue with trimethylsilane leads to the formation of dimeric compounds:

$$2\,RCH{=}CHCOOCH_3 + (CH_3)_3SiH \xrightarrow{H_2PtCl_6} (CH_3)_3SiOC(OCH_3){=}C(CH_2R)CH(R)CH_2COOCH_3 \quad (360)$$

In this case, when R = CH_3 the yield was 54%, and when R = C_2H_5 the yield was 25% [1321].

Hydrosilylation of α,β-unsaturated esters with triethylsilane catalysed by Wilkinson complex leads to saturated esters, whereas the reaction of fully conjugated diene esters were reduced to the dihydroderivatives to give β,γ or γ,β-unsaturated esters [678].

Much attention has been focused on the hydrosilylation of acryl and methacryl esters catalysed by H_2PtCl_6 and Pt metal [227, 300, 359, 423, 901, 985, 1127, 1133, U-182, U-187], as well as by nickel–phosphine [589, U-187] and rhodium–phosphine complexes [868, 869, 1321, U-187]. It has been shown that in the presence of platinum catalysts the mechanism of addition depends on the structure of the ester group [300, 423, 901, 985, 1127, 1133].

Acrylates react with methylchlorosilanes to form a mixture of α- and β-adducts [674, 1323] as well as the 1,4-addition product [985, 1178]. When triethylsilane is added, only β-triethylsilyl propionic ester is obtained [802]. Methacrylates in reaction with various hydrosilanes generally yield β-adducts [754, 795, 971, 1321, B-002, U-014] sometimes accompanied by α-adducts [359, 674, 1323] and 1,4-addition products [359]. Hydrosilylation of methyl methacrylate by triethylsilane catalysed by complexes of rhodium gives product of the 1,4-O-silylation, which is accompanied by β-adduct [E-023, U-187]. However, $Co_2(CO)_8$-catalysed reaction of acrylic acid esters (methyl acrylate, ethyl acrylate and butyl acrylate) with $HSi(C_2H_5)_2CH_3$ gave the corresponding (E)-3-silyl acrylate in high yields, along with only a small amount of β-adducts when the esters were used in excess.

In the presence of phosphine–nickel complexes, hydrosilylation of methyl acrylate gives a 97% yield of the α-adduct with no product of H/Cl exchange being observed [589]. The comprehensive work of Ojima *et al.* on the hydrosilylation of α,β-unsaturated esters generally in the presence of an $RhCl(PPh_3)_3$ catalyst has shed more light on the mechanism of this reaction which results in the selective formation of 1,2- and 1,4-adducts. Although the yields of these products are usually quite high, it is affected by the kind of ester and hydrosilane used [868]. Hydrosilylation of methyl acrylate with phenyldimethylsilane and ethyldimethylsilane leads exclusively to the β-adduct, which is accompanied by a disilyl product if the triethylsilane is used. The following mechanism has been taken into account on the formation of Rh(II) intermediate necessary for both products (Scheme 361).

On addition of chlorodimethylsilane, the α-adduct, is obtained selectively according to the following proposed mechanism [868] (Scheme 362).

Such a mechanism involves the nucleophilic attack of an Rh–H bond on the electron-deficient β-carbon atom of the acrylate or crotonate. If the silyl group migrates to the oxygen of the carbonyl group, the π-allyl intermediate is formed (Scheme 363); this decomposes subsequently to give the 1,4- or 1,2-adduct [868]. The methyl group at the olefin carbon apparently stabilizes the Π-allyl intermediate. Addition of trichlorosilane, methyldichlorosilane and trialkoxysilane and methyldialkoxysilane to the allyl esters of unsaturated carboxylic acids containing conjugated C=C and C=O bonds, e.g. the acrylate [A-004, D-011, E-029, G-001, G-018, J-016, J-027, U-016, U-157, U-176], methacrylate [E-007, J-074, U-176, U-157, U-183, J-070], maleinate [U-060, U-061, U-066, U-069, U-073, U-075] or allyl ester of 2,1-dicarboxycyclohexene-4 glycidic acid [D-053, G-020], always proceeds at the C=C bond of the allyl group. The addition of hydrosiloxanes to allyl methacrylate also gives 3-methacryloxypropylsiloxanes [J-073].

(Ph$_3$P)$_3$Rh^{I}Cl + R$_3$SiH $\xrightarrow{-Ph_3P}$ (Ph$_3$P)$_2$Rh^{III}(H)Cl(SiR$_3$) $\longrightarrow$ (Ph$_3$P)$_2$Rh^{II}(H)Cl

$\overset{\bullet}{\text{SiR}_3}$

$\xrightarrow{CH_2=CHCOOR}$

$$\left[\begin{array}{c} \text{R}_3\text{SiCH}_2\text{—CH}\cdots\text{COR} \\ \vdots\quad \text{O} \\ \text{(Ph}_3\text{P)}_2\text{Rh}^{II}\text{Cl(H)} \end{array}\right] \text{(V)}$$

R$_3$SiCH$_2$CH=C(OR)(OSiR$_3$)
+
(Ph$_3$P)$_2$Rh^{III}Cl(H)$_2$ $\xleftarrow{R_3SiH}$

(Ph$_3$P)$_2$Rh^{I}Cl + H$_2$
(IV)

(b)

R$_3$SiCH$_2$CH$_2$COOR + (Ph$_3$P)$_2$Rh^{I}Cl
(IV)

$\xrightarrow{Ph_3P}$ (Ph$_3$P)$_3$Rh^{I}Cl

(361)

(CH$_3$)$_2$SiHCl + (Ph$_3$P)$_3$Rh^{I}Cl $\xrightarrow{-Ph_3P}$ (Ph$_3$P)$_2$Rh^{II}Cl(H)[Si(CH$_3$)$_2$Cl]
(II)

$\downarrow$ RCH=CHCOOC$_2$H$_5$

(Ph$_3$P)$_2$Rh^{I}Cl
(IV)
+
RCH$_2$CHCOOC$_2$H$_5$
   |
   Si(CH$_3$)$_2$Cl

$$\left[\begin{array}{c} \overset{\delta\ominus}{\text{H}}\text{-----}\overset{\delta\pm}{\text{Rh}}[\text{Si(CH}_3)_2\text{Cl}]\text{Cl(PPh}_3)_2 \\ \vdots\qquad\vdots \\ \text{R—CH====CHCOOC}_2\text{H}_5 \end{array}\right]$$

Cl(CH$_3$)$_2$Si—Rh^{III}Cl(PPh$_3$)$_2$
   |
RCH$_2$CHCOOC$_2$H$_5$

(362)

(Ph$_3$P)$_3$Rh^{I}Cl + R$_3$SiH $\xrightarrow{-Ph_3P}$ [(Ph$_3$P)$_2$Rh^{III}(H)Cl(SiR$_3$) $\rightleftharpoons$ (Ph$_3$P)$_2$Rh^{II}(H)Cl-----•SiR$_3$]

$\downarrow$ R^{1}CH=CR^{2}COOR^3

$$\left[\begin{array}{c} \text{R}^2 \\ \text{C} \\ \text{R}^1\text{HC} \quad \text{C(OR}^3)\text{(OSiR}_3) \\ \text{(Ph}_3\text{P)}_2\text{Rh}^{III}\text{(H)Cl} \end{array}\right] \rightleftharpoons \left[\begin{array}{c} \text{R}^2 \\ \text{C} \\ \text{R}^1\text{HC} \quad \text{C(OR}^3)\text{(OSiR}_3) \\ \text{(Ph}_3\text{P)}_2\text{Rh}^{II}\text{(H)Cl} \end{array}\right]$$
(VII)

R^{1}CH$_2$-C(R^2)=C(OR^3)(OSiR$_3$)
+
(Ph$_3$P)$_2$Rh^{I}Cl
(IV)

R^{1}CH=C(R^2)-CH(OR^3)(OSiR$_3$)
+
(Ph$_3$P)$_2$Rh^{I}Cl
(IV)

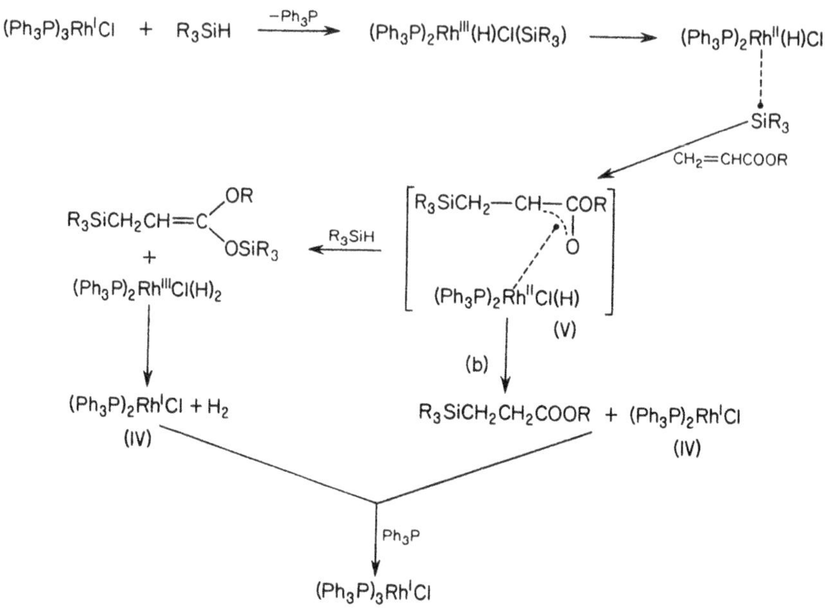

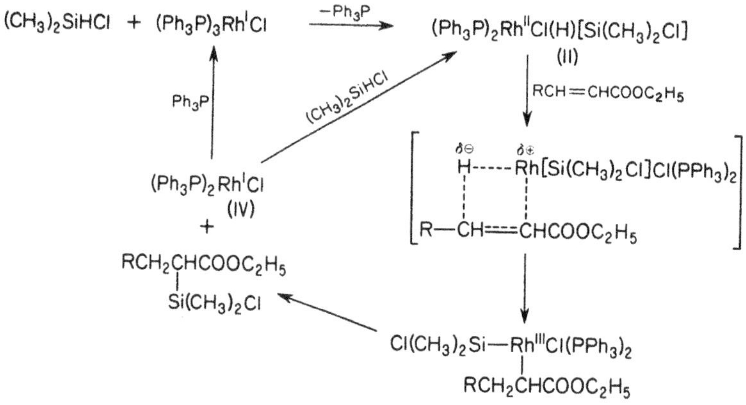

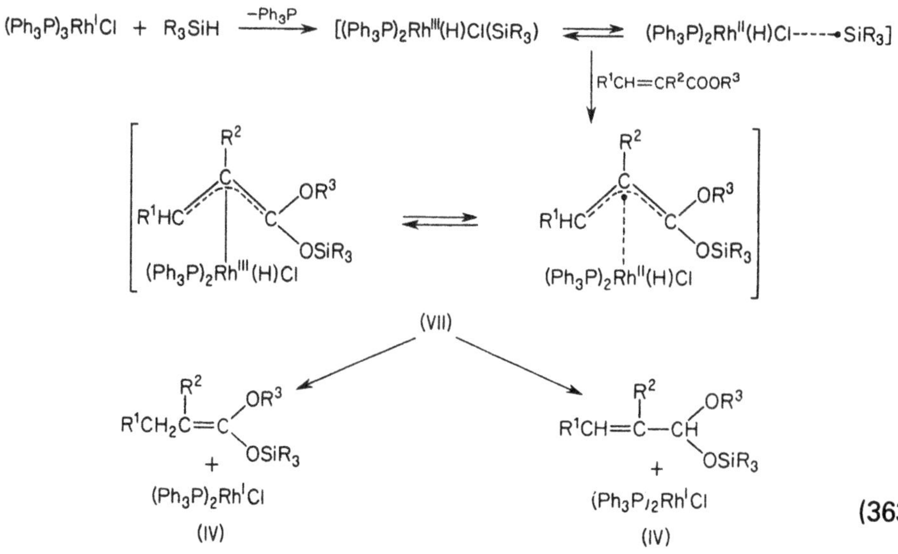

(363)

3.9 Compounds with Carbon–Nitrogen Multiple Bonds

Silicon hydrides add to the carbon–nitrogen double bonds of enamines and Schiff bases [Rev.26]. Thus, hydrosilylation of an N=C bond occurs via addition of the silyl group to the nitrogen atom and of the silane hydrogen to the carbon atom. Hydrosilylation of N-(benzylidene)aniline is only possible in the presence of $ZnCl_2$ as a Lewis acid, and is accompanied by reduction and reductive silylation connected with C=N bond cleavage [17]:

$$C_6H_5CH{=}NC_6H_5 + (C_2H_5)_3SiH \longrightarrow \begin{cases} C_6H_5CH_2N(C_6H_5)Si(C_2H_5)_3 & (I) \\ C_6H_5CH_2NHC_6H_5 & (II) \\ C_6H_5CH_3 + C_6H_5NH_2 + C_6H_5NHSi(C_2H_5)_3 & (III) \end{cases} \qquad (364)$$

The yield of the adduct increases as the catalyst concentration increases [17]. This competitive reaction also occurs in the presence of other catalysts such as $[(C_8H_{17})_3NH]_2RuCl_6$, $[(C_8H_{17})_3NH]_2IrCl_6$, $[(C_8H_{17})_3NH]_2PtCl_6$, and K_2IrCl_6. However, complexes such as $[(C_{10}H_{21})_4P]_2PtCl_6$ and $[(C_8H_{17})_3NH]_2PdCl_6$, when used as catalysts only enable the formation of the products (II) and (III), while the use of $RhCl(PPh_3)_3$ as a catalyst gives (I) (60% yield) and (III) as the products. In the presence of the latter catalyst, di-substituted silanes are readily added to N-(benzylidene)aniline [862, J-010], thus when diethylsilane is added (50°C, 2 h), the yield of product is 96%. Substitution of hydrogen at the carbon atom of C=N bond causes a reduction in the rate of hydrosilylation. Hydrosilylation of $C_6H_5C(CH_3){=}NCH_3$ with diethylsilane in the presence of $RhCl(PPh_3)_3$ (55°C, 72 h), results in an 85% yield of the adduct [862], while the hydrosilylation of $C_6H_5CH{=}NCH_3$ gives a 95% yield of the expected adduct even at room temperature (0.5 h) [862, J-010].

In the presence of $RhCl(PPh_3)(diop)(S)$, $RhCl(PPh_3)_3$ or $RhCl(diop)$, di-substituted silanes are satisfactorily added to enamines such as $C_6H_5CH{=}NCH_3$ [697, J-010], $C_6H_5C(CH_3){=}NCH_3$ [862], $C_6H_5CH{=}N(CH_2)_3CH_3$ [862], $C_6H_5CH_2C(CH_3){=}NCH_3$ [642], $C_6H_5C(CH_3){=}NC_6H_5$ [529, 642], $C_6H_5C(CH_3){=}NCH_2C_6H_5$ [529, 642] and $C_6H_5CH_2C(CH_3){=}NCH_2C_6H_5$ [659], and to

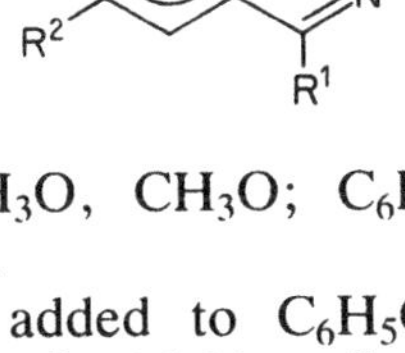

where R^1,R^2,R^3 = CH_3, CH_3O, CH_3O; $C_6H_5CH_2$, H, H; 4–CH_3O–3–$CH_3OC_6H_3CH_2$, CH_3O, CH_3O.

Polyhydrosiloxane may be added to $C_6H_5CH_2C(CH_3){=}NCH_3$ [642] and $C_6H_5C(CH_3){=}NCH_2C_6H_5$ [D-074]. Trichlorosilane is added to N-(benzylidene) anilines and their derivatives, $CH_3CH_2CH_2CH{=}N(CH_2)_3CH_3$ and $C_6H_5CH{=}NCH(CH_2)_5$, which can also function as basic catalysts in the system [115].

Alkyl aryl ketoximes $R^1R^2C{=}NOH$ were reported to undergo hydrosilylation with diphenylsilane giving, in the presence of phosphine rhodium precursor, silylamines $R^1R^2CHNHSiHPh_2$ and $Ph_2HSiOSiHPh_2$. Fourteen prochiral ketoximes were studied in the presence of optically active phosphines (diop, norphos) to reach optical inductions up to 36% e.e. [186]. On the other hand, Rh-catalysed enantioselective hydrosilylation of the five-membered ring imines with up to 64% optical induction allows isolation of products in chemical yields between 80% and 90% [85].

If the molecule in question also includes a C=C bond in addition to a C=N bond, e.g. *o*-alkylethethoximes, $(CH_2{=}CHCH_2ON{=}CR_2)$ [691, S-015], alkenylideneazynes, $[-N{=}C(CH_3)CH_2CH_2CH{=}CH_2]_2$ [U–061, U–126] and N-alkenyl Schiff bases, and $CH_2{=}CHCH_2N{=}C(CH_3)_2$ [J-040], addition occurs at the C=C bond when H_2PtCl_6 is used as the catalyst. Under mild conditions the hydrosilylation of 1,4-diaza-1,3-dienes and 1-monoazadienes occurred with rhodium (I) complexes) leading to *N*-silylated enamines [165], e.g.:

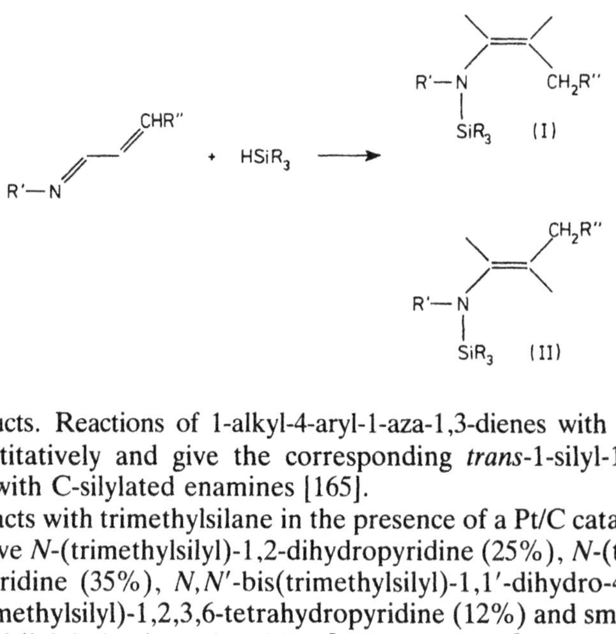

(365)

as main products. Reactions of 1-alkyl-4-aryl-1-aza-1,3-dienes with triethoxysilane proceed quantitatively and give the corresponding *trans*-1-silyl-1-aza-2-butene (II) together with C-silylated enamines [165].

Pyridine reacts with trimethylsilane in the presence of a Pt/C catalyst (30–42°C, 25–48 h) to give *N*-(trimethylsilyl)-1,2-dihydropyridine (25%), *N*-(trimethylsilyl)-1,4-dihydropyridine (35%), *N,N'*-bis(trimethylsilyl)-1,1'-dihydro-4,4'-dipyridine (25%), *N*-(trimethylsilyl)-1,2,3,6-tetrahydropyridine (12%) and small amounts of other trimethylsilyl derivatives of pyridine [268, 169, F-014]. In this instance, Pd/C appears to be a more active catalyst than Rh/C, $PdCl_2$, Pt/C, Ru/C or Raney nickel. However, Ru/C is more selective, yielding *N*-trimethylsilyl-1,4-dihydropyridine. In the presence of a Pd/C catalyst, 2-methylpyridine forms two isomers, [F-014], i.e.

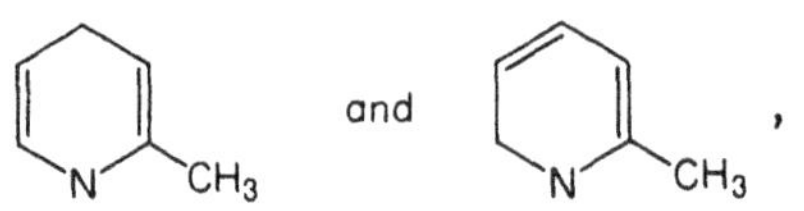

whereas 3-methylpyridine reacts with trimethylsilane to give three products [268, F-014], i.e.

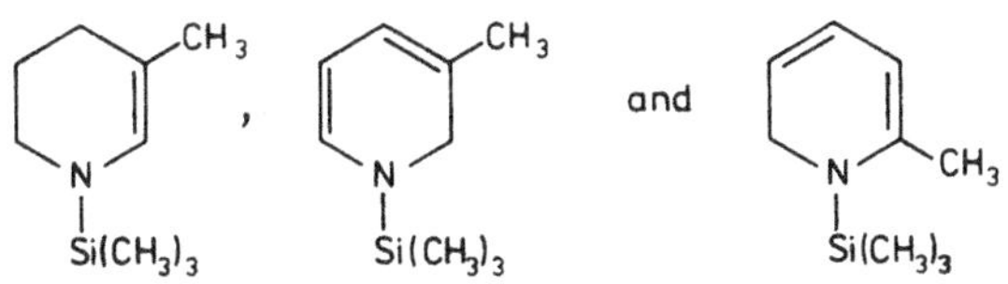

Compounds containing two C=N bonds in the molecule, such as carbodiimides, can also react with trisubstituted silanes. *N,N'*-Dialkylcarbodidimides, e.g. *N',N'*-diisopropyl- and N,N'-dicyclohexyl-carbodidimides, add triorganosilanes such as triethylsilane, dimethylphenylsilane and ethyldimethylsilane to one of the carbon–nitrogen double bonds to form *N*-silylformamidynes [J-015]:

$$(CH_3)_2CHN{=}C{=}NCH(CH_3)_2 + (C_2H_5)_3SiH \xrightarrow[140°C,\ 15\ h]{catalyst} (CH_3)_2CHN[Si(C_2H_5)_3]CH{=}NCH(CH_3)_2 \qquad (366)$$

When the catalyst employed in this reaction is $RhCl(PPh_3)_3$, a 60% yield was obtained after 15 h at 140°C, whereas $PdCl_2$ as the catalyst gave an 85–97% yield after 15–48 h at 140–200°C [849, 860].

Isocyanates containing saturated substituents undergo addition at the N=C bond [860, J-014], with any other multiple bonds present remaining intact [311, 862]:

$$C_6H_5SO_2N{=}C{=}O + (C_2H_5)_3SiH \xrightarrow[120\text{–}130°C,\ 3\ h]{Pd/C,\ C_6H_5CH_3} C_6H_5SO_2N(CHO)Si(C_2H_5)_3 \quad 89\% \qquad (367)$$

As discussed above, the reaction normally leads to the formation of *N*-silyl-*N*-organoformamide, but this is sometimes accompanied by C-silylamides which arise as the products of the hydrosilylation of the C=O bond [859]:

$$RN{=}C{=}O + (C_2H_5)_3SiH \xrightarrow[or\ Pt/C]{PdCl_2} \begin{cases} RN(CHO)Si(C_2H_5)_3 \\ RNHCOSi(C_2H_5)_3 \end{cases} \qquad (368)$$

where R = C_6H_5, $\overline{(CH_2)_5CH}$, $1-C_{10}H_7$.

Organic cyanides undergo hydrosilylation under special conditions. Thus, silicon hydrides which contain two Si–H bonds in the molecule, e.g. *o*-bis(dimethylsilyl)-benzene, may be added to alkyl cyanides in the presence of rhodium complexes with the formation of an additional ring (Scheme 369).

In this case, with $[RhCl(c-C_8H_{14})_2]_2$ as the catalyst and R = CH_3, R' = C_2H_5, the yield was 40%; with $RhCl(PPh_3)_3$ as the catalyst and R = C_2H_5, R' = C_2H_5 the yield was 15%, but when R' = $CH_2{=}CH$ the yield increased to 50%. Cobalt carbonyl catalysed reactions of $HSi(CH_3)_3$ with aromatic nitriles having various

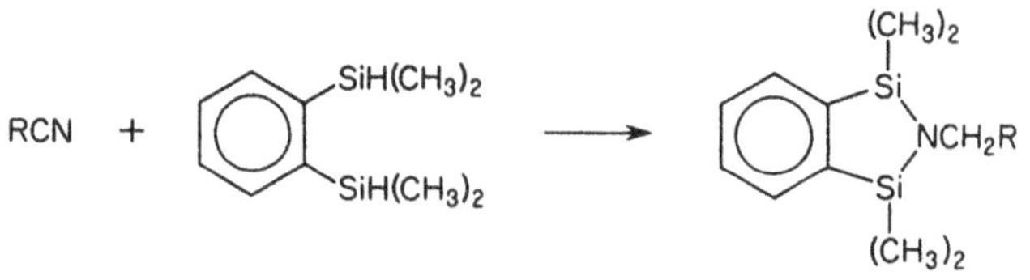

(369)

functional groups were reported recently to provide quite novel route to *N,N*-disilylamines [802]

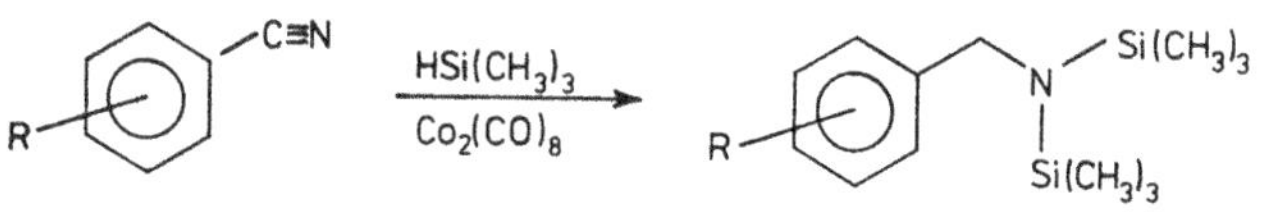

(370)

Using ethyl cyanide in the presence of Wilkinson's catalyst led to the formation of a di-adduct when R = C_3H_7 (20%) and the di-addition product of dehydrogenation (44%) when R = $CH_3CH{=}CH$ [283]. Dicyanides, e.g. $NC(CH_2)_4CN$, undergo di-addition reactions associated with one of the C≡N groups. However, if, in addition to a C–N multiple bond, the compound also includes a C=C bond, hydrosilylation at the latter is always favoured [D-024, U-008, U-039, J-024].

3.10 Compounds with Other Multiple Bonds

As already mentioned (Section 2.1.1), little information exists on the hydrosilylation of N=N and N=O bonds. The addition of triphenylsilane to azobenzene takes place according to the following equation [675]:

$$C_6H_5N{=}NC_6H_5 + (C_6H_5)_3SiH \xrightarrow{140^\circ C} C_6H_5NHN(C_6H_5)Si(C_6H_5)_3 \quad (371)$$

Radical-initiated addition to the N=N bond in diethylazodicarboxylane gives a high yield of product:

$$C_2H_5OCON{=}NCOOC_2H_5 + (C_6H_5)_3SiH \xrightarrow[3.5\ h]{UV} \quad (372)$$

$$(C_6H_5)_3SiN(COOC_2H_5)NHCOOC_2H_5$$
$$77\%$$

When the reaction is initiated by tert-butylperoxide it occurs only slowly (11 days, benzene, reflux 70%) [675, 676]. An unusual effect has been reported in connection with the addition of 1,1,2,2-tetraphenyldisilane, a disilane containing two SiH bonds, to dialkylazodicarboxylane [675]:

$$ROCON{=}NCOOR + H(C_6H_5)_2SiSi(C_6H_5)_2H \xrightarrow{(CH_3CO)_2,reflux} \quad (373)$$

$$[ROCONHN(COOR)Si(C_6H_5)_2]_2$$

the yield being 55% when R = CH_3, and 30% when R = C_2H_5.

As shown earlier, trimethylsilane is added to trifluoronitrosomethane according to the equation:

$$CF_3N{=}O + (CH_3)_3SiH \xrightarrow{UV} \underset{(I)}{CF_3NHOSi(CH_3)_3} + \underset{(II)}{(CF_3)_2NOSi(CH_3)_3} \qquad (374)$$

Irradiation with UV light gives a 43% yield of adduct (I) and a 28% yield of trimethylsilyl-*N,N*-bis(trifluoromethyl)hydroxyamine (II), whereas in the use of H_2PtCl_6 as a catalyst gives 74% and 10% of these two products, respectively [305]. Whilst the UV-initiated addition of trichlorosilane proceeds slowly (20°C, 15 h, 17%), trifluorosilane is readily added to trifluoronitrosomethane (80% yield) [587].

In the presence of H_2PtCl_6, nitrobenzene reacts with phenylmethylsilanes to form aniline rather than addition products [44].

CHAPTER 4

The Effect of Substituents at Silicon on the Reactivity of the Si–H Bond in Hydrosilylation

As already stated, the reactivity of the Si–H bond in hydrosilylation is also influenced by the physicochemical properties of silicon hydrides and the nature of the silicon atom, although the type of catalyst used actually determines the general, polar and/or radical mechanism of the addition process. Although abundant information exists on the process of hydrosilylation, little of this is useful as a means of classifying the reactivity of silicon hydrides. However, even in early studies, Meals found that the direction of the reaction is mostly determined by the nature of the silicon hydride employed [Rev.27]. The same applies to the formation of a given α- or β-isomer, but the reaction rate is affected only to a limited extent by this factor.

In spite of the paucity of relevant experimental data, later reviews have included material on this topic [Rev.21, Rev.22, Rev.38]. This chapter continues this trend and summarizes reports in the literature over the past 25 years on the effect of the substituent at the silicon atom on the rate, as well as on the yield and selectivity, of hydrosilylation reactions. In addition, an analysis of the practical applications of various silicon hydrides in the synthesis of useful organosilicon monomers is also attempted.

4.1 Effect of the Structure of Silicon Hydrides on the Rate and Yield in Hydrosilylation Processes

Although hydrosilylation reactions have been extensively studied, little kinetic information exists enabling a quantitative determination of the relative reactivities of silicon hydrides, particularly with respect to the effect of the substituent at the silicon atom. The reactivity of silanes towards addition at multiple bonds during hydrosilylation appears to depend on the inductive and steric effects of such substituents. Consequently, the following Taft correlation should provide a reasonable reflection of the relationship between the reactivity and the molecular structure of silicon hydrides:

$$\log k/k_0 = \rho^*\Sigma\sigma^* + \delta^*\Sigma E^*_s \quad (375)$$

where k_0 and k are the rate constants for the reference silane (usually $(CH_3)_3SiH$) and the silane under study, respectively; δ^* and E^*_s depict factors relevant to the inductive and steric effects of the substituents, respectively; and ρ^* and δ^* are the Taft inductive and steric constants, respectively.

Most reactivity sequences for hydrosilanes, hydrosiloxanes, etc. in hydrosilylation reactions are, however, based on the adduct yields, and only a few reports give relative rate constants. All the reactivity data described and presented in this chapter fit in with the catalyst criterion discussed in Chapter 2.

It is well known that $HSiCl_3$, $HSiBr_3$ and $(C_6H_5)_3SiH$ [Rev.13, Rev.36] are the most reactive of all the hydrosilanes as far as free-radical hydrosilylation reactions are concerned. Apparently, electron-withdrawing substituents such as Cl, when attached to silicon, reduce the polarity of the Si–H bond, and facilitate its homolytic cleavage to yield the silyl radical. The replacement of one chloro group by an alkyl substituent in a trichlorosilane markedly increases the hydride character of the hydrogen atom bonded to silicon, and leads to a decrease in the yield arising from Si–H addition to an alkene; thus in the case of 1-pentene [192] the reactivity follows the sequence:

silane	$SiHCl_3 >$	$CH_3SiHCl_2 >$	$CH_3(C_3H_7)_2SiH$	(376)
adduct yield [%]	44	10	5	

With trisubstituted silanes the general sequence of reactivity is as follows:

$$SiHCl_3 > RSiHCl_2 > R_2SiHCl > R_3SiH \quad (377)$$

Di- and mono-substituted silanes, however, are more reactive [Rev.36]:

$$RSiH_3 > R_2SiH_2 > R_3SiH \quad (378)$$

This latter order can be explained by a decrease in the steric effect of the silane brought about by R/H replacement.

The reactivity of a hydrosilane is virtually dependent on the stability of the silyl radical initiated during the first stage of the free-radical reaction. Thus, the extremely high activity of trichlorosilane in comparison with triethylsilane is due to the high stability of the trichlorosilyl radical. A kinetic study of the competitive addition of trichlorosilane and triethylsilane to 1-octene has enabled the relative rate constants to be determined as a measure of the activity [347]. Hence, $k_{SiHCl_3}/k_{(C_2H_5)_3SiH} = 3.1$–$2.3$ for the following step:

$$R_3SiCH_2\dot{C}(C_3H_7)_2 + X_3SiH \rightarrow R_3SiCH_2CH(C_3H_7)_2 + X_3Si^\bullet \quad (379)$$

Similarly, the reactivity of the following silanes towards 1-hexene (γ-irradiation) arises from the reactivity of the $X_3SiCH_2\dot{C}HR$ radical generated during the propagation step of the reaction:

$$\begin{array}{llll} & SiHCl_3 > & (C_6H_5)_3SiH > & (C_2H_5)_3SiH \\ V_0\,10^5 & 110 & 12 & 1.2 \end{array} \tag{380}$$

However, the replacement of a phenyl by a chloro substituent results in a substantial reduction in the reactivity of the silane [958]:

$$\begin{array}{llll} & (C_6H_5)_3SiH > & (C_6H_5)_2SiHCl > & C_6H_5SiHCl_2 \\ V_0\,10^5 & 12 & 1.7 & 0.69 \end{array} \tag{381}$$

In the free-radical hydrosilylation of olefin derivatives, e.g. halogeno-substituted alkenes, by (methyl, chloro)silanes (employing $\gamma^{60}Co$ as the initiator), the following reactivity sequence, which is similar to that for the hydrosilylation of unsubstituted alkenes, has been established [1360]:

$$SiHCl_3 > CH_3SiHCl_2 > (CH_3)_2SiHCl > (CH_3)_3SiH \tag{382}$$

Few reports have been published regarding the reactivity of hydrosilanes in hydrosilylation reactions using donor–acceptor catalysts. One example is the reactivity sequence based on the addition yields of the following tri-substituted silanes to acetylene:

$$\begin{gathered}(C_8H_{17})_3SiH < (C_7H_{15})_3SiH < (C_6H_{13})_3SiH < (C_5H_{11})_3SiH < \\ (C_6H_{13})_2CH_3SiH < (C_5H_{11})_2CH_3SiH < (CH_3)_2C_3H_7SiH < (C_2H_5)_3SiH\end{gathered} \tag{383}$$

This provides a neat illustration of the steric effect on the reactivity of substituents attached to the silicon atom. Functionalized silanes in the presence of Cu_2O–$[(CH_3)_2NCH_2]_2$ catalytic system can add to acrylonitrile in good yield, and the reactivity of silanes is in the order of

$$HSiCl_3 > HSi(C_6H_5)Cl_2 > HSiCl_2CH_3 \gg HSiCl(C_6H_5)_2 \tag{384}$$

$HSiCl_3$ is three times more reactive than $HSiCl_2(C_6H_5)$ and four times as reactive as $HSiCH_3Cl_2$ based on times required for completion of reaction [959]. Exclusively β-hydrosilylation of acrylonitrile in high yields (under mild conditions) is observed; ultrasonic waves significantly increase the rate of hydrosilylation.

The following order of reactivity has been observed for silanes in reactions with esters, e.g. with $C_4H_9COOC_2H_5$, subjected to heterogeneous catalysis by CsF:

$$H_2Si(C_6H_5)_2 < HSiCH_3(OC_2H_5)_2 < HSi(OC_2H_5)_3 \tag{385}$$

Under such circumstances, diphenylsilane undergoes addition with high yield at 140°C over period of 4 h, whereas triethoxysilane does so at room temperature over 1 min [154]. Unfortunately, few examples exist of the effect of the substituent attached to the silicon atom on the reactivity of silanes in reactions catalysed by supported metals, illustrating that the various reactivity sequences for such reactions are quite different from those for reactions proceeding via a free-radical mechanism [28, 897, 899, 900]. However, alkylchlorosilanes react with allyl chloride in the presence of a 1% Pt/C catalyst to give the following sequence of yields:

$$\underset{49\%}{CH_3Cl_2SiH} > \underset{41\%}{(C_2H_5)_2ClSiH} > \underset{2\%}{CH_3(C_2H_5)_2SiH} > \underset{trace}{Cl_3SiH} \qquad (386)$$

An analogous sequence for the reaction with ethylene catalysed by 5% Pt/C is:

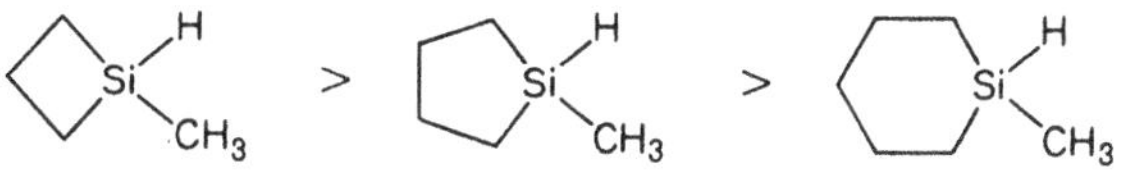

(387)

In general, it is found that silacycloalkanes are more reactive than their acyclic analogues [822].

The reactivity of silanes in reactions catalysed by heterogeneous systems is influenced by their various adsorptive capacities and by side-reactions. Thus, chlorosilyl radicals may be added to multiple bonds on the platinum catalyst, but they are also active in the abstraction of hydrogen from alkylsilanes. Hence, the overall data on the reactivities of particular silanes may be different depending on the method used in the catalytic study. This is probably the reason why the activity sequence observed in the competitive addition of hydrosilanes is:

$$CH_3(C_2H_5)_2SiH > CH_3(C_2H_5)SiHCl > C_2H_5SiHCl_2 > SiHCl_3 \qquad (388)$$

whereas a comparison of the reactivity of the same silanes on the basis of separate experiments gives the following order:

$$C_2H_5SiHCl_2 > CH_3(C_2H_5)SiHCl > SiHCl_3 > CH_3(C_2H_5)_2SiH \qquad (389)$$

The relative activities of hydrosilanes were measured for the hydrosilylation reaction between $CH_2{=}CHSi(CH_3)_3$ and R_3SiH, R_2SiH_2, $RSiH_3$ (R = ethyl, hexyl) in the presence of metal colloids (Pt, Rh) [668]. The relative rate of addition to olefins (catalysed by Pt) is $R_3SiH > R_2SiH_2 > RSiH_3$. For Rh catalyst three relative rates were thus opposite those for Pt. Authors conclude that R_2SiH and $RSiH_3$ poison Pt towards catalysis while no poisoning occurs for Rh [668].

In hydrosilylation processes catalysed by transition metal complexes the substituents attached to the silicon atom affect the various elementary steps in the catalytic process. For this reason it appears that the nature of the silicon hydride plays a complex role in such reactions. However, Chalk and Harrod's general scheme suggests that the substituents at the silicon atom may, in fact, exert their greatest influence on intermediate C (see Scheme 81), formed during the oxidative addition of the silicon hydride to the metal centre. Consequently, the inductive, steric and mesomeric effects of these substituents, and above all their influence on metal–silicon bonding, seems to be extremely important in hydrosilylation processes catalysed by transition metal complexes. Obviously, different substituents affect such factors as the reduction of the initial complex with the hydrosilane or the catalysis of the side-reactions to different extents, and this in turn alters the rate of hydrosilylation.

Generally, electron-withdrawing substituents attached to silicon stabilize the Si–metal intermediate. This is illustrated by the following reactivity observed in the hydrosilylation of 1-heptene [214]:

$$\begin{array}{lcccc} \text{silane} & SiHCl_3 > & C_2H_5Cl_2SiH > & (C_2H_5)_2ClSiH > & (C_2H_5)_3SiH \\ k_{rel} & 3.00 & 2.64 & 1.94 & 1.00 \end{array} \qquad (390)$$

and for 1-octene [341]:

$$\begin{array}{lccc} \text{silane} & CH_3Cl_2SiH > & C_2H_5Cl_2SiH > & (C_2H_5)_2ClSiH \\ k_{rel} & 1.64 & 1.36 & 1.00 \end{array} \qquad (391)$$

A kinetic study of the addition of isosteric trisubstituted silanes of general formula $(C_3H_7)_n(C_2H_5O)_{3-n}SiH$ to 1-heptene has enabled steric effects to be eliminated, and produced the following reactivity sequence:

$$\begin{array}{lccc} \text{silane} & (C_2H_5)_2C_3H_7SiH > & (C_2H_5O)_3SiH > & (C_2H_5O)(C_3H_7)_2SiH > \\ k_{rel} & 15.26 & 11.25 & 4.36 \\ & (C_3H_7)_3SiH & & \\ & 1.00 & & \end{array} \qquad (392)$$

The data indicate that the inductive effect of a given substituent is the controlling factor in the process, although no linear correlation exists with Taft's inductive constant ρ^*. Apparently, d_π–p_π interaction between the silicon atom and the oxygen of the alkoxy group can affect the stability of silyl–metal complexes. Addition of (chloro,alkoxy)silanes to $[CH_2{=}CHSi(CH_3)O]_4$ in the presence of H_2PtCl_6 shows that both effects discussed above are present [606], and the reactivity sequence is as follows:

$$CH_3SiHCl_2 > C_2H_5SiHCl_2 > C_2H_5(C_2H_5O)_2SiH > (C_2H_5O)_3SiH \qquad (393)$$

All the various effects mentioned (inductive, steric and mesomeric) lead to the differentiation of tri-substituted silane reactivity in the hydrosilylation of ethylene catalysed by chloroplatinic acid; hence the following complex sequence is obtained [554]:

$$CH_3Cl_2SiH > C_2H_5CH_3ClSiH > C_2H_5Cl_2SiH > C_3H_7Cl_2SiH > CH_3(C_2H_5)_2SiH > SiHCl_3 > (C_2H_5)_3SiH \qquad (394)$$

The position of trichlorosilane in this order is, however, somewhat anomalous. However, it is known that in the H_2PtCl_6 catalysed hydrosilylation of allyldichloromethylsilane introducing the methyl group instead of chlorine one at silicon atom increases reactivity (based on chemical yield) more than twice and the phenyl group causes the opposite effect [1359]:

	ratio of yields, mol/mol
$H(CH_3)SiCl_2/HSiCl_3$	2.89
$HSi(C_6H_5)_3/HSiCl_3$	0.09
$H(CH_3)Si(C_6H_5)_2/HSiCl_3$	0.25
$H(CH_3)_2SiC_6H_5/HSiCl_3$	2.46

Only slight differences are observed in the relative rate constants for the addition of the following aryldiethylsilane to ethylene [214]:

$$\begin{array}{llll} \text{silane} & p\text{-}ClC_6H_4(C_2H_5)_2SiH > & C_6H_5(C_2H_5)_2SiH > & p\text{-}CH_3OC_6H_4(C_2H_5)_2SiH \\ k_{rel} & 1.06 & 1.00 & 0.85 \end{array} \quad (395)$$

This order results both from the inductive effect of the chloro substituent and the predominant mesomeric effect of the p-methoxy substituent. The relative reactivities of methyldioctylsilane, triethylsilane and trioctylsilane toward 1-octene, were determined by a competitive procedure in the presence of chloroplatinic acid [876]. The order of reactivities with platinum catalyst followed the sequence

$$\begin{array}{llll} \text{silane} & CH_3Si(C_8H_{17})_2H > & (C_2H_5)_3SiH \approx & (C_8H_{17})_3SiH \\ k_{rel} & 2.9 & 1.0 & 0.8 \end{array} \quad (396)$$

The inductive effect appears to be rate-limiting in the hydrosilylation of 1-hexene by $RSiH_3$ as illustrated by the following reactivity sequence [614, 764]:

$$\begin{array}{lllll} R & ClC_6H_4 > & C_6H_5 > & c\text{-}C_6H_{11} > & C_4H_9 \\ k_{rel} & 3.78 & 3.22 & 1.79 & 1.00 \end{array} \quad (397)$$

The reactivity of chlorohydrosilanes towards the addition to a C=C bond, e.g. that of 1-octene, decreases markedly as the hydrogen atoms are successively substituted by chlorine [113]:

$$\begin{array}{llll} \text{silane} & ClSiH_3 > & Cl_2SiH_2 > & Cl_3SiH \\ k_{rel} & 100 & 65 & 1 \end{array} \quad (398)$$

However, the competitive reaction of equimolar amounts of trichlorosilane and dichlorosilane with 1-octene leads to relative product (β-adduct) yields of 86% and 3%, respectively; somewhat unexpectedly, this indicates the reverse order for reactivity of chlorosilanes [113]. Studies of substituent effects have also indicated the high hydrosilylation reactivity of silanes containing more than one Si–H bond [Rev.36]. The complex effect of the substituent (involving both inductive and d_π–p_π interactions) is demonstrated by the addition of the following silazanes and sililoamines to olefins [193]:

$$[(CH_3)_2SiH]_2NH > [(CH_3)_2SiH]_3N > (CH_3)_2SiHNR_2 \qquad (399)$$

When the reactivity of cyclic siloxanes, silazanes and silathianes towards the hydrosilylation of 1-methylstyrene is compared, the distinctive influence of heteroatom electronegativity is clearly shown by the following observed order [51]:

$$[CH_3HSiO]_4 > [C_2H_5HSiO]_4 > [CH_3HSiNH]_4 > [C_2H_5HSiNH]_4 > [C_2H_5HSiS]_3 \qquad (400)$$

With linear siloxanes the reactivity towards olefin hydrosilylation decreases significantly as the extent of steric hindrance increases [51]:

$$(CH_3)_2[(CH_3)_3SiO]SiH > CH_3[(CH_3)_3SiO]_2SiH > [(CH_3)_3SiO]_3SiH \qquad (401)$$

The presence of the following electron-withdrawing chloro-substituents at the silicon atom also leads to an increase in the extent of tri-substituted silane addition to allyl maleate and allyl formate, as illustrated by the following observed orders [1008]:

silane	$SiHCl_3 >$	$CH_3Cl_2SiiH >$	$(C_2H_5)_3SiH$	(402)
yield (%)	60	53	42	

silane	$SiHCl_3 >$	$CH_3Cl_2SiH >$	$(CH_3)_2ClSiH >$	$(C_2H_5O)_3SiH >$	$(C_2H_5)_3SiH$	(403)
yield (%)	74	68	52	49	48	

In the second series listed, the inductive effect is accompanied by d_{Π}–p_{Π} conjugation of the alkoxy groups.

A comparison of the reactivity of hydroorganosilanes, hydroaryloxysilanes, hydroorganoxysilazanes and silylamines during the hydrosilylation of 1,3,5-trimethyl- and 1,3,5-trivinylcyclotrisilazane, and of 1,3,5,7-tetramethyl and 1,3,5,7-tetravinylcyclotetrasiloxane leads to the following identical reactivity sequence [41]:

$$[(CH_3)_2HSi]_3N > [(CH_3)_2HSiO]_2C_6H_4 \approx [(CH_3)_2HSiOC_6H_4]_2C(CH_3)_2 > (C_6H_5SiHCH_3)_2NH > C_6H_5SiH_2CH_3 > (CH_3)_2HSiN[Si(CH_3)_3]_2 > (C_6H_5)_2SiH_2 \qquad (404)$$

However, the kinetic course of the different reactions varies for different silanes. An experimental order for silane reactivity during hydrosilylation of phenylacetylene catalysed by chloroplatinic acid has been reported [172], i.e. Scheme (405).

During the hydrosilylation of diphenylacetylene [837], the reactivity of the silane decreases as the size of the alkyl substituent increases, but increases as successively more electronegative aromatic substituents are introduced into the

$$\begin{array}{llll} \text{silane} & (CH_3)_2(CH_3C_6H_4)SiH < & CH_3(CH_3O)_2SiH < & (CH_3)_2ClSiH < \\ \text{yield (\%)} & 15 & 40 & 50 \\ & CH_3SiHCl_2 & & \\ & 90 & & \end{array} \quad (405)$$

molecule. According to Tsipis' studies, the effect of substituents at the silicon atom is virtually insignificant in reactions of substituted octylamines with various hydrosilanes catalysed by a phosphine platinum complex, whereas with phosphine ligands their basicity turns out to be decisive [1205, 1206].

However, in most hydrosilylation reactions catalysed by chloroplatinic acid it is the inductive effect of the substituent at silicon which exercises predominant kinetic control of the reaction; this mainly depends on the stability of the Si–Pt bond in the system, although steric effects and, occasionally, d_{Π}–p_{Π} conjugation also influence the rate and yield of the reactions. As a result, when other metal complexes are employed in hydrosilylation their reactivity is mainly determined by the stability of the metal–silyl bond involved, which is also influenced by the substituents attached to the silicon atom.

The order of reactivity observed in the hydrosilylation of 1-hexene by $RhX(PPh_3)_3$, on the basis of the adduct yield, varies in the opposite manner to that observed for H_2PtCl_6-catalysed reactions [476, 981]:

$$\begin{array}{lllll} \text{silane} & & (CH_3)_3SiH > & (C_2H_5O)_3SiH \geqslant & SiHCl_3 \\ \text{yield (\%)} & X = Cl & 82 & 64 & 68 \\ & X = Br & 71 & 52 & 52 \end{array} \quad (406.1)$$

$$\begin{array}{lllll} \text{silane} & & (C_6H_5)_3SiH > & (C_2H_5)_3SiH > & SiHCl_3 \\ \text{yield (\%)} & X = Cl & 100 & 60 & 8 \end{array} \quad (406.2)$$

However, in the hydrosilylation of 1-heptene by silanes with the general formula $(C_2H_5)_{3-n}Cl_nSiH$, in the presence of $RhCl(PPh_3)_3$ as the catalyst, the relative reaction rates follow an inverse sequence:

$$\begin{array}{lllll} \text{silane} & SiHCl_3 > & C_2H_5SiHCl_2 > & (C_2H_5)_2SiHCl > & (C_2H_5)_3SiH \\ k_{rel} & 8\times10^6 & 4\times10^4 & 2\times10^2 & 1 \end{array} \quad (407)$$

The unusual positive effect of chloro-substituents at the silicon atom on the reaction rate may be attributed to the structure of the pentacovalent intermediate in which the Rh–Si bond is stabilized by electron-withdrawing substituents at silicon.

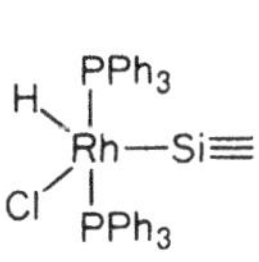

(408)

In the addition of isosteric tri(alkyl,alkoxy)silanes to 1-hexene, in the presence of $RhCl(PPh_3)_3$, the reactivity of the silanes decreases in the following order:

silane	$(C_2H_5O)_3SiH >$	$C_3H_7(C_2H_5O)_2SiH >$	$(C_3H_7)_2(C_2H_5O)SiH >$	$(C_3H_7)_3SiH$
k_{rel}	42.8	28.4	5.4	1.0

(409)

which is similar to that for the same reaction catalysed by $Co_2(CO)_8$:

silane	$(C_2H_5O)_3SiH >$	$C_3H_7(C_2H_5O)_2SiH >$	$(C_3H_7)_2(C_2H_5O)SiH >$	$(C_3H_7)_3SiH$
k_{rel}	119.0	59.5	27.5	1.0

(410)

On the other hand, the yield of 1-octylsilane catalysed by the system $RhCl(ArNC)_2$ at 20°C falls in the order [1]

silane	$HSi(CH_3)_2C_6H_5 >$	$HSi(C_2H_5)_3 >$	$HSi(OC_2H_5)_3$
yield (%)	81	66	40

(411)

Alkylsilanes give higher yields of octylalkylsilanes than alkoxysilanes [1].

Competitive kinetic studies of the addition of linear α,ω-dihydropolymethylsiloxanes of the general formula $(CH_3)_2HSi[OSi(CH_3)_2]_nOSiH(CH_3)_2$ to 1-heptene (in the presence of $RhCl(PPh_3)_3$ as the catalyst) have shown that the relative reaction rates decrease as the value of n increases [1131]:

n	0	1	2	3
k_{rel}	1.80	1.00	0.34	0.26

(412)

However, no correlation exists between the observed reactivity order and the electronic effect of the substituent. Apparently, in this case, the stability of the rhodium–siloxane intermediate is influenced by the size of the siloxane used.

In contrast to the results arising from the hydrosilylation of a C=C bond in the presence of Wilkinson's and related catalysts, the hydrosilylation yield arising from the hydrosilylation of a C≡C bond catalysed by rhodium and platinum complexes increases on replacement of ethoxy substituents by alkyl ones, e.g. in the hydrosilylation of n-butyne [947] with H_2PtCl_6 as catalyst:

silane	$(C_2H_5)_3SiH >$	$CH_3(C_2H_5)_2SiH >$	$(C_2H_5O)_3SiH$
yield (%)	84	72	49

(413)

and with $RhX(PPh)_3)_3$ as the catalyst:

silane		$(C_2H_5)_3SiH >$	$CH_3(C_2H_5)_2SiH >$	$(C_2H_5O)_3SiH$
yield (%)	X = Cl	74	53	57
	X = I	79	62	61

(414)

However, hydrosilylation of acetylene catalysed by $[RhCl(CO)_2]_2$ leads to the formation of the di-adduct, $\equiv SiCH_2CH_2Si\equiv$, accompanied by vinylsilane. The total product yield increased according to the following reactivity sequence of the silanes employed [1257]:

$$(C_2H_5O)_3SiH < SiHCl_3 < C_2H_5SiHCl_2 < CH_3SiHCl_2 < (C_2H_5)_3SiH \quad (415)$$

A comparison of the reactivity of mono-, di- and tri-substituted silanes during their addition to tert-butyl-substituted cyclohexanone catalysed by $RhClPPh_3$ indicates that, irrespective of the position of the tert-butyl group, diphenylsilane has the greatest reactivity (Scheme 416) [519]:

$$(C_6H_5)_2SiH_2 > (C_2H_5)_3SiH > C_6H_5SiH_3 \quad (416)$$

It was found recently that the reactivity of one Si–H bond of 1,2-bis(dimethylsilyl)ethane is somehow greatly enhanced relative to trialkylsilanes in the $RhCl(PPh_3)_3$ catalysed hydrosilylation of ketones [812]. Rate differences between the two Si–H bonds can be interpreted as the enhancement of the reactivity of one Si–H bond terminus of $HSi(CH_3)_2CH_2CH_2Si(CH_3)_2H$. This enhancement was diminished after one Si–H bond was converted.

In the presence of selected alkoxy- and other oxy-substituents at the silicon atom, catalysis by metal complexes of, for example, ruthenium [733, 735] and cobalt [61], enables the hydrosilylation process to be controlled kinetically and thermodynamically. Thus, the hydrosilylation of 1-alkenes catalysed by ruthenium phosphine complexes occurs only when trialkoxysilanes rather than chloro- and alkyl-substituted silanes are used as substrates [733, 735]. Exactly the same situation occurs for cobalt–phosphine complexes [61]. Attempts have been made to explain this unusual effect by advancing differences in the stabilities of the metal–silyl group intermediates and in particular (see Section 2.4.7), by simultaneous dynamic interaction between the d-orbitals of silicon, the d-orbitals of the metal and the p-orbitals of oxygen in the alkoxy group, which leads to the formation of a labile intermediate in one of the hydrosilylation steps [Rev.24].

The most extensive studies of hydrosilylation catalysed by supported complexes have been carried out with those containing $PtCl_6^{-2}$ anions. A preliminary study of the effect of the substituent at the silicon atom on the adduct yield obtained in such reactions with 1-alkenes has led to the following activity order (Scheme 417) [204]:

$$SiHCl_3 > (C_2H_5O)_3SiH > (C_2H_5)_3SiH \quad (417)$$

which is the same as for the reaction catalysed by soluble chloroplatinic acid. A quantitative study on the substituent effect at the silicon atom on the hydrosilylation of 1-hexene by tri-substituted silanes has been undertaken by Reikhsfel'd *et al.*, the use of $R(C_6H_5)CH_3SiH$ as a catalyst enabling the rate constant ω to be correlated precisely with the Taft constant δ^*. This correlation involves the steric factor E_s^*, and the experimental spectroscopic data, ν_{SiH} (IR) and δ_H (^1H NMR), and may be expressed by the equation [1345]:

$$\lg \omega = -732.6 - 8.87\delta^* + 3.2\, E_S^* + 0.34\, \nu + 0.04\delta \qquad (418)$$
$$r = 0.996,\ s_0 = 0.122$$

It has been found that this equation holds well for the eight silanes examined by Reikhsfel'd *et al.*, the d_π–p_π contribution to the interactions being reflected in the spectroscopic parameters. The larger steric effect in this heterogeneously catalysed reaction has been confirmed by the observed sequence of decreasing reaction rate [975]:

$$(CH_3)_2C_6H_5SiH > (C_6H_5)_2CH_3SiH > (1{-}C_{10}H_7)C_6H_5CH_3SiH \qquad (419)$$

In addition the reaction involving phenyldimethylsilane has been found to proceed with the lowest activation energy (18–20 kcal/mole).

An attempt to evaluate the effect of silane structure has been made by studying the hydrosilylation of styrene and phenylacetylene catalysed by chloroplatinic acid supported on phosphinated silica [172]. When rhodium complexes supported on silica, alumina and molecular sieves were used to catalyse the hydrosilylation of 1-heptene, the same reactivity order was observed for the silane as for the homogeneous catalyst [204]. This provides probably the most convincing proof that identical mechanisms are involved when reactions are catalysed by supported or homogeneous metal complexes. Not only do electronic and steric effects change the yield of specific cyclic products in intramolecular hydrosilylations (see Section 4.2) but they also influence the overall yield in the reaction. The yields obtained in the hydrosilylation of pentenyldiorganosilanes, $CH_2{=}CH(CH_2)_3SiHR_2$, when catalysed by H_2PtCl_6 were as follows [1012]:

R	C_6H_5 >	CH_3 ≈	Cl
yield (%)	91	55	54

(420)

Since experimental information on quantitative correlations between the rate/yield of a given hydrosilylation reaction and the structure of the silicon hydride involved are scarce, precise evaluation of the direct influence of substituents attached to the silicon atom on the catalyst centre (particularly for metal complex catalysts) is not possible. Irrespective of the catalyst employed, however, what is observed is the combined effect of inductive, steric and mesomeric influences arising from the substituents. It should be added that kinetic data obtained via competitive reactions of silicon hydrides with unsaturated compounds have often been regarded as suspect, since silanes may react specifically under such conditions [113].

4.2 Influence of the Structure of the Silicon Hydride on Regio- and Stereoselectivity in Hydrosilylation

As mentioned above, the influence that the substituent attached to the silicon atom of the hydrosilane has on the selectivity of a given hydrosilylation process depends strongly on the type of catalyst employed. Thus, the free-radical addition of silicon hydride to a C=C bond generally results in adducts possessing a terminal silyl group (see Section 2.1.2). But even in this case the nature of the silicon hydride can influence the regioselectivity of hydrosilylation. Thus, hydrosilylation of $CF_2{=}CHF$ by trichlorosilane leads selectively to the β-adduct, $CF_2HCHFSiCl_3$, while reaction with trimethylsilane leads to the formation of both the α- and β-adducts in a 1:1 ratio. However, hydrosilylation of $CF_2{=}CFCl$ and $CH_3CH{=}CH_2$ always gives β-adducts, regardless of the hydrosilane used [1360]. In the presence of $Ru_3(CO)_{12}$ as the catalyst, addition of trichlorosilane to $CF_3CH{=}CH_2$ yields the β-adduct in most cases, whereas the hydrosilylation of $CF_3CH{=}CH_2$ by methyldichlorosilane gives a mixture of α- and β-adducts [1033, 1314].

Hydrosilylation of the C=C bond in the presence of hexachloroplatinic acid generally leads to a product containing a terminal silyl group, although when styrene and its derivatives are involved in the hydrosilylation α- and β-adducts are formed. The overall yield of these products decreases in the sequence [172]:

$$CH_3Cl_2SiH > (CH_3)_2CH_3C_6H_4SiH > (CH_3)_2ClSiH \approx (C_2H_5)_3SiH \qquad (421)$$

with the yield of the β-adduct diminishing in the following sequence:

$$(CH_3)_2ClSiH > CH_3Cl_2SiH > Cl_3SiH \qquad (422)$$

Furyl- and naphthyl-substituted silanes are usually more reactive than chloro(organo) silanes in the hydrosilylation of styrene [257], with arylfluorosilanes forming mostly α-adducts. In the addition of tri(phenyl,methyl)silanes to p-nitrostyrene, the yield increases in the order:

$$C_6H_5(CH_3)_2SiH < (C_6H_5)_2CH_3SiH < (C_6H_5)_3SiH \qquad (423)$$

while the yield of the by-product, $p{-}NH_2C_6H_4CH{=}CH_2$, decreases in the same order [48].

Hydrosilylation is often accompanied by the reduction of unsaturated compounds, e.g. allyl compounds by hydrosilanes. Hence an extensive side-reduction has been observed in the reaction of allyl chloride, particularly with phenylmethyl-substituted silanes catalysed by chloroplatinic acid. For $X(CH_3)C_6H_5SiH$, the ratio of the addition product (β-adduct) formed to the reduction product (propene) varies as follows [98, 99]:

X	Cl	$ClCH_2$	C_6H_5	ClC_6H_4
[β-adduct] / [propene]	30/26	16/61	23/58	8/73

(424)

Since this ratio is equal to 0.07 in the case of triethylsilane, it seems probable that the inductive effect of the substituents at the silicon atom and their steric influence largely control the yields of the two products. A study of the effect of $d_\pi{-}p_\pi$ interaction of a furan or thiophene ring bound to the silicon atom on the reactivity of the Si–H bond during the hydrosilylation of acetylene derivatives (alcohols, ethers, amines, esters) has been reported by Lukevics [692, 695]. The reactions were undertaken in the presence of chloroplatinic acid and various other metal complexes to yield *trans*-β and α-isomers.

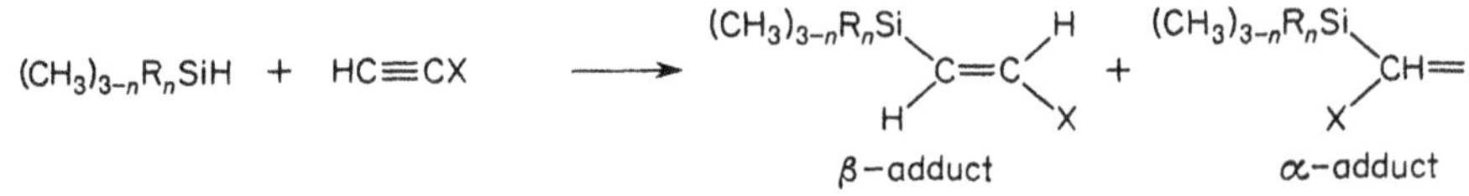

$$(CH_3)_{3-n}R_nSiH \ + \ HC{\equiv}CX \ \longrightarrow \ \underset{\beta-\text{adduct}}{\underset{H}{(CH_3)_{3-n}R_nSi}} C{=}C \underset{X}{\overset{H}{\diagup}} \ + \ \underset{\alpha-\text{adduct}}{\underset{X}{(CH_3)_{3-n}R_nSi}} CH{=}CH_2 \qquad (425)$$

where R is a furyl or thienyl group. It was found that the isomer ratio obtained depends on the structure of the acetylene compound used, but also to a smaller degree on the substituent at the silicon atom. No simple relationship exists between the structure of the tri-substituted silanes (electronic and/or steric properties) and the ratio of α- and β-isomers obtained in their reaction with 1-hexyne [266, 1252]:

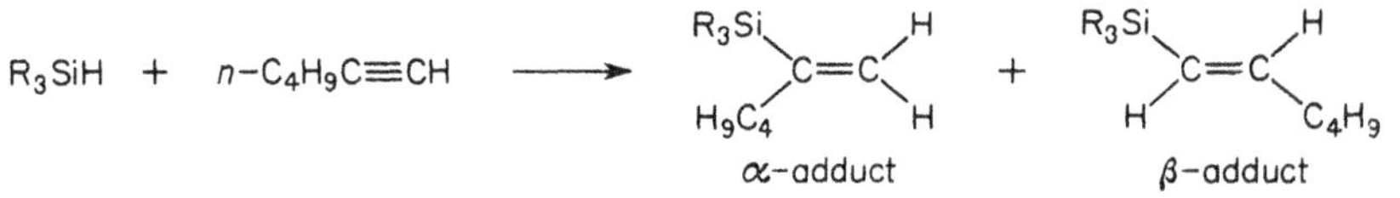

$$R_3SiH \ + \ n{-}C_4H_9C{\equiv}CH \ \longrightarrow \ \underset{\alpha-\text{adduct}}{\underset{H_9C_4}{R_3Si}} C{=}C\underset{H}{\overset{H}{\diagup}} \ + \ \underset{\beta-\text{adduct}}{\underset{H}{R_3Si}} C{=}C\underset{C_4H_9}{\overset{H}{\diagup}} \qquad (426)$$

The regioselectivity of the reaction increases from 70% to 92% (of the β-adduct) in the following sequence:

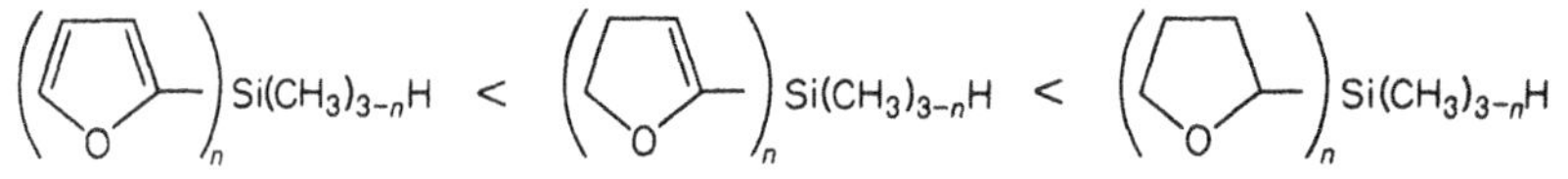

$$\left(\!\!\left(\underset{O}{\diagdown}\!\!\diagup\right)\!\!\right)_{\!\!n}\!\!Si(CH_3)_{3-n}H \ < \ \left(\!\!\left(\underset{O}{\diagdown}\!\!\diagup\right)\!\!\right)_{\!\!n}\!\!Si(CH_3)_{3-n}H \ < \ \left(\!\!\left(\underset{O}{\diagdown}\!\!\diagup\right)\!\!\right)_{\!\!n}\!\!Si(CH_3)_{3-n}H \qquad (427)$$

and with an increase in the number (n) of heteryl (furyl, thienyl) groups in the silane. The latter effect may be accounted for by steric hindrance during the formation of the α-isomer.

The hydrosilylation of styrene by trialkoxysilanes proceeds in the presence of Wilkinson's catalyst, $RhCl(PPh_3)_3$, in benzene solution to yield α- and β-adducts, as well as the products of dehydrogenative silylation (substitution) ($C_6H_5CH{=}CHSi{\equiv}$) or other reactions [630]. Branching of the alkyl group bound to the silicon atom has a substantial influence on the yield and composition of the organosilicon products obtained. Thus, branching of the alkyl groups attached to the α-carbon atom relative to silicon, e.g. $(CH_3)(iso{-}C_3H_7)_2SiH$, leads predominantly to dehydrogenative (substitution) products (D) (α:β:D = 1.7:27.8:50.5). On the other hand, if the branched alkyl group is attached to the β-carbon atom, e.g. $CH_3(iso{-}C_4H_9)_2SiH$, the β-adduct/D ratio remains similar (69.1:18.4) as for the reaction of trialkylsilanes containing unbranched alkyl groups, e.g. $CH_3(C_2H_5)_2SiH$ (56.9:14.1). Triisopropylsilane reacts with styrene only in the absence of the solvent, but no reaction occurs between tri-tert-butylsilane and with styrene in the absence of solvent, the presence of a high catalyst concentration, or at an elevated temperature (100°C). Under the latter conditions no α-adduct was observed amongst the products derived from all the triorganosilanes studied, and no β-adduct was formed with $CH_3(iso\text{-}C_3H_7)_2SiH$ and $(iso\text{-}C_3H_7)_3SiH$. The formation of the substitution product is mainly accompanied by the evolution of hydrogen.

The general decrease in the yield of the α-adduct during hydrosilylation of styrene by tri-substituted silanes follows the following sequence:

$$CH_3(C_2H_5)_2SiH > CH_3(iso\text{-}C_4H_9)_2SiH > CH_3(iso\text{-}C_3H_7)_2SiH \quad (428)$$

enabling the mechanism of the reaction to be explained on the basis of the scheme of Chalk and Harrod. These results also indicate that the Π–σ equilibrium depicted below shifts to the right with increasing bulkiness of the substituents attached to the silicon atom [630]:

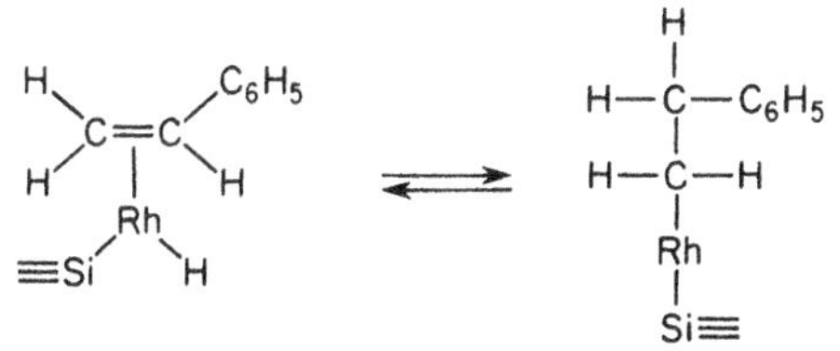

(429)

The substitution product (D) is formed by β-elimination of hydrogen from the α-complex, and is accompanied by β-adduct formation. Mechanistic considerations of this reaction have been given in Chapter 2 (Scheme 81).

Much attention has been devoted to the study of the hydrosilylation of 1,3-butadiene. Several products may be obtained from this reaction, depending on the catalyst employed. Thus, in the reaction of butadiene with trimethylsilane catalysed by Wilkinson's complex, 1:1 silylbutenes and 1:2 silyloctadienes are obtained as adducts (four products in all, amounting to 80% of the total yield), whereas in the hydrosilylation by triethylsilane and trichlorosilane only one adduct (*cis*-(1-silyl)-2-butene) is selectively obtained, but at a much reduced yield. The sequence of the silane reactivity observed is as follows:

$$(CH_3)_3SiH > (C_2H_5O)_3SiH > Cl_3SiH \quad (430)$$

However, in contrast to the nickel- and palladium catalysed hydrosilylation of 1,3-butadiene, the reaction of trimethyl and triethoxysilane does not lead to coupled (1:2) addition products, e.g. silyl-substituted octadienes [982]. When this reaction is catalysed by $Ni(cod)_2$, the yield of 1-silyl-2-butene (the 1:1 adduct), which is produced as the major product, is of a different order of magnitude [205].

$$\underset{93\%}{(C_2H_5O)_3SiH} > \underset{93\%}{CH_3Cl_2SiH} > \underset{76\%}{Cl_3SiH} > \underset{75\%}{(C_2H_5O)_2C_3H_7SiH} >$$
$$\underset{67\%}{(C_2H_5O)(C_3H_7)_2SiH} > \underset{65\%}{(C_2H_5)_3SiH} > \underset{52\%}{(C_3H_7)_3SiH} > \underset{51\%}{(CH_3)_3SiH} \quad (431)$$

In spite of the apparent similarity with Pd complexes [644], no simple correlation has been found between the substituent effect and the product distribution. Addition of triethylsilane to 1,3-butadiene gives *cis*-(2-silyl)-2-butene (52% yield) and 1-silyl-2,6-octadiene (24% yield). In contrast, hydrosilylation by trichlorosilane results in the selective formation of the 1:1 adduct (76%). To a rough approximation, the distribution of the two products appears to be controlled more by the electronegativity of the substituents at the silicon atom than by their steric influence [205]. As mentioned above (Section 2.4.4), the presence of a rhodium catalyst always leads to a *cis* product which then isomerizes to the *trans* product. The reaction of alkynes with triethylsilane catalysed by an Rh complex confirms the slight influence of the silyl group on the stereochemistry [160]. It also shows that no simple relationship exists between the stereoselectivity and the reactivity observed in the addition of tri-substituted alkoxysilanes to 1-butadiene, although the reaction occurs stereospecifically when H_2PtCl_6 is used as the catalyst. Wilkinson's complex, $RhCl(PPh_3)_3$, also catalyses the addition of mono-, di- and tri-substituted silanes to the C=C bond of cyclohexenone, leading to the different products, however, according to the following scheme:

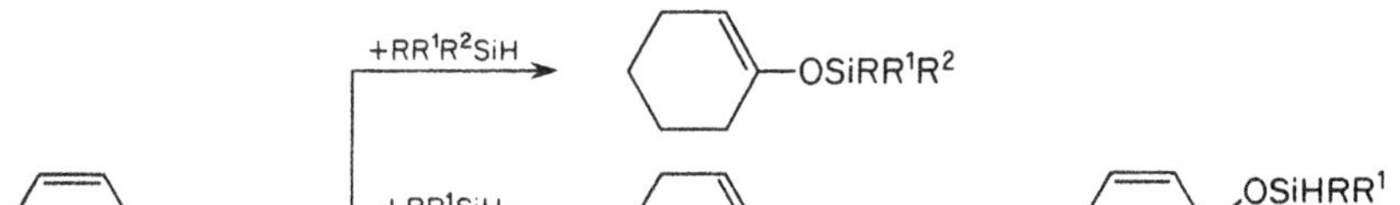

(432)

In the presence of Rh*(diop), the prochiral ketone $C_6H_5(CH_3)CO$ readily undergoes addition to di-substituted silanes with a high (90–95%) yield, regardless of the structure of the silicon hydride. However, the value of an enantiomeric excess, e.e., of the adduct, depends strongly on the substituent at the silicon atom [478]:

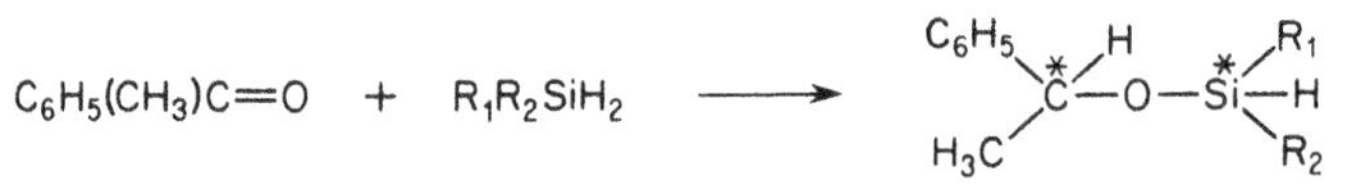

(433)

	e.e. [%]
$R_1 = R_2 = C_6H_5$	13
$R_1 = CH_3$, $R_2 = C_6H_5$	27
$R_1 = C_6H_5$, $R_2 = \alpha$-Np	58

The alkoxydiorganosilane obtained exhibits asymmetry on the carbon atom as well as on the silicon. Furthermore, as stated earlier (Section 2.4.4), the optical yield is differentiated by the steric hindrance of the substituents on the ketones and by the chiral phosphine, as well as by the bulkiness of the siloxy group [865]. When the first two ligands are relatively small, the third assumes control over the reaction. The regio- and stereoselectivity of the hydrosilylation of unsaturated compounds catalysed by catalysts other than platinum and rhodium is only marginally dependent on the substituents at the silicon atom. The distribution of α- and β-adducts in reactions occurring in the presence of nickel complexes depends more on the phosphine ligands than on the structure of the silicon hydride, although H/Cl exchange at silicon is a characteristic side-reaction which accompanies hydrosilylation catalysed by nickel complexes (see Section 2.4.8).

Most hydrosilylation reactions of the C=C bond catalysed by palladium phosphine complexes lead to β-adducts. Yet the reactions of trichlorosilane with alkenes involving electron-withdrawing functional groups, e.g. vinyltrichlorosilane [747], acrylonitrile and styrene [1208], lead to the production of α-adducts (see Section 2.4.8). In contrast to trichlorosilane, the reactions of tri(methyl,chloro)silane with vinyltri(chloro,methyl)silanes always give adducts with terminal silyl groups [740, 749]. The nature of the H–Pd–Si bond in palladium–silyl complexes involved in the catalytic cycle of the discussed hydrosilylation process, which depends on other substituents at silicon and on the ligands at the palladium atom, seems to be crucial in determining this unusual regioselectivity of the reaction examined [738].

Studies of intramolecular hydrosilylation indicate that various isomeric products can also be formed. In the reaction of $CH_2=CH(CH_2)_3SiHR_2$ catalysed by chloroplatinic acid, two cyclic adducts can be produced, i.e. the five-membered $(CH_2)_3CH(CH_3)SiR_2$ and the six-membered $(CH_2)_5SiR_2$, in ratios of 87:4 (for R = C_6H_5), 49:5 (R = CH_3) and 44:1 (R = Cl) [1012]. The mechanism of the reaction may be derived from the general scheme, and involves the simultaneous formation of Si–Pt and Pt–C bonds (Scheme 434).

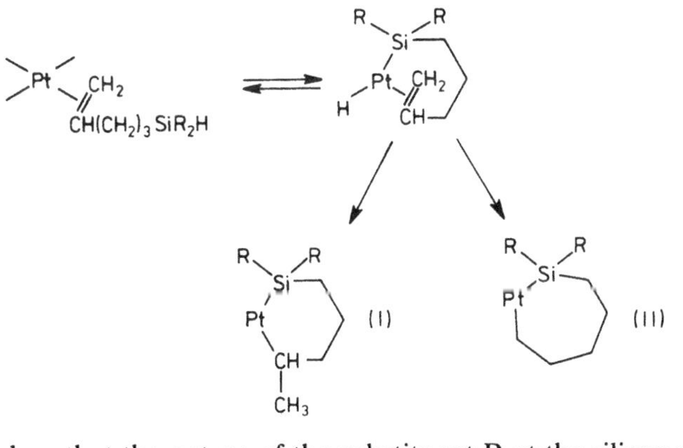

(434)

From this it is clear that the nature of the substituent R at the silicon atom can influence the yield of the various adducts formed. The tendency to form cyclic products from alkenyldimethylsilanes of the general formula $CH_2=CH(CH_2)_nSiH(CH_3)_2$ (n = 0–6) is generated mainly by the length of the alkenyl chain, as illustrated by the following results:

n	0	1	2	3	4	5	6
yield (%)	–	–	46	58	70	16	2

(435)

which indicate that the formation of hexenyldimethylsilane is favoured in intrahydrosilylation catalysed by H_2PtCl_6 [1172]. Moreover, the structure of the unsaturated organosilicon hydride can effect the reaction through promoting cyclization (intramolecular hydrosilylation) and/or polymerization (intermolecular hydrosilylation) [779, 780].

On the other hand, some unsaturated hydrosilanes do not undergo intra- and/or intermolecular hydrosilylation, but react readily with other unsaturated compound to yield product(s) with multiple bonds (see Sections 5.2 and 5.5), e.g. vinyldimethylsilane reacts with 1-hexene, vinyltrimethylsilane [272] or $HC{\equiv}CC(CH_3){=}CH_2$ [464] yielding various products which incorporate C=C bonds. Presumably, in this case coordination of the unsaturated compounds mentioned is preferred to that of vinyldimethylsilane, preventing intermolecular hydrosilylation of the latter.

In conclusion, the nature of the silicon hydrides employed has a slight influence on the regio- and stereoselectivity of their addition to the multiple bonds of organic compounds, although it only affects the final product composition because of the involvement of side-reactions, e.g. redistribution of the silane H/Cl exchange at the silicon atom [301], dehydrogenative silylation [Rev.38], etc.

4.3 Practical Choice of Silicon Hydride in the Hydrosilylation Reaction

The actual choice of silicon hydride employed in a given reaction almost always depends on the synthetic purpose of the reaction, i.e. the product desired, although as we have seen, mechanistic considerations associated with a reaction also often dictate the use of various types of silane substrates.

Organochlorosilanes are important compounds which provide precursors for other organosilicon compounds, i.e. monomers and polymers. The latter which contain siloxane chains are almost invariably prepared by the hydrolysis of organochlorosilanes and the condensation of the resulting organohydroxysilanes [Rev.39]. Diorganodichlorosilanes are essential monomers for the production of linear polyorganosiloxanes, and have been synthesized by the reaction of dichloroorgano (mainly methyl, ethyl, phenyl)silanes with a carbon–carbon unsaturated bond. Similarly, organotrichlorosilane and triorganochlorosilane may be obtained by the respective reactions of trichlorosilane and diorganochlorosilane with the C=C and C≡C bonds in organic compounds.

The replacement of the chloro-substituents on the alkoxy (mainly ethoxy) group enables the corresponding alkoxy-substituted silanes to be produced by hydrosilylation: these include diethoxy(diorgano)silanes, organotriethoxysilanes and triorganoethoxysilanes, all of which can be used as monomers. The last group of silanes, however, constitute a separate class of organosilicon compounds of general formula $R'Si(OR)_3$, which are well known as silane coupling agents (see Section 6.1). They are synthesized either by the reaction of trichlorosilane with a given alkenyl(alkynyl)compound followed by alcoholysis (predominantly ethanolysis and methanolysis of the Si–Cl bonds), or by the direct addition of $(RO)_3SiH$ to the unsaturated compounds.

Thus, although several hundred substituted silanes can be involved in hydrosilylation processes, most of the reports in fact refer to very limited groups of silanes. The first group consists of trichlorosilane, methyldichlorosilane and other organochlorosilanes; the second involves triethoxysilane, methyldiethoxysilane and other (organo)alkoxy-substituted silanes; whereas the third group of compounds is derived from triorganosilanes, mainly triethylsilane. Furthermore, $SiHCl_3$, $HSi(OC_2H_5)_3$ and $HSi(C_2H_5)_3$ may be regarded as typical representatives of the three classes of silicon hydrides in all dynamic studies, particularly when the effect of the substituent at silicon is under consideration. Whilst chloro- and ethoxy(alkoxy)silanes are used for the hydrosilylation of mainly C=C and C≡C bonds, triorganosilanes (as well as diorganosilanes) are commonly applied in the hydrosilylation of the C=O bond to yield the corresponding alkoxysilanes. Following hydrolysis these are useful in the synthesis of special organic alcohols (and also in asymmetric syntheses).

As we have seen, the available literature data reflect the industrial and research importance of hydrosilylation by silicon hydrides, particularly by substituted silanes. The selection of silicon hydrides for the hydrosilylation of polymers or alternatively polymeric organosilicon compounds containing Si–H bonds is a reflection of the synthetic requirements necessary for the production of a given polymer. This is discussed in detail in Sections 5.3–5.5 and 6.3, where consideration is given to the synthesis and modification of organic and organosilicon oligomers and polymers by hydrosilylation.

CHAPTER 5

The Hydrosilylation of Unsaturated Organosilicon Compounds

In comparison with the hydrosilylation of unsaturated organic compounds, reactions with unsaturated organosilicon substrates have received little attention. However, several applications of such processes have recently been reported, and they have now become one of the most frequently studied processes involving the addition of Si–H in organosilicon chemistry.

Despite its undoubted synthetic utility, the hydrosilylation of unsaturated silanes (mainly vinyl-substituted but to a lesser extent allyl-substituted silanes) may be regarded as a simplified model of the activated curing of polydimethylsiloxane chain containing unsaturated groups by polyfunctional silicon hydrides. Increasing the bulkiness of the substituents at the silicon atom in both the substrates involved in the hydrosilylation process, and its effect on the reactions, enables an approach to be made to the actual mechanism of the cure of organosilicon polymers by hydrosilylation. Hydrosilylation of the C≡C bond under mild conditions provides an efficient method for synthesizing silyl derivatives of ethene and other alkenes. In contrast, the polyaddition of difunctional organosilicon hydrides to difunctional unsaturated organosilicon compounds, and the intermolecular polyaddition of hydrosilanes and siloxanes containing unsaturated groups, leads to the synthesis of carbosiloxane oligomers and polymers.

5.1 Addition of Substituted Hydrosilanes to Carbon–Carbon Double Bonds in Unsaturated Silanes

The reaction of hydrosilanes (hydrosiloxanes) with alkenylsilanes (alkenylsiloxanes) usually takes place according to the following scheme:

$$\equiv SiH + CH_2{=}CH(CH_2)_nSi\equiv \longrightarrow \begin{cases} \equiv SiCH(CH_3)(CH_2)_nSi\equiv & \alpha\text{-adduct} \\ \equiv SiCH_2CH_2(CH_2)_nSi\equiv & \beta\text{-adduct} \end{cases} \quad (436)$$

where n is mainly equal to 0 or 1. The regioselectivity of the reaction arises from the substituent effect at the silicon atom in the alkenyl- and hydrosilanes (and siloxanes), as well as on the catalyst used.

Although vinyl-substituted organic compounds are generally less susceptible to hydrosilylation than allyl derivatives, a vinyl group adjacent to the silicon atom is more reactive than the allyl group in allylsilanes. The electron-releasing character of the silicon atom relative to the carbon atom, coupled to electron withdrawal by the silyl group attached to the vinyl substituent as a result of (p–d)Π back-donation, leads to the formation of α- or β-adducts when vinylsilanes undergo hydrosilylation.

When chloro(methyl,phenyl)-substituted silanes are hydrosilylated, the β-adduct is mainly formed if the reaction is catalysed by chloroplatinic acid in isopropanol (Speier's catalyst) [37, 1058, 1229, 1279], cyclohexanone [747] or $CH_3OCH_2CH_2OCH_3$ [466], Wilkinson's complex [600], or heterogenized Pt complexes [D-056]. In the presence of Speier's catalyst the yield of α-adducts increases as the number of electronegative groups ($-Cl$, $-CH{=}CH_2$) at the silicon atom increases, and also as the number of methyl groups at the silicon atom in the hydrosilane increases. Hydrosilylation of vinyltrimethylsilane leads to the β-adduct (with the amounts of α-adducts formed not exceeding 5–10% for all the hydrosilanes studied), whereas in the reaction of vinyltrichlorosilane (with trimethylsilane) and tetravinylsilane (with methyldichlorosilane), yields of β-adducts as high as 40% and 45%, respectively, have been observed. However, in the hydrosilylation of tetravinylsilane the selectivity of the reaction is drastically reduced when the number of methyl groups attached to the silicon atom of the hydrosilane is increased [1058]. In general, hydrosilylation of vinyltriorganosilanes by tri(organo,chloro)silanes always yields β-adducts in the presence of chloroplatinic acid [Rev.25, 552, 879, 921, 924, D-142].

The influence of the organic substituent attached to the silicon atom of the hydrosilane on the extent of reaction with vinyltriorganosilanes can be illustrated by the following examples:

$$\begin{aligned}&CH_3(C_5H_{11})_2SiH + CH_2{=}CHSi(CH_3)(C_6H_5)_2 \xrightarrow{96\%}\\&CH_3(C_6H_5)_2SiH + CH_2{=}CHSi(CH_3)(C_5H_{11})_2 \xrightarrow{60\%}\\&\rightarrow (C_5H_{11})_2(CH_3)Si(CH_2)_2Si(CH_3)(C_6H_5)_2\end{aligned} \tag{437.1}$$

$$\begin{aligned}&CH_3(C_6H_5CH_2)_2SiH + CH_2{=}CHSi(CH_3)(C_6H_5)_2 \xrightarrow{86\%}\\&CH_3(C_6H_5)_2SiH + CH_2{=}CHSi(CH_3)(CH_2C_6H_5)_2 \xrightarrow{38\%}\\&\rightarrow CH_3(C_6H_5CH_2)_2Si(CH_2)_2Si(CH_3)(C_6H_5)_2\end{aligned} \tag{437.2}$$

In all cases the reactivity of the substituents at the silicon atom of the hydrosilanes follows the sequence [921]:

$$C_5H_{11} > C_6H_5CH_2 > C_6H_5 \tag{437.3}$$

In addition, vinylsilanes appear to be more reactive than vinyl-substituted organic compounds [922].

The relative reactivity of the vinyl group attached to silicon has been evaluated by studying addition of methyldichlorosilane to unsaturated organosilicon compounds containing not only the vinyl group but also other unsaturated groups. From such studies the following reactivity sequences for reactions in the presence of chloroplatinic acid [402] have been obtained:

$$HC{\equiv}C- > CH_2{=}CH- > CH_2{=}CHCH_2- > CH_2{=}C(CH_3)CH_2-$$

and

$$CH_2{=}CH- > CH_2{=}CHC_6H_4- \tag{438}$$

In all cases, addition of the hydrosilanes usually occurs across only one of the unsaturated bonds (yield 50–70%) [402]. Exemplary study has shown that activity of Speier's catalyst in the hydrosilylation of vinyl- and butenylsilanes is 3–10 times higher than that of allyl-substituted silanes [1359].

As mentioned above, allylsilanes are less reactive than vinylsilanes. However, attachment of an allyloxy group at the silicon atom appears to be more effective than for the vinyl group [796]:

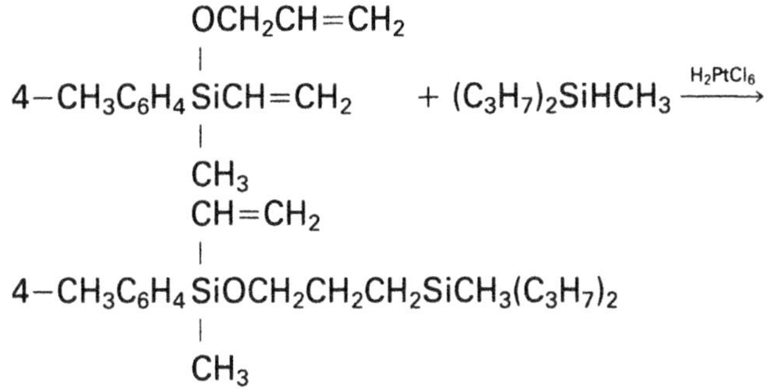

$$
\begin{array}{c}
OCH_2CH{=}CH_2 \\
| \\
4{-}CH_3C_6H_4\,SiCH{=}CH_2 \quad + (C_3H_7)_2SiHCH_3 \xrightarrow{H_2PtCl_6} \\
| \\
CH_3 \\
CH{=}CH_2 \\
| \\
4{-}CH_3C_6H_4\,SiOCH_2CH_2CH_2SiCH_3(C_3H_7)_2 \\
| \\
CH_3
\end{array}
\tag{439}
$$

On the other hand, the vinyloxy group is much less active than the vinyl one [J-103].

Hydrosilylation of β-chlorovinylsilanes occurs via preliminary reduction to vinylsilane followed by their hydrosilylation giving finally α and β bis(silyl)ethanes [1055]:

$$R_3SiCH{=}CHCl + HSiR_3' \rightarrow R_3SiCH_2CH_2SiR_3' + (R_3Si)(R_3'Si)CHCH_3 + R_3'SiCl \tag{440}$$

where $R_3 = (CH_3)_nCl_{3-n}$ ($n = 0$–3), $R_3' = (CH_3)_mCl_{3-m}$ ($m = 0$–3).

Very attractive 1,1,4,4-tetrakis(trimethylsilyl)butatriene readily undergoes hydrosilylation by trimethylsilane in the presence of RhCl(PPh$_3$)$_3$ [634]:

$$
\begin{array}{c}
(CH_3)_3Si \qquad\qquad Si(CH_3)_3 \\
\diagdown C{=}C{=}C{=}C\diagup \qquad + (CH_3)_3SiH \xrightarrow[80°C,\,90\%]{RhCl(PPh_3)_3} \\
(CH_3)_3Si \qquad\qquad Si(CH_3)_3 \\
[(CH_3)_3Si]_2CHC[Si(CH_3)_3]{=}C{=}C[Si(CH_3)_3]_2
\end{array}
$$

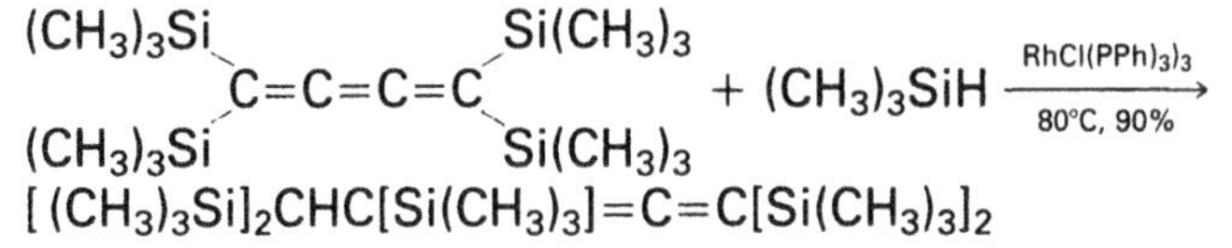

Catalysis of the hydrosilylation of vinylchloromethylsilanes by metal complexes other than H_2PtCl_6 is characterized by the high efficiency of the reaction and its regioselectivity (to give β-adducts), particularly in the presence of $Co_2(CO)_8$, $Rh_4(CO)_{12}$ and $Ir_4(CO)_{12}$ under relatively mild conditions (35–100°C, 2–4 h) [30, 575, U-019], but also in the presence of nickel–vinylsilane [1337] and palladium–phosphine complexes [738, 740, 749]. A series of studies on the hydrosilylation of vinyltri(chloro)methylsilanes by trichloro(methyl)hydrosilanes catalysed by $Pt(PPh_3)_4$ has revealed the following reactivity order for the hydrosilanes:

$$SiHCl_3 > CH_3SiHCl_2 \gg (CH_3)_2SiHCl \quad (442)$$

while the yield in the hydrosilylation of $CH_2{=}CHSi(CH_3)_nCl_{3-n}$ (n = 0–3) by trichlorosilane was found to be constant at 90% of the adduct. However, in the presence of Pd complexes the hydrosilylation of vinyltrichlorosilane by trichlorosilane led to the selective formation of the α-adduct, 1,1-bis(trichlorosilyl)ethane [747]. The mechanism of this reaction and its unusual regioselectivity has been discussed above (Section 2.4) [738, 740, 749].

In the presence of phosphino–nickel complexes a mixture of α- and β-adducts was found in hydrosilylations involving silane [269] and vinyltrichlorosilane [745]. Alkoxy-substituted organosilicon substrates have no effect on the significant regioselectivity of such hydrosilylations.

Vinyltrimethylsilane and vinyltrialkoxysilanes are hydrosilylated by trialkoxysilanes in the presence of chloroplatinic acid [390, 575, 745], Wilkinson's catalyst [745], $Pt(PPh_3)_2(C_2H_4)$ [1300], a Pt catalyst [D-071] and cobalt and rhodium carbonyls [30, 31, 713, 720]. The hydrosilylation of vinyl-substituted silanes by disubstituted silanes and dihydrosiloxanes did not lead to monohydro-functional products. The formation of the two adducts is usually accompanied by the products arising from side-reactions [846, 847].

One of the characteristic features of the hydrosilylation of vinylsilanes is substituent exchange at the silicon atom. The data reported by Speier indicate that vinyl/H exchange in this reaction, which proceeds in the presence of hexachloroplatinic acid

$$R_3SiH + R'_3SiCH{=}CH_2 \rightarrow R_3SiCH{=}CH_2 + R'_3SiH \quad (443)$$

leads to the formation of at least two, and sometimes four, hydrosilylation products [Rev.25, Rev.38, 1056].

The reaction of $(RO)_3SiH$ with $(R'O)_3SiCH{=}CH_2$, where R ≠ R′ (e.g. CH_3, C_2H_5) which occurs as a result of the preliminary exchange of successive alkoxy-substituents at silicon between substrates, followed by their hydrosilylation in the presence of ruthenium complexes, yields at least eight products which are mainly β-adducts. No vinyl/H exchange was observed between the silanes studied, however [745]. Exchanges of the type C_2H_5O/Cl occur readily (Scheme 444), leading to the formation of several hydrosilylation products [1163].

Numerous transition metal complexes form adducts with vinylsilanes. Some of them, such as $Fe(CO)_4[CH_2{=}CHSi(CH_3)_3]$, catalyse the hydrosilylation of vinyltrimethylsilanes by various chloro(methyl)hydrosilanes (70–80°C, 3 h, under increased pressure) and provide α-and β-adducts, and in addition the products of

$$(CH_3)_2ClSiH + CH_2{=}CH\underset{CH_3}{\overset{CH_3}{Si}}OC_2H_5 \rightleftharpoons C_2H_5O\underset{CH_3}{\overset{CH_3}{Si}}H + CH_2{=}CH\underset{CH_3}{\overset{CH_3}{Si}}Cl \qquad (444)$$

dehydrogenative silylation according to the following equation [398, 399, 844]:

$$\equiv Si{-}H + 2\ CH_2{=}CHSi\equiv \rightarrow \equiv SiCH{=}CHSi\equiv + CH_3CH_2Si\equiv \qquad (445)$$

Hydrosilylation of vinyltrimethylsilane by trimethylsilane results in bis-1,2(trimethylsilyl)ethene (60% yield) and ethyltrimethylsilanes (34% yield) but only a 6% yield of α- and β-adducts. Such a pathway was established by Speier as a result of studies on the addition of hydrosiloxanes to vinylsiloxanes in the presence of various catalysts [Rev.38].

An exceptionally strong tendency to yield the unsaturated organosilicon adducts arises in reactions catalysed by complexes of the iron triad; by iron carbonyl–vinylsilane and by hexachloroosmic acid [Rev.38], as well as by ruthenium phoshine and acetonate complexes [733, 745].

The latter group of catalysts has been mainly examined in the hydrosilylation of vinylalkoxysilanes by trialkoxysilanes, the yield of 1,2-bis(triethoxysilyl)ethene increase markedly when the reaction is carried out in the excess of vinylsilane. In the presence of ruthenium(II) and (III), the suggested mechanism for this reaction involves the existence of a vinylsilane–ruthenium intermediate which is stabilized by the presence of electron-withdrawing alkoxy groups at the silicon atom. The simplified scheme for the hydrosilylation and dehydrogenative double hydrosilylation had been written as follows [745]:

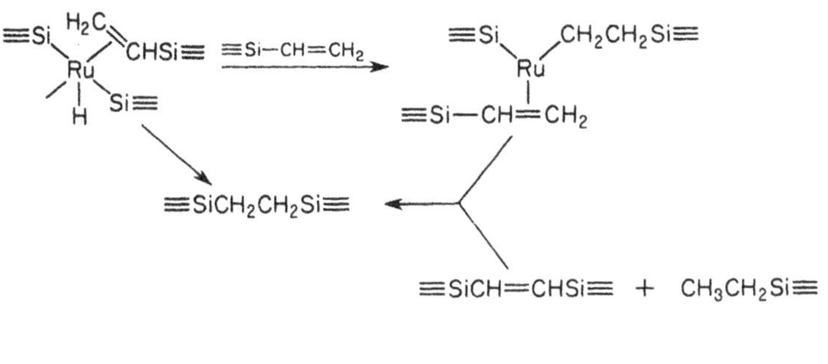

(446)

A more detailed mechanism has been recently reported for the catalysis [733, 740] of this process on the basis of the stoichiometry and kinetics of the reaction, as well as on catalytic studies of the active intermediates. A purely mechanistic discussion has suggested the dimeric and monomeric ruthenium–vinylsilane and ruthenium–carbene complexes as intermediates produced during fast equilibrium steps, in contrast to the situation which arises when the reaction is catalysed by other metal complexes. The presence of ruthenium complexes aids the formation of 1,2-bis(triethoxysilyl)ethene mainly as a result of the metathesis of excess vinyltriethoxysilane, a reaction which occurs competitively with hydrosilylation (Scheme 95). The metal–carbenes which can be responsible for this new reaction may be generated according to the following equation [736, 740]:

$$\begin{array}{ccccc} CH_2{=}CHSi{\equiv} & & CH{=}CHSi{\equiv} & & CH{-}CH_2Si{\equiv} \\ | & \rightarrow & \| & \rightarrow & \| \\ [Ru] & & [Ru{+}H & & [Ru] \end{array} \quad (447.1)$$

and/or

$$\begin{array}{ccccc} CH_2CH_2Si{\equiv} & \rightarrow & CHCH_2Si{\equiv} & \overset{}{\underset{\equiv SiH}{\rightleftharpoons}} & CHCH_2Si{\equiv} \\ | & & \| & & \| \\ {\equiv}Si{+}Ru] & & {\equiv}Si{+}Ru{+}H & & [Ru] \end{array} \quad (447.2)$$

The vinyl and allyl derivatives of α-aminosilanes, silanols and silanothiols constitute a separate class of organosilicon compounds containing unsaturated groups. Hydrosilylation of the C=C bond in hydrocarbons which have their chains bonded to an N, O or S heteroatom allows the synthesis of the respective unsaturated silyl derivatives also containing an alkylsilyl chain:

$$(CH_3)_3SiNHCH_2CH{=}CH_2 + H(CH_3)Si[OSi(CH_3)_3]_2 \xrightarrow[H_2PtCl_6]{reflux} (CH_3)_3SiNHCH_2CH_2CH_2Si(CH_3)[OSi(CH_3)_3]_2 \; (54\%) \quad (448.1) \; [585]$$

$$[(CH_3)_2C{=}NO]_2(CH_3)SiH + CH_2{=}CHCH_2COOSi(CH_3)_3 \xrightarrow[120°C]{H_2PtCl_6} (CH_3)_3SiOOC(CH_2)_3Si(CH_3)[ON{=}C(CH_3)_2]_2 \; 75\% \quad (448.2) \; [356]$$

$$(CH_3)_3SiOCH_2CH{=}CH_2 + (CH_3)_2(C_2H_5O)SiH \xrightarrow[H_2PtCl_6]{60°C,\ 4\ h} (CH_3)_3SiOCH_2CH_2CH_2Si(CH_3)_2OC_2H_5 \; 70\% \quad (448.3) \; [409]$$

$$(CH_3)_3SiCH_2SCH{=}CH_2 + (C_2H_5)_3SiH \xrightarrow{H_2PtCl_6} \underset{(\beta\text{-adduct})}{(CH_3)_3SiCH_2SCH_2CH_2Si(C_2H_5)_3} + \underset{(\alpha\text{-adduct})}{(CH_3)_3SiCH_2SCH(CH_3)Si(C_2H_5)_3} \quad (448.4) \; [1246]$$

Trialkylsilanes are added to 1-vinyl- and 1-allylsilatranes to form β-adducts [1237] of α- and β-adducts in the ratio I/II = 1:6 [1236]:

$$CH_3N(CH_2CH_2O)_2Si(CH_3)CH{=}CH_2 + R_3SiH \xrightarrow[100°C,\ 8\ h,\ 72\%]{Rh(acac)(CO)_2} \underset{(I)}{R_3SiCH_2CH_2(CH_3)Si(OCH_2CH_2)_2NCH_3} + \underset{(II)}{CH_3(R_3Si)CHSi(OCH_2CH_2)_2NCH_3} \quad (449)$$

Hydrosilylation of dimethylvinyldisilazane by triethoxysilane in the presence of platinum complex gives (triethoxysilyl)ethyl(dimethyl)disilazane [E-050]. The reaction of vinyl(methyl-1-pyrenyl)silane with hydrosilanes produces silyl derivatives of dyes [U-184].

Hydrosilylation of bis(silyl)ethenes by tri(methyl,chloro)silanes occurs under elevated temperatures and the yield of trisilylethanes depends strongly on the number of methyl groups at silicon. Trichlorosilane is not added to 1,2-bis(silyl)ethenes [1053].

$$(CH_3)_nCl_{3-n}SiCH{=}CHSi(CH_3)_nCl_{3-n} + HSi(CH_3)_mCl_{3-m} \rightarrow$$

$$(CH_3)_nCl_{3-n}SiCH_2CH\begin{array}{l}\diagup Si(CH_3)_nCl_{3-n}\\ \diagdown Si(CH_3)_mCl_{3-m}\end{array} \qquad (450)$$

A novel class of synthetic lubricants consisting of trisilahydrocarbons of the general structure $R_1R_2Si[\,(CH_2)_nSi(R_3R_4R_5)\,]_2$ (where R_1-R_5 = alkyl groups) can be synthesized by final hydrosilylation of bis(alkenyl)dichlorosilane with trichlorosilane followed by methylation (alkylation) of the product [U-188].

5.2 Addition of Substituted Hydrosilanes and Hydrosiloxanes to Unsaturated Silanes and Siloxanes to Yield Carbosiloxane Derivatives

The considerable number of reports on the hydrosilylation of unsaturated organosilicon compounds involving alkenyl (mainly vinyl)siloxane and/or hydro(cyclo)siloxane suggest that this problem has been studied extensively.

Vinyldisiloxanes and other linear multisiloxanes, as well as vinylcyclosiloxanes, have been most frequently used as unsaturated organosilicon monomers, although the corresponding derivatives containing more than one vinyl group have also been the subject of the synthetic studies. Most of the reactions yield monomeric or oligomeric β-adducts containing $\equiv SiCH_2CH_2SiO-$ or $-OSiCH_2CH_2SiO-$ sequences. Some of these are derived from vinyl- and/or hydro-siloxane substrates containing more than one functional group. Only a few reports have described synthetic methods involving the hydrosilylation of other alkenylsiloxanes or unsaturated organosilicon derivatives. The modification of the siloxane chain, for example in cyclosiloxanes, to yield new carbosiloxane monomers for ionic polymerization or polycondensation seems to be the main aim of the synthetic reactions described below.

The hydrosilylation of vinylheptaorganocyclotetrasiloxane by hydroheptaorganocyclotetrasiloxane provides a spectacular illustration of such a synthetic method which, in the presence of H_2PtCl_6 as the catalyst, proceeds according to the following equation [36]:

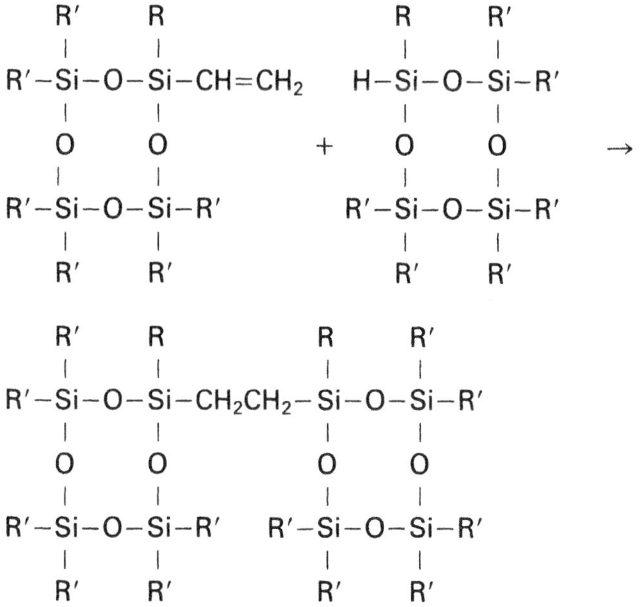

$$
\begin{array}{cc}
\text{R}' & \text{R} \\
| & | \\
\text{R}'-\text{Si}-\text{O}-\text{Si}-\text{CH}=\text{CH}_2 \\
| & | \\
\text{O} & \text{O} \\
| & | \\
\text{R}'-\text{Si}-\text{O}-\text{Si}-\text{R}' \\
| & | \\
\text{R}' & \text{R}'
\end{array}
\quad + \quad
\begin{array}{cc}
\text{R} & \text{R}' \\
| & | \\
\text{H}-\text{Si}-\text{O}-\text{Si}-\text{R}' \\
| & | \\
\text{O} & \text{O} \\
| & | \\
\text{R}'-\text{Si}-\text{O}-\text{Si}-\text{R}' \\
| & | \\
\text{R}' & \text{R}'
\end{array}
\quad \rightarrow
$$

(451)

$$
\begin{array}{cccc}
\text{R}' & \text{R} & & \text{R} & \text{R}' \\
| & | & & | & | \\
\text{R}'-\text{Si}-\text{O}-\text{Si}-\text{CH}_2\text{CH}_2- & \text{Si}-\text{O}-\text{Si}-\text{R}' \\
| & | & & | & | \\
\text{O} & \text{O} & & \text{O} & \text{O} \\
| & | & & | & | \\
\text{R}'-\text{Si}-\text{O}-\text{Si}-\text{R}' & \text{R}'-\text{Si}-\text{O}-\text{Si}-\text{R}' \\
| & | & & | & | \\
\text{R}' & \text{R}' & & \text{R}' & \text{R}'
\end{array}
$$

where R = CH_3, C_2H_5, R' = CH_3.

A similar modification occurs during the hydrosilylation of vinylcyclosiloxane by hydroorganosilane and/or the hydrosilylation of vinylorganosilane by hydrocyclosiloxane, e.g.; using tetrasiloxane substrates in the presence of $Rh_4(CO)_{12}$ or $Ir_4(CO)_{12}$ as catalysts:

$$
\begin{array}{cc}
\text{CH}_3 & \text{CH}_3 \\
| & | \\
\text{CH}_3-\text{Si}-\text{O}-\text{Si}-\text{CH}=\text{CH}_2 \\
| & | \\
\text{O} & \text{O} \\
| & | \\
\text{CH}_3-\text{Si}-\text{O}-\text{Si}-\text{CH}_3 \\
| & | \\
\text{CH}_3 & \text{CH}_3
\end{array}
\quad + \text{HSi}\equiv \xrightarrow{\text{[S-027, S-936]}}
$$

$$
\begin{array}{cc}
\text{CH}_3 & \text{CH}_3 \\
| & | \\
\text{CH}_3-\text{Si}-\text{O}-\text{Si}-\text{H} \\
| & | \\
\text{O} & \text{O} \\
| & | \\
\text{CH}_3-\text{Si}-\text{O}-\text{Si}-\text{CH}_3 \\
| & | \\
\text{CH}_3 & \text{CH}_3
\end{array}
\quad + \text{CH}_2=\text{CHSi}\equiv \xrightarrow{\text{[S-028]}}
$$

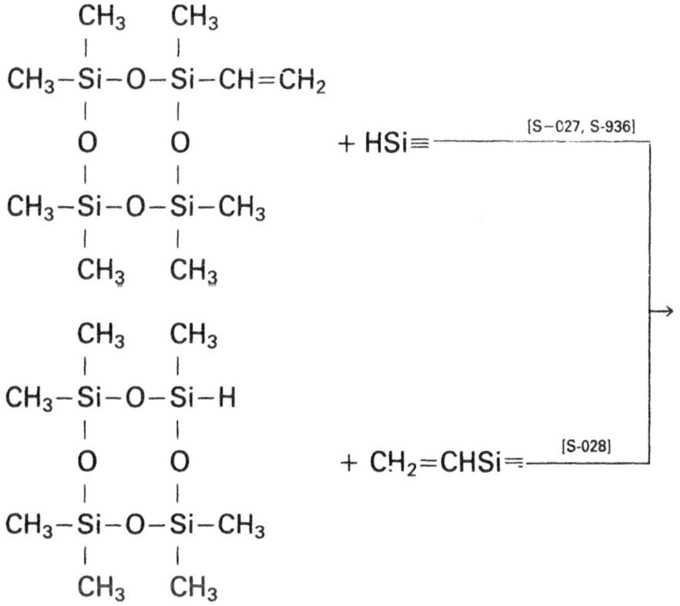

{452}

cont.

$$CH_3-\underset{\underset{\underset{\underset{CH_3}{|}}{\overset{|}{Si}-CH_3}}{\overset{|}{O}}}{\overset{\overset{CH_3}{|}}{Si}}-O-\underset{\underset{\underset{\underset{CH_3}{|}}{\overset{|}{Si}-CH_3}}{\overset{|}{O}}}{\overset{\overset{CH_3}{|}}{Si}}-CH_2CH_2Si\equiv$$

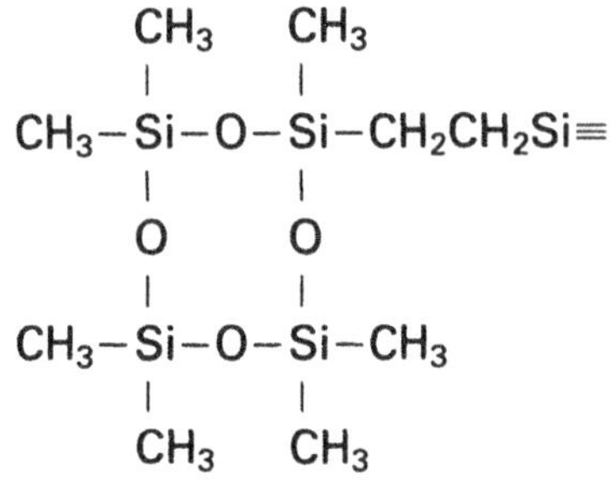

Linear hydrosiloxanes and hydrosilanes can also be readily added to vinylsilanes and vinylsiloxanes. Both methods provide characteristic derivatives (e.g. Scheme 453) in the case of use H_2PtCl_6 as the catalyst:

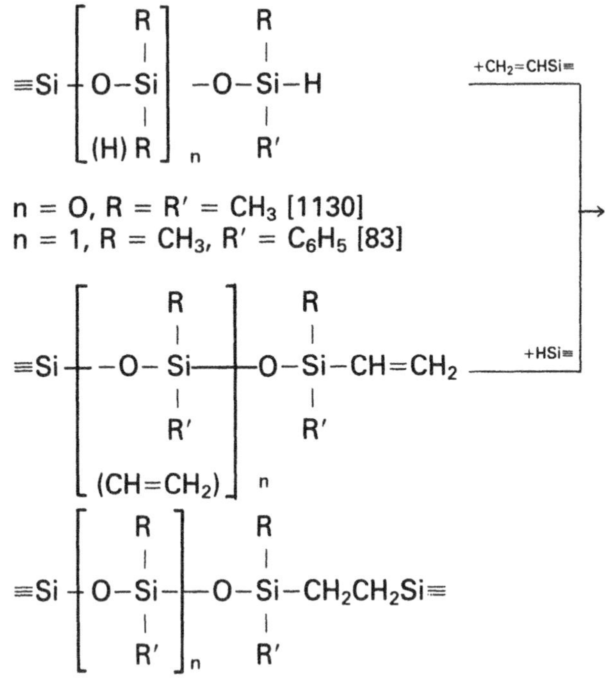

$$\equiv Si \left[O-\underset{\underset{(H)R}{|}}{\overset{\overset{R}{|}}{Si}} \right]_n -O-\underset{\underset{R'}{|}}{\overset{\overset{R}{|}}{Si}}-H \xrightarrow{+CH_2=CHSi\equiv} \qquad (453)$$

$$n = 0,\ R = R' = CH_3\ [1130]$$
$$n = 1,\ R = CH_3,\ R' = C_6H_5\ [83]$$

$$\equiv Si \left[-O-\underset{\underset{R'}{|}}{\overset{\overset{R}{|}}{Si}} \atop (CH=CH_2) \right]_n -O-\underset{\underset{R'}{|}}{\overset{\overset{R}{|}}{Si}}-CH=CH_2 \xrightarrow{+HSi\equiv}$$

$$\equiv Si \left[O-\underset{\underset{R'}{|}}{\overset{\overset{R}{|}}{Si}} \right]_n -O-\underset{\underset{R'}{|}}{\overset{\overset{R}{|}}{Si}}-CH_2CH_2Si\equiv$$

$n = 0,\ R = R' = CH_3\ [606,\ D-059,\ G-026,\ G-027,\ G-031]$
$n = 1,\ R = R' = CH_3\ [G-031,\ U-027]$
$n = 20,\ R = R' = CH_3\ [606].$

Isomeric linear vinylsiloxanes can also be used as the initial compounds [1130]:

$$\begin{array}{c}(CH_3)_3Si-O\\(CH_3)_3Si-O-SiCH=CH_2\\(CH_3)_3Si-O\end{array} + HSi\equiv \xrightarrow{H_2PtCl_6} \begin{array}{c}(CH_3)_3Si-O\\(CH_3)_3Si-O-Si-CH_2CH_2Si\equiv\\(CH_3)_3Si-O\end{array} \qquad (454)$$

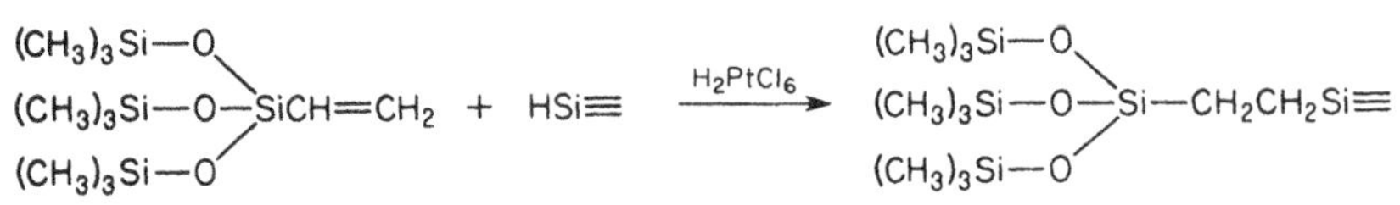

Instead vinylsilanes and alkenylsilanes, also alkenyloxysilanes [161] or alkenyl-oxysiloxanes [J-087, J-088], can be used.

Miscellaneous cyclo-linear siloxanes may be formed as a result of the following reactions (e.g. using cyclotetrasiloxane as the substrate and H_2PtCl_6 as the catalyst):

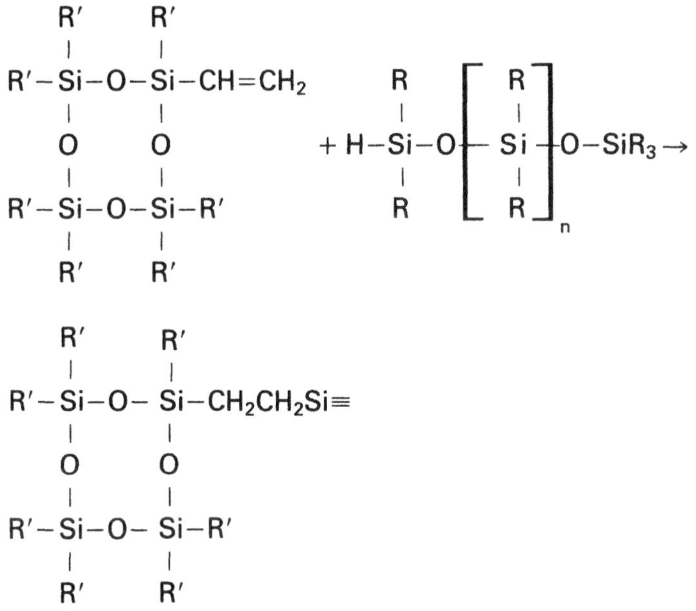

$$
\begin{array}{cc}
R' & R' \\
| & | \\
R'-Si-O-Si-CH{=}CH_2 \\
| & | \\
O & O \\
| & | \\
R'-Si-O-Si-R' \\
| & | \\
R' & R'
\end{array}
\quad + \;
H-Si-O\!\!\left[\begin{array}{c} R \\ | \\ Si \\ | \\ R \end{array}\right]_n\!\!O-SiR_3 \;\rightarrow
$$

$$
\begin{array}{cc}
R' & R' \\
| & | \\
R'-Si-O- & Si-CH_2CH_2Si{\equiv} \\
| & | \\
O & O \\
| & | \\
R'-Si-O- & Si-R' \\
| & | \\
R' & R'
\end{array}
\tag{455}
$$

$n = 3$, R = R′ = CH_3 [29]; $n = 2$, R = R′ = CH_3 [29]; $n = 1$, R = R′ = CH_3 [51, 1355].

The hydrosilylation reactions of divinylsiloxanes by monohydrosilanes, siloxanes and dihydro(cyclo)siloxanes lead to a similar type of adduct:

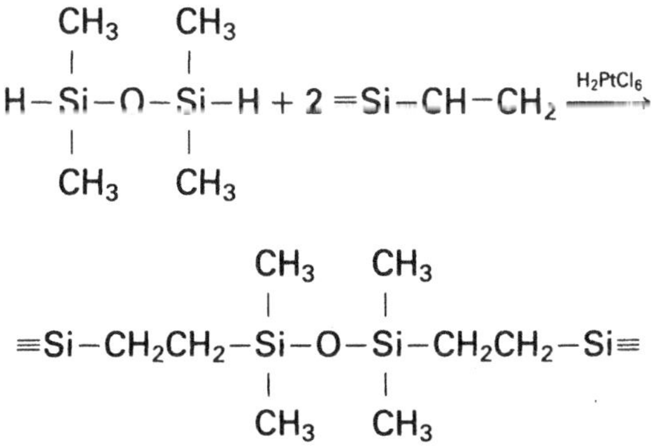

$$
\begin{array}{cc}
CH_3 & CH_3 \\
| & | \\
H-Si-O-Si-H & + \; 2 \; {=}Si-CH-CH_2 \xrightarrow{H_2PtCl_6} \\
| & | \\
CH_3 & CH_3
\end{array}
$$

$$
\begin{array}{cc}
 & CH_3 \quad CH_3 \\
 & | \qquad | \\
{\equiv}Si-CH_2CH_2-Si-O-Si-CH_2CH_2-Si{\equiv} \\
 & | \qquad | \\
 & CH_3 \quad CH_3
\end{array}
\tag*{(456.1) [390]}
$$

Hydrosilylation of vinyl-substituted (cyclo)siloxanes containing more than two vinyl groups per molecule always results in the production of cured polymers with various structures.

$$
\begin{array}{cc}
\quad|\quad\quad| & \quad\quad|\quad\quad| \\
-Si-O-Si-CH=CH_2 & -Si-O-Si-CH_2-CH_2-Si\equiv \\
\quad|\quad\quad| & \quad\quad|\quad\quad| \\
\;O\quad\;\;O \quad +\;2\,HSi\equiv\;\xrightarrow{H_2PtCl_6}\; & \;O\quad\;\;O \\
\quad|\quad\quad| & \quad\quad|\quad\quad| \\
-Si-O-Si-CH=CH_2 & -Si-O-Si-CH_2-CH_2-Si\equiv \\
\quad|\quad\quad| & \quad\quad|\quad\quad|
\end{array}
$$

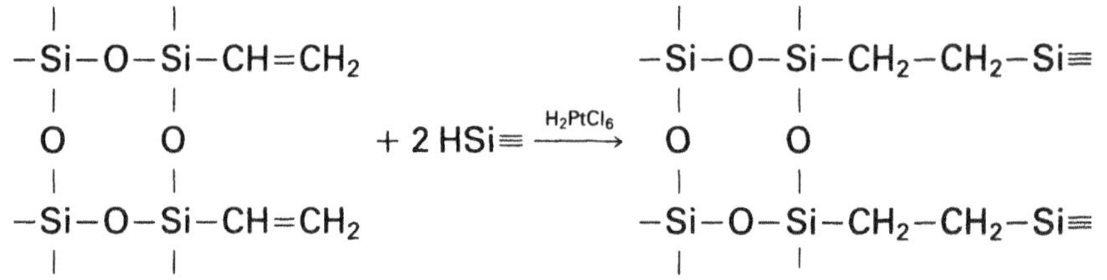

(456.2)

[50, 51, 1130, C-009]

5.3 Curing Organosilicon Polymers by Hydrosilylation

The addition of polyfunctional silicon hydrides to poly(vinyl)organosiloxane provides an activated cure for silicone rubber. Hydrolytic condensation of miscellaneous monomers under suitable conditions gives a method of synthesizing linear polyorganosiloxanes containing unsaturated (mainly vinyl) and hydrogen groups bonded to silicon:

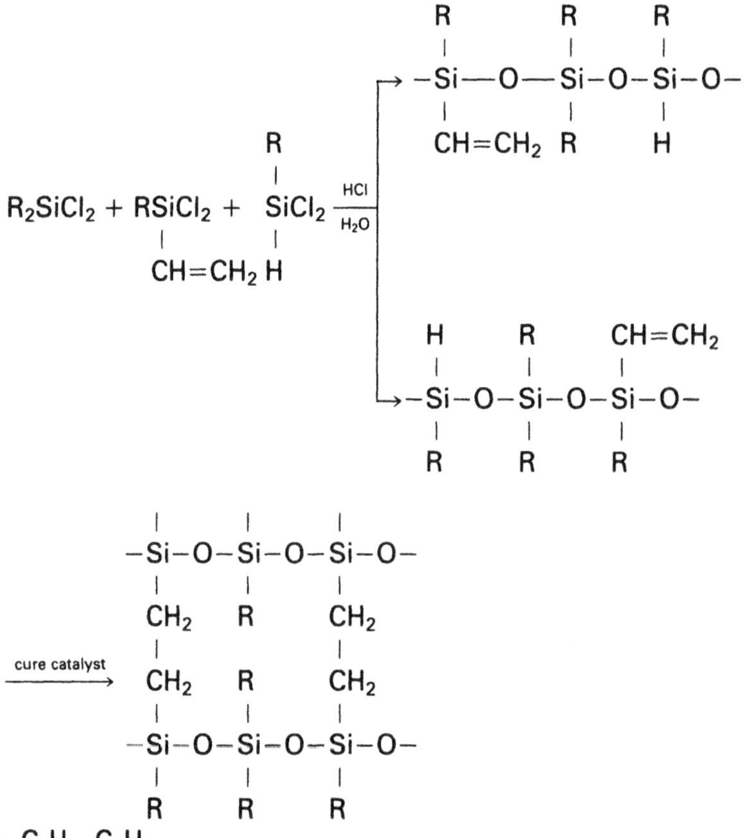

(457)

R = CH_3, C_2H_5, C_6H_5

Such polysiloxanes may be cured via the polyaddition of Si–H bonds to the vinyl (unsaturated) groups, a process which results in the crosslinking of the silicone polymer (Scheme 457). The commonest activated cure method is based on the hydrosilylation reactions of polydimethylsiloxanes, or generally polydiorgano-siloxanes containing small numbers of vinyl groups, e.g. polydimethylsiloxane whose polymer chain is end-blocked at one end by vinyldiorganosiloxy group and the other through the Si–H bond a polyalkylhydrosiloxane [1116, D-050].

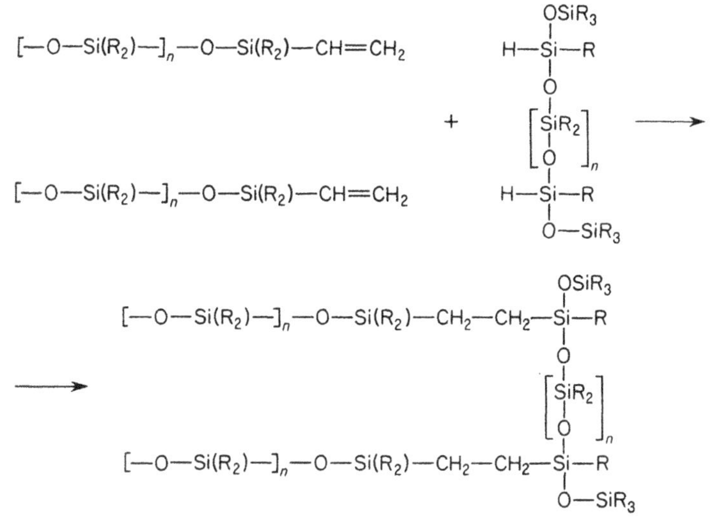

$$[-O-Si(R_2)-]_n-O-Si(R_2)-CH=CH_2 \qquad H-\underset{\displaystyle O}{\overset{\displaystyle OSiR_3}{Si}}-R$$

$$+ \quad \begin{bmatrix} SiR_2 \\ O \end{bmatrix}_n \longrightarrow$$

$$[-O-Si(R_2)-]_n-O-Si(R_2)-CH=CH_2 \qquad H-\underset{\displaystyle O-SiR_3}{Si}-R$$

$$\longrightarrow \qquad [-O-Si(R_2)-]_n-O-Si(R_2)-CH_2-CH_2-\underset{\displaystyle O}{\overset{\displaystyle OSiR_3}{Si}}-R$$

$$\begin{bmatrix} SiR_2 \\ O \end{bmatrix}_n$$

$$[-O-Si(R_2)-]_n-O-Si(R_2)-CH_2-CH_2-\underset{\displaystyle O-SiR_3}{Si}-R \qquad (458)$$

The viscosity of the polydimethylsiloxane employed is generally in the range 1000–100,000 cS at room temperature.

Since this reaction results in a high conversion and is relatively free from side-reactions, this cure method can be used as a model application of network topology theories. A silicone elastomer obtained by such a curing method is characterized by an exceptionally high tensile strength, toughness and low porosity [1287]. It may be employed in a composite consisting of a cured silicone elastomer

containing at least one hydrolyzable unit, e.g. $(RO)_3Si(CH_2)_x(CH_3)SiO-$ [U-062]

or $(RO)_3Si(CH_2)_3COOCH(CH_3)CH_2CH_2SiO-$ and a reinforcing silica filler [671]

which provides adhesion without the need to apply a primer at the interface between the silicone elastomer and a metal substrate.

Numerous papers have appeared containing mechanistic considerations based mainly on kinetic measurements, but also taking into account the physico-mechanical properties of silicone elastomers [24, 25, 222, 491, 723, 928, 1226, 1316, 1317]. Hexachloroplatinic acid and other platinum complexes are mainly used as soluble catalysts for the addition cure of such polymers [724], although H_2PtCl_6 does not catalyse the polymerization of vinylsiloxanes at 70–140°C even after 70 h [1354]. Catalytic systems used in the vulcanization of siloxane elastomers have recently been discussed in detail [Rev.15]. The most active catalyst used recently for vulcanization of silicon rubber based on polysiloxanes containing vinyl and Si–H groups is the platinum–alkenylsiloxanes complex, mainly the platinum–vinylsiloxane complex [e.g. U-064] (Karstedt catalyst). One important approach to the activated cure of silicone rubber makes use of various inhibitors which are added to the platinum catalyst to reduce, or temporarily inhibit, its catalytic activity in the presence of the alkenyl- and hydro-polysiloxanes. Generally, such inhibitors stop the vulcanization process but allow the platinum catalysts to be reactivated at elevated temperatures or by radiation ways. The inhibitors act as ligands which block the hydrosilylation reaction by coordination to the metal centre but, under the conditions employed for cure, release the active catalyst. The catalyst is usually added to the reaction mixture in quantities related to the number of unsaturated (e.g. vinyl) substituents in the polysiloxane. The temporary deactivation of the catalyst can be regulated not only by its concentration but also through the type and concentration of the inhibitor.

The following principal types of compounds have been reported as platinum catalyst inhibitors during the cure of silicone rubber through addition processes:

1. Alkenyl derivatives, e.g. $CH_2{=}CHCH_2OCOCH$ [D-112], $ClCH_2CH{=}CHCH_3$ [E-002] and allyl esters [D-112], esters of various unsaturated 1,4-dicarboxylic acids [U-148], allene derivatives [F-073].
2. Alkynyl derivatives [G-010], e.g. $ClCH_2C{\equiv}CH$ [E-002], $HC{\equiv}C(CH_2)_nOH$ [U-044, U-088, U-130, U-196], $HC{\equiv}CCH(CH_3)OH$ [F-029, U-175, J-108], phenylacetylene [U-093], 2-butyne [U-034, U-093], $C_6H_5Si(OCH_2C{\equiv}CH)_3$ [U-034] and $ROOCC{\equiv}CCOOR$ [D-144, U-133].
3. Cyclic crown ethers [920, S-032].
4. Organic nitrogen compounds, e.g. nitrile [U-024], heterocyclic nitrogen compounds (pyrazine) [U-012, U-085] and hydrazine derivatives [U-095].
5. Organophosphorous compounds, e.g. triphenylphosphine and triphenylphosphite [G-005, U-013].
6. Organic sulphur compounds, e.g. sulphoxides and organosilicon sulphoxides [U-036], thioacetamine [G-005] and phenothiazine [U-085].
7. Peroxides, e.g. tetra-butylhydroperoxide [N-006, U-103] and methylethylketoneperoxide [N-006].
8. Salts of Sn(II), Hg(II), Cu(II), Cu(I) and Bi(III) [D-020, G-014].
9. Linear and cyclic vinylsiloxanes [D-063, D-101,D-133, E-033, F-030, U-029, U-089, U-094, U-185].
10. Poly(vinyl)siloxanes [E-030].

Instead of the catalyst inhibitors the method of microcapsulated platinum group catalyst was effectively used. The catalyst is encapsulated within one or two layers of thermoplastic organic polymers. The microcapsules can be incorporated into storage-stable one-part polysiloxane compositions that cure at the given temperature [E-047, U-186].

Other platinum complexes also exhibit a high activity during the curing of polysiloxanes containing vinyl groups at elevated temperatures, e.g. $Pt(PPh_3)_2Cl_2$ [G-005], $PtCl_2(C_2H_4)_2$ [F-020], $Pt(Et_2S)_2Cl_2$ [1226, G-037] and $[(PhO)_3P]_4Pt$ [J-031, U-117], $[Pt(CH_3)_3]_4$ [U-149], $Pt[HC{\equiv}CC(CH_3)_2OH]_2$ [E-014], Pt(vinylcyclohexene) complex [F-036], (diolefin)Pt(α-aryl)$_2$ [E-006], Pt(phosphine)$_n$(vinylsiloxane)$_n$ (n = 1, 2) [E-017], $CpPtR_3$ [U-149], $(NR_4)_2[PtCl_n]$, where n = 4 or 6 [426]. Dicobalt octacarbonyl, rhodium phosphine [491, 492, U-127] and other rhodium(I) and rhodium(III) complexes [491, J-59, J-080, J-081] as well as palladium complexes [J-059, J-080, J-081] are also effective catalysts for the addition of polyhydroorganosiloxanes to unsaturated groups in polydimethylsiloxanes.

The addition of hydrosiloxanes as a crosslinking process in combination with curing can also be effected by accessible radiation methods, e.g. UV and electron beam methods [523, 1287, E-010, E-057] or microwaves [U-099] or actinic radiation [E-006, U-149]. Such methods of curing are fast, taking only a few seconds [1287].

The replacement of methyl groups by ethyl or phenyl in α,ω-divinylpolydimethylsiloxane and α,ω-dihydropolysiloxane leads to a marked decrease in the rate of vulcanization by hydrosilylation [24]. Phenyl substituents, not only in polysiloxanes but also in polycarboorganosilicon chains of the following general formula, also lead to a substantial decrease in the extent of the polyaddition reaction rate in comparison with a methyl group [918]:

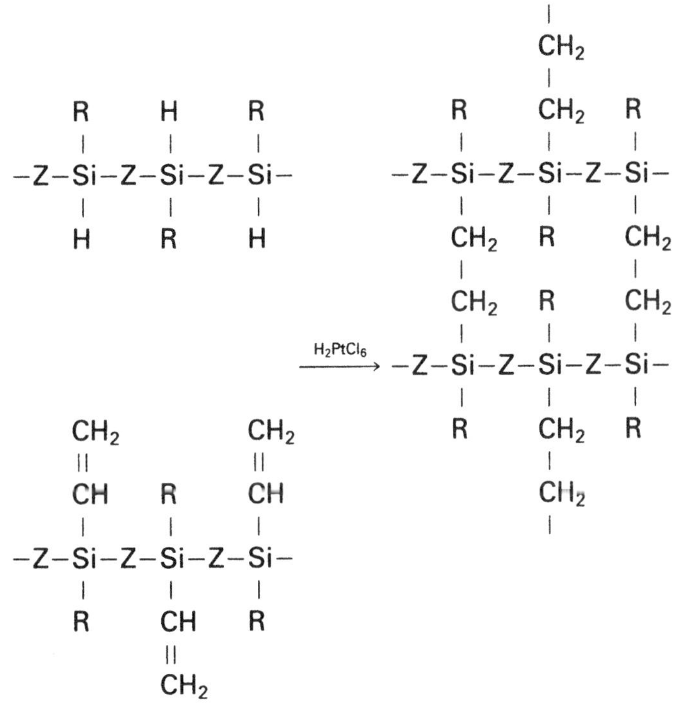

(459)

R = CH_3, C_6H_5;

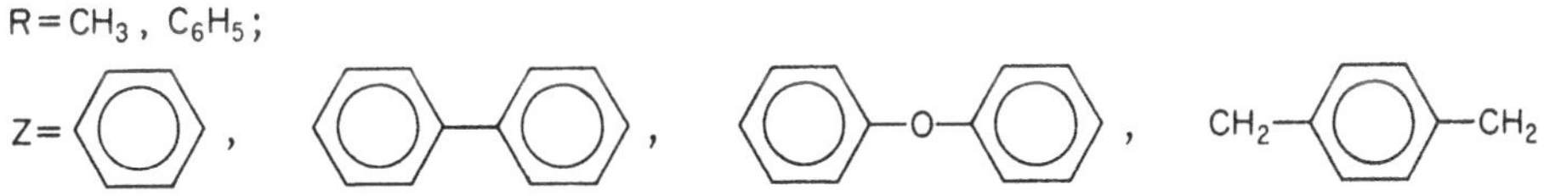

Various polyhydrosiloxanes and cyclosiloxanes have been employed as crosslinking agents in the vulcanization of vinyl-terminated polydimethylsiloxanes, e.g. $H(CH_3)_2Si(OSi(CH_3))_{2n}OSi(CH_3)_2H$ [U-137], tri- and tetra-functional hydrosiloxanes [1225, 1226], $[HSi(CH_3)_2O]_4Si$ [404, 405], $[HSi(CH_3)_2O]_3SiC_6H_5$, $[HSi(CH_3)_2]_2O$ and $(CH_3)_3SiO(CH_3HSiO)_n[(CH_3)_2SiO]_mSi(CH_3)_3$.

Hydrosilylation crosslinking of silicone rubber has been increasingly used in rubber manufacture since it has many technological and economic advantages. The composition and proportion of both polysiloxane components, as well as the catalyst used, are deciding factors of the quality of the vulcanizates and particularly allow to reach materials of the special properties and of the special applications. Among the recent patents some describe procedures for preparation of silicon rubber used for dental purposes [E-025], reducing and controlling density of silicone foam [U-167, U-195, U-197], improving adhesion to metal, glass and plastics [677, 1199, E-058, E-059, J-060, J-066, U-169, U-175] and also the release of adhesive materials therefrom [E-024, E-056, J-071, J-096, J-108], as well as being used as discoloration-resistant compounds [J-075, J-086], gas-separating siloxane membranes [J-063, J-089], electric insulators [J-058], gasoline and oil-resistant fluorinated silicone elastomer [E-039], and those exhibiting generally better physical properties [888, J-107, U-185, U-198] and also preventing premature deterioration due to exposure to chemical reagents [U-171]. Silicone polymers useful as optical materials are also synthesized via hydrosilylation [E-020, E-021, J-077, J-097]. Crosslinked polyorganosiloxane networks can be prepared by dispersing vinylsiloxane and silicon hydride components in a liquid monomer, crosslinking the siloxane components by hydrosilylation and polymerizing the organic monomers [E-055].

5.4 Addition of Substituted Hydrosilanes to Carbon–Carbon Triple Bonds in Unsaturated Silanes

In contrast to the hydrosilylation of alkenylsilanes the corresponding process with alkynylsilanes has rarely been studied, and even then has been predominantly directed towards the synthesis of disilylethene derivatives. The reaction of trisubstituted silanes with ethynyl trisubstituted silanes occurs according to the following equation:

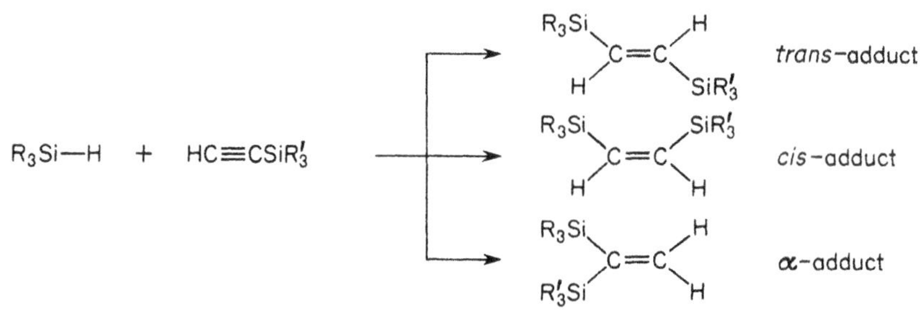

$$R_3Si\text{—}H \ + \ HC{\equiv}CSiR'_3 \ \longrightarrow$$

$$\begin{array}{ll}
\underset{H}{\overset{R_3Si}{>}}C{=}C\underset{SiR'_3}{\overset{H}{<}} & trans\text{–adduct} \\[6pt]
\underset{H}{\overset{R_3Si}{>}}C{=}C\underset{H}{\overset{SiR'_3}{<}} & cis\text{–adduct} \\[6pt]
\underset{R'_3Si}{\overset{R_3Si}{>}}C{=}C\underset{H}{\overset{H}{<}} & \alpha\text{–adduct}
\end{array} \qquad (460)$$

yielding 1,2-bis(silyl)ethenes (*trans* and *cis* adducts) and 1,1-bis(silyl)ethenes in most reactions with alkyl-, alkoxy- and chloro-substituents at the silicon atom; however, the *trans* adduct is the only product:

$R_3 = (C_2H_5)_2(CH_3O)$	$R'_3 = Cl_2C_2H_5$	[1278]
$(CH_3)_2CH_3O$	$CH_3(C_6H_5)Cl$	[1261]
Cl_3	$(CH_3O)_2CH_3$	[1281]
$Cl(CH_3)_2$	$(CH_3)_3$	[134]
Cl_3	$(CH_3O)_3$	[1278]
Cl_2CH_3	$(CH_3)_2CH_3O$	[1281]
$(CH_3)_2Cl$	CH_3Cl_2	[759]
Cl_3	$ClCH_2(CH_3)_2$	[1240]

Most reactions proceed in the presence of chloroplatinic acid as the catalyst, although $Pt(PPh_3)_4$ and $RhCl(PPh_3)_3$ have also been employed [759]. The hydrosilylation of ethynylmethyldichlorosilane by dimethylchlorosilane catalysed by chloroplatinic acid gives the *trans*-β-adduct (94%), while in the presence of $Pt(PPh_3)_4$ the α-adduct ((II), 90% yield) is formed and with Wilkinson's catalyst a mixture of both adducts (I) + (II) (85% yield) is obtained.

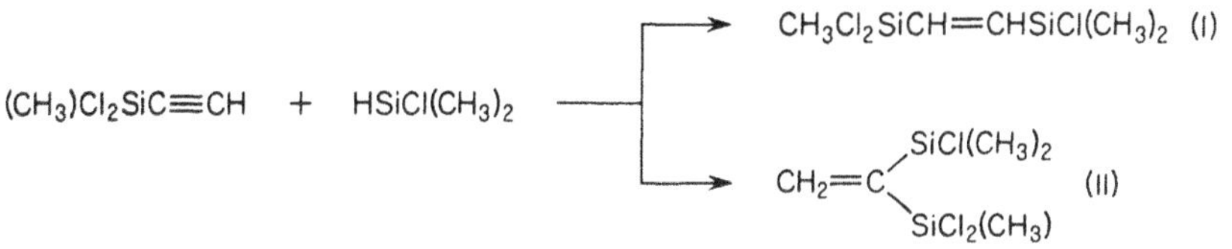

(461)

An exception to this rule is the formation of *cis* adducts established in the following reactions by ^1H NMR and IR spectroscopic methods:

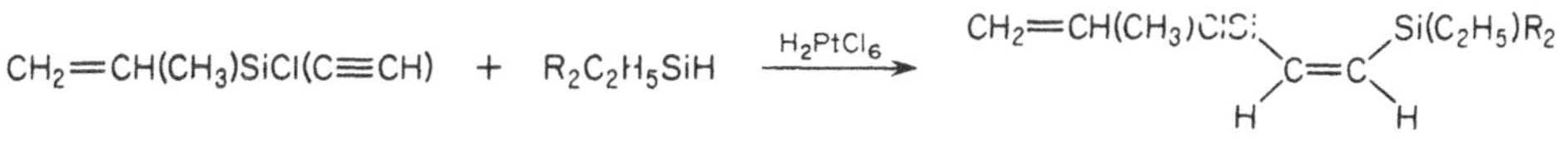

(462)

R = C_2H_5, Cl.

In the hydrosilylation of ethynyl tri(methyl,alkoxy)silane, 1,1-bis(silyl)ethene was also formed in addition to the *trans* adduct [1309]:

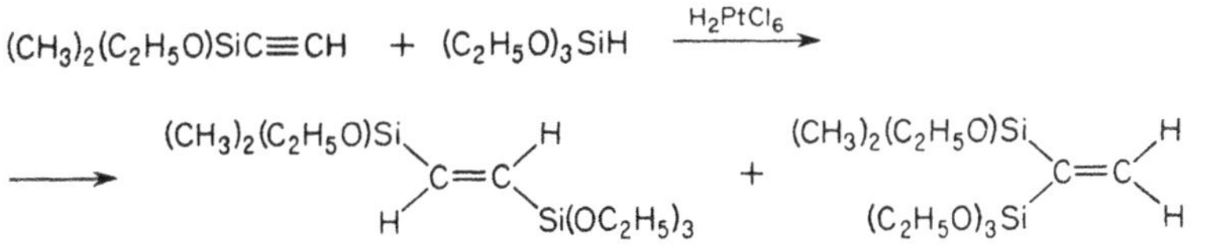

(463)

The reaction of the silyl derivatives of ferrocenylacetylene with triorganosilanes (Scheme 464) leads to the *trans* adduct and 1,1-bis(silyl)ethene as the two products formed [62]. With trialkyl substituents attached to the silicon atom of the ferrocenylacetylene derivative more of the trans adduct is formed, whereas the triphenyl substituents in the same position more of the 1,1-bis(silyl)ethene is produced. The processes are accompanied by side-reactions in which the ≡Si–C≡

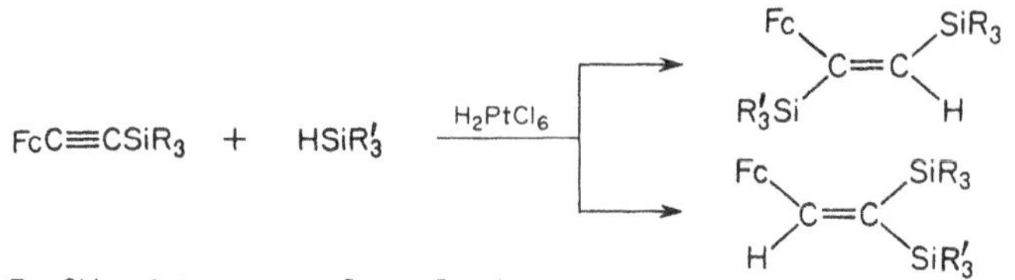

(464)

$R=CH_3$, C_2H_5, C_6H_5 [54, 55]; $R'=C_2H_5$

bond is cleaved giving, for example, $FeC(=CH_2)Si(C_2H_5)_3$ among the reaction products [62].

The hydrosilylation of silylacetylenes [436, 461, 462, 989, 1140, 1312] and particularly silyldiacetylenes [340, 459, 461, 462, 989, 1100, 1182, 1261, 1312] has been employed in the synthesis of more complex monomeric as well as polymeric products [462] (Scheme 465):

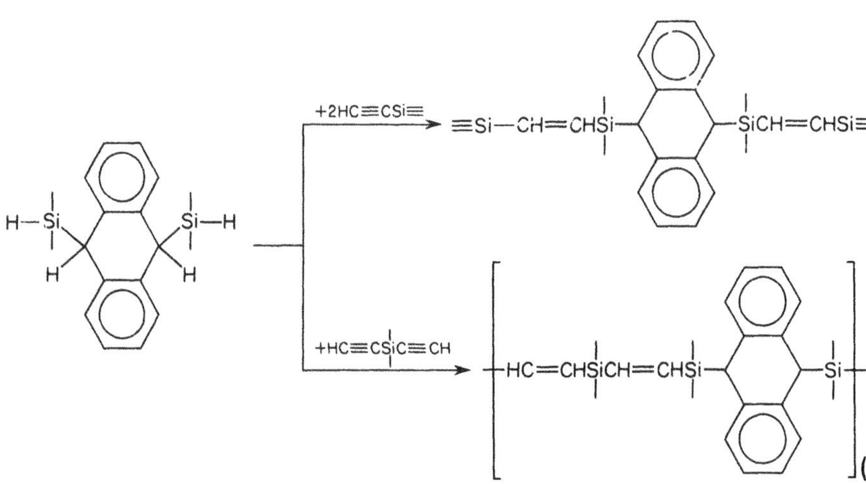

(465)

The ratio of products (I)/(II) depends strongly on the solvent used. 1,3,5-trisilacyclohexenes containing a larger number of Si–H groups tend to polymerize [389] (Scheme 466).

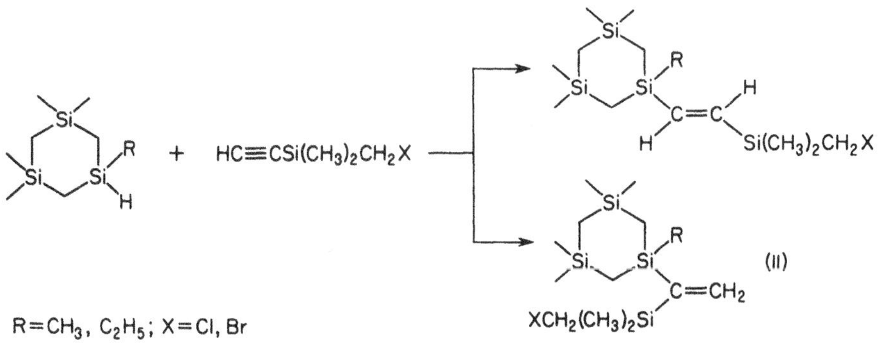

$R=CH_3$, C_2H_5; $X=Cl$, Br

(466)

The products which are similar to those obtained from 1,4-bis(methylphenylsilyl)-benzene [461], act as fluorescent ligands, whereas the oligomers are solid compounds soluble in organic solvents. Silyl acetylene ethers may be regarded as a separate class of organosilicon compounds which undergo hydrosilylation of the C≡C bond to give mainly silyl alkenyl ethers.

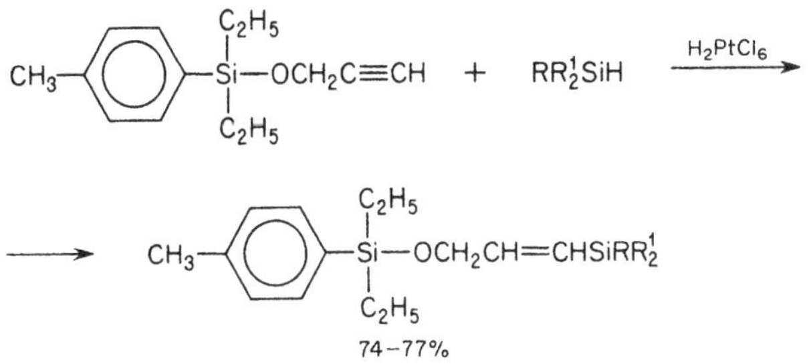

$$CH_3-\bigcirc-\underset{\underset{C_2H_5}{|}}{\overset{\overset{C_2H_5}{|}}{Si}}-OCH_2C\equiv CH \ + \ RR_2^1SiH \ \xrightarrow{H_2PtCl_6}$$

$$\longrightarrow \ CH_3-\bigcirc-\underset{\underset{C_2H_5}{|}}{\overset{\overset{C_2H_5}{|}}{Si}}-OCH_2CH=CHSiRR_2^1$$

74–77%

R = C₂H₅, C₃H₇, C₄H₉, R¹ = CH₃, C₂H₅   (467)

Hydrosilylation of ethynylsilanes RC≡CH (R = (CH₃)₃Si, Cl₃Si, (C₂H₅O)₃Si, N(CH₂CH₂O)₃Si) in the presence of Pt and Rh catalysts by phenyl- and thienylhydrosilanes, recently reported [704], yields predominantly β-*trans*-isomer which in the case of ethynylsilatrane is the only product formed.

Hydrosilylation of 1,4-bis(trimethylsiloxy)2-butynes yields 2-triorganosilyl-1,4-bis(trimethylsiloxy)-2-butenes in 68–80% yield [698, Dd-010]:

$$(CH_3)_3SiOCH_2C\equiv CCH_2OSi(CH_3)_3 + HSi(CH_3)_3 \xrightarrow[90-100°C]{H_2PtCl_6}$$

$$(CH_3)_3SiOCH_2\underset{\underset{Si(CH_3)_3}{|}}{C}=CHCH_2OSi(CH_3)_3 \qquad (468)$$

1-Methyl-1-ethynylsilacyclopentane is hydrosilylated with trichlorosilane and methyldichlorosilane giving *trans*-1,2-bis(silyl)ethene:

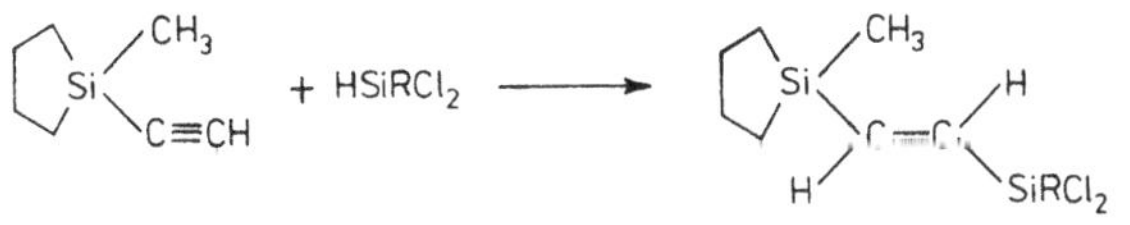

$$\underset{}{\overset{}{Si}}\begin{matrix}CH_3\\C\equiv CH\end{matrix} \ + \ HSiRCl_2 \longrightarrow \ \underset{}{\overset{}{Si}}\begin{matrix}CH_3\\ \\ \end{matrix} \ C=C \begin{matrix}H\\SiRCl_2\end{matrix} \qquad (469)$$

where R = Cl, CH₃.

Concurrent hydrosilylation of vinyl, ethynyl and propargyloxy groups attached to silicon results in the reactivity of propargyloxy-substituent reacting as:

$$R(CH_3)_2SiOCH_2C=CH + HSi(C_2H_5)_3 \xrightarrow{H_2PtCl_6}$$
$$R(CH_3)_2SiOCH_2CH=CHSi(C_2H_5)_3 \qquad (470)$$

(R = CH₂=CH−, −C≡CH)

Virtually the only characteristic feature of the hydrosilylation of the C≡C bond in unsaturated organosilicon compounds is its exceptional ability to undergo double hydrosilylation giving saturated products when the initial equimolar ratio of substrates in the system is unit.

Quite novel reaction of hydrosilylation of bis(silyl)acetylenes has been recently discovered [1024, 1053, 1310] to provide tris(silyl)ethenes:

$$R_3SiC{\equiv}CSiR_3 + HSiX_3 \xrightarrow[36-90\%]{H_2PtCl_6} R_3SiCH{=}C(SiR_3)SiX_3 \qquad (471)$$

where $R_3 = (CH_3)_3$; $X_3 = Cl_3$ [1024]
$R_3 = (CH_3)_3$; $X_3 = (CH_3)_{3-n}Cl_n$ ($n = 1,3$) [1310]
$R_3 = (CH_3)_nCl_{3-n}$ ($n = 0$–3); $X_3 = (CH_3)_mCl_{3-m}$ ($m = 0$–3) [1053].

5.5 Polyaddition of Di-, Tri- and Tetra-functional Silicon Hydride Derivatives to Di-, Tri- and Tetra-functional Unsaturated Organosilicon Monomers

The search for new thermoplastic elastomers has stimulated considerable interest in the synthesis of block copolymers characterized by quite unique mechanical and physiocochemical properties. Hydrosilylation of the vinyl group at silicon seems to provide a suitable synthetic method for the preparation of di-functional (hard segment) oligomers and polymers with a linear and cyclo-linear structure. A review by Andrianov *et al.* has shown notable progress in the development of polymeric structures by polyhydrosilylation [Rev.2]. The following examples illustrate the types of polyaddition which result in the formation of ethylene bridges in a polymeric chain:

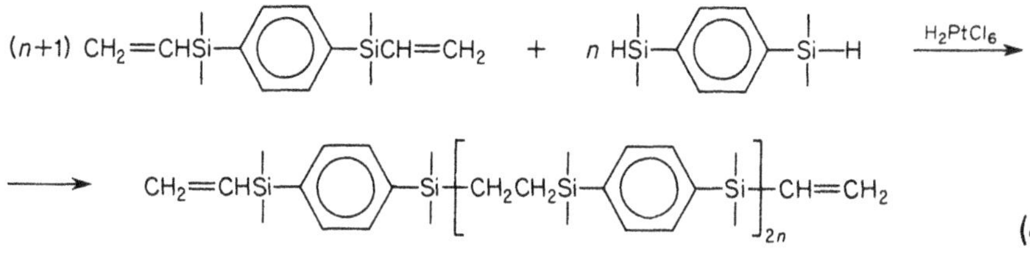

(472)

When this oligomer is coupled to an α,ω-dihydrogenopolydimethylsiloxane, high-molecular-weight, semicrystalline block copolymers are obtained:

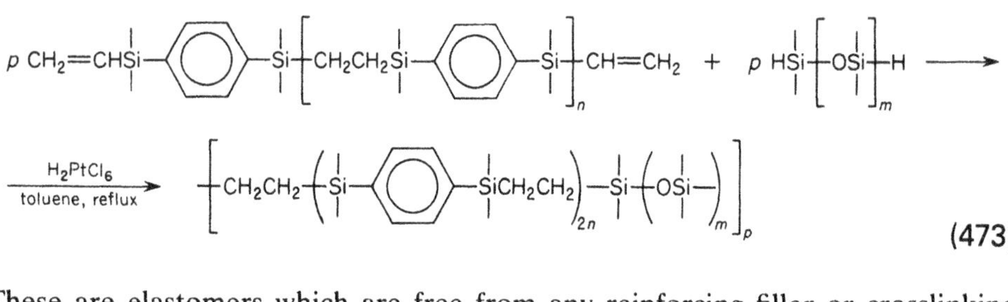

(473)

These are elastomers which are free from any reinforcing filler or crosslinking [940]. Siladialkenes in their reaction with bi-functional silanes exhibit exceptional reactivity. This leads to ring closure to give silamacrocycles according to Scheme (474) [276]:

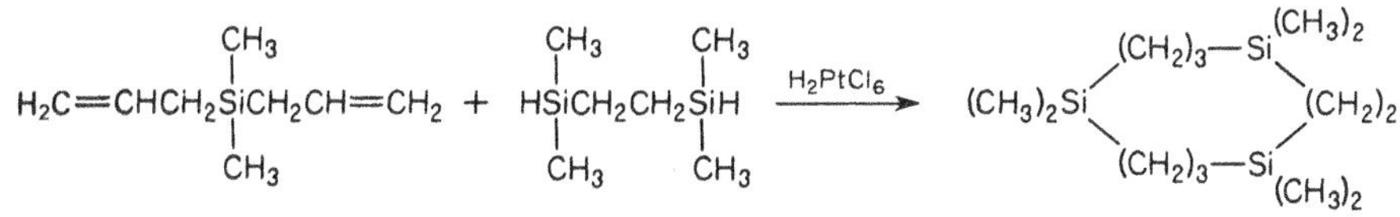

(474)

Linear polyorganosiloxanes can generally be obtained by polyaddition of linear substrates [1040, C-021, G-035]:

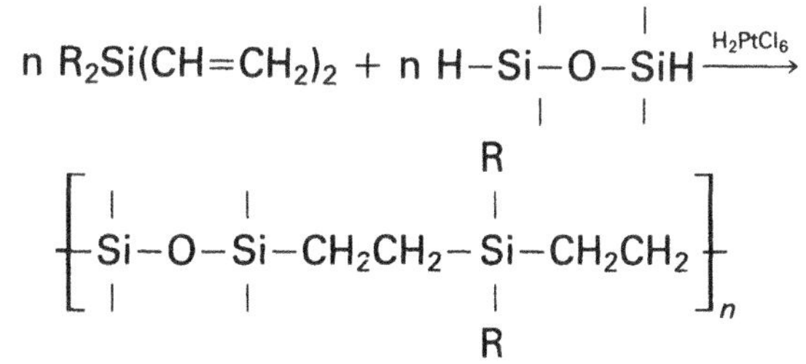

(475) [390]

$R_2 = (C_2H_5)_2, (C_6H_5)_2$

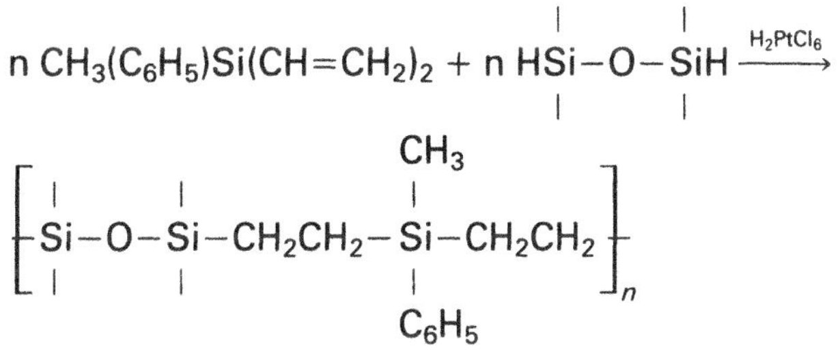

(476)

Linear multiblock copolymers have been synthesized by hydrosilylating α,ω-dihydrogenopolydimethylsiloxane with α,ω-divinylsilylated polystyrene derivatives [233, 234]. The synthesis of cyclo-linear polyorganosiloxanes occurs readily via the polyaddition of dihydropolyorganosiloxanes to divinylcyclosiloxanes:

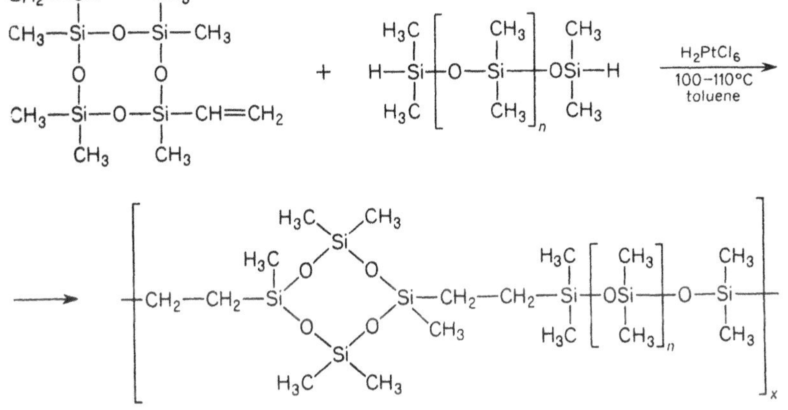

(477)

where $n = 0, 1, 4, 5, 6, 10, 20, 27, 34, 57, 94, 150, 200$ [810].

A combined procedure can be applied to synthesize carbosiloxane polymers. At first a vinyl-containing organopolysiloxane is hydrosilylated by chloro-substituted silanes (e.g. dichlorosilane). Then the silicon-bonded chlorine atoms are subjected to a dehydrochlorination reaction to yield silanol groups reacted immediately with other silanol groups, e.g. from acryloxymethyldimethyl silanol [E-036].

Carbosiloxane polymers with a cyclo-linear structure can be obtained by hydrosilylating divinyl cyclotetrasiloxane, for example with dihydrocyclotetrasiloxanes. The structure of these polymers varies depending on the arrangement of the reactive groups in the ring. Rings with reactive groups at positions 1 and 5 are the most reactive (the reaction proceeds smoothly at 60°C), to yield polymers with molecular weights up to 25×10^4. In contrast, rings with reactive groups at positions 1 and 3 exhibit a low reactivity [1353]:

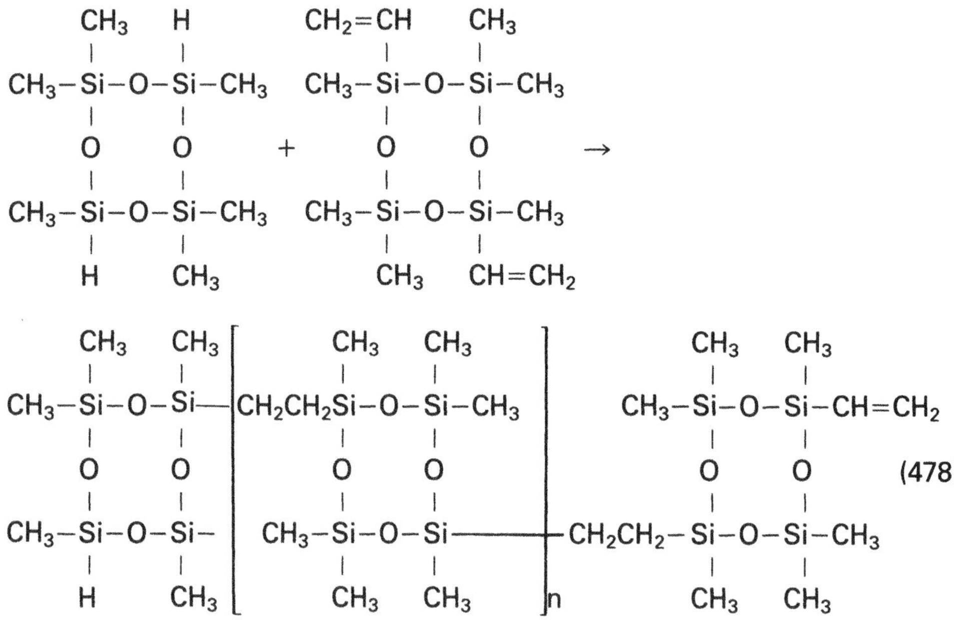

(478)

With substrates where the reactive sites are at positions 1 and 3 it is possible to obtain a crosslinking polymer at temperatures of 100–150°C, especially with substrates having methyl groups attached to silicon.

The extensive experiments of Andrianov, Zhdanov and their co-workers have led to the synthesis of organocarbosiloxane polymers through the polyhydrosilylation of cyclolinear and crosslinked structures, in particular [41, 52, 666, 1354] (see Scheme 479).

As a rule the poly-addition of cyclosiloxanes and cyclosilazanes is carried out at 120°C in the presence of H_2PtCl_6 in THF as the catalyst. The Si–H reactivity sequence obtained is as follows: No dehydrogenative side-reaction was observed in the presence of cyclosilazanes. It was found that 100% conversion could be achieved for all the substrates examined when the reaction mixture was heated at 200°C for 4 h [40]. There is now a great deal of interest in preparation of

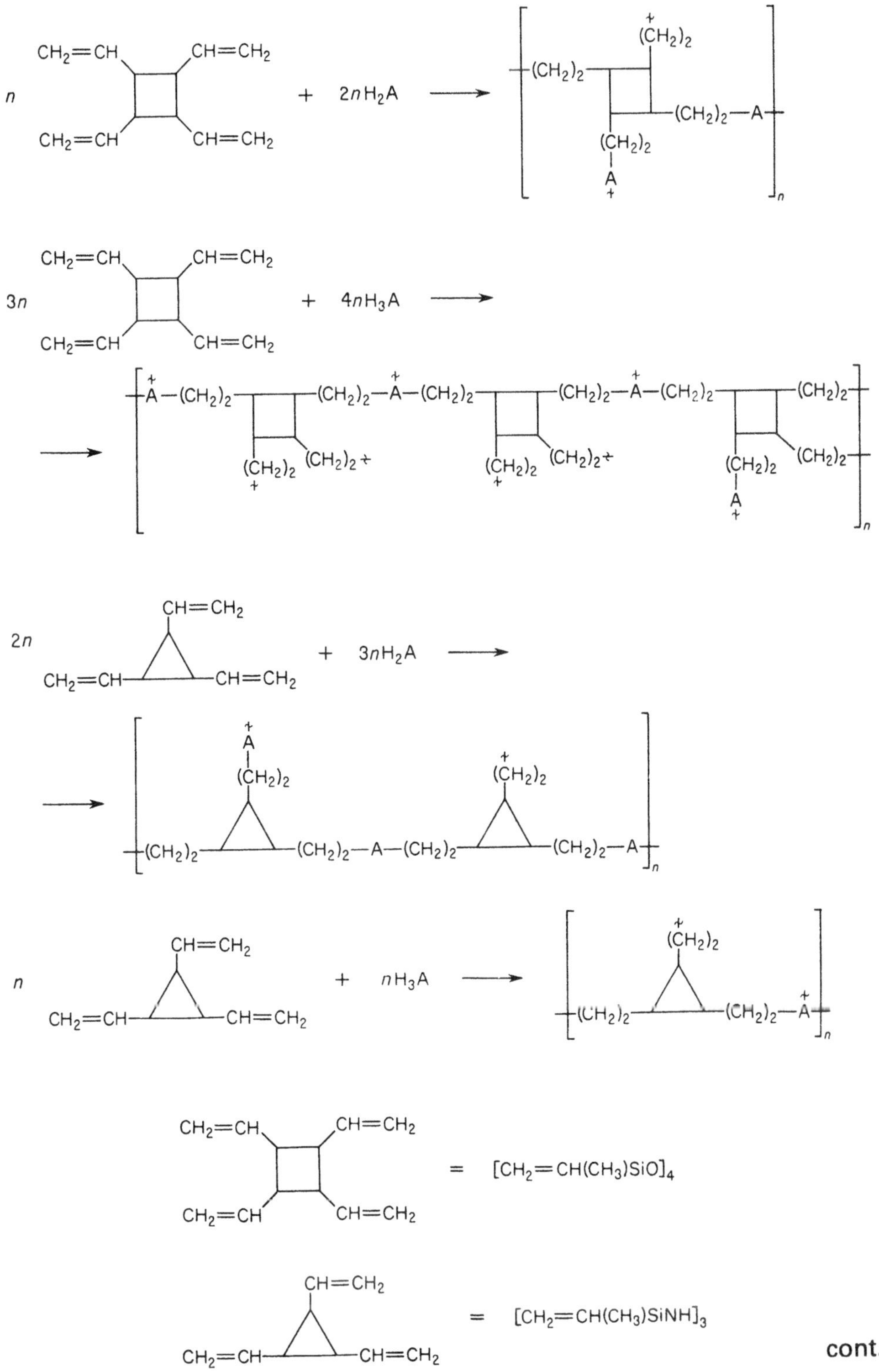
n
+ 2nH2A
3n
+ 4nH3A
2n
+ 3nH2A
n
+ nH3A
= [CH2=CH(CH3)SiO]4
= [CH2=CH(CH3)SiNH]3

cont.

H_2A = $CH_3C_6H_5SiH_2$, $(C_6H_5)_2SiH_2$, $[(CH_3)_2HSiO]_2C_6H_4$,

$[(CH_3)_2HSiOC_6H_4]_2C(CH_3)_2$, $[(CH_3)_2HSi]_2NH$,

$[CH_3(C_6H_5)HSi]_2NH$, $[(CH_3)_2HSiNSi(CH_3)_2]_2$,

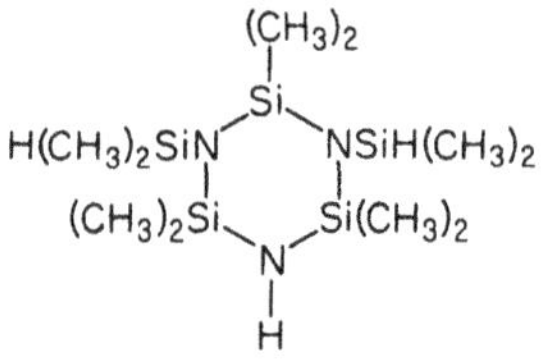

H_3A = $[H(CH_3)_2Si]_3N$, $[H(CH_3)_2SiO]_3C_6H_3$ (479)

$[H(CH_3)_2Si]_3N$ > $[H(CH_3)_2SiO]_2C_6H_4$ ≈ $[H(CH_3)_2SiOC_6H_4]_2C(CH_3)_2$

> $[HCH_3C_6H_5Si]_2NH$ > $CH_3C_6H_5SiH_2$ > $[H(CH_3)_2SiNSi(CH_3)_2]_2$

> $(C_6H_5)_2SiH_2$ ≫ 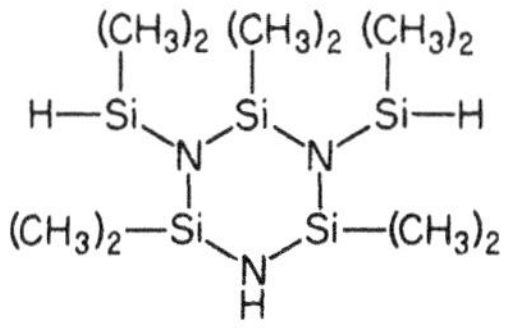 (480)

preceramic polymer materials which can be pyrolysed to yield SiC, Si_3N_4 and other silicon-based ceramic materials. Such polymers; polycarbosilane, polycarbosilazane, polycarbosiloxane are formed by hydrosilylation, preferably polyvinyl silicon compounds including those of the formula $[R'_2(CH_2{=}CH)Si]_nX$, $[R'(CH_2{=}CH)_2Si]_nX$ or $[(CH_2{=}CH)_3Si]_nX$ where X = O, S, NH, NR_2, CH_2, CH_2CH_2, C_6H_4 and others $n = 2, 3 \ldots$) by organopolysilanes of the formula $[(RSiH)_x(R_2Si)_y]_n$ (where $x + y = 1$, R = CH_3,Ph) or by polycarbosilane involving Si–H bond.

The reaction is preferably carried out in the presence of free-radical (e.g. AIBN) or transition metal catalyst (e.g. Pt complex) [E-017, E-028, U-170].

A discussion of the polyadditive copolymerization of dihydropolysiloxanes with di-functional unsaturated organic monomers and polymers is given in Section 6.3 below.

5.6 Intra- and Intermolecular Hydrosilylation

The hydrosilylation of an unsaturated organosilicon compound which contain an Si–H bond leads, in most cases, to two different cyclic and/or polymeric adducts according to the general scheme:

$$\equiv SiH(CH_2)_nCH=CHR \longrightarrow \begin{cases} \rightarrow \equiv Si(CH_2)_nCH_2CHR + \equiv Si(CH_2)_nCHCH_2R \\ \qquad\quad \text{β-adduct} \qquad\qquad\qquad \text{α-adduct} \\ \\ \\ \underset{|}{\overset{|}{\rightarrow}} [-Si-(CH_2)_n-CH_2CHR-]_x \\ \qquad\qquad |\qquad\qquad \text{polymer} \end{cases}$$

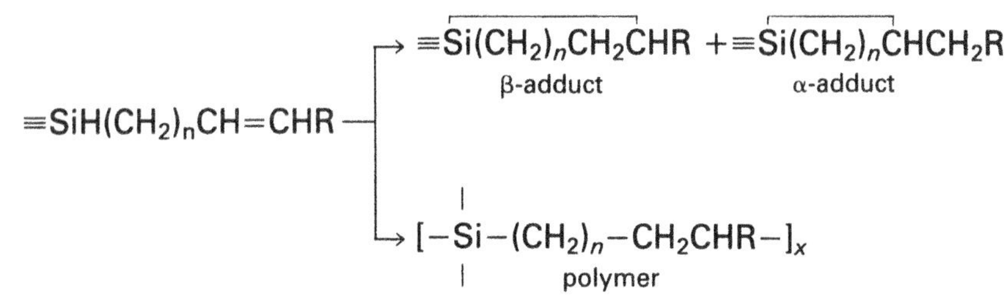

(481)

where R = H, C_mH_{2m+1}, C_6H_5.

In this section we mention only the main trends observed during the study of these reactions.

Silacycloalkanes are readily prepared by the intramolecular hydrosilylation of alkenyl-disubstituted silanes (when n = 2–8) (Scheme 481 [1172]), although the reaction is often accompanied by polymeric by-products. A general rule may be applied in such syntheses, i.e. alkenylsilanes with an internal C=C bond are inert towards intermolecular hydrosilylation under mild conditions [1186]. The intramolecular hydrosilylation of alkenyloxysilanes, silyl esters of unsaturated acids, sililoxyketones, α-hydroxy enol ethers, alkenylalcohols, and of compounds with other heteroatoms in unsaturated groups bonded to silicon and containing a Si–H bond also provides a recommended method for the preparation of various silacyclic organic compounds [57, 1185, 1187, 1190, 1191, U-165]. After hydrolysis of the Si–X–C bond (where X = O, N, S), new organic products, e.g. 1,3-diols (Scheme 482), may be synthesized from these compounds [1186].

$$CH_3CH_2CH=CHCH_2CH_2OSiH(CH_3)_2 \xrightarrow[60°C]{H_2PtCl_6}$$

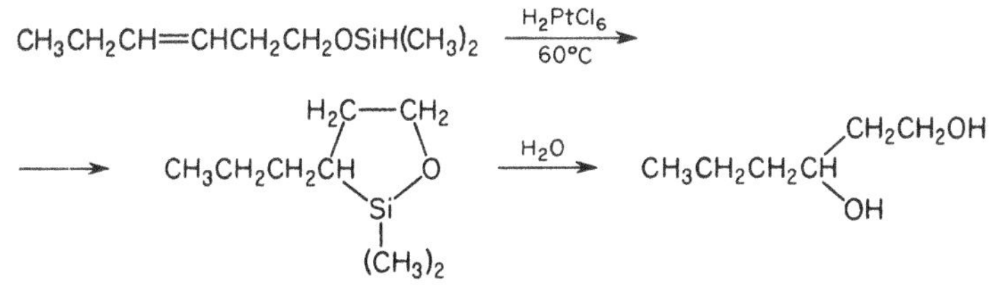

(482)

Intramolecular hydrosilylation of cyclic compounds, e.g. benzene or cycloalkene silyl derivatives, leads to silabicycloderivatives [781]:

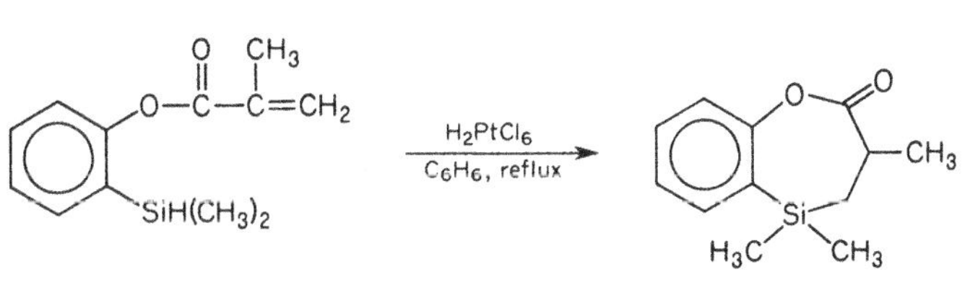

(483)

The reaction of an organosilicon substrate containing C≡C may be utilized for the synthesis of (sila)unsaturated rings [35]:

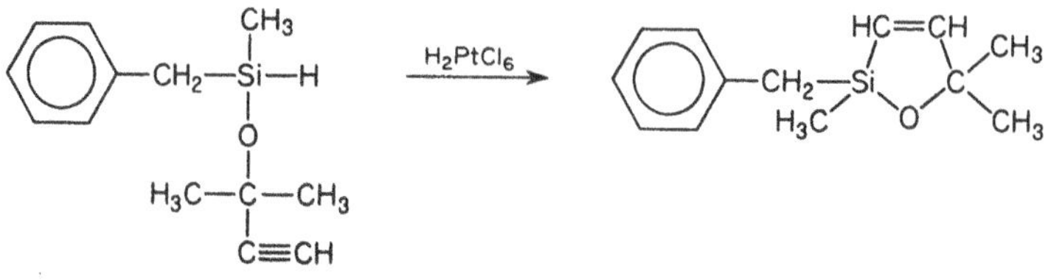

(484)

Intramolecular hydrosilylation of silyl acetylenes constitutes as novel method for synthesizing predominantly cyclic products having an exocyclic rather than an endocyclic double bond [1139, 1140, 1183]. Chloroplatinic acid is the most-used catalyst but the reaction can also be effected photochemically [1140]. Intramolecular hydrosilylation of disilalkynes leads regioselectively to exocyclic adducts [1139].

No ring closure occurs when a short alkenyl substituent is present at the silicon atom ($n = 0$, 1 in Scheme 485) of the hydrosilylation substrates, and this has led to this class of compounds being applied with a good effect in intermolecular hydrosilylation. Such reactions yield oligomers [16, 782].

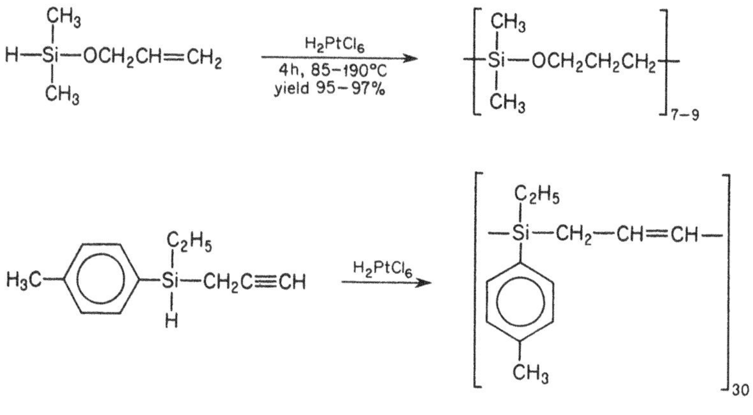

(485)

The addition of dihydrocarbosilanes and siloxanes to divinylsilanes and siloxanes mentioned above (Section 5.4) can be replaced by the intermolecular hydrosilylation of carbosilanes, siloxanes and silazanes containing an unsaturated substituent and a hydrogen substituent in the same molecule, although not attached to the same silicon atom:

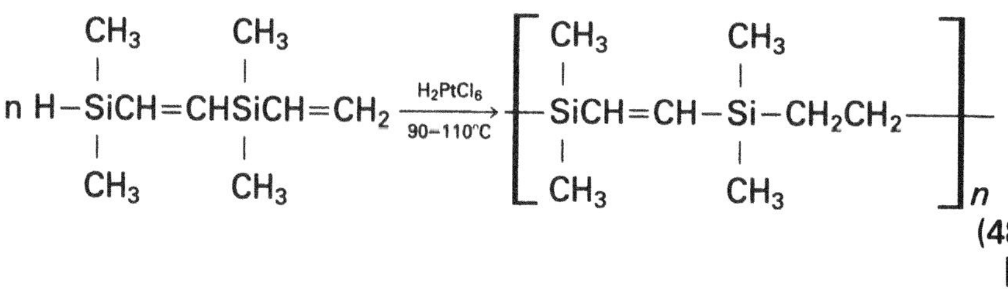

(486.1)
[776]

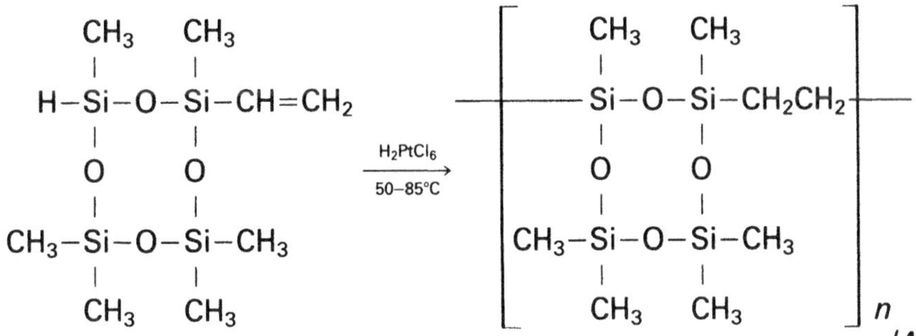
CH3
CH3
H-Si-O-Si-CH=CH2
O
O
CH3-Si-O-Si-CH3
CH3
CH3
H2PtCl6
50-85°C
CH3
CH3
Si-O-Si-CH2CH2
O
O
CH3-Si-O-Si-CH3
CH3
CH3
n

(486.2)
[49]

CHAPTER 6

The Application of Hydrosilylation Processes

6.1 Silanes as Coupling Agents

Silane coupling agents are compounds which combine the organic chemistry of organofunctional groups with the inorganic chemistry of silicates or other surfaces with siliceous properties to bridge the hydrophilic interface between mineral substrates and organic molecule and polymers [916]. The general formula of such organosilanes is as follows:

$$R_nX_{3-n}Si(CH_2)_mY \tag{487}$$

where $n = 0, 1$; $m = 0, 1, 2, 3, 4 \ldots 18$;

Y = Cl, NH_2, SH, $CH{=}CH_2$, CN, NCO, $OCH_2\overline{CHCH_2O}$, OCOR′, NR_1R_2;

X = halogen, OR′, OCOR′.

Compounds which are commercially available are mainly those in which $m = 3$ and $n = 0$, and are listed in Table 1 (p. 206).

These organosilanes exhibit two types of functionality. If X is a hydrolysable group, such as a halogen, alkoxy or acyloxy group, these materials when hydrolysed yield silanol groups for bonding to mineral surface, e.g. siliceous fillers to form siloxane linkages. Stable products can also be produced with other oxides such as those of aluminium and titanium, but in this case less stable bonds are formed with hydrophilic surfaces.

The non-hydrolysable organofunctional groups Y are generally chosen for their reactivity or compatibility with organic molecules or polymer substrates. Thus, the addition of traces of appropriate reactive silanes at the interface permits a marked improvement in the adhesion of plastics and organic or bio-organic compounds to minerals.

An excellent monograph has recently been written by Plueddemann of the Dow Corning Corporation, who has made a substantial contribution to the application of silanes as coupling agents [917]. The laboratory and industrial syntheses of these compounds always involve a hydrosilylation step.

The general method employed is based either on the addition of trichlorosilane (or methyldichlorosilane, where $n = 1$) mainly to allyl compounds followed by alcoholysis of the Si–Cl bond (together with the introduction of a desired functional group at C3), or on the direct addition of triethoxysilane (or methyldiethoxysilane, where $n = 1$; other trialkoxysilanes are rarely used) to the allyl derivatives.

One of the fundamental reactions involved in the production of silane coupling agents is the addition of trichlorosilane to allyl chloride, resulting in the formation of 3-chloropropyltrichlorosilane as the main product (Scheme 252). Various homogeneous and heterogenizing platinum complexes have been used as effective catalysts of this reaction (see Table 1.3, Part II). Alcoholysis of this product occurs very readily according to the following equation, giving a 90% yield or greater:

$$Cl_3SiCH_2CH_2CH_2Cl + 3\ ROH \rightarrow (RO)_3SiCH_2CH_2CH_2Cl + 3\ HCl \quad (488)$$

where R = CH_3, C_2H_5.

If secondary alcohols are used the yield of this reaction is diminished. Although direct synthesis of 3-chloropropyltriethoxysilane by the addition of triethoxysilane to allyl chloride has also been reported [D-071, U-172] as occurring with a relatively high yield (70%). This pathway appears to be of less importance since the addition reaction is accompanied by a condensation reaction yielding chlorotriethoxysilane and propene under the same conditions [613].

3-Chloropropyltriethoxysilane and 3-chloropropyltrimethoxysilane are important substrates which are the basis of various silane coupling agents which can be prepared with the following functional groups attached to the γ-carbon atom: NH_2, SH, OC_6H_4X, OCH_3, OC_2H_5, CN, S_x, S, NCO, $SO_2C_6H_5$, NCS, SR, OCOR, $NHCH_2CH_2NH_2$, NHR, NR_2, etc.

The products are generated as a result of chlorine substitution which occurs with HCl elimination. The reaction usually proceeds in non-aqueous polar media, mainly alcohols, DMF and DMSO. Excess of hydrogen chloride may be removed by alcoholates, tertiary amines, ammonia or epoxy compounds. The following are examples of possible reactions [82, 313a]:

$$(C_2H_5O)_3Si(CH_2)_3Cl \xrightarrow[\text{(2) EtONa or } NH_3,\ 20\text{ h, reflux}]{\text{(1) } NH_2CSNH_2,\text{ EtOH}} (C_2H_5O)_3Si(CH_2)_3SH + HCl \quad (489.1)$$
[D-004]

$$(C_2H_5O)_3Si(CH_2)_3Cl \xrightarrow[\text{6 h, 155°C}]{\text{NaCN, DMF}} (C_2H_5O)_3Si(CH_2)_3CN + NaCl \quad (489.2)$$
[U-009]

$$(C_2H_5)O_3Si(CH_2)_3Cl \xrightarrow{Na_2S_4, C_2H_5OH} (C_2H_5O)_3Si(CH_2)_3S_4(CH_2)_3Si(OC_2H_5)_3 \quad (489.3)$$
[D-051]

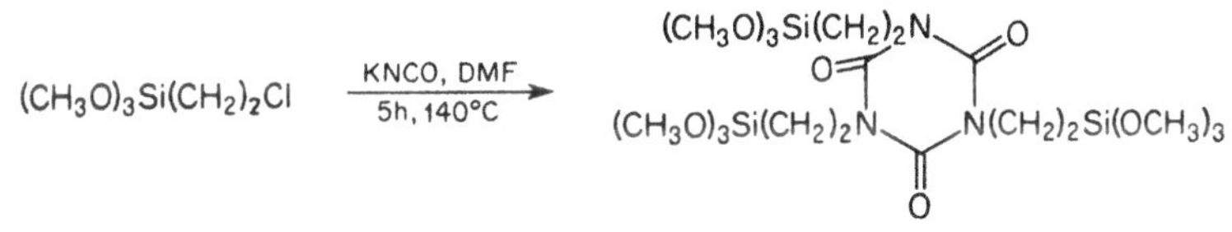

(489.4)
[U-041]

$$(CH_3O)_3Si(CH_2)_3Cl + H_2NCH_2CH_2NH_2 + H_2S \rightarrow (CH_3O)_3Si(CH_2)_3SH + H_2NCH_2CH_2NH_2\ HCl \qquad (489.5)$$
[120]

$$(CH_3O)_3Si(CH_2)_3Cl \xrightarrow{NH_2CH_2CH_2NH_2} (CH_3O)_3Si(CH_2)_3NHCH_2CH_2NH_2 \qquad (489.6)$$
[C-040, PI-024]

$$(RO)_3SiCH_2CH_2CH_2Cl + HNR_1R_2 \rightarrow (RO)_3SiCH_2CH_2CH_2NR_1R_2$$

R_1,R_2, saturated and unsaturated hydrocarbon substituents (489.7)
[313a]

$$(C_2H_5O)_3Si(CH_2)_3Cl \xrightarrow[75°C,\ 36\ bar]{NH_3} (C_2H_5O)_3Si(CH_2)_3NH_2 \qquad (489.8)$$
[D-091]

The last product, 3-aminopropyltriethoxysilane, is also prepared through the hydrosilylation of allylamine (Table 9.3). 3-Aminopropyltrialkoxysilanes can be also obtained by hydrosilylating of acrylonitrile (Table 1.3) and then via hydrogenating the CN group in the presence of a Raney nickel catalyst or palladium supported on carbon or other catalysts [U-151]. Other aminoalkyltrialkoxysilanes can be prepared by hydrosilylating *N*-(2-aminoethyl)3-aminopropene or alkenylsilylamines. The latter reaction, which produces cyclic silazane, may be followed by hydrolysis in an alcoholic medium to give 3-aminopropyltrialkoxysilane (or 3-aminopropylalkyldialkoxysilane) [D-073]. The reaction of cyclic silazane with phosgene leads to 3-isocyanatopropyldichlorosilane [S-001]:

$$CH_2{=}CHCH_2NHSiH(CH_3)_2 \xrightarrow[100\text{–}215°C,\ 10\text{–}15\ h]{H_2PtCl_6\ (iso\text{-}PrOH)} [(CH_3)_2Si(CH_2)_3NH]_n \qquad (490.1)$$

$$[(CH_3)_2Si(CH_2)_3NH]_n \rightarrow \begin{cases} \xrightarrow{+ROH} CH_3(RO)_2Si(CH_2)_3NH_2 \\ \xrightarrow[(-20)\ -10°C]{+\ COCl_2,\ C_6H_5CH_3} ClSi(CH_3)_2(CH_2)_3NCO \end{cases} \qquad (490.2)$$

N,N'-saturated or unsaturated hydrocarbon-substituted 3-propyltrialkoxysilanes can also be an increasingly common group of silane coupling agents produced either by the method of HCl elimination from 3-chloropropyltrialkoxysilane (equations 489) or by direct hydrosilylation of the corresponding allylamines by trialkoxysilane, e.g. [J-082].

Aminopropyltrialkoxysilanes may act as initial substrates for other silane coupling agents. They undergo addition to α,β-unsaturated compounds, e.g. acrylates, methacrylates and vinylcyanides [82]:

$$(CH_3O)_3Si(CH_2)_3NH_2 + CH_2{=}C(CH_3)COOCH_3 \rightarrow (CH_3O)_3Si(CH_2)_3NHCH_2CH(CH_3)COOCH_3 \quad (491)$$

as well as reacting with ethylene oxide, ethylene sulphide and their derivatives:

$$(C_2H_5O)_3Si(CH_2)_3NH_2 + \overline{CH_2CH_2S} \xrightarrow[100°C,\ 20\ h]{N_2} (C_2H_5O)_3Si(CH_2)_3NHCH_2CH_2SH \quad (492)$$

The above reactions lead to novel reactive organosilicon compounds of higher molecular weight which sometimes involve further functional groups. They may be regarded as silane coupling agents with a potentially greater efficiency.

Organofunctional epoxy silanes may be prepared by hydrosilylating unsaturated epoxides or epoxidizing unsaturated silanes. A commercial silane, γ-glycydoxytrimethoxysilane, may be obtained as follows, e.g. [E-042]:

$$(CH_3O)_3SiH + CH_2{=}CHCH_2OCH_2\overline{CHCH_2O} \xrightarrow{Pt} (CH_3O)_3Si(CH_2)_3OCH_2\overline{CHCH_2O} \quad (493)$$

Siloxane-containing oxiranes can be prepared by hydrosilylation of allylglycidyl ether with $(CH_3SiHR)_2O$ [1147].

Mercaptofunctional silanes are usually prepared from unsaturated silanes of chloroalkylsilanes (see Scheme 488).

The addition of trichlorosilane to acrylonitrile, i.e.

$$HSiCl_3 + CH_2{=}CHCN \rightarrow Cl_3SiCH_2CH_2CN \quad (494)$$

produces 2-cyanoethyltrichlorosilane as well as siliconates with carboxy functional groups [917]:

$$Cl_3SiCH_2CH_2CN + 4\ NaOH \xrightarrow{H_2O} O_{1.5}SiCH_2CH_2COONa + 3NaCl + \uparrow NH_3 \quad (495)$$

Since the direct addition of tri-substituted silanes to alcohols is accompanied by solvolysis of the silane hydrogen, hydroxy-functional propylsilanes may be obtained by hydrosilylating allyl esters and then alcoholysing the product to recover the hydroxypropylsilanes [917].

N-[(3-trimethoxysilyl)propyl]maleimide was synthesized as coupling agent by hydrosilylation of *N*-allylmaleimide with trimethoxysilane [J-109].

$$(CH_3O)_3SiH + CH_2{=}CHCH_2OCOCH_3 \xrightarrow{H_2PtCl_6} (CH_3O)_3Si(CH_2)_3OCOCH_3 \xrightarrow[H^+]{+\ CH_3OH} (CH_3O)_3Si(CH_2)_3OH + CH_3OCOCH_3 \quad (496)$$

Generally, unsaturated silane coupling agents are prepared by the addition of hydrochloro- and hydroalkoxy-silanes to: alkynes (acetylene, phenylacetylene and propargyl derivatives); alkadienes (butadiene, allene, 4-vinylcyclohexene, α,ω-dienes); allyl esters of unsaturated carboxylic acids (acrylate, methacrylate, maleate, fumarate, tetrahydrophthalate); propargyl esters of saturated and unsaturated carboxylic acids; bis-unsaturated ethers (allylvinyl, diallyl ethers); bis-unsaturated thioethers (diallyl thioethers); and bis-unsaturated ketones (allylvinyl ketone, α-methylvinyl- and γ-methylallyl ketone).

Vinylsilanes (vinyltrichlorosilane, vinyltrialkoxysilanes and vinyltriacetoxysilane), which were the first commercially used silane coupling agents, are mainly produced by the addition of trichlorosilane to acetylene in a liquid or gas-phase reaction catalysed by various homogeneous and heterogenizing complexes (see Table 1.4, Part II) followed, if necessary, by alcoholysis or acylation of the Si–Cl bonds. All the other above-mentioned unsaturated silanes may be prepared mainly by hydrosilylation with trichlorosilane or occasionally trimethoxy- and triethoxysilanes (see corresponding Tables 1, 5, 9, Part II). Alkyltri(chloro, alkoxy)silanes can also be considered as coupling agents improving, as additives, of composites mechanical and other physicochemical and physical properties of the final products. A novel class of silane coupling agents has been extended by introduction of new functional groups predominantly at the γ-position of propyl-substituted trialkoxysilanes, e.g. [J-079, J-110, E-016, J-065, C-037, C-038, U-194a, E-014, 156, 552] or corresponding derivatives of (poly)siloxanes [552, 597, U-153]. Silane coupling agents $(RO)_3Si(CH_2)_nX$ with a higher number of methylene chains (predominantly n = 8, 12, 16 and 18) have been synthesized by methods similar to these for γ-propyl derivatives e.g. 11-triethoxysilylundecanal [1060], and octyltri(chloro,alkoxy)silanes [Pl-030] and 8-(trichlorosilyl)octyl-methacrylate [J-112]. Such compounds are frequently applied in stationary phases preferentially to high-performance liquid chromatography (HPLC) or also as coupling agents and water resistants.

The role of the difunctional (X,Y) group of silanes as coupling agents, and particularly those introduced to a polymer filler system, can be substituted by the preliminary preparation of a copolymer with a tri(alkoxy, chloro, acetoxy)silyl group which reacts directly with a mineral substrate under hydrolytic conditions, e.g. [E-029] (see also Section 6.3).

6.2 Reduction of Organic Compounds by Hydrosilylation

The synthesis of organic compounds via hydrosilylation processes continues to enjoy widespread application. An organosilicon compound synthesized by hydrosilylation may be regarded as an intermediate which, when suitably reacted (usually via solvolysis), leads to the desired organic product. The best-known example of such a procedure is the addition of silicon hydrides to the carbonyl group of aldehydes and ketones, resulting in the formation of alkoxysilanes. When hydrolysed, these give the corresponding primary and secondary alcohols, respectively (see Section 3.7.1):

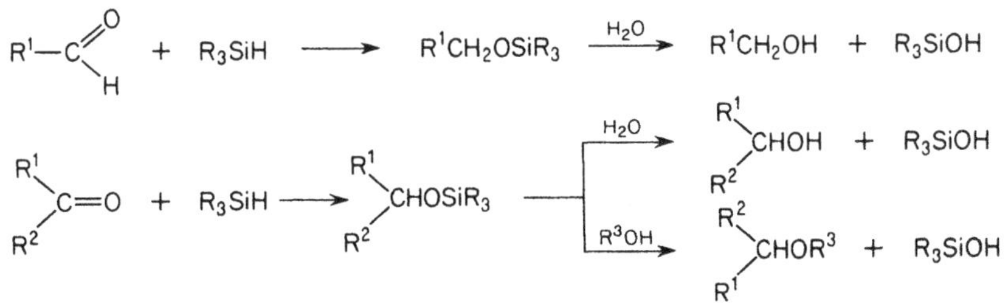

$$R^1-C\overset{O}{\underset{H}{\diagdown}} + R_3SiH \longrightarrow R^1CH_2OSiR_3 \xrightarrow{H_2O} R^1CH_2OH + R_3SiOH$$

$$R^1_{\diagdown} \atop R^2_{\diagup}C=O + R_3SiH \longrightarrow {R^1_{\diagdown} \atop R^2_{\diagup}}CHOSiR_3 \left\{ \begin{array}{l} \xrightarrow{H_2O} {R^1_{\diagdown} \atop R^2_{\diagup}}CHOH + R_3SiOH \\ \xrightarrow{R^3OH} {R^2_{\diagdown} \atop R^1_{\diagup}}CHOR^3 + R_3SiOH \end{array} \right. \quad (497)$$

Alcoholysis of alkoxysilanes results in the formation of the respective ethers (Scheme 497).

When prochiral ketones are used in the presence of a catalyst (mainly rhodium complexes) with chiral ligands, hydrosilylation followed by hydrolysis finally leads to an R- or S-alcohol (see Section 2.4.4). In the laboratory, polymethylhydrosiloxane appears to be a convenient reagent for selective reduction of aldehydes and ketones; it is active in the presence of organotin catalysts and in protic solvents [857]:

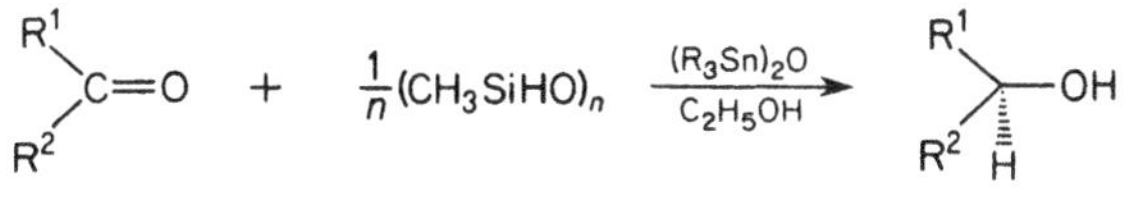

$$R^1_{\diagdown} \atop R^2_{\diagup}C=O + \frac{1}{n}(CH_3SiHO)_n \xrightarrow[C_2H_5OH]{(R_3Sn)_2O} {R^1_{\diagdown} \atop R^2_{\diagup}}\overset{|}{\underset{\underline{H}}{C}}-OH$$

$$R^2 = H, \text{alkyl} \quad (498)$$

As already discussed, hydrosilylation of a C=O group is often accompanied by dehydrocondensation. This may be a desirable reaction (see Section 3.7.1), as it leads to alkenoxysilanes which yield unsaturated alcohols on hydrolysis.

α,β-Unsaturated carbonyl compounds react with triorganosilanes to yield 1,4-addition products, whereas with mono- or diorganosilanes these compounds react selectively to form 1,2-adducts. The respective 1,2- or 1,4-reduction of α,β-unsaturated substrates by hydrosilylation (followed by hydrolysis) consequently leads to alcohols or aldehydes [857]. Selective 1,4-addition of hydrosilanes to crotonates and methacrylates gives silyl acetals:

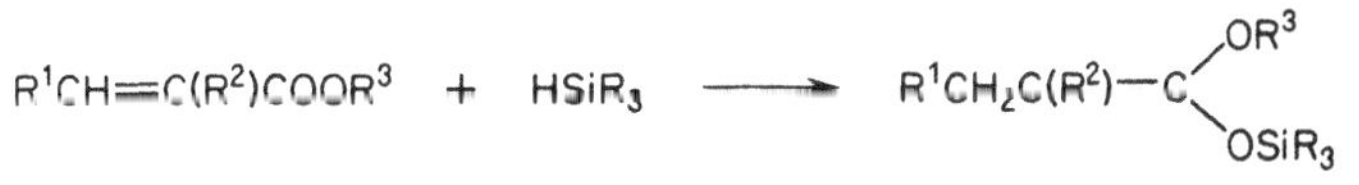

$$R^1CH=C(R^2)COOR^3 + HSiR_3 \longrightarrow R^1CH_2C(R^2)-C\overset{OR^3}{\underset{OSiR_3}{\diagdown}} \quad (499)$$

which after hydrolysis produce saturated esters:

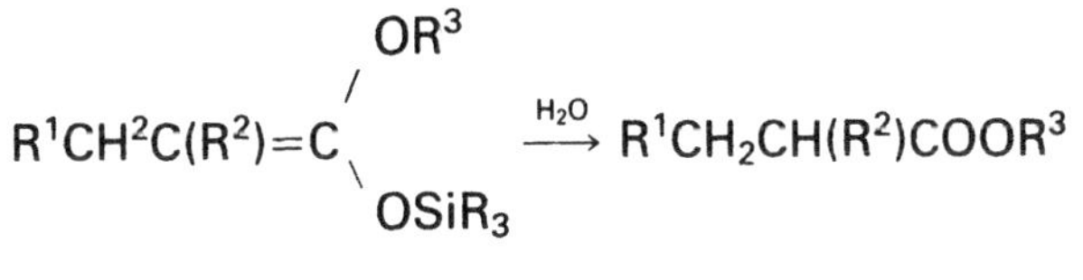

$$R^1CH^2C(R^2)=C\overset{\diagup OR^3}{\underset{\diagdown OSiR_3}{}} \xrightarrow{H_2O} R^1CH_2CH(R^2)COOR^3 \quad (500)$$

Carboxylic acid chlorides are reduced to aldehydes via a reaction catalysed by palladium complexes, which provides an alternative to the Rosenmund reduction [28]:

$$RCOCl \xrightarrow[\text{Pd catalyst}]{(C_2H_5)_3SiH} RCHO \qquad (501)$$

Trichlorosilane reduction of acid chloride induced by γ-rays occurs in two ways. When the alkyl moiety (R) is tertiary the acyl radical undergoes decomposition into carbon monoxide and an alkyl radical which finally yields an alkane:

$$RCOCl + HSiCl_3 \rightarrow RH + SiCl_4 + CO \qquad (502)$$

When R is primary, the acyl radical abstracts a hydrogen atom from trichlorosilane producing aldehyde, which, by subsequent hydrosilylation, gives an alkoxysilane [874]:

$$RCOCl + 2\ HSiCl_3 \rightarrow RCH_2OSiCl_3 + SiCl_4 \qquad (503)$$

Aldehydes are also formed from nitriles. Thus, *N*-ethylnitrilium tetrafluoroborate (prepared from nitrile and triehtyloxonium tetrafluoroborate) react with triethylsilane in dichloromethane solution at room temperature to give, after hydrolysis, good yields of aldehydes [391]:

$$RCN \xrightarrow{C_2H_5BF_4} RCNC_2H_5BF_4^- \xrightarrow{(C_2H_5)_3SiH} RCH{=}NC_2H_5 \xrightarrow{H_2O} RCHO \qquad (504)$$

The following yields (%) have been observed: R = C_4H_9 (71); R = $(CH_3)_2CH$ (85); R = $C_6H_5CH_2$ (41); R = C_6H_5 (90); R = 1−$C_{10}H_7$ (84); R = 1-adamantyl (83); R = cyclopropyl (79).

Chloroformates can be reduced by using tripropylsilane in the presence of tert-butyl peroxide [80, 131]:

$$ROH + COCl_2 \rightarrow ROCOCl \xrightarrow[[(CH_3)_3CO]_2]{(C_3H_7)_3SiH} RH + CO_2 + ClSi(C_3H_7)_3 \qquad (505)$$

Aromatic carbonyl compounds, and carboxylic acids in particular, react with triethylsilane in the presence of triethylamine to yield benzylsilane. Hydrolysis converts this into the corresponding hydrocarbon derivative [108, 109]:

$$C_2H_5OCOC_6H_4COOH \xrightarrow[(C_2H_5)_3N]{HSiCl_3} C_2H_5OCOC_6H_4CH_2SiCl_3 \xrightarrow[C_2H_5OH]{KOH} C_2H_5OCOC_6H_4CH_3 \qquad (506)$$

$RhCl(PPh_3)_3$ appeared to be an effective catalyst for reductive hydrosilylation of quinones, e.g. benzoquinone, and naphthalenediones to give the corresponding bis-silyl ethers [76].

Facile synthesis of symmetrical ethers is achieved by either trimethylsilyl triflate or trimethylsilyl iodide catalysed reductive coupling of carbonyl compounds (aldehydes and ketones) with trialkylsilanes [1021].

The hydrogenation of alkenes to alkanes, or of alkynes or alkatrienes to alkenes, via silicon hydrides may proceed in two ways. The first route involves hydrosilylation followed by desilylation in the presence of hydrogen iodide (elimination of silyl iodide) or potassium fluoride [1192]:

$$-C{\equiv}C- \; + \; R_3SiH \longrightarrow -HC{=}C(SiR_3)- \xrightarrow[-ISiR_3]{HI} -CH{=}CH-$$

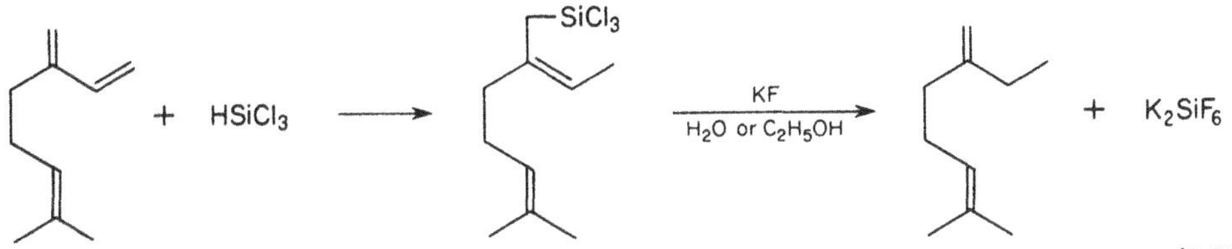

(507)

The second route is "ionic hydrogenation", which involves the protonations of the double bond to form a carbonium ion, followed by reaction with a hydride donor to give the hydrogenated product [329]. This reaction is selective; e.g. reducing a mixture of 1-methylcyclohexene and cyclohexene with triethylsilane yields methylcyclohexane, while cyclohexene remains intact [632].

Organic compounds containing a functional group, such as hydroxyl, epoxy, acetal, ketal, semiacetal and semiketal, react with hydrosilanes such as triethylsilane, R(+)-1-naphthyl-phenylmethylsilane, dimethylethylsilane, phenylneopentylsilane and tri-n-hexylsilane in the presence of compounds generating carbocations, e.g. BF_3, to yield finally the corresponding hydrocarbons [U-112].

Since they are stable in air, organopentafluorosilicates have become very useful reagents in organic syntheses. They are obtained by the reaction of organotrichlorosilanes (the products of the addition of trichlorosilane to the unsaturated compound) with potassium fluoride:

$$RSiCl_3 + KF \xrightarrow[0°C\text{–room temp.}]{H_2O} K_2[RSiF_5] \qquad (508)$$

Alkylpentafluorosilicates react exothermally with halogens such as chloride, bromide, iodine, iodine monobromide and N-bromosuccinimide (NBS) in carbon tetrachloride to give the corresponding alkyl halides in good yields [U-112]. Thus, *exo*-1-bromonorbornane may be obtained by this method as follows:

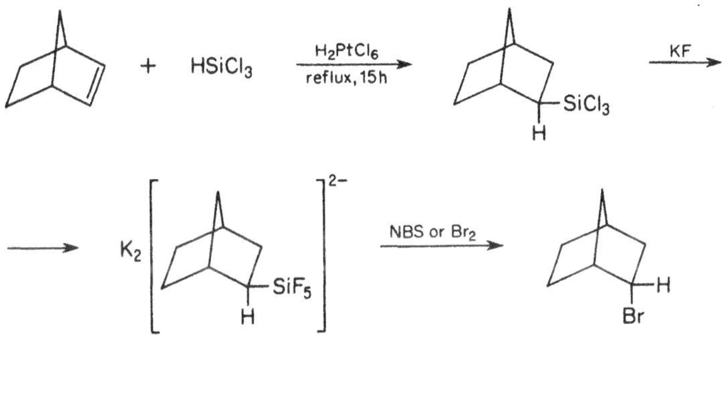

(509)

Difluorotrimethylsilicate $[(CH_3)_3SiF_2]^{2-}$ can also play a role of reagent in an exemplary stereoselective reduction of α-substituted β-ketoamides [392]. In contrast with the common use of fluorosilicate alkoxy-substituted pentacoordinate silicon compounds, generated *in situ* from trialkoxysilane and metal alkoxides have been applied in organic synthesis only very recently [501, J-093].

The hydrosilylation products of isocyanates (see Section 3.9) yield formamides on hydrolysis. Hydrosilylation can occur via two intermediates, N-silylformamides or C-silylamides [860, J-014].

$$R{-}N{=}C{=}O + (C_2H_5)_3SiH \rightarrow \left[\begin{array}{l} C_6H_5N(CHO)Si(C_2H_5)_3 \\ \overline{(CH_2)_5CH}NHCOSi(C_2H_5)_3 \end{array}\right] \xrightarrow{CH_3OH} RNHCHO \tag{510}$$

Although the hydrosilylation of carbodiimides does not proceed by addition of a silyl group to the carbon atom, the products obtained are N-silylformamides. These compounds react exothermally with methanol or water to give formamides in quantitative amounts [849, 860].

The hydrosilylation of Schiff bases (imines) catalysed by Wilkinson's complex results in an almost quantitative yield of silylated amines [23, 862]. Distillation of this product by methanol or acid chloride finally results in the formation of amines or amidines, as already mentioned. The hydrosilylation of Schiff bases can also be catalysed by Lewis acids or other complexes [43, 379] (see also Section 3.9):

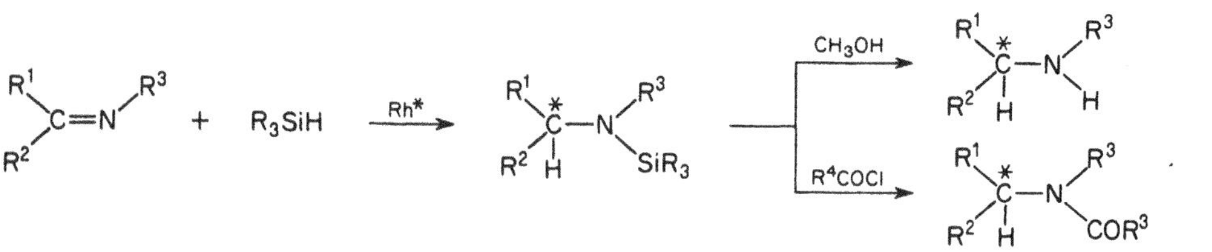

(511)

It is interesting that divinylsulphoxide reacts with triethylsilane to give instead of the C=C hydrosilylation product, divinylsulphides [1273]

$$CH_2{=}CHS(O)CH{=}CH_2 + 2\ HSi(C_2H_5)_3 \xrightarrow{H_2PtCl_6, NiCl_2, Ni(acac)_2} CH_2{=}CHSCH{=}CH_2 + (C_2H_5)_3SiOSi(C_2H_5)_3 + H_2 \tag{512}$$

On the other hand, the stereoselective reduction of acetylenic sulphides to *cis* vinyl sulphones by hydrosilane was observed in the presence of copper(II) salts [993].

Applications of silicon hydrides as reducing agents and of hydrosilylation reactions in organic syntheses have been reviewed by Nagai [Rev.28]. Separate but increasing application of the hydrosilylation is observed in the synthesis of various stationary phases in capillary column GLC [4, 58, 128, 159] and HPLC [910, 1362, J-101]. The above-mentioned examples of reactions are intended only as indications of the possibilities of hydrosilylation in organic chemistry. This field is still continuing to expand.

6.3 Modification of Polymers via Hydrosilylation

The application of hydrosilylation in the polymer chemistry of organosilicon chains has already been discussed in Chapter 5, particularly with regard to the curing of polyorganosiloxanes and the poly-addition of silicon monomers involving Si–H and unsaturated carbon–carbon bonds. Both substrates in these reactions are organosilicon compounds.

Hydrosilylation can also be used to modify unsaturated organic polymer through the use of organosilicon hydrosilanes and hydrosiloxanes in the modification of polyorganosiloxanes with Si–H bonds by unsaturated compounds, especially those involving functional groups, and lastly to co-polymerize and co-oligomerize of dihydropolysiloxanes with organic monomers and polymers.

The present considerations are limited simply to the basic features of these reactions. The most advanced concept for modifying organic polymers by hydrosilanes is connected with the replacement of the silane coupling agents. The hydrosilylation of unsaturated polymers, e.g. polybutadiene [203, 166, 384, J-030, J-085], polypiperylene, polyisoprene [384, 908], polyesters [J-083, J-091, J-106, J-078], polycarbonate [J-113], phenolic resins [906], polyphosphazenes [U-177], polyimines [263], oligophenylenes with $-C{\equiv}CH$ groups [770], prepolymer diallylphthalate [U-120] and other polyenes [E-031, E-043, F-032, J-100, U-191], polyoxyenes [J-057, J-067, J-076] and polyoxy-ynes [Dd-011] with a silane having hydrolysable substituents (e.g. trichlorosilane, triethoxysilane, methyldichlorosilane, methylalkoxysilanes) or dihydroorganosilanes as well as hydrosiloxanes and others in the presence of free-radical initiators or chloroplatinic acid, leads to the formation of moisture-curable polymers [U-046]. Such modification also enhances their activity towards mineral fillers:

$$(CH_2CH{=}CHCH_2)^n + {\equiv}SiH \rightarrow (CH_2CH_2\underset{\displaystyle \overset{|}{Si}{\equiv}}{CH}CH_2)_n \qquad (513)$$

Modified polymers have been used, for example, as compositions for the treatment of glass fibres to improve their processing and performance characteristics [U-091].

The C=O group in polymers is also capable of hydrosilylation [756]:

$$\begin{array}{ccc} (CH_2{-}CH)_n & & (CH_2{-}CH)_n \\ | & & | \\ C{=}O & \xrightarrow[\text{RhCl(DIOP)S}]{(C_6H_5)_2SiH_2} & H{-}C{-}OSiH(C_6H_5)_2 \\ | & & | \\ CH_3 & & CH_3 \end{array} \qquad (514)$$

Generally, silyl substituted polydienes enhance the bonding strength of filled and reinforced polymer products, and also of conventional adhesives [327, U-059, U-082, U-120].

Linear and cyclic polyorganohydrosiloxanes effectively undergo addition to carbon–carbon bonds in various unsaturated organic compounds which usually contain functional groups such as esters [58, 572, U-034, U-070], ethers and epoxyethers [1209, D-127, D-129, D-136, J-094, U-083, U-084], carbonates and thiocarbonates [F-040], amines [56, 1117, U-189], nitriles [C-043, steroids [2], imides [F-041, U-194] of phospholipid-like structure [E-044] as well as alkenes [D-137], chiral alkenes [4] and chromopheric alkenes [G-038].

Polyorganosiloxanes with terminal Si–H bonds may be applied to introduce corresponding end-groups for various purposes [19, 1046, F-041, J-094]. The modification of cyclo-linear polymethylhydrosiloxane by acrylonitrile can also lead to the polyaddition of Si–H bonds to C≡N bonds [798]:

$$\equiv Si-H + CH_2{=}CHC{\equiv}N \xrightarrow{H_2PtCl_6} CH_2{=}CH-CH{=}NSi\equiv \qquad (515)$$

The co-polymerization of dihydro(poly)siloxanes with diolefins [36, E-021, F-038, F-039, U-190], polyethers [D-105, E-054, U-114, U-150], oligophenylenes [1039, J-092], fluoropolymers [J-111] and other monomers [292, 711, J-102], polyoxazaline [264], polyamides [J-113] has also been the subject of extensive investigations:

$$HSi{\sim}SiH + CH_2{=}CHCH_2O{\sim}CH_2CH{=}CH_2 \xrightarrow{H_2PtCl_6} [Si{\sim}Si(CH_2)_3O{\sim}(CH_2)_3]_n \qquad (516)$$

Kennedy *et al.* have applied hydrosilylation process to the synthesis of block co-polymers [351]. The hydrosilylation of polyisobutylene telomers by various chlorohydrosilanes proceeds quantitatively to finally yield a variety of glassy rubber block co-polymers. Hydrosilylation of living polystyrene anions PSt^- with an H–Si–Cl bond may be followed by coupling:

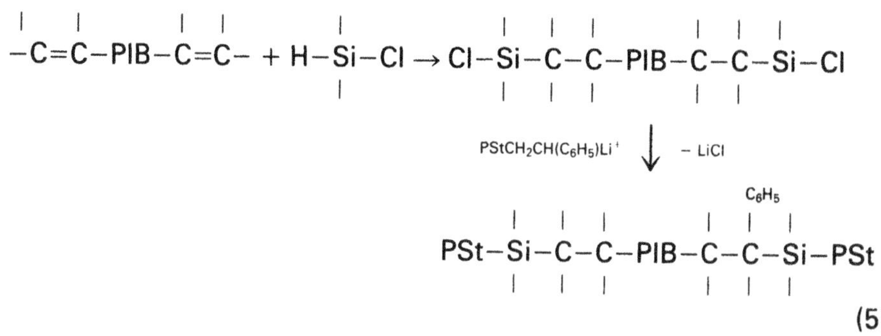

(517)

where PIB depicts polyisobutylene and PSt depicts polystyrene. In this way block copolymers with the following structures may be synthesized [351, 565]. (See Scheme 518.)

Statistical polydiene networks have also been prepared by reacting ligand polymers, e.g. polybutadiene, polyisoprene with telomeric siloxanes, e.g. $HSi(CH_3)_2OSi(CH_3)_2H$ and $[Si(CH_3)_2O]_5Si(CH_3)_2H$ [385] and others [D-143].

$$\mathrm{PSt - Si \sim PIB \sim Si - PSt;} \qquad \begin{array}{ccccc} \mathrm{PSt} & & & & \mathrm{PSt} \\ & \backslash & & / & \\ & \mathrm{Si} & \mathrm{\sim PIB \sim} & \mathrm{Si} & \\ & / & & \backslash & \\ \mathrm{PSt} & & & & \mathrm{St} \end{array} \qquad (518)$$

Block copolymers of aromatic polyethersulphone (PSU) and polydimethylsiloxane (PDMS) have been recently prepared by Pt-catalysed hydrosilylation of α,ω-di(vinylbenzyl)PSU or α,ω-di(allylether)PSU with α,ω-di(silane)PDMS [70]. The hydrosilylation reaction is commonly applied as a method of crosslinking vinyl- and allyl-containing polymers by multi-Si–H containing siloxanes and polysiloxanes [665, E-051, E-052, E-053].

Very recently preceramic polymers were reported to be synthesized by the reaction of polycarbosilanes containing Si–H groups with unsaturated compounds selected from the group consisting of reactive diolefins, reactive alkynes, polyolefins and vinyl-containing silicon derivatives [E-028]. Fine and uniform SiC (β-form) powders were synthesized by laser-driven gas-phase reactions of SiH_4 and C_2H_4 [1164].

Part II

A COMPILATION OF INFORMATION ON HYDROSILYLATION STUDIES OVER THE PERIOD FROM 1965 TO 1990

The following tables contain detailed information on hydrosilylation processes, of selected tri-substituted silanes, which are most frequently referred to in the literature, i.e. initial reagents, catalysts, reaction conditions (temperature, time, solvents, etc.), reaction products and their yields. They also provide references covering the bibliography from 1965 to the beginning of 1990.

The basic criterion whereby all the data are presented in the tables is the structure of the silicon hydride. The data are presented in order of increasing number of particular atoms such as Si, C, H and others (using the alphabetical order of the element symbol) in the silicon hydride molecule.

Each table which lists the reactions of a given silicon hydride, particularly those with many references, has been subdivided to show unsaturated compounds in separate subtables, mainly in the following order (similar to that discussed in Chapter 3):

1. Compounds containing a C=C bond (alkenes and arylalkenes, dienes, trienes and polyenes and other substituted derivatives).
2. Compounds containing a C≡C bond (alkynes and arylalkynes and their substituted derivatives).
3. Compounds containing a C=O bond (aldehydes, ketones, esters).
4. Compounds containing C=N, C≡N and N=N bonds.

The above listed order of unsaturated substrates is maintained or only slightly modified within a given table for those reactions where less information is available. For a given group of unsaturated compounds the data are arranged in order of increasing number of C and H atoms in the molecule. In order that an unsaturated substrate for a given reaction can be readily found in a table, discriminants C_xH_y have been placed to the left of that table. In a given table the columns are arranged in the following order: "Catalyst", "Temperature", "Time", "Addition Products", "Reaction Yield" and "References".

"Catalyst" is mainly listed in accordance with the order suggested in Chapter 2, i.e. after reactions in the absence of catalyst the following catalysts are tabulated: free-radical initiators, nucleophilic and electrophilic catalysts, supported metal catalysts and homogeneous and heterogeneous complex catalysts. Within a given catalyst group the information is classified according to the alphabetical order of element in the catalyst, i.e. amines are listed before phosphines, cobalt complexes before iron complexes, etc., and according to increasing complexity of the organic group, anions, ligands and the like. Where identical catalysts occur, particularly metal complexes, reaction data are successively listed in accordance with the increasing complexity of the catalyst solvent, e.g. chloroplatinic acid and/or co-catalyst. If the complexity is also identical in both cases, then the increasing reaction temperature (in °C) becomes a factor in the table listing. The reaction time is usually given in hours (h) unless it is necessary to denote it differently, i.e. in seconds (s) or days. The reaction solvent (if present or known) is noted in the column for the reaction time. The subsequent columns list product(s) and its (their) percentage yield and/or the product ratio. Unfortunately, optical yields are not listed.

The final column gives the reference numbers listed in alphabetical order according to the first author's name (see "Bibliography"). Russian names and journal title abbreviations have been transcribed in accordance with the rules laid down by *Chemical Abstracts*. Where references were commonly unavailable, the appropriate number of the abstract from *Chemical Abstracts* or *Ref. Zh. Khim.* is given. Patents are listed in the international notation, and those for given countries are in increasing patent number order. If the particular information has been patented in many countries, predominantly, only the number of the original patent is given in the tables. The list of patents includes mainly original patents.

In order to make the reader's task easier in finding information of interest, the following annexes follow:

1. "List of Tables", including chemical names and formulae of silicon hydrides.
2. "Abbreviations" for ligands, solvents, co-catalysts and other chemical compounds included in the tables and the text.

List of Tables

Abbreviations

a	Acetone
Ac	Acetyl
acac	Acetylacetonate
ACDVB	Allyl chloride–divinylbenzene co-polymer
AcO	Acetate
AIBN	Azoisobutyronitrile
All	Allyl
Π-All	Π-Allyl
b	Benzene
BCHD	Bicyclo[2.2.1]heptadiene-2,5
BDFB	1,2-Bis(diphenylphosphino)benzene
bipy	2,2′-Bipyridyl
BMPP	Benzylmethylphenylphosphine
Born	Bornyl
i-Born	iso-Bornyl
DPPFA	(2,2′-Diphenylphosphinoferrocenyl)ethyldimethylamine
Bu	Butyl
Bz	Benzyl
c	$CHCl_3$
c-Hex	Cyclohexyl
cod	*cis,cis*-1,5-Cyclooctadiene
cp	Cyclopentadienyl
cpd	Cyclopentadiene
cy	Cyclohexane
d	1,4-Dioxane
dc	Dichloroethane
DCE	Dichloroethylene
dcm	Dichloromethane
DCIUB	7,8-Dicarbo-indo-undecarborate
DESO	Diethylsulphoxide
dg	diglyme
diarsop	2,3-iso-Propylidenedioxy-,4 bis(diphenylarsine)butane
diop	2,3-iso-Propylidenedioxy-,4-bis(diphenylphosphine)butane
diphos	1,2-Bis(diphenylphosphine)ethane
DIPPC	2,6-Di(isopropyl)phenylisocyanide

dmg	Dimethylglyoxime
DMF	Dimethylformamide
DMPC	1,2-Bis(dimethylphenyl)carborane
DMPF	Dimethylphosphinoferrocene
DMPIC	2,6-Dimethylphenylphosphine
DMPP	Dimethylphenylphosphine
DMSO	Dimethylsulphoxide
e	Diethylether
EDM	Ethylenedimethacrylate
Et	Ethyl
Fc	Ferrocenyl
h	Hexane
Hept	Heptyl
Hex	Hexyl
ht	Heptane
kin.	Kinetic study
L	Ligand
LAN	Lauric acid nitrile
MCE	Cyanoethyl methacrylate
MDPP	Menthyldiphenylphosphine
Me	Methyl
MEE	2-Dimethylaminoethyl methacrylate
Ment	Menthyl
n-Ment	Neomenthyl
Mes	Mesityl
m.i.	Isomer mixture
MPFA	(2-Dimethylphosphinoferrocenyl)ethyldimethylamine
MPPP	Methylphenylpropylphosphine
NMDPP	Neomenthyldiphenylphosphine
Np	Naphtyl
Oct	Octyl
p	Pentane
PDPPS	Poly-p-diphenylphosphinostyrene
Pent	Pentyl
Ph	Phenyl
PMC	Polymethacrylate
PMM	Polymethylmethacrylate
PPFA	*N,N*-Dimethyl(isodiphenylphosphine)(ferrocenyl)-ethylamine
Pr	Propyl
PS	Polystyrene
py	Pyridine
r.t.	Room temperature
s	Second
SDVB	Styrene-divinylbenzene copolymer
SG	Silica gel
solv.	Solvent(s)
t	Toluene

TCE	Trichloroethylene
te	1,1,2-Trichloroethane
THF	Tetrahydrofuran
X	Anion
xy	Xylene

Table 1.1

= $HSiCl_3$ =

=1=

C_2H_4	$CH_2=CH_2$				
	UV	130-275	-	$C_2H_5SiCl_3$ (kin.)	320
	UV + Me_2CO	100-275		$C_2H_5SiCl_3$ (-)	325
	none	325	2	$C_2H_5SiCl_3$ (64) $C_4H_9SiCl_3$ (9) $C_6H_{13}SiCl_3$ (2)	D -097
	$Co_2(CO)_8PtCl_2$	reflux	2(THF)	$C_2H_5SiCl_3$ (60)	U -053
	$Co_2(CO)_8PtCl_2$	reflux	2(THF)	$C_2H_5SiCl_3$ (60)	U -102
	Pd + $AsPh_2Cl$	120	6	$C_2H_5SiCl_3$ (40)	D -027
	Pd +PPh_3	120	5	$C_2H_5SiCl_3$ (93)	D -027
	$Pd(PPh_3)_4$	110	6	$C_2H_5SiCl_3$ (90)	1208
	$PdCl_2$+ PPh_3	100	5	$C_2H_5SiCl_3$ (89)	D -027
	H_2PtCl_6	-38	37	$C_2H_5SiCl_3$ (17) $C_4H_9SiCl_3$ (7) $C_6H_{13}SiCl_3$ (trace)	D -097
	H_2PtCl_6 ($MeOC_2H_4OMe$ + EtOH + $\overline{C_6H_4NHC_6H_4S}$)	-	2-6	$C_2H_5SiCl_3$ (5-14)	U -085
	H_2PtCl_6 + ($MeOC_2H_4OMe$ + EtOH)	-	2-19	$C_2H_5SiCl_3$ (2-12)	U -085
	H_2PtCl_6 [$(\overline{CH_2)_5CO}$]	reflux	-	$C_2H_5SiCl_3$ (100)	D -041
	Pt/C	110-120	1	$C_2H_5SiCl_3$ (92) $C_4H_9SiCl_3$ (8)	D -097
	$[RhCl(CO)_2]_2$	20	65	$C_2H_5SiCl_3$ (95)	227
	$[RhCl(CO)_2]_2$	200	4 (xy)	$C_2H_5SiCl_3$ (-) $CH_2=CHSiCl_3$ (80)	U -173
C_3H_6	$CH_3CH=CH_2$				
	UV	130-275	-	$CH_3CH_2CH_2SiCl_3$ (kin.)	320
	UV + Me_2CO	103-250	-	$CH_3CH_2CH_2SiCl_3$ + $(CH_3)_2CHOSiCl_3$ (kin.)	322
	gamma(^{60}Co)	80	-	$CH_3CH_2CH_2SiCl_3$ (94)	1360
	none	280	2	$CH_3CH_2CH_2SiCl_3$ (25)	1085
	Pd +PPh_3	120	8	$C_3H_7SiCl_3$ (94)	D -027
	$Pd(PBu_3)_4$	120	5	$C_3H_7SiCl_3$ (86)	D -027

	$Pd(PPh_3)_4$	120	5	$C_3H_7SiCl_3$ (92)	1208
	H_2PtCl_6 (i-PrOH)	-	-	$C_3H_7SiCl_3$ (-)	141
	H_2PtCl_6 (i-PrOH)	70-122	2	$C_3H_7SiCl_3$ (95)	D -013
	H_2PtCl_6 [$(\overline{CH_2)_5CO}$]	-	-	$C_3H_7SiCl_3$ (100)	D -041
	Pt/C	110-120	1	$C_3H_7SiCl_3$ (92)	D -096
	$PtCl_2(acac)_2$	-	flow	$C_3H_7SiCl_3$ (-)	D -013
	$PtCl_2(acac)_2$ (C_6H_6)	70	-	$C_3H_7SiCl_3$ (96)	J -011
C_4H_8	C_4H_8				
	H_2PtCl_6	145	-	$C_4H_9SiCl_3$ (96)	D -013
C_4H_8	$(CH_3)_2C=CH_2$				
	UV + Me_2CO	103-250	-	$CH_3(CH_2)_3SiCl_3$ + $(CH_3)_2CHOSiCl_3$ (kin.)	322
	$Mn(CO)_3(cpd)$	240	2	$C_4H_9SiCl_3$ (65)	38
	$Pd[P(c\text{-}Hex)_3]_4$	120	10	$(CH_3)_2CHCH_2SiCl_3$ (61)	D -027
C_4H_8	$(CH_3)_2C=CH_2$ + $CH_3CH_2CH=CH_2$				
	H_2PtCl_6 [$(\overline{CH_2)_5CO}$]	30-120	16	$CH_3(CH_2)_3SiCl_3$ + $(CH_3)_2CHCH_2SiCl_3$ (-)	Pl-013
	H_2PtCl_6 [$(\overline{CH_2)_5CO}$]	76-120	5.5(prod.)	$CH_3(CH_2)_3SiCl_3$ + $(CH_3)_2CHCH_2SiCl_3$ (-)	Pl-012
	SiO_2/Pt(II)complex	40-90	28(prod.)	$CH_3(CH_2)_3SiCl_3$ + $(CH_3)_2CHCH_2SiCl_3$ (-)	Pl-014
	$RhCl(PPh_3)_3$	60-120	7(prod.)	$CH_3(CH_2)_3SiCl_3$ + $(CH_3)_2CHCH_2SiCl_3$ (-)	Pl-012
C_4H_8	$CH_3CH_2CH=CH_2$				
	UV	130-275	-	$CH_3(CH_2)_3SiCl_3$ (kin.)	320
	UV + Me_2CO	103-250	-	$C_4H_9SiCl_3$ + $(CH_3)_2CHOSiCl_3$ (kin.)	322
	UV + Me_2CO	78-272	-	$C_4H_9SiCl_3$ + $(CH_3)_2CHOSiCl_3$ (kin.)	323
	$Pd[P(4\text{-}C_6H_{10}Me)_3]_4$	120	7	$CH_3(CH_2)_3SiCl_3$ (89)	D -038
	H_2PtCl_6 (i-PrOH)	145	4	$C_4H_9SiCl_3$ (96)	D -013
	H_2PtCl_6 [$(\overline{CH_2)_5CO}$]	30-145	20	$CH_3(CH_2)_3SiCl_3$ (-)	Pl-013
	SiO_2/Pt(II)complex	40-90	6(prod.)	$CH_3(CH_2)_3SiCl_3$ (-)	Pl-014
C_4H_8	$CH_3CH=CHCH_3$				

UV (cis)	152-252	-	$CH_3CH_2CH(CH_3)SiCl_3$ (kin.)	318
UV + Me_2CO (cis+trans)	167	-	$CH_3CH_2CH(CH_3)SiCl_3$ + $(CH_3)_2CHOSiCl_3$ (kin.)	319
C_5H_8 $\overline{(CH_2)_3CH=CH}$				
UV + Me_2CO	167	-	$\overline{(CH_2)_4CH}SiCl_3$ + $(CH_3)_2CHOSiCl_3$ (kin.)	321
H_2PtCl_6(i-PrOH)	120	3	$\overline{(CH_2)_4CH}SiCl_3$ (53)	784
C_5H_{10} $(CH_3)_2C=CHCH_3$				
UV + Me_2CO	167	-	$(CH_3)_2CHCH(CH_3)SiCl_3$ + $CH_3CH_2C(CH_3)_2SiCl_3$ + $(CH_3)_2CHOSiCl_3$ (kin.)	319
gamma(^{60}Co)	16	19	$(CH_3)_2CHCH(CH_3)SiCl_3$ (99)	958
H_2PtCl_6	120	2	$(CH_3)_2CHCH(CH_3)SiCl_3$ (-)	215
C_5H_{10} $CH_3CH_2C(CH_3)=CH_2$				
gamma(^{60}Co)	16	14	$C_5H_{11}SiCl_3$ (99)	958
H_2PtCl_6	100	20	$CH_3CH_2CH(CH_3)CH_2SiCl_3$ (93)	1136
Pt/C	100	20	$CH_3CH_2CH(CH_3)CH_2SiCl_3$ (55)	1136
cis-$PtCl_2$(BMPP)(C_2H_4)	-	-	$CH_3CH_2CH(CH_3)SiCl_3$ (54)	1303
C_5H_{10} $CH_3CH_2C(CH_3)=CH_2$ + $CH_3CH=C(CH_3)_2$				
H_2PtCl_6 [$\overline{(CH_2)_5CO}$]	120	-	$C_5H_{11}SiCl_3$ (-)	D -103
C_5H_{10} $CH_3CH_2CH_2CH=CH_2$				
UV	130-275	-	$CH_3(CH_2)_4SiCl_3$ (kin.)	320
UV + Me_2CO	103-250	-	$C_5H_{11}SiCl_3$ + $(CH_3)_2CHOSiCl_3$ (kin.)	322
gamma(^{60}Co)	16	15	$C_5H_{11}SiCl_3$ (99)	958
none	110-115	1.16	$CH_3(CH_2)_4SiCl_3$ (95)	D -096
$Ir_4(CO)_{12}$	30-35	1	$C_5H_{11}SiCl_3$ (98)	S -037
$PdCl_2(PPh_3)_2$	120	5	$C_5H_{11}SiCl_3$ (91)	D -027
H_2PtCl_6	reflux	0.67	$C_5H_{11}SiCl_3$ (100)	F -026
H_2PtCl_6	reflux	0.67	$C_5H_{11}SiCl_3$ (100)	U -074
H_2PtCl_6 [$\overline{(CH_2)_5CO}$]	reflux	-	$CH_3(CH_2)_4SiCl_3$ (100)	D -041
$[HPt(SiCl_3)P(c\text{-}Hex)_3]_2$	20	4(t)	$C_5H_{11}SiCl_3$ (70)	429

Table 1.1

= $HSiCl_3$ = =4=

	$[PtCl_2(c\text{-}C_3H_6)]_2$	70-170	6	$CH_3(CH_2)_4SiCl_3$ (92)	U -005
	$Rh_4(CO)_{12}$	30-35	1	$C_5H_{11}SiCl_3$ (96-98)	S -037
C_5H_{10}	$CH_3CH_2CH{=}CHCH_3$				
	UV + Me_2CO (cis)	167	-	$CH_3CH_2CH_2CH(CH_3)SiCl_3$ + $CH_3CH_2CH(CH_2CH_3)SiCl_3$ + $(CH_3)_2CHOSiCl_3$ (kin.)	319
	gamma(^{60}Co)	16	20	$CH_3CH_2CH_2CH(CH_3)SiCl_3$ (99)	958
	H_2PtCl_6	reflux	0.67(p)	$CH_3(CH_2)_4SiCl_3$ (95)	F -026
	H_2PtCl_6	reflux	0.67(p)	$CH_3(CH_2)_4SiCl_3$ (95)	U -074
	H_2PtCl_6 (i-PrOH)	130	0.5	$CH_3(CH_2)_4SiCl_3$ (97)	113
	H_2PtCl_6 (i-PrOH)	reflux	0.67	$CH_3(CH_2)_4SiCl_3$ (95)	U -100
C_6H_{10}	$(\overline{CH_2)_4CH{=}CH}$				
	UV + $(t\text{-}BuO)_2$	40	12	$(\overline{CH_2)_5CH}SiCl_3$ (84)	126
	UV + $(t\text{-}BuO)_2$	40	14	$(\overline{CH_2)_5CH}SiCl_3$ (62)	338
	UV + $Hg(SiMe_3)_2$	40	12	$(\overline{CH_2)_5CH}SiCl_3$ (80)	126
	gamma(^{60}Co)	16	15	$(\overline{CH_2)_5CH}SiCl_3$ (98)	958
	H_2PtCl_6	120	2	$(\overline{CH_2)_5CH}SiCl_3$ (-)	215
	H_2PtCl_6 (i-PrOH)	100	20	$(\overline{CH_2)_5CH}SiCl_3$ (78)	1192
	SDVB-NMe_2/H_2PtCl_6	50	3	$(\overline{CH_2)_5CH}SiCl_3$ (82)	C -002
	$Rh_4(CO)_{12}$	30-35	1	$(\overline{CH_2)_5CH}SiCl_3$ (96-99)	S -037
C_6H_{12}	$(CH_3)_2CHC(CH_3){=}CH_2$				
	cis-$[PtCl_2(BMPP)(C_2H_4)]$	90	40(b)	$(CH_3)_2CHCH(CH_3)CH_2SiCl_3$ (70)	1303
C_6H_{12}	$(CH_3)_2CHCH_2CH{=}CH_2$				
	gamma(^{60}Co)	16	19	$(CH_3)_2CH(CH_2)_3SiCl_3$ (90)	958
	Pt/C + sonic waves	30	1	$(CH_3)_2CH(CH_2)_3SiCl_3$ (94)	467
C_6H_{12}	$(CH_3)_3CCH{=}CH_2$				
	H_2PtCl_6	120	2	$C_6H_{13}SiCl_3$ (-)	215
	H_2PtCl_6 + Et_3SiH	57	0.06	$(CH_3)_3CCH_2CH_2SiCl_3$ (8) $(CH_3)_3CCH_2CH_2Si(C_2H_5)_3$ (60)	U -168

	H_2PtCl_6(i-PrOH)	r.t.	12	$(CH_3)_3CCH_2CH_2SiCl_3$ (71)	1192
C_6H_{12}	$CH_3(CH_2)_3CH{=}CH_2$				
	UV + $(t\text{-}BuO)_2$	0-40	9-14	$C_6H_{13}SiCl_3$ (61-88)	338
	gamma(^{60}Co)	16	23	$C_6H_{13}SiCl_3$ (97)	958
	gamma(^{60}Co)	52	-	$C_6H_{13}SiCl_3$ (kin.)	958
	Ni(THF)+ PPh_3	0-25	20	$CH_3(CH_2)_5SiCl_3$ (75)	E -049
	Ni(THF) + PPh_3 + sonic waves	0-25	2-3	$CH_3(CH_2)_5SiCl_3$ (62-81)	E -049
	polymer/Ni complex	100	2	$C_6H_{13}SiCl_3$ (5)	1220
	$K_2[Pd(C{\equiv}CH)_2]_2$ + PBu_3	120	10	$CH_3(CH_2)_5SiCl_3$ (57)	D -027
	Na_2PdCl_4+ P(c-Hex)$_3$	130	10	$CH_3(CH_2)_5SiCl_3$ (58)	D -027
	Pd + $AsHPh_2$	130	10	$CH_3(CH_2)_5SiCl_3$ (-)	F -019
	Pd + $AsPh_3$	120	10	$CH_3(CH_2)_5SiCl_3$ (51)	F -019
	Pd +PPh_3	120	10	$C_6H_{13}SiCl_3$ (92)	468
	Pd + $SbPh_3$	130	10	$CH_3(CH_2)_5SiCl_3$ (40)	F -019
	$Pd(AcO)_2$ + PBu_3	80	10	$C_6H_{13}SiCl_3$ (76)	468
	$Pd(CN)_2$ + PPh_3	120	8	$CH_3(CH_2)_5SiCl_3$ (70)	D -027
	$Pd(HCOO)_2$ + PPh_3	110	5	$CH_3(CH_2)_5SiCl_3$ (59)	D -027
	$Pd(NO_3)_2$ + PPh_3	100	8	$CH_3(CH_2)_5SiCl_3$ (82)	D -027
	$Pd(O)_2(PPh_3)_2$	80	8	$CH_3(CH_2)_5SiCl_3$ (96)	D -027
	Pd($\overline{OCOCH{=}CHCO}$)$(PPh_3)_2$	120	5	$CH_3(CH_2)_5SiCl_3$ (95)	D -027
	$Pd(OPh_2)_2(PPh_3)_2$	120	5	$CH_3(CH_2)_5SiCl_3$ (80)	D -027
	$Pd(PPh_3)_2L$ L=p-Benzoquinone;1,4-Naphtoquinone; $Cl_2C{=}CHCl$; $(NC)_2C{=}C(CN)_2$	110-130	5-10	$CH_3(CH_2)_5SiCl_3$ (60-95)	D -027
	$Pd(PPh_3)_4$	100	5	$C_6H_{13}SiCl_3$ (90)	468
	$Pd(PPh_3)_4$	110	6	$CH_3(CH_2)_5SiCl_3$ (90)	1208
	$Pd(PPh_3)_4$	120	5	$CH_3(CH_2)_5SiCl_3$ (90)	J -002
	$Pd(PPh_3)_4$	120	5	$CH_3(CH_2)_5SiCl_3$ (90)	D -027
	$Pd(PhCOO)_2$ + PPh_3	120	8	$CH_3(CH_2)_5SiCl_3$ (75)	D -027

Table 1.1

= $HSiCl_3$ = =6=

$Pd(acac)_2 + PPh_3$	120	10	$CH_3(CH_2)_5SiCl_3$ (83)	D -027
$Pd(acac)_2(\pi\text{-All}) + PPh_3$	120	5	$CH_3(CH_2)_5SiCl_3$ (76)	D -027
$Pd/C + PPh_3$	120	5	$CH_3(CH_2)_5SiCl_3$ (81)	D -027
$PdBr_2(PEt_3)_2$	80	10	$CH_3(CH_2)_5SiCl_3$ (65)	D -027
$PdCl_2 + PPh_3$	110-120	5-6	$CH_3(CH_2)_5SiCl_3$ (66-82)	D -027
$PdCl_2(\pi\text{-All})_2$	110	8	$CH_3(CH_2)_5SiCl_3$ (70)	D -027
$PdCl_2(BCHD) + AsPh_3$	150	4	$CH_3(CH_2)_5SiCl_3$ (56)	D -027
$PdCl_2(NH_2CH_2COOH)_2 + PEt_3$	120	8	$CH_3(CH_2)_5SiCl_3$ (75)	D -027
$PdCl_2(PPh_3)_2$	120	5	$C_6H_{13}SiCl_3$ (91)	468
$PdCl_2(cod) + PBu_3$	120	7	$CH_3(CH_2)_5SiCl_3$ (86)	D -027
$PdCl_2L_2 + PPh_3$ L=$PhNH_2$, $(\overline{CH_2)_5C}$=NOH, L_2 = HON=C(Me)C(Me)=NOH	120-130	6-8	$CH_3(CH_2)_5SiCl_3$ (68-76)	D -027
$PdCl_2[P(OBu)_3]_2$	120	5	$CH_3(CH_2)_5SiCl_3$ (77)	D -027
$PdI_2(PEt_3)_2$	100	10	$CH_3(CH_2)_5SiCl_3$ (61)	D -027
$PdO + PPh_3$	110	6	$CH_3(CH_2)_5SiCl_3$ (80)	D -027
$[Pd(PPh_3)_2]_2$	120	5	$CH_3(CH_2)_5SiCl_3$ (92)	D -027
polymer-Pd	70	24	$CH_3(CH_2)_5SiCl_3$ (0-26)	293
$Al_2O_3\text{-}PPh_2/H_2PtCl_6$	80	2-3	$C_6H_{13}SiCl_3$ (84-87)	S -006
H_2PtCl_6 (i-PrOH) + CH_2=CHOBu	90-120	1-18	$C_6H_{13}SiCl_3$ (98)	S -012
$H_2PtCl_6 + Me_2SiHN(SiMe_3)_2$	reflux	2	$C_6H_{13}SiCl_3$ (90)	193
Pt/C + sonic waves	30	1	$CH_3(CH_2)_5SiCl_3$ (90)	467
$PtIMe_3$	125	3	$C_6H_{13}SiCl_3$ (-)	U -021
$[HPt(SiCl_3)P(c\text{-}Hex)_3]_2$	20	7(t)	$C_6H_{13}SiCl_3$ (70)	429
polymer/Pt complex	100	2	$C_6H_{13}SiCl_3$ (70)	1220
polymer/Pt complex	25	20	$C_6H_{13}SiCl_3$ (81)	1220
polymer-Pt	70	24	$CH_3(CH_2)_5SiCl_3$ (94-95)	293
$Al_2O_3\text{-}CN/[RhCl(CO)_2]_2$	120	3	$CH_3(CH_2)_5SiCl_3$ (61)	Dd-003
C + $C_6H_8(OC_2H_4CN)_6/[RhCl(CO)_2]_2$	120	3	$CH_3(CH_2)_5SiCl_3$ (57)	Dd-003

C/$RhCl_3$ (c-$C_{12}H_{18}$ + EtOH)	130	3	$CH_3(CH_2)_5SiCl_3$ (10)	Dd-003
C-CN/$[RhCl(CO)_2]_2$	150	3	$CH_3(CH_2)_5SiCl_3$ (57)	Dd-003
$CaCO_3$ + $(CH_2OC_2H_4CN)_2$/$[RhCl(CO)_2]_2$	120	3	$CH_3(CH_2)_5SiCl_3$ (77)	Dd-003
$Rh_4(CO)_{12}$	30-35	1	$C_6H_{13}SiCl_3$(95-99)	S -037
$RhBr(PPh_3)_3$	120	2(b)	$CH_3(CH_2)_5SiCl_3$ (52)	981
$RhCl(CO)(PEt_3)_2$	30	55(b)	$C_6H_{13}SiCl_3$(2)	232
$RhCl(CO)(PEt_3)_2$	55	30	$C_6H_{13}SiCl_3$ (2)	496
$RhCl(CO)(PPh_3)_2$	120	2(b)	$CH_3(CH_2)_5SiCl_3$(41)	981
$RhCl(CO)(PPh_3)_2$	15	75	$C_6H_{13}SiCl_3$(90)	232
$RhCl(CO)(PPh_3)_2$	3	50(b)	$C_6H_{13}SiCl_3$(5)	232
$RhCl(CO)(PPh_3)_2$	75	15	$C_6H_{13}SiCl_3$ (90)	496
$RhCl(PPh_3)_3$	120	2(b)	$CH_3(CH_2)_5SiCl_3$ (68)	981
$RhCl(PPh_3)_3$	15	75	$C_6H_{13}SiCl_3$ (40)	232
$RhCl(PPh_3)_3$	3	50 (b)	$C_6H_{13}SiCl_3$ (5)	232
$RhCl(PPh_3)_3$	60	6 days	$C_6H_{13}SiCl_3$ (8)	476
$RhCl_3[(EtO)_3SiH]$	reflux	20	$CH_3(CH_2)_5SiCl_3$ (-)	F -004
$RhH(CO)(PPh_3)_3$	120	2	$CH_3(CH_2)_5SiCl_3$ (39)	981
$RhI(PPh_3)_3$	120	2(b)	$CH_3(CH_2)_5SiCl_3$ (44)	981
$[RhCl(CO)_2]_2$	15	20	$C_6H_{13}SiCl_3$ (45)	232
$[RhCl(CO)_2]_2$	2	70	$C_6H_{13}SiCl_3$ (80)	232
porolith-$(CH_2OC_2H_4CN)_2$/$[RhCl(CO)_2]_2$	120	3	$CH_3(CH_2)_5SiCl_3$ (31)	Dd-003
zeolite-CN/$RhCl(CO)(PPh_3)_2$	120	3	$CH_3(CH_2)_5SiCl_3$ (51)	Dd-003

C_6H_{12} $CH_3CH_2CH_2C(CH_3)=CH_2$

gamma(^{60}Co)	16	24	$CH_3CH_2CH_2C(CH_3)_2SiCl_3$ (55)	958
none	260	5	$C_6H_{13}SiCl_3$ (10-16)	38
$Mn(CO)_3$(cpd)	260	5	$CH_3CH_2CH_2C(CH_3)_2SiCl_3$ (60-70)	38
$Pd(PPh_3)_4$	120	10	$CH_3CH_2CH_2CH(CH_3)CH_2SiCl_3$ (55)	D -027

	Pt/C + sonic waves	30	2	$CH_3CH_2CH_2CH(CH_3)CH_2SiCl_3$ (71)	467
C_6H_{12}	$CH_3CH_2CH{=}C(CH_3)_2$				
	none	260	5	$C_6H_{13}SiCl_3$ (45-48)	38
C_6H_{12}	$CH_3CH{=}CH_2$ dimer				
	$Rh_4(CO)_{12}$	30-35	1	$C_6H_{13}SiCl_3$ (96-98)	S -037
C_7H_{10}	Fig.1.				
	$PdCl_2$(PPFA)	70	40	Fig.4. (53)	485
	H_2PtCl_6(i-PrOH)	reflux	40	Fig.4. (79)	1192
	$[HPt(SiCl_3)P(c\text{-}Hex)_3]_2$	20	72	Fig.4. (58)	429
	$[HPt(SiCl_3)P(c\text{-}Hex)_3]_2$	65	8	Fig.4. (76)	429
C_7H_{12}	Fig.2.				
	gamma(^{60}Co)	16	16	Fig.5. (97)	958
C_7H_{12}	Fig.3.				
	$PtCl_2(C_2H_4)$	-	6 days	Fig.6. (97)	E -004
C_7H_{14}	$(CH_3)_3CCH_2CH{=}CH_2$				
	H_2PtCl_6	120	2	$C_7H_{15}SiCl_3$ (-)	215
C_7H_{14}	$(CH_3CH_2)_2CHCH{=}CH_2$				
	H_2PtCl_6 (THF)	110 (b)	-	$(CH_3CH_2)_2CHCH_2CH_2SiCl_3$ (-)	907
C_7H_{14}	$CH_3(CH_2)_4CH{=}CH_2$				
	gamma(^{60}Co)	16	18	$C_7H_{15}SiCl_3$ (92)	958
	$Co_2(CO)_8$	120	-	$C_7H_{15}SiCl_3$ (kin.)	1169
	ACDVB-PPh_2/H_2PtCl_6	80	2	$C_7H_{15}SiCl_3$ (93)	217
	H_2PtCl_6	120	2	$C_7H_{15}SiCl_3$ (-)	215
	H_2PtCl_6 (THF)	100	4	$C_7H_{15}SiCl_3$ (-)	214
	H_2PtCl_6 (i-PrOH)	-	0.1	$C_7H_{15}SiCl_3$ (80)	96
	H_2PtCl_6 (i-PrOH) + DCIUB	20	24	$C_7H_{15}SiCl_3$ (80)	96
	PMC-CN/H_2PtCl_6	24	24	$C_7H_{15}SiCl_3$ (28)	217

Table 1.1
= $HSiCl_3$ = =9=

PMC-CN/H_2PtCl_6	80	2	$C_7H_{15}SiCl_3$ (88)	217
PMC-NMe_2/H_2PtCl_6	24	24	$C_7H_{15}SiCl_3$ (87)	217
PMC-NMe_2/H_2PtCl_6	80	2	$C_7H_{15}SiCl_3$ (90)	217
PMC-PPh_2/H_2PtCl_6	80	2	$C_7H_{15}SiCl_3$ (3-8)	217
SDVB-CN/H_2PtCl_6	80	2	$C_7H_{15}SiCl_3$ (79)	217
SDVB-NMe_2/H_2PtCl_6	24	24	$C_7H_{15}SiCl_3$ (95)	217
SDVB-NMe_2/H_2PtCl_6	80	2	$C_7H_{15}SiCl_3$ (97)	217
SDVB-PPh_2/H_2PtCl_6	24	24	$C_7H_{15}SiCl_3$ (69)	217
SDVB-PPh_2/H_2PtCl_6	80	2	$C_7H_{15}SiCl_3$ (73)	217
polymer-NMe_2/H_2PtCl_6	80	2	$C_7H_{15}SiCl_3$ (79)	217
zeolite-PPh_2/H_2PtCl_6	70	4	$CH_3(CH_2)_6SiCl_3$ (87)	C -007
ACDVB-PPh_2/$RhCl_3$	80	2	$C_7H_{15}SiCl_3$ (7)	217
Al_2O_3-CN/$[RhCl(CO)_2]_2$	120	3	$CH_3(CH_2)_6SiCl_3$ (61)	C -030
C + $C_6H_8(OC_2H_4CN)_6$/$[RhCl(CO)_2]_2$	120	3	$CH_3(CH_2)_6SiCl_3$ (57)	C -030
C-CN/$RhCl_3$ + c-$C_{12}H_{18}$	150	3	$CH_3(CH_2)_6SiCl_3$ (16)	C -030
C-CN/$[RhCl(CO)_2]_2$	150	3	$C_7H_{15}SiCl_3$ (57)	C -030
$CaCO_3$ + $(CH_2OC_2H_4CN)_2$/$[RhCl(CO)_2]_2$	120	3	$CH_3(CH_2)_6SiCl_3$ (77)	C -030
PMC-NMe_2/$RhCl_3$	80	2	$C_7H_{15}SiCl_3$ (24)	217
$RhCl(PHPh_2)_3$	85	2	$CH_3(CH_2)_6SiCl_3$ (10)	1166
SDVB-CH_2Cl/$RhCl(CO)(PPh_3)_2$	150	3	$CH_3(CH_2)_6SiCl_3$ (71)	C -022
SDVB-CH_2Cl/$RhCl(CO)(PPh_3)_2$	80	3	$CH_3(CH_2)_6SiCl_3$ (1)	C -022
SDVB-CH_2Cl/$RhCl_3$(THF)	120	3	$CH_3(CH_2)_6SiCl_3$ (29)	C -022
SDVB-NMe_2/$[RhCl(CO)_2]_2$	100	2	$C_7H_{15}SiCl_3$ (3-72)	316
$[PS-PPh_2RhCl(PPh_3)_2]_n$	120	2	$CH_3(CH_2)_6SiCl_3$ (44)	C -008
$[RhCl(C_2H_4)_2]_2$	85	2 (b)	$CH_3(CH_2)_6SiCl_3$ (54)	1167
glass + $(CH_2OC_2H_4CN)_2$/$[RhCl(CO)_2]_2$	120	3	$CH_3(CH_2)_6SiCl_3$ (31)	C -030
support/Rh complex	80	2	$C_7H_{15}SiCl_3$ (6-47)	204

	zeolite-CN/RhCl(CO)(PPh$_3$)$_2$	120	3	CH$_3$(CH$_2$)$_6$SiCl$_3$ (51)	C -030
C$_7$H$_{14}$	CH$_3$CH$_2$CH=CHCH$_2$CH$_2$CH$_3$ (cis,trans)				
	H$_2$PtCl$_6$	120	3	CH$_3$(CH$_2$)$_6$SiCl$_3$ (-)	215
C$_7$H$_{14}$	CH$_3$CH=CH(CH$_2$)$_3$CH$_3$				
	H$_2$PtCl$_6$ (i-PrOH)	110	8	CH$_3$(CH$_2$)$_6$SiCl$_3$ (87)	114
C$_7$H$_{14}$	CH$_3$CH=CH(CH$_2$)$_3$CH$_3$ (cis)				
	[RhCl(C$_2$H$_4$)]$_2$	85	2 (b)	CH$_3$(CH$_2$)$_6$SiCl$_3$ (10)	1167
C$_7$H$_{14}$	CH$_3$CH=CH(CH$_2$)$_3$CH$_3$ (trans)				
	H$_2$PtCl$_6$	120	3	CH$_3$(CH$_2$)$_6$SiCl$_3$ (-)	215
	[RhCl(C$_2$H$_4$)]$_2$	85	2 (b)	CH$_3$(CH$_2$)$_6$SiCl$_3$ (1)	1167
C$_8$H$_8$	C$_6$H$_5$CH=CH$_2$				
	Co$_2$(CO)$_8$	20-30	70	C$_6$H$_5$CH$_2$CH$_2$SiCl$_3$ (-)	715
	Co$_2$(CO)$_8$	30-40	1	C$_6$H$_5$CH$_2$CH$_2$SiCl$_3$ (91)	719
	Ir$_4$(CO)$_{12}$	30-35	1	C$_6$H$_5$CH$_2$CH$_2$SiCl$_3$ (90)	S -023
	Ni(AcO)$_2$ +PPh$_3$ + CuCl	25-150	3	C$_6$H$_5$CH$_2$CH$_2$SiCl$_3$ (4) C$_6$H$_5$CH(CH$_3$)SiCl$_3$ (31)	125
	Ni(CO)$_4$	80	2	C$_6$H$_5$CH(CH$_3$)SiCl$_3$ (97)	213
	Ni(THF)+ PPh$_3$	0-25	20	C$_6$H$_5$CH$_2$CH$_2$SiCl$_3$ (58)	E -049
	Ni(THF) + PPh$_3$, sonic waves	0-25	2	C$_6$H$_5$CH$_2$CH$_2$SiCl$_3$ (88)	E -049
	Ni(acac)$_2$ + PPh$_3$ + CuCl	25-175	3	C$_6$H$_5$CH$_2$CH$_2$SiCl$_3$ (4) C$_6$H$_5$CH(CH$_3$)SiCl$_3$ (31)	125
	NiBr$_2$+ PPh$_3$	25-150	16	C$_6$H$_5$CH$_2$CH$_2$SiCl$_3$ (5) C$_6$H$_5$CH(CH$_3$)SiCl$_3$ (12)	125
	NiBr$_2$ + PPh$_3$ + CuBr	25-145	3	C$_6$H$_5$CH$_2$CH$_2$SiCl$_3$ (4) C$_6$H$_5$CH(CH$_3$)SiCl$_3$ (16)	125
	NiBr$_2$ + PPh$_3$ + CuCl	25-145	3	C$_6$H$_5$CH$_2$CH$_2$SiCl$_3$ (4) C$_6$H$_5$CH(CH$_3$)SiCl$_3$ (66)	125
	NiBr$_2$ + PPh$_3$ + CuI	25-145	3	C$_6$H$_5$CH$_2$CH$_2$SiCl$_3$ (5) C$_6$H$_5$CH(CH$_3$)SiCl$_3$ (10)	125
	NiCl$_2$ + C$_5$H$_5$N	155-170	4	C$_6$H$_5$CH$_2$CH$_2$SiCl$_3$ +C$_6$H$_5$CH(CH$_3$)SiCl$_3$ (30)	125

Table 1.1

= $HSiCl_3$ = =11=

$NiCl_2$ + NBu_3 + CuCl	25-145	3	$C_6H_5CH_2CH_2SiCl_3$ (3) $C_6H_5CH(CH_3)SiCl_3$ (17)	125
$NiCl_2$+ PBu_3	25-150	3	$C_6H_5CH_2CH_2SiCl_3$ (1) $C_6H_5CH(CH_3)SiCl_3$ (19)	125
$NiCl_2$ + PBu_3 + CuCl	25-175	3	$C_6H_5CH_2CH_2SiCl_3$ (6) $C_6H_5CH(CH_3)SiCl_3$ (54)	125
$NiCl_2$+ PPh_3	25-150	3	$C_6H_5CH_2CH_2SiCl_3$ (2) $C_6H_5CH(CH_3)SiCl_3$ (8)	125
$NiCl_2$ + PPh_3 + CuCl	25-150	3	$C_6H_5CH_2CH_2SiCl_3$ (4) $C_6H_5CH(CH_3)SiCl_3$ (71)	125
$NiCl_2$ + diphos + CuCl	-	-	$C_6H_5CH_2CH_2SiCl_3$ (10) $C_6H_5CH(CH_3)SiCl_3$ (60)	125
NiO + PPh_3 + CuCl	25-160	3	$C_6H_5CH_2CH_2SiCl_3$ (6) $C_6H_5CH(CH_3)SiCl_3$ (59)	125
$NiSO_4$ + PPh_3 + CuCl	25-150	3	$C_6H_5CH_2CH_2SiCl_3$ (6) $C_6H_5CH(CH_3)SiCl_3$ (59)	125
$Ni[P(OPh)_3]_4$	25-150	18	$C_6H_5CH_2CH_2SiCl_3$ (15) $C_6H_5CH(CH_3)SiCl_3$ (15)	125
$[NiCO(\pi\text{-}C_5H_5)_2]_2$	r.t.	2.5-3	$C_6H_5CH(CH_3)SiCl_3$ (82)	1171
Pd +PPh_3	100	5	$C_6H_5CH(CH_3)SiCl_3$ (92)	D -027
$Pd(PPh_3)_4$	100	5	$C_6H_5CH(CH_3)SiCl_3$ (95)	1208
$PdBr_2$+ PPh_3	110	5	$C_6H_5CH(CH_3)SiCl_3$ (96)	D -027
$PdBr_2(PPh_3)_2$	110	5	$C_6H_5CH(CH_3)SiCl_3$ (90)	D -027
$PdCl_2$ + PPFA	70	40	$C_6H_5CH(CH_3)SiCl_3$ (95)	485
$PdCl_2(PhCN)_2$ + BMPP	120	12	$C_6H_5CH(CH_3)SiCl_3$ (70)	1304
$PdCl_2(PhCN)_2$ + BMPP	120	12	$C_6H_5CH(CH_3)SiCl_3$ (70)	590
$PdCl_2(PhCN)_2$ + MDPP	20	5	$C_6H_5CH(CH_3)SiCl_3$ (87)	1304
$PdCl_2(PhCN)_2$ + MDPP	20	5	$C_6H_5CH(CH_3)SiCl_3$ (87)	590
$PdCl_2(PhCN)_2$ + NMDP	20	5	$C_6H_5CH(CH_3)SiCl_3$ (87)	1304
$PdCl_2(PhCN)_2$ + NMDP	20	5	$C_6H_5CH(CH_3)SiCl_3$ (87)	590
$PdCl_2(PhCN)_2$ + PPh_3	110	5 (b)	$C_6H_5CH(CH_3)SiCl_3$ (90)	D -027
polymer/Pd complex	100	2	$C_6H_5CH(CH_3)SiCl_3$ (19)	1220

Table 1.1

= $HSiCl_3$ =

polymer-Pd	40-90	24-28	$C_6H_5CH_2CH_2SiCl_3$ + $C_6H_5CH(CH_3)SiCl_3$ (45-100)	293
polymer-Pd	70	48	$C_6H_5CH_2CH_2SiCl_3$ + $C_6H_5CH(CH_3)SiCl_3$ (kin.)	293
$(CH_2CMeCOOCH_2CH_2NMe_2)_n/H_2PtCl_6$	70	1	$C_6H_5CH_2CH_2SiCl_3$ +$C_6H_5CH(CH_3)SiCl_3$ (91)	C -002
$C_6H_8(OC_3H_6CN)_6/PtCl_2(PPh_3)_2$	80	3	$C_6H_5CH_2CH_2SiCl_3$ (84)	C -013
C-CN/$Pt(PPh_3)_4$	80	3	$C_6H_5CH_2CH_2SiCl_3$ (76)	C -013
C-CN/$PtCl_2(PPh_3)_2$	80	3	$C_6H_5CH_2CH_2SiCl_3$ (59)	C -013
H_2PtCl_6	-	-	$C_6H_5CH_2CH_2SiCl_3$ +$C_6H_5CH(CH_3)SiCl_3$ (-)	571
H_2PtCl_6	120	3	$C_6H_5CH_2CH_2SiCl_3$ +$C_6H_5CH(CH_3)SiCl_3$ (-)	215
H_2PtCl_6	25-150	1.5	$C_6H_5CH_2CH_2SiCl_3/C_6H_5CH(CH_3)SiCl_3$ (65/35)	125
H_2PtCl_6	5-20	24	$C_6H_5CH_2CH_2SiCl_3$ (kin.)	966
H_2PtCl_6	reflux	0.67 (p)	$C_6H_5CH_2CH_2SiCl_3$ (80)	F -026
H_2PtCl_6	reflux	0.67 (p)	$C_6H_5CH_2CH_2SiCl_3$ (80)	U -074
H_2PtCl_6 (Me_2CO) + $\overline{C_6H_4NHC_6H_4S}$	72	6	$C_6H_5CH_2CH_2SiCl_3$ (97)	D -123
H_2PtCl_6 ($MeOCH_2CH_2OMe$)	90-95	1.5	$C_6H_5CH_2CH_2SiCl_3$ (94)	U -092
H_2PtCl_6 ($MeOCH_2CH_2OMe$)	90-95	1.5	$C_6H_5CH_2CH_2SiCl_3$ (94)	D -084
H_2PtCl_6 (i-PrOH)	-	-	$C_6H_5CH_2CH_2SiCl_3$ (70)	751
H_2PtCl_6 (i-PrOH)	80	0.5	$C_6H_5CH_2CH_2SiCl_3$ +$C_6H_5CH(CH_3)SiCl_3$ (-)	1073
H_2PtCl_6 (i-PrOH)	reflux	3 (THF)	$C_6H_5CH_2CH_2SiCl_3$ (92)	1192
H_2PtCl_6 (i-PrOH)	reflux	0.67 (p)	$C_6H_5CH_2CH_2SiCl_3$ (-)	U -100
H_2PtCl_6 (i-PrOH) + C_5H_5N	80	2	$C_6H_5CH_2CH_2SiCl_3$ (90-92)	213
H_2PtCl_6 (i-PrOH) + PPh_3	80	2	$C_6H_5CH_2CH_2SiCl_3$ (88-95)	213
H_2PtCl_6 (i-PrOH) + PPh_3	80-150	1	$C_6H_5CH_2CH_2SiCl_3$ (81)	213
H_2PtCl_6 (i-PrOH) + $PhCH_2CN$	80	2	$C_6H_5CH_2CH_2SiCl_3$ (64-73) $C_6H_5CH(CH_3)SiCl_3$ (27-36)	213
H_2PtCl_6 (i-PrOH) + $PhCH_2NMe_2$	80	2	$C_6H_5CH_2CH_2SiCl_3$ (80-85)	213
H_2PtCl_6 + ($MeOCH_2CH_2OMe$ + EtOH)	-	-	$C_6H_5CH_2CH_2SiCl_3/C_6H_5CH(CH_3)SiCl_3$ (5/95)	U -085
H_2PtCl_6 + $\overline{C_6H_4NHC_6H_4S}$	-	-	$C_6H_5CH_2CH_2SiCl_3$ (100)	U -085
$KPtCl_3(CH_2{=}CH_2)$	5	-	$C_6H_5CH_2CH_2SiCl_3$ (kin.)	966

	Pt catalyst + Bu_3N	80-100	2	$C_6H_5CH_2CH_2SiCl_3$ (99)	U -141
	Pt/C + sonic waves	30	1.5	$C_6H_5CH_2CH_2SiCl_3$ (94)	467
	SiO_2-PPh_2/H_2PtCl_6	90	2	$C_6H_5CH_2CH_2SiCl_3$ + $C_6H_5CH(CH_3)SiCl_3$ (57)	C -007
	$[HPt(SiCl_3)P(c\text{-}Hex)_3]_2$	20	6	$C_6H_5CH_2CH_2SiCl_3$ (94)	429
	$[PtCl_2(PhCH=CH_2)]_2$	5-20	24	$C_6H_5C_2H_4SiCl_3$ (88-95)	965
	polymer/Pt complex	100	2	$C_6H_5CH_2CH_2SiCl_3$ (92)	1220
	polymer-Pt	70-90	24	$C_6H_5CH_2CH_2SiCl_3$ + $C_6H_5CH(CH_3)SiCl_3$ (95-100)	293
	support/Pt complex	80	2	$C_6H_5CH_2CH_2SiCl_3$ (2-89) $C_6H_5CH(CH_3)SiCl_3$ (0-9)	204
	$Rh_4(CO)_{12}$	20-30	70	$C_6H_5CH_2CH_2SiCl_3$ (-)	716
	$Rh_4(CO)_{12}$	30-35	1	$C_6H_5CH_2CH_2SiCl_3$ (95-98)	S -023
	$Rh_4(CO)_{12}$	30-40	1	$C_6H_5CH_2CH_2SiCl_3$ (96)	719
	$RhCl(PPh_3)_3$	100	3.5 (t)	$C_6H_5CH_2CH_2SiCl_3$ (20) $C_6H_5CH(CH_3)SiCl_3$ (6) $C_6H_5CH=CHSiCl_3$ (trace)	878
	$RhCl(acac)(CO)_2$	100	2	$C_6H_5CH_2CH_2SiCl_3$ + $C_6H_5CH(CH_3)SiCl_3$ (60)	C -009
	$RhCl(acac)(CO)_2$ + PPh_3	100	2	$C_6H_5CH_2CH_2SiCl_3$ + $C_6H_5CH(CH_3)SiCl_3$ (20)	C -009
	SDVB-NMe_2/$[RhCl(CO)_2]_2$	100	2	$C_6H_5CH_2CH_2SiCl_3$ + $C_6H_5CH(CH_3)SiCl_3$ (1-42)	316
	polymer/Rh complex	100	2	$C_6H_5CH_2CH_2SiCl_3$ (95)	1220
C_8H_{12}	Fig.7.				
	$(PhCOO)_2$	reflux	64 (ht)	Fig.9. (-)	1297
	H_2PtCl_6 (i-PrOH)	120	120	Fig.9. (-)	1297
C_8H_{12}	Fig.8.				
	$Ni(acac)_2$ + PPh_3	100	7	Fig.10. (98)	1329
C_8H_{14}	$\overline{(CH_2)_4CH=C}CH_2CH_3$				
	H_2PtCl_6 (i-PrOH)	reflux	3.5-23	$\overline{(CH_2)_5CH}C_2H_4SiCl_3$ (17-74)	119
	Pt/C	reflux	164	$\overline{(CH_2)_5CH}C_2H_4SiCl_3$ (trace)	119
	$[PtCl_2(CH_2=CH_2)]_2$	reflux	1	$\overline{(CH_2)_5CH}C_2H_4SiCl_3$ (80)	119
	$[PtCl_2(CH_2CH=CEt)]_2$	reflux	1.17	$\overline{(CH_2)_5CH}C_2H_4SiCl_3$ (75)	119
C_8H_{14}	$\overline{(CH_2)_5CH}CH=CH_2$				

	H_2PtCl_6 (i-PrOH)	reflux	0.5	$(CH_2)_5CHCH_2CH_2SiCl_3$ (68)	119
C_8H_{14}	$(CH_2)_6CH=CH$				
	Pd + PPh_3	120	5	$(CH_2)_7CHSiCl_3$ (20)	D -027
	$Pd(PPh_3)_4$	110	8	$(CH_2)_7CHSiCl_3$ (30)	D -027
	$Pd(PPh_3)_4$	120	15	$(CH_2)_7CHSiCl_3$ (30)	1208
C_8H_{14}	Fig.11.				
	H_2PtCl_6 (i-PrOH)	reflux	6-12	$(CH_2)_5CHC_2H_4SiCl_3$ (59-88)	119
C_8H_{14}	Fig.12.				
	H_2PtCl_6 (i-PrOH)	reflux	3.5-96	$(CH_2)_5CHC_2H_4SiCl_3$ (19-83)	119
C_8H_{16}	$(CH_3)_2C=CHC(CH_3)_3$				
	gamma(^{60}Co)	16	23	$(CH_3)_2CHCH(SiCl_3)C(CH_3)_3$ (20)	958
C_8H_{16}	$(CH_3)_3CCH_2C(CH_3)=CH_2$				
	gamma(^{60}Co)	16	15	$C_8H_{17}SiCl_3$ (50)	958
	H_2PtCl_6 ($MeOC_2H_4OMe$)	60	- (c)	$(CH_3)_3CCH_2CH(CH_3)CH_2SiCl_3$ (-)	351
C_8H_{16}	$(CH_3)_3CCH_2C(CH_3)=CH_2$ + $(CH_3)_3CCH=C(CH_3)_2$				
	H_2PtCl_6 (i-PrOH)	140	9	$(CH_3)_3CCH_2CH(CH_3)CH_2SiCl_3$ (82)	1192
	H_2PtCl_6 (i-PrOH)	20-78	1-2 (b)	$(CH_3)_3CCH_2CH(CH_3)CH_2SiCl_3$ (-)	G -002
	H_2PtCl_6 (i-PrOH)	reflux	2 (b)	$(CH_3)_3CCH_2CH(CH_3)CH_2SiCl_3$ (-)	G -002
	H_2PtCl_6 [$(CH_2)_5CO$]	120	-	$(CH_3)_3CCH_2CH(CH_3)CH_2SiCl_3$ (-)	D -103
C_8H_{16}	$(CH_3CH_2)_2CHC(CH_3)=CH_2$				
	H_2PtCl_6 (THF)	110	- (b)	$(CH_3CH_2)_2CHCH(CH_3)CH_2SiCl_3$ + $(CH_3CH_2)_2CHCH(CH_2CH_3)CH_2CH_2SiCl_3$ (-)	907
C_8H_{16}	$^{13}CH_3(CH_2)_5CH=CH_2$				
	H_2PtCl_6	-	-	$^{13}CH_3(CH_2)_7SiCl_3$ (-)	401
C_8H_{16}	$CH_3(CH_2)_4CH=CHCH_3$				
	UV + $(t\text{-}BuO)_2$	40	14	$CH_3CH_2CH(SiCl_3)(CH_2)_4CH_3$ (6)	338
	gamma(^{60}Co)	16	21	$C_6H_{13}CH(CH_3)SiCl_3$ (99)	958
C_8H_{16}	$CH_3(CH_2)_5CH=CH_2$				

Table 1.1
= $HSiCl_3$ =

UV	40	12	$CH_3(CH_2)_7SiCl_3$ (2)	126
UV + $(t\text{-}BuO)_2$	30-40	8-14	$C_8H_{17}SiCl_3$ (63-78)	338
UV + $(t\text{-}BuO)_2$	40	0.2-12	$CH_3(CH_2)_7SiCl_3$ (16-95)	126
UV + $Hg(SiMe_3)_2$	40	0.2-12	$CH_3(CH_2)_7SiCl_3$ (32-97)	126
gamma(^{60}Co)	16	17	$C_8H_{17}SiCl_3$ (98)	958
Ni(diphos)($CH_2=CH_2$)	120	20	$C_8H_{17}SiCl_3$ (98)	588
$NiCl_2$(DMPC)	120	20	$CH_3(CH_2)_7SiCl_3/C_8H_{17}SiCl_3$ (15/85)	337
$NiCl_2$(DMPF)	120	20	$C_8H_{17}SiCl_3/C_8H_{17}SiHCl_2$ (50/50)	588
$NiCl_2$(DMPF)	120	20	$C_8H_{17}SiCl_3/C_8H_{17}SiHCl_2$ (50/50)	628
$NiCl_2(PEt_3)_2$	120	20	$C_8H_{17}SiCl_3$ + $C_8H_{17}SiHCl_2$ (90)	589
$NiCl_2$(diphos)	120	20	$C_8H_{17}SiCl_3$ (90)	588
Pd + PBu_3	120	10	$CH_3(CH_2)_7SiCl_3$ (54)	1208
Pd + PPh_3	120	5	$CH_3(CH_2)_7SiCl_3$ (70-88)	1208
Pd + PPh_3	120	5	$CH_3(CH_2)_7SiCl_3$ (70-88)	D -027
$Pd(PPh_3)_4$	100	5	$CH_3(CH_2)_7SiCl_3$ (85)	1208
$Pd(PPh_3)_4$	120	5	$CH_3(CH_2)_7SiCl_3$ (93)	D -027
$Pd(PPh_3)_4$	80	10	$CH_3(CH_2)_7SiCl_3$ (52)	D -027
$Pd(acac)_2$ + PBu_3	120	10	$CH_3(CH_2)_7SiCl_3$ (76)	1208
$PdBr_2(PEt_3)_2$	80	10	$CH_3(CH_2)_7SiCl_3$ (65)	1208
$PdCl_2(AsPh_3)_2$	120	10	$CH_3(CH_2)_7SiCl_3$ (30)	D -027
$PdCl_2(PPh_3)_2$	120	5	$CH_3(CH_2)_7SiCl_3$ (90)	1208
$PdCl_2(SbPh_3)_2$	130	8	$CH_3(CH_2)_7SiCl_3$ (20)	D -027
$PdCl_2[(\overline{CH_2)_5C}=NOH]$	120	5	$CH_3(CH_2)_7SiCl_3$ (85)	D -027
$PdCl_2[P(OBu)_3]_2$	120	10	$CH_3(CH_2)_7SiCl_3$ (20)	1208
$PdCl_2[P(OPh)_3]_2$	120	15	$CH_3(CH_2)_7SiCl_3$ (16)	1208
$PdCl_2[P(OPh)_3]_2$	120	10	$CH_3(CH_2)_7SiCl_3$ (61)	D -027
$PdCl_2[P(c\text{-}Hex)_3]_2$	120	8	$CH_3(CH_2)_7SiCl_3$ (80)	D -027

	$PdI_2(PEt_3)_2$	100	10	$CH_3(CH_2)_7SiCl_3$ (61)	1208
	H_2PtCl_6	-	-	$CH_3(CH_2)_7SiCl_3$ (-)	J -035
	H_2PtCl_6 (THF)	reflux	4	$CH_3(CH_2)_7SiCl_3$ (92)	C -041
	H_2PtCl_6 (THF) + Ph_2SiH_2	reflux	4	$CH_3(CH_2)_7SiCl_3$ (93)	C -041
	H_2PtCl_6 (i-PrOH)	20	3	$CH_3(CH_2)_7SiCl_3$ (86)	1192
	H_2PtCl_6 (i-PrOH)	25	15	$CH_3(CH_2)_7SiCl_3$ (78)	748
	H_2PtCl_6 (i-PrOH)	25	6	$CH_3(CH_2)_7SiCl_3$ (84)	748
	H_2PtCl_6 (i-PrOH)	50	4	$CH_3(CH_2)_7SiCl_3$ (72)	748
	H_2PtCl_6 (i-PrOH)	50	6	$CH_3(CH_2)_7SiCl_3$ (60)	748
	H_2PtCl_6 (i-PrOH)	50	2	$CH_3(CH_2)_7SiCl_3$ (59)	748
	H_2PtCl_6 (i-PrOH)	64	1	$C_8H_{17}SiCl_3$ (86)	113
	$Pt(PPh_3)_2(CH_2{=}CH_2)$	25	15	$CH_3(CH_2)_7SiCl_3$ (91)	748
	$Pt(PPh_3)_2(CH_2{=}CH_2)$	25	6	$CH_3(CH_2)_7SiCl_3$ (90)	748
	$Pt(PPh_3)_2(CH_2{=}CH_2)$	25	6	$CH_3(CH_2)_7SiCl_3$ (90)	Pl-030
	$Pt(PPh_3)_2(CH_2{=}CH_2)$	25	30	$CH_3(CH_2)_7SiCl_3$ (89)	748
	$Pt(PPh_3)_2(CH_2{=}CH_2)$	50	6	$CH_3(CH_2)_7SiCl_3$ (80)	748
	$Pt(PPh_3)_2(CH_2{=}CH_2)$	50	4	$CH_3(CH_2)_7SiCl_3$ (82)	748
	$Pt(PPh_3)_2(CH_2{=}CH_2)$	50	2	$CH_3(CH_2)_7SiCl_3$ (89)	748
	$Pt(PPh_3)_2(CH_2{=}CH_2)$	50	2	$CH_3(CH_2)_7SiCl_3$ (76)	Pl-030
	$Pt(PPh_3)_2[CH_2{=}CHSi(OEt)_3]$	50	2	$CH_3(CH_2)_7SiCl_3$ (78)	Pl-030
	$PtCl_2(PPh_3)_2$	80	2	$CH_3(CH_2)_7SiCl_3$ (83)	Pl-031
	$PtCl_2(PhCN)_2$	50	2	$CH_3(CH_2)_7SiCl_3$ (68)	Pl-031
	RhCl(cod) + diphos	-	-	$CH_3(CH_2)_7SiCl_3$ (-)	J -020
C_8H_{16}	$CH_3CH_2CH_2CH{=}CHCH_2CH_2CH_3$				
	H_2PtCl_6 (THF)	110	- (h)	$C_8H_{17}SiCl_3$ (-)	907
C_9H_{10}	$C_6H_5C(CH_3){=}CH_2$				
	$[PtCl_2(BMPP)]_2$	90	24	$C_6H_5C(CH_3)_2CH_2CH(C_6H_5)CH_2SiCl_3$ (52)	1303

	$Rh_4(CO)_{12}$	30-35	1	$C_6H_5CH(CH_3)CH_2SiCl_3$ (96-99)	S -023
C_9H_{10}	$C_6H_5CH_2CH=CH_2$				
	H_2PtCl_6	120	2	$C_6H_5C_3H_6SiCl_3$ (-)	215
	H_2PtCl_6 (i-PrOH) + PPh_3	reflux	3 (b)	$C_6H_5(CH_2)_3SiCl_3$ (75)	1218
C_9H_{16}	$\overline{(CH_2)_4CH=C}CH_2CH_2CH_3$				
	H_2PtCl_6 (i-PrOH)	reflux	162	$\overline{(CH_2)_5C}HC_3H_6SiCl_3$ (30)	119
C_9H_{16}	$\overline{(CH_2)_5C}HCH_2CH=CH_2$				
	$(PhCOO)_2$	reflux	72	$\overline{(CH_2)_5C}H(CH_2)_3SiCl_3$ (30)	119
	H_2PtCl_6 (i-PrOH)	reflux	-	$\overline{(CH_2)_5C}H(CH_2)_3SiCl_3$ (44)	119
	Pt/C	reflux	20	$\overline{(CH_2)_5C}H(CH_2)_3SiCl_3$ (60)	119
	$[PtCl_2(CH_2=CH_2)_2]_2$	reflux	48	$\overline{(CH_2)_5C}H(CH_2)_3SiCl_3$ (81)	119
	$\{PtCl_2[(\overline{(CH_2)_4CH=C}Et]_2\}_2$	reflux	48	$\overline{(CH_2)_5C}H(CH_2)_3SiCl_3$ (78)	119
C_9H_{18}	$CH_3CH_2CH_2CH=C(CH_3)CH_2CH_2CH_3$				
	H_2PtCl_6 (THF)	110	- (b)	$CH_3CH_2CH_2CH(SiCl_3)CH(CH_3)CH_2CH_2CH_3$ (-)	907
C_9H_{18}	$CH_3CH=CH_2$ trimer				
	$Rh_4(CO)_{12}$	30-35	1	$C_9H_{19}SiCl_3$ (97-99)	S -037
$C_{10}H_{10}$	$C_6H_4(CH=CH_2)_2$				
	H_2PtCl_6 (i-PrOH) + PPh_3 (CH_2Cl_2)	70-75	3	$C_6H_4[CH_2CH_2SiCl_3]_2$ (72) $C_6H_4[CH_2CH_2SiCl_3][CH(CH_3)SiCl_3]$ + $C_6H_4[CH(CH_3)SiCl_3]_2$ (88)	D -131
$C_{10}H_{12}$	$(C_2H_5)C_6H_4(CH=CH_2)$				
	H_2PtCl_6 (i-PrOH) + PPh_3 (CH_2Cl_2)	70-75	3	$(C_2H_5)C_6H_4(CH_2CH_2SiCl_3)$ (90)	D -131
$C_{10}H_{16}$	Fig.15.				
	$Ni(acac)_2$ + PCl_2(i-Bu)	120	1	Fig.16. (63)	557
	$Ni(acac)_2$ + PPh_3	120	10	Fig.17. (50)	1335
	$Ni(acac)_2$ + PPh_3	120	1	Fig.16. + Fig.17. (63)	557
$C_{10}H_{16}$	Fig.18.				
	$Ni(acac)_2$ + PPh_3	100	7	Fig.19. (98)	1329
$C_{10}H_{18}$	Fig.20.				

	$[PtCl_2(CH_2=CH_2)]_2$	140	92	Fig.21. + Fig.22. (14)	106
$C_{10}H_{18}$	Fig.23.				
	$[PtCl_2(CH_2=CH_2)]_2$	140	27	Fig.21. + Fig.22. (69)	106
	$[PtCl_2(CH_2=CH_2)]_2$	reflux	114	Fig.21. + Fig.22. (62)	106
$C_{10}H_{18}$	Fig.24.				
	$(PhCOO)_2$	reflux	20 (ht)	Fig.22. (35)	106
	$[PtCl_2(CH_2=CH_2)]_2$	reflux	19	Fig.21. + Fig.22. (49)	106
$C_{10}H_{20}$	$CH_3(CH_2)_7CH=CH_2$				
	Pd/support	80	2	$C_{10}H_{21}SiCl_3$ (71)	204
	$PdCl_2(PEt_3)_2$	120	8	$CH_3(CH_2)_9SiCl_3$ (84)	D -027
	Al_2O_3-CN/$Pt(PPh_3)_4$	80	3	$C_{10}H_{21}SiCl_3$ (64)	C -013
	Al_2O_3-CN/$PtCl_2(PPh_3)_2$	80	3	$C_{10}H_{21}SiCl_3$ (54)	C -013
	$C_6H_8[O(CH_2)_3CN]_6/Pt(PPh_3)_4$	80	3 (b,THF,e)	$C_{10}H_{21}SiCl_3$ (25-61)	C -013
	C-CN/$Pt(PPh_3)_4$	80	3	$C_{10}H_{21}SiCl_3$ (58)	C -013
	C-CN/$PtCl_2(PPh_3)_2$	80	3	$C_{10}H_{21}SiCl_3$ (52)	C -013
	H_2PtCl_6	120	2	$C_{10}H_{21}SiCl_3$ (-)	215
	H_2PtCl_6	80	3	$C_{10}H_{21}SiCl_3$ (77)	204
	Pt/support	80	2	$C_{10}H_{21}SiCl_3$ (22-85)	204
	siloxane-NH_2/H_2PtCl_6	60-85	3	$C_{10}H_{21}SiCl_3$ (89)	D -071
$C_{12}H_{24}$	$CH_3(CH_2)_9CH=CH_2$				
	H_2PtCl_6 (i-PrOH)	60-70	6.5	$CH_3(CH_2)_{11}SiCl_3$ (73)	1192
$C_{13}H_{26}$	$CH_3(CH_2)_{10}CH=CH_2$				
	$(PhCOO)_2$	130-200	2.5	$CH_3(CH_2)_{12}SiCl_3$ (88-91)	F -026
	$(PhCOO)_2$	130-200	2.5	$CH_3(CH_2)_{12}SiCl_3$ (88-91)	U -074
	$(PhCOO)_2$	130-200	2.5	$CH_3(CH_2)_{12}SiCl_3$ (88-91)	U -098
$C_{14}H_{12}$	$(C_6H_5)_2C=CH_2$				
	H_2PtCl_6	reflux	- (h)	$(C_6H_5)_2CHCH_2SiCl_3$ (78)	833

	$Rh_4(CO)_{12}$	35-40	1	$(C_6H_5)_2CHCH_2SiCl_3$ (95-98)	S -023
$C_{14}H_{12}$	$C_6H_5CH=CHC_6H_5$				
	H_2PtCl_6	20	40 days	$C_6H_5CH_2CH(C_6H_5)SiCl_3$ (76)	833
$C_{16}H_{32}$	$CH_2=CH(CH_2)_{13}CH_3$				
	$Co_2(CO)_8PtCl_2$	reflux	3	$C_{16}H_{33}SiCl_3$ (90)	U -102
	$Co_2(CO)_8PtCl_2$	reflux	3	$C_{16}H_{33}SiCl_3$ (90)	U -053
$C_{18}H_{36}$	$CH_2=CH(CH_2)_{15}CH_3$				
	H_2PtCl_6 (i-PrOH)	110	12	$C_{18}H_{37}SiCl_3$ (84)	C -032
	H_2PtCl_6 (i-PrOH)	50-60	5	$CH_3(CH_2)_{17}SiCl_3$ (90)	1192
	H_2PtCl_6 (i-PrOH)	80	6	$CH_3(CH_2)_{17}SiCl_3$ (45)	748
	$Pt(PPh_3)_2(CH_2=CH_2)$	80	6	$CH_3(CH_2)_{17}SiCl_3$ (50)	748
	$Pt(PPh_3)_2[CH_2=CHSi(OEt)_3]$	110	8	$CH_3(CH_2)_{17}SiCl_3$ (75)	Pl-030
	$Rh_4(CO)_{12}$	30-35	1	$C_{18}H_{37}SiCl_3$ (97-100)	S -037
	RhCl(cod) + diphos	-	-	$CH_3(CH_2)_{17}SiCl_3$ (-)	J -020

Table 1.2

= $HSiCl_3$ = =1=

C_3H_4	$CH_2=C=CH_2$				
	Pd + PPh_3	120	5	$CH_2=CHCH_2SiCl_3$ (78)	D -027
	$Pd(PPh_3)_4$	120	5	$CH_2=CHCH_2SiCl_3$ (71)	1208
	$PdCl_2(PPh_3)_2$	120	5	$CH_2=CHCH_2SiCl_3$ (71)	D -027
C_4H_6	C_4 + C_7 fraction				
	$NiBr_2$ + P(c-Hex)$_3$	120	3	$C_4H_7SiCl_3$ (8)	C -018
	$PdCl_2(PPh_3)_2$	80	5	$C_4H_7SiCl_3$ (9)	C -017
C_4H_6	C_4 fraction				
	Ni(acac)$_2$ + PPh_3	120	3	$C_4H_7SiCl_3$ (63)	C -018
	$NiBr_2$ + P(c-Hex)$_3$	120	3	$C_4H_7SiCl_3$ (90)	C -018
	$NiCl_2$ + PBu_3	120	3	$C_4H_7SiCl_3$ (92)	C -018
	$NiCl_2(PPh_3)_2$	120	3	$C_4H_7SiCl_3$ (33)	C -018
	$Pd(PPh_3)_4$	80	2	$C_4H_7SiCl_3$ (91)	C -017
	$PdCl_2$ + PPh_3	120	3	$C_4H_7SiCl_3$ (35)	C -017
	$PdCl_2$(π-All)$_2$	120	3	$C_4H_7SiCl_3$ (61)	C -017
	$PdCl_2(PPh_3)_2$	125	4	$C_4H_7SiCl_3$ (88)	C -017
	$PdCl_2(PPh_3)_2$	45-120	3-8	$C_4H_7SiCl_3$ (4-79)	C -017
	$PdCl_2(PhCN)_2$	120	3	$C_4H_7SiCl_3$ (58)	C -017
C_4H_6	$CH_2=CHCH=CH_2$				
	$Ni(CN)_2(PPh_3)_2$	120	3	$C_4H_7SiCl_3$ (47)	D -070
	$Ni(NO_3)_2 \cdot 6\ H_2O$ + PBu_3	120	3	$C_4H_7SiCl_3$ (61)	D -070
	$Ni(NO_3)_2 \cdot 6\ H_2O$ + PPh_3	120	3	$C_4H_7SiCl_3$ (53)	D -070
	$Ni(NO_3)_2 \cdot 6\ H_2O$ + diphos	120	3	$C_4H_7SiCl_3$ (12)	D -070
	$Ni(NO_3)_2$ + phosphine; phosphine: PBu_3, P(c-Hex)$_3$, PPh_3, diphos	120	3	$CH_3CH=CHCH_2SiCl_3$ (0-56) $CH_2=CHCH_2CH_2SiCl_3$ (0-20) $Cl_3SiCH_2CH=CHCH_2SiCl_3$ (0-14)	206
	$Ni(PPh_3)_2(CO)_2$	120	3	$C_4H_7SiCl_3$ (57)	D -070
	$Ni(PPh_3)_2(CO)_2$ + $CH_2=CHBu$	120	4	$C_4H_7SiCl_3$ (37)	C -018

Table 1.2

= $HSiCl_3$ = =2=

$Ni(PPh_3)_4$	120	3	cis,trans-$CH_3CH=CHCH_2SiCl_3$ (33)	210
$Ni(PPh_3)_4$	120	3	$C_4H_9SiCl_3$ (91)	D -070
$Ni(PPh_3)_4$	120	3	$CH_3CH=CHCH_2SiCl_3$ (53) $CH_2=CHCH_2CH_2SiCl_3$ (8) $Cl_3SiCH_2CH=CHCH_2SiCl_3$ (8)	206
$Ni(PPh_3)_4$ + $CH_2=CHBu$	120	3	$C_4H_9SiCl_3$ (33)	C -018
$Ni(PPh_3)_4$ + phosphine; phosphine: PBu_3, $P(c\text{-}Hex)_3$, PPh_3, $P(OPh)_3$, diphos	120	3	$CH_3CH=CHCH_2SiCl_3$ (10-76) $CH_2=CHCH_2CH_2SiCl_3$ (0-24) $Cl_3SiCH_2CH=CHCH_2SiCl_3$ (0-26)	206
$Ni(acac)_2$	120	3	$C_4H_7SiCl_3$ (17)	D -070
$Ni(acac)_2$ + BDFB	120	3	$C_4H_7SiCl_3$ (44)	D -070
$Ni(acac)_2$ + NEt_3	120	3	$C_4H_7SiCl_3$ (68)	D -070
$Ni(acac)_2$ + $Na[AlH_2(OC_2H_4OMe)_2]$	40	6	$C_4H_7SiCl_3$ (14)	C -010
$Ni(acac)_2$ + PBu_3 + $Na[AlH_2(OC_2H_4OMe)_2]$	until 78	4	$C_4H_7SiCl_3$ (34)	C -010
$Ni(acac)_2$ + PPh_3	120	3	cis-$CH_3CH=CHCH_2SiCl_3$ + trans-$CH_3CH=CHCH_2SiCl_3$ (64)	210
$Ni(acac)_2$ + PPh_3 + $Na[AlH_2(OC_2H_4OMe)_2]$	until 78	4	$C_4H_7SiCl_3$ (49)	C -010
$Ni(acac)_2$ + PPh_3 +$Na[AlH_2(OC_2H_4OMe)_2]$	70	6	$C_4H_7SiCl_3$ (58) trans-$C_8H_{13}SiCl_3$ (2)	C -010
$Ni(acac)_2$ + diphos	120	3	cis-$CH_3CH=CHCH_2SiCl_3$ + trans-$CH_3CH=CHCH_2SiCl_3$ (44)	210
$Ni(acac)_2$ + phosphine, phosphine:PBu_3, $P(c\text{-}Hex)_3$,PPh_3, diphos	120	3	$CH_3CH=CHCH_2SiCl_3$ (3-75) $CH_2=CHCH_2CH_2SiCl_3$ (0-25) $Cl_3SiCH_2CH=CHCH_2SiCl_3$ (0-25)	206
$Ni(acac)_2$ +diphos+$Na[AlH_2(OC_2H_4OMe)_2]$	70	6	$C_4H_7SiCl_3$ (12)	C -010
$Ni(cod)_2$	120	3	$CH_3CH=CHCH_2SiCl_3$ (86) $CH_2=CHCH_2CH_2SiCl_3$ (8) $Cl_3SiCH_2CH=CHCH_2SiCl_3$ (6)	206
$Ni(cod)_2$	70	3	$CH_3CH=CHCH_2SiCl_3$ (60) $CH_2=CHCH_2CH_2SiCl_3$ (2)	207
$Ni(cod)_2$	80	1.5 (b)	$CH_3CH=CHCH_2SiCl_3$ (76)	205
$Ni(cod)_2$	80-120	3-4	$CH_3CH=CHCH_2SiCl_3$ (54-87)	210
$Ni(cod)_2$+ $CH_2=CHBu$	80	4	$C_4H_7SiCl_3$ (85)	C -018

Table 1.2
= $HSiCl_3$ = =3=

$Ni(cod)_2 + PPh_3 + Na[AlH_2(OC_2H_4OMe)_2]$	70	3	$CH_3CH=CHCH_2SiCl_3$ (60) $CH_2=CHCH_2CH_2SiCl_3$ (2)	207
$Ni(cod)_2$ + phosphine; phosphine: PBu_3, $P(c\text{-}Hex)_3$,PPh_3, $P(OPh)_3$, diphos	120	3	$CH_3CH=CHCH_2SiCl_3$ (8-75) $CH_2=CHCH_2CH_2SiCl_3$ (1-27) $Cl_3SiCH_2CH=CHCH_2SiCl_3$ (0-20)	206
$NiBr_2 + P(c\text{-}Hex)_3$	120	3	cis-$CH_3CH=CHCH_2SiCl_3$ + trans-$CH_3CH=CHCH_2SiCl_3$ (85)	210
$NiBr_2 + PPh_3 + Na[AlH_2(OC_2H_4OMe)_2]$	70	3	$CH_3CH=CHCH_2SiCl_3$ (60) $CH_2=CHCH_2CH_2SiCl_3$ (2)	207
$NiBr_2 + PPh_3 + Na[AlH_2(OC_2H_4OMe)_2]$	until 78	4	$C_4H_7SiCl_3$ (53)	C -010
$NiBr_2$ + phosphine; phosphine: PBu_3, $P(c\text{-}Hex)_3$,PPh_3, diphos	120	3	$CH_3CH=CHCH_2SiCl_3$ (3-76) $CH_2=CHCH_2CH_2SiCl_3$ (1-30) $Cl_3SiCH_2CH=CHCH_2SiCl_3$ (0-11)	206
$NiBr_2(PPh_3)_2$	120	3	$C_4H_7SiCl_3$ (67)	D -070
$NiCl_2 + PBu_3$	120	3	cis-$CH_3CH=CHCH_2SiCl_3$ + trans-$CH_3CH=CHCH_2SiCl_3$ (85)	210
$NiCl_2 + PMe_2Ph + Na[AlH_2(OC_2H_4OMe)_2]$	until 78	4	$C_4H_7SiCl_3$ (62)	C -010
$NiCl_2 + PPh_3 + Na[AlH_2(OC_2H_4OMe)_2]$	70	3	$CH_3CH=CHCH_2SiCl_3$ (54) $CH_2=CHCH_2CH_2SiCl_3$ (4)	207
$NiCl_2$ + phosphine; phosphine: PBu_3, $P(c\text{-}Hex)_3$,PPh_3, diphos	120	3	$CH_3CH=CHCH_2SiCl_3$ (4-79) $CH_2=CHCH_2CH_2SiCl_3$ (0-29) $Cl_3SiCH_2CH=CHCH_2SiCl_3$ (0-12)	206
$NiCl_2 \cdot 6\ H_2O + PPh_3$	120	3	$C_4H_7SiCl_3$ (4)	D -070
$NiCl_2 \cdot 6\ H_2O$ + diphos	120	3	$C_4H_7SiCl_3$ (71)	D -070
$NiCl_2(PPh_3)_2$	120	3 (b)	$C_4H_7SiCl_3$ (69)	D -070
$NiCl_2(PPh_3)_2$	120	3	$C_4H_7SiCl_3$ (64)	D -070
$NiCl_2(PPh_3)_2$	120	3	cis-$CH_3CH=CHCH_2SiCl_3$ + trans-$CH_3CH=CHCH_2SiCl_3$ (20)	210
$NiCl_2(PPh_3)_2$	150-200	3	$C_4H_7SiCl_3$ (53-83)	D -070
$NiCl_2(PPh_3)_2 + NEt_3$	120	3 (b)	$C_4H_7SiCl_3$ (47)	D -070
NiF_2 + phosphine; phosphine: PBu_3, $P(c\text{-}Hex)_3$,PPh_3, diphos	120	3	$CH_3CH=CHCH_2SiCl_3$ (4-56) $CH_2=CHCH_2CH_2SiCl_3$ (0-32) $Cl_3SiCH_2CH=CHCH_2SiCl_3$ (0-10)	206
NiI_2 + phosphine; phosphine: PBu_3, $P(c\text{-}Hex)_3$,PPh_3, diphos	120	3	$CH_3CH=CHCH_2SiCl_3$ (4-52) $CH_2=CHCH_2CH_2SiCl_3$ (0-24) $Cl_3SiCH_2CH=CHCH_2SiCl_3$ (0-8)	206

Table 1.2
= $HSiCl_3$ =   =4=

$NiI_2(PPh_3)_2$	120	3	$C_4H_7SiCl_3$ (54)	D -070
$Ni[MeCH(OH)COO]_2$ + PPh_3	120	3	$C_4H_7SiCl_3$ (31)	D -070
$Ni[MeCH(OH)COO]_2$ + PPh_3 + $Na[AlH_2(OC_2H_4OMe)_2]$	until 78	4	$C_4H_7SiCl_3$ (42)	C -010
$Ni[MeCH(OH)COO]_2$ + diphos	120	3	trans-$CH_3CH=CHCH_2SiCl_3$/cis-$CH_3CH=CHCH_2SiCl_3$ (80/20)	983
$Ni[MeCH(OH)COO]_2$ + phosphine;phosphine: PBu_3, $P(c\text{-}Hex)_3$, PPh_3, diphos	120	3	$CH_3CH=CHCH_2SiCl_3$ (3-74) $CH_2=CHCH_2CH_2SiCl_3$ (1-31) $Cl_3SiCH_2CH=CHCH_2SiCl_3$ (0-9)	206
$Ni[P(OPh)_3]_4$	120	3	trans-$CH_3CH=CHCH_2SiCl_3$ (8) cis-$CH_3CH=CHCH_2SiCl_3$ (67)	206
$Ni[P(OPh)_3]_4$ + phosphine; phosphine: PBu_3, $P(c\text{-}Hex)_3$, PPh_3, $P(OPh)_3$, diphos	120	3	$CH_3CH=CHCH_2SiCl_3$ (0-73) $CH_2=CHCH_2CH_2SiCl_3$ (0-2)	206
SDVB-NMe_2/$NiCl_2$(EtOH)	120	3	$C_4H_7SiCl_3$ (28)	D -070
SDVB-PPh_2/$Ni(NO_3)_2$(EtOH)	120	3	$C_4H_7SiCl_3$ (28)	D -070
SDVB-PPh_2/$NiCl_2$(EtOH)	120	3	$C_4H_7SiCl_3$ (53)	D -070
SiO_2-CN/$NiCl_2$(EtOH) + PPh_3	120	3	$C_4H_7SiCl_3$ (59)	D -070
Al_2O_3-CN/$PdCl_2$	25	24	$CH_3CH=CHCH_2SiCl_3$ (80)	C -007
MCE-EDM/$PdCl_2$	80	2	$C_4H_7SiCl_3$ (56)	D -060
MCE-EDM/$PdCl_2$	80	2	$C_4H_7SiCl_3$ (56)	U -080
Pd + $AsPh_3$	120	5	$CH_3CH=CHCH_2SiCl_3$ (69)	D -027
Pd + $P(OBu)_3$	120	4	$CH_3CH=CHCH_2SiCl_3$ (78)	D -027
Pd + $P(OMe)_3$	110	6	$CH_3CH=CHCH_2SiCl_3$ (61)	D -027
Pd + $P(c\text{-}Hex)_3$	100	4	$CH_3CH=CHCH_2SiCl_3$ (54)	D -027
Pd +PBu_3	100	4	$CH_3CH=CHCH_2SiCl_3$ (80)	D -027
Pd +PEt_3	110	5	$CH_3CH=CHCH_2SiCl_3$ (75)	D -027
Pd + PMe_2Ph	110	5	$CH_3CH=CHCH_2SiCl_3$ (74)	D -027
Pd + $PMePh_2$	100	6	$CH_3CH=CHCH_2SiCl_3$ (82)	D -027
Pd +PPh_3	100	6	$CH_3CH=CHCH_2SiCl_3$ (94)	468
Pd +PPh_3	120	8	$CH_3CH=CHCH_2SiCl_3$ (76)	D -027

Table 1.2

= $HSiCl_3$ = =5=

Pd + PPh_3	80	5	$CH_3CH=CHCH_2SiCl_3$ (-)	J -001
Pd + $SbPh_3$	120	6	$CH_3CH=CHCH_2SiCl_3$ (58)	D -027
$Pd(AsPh_3)_4$	120	5	$CH_3CH=CHCH_2SiCl_3$ (60)	D -027
$Pd(NO_3)_2$ + PPh_3	120	5	$CH_3CH=CHCH_2SiCl_3$ (86)	D -027
$Pd(OAc)_2$ + PPh_3	110	5	$CH_3CH=CHCH_2SiCl_3$ (86)	D -027
$Pd(\overline{OCOCH=CHCO})(PPh_3)_2$	-	-	$CH_2=CHCH_2CH_2SiCl_3$ (-)	1176
$Pd(PBu_3)_4$	110	6	$CH_3CH=CHCH_2SiCl_3$ (79)	D -027
$Pd(PPh_3)_2(C_2H_4)$	80	4	$C_4H_7SiCl_3$ (49)	C -017
$Pd(PPh_3)_4$	-	-	$C_4H_7SiCl_3$ (-)	1176
$Pd(PPh_3)_4$	100	2	$CH_3CH=CHCH_2SiCl_3$ (-)	D -027
$Pd(PPh_3)_4$	100	2	$CH_3CH=CHCH_2SiCl_3$ (-)	J -003
$Pd(PPh_3)_4$	100	5	$CH_3CH=CHCH_2SiCl_3$ (85)	1208
$Pd(PPh_3)_4$	80	2	cis-$CH_3CH=CHCH_2SiCl_3$ (100)	210
$Pd(SbPh_3)_4$	130	5	$CH_3CH=CHCH_2SiCl_3$ (52)	D -027
$PdBr_2$ + PPh_3	110	6	$CH_3CH=CHCH_2SiCl_3$ (91)	D -027
$PdBr_2(PPh_3)_2$	110	5	$CH_3CH=CHCH_2SiCl_3$ (92)	D -027
$PdBr_2(PPh_3)_2$	80	2	$C_4H_7SiCl_3$ (75)	U -080
$PdBr_2(PPh_3)_2$	80	2	$C_4H_7SiCl_3$ (75)	D -060
$PdBr_2$(diphos)	100	6	$CH_3CH=CHCH_2SiCl_3$ (75)	D -027
$PdBr_2(p\text{-}Me_2NC_6H_4CN)_2$	25	2(b or THF)	$CH_3CH=CHCH_2SiCl_3$ (33-91)	1221
$PdBr_2[P(OEt)_3]_2$	120	5	$CH_3CH=CHCH_2SiCl_3$ (71)	D -027
$PdCl_2$	120	3	$C_4H_7SiCl_3$ (78)	D -060
$PdCl_2$	120	3	$C_4H_7SiCl_3$ (78)	U -080
$PdCl_2$ + PPh_3	80	5	$CH_3CH=CHCH_2SiCl_3$ (-)	D -027
$PdCl_2(PBu_3)_2$	120	5	$CH_3CH=CHCH_2SiCl_3$ (90)	D -027
$PdCl_2(PEt_3)_2$	110	5	$CH_3CH=CHCH_2SiCl_3$ (86)	D -027
$PdCl_2(PPh_3)_2$	100	5	$CH_3CH=CHCH_2SiCl_3$ (87)	D -027

$PdCl_2(PPh_3)_2$	80	2	$CH_3CH=CHCH_2SiCl_3$ (99)	210
$PdCl_2(Ph_2PCH_2CH_2SPh)_2$	130	6	$CH_3CH=CHCH_2SiCl_3$ (61)	D -027
$PdCl_2(PhCN)_2$	(-15)-10	-	cis-$CH_3CH=CHCH_2SiCl_3$ + $CH_3CH=CHCH_2SiCl_3$ (82)	D -060
$PdCl_2(PhCN)_2$	-	-	$CH_3CH=CHCH_2SiCl_3$ (84)	1219
$PdCl_2(PhCN)_2$	0-20	- (t)	$CH_3CH=CHCH_2SiCl_3$ (82)	U -080
$PdCl_2(PhCN)_2$	100	2	$C_4H_7SiCl_3$ (92)	U -080
$PdCl_2(PhCN)_2$	100	2	$C_4H_7SiCl_3$ (92)	D -060
$PdCl_2(PhCN)_2$	25	2(b or THF)	$CH_3CH=CHCH_2SiCl_3$ (43-93)	1221
$PdCl_2(PhCN)_2$	80	2	cis-$CH_3CH=CHCH_2SiCl_3$ (74)	210
$PdCl_2(PhCN)_2$	80	2	$CH_3CH=CHCH_2SiCl_3$ (87-92)	U -080
$PdCl_2(PhCN)_2$ + $CH_2=CHBu$	120	3	$C_4H_7SiCl_3$ (66)	C -017
$PdCl_2$(diphos)	130	5	$CH_3CH=CHCH_2SiCl_3$ (58)	D -027
$PdCl_2(o\text{-}Ph_2PC_6H_4SMe)_2$	100	6	$CH_3CH=CHCH_2SiCl_3$ (72)	D -027
$PdCl_2(p\text{-}Me_2NC_6H_4CN)_2$	25	2(b or THF)	$CH_3CH=CHCH_2SiCl_3$ (36-93)	1221
$PdCl_2(p\text{-}MeC_6H_4CN)_2$	25	2(b or THF)	$CH_3CH=CHCH_2SiCl_3$ (37-98)	1221
$PdCl_2(p\text{-}MeOC_6H_4CN)_2$	25	2(b or THF)	$CH_3CH=CHCH_2SiCl_3$ (45-98)	1221
$PdCl_2[P(OBu)_3]_2$	120	6	$CH_3CH=CHCH_2SiCl_3$ (73)	D -027
$PdCl_2[P(OPh)_3]_2$	120	4	$CH_3CH=CHCH_2SiCl_3$ (66)	D -027
$PdCl_2[P(c\text{-}Hex)_3]_4$	120	5	$CH_3CH=CHCH_2SiCl_3$ (83)	D -027
$PdCl_2[o\text{-}C_6H_4(CN)_2]$	80	2	$C_4H_7SiCl_3$ (92)	U -000
$PdCl_2[o\text{-}C_6H_4(CN)_2]$	80	2	$C_4H_7SiCl_3$ (92)	D -060
PdI_2+ PPh_3	120	5	$CH_3CH=CHCH_2SiCl_3$ (92)	D -027
$PdI_2(PPh_3)_2$	100	5	$CH_3CH=CHCH_2SiCl_3$ (91)	D -027
PdO + PPh_3	110	6	$CH_3CH=CHCH_2SiCl_3$ (90)	D -027
$Pd[P(OPh)_3]_4$	110	5	$CH_3CH=CHCH_2SiCl_3$ (84)	D -027
$Pd[P(c\text{-}Hex)_3]_4$	100	5	$CH_3CH=CHCH_2SiCl_3$ (87)	D -027
SDVB-CN/$PdCl_2$	100	- (flow)	$C_4H_7SiCl_3$ (80)	D -060

Table 1.2
= $HSiCl_3$ = =7=

SDVB-CN/$PdCl_2$	100	- (flow)	$C_4H_7SiCl_3$ (80)	U -080
SDVB-CN/$PdCl_2$	23	24	$C_4H_7SiCl_3$ (85)	U -080
SDVB-CN/$PdCl_2$	23	24	$C_4H_7SiCl_3$ (85)	D -060
SDVB-NMe_2/$PdCl_2$	100	2	$C_4H_7SiCl_3$ (76)	U -080
SDVB-NMe_2/$PdCl_2$	100	2	$C_4H_7SiCl_3$ (76)	D -060
SDVB-PPh_2/$PdCl_2$	90	3	$C_4H_7SiCl_3$ (73)	U -080
SDVB-PPh_2/$PdCl_2$	90	3	$C_4H_7SiCl_3$ (73)	D -060
[PdCl(π-All)]$_2$	(-15)-10	-	$CH_3CH=CHCH_2SiCl_3$ (76)	D -060
[PdCl(π-All)]$_2$	0-20	- (t)	$CH_2=CHCH_2CH_2SiCl_3$ (76)	U -080
[PdCl(π-All)]$_2$	23	30	$C_4H_7SiCl_3$ (67-87)	U -080
[PdCl(π-All)]$_2$	23	30	$C_4H_7SiCl_3$ (67-87)	D -060
[PdCl(π-All)]$_2$	80	2	cis-$CH_3CH=CHCH_2SiCl_3$ (69)	210
[PdCl(π-All)]$_2$	80	2	$C_4H_7SiCl_3$ (84-91)	D -060
[PdCl(π-All)]$_2$	80	2	$C_4H_7SiCl_3$ (84-91)	U -080
polymer/Pd complex	100	2	cis-$CH_3CH=CHCH_2SiCl_3$ (86)	1220
polymer/Pd complex	25	20	cis-$CH_3CH=CHCH_2SiCl_3$ (93)	1220
Al_2O_3-NMe_2/H_2PtCl_6	80	2	$C_4H_7SiCl_3$ (56)	C -007
$Pt(PPh_3)_4$	-	-	$C_4H_7SiCl_3$ (-)	1176
$Pt(PPh_3)_4$	95	5	cis-$CH_3CH=CHCH_2SiCl_3$/$CH_2=CHCH_2CH_2SiCl_3$ (14/86)	983
SDVB-NMe_2/H_2PtCl_6	110	- (flow)	$Cl_3Si(CH_2)_4SiCl_3$ (trace) $C_4H_7SiCl_3$ (70)	C -003
polymer/Pt complex	100	2	$CH_3CH=CHCH_2SiCl_3$ (53) $CH_2=CHCH_2CH_2SiCl_3$ (14) $Cl_3Si(CH_2)_4SiCl_3$ (4)	1220
$RhBr(PPh_3)_3$	100	6 (b)	$C_4H_7SiCl_3$ (5)	982
$RhCl(CO)(PPh_3)_3$	100	6 (b)	$CH_3CH=CHCH_2SiCl_3$ (2)	982
$RhCl(PPh_3)_3$	100	6 (b)	$C_4H_7SiCl_3$ (80)	982
$RhH(CO)(PPh_3)_3$	100	6 (b)	$CH_3CH=CHCH_2SiCl_3$ (2)	982
$RhI(PPh_3)_3$	100	6 (b)	$C_4H_7SiCl_3$ (2)	982

	polymer/Rh complex	100	2	cis-$CH_3CH=CHCH_2SiCl_3$ (23) trans-$CH_3CH=CHCH_2SiCl_3$ (10) $Cl_3Si(CH_2)_4SiCl_3$ (5)	1220
	polymer/$RhCl_3$ + THF	100	3	$CH_3CH=CHCH_2SiCl_3$ (70)	S -029
C_5H_6	Fig.25.				
	$Ni(CO)_3PPh_3$	120	8 (xy)	Fig.26. (7) Fig.27. (3) Fig.28. (72)	C -015
	$Ni(CO)_3PPh_3$	120	8	Fig.26. (7) Fig.27. (3) Fig.28. (72)	1217
	$Ni(CO)_4$	120	4 (xy)	Fig.26. (76)	C -015
	$Ni(CO)_4$	120	4	Fig.26. (76) Fig.27. (5)	C -015
	$Ni(CO)_4$	120	30 (xy)	Fig.26. (91)	C -015
	$Ni(CO)_4$	120	8	Fig.26. (90) Fig.27. (5)	1217
	$Ni(CO)_4$ + phosphine; phosphine: PBu_3, PPh_3,$P(OPh)_3$	120	8	Fig.26. (0-55) Fig.27. (4-44) Fig.28.(0-55)	1217
	$Ni(PPh_3)_4$	120	8 (xy)	Fig.26. (4) Fig.28. (62)	C -015
	$Ni(PPh_3)_4$	120	8	Fig.27. (5) Fig.28. (61)	1217
	$Ni(cod)_2$	120	8	Fig.26. (9) Fig.28. (67)	1217
	$NiCl_2(PPh_3)_2$	120	8	Fig.27. (5) Fig.28. (67)	1217
	$[Ni(\pi\text{-}C_5H_5)CO]_2$	120	8	Fig.26. (7) Fig.28. (63)	1217
	$[Ni(CO)_2PPh_3]_2$	120	8	Fig.27. (2) Fig.28. (65)	1217
	$Pd(PPh_3)_4$	100	12	Fig.26./Fig.29. (80/20)	C -014
	$Pd(PPh_3)_4$	120	7 (xy)	Fig.26. (85) Fig.27. (7) Fig.28. (3)	C -014

	Pd(PPh$_3$)$_4$	120	8	Fig.26. (74) Fig.27. (4) Fig.28. (8)	1217
	Pd(PPh$_3$)$_4$	reflux	-	Fig.26. (91)	C -014
	PdCl$_2$(PPh$_3$)$_2$	120	8	Fig.26. (67) Fig.28. (12)	1217
	PdCl$_2$(PhCN)$_2$	120	8	Fig.26. (74) Fig.27. (4) Fig.28. (10)	1217
	PdCl$_2$(PhCN)$_2$ + BMPP	120	48	Fig.26. (41)	1304
	PdCl$_2$(PhCN)$_2$ + MDPP	120	44	Fig.26. (69)	590
	PdCl$_2$(PhCN)$_2$ + MDPP	120	44	Fig.26. (69)	1304
	PdCl$_2$(PhCN)$_2$ + NMDPP	120	58	Fig.26. (81)	1304
	PdCl$_2$(PhCN)$_2$ + NMDPP	120	58	Fig.26. (81)	590
	PdCl$_2$(PhCN)$_2$ + PBu$_3$	100	48	Fig.27. (15)	933
	PdCl$_2$(PhCN)$_2$ + PPh$_3$	100	48	Fig.27. (60)	933
	PdCl$_2$(PhCN)$_2$ + PPh$_3$	90	6-24	Fig.26. (83)	867
	PdCl$_2$(c-HexNC)(PPh$_3$)	120	8	Fig.26. (86) Fig.28. (12)	1217
	PdCl$_2$(c-HexNC)$_2$	120	8	Fig.26. (78) Fig.27. (4) Fig.28. (16)	1217
	[PdCl(π-All)]$_2$	120	8	Fig.26. (60) Fig.28. (8)	1217
	H$_2$PtCl$_6$ (i-PrOH)	240-250	5	Fig.26. (52)	784
	H$_2$PtCl$_6$ (i-PrOH)	550	8	Fig.26. (30)	784
C_5H_8	$CH_2=C(CH_3)CH=CH_2$				
	Ni(CO)$_2$(PPh$_3$)$_2$	130	4	$CH_3CH=C(CH_3)CH_2SiCl_3$ (87)	1222
	Ni(CO)$_3$PPh$_3$	130	4	$CH_3CH=C(CH_3)CH_2SiCl_3$ (96)	1222
	Ni(PPh$_3$)$_4$	120	3 (THF)	$C_5H_9SiCl_3$ (86)	D -070
	Ni(PPh$_3$)$_4$	130	4	$CH_3CH=C(CH_3)CH_2SiCl_3$ (74)	1222
	Ni(acac)$_2$ + PPh$_3$	80	6	$(CH_3)_2C=CHCH_2SiCl_3$ (89)	1335

$Ni(acac)_2$ + PPh_3 + $Na[AlH_2(OC_2H_4OMe)_2]$	until 82	5	$C_5H_9SiCl_3$ (61)	C -010
$Ni(cod)_2$ + diphos	130	4	$CH_3CH=C(CH_3)CH_2SiCl_3$ (81)	1222
$NiCl_2(PPh_3)_2$	130	4	$CH_3CH=C(CH_3)CH_2SiCl_3$ (80)	1222
ACDVB/$PdCl_2(PPh_3)_2$	110	3	$C_5H_9SiCl_3$ (86)	U -080
ACDVB-PPh_2/$PdCl_2(PPh_3)_2$	110	3	$C_5H_9SiCl_3$ (86)	D -060
MEE-EDM/$PdCl_2$	80	2	$C_5H_9SiCl_3$ (47)	D -060
MEE-EDM/$PdCl_2$	80	2	$C_5H_9SiCl_3$ (47)	U -080
Pd +PPh_3	100	5	$CH_3CH=C(CH_3)CH_2SiCl_3$ (95)	D -027
$Pd(PPh_3)_4$	100	2	$CH_3CH=C(CH_3)CH_2SiCl_3$ (92)	D -027
$Pd(PPh_3)_4$	110	6	$CH_3CH=C(CH_3)CH_2SiCl_3$ (82)	1208
$PdCl_2(PPh_3)_2$	110	3	$CH_3CH=C(CH_3)CH_2SiCl_3$ (-)	D -027
$PdCl_2(PPh_3)_2$	80	4	$CH_3CH=C(CH_3)CH_2SiCl_3$ (73-100)	1222
$PdCl_2(PhCN)_2$	-	-	$CH_3CH=C(CH_3)CH_2SiCl_3$ (78)	1219
$PdCl_2(PhCN)_2$	125	8	$C_5H_9SiCl_3$ (-)	C -017
$PdCl_2(PhCN)_2$	80	2	$C_5H_9SiCl_3$ (87)	D -060
$PdCl_2(PhCN)_2$	80	2	$C_5H_9SiCl_3$ (87)	U -080
$PdCl_2(PhCN)_2$	80	4	$CH_3CH=C(CH_3)CH_2SiCl_3$ (73)	1222
$PdCl_2(PhCN)_2$ + PPh_3	60-80	6	$CH_3CH=C(CH_3)CH_2SiCl_3$ (95)	855
$PdCl_2(PhCN)_2$ + PPh_3	70	6	$CH_3CH=C(CH_3)CH_2SiCl_3$ (100)	851
SDVB-PPh_2/$PdCl_2$ (EtOH)	80	2	$C_5H_9SiCl_3$ (82-87)	U -080
SDVB-PPh_2/$PdCl_2$ (EtOH)	80	2	$C_5H_9SiCl_3$ (82-87)	D -060
H_2PtCl_6 (i-PrOH)	165	17	$(CH_3)_2C=CHCH_2SiCl_3$ (72) $Cl_3SiC_5H_{10}SiCl_3$ (6)	122
H_2PtCl_6 (i-PrOH)	80	4	$CH_3CH=C(CH_3)CH_2SiCl_3$ (95)	1222
$PtCl_2(acac)_2$	until 95	9 (a)	$C_5H_9SiCl_3$ (41) $Cl_3SiC_5H_{10}SiCl_3$ (20)	D -010
$[HPt(SiCl_3)P(c\text{-}Hex)_3]_2$	65	23	$Cl_3SiC_5H_{10}SiCl_3$ (97)	G -036
$[HPt(SiCl_3)P(c\text{-}Hex)_3]_2$	65	23	$C_5H_9SiCl_3$/$Cl_3SiC_5H_{10}SiCl_3$ (30/70)	G -036

C_5H_8 $CH_2=CHCH_2CH=CH_2$

	$Ni(PPh_3)_4$	120	3 (THF)	$C_5H_9SiCl_3$ (69)	D -070
	$Ni(cod)_2$	120	3	$C_5H_9SiCl_3$ (85)	C -018
	ACDVB-PPh_2/$PdCl_2(PPh_3)_2$	80	3	$C_5H_9SiCl_3$ (83)	D -060
	$PdCl_2(PhCN)_2$	125	8	$C_5H_9SiCl_3$ (56)	C -017
	$[PdCl(\pi\text{-All})]_2$	90	4	$C_5H_9SiCl_3$ (72)	D -060
C_5H_8	$CH_2=CHCH=CHCH_3$				
	$Ni(acac)_2 + PPh_3$	80	6	$CH_3CH_2CH=CHCH_2SiCl_3$ (85)	1335
	$Ni(acac)_2 + PPh_3 + Na[AlH_2(OC_2H_4OMe)_2]$	until 82	5	$C_5H_9SiCl_3$ (48)	C -010
	ACDVB/$PdCl_2(PPh_3)_2$	80	3	$C_5H_9SiCl_3$ (83)	U -080
	Pd +PPh_3	110	6	$CH_3CH=CHCH(CH_3)SiCl_3$ (82)	D -027
	$Pd(PPh_3)_4$	110	6	$CH_3CH=CHCH(CH_3)SiCl_3$ (81)	1208
	$Pd(PPh_3)_4$	80	4	$CH_3CH=CHCH(CH_3)SiCl_3$ (90)	D -027
	$[PdCl(\pi\text{-All})]_2$	90	4	$C_5H_9SiCl_3$ (72)	U -080
C_5H_8	$CH_2C(CH_3)CH=CH_2$				
	$Ni(cod)_2$	120	3	$C_5H_9SiCl_3$ (82)	C -018
C_6H_8	Fig.30.				
	$PdCl_2(PhCN)_2$ + MDPP	120	58-63	Fig.31. + Fig.32. (64)	1302
	$PdCl_2(PhCN)_2$ + MDPP	120	58-63	Fig.31. + Fig.32. (64)	590
	$PdCl_2(PhCN)_2$ + MDPP	120	58-63	Fig.31. + Fig.32. (64)	1304
	$PdCl_2(PhCN)_2$ + NMDPP	120	58-63	Fig.31. (56)	1304
	$PdCl_2(PhCN)_2$ + NMDPP	120	58-63	Fig.31. (56)	590
	H_2PtCl_6 (i-PrOH)	-	-	1,3-$C_6H_{10}(SiCl_3)_2$ (-)	239
C_6H_8	Fig.33.				
	$PdCl_2(PhCN)_2$ + MDPP	135	58	Fig.31. (58)	1304
	$PdCl_2(PhCN)_2$ + MDPP	135	67	Fig.31. + Fig.32. (80)	1302
	$PdCl_2(PhCN)_2$ + NMDPP	135	66	Fig.31. (80)	1304
C_6H_{10}	$CH_2=C(CH_3)C(CH_3)=CH_2$				

	$NiCl_2(PPh_3)_2$	130	4	$(CH_3)_2C=C(CH_3)CH_2SiCl_3$ (48)	1222
	Pd + PPh_3	90	5	$(CH_3)_2C=C(CH_3)CH_2SiCl_3$ (87)	D -027
	$Pd(PPh_3)_4$	50	4	$(CH_3)_2C=C(CH_3)CH_2SiCl_3$ (100)	D -027
	$PdCl_2(PhCN)_2$	80	4	$(CH_3)_2C=C(CH_3)CH_2SiCl_3$ (64-97)	1222
	SDVB-CN/$PdCl_2$	90	3	$(CH_3)_2C=C(CH_3)CH_2SiCl_3$ (43)	D -060
	SDVB-CN/$PdCl_2$	90	3	$(CH_3)_2C=C(CH_3)CH_2SiCl_3$ (43)	U -080
	H_2PtCl_6 (i-PrOH)	130	4	$(CH_3)_2C=C(CH_3)CH_2SiCl_3$ (61) $CH_2=C(CH_3)CH(CH_3)CH_2SiCl_3$ (5)	1222
	$[HPt(SiCl_3)P(c\text{-}Hex)_3]_2$	65	16	$(CH_3)_2C=C(CH_3)CH_2SiCl_3$ (92)	G -036
C_6H_{10}	$CH_2=CHCH_2CH_2CH=CH_2$				
	H_2PtCl_6 (EtOH)	50	65	$CH_2=CH(CH_2)_4SiCl_3$ (90) $Cl_3Si(CH_2)_6SiCl_3$ (4)	U -109
	H_2PtCl_6 (EtOH)	50	65	$CH_2=CH(CH_2)_4SiCl_3$ (90) $Cl_3Si(CH_2)_6SiCl_3$ (4)	D -068
	H_2PtCl_6 (EtOH)	50	65	$CH_2=CH(CH_2)_4SiCl_3$ (90) $Cl_3Si(CH_2)_6SiCl_3$ (4)	U -118
	H_2PtCl_6 (i-PrOH)	reflux	8.5	$Cl_3Si(CH_2)_6SiCl_3$ (27)	346
C_6H_{10}	$CH_2=CHCH_2CH=CHCH_3$				
	$Pd(PPh_3)_4$	110	8	$CH_3CH=CH(CH_2)_3SiCl_3$ + $CH_3CH_2CH=CHCH_2CH_2SiCl_3$ (92)	1208
	H_2PtCl_6 (EtOH)	55	18	$CH_3CH=CH(CH_2)_3SiCl_3$ (57)	U -118
	H_2PtCl_6 (EtOH)	55	18	$CH_3CH=CH(CH_2)_3SiCl_3$ (57)	D -068
	H_2PtCl_6 (EtOH)	55	18	$CH_3CH=CH(CH_2)_3SiCl_3$ (57)	U -109
C_8H_{12}	$CH_2=CHCH(CH_3)CH=CHCH=CH_2$				
	$Ni(acac)_2$ + PPh_3	80	6	$CH_2=CHCH(CH_3)CH_2CH=CHCH_2SiCl_3$ (87)	1335
C_8H_{12}	$CH_2=CHCH=CHCH_2CH_2CH=CH_2$				
	Pd + PPh_3	120	5	$CH_3CH=CHCH(SiCl_3)CH_2CH_2CH=CH_2$ (81)	D -027
	$Pd(PPh_3)_4$	100	3	$CH_3CH=CHCH(SiCl_3)CH_2CH_2CH=CH_2$ (82)	D -027
	$Pd(PPh_3)_4$	100	3	$CH_3CH=CHCH(SiCl_3)CH_2CH_2CH=CH_2$ (82)	468
	$Pd(PPh_3)_4$	110	5	$CH_3CH=CHCH(SiCl_3)CH_2CH_2CH=CH_2$ (82)	1208
C_8H_{12}	Fig.34.				

	$Ni(acac)_2 + PPh_3 + Na[AlH_2(OC_2H_4OMe)_2]$	until 82	5	Fig.38. (-)	C -010
	Pd +PPh_3	130	8	Fig.37. (45)	D -027
C_8H_{12}	Fig.35.				
	$(PhCOO)_2$	reflux	48 (ht)	Fig.40. (-)	1297
	$Ni(acac)_2$	120	10	Fig.39. (70)	1335
	H_2PtCl_6 (THF)	110	- (b)	Fig.39. (-)	907
	H_2PtCl_6 [$\overline{(CH_2)_5CO}$]	120	-	Fig.41. (-)	D -103
C_8H_{12}	Fig.36.				
	Ni(II)salt + $AlEt_3$ + PPh_3	120	4	Fig.41. (70)	1327
	$Ni(acac)_2 + PPh_3$	120	10	Fig.41. (70)	1335
	$PdBr_2(PPh_3)_2$	120	10	Fig.41. (82)	D -027
	H_2PtCl_6	63	30	Fig.41. (62)	C -028
	H_2PtCl_6 (THF)	110	- (b)	Fig.41. (-)	907
	H_2PtCl_6 (i-PrOH)	120	1	Fig.41. (96)	D -085
	H_2PtCl_6 (i-PrOH)	r.t.	22	Fig.41. (88)	1192
	H_2PtCl_6 (i-PrOH)	reflux	72	Fig.41. (-)	D -085
	H_2PtCl_6 (i-PrOH)	reflux	1	Fig.41. (97)	D -085
	H_2PtCl_6 (i-PrOH) + Cl_2SiH_2 + Cl_4Si	105-130	-	Fig.41. (3)	C -042
	H_2PtCl_6 (i-PrOH) + P-donor (PhH)	105-115	0.2-2	Fig.41. (92-98)	C -039
	H_2PtCl_6 (i-PrOH) + PPh_3(PhH) + Cl_2SiH_2 + Cl_4Si	105-130	-	Fig.41. (95)	C -042
	H_2PtCl_6 + CF_3COOH	100-115	5	Fig.41. (87)	C -028
	H_2PtCl_6 + PrCOOH	reflux	2	Fig.41. (96)	C -028
	H_2PtCl_6 + o-HOC_6H_4COOH	100	6	Fig.41. (95)	C -028
C_8H_{14}	$CH_2=CH(CH_2)_4CH=CH_2$				
	-	-	-	$Cl_3Si(CH_2)_6CH=CH_2$ + $Cl_3Si(CH_2)_8SiCl_3$ (-)	881
	H_2PtCl_6 (EtOH)	35-50	24	$Cl_3Si(CH_2)_6CH=CH_2$ (82) $Cl_3Si(CH_2)_8SiCl_3$ (12)	D -068

	H_2PtCl_6 (EtOH)	35-50	24	$Cl_3Si(CH_2)_6CH=CH_2$ (82) $Cl_3Si(CH_2)_8SiCl_3$ (12)	U -109
	H_2PtCl_6 (EtOH)	50	44	$Cl_3Si(CH_2)_6CH=CH_2$ (80) $Cl_3Si(CH_2)_8SiCl_3$ (10)	D -068
	H_2PtCl_6 (EtOH)	50	44	$Cl_3Si(CH_2)_6CH=CH_2$ (80) $Cl_3Si(CH_2)_8SiCl_3$ (10)	U -109
	H_2PtCl_6 (EtOH)	50	44	$Cl_3Si(CH_2)_6CH=CH_2$ (80) $Cl_3Si(CH_2)_8SiCl_3$ (10)	U -118
	H_2PtCl_6 (EtOH)	reflux	28	$Cl_3Si(CH_2)_6CH=CH_2$ (70) $Cl_3Si(CH_2)_8SiCl_3$ (trace)	U -109
	H_2PtCl_6 (EtOH)	reflux	28	$Cl_3Si(CH_2)_6CH=CH_2$ (70) $Cl_3Si(CH_2)_8SiCl_3$ (trace)	U -118
	H_2PtCl_6 (EtOH)	reflux	28	$Cl_3Si(CH_2)_6CH=CH_2$ (70) $Cl_3Si(CH_2)_8SiCl_3$ (trace)	D -068
	$[PtH(SiCl_3)P(c\text{-}Hex)_3]_2$	20	2	$Cl_3Si(CH_2)_8SiCl_3$ (76)	429
C_9H_{12}	Fig.107.				
	H_2PtCl_6	80	50	Fig.14. (49)	E -019
C_9H_{12}	Fig.166.				
	$PdCl_2(PPh_3)_2$	90	5	Fig.13. (89)	U -192
$C_{10}H_{10}$	$C_6H_5CH=CHCH=CH_2$				
	$PdCl_2$(PPFA)	80	16	$C_6H_5CH(SiCl_3)CH=CHCH_3+C_6H_5CH=CHCH(SiCl_3)CH_3$ (-)	482
	$PdCl_2$(PPFA)	80	16	$C_6H_5CH(SiCl_3)CH=CHCH_3+C_6H_5CH=CHCH(SiCl_3)CH_3$ (-)	483
$C_{10}H_{10}$	$CH_2=CHC_6H_4CH=CH_2$				
	H_2PtCl_6 (i-PrOH)	reflux	16	$CH_2=CHC_6H_4CH_2CH_2SiCl_3$ (56)	U -074
	H_2PtCl_6 (i-PrOH)	reflux	16	$CH_2=CHC_6H_4CH_2CH_2SiCl_3$ (56)	F -026
$C_{10}H_{12}$	Fig.42.				
	$Co_2(CO)_8$	80	3 (ht)	Fig.43. (65)	721
	H_2PtCl_6 (i-PrOH)	-	-	Fig.43. +Fig.44. (-)	260
	H_2PtCl_6 (i-PrOH)	20	-	Fig.43. (86)	S -042
$C_{10}H_{16}$	$(CH_3)_2C=CHCH_2CH_2C(=CH_2)CH=CH_2$				
	$PdCl_2(PhCN)_2$ + PPh_3	100	24	$(CH_3)_2C=CHCH_2CH_2C(CH_2SiCl_3)=CHCH_3$ (90)	851

H_2PtCl_6 (i-PrOH)	reflux	4	$(CH_3)_2C{=}CHCH_2CH_2C({=}CH_2)CH_2CH_2SiCl_3$ (39-41) $(CH_3)_2C{=}CHCH_2CH_2CH(CH_2SiCl_3)CH_2CH_2SiCl_3$ (17-19)	823
H_2PtCl_6 (i-PrOH)	reflux	35	$(CH_3)_2C{=}CHCH_2CH_2CH(CH_2SiCl_3)CH_2CH_2SiCl_3$ (44-46)	823
H_2PtCl_6 (i-PrOH)	reflux	6	$(CH_3)_2C{=}CHCH_2CH_2CH(CH_2SiCl_3)CH_2CH_2SiCl_3$ (48)	823
$C_{10}H_{16}$ $(CH_3)_2C{=}CHCH_2CH{=}C(CH_3)CH{=}CH_2$				
$PdCl_2(PhCN)_2 + PPh_3$	100	24	$(CH_3)_2C{=}CHCH_2CH(SiCl_3)C(CH_3){=}CHCH_3$/ $(CH_3)_2C{=}CHCH_2CH_2CH(CH_3)CH(CH_3)SiCl_3$ (42/58)	851
$C_{10}H_{16}$ Fig.45.				
H_2PtCl_6 (i-PrOH)	reflux	72	Fig.46. (-)	D -080
$C_{10}H_{16}$ trans-$CH_2{=}CHCH_2CH_2CH{=}CHCH_2CH_2CH{=}CH_2$				
H_2PtCl_6	130	-	trans-$Cl_3Si(CH_2)_4CH{=}CH(CH_2)_4SiCl_3$ (-)	D -046
$C_{10}H_{18}$ $CH_2{=}CH(CH_2)_6CH{=}CH_2$				
H_2PtCl_6 (EtOH)	till 130	6.5	$CH_2{=}CH(CH_2)_8SiCl_3$ (41) $Cl_3Si(CH_2)_{10}SiCl_3$ (10)	U -118
H_2PtCl_6 (EtOH)	till 130	6.5	$CH_2{=}CH(CH_2)_8SiCl_3$ (41) $Cl_3Si(CH_2)_{10}SiCl_3$ (10)	D -068
H_2PtCl_6 (EtOH)	till 130	6.5	$CH_2{=}CH(CH_2)_8SiCl_3$ (41) $Cl_3Si(CH_2)_{10}SiCl_3$ (10)	U -109
$C_{11}H_{12}$ $C_6H_5CH{=}C(CH_3)CH{=}CH_2$				
$PdCl_2$(PPFA)	80	16	$C_6H_5CH{=}C(CH_3)CH(CH_3)SiCl_3$ + $CH_2{=}CHCH(CH_3)CH(C_6H_5)SiCl_3$ (-)	483
$C_{11}H_{12}$ $C_6H_5CH{=}CHC(CH_3){=}CH_2$				
$PdCl_2$(PPFA)	80	16	$C_6H_5CH{=}CHC(CH_3)_2SiCl_3$ + $(CH_3)_2C{=}CHCH(C_6H_5)SiCl_3$ (-)	483
$C_{11}H_{12}$ $C_6H_5CH{=}CHCH{=}CHCH_3$				
$PdCl_2$(PPFA)	80	16	$C_6H_5CH(SiCl_3)CH{=}CHCH_2CH_3$ + $C_6H_5CH{=}CHCH(SiCl_3)CH_2CH_3$ (-)	482
$PdCl_2$(PPFA)	80	16	$C_6H_5CH{=}CHCH(CH_3)SiCl_3 + CH_3CH{=}CHCH(C_6H_5)SiCl_3$ (-)	483
$C_{12}H_{18}$ Fig.47.				
H_2PtCl_6 (i-PrOH)	160	72	Fig.48. (-)	1175
H_2PtCl_6 (i-PrOH)	180	80	Fig.49. (-)	1175
$C_{12}H_{22}$ $CH_2{=}CH(CH_2)_8CH{=}CH_2$				

	-	-	-	$Cl_3Si(CH_2)_{10}CH=CH_2$ + $Cl_3Si(CH_2)_{12}SiCl_3$ (-)	881
$C_{14}H_{12}$	1-$C_{10}H_7CH=CHCH=CH_2$				
	$PdCl_2$(PPFA)	80	16	$CH_3CH=CHCH(1\text{-}C_{10}H_7)SiCl_3$ + 1-$C_{10}H_7CH=CHCH(CH_3)$ $SiCl_3$ (-)	483
	$PdCl_2$(PPFA)	80	16	1-$C_{10}H_7CH=CHCH(SiCl_3)CH_3$ + 1-$C_{10}H_7CH(SiCl_3)CH=CHCH_3$ (-)	482
$C_{14}H_{26}$	$CH_2=CH(CH_2)_{10}CH=CH_2$				
	UV	45	72	$Cl_3Si(CH_2)_{12}CH=CH_2$ (25)	U -118
	UV	45	72	$Cl_3Si(CH_2)_{12}CH=CH_2$ (25)	D -068
	UV	45	72	$Cl_3Si(CH_2)_{12}CH=CH_2$ (25)	U -109
	H_2PtCl_6 (EtOH)	30	2	$Cl_3Si(CH_2)_{12}CH=CH_2$ (41)	U -118
	H_2PtCl_6 (EtOH)	30	2	$Cl_3Si(CH_2)_{12}CH=CH_2$ (41)	U -109
	H_2PtCl_6 (EtOH)	50	24	$Cl_3Si(CH_2)_{12}CH=CH_2/Cl_3Si(CH_2)_{14}SiCl_3$ (70/30)	D -068
	H_2PtCl_6 (EtOH)	till 70	1	$Cl_3Si(CH_2)_{12}CH=CH_2/Cl_3Si(CH_2)_{14}SiCl_3$ (70/30)	U -118
	H_2PtCl_6 (EtOH)	till 70	1	$Cl_3Si(CH_2)_{12}CH=CH_2/Cl_3Si(CH_2)_{14}SiCl_3$ (70/30)	U -109
	H_2PtCl_6 (i-PrOH)	25-30	3	$Cl_3Si(CH_2)_{12}CH=CH_2$ (41)	D -068

C_2	ClFC=CClF					
	UV	20	60	$CHClFCClFSiCl_3$ (16) $CHClFCHFSiCl_3$ (41)	130
C_2	ClFC=CF_2					
	gamma(^{60}Co)	20-30	-	$CHClFCF_2SiCl_3$ (58) $CH_2FCF_2SiCl_3$ (9)	1360
C_2H_1	BrHC=CF_2					
	UV	-	100	$Cl_3SiCH_2CBrF_2$ (17) $Cl_3SiCH_2CHF_2$ (47)	129
C_2H_1	ClHC=CF_2					
	UV	-	8-120	$Cl_3SiCH_2CHF_2$ (10-56) $Cl_3SiCF_2CH_3$ (0-21) $Cl_3SiCH_2CClF_2$ (7-62) $Cl_3SiCF_2CH_2Cl$ (0-9)	129
C_2H_1	F_2C=CHF					
	gamma(^{60}Co)	20-30	-	$CHF_2CHFSiCl_3$ (100)	1360
C_2H_2	ClHC=CHF					
	UV	-	100-116	$CH_2ClCHFSiCl_3$ (8-79) $CH_2FCHClSiCl_3$ (8-79) $CH_3CHFSiCl_3$ (3-73) $CH_2FCH_2SiCl_3$ (3-75)	270
	UV	20	100	$CH_2ClCHFSiCl_3$ (34) $CH_2FCHClSiCl_3$ (17)	480
C_2H_2	F_2C=CH_2					
	UV	-	24	$CHF_2CH_2SiCl_3$ (59)	86
C_2H_2	cis-FHC=CHF					
	UV	20	70	$CH_2FCHFSiCl_3$ (85)	480
C_2H_3	CH_2=$CHSiCl_3$					
	$Ni(PHPh_2)_4$	120-140	8	$Cl_3SiCH_2CH_2SiCl_3$ (91)	D-111
	$Ni(PPh_3)_4$	120-140	8	$Cl_3SiCH_2CH_2SiCl_3$ (91)	D-111
	$NiCl_2(PPh_3)_2$	120	6	$Cl_3SiCH_2CH_2SiCl_3$ (49-50) $Cl_3SiCH(CH_3)SiCl_3$ (28-30)	Pl-010
	$NiCl_2(PPh_3)_2$	120	6	$Cl_3SiCH_2CH_2SiCl_3$ (49-50) $Cl_3SiCH(CH_3)SiCl_3$ (28-30)	747

Table 1.3

= $HSiCl_3$ =

=2=

Pd + PPh_3	120	6	$Cl_3SiCH_2CH_2SiCl_3$ (10) $Cl_3SiCH(CH_3)SiCl_3$ (58)	749
Pd black	120	6	$Cl_3SiCH_2CH_2SiCl_3$ (trace)	749
$Pd(PPh_3)_4$	120	6	$Cl_3SiCH_2CH_2SiCl_3$ (2-4) $Cl_3SiCH(CH_3)SiCl_3$ (81-90)	749
$Pd(PPh_3)_4$	120	6	$Cl_3SiCH_2CH_2SiCl_3$ (2-4) $Cl_3SiCH(CH_3)SiCl_3$ (81-90)	747
$Pd(PPh_3)_4$	120	2	$Cl_3SiCH(CH_3)SiCl_3$ (70)	738
$Pd(PPh_3)_4$	60	6	$Cl_3SiCH_2CH_2SiCl_3$ (6) $Cl_3SiCH(CH_3)SiCl_3$ (70)	747
$Pd(PPh_3)_4$	80	1	$Cl_3SiCH(CH_3)SiCl_3$ (20)	738
$Pd(PPh_3)_4$	80-120	6 (solv.)	$Cl_3SiCH(CH_3)SiCl_3$ (54-94)	749
$Pd(PPh_3)_4$	reflux	2-5	$Cl_3SiCH_2CH_2SiCl_3$ (2) $Cl_3SiCH(CH_3)SiCl_3$ (96)	747
$Pd(PPh_3)_4$	reflux	3	$Cl_3SiCH(CH_3)SiCl_3$ (95)	Pl-010
$Pd(PPh_3)_4$ + $CH_2{=}CHSiCl_3$	80-120	1	$Cl_3SiCH(CH_3)SiCl_3$ (10-96)	738
$Pd(PPh_3)_4$ + $HSiCl_3$	45-120	0.08-12	$Cl_3SiCH(CH_3)SiCl_3$ (52-95)	738
$Pd(PPh_3)_4$ + $HSiCl_3$ ($PhCH_3$)	90-105	-	$Cl_3SiCH(CH_3)SiCl_3$ (kin.)	738
$PdCl_2$	120	6	$Cl_3SiCH_2CH_2SiCl_3$ (trace)	749
$PdCl_2$ + PPh_3	120	6	$Cl_3SiCH_2CH_2SiCl_3$ (3) $Cl_3SiCH(CH_3)SiCl_3$ (90)	749
$PdCl_2(PPh_3)_2$	120	6	$Cl_3SiCH_2CH_2SiCl_3$ (3) $Cl_3SiCH(CH_3)SiCl_3$ (90)	749
$PdCl_2(PPh_3)_2$	120	6	$Cl_3SiCH(CH_3)SiCl_3$ (96)	738
$PdCl_2(PPh_3)_2$	120	2	$Cl_3SiCH(CH_3)SiCl_3$ (62)	738
$PdCl_2(PPh_3)_2$ + $CH_2{=}CHSiCl_3$	120	0.08	$Cl_3SiCH(CH_3)SiCl_3$ (94)	738
$PdCl_2(PPh_3)_2$ + $CH_2{=}CHSiCl_3$	80-120	0.08-1	$Cl_3SiCH(CH_3)SiCl_3$ (5-94)	738
$PdCl_2(PPh_3)_2$ + $CH_2{=}CHSiCl_3$	95	1	$Cl_3SiCH(CH_3)SiCl_3$ (92)	738
$PdCl_2(PPh_3)_2$ + $HSiCl_3$	120	6	$Cl_3SiCH(CH_3)SiCl_3$ (70)	738
$PdCl_2(PPh_3)_2$ + $HSiCl_3$	45-120	0.08-4	$Cl_3SiCH(CH_3)SiCl_3$ (60-95)	738
$PdCl_2(PPh_3)_2$ + $HSiCl_3$	60-120	0.08-4	$Cl_3SiCH(CH_3)SiCl_3$ (65-95)	738

Table 1.3
= $HSiCl_3$ =

=3=

	$PdCl_2(PPh_3)_2$+ $HSiCl_3$ ($PhCH_3$)	90-105	-	$Cl_3SiCH(CH_3)SiCl_3$ (kin.)	738
	$(CH_2CMeCOOCH_2CH_2NMe_2)_n/H_2PtCl_6$	100	1	$Cl_3SiCH_2CH_2SiCl_3$ (85)	C -002
	H_2PtCl_6	26-98	12	$Cl_3SiCH_2CH_2SiCl_3$ (8) $Cl_3SiCH(CH_3)SiCl_3$ (2)	1058
	H_2PtCl_6	until 139	6	$Cl_3SiCH_2CH_2SiCl_3$ (43) $Cl_3SiCH(CH_3)SiCl_3$ (29)	1058
	H_2PtCl_6 (MeCOMe)	reflux		$Cl_3SiCH_2CH_2SiCl_3$ (100)	D -041
	H_2PtCl_6 (i-PrOH)	reflux	72	$Cl_3SiCH_2CH_2SiCl_3$ (-)	D -080
	H_2PtCl_6 (i-PrOH)	reflux	4	$Cl_3SiCH_2CH_2SiCl_3/Cl_3SiCH(CH_3)SiCl_3$ (60/40)	1059
	H_2PtCl_6 [$\overline{(CH_2)_5CO}$]	120	6	$Cl_3SiCH_2CH_2SiCl_3$ (93)	747
	H_2PtCl_6 [$\overline{(CH_2)_5CO}$]	60	6	$Cl_3SiCH_2CH_2SiCl_3$ (65)	747
	H_2PtCl_6 [$\overline{(CH_2)_5CO}$]	reflux	2	$Cl_3SiCH_2CH_2SiCl_3$ (97)	747
	$Rh_4(CO)_{12}$	35-40	1	$Cl_3SiCH_2CH_2SiCl_3$ (96-98)	S -024
	$RhCl(PPh_3)_3$	120	6	$Cl_3SiCH_2CH_2SiCl_3$ (79)	Pl-010
	$RhCl(PPh_3)_3$	120	6	$Cl_3SiCH_2CH_2SiCl_3$ (90) $Cl_3SiCH(CH_3)SiCl_3$ (6)	747
	$RhCl(PPh_3)_3$	60	6	$Cl_3SiCH_2CH_2SiCl_3$ (8)	747
	$RhCl(PPh_3)_3$	reflux	2	$Cl_3SiCH_2CH_2SiCl_3$ (8)	747
C_2H_3	$CHCl=CH_2$				
	gamma(^{60}Co)	100-180	-	$CH_3CH_2SiCl_3$ (21) $CH_3(CH_2)_3SiCl_3$ (7)	1360
	$PdBr_2[P(C_6H_4OMe)_3]_2$	120	5	$ClCH_2CH_2SiCl_3$ (75)	D -027
	Pt/C	110-120	0.5	$C_2H_5SiCl_3$ (99)	D -097
C_2H_3	$CHF=CH_2$				
	UV	-	100-150	$FCH_2CH_2SiCl_3$ (40-59)	270
	none	80-90	5	$FCH_2CH_2SiCl_3$ (34)	1228
C_3	$CF_3FC=CF_2$				
	gamma(^{60}Co)	20-30	-	$CF_3CHFCF_2SiCl_3/CF_3CF(SiCl_3)CHF_2$ (35/65)	1360
C_3H_1	$CF_3CH=CF_2$				

	gamma(^{60}Co)	40-60		$CF_3CH(SiCl_3)CHF_2$ (100)	1360
C_3H_1	$CF_3ClC=CHF$				
	UV	20	300	$CF_3CHFCH_2SiCl_3$ (65) $CF_3CH_2CHFSiCl_3$ (29)	479
C_3H_2	$CF_3CH=CHF$				
	UV	20	360	$CF_3CH_2CHFSiCl_3$ (16) $CF_3CHFCH_2SiCl_3$ (trace) $CF_3CH(SiCl_3)CH_2F$ (46)	479
C_3H_3	$CF_3CH=CH_2$				
	UV	-	100	$CF_3CH_2CH_2SiCl_3$ (66) $CF_3CH_2CH_2CH(CF_3)CH_2SiCl_3$ (13)	86
	$PdCl_2(PhCN)_2$	120	27	$CF_3CH_2CH_2SiCl_3$ (40) $CF_3CH(CH_3)SiCl_3$ (3)	858
	H_2PtCl_6 + $SnCl_2$ (i-PrOH)	50	3	$CF_3CH_2CH_2SiCl_3$ (95)	D -079
	$RhCl(PPh_3)_3$	120	36	$CF_3CH_2CH_2SiCl_3$ (75)	858
C_3H_3	$CH_2=CHCN$				
	none	200	6	$Cl_3SiCH(CN)CH_3$ (-)	U -011
	none	37-60	25	$Cl_3SiCH_2CH_2CN$ (13)	S -009
	$(EtO)_3Si(CH_2)_3NAll_2$	reflux	33	$Cl_3SiCH_2CH_2CN$ (40)	S -009
	$(Me_2N)_3PO$	85-90	1	$Cl_3SiCH_2CH_2CN$ (96)	J -008
	$AllN[Si(OEt)_3]_2$	40-110	33	$Cl_3SiCH_2CH_2CN$ (49)	S -009
	Antipirine	120-170	-	$Cl_3SiCH_2CH_2CN$ (72-76)	F -013
	$AsPh_3$	200	2	$Cl_3SiCH_2CH_2CN$ (26)	U -007
	$AsPh_3$	200	2	$Cl_3SiCH_2CH_2CN$ (26)	U -015
	BCl_3	120-140	4	$Cl_3SiCH(CN)CH_3$ (18)	95
	C_5H_5N	100	5	$Cl_3SiCH_2CH_2CN$ (54)	1192
	Et-4-C_5H_4N	200	2	$Cl_3SiCH_2CH_2CN$ (41)	U -010
	Me-4-C_5H_4N	200	2	$Cl_3SiCH_2CH_2CN$ (32)	U -010
	$N(All)_3$	reflux	21	$Cl_3SiCH_2CH_2CN$ (82)	S -009
	$N(All)_3$	reflux	34	$Cl_3SiCH_2CH_2CN$ (97)	S -009

Table 1.3
= $HSiCl_3$ =

=5=

$N(All)_3$	reflux	25	$Cl_3SiCH_2CH_2CN$ (71)	S -009
NBu_3	100	2	$Cl_3SiCH_2CH_2$ (80)	95
NBu_3	reflux	24	$Cl_3SiCH_2CH_2CN$ (56)	U -011
NBu_3	until 97	112	$Cl_3SiCH(CN)CH_3$ (44)	U -011
NEt_3	200	2	$Cl_3SiCH_2CH_2CN$ (27)	U -015
NEt_3	200	2	$Cl_3SiCH_2CH_2CN$ (27)	U -007
$P(OEt)_3$	130	4	$Cl_3SiCH_2CH_2CN$ (-)	J -029
PBu_3	200	2	$Cl_3SiCH_2CH_2CN$ (56)	U -007
PBu_3	200	2	$Cl_3SiCH_2CH_2CN$ (56)	U -015
PEt_3	150	2	$Cl_3SiCH_2CH_2CN$ (55)	U -015
PPh_3	100	0.5	$Cl_3SiCH_2CH_2CN$ (66)	U -015
PPh_3	150	2	$Cl_3SiCH_2CH_2CN$ (59-67)	U -015
PPh_3	175-179	-	$Cl_3SiCH_2CH_2CN$ (60)	U -015
PPh_3	200	2	$Cl_3SiCH_2CH_2CN$ (56)	U -015
PPh_3	200	2	$Cl_3SiCH_2CH_2CN$ (56)	U -007
PPh_3	200	3	$Cl_3SiCH_2CH_2CN$ (-)	J -044
PPh_3	208-225	-	$Cl_3SiCH_2CH_2CN$ (74)	U -015
PPh_3	223	-	$Cl_3SiCH_2CH_2CN$ (71)	U -015
PPh_3	230-231	-	$Cl_3SiCH_2CH_2CN$ (78)	U -015
PPh_3	40	18	$Cl_3SiCH_2CH_2CN$ (8)	U -015
PPh_3	50	20	$Cl_3SiCH_2CH_2CN$ (24)	U -015
PPh_3	60	26	$Cl_3SiCH_2CH_2CN$ (66)	U -015
PPh_3	75	2	$Cl_3SiCH_2CH_2CN$ (13)	U -015
Ph_2NSH	200	2	$Cl_3SiCH_2CH_2CN$ (25)	U -010
ion exchanger	110-120	2	$Cl_3SiCH_2CH_2CN$ (24-45)	95
polyalkylpyridine	200	2	$Cl_3SiCH_2CH_2CN$ (21)	U -010
$Cu(acac)_2$ + c-HexNC	120	2	$Cl_3SiCH_2CH_2CN$ (75)	1165

Table 1.3
= $HSiCl_3$ =

=6=

$Cu(acac)_2$ + c-HexNC	120	2	$Cl_3SiCH_2CH_2CN$ (75)	C -005
$Cu(acac)_2$ + c-HexNC	60	2	$Cl_3SiCH_2CH_2CN$ (10)	1165
$Cu(acac)_2$ + c-HexNC	80	2	$Cl_3SiCH_2CH_2CN$ (45)	1165
$Cu(acac)_2$ + c-HexNC	80	2	$Cl_3SiCH_2CH_2CN$ (45)	C -005
$Cu(acac)_2$ + t-BuNC	100	1.5	$Cl_3SiCH_2CH_2CN$ (51)	C -005
Cu_2O + $Me_2NCH_2CH_2NMe_2$	30	4	$Cl_3SiCH_2CH_2CN$ (30)	959
Cu_2O + $Me_2NCH_2CH_2NMe_2$	reflux	0.16	$Cl_3SiCH_2CH_2CN$ (95-100)	959
Cu_2O + $Me_2NCH_2CH_2NMe_2$ + ultrasound	30	2	$Cl_3SiCH_2CH_2CN$ (80)	959
Cu_2O + c-HexNC	100	2	$Cl_3SiCH_2CH_2CN$ (70)	1165
Cu_2O + c-HexNC	100	- (b)	$Cl_3SiCH_2CH_2CN$ (67)	C -005
Cu_2O + c-HexNC	120	2	$Cl_3SiCH_2CH_2CN$ (66)	C -005
Cu_2O+ t-BuNC	120	2	$Cl_3SiCH_2CH_2CN$ (42)	C -005
Cu_2O+ t-BuNC	120	2	$Cl_3SiCH_2CH_2CN$ (42)	1165
CuCl + NEt_3 + $Me_2NC_2H_4NMe_2$	60	1	$Cl_3SiCH_2CH_2CN$ (91)	1103
CuCl +c-HexNC	110	2.5	$Cl_3SiCH_2CH_2CN$ (78)	C -005
CuCl +c-HexNC	120	2	$Cl_3SiCH_2CH_2CN$ (78)	1165
CuCl + t-BuNC	120	2	$Cl_3SiCH_2CH_2CN$ (56)	1165
$CuCl_2$+ t-BuNC	120	2	$Cl_3SiCH_2CH_2CN$ (56)	C -005
Ni salt+ AlR_3	-	-	$Cl_3SiCH(CN)CH_3$ (-)	561
Ni(THF)	0-25	2	$Cl_3SiCH(CN)CH_3$ (75)	E -049
Ni(THF) +sonic waves	0-25	1	$Cl_3SiCH(CN)CH_3$ (70)	E -049
$Ni(acac)_2$ + PPh_3	100	5	$Cl_3SiCH(CN)CH_3$ (70)	560
$Ni(acac)_2$ + PPh_3	60	7	$Cl_3SiCH(CN)CH_3$ (70)	1325
$NiCl_2(DMPF)$	100	156	$Cl_3SiCH(CN)CH_3$ (59)	589
Pd +PBu_3	100	5	$Cl_3SiCH(CN)CH_3$ (40)	D -027
$Pd(PBu_3)_4$	120	8	$Cl_3SiCH(CN)CH_3$ (40)	1208
$Pd(PBu_3)_4$	120	8	$Cl_3SiCH(CN)CH_3$ (40)	D -027

Table 1.3
= $HSiCl_3$ = =7=

	$RhCl(PPh_3)_3$	140	24	$Cl_3SiCH_2CH_2CN$ (80)	868
	polymer/(Rh complex)	100	2	$Cl_3SiCH(CN)CH_3$ (8-12)	1220
	polymer/$RhCl_3$(THF)	reflux	3	$Cl_3SiCH_2CH_2CN$ (60)	S -029
	$RuCl_2(PPh_3)_3$	120	2	$Cl_3SiCH_2CH_2CN$ (93)	1170
C_3H_5	$CH_2=CHCH_2Br$				
	H_2PtCl_6	42-80	19.5-20	$Cl_3Si(CH_2)_3Br$ (64)	U -068
	H_2PtCl_6	reflux	72	$Cl_3Si(CH_2)_3Br$ (60)	637
	H_2PtCl_6 (i-PrOH)	-	-	$Cl_3Si(CH_2)_3Br$ (60)	751
	H_2PtCl_6(i-PrOH) + Et_3N(PhMe)+Ph_2SiH_2	70-reflux	2	$Cl_3Si(CH_2)_3Br$ (72)	C -036
C_3H_5	$CH_2=CHCH_2Cl$				
	gamma(^{60}Co)	110-160	0.9-5.4	$Cl_3Si(CH_2)_3Cl$ (57-81) $CH_2=CHCH_2SiCl_3$ (2-5) $Cl_3Si(CH_2)_3SiCl_3$ (1-19) $CH_3CH_2CH_2SiCl_3$ (0-1)	D -012
	gamma(^{60}Co)	20-80	5-62	$Cl_3Si(CH_2)_3Cl$ (40-78)	1034
	gamma(^{60}Co)	80	-	$Cl_3Si(CH_2)_3Cl$ (100)	1360
	$0_{1,5}Si-NR_3/PtCl_4^{-2}$	-	-	$Cl_3Si(CH_2)_3Cl$ (74-75)	D -124
	$0_{1,5}Si-PPh_2$/Pt(II) complex	70	8	$Cl_3Si(CH_2)_3Cl$ (-)	D -067
	$Al_2O_3-NMe_2/H_2PtCl_6$	80	3	$Cl_3Si(CH_2)_3Cl$ (85)	C -007
	$C_6H_8(OC_3H_6)_6/Pt(PPh_3)_4$	80	3	$Cl_3Si(CH_2)_3Cl$ (41)	C -013
	H_2PtCl_6	-	-	$Cl_3Si(CH_2)_3Cl$ (44)	926
	H_2PtCl_6	-	-	$Cl_3Si(CH_2)_3Cl$ (59)	507
	H_2PtCl_6	-	-	$Cl_3Si(CH_2)_3Cl$ (58)	508
	H_2PtCl_6	100	8	$Cl_3Si(CH_2)_3Cl$ (47)	U -121
	H_2PtCl_6	35-80	9	$Cl_3Si(CH_2)_3Cl$ (65)	U -068
	H_2PtCl_6	40	1	$Cl_3Si(CH_2)_3Cl$ (66)	E -037
	H_2PtCl_6	70	-	$Cl_3Si(CH_2)_3Cl$ (60)	505
	H_2PtCl_6	reflux	4	$Cl_3Si(CH_2)_3Cl$ (63)	U -074
	H_2PtCl_6	reflux	4	$Cl_3Si(CH_2)_3Cl$ (63)	F -026

Table 1.3

= $HSiCl_3$ = =8=

H_2PtCl_6 (AcOAll)	44-78	5	$Cl_3Si(CH_2)_3Cl$ (53)	Pl-004
H_2PtCl_6 (AcOAll)	44-78	5	$Cl_3Si(CH_2)_3Cl$ (53)	Pl-009
H_2PtCl_6 ($AllOCH_2COOAll$)	until 78	4.33	$Cl_3Si(CH_2)_3Cl$ (71)	Pl-008
H_2PtCl_6 (AllOH)	44-75	5	$Cl_3Si(CH_2)_3Cl$ (56)	Pl-004
H_2PtCl_6 ($MeOC_2H_4OMe$ + $\overline{C_6H_4NHC_6H_4S}$)	40-60	0.75	$Cl_3Si(CH_2)_3Cl$ (68) $CH_3CH_2CH_2SiCl_3$ (6)	D -104
H_2PtCl_6 ($MeOC_2H_4OMe$ + $\overline{C_6H_4NHC_6H_4S}$)	70	-	$Cl_3Si(CH_2)_3Cl$ (58) $CH_3CH_2CH_2SiCl_3$ (12)	D -104
H_2PtCl_6 ($MeOC_2H_4OMe$ + EtOH)	70	1	$Cl_3Si(CH_2)_3Cl$ (54) $CH_3CH_2CH_2SiCl_3$ (13)	D -104
H_2PtCl_6 (THF)	40	1	$Cl_3Si(CH_2)_3Cl$ (68)	E -037
H_2PtCl_6 (THF)	until 110	-	$Cl_3Si(CH_2)_3Cl$ (43)	955
H_2PtCl_6 (i-PrOH)	40	1	$Cl_3Si(CH_2)_3Cl$ (23)	E -037
H_2PtCl_6 (i-PrOH)	reflux	12	$Cl_3Si(CH_2)_3Cl$ (47-57)	C -020
H_2PtCl_6 (i-PrOH)	reflux	12	$Cl_3Si(CH_2)_3Cl$ (47-57)	C -019
H_2PtCl_6 (i-PrOH)	reflux	1	$Cl_3Si(CH_2)_3SiCl_3$ (12)	748
H_2PtCl_6 (i-PrOH)	reflux	- (solv.)	$Cl_3Si(CH_2)_3Cl$ (78)	U -178
H_2PtCl_6 (i-PrOH) + $AsPh_3$	reflux	3	$Cl_3Si(CH_2)_3Cl$ (60)	C -020
H_2PtCl_6 (i-PrOH) + $AsPh_3$ + Ph_2SiH_2	reflux	1.75	$Cl_3Si(CH_2)_3Cl$ (78)	C -024
H_2PtCl_6 (i-PrOH) + $CH_2{=}CHOBu$	until 115	5	$Cl_3Si(CH_2)_3Cl$ (67)	S -012
H_2PtCl_6 (i-PrOH) + Et_2NH	reflux	3-4	$Cl_3Si(CH_2)_3Cl$ (73)	C -019
H_2PtCl_6 (i-PrOH) + FCH_2COOH	reflux	6	$Cl_3Si(CH_2)_3Cl$ (65)	C -029
H_2PtCl_6 (i-PrOH) + $MeCOONMe_2$	reflux	- (solv.)	$Cl_3Si(CH_2)_3Cl$ (85)	U -178
H_2PtCl_6 (i-PrOH) + NBu_3	reflux	3-4	$Cl_3Si(CH_2)_3Cl$ (79)	C -019
H_2PtCl_6 (i-PrOH) + $NpPPh_2$ + Ph_2SiH_2	reflux	1.6	$Cl_3Si(CH_2)_3Cl$ (79)	C -024
H_2PtCl_6 (i-PrOH) + PPh_3	reflux	3	$Cl_3Si(CH_2)_3Cl$ (58-81)	C -020
H_2PtCl_6 (i-PrOH) + Ph_2PH	reflux	3	$Cl_3Si(CH_2)_3Cl$ (56)	C -020
H_2PtCl_6 (i-PrOH) + $PhC_6H_4PMe_2$	reflux	3	$Cl_3Si(CH_2)_3Cl$ (57)	C -020
H_2PtCl_6 (i-PrOH) + $PhNMe_2$	46-97	3.25	$Cl_3Si(CH_2)_3Cl$ (-)	C -024

Table 1.3

= $HSiCl_3$ =

H_2PtCl_6 (i-PrOH) + $PhNMe_2$	reflux	3-4	$Cl_3Si(CH_2)_3Cl$ (60-82)	C -019
H_2PtCl_6 (i-PrOH) + $PhNMe_2$ + $(EtO)_3SiH$	reflux	1.59	$Cl_3Si(CH_2)_3Cl$ (83)	C -024
H_2PtCl_6 (i-PrOH) + $PhNMe_2$ + $C_{10}H_{21}SiH_3$	reflux	1.33	$Cl_3Si(CH_2)_3Cl$ (82)	C -024
H_2PtCl_6 (i-PrOH)+ $PhNMe_2$ + $LiAlH_4$	reflux	24	$Cl_3Si(CH_2)_3Cl$ (52)	C -024
H_2PtCl_6 (i-PrOH) + $PhNMe_2$ + $Na[AlH_2$-$(OC_2H_4OMe)_2]$	reflux	24	$Cl_3Si(CH_2)_3Cl$ (54)	C -024
H_2PtCl_6 (i-PrOH) + $PhNMe_2$ + $NpSiH_2Me$	reflux	1.5	$Cl_3Si(CH_2)_3Cl$ (80)	C -024
H_2PtCl_6 (i-PrOH) + $PhNMe_2$ + Ph_2SiH_2	46-98	1.75	$Cl_3Si(CH_2)_3Cl$ (81)	C -024
H_2PtCl_6 (i-PrOH) + $SbPh_3$	reflux	3	$Cl_3Si(CH_2)_3Cl$ (61)	C -020
H_2PtCl_6 (i-PrOH) + $[CH(OH)COOH]_2$	reflux	6	$Cl_3Si(CH_2)_3Cl$ (65)	C -029
H_2PtCl_6 (i-PrOH) + c-$C_5H_{10}NH$	reflux	3-4	$Cl_3Si(CH_2)_3Cl$ (64)	C -019
H_2PtCl_6 (i-PrOH) + c-$HexNH_2$	reflux	3-4	$Cl_3Si(CH_2)_3Cl$ (66)	C -019
H_2PtCl_6 (i-PrOH) +c-$HexNH_2$ + Ph_2SiH_2	reflux	1.66	$Cl_3Si(CH_2)_3Cl$ (82)	C -024
H_2PtCl_6 (i-PrOH) + diphos	reflux	3	$Cl_3Si(CH_2)_3Cl$ (79)	C -020
H_2PtCl_6 (i-PrOH) + o-HOC_6H_4COOH	reflux	7	$Cl_3Si(CH_2)_3Cl$ (66)	C -029
H_2PtCl_6 + $BuNH_2$	-	-	$Cl_3Si(CH_2)_3Cl$ (85)	507
H_2PtCl_6 + Et_3SiH	19-66	2	$(C_2H_5)_3Si(CH_2)_3Cl$ (4) $Cl_3Si(CH_2)_3Cl$ (54)	U -168
H_2PtCl_6+ H-siloxane	23-96	0.05	$Cl_3Si(CH_2)_3Cl$ (traces)	U -168
H_2PtCl_6+ H-siloxane	83	0.8	$Cl_3Si(CH_2)_3Cl$ (29)	U -168
H_2PtCl_6 + PPh_3	r.t.	-	$Cl_3Si(CH_2)_3Cl$ (85)	508
H_2PtCl_6 + amines	-	-	$Cl_3Si(CH_2)_3Cl$ (-)	507
H_2PtCl_6 $[(\overline{CH_2)_5C}O$, $CH_2=CHSi(OEt)_3]$	r.t.	2-5	$Cl_3Si(CH_2)_3Cl$ (71-78)	Pl-023
H_2PtCl_6 $[(\overline{CH_2)_5C}O$, $CH_2=CHSi(OEt)_3]$	r.t.	5	$Cl_3Si(CH_2)_3Cl$ (71)	Pl-024
H_2PtCl_6 $[(\overline{CH_2)_5C}O$, $CH_2=CHSi(OMe)_3]$	r.t.	5	$Cl_3Si(CH_2)_3Cl$ (73-75)	Pl-023
H_2PtCl_6 $[(\overline{CH_2)_5C}O$, $CH_2=CHSi(iso\text{-}OPr)_3]$	r.t.	5	$Cl_3Si(CH_2)_3Cl$ (76)	Pl-023
H_2PtCl_6 $[(\overline{CH_2)_5C}O]$	120	-	$Cl_3Si(CH_2)_3Cl/CH_3CH_2CH_2SiCl_3$ (78/22)	D -103
H_2PtCl_6 $[(\overline{CH_2)_5C}O]$	reflux	1	$Cl_3Si(CH_2)_3SiCl_3$ (60)	748

Table 1.3

= $HSiCl_3$ = =10=

H_2PtCl_6 [$\overline{(CH_2)_5CO}$]	until 76	-	$Cl_3Si(CH_2)_3Cl$ (78) $CH_2=CHCH_2SiCl_3$ (1) $CH_3CH_2CH_2SiCl_3$ (5)	D -041
H_2PtCl_6 [$\overline{CH_2(CH_2)_2COO}$]	40	1	$Cl_3Si(CH_2)_3Cl$ (55)	E -037
Pt catalyst	-	-	$Cl_3Si(CH_2)_3SiCl_3$ (-)	313
Pt(IV)complex	-	- (flow)	$Cl_3Si(CH_2)_3Cl$ (-)	J -011
$Pt(PPh_3)_2(CH_2=CH_2)$	reflux	1.5	$Cl_3Si(CH_2)_3Cl$ (72)	Pl-032
$Pt(PPh_3)_2(CH_2=CH_2)$	reflux	1	$Cl_3Si(CH_2)_3SiCl_3$ (75)	748
$Pt(PPh_3)_2(CH_2=CHSiMe_3)$	reflux	3.5	$Cl_3Si(CH_2)_3Cl$ (68)	Pl-032
$Pt(PPh_3)_2[CH_2=CHSi(OEt)_3]$	reflux	3	$Cl_3Si(CH_2)_3Cl$ (71)	Pl-032
$Pt(PPh_3)_4$	100	8	$Cl_3Si(CH_2)_3Cl$ (44)	U -121
$Pt(PPh_3)_4$	100	8	$Cl_3Si(CH_2)_3Cl$ (82)	J -038
$Pt(PPh_3)_4$	30-73	4.3	$Cl_3Si(CH_2)_3Cl$ (65-66)	Pl-005
$Pt(PPh_3)_4$	reflux	1	$Cl_3Si(CH_2)_3SiCl_3$ (65)	748
Pt/C	80-90	14-27	$Cl_3Si(CH_2)_3Cl$ (-)	D -033
Pt/C	95	4	$Cl_3Si(CH_2)_3Cl$ (83)	D -096
Pt/C	until 85	8.5-27	$Cl_3Si(CH_2)_3Cl$ (64-65)	D -096
$PtCl_2(PPh_3)_2$	100	8	$Cl_3Si(CH_2)_3Cl$ (82-84)	U -121
$PtCl_2(PPh_3)_2$	reflux	1	$Cl_3Si(CH_2)_3SiCl_3$ (32)	748
$PtCl_4$ + $CH_2=CHHex$ + sec-$BuNH_2$	20-87	0.1	$Cl_3Si(CH_2)_3Cl$ (76)	D -110
SDVB-NMe_2/H_2PtCl_6 (EtOH)	80	2	$Cl_3Si(CH_2)_3Cl$ (93)	C -002
SiO_2-$\overline{N(CH_2)_4}$/H_2PtCl_6 (MeOH)	20-83	6	$Cl_3Si(CH_2)_3Cl$ (60)	Pl-007
SiO_2-$\overline{N(CH_2)_4}$/H_2PtCl_6 (MeOH)	20-83	6	$Cl_3Si(CH_2)_3Cl$ (60)	Pl-003
SiO_2-$\overline{N(CH_2)_4}$/$Pt(PPh_3)_4$	20-83	6	$Cl_3Si(CH_2)_3Cl$ (13)	Pl-003
SiO_2-$\overline{N(CH_2)_4}$/$Pt(PPh_3)_4$	20-83	6	$Cl_3Si(CH_2)_3Cl$ (13)	Pl-007
SiO_2-$\overline{N(CH_2)_5}$/H_2PtCl_6 (MeOH)	20-83	6	$Cl_3Si(CH_2)_3Cl$ (61-64)	Pl-003
SiO_2-$\overline{N(CH_2)_5}$/H_2PtCl_6 (MeOH)	20-83	6	$Cl_3Si(CH_2)_3Cl$ (61-64)	Pl-007
SiO_2-$\overline{N(CH_2)_5}$/$Pt(PPh_3)_4$	20-83	6	$Cl_3Si(CH_2)_3Cl$ (5-43)	Pl-003

Table 1.3

= $HSiCl_3$ = =11=

	SiO_2-$N(CH_2)_5$/$Pt(PPh_3)_4$	20-83	6	$Cl_3Si(CH_2)_3Cl$ (5-43)	Pl-007
	SiO_2-$N(CH_2CH_2)_2O$/H_2PtCl_6 (MeOH)	20-83	6	$Cl_3Si(CH_2)_3Cl$ (62-63)	Pl-003
	SiO_2-$N(CH_2CH_2)_2O$/H_2PtCl_6 (MeOH)	20-83	6	$Cl_3Si(CH_2)_3Cl$ (62-63)	Pl-007
	SiO_2-$N(CH_2CH_2)_2O$/$Pt(PPh_3)_4$	20-83	6	$Cl_3Si(CH_2)_3Cl$ (32-54)	Pl-007
	SiO_2-$N(CH_2CH_2)_2O$/$Pt(PPh_3)_4$	20-83	6	$Cl_3Si(CH_2)_3Cl$ (32-54)	Pl-003
	SiO_2-$OSi(CH_2)_3SH$/Pt	60	16	$Cl_3Si(CH_2)_3Cl$ (92)	U -146
	$[HPt(SiCl_3)P(c\text{-}Hex)_3]_2$	20	2 (t)	$Cl_3Si(CH_2)_3Cl$ (70)	429
	$[HPt(SiCl_3)P(c\text{-}Hex)_3]_2$	20	-	$Cl_3Si(CH_2)_3Cl$ (70)	428
	$[HPt(SiCl_3)P(c\text{-}Hex)_3]_2$	r.t.	0.5	$Cl_3Si(CH_2)_3Cl$ (70)	G -036
	polymer-$(CH_2)_nS$/$PtCl_x$	70	-	$Cl_3Si(CH_2)_3Cl$ (84)	505
	siloxane-ammonium-Pt complex	-	- (flow)	$Cl_3Si(CH_2)_3Cl$ (75)	D -124
	C/$RhCl_3 \cdot 3H_2O$ +c-$C_{12}H_{18}$ (EtOH)	150	5	$Cl_3Si(CH_2)_3Cl$ (10)	Dd-003
	C-CN/$RhCl_3$ + c-$C_{12}H_{18}$	80	5	$Cl_3Si(CH_2)_3Cl$ (10)	C -030
	SDVB-CH_2Cl/$RhCl(CO)(PPh_3)_2$	150	5	$Cl_3Si(CH_2)_3Cl$ (8)	C -022
C_3H_5	$CH_2=CHCH_2F$				
	H_2PtCl_6 (THF)	reflux	3	$Cl_3Si(CH_2)_3F$ (93)	955
C_3H_5	$CH_2=CHCH_2SiCl_3$				
	H_2PtCl_6 (i-PrOH)	150-180	-	$Cl_3Si(CH_2)_3SiCl_3$ (79)	1359
C_3H_5	$ClCH_2SiCl_2CH=CH_2$				
	H_2PtCl_6 (i-PrOH)	80-100	-	$ClCH_2SiCl_2CH_2CH_2SiCl_3$ (73)	387
C_3H_6	$CH_3SiCl_2CH=CH_2$				
	$Co_2PtCl_2(CO)_8$	reflux	1.2	$CH_3SiCl_2CH_2CH_2CH_2SiCl_3$ (75)	U -053
	$Co_2PtCl_2(CO)_8$	reflux	1.2	$CH_3SiCl_2CH_2CH_2CH_2SiCl_3$ (75)	U -102
	$Pd(PPh_3)_4$	120	6	$CH_3SiCl_2CH_2CH_2CH_2SiCl_3$ (1) $CH_3SiCl_2CH(CH_3)SiCl_3$ (89)	749
	H_2PtCl_6 (i-PrOH)	reflux	4	$CH_3SiCl_2CH_2CH_2CH_2SiCl_3$/$CH_3SiCl_2CH(CH_3)SiCl_3$ (97/3)	1059
C_4	$CF_2CF_2CF=CF$				

	UV	-	150	trans-$\overline{CF_2CF_2CHFC}HSiCl_3$ (80)	69
C_4	$CF_3CF=CFCF_3$				
	UV	-	150	$CF_3CHFCF(SiCl_3)CF_3$ (91)	69
C_4H_1	$\overline{CF_2CF_2ClC=C}H$				
	gamma	-	-	trans-$\overline{CF_2CF_2CHClC}HSiCl_3$ + cis-$\overline{CF_2CF_2CHClC}HSiCl_3$ + $\overline{CF_2CF_2CH_2C}HSiCl_3$ +$\overline{CF_2CF_2CH_2C}ClSiCl_3$ (-)	1214
	gamma	-	-	trans-$\overline{CF_2CF_2CHClC}HSiCl_3$ + cis-$\overline{CF_2CF_2CHClC}HSiCl_3$ + $\overline{CF_2CF_2CH_2C}HSiCl_3$ +$\overline{CF_2CF_2CH_2C}ClSiCl_3$ (-)	J -007
	gamma	-	-	trans-$\overline{CF_2CF_2CHClC}HSiCl_3$ (31) cis-$\overline{CF_2CF_2CHClC}HSiCl_3$ (38)	1213
C_4H_1	$\overline{CF_2CF_2FC=C}H$				
	gamma	-	-	trans-$\overline{CF_2CF_2CHFC}HSiCl_3$ + cis-$\overline{CF_2CF_2CHFC}HSiCl_3$ (-)	1213
C_4H_3	$(CF_3)_2NCH=CH_2$				
	UV	-	40	$(CF_3)_2NCH_2CH_2SiCl_3$ (93)	D -038
	UV	-	48	$(CF_3)_2NCH_2CH_2SiCl_3$ (100)	11
C_4H_5	$CH_2=C(CH_3)CN$				
	NBu_3	until 97	112	$Cl_3SiCH_2CH(CH_3)CN$ (93)	U -011
	PPh_3	150	2	$Cl_3SiCH_2CH(CH_3)CN$ (35)	U -015
	$NiCl_2(PhPMe_2)_2$	120-140	8	$Cl_3SiC(CH_3)_2CN$ (70)	D -111
	$NiCl_3(HPBu_3)(PBu_3)$	until 120	36	$Cl_3SiC(CH_3)_2CN$ (51)	D -111
C_4H_5	$CH_2=CHCH_2CN$				
	BCl_3	120-140	4	$Cl_3Si(CH_2)_3CN$ (18)	95
	Me_2NHCO	100	2.5	$Cl_3Si(CH_2)_3CN$ (7)	95
	PPh_3	150	2	$Cl_3Si(CH_2)_3CN$ (31)	U -007
	PPh_3	150	2	$Cl_3Si(CH_2)_3CN$ (31)	U -015
	H_2PtCl_6	80-136	12.5	$Cl_3Si(CH_2)_3CN$ (99)	95
	Pt/Al_2O_3	200	2	$Cl_3Si(CH_2)_3CN$ (82)	U -007
C_4H_5	$CH_2=CHCH_2NCO$				

	H_2PtCl_6 (EtOH)	46-49	1	$Cl_3Si(CH_2)_3NCO$ (73)	D -024
	H_2PtCl_6 (EtOH)	until 49	2	$Cl_3Si(CH_2)_3NCO$ (73)	U -039
C_4H_5	$ClCOOCH_2CH=CH_2$				
	H_2PtCl_6 (i-PrOH)	20-25	45	$ClCOO(CH_2)_3SiCl_3$ (-)	G -016
	H_2PtCl_6 (i-PrOH)	60-200	-	$ClCOO(CH_2)_3SiCl_3$ (55)	S -026
	H_2PtCl_6 (i-PrOH)	90-105	-	$ClCOO(CH_2)_3SiCl_3$ (22)	783
	H_2PtCl_6 (i-PrOH)	90-105	-	$ClCOO(CH_2)_3SiCl_3$ (22)	786
C_4H_5	$N_3COOCH_2CH=CH_2$				
	H_2PtCl_6 (i-PrOH)	50	16	$N_3COO(CH_2)_3SiCl_3$ (76)	G -019
C_4H_6	$CH_2=CHCOOCH_3$				
	$Ni(acac)_2 + PPh_3$	100	5	$CH_3CH(SiCl_3)COOCH_3$ (80)	560
	$Ni(acac)_2 + PPh_3$	120	10	$CH_3CH(SiCl_3)COOCH_3$ (80)	1335
	$Ni(acac)_2 + PPh_3$	60	7	$CH_3CH(SiCl_3)COOCH_3$ (80)	1325
C_4H_6	$CH_2=CHOCH=CH_2$				
	$Ni(acac)_2 + PPh_3$	80	6 (THF)	$Cl_3SiCH_2CH_2OCH_2CH_2SiCl_3$ (75)	560
C_4H_6	$CH_3COOCH=CH_2$				
	gamma(^{60}Co)	20	-	$CH_3COOCH_2CH_2SiCl_3 + CH_3CH_2OC_2H_3SiCl_3$ (30)	1210
	Ni(THF)	0-25	12	$CH_3COOCH_2CH_2SiCl_3$ (10)	E -049
	Ni(THF)	0-25	2.5	$CH_3COOCH(CH_3)SiCl_3 + CH_3CH=C(OCH_3)OSiCl_3$ (65)	E -049
	Ni(THF) +sonic waves	0-25	3	$CH_3COOCH_2CH_2SiCl_3$ (42)	E -049
	Ni(THF) +sonic waves	0-25	2	$CH_3COOCH(CH_3)SiCl_3 + CH_3CH=C(OCH_3)OSiCl_3$ (74)	E -049
	$Ni(acac)_2 + PPh_3$	120	7	$CH_3COOCH_2CH_2SiCl_3$ (45)	560
	$Ni(acac)_2 + PPh_3$	120	7	$CH_3COOCH_2CH_2SiCl_3$ (45)	1325
	$Pd[P(C_6H_4Me)_3]_4$	130	5	$CH_3COOCH_2CH_2SiCl_3$ (88)	D -027
	$C_6H_8(OC_3H_6CN)_6/Pt(PPh_3)_4$	80	3	$CH_3COOCH_2CH_2SiCl_3 + CH_3COOCH(CH_3)SiCl_3$ (34)	C -013
	H_2PtCl_6 ($AllOCH_2COOAll$)	43-116	18	$CH_3COOCH_2CH_2SiCl_3$ (54)	Pl-008
	H_2PtCl_6 (OctOH)	until 118	4	$CH_3COOCH_2CH_2SiCl_3$ (90)	D -040

	SiO_2-$\overline{N(CH_2CH_2)_2O}$/$H_2PtCl_6$ (EtOH)	73-108	11.5	$CH_3COOCH_2CH_2SiCl_3$ (50)	PI-007
	[PS-$PPh_2RhCl(PPh_3)_2]_n$	80	3	$CH_3COOCH_2CH_2SiCl_3$ (28)	C -008
C_4H_6	$Cl_2Si(CH{=}CH_2)_2$				
	H_2PtCl_6 (i-PrOH)	reflux	4	$Cl_2Si(CH{=}CH_2)C_2H_4SiCl_3$ (-) $Cl_2Si(C_2H_4SiCl_3)_2$ (-)	1059
C_4H_6	$ClCH_2CHClCH{=}CH_2$				
	H_2PtCl_6 (i-PrOH)	90	1	$ClCH_2CHClCH_2CH_2SiCl_3$ (52)	730
C_4H_6	$HCOOCH_2CH{=}CH_2$				
	H_2PtCl_6 (i-PrOH)	45-150	2	$HCCOO(CH_2)_3SiCl_3$ (74)	1052
	H_2PtCl_6 (i-PrOH)	45-150	2	$HCCOO(CH_2)_3SiCl_3$ (74)	S -026
C_4H_7	$BrCH_2CH_2CH{=}CH_2$				
	H_2PtCl_6 (THF)	reflux	7	$Br(CH_2)_4SiCl_3$ (84)	212
C_4H_7	$CH_2{=}C(CH_3)CH_2Cl$				
	H_2PtCl_6	reflux	6	$Cl_3SiCH_2CH(CH_3)CH_2Cl$ (67)	U -168
C_4H_7	$ClCH_2C(CH_3){=}CH_2$				
	H_2PtCl_6 + Et_3SiH	100	3	$ClCH_2CH(CH_3)CH_2SiCl_3$ (87)	U -168
	H_2PtCl_6+ Me_2ClSiH	31-60	0.5	$(CH_3)_2SiClCH_2CH(CH_3)CH_2Cl$ (74) $Cl_3SiCH_2CH(CH_3)CH_2Cl$ (3)	U -168
	H_2PtCl_6+ $MeCl_2SiH$	21-52	0.35	$CH_3Cl_2SiCH_2CH(CH_3)CH_2Cl$ (69) $Cl_3SiCH_2CH(CH_3)CH_2Cl$ (22)	U -168
	Pt/Al_2O_3	until 175	4 (TCE)	$ClCH_2CH(CH_3)CH_2SiCl_3$ (76)	D -002
C_4H_7	$ClCH_2OCH_2CH{=}CH_2$				
	H_2PtCl_6 (i-PrOH)	80	3	$ClCH_2O(CH_2)_3SiCl_3$ (18)	G -006
C_4H_8	$(CH_3)_2ClSiCH{=}CHCl$				
	H_2PtCl_6 (i-PrOH)	120-130	70-80	$(CH_3)_2ClSiC_2H_2ClSiCl_3$ (9)	1055
C_4H_8	$CH_3CH_2OCH{=}CH_2$				
	$PdCl_2[P(OBu)_3]_2$	120	8	$CH_3CH_2OCH_2CH_2SiCl_3$ (56)	D -027
	H_2PtCl_6	120	3	$CH_3CH_2OCH_2CH_2SiCl_3$ (-)	215
C_4H_8	$CH_3OCH_2CH{=}CH_2$				

Table 1.3
= HSiCl$_3$ =   =15=

	H$_2$PtCl$_6$	reflux	8.5	CH$_3$O(CH$_2$)$_3$SiCl$_3$ (-)	88
C$_4$H$_8$	CH$_3$SO$_2$CH$_2$CH=CH$_2$				
	[PtCl$_2$(C$_2$H$_4$)]$_2$	110	2	CH$_3$SO$_2$(CH$_2$)$_3$SiCl$_3$ (85)	U -037
C$_4$H$_8$	ClCH$_2$Si(CH$_3$)ClCH=CH$_2$ +ClCH$_2$Si(CH$_3$)(CH=CH$_2$)$_2$				
	H$_2$PtCl$_6$ (i-PrOH)	80-100	-	ClCH$_2$Si(CH$_3$)ClCH$_2$CH$_2$SiCl$_3$ + ClCH$_2$Si(CH$_3$)(CH$_2$CH$_2$SiCl$_3$)$_2$ (-)	387
C$_4$H$_9$	CH$_2$=CHSi(CH$_3$)$_2$Cl				
	Pd(PPh$_3$)$_4$	120	6	Cl$_3$SiCH$_2$CH$_2$Si(CH$_3$)$_2$Cl (80)	749
	H$_2$PtCl$_6$ (i-PrOH)	-	-	Cl$_3$SiCH$_2$CH$_2$Si(CH$_3$)$_2$Cl (62)	1056
	H$_2$PtCl$_6$ (i-PrOH)	reflux	4	Cl$_3$SiCH$_2$CH$_2$Si(CH$_3$)$_2$Cl/Cl$_3$SiCH(CH$_3$)Si(CH$_3$)$_2$Cl (96/4)	1059
C$_4$H$_{14}$	$\overline{CHB_{10}H_{10}}$CCH=CH$_2$				
	none	260	-	$\overline{CHB_{10}H_{10}}CC_2H_4$SiCl$_3$ (-)	1045
	Co$_2$(CO)$_8$	20-30	60	$\overline{CHB_{10}H_{10}}CCH_2CH_2$SiCl$_3$ (-)	715
	Rh$_4$(CO)$_{12}$	20-30	60	$\overline{CHB_{10}H_{10}}CCH_2CH_2$SiCl$_3$ (-)	715
C$_5$H$_1$	$(\overline{CF_2)_3CH=CCl}$				
	gamma	-	-	$(\overline{CF_2)_3CH_2}$CClSiCl$_3$ + trans-$\overline{CHCl(CF_2)_3}$CHSiCl$_3$ + cis-$\overline{CHCl(CF_2)_3}$CHSiCl$_3$ (-)	1214
C$_5$H$_3$	(CF$_3$)$_2$CFOCH=CH$_2$				
	H$_2$PtCl$_6$ (i-PrOH)	90	6	(CF$_3$)$_3$CFOCH$_2$CH$_2$SiCl$_3$ (84)	U -022
	H$_2$PtCl$_6$ (i-PrOH)	90	6	(CF$_3$)$_3$CFOCH$_2$CH$_2$SiCl$_3$ (84)	U -029
	H$_2$PtCl$_6$ (i-PrOH)	90	6	(CF$_3$)$_3$CFOCH$_2$CH$_2$SiCl$_3$ (84)	U -033
C$_5$H$_5$	CF$_3$COOCH$_2$CH=CH$_2$				
	H$_2$PtCl$_6$ (i-PrOH)	until 105	3	CF$_3$COO(CH$_2$)$_3$SiCl$_3$ (62)	103
C$_5$H$_6$	$\overline{CCl_2CH_2}$CHCH=CH$_2$				
	H$_2$PtCl$_6$ (i-PrOH)	60	1	$\overline{CCl_2CH_2}$CHCH$_2$CH$_2$SiCl$_3$ (64)	1044
C$_5$H$_6$	CHF$_2$CF$_2$OCH$_2$CH=CH$_2$				
	[PtCl$_2$(CH$_2$=CHC$_6$H$_{13}$)$_2$]	100	flow	CHF$_2$CF$_2$O(CH$_2$)$_3$SiCl$_3$ (92)	D -117
C$_5$H$_7$	CH$_2$=C(CH$_3$)CH$_2$CN				

	Pt/Al_2O_3	200	2	$Cl_3SiCH_2CH(CH_3)CH_2CN$ (5)	U -007
C_5H_7	$ClCH_2COOCH_2CH=CH_2$				
	H_2PtCl_6 (AllOCH$_2$COOAll)	43-134	7	$ClCH_2COO(CH_2)_3SiCl_3$ (79)	Pl-008
	H_2PtCl_6 [O(CH$_2$COOAll)$_2$]	49-136	4.25	$ClCH_2COO(CH_2)_3SiCl_3$ (65)	Pl-008
	H_2PtCl_6 [O(CH$_2$COOAll)$_2$]	49-136	4.25	$ClCH_2COO(CH_2)_3SiCl_3$ (65)	Pl-002
	SiO_2-$\overline{N(CH_2CH_2)_2O}$/$H_2PtCl_6$ (EtOH)	90-135	5.33	$ClCH_2COO(CH_2)_3SiCl_3$ (83-86)	Pl-003
	SiO_2-$\overline{N(CH_2CH_2)_2O}$/$H_2PtCl_6$ (EtOH)	90-135	5.33	$ClCH_2COO(CH_2)_3SiCl_3$ (83-86)	Pl-008
C_5H_8	$CH_2=C(CH_3)COOCH_3$				
	$Ni(acac)_2 + PPh_3$	100	5-9	$Cl_3SiCH_2CH(CH_3)COOCH_3$ (80) $Cl_3SiC(CH_3)_2COOCH_3$ (20)	560
	$Ni(acac)_2 + PPh_3$	120	10	$Cl_3SiC(CH_3)_2COOCH_3$ (80)	1335
	H_2PtCl_6 (i-PrOH)	reflux	1	$Cl_3SiCH_2CH(CH_3)COOCH_3$ (-)	B -002
C_5H_8	$CH_3COOC(CH_3)=CH_2$				
	$Ni(acac)_2 + PPh_3$	120	7	$CH_3COOCH(CH_3)CH_2SiCl_3$ (40)	560
C_5H_8	$CH_3COOCH_2CH=CH_2$				
	H_2PtCl_6 (AllOCH$_2$COOAll)	20-82	4	$CH_3COO(CH_2)_3SiCl_3$ (84)	Pl-002
	H_2PtCl_6 (AllOCH$_2$COOAll)	20-82	4	$CH_3COO(CH_2)_3SiCl_3$ (84)	Pl-008
	H_2PtCl_6 (OctOH)	100-110	2	$CH_3COO(CH_2)_3SiCl_3$ (92)	D -040
	H_2PtCl_6 (i-PrOH)	60-100	6	$CH_3COO(CH_2)_3SiCl_3$ (69)	98
	H_2PtCl_6 [(CH$_2$=CHSiMe$_2$)$_2$O]	75-95	-	$CH_3COO(CH_2)_3SiCl_3$ (70)	U -134
	$[PtCl_2(Me_2C=CHAc)]_2$	65-105	-	$CH_3COO(CH_2)_3SiCl_3$ (-)	D -023
C_5H_8	$CH_3OCOOCH_2CH=CH_2$				
	H_2PtCl_6 (i-PrOH)	38-148	0.5	$CH_3OCOO(CH_2)_3SiCl_3$ (64)	S -026
C_5H_8	$ClCH_2SiCl(CH=CH_2)_2$				
	H_2PtCl_6 (i-PrOH)	80-100	-	$ClCH_2SiCl(CH_2CH_2SiCl_3)_2$ (-)	387
C_5H_9	$ClSiCH_3(CH=CH_2)_2$				
	H_2PtCl_6 (i-PrOH)	reflux	4	$ClSiCH_3(CH=CH_2)CH_2CH_2SiCl_3$ (-) $ClSiCH_3(CH_2CH_2SiCl_3)_2$ (-)	1059
C_5H_{10}	$CH_3CH_2OCH_2CH=CH_2$				

	$Pd[P(C_6H_4Cl)_3]_4$	120	10	$CH_3CH_2O(CH_2)_3SiCl_3$ (77)	D -027
C_5H_{11}	$(CH_3)_3SiCH=CHCl$				
	H_2PtCl_6 (i-PrOH)	120-130	70-80	$(CH_3)_3SiC_2H_2Cl(SiCl_3)$ (4)	1055
C_5H_{11}	$BrCH_2Si(CH_3)_2CH=CH_2$				
	H_2PtCl_6 (i-PrOH)	80-100	-	$BrCH_2Si(CH_3)_2CH_2CH_2SiCl_3$ (88)	387
C_5H_{12}	$(CH_3)_3SiCH=CH_2$				
	$Co_2(CO)_8$	30-40	1	$(CH_3)_3SiCH_2CH_2SiCl_3$ (-)	715
	$Co_2(CO)_8$	65-70	1	$(CH_3)_3SiCH_2CH_2SiCl_3$ (73)	31
	$Fe(CO)_4(CH_2=CHSiMe_3)$	-	-	$(CH_3)_3SiCH_2CH_2SiCl_3$ + $(CH_3)_3SiCH(CH_3)SiCl_3$ + $(CH_3)_3SiCH=CHSiCl_3$ (-)	398
	$Fe(CO)_4(CH_2=CHSiMe_3)$	70-80	3	$(CH_3)_3SiCH_2CH_2SiCl_3$ (7) $(CH_3)_3SiCH(CH_3)SiCl_3$ (68) $(CH_3)_3SiCH=CHSiCl_3$ (12)	399
	$Fe(CO)_4(CH_2=CHSiMe_3)$	80	4	$(CH_3)_3SiCH_2CH_2SiCl_3$ (4) $(CH_3)_3SiCH(CH_3)SiCl_3$ (36) $(CH_3)_3SiCH=CHSiCl_3$ (12)	844
	$Ni(PPh_3)_2(CH_2=CHSiMe_3)$	-	-	$(CH_3)_3SiCH_2CH_2SiCl_3$ (78-90)	1337
	$Ni(PPh_3)_2[(CH_2=CH)_4Si]$	-	-	$(CH_3)_3SiCH_2CH_2SiCl_3$ (78-90)	1337
	$Ni(PPh_3)_2[CH_2=CHSi(OMe)_3]$	-	-	$(CH_3)_3SiCH_2CH_2SiCl_3$ (78-90)	1337
	$Pd(PPh_3)_4$	120	6	$(CH_3)_3SiCH_2CH_2SiCl_3$ (88)	749
	H_2PtCl_6	until 127	6	$(CH_3)_3SiCH_2CH_2SiCl_3$ (87)	1058
	H_2PtCl_6 (i-PrOH)	-	-	$(CH_3)_3SiCH_2CH_2SiCl_3$ (87)	1056
	H_2PtCl_6 (i-PrOH)	reflux	4	$(CH_3)_3SiCH_2CH_2SiCl_3/(CH_3)_3SiCH(CH_3)SiCl_3$(97/3)	1059
	$Rh_4(CO)_{12}$	30-40	1	$(CH_3)_3SiCH_2CH_2SiCl_3$ (-)	715
	$Rh_4(CO)_{12}$	35-40	1	$(CH_3)_3SiCH_2CH_2SiCl_3$ (95-98)	148
	$Rh_4(CO)_{12}$	65-70	1	$(CH_3)_3SiCH_2CH_2SiCl_3$ (91)	31
C_5H_{16}	$\overline{CHB_{10}H_{10}C}C(CH_3)=CH_2$				
	none	260	12	$\overline{CHB_{10}H_{10}C}CH(CH_3)CH_2SiCl_3$ (-)	1045
	$Co_2(CO)_8$	20-30	60	$\overline{CHB_{10}H_{10}C}CH(CH_3)CH_2SiCl_3$ (-)	715
	$Rh_4(CO)_{12}$	20-30	60	$\overline{CHB_{10}H_{10}C}CH(CH_3)CH_2SiCl_3$ (-)	715
C_6H_3	$(CF_3)_3CCH=CH_2$				

	gamma(^{60}Co)	135	-	$(CF_3)_3CCH_2CH_2SiCl_3$ (100)	1360
C_6H_5	$(CF_3)_2CFOCH_2CH=CH_2$				
	H_2PtCl_6 (i-PrOH)	90	6	$(CF_3)_2CFO(CH_2)_3SiCl_3$ (85)	U -033
	H_2PtCl_6 (i-PrOH)	90	6	$(CF_3)_2CFO(CH_2)_3SiCl_3$ (85)	U -022
	H_2PtCl_6 (i-PrOH)	90	6	$(CF_3)_2CFO(CH_2)_3SiCl_3$ (85)	U -029
C_6H_5	$ClCF_2CF(CF_3)OCH_2CH=CH_2$				
	H_2PtCl_6 (i-PrOH)	90	6	$ClCF_2CF(CF_3)O(CH_2)_3SiCl_3$ (60)	U -022
	H_2PtCl_6 (i-PrOH)	90	6	$ClCF_2CF(CF_3)O(CH_2)_3SiCl_3$ (60)	U -029
	H_2PtCl_6 (i-PrOH)	90	6	$ClCF_2CF(CF_3)O(CH_2)_3SiCl_3$ (60)	U -033
C_6H_6	$CF_3CHFCF_2OCH_2CH=CH_2$				
	$[PtCl_2(CH_2=CHHex)]_2$	until 100	3	$CF_3CHFCF_2O(CH_2)_3SiCl_3$ (93)	D -116
C_6H_6	Fig.50.				
	H_2PtCl_6	80-85	18 (THF)	Fig.51. (60)	693
C_6H_9	$ClSi(CH=CH_2)_3$				
	H_2PtCl_6 (i-PrOH)	reflux	4	$ClSi(CH=CH_2)_2C_2H_4SiCl_3$ (-) $ClSiCH=CH_2(C_2H_4SiCl_3)_2$ (-) $ClSi(C_2H_4SiCl_3)_3$ (-)	1059
C_6H_9	$N_3COOCH_2CH_2OCH_2CH=CH_2$				
	H_2PtCl_6 (i-PrOH)	50	20	$N_3COOCH_2CH_2O(CH_2)_3SiCl_3$ (88)	G -019
C_6H_{10}	$(CH_2=CHCH_2)_2O$				
	$Ni(acac)_2$	80	6 (THF)	$CH_2=CHCH_2O(CH_2)_3SiCl_3$ (28)	560
	H_2PtCl_6 (i-PrOH)	reflux	2	$CH_2=CHCH_2O(CH_2)_3SiCl_3$ (28)	1192
C_6H_{10}	$(CH_2=CHCH_2)_2S$				
	$Ni(acac)_2$ + PCl_2(i-Bu)	100	1	$CH_2=CHCH_2S(CH_2)_3SiCl_3$ (40)	558
	$Ni(acac)_2$ + PPh_3	100	1	$CH_2=CHCH_2S(CH_2)_3SiCl_3$ (15-29) $[Cl_3Si(CH_2)_3]_2S$ (0-4) $CH_3CH_2CH_2S(CH_2)_3SiCl_3$ (0-18)	558
	$Ni(acac)_2$ + PPh_3	120	3	$[Cl_3Si(CH_2)_3]_2S$ (14) $CH_3CH_2CH_2S(CH_2)_3SiCl_3$ (15)	558
C_6H_{10}	$CH_3COCH_2CH_2CH=CH_2$				

	H_2PtCl_6	130-140	1.5-2	$CH_3CO(CH_2)_4SiCl_3$ (82)	602
	H_2PtCl_6 (i-PrOH)	50-60	4	$CH_3CO(CH_2)_4SiCl_3$ (85)	1192
C_6H_{10}	$CH_3OCH_2COOCH_2CH=CH_2$				
	H_2PtCl_6 (AllOCH$_2$COOAll)	34-145	5.75	$CH_3OCH_2COO(CH_2)_3SiCl_3$ (69)	Pl-008
C_6H_{11}	$CH_3CHClCH_2CH_2CH=CH_2$				
	$(PhCOO)_2$	65	90 (cy)	$CH_3CHCl(CH_2)_4SiCl_3$ (87)	362
C_6H_{11}	$ClCH_2Si(CH_3)(CH=CH_2)_2$ + $ClCH_2Si(CH_3)ClCH=CH_2$				
	H_2PtCl_6 (i-PrOH)	80-100	-	$ClCH_2Si(CH_3)(CH_2CH_2SiCl_3)_2$ + $ClCH_2SiCl(CH_3)CH_2CH_2SiCl_3$ (-)	387
C_6H_{12}	$(CH_3)_2Si(CH=CH_2)_2$				
	H_2PtCl_6 (i-PrOH)	reflux	4	$(CH_3)_2Si(CH=CH_2)C_2H_4SiCl_3$ (-) $(CH_3)_2Si(C_2H_4SiCl_3)_2$ (-)	1059
C_6H_{12}	$C_4H_9OCH=CH_2$				
	$Ni(acac)_2$ + PPh_3	120	7 (THF)	$C_4H_9OCH_2CH_2SiCl_3$ (60)	560
C_6H_{12}	$CH_3(CH_2)_3OCH=CH_2$				
	Ni(THF)	0-25	20	$CH_3(CH_2)_3OCH_2CH_2SiCl_3$ (53)	E -049
	Ni(THF) +sonic waves	0-25	4	$CH_3(CH_2)_3OCH_2CH_2SiCl_3$ (78)	E -049
C_6H_{14}	$(CH_3)_3SiOCH_2CH=CH_2$				
	H_2PtCl_6 (i-PrOH)	reflux	-	$\overline{(CH_2)_3OSi}Cl_2$ (62)	777
C_7H_5	$C_3F_7COOCH_2CH=CH_2$				
	H_2PtCl_6 (i-PrOH)	until 130	14	$C_3F_7COO(CH_2)_3SiCl_3$ (37)	103
C_7H_5	$CH_2=CHCOOCH_2CF_2CF_2CF_3$				
	CuCl + $(CMe_2NH_2)_2$ + NBu_3	reflux	2	$Cl_3SiCH_2CH_2COOCH_2CF_2CF_2CF_3$ (49)	419
C_7H_7	2-$CH_2=CHC_5H_4N$				
	-	20	-	2-$(Cl_3SiCH_2CH_2)C_5H_4N$ (-)	710
	none or H_2PtCl_6 (i-PrOH)	150	5	2-$(Cl_3SiCH_2CH_2)C_5H_4N$ (90)	819
C_7H_7	4-$CH_2=CHC_5H_4N$				
	-	150	5	4-$(Cl_3SiCH_2CH_2)C_5H_4N$ (-)	710

Table 1.3

= $HSiCl_3$ =

	NPr_3	reflux	24 (a)	4-($Cl_3SiCH_2CH_2$)C_5H_4N (52)	433
	none or H_2PtCl_6 (i-PrOH)	150	5	4-($Cl_3SiCH_2CH_2$)C_5H_4N (78)	818
C_7H_7	$CF_3CF_2OCF_2CF_2SO_2NHCH_2CH=CH_2$				
	H_2PtCl_6	-		$CF_3CF_2OCF_2CF_2SO_2NH(CH_2)_3SiCl_3$ (-)	238
C_7H_8	$CH_2=CHCH_2\overline{CHCH_2COOCO}$				
	H_2PtCl_6	110-115	5	$Cl_3Si(CH_2)_3\overline{CHCH_2COOCO}$ (-)	D -093
C_7H_8	$CH_3COOCH_2CF_2CF_2CH=CH_2$				
	H_2PtCl_6 (i-PrOH)	110	17	$CH_3COOCH_2CF_2CF_2CH_2CH_2SiCl_3$ (-)	U -089
	H_2PtCl_6 (i-PrOH)	reflux	17	$CH_3COOCH_2CF_2CF_2CH_2CH_2SiCl_3$ (-)	U -054
C_7H_8	$CHF_2CF_2CH_2COOCH_2CH=CH_2$				
	H_2PtCl_6 (i-PrOH)	until 190	4	$CHF_2CF_2CH_2COO(CH_2)_3SiCl_3$ (48)	103
C_7H_9	$CH_2=C(CH_3)COOCH_2CH_2CN$				
	H_2PtCl_6 (OctOH)	115-130	24	$Cl_3SiCH_2CH(CH_3)COOCH_2CH_2CN$ (-)	U -047
C_7H_{10}	$CH_2=C(CH_3)COOCH_2CH=CH_2$				
	H_2PtCl_6	70-80	1	$CH_2=C(CH_3)COO(CH_2)_3SiCl_3$ (-)	U -016
	H_2PtCl_6	70-80	1	$CH_2=C(CH_3)COO(CH_2)_3SiCl_3$ (-)	G -001
	H_2PtCl_6 (OctOH)	95		$CH_2=C(CH_3)COO(CH_2)_3SiCl_3$ (-)	J -074
	H_2PtCl_6 (i-PrOH)	-	4	$CH_2=C(CH_3)COO(CH_2)_3SiCl_3$ (87)	748
	H_2PtCl_6 (i-PrOH)	120	2	$CH_2=C(CH_3)COO(CH_2)_3SiCl_3$ (90)	U -052
	H_2PtCl_6 [$\overline{(CH_2)_5CO}$]	until 112	0.5	$CH_2=C(CH_3)COO(CH_2)_3SiCl_3$ (-)	D -110
	H_2PtCl_6 [$\overline{(CH_2)_5CO}$, $CH_2=CHSi(OEt)_3$]	up to 104	1	$CH_2=C(CH_3)COO(CH_2)_3SiCl_3$ (97)	Pl-034
	H_2PtCl_6 [$\overline{(CH_2)_5CO}$]	140	-	$CH_2=C(CH_3)COO(CH_2)_3SiCl_3$ (98)	D -103
	$Pt(PPh_3)_2(CH_2=CH_2)$	-	2	$CH_2=C(CH_3)COO(CH_2)_3SiCl_3$ (98)	748
	$Pt(PPh_3)_2(CH_2=CH_2)$	up to 120	3.5	$CH_2=C(CH_3)COO(CH_2)_3SiCl_3$ (98)	Pl-034
	$Pt(PPh_3)_2[CH_2=CHSi(OEt)_3]$	up to 117	2.5	$CH_2=C(CH_3)COO(CH_2)_3SiCl_3$ (96)	Pl-034
	$PtCl_2(PPh_3)_2$	-	6.5	$CH_2=C(CH_3)COO(CH_2)_3SiCl_3$ (93)	748
	$PtCl_2(PPh_3)_2$	up to 80	4	$CH_2=C(CH_3)COO(CH_2)_3SiCl_3$ (93)	Pl-034

	$PtCl_2(PhCN)_2$	up to 120	3.5	$CH_2=C(CH_3)COO(CH_2)_3SiCl_3$ (96)	Pl-034
	$PtCl_2(acac)_2$	32-r.t	4	$CH_2=C(CH_3)COO(CH_2)_3SiCl_3$ (96)	G -011
	$PtCl_2(acac)_2$	32-r.t	4	$CH_2=C(CH_3)COO(CH_2)_3SiCl_3$ (96)	D -011
	$PtCl_4(HexCH=CH_2$ + sec-$BuNH_2)$	until 139	0.05	$CH_2=C(CH_3)COO(CH_2)_3SiCl_3$ (-)	D -110
	$PtCl_4(HexCH=CH_2)$	until 24	0.35	$CH_2=C(CH_3)COO(CH_2)_3SiCl_3$ (-)	D -110
	SiO_2-$OSi(CH_2)_3SH/Pt$	reflux	2.5	$CH_2=C(CH_3)COO(CH_2)_3SiCl_3$ (98)	U -146
	$[PtCl_2(Me_2C=CHAc)]_2$	60	0.3	$CH_2=C(CH_3)COO(CH_2)_3SiCl_3$ (97)	D -023
C_7H_{10}	$CO(OCH_2CH=CH_2)_2$				
	H_2PtCl_6	40-105	1-2	$CO[O(CH_2)_3SiCl_3]_2$ (76)	1051
C_7H_{12}	$CH_3Si(CH=CH_2)_3$				
	H_2PtCl_6 (i-PrOH)	reflux	4	$CH_3Si(CH=CH_2)_2C_2H_4SiCl_3$ (-) $CH_3SiCH=CH_2(C_2H_4SiCl_3)_2$ (-) $CH_3Si(C_2H_4SiCl_3)_3$ (-)	1059
C_7H_{14}	$CH_2=CHCH_2COOSi(CH_3)_3$				
	H_2PtCl_6 (i-PrOH)	reflux	2	$Cl_2\overline{Si(CH_2)_3COO}$ (70)	774
C_7H_{14}	$CH_2=CHSi(CH_3)_2CH_2CH=CH_2$				
	H_2PtCl_6	110-114	0.5	$Cl_3SiCH_2CH_2Si(CH_3)_2CH_2CH=CH_2$ + $CH_2=CHSi(CH_3)_2$-$(CH_2)_3SiCl_3$ (64)	457
C_8H_3	$C_6F_5CH=CH_2$				
	H_2PtCl_6 (i-PrOH)	100	1	$C_6F_5CH_2CH_2SiCl_3$ (78)	132
	$RhCl(PPh_3)_3$	120	24	$C_6F_5CH_2CH_2SiCl_3$ (84)	858
	$Ru_3(CO)_{12}$	120	70	$C_6F_5CH_2CH_2SiCl_3$ (27) $C_6F_5CH=CHSiCl_3$ (24)	858
C_8H_6	$O(CF_2CF_2CH=CH_2)_2$				
	$(t$-$BuO)_2$	reflux	72	$O(CF_2CF_2CH_2CH_2SiCl_3)_2$ (-)	D -025
C_8H_7	$(CF_3)_2CFOCH_2COOCH_2CH=CH_2$				
	H_2PtCl_6 (i-PrOH)	80	1	$(CF_3)_2CFOCH_2COO(CH_2)_3SiCl_3$ (-)	D -043
C_8H_7	$C_6H_5CH=CHSiCl_3$				
	NBu_3	-	-	$Cl_3SiCH(C_6H_5)CH_2SiCl_3$ (-)	118
C_8H_7	$ClC_6H_4CH=CH_2$				

	$KPtCl_3(C_2H_4)$	5	-	$ClC_6H_4CH_2CH_2SiCl_3$ (kin.)	966
C_8H_8	$CF_3CF_2OCF_2CF_2SO_2N(CH_3)CH_2CH=CH_2$				
	H_2PtCl_6	-	-	$CF_3CF_2OCF_2CF_2SO_2N(CH_3)(CH_2)_3SiCl_3$ (-)	238
C_8H_9	Fig.52.				
	none or H_2PtCl_6 (i-PrOH)	150	5	Fig.53. (70)	819
C_8H_9	Fig.54.				
	$Ni(acac)_2$ + PPh_3	100	7	Fig.55. (98)	1329
C_8H_{11}	Fig.56.				
	H_2PtCl_6 (i-PrOH)	reflux	2	Fig.57 (60)	728
C_8H_{11}	Fig.58.				
	H_2PtCl_6 (i-PrOH)	215	2	Fig.59. (80)	731
	H_2PtCl_6 (i-PrOH)	reflux	2	Fig.59. (49)	728
C_8H_{12}	$CH_2=CHCH_2OCH_2COOCH_2CH=CH_2$				
	H_2PtCl_6 (AllOCH$_2$COOAll)	70-157	5.5	$Cl_3Si(CH_2)_3OCH_2COO(CH_2)_3SiCl_3$ (65)	Pl-008
	H_2PtCl_6 (AllOCH$_2$COOAll)	80-137	0.83	$Cl_3Si(CH_2)_3OCH_2COOCH_2CH=CH_2$ + $CH_2=CHCH_2OCH_2COO(CH_2)_3SiCl_3$ (37)	Pl-008
C_8H_{12}	$Si(CH=CH_2)_4$				
	$Ni(acac)_2$ + PPh_3	120-150	4	$(CH_2=CH)_3SiCH_2CH_2SiCl_3+(CH_2=CH)_3SiCH(CH_3)SiCl_3$ + $(CH_2=CH)_2Si(C_2H_4SiCl_3)_2$ (18)	1059
	$Ni(acac)_2$+ i-BuPCl$_2$	120-150	4	$(CH_2=CH)_3SiCH_2CH_2SiCl_3+(CH_2=CH)_3SiCH(CH_3)SiCl_3$ + $(CH_2=CH)_2Si(C_2H_4SiCl_3)_2+CH_2=CHSi(C_2H_4SiCl_3)_3$ (48)	1059
	H_2PtCl_6	-	33-126	$(CH_2=CH)_3SiCH_2CH_2SiCl_3+(CH_2=CH)_3SiCH(CH_3)SiCl_3$ (65) $(CH_2=CH)_2Si(C_2H_4SiCl_3)_2$ (17)	1058
	H_2PtCl_6 (MeOC$_2$H$_4$OMe)	40	-	$Si(CH_2CH_2SiCl_3)_4$ (-)	466
	H_2PtCl_6 (i-PrOH)	reflux	4	$(CH_2=CH)_3SiCH_2CH_2SiCl_3+(CH_2=CH)_3SiCH(CH_3)SiCl_3$ + $(CH_2=CH)_2Si(C_2H_4SiCl_3)_2+CH_2=CHSi(C_2H_4SiCl_3)_3$ (-)	1059
C_8H_{14}	$CH_2=CHCOO(CH_2CH_2O)_2CH_3$				
	H_2PtCl_6 (i-PrOH)	-	-	$Cl_3SiCH(CH_3)COO(CH_2CH_2O)_2CH_3$ (-)	U -045

	-	reflux	2.5(MeCN)	$(CH_2)_4NC(SiCl_3)(CH_2)_4$ (28)	1114
C_9H_{17}	$CH_3CH_2C[N(CH_2)_4]=CHCH_3$				
	-	reflux	2.5(MeCN)	$CH_3CH_2C(SiCl_3)[N(CH_2)_4]CH_2CH_3$ (8)	1114
$C_{10}H_6$	$CH_2=CH(CF_2)_6CH=CH_2$				
	$(t\text{-}BuO)_2$	reflux	72	$Cl_3SiCH_2CH_2(CF_2)_6CH_2CH_2SiCl_3$ (-)	D -025
$C_{10}H_8$	$CF_3(CF_2)_3OCF_2CF_2SO_2N(CH_3)CH_2CH=CH_2$				
	H_2PtCl_6	-	-	$CF_3(CF_2)_3OCF_2CF_2SO_2N(CH_3)(CH_2)_3SiCl_3$ (-)	238
$C_{10}H_9$	m-$O_2NC_6H_4COOCH_2CH=CH_2$				
	H_2PtCl_6	30-90	2.5 (t)	m-$O_2NC_6H_4COO(CH_2)_3SiCl_3$ (90)	D -052
$C_{10}H_{12}$	cis-$CH_2=CHCH_2OCOCH=CHCOOCH_2CH=CH_2$				
	H_2PtCl_6 (OctOH)	100	3	$CH_2=CHCH_2OCOCH=CHCOO(CH_2)_3SiCl_3$ + $Cl_3Si(CH_2)_3OCOCH=CHCOO(CH_2)_3SiCl_3$ (-)	U -061
	H_2PtCl_6 (OctOH)	100	3	$CH_2=CHCH_2OCOCH=CHCOO(CH_2)_3SiCl_3$ + $Cl_3Si(CH_2)_3OCOCH=CHCOO(CH_2)_3SiCl_3$ (-)	U -060
	H_2PtCl_6 (OctOH)	100	3	$CH_2=CHCH_2OCOCH=CHCOO(CH_2)_3SiCl_3$ + $Cl_3Si(CH_2)_3OCOCH=CHCOO(CH_2)_3SiCl_3$ (-)	U -066
	H_2PtCl_6 (OctOH)	100	3	$CH_2=CHCH_2OCOCH=CHCOO(CH_2)_3SiCl_3$ + $Cl_3Si(CH_2)_3OCOCH=CHCOO(CH_2)_3SiCl_3$ (-)	U -067
	H_2PtCl_6 (OctOH)	100	3	$CH_2=CHCH_2OCOCH=CHCOO(CH_2)_3SiCl_3$ + $Cl_3Si(CH_2)_3OCOCH=CHCOO(CH_2)_3SiCl_3$ (-)	U -069
	H_2PtCl_6 (OctOH)	100	3	$CH_2=CHCH_2OCOCH=CHCOO(CH_2)_3SiCl_3$ + $Cl_3Si(CH_2)_3OCOCH=CHCOO(CH_2)_3SiCl_3$ (-)	U -073
	H_2PtCl_6 (OctOH)	100	3	$CH_2=CHCH_2OCOCH=CHCOO(CH_2)_3SiCl_3$ + $Cl_3Si(CH_2)_3OCOCH=CHCOO(CH_2)_3SiCl_3$ (-)	U -075
$C_{10}H_{12}$	m-$CH_3OCH_2C_6H_4CH=CH_2$				
	H_2PtCl_6 (i-PrOH + $BuNH_2$)	100-120	1	m-$CH_3OCH_2C_6H_4CH_2CH_2SiCl_3$ (-)	F -015
$C_{10}H_{14}$	$CH_2=CHCH_2OCOCH_2CH_2COOCH_2CH=CH_2$				
	H_2PtCl_6 (i-PrOH)	100	0.25	$CH_2=CHCH_2OCOCH_2CH_2COO(CH_2)_3SiCl_3$ (14)	32
$C_{10}H_{14}$	Fig.60.				
	H_2PtCl_6 (i-PrOH)	160	60	Fig.61. (88)	S -042
$C_{10}H_{14}$	Fig.62.				

	H_2PtCl_6 (i-PrOH)	reflux	1	Fig.63. (-)	B -002
$C_{10}H_{16}$	$CH_2=CHCH_2CH(COOCH_2CH_3)_2$				
	H_2PtCl_6 (i-PrOH)	reflux	-	$Cl_3Si(CH_2)_3CH(COOCH_2CH_3)_2$ (60)	1008
	H_2PtCl_6 (i-PrOH)	reflux	8 (h)	$Cl_3Si(CH_2)_3CH(COOCH_2CH_3)_2$ (60)	J -052
$C_{10}H_{16}$	$CH_3COOCH_2CH=CH(CH_2)_3CH=CH_2$				
	H_2PtCl_6 (i-PrOH)	reflux	2	$CH_3COOCH_2CH=CH(CH_2)_5SiCl_3$ (32)	D -055
$C_{10}H_{17}$	$(CH_2)_4N\text{-}C=CH(CH_2)_4$				
	-	reflux	2.5(MeCN)	$(CH_2)_4NC(SiCl_3)(CH_2)_5$ (58)	1114
$C_{10}H_{18}$	$(CH_2)_5N\text{-}C=CH(CH_2)_3$				
	-	reflux	2.5(MeCN)	$(CH_2)_5NC(SiCl_3)(CH_2)_4$ (60)	1114
$C_{10}H_{19}$	$CH_3CH_2C[N(CH_2)_5]=CHCH_3$				
	-	reflux	2.5(MeCN)	$CH_3CH_2C(SiCl_3)[N(CH_2)_5]CH_2CH_3$ (62)	1114
$C_{10}H_{21}$	p-$(C_6H_5)_2PC_6H_4CH_2CH_2CH=CH_2$				
	UV	-	120	p-$(C_6H_5)_2PC_6H_4(CH_2)_4SiCl_3$ (30)	14
$C_{11}H_5$	Fig.70.				
	H_2PtCl_6 (i-PrOH)	reflux	24 (dcm)	Fig.71. (-)	522
$C_{11}H_6$	$CF_3(CF_2)_5OCF_2CF_2SO_2NHCH_2CH=CH_2$				
	H_2PtCl_6	-	-	$CF_3(CF_2)_5OCF_2CF_2SO_2NH(CH_2)_3SiCl_3$ (-)	238
$C_{11}H_{11}$	Fig.64.				
	H_2PtCl_6 (i-PrOH)	50	5	Fig.65. (90)	G -019
$C_{11}H_{12}$	$C_6H_5OCH_2COOCH_2CH=CH_2$				
	H_2PtCl_6 (AllOCH$_2$COOAll)	30-166	6.33	$C_6H_5OCH_2COO(CH_2)_3SiCl_3$ (84)	Pl-008
$C_{11}H_{16}$	Fig.66.				
	H_2PtCl_6	reflux	5 (h)	Fig.67. (-)	J 055
$C_{11}H_{17}$	Fig.68.				
	H_2PtCl_6 (i-PrOH)	160	62	Fig.69. (85)	S -042
$C_{11}H_{19}$	$(CH_2)_4N\text{-}C=CH(CH_2)_5$				

	-	reflux	2.5(MeCN)	$(CH_2)_4NC(SiCl_3)(CH_2)_6$ (9)	1114
$C_{11}H_{19}$	$(CH_2)_5N$-$C=CH(CH_2)_4$				
	-	reflux	2.5(MeCN)	$(CH_2)_5NC(SiCl_3)(CH_2)_5$ (77)	1114
$C_{11}H_{19}$	$CH_2=CH(CH_2)_8CN$				
	H_2PtCl_6 (i-PrOH)	120	8	$Cl_3Si(CH_2)_{10}CN$ (85)	1138
$C_{11}H_{21}$	$Br(CH_2)_9CH=CH_2$				
	H_2PtCl_6 (i-PrOH)	50	11	$Br(CH_2)_{11}SiCl_3$ (77)	1192
$C_{11}H_{21}$	$Cl(CH_2)_9CH=CH_2$				
	H_2PtCl_6 (i-PrOH)	120	8	$Cl(CH_2)_{11}SiCl_3$ (95)	1138
$C_{11}H_{24}$	$CH_2=CHSi(C_3H_7)_3$				
	H_2PtCl_6 (i-PrOH)	80	96	$Cl_3SiCH_2CH_2Si(C_3H_7)_3$ (88)	135
$C_{12}H_5$	$(CF_3)_2C=C[CF(CF_3)_2]CF(CF_3)OCH_2CH=CH_2$				
	H_2PtCl_6	100	8	$(CF_3)_2C=C[CF(CF_3)_2]CF(CF_3)O(CH_2)_3SiCl_3$ (-)	J -025
	H_2PtCl_6 (Me_2CO)	90-130	1	$(CF_3)_2C=C[CF(CF_3)_2]CF(CF_3)O(CH_2)_3SiCl_3$ (99)	J -032
$C_{12}H_8$	$CF_3(CF_2)_5OCF_2CF_2SO_2N(CH_3)CH_2CH=CH_2$				
	H_2PtCl_6	-	-	$CF_3(CF_2)_5OCF_2CF_2SO_2N(CH_3)(CH_2)_3SiCl_3$ (-)	238
$C_{12}H_{18}$	$CH_2=CHCH_2OCO(CH_2)_4COOCH_2CH=CH_2$				
	H_2PtCl_6 (i-PrOH)	100	0.25	$CH_2=CHCH_2OCO(CH_2)_4COO(CH_2)_3SiCl_3$ (11)	32
$C_{12}H_{18}$	$CH_2=CHCH_2OCOO(CH_2CH_2O)_2COOCH_2CH=CH_2$				
	H_2PtCl_6	reflux	8 (h)	$Cl_3Si(CH_2)_3OCOO(CH_2CH_2O)_2COO(CH_2)_3SiCl_3$ (97)	J -046
$C_{12}H_{18}$	$Cl_3SiCH_2CH_2CH(CH_2SiCl_3)CH_2CH_2CH=C(CH_3)_2$				
	H_2PtCl_6 (i-PrOH)	60-105	6	$Cl_3SiCH_2CH_2CH(CH_2SiCl_3)CH_2CH_2CH(SiCl_3)CH(CH_3)_2$ (49)	823
$C_{12}H_{20}$	$CH_2=C(CH_3)COO(CH_2)_6CH=CH_2$				
	H_2PtCl_6 (THF)	0-50	28	$CH_2=C(CH_3)COO(CH_2)_8SiCl_3$ (-)	J -112
$C_{12}H_{21}$	$(CH_2)_5N$-$C=CH(CH_2)_5$				
	-	reflux	2.5(MeCN)	$(CH_2)_5NC(SiCl_3)(CH_2)_6$ (13)	1114
$C_{12}H_{22}$	$CH_2=CH(CH_2)_8COOCH_3$				

H_2PtCl_6 (AllOCH$_2$COOAll)	67-181	6.3	$Cl_3Si(CH_2)_{10}COOCH_3$ (67)	Pl-008
H_2PtCl_6 (i-PrOH)	120	8	$Cl_3Si(CH_2)_{10}COOCH_3$ (95)	1138
H_2PtCl_6 (i-PrOH)	60	7	$Cl_3Si(CH_2)_{10}COOCH_3$ (91)	105
H_2PtCl_6 (i-PrOH)	85	1	$Cl_3Si(CH_2)_{10}COOCH_3$ (-)	B -002
H_2PtCl_6 (i-PrOH)	85	1	$Cl_3Si(CH_2)_{10}COOCH_3$ (-)	U -038
H_2PtCl_6 (i-PrOH)	reflux	5	$Cl_3Si(CH_2)_{10}COOCH_3$ (95)	1011
SiO_2-$\overline{N(CH_2CH_2)_2O}$/$H_2PtCl_6$ (EtOH)	100-165	3.5	$Cl_3Si(CH_2)_{10}COOCH_3$ (60)	Pl-007
$C_{12}H_{22}$ $CH_2=CHCH_2O(CH_2CH_2O)_3CH_2CH=CH_2$				
PhCOOBu-t	100	overnight	$Cl_3Si(CH_2)_3O(CH_2CH_2O)_3(CH_2)_3SiCl_3$ (-)	U -045
PhCOOBu-t	100	overnight	$Cl_3Si(CH_2)_3O(CH_2CH_2O)_3(CH_2)_3SiCl_3$ (-)	U -007
PhCOOBu-t	100	overnight	$Cl_3Si(CH_2)_3O(CH_2CH_2O)_3(CH_2)_3SiCl_3$ (-)	U -077
$C_{13}H_5$ $CF_3(CF_2)_9CH_2CH=CH_2$				
-	-	-	$CF_3(CF_2)_9(CH_2)_3SiCl_3$ (-)	465
$C_{13}H_7$ $CF_3(CF_2)_7OCF_2CF_2SO_2NHCH_2CH=CH_2$				
H_2PtCl_6	-	-	$CF_3(CF_2)_7OCF_2CF_2SO_2NH(CH_2)_3SiCl_3$ (-)	238
$C_{13}H_{20}$ $CH_2=CH(CH_2)_3CH=CHCH_2CH(COOCH_3)_2$				
H_2PtCl_6 (i-PrOH)	reflux	2	$Cl_3Si(CH_2)_5CH=CHCH_2CH(COOCH_3)_2$ (42)	D -055
$C_{13}H_{22}$ $CH_2=CHCOOC_{10}H_{19}$				
H_2PtCl_6 (i-PrOH)	120	1.5	$Cl_3SiCH_2CH_2COOC_{10}H_{19}$ (30)	34
$C_{13}H_{24}$ $C_{10}H_{19}OCH_2CH=CH_2$				
H_2PtCl_6 (i-PrOH)	140	0.66	$C_{10}H_{19}O(CH_2)_3SiCl_3$ (74)	33
$C_{13}H_{24}$ $CH_3COO(CH_2)_9CH=CH_2$				
H_2PtCl_6 (i-PrOH)	100-150	1	$CH_3COO(CH_2)_{11}SiCl_3$ (-)	A -001
H_2PtCl_6 (i-PrOH)	120	8	$CH_3COO(CH_2)_{11}SiCl_3$ (93)	1138
H_2PtCl_6 (i-PrOH)	reflux	5	$CH_3COO(CH_2)_{11}SiCl_3$ (93)	1011
$C_{14}H_8$ $CF_3(CF_2)_7OCF_2CF_2SO_2N(CH_3)CH_2CH=CH_2$				
H_2PtCl_6	-	-	$CF_3(CF_2)_7OCF_2CF_2SO_2N(CH_3)(CH_2)_3SiCl_3$ (-)	238
$C_{14}H_{24}$ $CH_2=C(CH_3)COOC_{10}H_{19}$				

	H_2PtCl_6 (i-PrOH)	120	1.5	$Cl_3SiCH_2CH(CH_3)COOC_{10}H_{19}$ (44)	34
$C_{15}H_{24}$	$CH_2{=}C(COOCH_3)CH_2CH_2CH{=}CH(CH_2)_3CH{=}CHCH_2CH_3$				
	H_2PtCl_6 (i-PrOH)	80	1	$CH_2{=}C(COOCH_3)CH_2CH_2CH{=}CH(CH_2)_4CH(SiCl_3)CH_2CH_3$ (70)	D -055
$C_{15}H_{26}$	$CH_2{=}CHCH_2CH[COO(CH_2)_3CH_3]CH_2COO(CH_2)_3CH_3$				
	H_2PtCl_6 [$(CH_2)_5CO$]	60-100	6 (t)	$Cl_3Si(CH_2)_3CH[COO(CH_2)_3CH_3]CH_2COO(CH_2)_3CH_3$ (-)	D -093
$C_{15}H_{28}$	$CH_2{=}CH(CH_2)_8COOOC(CH_3)_3$				
	$PtCl_2(p\text{-}ClC_6H_4CN)_2$	25	94 (b)	$Cl_3Si(CH_2)_{10}COOOC(CH_3)_3$ (78)	D -045
$C_{16}H_{11}$	$CF_3(CF_2)_9(CH_2)_4CH{=}CH_2$				
	H_2PtCl_6 (THF)	reflux	20 (d)	$CF_3(CF_2)_9(CH_2)_6SiCl_3$ (-)	N -002
$C_{17}H_3$	$CF_3(CF_2)_{14}CH{=}CH_2$				
	-	-	-	$CF_3(CF_2)_{14}CH_2CH_2SiCl_3$ (-)	465
$C_{17}H_{26}$	Fig.72.				
	H_2PtCl_6 (i-PrOH)	60	1	Fig.73. (-)	J -068
$C_{18}H_{26}$	$CH_3CH_2CH{=}CH(CH_2)_3CH{=}CHCH_2CH_2C(COOCH_3){=}CHCOOCH_3$				
	H_2PtCl_6 (i-PrOH)	140	2	$CH_3CH_2CH(SiCl_3)(CH_2)_4CH{=}CHCH_2CH_2C(COOCH_3){=}CH\text{-}COOCH_3$ (46)	D -055
$C_{18}H_{26}$	cis-1,4-$CH_3CH_2CH_2C_6H_{10}C_6H_4OCH_2CH{=}CH_2$-p				
	H_2PtCl_6 (i-PrOH)	-	- (dcm)	cis-1,4-$CH_3CH_2CH_2C_6H_{10}C_6H_4O(CH_2)_3SiCl_3$-p (-)	D -122
$C_{19}H_9$	c-$C_{10}H_{19}OC_6H_4CH_2CH{=}CH_2$				
	$PtCl_2(Et_2S)_2$	until 120	0.5	o-$C_{10}H_{19}OC_6H_4(CH_2)_3SiCl_3$ (-)	G -002
$C_{19}H_{14}$	Fig.74.				
	H_2PtCl_6	85	48	Fig.75. (-)	D -018
$C_{19}H_{28}$	3-$CH_2{=}CHCH_2C_6H_4$-15-crown-5				
	H_2PtCl_6 (OctOH)	-	-	3-$Cl_3Si(CH_2)_3C_6H_4$-15-crown-5 (-)	241
$C_{19}H_{36}$	$CH_2{=}CH(CH_2)_{15}COOCH_3$				
	$(PhCOO)_2$	reflux	5	$Cl_3Si(CH_2)_{17}COOCH_3$ (32)	1011
	H_2PtCl_6 (i-PrOH)	90	5	$Cl_3Si(CH_2)_{17}COOCH_3$ (75)	1011

	Pt/C	reflux	5	$Cl_3Si(CH_2)_{17}COOCH_3$ (70)	1011
$C_{19}H_{36}$	$CH_3(CH_2)_{14}COOCH_2CH=CH_2$				
	-	-	-	$CH_3(CH_2)_{14}COO(CH_2)_3SiCl_3$ (-)	E -003
$C_{19}H_{36}$	$CH_3(CH_2)_7CH=CH(CH_2)_7COOCH_3$				
	$(PhCOO)_2$	70-150	2	$CH_3(CH_2)_7CH(SiCl_3)(CH_2)_8COOCH_3$ + $CH_3(CH_2)_8CH(SiCl_3)(CH_2)_7COOCH_3$ (62)	D -090
	$(PhCOO)_2$	70-150	2	$CH_3(CH_2)_7CH(SiCl_3)(CH_2)_8COOCH_3$ + $CH_3(CH_2)_8CH(SiCl_3)(CH_2)_7COOCH_3$ (62)	D -099
	$(PhCOO)_2$	85	11	$CH_3(CH_2)_7CH(SiCl_3)(CH_2)_8COOCH_3$ + $CH_3(CH_2)_8CH(SiCl_3)(CH_2)_7COOCH_3$ (89-92)	D -099
	$(PhCOO)_2$	85	11	$CH_3(CH_2)_7CH(SiCl_3)(CH_2)_8COOCH_3$ + $CH_3(CH_2)_8CH(SiCl_3)(CH_2)_7COOCH_3$ (89-92)	D -090
	-	-	-	$CH_3(CH_2)_7CH(SiCl_3)(CH_2)_8COOCH_3$ + $CH_3(CH_2)_8CH(SiCl_3)(CH_2)_7COOCH_3$ (-)	E -003
	H_2PtCl_6 (i-PrOH)	reflux	5	$CH_3(CH_2)_7CH(SiCl_3)(CH_2)_8COOCH_3$ + $CH_3(CH_2)_8CH(SiCl_3)(CH_2)_7COOCH_3$ (65)	1011
$C_{21}H_{32}$	$CH_2=CH(CH_2)_3CH=CHCH_2CH_2CH=C(COOCH_3)CH_2CH=CH(CH_2)_3CH=CH_2$				
	H_2PtCl_6	80-90	1	$Cl_3Si(CH_2)_5CH=CHCH_2CH_2CH=C(COOCH_3)CH_2CH=CH$-$(CH_2)_5SiCl_3$ (48)	D -055
$C_{21}H_{40}$	$CH_3(CH_2)_{16}COOCH_2CH=CH_2$				
	-	-	-	$CH_3(CH_2)_{16}COO(CH_2)_3SiCl_3$ (-)	E -003
$C_{21}H_{40}$	$CH_3CH_2COO(CH_2)_{16}CH=CH_2$				
	H_2PtCl_6 (i-PrOH)	150	1	$CH_3CH_2COO(CH_2)_{18}SiCl_3$ (-)	A -001
$C_{22}H_{38}$	$[CH_2=CH(CH_2)_8COO]_2$				
	$Pt_2Cl_4(PhCH=CH_2)_2$	25-18	114	$[Cl_3Si(CH_2)_{10}COO]_2$ (89)	D -045
$C_{23}H_{44}$	$CH_3(CH_2)_7CH=CH(CH_2)_{11}COOCH_3$				
	-	-	-	$CH_3(CH_2)_7CH(SiCl_3)(CH_2)_{12}COOCH_3$ + $CH_3(CH_2)_8CH(SiCl_3)(CH_2)_{11}COOCH_3$ (-)	E -003
$C_{26}H_{22}$	$(C_6H_5)_3SiC(C_6H_5)=CH_2$				
	H_2PtCl_6	reflux	- (cy)	$(C_6H_5)_3SiCH(C_6H_5)CH_2SiCl_3$ (-)	118
$C_{26}H_{37}$	$1\text{-}C_{10}H_7NHCH(i\text{-}C_3H_7)COO(CH_2)_9CH=CH_2$				

	H_2PtCl_6 (i-PrOH)	reflux	5	1-$C_{10}H_7NHCH(i\text{-}C_3H_7)COO(CH_2)_{11}SiCl_3$ (-)	910
$C_{26}H_{50}$	$CH_3(CH_2)_7CH=CH(CH_2)_7COO(CH_2)_5CH(CH_3)_2$				
	-	-	-	$CH_3(CH_2)_7CH(SiCl_3)(CH_2)_8COO(CH_2)_5CH(CH_3)_2$ + $CH_3(CH_2)_8CH(SiCl_3)(CH_2)_7COO(CH_2)_5CH(CH_3)_2$ (-)	E -003
$C_{28}H_{56}$	$CH_3(OCH_2CH_2)_{12}OCH_2CH=CH_2$				
	PhCOOOBu-t	100	overnight	$CH_3(OCH_2CH_2)_{12}O(CH_2)_3SiCl_3$ (-)	U -045
	PhCOOOBu-t	100	overnight	$CH_3(OCH_2CH_2)_{12}O(CH_2)_3SiCl_3$ (-)	U -076
	PhCOOOBu-t	100	overnight	$CH_3(OCH_2CH_2)_{12}O(CH_2)_3SiCl_3$ (-)	U -077
$C_{29}H_{45}$	Fig.76.				
	H_2PtCl_6 (i-PrOH)	reflux	0.5	Fig.77. (73)	911
$C_{33}H_{28}$	Fig.78.				
	Et_3N	155-160	5 (MeCN)	Fig.79. (-)	410
$C_{38}H_{26}$	Fig.80.				
	Et_3N	155-160	5 (MeCN)	Fig.81. (-)	410

Table 1.4

= $HSiCl_3$ = =1=

C_2H_2 HC≡CH

-	20	2	CH_2=$CHSiCl_3$ (19) $Cl_3SiCH_2CH_2SiCl_3$ (67)	956
none	500	45s	CH_2=$CHSiCl_3$ (20) $Cl_3SiCH_2CH_2SiCl_3$ (5)	258
$Co_2(CO)_8$ + Et_2O	15-20	1.3	CH_2=$CHSiCl_3$ (95-97)	S -043
$Co_2(CO)_8$ + KI	15-20	1.2-1.5	CH_2=$CHSiCl_3$ (94-95)	S -043
$Co_2(CO)_8$ + THF	15-20	1.5-1.7	CH_2=$CHSiCl_3$ (91-94)	S -043
H_2PtCl_6	-	- (p)	CH_2=$CHSiCl_3$ (100)	F -026
H_2PtCl_6	-	- (p)	CH_2=$CHSiCl_3$ (100)	U -074
H_2PtCl_6	105	2.2	CH_2=$CHSiCl_3$ (87) $Cl_3SiCH_2CH_2SiCl_3$ (trace)	F -007
H_2PtCl_6	until 150	flow	CH_2=$CHSiCl_3$ (95)	D -031
H_2PtCl_6 ($BuEtCHCH_2OH$)	115-120	-	CH_2=$CHSiCl_3$ (-)	J -039
H_2PtCl_6 (BzOH)	130-150	-	$Cl_3SiCH_2CH_2SiCl_3$ (95)	D -042
H_2PtCl_6 (THF) + $AlCl_3$	30	- (dc)	CH_2=$CHSiCl_3$ (70)	1234
H_2PtCl_6 (THF) + $AlCl_3$	r.t.	2(BuOBu)	CH_2=$CHSiCl_3$ (98)	S -041
H_2PtCl_6 (i-PrOH)	-	- (p)	CH_2=$CHSiCl_3$ (100)	U -100
H_2PtCl_6 (i-PrOH)	60	3 (xy)	CH_2=$CHSiCl_3$ (90)	1290
H_2PtCl_6 (i-PrOH)	60	3 (xy)	CH_2=$CHSiCl_3$ (90)	1288
H_2PtCl_6 [$(CH_2)_5CO$]	-	-	CH_2=$CHSiCl_3$ (93-97) $Cl_3SiCH_2CH_2SiCl_3$ (1-3)	D -041
H_2PtCl_6 [$(CH_2)_5CO$]	100-140	1	CH_2=$CHSiCl_3$ (93-97) $Cl_3SiCH_2CH_2SiCl_3$ (1-3)	D -042
Pt	105	3	CH_2=$CHSiCl_3$ (87) $Cl_3SiCH_2CH_2SiCl_3$ (trace)	F -007
$PtCl_2(PPh_3)_2$	90	3 (xy)	CH_2=$CHSiCl_3$ (54) $Cl_3SiCH_2CH_2SiCl_3$ (15)	1290
$PtCl_2(PPh_3)_2$	90	3 (xy)	CH_2=$CHSiCl_3$ (54) $Cl_3SiCH_2CH_2SiCl_3$ (15)	1288
SDVB-CN/H_2PtCl_6	100	-	CH_2=$CHSiCl_3$ (kin.)	621

Table 1.4

= $HSiCl_3$ = =2=

SDVB-CN/H_2PtCl_6	100	flow	$CH_2=CHSiCl_3$ (95)	C -003
SDVB-Me_2N/H_2PtCl_6	100	-	$CH_2=CHSiCl_3$ (kin.)	764
SiO_2-CN/Pt	110	1	$CH_2=CHSiCl_3$ (74)	C -013
$RhCl(PPh_3)_3$	75-80	10 (xy)	$CH_2=CHSiCl_3$ (45) $Cl_3SiCH_2CH_2SiCl_3$ (11)	1288
$RhCl(PPh_3)_3$	75-80	10 (xy)	$CH_2=CHSiCl_3$ (45) $Cl_3SiCH_2CH_2SiCl_3$ (11)	1290
$RhH(PPh_3)_4$	70-100	6 (xy)	$CH_2=CHSiCl_3$ (50) $Cl_3SiCH_2CH_2SiCl_3$ (12)	1290
SiO_2-CN/$RhCl(PPh_3)_3$	140	flow	$CH_2=CHSiCl_3$ (63) $Cl_3SiCH_2CH_2SiCl_3$ (3)	Pl-019
SiO_2-CN/$RhCl(PPh_3)_3$	140	flow	$CH_2=CHSiCl_3$ (63) $Cl_3SiCH_2CH_2SiCl_3$ (3)	744
SiO_2-$NHCH_2CH_2NH_2$/$RhCl(PPh_3)_3$	140	flow	$CH_2=CHSiCl_3$ (49) $Cl_3SiCH_2CH_2SiCl_3$ (16)	744
SiO_2-$NHCH_2CH_2NH_2$/$RhCl(PPh_3)_3$	140	flow	$CH_2=CHSiCl_3$ (49) $Cl_3SiCH_2CH_2SiCl_3$ (16)	Pl-017
SiO_2-$NHCH_2CH_2NH_2$/$RhCl(PPh_3)_3$	140	flow	$CH_2=CHSiCl_3$ (49) $Cl_3SiCH_2CH_2SiCl_3$ (16)	Pl-019
SiO_2-PPh_2/$RhCl(PPh_3)_3$	115	flow	$CH_2=CHSiCl_3$ (50-68) $Cl_3SiCH_2CH_2SiCl_3$ (5-20)	Pl-017
SiO_2-PPh_2/$RhCl(PPh_3)_3$	115	flow	$CH_2=CHSiCl_3$ (50-68) $Cl_3SiCH_2CH_2SiCl_3$ (5-20)	Pl-019
SiO_2-PPh_2/$RhCl(PPh_3)_3$	115	flow	$CH_2=CHSiCl_3$ (50-68) $Cl_3SiCH_2CH_2SiCl_3$ (5-20)	744
SiO_2-SH/$RhCl(PPh_3)_3$	140	flow	$CH_2=CHSiCl_3$ (62) $Cl_3SiCH_2CH_2SiCl_3$ (3)	Pl-019
SiO_2-SH/$RhCl(PPh_3)_3$	140	flow	$CH_2=CHSiCl_3$ (62) $Cl_3SiCH_2CH_2SiCl_3$ (3)	744
$[RhCl(CO)(C_8H_{14})]_2$	40	4	$CH_2=CHSiCl_3$ (12)	1257
$[RhCl(CO)_2]_2$	20-25	2	$Cl_3SiCH_2CH_2SiCl_3$ (51)	S -045
asbestos-PPh_2/$RhCl(PPh_3)_3$	140-150	flow	$CH_2=CHSiCl_3$ (38-75)	744
asbestos-PPh_2/$RhCl(PPh_3)_3$	140-150	flow	$CH_2=CHSiCl_3$ (38-75)	Pl-016
asbestos-PPh_2/$RhCl(PPh_3)_3$	140-150	flow	$CH_2=CHSiCl_3$ (38-75)	Pl-020

Table 1.4

= $HSiCl_3$ = =3=

	siloxane-$(CH_2)_n[RhCl(C_8H_{14})_2]_2$ n=1,3	20	1	$CH_2=CHSiCl_3$ (1-2)	941
	siloxane-$PPh_2/RhCl(PPh_3)_3$	140	flow	$CH_2=CHSiCl_3$ + $Cl_3SiCH_2CH_2SiCl_3$ (38-66)	742
	$RuCl_2(PPh_3)_3$	20	2.5 (b)	$CH_2=CHSiCl_3$ (85)	1290
	$RuCl_2(PPh_3)_3$	20	2.5 (b)	$CH_2=CHSiCl_3$ (85)	1288
	$RuCl_2(PPh_3)_3$	80	3 (xy)	$CH_2=CHSiCl_3$ (90)	1290
	$RuCl_2(PPh_3)_3$	80	3 (xy)	$CH_2=CHSiCl_3$ (90)	1288
	SiO_2-$PPh_2/RuCl_2(PPh_3)_3$	130	flow	$CH_2=CHSiCl_3$ (46) $Cl_3SiCH_2CH_2SiCl_3$ (13-16)	PI-017
	SiO_2-$PPh_2/RuCl_2(PPh_3)_3$	130	flow	$CH_2=CHSiCl_3$ (46) $Cl_3SiCH_2CH_2SiCl_3$ (13-16)	PI-018
	SiO_2-$PPh_2/RuCl_3(PPh_3)_3$	130	flow	$CH_2=CHSiCl_3$ (52-63)	PI-018
	SiO_2-SH/$RuCl_3(PPh_3)_3$	140	flow	$CH_2=CHSiCl_3$ (28) $Cl_3SiCH_2CH_2SiCl_3$ (9-16)	PI-018
	SiO_2-SH/$RuCl_3(PPh_3)_3$	140	flow	$CH_2=CHSiCl_3$ (28) $Cl_3SiCH_2CH_2SiCl_3$ (9-16)	PI-017
	Metal/Al_2O_3 or C; Metal: Ru,Rh,Pd,Ir,Pt	150	7s	$CH_2=CHSiCl_3$ + $Cl_3SiCH_2CH_2SiCl_3$ (-)	435
C_4H_6	$CH_3C\equiv CCH_3$				
	$[HPt(SiCl_3)P(c\text{-}Hex)_3]_2$	100	0.25	cis-$CH_3CH=C(CH_3)SiCl_3$ (87)	430
C_4H_6	$CH_3CH_2C\equiv CH$				
	$[HPt(SiCl_3)P(c\text{-}Hex)_3]_2$	65	3	trans-$CH_3CH_2CH=CHSiCl_3$ + $CH_3CH_2C(SiCl_3)=CH_2$ (88)	430
C_5H_6	$CH_2=C(CH_3)C\equiv CH$				
	H_2PtCl_6	80	6	$CH_2=C(CH_3)CH=CHSiCl_3$ (65)	672
	H_2PtCl_6 + $HSiMe_3Cl_2$	80	6	$CH_2=C(CH_3)CH=CHSiCl_3$ (14) $CH_2=C(CH_3)CH=CHSiCH_3Cl_2$ (56)	672
C_5H_8	$CH_3C\equiv CCH_2CH_3$				
	$[HPt(SiMe_2Bz)P(c\text{-}Hex)_3]_2$	65	3	$Cl_3SiC(CH_3)=CHCH_2CH_3$ (66) $CH_3CH=C(SiCl_3)CH_2CH_3$ (28)	1205
C_5H_8	$CH_3CH_2CH_2C\equiv CH$				
	$[HPt(SiMe_2Bz)P(c\text{-}Hex)_3]_2$	65	1.25	trans-$CH_3CH_2CH_2CH=CHSiCl_3$ (84) $CH_3CH_2CH_2C(SiCl_3)=CH_2$ (4)	1205
C_6H_{10}	$(CH_3)_3CC\equiv CH$				

	H_2PtCl_6 (i-PrOH)	r.t.	12	$(CH_3)_3CCH=CHSiCl_3$ (72)	1192
C_6H_{10}	$CH_3(CH_2)_3C\equiv CH$				
	e + $CH_2=CHSiMe(OOBu)_2$+$(EtO)_3SiCH_2CH_2$-$SiMe(OOBu)_2$	270	-	Z-$CH_3(CH_2)_3CH=CHSiCl_3$ (65)	950
	H_2PtCl_6 (i-PrOH)	r.t.	15	$CH_3(CH_2)_3CH=CHSiCl_3$ (75)	1192
	H_2PtCl_6 (i-PrOH)	reflux	0.5-110	Z-$CH_3(CH_2)_3CH=CHSiCl_3$ + E-$CH_3(CH_2)_3CH=CHSiCl_3$ + $CH_3(CH_2)_3C(SiCl_3)=CH_2$ + $Cl_3Si(CH_2)_6SiCl_3$ + $CH_3(CH_2)_3CH(SiCl_3)CH_2SiCl_3$ (79-98)	117
	$[HPt(SiMe_2Bz)P(c\text{-}Hex)_3]_2$	65	1.5	$CH_3(CH_2)_3CH=CHSiCl_3$ (89) $CH_3(CH_2)_3C(SiCl_3)=CH_2$ (5)	1205
	$RhCl_3$•4 H_2O	100	2 (THF)	$CH_3(CH_2)_5SiCl_3$ (44)	956
	$RhCl_3$•4 H_2O	100	2 (THF)	cis-$CH_3(CH_2)_5CH=CHSiCl_3$/ trans-$CH_3(CH_2)_5CH=CHSiCl_3$ (62/38)	956
	$[(RhCOCl)_2C_4H_6]_n$	100	2 (THF)	$CH_3(CH_2)_5SiCl_3$ (84)	956
	$[(RhCOCl)_2C_4H_6]_n$	100	2 (THF)	cis-$CH_3(CH_2)_5CH=CHSiCl_3$ / trans-$CH_3(CH_2)_5CH=CHSiCl_3$ (44/56)	956
	$[Rh(CO)_2Cl]_2$	100	2 (THF)	$CH_3(CH_2)_5SiCl_3$ (56)	956
	$[Rh(CO)_2Cl]_2$	100	2 (THF)	cis-$CH_3(CH_2)_5CH=CHSiCl_3$ / trans-$CH_3(CH_2)_5CH=CHSiCl_3$ (62/38)	956
C_6H_{10}	$CH_3CH_2C\equiv CCH_2CH_3$				
	$Ni(bpy)(Et)_2$	20	4	$Cl_3SiC(CH_2CH_3)=C(CH_2CH_3)SiCl_3$ (62)	1184
	$[HPt(SiBzMe_2)P(c\text{-}Hex)_3]_2$	65	1.5	$CH_3CH_2CH=C(CH_2CH_3)SiCl_3$ (93)	1205
C_6H_{10}	$CH_3CH_2CH_2C\equiv CCH_3$				
	$[HPt(SiMe_2Bz)P(c\text{-}Hex)_3]_2$	65	3.5	$CH_3CH_2CH_2CH=C(CH_3)SiCl_3$ (62) $CH_3CH_2CH_2C(SiCl_3)=CHCH_3$ (30)	1205
C_7H_{10}	$(\overline{CH_2)_4C}HC\equiv CH$				
	$[HPt(SiBzMe_2)P(c\text{-}Hex)_3]_2$	65	1.5	$(\overline{CH_2)_4C}HCH=CHSiCl_3$ (90)	1205
C_7H_{12}	$CH_3(CH_2)_4C\equiv CH$				
	$C_6H_8(OC_3H_6CN)_6/Pt(PPh_3)_4$	80	3	$C_7H_{13}SiCl_3$ (72)	C -013
	$[HPt(SiBzMe_2)P(c\text{-}Hex)_3]_2$	65	1.5	$CH_3(CH_2)_4CH=CHSiCl_3$ (89) $CH_3(CH_2)_4C(SiCl_3)=CH_2$ (4)	1205
C_8H_6	$C_6H_5C\equiv CH$				

Table 1.4
= $HSiCl_3$ = =5=

NBu_3	reflux	- (a)	cis-$C_6H_5CH=CHSiCl_3$ +trans-$C_6H_5CH=CHSiCl_3$ + $C_6H_5C(SiCl_3)=CH_2$ + $C_6H_5CH(SiCl_3)CH_2SiCl_3$ (38)	118
H_2OsCl_6 (THF)	110	4	$C_6H_5CH=CHSiCl_3$ (86) $C_6H_5C(SiCl_3)=CH_2$ (trace)	608
$(Bu_4N)_2PtCl_6$	80	1-3	$C_6H_5CH=CHSiCl_3$ + $C_6H_5C(SiCl_3)=CH_2$ (-)	673
H_2PtCl_6	100	5	cis-$C_6H_5CH=CHSiCl_3$ +trans-$C_6H_5CH=CHSiCl_3$ + $C_6H_5C(SiCl_3)=CH_2$ (96)	942
H_2PtCl_6 (CH_2=CH-N donor)	100	5	$C_6H_5CH=CHSiCl_3$ + $C_6H_5C(SiCl_3)=CH_2$ (98)	612
H_2PtCl_6 (THF)	110	4	$C_6H_5CH=CHSiCl_3$ (75) $C_6H_5C(SiCl_3)=CH_2$ (17)	608
H_2PtCl_6 (i-PrOH)	80	0.5	$C_6H_5CH=CHSiCl_3$ (68) $C_6H_5C(SiCl_3)=CH_2$ (22)	1280
H_2PtCl_6 (i-PrOH)	r.t.	20	$C_6H_5CH=CHSiCl_3$ (88)	1192
H_2PtCl_6 (i-PrOH)	reflux	6-12	$C_6H_5CH=CHSiCl_3$ (78)	164
H_2PtCl_6 + LiI	20	- (h)	$C_6H_5CH=CHSiCl_3$ + $C_6H_5C(SiCl_3)=CH_2$ (70)	1259
$PtCl_2$ (CH_2=CH-N donor)	100	5	$C_6H_5CH=CHSiCl_3$ + $C_6H_5C(SiCl_3)=CH_2$ (93-98)	612
$PtCl_4$ (CH_2=CH-N donor)	100	5	$C_6H_5CH=CHSiCl_3$ + $C_6H_5C(SiCl_3)=CH_2$ (92-98)	612
$[HPt(SiCl_3)P(c\text{-}Hex)_3]_2$	25	23 (t)	trans-$C_6H_5CH=CHSiCl_3$ (80)	G -036
$[HPt(SiCl_3)P(c\text{-}Hex)_3]_2$	25	23 (t)	trans-$C_6H_5CH=CHSiCl_3$ (80)	430
$Rh(CO)_2(acac)$	100	5	cis-$C_6H_5CH=CHSiCl_3$ +trans-$C_6H_5CH=CHSiCl_3$ + $C_6H_5C(SiCl_3)=CH_2$ (71)	942
$RhCl(PBz_3)_3$	100	5	cis-$C_6H_5CH=CHSiCl_3$ +trans-$C_6H_5CH=CHSiCl_3$ + $C_6H_5C(SiCl_3)=CH_2$ (53)	942
$RhCl(PPh_3)_3$	100	5	cis-$C_6H_5CH=CHSiCl_3$ +trans-$C_6H_5CH=CHSiCl_3$ + $C_6H_5C(SiCl_3)=CH_2$ (60)	942
$RhCl(SbPh_3)_3$	100	5	cis-$C_6H_5CH=CHSiCl_3$ +trans-$C_6H_5CH=CHSiCl_3$ + $C_6H_5C(SiCl_3)=CH_2$ (42)	942
$RhCl_3{\cdot}4\ H_2O$	100	2 (THF)	$C_6H_5C(SiCl_3)=CH_2$ (10) $C_6H_5CH=CHSiCl_3$ (43)	956
$RhCl_3{\cdot}4\ H_2O$	100	2 (THF)	cis-$C_6H_5CH=CHSiCl_3$/trans-$C_6H_5CH=CHSiCl_3$ (32/68)	956
$RhClCO(PPh_3)_2$	100	5	cis-$C_6H_5CH=CHSiCl_3$ +trans-$C_6H_5CH=CHSiCl_3$ + $C_6H_5C(SiCl_3)=CH_2$ (88)	942
$RhHCO(PPh_3)_3$	100	5	cis-$C_6H_5CH=CHSiCl_3$ +trans-$C_6H_5CH=CHSiCl_3$ + $C_6H_5C(SiCl_3)=CH_2$ (88)	942

Table 1.4
= $HSiCl_3$ =

=6=

$RhI(PPh_3)_3$	100	5	cis-$C_6H_5CH{=}CHSiCl_3$ +trans-$C_6H_5CH{=}CHSiCl_3$ + $C_6H_5C(SiCl_3){=}CH_2$ (92)	942
$[(RhCOCl)_2C_4H_6]_n$	100	2 (THF)	$C_6H_5C(SiCl_3){=}CH_2$ (10) $C_6H_5CH{=}CHSiCl_3$ (43)	956
$[(RhCOCl)_2C_4H_6]_n$	100	2 (THF)	cis-$C_6H_5CH{=}CHSiCl_3$/trans-$C_6H_5CH{=}CHSiCl_3$ (26/74)	956
$[Rh(CO)_2Cl]_2$	100	2 (THF)	cis-$C_6H_5CH{=}CHSiCl_3$/trans-$C_6H_5CH{=}CHSiCl_3$ (35/65)	956
$[Rh(CO)_2Cl]_2$	100	2 (THF)	$C_6H_5C(SiCl_3){=}CH_2$ (10) $C_6H_5CH{=}CHSiCl_3$ (43)	956
C_8H_{12} $\overline{(CH_2)_5C}HC\equiv CH$				
H_2PtCl_6 (i-PrOH)	reflux	75	$\overline{(CH_2)_5C}HCH{=}CHSiCl_3$ (69)	119
C_8H_{14} $CH_3(CH_2)_5C\equiv CH$				
H_2PtCl_6 (i-PrOH)	80	20	$CH_3(CH_2)_5CH{=}CHSiCl_3+CH_3(CH_2)_5C(SiCl_3){=}CH_2$ (92)	1192
H_2PtCl_6 (i-PrOH)	reflux	3	$CH_3(CH_2)_5CH{=}CHSiCl_3$ (92)	342
$C_{10}H_{18}$ $CH_3(CH_2)_3C\equiv C(CH_2)_3CH_3$				
$Ni(bpy)(C_2H_5)_2$	20	2-4	$CH_3(CH_2)_3C(SiCl_3){=}C(SiCl_3)(CH_2)_3CH_3$ (74)	1184
H_2PtCl_6 (i-PrOH)	100	24	$CH_3(CH_2)_3CH{=}C(SiCl_3)(CH_2)_3CH_3$ (90)	1192
$C_{14}H_{10}$ $C_6H_5C\equiv CC_6H_5$				
H_2PtCl_6 (i-PrOH)	reflux	2 (t)	$C_6H_5CH{=}C(SiCl_3)C_6H_5$ (87)	837
$[HPt(SiBzMe_2)P(c\text{-}Hex)_3]_2$	65	8	cis-$C_6H_5CH{=}C(SiCl_3)C_6H_5$ (76)	430
$C_{14}H_{26}$ $CH_3(CH_2)_5C\equiv C(CH_2)_5CH_3$				
H_2PtCl_6 (i-PrOH)	100	43	$CH_3(CH_2)_5CH{=}C(SiCl_3)(CH_2)_5CH_3$ (89)	1192
$C_{18}H_{34}$ $CH_3(CH_2)_{15}C\equiv CH$				
H_2PtCl_6 (i-PrOH)	r.t.	48	$CH_3(CH_2)_{15}CH{=}CHSiCl_3$ (48)	1192

Table 1.5

= $HSiCl_3$ = =1=

C_2	$Cl_3SiC{\equiv}CSiCl_3$				
	H_2PtCl_6 (i-PrOH)	120-130	-	$Cl_3SiCH{=}C(SiCl_3)_2$ (2)	1053
	H_2PtCl_6 (i-PrOH)	until 90	6 (e)	$(Cl_3Si)_2C{=}CHSiCl_3$ (80)	1024
C_2H_1	$HC{\equiv}CSiCl_3$				
	none	520	30s	$Cl_3SiCH{=}CHSiCl_3$ (35)	1057
	H_2PtCl_6 (i-PrOH)	130	5	$Cl_3SiCH{=}CHSiCl_3$ (74)	1057
C_3	$FC{\equiv}CCF_3$				
	UV	-	97	$Cl_3SiCF{=}CHCF_3$ (45) $CHF{=}C(SiCl_3)CF_3$ (45)	481
C_3H_1	$HC{\equiv}CCF_3$				
	UV	-	97	$Cl_3SiCH{=}CHCF_3$ (27) $CH_2{=}C(SiCl_3)CF_3$ (70)	481
C_3H_3	$HC{\equiv}CCH_2Cl$				
	$Cu_2Cl_2 + NEt_3$	50-52	2.2-15.5	$CH_2{=}C(SiCl_3)CH_2Cl$ (6-13)	403
	Pt/polysiloxane	50-118	6 (t)	$Cl_3SiCH{=}CHCH_2Cl$ (95)	E -009
	$PtCl_2(acac)_2$	until 89	4 (a)	$Cl_3SiC_2H_2CH_2Cl$ (76)	D -011
C_4H_4	$ClC{\equiv}CCH_2CH_2Cl$				
	H_2PtCl_6 (i-PrOH)	-	8	$ClCH_2C(SiCl_3){=}CHCH_2Cl$ (94)	E -015
C_4H_4	$ClCH_2C{\equiv}CCH_2Cl$				
	H_2PtCl_6	-	-	$ClCH_2CH{=}C(SiCl_3)CH_2Cl$ (-)	1022
	H_2PtCl_6	90	1	$ClCH_2CH{=}C(SiCl_3)CH_2Cl$ (31-58)	730
	H_2PtCl_6 (i-PrOH)	reflux	8	$ClCH_2CH{=}C(SiCl_3)CH_2Cl$ (94)	F -009
C_4H_6	$CH_3Cl_2SiC{\equiv}CSiCl_2CH_3$				
	H_2PtCl_6 (i-PrOH)	120-130	-	$CH_3Cl_2SiCH{=}C(SiCl_2CH_3)SiCl_3$ (58)	1053
C_4H_6	$HC{\equiv}CCH_2OCH_3$				
	H_2PtCl_6 (i-PrOH)	55	8	$Cl_3SiCH{=}CHCH_2OCH_3 + CH_2{=}C(SiCl_3)CH_2OCH_3$ (85)	1192
C_5H_6	$CH_3COOCH_2C{\equiv}CH$				
	H_2PtCl_6 (OctOH)	60-115	20	$CH_3COOCH_2CH{=}CHSiCl_3 + CH_3COOCH_2C(SiCl_3){=}CH_2$ (-)	U -051
C_5H_6	$HC{\equiv}CCOOCH_2CH_3$				

	H_2PtCl_6 (i-PrOH)	reflux	6	$Cl_3SiCH{=}CHCOOCH_2CH_3$ (-)	B -002
C_5H_7	$HC{\equiv}C(CH_2)_3Cl$				
	$(PhCOO)_2$ [$\overline{(CH_2)_5CO}$]	65	70	$Cl_3SiCH{=}CH(CH_2)_3Cl$ (52)	168
	$(PhCOO)_2$ [$\overline{(CH_2)_5CO}$]	reflux	70	cis-$Cl_3SiCH{=}CH(CH_2)_3Cl$+trans-$Cl_3SiCH{=}CH(CH_2)_3Cl$ (54)	107
C_5H_9	$HC{\equiv}CSi(CH_3)_2CH_2Br$				
	$(PhCOO)_2$	40-60	21days (cy)	trans-$Cl_3SiCH{=}CHSi(CH_3)_2CH_2Br$ (15)	387
C_5H_9	$HC{\equiv}CSi(CH_3)_2CH_2Cl$				
	$(PhCOO)_2$	40-60	60 (cy)	cis-$Cl_3SiCH{=}CHSi(CH_3)_2CH_2Cl$ + trans-$Cl_3SiCH{=}CHSi(CH_3)_2CH_2Cl$ (33)	387
	H_2PtCl_6 (i-PrOH)	80-100	-	trans-$Cl_3SiCH{=}CHSi(CH_3)_2CH_2Cl$ (94)	387
	H_2PtCl_6 (i-PrOH)	reflux	1	$Cl_3SiCH{=}CHSi(CH_3)_2CH_2Cl$ (86)	1240
C_5H_{10}	$HC{\equiv}CCH_3Si(OCH_3)_2$				
	H_2PtCl_6 (i-PrOH)	90-100	1	$Cl_3SiCH{=}CHSiCH_3(OCH_3)_2$ (74)	1281
C_5H_{10}	$HC{\equiv}CSi(CH_3)_2OCH_3$				
	H_2PtCl_6 (i-PrOH)	90-100	1	$Cl_3SiCH{=}CHSi(CH_3)_2OCH_3$ (76)	1281
C_5H_{10}	$HC{\equiv}CSi(CH_3)_3$				
	H_2PtCl_6 (i-PrOH)	90-100	1	$Cl_3SiCH{=}CHSi(CH_3)_3$ (82) $(Cl_3Si)_2CHCH_2Si(CH_3)_3$ (trace)	1100
C_5H_{10}	$HC{\equiv}CSi(OCH_3)_3$				
	H_2PtCl_6 (i-PrOH)	80-90	0.5	$Cl_3SiCH{=}CHSi(OCH_3)_3$ (76)	1278
C_6H_7	$HC{\equiv}CCH_2OCH_2CH_2CN$				
	H_2PtCl_6 (i-PrOH)	20	20	$Cl_3SiCH{=}CHCH_2OCH_2CH_2CN$ + $CH_2{=}C(SiCl_3)CH_2OCH_2CH_2CN$ (93)	1192
C_6H_8	$(HC{\equiv}C)_2Si(CH_3)_2$				
	H_2PtCl_6	-	- (HF)	$(CH_3)_2Si(C{\equiv}CH)(CH{=}CHSiF_3)$ + $(CH_3)_2Si(CH{=}CHSiF_3)_2$ (-)	1308
C_6H_8	$(HC{\equiv}C)_2Si(OCH_3)CH_3$				
	H_2PtCl_6 (i-PrOH)	80-90	0.5	$Cl_3SiCH{=}CHSi(OCH_3)(CH_3)C{\equiv}CH$ (66) $(Cl_3SiCH{=}CH)_2Si(OCH_3)CH_3$ (trace)	1261
C_6H_{10}	$CH_3C{\equiv}CCH_2OCH_2CH_3$				

Table 1.5
= HSiCl$_3$ =   =3=

	-	-	-	Cl$_3$SiC(CH$_3$)=CHCH$_2$OCH$_2$CH$_3$ + CH$_3$CH=C(SiCl$_3$)CH$_2$OCH$_2$CH$_3$ (-)	805
C$_7$H$_6$	(HC≡C)$_3$SiCH$_3$				
	H$_2$PtCl$_6$ (i-PrOH)	80-90	0.5	(CH≡C)$_2$Si(CH$_3$)CH=CHSiCl$_3$ (44) HC≡CSi(CH$_3$)(CH=CHSiCl$_3$)$_2$ (44)	1278
C$_7$H$_8$	CH$_2$=C(CH$_3$)COOCH$_2$C≡CH				
	H$_2$PtCl$_6$ (OctOH)	60-65	12	CH$_2$=C(CH$_3$)COOCH$_2$CH=CHSiCl$_3$ (56) CH$_2$=C(CH$_3$)COOCH$_2$C(SiCl$_3$)=CH$_2$ (38)	D -062
	H$_2$PtCl$_6$ (OctOH)	60-65	9	CH$_2$=C(CH$_3$)COOCH$_2$CH=CHSiCl$_3$ (56) CH$_2$=C(CH$_3$)COOCH$_2$C≡CSiCl$_3$ (38)	U -057
C$_7$H$_{10}$	CH$_3$COOC(CH$_3$)$_2$C≡CH				
	H$_2$PtCl$_6$ (OctOH)	reflux	8	CH$_3$COOC(CH$_3$)$_2$CH=CHSiCl$_3$ (-)	U -079
C$_7$H$_{10}$	HC≡CCOO(CH$_2$)$_3$CH$_3$				
	H$_2$PtCl$_6$ (i-PrOH)	50	6	Cl$_3$SiCH=CHCOO(CH$_2$)$_3$CH$_3$ (-)	B -002
C$_7$H$_{12}$	CH$_3$Si(C≡CH)(CH$_2$)$_4$				
	H$_2$PtCl$_6$ (i-PrOH)	1	90	CH$_3$Si(CH=CHSiCl$_3$)(CH$_2$)$_4$ (90)	1276
C$_7$H$_{14}$	HC≡CSi(OCH$_3$)(CH$_2$CH$_3$)$_2$				
	H$_2$PtCl$_6$ (i-PrOH)	90-100	1	Cl$_3$SiCH=CHSi(OCH$_3$)(CH$_2$CH$_3$)$_2$ (76)	1281
C$_8$H$_4$	(HC≡C)$_3$SiCH$_3$				
	H$_2$PtCl$_6$	-	- (HF)	CH$_3$Si(CH=CHSiF$_3$)$_3$ (-)	1308
C$_8$H$_4$	(HC≡C)$_4$Si				
	H$_2$PtCl$_6$	-	- (HF)	(HC≡C)$_2$Si(CH=CHSiF$_3$)$_2$ (-)	1308
C$_8$H$_{10}$	CH$_3$COOCH$_2$C≡CCH$_2$OCOCH$_3$				
	H$_2$PtCl$_6$ (OctOH)	reflux	2	CH$_3$COOCH$_2$CH=C(SiCl$_3$)CH$_2$OCOCH$_3$ (-)	U -051
C$_8$H$_{16}$	HC≡CSi(CH$_2$CH$_3$)$_3$				
	H$_2$PtCl$_6$ (i-PrOH)	50	-	Cl$_3$SiCH=CHSi(CH$_2$CH$_3$)$_3$ (70)	1279
C$_8$H$_{18}$	(CH$_3$)$_3$SiC≡CSi(CH$_3$)$_3$				
	H$_2$PtCl$_6$ (i-PrOH)	120-130	-	(CH$_3$)$_3$SiCH=C[Si(CH$_3$)$_3$]SiCl$_3$ (3)	1053
	H$_2$PtCl$_6$ (i-PrOH)	150-180	8	(CH$_3$)$_3$SiCH=C[Si(CH$_3$)$_3$]SiCl$_3$ (95)	1310

Table 1.5

= $HSiCl_3$ = =4=

	H_2PtCl_6 (i-PrOH)	reflux	-	$(CH_3)_3SiCH{=}C[Si(CH_3)_3]SiCl_3$ (23)	1179
	H_2PtCl_6 (i-PrOH)	until 90	166 (h)	$(CH_3)_3SiCH{=}C[Si(CH_3)_3]SiCl_3$ (68)	1024
C_9H_{18}	$CH_3(CH_2)_3C{\equiv}CSi(CH_3)_3$				
	H_2PtCl_6 (i-PrOH)	reflux	6	$CH_3(CH_2)_3CH{=}C[Si(CH_3)_3]SiCl_3$ (97)	1179
C_9H_{18}	$HC{\equiv}CSi(CH_3)_2CH{=}CHSi(CH_3)_3$				
	H_2PtCl_6 (i-PrOH)	90	1	$Cl_3SiCH{=}CHSi(CH_3)_2CH{=}CHSi(CH_3)_3$ (72)	1311
$C_{10}H_{12}$	$HC{\equiv}CSi(CH_3)_2C_6H_5$				
	H_2PtCl_6 (i-PrOH)	80-90	0.5	$Cl_3SiCH{=}CHSi(CH_3)_2C_6H_5$ (78)	1278
$C_{10}H_{14}$	$(CH_3CH_2COO)_2CHCH_2C{\equiv}CH$				
	H_2PtCl_6	reflux	-	$(CH_3CH_2COO)_2CHCH_2CH{=}CHSiCl_3$ (67)	1009
$C_{10}H_{16}$	$HC{\equiv}C(CH_2)_6COOCH_3$				
	H_2PtCl_6	-	-	$Cl_3SiCH{=}CH(CH_2)_6COOCH_3$ (-)	1318
$C_{10}H_{18}$	$CH_3CH_2CH_2OCH_2C{\equiv}CCH_2OCH_2CH_2CH_3$				
	H_2PtCl_6 (i-PrOH)	70-90	2	$CH_3CH_2CH_2OCH_2C(SiCl_3){=}CHCH_2OCH_2CH_2CH_3$ (49)	569
$C_{11}H_{14}$	$C_6H_5C{\equiv}CSi(CH_3)_3$				
	H_2PtCl_6 (i-PrOH)	reflux	11	$C_6H_5C(SiCl_3){=}CHSi(CH_3)_3$ (91)	1179
$C_{12}H_{20}$	$HC{\equiv}C(CH_2)_8COOCH_3$				
	H_2PtCl_6 (i-PrOH)	20	20	$Cl_3SiCH{=}CH(CH_2)_8COOCH_3 + CH_2{=}C(SiCl_3)(CH_2)_8COOCH_3$ (93)	1192

Table 1.6

= $HSiCl_3$ = =1=

C_1	CF_3NO				
	UV	20	15	$CF_3NHOSiCl_3$ (17)	587
C_3	$(CF_3)_2CO$				
	UV	22	-	$(CF_3)_2CHOSiCl_3$ (-)	502
	none	20	13	$(CF_3)_2CHOSiCl_3$ (68)	526
	none	50	70	$(CF_3)_2C(Cl)OSiHCl_2$ + $[(CF_3)_2C(Cl)O]_2SiHCl$ (63)	304
C_3H_5	CH_3CH_2COCl				
	gamma	-	-	$CH_3CH_2CH_2OSiCl_3$ (71) C_2H_6 (15)	874
C_3H_6	$(CH_3)_2CO$				
	UV	78-253	- (h)	$(CH_3)_2CHOSiCl_3$ (kin.)	326
	UV + $CH_2{=}C(CH_3)_2$	103-250	-	$(CH_3)_2CHOSiCl_3$ + $C_4H_9SiCl_3$ (kin.)	323
	UV + $CH_2{=}C(CH_3)_2$	78-272	-	$(CH_3)_2CHOSiCl_3$ + $C_4H_9SiCl_3$ (kin.)	323
	UV + $CH_2{=}C(CH_3)CH_2CH_3$	78-272	-	$(CH_3)_2CHOSiCl_3$ +$C_5H_{11}SiCl_3$ (kin.)	323
	UV + $CH_2{=}CH_2$	150-275	-	$(CH_3)_2CHOSiCl_3$ +$CH_3CH_2SiCl_3$ (kin.)	324
	UV + $CH_2{=}CH_2$	78-272	-	$(CH_3)_2CHOSiCl_3$ +$CH_3CH_2SiCl_3$ (kin.)	323
	UV + $CH_2{=}CHCH_2CH_2CH_3$	103-250	-	$(CH_3)_2CHOSiCl_3$ +$C_5H_{11}SiCl_3$ (kin.)	322
	UV + $CH_2{=}CHCH_2CH_3$	103-250	-	$(CH_3)_2CHOSiCl_3$ + $C_4H_9SiCl_3$ (kin.)	322
	UV + $CH_2{=}CHCH_2CH_3$	78-272	-	$(CH_3)_2CHOSiCl_3$ + $C_4H_9SiCl_3$ (kin.)	323
	UV + $CH_2{=}CHCH_3$	103-250	-	$(CH_3)_2CHOSiCl_3$ + $C_3H_7SiCl_3$ (kin.)	322
	UV + $CH_2{=}CHCH_3$	78-272	-	$(CH_3)_2CHOSiCl_3$ + $C_3H_7SiCl_3$ (kin.)	323
	UV + $CH_3CH{=}CHCH_2CH_3$ (cis,trans)	30-136	-	$(CH_3)_2CHOSiCl_3$ +$C_5H_{11}SiCl_3$ (kin.)	322
C_3H_6	CH_3COOCH_3				
	gamma(^{60}Co)	-	-	$CH_3CH(OCH_3)OSiCl_3$ (kin.)	816
	gamma(^{60}Co)	-	-	$CH_3CH(OCH_3)OSiCl_3$ (-)	813
C_4H_7	$(CH_3)_2CHCOCl$				
	gamma	-	-	$(CH_3)_2CHCH_2OSiCl_3$ (30) C_3H_8 (52)	874
C_5H_9	C_4H_9COCl				

	gamma	-	-	$C_4H_9CH_2OSiCl_3$ (81) C_4H_{10} (2)	874
C_6H_{11}	$(CH_3CH_2)_2CHCOCl$				
	gamma	-	-	$(CH_3CH_2)_2CHCH_2OSiCl_3$ (37) C_5H_{12} (42)	874
C_6H_{11}	$C_5H_{11}COCl$				
	gamma	-	-	$C_5H_{11}CH_2OSiCl_3$ (84) C_5H_{12} (7)	874
C_7H_{11}	c-$C_6H_{11}COCl$				
	gamma	-	-	c-$C_6H_{11}CH_2OSiCl_3$ (84) c-C_6H_{12} (10)	874
C_8H_{15}	$C_7H_{15}COCl$				
	gamma	-	-	$C_7H_{15}CH_2OSiCl_3$ (84) C_7H_{16} (12)	874
C_8H_{17}	$CH_3CH_2CH_2CH=N(CH_2)_3CH_3$				
	none	reflux	4 (a)	$[CH_3(CH_2)_3]_2NSiCl_3$ (-)	115
$C_{10}H_{16}$	Fig.82.				
	UV	reflux	100	Fig.83. (83)	371
$C_{10}H_{16}$	Fig.84.				
	UV	reflux	100	Fig.85. (84)	371
$C_{10}H_{18}$	Fig.86.				
	$RhCl(PPh_3)_3$	80	18 (b)	Fig.87. (-)	1037
$C_{10}H_{19}$	$C_9H_{19}COCl$				
	gamma	-	-	$C_9H_{19}CH_2OSiCl_3$ (81) C_9H_{20} (10)	874
$C_{11}H_{15}$	$C_6H_5CH=N(CH_2)_3CH_3$				
	none	reflux	4 (a)	$C_6H_5CH_2N(SiCl_3)(CH_2)_3CH_3$ (-)	115
$C_{12}H_{15}$	$(\overline{CH_2)_5C}=NC_6H_5$				
	none	reflux	4 (a)	$(\overline{CH_2)_5C}HN(C_6H_5)SiCl_3$ (-)	115
$C_{12}H_{21}$	$(\overline{CH_2)_5C}=N\overline{CH(C}H_2)_5$				

	none	reflux	4 (a)	$[(\overline{CH_2)_5CH}]_2NSiCl_3$ (-)	115
$C_{13}H_{11}$	$C_6H_5CH=NC_6H_5$				
	none	30-35	0.67-4 (a)	$C_6H_5CH_2N(C_6H_5)SiCl_3$ (38)	115
$C_{13}H_{17}$	$C_6H_5CH=N\overline{CH(CH_2)_5}$				
	none	reflux	4 (a)	$C_6H_5CH_2N(SiCl_3)\overline{CH(CH_2)_5}$ (-)	115
$C_{13}H_{17}$	$XC_6H_4CH=NC_6H_5$				
	none	reflux	4 (a)	$XC_6H_4CH_2N(C_6H_5)SiCl_3$;X = o,m,p,Cl,$CH_3O$,$C_3H_7$ (-)	115
$C_{14}H_{27}$	$C_{13}H_{27}COCl$				
	gamma	-	-	$C_{13}H_{27}CH_2OSiCl_3$ (89) $C_{13}H_{28}$ (-)	874

Table 2.1

= CH_3SiHCl_2 = =1=

C_2H_4 $CH_2{=}CH_2$				
$NiCl_2$(DMPF)	90	6.5 days	$CH_3CH_2SiCl_2CH_3$ + $CH_3CH_2SiHClCH_3$ (100)	589
H_2PtCl_6 (Me_2CO)	-	-	$CH_3CH_2SiCl_2CH_3$ + $CH_3CH_2SiCl_3$ (97)	D -128
H_2PtCl_6 (Me_2CO) + benzothiazole	130	4	$CH_3CH_2SiCl_2CH_3$ (98)	D -128
H_2PtCl_6 (Me_2CO) + benzothiazole	130	1.7	$CH_3CH_2SiCl_2CH_3$ (97-99)	D -128
H_2PtCl_6 [$(CH_2)_5CO$]	reflux	-	$CH_3CH_2SiCl_2CH_3$ (100)	D -041
Pt/Al_2O_3	130	8	$CH_3CH_2SiCl_2CH_3$ + $CH_3CH_2SiCl_3$ (81)	D -128
Pt/C	130	8	$CH_3CH_2SiCl_2CH_3$ (84)	D -128
C_3H_6 $CH_3CH{=}CH_2$				
gamma (^{60}Co)	80	-	$CH_3CH_2CH_2SiCl_2CH_3$ (100)	1360
$NiCl_2$(DMPC)	120	20	$CH_3CH_2CH_2SiCl_2CH_3$ + $CH_3CH(CH_3)SiCl_2CH_3$ (40)	337
$NiCl_2$(DMPF)	120	20	$C_3H_7SiCl_2CH_3$ + $C_3H_7SiHClCH_3$(100)	589
H_2PtCl_6 (Me_2CO) + benzothiazole	130	4.5	$CH_3CH_2CH_2SiCl_2CH_3$ (98)	D -128
H_2PtCl_6 (i-PrOH)	33-120	-	$C_3H_7SiCl_2CH_3$ (75)	141
H_2PtCl_6 [$(CH_2)_5CO$]	reflux	-	$CH_3CH_2CH_2SiCl_2CH_3$ (100)	D -041
Pt catalyst	-	-	$CH_3CH_2CH_2SiCl_2CH_3$ (kin.)	528
C_4H_8 $(CH_3)_2C{=}CH_2$				
$NiCl_2$(DMPC)	120	40	$(CH_3)_2CHCH_2SiCl_2CH_3$ (-) $(CH_3)_3CSiCl_2CH_3$ (20)	1360
H_2PtCl_6 (i-PrOH)	60	8	$(CH_3)_2CHCH_2SiCl_2CH_3$ (-)	J -045
C_4H_8 $(CH_3)_2C{=}CH_2$ + $CH_3CH_2CH{=}CH_2$				
H_2PtCl_6 [$(CH_2)_5CO$]	30-120	19	$CH_3(CH_2)_3SiCl_2CH_3$ + $(CH_3)_2CHCH_2SiCl_2CH_3$ (-)	Pl-013
H_2PtCl_6 [$(CH_2)_5CO$]	40-60	9	$CH_3(CH_2)_3SiCl_2CH_3$ + $(CH_3)_2CHCH_2SiCl_2CH_3$ (-)	Pl-014
$RhCl(PPh_3)_3$	60-120	8	$CH_3(CH_2)_3SiCl_2CH_3$ + $(CH_3)_2CHCH_2SiCl_2CH_3$ (-)	Pl-012
C_4H_8 $CH_3CH_2CH{=}CH_2$				
$NiCl_2$(DMPF)	120	20	$C_4H_9SiCl_2CH_3$ + $C_4H_9SiHClCH_3$ (80)	589
H_2PtCl_6 (Me_2CO) + benzothiazole	130	5	$CH_3(CH_2)_3SiCl_2CH_3$ (97)	D -128
C_5H_8 $(CH_2)_3CH{=}CH$				

	H_2PtCl_6 (2-EtC_6H_{10}OH) + UV light	40	6	$\overline{(CH_2)_4C}HSiCl_2CH_3$ (93)	E -038
	H_2PtCl_6 (i-PrOH)	120	3	$\overline{(CH_2)_4C}HSiCl_2CH_3$ (55)	784
C_5H_{10}	$CH_3C(CH_3)=CHCH_3$				
	H_2PtCl_6 +$(ViMe_2Si)_2O$	25-80	0.25	$(CH_3)_2CHCH(CH_3)SiCl_2CH_3$ (-)	U -180
C_5H_{10}	$CH_3CH_2C(CH_3)=CH_2$				
	gamma (^{60}Co)	16	60	$CH_3CH_2CH(CH_3)CH_2SiCl_2CH_3$ (75)	958
	gamma (^{60}Co)	16	60	$CH_3CH_2CH(CH_3)CH_2SiCl_2CH_3$ (75)	958
	$NiCl_2(BMPP)_2$	90	60	$CH_3CH_2CH(CH_3)CH_2SiCl_2CH_3$(17) $CH_3CH_2CH(CH_3)CH_2SiHClCH_3$(19)	1302
	$PtCl_2(C_2H_4)(BMPP)$	40	24-40	$CH_3CH_2CH(CH_3)CH_2SiCl_2CH_3$ (69)	1301
	$PtCl_2(C_2H_4)(BMPP)$	40	24-40	$CH_3CH_2CH(CH_3)CH_2SiCl_2CH_3$ (69)	1303
	$PtCl_2(C_2H_4)(MPPP)$	40	40	$CH_3CH_2CH(CH_3)CH_2SiCl_2CH_3$ (69)	1301
	$[PtCl_2(BMPP)]_2$	40	24	$CH_3CH_2CH(CH_3)CH_2SiCl_2CH_3$ (68)	1303
	$[PtCl_2(MDPP)]_2$	90	40	$CH_3CH_2CH(CH_3)CH_2SiCl_2CH_3$ (35)	1303
	$[PtCl_2(MPPP)]_2$	40	24-40	$CH_3CH_2CH(CH_3)CH_2SiCl_2CH_3$ (55-69)	1303
	$[PtCl_2(MPPP)]_2$	40	24-40	$CH_3CH_2CH(CH_3)CH_2SiCl_2CH_3$ (55-69)	1301
C_5H_{10}	$CH_3CH_2CH_2CH=CH_2$				
	$(PhCOO)_2$	100	8	$C_5H_{11}SiCl_2CH_3$ (-)	265
	gamma (^{60}Co)	16	60	$C_5H_{11}SiCl_2CH_3$ (72)	958
	$Fe(CO)_5$	145	8	$C_5H_{11}SiCl_2CH_3$ (35-41)	265
	$NiCl_2(DMPC)$	120	20	$CH_3(CH_2)_4SiCl_2CH_3 + CH_3CH_2CH_2CH(CH_3)SiCl_2CH_3$ (84)	337
	$NiCl_2(DMPF)$	120	20	$C_5H_{11}SiCl_2CH_3$ (18) $C_5H_{11}SiHClCH_3$ (80)	589
	$NiCl_2(DMPF)$	120	20	$C_5H_{11}SiCl_2CH_3$ (18) $C_5H_{11}SiHClCH_3$ (80)	628
	H_2PtCl_6	25	-(ht)	$CH_3(CH_2)_4SiCl_2CH_3$ (-)	845
	H_2PtCl_6 (EtOH)	50	2	$CH_3(CH_2)_4SiCl_2CH_3$ (-)	938
	H_2PtCl_6 [$\overline{(CH_2)_5C}O$]	reflux	-	$CH_3(CH_2)_4SiCl_2CH_3$ (100)	D -041
	$PtCl_2(PBu_3)_2$	120	2	$CH_3(CH_2)_4SiCl_2CH_3$ (70)	G -008

Table 2.1

= CH_3SiHCl_2 = =3=

$PtCl_2(PPh_3)_2$	125	17	$CH_3(CH_2)_4SiCl_2CH_3$ (95)	G -008
$RhCl_3$/polyolefin	80	2	$CH_3(CH_2)_4SiCl_2CH_3$ (70)	S -029
C_5H_{10} $CH_3CH_2CH{=}CHCH_3$				
gamma (^{60}Co)	16	60	$CH_3CH_2CH_2CH(CH_3)SiCl_2CH_3$ (80)	958
$NiCl_2$(DMPC)	120	40	$CH_3CH_2CH_2CH(CH_3)SiCl_2CH_3+(CH_3CH_2)_2CHSiCl_2CH_3$ (47)	337
$NiCl_2$(DMPF)	120	20	$C_5H_{11}SiCl_2CH_3$ $+C_5H_{11}SiHClCH_3$ (90)	589
SiO_2-$L_2Pt(C_2O_4)$	40	48	$CH_3(CH_2)_4SiCl_2CH_3$ (90)	938
C_5H_{10} $CH_3CH{=}C(CH_3)_2$				
gamma (^{60}Co)	16	60	$CH_3CH_2C(CH_3)_2SiCl_2CH_3$ (71)	958
C_6H_{10} $(\overline{CH_2)_4CH{=}C}H$				
UV + $(t\text{-}BuO)_2$	40	14	$(\overline{CH_2)_5C}HSiCl_2CH_3$ (10)	338
$NiCl_2$(DMPF)	120	20	$(\overline{CH_2)_5C}HSiCl_2CH_3$ (34) $(\overline{CH_2)_5C}HSiHClCH_3$ (41)	589
H_2PtCl_6	100	7	$(\overline{CH_2)_5C}HSiCl_2CH_3$ (32)	E -038
H_2PtCl_6 (2-EtC_6H_{10}OH)	55-71	20	$(\overline{CH_2)_5C}HSiCl_2CH_3$ (9)	E -038
H_2PtCl_6 (2-EtC_6H_{10}OH) + UV light	20-26	40	$(\overline{CH_2)_5C}HSiCl_2CH_3$ (90)	E -038
H_2PtCl_6 (2-EtC_6H_{10}OH) + sunlight	r.t.	70	$(\overline{CH_2)_5C}HSiCl_2CH_3$ (91)	E -038
H_2PtCl_6 (OctOH)	100	16 (t)	$(\overline{CH_2)_5C}HSiCl_2CH_3$ (50)	U -014
$Rh_4(CO)_{12}$	40-45	1	$(\overline{CH_2)_5C}HSiCl_2CH_3$ (96-99)	S -037
C_6H_{12} $(CH_3)_2CHC(CH_3){=}CH_2$				
$NiCl_2$(BMPP)	90	60	$(CH_3)_2CHCH(CH_3)CH_2SiCl_2CH_3$ (21) $(CH_3)_2CHCH(CH_3)CH_2SiHClCH_3$ (26)	1302
$PtCl_2(C_2H_4)$(BMPP)	40	24	$(CH_3)_2CHCH(CH_3)CH_2SiCl_2CH_3$ (76)	1303
$[PtCl_2(BMPP)]_2$	40	24	$(CH_3)_2CHCH(CH_3)CH_2SiCl_2CH_3$ (83)	1303
$[PtCl_2(MDPP)]_2$	40	24	$(CH_3)_2CHCH(CH_3)CH_2SiCl_2CH_3$ (70)	1303
$[PtCl_2(MPPP)]_2$	40	24	$(CH_3)_2CHCH(CH_3)CH_2SiCl_2CH_3$ (86)	1303
C_6H_{12} $(CH_3)_2CHCH_2CH{=}CH_2$				
gamma (^{60}Co)	16	21	$(CH_3)_2CH(CH_2)_3SiCl_2CH_3$ (74)	958

Table 2.1

= CH_3SiHCl_2 = =4=

	H_2PtCl_6 + Me_3Cl_2SiH	55	1.5	$(CH_3)_2CH(CH_2)_3SiCl_2CH_3$ (-)	U -168
	Pt/C + sonic waves	30	1	$(CH_3)_2CH(CH_2)_3SiCl_2CH_3$ (93)	467
C_6H_{12}	$CH_3(CH_2)_3CH{=}CH_2$				
	Fig.88.	120	5	$CH_3(CH_2)_5SiCl_2CH_3$ (88)	D -027
	$SiO_2[OSi(CH_2)_3ER_3]_2PtCl_6$;$ER_3{=}NMe_3$,$NEt_3$, PPh_3, $NEt_2(C_{10}H_{21})$, $NMeEt_2$	18-40	-	$C_6H_{13}SiCl_2CH_3$ (kin.)	1106
	UV + $(t\text{-}BuO)_2$	30	40	$CH_3(CH_2)_5SiCl_2CH_3$ (27-42)	338
	gamma	0-55	-	$CH_3(CH_2)_5SiCl_2CH_3$ (kin.)	189
	gamma (^{60}Co)	16	60	$C_6H_{13}SiCl_2CH_3$ (81)	958
	$Ni(acac)_2(Bz_3P)_2$	120	20	$CH_3(CH_2)_5SiCl_2CH_3$ (44)	1135
	$Ni(acac)_2(Et_2PhP)_2$	120	20	$CH_3(CH_2)_5SiCl_2CH_3$ (74)	1135
	$Ni(acac)_2(Ph_2PCH_2CH_2PPh_2)$	120	20	$CH_3(CH_2)_5SiCl_2CH_3$ (39)	1135
	$NiBr_2(Me_3SiCH_2PBu_2)_2$	120	16	$CH_3(CH_2)_3CH(CH_3)SiCl_2CH_3$ (7) $C_6H_{13}SiCl_2CH_3$ (48) $C_6H_{13}SiHClCH_3$ (15) $C_4H_9CH(CH_3)SiHClCH_3$ (4)	D -061
	$NiBr_2[Me_2Si(CH_2PPh_2)_2]$	120	16	$CH_3(CH_2)_5SiCl_2CH_3$ (24) $CH_3(CH_2)_3CH(CH_3)SiCl_2CH_3$ (17) $C_6H_{13}SiHClCH_3$ (2) $C_4H_9CH(CH_3)SiHClCH_3$ (1)	G -024
	$NiCl_2(DMPC)$	120	20	$CH_3(CH_2)_5SiCl_2CH_3$ + $CH_3(CH_2)_3CH(CH_3)SiCl_2CH_3$ (98)	337
	$NiCl_2(DMPF)$	120	20	$C_6H_{13}SiCl_2CH_3$ + $C_6H_{13}SiHClCH_3$ (100)	589
	$NiCl_2(DMPF)$	120	20	$C_6H_{13}SiCl_2CH_3$ (8) $C_6H_{13}SiHClCH_3$ (87)	628
	$K_2[Pd(C{\equiv}CMe)_2]$ + PPh_3	100	8	$CH_3(CH_2)_5SiCl_2CH_3$ (57)	D -027
	Pd + PEt_3	120	6-8	$C_6H_{13}SiCl_2CH_3$ (57-68)	D -027
	$Pd(PPh_3)_2(MeOCOCH{=}CHCOOMe)$	110	8	$CH_3(CH_2)_5SiCl_2CH_3$ (89)	D -027
	$Pd(PPh_3)_2[\overline{O(O)CC_2H_2C}O]$	110	6	$CH_3(CH_2)_5SiCl_2CH_3$ (90)	D -027
	$PdBr_2(PPh_3)_2$	110	5	$CH_3(CH_2)_5SiCl_2CH_3$ (80)	D -027
	$PdCl_2[Ph\overline{C{=}C(Ph)C(Ph){=}C}Ph]$ + PPh_3	110	6	$CH_3(CH_2)_5SiCl_2CH_3$ (85)	D -027
	H_2PtCl_6	25	- (ht)	$CH_3(CH_2)_5SiCl_2CH_3$ (-)	845

	H_2PtCl_6 (i-PrOH)	150	1	$C_6H_{13}SiCl_2CH_3$ (-)	D -071
	H_2PtCl_6 [$(\overline{CH_2)_5C}O$]	reflux	2	$C_6H_{13}SiCl_2CH_3$ (85)	Pl-011
	K_2PtCl_4+ crown ether	reflux	-	$CH_3(CH_2)_5SiCl_2CH_3$ (100)	J -027
	$Pt(PPh_3)_4$	45-52	0.5	$CH_3(CH_2)_3CH(CH_3)SiCl_2CH_3$ (94)	364
	Pt/C + sonic waves	30	1	$CH_3(CH_2)_5SiCl_2CH_3$ (95)	467
	$PtL_2(C_2H_4)$	-	-	$CH_3(CH_2)_5SiCl_2CH_3$ (-)	1300
	siloxane-$NHCH_2CH_2NH_2/H_2PtCl_6$	150	1	$C_6H_{13}SiCl_2CH_3$ (-)	D -071
	$Rh_4(CO)_{12}$	40-45	1	$C_6H_{13}SiCl_2CH_3$ (99)	S -037
	$RhCl(PPh_3)_3$	120	6	$C_6H_{13}SiCl_2CH_3$ (81)	Pl-011
C_6H_{12}	$CH_3(CH_2)_3CH{=}CH_2$ + $CH_3COCH(CH_3)_2$				
	gamma (^{60}Co)	-	-	$CH_3(CH_2)_5SiCl_2CH_3$ + $CH_3Cl_2SiOCH(CH_3)_2$ (kin.)	303
C_6H_{12}	$CH_3(CH_2)_3CH{=}CH_2$ + CH_3COCH_3				
	gamma (^{60}Co)	-	-	$CH_3(CH_2)_5SiCl_2CH_3$ + $(CH_3)_2CHOSiCl_2CH_3$ (kin.)	687
C_6H_{12}	$CH_3CH_2CH_2C(CH_3){=}CH_2$				
	gamma (^{60}Co)	16	24	$CH_3CH_2CH_2CH(CH_3)CH_2SiCl_2CH_3$ (73)	958
	Pt(II)complex	-		$CH_3CH_2CH_2CH(CH_3)CH_2SiCl_2CH_3$ (-)	J -009
	Pt/C + sonic waves	30	2	$CH_3CH_2CH_2CH(CH_3)CH_2SiCl_2CH_3$ (30)	467
C_6H_{12}	$CH_3CH_2CH_2CH{=}CHCH_3$				
	$NiCl_2$(DMPC)	120	40	$CH_3CH_2CH_2CH(CH_2CH_3)SiCl_2CH_3$ + $CH_3(CH_2)_3CH(CH_3)SiCl_2CH_3$ (38)	337
C_6H_{12}	$CH_3CH_2CH{=}CHCH_2CH_3$				
	$Co(CO)_8PtCl_2$	reflux	2 (THF)	$C_6H_{13}SiCl_2CH_3$ (64)	U -053
	$Co(CO)_8PtCl_2$	reflux	2 (THF)	$C_6H_{13}SiCl_2CH_3$ (64)	U -102
C_7H_{10}	Fig.1.				
	$[HPt(SiMe_2Bz)P(c\text{-}Hex)_3]_2$	20	5	Fig.89. (85)	428
	$[HPt(SiMe_2Bz)P(c\text{-}Hex)_3]_2$	20	5	Fig.89. (85)	429
C_7H_{14}	$CH_3(CH_2)_4CH{=}CH_2$				
	gamma (^{60}Co)	16	20	$CH_3(CH_2)_6SiCl_2CH_3$ (69)	958

$Ni(NCS)_2(PBu_3)_2$	120	-	$C_7H_{15}SiCl_2CH_3 + C_7H_{15}SiHClCH_3$ (kin.)	1109
$NiBr_2(PBu_3)_2$	120	-	$C_7H_{15}SiCl_2CH_3 + C_7H_{15}SiHClCH_3$ (kin.)	1109
$NiCl_2 + (Me_3SiCH_2)_3PO$	120	10	$CH_3(CH_2)_6SiCl_2CH_3 + CH_3(CH_2)_4CH(CH_3)SiCl_2CH_3$ (82)	1108
$NiCl_2 + Me(MeO)_2Si(CH_2)_nP(O)R_2$	120	10	$CH_3(CH_2)_6SiCl_2CH_3 + CH_3(CH_2)_4CH(CH_3)SiCl_2CH_3$ (53-67)	1108
$NiCl_2 + Me_2P(O)C{\equiv}CPh$	120	10	$CH_3(CH_2)_6SiCl_2CH_3$ (89) $C_7H_{15}SiHClCH_3$ (3)	S -048
$NiCl_2 + Me_3Si(CH_2)_3OP(O)$ (PhMe)	120	10	$CH_3(CH_2)_6SiCl_2CH_3 + CH_3(CH_2)_4CH(CH_3)SiCl_2CH_3$ (70)	1108
$NiCl_2 + Me_3Si(CH_2)_3P(O)Me_2$	120	10	$CH_3(CH_2)_6SiCl_2CH_3 + CH_3(CH_2)_4CH(CH_3)SiCl_2CH_3$ (76)	1108
$NiCl_2 + R_3PO$	-	-	$CH_3(CH_2)_6SiCl_2CH_3 + C_7H_{15}SiHClCH_3$ (14-84)	793
$NiCl_2(DMPF)$	120	20	$C_7H_{15}SiCl_2CH_3$ (28) $C_7H_{15}SiHClCH_3$ (55)	628
$NiCl_2(OPCy_3)_2$	120	10	$C_7H_{15}SiCl_2CH_3$ (25) $C_7H_{15}SiHClCH_3$ (10)	1109
$NiCl_2(PBu_3)_2$	120	-	$C_7H_{15}SiCl_2CH_3 + C_7H_{15}SiHClCH_3$ (kin.)	1109
$NiCl_2(PCy_3)_2$	120	10	$C_7H_{15}SiCl_2CH_3$ (18) $C_7H_{15}SiHClCH_3$ (12)	1109
$NiCl_2(PPh_3)_2$	120	10	$C_7H_{15}SiCl_2CH_3$ (5)	1109
$NiCl_2L_2$	120	10	$CH_3(CH_2)_6SiCl_2CH_3$ (9-74) $C_7H_{15}SiHClCH_3$ (2-6)	792
$NiCl_2[(Me_3SiCH_2)_3PO]_2$	120	10	$C_7H_{15}SiCl_2CH_3$ (37) $C_7H_{15}SiHClCH_3$ (37)	1109
$NiCl_2[OP(C_6H_4)_2Ph]_2$	120	10	$C_7H_{15}SiCl_2CH_3$ (42)	1109
$NiCl_2[OP(C_6H_4NMe_2\text{-}p)_3]_2$	120	10	$C_7H_{15}SiCl_2CH_3$ (18) $C_7H_{15}SiHClCH_3$ (11)	1109
$NiCl_2[P(C_6H_4NMe_2\text{-}p)_3]_2$	120	10	$C_7H_{15}SiCl_2CH_3$ (17) $C_7H_{15}SiHClCH_3$ (13)	1109
NiX_2L_2 L=PR_3,OPR_3, X=CN,NCO,NCS,NO_2	-	-	$CH_3(CH_2)_6SiCl_2CH_3 + C_7H_{15}SiHClCH_3$ (-)	790
NiX_2L_2 L=PR_3,OPR_3, X=CN,NCO,NCS,NO_2	-	-	$CH_3(CH_2)_6SiCl_2CH_3 + C_7H_{15}SiHClCH_3$ (-)	794
H_2PtCl_6	25	- (ht)	$CH_3(CH_2)_6SiCl_2H_3$ (-)	845
H_2PtCl_6 (EtOH)	50	2	$CH_3(CH_2)_6SiCl_2CH_3$ (-)	938
$Pt(C_2O_4)(PEt_3)_2$	-	0.7(MeCN)	$CH_3(CH_2)_6SiCl_2CH_3$ (kin.)	938

= CH$_3$SiHCl$_2$ =

=7=

Pt(C$_2$O$_4$)(PEt$_3$)$_2$	30	- (MeCN)	CH$_3$(CH$_2$)$_6$SiCl$_2$CH$_3$ (kin.)	938
Pt(PPh$_3$)$_4$	45-52	0.25	CH$_3$(CH$_2$)$_6$SiCl$_2$CH$_3$ (97)	364
PtCl$_3$(CH$_2$C≡CH)PPh$_3$	20-60	1-3	C$_7$H$_{15}$SiCl$_2$CH$_3$ (-)	171
PtCl$_3$(CH$_2$CH=CH$_2$)PPh$_3$	20-60	1-3	C$_7$H$_{15}$SiCl$_2$CH$_3$ (-)	171
PtCl$_4$(PPh$_2$C$_{12}$H$_{25}$)$_3$	20-60	1-3	C$_7$H$_{15}$SiCl$_2$CH$_3$ (-)	171
SiO$_2$-(CH$_2$)$_3$S/H$_2$PtCl$_6$	-	-	CH$_3$(CH$_2$)$_6$SiCl$_2$CH$_3$ (91)	506
SiO$_2$-L$_2$Pt(C$_2$O$_4$)	30	48	CH$_3$(CH$_2$)$_6$SiCl$_2$CH$_3$ (90)	938
SiO$_2$-L$_2$Pt(C$_2$O$_4$)	30	5 (h)	CH$_3$(CH$_2$)$_6$SiCl$_2$CH$_3$ (95)	938
SiO$_2$-L$_2$Pt(C$_2$O$_4$)	30	- (MeCN)	CH$_3$(CH$_2$)$_6$SiCl$_2$CH$_3$ (kin.)	938
SiO$_2$-L$_2$Pt(C$_2$O$_4$)	30	2	CH$_3$(CH$_2$)$_6$SiCl$_2$CH$_3$ (45)	938
SiO$_2$-L$_2$Pt(C$_2$O$_4$)	r.t.	15	CH$_3$(CH$_2$)$_6$SiCl$_2$CH$_3$ (100)	938
SiO$_2$-L$_2$Pt(C$_2$O$_4$)	r.t.	- (h)	CH$_3$(CH$_2$)$_6$SiCl$_2$CH$_3$ (-)	1203
SiO$_2$-PEt$_2$/Pt(C$_2$O$_4$)	-	- (h)	CH$_3$(CH$_2$)$_6$SiCl$_2$CH$_3$ (-)	939
SiO$_2$[OSi(CH$_2$)$_3$ER$_3$]$_2$PtCl$_6$;ER$_3$=NMe$_3$,PPh$_3$	18-40	-	C$_7$H$_{15}$SiHClCH$_3$ (kin.)	1106
polymer-(CH$_2$)$_n$S/PtCl$_x$	60	10	CH$_3$(CH$_2$)$_6$SiCH$_3$Cl$_2$ (91)	505

C$_8$H$_8$ C$_6$H$_5$CH=CH$_2$

Co$_2$(CO)$_8$	20-30	70	C$_6$H$_5$CH$_2$CH$_2$SiCl$_2$CH$_3$ (-)	715
anionite- Co salt	60	20	C$_6$H$_5$CH$_2$CH$_2$SiCl$_2$CH$_3$ (8-14)	828
anionite/[Co(NH$_3$)$_5$Cl]$^{+2}$	90	10	C$_6$H$_5$CH$_2$CH$_2$SiCl$_2$CH$_3$ (43) C$_6$H$_5$CH(CH$_3$)SiCl$_2$CH$_3$ (47)	S -033
anionite- Ir salt	60	10	C$_6$H$_5$CH$_2$CH$_2$SiCl$_2$CH$_3$ (41-71)	828
Ni(CO)$_4$	reflux	4	C$_6$H$_5$CH(CH$_3$)SiCl$_2$CH$_3$ (88)	213
Ni(cpd)$_2$	20-50	-	C$_6$H$_5$CH(CH$_3$)SiCl$_2$CH$_3$ (89)	978
Ni(cpd)$_2$ + PPh$_3$	50	-	C$_6$H$_5$CH$_2$CH$_2$SiCl$_2$CH$_3$ + C$_6$H$_5$CH(CH$_3$)SiCl$_2$CH$_3$ (62-74)	791
NiCl$_2$ + (Me$_3$SiCH$_2$)$_3$PO	100	10	C$_6$H$_5$CH$_2$CH$_2$SiCl$_2$CH$_3$ + C$_6$H$_5$CH(CH$_3$)SiCl$_2$CH$_3$ + C$_6$H$_5$CH$_2$CH$_2$SiHClCH$_3$ (100)	S -048
NiCl$_2$(BMPP)	120	12	C$_6$H$_5$CH$_2$CH$_2$SiCl$_2$CH$_3$ +C$_6$H$_5$CH(CH$_3$)SiCl$_2$CH$_3$ (24)	1302
NiCl$_2$(DMPC)	120	40	C$_6$H$_5$CH$_2$CH$_2$SiCl$_2$CH$_3$ +C$_6$H$_5$CH(CH$_3$)SiCl$_2$CH$_3$ (54)	337

Table 2.1
= CH_3SiHCl_2 = =8=

$NiCl_2$(DMPF)	120	9 days	$C_6H_5CH_2CH_2SiCl_2CH_3$ (6) $C_6H_5CH(CH_3)SiCl_2CH_3$ (32) $C_6H_5CH_2CH_2SiHClCH_3$ (17)	589
$[Ni(CO)(cpd)]_2$	20	2.5-3	$C_6H_5CH(CH_3)SiCl_2CH_3$ (3)	565
$[NiCl(\pi\text{-}C_4H_7)]_2$	20	-	$C_6H_5CH(CH_3)SiCl_2CH_3$ (3-11)	791
$[NiCl(\pi\text{-}C_5H_9)]_2$	20-60	-	$C_6H_5CH(CH_3)SiCl_2CH_3$ (2-23)	791
$[NiCl(\pi\text{-}C_5H_9)]_2 + PPh_3$	20-60	-	$C_6H_5CH_2CH_2SiCl_2CH_3$ + $C_6H_5CH(CH_3)SiCl_2CH_3$ (9-34)	791
anionite- Ni salt	60	20	$C_6H_5CH_2CH_2SiCl_2CH_3$ (16)	828
anionite/Ni^{+2}	70	15	$C_6H_5CH_2CH_2SiCl_2CH_3$ (41) $C_6H_5CH(CH_3)SiCl_2CH_3$ (44)	S -033
anionite- Os salt	60	20	$C_6H_5CH_2CH_2SiCl_2CH_3$ (2)	828
anionite- Pd salt	60	5-10	$C_6H_5CH_2CH_2SiCl_2CH_3$ (3-54)	828
SiO_2-$SiMe_2(CH_2)_3NMe_3/PtCl_4^{-2}$	27	1	$C_6H_5CH_2CH_2SiCl_2CH_3$ (kin.)	172
SiO_2-$SiMe_2(CH_2)_3PPh_3/PtCl_6^{-2}$	27	3	$C_6H_5CH_2CH_2SiCl_2CH_3$ (kin.)	172
SiO_2-$^+PPh_3/PtBr_4^-$	20	10	$C_6H_5CH_2CH_2SiCl_2CH_3$ (80)	S -036
$[PtCl_2(XC_6H_4CH{=}CH_2)]_2$	-	-	$C_6H_5C_2H_4SiCl_2CH_3$ (kin.)	972
$(Bu_4N)[PtCl_3(PhCH{=}CH_2)]$	r.t	0.22	$C_6H_5CH_2CH_2SiCl_2CH_3$ (96) $C_6H_5CH(CH_3)CH_2SiCl_2CH_3$ (-)	220
$(HexMe_3N)_2PtCl_4$	60	1	$C_6H_5CH_2CH_2SiCl_2CH_3$ (93)	172
$(MePh_3P)_2PtCl_4$	20	20	$C_6H_5CH_2CH_2SiCl_2CH_3$ (95)	172
H_2PtCl_6	-	-	$C_6H_5CH_2CH_2SiCl_2CH_3$ (-)	971
H_2PtCl_6	20	-	$C_6H_5CH_2CH_2SiCl_2CH_3$ (56) $C_6H_5CH(CH_3)SiCl_2CH_3$ (24)	172
H_2PtCl_6	30	-	$C_6H_5C_2H_4SiCl_2CH_3$ (kin.)	795
H_2PtCl_6	5-20	- (solv.)	$C_6H_5C_2H_4SiCl_2CH_3$ (kin.)	966
H_2PtCl_6	80	2	$C_6H_5CH_2CH_2SiCl_2CH_3$ (64) $C_6H_5CH(CH_3)SiCl_2CH_3$ (35)	172
H_2PtCl_6 ($MeOC_2H_4OMe$ + EtOH + $\overline{C_6H_4NHC_6H_4S}$)	20	1-2 days	$C_6H_5CH_2CH_2SiCl_2CH_3$ (5-65)	U -085
H_2PtCl_6 ($MeOC_2H_4OMe$ + EtOH + $\overline{C_6H_4NHC_6H_4S}$)	20	300 s	$C_6H_5CH_2CH_2SiCl_2CH_3/C_6H_5CH(CH_3)SiCl_2CH_3$ (94/6)	U -085

Table 2.1

= CH_3SiHCl_2 =

=9=

H_2PtCl_6 ($MeOC_2H_4OMe$ + EtOH + $Ph_2NC_6H_4NH_2$-p)	20	1-2	$C_6H_5CH_2CH_2SiCl_2CH_3$ (15-86)	U -085
H_2PtCl_6 ($MeOC_2H_4OMe$ + EtOH + $Ph_2NC_6H_4NH_2$-p)	45	1-2	$C_6H_5CH_2CH_2SiCl_2CH_3$ (99)	U -085
H_2PtCl_6 ($MeOC_2H_4OMe$ + EtOH + $Ph_2NC_6H_4NH_2$-p)	60	1-2	$C_6H_5CH_2CH_2SiCl_2CH_3$ (96) $C_6H_5CH(CH_3)SiCl_2CH_3$ (4)	U -085
H_2PtCl_6 ($MeOC_2H_4OMe$ + EtOH)	20	240 s	$C_6H_5CH_2CH_2SiCl_2CH_3$/$C_6H_5CH(CH_3)SiCl_2CH_3$ (75/25)	U -085
H_2PtCl_6 ($MeOC_2H_4OMe$ + EtOH)	20	2 days	$C_6H_5CH_2CH_2SiCl_2CH_3$ +$C_6H_5CH(CH_3)SiCl_2CH_3$ (20)	U -085
H_2PtCl_6 ($MeOC_2H_4OMe$ + EtOH)	70	240 s	$C_6H_5CH_2CH_2SiCl_2CH_3$/$C_6H_5CH(CH_3)SiCl_2CH_3$ (65/35)	U -085
H_2PtCl_6 (i-PrOH)	80	2	$C_6H_5CH_2CH_2SiCl_2CH_3$ (64) $C_6H_5CH(CH_3)SiCl_2CH_3$ (34)	213
H_2PtCl_6 (i-PrOH)	reflux	-	$C_6H_5CH_2CH_2SiCl_2CH_3$ (6) $C_6H_5CH(CH_3)SiCl_2CH_3$ (4)	U -178
H_2PtCl_6 (i-PrOH) + BzCN	80	2	$C_6H_5CH_2CH_2SiCl_2CH_3$ (66-67) $C_6H_5CH(CH_3)SiCl_2CH_3$ (32-33)	213
H_2PtCl_6 (i-PrOH) + C_5H_5N	80	2	$C_6H_5CH_2CH_2SiCl_2CH_3$ (69-72) $C_6H_5CH(CH_3)SiCl_2CH_3$ (27-30)	213
H_2PtCl_6 (i-PrOH) + $MeCOONMe_2$	reflux	-	$C_6H_5CH_2CH_2SiCl_2CH_3$ (9) $C_6H_5CH(CH_3)SiCl_2CH_3$ (1)	U -178
H_2PtCl_6 (i-PrOH) + NMe_2Bz	80	2	$C_6H_5CH_2CH_2SiCl_2CH_3$ (89-95) $C_6H_5CH(CH_3)SiCl_2CH_3$ (4-5)	213
H_2PtCl_6 (i-PrOH) + PPh_3	80	2	$C_6H_5CH_2CH_2SiCl_2CH_3$ (69)	213
H_2PtCl_6 (i-PrOH) + PPh_3	80	2	$C_6H_5CH_2CH_2SiCl_2CH_3$ (93-99) $C_6H_5CH(CH_3)SiCl_2CH_3$ (0-5)	213
$KPtCl_3(C_2H_4)$	5	- (silane)	$C_6H_5CH_2CH_2SiCl_2CH_3$ (kin.)	966
$KPtCl_3$(DMSO)	20	-	$C_6H_5C_2H_4SiCl_2CH_3$ (kin.)	68
$K[PtCl_3(CH_2{=}CH_2)]$	r.t	0.05	$C_6H_5CH_2CH_2SiCl_2CH_3$ (63) $C_6H_5CH(CH_3)CH_2SiCl_2CH_3$ (37)	220
$Pt(PPh_3)_4$	15	- (solv.)	$C_6H_5C_2H_4SiCl_2CH_3$ (kin.)	977
$Pt(PPh_3)_4$	15	1.5-4	$C_6H_5C_2H_4SiCl_2CH_3$ (kin.)	576
$Pt(PPh_3)_4$	45-52	12	$C_6H_5CH_2CH_2SiCl_2CH_3$ (97)	364
$Pt(PPh_3)_4$ + PPh_3	15	4	$C_6H_5C_2H_4SiCl_2CH_3$ (kin.)	576
Pt/C + sonic waves	30	1.5	$C_6H_5CH_2CH_2SiCl_2CH_3$ (94)	467

Table 2.1
= CH_3SiHCl_2 =

=10=

$PtCl_2(PBu_3)_2$	120	2	$C_6H_5CH_2CH_2SiCl_2CH_3/C_6H_5CH(CH_3)SiCl_2CH_3$ (97/3)	G -008
$PtCl_2(PPh_3)_2$	50	1.5	$C_6H_5C_2H_4SiCl_2CH_3$ (kin.)	577
$PtCl_2(PPh_3)_2$	50	-(solv.)	$C_6H_5C_2H_4SiCl_2CH_3$ (kin.)	976
$PtCl_3[(CH_2C{\equiv}CH)PPh_3]$	22	0.5	$C_6H_5CH_2CH_2SiCl_2CH_3$ (90)	172
$PtCl_3[(CH_2CH{=}CH_2)PPh_3]$	22	1	$C_6H_5CH_2CH_2SiCl_2CH_3$ (90)	172
$PtCl_5[(CH_2CH{=}CH_2)PPh_3]$	22	20	$C_6H_5CH_2CH_2SiCl_2CH_3$ (90)	172
$PtH(PPh_3)_3$	50	-(solv.)	$C_6H_5C_2H_4SiCl_2CH_3$ (kin.)	976
$PtHCl(PPh_3)_2$	50	-(solv.)	$C_6H_5C_2H_4SiCl_2CH_3$ (kin.)	976
SiO_2-O-$[Si(CH_2)_3ER_3]_2PtCl_6$ $ER_3{=}NMe_3$, PBu_3, PPh_3, NEt_3, NBu_3	18-40	-	$C_6H_5C_2H_4SiCl_2CH_3$ (kin.)	1106
SiO_2-$PPh_3/PtCl_6^{-2}$	20	-	$C_6H_5CH_2CH_2SiCl_2CH_3$ (80)	172
$[Pt(\pi\text{-All})(PPh_3)_2](ClO_4)$	50	-(solv.)	$C_6H_5C_2H_4SiCl_2CH_3$ (kin.)	976
$[PtCl(NH_3)_2(C_2H_4)]NO_3$	20	-	$C_6H_5C_2H_4SiCl_2CH_3$ (kin.)	68
$[PtCl_2(C_2H_4)]_2$ + DMSO	-	-	$C_6H_5C_2H_4SiCl_2CH_3$ (kin.)	68
$[PtCl_2(C_8H_8)]_2$ + DMSO	85	-	$C_6H_5C_2H_4SiCl_2CH_3$ (kin.)	68
$[PtCl_2(PhCH{=}CH_2)]_2$	20	-	$C_6H_5C_2H_4SiCl_2CH_3$ (kin.)	68
$[PtCl_2(PhCH{=}CH_2)]_2$	5-20	until 24	$C_6H_5CH_2CH_2SiCl_2CH_3$ (kin.)	965
$[PtCl_2(PhCH{=}CH_2)]_2$	5-20	-	$C_6H_5C_2H_4SiCl_2CH_3$ (kin.)	967
$[PtCl_2(PhCH{=}CH_2)]_2$	5-20	-	$C_6H_5C_2H_4SiCl_2CH_3$ (kin.)	968
$[PtCl_2(PhCH{=}CH_2)]_2$ + C_5H_5N	20-45	-	$C_6H_5C_2H_4SiCl_2CH_3$ (kin.)	68
$[PtCl_2(PhCH{=}CH_2)]_2$ + PPh_3	20-45	-	$C_6H_5C_2H_4SiCl_2CH_3$ (kin.)	68
$[PtCl_2(XC_6H_4CH{=}CH_2)]_2$	-	-	$C_6H_5C_2H_4SiCl_2CH_3$ (kin.)	68
$[PtH(BzSiMe_2)P(c\text{-}Hex)_3]_2$	20	1	$C_6H_5CH_2CH_2SiCl_2CH_3$ (96)	429
$[PtH(CO)(PPh_3)_2][ClO_4]$	50	-(solv.)	$C_6H_5C_2H_4SiCl_2CH_3$ (kin.)	976
$[PtH(MeSiCl_2)(PPh)_3)]_2$	50	1.5	$C_6H_5C_2H_4SiCl_2CH_3$ (kin.)	577
anionite- Pt salt	60	2-10	$C_6H_5CH_2CH_2SiCl_2CH_3$ (67-99)	828
anionite/K_2PtCl_4	60	7	$C_6H_5CH_2CH_2SiCl_2CH_3$ (83) $C_6H_5CH(CH_3)SiCl_2CH_3$ (16)	172

	cis-$PtCl_2(PhCH{=}CH_2)$	r.t	0.05	$C_6H_5CH_2CH_2SiCl_2CH_3$ (60) $C_6H_5CH(CH_3)SiCl_2CH_3$ (40)	220
	glass-$^{+}NEt_3HPtCl_6^{-}$	20	10	$C_6H_5CH_2CH_2SiCl_2CH_3$ (91-98)	S -039
	glass-$^{+}NEt_3HPtCl_6^{-}$	20	10	$C_6H_5CH_2CH_2SiCl_2CH_3$ (72-76)	S -039
	glass-$^{+}NEt_3HPtCl_6^{-}$	20	10	$C_6H_5CH_2CH_2SiCl_2CH_3$ (72-76)	S -036
	polymer-$(CH_2)_nS/PtCl_x$	-	-	$C_6H_5CH_2CH_2SiCl_2CH_3$ (52) $C_6H_5CH(CH_3)SiCl_2CH_3$ (24)	505
	$Rh_4(CO)_{12}$	20-30	70	$C_6H_5CH_2CH_2SiCl_2CH_3$ (-)	715
	anionite- Rh salt	60	10	$C_6H_5CH_2CH_2SiCl_2CH_3$ (63-93)	828
	anionite- Ru salt	60	20	$C_6H_5CH_2CH_2SiCl_2CH_3$ (4)	828
C_8H_{14}	$\overline{(CH_2)_6CH{=}CH}$				
	H_2PtCl_6 (2-EtC_6H_{10}OH) + UV light	70	10	$\overline{(CH_2)_7CH}SiCl_2CH_3$ (90)	E -038
C_8H_{16}	$(CH_3)_3CCH_2C(CH_3){=}CH_2$				
	H_2PtCl_6 ($MeOC_2H_4OMe$)	60	-	$(CH_3)_3CCH_2CH(CH_3)CH_2SiCl_2CH_3$ (-)	351
C_8H_{16}	$(CH_3)_3CCH_2C(CH_3){=}CH_2$ + $(CH_3)_3CCH{=}C(CH_3)_2$				
	H_2PtCl_6 (i-PrOH)	reflux	4.5(b)	$(CH_3)_3CCH_2CH(CH_3)CH_2SiCl_2CH_3$ (-)	G -002
	$[PtCl_2(c\text{-}C_6H_{10})]_2$	reflux	4.5	$(CH_3)_3CCH_2CH(CH_3)CH_2SiCl_2CH_3$ (-)	G -002
C_8H_{16}	$CH_3(CH_2)_5CH{=}CH_2$				
	UV + $(t\text{-}BuO)_2$	40	14	$C_8H_{17}SiCl_2CH_3$ (23)	338
	anionite/Fe^{+2}	110	30	$CH_3(CH_2)_7SiCl_2CH_3$ (48)	S -033
	$Ni(C_2H_4)(PPh_3)_2$	120	20	$CH_3(CH_2)_5CH(CH_3)SiCl_2CH_3$ (5)	588
	$Ni(C_2H_4)$(diphos)	120-135	20-40	$CH_3(CH_2)_5CH(CH_3)SiCl_2CH_3$ + $CH_3(CH_2)_7SiHClCH_3$ (72-79)	588
	$Ni(C_2H_4)$(diphos)	90	65	$CH_3(CH_2)_5CH(CH_3)SiCl_2CH_3$ + $CH_3(CH_2)_7SiHClCH_3$ (84)	588
	Ni(cpd)(bipy) + PPh_3	120	10	$C_8H_{17}SiCl_2CH_3$ + $CH_3(CH_2)_7SiHClCH_3$ (40-83)	791
	$Ni(cpd)_2$	20-50	-	$CH_3(CH_2)_7SiCl_2CH_3$ (26)	978
	$NiBr_2(PEt_3)_2$	120	20	$C_8H_{17}SiCl_2CH_3$ + $CH_3(CH_2)_7SiHClCH_3$ (59)	589
	$NiBr_2(PPhEt_2)_2$	120	-	$C_8H_{17}SiCl_2CH_3$ + $CH_3(CH_2)_7SiHClCH_3$ (kin.)	1109
	$NiBr_2(PhBu_2PO)_2$	120	10	$C_8H_{17}SiCl_2CH_3$ + $CH_3(CH_2)_7SiHClCH_3$ (96)	S -048

$NiBr_2$(diphos)	120	-	$C_8H_{17}SiCl_2CH_3$ + $CH_3(CH_2)_7SiHClCH_3$ (kin.)	1109
$NiCl_2$	100	20	$CH_3(CH_2)_7SiCl_2CH_3$ + $CH_3(CH_2)_5CH(CH_3)SiCl_2CH_3$ (8)	S -033
$NiCl_2(Bu_3PO)_2$	120	10	$C_8H_{17}SiCl_2CH_3$ + $CH_3(CH_2)_7SiHClCH_3$ (94)	589
$NiCl_2$(DMPC)	120	20	$CH_3(CH_2)_5CH(CH_3)SiCl_2CH_3$ + $C_8H_{17}SiCl_2CH_3$ (59-97)	337
$NiCl_2$(DMPF)	120	20	$C_8H_{17}SiCl_2CH_3$ + $CH_3(CH_2)_7SiHClCH_3$ (95)	628
$NiCl_2$(DMPF)	120	20	$C_8H_{17}SiCl_2CH_3$ + $CH_3(CH_2)_7SiHClCH_3$ (95)	589
$NiCl_2(Me_2PCH_2CH_2PMe_2)$	120-135	20-40	$C_8H_{17}SiCl_2CH_3$ + $CH_3(CH_2)_7SiHClCH_3$ (79-87)	589
$NiCl_2(PBu_3)_2$	120	20	$C_8H_{17}SiCl_2CH_3$ + $CH_3(CH_2)_7SiHClCH_3$ (44)	589
$NiCl_2(PEt_3)_2$	120	20	$C_8H_{17}SiCl_2CH_3$ + $CH_3(CH_2)_7SiHClCH_3$ (61)	589
$NiCl_2(PMe_2Ph)_2$	120	20	$C_8H_{17}SiCl_2CH_3$ + $CH_3(CH_2)_7SiHClCH_3$ (77)	589
$NiCl_2(PMe_3)_2$	120	20	$C_8H_{17}SiCl_2CH_3$ + $CH_3(CH_2)_7SiHClCH_3$ (78)	589
$NiCl_2(PPh_3)_2$	120	20	$CH_3(CH_2)_5CH(CH_3)SiCl_2CH_3$ (3-5)	589
$NiCl_2(Ph_2PCH{=}CHPPh_2)$	135	20	$C_8H_{17}SiCl_2CH_3$ + $CH_3(CH_2)_7SiHClCH_3$ (84)	589
$NiCl_2$(diphos)	-	-	$C_8H_{17}SiCl_2CH_3$ (-)	577
$NiCl_2$(diphos)	120	-	$C_8H_{17}SiCl_2CH_3$ + $CH_3(CH_2)_7SiHClCH_3$ (kin.)	1109
$NiCl_2$(diphos)	120	40	$CH_3(CH_2)_5CH(CH_3)SiCl_2CH_3$ + $CH_3(CH_2)_7SiHClCH_3$ (75)	588
$NiCl_2$(diphos)	120-135	20-40	$C_8H_{17}SiCl_2CH_3$ + $CH_3(CH_2)_7SiHClCH_3$ (75-98)	589
$NiCl_2[(Me_3SiCH_2)_3PO]_2$	120	10	$C_8H_{17}SiCl_2CH_3$ + $CH_3(CH_2)_7SiHClCH_3$ (95)	S -048
$NiCl_2[(Me_3SiCH_2)_3PO]_2$	120	10	$C_8H_{17}SiCl_2CH_3$(50) $CH_3(CH_2)_7SiHClCH_3$ (17)	1109
$NiCl_2[Ph_2P(CH_2)_3PPh_2]$	135	20	$C_8H_{17}SiCl_2CH_3$ + $CH_3(CH_2)_7SiHClCH_3$ (94)	589
NiI_2(diphos)	-	-	$C_8H_{17}SiCl_2CH_3$ (-)	577
NiI_2(diphos)	120	-	$C_8H_{17}SiCl_2CH_3$ + $CH_3(CH_2)_7SiHClCH_3$ (kin.)	1109
$NiX_2(PR_3)_2$	-	-	$CH_3(CH_2)_7SiCl_2CH_3$ + $CH_3(CH_2)_7SiHClCH_3$ (-)	J -006
NiX_2L_2	120	10	$CH_3(CH_2)_7SiCl_2CH_3$ + $CH_3(CH_2)_7SiHClCH_3$ (2-96)	793
anionite/Ni^{+2}	110	20	$CH_3(CH_2)_7SiCl_2CH_3$ (50)	S -033
$Pd(PPh_3)_4$	120	6	$CH_3(CH_2)_7SiCl_2CH_3$ (90)	1208

	H_2PtCl_6 (THF) + Ph_2SiH_2	reflux	4	$CH_3(CH_2)_7SiCl_2CH_3$ (89)	C -041
	H_2PtCl_6 (i-PrOH)	25	15	$CH_3(CH_2)_7SiCl_2CH_3$ (75)	748
	H_2PtCl_6 (i-PrOH)	25	6	$CH_3(CH_2)_7SiCl_2CH_3$ (85)	748
	H_2PtCl_6 (i-PrOH)	50	4	$CH_3(CH_2)_7SiCl_2CH_3$ (62)	748
	H_2PtCl_6 (i-PrOH)	50	2	$CH_3(CH_2)_7SiCl_2CH_3$ (65)	748
	H_2PtCl_6 (i-PrOH)	50	6	$CH_3(CH_2)_7SiCl_2CH_3$ (71)	748
	$Pt(CH_2{=}CH_2)(PPh_3)_2$	25	15	$CH_3(CH_2)_7SiCl_2CH_3$ (24)	748
	$Pt(CH_2{=}CH_2)(PPh_3)_2$	25	30	$CH_3(CH_2)_7SiCl_2CH_3$ (63)	748
	$Pt(CH_2{=}CH_2)(PPh_3)_2$	25	30	$CH_3(CH_2)_7SiCl_2CH_3$ (63)	Pl-030
	$Pt(CH_2{=}CH_2)(PPh_3)_2$	50	6	$CH_3(CH_2)_7SiCl_2CH_3$ (79)	Pl-030
	$Pt(CH_2{=}CH_2)(PPh_3)_2$	50	4	$CH_3(CH_2)_7SiCl_2CH_3$ (75)	748
	$Pt(CH_2{=}CH_2)(PPh_3)_2$	50	6	$CH_3(CH_2)_7SiCl_2CH_3$ (79)	748
	$Pt(CH_2{=}CH_2)(PPh_3)_2$	50	2	$CH_3(CH_2)_7SiCl_2CH_3$ (72)	748
	$Pt(CH_2{=}CH_2)(PPh_3)_2$	80	4	$CH_3(CH_2)_7SiCl_2CH_3$ (80)	Pl-030
	Pt/C	-	-	$CH_3(CH_2)_7SiCl_2CH_3$ (-)	877
	$PtCl_2(PBu_3)_2$	25-150	15	$C_8H_{17}SiCl_2CH_3$ (1-100)	G -008
	$PtCl_2(PPh_3)_2$	80	2	$CH_3(CH_2)_7SiCl_2CH_3$ (76)	Pl-031
	$PtCl_2(PhCN)_2$	25	15	$CH_3(CH_2)_7SiCl_2CH_3$ (75)	Pl-031
	$PtCl_2(SEt_2)_2$	reflux	1 (t)	$CH_3(CH_2)_7SiCl_2CH_3$ (-)	F -002
	$Pt[PhC{\equiv}CC(Me)Ph(OH)]_2$	110	0.5	$CH_3(CH_2)_7SiCl_2CH_3$ (84)	E -014
	$RhCl(PPh_3)_3$	120	6	$C_8H_{17}SiCl_2CH_3$ (42)	Pl-011
	$RhH(PPh_3)_4$	-	-	$CH_3(CH_2)_7SiCl_2CH_3$ (-)	J -019
C_9H_{10}	$C_6H_5C(CH_3){=}CH_2$				
	$NiCl_2(BMPP)_2$	90	60	$C_6H_5CH(CH_3)CH_2SiCl_2CH_3$ (31) $C_6H_5CH(CH_3)CH_2SiHClCH_3$ (8)	1302
	$NiCl_2(BMPP)_2$	90	60	$C_6H_5CH(CH_3)CH_2SiCl_2CH_3$ (31) $C_6H_5CH(CH_3)CH_2SiHClCH_3$ (8)	1306
	$(PtCl_2BMPP)_2$	40	24	$C_6H_5CH(CH_3)CH_2SiCl_2CH_3$ (56)	1303

	$(PtCl_2BMPP)_2$	r.t.	-	$C_6H_5CH(CH_3)CH_2SiCl_2CH_3$ (-)	1299
	$(PtCl_2MDPP)_2$	120	60	$C_6H_5CH(CH_3)CH_2SiCl_2CH_3$ (33)	1303
	$(PtCl_2MPPP)_2$	40	24	$C_6H_5CH(CH_3)CH_2SiCl_2CH_3$ (64)	1303
	$(PtCl_2PMePrPh)_2$	40	40	$C_6H_5CH(CH_3)CH_2SiCl_2CH_3$ (47)	1301
	$H_2PtCl_4(L)(MeCH=CHMe)$	135	49	$C_6H_5CH(CH_3)CH_2SiCl_2CH_3$ (-)	J -009
	H_2PtCl_6	-	-	$C_6H_5CH(CH_3)CH_2SiCl_2CH_3$ (-)	971
	$H_2PtCl_6[(\overline{CH_2)_5CO}]$	120-140	-	$C_6H_5CH(CH_3)CH_2SiCl_2CH_3$ (-)	D -103
	$PtCl_2(BMPP)(C_2H_4)$	40	40	$C_6H_5CH(CH_3)CH_2SiCl_2CH_3$ (43)	1301
C_9H_{10}	$C_6H_5CH_2CH=CH_2$				
	$NiCl_2 + [(Me_3SiCH_2)_3PO]$	120	20	$C_6H_5(CH_2)_3SiCl_2CH_3$ (48) $C_6H_5(CH_2)_3SiHClCH_3$ (23)	S -048
	$NiCl_2[(Me_3SiCH_2)_3PO]_2$	120	10	$C_6H_5(CH_2)_3SiCl_2CH_3$ (35) $C_6H_5(CH_2)_3SiHClCH_3$ (25)	1109
	H_2PtCl_6 (i-PrOH) + PPh_3	reflux	3 (b)	$C_6H_5(CH_2)_3SiCl_2CH_3$ (72)	1218
C_9H_{10}	$p-CH_3C_6H_4CH=CH_2$				
	H_2PtCl_6	5-20	24	$p-CH_3C_6H_4C_2H_4SiCl_2CH_3$ (85-90)	966
	$KPtCl_3(CH_2=CH_2)$	5	-	$p-CH_3C_6H_4CH_2CH_2SiCl_2CH_3$ (kin.)	966
	$Pt(PPh_3)_4$	15	4	$p-CH_3C_6H_4CH_2CH_2SiCl_2CH_3$ + $p-CH_3C_6H_4CH(CH_3)SiCl_2CH_3$ (kin.)	576
	$PtCl_2(PhCH=CH_2)_2$	5-20	24	$p-CH_3C_6H_4C_2H_4SiCl_2CH_3$ (kin.)	965
C_9H_{18}	$CH_3(CH_2)_6CH=CH_2$				
	H_2PtCl_6	20-40	2	$CH_3(CH_2)_8SiCl_2CH_3$ (kin.)	341
$C_{10}H_{16}$	Fig.15.				
	H_2PtCl_6	-	-	Fig.90. (-)	1283
$C_{10}H_{16}$	Fig.91.				
	H_2PtCl_6	-	-	Fig.92. (75)	1181
$C_{10}H_{20}$	$CH_3(CH_2)_7CH=CH_2$				
	Pt complex	-	-	$CH_3(CH_2)_9SiCl_2CH_3$ (-)	J -011
	H_2PtCl_6 (i-PrOH)	100-120	6	$CH_3(CH_2)_9SiCl_2CH_3$ (94)	D -118

	Pt/C	-	-	$CH_3(CH_2)_9SiCl_2CH_3$ (-)	877
$C_{12}H_{10}$	Fig.93.				
	H_2PtCl_6	reflux	2 (b)	Fig.94. (80)	638
$C_{12}H_{24}$	$CH_3(CH_2)_9CH=CH_2$				
	polymer-$(CH_2)_nS/PtCl_x$	60	10	$CH_3(CH_2)_{11}SiCl_2CH_3$ (93)	505
$C_{14}H_{12}$	$(C_6H_5)_2C=CH_2$				
	H_2PtCl_6	reflux	-	$(C_6H_5)_2CHCH_2SiCl_2CH_3$ (63)	833
$C_{14}H_{12}$	$C_6H_5CH=CHC_6H_5$				
	H_2PtCl_6	20	60 days	$C_6H_5CH_2CH(C_6H_5)SiCl_2CH_3$ (53)	833
	$Rh_4(CO)_{12}$	30-40	1	$C_6H_5CH_2CH(C_6H_5)SiCl_2CH_3$ (96-98)	S -023
$C_{14}H_{28}$	$C_{12}H_{25}CH=CH_2$				
	$PtCl_2$ (c-C_6H_{10})	reflux	2	$C_{14}H_{29}SiCl_2CH_3$ (-)	G -012
$C_{18}H_{36}$	$C_{16}H_{33}CH=CH_2$				
	H_2PtCl_6	100-120	3.5	$C_{18}H_{37}SiCl_2CH_3$ (90)	D -021
	H_2PtCl_6 (i-PrOH)	100	14	$C_{18}H_{37}SiCl_2CH_3$ (83)	C -032
$C_{18}H_{36}$	$CH_3(CH_2)_{15}CH=CH_2$				
	H_2PtCl_6 (i-PrOH)	80	6	$CH_3(CH_2)_{17}SiCl_2CH_3$ (52)	748
	$Pt(PPh_3)_2(CH_2=CH_2)$	80	6	$CH_3(CH_2)_{17}SiCl_2CH_3$ (77)	748
	$Pt(PPh_3)_4$	80	6	$CH_3(CH_2)_{17}SiCl_2CH_3$ (54)	748
	$PtCl_2(PhCN)_2$	80	6	$CH_3(CH_2)_{17}SiCl_2CH_3$ (81)	Pl-031

Table 2.2

= CH_3SiHCl_2 = =1=

C_4H_6	$CH_2=CHCH=CH_2$				
	M + $Al(i\text{-}Bu)_3$	-	-	$CH_3CH=CHCH_2SiCl_2CH_3$ + $CH_3CH_2CH=CHSiCl_2CH_3$ + $CH_3Cl_2SiCH_2CH=CHCH_2SiCl_2CH_3$ (-)	7
	M + $AlEt_3$	-	-	$CH_3CH=CHCH_2SiCl_2CH_3$ + $CH_3CH_2CH=CHSiCl_2CH_3$ + $CH_3Cl_2SiCH_2CH=CHCH_2SiCl_2CH_3$ (-)	7
	$Ni(PPh_3)_4$	120	3	$C_4H_7SiCl_2CH_3$ (38)	C -018
	$Ni(acac)_2$ + PPh_3	100	6	$CH_3CH=CHCH_2SiCl_2CH_3$ (63) $CH_3Cl_2SiCH_2CH=CHCH_2SiCl_2CH_3$ (7)	1018
	$Ni(acac)_2$ + PPh_3 + $Al(i\text{-}Bu)_3$	100	6	$CH_3CH=CHCH_2SiCl_2CH_3$ (18) $CH_3Cl_2SiCH_2CH=CHCH_2SiCl_2CH_3$ (42)	1018
	$Ni(acac)_2$ + PPh_3 + $Al(i\text{-}Bu)_3$	20	12	$CH_3CH=CHCH_2SiCl_2CH_3$ (54) $CH_3Cl_2SiCH_2CH=CHCH_2SiCl_2CH_3$ (6)	1018
	$Ni(acac)_2$ + PPh_3 + $AlEt_3$	100	5	$CH_3CH=CHCH_2SiCl_2CH_3$ (14) $CH_3Cl_2SiCH_2CH=CHCH_2SiCl_2CH_3$ (20)	1018
	$Ni(acac)_2$ + PPh_3 + $AlEt_3$	20	12	$CH_3CH=CHCH_2SiCl_2CH_3$ (40) $CH_3Cl_2SiCH_2CH=CHCH_2SiCl_2CH_3$ (10)	1018
	$Ni(acac)_2$ + PPh_3 + $Na[AlH_2(OC_2H_4OMe)_2]$	82	5	$C_4H_7SiCl_2CH_3$ (57)	C -010
	$Ni(cod)_2$	80	1.5	$CH_3CH=CHCH_2SiCl_2CH_3$ (93)	205
	$NiCl_2(PPh_3)_2$	120	3	$C_4H_7SiCl_2CH_3$ (65)	D -070
	Pd + PPh_3	100	6	$CH_3CH=CHCH_2SiCl_2CH_3$ (89)	D -027
	$Pd(PPh_3)_4$	110	6	$CH_3CH=CHCH_2SiCl_2CH_3$ (83) $CH_3CH=CHCH_2CH_2CH=CHCH_2SiCl_2CH_3$ (5)	1208
	$Pd(PPh_3)_4$	80	2	$C_4H_7SiCl_2CH_3$ (8)	C -017
	$PdCl_2(PhCN)_2$	(-15)-10	-	$CH_3CH=CHCH_2SiCl_2CH_3$ (62)	D -060
	$PdCl_2(PhCN)_2$	0-20	- (t)	$CH_3CH=CHCH_2SiCl_2CH_3$ (62)	U -080
	$[PdCl(\pi\text{-}All)]_2$	80	2	$C_4H_7SiCl_2CH_3$ (67)	D -060
	$[PdCl(\pi\text{-}All)]_2$	80	2	$C_4H_7SiCl_2CH_3$ (67)	U -080
	H_2PtCl_6 (i-PrOH)	200	4	$CH_3CH=CHCH_2SiCl_2CH_3$ (18-47) $CH_3CH=CHCH_2CH_2CH=CHCH_2SiCl_2CH_3$(27-46)	90
C_5H_6	Fig.25.				
	$PdCl_2$ + PPh_3	60	6	Fig.95. (91)	J -028
	$PdCl_2(PPFA)$	30	20	Fig.95. (87)	484

	$PdCl_2(PhCN)_2$ + PPh_3	90	6-24	Fig.95. (90)	867
C_5H_8	$CH_2=C(CH_3)CH=CH_2$				
	$Ni(acac)_2$ + PPh_3	100	6	$(CH_3)_2C=CHCH_2SiCl_2CH_3$ (63) $CH_3Cl_2SiCH_2C(CH_3)=CHCH_2SiCl_2CH_3$ (7)	1018
	$Ni(acac)_2$ + PPh_3 + $Al(i\text{-}Bu)_3$	100	6	$(CH_3)_2C=CHCH_2SiCl_2CH_3$ (25) $CH_3Cl_2SiCH_2C(CH_3)=CHCH_2SiCl_2CH_3$ (45)	1018
	$Ni(acac)_2$ + PPh_3 + $Al(i\text{-}Bu)_3$	20	12	$(CH_3)_2C=CHCH_2SiCl_2CH_3$ (60) $CH_3Cl_2SiCH_2C(CH_3)=CHCH_2SiCl_2CH_3$ (10)	1018
	$Ni(acac)_2$ + PPh_3 + $AlEt_3$	20	12	$(CH_3)_2C=CHCH_2SiCl_2CH_3$ (60) $CH_3Cl_2SiCH_2C(CH_3)=CHCH_2SiCl_2CH_3$ (15)	1018
	$NiCl_2(DMPF)$	105	8 days	$(CH_3)_2C=CHCH_2SiCl_2CH_3$ (33) $CH_3CH=C(CH_3)CH_2SiCl_2CH_3$ (41)	589
	Al_2O_3-PPh_2/$PdCl_2(PhCN)_2$	100	1.5	$CH_3CH=C(CH_3)CH_2SiCl_2CH_3$ (64)	C -007
	$PdCl_2(PhCN)_2$	60	6	$CH_3CH=C(CH_3)CH_2SiCl_2CH_3$ (95)	855
	$PdCl_2(PhCN)_2$ + PPh_3	70	6	$CH_3CH=C(CH_3)CH_2SiCl_2CH_3$ (95)	851
	SDVB-PPh_2/$PdCl_2$(EtOH)	80	2	$C_5H_9SiCl_2CH_3$ (91)	U -080
	SDVB-PPh_2/$PdCl_2$(EtOH)	80	2	$C_5H_9SiCl_2CH_3$ (91)	D -060
	$[PdCl(\pi\text{-All})]_2$	80	2	$C_5H_9SiCl_2CH_3$ (64)	U -080
	$[PdCl(\pi\text{-All})]_2$	80	2	$C_5H_9SiCl_2CH_3$ (64)	D -060
	H_2PtCl_6 (i-PrOH)	165	17	$CH_2=C(CH_3)CH_2CH_2SiCl_2CH_3$ (3-31) $(CH_3)_2C=CHCH_2SiCl_2CH_3$ (51-57) $CH_3Cl_2SiCH_2CH(CH_3)CH_2CH_2SiCl_2CH_3$ (9-25)	122
	$Pt(PPh_3)_4$	45-52	25	$CH_2C(CH_3)CH_2CH_2SiCl_2CH_3$ (37) $C_5H_9SiCl_2CH_3$ (37)	364
	siloxane/$PtCl_2(CO)_2$	reflux	36	$C_5H_9SiCl_2CH_3$ (-)	D -059
	$RhCl(PPh_3)_3$	130	15	$(CH_3)_2C=CHCH_2SiCl_2CH_3$ + $CH_3CH=C(CH_3)CH_2SiCl_2CH_3$ (60)	852
C_5H_8	$CH_2=CHCH_2CH=CH_2$				
	$Ni(acac)_2$ + PPh_3	100	6	$(CH_3)_2C=CHCH_2SiCl_2CH_3$ (45) $CH_3CH_2SiCH_2C(CH_3)=CHCH_2SiCl_2CH_3$ (5)	1018
	$Ni(acac)_2$ + PPh_3 + $Al(i\text{-}Bu)_3$	100	6	$(CH_3)_2C=CHCH_2SiCl_2CH_3$ (24) $CH_3CH_2SiCH_2C(CH_3)=CHCH_2SiCl_2CH_3$ (46)	1018
	$Ni(acac)_2$ + PPh_3 + $Al(i\text{-}Bu)_3$	20	12	$(CH_3)_2C=CHCH_2SiCl_2CH_3$ (60) $CH_3CH_2SiCH_2C(CH_3)=CHCH_2SiCl_2CH_3$ (10)	1018

Table 2.2
= CH_3SiHCl_2 = =3=

	$Ni(acac)_2 + PPh_3 + AlEt_3$	20	12	$(CH_3)_2C=CHCH_2SiCl_2CH_3$ (32) $CH_3CH_2SiCH_2C(CH_3)=CHCH_2SiCl_2CH_3$ (8)	1018
	H_2PtCl_6 (i-PrOH)	93	24 (cy)	$CH_2=CH(CH_2)_3SiCl_2CH_3$ + $CH_3CH_2CH=CHCH_2SiCl_2CH_3$ + $CH_3CH=CH_2CH_2SiCl_2CH_3$ (75)	1173
C_5H_8	$CH_2=CHCH=CHCH_3$				
	$Ni(acac)_2 + AlEt_3$	20	3	$CH_3CH_2CH=CHCH_2SiCl_2CH_3$ (96)	662
	$Ni(acac)_2 + AlEt_3$	20	3	$CH_3CH_2CH=CHCH_2SiCl_2CH_3$ (96)	G -030
	$Ni(acac)_2 + AlEt_3$	20	3	$CH_3CH_2CH=CHCH_2SiCl_2CH_3$ (96)	663
	H_2PtCl_6 (i-PrOH) (cis)	95	24 (cy)	$CH_3CH=CHCH_2CH_2SiCl_2CH_3$ + cis,trans-$CH_3CH=CHCH(CH_3)SiCl_2CH_3$ + $CH_3CH_2CH=CHCH_2SiCl_2CH_3$ (82)	1173
	H_2PtCl_6 (i-PrOH) (trans)	95	24 (cy)	cis,trans-$CH_3CH=CHCH(CH_3)SiCl_2CH_3$ + $CH_3CH_2CH=CHCH_2SiCl_2CH_3$ (75)	1173
	$Pt(PPh_3)_4$ (cis)	45-52	17	cis,trans-$CH_3CH=CHCH_2CH_2SiCl_2CH_3$ (92)	364
	$Pt(PPh_3)_4$ (trans)	45-52	18	cis,trans-$CH_3CH=CHCH_2CH_2SiCl_2CH_3$ (86)	364
C_6H_8	Fig.30.				
	$NiCl_2(BMPP)$	120	40	Fig.96. (38)	1302
C_6H_8	Fig.33.				
	$NiCl_2(BMPP)$	90	40	Fig.97. + Fig.98. + Fig.99. (37)	1302
	$NiCl_2(DMPP)$	120	3.8 days	Fig.97. (55)	589
	$PtCl_2(BMPP)(C_2H_4)$	90	28	Fig.97. (83)	1302
	$PtCl_2(PhMeCHNH_2)$	90	43	Fig.97. + Fig.98. (90)	1302
C_6H_{10}	$CH_2=C(CH_3)C(CH_3)=CH_2$				
	$PdCl_2(PhCN)_2$	80	2	$C_6H_{11}SiCl_2CH_3$ (27)	D -060
	$PdCl_2(PhCN)_2$	80	2	$C_6H_{11}SiCl_2CH_3$ (27)	U -080
C_6H_{10}	$CH_2=CHCH_2CH_2CH=CH_2$				
	H_2PtCl_6 (AcOEt)	30-55	3.25	$CH_3Cl_2Si(CH_2)_6SiCl_2CH_3$ (-)	G -004
	H_2PtCl_6 (t-BuOH)	15-60	2	$CH_3Cl_2Si(CH_2)_6SiCl_2CH_3$ (74)	G -004
	$Pt(PPh_3)_4$	45-52	0.25	$CH_3Cl_2Si(CH_2)_6SiCl_2CH_3$ (96)	364
	$Pt[PhC\equiv CCPh(OH)Me]_2$	55	-	$CH_2=CH(CH_2)_4SiCl_2CH_3$ (88)	E -024
C_6H_{10}	$CH_2=CHCH_2CH=CHCH_3$				

Table 2.2

= CH_3SiHCl_2 = =4=

$Pt(PPh_3)_4$	45-52	0.25	trans-$CH_3CH{=}CH(CH_2)_3SiCl_2CH_3$ (95)	364
C_7H_{10} $CH_2{=}CHCH_2CH{=}CHCH{=}CH_2$				
M + AlR_3	-	-	$C_7H_{11}SiCl_2CH_3$ (-)	7
C_8H_{10} Fig.112.				
H_2PtCl_6 (i-PrOH)	reflux	4	Fig.113. (-)	725
C_8H_{12} $CH_2{=}CHCH(CH_2)_2CH{=}CHCH_2$				
H_2PtCl_6 (i-PrOH) + PPh_3(PhH)	110-115	0.66	$CH_2CH{=}CH(CH_2)_2CHCH_2CH_2SiCl_2CH_3$ (87)	C -039
C_8H_{12} $CH_2{=}CHCH(CH_3)CH{=}CHCH{=}CH_2$				
$Ni(acac)_2$ + PPh_3	100	6	$CH_2{=}CHCH(CH_3)CH{=}CHCH_2CH_2SiCl_2CH_3$ (49) $CH_3CH{=}C(CH_3)CH_2CH{=}CHCH_2SiCl_2CH_3$ (1)	1018
$Ni(acac)_2$ + PPh_3 + $Al(i\text{-}Bu)_3$	20	24	$CH_2{=}CHCH(CH_3)CH{=}CHCH_2CH_2SiCl_2CH_3$ (9) $CH_3CH{=}C(CH_3)CH_2CH{=}CHCH_2SiCl_2CH_3$ (51)	1018
$Ni(acac)_2$ + PPh_3 + $Al(i\text{-}Bu)_3$	20	12	$CH_2{=}CHCH(CH_3)CH{=}CHCH_2CH_2SiCl_2CH_3$ (59) $CH_3CH{=}C(CH_3)CH_2CH{=}CHCH_2SiCl_2CH_3$ (6)	1018
$Ni(acac)_2$ + PPh_3 + $AlEt_3$	20	12	$CH_2{=}CHCH(CH_3)CH{=}CHCH_2CH_2SiCl_2CH_3$ (54) $CH_3CH{=}C(CH_3)CH_2CH{=}CHCH_2SiCl_2CH_3$ (6)	1018
C_8H_{12} Fig.106.				
H_2PtCl_6 (i-PrOH)	70	1	Fig.105. (-)	U -193
C_8H_{12} Fig.34.				
$NiCl_2$ (DMPF)	120	20	Fig.100. (6)	589
C_8H_{12} Fig.35.				
$NiCl_2$ (DMPF)	-	-	Fig.101. + Fig.102. + Fig.103. (-)	628
$NiCl_2$ (DMPF)	-	-	Fig.101. (-)	556
$NiCl_2$ (DMPF)	120	20	Fig.101. (40) Fig.102. (31)	589
NiX_2 + PPh_3 + $AlEt_3$	120	4	Fig.101. (-)	1325
C_8H_{12} Fig.36.				
NiX_2 + PPh_3 + $AlEt_3$	120	4	Fig.104. (60)	1325
NiX_2 + PPh_3 + $HSiCl_3$	150	8	Fig.104. (24) Fig.105. (5)	559

Table 2.2

= CH_3SiHCl_2 = =5=

	H_2PtCl_6 (i-PrOH)	80-170	-	Fig.104. (85)	Dd-007
	H_2PtCl_6 (i-PrOH)	reflux	0.83	Fig.104. (77)	D -085
	H_2PtCl_6 (i-PrOH)	until 180	1	Fig.104. (78)	Dd-007
	$[HPt(SiMe_2Et)P(c\text{-}Hex)_3]_2$	20	0.5	Fig.104. (82)	429
C_8H_{14}	$CH_2=CH(CH_2)_4CH=CH_2$				
	-	-	-	$CH_3SiCl_2(CH_2)_8SiCl_2CH_3$ (83)	622
	-	-	-	$CH_3SiCl_2(CH_2)_8SiCl_2CH_3$ (-)	306
	H_2PtCl_6 (EtOH)	until 130	1	$CH_2=CH(CH_2)_6SiCl_2CH_3$ (70) $CH_3SiCl_2(CH_2)_8SiCl_2CH_3$ (14)	U -118
	H_2PtCl_6 (EtOH)	until 130	1	$CH_2=CH(CH_2)_6SiCl_2CH_3$ (70) $CH_3SiCl_2(CH_2)_8SiCl_2CH_3$ (14)	D -068
	H_2PtCl_6 (EtOH)	until 130	1	$CH_2=CH(CH_2)_6SiCl_2CH_3$ (70) $CH_3SiCl_2(CH_2)_8SiCl_2CH_3$ (14)	U -109
	$[HPt(SiBzMe_2)P(c\text{-}Hex)_3]_2$	20	1	$CH_3SiCl_2(CH_2)_8SiCl_2CH_3$ (74)	429
C_8H_{14}	$CH_2=CHCH_2CH=CHCH_2CH_2CH_3$				
	$Pt(PPh_3)_4$	45-52	4	$CH_3CH_2CH_2CH=CH(CH_2)_3SiCl_2CH_3$ (92)	364
C_9H_{12}	Fig.107.				
	H_2PtCl_6	80	20	Fig.108. + Fig.109. (79)	E -019
$C_{10}H_{10}$	$CH_2=CHC_6H_4CH=CH_2$				
	H_2PtCl_6 (i-PrOH)	60	4	$CH_2=CHC_6H_4CH_2CH_2SiCl_2CH_3$ (-)	E -029
$C_{10}H_{12}$	Fig.42.				
	$Co_2(CO)_8$	80	3 (ht)	Fig.110. (70)	721
	H_2PtCl_6 (i-PrOH)	-	-	Fig.110. (80)	836
	H_2PtCl_6 (i-PrOH)	-	-	Fig.110. +Fig.111. (-)	260
	H_2PtCl_6 (i-PrOH)	70	-	Fig.110. (82)	S -042
$C_{10}H_{16}$	$(CH_3)_2C=CHCH_2CH_2C(=CH_2)CH=CH_2$				
	$PdCl_2(PhCN)_2$ + PPh_3	100	24	$(CH_3)_2C=CHCH_2CH_2C(CH_2SiCl_2CH_3)=CHCH_3$ (90)	851
$C_{10}H_{16}$	$(CH_3)_2C=CHCH_2CH=C(CH_3)CH=CH_2$				
	$PdCl_2(PhCN)_2$ + PPh_3	100	42	$(CH_3)_2C=CHCH_2CH(SiCl_2CH_3)C(CH_3)=CHCH_3$ (77)	851
$C_{10}H_{18}$	$CH_2=CH(CH_2)_6CH=CH_2$				

Table 2.2

= CH_3SiHCl_2 = =6=

	-	-	-	$CH_3SiCl_2(CH_2)_8CH{=}CH_2$ + $CH_3SiCl_2(CH_2)_{10}SiCl_2CH_3$ (-)	881
	-	-	-	$CH_3SiCl_2(CH_2)_{10}SiCl_2CH_3$ (79)	622
$C_{12}H_{18}$ Fig.47.					
	H_2PtCl_6 (i-PrOH)	185	80	Fig.114. (-)	1175

Table 2.3

= CH_3SiHCl_2 = =1=

C_2H_1	$CHF=CF_2$				
	gamma (^{60}Co)	20-30	-	$CH_3SiCl_2CHFCHF_2$ (75) $CH_2FCF_2SiCl_2CH_3$ (25)	1360
C_2H_2	$Cl_3SiCH=CHCl$				
	H_2PtCl_6 (i-PrOH)	120-130	70-80	$Cl_3SiC_2H_2Cl(SiCl_2CH_3)$ (2)	1055
C_2H_3	$CH_2=CHCl$				
	gamma	0-35	-	$ClCH_2CH_2SiCl_2CH_3$ (kin.)	189
	Pt/C	110-120	1	$CH_3SiCl_2CH_2CH_3$ (98)	D -097
C_2H_3	$CH_2=CHF$				
	UV	-	100	$CH_3SiCl_2CH_2CH_2F$ (95)	270
C_2H_3	$CH_2=CHSiCl_3$				
	$NiCl_2(DMPP)_2$	130-140	8	$CH_3SiCl_2CH_2CH_2SiCl_3$ (76)	D -111
	$Pd(PPh_3)_4$	120	6	$CH_3SiCl_2CH_2CH_2SiCl_3$ (90) $CH_3SiCl_2CH(CH_3)SiCl_3$ (3)	749
	H_2PtCl_6	64-173	4	$CH_3SiCl_2CH_2CH_2SiCl_3$ (66) $CH_3SiCl_2CH(CH_3)SiCl_3$ (16)	1058
	H_2PtCl_6 (i-PrOH)	150	4	$CH_3SiCl_2CH_2CH_2SiCl_3$ (58)	893
	H_2PtCl_6 (i-PrOH)	reflux	4	$CH_3SiCl_2CH_2CH_2SiCl_3$ + $CH_3SiCl_2CH(CH_3)SiCl_3$ (-)	1059
C_3	$CF_2=CFCF_3$				
	gamma (^{60}Co)	20-30	-	$CH_3SiCl_2CF_2CHFCF_3$ (56) $CHF_2CF(CF_3)SiCl_2CH_3$ (44)	1360
C_3H_3	$CCl_3CH=CH_2$				
	H_2PtCl_6 (i-PrOH)	65	48	$CCl_3CH_2CH_2SiCl_2CH_3$ + $CCl_2=CHCH_2SiCl_3$ (-)	D -037
C_3H_3	$CF_3CH=CH_2$				
	$(c\text{-}HexOCOO)_2$	55-60	10	$CF_3CH_2CH_2SiCl_2CH_3$ (15-61) $CF_3CH_2CH_2[CH(CF_3)CH_2]_nSiCl_2CH_3$ (20-21)	S -002
	$PdCl_2(PhCN)_2$ + PPh_3	100	14	$CF_3CH(CH_3)SiCl_2CH_3$ (88)	858
	H_2PtCl_6	20	-	$CF_3CH_2CH_2SiCl_2CH_3$ (kin.)	528
	H_2PtCl_6 (i-PrOH)	38-42	12	$CF_3CH_2CH_2SiCl_2CH_3$ (72)	D -079
	H_2PtCl_6 (i-PrOH) + $SnCl_2$	50	3	$CF_3CH_2CH_2SiCl_2CH_3$ (94)	D -079

	Pt catalyst	-	-	$CF_3CH_2CH_2SiCl_2CH_3$ (-)	U -115
	Pt/C	250	9	$CF_3CH_2CH_2SiCl_2CH_3$ (70)	D -079
	$RhCl(PPh_3)_3$	120	12	$CF_3CH_2CH_2SiCl_2CH_3$ (85)	858
	$Ru_3(CO)_{12}$	120	70	$CF_3CH_2CH_2SiCl_2CH_3$ + $CF_3CH{=}CHSiCl_2CH_3$ (11)	858
C_3H_3	$CH_2{=}CHCN$				
	PPh_3	150	2	$CH_3SiCl_2CH_2CH_2CN$ (-)	U -007
	PPh_3	150	2	$CH_3SiCl_2CH_2CH_2CN$ (-)	U -015
	PPh_3	150	5	$CH_3SiCl_2CH_2CH_2CN$ + $CH_3SiCl(CH_2CH_2CN)_2$ (-)	U -015
	$Cu(acac)_2$ + c-HexNC	-	-	$CH_3SiCl_2CH_2CH_2CN$ (80)	1165
	$Cu(acac)_2$ + c-HexNC	120	2	$CH_3SiCl_2CH_2CH_2CN$ (65)	C -005
	$Cu(acac)_2$ + t-BuNC	-	-	$CH_3SiCl_2CH_2CH_2CN$ (60)	1165
	Cu_2O + $(Me_2NCH_2)_2$	30	18	$CH_3SiCl_2CH_2CH_2CN$ (11)	959
	Cu_2O + $(Me_2NCH_2)_2$	reflux	0.75	$CH_3SiCl_2CH_2CH_2CN$ (98-100)	959
	Cu_2O + $(Me_2NCH_2)_2$ + ultrasonic	30	6	$CH_3SiCl_2CH_2CH_2CN$ (60)	959
	CuCl + $(Me_2NCH_2)_2$ + Et_3N	reflux	50	$CH_3SiCl_2CH_2CH_2CN$ (87)	J -095
	Ni(THF)	0-25	2	$CH_3SiCl_2CH(CH_3)CN$ (74)	E -049
	Ni(THF) +sonic waves	0-25	2	$CH_3SiCl_2CH(CH_3)CN$ (96)	E -049
	NiL_2 + PPh_3 + Cl_3SiH	100	5	$CH_3SiCl_2CH(CH_3)CN$ (62) $CH_3SiHClCH(CH_3)CN$ (7)	559
	H_2PtCl_6	30	-	$CH_3SiCl_2C_2H_4CN$ (kin.)	795
	H_2PtCl_6 (i-PrOH) + $AllOCH_2\overline{CHCH_2O}$	reflux	22	$CH_3SiCl_2CH(CH_3)CN$ (50)	S -012
	H_2PtCl_6 (i-PrOH) + $BuOCH{=}CH_2$	reflux	30	$CH_3SiCl_2CH(CH_3)CN$ (52)	S -012
	$RhCl(PPh_3)_3$	60	24	$CH_3SiCl_2CH(CH_3)CN$ (80)	868
C_3H_4	$CH_2{=}CHCHO$				
	H_2PtCl_6	30	-	$CH_3SiCl_2CH_2CH_2CHO$ (kin.)	795
C_3H_4	$CH_3Cl_2SiCH{=}CCl_2$				
	H_2PtCl_6 (i-PrOH)	120-130	70-80	$CH_3Cl_2SiC_2H_2Cl_2(SiCl_2CH_3)$ (4)	1055
C_3H_5	$CH_2{=}CHCH_2Br$				

	H_2PtCl_6	reflux	-	$Br(CH_2)_3SiCl_2CH_3$ (22)	1314
	H_2PtCl_6 (i-PrOH)+ amine + Ph_2SiH_2	70 - reflux	2	$Br(CH_2)_3SiCl_2CH_3$ (60-74)	C -036
	H_2PtCl_6 (i-PrOH)+Et_3N(PhMe)+Ph_2SiH_2	70 - reflux	2	$Br(CH_2)_3SiCl_2CH_3$ (68)	C -036
C_3H_5	$CH_2=CHCH_2Cl$				
	gamma (^{60}Co)	20-80	6-65	$CH_3SiCl_2(CH_2)_3Cl$ (4-70)	1034
	H_2PtCl_6	-	-	$CH_3SiCl_2(CH_2)_3Cl$ (12)	508
	H_2PtCl_6	30	-	$CH_3SiCl_2C_3H_6Cl$ (kin.)	795
	H_2PtCl_6	until 90	0.4	$CH_3SiCl_2(CH_2)_3Cl$ (61)	D -094
	H_2PtCl_6 (Me_2CO)	-	-	$CH_3SiCl_2(CH_2)_3Cl$ (48)	D -128
	H_2PtCl_6 (Me_2CO) + benzothiazole	139	3.5	$CH_3SiCl_2(CH_2)_3Cl$ (49)	D -128
	H_2PtCl_6 (i-PrOH)	120	0.5	$CH_3SiCl_2C_3H_6Cl$ (60)	96
	H_2PtCl_6 (i-PrOH)	reflux	-(solv.)	$CH_3SiCl_2(CH_2)_3Cl$ (59)	U -178
	H_2PtCl_6 (i-PrOH) + DClUB	80-105	3.5	$CH_3SiCl_2C_3H_6Cl$ (60)	96
	H_2PtCl_6 (i-PrOH) + $MeCOON(Me)_2$	reflux	-(solv.)	$CH_3SiCl_2(CH_2)_3Cl$ (79)	U -178
	H_2PtCl_6 (i-PrOH) + $PhNMe_2$	reflux	5	$CH_3SiCl_2(CH_2)_3Cl$ (61)	C -019
	H_2PtCl_6 (i-PrOH) + $SbPh_3$	reflux	5	$CH_3SiCl_2(CH_2)_3Cl$ (60)	C -020
	H_2PtCl_6 (i-PrOH) + ascorbic acid	reflux	8	$CH_3SiCl_2(CH_2)_3Cl$ (63)	C -029
	H_2PtCl_6 +$(CH_3)_2ClSiH$	32-57	0.16	$(CH_3)_2SiClCH_2CH_2CH_2Cl$ (54) $CH_3SiCl_2CH_2CH_2CH_2Cl$ (20)	U -168
	H_2PtCl_6+ H-siloxane	19-95	0.75	$CH_3SiCl_2(CH_2)_3Cl$ (22)	U -168
	H_2PtCl_6 + PPh_3	-	-	$CH_3SiCl_2(CH_2)_3Cl$ (68)	508
	H_2PtCl_6 + amines	-	-	$CH_3SiCl_2(CH_2)_3Cl$ (-)	507
	H_2PtCl_6 [$(CH_2)_5CO$]	120-170	-	$CH_3SiCl_2(CH_2)_3Cl$ (-)	D -103
	H_2PtCl_6 [$(CH_2)_5CO$]	until 90	0.75	$CH_3SiCl_2(CH_2)_3Cl$ (54)	D -094
	$PtCl_2(C_2H_4)(C_5H_5N)$	105-120	0.5	$CH_3SiCl_2(CH_2)_3Cl$ (62)	D -006
	$PtCl_2(C_3H_6)(2\text{-}MeC_5H_4N)$	until 90	1	$CH_3SiCl_2(CH_2)_3Cl$ (72)	D -094
	$PtCl_2(C_3H_6)(3\text{-}MeC_5H_4N)$	until 90	2	$CH_3SiCl_2(CH_2)_3Cl$ (71)	D -094
	$PtCl_2(C_3H_6)(4\text{-}MeC_5H_4N)$	until 90	1	$CH_3SiCl_2(CH_2)_3Cl$ (70)	D -094

Table 2.3

= CH_3SiHCl_2 =

	$PtCl_2(C_3H_6)(C_5H_5N)$	until 90	0.17	$CH_3SiCl_2(CH_2)_3Cl$ (72)	D -094
	SiO_2-NH_2/Na_2PtCl_4	36-90	0.45-12.5	$CH_3SiCl_2(CH_2)_3Cl$ (-)	D -089
	SiO_2-$NHCH_2CH_2NH_2$/Na_2PtCl_4	36-90	-	$CH_3SiCl_2(CH_2)_3Cl$ (-)	D -089
	SiO_2-SEt/Na_2PtCl_4	40-80	15-20	$CH_3SiCl_2(CH_2)_3Cl$ (50-56)	U -108
	SiO_2-SEt/Na_2PtCl_4	40-80	15-20	$CH_3SiCl_2(CH_2)_3Cl$ (50-56)	D -083
	SiO_2-SEt/$PtCl_2$	80	3.3	$CH_3SiCl_2(CH_2)_3Cl$ (56)	U -108
	SiO_2-SH/Na_2PtCl_4	reflux	-	$CH_3SiCl_2(CH_2)_3Cl$ (-)	D -082
	$[PtCl_2(C_2H_4)]_2$	93-120	0.5 (prod.)	$CH_3SiCl_2(CH_2)_3Cl$ (48)	D -006
	$[PtCl_2(PhCH=CH_2)]_2$	99-122	0.5	$CH_3SiCl_2(CH_2)_3Cl$ (51)	D -006
	$[PtCl_2(c\text{-}C_6H_{10})]_2$	102-121	0.5	$CH_3SiCl_2(CH_2)_3Cl$ (53)	D -006
	cis-$PtCl_2(PBu_3)_2$	120	2	$CH_3SiCl_2(CH_2)_3Cl$ (78) $CH_3SiCl_2C_3H_6Cl$ (22)	G -008
	polymer/Pt	reflux	12	$CH_3SiCl_2(CH_2)_3Cl$ (73)	S -046
	polymer/Pt	reflux	12	$CH_3SiCl_2(CH_2)_3Cl$ (73)	932
	polymer-$(CH_2)_nS/PtCl_x$	70	0.5	$CH_3SiCl_2(CH_2)_3Cl$ (69)	505
C_3H_5	$CH_2=CHCH_2SiCl_3$				
	H_2PtCl_6 (i-PrOH)	150-180	-	$Cl_3Si(CH_2)_3SiCl_2CH_3$ (83)	1359
C_3H_5	$CH_3Cl_2SiCH=CHCl$				
	H_2PtCl_6 (i-PrOH)	120-130	70-80	$CH_3Cl_2SiC_2H_2Cl(SiCl_2CH_3)$ (17)	1055
C_3H_6	$CH_3OCH=CH_2$				
	$PtCl_2(SEt_2)_2$ + NBu_3	90	2	$CH_3OCH_2CH_2SiCl_2CH_3$ (-)	U -086
C_3H_6	$CH_3SiCl_2CH=CH_2$				
	$NiCl_2$(DMPC)	120	40	$CH_3SiCl_2CH_2CH_2SiCl_2CH_3$ + $CH_3SiCl_2CH(CH_3)SiCl_2CH_3$ (19)	337
	$Pd(PPh_3)_4$	120	6	$CH_3SiCl_2CH_2CH_2SiCl_2CH_3$ (84)	749
	H_2PtCl_6 (i-PrOH)	150	4	$CH_3SiCl_2CH_2CH_2SiCl_2CH_3$ (76)	893
	H_2PtCl_6 (i-PrOH)	reflux	4	$CH_3SiCl_2CH_2CH_2SiCl_2CH_3$ + $CH_3SiCl_2CH(CH_3)SiCl_2CH_3$ (-)	1059
	$Rh_4(CO)_{12}$	35-40	1	$CH_3SiCl_2CH_2CH_2SiCl_2CH_3$ (97)	S -024
C_3H_7	$CH_2=CHCH_2NH_2$				

Table 2.3

= CH_3SiHCl_2 = =5=

	H_2PtCl_6	30		$CH_3SiCl_2C_3H_6NH_2$ (kin.)	795
C_4	Fig.115.				
	gamma	-	-	Fig.116. (-)	1214
C_4H_1	Fig.117.				
	gamma	-	-	Fig.118. (-)	1214
C_4H_3	$(CF_3)_2NCH{=}CH_2$				
	UV	-	48	$(CF_3)_2NCH_2CH_2SiCl_2CH_3$ (100)	11
	UV	-	48	$(CF_3)_2NCH_2CH_2SiCl_2CH_3$ (100)	D -038
	H_2PtCl_6	100	48	$(CF_3)_2NCH_2CH_2SiCl_2CH_3$ (91)	290
	H_2PtCl_6	100	48	$(CF_3)_2NCH_2CH_2SiCl_2CH_3$ (91)	D -038
C_4H_3	$CF_3CF_2CH{=}CH_2$				
	$PdCl_2(PhCN)_2$ + PPh_3	100	14	$CF_3CF_2CH(CH_3)SiCl_2CH_3$ (38)	858
	H_2PtCl_6-$SnCl_2$ (i-PrOH)	70	-	$CF_3CF_2CH_2CH_2SiCl_2CH_3$ (91)	D -079
C_4H_5	$CH_2{=}CClCH{=}CH_2$				
	$PdCl_2(PPh_3)_2$	90	7	$C_4H_6ClSiCl_2CH_3$ (37)	U -042
	$PdCl_2(PPh_3)_2$	90	7	$C_4H_6ClSiCl_2CH_3$ (37)	U -080
C_4H_5	$CH_2{=}CHCH_2CN$				
	H_2PtCl_6	-	-	$CH_3SiCl_2(CH_2)_3CN$ (-)	971
	H_2PtCl_6 ($MeOC_2H_4OMe$ + EtOH + $NHPh_2$)	70	0.25	$CH_3SiCl_2(CH_2)_3CN$ (95)	U -085
	H_2PtCl_6 (i-PrOH)	reflux	3	$CH_3SiCl_2(CH_2)_3CN$ (97)	935
	Pt/Al_2O_3	200	2	$CH_3SiCl_2(CH_2)_3CN$ (85)	U -007
	$PtCl_2(PhCN)_2$	97-151	1.2	$CH_3SiCl_2(CH_2)_3CN$ (89)	U -027
C_4H_5	$CH_2{=}CHCH_2NCO$				
	H_2PtCl_6	80-135	3	$CH_3SiCl_2(CH_2)_3NCO$ (78)	783
	H_2PtCl_6 ($MeOC_2H_4OMe$)	42-85	1.5	$CH_3SiCl_2(CH_2)_3NCO$ (96)	U -039
	H_2PtCl_6 ($MeOC_2H_4OMe$)	42-85	1.5	$CH_3SiCl_2(CH_2)_3NCO$ (96)	D -024
	Rh catalyst	-	-	$CH_3SiCl_2(CH_2)_3NCO$ (-)	J -024
C_4H_5	$CH_2{=}CHCH_2OCON_3$				

	H_2PtCl_6 (i-PrOH)	60	16	$CH_3SiCl_2(CH_2)_3OCON_3$ (57)	G -019
C_4H_5	$ClCOOCH_2CH=CH_2$				
	H_2PtCl_6 (i-PrOH)	120-200	-	$ClCOO(CH_2)_3SiCl_2CH_3$ (57)	S -026
	H_2PtCl_6 (i-PrOH)	60-85	-	$ClCOO(CH_2)_3SiCl_2CH_3$ (34)	783
	H_2PtCl_6 (i-PrOH)	60-85	-	$ClCOO(CH_2)_3SiCl_2CH_3$ (34)	786
C_4H_6	$CH_2=CHCHClCH_2Cl$				
	H_2PtCl_6 (i-PrOH)	90	1	$CH_3SiCl_2CH_2CH_2CHClCH_2Cl$ (40)	730
C_4H_6	$CH_2=CHCOOCH_3$				
	-	-	-	$CH_3SiCl_2CH_2CH_2COOCH_3$ + $CH_3SiCl_2C(CH_3)HCOOCH_3$ (-)	1323
	Ni(THF)	0-25	20	$CH_3SiCl_2CH(CH_3)COOCH_3$ (63)	E -049
	Ni(THF) +sonic waves	0-25	3	$CH_3SiCl_2CH(CH_3)COOCH_3$ (87)	E -049
	Ni(THF) +sonic waves	0-25	12	$CH_3SiCl_2CH_2CH_2COOCH_3$ (5)	E -049
	$NiCl_2$ (DMF)	120	20	$CH_3SiCl_2CH_2CH_2COOCH_3$ + $CH_3SiCl_2CH(CH_3)COOCH_3$ (77)	589
	NiX_2 + PPh_3 + $HSiCl_3$	100	8	$CH_3SiCl_2CH(CH_3)COOCH_3$ (29)	559
	$PtCl_2(PBu_3)_2$	120	2	$CH_3SiCl_2CH_2CH_2COOCH_3/CH_3SiCl_2CH(CH_3)COOCH_3$ (79/21)	G -008
C_4H_6	$CH_2=CHOCH=CH_2$				
	H_2PtCl_6 (i-PrOH)	75-80	4-5 (d)	$CH_2=CHOCH_2CH_2SiCl_2CH_3$ (28) $CH_3SiCl_2CH_2CH_2OCH_2CH_2SiCl_2CH_3$ (48)	152
C_4H_6	$CH_3COOCH=CH_2$				
	-	-	-	$CH_3COOCH_2CH_2SiCl_2CH_3$ + $CH_3COOCH(CH_3)SiCl_2CH_3$ (kin.)	674
	NiL_2 + PPh_3 + $HSiCl_3$	120	8	$CH_3COOCH_2CH_2SiCl_2CH_3$ (28)	559
	H_2PtCl_6	-	-	$CH_3COOCH_2CH_2SiCl_2CH_3$ (66)	926
	H_2PtCl_6 (OctOH)	reflux	3	$CH_3COOCH_2CH_2SiCl_2CH_3$ (100)	F -002
	H_2PtCl_6 (OctOH)	reflux	24 (t)	$CH_3COOCH_2CH_2SiCl_2CH_3$ (75)	U -014
	$PtCl_2(CO)_2$	60-98	16	$CH_3COOCH_2CH_2SiCl_2CH_3$ (-)	D -059
	$PtCl_2(SEt_2)_2$	65-94	7	$CH_3COOC_2H_4SiCl_2CH_3$ (-)	F -011
C_4H_6	$Cl_2Si(CH=CH_2)_2$				

	H_2PtCl_6 (i-PrOH)	reflux	4	$Cl_2Si(CH{=}CH_2)C_2H_4SiCl_2CH_3$ + $Cl_2Si(C_2H_4SiCl_2CH_3)_2$ (-)	1059
C_4H_6	$ClCH_2CH{=}CHCH_2Cl$				
	H_2PtCl_6 (i-PrOH)	78-125	-	$CH_3CH_2ClCH_2CH_2SiCl_2CH_3$ (20) $Cl(CH_2)_4SiCl_2CH_3$ (32)	730
C_4H_6	$HCOOCH_2CH{=}CH_2$				
	H_2PtCl_6 (i-PrOH)	55-126	1	$HCOO(CH_2)_3SiCl_2CH_3$ (68)	1052
	H_2PtCl_6 (i-PrOH)	55-126	1	$HCOO(CH_2)_3SiCl_2CH_3$ (68)	S -026
C_4H_7	$(CH_3)_2ClSiCH{=}CCl_2$				
	H_2PtCl_6 (i-PrOH)	120-130	70-80	$(CH_3)_2ClSiC_2H_2Cl_2(SiCl_2CH_3)$ (10)	1055
C_4H_7	$CH_2{=}C(CH_3)CH_2Cl$				
	Al_2O_3	80	3.5	$CH_3SiCl_2CH_2CH(CH_3)CH_2Cl$ (80)	D -002
	H_2PtCl_6	120	0.3	$CH_3SiCl_2CH_2CH(CH_3)CH_2Cl$ (98)	U -168
	H_2PtCl_6 + CH_3Cl_2SiH	21-52	0.35	$ClCH_2CH(CH_3)CH_2SiCl_2CH_3$ (69) $Cl_3SiCH_2CH(CH_3)CH_2Cl$ (22)	U -168
	H_2PtCl_6 + CH_3Cl_2SiH	84	0.07	$ClCH_2CH(CH_3)CH_2SiCl_2CH_3$ (67) $(C_2H_5)_3SiCH_2CH(CH_3)CH_2Cl$ (21)	U -168
C_4H_7	$CH_2{=}CHCH_2OCH_2Cl$				
	H_2PtCl_6 (i-PrOH)	80	6	$CH_3SiCl_2(CH_2)_3OCH_2Cl$ (64)	G -006
C_4H_7	$CH_2{=}CHOCH_2CH_2Cl$				
	H_2PtCl_6 (i-PrOH)	85-90	5-6	$CH_3SiCl_2CH_2CH_2OCH_2CH_2Cl$ (94)	600
C_4H_8	$CH_3Cl_2SiCH{=}CHSiCl_2CH_3$				
	H_2PtCl_6 (i-PrOH)	120-130	-	$CH_3Cl_2SiCH_2CH(SiCl_2CH_3)_2$ (6)	1053
C_4H_8	$(CH_3)_2ClSiCH{=}CHCl$				
	H_2PtCl_6 (i-PrOH)	120-130	70-80	$(CH_3)_2ClSiC_2H_2(Cl)SiCl_2CH_3$ (40)	1055
C_4H_8	$CH_2{=}CHCH_2SiCl_2CH_3$				
	H_2PtCl_6 (i-PrOH)	150-180	-	$CH_3Cl_2Si(CH_2)_3SiCl_2CH_3$ (66)	1359
C_4H_8	$CH_3OCH_2CH{=}CH_2$				
	H_2PtCl_6	80-90	-	$CH_3O(CH_2)_3SiCl_2CH_3$ (43)	1338

	H_2PtCl_6 (i-PrOH)	80	3	$CH_3O(CH_2)_3SiCl_2CH_3$ (41)	1115
C_4H_8	$CH_3SO_2CH_2CH=CH_2$				
	H_2PtCl_6	115	3	$CH_3SO_2(CH_2)_3SiCl_2CH_3$ (70)	U -037
C_4H_8	$ClCH_2SiCl(CH_3)CH=CH_2$ + $ClCH_2Si(CH_3)(CH=CH_2)_2$				
	H_2PtCl_6 (i-PrOH)	80-100		$ClCH_2SiCl(CH_3)CH_2CH_2SiCl_2CH_3$ + $ClCH_2Si(CH_3)(CH_2CH_2SiCl_2CH_3)_2$ (-)	387
C_4H_9	$ClSi(CH_3)_2CH=CH_2$				
	$Pd(PPh_3)_4$	120	6	$ClSi(CH_3)_2CH_2CH_2SiCl_2CH_3$ (18)	749
	H_2PtCl_6 (i-PrOH)	-	-	$ClSi(CH_3)_2CH_2CH_2SiCl_2CH_3$ (67) $(CH_3)_2Si(CH_2CH_2SiCl_2CH_3)_2$ (1)	1056
	H_2PtCl_6 (i-PrOH)	reflux	4	$ClSi(CH_3)_2C_2H_4SiCl_2CH_3$ (-)	1059
C_4H_{14}	$\overline{CHB_{10}H_{10}C}CH=CH_2$				
	none	260	-	$\overline{CHB_{10}H_{10}C}CH_2CH_2SiCl_2CH_3$ (-)	1045
	$Co_2(CO)_8$	20-30	60	$\overline{CHB_{10}H_{10}C}CH_2CH_2SiCl_2CH_3$ (-)	715
	$Rh_4(CO)_{12}$	20-30	60	$\overline{CHB_{10}H_{10}C}CH_2CH_2SiCl_2CH_3$ (-)	715
C_5	$(\overline{CF_2)_3CCl=C}Cl$				
	gamma	-	-	$(\overline{CF_2)_3CHClC}ClSiCl_2CH_3$ (-)	1214
C_5H_1	$(\overline{CF_2)_3CH=C}Cl$				
	gamma	-	-	$(\overline{CF_2)_3CHClC}HSiCl_2CH_3$ + $(\overline{CF_2)_3CH_2C}ClSiCl_2CH_3$ (-)	1214
C_5H_3	$(\overline{CF_2)_2CCl=C}SiCl_2CH_3$				
	gamma	-	-	$(\overline{CF_2)_2CCl(SiCl_2CH_3)C}HSiCl_2CH_3$ + $(\overline{CF_2)_2CHClC}(SiCl_2CH_3)_2$ (-)	1213
C_5H_3	$(CF_3)_2CFOCH=CH_2$				
	H_2PtCl_6 (i-PrOH)	90	6	$(CF_3)_2CFOCH_2CH_2SiCl_2CH_3$ (73)	U -033
	H_2PtCl_6 (i-PrOH)	90	6	$(CF_3)_2CFOCH_2CH_2SiCl_2CH_3$ (73)	U -029
	H_2PtCl_6 (i-PrOH)	90	6	$(CF_3)_2CFOCH_2CH_2SiCl_2CH_3$ (73)	U -022
C_5H_3	$CCl_3CF_2CF_2CH=CH_2$				
	H_2PtCl_6 (i-PrOH)	105-130	36	$CCl_3CF_2CF_2CH_2CH_2SiCl_2CH_3$ (-)	U -056
	H_2PtCl_6 (i-PrOH)	105-130	36	$CCl_3CF_2CF_2CH_2CH_2SiCl_2CH_3$ (-)	U -042
C_5H_3	$CF_2ClCF(CF_3)OCH=CH_2$				

	H_2PtCl_6 (i-PrOH)	90	6	$CF_2ClCF(CF_3)OCH_2CH_2SiCl_2CH_3$ (75)	U -033
	H_2PtCl_6 (i-PrOH)	90	6	$CF_2ClCF(CF_3)OCH_2CH_2SiCl_2CH_3$ (75)	U -029
	H_2PtCl_6 (i-PrOH)	90	6	$CF_2ClCF(CF_3)OCH_2CH_2SiCl_2CH_3$ (75)	U -022
C_5H_3	$CF_3CF_2CF_2CH{=}CH_2$				
	H_2PtCl_6 (OctOH)	100	16(t)	$CF_3CF_2CF_2CH_2CH_2SiCl_2CH_3$ (90)	U -014
	$H_2PtCl_6 + SnCl_2$ (i-PrOH)	70	-	$CF_3CF_2CF_2CH_2CH_2SiCl_2CH_3$ (90)	D -079
C_5H_5	$(CF_3)_2NCH_2CH{=}CH_2$				
	H_2PtCl_6	80	24	$(CF_3)_2N(CH_2)_3SiCl_2CH_3$ (85-96)	D -038
	H_2PtCl_6	80	24	$(CF_3)_2N(CH_2)_3SiCl_2CH_3$ (85-96)	290
C_5H_5	$CCl_3COOCH_2CH{=}CH_2$				
	H_2PtCl_6 (THF)	95-100	3.5	$CCl_3COO(CH_2)_3SiCl_2CH_3$ (50)	S -052
	Pt/C or H_2PtCl_6 (THF)	95-100	3	$CCl_3COO(CH_2)_3SiCl_2CH_3$ (50)	553
C_5H_5	$CF_3COOCH_2CH{=}CH_2$				
	H_2PtCl_6	100	3(solv)	$CF_3COO(CH_2)_3SiCl_2CH_3$ (20-52)	100
	H_2PtCl_6 (i-PrOH)	until 115	2.5	$CF_3COO(CH_2)_3SiCl_2CH_3$ (55)	103
C_5H_6	$CHF_2CF_2OCH_2CH{=}CH_2$				
	$[PtCl_2(CH_2{=}CHC_6H_{13})]_2$	100	flow	$CHF_2CF_2O(CH_2)_3SiCl_2CH_3$ (78)	D -117
C_5H_8	$CH_2{=}C(CH_3)COOCH_3$				
	-	-	-	$CH_3SiCl_2CH(CH_3)COOCH_3 + CH_3SiCl_2CH_2CH_2COOCH_3$ (-)	1323
	H_2PtCl_6	-	-	$CH_3SiCl_2CH_2CH(CH_3)COOCH_3$ (-)	971
	H_2PtCl_6	30	-	$CH_3SiCl_2CH_2CH(CH_3)COOCH_3$ (kin.)	795
	H_2PtCl_6 (OctOH)	reflux	8(t)	$CH_3SiCl_2CH_2CH(CH_3)COOCH_3$ (90)	U -014
	H_2PtCl_6 (i-PrOH)	85	1	$CH_3SiCl_2CH_2CH(CH_3)COOCH_3$ (-)	B -002
C_5H_8	$CH_2{=}CHCOOC_2H_5$				
	-	-	-	$CH_3SiCl_2CH_2CH_2COOC_2H_5$ + $CH_3SiCl_2CH(CH_3)COOC_2H_5$ (-)	1323
C_5H_8	$CH_3COOCH_2CH{=}CH_2$				

Table 2.3

= CH_3SiHCl_2 = =10=

	H_2PtCl_6	reflux	-(THF)	$CH_3COO(CH_2)_3SiCl_2CH_3$ (79)	754
	H_2PtCl_6 (i-PrOH)	60-100	6	$CH_3COO(CH_2)_3SiCl_2CH_3$ (84)	512
	$PtCl_2(CO)(C_5H_5N)$	90-120	0.5	$CH_3COO(CH_2)_3SiCl_2CH_3$ (80-83)	D -008
	$PtCl_2(CO)(PhNH_2)$	90-120	0.5	$CH_3COO(CH_2)_3SiCl_2CH_3$ (80-83)	D -008
	$[PtCl_2(C_2H_4)]_2$	90-120	0.5	$CH_3COO(CH_2)_3SiCl_2CH_3$ (80-83)	D -008
	$[PtCl_2(PhCH=CH_2)]_2$	90-120	0.5	$CH_3COO(CH_2)_3SiCl_2CH_3$ (80-83)	D -008
	$[PtCl_2(c\text{-}C_6H_{10})]_2$	90-120	0.5	$CH_3COO(CH_2)_3SiCl_2CH_3$ (80-83)	D -008
	polymer-$(CH_2)_nS/PtCl_x$	70	1	$CH_3COO(CH_3)CH_2CH_2SiCl_2CH_3$ (79)	505
	polymer-$(CH_2)_nS/PtCl_x$	70	10	$CH_3COO(CH_2)_3SiCl_2CH_3$ (95)	505
C_5H_8	$CH_3COOOCH_2CH=CH_2$				
	H_2PtCl_6 (i-PrOH)	40-132	0.6	$CH_3COOO(CH_2)_3SiCl_2CH_3$ (51)	1051
	H_2PtCl_6 (i-PrOH)	40-132	0.6	$CH_3COOO(CH_2)_3SiCl_2CH_3$ (51)	S -026
C_5H_9	$ClSiCH_3(CH=CH_2)_2$				
	H_2PtCl_6 (i-PrOH)	reflux	4	$ClSiCH_3(CH=CH_2)C_2H_4SiCl_2CH_3$ + $ClSiCH_3(C_2H_4SiCl_2CH_3)_2$ (-)	1059
C_5H_{10}	$CH_2=C(CH_3)CH_2OCH_3$				
	H_2PtCl_6 (i-PrOH)	80	3	$CH_3SiCl_2CH_2CH(CH_3)CH_2OCH_3$ (37)	1115
C_5H_{10}	$CH_3SO_2CH_2C(CH_3)=CH_2$				
	H_2PtCl_6 (OctOH)	110	2	$CH_3SO_2CH_2CH(CH_3)CH_2SiCl_2CH_3$ (75)	U -022
C_5H_{11}	$(CH_3)_2ClSiCH_2CH=CH_2$				
	H_2PtCl_6	-	-	$(CH_3)_2ClSi(CH_2)_3SiCl_2CH_3$ (80-89)	500
	H_2PtCl_6 (i-PrOH)	150-180	-	$(CH_3)_2ClSi(CH_2)_3SiCl_2CH_3$ (84)	1359
C_5H_{11}	$(CH_3)_3SiCH=CHCl$				
	H_2PtCl_6 (i-PrOH)	120-130	70-80	$(CH_3)_3SiC_2H_2Cl(SiCl_2CH_3)$ (27)	1055
C_5H_{11}	$ClCH_2Si(CH_3)_2CH=CH_2$				
	H_2PtCl_6 (i-PrOH)	80-100	-	$ClCH_2Si(CH_3)_2CH_2CH_2SiCl_2CH_3$ (86)	387
C_5H_{12}	$(CH_3)_3SiCH=CH_2$				
	$Co_2(CO)_8$	30-40	1	$(CH_3)_3SiCH_2CH_2SiCl_2CH_3$ (-)	715

Table 2.3

= CH_3SiHCl_2 = =11=

$Co_2(CO)_8$	65-70	1	$(CH_3)_3SiCH_2CH_2SiCl_2CH_3$ (85)	31
$Fe(CO)_4(Me_3SiCH=CH_2)$	70-80	3	$(CH_3)_3SiCH_2CH_2SiCl_2CH_3$ (8) $(CH_3)_3SiCH(CH_3)SiCl_2CH_3$ (61) $(CH_3)_3SiCH=CHSiCl_2CH_3$ (18)	399
$Fe(CO)_4(Me_3SiCH=CH_2)$	70-80	3	$(CH_3)_3SiCH_2CH_2SiCl_2CH_3$ (8) $(CH_3)_3SiCH(CH_3)SiCl_2CH_3$ (61) $(CH_3)_3SiCH=CHSiCl_2CH_3$ (18)	398
$Fe(CO)_4(Me_3SiCH=CH_2)$	80	4	$(CH_3)_3SiCH_2CH_2SiCl_2CH_3$ + $(CH_3)_3SiCH(CH_3)SiCl_2CH_3$ + $(CH_3)_3SiCH=CHSiCl_2CH_3$ (50)	844
$Ni(PPh_3)_2(CH_2=CHSiMe_3)$	-	-	$(CH_3)_3SiCH_2CH_2SiCl_2CH_3$ (78-90)	1337
$Ni(PPh_3)_2[CH_2=CHSi(OMe)_3]$	-	-	$(CH_3)_3SiCH_2CH_2SiCl_2CH_3$ (78-90)	1337
$NiCl_2(DMPC)$	120	40	$(CH_3)_3SiCH_2CH_2SiCl_2CH_3$ + $(CH_3)_3SiCH(CH_3)SiCl_2CH_3$ (100)	337
$[Ni(PPh_3)_2]_2[(CH_2=CH)_4Si]$	-	-	$(CH_3)_3SiCH_2CH_2SiCl_2CH_3$ (78-90)	1337
$Pd(PPh_3)_4$	120	6	$(CH_3)_3SiCH_2CH_2SiCl_2CH_3$ (5)	749
H_2PtCl_6	146-148	6	$(CH_3)_3SiCH_2CH_2SiCl_2CH_3$ (90-96)	1058
H_2PtCl_6 (i-PrOH)	reflux	4	$(CH_3)_3SiCH_2CH_2SiCl_2CH_3$ + $(CH_3)_3SiCH(CH_3)SiCl_2CH_3$ (-)	1059
$Rh_4(CO)_{12}$	30-40	1	$(CH_3)_3SiCH_2CH_2SiCl_2CH_3$ (-)	715
$Rh_4(CO)_{12}$	35-40	1	$(CH_3)_3SiCH_2CH_2SiCl_2CH_3$ (98)	S -024
$Rh_4(CO)_{12}$	65-70	1	$(CH_3)_3SiCH_2CH_2SiCl_2CH_3$ (90)	31
$RhCl(PPh_3)_3$	120	6	$(CH_3)_3SiCH_2CH_2SiCl_2CH_3$ (75)	Pl-010
C_5H_{12} $CH_2=CHSi(CH_3)_3$				
H_2PtCl_6 (i-PrOH)	-	-	$(CH_3)_3SiCH_2CH_2SiCl_2CH_3$ (60) $(CH_3)_3SiCH_2CH_2Si(CH_3)_3$ (1) $CH_3Cl_2SiCH_2CH_2SiCl_2CH_3$ (19)	1056
C_5H_{16} $HCB_{10}H_{10}CC(CH_3)=CH_2$				
none	200-260	12	$HCB_{10}H_{10}CCH(CH_3)CH_2SiCl_2CH_3$ (30)	1045
$Co_2(CO)_8$	20-30	60	$HCB_{10}H_{10}CCH(CH_3)CH_2SiCl_2CH_3$ (-)	715
$FeCl_3$	200-260	12-48	$HCB_{10}H_{10}CCH(CH_3)CH_2SiCl_2CH_3$ (43-84)	1045
H_2PtCl_6 (i-PrOH)	150-230	12	$HCB_{10}H_{10}CCH(CH_3)CH_2SiCl_2CH_3$ (48)	1045

Table 2.3

= CH_3SiHCl_2 =

	$Rh_4(CO)_{12}$	20-30	60	$HCB_{10}H_{10}CCH(CH_3)CH_2SiCl_2CH_3$ (-)	715
C_5H_{16}	$HCB_{10}H_{10}CCH_2CH{=}CH_2$				
	none	260	-	$HCB_{10}H_{10}CC_3H_6SiCl_2CH_3$ (-)	1045
	H_2PtCl_6	40-45	-	$HCB_{10}H_{10}CC_3H_6SiCl_2CH_3$ (92)	761
C_6H_3	$CF_3(CF_2)_3CH{=}CH_2$				
	H_2PtCl_6 (i-PrOH)	reflux	68	$CF_3(CF_2)_3CH_2CH_2SiCl_2CH_3$ (86)	581
C_6H_5	$(CClF_2)_2CFOCH_2CH{=}CH_2$				
	H_2PtCl_6 (i-PrOH)	80-100	6	$(CClF_2)_2CFO(CH_2)_3SiCl_2CH_3$ (82)	U -033
	H_2PtCl_6 (i-PrOH)	80-100	6	$(CClF_2)_2CFO(CH_2)_3SiCl_2CH_3$ (82)	U -022
	H_2PtCl_6 (i-PrOH)	80-100	6	$(CClF_2)_2CFO(CH_2)_3SiCl_2CH_3$ (82)	U -029
C_6H_5	$(CF_3)_2CFOCH_2CH{=}CH_2$				
	PhCOOOBu-t	90	10	$(CF_3)_2CFO(CH_2)_3SiCl_2CH_3$ (25)	U -022
	PhCOOOBu-t	90	10	$(CF_3)_2CFO(CH_2)_3SiCl_2CH_3$ (25)	U -033
	H_2PtCl_6 (i-PrOH)	80-100	6	$(CF_3)_2CFO(CH_2)_3SiCl_2CH_3$ (71)	U -033
	H_2PtCl_6 (i-PrOH)	80-100	6	$(CF_3)_2CFO(CH_2)_3SiCl_2CH_3$ (71)	U -022
	H_2PtCl_6 (i-PrOH)	80-100	6	$(CF_3)_2CFO(CH_2)_3SiCl_2CH_3$ (71)	U -029
	$PtCl_2$	reflux	5.5	$(CF_3)_2CFO(CH_2)_3SiCl_2CH_3$ (86)	U -081
C_6H_5	$CClF_2CF(CF_3)OCH_2CH{=}CH_2$				
	H_2PtCl_6	80-100	6	$CClF_2CF(CF_3)O(CH_2)_3SiCl_2CH_3$ (80)	U -022
	H_2PtCl_6	80-100	6	$CClF_2CF(CF_3)O(CH_2)_3SiCl_2CH_3$ (80)	U -029
	H_2PtCl_6	80-100	6	$CClF_2CF(CF_3)O(CH_2)_3SiCl_2CH_3$ (80)	U -033
C_6H_6	$CF_3CHFCF_2OCH_2CH{=}CH_2$				
	$PtCl_2(CH_2{=}CHC_6H_{13})_2$	reflux	3	$CF_3CHFCF_2O(CH_2)_3SiCl_2CH_3$ (92)	D -116
C_6H_6	$CH_2{=}CHCF_2CF_2CH{=}CH_2$				
	H_2PtCl_6 (i-PrOH)	55-70	overnight	$CH_3SiCl_2CH_2CH_2CF_2CF_2CH_2CH_2SiCl_2CH_3$ (-)	U -105
C_6H_6	$CH_2{=}CHCF_2SCF_2CH{=}CH_2$				
	$(t\text{-}BuO)_2$	reflux	72	$CH_3SiCl_2CH_2CH_2CF_2SCF_2CH_2CH_2SiCl_2CH_3$ (-)	D -025
C_6H_6	Fig.119.				

	H_2PtCl_6	50-70	20-25	Fig.120. (58-71)	693
C_6H_8	$CHF_2CF_2CH_2OCH_2H=CH_2$				
	H_2PtCl_6	-	-	$CHF_2CF_2CH_2O(CH_2)_3SiCl_2CH_3$ (75)	237
C_6H_9	$ClSi(CH=CH_2)_3$				
	H_2PtCl_6 (i-PrOH)	reflux	4	$ClSi(CH=CH_2)_2C_2H_4SiCl_2CH_3$ (-) $ClSi(CH=CH_2)(C_2H_4SiCl_2CH_3)_2$ (-) $ClSi(C_2H_4SiCl_2CH_3)_3$ (-)	1059
C_6H_{10}	$(CH_2=CHCH_2)_2O$				
	H_2PtCl_6	100	12	$CH_2=CHCH_2O(CH_2)_3SiCl_2CH_3$ (-)	U -098
	polymer-$(CH_2)_nS/PtCl_x$	70	0.5	$(Cl_2CH_3SiCH_2CH_2CH_2)_2O$ (87)	505
C_6H_{10}	$CH_2=C(CH_3)CH_2OCOCH_3$				
	H_2PtCl_6	reflux	- (THF)	$CH_3SiCl_2CH_2CH(CH_3)CH_2OCOCH_3$ (75)	754
C_6H_{10}	$CH_2=C(CH_3)COOC_2H_5$				
	-	-	-	$CH_3SiCl_2CH(CH_3)COOC_2H_5$ + $CH_3SiCl_2CH_2CH_2COOC_2H_5$ (-)	1323
C_6H_{10}	$CH_2=CHCH_2CH_2COCH_3$				
	H_2PtCl_6	120	1	$CH_3SiCl_2(CH_2)_4COCH_3$ (-)	431
	H_2PtCl_6	130-140	1.5 - 2	$CH_3SiCl_2(CH_2)_4COCH_3$ (84)	602
C_6H_{11}	$ClCH_2Si(CH_3)(CH=CH_2)_2$				
	H_2PtCl_6 (i-PrOH)	80-100	-	$ClCH_2Si(CH_3)(CH_2CH_2SiCl_2CH_3)_2$ (-)	387
C_6H_{12}	$(CH_3)_2Si(CH=CH_2)_2$				
	H_2PtCl_6 (i-PrOH)	reflux	4	$(CH_3)_2Si(CH=CH_2)C_2H_4SiCl_2CH_3$ $(CH_3)_2Si(C_2H_4SiCl_2CH_3)_2$ (-)	1059
C_6H_{12}	$CH_2=CHCH_2OCH(CH_3)_2$				
	H_2PtCl_6 (i-PrOH)	80	3	$CH_3SiCl_2(CH_2)_3OCH(CH_3)_2$ (21)	1115
C_6H_{12}	$CH_2=CHO(CH_2)_3CH_3$				
	Ni(THF)	0-25	20	$CH_3(CH_2)_3OCH_2CH_2SiCl_2CH_3$ (48)	E -049
	Ni(THF) +sonic waves	0-25	4	$CH_3(CH_2)_3OCH_2CH_2SiCl_2CH_3$ (71)	E -049
C_6H_{12}	$CH_2=CHOCH_2CH(CH_3)_2$				

	$PtCl_2(SEt_2)_2$	reflux	3	$CH_3SiCl_2CH_2CH_2OCH_2CH(CH_3)_2$ (-)	U -086
C_6H_{13}	$CH_2{=}CHCH_2SiCl(CH_3)CH_2CH_3$				
	H_2PtCl_6	-	-	$CH_3SiCl_2(CH_2)_3SiCl(CH_3)CH_2CH_3$ (80-81)	500
C_6H_{13}	$ClCH_2(CH_3)_2SiCH_2CH{=}CH_2$				
	H_2PtCl_6 (i-PrOH)	150-180	-	$ClCH_2(CH_3)_2Si(CH_2)_3SiCl_2CH_3$ (61)	1359
C_6H_{14}	$(CH_3)_3SiCH_2CH{=}CH_2$				
	H_2PtCl_6 (i-PrOH)	150-180	-	$(CH_3)_3Si(CH_2)_3SiCl_2CH_3$ (76)	1359
C_6H_{14}	$(CH_3)_3SiOCH_2CH{=}CH_2$				
	H_2PtCl_6 (i-PrOH)	reflux	-	$\overline{(CH_2)_3OSi}ClCH_3$ (80)	777
C_6H_{18}	$CH_3\overline{CB_{10}H_{10}C}CH_2CH{=}CH_2$				
	H_2PtCl_6	90-95	-	$CH_3\overline{CB_{10}H_{10}C}(CH_2)_3SiCl_2CH_3$ (92)	761
	H_2PtCl_6 (i-PrOH)	50	2	$CH_3\overline{CB_{10}H_{10}C}(CH_2)_3SiCl_2CH_3$ (-)	U -032
C_6H_{18}	$\overline{CHB_{10}H_{10}C}CH_2CH_2CH_2CH{=}CH_2$				
	none	260	-	$\overline{CHB_{10}H_{10}C}(CH_2)_4SiCl_2CH_3$ (-)	1045
	Pt/C	95	68	$\overline{CHB_{10}H_{10}C}(CH_2)_4SiCl_2CH_3$ (-)	U -025
C_7H_5	$CH_2{=}CHCH_2OCOC_3F_7$				
	H_2PtCl_6 (i-PrOH)	until 140	5	$CH_3SiCl_2(CH_2)_3OCOC_3H_7$ (51)	103
C_7H_5	$CH_2{=}CHCOOCH_2C_3F_7$				
	CuCl + $NH_2CMe_2CMe_2NH_2$ + NBu_3	80-140	2	$CH_3SiCl_2CH_2CH_2COOCH_2C_3H_7$ (67)	419
C_7H_7	$CH_2{=}CHCH_2OCH_2(CF_2)_2OCF_3$				
	H_2PtCl_6 (i-PrOH)	reflux	5	$CH_3SiCl_2(CH_2)_3OCH_2(CF_2)_2OCF_3$ (-)	419
C_7H_8	$CH_2{=}CHCH_2COOCH_2(CF_2)_2H$				
	H_2PtCl_6 (i-PrOH)	until 215	3	$CH_3SiCl_2(CH_2)_3COOCH_2(CF_2)_2H$ (74)	103
C_7H_8	$CH_3COOCH_2CF_2CF_2CH{=}CH_2$				
	H_2PtCl_6 (i-PrOH)	reflux	17	$CH_3COOCH_2CF_2CF_2CH_2CH_2SiCl_2CH_3$ (-)	U -048
	H_2PtCl_6 (i-PrOH)	reflux	17	$CH_3COOCH_2CF_2CF_2CH_2CH_2SiCl_2CH_3$ (-)	U -054
C_7H_9	$CH_2{=}CHCH_2COOCH_2CH_2CN$				

	H_2PtCl_6 (OctOH)	100-120	-	$CH_3SiCl_2(CH_2)_3COOCH_2CH_2CN$ (-)	U -047
C_7H_9	Fig.121.				
	H_2PtCl_6	140-160	-	Fig.122. (74)	729
C_7H_9	Fig.123.				
	H_2PtCl_6	140-160	-	Fig.124. (93)	729
C_7H_{10}	$(CH_2{=}CHCH_2O)_2CO$				
	H_2PtCl_6	70-107	1.3	$[CH_3SiCl_2(CH_2)_3O]_2CO$ (67)	1051
	H_2PtCl_6 (i-PrOH)	reflux	12 (h)	$[CH_3SiCl_2(CH_2)_3O]_2CO$ (79)	J -054
C_7H_{10}	$CH_2{=}C(CH_3)COOCH_2CH{=}CH_2$				
	$Pt(ViMe_2SiOSiMe_2Vi)_2$	70	1	$CH_2{=}C(CH_3)COO(CH_2)_3SiCl_2CH_3$ (80)	U -157
	$Pt(ViMe_2SiOSiMe_2Vi)_2$ + phenothiazine	60-80	15	$CH_2{=}C(CH_3)COO(CH_2)_3SiCl_2CH_3$ (92)	U -157
	$Pt(ViMe_2SiOSiMe_2Vi)_2$ + phenothiazine	70	-	$CH_2{=}C(CH_3)COO(CH_2)_3SiCl_2CH_3$ (68)	U -157
C_7H_{10}	$CH_2{=}CHCH(OCOCH_3)_2$				
	H_2PtCl_6 (EtOH) + NBu_3	120-150	-	$CH_3SiCl_2CH_2CH_2CH(OCOCH_3)_2$ (82)	D -003
C_7H_{11}	$\overline{CH_2CH_2CH(CH_2Cl)CH_2CH{=}CH}$				
	H_2PtCl_6 (i-PrOH)	reflux	2	$(\overline{CH_2)_3CH(CH_2Cl)CH_2C}HSiCl_2CH_3$ + $\overline{CH_2CH_2CH(CH_2Cl)CH_2CH_2C}HSiCl_2CH_3$ (12)	728
C_7H_{12}	$CH_2{=}C(CH_3)CH(CH_3)OCOCH_3$				
	H_2PtCl_6	reflux	- (THF)	$CH_3SiCl_2CH_2CH(CH_3)CH(CH_3)OCOCH_3$ (71)	754
C_7H_{12}	$CH_2{=}CHCOOC_4H_9$				
	-	-	-	$CH_3SiCl_2CH_2CH_2COOC_4H_9 + CH_3SiCl_2CH(CH_3)COOC_4H_9$ (-)	1323
C_7H_{12}	$CH_3Si(CH{=}CH_2)_3$				
	H_2PtCl_6 (i-PrOH)	reflux	4	$CH_3Si(CH{=}CH_2)_2C_2H_4SiCl_2CH_3$ + $CH_3SiCH{=}CH_2(C_2H_4SiCl_2CH_3)_2$ + $CH_3Si(C_2H_4SiCl_2CH_3)_3$ (-)	1059
C_7H_{13}	$CH_2{=}CH(CH_2)_5Br$				
	-	-	-	$CH_3SiCl_2(CH_2)_7Br$ (-)	979
C_7H_{14}	$(CH_3)_3SiOCOCH_2CH{=}CH_2$				

	H_2PtCl_6 (i-PrOH)	reflux	2	$CH_3\overline{SiCl(CH_2)_3CO}$ (88)	774
C_7H_{16}	$(CH_3)_3SiOCH_2C(CH_3)=CH_2$				
	H_2PtCl_6 (i-PrOH)	reflux	-	$\overline{CH_2CH(CH_3)CH_2OSi}ClCH_3$ (85)	777
C_7H_{16}	$CH_2=CH(CH_2)_2Si(CH_3)_3$				
	H_2PtCl_6 (i-PrOH)	150-180	-	$(CH_3)_3Si(CH_2)_4SiCl_2CH_3$ (89)	1359
C_7H_{18}	$CH_2=CHSi(CH_3)_2Si(CH_3)_3$				
	H_2PtCl_6 (i-PrOH)	60	10 (p)	$CH_3SiCl_2CH_2CH_2Si(CH_3)_2Si(CH_3)_3$ (60)	1305
C_8H_3	$C_6F_5CH=CH_2$				
	H_2PtCl_6 (i-PrOH)	100	1	$C_6F_5CH_2CH_2SiCl_2CH_3$ (77)	132
	$RhCl(PPh_3)_3$	120	24	$C_6F_5CH_2CH_2SiCl_2CH_3$ (72)	858
	$Ru_3(CO)_{12}$	120	48	$C_6F_5CH_2CH_2SiCl_2CH_3$ (25) $C_6F_5CH=CHSiCl_2CH_3$ (35)	858
C_8H_7	$C_3F_7CH_2OCOCH_2CH=CH_2$				
	H_2PtCl_6 (i-PrOH)	until 150	8.5	$C_3F_7CH_2OCO(CH_2)_3SiCl_2CH_3$ (68)	103
C_8H_7	$CClF_2CF(CF_3)OCH_2COOCH_2CH=CH_2$				
	H_2PtCl_6	80	1	$CClF_2CF(CF_3)OCH_2COO(CH_2)_3SiCl_2CH_3$ (65)	D -043
C_8H_7	m-$CH_2=CHC_6H_4NO_2$				
	H_2PtCl_6 (i-PrOH)	50-100	2.5	$CH_3SiCl_2CH_2CH_2C_6H_4NO_2$-m (53)	45
	H_2PtCl_6 (i-PrOH)	50-100	2.5	$CH_3SiCl_2CH_2CH_2C_6H_4NO_2$-m (53)	47
	$Rh_4(CO)_{12}$	30-40	1	$CH_3SiCl_2CH_2CH_2C_6H_4NO_2$-m (95-99)	S -023
C_8H_7	p-$CH_2=CHC_6H_4NO_2$				
	$[PtCl_2(PhCH=CH_2)]_2$	5-10	24	$CH_3SiCl_2C_2H_4C_6H_4NO_2$-p (kin.)	965
C_8H_7	p-$ClC_6H_4CH=CH_2$				
	$KPtCl_3(C_2H_4)$	5	-	p-$ClC_6H_4CH_2CH_2SiCl_2CH_3$ (kin.)	977
	$Pt(PPh_3)_4$	15	4 (solv.)	p-$ClC_6H_4C_2H_4SiCl_2CH_3$ (kin.)	977
	$Pt(PPh_3)_4$	15	4 (solv.)	p-$ClC_6H_4C_2H_4SiCl_2CH_3$ (kin.)	576
	$PtCl_2(PPh_3)_2$	50	-	p-$ClC_6H_4CH_2CH_2SiCl_2CH_3$ (kin.)	976
	$PtHCl(PPh_3)_2$	50	-	p-$ClC_6H_4CH_2CH_2SiCl_2CH_3$ (kin.)	976

	$[PtCl_2(PhCH=CH_2)]_2$	5-20	24	p-$ClC_6H_4C_2H_4SiCl_2CH_3$ (kin.)	965
C_8H_9	Fig.54.				
	NiX_2 + PPh_3 + $HSiCl_3$	100	7	Fig.125. (60)	559
C_8H_{11}	Fig.128.				
	H_2PtCl_6 (i-PrOH)	reflux	2	Fig.129. (79)	728
C_8H_{11}	Fig.130.				
	H_2PtCl_6 (i-PrOH)	reflux	2	Fig.131. (21)	728
C_8H_{11}	Fig.56.				
	H_2PtCl_6 (i-PrOH)	reflux	2	Fig.126. (57)	728
C_8H_{11}	Fig.58.				
	H_2PtCl_6 (i-PrOH)	215	2	Fig.127. (80)	731
	H_2PtCl_6 (i-PrOH)	reflux	2	Fig.127. (77)	728
C_8H_{12}	$Si(CH=CH_2)_4$				
	$(t\text{-}BuO)_2$	reflux	1	$(CH_2=CH)_3SiCH_2CH_2SiCl_2CH_3$ + $(CH_2=CH)_3SiCH(CH_3)SiCl_2CH_3$ + $(CH_2=CH)_2Si(C_2H_4SiCl_2CH_3)_2$ (4)	1059
	$Co_2(CO)_8$	reflux	1	$(CH_2=CH)_3SiCH_2CH_2SiCl_2CH_3$ + $(CH_2=CH)_3SiCH(CH_3)SiCl_2CH_3$ + $(CH_2=CH)_2Si(C_2H_4SiCl_2CH_3)_2$ (8)	1059
	$Fe(CO)_5$	reflux	1	$(CH_2=CH)_3SiCH_2CH_2SiCl_2CH_3$ + $(CH_2=CH)_3SiCH(CH_3)SiCl_2CH_3$ + $(CH_2=CH)_2Si(C_2H_4SiCl_2CH_3)_2$ (9)	1059
	$Ni(acac)_2$ + i-$BuPCl_2$	120-150	4	$(CH_2=CH)_3SiCH_2CH_2SiCl_2CH_3$ + $(CH_2=CH)_3SiCH(CH_3)SiCl_2CH_3$ + $(CH_2=CH)_2Si(C_2H_4SiCl_2CH_3)_2$ + $CH_2=CHSi(C_2H_4SiCl_2CH_3)_3$ (19)	1059
	H_2PtCl_6	66-105	6	$(CH_2=CH)_3SiCH_2CH_2SiCl_2CH_3$ (38) $(CH_2=CH)_3SiCH(CH_3)SiCl_2CH_3$ (13) $(CH_2=CH)_2Si(CH_2CH_2SiCl_2CH_3)_2$ (31)	1058
	H_2PtCl_6 ($MeOC_2H_4OMe$)	40	-	$Si(CH_2CH_2SiCl_2CH_3)_4$ (-)	466
	H_2PtCl_6 (i-PrOH)	reflux	1	$(CH_2=CH)_3SiCH_2CH_2SiCl_2CH_3$ + $(CH_2=CH)_3SiCH(CH_3)SiCl_2CH_3$ + $(CH_2=CH)_2Si(C_2H_4SiCl_2CH_3)_2$ (81)	1059
	H_2PtCl_6 (i-PrOH)	reflux	1 (solv.)	$(CH_2=CH)_3SiCH_2CH_2SiCl_2CH_3$ + $(CH_2=CH)_3SiCH(CH_3)SiCl_2CH_3$ + $(CH_2=CH)_2Si(C_2H_4SiCl_2CH_3)_2$ (37-83)	1059

Table 2.3

= CH_3SiHCl_2 = =18=

	H_2PtCl_6 (i-PrOH)	reflux	4	$(CH_2{=}CH)_3SiCH_2CH_2SiCl_2CH_3$ + $(CH_2{=}CH)_3SiCH(CH_3)SiCl_2CH_3$ + $(CH_2{=}CH)_2Si(C_2H_4SiCl_2CH_3)_2$ + $CH_2{=}CHSi(C_2H_4SiCl_2CH_3)_3$ (-)	1059
C_8H_{14}	$C_4H_9OCH{=}CHCH{=}CH_2$				
	H_2PtCl_6 (i-PrOH)	60	3	$C_4H_9OCH{=}CHCH_2CH_2SiCl_2CH_3$ (1)	1119
C_8H_{14}	$CH_2{=}C(CH_3)COOC_4H_9$				
	-	-	-	$CH_3SiCl_2CH_2CH_2COOC_4H_9$ + $CH_3SiCl_2CH(CH_3)COOC_4H_9$ (-)	1323
C_8H_{14}	$\overline{OCH_2CH_2OCH_2C}HCH_2OCH_2CH{=}CH_2$				
	H_2PtCl_6 (i-PrOH)	reflux	52	$\overline{OCH_2CH_2OCH_2C}HCH_2O(CH_2)_3SiCl_2CH_3$ (58)	554
C_8H_{16}	$(CH_3)_2Si(CH_2CH{=}CH_2)_2$				
	H_2PtCl_6 (i-PrOH)	150-180	-	$(CH_3)_2(CH_2{=}CHCH_2)Si(CH_2)_3SiCl_2CH_3$ (72) $(CH_3)_2Si[(CH_2)_3SiCl_2CH_3]_2$ (28)	1359
C_8H_{16}	$(CH_3SiCl_2CH{=}CH)_2Si(CH_3)_2$				
	H_2PtCl_6 (i-PrOH)	150-160	6	$CH_3SiCl_2CH{=}CHSi(CH_3)_2CH_2CH(SiCl_2CH_3)_2$ (88)	1100
C_8H_{18}	$[CH_2{=}CHSi(CH_3)_2]_2O$				
	H_2PtCl_6 (i-PrOH)	50-60	3	$[CH_3Cl_2SiCH_2CH_2Si(CH_3)_2]_2O$ (-)	E -036
C_8H_{20}	$(CH_3)_3SiCH{=}CHSi(CH_3)_3$				
	H_2PtCl_6 (i-PrOH)	150-180	8	$(CH_3)_3SiCH_2CH(SiCl_2CH_3)Si(CH_3)_3$ (85)	1310
C_9H_5	$C_3F_7OCF(CF_3)CF_2OCH_2CH{=}CH_2$				
	H_2PtCl_6	70	0.5	$C_3F_7OCF(CF_3)CF_2O(CH_2)_3SiCl_2CH_3$ (-)	J -104
C_9H_8	$CH_2{=}CHCH_2COOCH_2(CF_2)_4H$				
	H_2PtCl_6 (i-PrOH)	until 200	6	$CH_3SiCl_2(CH_2)_3COOCH_2(CF_2)_4H$ (63)	103
C_9H_9	$CH_2{=}CHC_6H_4CH_2Cl$-p				
	H_2PtCl_6	-	-	$CH_3SiCl_2CH_2CH_2C_6H_4CH_2Cl$-p (-)	566
	H_2PtCl_6 ($MeOC_2H_4OMe$)	45-50	3	$CH_3SiCl_2CH_2CH_2C_6H_4CH_2Cl$-p + $CH_3SiCl_2CH(CH_3)C_6H_4CH_2Cl$-p (93)	567
C_9H_9	$CH_2{=}CHCH_2OC_6H_4NO_2$-o				
	H_2PtCl_6	90-110	2	$CH_3SiCl_2(CH_2)_3OC_6H_4NO_2$-o (87)	D -075

	H_2PtCl_6 (i-PrOH)	reflux	7	$CH_3SiCl_2(CH_2)_3OC_6H_4NO_2$-o (78)	930
C_9H_9	$CH_2=CHCH_2OC_6H_4NO_2$-p				
	H_2PtCl_6	80-110	2	$CH_3SiCl_2(CH_2)_3OC_6H_4NO_2$-p (81)	D -075
	H_2PtCl_6 (i-PrOH)	reflux	7	$CH_3SiCl_2(CH_2)_3OC_6H_4NO_2$-p (79)	930
C_9H_9	o-$BrC_6H_4OCH_2CH=CH_2$				
	H_2PtCl_6	reflux	7	o-$BrC_6H_4O(CH_2)_3SiCl_2CH_3$ (60)	930
C_9H_9	p-$ClC_6H_4CH_2CH=CH_2$				
	H_2PtCl_6	reflux	7	p-$ClC_6H_4(CH_2)_3SiCl_2CH_3$ (67)	930
C_9H_{10}	$C_6H_5SO_2CH_2CH=CH_2$				
	H_2PtCl_6 (OctOH)	110-120	4-5	$C_6H_5SO_2(CH_2)_3SiCl_2CH_3$ (65)	U -037
C_9H_{10}	$CH_2=CHC_6H_4OCH_3$-p				
	$Pt(PPh_3)_4$	15	4 (solv.)	$CH_3SiCl_2C_2H_4C_6H_4OCH_3$-p (kin.)	977
	$PtCl_2(PPh_3)_2$	50	-	$CH_3SiCl_2C_2H_4C_6H_4OCH_3$-p (kin.)	976
	$PtHCl(PPh_3)_2$	50	-	$CH_3SiCl_2C_2H_4C_6H_4OCH_3$-p (kin.)	976
	$[PtCl_2(PhCH=CH_2)]_2$	5-20	24	$CH_3SiCl_2C_2H_4C_6H_4OCH_3$-p (kin.)	965
C_9H_{10}	$CH_2=CHCH_2C_6H_4OH$-o				
	H_2PtCl_6 (Me_2CO)	82-97	2.3 (TCE)	$CH_3SiCl_2(CH_2)_3C_6H_4OH$-o/ $CH_3SiCl_2(CH_2)_3C_6H_4OSiCl_2CH_3$-o (71/29)	D -121
C_9H_{10}	$CH_2=CHCH_2OC_6H_5$				
	H_2PtCl_6 (i-PrOH)	110-130	2	$CH_3SiCl_2(CH_2)_3OC_6H_5$ (47)	53
	H_2PtCl_6 (i-PrOH)	reflux	-	$CH_3SiCl_2(CH_2)_3OC_6H_5$ (67)	102
	H_2PtCl_6 (i-PrOH)	reflux	7	$CH_3SiCl_2(CH_2)_3OC_6H_5$ (58)	930
C_9H_{11}	$CH_3SiCl(C_6H_5)CH=CH_2$				
	H_2PtCl_6 (i-PrOH)	145-156	2	$CH_3SiCl(C_6H_5)CH_2CH_2SiCl_2CH_3$ (79)	37
	$Rh_4(CO)_{12}$	35-40	1	$CH_3SiCl(C_6H_5)CH_2CH_2SiCl_2CH_3$ (95-98)	S -024
C_9H_{11}	$\overline{COC(CH_3)=C(CH_3)CON}CH_2CH=CH_2$				
	H_2PtCl_6 ($MeOC_2H_4OMe$)	100-115	- (xy)	$\overline{COC(CH_3)=C(CH_3)CON}(CH_2)_3SiCl_2CH_3$ (54)	D -106
C_9H_{11}	Fig.132.				

Table 2.3

= CH_3SiHCl_2 = =20=

	H_2PtCl_6 (i-PrOH))	200	2.5	Fig.133. (55)	727
C_9H_{14}	$CH_2=CHCH(OCOCH_2CH_3)_2$				
	H_2PtCl_6 (EtOH) + NBu_3	reflux	-	$CH_3SiCl_2CH_2CH_2CH(OCOCH_2CH_3)_2$ (80)	D -003
C_9H_{21}	$(CH_3)_2CHN[Si(CH_3)_3]CH_2CH=CH_2$				
	$PtCl_2(C_2H_4)(C_5H_5N)$	120	4	$(CH_3)_2CH\overline{N(CH_2)_3Si}ClCH_3$ (23)	D -007
C_9H_{24}	$[\overline{OSi(CH_3)_2]_3OS}i(CH_3)CH=CH_2$				
	$Ir_4(CO)_{12}$	60-65	1	$[\overline{OSi(CH_3)_2]_3OSi}CH_3CH_2CH_2SiCl_2CH_3$ (99)	S -027
	$Rh_4(CO)_{12}$	60-65	1	$[\overline{OSi(CH_3)_2]_3OSi}CH_3CH_2CH_2SiCl_2CH_3$ (98-99)	S -027
$C_{10}H_5$	$CF_3(CF_2)_5COOCH_2CH=CH_2$				
	H_2PtCl_6 (i-PrOH)	reflux	6	$CF_3(CF_2)_5COO(CH_2)_3SiCl_2CH_3$ (60)	419
$C_{10}H_6$	$CH_2=CH(CF_2)_6CH=CH_2$				
	$(t\text{-}BuO)_2$	reflux	72	$CH_3SiCl_2CH_2CH_2(CF_2)_6CH_2CH_2SiCl_2CH_3$ (-)	D -025
$C_{10}H_{11}$	Fig.134.				
	H_2PtCl_6 (i-PrOH)	200	68	Fig.135. (87)	S -042
$C_{10}H_{12}$	$CH_2=CHC_6H_4CH_2OCH_3$-p				
	H_2PtCl_6 (CH_3CN) + PPh_3	reflux	2 days	$CH_3SiCl_2(CH_2)_3C_6H_4CH_2OCH_3$-p (92)	155
	H_2PtCl_6 (i-PrOH) + NH_2Bu	-	-	$CH_3SiCl_2CH_2CH_2C_6H_4CH_2OCH_3$-p (-)	F -015
$C_{10}H_{12}$	$CH_2=CHCH_2OC_6H_4CH_3$-o				
	H_2PtCl_6 (i-PrOH)	reflux	7	$CH_3SiCl_2(CH_2)_3OC_6H_4CH_3$-o (54)	930
$C_{10}H_{12}$	$CH_2=CHCH_2OC_6H_4CH_3$-p				
	H_2PtCl_6 (i-PrOH)	reflux	7	$CH_3SiCl_2(CH_2)_3OC_6H_4CH_3$-p (63)	930
$C_{10}H_{14}$	$CH_2(COOCH_2CH=CH_2)_2$				
	H_2PtCl_6 (i-PrOH)	100	0.25	$CH_2=CHCH_2OCOCH_2COO(CH_2)_3SiCl_2CH_3$ (23)	32
$C_{10}H_{16}$	$CH_2=CHCH_2CH(COOCH_2CH_3)_2$				
	H_2PtCl_6 (i-PrOH)	reflux	-	$CH_3SiCl_2(CH_2)_3CH(COOCH_2CH_3)_2$ (53)	1008
$C_{10}H_{18}$	$CH_3Si(CH_2CH=CH_2)_3$				
	H_2PtCl_6 (i-PrOH)	150-180	-	$CH_3Si[(CH_2)_3SiCl_2CH_3]_3$ (4) $CH_3Si(CH_2CH=CH_2)[(CH_2)_3SiCl_2CH_3]_2$ (31) $CH_3Si(CH_2CH=CH_2)_2(CH_2)_3SiCl_2CH_3$ (65)	1359
$C_{10}H_{24}$	$[OSi(CH_3)_2OSi(CH_3)CH=CH_2]_2$				

H_2PtCl_6 (i-PrOH)	160-180	2	$[OSi(CH_3)_2][OSi(CH_3)(CH=CH_2)][OSi(CH_3)_2][OSi(CH_3)CH_2CH_2SiCl_2CH_3]$ + $[OSi(CH_3)_2OSi(CH_3)CH_2CH_2SiCl_2CH_3]_2$ (58-60)	46
$C_{11}H_{12}$ $CH_2=C(CH_3)C_6H_4OCOCH_3$-p				
H_2PtCl_6 (i-PrOH)	120	23	$CH_3SiCl_2CH_2CH(CH_3)C_6H_4OCOCH_3$-p (32)	1038
$C_{11}H_{12}$ $CH_2=CHCH_2C_6H_4OCOCH_3$-o				
H_2PtCl_6 (i-PrOH)	120	23	$CH_3SiCl_2(CH_2)_3C_6H_4OCOCH_3$-o (60)	1038
$C_{11}H_{13}$ p-$(CH_3)_2NC_6H_4COOCH=CH_2$				
H_2PtCl_6 (i-PrOH)	reflux	1 (cy)	p-$(CH_3)_2NC_6H_4COOCH_2CH_2SiCl_2CH_3$ (-)	D -015
$C_{11}H_{14}$ Fig. 136.				
H_2PtCl_6 (i-PrOH)	200	70	Fig. 137. (86)	S -042
$C_{11}H_{16}$ $C_6H_5(CH_3)_2SiCH_2CH=CH_2$				
H_2PtCl_6 (i-PrOH)	150-180	-	$C_6H_5(CH_3)_2Si(CH_2)_3SiCl_2CH_3$ (77)	1359
$C_{12}H_{15}$ $CH_2=CHCH_2OCOC_6H_4N(CH_3)_2$-p				
H_2PtCl_6 (i-PrOH)	reflux	1.3 (cy)	$CH_3SiCl_2(CH_2)_3OCOC_6H_4N(CH_3)_2$-p (-)	Ch-001
H_2PtCl_6 (i-PrOH)	reflux	1.3 (cy)	$CH_3SiCl_2(CH_2)_3OCOC_6H_4N(CH_3)_2$-p (-)	Ch-002
H_2PtCl_6 (i-PrOH)	reflux	1.3 (cy)	$CH_3SiCl_2(CH_2)_3OCOC_6H_4N(CH_3)_2$-p (-)	D -015
$C_{12}H_{18}$ 2-$CH_2=CHCH_2C_6H_4OSi(CH_3)_3$				
H_2PtCl_6 (i-PrOH)	reflux	12	2-$(CH_3)_3SiOC_6H_4(CH_2)_3SiCl_2CH_3$ (-)	906
$C_{12}H_{18}$ $CH_2=CHCH_2OCO(CH_2)_4COOCH_2CH=CH_2$				
H_2PtCl_6 (i-PrOH)	100	0.25	$CH_2=CHCH_2OCO(CH_2)_4COO(CH_2)_3SiCl_2CH_3$ (23)	32
$C_{12}H_{18}$ $CH_2=CHCH_2OCOO(C_2H_4O)_2COOCH_2CH=CH_2$				
H_2PtCl_6 (C_6H_{14})	reflux	8	$CH_3SiCl_2(CH_2)_3OCOO(C_2H_4O)_2COO(CH_2)_3SiCl_2CH_3$ (96)	J -051
$C_{12}H_{20}$ $Si(CH_2CH=CH_2)_4$				
H_2PtCl_6 (i-PrOH)	150-180	-	$(CH_2=CHCH_2)_3Si(CH_2)_3SiCl_2CH_3$ (37) $(CH_2=CHCH_2)_2Si[(CH_2)_3SiCl_2CH_3]_2$ (20) $(CH_2=CHCH_2)Si[(CH_2)_3SiCl_2CH_3]_3$ (4)	1359
$C_{12}H_{20}$ $[CH_2=CHCH_2CH_2C(CH_3)=N]_2$				
H_2PtCl_6	25-74	1.5	$[CH_3SiCl_2(CH_2)_4C(CH_3)=N]_2$ (-)	U -063

	H_2PtCl_6	50-60	3	$[CH_3SiCl_2(CH_2)_4C(CH_3)=N]_2$ (-)	267
	H_2PtCl_6 (i-PrOH)	50-70	1.5	$[CH_3SiCl_2(CH_2)_4C(CH_3)=N]_2$ (-)	U -065
$C_{12}H_{22}$	$CH_2=CH(CH_2)_8COOCH_3$				
	H_2PtCl_6 (MeOH)	120	5	$CH_3SiCl_2(CH_2)_{10}COOCH_3$ (-)	D -108
	H_2PtCl_6 (i-PrOH)	reflux	5	$CH_3SiCl_2(CH_2)_{10}COOCH_3$ (97)	1011
$C_{12}H_{24}$	$[OSi(CH_3)CH=CH_2]_4$				
	H_2PtCl_6	50	-	$[OSi(CH_3)CH_2CH_2SiCl_2CH_3]_4$ (-)	606
	$PtCl_2(CO)_2$	80-110	-	$[OSi(CH_3)CH_2CH_2SiCl_2CH_3]_4$ (-)	D -059
$C_{13}H_9$	Fig.138.				
	H_2PtCl_6 ($MeOC_2H_4OMe$)	70-104	3 (t)	Fig.139. (-)	D -124
	H_2PtCl_6 (glycol ether)	reflux	3	Fig.140. (74)	D -126
$C_{13}H_{10}$	$C_8F_{17}SO_2N(CH_3)CH_2CH_2CH=CH_2$				
	Pt/C	reflux	16	$C_8F_{17}SO_2N(CH_3)(CH_2)_4SiCl_2CH_3$ (-)	U -030
$C_{13}H_{22}$	$CH_2=CHCOOC_{10}H_{19}$				
	H_2PtCl_6 (i-PrOH)	120	1.5	$CH_3SiCl_2CH_2CH_2COOC_{10}H_{19}$ (24) $CH_3SiCl_2CH(CH_3)COOC_{10}H_{19}$ (30)	34
$C_{13}H_{24}$	$CH_2=CH(CH_2)_8COOCH_2CH_3$				
	H_2PtCl_6 (i-PrOH)	reflux	1	$CH_3SiCl_2(CH_2)_{10}COOCH_2CH_3$ (-)	B -002
$C_{13}H_{24}$	$CH_2=CH(CH_2)_9OCOCH_3$				
	H_2PtCl_6 (MeOH)	120	5	$CH_3SiCl_2(CH_2)_{11}OCOCH_3$ (-)	D -108
	H_2PtCl_6 (i-PrOH)	reflux	5	$CH_3SiCl_2(CH_2)_{11}OCOCH_3$ (96)	1011
$C_{13}H_{24}$	$CH_2=CHCH_2OC_{10}H_{19}$				
	H_2PtCl_6 (i-PrOH)	140	0.7	$CH_3SiCl_2(CH_2)_3OC_{10}H_{19}$ (65)	33
$C_{14}H_{14}$	o-$C_6H_4(COOCH_2CH=CH_2)_2$				
	H_2PtCl_6	90	3	o-$C_6H_4[COO(CH_2)_3SiCl_2CH_3]_2$ (-)	J -033
	H_2PtCl_6	90	3	o-$C_6H_4[COO(CH_2)_3SiCl_2CH_3]_2$ (-)	D -100
	H_2PtCl_6	90	3	o-$C_6H_4[COO(CH_2)_3SiCl_2CH_3]_2$ (-)	U -120
	H_2PtCl_6 (i-PrOH)	100	0.25	o-$CH_2=CHCH_2OCOC_6H_4COO(CH_2)_3SiCl_2CH_3$ (32)	32
$C_{14}H_{21}$	Fig.141.				

Table 2.3

= CH_3SiHCl_2 = =23=

	H_2PtCl_6 (i-PrOH)	until 170	6	Fig.142. (63)	777
$C_{14}H_{24}$	$CH_2=C(CH_3)COOC_{10}H_{19}$				
	H_2PtCl_6 (i-PrOH)	120	1.5	$CH_3SiCl_2CH_2CH(CH_3)COOC_{10}H_{19}$ (52)	34
$C_{16}H_{18}$	$(C_6H_5)_2CH_3SiCH_2CH=CH_2$				
	H_2PtCl_6 (i-PrOH)	150-180	-	$(C_6H_5)_2CH_3Si(CH_2)_3SiCl_2CH_3$ (68)	1359
$C_{16}H_{24}$	Fig.143.				
	Pt/C	reflux	24 (e)	Fig.144. (80-90)	498
$C_{19}H_9$	o-$C_{10}F_{19}OC_6H_4CH_2CH=CH_2$				
	$PtCl_2(SEt_2)_2$	reflux	1.5	o-$C_{10}F_{19}OC_6H_4(CH_2)_3SiCl_2CH_3$ (-)	G -025
$C_{19}H_{14}$	Fig.145.				
	H_2PtCl_6	70-80	3-4	Fig.146. (-)	D -018
$C_{19}H_{36}$	$CH_2=CHC_{15}H_{30}COOCH_3$				
	$(PhCOO)_2$	reflux	5	$CH_3SiCl_2C_{17}H_{34}COOCH_3$ (23)	1011
	H_2PtCl_6 (i-PrOH)	reflux	5	$CH_3SiCl_2C_{17}H_{34}COOCH_3$ (75)	1011
	Pt/C	reflux	5	$CH_3SiCl_2C_{17}H_{34}COOCH_3$ (73)	1011
$C_{19}H_{36}$	$CH_3(CH_2)_7CH=CH(CH_2)_7COOCH_3$				
	H_2PtCl_6 (i-PrOH)	reflux	5	$CH_3(CH_2)_7CH(SiCl_2CH_3)(CH_2)_8COOCH_3$ + $CH_3(CH_2)_8CH(SiCl_2CH_3)(CH_2)_7COOCH_3$ (70)	1011
	H_2PtCl_6 (i-PrOH)	reflux	1	$CH_3(CH_2)_7CH(SiCl_2CH_3)(CH_2)_8COOCH_3$ + $CH_3(CH_2)_8CH(SiCl_2CH_3)(CH_2)_7COOCH_3$ (-)	B -002
$C_{20}H_{39}$	$CH_2=CHSi(CH_3)_2O[SiCH_3(CH_2CH_2CF_3)O]_3Si(CH_3)_2CH=CH_2$				
	H_2PtCl_6	110-125	1	$CH_2=CHSi(CH_3)_2O[SiCH_3(CH_2CH_2CF_3)O]_3Si(CH_3)_2$-$CH_2CH_2SiCl_2CH_3$ (74)	U -018
$C_{21}H_{20}$	$(C_6H_5)_3SiCH_2CH=CH_2$				
	H_2PtCl_6 (i-PrOH)	150-180	-	$(C_6H_5)_3Si(CH_2)_3SiCl_2CH_3$ (66)	1359
$C_{21}H_{20}$	$CH_2=CHCH_2P(C_6H_5)_3Br$				
	H_2PtCl_6	-	10	$CH_3SiCl_2(CH_2)_3P(C_6H_5)_3Br$ (-)	173
$C_{24}H_{33}$	Fig.147.				
	Pt/C	reflux	24 (e)	Fig.148. (80-90)	498
$C_{24}H_{38}$	m-$(CH_3)_3SiOC_6H_4(CH_2)_7CH=CHCH_2CH=CHCH_2CH=CH_2$				

Table 2.3

= CH_3SiHCl_2 = =24=

	H_2PtCl_6 (i-PrOH)	reflux	-	m-$(CH_3)_3SiOC_6H_4(CH_2)_7CH{=}CHCH_2CH{=}CH(CH_2)_3$-$SiCl_2CH_3$ (-)	906
$C_{24}H_{42}$	m-$(CH_3)_3SiOC_6H_4(CH_2)_7CH{=}CH(CH_2)_5CH_3$				
	H_2PtCl_6 (i-PrOH)	reflux	136	m-$(CH_3)_3SiOC_6H_4(CH_2)_7CH_2CH(SiCl_2CH_3)(CH_2)_5CH_3$ (-)	906
$C_{37}H_{66}$	$CH_2{=}C(CH_3)CH_2(OCH_2CH_2)_9OC_6H_4C_9H_{19}$				
	PhCOOOBu-t	-	-	$CH_3SiCl_2CH_2CH(CH_3)CH_2(OCH_2CH_2)_9OC_6H_4C_9H_{19}$ (-)	U -077
$C_{43}H_{78}$	$C_9H_{19}C_6H_4O(CH_2CH_2O)_{12}CH_2C(CH_3){=}CH_2$				
	PhCOOOBu-t	-	overnight	$C_9H_{19}C_6H_4O(CH_2CH_2O)_{12}CH_2CH(CH_3)CH_2SiCl_2CH_3$ (-)	U -045
	PhCOOOBu-t	-	overnight	$C_9H_{19}C_6H_4O(CH_2CH_2O)_{12}CH_2CH(CH_3)CH_2SiCl_2CH_3$ (-)	U -076

Table 2.4

= CH_3SiHCl_2 = =1=

C_2H_2 CH≡CH

$PdCl_2(PPh_3)_2$	70-80	3	CH_2=$CHSiCl_2CH_3$ (30) $CH_3SiCl_2CH_2CH_2SiCl_2CH_3$ (40)	1288
$PdCl_2(PPh_3)_2$	70-80	3	CH_2=$CHSiCl_2CH_3$ (30) $CH_3SiCl_2CH_2CH_2SiCl_2CH_3$ (40)	1290
H_2PtCl_6	15-160	flow	CH_2=$CHSiCl_2CH_3$ (88)	D -031
H_2PtCl_6 ($BuCHEtCH_2OH$)	115-120	-	CH_2=$CHSiCl_2CH_3$ (-)	J -039
H_2PtCl_6 (BzOH)	140		$CH_3SiCl_2CH_2CH_2SiCl_2CH_3$ (88)	D -042
H_2PtCl_6 (BzOH)	140-170	-	CH_2=$CHSiCl_2CH_3$ (87)	D -042
H_2PtCl_6 (THF) + $AlCl_3$	30	- (dcm)	CH_2=$CHSiCl_2CH_3$ (75)	1234
H_2PtCl_6 (i-PrOH)	20	0.5 (xy)	CH_2=$CHSiCl_2CH_3$ (8) $CH_3SiCl_2CH_2CH_2SiCl_2CH_3$ (80)	1290
H_2PtCl_6 (i-PrOH)	20	0.5 (xy)	CH_2=$CHSiCl_2CH_3$ (8) $CH_3SiCl_2CH_2CH_2SiCl_2CH_3$ (80)	1288
H_2PtCl_6 [$\overline{(CH_2)_5CO}$]	140	-	CH_2=$CHSiCl_2CH_3$ (92-96) $CH_3SiCl_2CH_2CH_2SiCl_2CH_3$ (3-7)	D -041
H_2PtCl_6 [$\overline{(CH_2)_5CO}$]	140	-	CH_2=$CHSiCl_2CH_3$ (92-96) $CH_3SiCl_2CH_2CH_2SiCl_2CH_3$ (3-7)	D -042
$Pt(PPh_3)_4$	70-80	11 (xy)	CH_2=$CHSiCl_2CH_3$ (45) $CH_3SiCl_2CH_2CH_2SiCl_2CH_3$ (23)	1288
$Pt(PPh_3)_4$	70-80	11 (xy)	CH_2=$CHSiCl_2CH_3$ (45) $CH_3SiCl_2CH_2CH_2SiCl_2CH_3$ (23)	1290
Pt/C (o-$C_6H_4Cl_2$)	105	2	CH_2=$CHSiCl_2CH_3$ (86) $CH_3SiCl_2CH_2CH_2SiCl_2CH_3$ (trace)	F -007
Pt/C (o-$C_6H_4Cl_2$)	114-116	3	CH_2=$CHSiCl_2CH_3$ (91)	F -007
$PtCl_2(PPh_3)_2$	72	7 (xy)	CH_2=$CHSiCl_2CH_3$ (-) $CH_3SiCl_2CH_2CH_2SiCl_2CH_3$ (12-36)	1290
$PtCl_2(PPh_3)_2$	72	7 (xy)	CH_2=$CHSiCl_2CH_3$ (-) $CH_3SiCl_2CH_2CH_2SiCl_2CH_3$ (12-36)	1288
SDVB-CN/H_2PtCl_6	70	flow	CH_2=$CHSiCl_2CH_3$ (90)	C -003
SDVB-PPh_2/H[$PtCl_3(C_2H_4)$]	20	18	CH_2=$CHSiCl_2CH_3$ (33)	C -002
SiO_2-$(CH_2)_3S$/H_2PtCl_6	20 (1 atm)	6.5	CH_2=$CHSiCl_2CH_3$ (35) $CH_3SiCl_2CH_2CH_2SiCl_2CH_3$ (42)	504

Table 2.4

= CH_3SiHCl_2 = =2=

SiO_2-$(CH_2)_3S/H_2PtCl_6$	20 (1 atm)	9	$CH_2=CHSiCl_2CH_3$ (25) $CH_3SiCl_2CH_2CH_2SiCl_2CH_3$ (63)	504
SiO_2-$(CH_2)_3S/H_2PtCl_6$	30 (1 atm)	6.5	$CH_2=CHSiCl_2CH_3$ (13) $CH_3SiCl_2CH_2CH_2SiCl_2CH_3$ (65)	504
polymer-$(CH_2)_nS/PtCl_x$	r.t.	-	$CH_2=CHSiCl_2CH_3$ + $CH_3Cl_2SiCH_2CH_2SiCl_2CH_3$ (-)	505
$RhCl(PPh_3)_3$	60-70	10 (xy)	$CH_2=CHSiCl_2CH_3$ (46) $CH_3SiCl_2CH_2CH_2SiCl_2CH_3$ (18)	1288
$RhCl(PPh_3)_3$	60-70	10 (xy)	$CH_2=CHSiCl_2CH_3$ (46) $CH_3SiCl_2CH_2CH_2SiCl_2CH_3$ (18)	1290
$RhH(PPh_3)_4$	55	3	$CH_2=CHSiCl_2CH_3$ (54) $CH_3SiCl_2CH_2CH_2SiCl_2CH_3$ (17)	1288
$RhH(PPh_3)_4$	55	3	$CH_2=CHSiCl_2CH_3$ (54) $CH_3SiCl_2CH_2CH_2SiCl_2CH_3$ (17)	1290
SiO_2-$PPh_2/RhCl(PPh_3)_3$	140	flow	$CH_2=CHSiCl_2CH_3$ (42-46)	744
SiO_2-$PPh_2/RhCl(PPh_3)_3$	140	flow	$CH_2=CHSiCl_2CH_3$ (44) $CH_3SiCl_2CH_2CH_2SiCl_2CH_3$ (-)	742
SiO_2-$PPh_2/RhCl(PPh_3)_3$	140	flow	$CH_2=CHSiCl_2CH_3$ (42-46)	Pl-019
$[RhCl(CO)_2]_2$	50	4	$CH_2=CHSiCl_2CH_3$ (11) $CH_3SiCl_2CH_2CH_2SiCl_2CH_3$ (60)	1257
$[RhCl(CO)_2]_2[(MeOC_2H_4)_2O]$	r.t.	6	$CH_3SiCl_2CH_2CH_2SiCl_2CH_3$ (62)	S -045
asbestos-$PPh_2/RhCl(PPh_3)_3$	140	flow	$CH_2=CHSiCl_2CH_3$ (68)	744
asbestos-$PPh_2/RhCl(PPh_3)_3$	140	flow	$CH_2=CHSiCl_2CH_3$ (68)	Pl-016
siloxane-$(CH_2)_2PPh_2/[RhCl(C_8H_{14})_2]_2$	40	1	$CH_2=CHSiCl_2CH_3$ (5)	941
siloxane-$(CH_2)_3PPh_2/[RhCl(C_8H_{14})_2]_2$	40	1	$CH_2=CHSiCl_2CH_3$ (9)	941
siloxane-$PPh_2/RhCl(PPh_3)_3$	140	flow	$CH_2=CHSiCl_2CH_3$ (68-79) $CH_3SiCl_2CH_2CH_2SiCl_2$ (-)	742
$RuCl_2(PPh_3)_3$	65	6 (xy)	$CH_2=CHSiCl_2CH_3$ (28) $CH_3SiCl_2CH_2CH_2SiCl_2CH_3$ (5)	1288
$RuCl_2(PPh_3)_3$	65	6 (xy)	$CH_2=CHSiCl_2CH_3$ (28) $CH_3SiCl_2CH_2CH_2SiCl_2CH_3$ (5)	1290
SiO_2-$PPh_2/RuCl_2(PPh_3)_3$	115	flow	$CH_2=CHSiCl_2CH_3$ (49) $CH_3SiCl_2CH_2CH_2SiCl_2CH_3$ (11-16)	Pl-018
SiO_2-$PPh_2/RuCl_2(PPh_3)_3$	115	flow	$CH_2=CHSiCl_2CH_3$ (49) $CH_3SiCl_2CH_2CH_2SiCl_2CH_3$ (11-16)	744

	SiO_2-PPh_2/$RuCl_2(PPh_3)_3$	115	flow	CH_2=$CHSiCl_2CH_3$ (49) $CH_3SiCl_2CH_2CH_2SiCl_2CH_3$ (11-16)	PI-017
	SiO_2-PPh_2/$RuCl_2(PPh_3)_3$	140	flow	CH_2=$CHSiCl_2CH_3$ (41-45)	744
	SiO_2-PPh_2/$RuCl_2(PPh_3)_3$	140	flow	CH_2=$CHSiCl_2CH_3$ (41-45)	PI-018
C_4H_6	$CH_3C{\equiv}CCH_3$				
	H_2PtCl_6	25-30	-	cis-CH_3CH=$C(CH_3)SiCl_2CH_3$ (47)	991
	[HPt(SiBzMe$_2$)P(c-Hex)$_3$]$_2$	65	0.5	cis-CH_3CH=$C(CH_3)SiCl_2CH_3$ (87)	430
C_4H_6	$CH_3CH_2C{\equiv}CH$				
	[HPt(SiBzMe$_2$)P(c-Hex)$_3$]$_2$	75	0.5	trans-CH_3CH_2CH=$CHSiCl_2CH_3$+CH_3CH_2C(=CH_2)$SiCl_2CH_3$ (85)	430
C_5H_6	CH_2=$C(CH_3)C{\equiv}CH$				
	H_2PtCl_6 +$(CH_3)_2ClSiH$	80	6	$(CH_3)_2C$=$CHCH(SiCl_2CH_3)_2$ (18) $(CH_3)_2C$=$CHCH[SiCl(CH_3)_2]_2$ (29) $(CH_3)_2C$=$CHCH(SiCl_2CH_3)[SiCl(CH_3)_2]$ (53)	672
	H_2PtCl_6+ $(PhCOO)_2$	80	18	CH_2=$C(CH_3)CH$=$CHSiCl_2CH_3$ (75)	672
	H_2PtCl_6 + $HSiCl_3$	80	6	CH_2=$C(CH_3)CH$=$CHSiCl_2CH_3$ (56) CH_2=$C(CH_3)CH$=$CHSiCl_3$ (14)	672
	H_2PtCl_6 + $HSiCl_3$	80	6	CH_2=$C(CH_3)CH$=$CHSiCl_2CH_3$ (-)	672
C_5H_8	$CH_3CH_2C{\equiv}CCH_3$				
	H_2PtCl_6 (i-PrOH)	95	24 (cy)	CH_3CH_2CH=$C(CH_3)SiCl_2CH_3$+$CH_3CH_2C(SiCl_2CH_3)$=$CHCH_3$ (40)	1173
	[HPt(SiBzMe$_2$)P(c-Hex)$_3$]$_2$	65	1	CH_3CH_2CH=$C(CH_3)SiCl_2CH_3$ (62) $CH_3CH_2C(SiCl_2CH_3)$=$CHCH_3$ (34)	1205
C_5H_8	$CH_3CH_2CH_2C{\equiv}CH$				
	[HPt(SiBzMe$_2$)P(c-Hex)$_3$]$_2$	65	0.5	trans-$CH_3CH_2CH_2CH$=$CHSiCl_2CH_3$ (84) $CH_3CH_2CH_2C$(=CH_2)$SiCl_2CH_3$ (4)	1205
C_6H_{10}	$(CH_3)_3CC{\equiv}CH$				
	e	120 270	0.17	$(CH_3)_3CCH$=$CHSiCl_2CH_3$ (63-67)	1270
C_6H_{10}	$CH_3(CH_2)_3C{\equiv}CH$				
	e	120-270	0.17	$CH_3(CH_2)_3CH$=$CHSiCl_2CH_3$ (83-87)	1270
	e	270	-	$CH_3(CH_2)_3CH$=$CHSiCl_2CH_3$ (-)	950

Table 2.4
= CH_3SiHCl_2 = =4=

	$[HPt(SiBzMe_2)P(c\text{-}Hex)_3]_2$	65	1.5	$CH_3(CH_2)_3CH{=}CHSiCl_2CH_3$ (91) $CH_3(CH_2)_3C(SiCl_2CH_3){=}CH_2$ (3)	1205
C_6H_{10}	$CH_3CH_2C{\equiv}CCH_2CH_3$				
	$Ni(bipy)(C_2H_5)_2$	20	70	$CH_3Cl_2SiC(CH_2CH_3){=}CHCH_2CH_3$ (11) $CH_3Cl_2SiC(CH_2CH_3){=}C(CH_2CH_3)SiCl_2CH_3$ (56)	1184
	$[HPt(SiBzMe_2)P(c\text{-}Hex)_3]_2$	65	0.5	cis-$CH_3Cl_2SiC(CH_2CH_3){=}CHCH_2CH_3$ (95)	1205
C_6H_{10}	$CH_3CH_2CH_2C{\equiv}CCH_3$				
	$[HPt(SiBzMe_2)P(c\text{-}Hex)_3]_2$	65	1	$CH_3CH_2CH_2CH{=}C(CH_3)SiCl_2CH_3$ (68) $CH_3CH_2CH_2(CH_3Cl_2Si)C{=}CHCH_3$ (27)	1205
C_7H_{10}	$\overline{(CH_2)_4C}HC{\equiv}CH$				
	$[HPt(SiBzMe_2)P(c\text{-}Hex)_3]_2$	65	1.5	$\overline{(CH_2)_4C}HCH{=}CHSiCl_2CH_3$ (92)	1205
C_7H_{12}	$CH_3(CH_2)_4C{\equiv}CH$				
	$[HPt(SiBzMe_2)P(c\text{-}Hex)_3]_2$	65	1.5	$CH_3(CH_2)_4CH{=}CHSiCl_2CH_3$ (91) $CH_3(CH_2)_4C(SiCl_2CH_3){=}CH_2$ (4)	1205
C_8H_6	$C_6H_5C{\equiv}CH$				
	e	120-270	0.17	$C_6H_5CH{=}CHSiCl_2CH_3$ (0-19)	1270
	H_2OsCl_6 (THF)	110	4	$C_6H_5CH{=}CHSiCl_2CH_3$ (88) $C_6H_5CH(SiCl_2CH_3){=}CH_2$ (trace)	608
	$(Bu_4N)_2(PtCl_6)$	80	1-3	$C_6H_5CH{=}CHSiCl_2CH_3$ (51)	673
	H_2PtCl_6	100	5	$C_6H_5CH{=}CHSiCl_2CH_3$ + $C_6H_5CH(SiCl_2CH_3){=}CH_2$ (97)	942
	H_2PtCl_6	20	-	$C_6H_5CH{=}CHSiCl_2CH_3$ (81) $C_6H_5CH(SiCl_2CH_3){=}CH_2$ (10)	172
	H_2PtCl_6 ($CH_2{=}CH$-N donor)	100	5	$C_6H_5CH{=}CHSiCl_2CH_3$ + $C_6H_5C(SiCl_2CH_3){=}CH_2$ (95)	612
	H_2PtCl_6 (THF)	110	4	$C_6H_5CH{=}CHSiCl_2CH_3$ (63) $C_6H_5CH(SiCl_2CH_3){=}CH_2$ (27)	608
	H_2PtCl_6 (i-PrOH)	reflux	6-12	$C_6H_5CH{=}CHSiCl_2CH_3$ (96)	164
	$PtCl_2$ ($CH_2{=}CH$-N donor)	100	5	$C_6H_5CH{=}CHSiCl_2CH_3$+$C_6H_5C(SiCl_2CH_3){=}CH_2$ (91-98)	612
	$PtCl_3(CH_2CH{=}CH_2)PPh_3$	-	-	$C_6H_5CH{=}CHSiCl_2CH_3$ (95)	172
	$PtCl_4$ ($CH_2{=}CH$-N donor)	100	5	$C_6H_5CH{=}CHSiCl_2CH_3$+$C_6H_5C(SiCl_2CH_3){=}CH_2$ (90-96)	612
	SiO_2-PPh_3/$PtCl_6^{-2}$	-	-	$C_6H_5CH{=}CHSiCl_2CH_3$ (75)	172
	$(RhLPPh_3)_2$ (C_6H_6)	80	3	$C_6H_5CH{=}CHSiCl_2CH_3$ (51)	886

Table 2.4

= CH_3SiHCl_2 = =5=

	$(RhLPPh_3)_2(C_6H_6)$	80	3	$C_6H_5CH{=}CHSiCl_2CH_3$ (51)	S -003
	$Rh(acac)(CO)_2$	100	5	$C_6H_5CH{=}CHSiCl_2CH_3 + C_6H_5CH(SiCl_2CH_3){=}CH_2$ (80)	942
	$RhCl(CO)(PPh_3)_2$	100	5	$C_6H_5CH{=}CHSiCl_2CH_3 + C_6H_5CH(SiCl_2CH_3){=}CH_2$ (85)	942
	$RhCl(PPh_2CH_3)_3$	100	5	$C_6H_5CH{=}CHSiCl_2CH_3 + C_6H_5CH(SiCl_2CH_3){=}CH_2$ (81)	942
	$RhCl(PPh_3)_3$	100	5	$C_6H_5CH{=}CHSiCl_2CH_3 + C_6H_5CH(SiCl_2CH_3){=}CH_2$ (98)	942
	$RhCl(SbPh_3)_3$	100	5	$C_6H_5CH{=}CHSiCl_2CH_3 + C_6H_5CH(SiCl_2CH_3){=}CH_2$ (74)	942
	$RhH(CO)(PPh_3)_3$	100	5	$C_6H_5CH{=}CHSiCl_2CH_3 + C_6H_5CH(SiCl_2CH_3){=}CH_2$ (86)	942
	$RhI(PPh_3)_3$	100	5	$C_6H_5CH{=}CHSiCl_2CH_3 + C_6H_5CH(SiCl_2CH_3){=}CH_2$ (87)	942
C_8H_{14}	$CH_3(CH_2)_5C{\equiv}CH$				
	$Pt[(CH_2{=}CHSiMe_2)_2O]_2(xy)$	-	-	$CH_3Cl_2SiCH{=}CH(CH_2)_5CH_3$ (-)	1188
C_9H_8	$C_6H_5C{\equiv}CCH_3$				
	$Ni(bipy)(C_2H_5)_2$	20	49	$C_6H_5CH{=}C(CH_3)SiCl_2CH_3$ (52) $C_6H_5C(SiCl_2CH_3){=}C(CH_3)iCl_2CH_3$ (27)	1184
$C_{10}H_{18}$	$(CH_3)_3CC{\equiv}CC(CH_3)_3$				
	H_2PtCl_6 (i-PrOH)	reflux	20	$(CH_3)_3CCH{=}C(SiCl_2CH_3)C(CH_3)_3$ (90)	137
$C_{10}H_{18}$	$CH_3(CH_2)_3C{\equiv}C(CH_2)_3CH_3$				
	$Ni(bipy)(C_2H_5)_2$	20	14-64	$CH_3(CH_2)_3CH{=}C(SiCl_2CH_3)(CH_2)_3CH_3$ (22-24) $CH_3(CH_2)_3C(SiCl_2CH_3){=}C(SiCl_2CH_3)(CH_2)_3CH_3$ (69-72)	1184
$C_{14}H_{10}$	$C_6H_5C{\equiv}CC_6H_5$				
	$Ni(bipy)(C_2H_5)_2$	20	24	$C_6H_5CH{=}C(SiCl_2CH_3)C_6H_5$ (60) $CH_3Cl_2SiC(C_6H_5){=}C(C_6H_5)SiCl_2CH_3$ (3)	1184
	$NiCl_2(DMPF)$	120	2 days	$C_6H_5CH{=}C(SiCl_2CH_3)C_6H_5$ (78)	589
	H_2PtCl_6 (i-PrOH)	reflux	2 (t)	$C_6H_5CH{=}C(SiCl_2CH_3)C_6H_5$ (83)	837
	$[HPt(SiBzMe_2)P(c\text{-}Hex)_3]_2$	60	3	cis-$C_6H_5CH{=}C(SiCl_2CH_3)C_6H_5$ (82)	430

Table 2.5

= CH_3SiHCl_2 = =1=

C_2	$Cl_3SiC{\equiv}CSiCl_3$				
	H_2PtCl_6 (i-PrOH)	120-130	-	$Cl_3SiCH{=}C(SiCl_3)SiCl_2CH_3$ (35)	1053
C_2H_1	$CH{\equiv}CSiCl_3$				
	none	520	30 s	$CH_3SiCl_2CH{=}CHSiCl_3$ (16)	1057
	H_2PtCl_6 (i-PrOH)	130	4.5	$CH_3SiCl_2CH{=}CHSiCl_3$ (83)	1057
C_3H_3	$CH{\equiv}CCH_2Cl$				
	$SiO_2[OSi(CH_2)_3ER_3]_2PtCl_6$;$ER_3{=}NMe_3,PBu_3$, PPh_3	37	-	$CH_3SiCl_2C_3H_4Cl$ (kin.)	1106
C_4H_4	$ClCH_2C{\equiv}CCH_2Cl$				
	H_2PtCl_6	-	-	$ClCH_2CH{=}C(CH_2Cl)SiCl_2CH_3$ (-)	1022
	H_2PtCl_6	90	1	$ClCH_2CH{=}C(CH_2Cl)SiCl_2CH_3$ (60)	730
	H_2PtCl_6 (i-PrOH)	reflux	8	$ClCH_2CH{=}C(CH_2Cl)SiCl_2CH_3$ (87)	E -011
C_4H_6	$CH_3Cl_2SiC{\equiv}CSiCl_2CH_3$				
	H_2PtCl_6 (i-PrOH)	120-130	-	$CH_3Cl_2SiCH{=}C(SiCl_2CH_3)_2$ (65)	1053
C_5H_6	$CH_3COOCH_2C{\equiv}CH$				
	H_2PtCl_6 (OctOH)	reflux	10	$CH_3COOCH_2CH{=}CHSiCl_2CH_3/CH_3COOCH_2C({=}CH_2)SiCl_2CH_3$ (60/40)	U -051
C_5H_{10}	$CH{\equiv}CSi(CH_3)_2OCH_3$				
	H_2PtCl_6 (i-PrOH)	90-100	1	$CH_3SiCl_2CH{=}CHSi(CH_3)_2OCH_3$ (78)	1281
C_5H_{10}	$CH{\equiv}CSi(CH_3)_3$				
	H_2PtCl_6	40-100	1	$CH_3SiCl_2CH{=}CHSi(CH_3)_3$ (92) $(CH_3SiCl_2)_2CHCH_2Si(CH_3)_3$ (6)	1100
C_5H_{10}	$CH{\equiv}CSi(OCH_3)_2CH_3$				
	H_2PtCl_6 (i-PrOH)	80	0.5	$CH_3SiCl_2CH{=}CHSi(OCH_3)_2CH_3$ (74)	1309
C_5H_{10}	$CH{\equiv}CSi(OCH_3)_3$				
	H_2PtCl_6	80-90	0.5	$CH_3SiCl_2CH{=}CHSi(OCH_3)_3$ (68)	1261
C_6H_8	$(CH{\equiv}C)_2Si(CH_3)_2$				
	H_2PtCl_6 (i-PrOH)	40-100	1	$CH{\equiv}CSi(CH_3)_2CH{=}CHSiCl_2CH_3$ + $CH_3SiCl_2CH{=}CH)_2Si(CH_3)_2$ (-)	1100
C_6H_8	$(CH{\equiv}C)_2Si(OCH_3)CH_3$				

Table 2.5

= CH_3SiHCl_2 = =2=

	H_2PtCl_6 (i-PrOH)	80-90	0.5	$CH{\equiv}CSi(OCH_3)(CH_3)CH{=}CHSiCl_2CH_3$ (67)	1261
C_6H_{12}	$(CH_3)_2ClSiC{\equiv}CSiCl(CH_3)_2$				
	H_2PtCl_6 (i-PrOH)	120-130	-	$(CH_3)_2ClSiCH{=}C(SiCl_2CH_3)SiCl(CH_3)_2$ (23)	1053
C_7H_6	$(CH{\equiv}C)_3SiCH_3$				
	H_2PtCl_6 (i-PrOH)	80-90	0.5	$(CH{\equiv}C)_2Si(CH_3)CH{=}CHSiCl_2CH_3$ (45) $CH{\equiv}CSi(CH_3)(CH{=}CHSiCl_2CH_3)_2$ (44)	1278
C_7H_{10}	$CH{\equiv}CCOO(CH_2)_3CH_3$				
	H_2PtCl_6 (i-PrOH)	reflux	1	$CH_3SiCl_2CH{=}CHCOO(CH_2)_3CH_3$ (-)	B -002
C_7H_{12}	$(CH_3)_3SiC{\equiv}CCH{=}CH_2$				
	H_2PtCl_6 (i-PrOH)	-	-	$(CH_3)_3SiCH{=}C(SiCl_2CH_3)CH{=}CH_2$ (-)	139
C_7H_{12}	$CH_3Si(C{\equiv}CH)(CH_2)_4$				
	H_2PtCl_6 (i-PrOH)	90	1	$CH_3Si(CH{=}CHSiCl_2CH_3)(CH_2)_4$ (92)	1276
C_7H_{14}	$CH{\equiv}CSi(OCH_3)(CH_2CH_3)_2$				
	H_2PtCl_6 (i-PrOH)	90-100	1	$CH_3SiCl_2CH{=}CHSi(OCH_3)(CH_2CH_3)_2$ (74)	1281
C_8H_{14}	$CH{\equiv}CSi(CH_3)_2OSi(CH_3)_2C{\equiv}CH$				
	H_2PtCl_6 (i-PrOH)	80	0.5	$CH{\equiv}CSi(CH_3)_2OSi(CH_3)_2CH{=}CHSiCl_2CH_3$ (62) $[CH_3SiCl_2CH{=}CHSi(CH_3)_2]_2O$ (trace)	1309
C_8H_{14}	$CH_3CH_2OCH_2C{\equiv}CCH_2OCH_2CH_3$				
	H_2PtCl_6 (i-PrOH)	70-90	2	$CH_3CH_2OCH_2C(SiCl_2CH_3){=}CHCH_2OCH_2CH_3$ (-)	569
C_8H_{18}	$(CH_3)_3SiC{\equiv}CSi(CH_3)_3$				
	H_2PtCl_6 (i-PrOH)	120-130	-	$(CH_3)_3SiCH{=}C(SiCl_2CH_3)Si(CH_3)_3$ (3)	1053
	H_2PtCl_6 (i-PrOH)	150-180	8	$(CH_3)_3SiCH{=}C(SiCl_2CH_3)Si(CH_3)_3$ (87)	1310
	H_2PtCl_6 (i-PrOH)	reflux	20	$(CH_3)_3SiCH{=}C(SiCl_2CH_3)Si(CH_3)_3$ (13)	138
	H_2PtCl_6 (i-PrOH)	reflux	20	$(CH_3)_3SiCH{=}C(SiCl_2CH_3)Si(CH_3)_3$ (13)	137
C_9H_{18}	$(CH_3)_3CC{\equiv}CSi(CH_3)_3$				
	H_2PtCl_6 (i-PrOH)	reflux	20	$(CH_3)_3CCH{=}C(SiCl_2CH_3)Si(CH_3)_3$ (77)	137
$C_{10}H_{12}$	$CH{\equiv}CSi(CH_3)_2C_6H_5$				
	H_2PtCl_6 (i-PrOH)	80-90	0.5	$CH_3SiCl_2CH{=}CHSi(CH_3)_2C_6H_5$ (92)	1278
$C_{10}H_{14}$	$CH{\equiv}CCH_2CH(COOCH_2CH_3)_2$				

Table 2.5

= CH_3SiHCl_2 = =3=

	H_2PtCl_6	reflux	-	$CH_3SiCl_2CH{=}CHCH_2CH(COOCH_2CH_3)_2$ (66)	1009
$C_{10}H_{18}$	$(CH_3)_3SiC{\equiv}CC{\equiv}CSi(CH_3)_3$				
	H_2PtCl_6 (i-PrOH)	reflux	3.5 (h)	$C_{10}Si_2H_{19}SiCl_2CH_3$ (-)	139
$C_{10}H_{18}$	$CH_3CH_2CH_2OCH_2C{\equiv}CCH_2OCH_2CH_2CH_3$				
	H_2PtCl_6 (i-PrOH)	70-90	2	$CH_3CH_2CH_2OCH_2C(SiCl_2CH_3){=}CHCH_2OCH_2CH_2CH_3$ (42)	569
$C_{11}H_{22}$	$CH_3(CH_2)_5C{\equiv}CSi(CH_3)_3$				
	H_2PtCl_6	20-50	26	$CH_3(CH_2)_5C(SiCl_2CH_3){=}CHSi(CH_3)_3$ (-)	510

Table 2.6

= CH_3SiHCl_2 = =1=

C_3	$(CF_3)_2CO$				
	UV	-	16	$(CF_3)_2CHOSiCl_2CH_3$ (-)	502
C_3H_6	CH_3COCH_3				
	cis-$PtCl_2(PhCH{=}CH_2)_2$ + 4-$NH_2C_5H_4N$	r.t	28	$(CH_3)_2CHOSiCl_2CH_3$ (88)	220
C_3H_6	CH_3COCH_3 + $CH_2{=}CH(CH_2)_3CH_3$				
	gamma (^{60}Co)	-	-	$CH_3(CH_2)_5SiCl_2CH_3$ + $(CH_3)_2CHOSiCl_2CH_3$ (kin.)	687
C_3H_6	CH_3COOCH_3				
	gamma (^{60}Co)	-	-	$C_3H_7O_2SiCl_2CH_3$ (kin.)	816
C_4H_8	$CH_3COCH_2CH_3$				
	cis-$PtCl_2(PhCH{=}CH_2)_2$ + 4-$NH_2C_5H_4N$	r.t	28	$CH_3(CH_3CH_2)CHOSiCl_2CH_3$ (80)	220
C_5H_8	$\overline{(CH_2)_4C}O$				
	cis-$PtCl_2(PhCH{=}CH_2)_2$ + 4-$NH_2C_5H_4N$	r.t	28	$\overline{(CH_2)_4C}HOSiCl_2CH_3$ (84)	220
C_5H_8	$CH_2{=}C(CH_3)COOCH_3$				
	Rh/C	110-115	1	$(CH_3)_2C{=}C(OCH_3)OSiCl_2CH_3$ (-) $CH_3SiOC(CH_3)_2COOCH_3$ (-)	U -187
C_6H_{12}	$CH_3COC(CH_3)_3$				
	cis-$PtCl_2(PhCH{=}CH_2)_2$ + 4-$NH_2C_5H_4N$	r.t	32	$CH_3[(CH_3)_3C]CHOSiCl_2CH_3$ (69)	220
C_6H_{12}	$CH_3COCH(CH_3)_2$ + $CH_3(CH_2)_3CH{=}CH_2$				
	gamma (^{60}Co)	-	-	$CH_3(CH_2)_5SiCl_2CH_3$ + $CH_3Cl_2SiOCH(CH_3)_2$ (kin.)	303
C_7H_6	C_6H_5CHO				
	cis-$PtCl_2(PhCH{=}CH_2)_2$ + 4-$NH_2C_5H_4N$	r.t	28	$C_6H_5CH_2OSiCl_2CH_3$ (59)	220
C_8H_8	$C_6H_5COCH_3$				
	$K[PtCl_3(C_2H_4)]$	0-20	12	$C_6H_5CH(CH_3)OSiCl_2CH_3$ (70)	177
	$K[PtCl_3(C_2H_4)]$ + N,P donors	20	20-62	$C_6H_5CH(CH_3)OSiCl_2CH_3$ (56-100)	177
	$PtCl_2(PhMeCHNH_2)(C_2H_4)$	20	40	$C_6H_5CH(CH_3)OSiCl_2CH_3$ (51)	490
	$[PtCl_2(BMPP)]_2$	r.t.	-	$C_6H_5CH(CH_3)OSiCl_2CH_3$ (81)	1299
	$[PtCl_2(BMPP)]_2$	r.t.	-	$C_6H_5CH(CH_3)OSiCl_2CH_3$ (81)	490
	$[PtCl_2(DMPP)]_2$	20	40	$C_6H_5CH(CH_3)OSiCl_2CH_3$ (80)	490

Table 2.6

= CH_3SiHCl_2 = 	=2=

$[PtCl_2(MPPP)]_2$	r.t.	-	$C_6H_5CH(CH_3)OSiCl_2CH_3$ (71)	1299
$[PtCl_2(MPPP)]_2$	r.t.	-	$C_6H_5CH(CH_3)OSiCl_2CH_3$ (71)	490
cis-$PtCl_2(PPh_3)_2$	20	23	$C_6H_5CH(CH_3)OSiCl_2CH_3$ (10)	177
cis-$PtCl_2(PhCH{=}CH_2)_2$	r.t	26	$C_6H_5CH(CH_3)OSiCl_2CH_3$ (4)	220
cis-$PtCl_2(PhCH{=}CH_2)_2$ + $AsPh_3$	r.t	32	$C_6H_5CH(CH_3)OSiCl_2CH_3$ (75)	220
cis-$PtCl_2(PhCH{=}CH_2)_2$ + N donor	r.t	7-41	$C_6H_5CH(CH_3)OSiCl_2CH_3$ (72-98)	220
cis-$PtCl_2(PhCH{=}CH_2)_2$ + PPh_3	r.t	26	$C_6H_5CH(CH_3)OSiCl_2CH_3$ (51)	220
cis-$PtCl_2(PhCH{=}CH_2)_2$ + $SbPh_3$	r.t	32	$C_6H_5CH(CH_3)OSiCl_2CH_3$ (89)	220
C_9H_{10} $C_6H_5COCH_2CH_3$				
$[PtCl_2(BMPP)]_2$	r.t.	-	$C_6H_5CH(CH_2CH_3)OSiCl_2CH_3$ (81)	490
$[PtCl_2(BMPP)]_2$	r.t.	-	$C_6H_5CH(CH_2CH_3)OSiCl_2CH_3$ (81)	1299
$[PtCl_2(MPPP)]_2$	r.t.	-	$C_6H_5CH(CH_2CH_3)OSiCl_2CH_3$ (83)	490
$[PtCl_2(MPPP)]_2$	r.t.	-	$C_6H_5CH(CH_2CH_3)OSiCl_2CH_3$ (83)	1299
$C_{10}H_{12}$ $C_6H_5COCH(CH_3)_2$				
$[PtCl_2(BMPP)]_2$	r.t.	-	$(CH_3)_2CHCH(C_6H_5)OSiCl_2CH_3$ (55) $C_6H_5C(OSiCl_2CH_3){=}C(CH_3)_2$ (10)	1299
$[PtCl_2(BMPP)]_2$	r.t.	-	$(CH_3)_2CHCH(C_6H_5)OSiCl_2CH_3$ (55) $C_6H_5C(OSiCl_2CH_3){=}C(CH_3)_2$ (10)	490
$[PtCl_2(MPPP)]_2$	r.t.	-	$(CH_3)_2CHCH(C_6H_5)OSiCl_2CH_3$ (57) $C_6H_5C(OSiCl_2CH_3){=}C(CH_3)_2$ (10)	490
$[PtCl_2(MPPP)]_2$	r.t.	-	$(CH_3)_2CHCH(C_6H_5)OSiCl_2CH_3$ (57) $C_6H_5C(OSiCl_2CH_3){=}C(CH_3)_2$ (10)	1299
$C_{10}H_{12}$ $C_6H_5COCH_2CH_2CH_3$				
$[PtCl_2(BMPP)]_2$	r.t.	-	$C_6H_5CH(OSiCl_2CH_3)CH_2CH_2CH_3$ (77)	490
$[PtCl_2(BMPP)]_2$	r.t.	-	$C_6H_5CH(OSiCl_2CH_3)CH_2CH_2CH_3$ (77)	1299
$[PtCl_2(MPPP)]_2$	r.t.	-	$C_6H_5CH(OSiCl_2CH_3)CH_2CH_2CH_3$ (74)	490
$[PtCl_2(MPPP)]_2$	r.t.	-	$C_6H_5CH(OSiCl_2CH_3)CH_2CH_2CH_3$ (74)	1299
$C_{11}H_{14}$ $C_6H_5COC(CH_3)_3$				
$[PtCl_2(BMPP)]_2$	90	10 days	$C_6H_5CH(OSiCl_2CH_3)C(CH_3)_3$ (33)	1299

Table 2.6

= CH_3SiHCl_2 = =3=

	$[PtCl_2(BMPP)]_2$	90	10 days	$C_6H_5CH(OSiCl_2CH_3)C(CH_3)_3$ (33)	490
	$[PtCl_2(MPPP)]_2$	90	10 days	$C_6H_5CH(OSiCl_2CH_3)C(CH_3)_3$ (24)	490
	$[PtCl_2(MPPP)]_2$	90	10 days	$C_6H_5CH(OSiCl_2CH_3)C(CH_3)_3$ (24)	1299
$C_{23}H_{30}$	Fig.149.				
	H_2PtCl_6 (i-PrOH)	60	0.5 (t)	Fig.150. (100)	1320
	H_2PtCl_6 (i-PrOH)	60	0.5 (t)	Fig.150. (100)	1322

Table 3.1

= $(CH_3)_2SiHCl$ = =1=

C_3H_6	$CH_3CH=CH_2$				
	$AlCl_3$	28-35	24	$CH_3CH_2CH_2SiCl(CH_3)_2$ (70)	D -092
	gamma(^{60}Co)	80	-	$CH_3CH_2CH_2SiCl(CH_3)_2$ (100)	1360
	H_2PtCl_6 (i-PrOH)	-	-	$CH_3CH_2CH_2SiCl(CH_3)_2$ (-)	141
C_4H_8	$CH_3CH_2CH=CH_2$				
	H_2PtCl_6	-	-	$C_4H_9SiCl(CH_3)_2$ (-)	330
C_4H_8	$CH_3CH_2CH=CH_2 + (CH_3)_2C=CH_2$				
	H_2PtCl_6 [$\overline{(CH_2)_5C}O$]	30-120	8.5	$CH_3(CH_2)_3SiCl(CH_3)_2 + (CH_3)_2CHCH_2SiCl(CH_3)_2$ (-)	Pl-013
C_5H_8	$(CH_2)_3C=CH_2$				
	H_2PtCl_6 (THF)	until 110	10	$(CH_2)_3CHCH_2SiCl(CH_3)_2$ (51)	817
C_5H_9	$CH_3CHBrCH_2CH=CH_2$				
	H_2PtCl_6 (i-PrOH)	160	4	$CH_3CHBr(CH_2)_3SiCl(CH_3)_2$ (82)	800
C_5H_{10}	$(CH_3)_2C=CHCH_3$				
	Al(O)Cl	28-35	24	$(CH_3)_2CHCH(CH_3)SiCl(CH_3)_2$ (80)	D -092
	$AlBr_3$	28-35	24	$(CH_3)_2CHCH(CH_3)SiCl(CH_3)_2$ (90)	D -092
	$AlCl_3$	28-35	24	$(CH_3)_2CHCH(CH_3)SiCl(CH_3)_2$ (95)	D -092
	$AlCl_3(SiMe_4)$	0-10	- (p)	$(CH_3)_2CHCH(CH_3)SiCl(CH_3)_2$ (2-98)	D -109
C_5H_{10}	$CH_3CH_2CH_2CH=CH_2$				
	$[HPt(SiMe_2Et)P(c\text{-}Hex)_3]_2$	20	1 (h)	$CH_3(CH_2)_4SiCl(CH_3)_2$ (75)	429
C_6H_{10}	$\overline{(CH_2)_4C}=CH_2$				
	$AlCl_3$	28-35	24	$\overline{(CH_2)_4C}HCH_2SiCl(CH_3)_2$ (72)	D -092
C_6H_{12}	$(CH_3)_2C=C(CH_3)_2$				
	$AlCl_3$	25	-	$(CH_3)_2ClSiC(CH_3)_2CH(CH_3)_2$ (93)	1362
C_6H_{12}	$CH_3(CH_2)_3CH=CH_2$				
	$AlCl_3$	28-35	24	$CH_3(CH_2)_5SiCl(CH_3)_2$ (81)	D -092
	Fig.151.	110	6	$CH_3(CH_2)_5SiCl(CH_3)_2$ (65)	D -027
	Pd + $P(NMe_2)_3$	100	10	$CH_3(CH_2)_5SiCl(CH_3)_2$ (63)	D -027

Table 3.1

= $(CH_3)_2SiHCl$ = =2=

	Pd + PCl_3	130	6	$CH_3(CH_2)_5SiCl(CH_3)_2$ (36)	D -027
	$Pd(AsPh_3)_4$	120	6	$CH_3(CH_2)_5SiCl(CH_3)_2$ (77)	D -027
	$Pd(NO_2)_2(NH_3)_2$ + PPh_3	130	5	$CH_3(CH_2)_5SiCl(CH_3)_2$ (76)	D -027
	$Pd(PPh_3)_4$	110	6	$CH_3(CH_2)_5SiCl(CH_3)_2$ (90)	D -027
	$PdCl_2(PPh_3)_2$	120	5	$CH_3(CH_2)_5SiCl(CH_3)_2$ (86)	D -027
	H_2PtCl_6	-	-	$C_6H_{13}SiCl(CH_3)_2$ (-)	330
	$PtCl_2(py)_2$	-	2.5	$CH_3(CH_2)_5SiCl(CH_3)_2$ (90)	987
	$[HPt(SiBzMe_2)P(c\text{-}Hex)_3]_2$	20	0.75	$CH_3(CH_2)_5SiCl(CH_3)_2$ (83)	429
	$[HPt(SiEtMe_2)P(c\text{-}Hex)_3]_2$	20	2 (h)	$CH_3(CH_2)_5SiCl(CH_3)_2$ (85)	429
	glass-NEt_3/H_2PtCl_6	100	8	$CH_3(CH_2)_5SiCl(CH_3)_2$ (60)	S -036
C_7H_{10}	Fig.1.				
	$[HPt(SiBzMe)P(c\text{-}Hex)_3]_2$	20	0.5	Fig.152. (80)	429
C_7H_{14}	$CH_3(CH_2)_4CH=CH_2$				
	$PtCl_4(C_{12}H_{25}PPh_2)_3$	20-60	1-3	$CH_3(CH_2)_6SiCl(CH_3)_2$ (kin.)	171
C_8H_8	$C_6H_5CH=CH_2$				
	$AlCl_3$	28-35	24	$C_6H_5CH_2CH_2SiCl(CH_3)_2$ (80)	D -092
	$Co_2(CO)_8$	20-30	70	$C_6H_5CH_2CH_2SiCl(CH_3)_2$ (-)	715
	$Bu_4N^+[PtCl_3(PhCH=CH_2)]^-$	r.t.	4	$C_6H_5CH_2CH_2SiCl(CH_3)_2$ (93) $C_6H_5CH(CH_3)SiCl(CH_3)_2$ (-)	220
	H_2PtCl_6	20	-	$C_6H_5CH_2CH_2SiCl(CH_3)_2$ (51) $C_6H_5CH(CH_3)SiCl(CH_3)_2$ (9)	172
	H_2PtCl_6	5-20	24	$C_6H_5C_2H_4SiCl(CH_3)_2$ (kin.)	966
	$K[PtCl_3(CH_2=CH_2)]$	r.t.	1.25	$C_6H_5CH_2CH_2SiCl(CH_3)_2$ (82) $C_6H_5CH(CH_3)SiCl(CH_3)_2$ (17)	220
	$Pt(PPh_3)_4$	15	-	$C_6H_5C_2H_4SiCl(CH_3)_2$ (kin.)	977
	$PtCl_2(PhCH=CH_2)_2$	5-20	24	$C_6H_5C_2H_4SiCl(CH_3)_2$ (kin.)	965
	$PtCl_2(py)_2$	-	1.5-24	$C_6H_5CH_2CH_2SiCl(CH_3)_2$ (59-85) $C_6H_5CH(CH_3)SiCl(CH_3)_2$ (11-15)	987
	$PtCl_3[(CH_2CH=CH_2)PPh_3]$	20	-	$C_6H_5CH_2CH_2SiCl(CH_3)_2$ (80)	171

Table 3.1

= $(CH_3)_2SiHCl$ = =5=

	H_2PtCl_6	reflux	10	$C_{22}H_{45}SiCl(CH_3)_2$ (68)	140
C_mH_{2m}	$C_nH_{2n+1}CH{=}CH_2$				
	$Co_2(CO)_8$	40-50	1	$C_nH_{2n+1}CH_2CH_2SiCl(CH_3)_2$ (-)	715
	$Rh_4(CO)_{12}$	40-50	1	$C_nH_{2n+1}CH_2CH_2SiCl(CH_3)_2$ (-)	715

Table 3.2
= $(CH_3)_2SiHCl$ = =1=

C_4H_6	$CH_2=CHCH=CH_2$				
	$Ni(acac)_2$	80-100	3	cis-$CH_3CH=CHCH_2SiCl(CH_3)_2$ (69)	1019
	Pd + PPh_3	110	6	$CH_3CH=CHCH_2SiCl(CH_3)_2$ (83)	D -027
	H_2PtCl_6	200	4	$CH_3CH=CHCH_2SiCl(CH_3)_2$ (47) $(CH_3)_2SiCl(CH_2)_4SiCl(CH_3)_2$ (46)	90
C_5H_6	Fig.25.				
	$Co_2(CO)_8$	30-35	1	Fig.160. (91-93)	715
	$Rh_4(CO)_{12}$	30-35	1	Fig.160. (91-93)	715
C_5H_8	$CH_2=C(CH_3)CH=CH_2$				
	$Ni(acac)_2$	80-100	3	$CH_3CH=C(CH_3)CH_2SiCl(CH_3)_2$ + $(CH_3)_2C=CHCH_2SiCl(CH_3)_2$ (60)	1019
	$PdCl_2(PPh_3)_3$	130	15	$CH_3CH=C(CH_3)CH_2SiCl(CH_3)_2$ (18) $(CH_3)_2C=CHCH_2SiCl(CH_3)_2$ (42)	851
	$PdCl_2(PhCN)_2$ + PPh_3	60-80	6	$CH_3CH=C(CH_3)CH_2SiCl(CH_3)_2$ (95)	855
	$PdCl_2(PhCN)_2$ + PPh_3	70	6	$CH_3CH=C(CH_3)CH_2SiCl(CH_3)_2$ (90)	851
C_5H_8	$CH_2=CHCH_2CH=CH_2$				
	H_2PtCl_6	-	-	$CH_2=CH(CH_2)_3SiCl(CH_3)_2$ (70)	594
C_5H_8	$CH_3CH=CHCH=CH_2$				
	$Ni(acac)_2$	80-100	3	$CH_3CH_2CH=CHCH_2SiCl(CH_3)_2$ (50)	1019
C_6H_{10}	$CH_2=C(CH_3)C(CH_3)=CH_2$				
	$Ni(acac)_2$	80-100	3	$(CH_3)_2C=C(CH_3)CH_2SiCl(CH_3)_2$ (60)	1019
C_6H_{10}	$CH_2=CHCH_2CH_2CH=CH_2$				
	H_2PtCl_6	-	-	$CH_2=CH(CH_2)_4SiCl(CH_3)_2$ (72)	593
	$[HPt(SiBzMe_2)P(c\text{-}Hex_3)]_2$	20	0.2	$(CH_3)_2SiCl(CH_2)_6SiCl(CH_3)_2$ (96)	429
C_8H_{12}	$CH_2=CHCH(CH_3)CH=CHCH=CH_2$				
	$Ni(acac)_2$	80-100	3	$CH_2=CHCH(CH_3)CH_2CH=CHCH_2SiCl(CH_3)_2$ (62)	1019
C_8H_{12}	Fig.161.				
	-	20	-	Fig.164. (83)	622
	$[HPt(SiEtMe_2)P(c\text{-}Hex_3)]_2$	20	0.5	Fig.162. (89)	429
C_8H_{12}	Fig.36.				

Table 3.3

= $(CH_3)_2SiHCl$ =

	H_2PtCl_6 (i-PrOH)	120-130	70-80	$CH_3Cl_2SiC_2H_2Cl_2[SiCl(CH_3)_2]$ (12)	1055
C_3H_5	$CH_2=CHCH_2Br$				
	H_2PtCl_6 (i-PrOH)	reflux	-	$(CH_3)_2SiCl(CH_2)_3Br$ (10)	565
C_3H_5	$CH_2=CHCH_2Cl$				
	-	-	-	$(CH_3)_2SiCl(CH_2)_3Cl$ (-)	65
	H_2PtCl_6 + H-siloxane	42-98	2.5	$(CH_3)_2SiCl(CH_2)_3Cl$ (23)	U -168
	H_2PtCl_6 + $MeCl_2SiH$	32-57	0.16	$(CH_3)_2SiCl(CH_2)_3Cl$ (54) $CH_3SiCl_2(CH_2)_3Cl$ (20)	U -168
	$[HPt(SiMe_2Et)P(c\text{-}Hex)_3]_2$	65	17 (h)	$(CH_3)_2SiCl(CH_2)_3Cl$ (32)	429
C_3H_5	$CH_2=CHCONH_2$				
	H_2PtCl_6 (THF)	60-80	2-3	$(CH_3)_2SiCH_2CH_2CONH$ (61) $(CH_3)_2Si(NHOCCH_2CH_2)(CH_2CH_2CONH)Si(CH_3)_2$ (7)	570
C_3H_5	$CH_3Cl_2SiCH=CHCl$				
	H_2PtCl_6 (i-PrOH)	120-130	70-80	$CH_3Cl_2SiC_2H_2Cl[SiCl(CH_3)_2]$ (47)	1055
C_3H_6	$CH_2=CHSiCl_2CH_3$				
	H_2PtCl_6 (i-PrOH)	-	-	$(CH_3)_2SiClCH_2CH_2SiCl_2CH_3$ (86) $(CH_3)_2SiClCH_2CH_2Si(CH_3)_2Cl$ (1)	1056
	H_2PtCl_6 (i-PrOH)	150	4	$(CH_3)_2SiClCH_2CH_2SiCl_2CH_3$ (87)	893
	H_2PtCl_6 (i-PrOH)	reflux	4	$(CH_3)_2SiClC_2H_4SiCl_2CH_3$ (-)	1059
C_4	$CF_2CF_2CCl=CCl$				
	gamma	-	-	$CF_2CF_2CH(Cl)C(Cl)SiCl_2CH_3$ + $CF_2CF_2C(Cl)[SiCl(CH_3)_2]CH(Cl)$ (-)	1214
C_4H_1	$CF_2CF_2CH=CCl$				
	gamma	-	-	$CF_2CF_2CH_2C(Cl)SiCl_2CH_3$ + $CF_2CF_2CH[SiCl(CH_3)_2]CH(Cl)$ (-)	1214
C_4H_5	$CH_2=C(CH_3)CN$				
	$Co_2(CO)_8$	25	6 days	$(CH_3)_2SiClCH_2CH(CH_3)CN$ + $(CH_3)_2SiClC(CH_3)_2CN$(20)	227
C_4H_5	$CH_2=CHCH_2CN$				
	H_2PtCl_6 (i-PrOH)	130-140	2	$(CH_3)_2SiCl(CH_2)_3CN$ (92)	1099
C_4H_5	$CH_2=CHCH_2NCO$				

Table 3.3

= $(CH_3)_2SiHCl$ = =3=

	H_2PtCl_6 ($MeOC_2H_4OMe$)	57-68	4-6	$(CH_3)_2SiCl(CH_2)_3NCO$ (42)	U -039
	H_2PtCl_6 ($MeOC_2H_4OMe$)	57-68	4-6	$(CH_3)_2SiCl(CH_2)_3NCO$ (42)	D -024
	H_2PtCl_6 (i-PrOH)	-	-	$(CH_3)_2SiCl(CH_2)_3NCO$ (-)	1230
C_4H_5	$CH_3CH{=}CHCN$				
	$RhCl(PPh_3)_3$	80	60	$CH_3CH_2CH(CN)SiCl(CH_3)_2$ (90)	868
C_4H_5	$ClCOOCH_2CH{=}CH_2$				
	$(t\text{-}BuO)_2$	105	15	$ClCOO(CH_2)_3SiCl(CH_3)_2$ (54)	786
	H_2PtCl_6	60-100	4	$ClCOO(CH_2)_3SiCl(CH_3)_2$ (24)	786
	H_2PtCl_6	60-100	4	$ClCOO(CH_2)_3SiCl(CH_3)_2$ (24)	783
	H_2PtCl_6 (i-PrOH)	60-83	-	$ClCOO(CH_2)_3SiCl(CH_3)_2$ (54)	S -026
C_4H_6	$CH_2{=}CHCOOCH_3$				
	Ni(THF)	0-25	20	$CH_3CH(COOCH_3)SiCl(CH_3)_2$ (3)	E -049
	Ni(THF) +sonic waves	0-25	8	$CH_3CH(COOCH_3)SiCl(CH_3)_2$ (16)	E -049
C_4H_6	$CH_2{=}CHOCH{=}CH_2$				
	H_2PtCl_6 (i-PrOH)	150	4	$(CH_3)_2SiClCH_2CH_2OCH_2CH_2SiCl(CH_3)_2$ (-)	1266
C_4H_6	$Cl_2Si(CH{=}CH_2)_2$				
	H_2PtCl_6 (i-PrOH)	reflux	4	$Cl_2Si(CH{=}CH_2)C_2H_4SiCl(CH_3)_2$ + $Cl_2Si[C_2H_4SiCl(CH_3)_2]_2$ (-)	1059
C_4H_6	$HCOOCH_2CH{=}CH_2$				
	H_2PtCl_6 (i-PrOH)	40-115	1.5	$HCOO(CH_2)_3SiCl(CH_3)_2$ (52)	1052
	H_2PtCl_6 (i-PrOH)	40-115	1.5	$HCOO(CH_2)_3SiCl(CH_3)_2$ (52)	S -026
C_4H_7	$(CH_3)_2ClSiCH{=}CCl_2$				
	H_2PtCl_6 (i-PrOH)	120-130	70-80	$CH_3Cl_2SiC_2H_2Cl_2[SiCl(CH_3)_2]$ (2)	1055
C_4H_7	$CH_2{=}C(CH_3)CH_2Cl$				
	H_2PtCl_6+ H-siloxane	95	1	$(CH_3)_2SiClCH_2CH(CH_3)CH_2Cl$ (42)	U -168
	H_2PtCl_6+ H-siloxane	96	0.1	$(CH_3)_2SiClCH_2CH(CH_3)CH_2Cl$ (72)	U -168
	H_2PtCl_6 + $HSiCl_3$	31-60	0.05	$(CH_3)_2SiClCH_2CH(CH_3)CH_2Cl$ (74) $Cl_3SiCH_2CH(CH_3)CH_2Cl$ (3)	U -168
C_4H_7	$CH_2{=}CHCH_2CH_2Br$				

Table 3.3

= $(CH_3)_2SiHCl$ = =6=

	H_2PtCl_6	76-132	6	$(CH_3)_2SiClCH_2CH_2Si(CH_3)_3$ (92)	1058
	H_2PtCl_6 (i-PrOH)	-	-	$(CH_3)_2SiClCH_2CH_2Si(CH_3)_3$ (92) $(CH_3)_3SiCH_2CH_2Si(CH_3)_3$ (traces) $(CH_3)_2SiClCH_2CH_2SiCl(CH_3)_2$ (1) $(CH_3)_2Si[CH_2CH_2SiCl(CH_3)_2]_2$ (1)	1056
	H_2PtCl_6 (i-PrOH)	reflux	4	$(CH_3)_2SiClCH_2CH_2Si(CH_3)_3$ + $(CH_3)_2SiClCH(CH_3)Si(CH_3)_3$ (-)	1059
	$Rh_4(CO)_{12}$	30-40	1	$(CH_3)_2SiClCH_2CH_2Si(CH_3)_3$ (-)	715
C_5H_{16}	$H\overline{CB_{10}H_{10}C}C(CH_3)=CH_2$				
	none	260	-	$H\overline{CB_{10}H_{10}C}CH(CH_3)CH_2CH_2SiCl(CH_3)_2$ (-)	1045
	$Co_2(CO)_8$	20-30	60	$H\overline{CB_{10}H_{10}C}CH(CH_3)CH_2CH_2SiCl(CH_3)_2$ (-)	715
	$Rh_4(CO)_{12}$	20-30	60	$H\overline{CB_{10}H_{10}C}CH(CH_3)CH_2CH_2SiCl(CH_3)_2$ (-)	715
C_6H_3	$(CF_3)_3CCH=CH_2$				
	gamma(^{60}Co)	135	-	$(CF_3)_3CCH_2CH_2SiCl(CH_3)_2$ (100)	1360
C_6H_6	$CH_2=CHCH_2COOCF_2CF_2H$				
	H_2PtCl_6 (i-PrOH)	reflux	3.5	$(CH_3)_2SiCl(CH_2)_3COOCF_2CF_2H$ (70)	103
C_6H_9	$ClSi(CH=CH_2)_3$				
	H_2PtCl_6 (i-PrOH)	reflux	4	$ClSi(CH=CH_2)_2C_2H_4SiCl(CH_3)_2$ + $ClSi(CH=CH_2)[C_2H_4SiCl(CH_3)_2]_2$ + $ClSi[C_2H_4SiCl(CH_3)_2]_3$ (-)	1059
C_6H_{10}	$(CH_2=CHCH_2)_2S$				
	$RhCl(PPh_3)_3$ or H_2PtCl_6	-	-	$CH_2=CHCH_2S(CH_2)_3SiCl(CH_3)_2$ + $CH_2=CHCH_2SCH_2CH(CH_3)SiCl(CH_3)_2$ (-)	586
C_6H_{10}	$CH_2=C(CH_3)CH_2OCOCH_3$				
	H_2PtCl_6	reflux	- (THF)	$(CH_3)_2SiClCH_2CH(CH_3)CH_2OCOCH_3$ (74)	754
C_6H_{10}	$CH_3CH=CHCOOCH_2CH_3$				
	$RhCl(PPh_3)_3$	-	-	$CH_3CH_2CH[SiCl(CH_3)_2]COOCH_2CH_3$ (-)	J -021
	$RhCl(PPh_3)_3$	110	6	$CH_3CH_2CH[SiCl(CH_3)_2]COOCH_2CH_3$ (65)	868
C_6H_{11}	$CH_2=CHCH_2NHCOOCH_2CH_3$				
	H_2PtCl_6	100-125	-	$(CH_3)_2SiCl(CH_2)_3NHCOOCH_2CH_3$ (29)	783
C_6H_{11}	$CH_2=CHCH_2OCON(CH_3)_2$				

	H_2PtCl_6 (i-PrOH)	100-165	2	$(CH_3)_2SiCl(CH_2)_3OCON(CH_3)_2$ (68)	783
C_6H_{12}	$(CH_3)_2Si(CH=CH_2)_2$				
	H_2PtCl_6 (i-PrOH)	reflux	4	$(CH_3)_2Si(CH=CH_2)C_2H_4SiCl(CH_3)_2$ + $(CH_3)_2Si[C_2H_4SiCl(CH_3)_2]_2$ (-)	1059
	H_2PtCl_6 (i-PrOH)	until 140	0.5	$(CH_3)_2SiCl(CH=CH_2)CH_2CH_2SiCl(CH_3)_2$ (72)	594
C_6H_{12}	$CH_2=CHCH_2OCH_2CH_2OCH_3$				
	H_2PtCl_6 (i-PrOH)	reflux	3	$(CH_3)_2SiCl(CH_2)_3OCH_2CH_2OCH_3$ (35)	1141
C_6H_{12}	$CH_3O(CH_2)_3CH=CH_2$				
	H_2PtCl_6	40	18	$CH_3O(CH_2)_5SiCl(CH_3)_2$ (87)	140
C_6H_{14}	$(CH_3)_2ClSiCH=CHSiCl(CH_3)_2$				
	H_2PtCl_6 (i-PrOH)	120-130	-	$(CH_3)_2ClSiCH_2CH[(SiCl(CH_3)_2]_2$ (9)	1053
C_6H_{14}	$(CH_3)_3SiCH_2CH=CH_2$				
	H_2PtCl_6 (i-PrOH)	-	-	$(CH_3)_3Si(CH_2)_3SiCl(CH_3)_2$ (-)	525
C_6H_{14}	$(CH_3)_3SiOCH_2CH=CH_2$				
	H_2PtCl_6	60	-	$(CH_3)_3SiO(CH_2)_3SiCl(CH_3)_2$ (45)	574
	H_2PtCl_6 (i-PrOH)	reflux	-	$\overline{O(CH_2)_3Si}(CH_3)_2$ (78)	777
C_6H_{14}	$CH_3CH_2OSi(CH_3)_2CH=CH_2$				
	H_2PtCl_6	120	-	$CH_3CH_2OSi(CH_3)_2CH_2CH_2SiCl(CH_3)_2$ (-)	1076
C_6H_{14}	$F(CH_3)_2SiCH=CHSiCl(CH_3)_2$				
	H_2PtCl_6 (i-PrOH)	90	0.5	$F(CH_3)_2SiCH[SiCl(CH_3)_2]CH_2SiCl(CH_3)_2$ (-) $F(CH_3)_2SiCH_2CH[SiCl(CH_3)_2]_2$ (-)	1312
C_7H_5	$C_3F_7COOCH_2CH=CH_2$				
	H_2PtCl_6 (i-PrOH)	reflux	6	$C_3F_7COO(CH_2)_3SiCl(CH_3)_2$ (22)	103
C_7H_7	$\overline{COCH=CHCON}CH_2CH=CH_2$				
	H_2PtCl_6 (OctOH)	80-90	-	$\overline{COCH=CHCON}(CH_2)_3SiCl(CH_3)_2$ (100)	U -058
C_7H_{10}	$(CH_2=CHCH_2O)_2CO$				
	H_2PtCl_6	45-115	1.1	$[(CH_3)_2SiCl(CH_2)_3O]_2CO$ (56)	1051
C_7H_{12}	$CH_3COOCH(CH_3)C(CH_3)=CH_2$				

	H$_2$PtCl$_6$	reflux	- (THF)	CH$_3$COOCH(CH$_3$)CH(CH$_3$)CH$_2$SiCl(CH$_3$)$_2$ (71)	754
C$_7$H$_{12}$	CH$_3$Si(CH=CH$_2$)$_3$				
	H$_2$PtCl$_6$ (i-PrOH)	reflux	4	CH$_3$Si(CH=CH$_2$)$_2$C$_2$H$_4$SiCl(CH$_3$)$_2$ + CH$_3$Si(CH=CH$_2$)[C$_2$H$_4$SiCl(CH$_3$)$_2$]$_2$ + CH$_3$Si[C$_2$H$_4$SiCl(CH$_3$)$_2$]$_3$ (-)	1059
C$_7$H$_{13}$	CH$_2$=CH(CH$_2$)$_5$Cl				
	H$_2$PtCl$_6$ (i-PrOH)	-	-	(CH$_3$)$_2$SiCl(CH$_2$)$_7$Cl (-)	1230
C$_7$H$_{14}$	CH$_2$=CHCH$_2$COOSi(CH$_3$)$_3$				
	H$_2$PtCl$_6$ (i-PrOH)	reflux	2	$\overline{(\text{CH}_2)_3\text{COOS}}$i(CH$_3$)$_2$ (77)	774
C$_7$H$_{14}$	CH$_2$=CHSi(CH$_3$)$_2$CH$_2$CH=CH$_2$				
	H$_2$PtCl$_6$ (i-PrOH)	until 140	0.5	(CH$_3$)$_2$SiClCH$_2$CH$_2$Si(CH$_3$)$_2$CH$_2$CH=CH$_2$ (61)	402
C$_7$H$_{15}$	CH$_3$CH$_2$CH$_2$SiCl(CH$_3$)CH$_2$CH=CH$_2$				
	H$_2$PtCl$_6$ (i-PrOH)	reflux	3	CH$_3$CH$_2$CH$_2$SiCl(CH$_3$)(CH$_2$)$_3$SiCl(CH$_3$)$_2$ (61)	22
C$_7$H$_{16}$	(CH$_3$)$_3$SiOCH(CH$_3$)CH=CH$_2$				
	H$_2$PtCl$_6$ (i-PrOH)	reflux	0.5	$\overline{\text{OCH}_2\text{CH(CH}_3)\text{CH}_2}$Si(CH$_3$)$_2$ (65)	619
C$_7$H$_{16}$	(CH$_3$)$_3$SiOCH$_2$CH$_2$CH=CH$_2$				
	H$_2$PtCl$_6$ (i-PrOH)	reflux	2	(CH$_3$)$_3$SiO(CH$_2$)$_4$SiCl(CH$_3$)$_2$ (56) $\overline{\text{O(CH}_2)_4}$Si(CH$_3$)$_2$ (28)	619
C$_8$H$_3$	C$_6$F$_5$CH=CH$_2$				
	RhCl(PPh$_3$)$_3$	120	24	C$_6$F$_5$CH$_2$CH$_2$SiCl(CH$_3$)$_2$ (44)	858
	Ru$_3$(CO)$_{12}$	120	76	C$_6$F$_5$CH$_2$CH$_2$SiCl(CH$_3$)$_2$ (8) C$_6$H$_5$CH=CHSiCl(CH$_3$)$_2$ (67)	858
C$_8$H$_7$	(CF$_3$)$_2$CFOCH$_2$COOCH$_2$CH=CH$_2$				
	H$_2$PtCl$_6$ (i-PrOH)	80	1	(CF$_3$)$_2$CFOCH$_2$COO(CH$_2$)$_3$SiCl(CH$_3$)$_2$ (60)	D -042
C$_8$H$_7$	m-NO$_2$C$_6$H$_4$CH=CH$_2$				
	H$_2$PtCl$_6$ (i-PrOH)	120	2.5	m-NO$_2$C$_6$H$_4$CH$_2$CH$_2$SiCl(CH$_3$)$_2$ (50)	45
	H$_2$PtCl$_6$ (i-PrOH)	50-100	2.5	m-NO$_2$C$_6$H$_4$CH$_2$CH$_2$SiCl(CH$_3$)$_2$ (53)	47
C$_8$H$_7$	p-ClC$_6$H$_4$CH=CH$_2$				
	H$_2$PtCl$_6$	5-20	24	p-ClC$_6$H$_4$C$_2$H$_4$SiCl(CH$_3$)$_2$ (kin.)	966

Table 3.3

= $(CH_3)_2SiHCl$ = =9=

	$Pt(PPh_3)_4$	15	- (solv.)	p-$ClC_6H_4C_2H_4SiCl(CH_3)_2$ (kin.)	977
C_8H_8	$C_6H_5SiCl_2CH=CH_2$				
	H_2PtCl_6 (i-PrOH)	150	4	$C_6H_5SiCl_2CH_2CH_2SiCl(CH_3)_2$ (68)	893
C_8H_{12}	$Si(CH=CH_2)_4$				
	H_2PtCl_6	50-137	6	$(CH_2=CH)_3SiCH_2CH_2SiCl(CH_3)_2$ (53) $(CH_2=CH)_2Si[CH_2CH_2SiCl(CH_3)_2]_2$ (25)	1058
	H_2PtCl_6 (i-PrOH)	reflux	4	$(CH_2=CH)_3SiC_2H_4SiCl(CH_3)_2$ + $(CH_2=CH)_2Si[C_2H_4SiCl(CH_3)_2]_2$ + $CH_2=CHSi[C_2H_4SiCl(CH_3)_2]_3$ + $Si[C_2H_4SiCl(CH_3)_2]_4$ (-)	1059
C_8H_{14}	$(CH_2=CHOCH_2CH_2)_2O$				
	H_2PtCl_6	110-135	0.5	$[(CH_3)_2SiClCH_2CH_2OCH_2CH_2]_2O$ (-)	D -021
C_8H_{14}	$CH_3CH(OCH_2CH=CH_2)_2$				
	H_2PtCl_6 (i-PrOH)	reflux	10	$CH_3CH[O(CH_2)_3SiCl(CH_3)_2]_2$ (34)	1115
C_8H_{16}	$CH_3OCH_2CH_2O(CH_2)_3CH=CH_2$				
	H_2PtCl_6	100	5	$CH_3OCH_2CH_2O(CH_2)_5SiCl(CH_3)_2$ (67)	140
C_8H_{17}	$CH_3(CH_2)_3SiCl(CH_3)CH_2CH=CH_2$				
	H_2PtCl_6 (i-PrOH)	reflux	3	$CH_3(CH_2)_3SiCl(CH_3)(CH_2)_3SiCl(CH_3)_2$ (78)	22
C_8H_{18}	$(CH_3)_3SiO(CH_2)_3CH=CH_2$				
	H_2PtCl_6 (i-PrOH)	reflux	0.5	$\overline{O(CH_2)_5Si}(CH_3)_2$ (10)	619
C_8H_{18}	$(CH_3)_3SiOCH_2CH_2C(CH_3)=CH_2$				
	H_2PtCl_6 (i-PrOH)	reflux	2	$(CH_3)_3SiOCH_2CH_2C(CH_3)CH_2SiCl(CH_3)_2$ (26) $\overline{OCH_2CH_2CH(CH_3)CH_2Si}(CH_3)_2$ (60)	619
	H_2PtCl_6 (i-PrOH)	reflux	-	$\overline{OCH_2CH_2CH(CH_3)CH_2Si}(CH_3)_2$ (80)	777
C_8H_{20}	$(CH_3)_3SiCH=CHSi(CH_3)_3$				
	H_2PtCl_6 (i-PrOH)	120-130	-	$(CH_3)_3SiCH_2CH[Si(CH_3)_3]SiCl(CH_3)_2$ (5)	1053
	H_2PtCl_6 (i-PrOH)	150-180	8	$(CH_3)_3SiCH_2CH[Si(CH_3)_3]SiCl(CH_3)_2$ (35)	1310
C_8H_{20}	$CH_2=CHCH_2\overline{CB_{10}H_{10}C}CH_2CH-CH_2$				
	Pt/C	reflux	overnight(e)	$(CH_3)_2SiCl(CH_2)_3\overline{CB_{10}H_{10}C}(CH_2)_3SiCl(CH_3)_2$ (85)	761
C_9H_8	$CH_2=CHCH_2COOCH_2(CF_2)_3CHF_2$				

	H_2PtCl_6 (i-PrOH)	reflux	4	$(CH_3)_2SiCl(CH_2)_3COOCH_2(CF_2)_3CHF_2$ (54)	103
C_9H_8	Fig.172.				
	Pt catalyst	-	-	Fig.173. (98)	U -154
	Pt complex	80	- (t)	Fig.173 (-)	D -132
C_9H_9	$ClCH_2C_6H_4CH=CH_2$				
	H_2PtCl_6	-	-	$ClCH_2C_6H_4CH_2CH_2SiCl(CH_3)_2$ (-)	566
	H_2PtCl_6 (i-PrOH)	30	- (dg)	$ClCH_2C_6H_4CH_2CH_2SiCl(CH_3)_2$ (-)	E -026
	H_2PtCl_6 (i-PrOH)	reflux	2	$ClCH_2C_6H_4CH_2CH_2SiCl(CH_3)_2$ (92)	U -131
C_9H_9	o-$NO_2C_6H_4OCH_2CH=CH_2$				
	H_2PtCl_6	90-110	2	o-$NO_2C_6H_4O(CH_2)_3SiCl(CH_3)_2$ (85)	D -075
C_9H_9	p-$ClCH_2C_6H_4CH=CH_2$				
	H_2PtCl_6 ($MeOC_2H_4OMe$)	45-50	3	p-$ClCH_2C_6H_4CH_2CH_2SiCl(CH_3)_2$ (95)	567
C_9H_9	p-$NO_2C_6H_4OCH_2CH=CH_2$				
	H_2PtCl_6	90-110	2	p-$NO_2C_6H_4O(CH_2)_3SiCl(CH_3)_2$ (90)	D -075
C_9H_{10}	o-$HOC_6H_4CH_2CH=CH_2$				
	H_2PtCl_6 ($MeOC_2H_4OMe$)	reflux	- (b)	o-$HOC_6H_4(CH_2)_3Si(CH_3)_2$ (-)	F -006
C_9H_{10}	p-$CH_3OC_6H_4CH=CH_2$				
	$Pt(PPh_3)_4$	15	- (solv.)	p-$CH_3OC_6H_4C_2H_4SiCl(CH_3)_2$ (kin.)	977
C_9H_{14}	$CH_2=C(CH_3)COO(CH_2)_3CH=CH_2$				
	-	-	-	$CH_2=C(CH_3)COO(CH_2)_5SiCl(CH_3)_2$ (-)	562
C_9H_{17}	$CH_2=CH(CH_2)_7Cl$				
	H_2PtCl_6 (i-PrOH)	-	-	$(CH_3)_2SiCl(CH_2)_9Cl$ (-)	1230
C_9H_{18}	$CH_2=CHCH_2Si(CH_3)_2CH_2C(CH_3)=CH_2$				
	H_2PtCl_6 (i-PrOH)	85	0.5	$(CH_3)_2SiCl(CH_2)_3Si(CH_3)_2CH_2C(CH_3)=CH_2$ (62)	402
C_9H_{18}	$CH_3O(CH_2)_5OCH_2CH=CH_2$				
	H_2PtCl_6	100	5	$CH_3O(CH_2)_5O(CH_2)_3SiCl(CH_3)_2$ (69)	140
C_9H_{20}	$(CH_3)_3SiOCH_2CH_2OCH_2CH_2CH=CH_2$				

	H_2PtCl_6 (OctOH)	90-110	-	$OCH_2CH_2O(CH_2)_3Si(CH_3)_2$ (98)	F -017
C_9H_{24}	$[OSi(CH_3)_2]_3OSi(CH_3)CH=CH_2$				
	$Ir_4(CO)_{12}$	65-70	1	$[OSi(CH_3)_2]_3OSi(CH_3)CH_2CH_2SiCl(CH_3)_2$ (98)	S -027
	H_2PtCl_6 (THF)	70-80	4	$[OSi(CH_3)_2]_3OSi(CH_3)CH_2CH_2SiCl(CH_3)_2$ (55)	55
	H_2PtCl_6 (i-PrOH)	160-180	2	$[OSi(CH_3)_2]_3OSi(CH_3)CH_2CH_2SiCl(CH_3)_2$ (69)	46
$C_{10}H_9$	o-$ClCOOC_6H_4CH_2CH=CH_2$				
	H_2PtCl_6	65-140	3.5	o-$ClCOOC_6H_4(CH_2)_3SiCl(CH_3)_2$ (40)	574
$C_{10}H_{14}$	$(CH_2COOCH_2CH=CH_2)_2$				
	H_2PtCl_6 (i-PrOH)	100	0.25	$CH_2=CHCH_2OCOCH_2CH_2COO(CH_2)_3SiCl(CH_3)_2$ (42)	32
$C_{10}H_{15}$	Fig.174.				
	H_2PtCl_6 (i-PrOH)	125	6	Fig.175. (70)	S -042
$C_{10}H_{18}$	$CH_2=CH(CH_2)_6SCOCH_3$				
	Pt catalyst	90	50	$(CH_3)_2SiCl(CH_2)_8SCOCH_3$ (-)	U -040
$C_{10}H_{24}$	$[(CH_3)_2SiO]_2[OSi(CH_3)CH=CH_2]_2$				
	H_2PtCl_6 (THF)	80	2	$[(CH_3)_2SiO]_2[OSi(CH_3)CH_2CH_2SiCl(CH_3)_2]_2$ (80)	26
	H_2PtCl_6 (THF)	80	2	$[(CH_3)_2SiO]_2[OSi(CH_3)CH_2CH_2SiCl(CH_3)_2]_2$ (80)	41
	H_2PtCl_6 (THF)	80	2	$[(CH_3)_2SiO]_2[OSi(CH_3)CH_2CH_2SiCl(CH_3)_2]_2$ (80)	52
	H_2PtCl_6 (i-PrOH)	160-180	2	$[(CH_3)_2SiO]_2[OSi(CH_3)CH=CH_2][OSi(CH_3)CH_2CH_2$-$SiCl(CH_3)_2]$ (21) $[(CH_3)_2SiO]_2[OSi(CH_3)CH_2CH_2SiCl(CH_3)_2]_2$ (49)	46
$C_{11}H_{14}$	Fig.176.				
	H_2PtCl_6 (i-PrOH)	170	65	Fig.177. (-)	S -042
$C_{11}H_{19}$	$CH_2=CH(CH_2)_8CN$				
	H_2PtCl_6	reflux	16	$(CH_3)_2SiCl(CH_2)_{10}CN$ (60)	354
	H_2PtCl_6 (i-PrOH)	reflux	24	$(CH_3)_2SiCl(CH_2)_{10}CN$ (82)	354
$C_{12}H_{12}$	Fig.178.				
	H_2PtCl_6	34-38	5	Fig.179. (-)	F -035
$C_{12}H_{15}$	p-$(CH_3)_2NC_6H_4COOCH_2CH=CH_2$				

Table 3.3

= $(CH_3)_2SiHCl$ = =12=

	H_2PtCl_6 (i-PrOH)	reflux	1.6 (cy)	p-$(CH_3)_2NC_6H_4COO(CH_2)_3SiCl(CH_3)_2$ (-)	Ch-001
	H_2PtCl_6 (i-PrOH)	reflux	1.6 (cy)	p-$(CH_3)_2NC_6H_4COO(CH_2)_3SiCl(CH_3)_2$ (-)	Ch-002
	H_2PtCl_6 (i-PrOH)	reflux	1.6 (cy)	p-$(CH_3)_2NC_6H_4COO(CH_2)_3SiCl(CH_3)_2$ (-)	D -015
$C_{12}H_{16}$	p-$CH_2=CHC_6H_4Si(CH_3)_2CH=CH_2$				
	H_2PtCl_6 (i-PrOH)	until 140	0.5	p-$CH_2=CHC_6H_4Si(CH_3)_2CH_2CH_2SiCl(CH_3)_2$ (62)	402
$C_{12}H_{18}$	$(CH_2CH_2COOCH_2CH=CH_2)_2$				
	H_2PtCl_6 (i-PrOH)	100	0.25	$CH_2=CHCH_2OCO(CH_2)_4COO(CH_2)_3SiCl(CH_3)_2$ (32)	32
	H_2PtCl_6 (i-PrOH)	reflux	10	$[CH_2CH_2COO(CH_2)_3SiCl(CH_3)_2]_2$ (47)	1115
$C_{12}H_{20}$	$CH_2=CHCH_2CH_2C(CH_3)=N\text{-}N=C(CH_3)CH_2CH_2CH=CH_2$				
	H_2PtCl_6 (i-PrOH)	110	5	$(CH_3)_2SiCl(CH_2)_4C(CH_3)=N\text{-}N=C(CH_3)(CH_2)_4$- $SiCl(CH_3)_2$ (75)	U -065
$C_{12}H_{20}$	Fig.180.				
	gamma	160	120	Fig.181. (63)	S -042
$C_{12}H_{22}$	$CH_2=CHC_8H_{16}COOCH_3$				
	H_2PtCl_6 (i-PrOH)	reflux	5	$(CH_3)_2SiClC_{10}H_{20}COOCH_3$ (90)	1011
$C_{13}H_{12}$	Fig.182.				
	H_2PtCl_6 (THF)	140	15	Fig.183. (70)	U -200
$C_{13}H_{14}$	Fig.184.				
	H_2PtCl_6	34	5	Fig.185. (-)	F -035
$C_{13}H_{22}$	$CH_2=CHCOOC_{10}H_{19}$				
	H_2PtCl_6 (i-PrOH)	120	1.5	$(CH_3)_2SiClCH_2CH_2COOC_{10}H_{19}$ (29)	34
$C_{13}H_{24}$	$CH_2=CHCH_2OC_{10}H_{19}$				
	H_2PtCl_6 (i-PrOH)	140	0.5	$(CH_3)_2SiCl(CH_2)_3OC_{10}H_{19}$ (33)	637
$C_{13}H_{24}$	$CH_3COO(CH_2)_9CH=CH_2$				
	H_2PtCl_6 (i-PrOH)	reflux	5	$CH_3COO(CH_2)_{11}SiCl(CH_3)_2$ (95)	1011
$C_{13}H_{26}$	$CH_3(OCH_2CH_2)_5CH=CH_2$				
	H_2PtCl_6	80	6	$CH_3(OCH_2CH_2)_5CH_2CH_2SiCl(CH_3)_2$ (-)	140
$C_{14}H_{14}$	$C_6H_4[COOCH_2CH=CH_2]_2$				

Table 3.3

= $(CH_3)_2SiHCl$ = =13=

	H_2PtCl_6 (i-PrOH)	100	0.25	$CH_2=CHCH_2OCOC_6H_4COO(CH_2)_3SiCl(CH_3)_2$ (32)	32
	H_2PtCl_6 (i-PrOH)	reflux	10	$C_6H_4[COO(CH_2)_3SiCl(CH_3)_2]_2$ (52)	1115
$C_{14}H_{14}$	Fig.186.				
	H_2PtCl_6	34	5	Fig.187. (-)	F -035
$C_{14}H_{16}$	Fig.188.				
	H_2PtCl_6	34	5	Fig.189. (-)	F -035
$C_{14}H_{18}$	$CH_2=CHCONHCH(C_6H_5)CONHC_3H_7$				
	H_2PtCl_6	reflux	2	$(CH_3)_2ClSiCH_2CH_2CONHCH(C_6H_5)CONHC_3H_7$ (-)	631
$C_{14}H_{24}$	$CH_2=C(CH_3)COOC_{10}H_{19}$				
	H_2PtCl_6 (i-PrOH)	120	1.5	$(CH_3)_2SiClCH_2CH(CH_3)COOC_{10}H_{19}$ (49)	34
$C_{14}H_{28}$	$CH_2=CH(CH_2)_8COOSi(CH_3)_3$				
	H_2PtCl_6 (i-PrOH)	reflux	2.5	$(CH_2)_{10}COOSi(CH_3)_2$ (45)	778
$C_{16}H_6$	m-$(C_6F_4CH=CH_2)_2$				
	H_2PtCl_6 (i-PrOH)	-	-	m-$(CH_3)_2SiClCH_2CH_2C_6F_4CH_2CH_2SiCl(CH_3)_2$ (-)	U -043
$C_{16}H_{24}$	Fig.143.				
	Pt/C	reflux	24	Fig.190. (80-90)	498
$C_{19}H_{32}$	$C_4H_9(CH=CH)_3(CH_2)_7COOCH_3$				
	-	-	-	$C_4H_9CH_2CH[SiCl(CH_3)_2](CH=CH)_2(CH_2)_7COOCH_3$ (-)	1198
$C_{19}H_{36}$	$CH_2=CHC_{15}H_{30}COOCH_3$				
	H_2PtCl_6 (i-PrOH)	reflux	5	$(CH_3)_2SiClC_{17}H_{34}COOCH_3$ (47-57)	1011
	Pt/C	reflux	5	$(CH_3)_2SiClC_{17}H_{34}COOCH_3$ (40)	1011
$C_{19}H_{36}$	$CH_3(CH_2)_7CH=CH(CH_2)_7COOCH_3$				
	H_2PtCl_6 (i-PrOH)	reflux	5	$(CH_3)_2SiClC_{17}H_{34}COOCH_3$ (50)	1011
$C_{20}H_{28}$	$[OSi(C_6H_5)CH_3]_2[OSi(CH_3)CH=CH_2]_2$				
	H_2PtCl_6 (THF)	80	2	$[OSi(C_6H_5)CH_3]_2[OSi(CH_3)CH_2CH_2SiCl(CH_3)_2]_2$ (-)	26
$C_{20}H_{30}$	Fig.191.				
	H_2PtCl_6 (i-PrOH)	20-80	4	Fig.192. (100)	E -013
$C_{21}H_{28}$	Fig.193.				

Table 3.3

= $(CH_3)_2SiHCl$ = =14=

	H_2PtCl_6	34	5	Fig.194. (-)	F -035
$C_{24}H_{36}$	Fig.147.				
	Pt/C	reflux	24	Fig.195. (80-90)	498
$C_{36}H_{48}$	Fig.196.				
	H_2PtCl_6	reflux	2	Fig.197. (-)	1362

Table 3.4

= $(CH_3)_2SiHCl$ = =1=

C_2	$Cl_3SiC\equiv CSiCl_3$				
	H_2PtCl_6 (i-PrOH)	120-130	-	$Cl_3SiCH=C(SiCl_3)SiCl(CH_3)_2$ (42)	1053
C_2H_1	$CH\equiv CSiCl_3$				
	none	520	30s	$(CH_3)_2SiClCH=CHSiCl_3$ (10)	1057
	H_2PtCl_6 (i-PrOH)	130	3	$(CH_3)_2SiClCH=CHSiCl_3$ (82)	1057
C_2H_2	$CH\equiv CH$				
	$AlCl_3$	28-35	24	$CH_2=CHSiCl(CH_3)_2$ (70)	D -092
	H_2PtCl_6 $[(\overline{CH_2)_5C}O]$	-	-	$CH_2=CHSiCl(CH_3)_2$ (74)	D -041
	H_2PtCl_6 $[(\overline{CH_2)_5C}O]$	150	-	$CH_2=CHSiCl(CH_3)_2$ (74) $(CH_3)_2SiClCH_2CH_2SiCl(CH_3)_2$ (6-12)	D -042
	Pt (o-$C_6H_4Cl_2$)	86-88	flow	$CH_2=CHSiCl(CH_3)_2$ (81)	F -007
C_3H_4	$CH\equiv CSiCl_2CH_3$				
	H_2PtCl_6 (i-PrOH)	120	10	$(CH_3)_2SiClCH=CHSiCl_2CH_3$ (94)	759
	$Pt(PPh_3)_4$	120	10 (t)	$(CH_3)_2SiClCH=CHSiCl_2CH_3$ (8) $CH_2=C[SiCl(CH_3)_2]SiCl_2CH_3$ (87)	J -056
	$Pt(PPh_3)_4$	120	10	$CH_2=C[SiCl(CH_3)_2]SiCl_2CH_3$ (90)	759
	$RhCl(PPh_3)_3$	120	10	$(CH_3)_2SiClCH=CHSiCl_2CH_3$ + $CH_2=C[SiCl(CH_3)_2]SiCl_2CH_3$ (85)	759
C_4H_4	$ClCH_2C\equiv CCH_2Cl$				
	H_2PtCl_6	-	-	$ClCH_2CH=C(CH_2Cl)SiCl(CH_3)_2$ (-)	1022
	H_2PtCl_6 (i-PrOH)	reflux	8	$ClCH_2CH=C(CH_2Cl)SiCl(CH_3)_2$ (91)	E -011
C_4H_6	$CH_3C\equiv CCH_3$				
	$[HPt(SiBzMe_2)P(c\text{-}Hex)_3]_2$	55	0.2	cis-$CH_3CH=C(CH_3)SiCl(CH_3)_2$ (94)	430
C_4H_6	$CH_3CH_2C\equiv CH$				
	$[HPt(SiBzMe_2)P(c\text{-}Hex)_3]_2$	60	0.2	trans-$CH_3CH_2CH=CHSiCl(CH_3)_2$ + $CH_3CH_2C[SiCl(CH_3)_2]-CH_2$ (85)	430
C_4H_6	$CH_3Cl_2SiC\equiv CSiCl_2CH_3$				
	H_2PtCl_6 (i-PrOH)	120-130	-	$CH_3Cl_2SiCH=C(SiCl_2CH_3)SiCl(CH_3)_2$ (76)	1053
C_4H_7	$CH\equiv CSi(CH_3)_2F$				

Table 3.4

$= (CH_3)_2SiHCl =$ =2=

	H_2PtCl_6 (i-PrOH)	90	0.5	$F(CH_3)_2SiCH{=}CHSiCl(CH_3)_2$ (76)	1312
C_5H_6	$CH_2{=}C(CH_3)C{\equiv}CH$				
	H_2PtCl_6	80	6	$CH_2{=}C(CH_3)CH{=}CHSiCl(CH_3)_2$ (80)	672
	H_2PtCl_6	80	6	$CH_2{=}C(CH_3)CH{=}CHSiCl(CH_3)_2$ (80)	672
	H_2PtCl_6 + Me_3Cl_2SiH	80	6	$(CH_3)_2C{=}CHCH(SiCl_2CH_3)_2$ (18) $(CH_3)_2C{=}CHCH[Si(CH_3)_2Cl]_2$ (29) $(CH_3)_2C{=}CHCH(SiCH_3Cl_2)[Si(CH_3)_2Cl]$ (53)	672
C_5H_9	$CH{\equiv}CSi(CH_3)_2CH_2Cl$				
	H_2PtCl_6 (i-PrOH)	80-110	-	trans-$(CH_3)_2SiClCH{=}CHSi(CH_3)_2CH_2Cl$ (68)	387
	H_2PtCl_6 (i-PrOH)	90-110	- (h)	$(CH_3)_2SiClCH{=}CHSi(CH_3)_2CH_2Cl$ (89)	776
C_5H_{10}	$CH{\equiv}CSi(CH_3)_3$				
	H_2PtCl_6 (i-PrOH)	50-60	1.17	$(CH_3)_2SiClCH{=}CHSi(CH_3)_3$ (66)	134
	H_2PtCl_6 (i-PrOH)	90	0.5	$(CH_3)_2SiClCH{=}CHSiCl(CH_3)_3$ (80)	1312
C_6H_8	$(CH{\equiv}C)_2Si(CH_3)_2$				
	H_2PtCl_6 (i-PrOH)	90	0.5	$(CH_3)_2(CH{\equiv}C)SiCH{=}CHSiCl(CH_3)_2$ (25) $(CH_3)_2Si[CH{=}CHSiCl(CH_3)_2]_2$ (48)	1312
C_6H_{10}	$(CH_3)_3CC{\equiv}CH$				
	H_2PtCl_6	reflux	-	$(CH_3)_3CCH{=}CHSiCl(CH_3)_2$ (93)	773
C_6H_{10}	$CH{\equiv}CSi(CH_3)_2CH{=}CH_2$				
	H_2PtCl_6 (i-PrOH)	100-140	0.5	$(CH_3)_2SiClCH{=}CHSi(CH_3)_2CH{=}CH_2$ (-)	402
C_6H_{10}	$CH_3(CH_2)_3C{\equiv}CH$				
	H_2PtCl_6	45-50	overnight	$CH_3(CH_2)_3CH{=}CHSiCl(CH_3)_2$+ $CH_3(CH_2)_3C[SiCl(CH_3)_2]{=}$ $={CH_2}$ (84)	511
	H_2PtCl_6	reflux	-	$CH_3(CH_2)_3CH{=}CHSiCl(CH_3)_2$+ $CH_3(CH_2)_3C[SiCl(CH_3)_2]{=}$ $={CH_2}$ (79)	773
C_6H_{12}	$(CH_3)_2ClSiC{\equiv}CSiCl(CH_3)_2$				
	H_2PtCl_6 (i-PrOH)	120-130	-	$(CH_3)_2ClSiCH{=}C[SiCl(CH_3)_2]SiCl(CH_3)_2$ (67)	1053
C_6H_{12}	$CH{\equiv}CCH_2OSi(CH_3)_3$				
	H_2PtCl_6 (i-PrOH)	reflux	2	$(CH_3)_2SiClCH{=}CHCH_2OSi(CH_3)_3$ (72)	619
C_7H_6	$(CH{\equiv}C)_3SiCH_3$				

Table 3.4
= $(CH_3)_2SiHCl$ = =3=

	H_2PtCl_6 (i-PrOH)	90	0.5	$CH_3(CH\equiv C)_2SiCH=CHSiCl(CH_3)_2$ (24) $CH_3(CH\equiv C)Si[CH=CHSiCl(CH_3)_2]_2$ (43) $CH_3Si[CH=CHSiCl(CH_3)_2]_3$ (-)	1312
C_7H_8	$(CH\equiv C)_2Si(CH_3)CH=CH_2$				
	H_2PtCl_6 (i-PrOH)	90	0.5	$(CH_2=CH)(CH_3)(CH\equiv C)SiCH=CHSiCl(CH_3)_2$ (51) $(CH_2=CH)(CH_3)Si[CH=CHSiCl(CH_3)_2]_2$ (30)	1312
C_8H_6	$C_6H_5C\equiv CH$				
	$(Bu_4N)_2PtCl_6$	80	1-3	$C_6H_5CH=CHSiCl(CH_3)_2$ + $C_6H_5C[SiCl(CH_3)_2]=CH_2$ (-)	673
	H_2PtCl_6	20	-	$C_6H_5CH=CHSiCl(CH_3)_2$ (45) $C_6H_5C[SiCl(CH_3)_2]=CH_2$ (5)	172
	H_2PtCl_6	reflux	-	$C_6H_5CH=CHSiCl(CH_3)_2$ (80)	773
	H_2PtCl_6 (i-PrOH)	reflux	6-12	$C_6H_5CH=CHSiCl(CH_3)_2$ (94)	164
	$PtCl_3[(CH_2CH=CH_2)PPh_3]$	20	-	$C_6H_5CH=CHSiCl(CH_3)_2$ (49) $C_6H_5C[SiCl(CH_3)_2]=CH_2$ (1)	172
	$PtCl_3[(CH_2CH=CH_2)PPh_3]$	20	-	$C_6H_5CH=CHSiCl(CH_3)_2$ (49) $C_6H_5C[SiCl(CH_3)_2]=CH_2$ (1)	171
	$PtCl_4(NC_{12}H_{25}PPh_2)_3$	20-60	1-3	$C_6H_5CH=CHSiCl(CH_3)_2$ + $C_6H_5C[SiCl(CH_3)_2]=CH_2$ (-)	171
	SiO_2-$PPh_3/PtCl_6^{-2}$	20	-	$C_6H_5CH=CHSiCl(CH_3)_2$ (65) $C_6H_5C[SiCl(CH_3)_2]=CH_2$ (7)	172
	$[HPt(SiBzMe_2)P(c\text{-}Hex)_3]_2$	25	20	trans-$C_6H_5CH=CHSiCl(CH_3)_2$ (92)	430
	$[HPt(SiBzMe_2)P(c\text{-}Hex)_3]_2$	25	20	trans-$C_6H_5CH=CHSiCl(CH_3)_2$ (92)	428
C_8H_{12}	$\overline{(CH_2)_5C}HC\equiv CH$				
	H_2PtCl_6	reflux		$\overline{(CH_2)_5C}HCH=CHSiCl(CH_3)_2$ + $\overline{(CH_2)_5C}HC[SiCl(CH_3)_2]=CH_2$ (93)	773
C_8H_{14}	$CH_3(CH_2)_5C\equiv CH$				
	$Pt[(ViSiMe_2)_2O]_2$	r.t.	4	$CH_3(CH_2)_5CH=CHSiCl(CH_3)_2$ (-)	1188
C_8H_{18}	$(CH_3)_3SiC\equiv CSi(CH_3)_3$				
	H_2PtCl_6 (i-PrOH)	120-130	-	$(CH_3)_3SiCH=C[Si(CH_3)_2]SiCl(CH_3)_3$ (15)	1053
	H_2PtCl_6 (i-PrOH)	150-180	8	$(CH_3)_3SiCH=C[SiCl(CH_3)_2]Si(CH_3)_3$ (62)	1310
$C_{10}H_6$	$C_6H_5C\equiv CC_6H_5$				
	H_2PtCl_6 (i-PrOH)	100	2	$C_6H_5CH=C(C_6H_5)SiCl(CH_3)_2$ (79)	837

Table 3.4

= $(CH_3)_2SiHCl$ = =4=

	$[HPt(SiBzMe_2)P(c\text{-}Hex)_3]_2$	50	3	cis-$C_6H_5CH{=}C(C_6H_5)SiCl(CH_3)_2$ (84)	430
$C_{10}H_6$	$CH{\equiv}CC_6H_4C{\equiv}CH$				
	H_2PtCl_6 (THF)	34-40	1.5 (b)	$CH{\equiv}CC_6H_4CH{=}CHSiCl(CH_3)_2$ (10-73) $(CH_3)_2SiClCH{=}CHC_6H_4CH{=}CHSiCl(CH_3)_2$ (8-71)	770
$C_{12}H_{16}$	Fig.198.				
	H_2PtCl_6 (i-PrOH)	150	6	Fig.199. (66)	708

Table 3.5

= $(CH_3)_2SiHCl$ = =1=

C_3H_6	CH_3COOCH_3				
	gamma (^{60}Co)	-	-	$CH_3CH[OSiCl(CH_3)_2]OCH_3$ (-)	816
C_5H_8	$CH_2=C(CH_3)COOCH_3$				
	Rh/C (5%)	100-110	1	$(CH_3)_2C=C(OCH_3)OSi(CH_3)_2Cl$ (-)	U -187
C_7H_6	C_6H_5CHO				
	UV	-	20-45	$C_6H_5CH_2OSiCl(CH_3)_2$ (23-29)	415
	Ni	110-135	3-12	$C_6H_5CH_2OSiCl(CH_3)_2$ (25-28) $C_6H_5CH[OSiCl(CH_3)_2]CH(C_6H_5)OSiCl(CH_3)_2$ (41-65)	415
C_9H_8	$C_6H_5CH=CHCHO$				
	$PtCl_2(py)_2$	-	68	$C_6H_5CH_2CH=CHOSiCl(CH_3)_2$ (100)	987
$C_{10}H_{10}$	$C_6H_5CH=CHCOCH_3$				
	$PtCl_2(py)_2$	-	68	$C_6H_5CH_2CH=C(CH_3)OSiCl(CH_3)_2$ (75)	987
$C_{23}H_{30}$	Fig.149.				
	H_2PtCl_6 (i-PrOH)	20-60	0.5 (t)	Fig.200. (-)	1320

Table 4.1

= $(CH_3)_3SiH$ = =1=

C_2H_4	$CH_2=CH_2$				
	UV + $(t\text{-}BuO)_2$	(-74)-20	-	$CH_3CH_2Si(CH_3)_3$ (kin.)	261
	$HCo(CO)_4$	20	0.2	$CH_3CH_2Si(CH_3)_3$ (81)	74
C_3H_6	$CH_3CH=CH_2$				
	$SiO_2 + Al_2O_3$	104	flow	$C_3H_7Si(CH_3)_3$ (-)	U -017
	UV + $Fe(CO)_5$	25-35	12	$C_3H_7Si(CH_3)_3$ (81)	1028
C_4H_8	$(CH_3)_2C=CH_2$				
	$SiO_2 + Al_2O_3$	143	flow	$(CH_3)_2CHCH_2Si(CH_3)_3 + C_8H_{17}Si(CH_3)_3$ (-)	U -017
	$SiO_2 + Al_2O_3$	99-104	flow	$(CH_3)_2CHCH_2Si(CH_3)_3$ (-)	U -017
	UV + $Fe(CO)_5$	25-35	168	$C_4H_9Si(CH_3)_3$ (64)	1028
C_4H_8	$CH_3CH_2CH=CH_2$				
	UV + $Fe(CO)_5$	25-35	12	$C_4H_9Si(CH_3)_3$ (18)	1028
C_4H_8	$CH_3CH=CHCH_3$				
	$SiO_2 + Al_2O_3$	104	flow	$C_4H_9Si(CH_3)_3$ (-)	U -017
C_5H_8	$(\overline{CH_2)_3CH=C}H$				
	$Fe(CO)_5$	25-35	168	$(\overline{CH_2)_4C}HSi(CH_3)_3$ (3)	1028
C_5H_{10}	$CH_3CH_2CH_2CH=CH_2$				
	UV + $Fe(CO)_5$	25-35	12	$C_5H_{11}Si(CH_3)_3$ (18)	1028
C_6H_{10}	$(\overline{CH_2)_4CH=C}H$				
	UV + $(t\text{-}BuO)_2$	40	12	$(\overline{CH_2)_5C}HSi(CH_3)_3$ (1)	126
	UV + $Hg(SiMe_3)_2$	40	12	$(\overline{CH_2)_5C}HSi(CH_3)_3$ (2)	126
C_6H_{12}	$(CH_3)_3CCH=CH_2$				
	UV + $Fe(CO)_5$	25-35	96	$(CH_3)_3CCH_2CH_2Si(CH_3)_3$ (18)	1028
C_6H_{12}	$CH_3(CH_2)_3CH=CH_2$				
	$[HPt(SiMe_2Et)P(c\text{-}Hex)_3]_2$	20	2 (h)	$C_6H_{13}Si(CH_3)_3$ (90)	429
	$RhBr(PPh_3)_3$	50	0.25 (b)	$C_6H_{13}Si(CH_3)_3$ (71)	981
	$RhCl(CO)(PPh_3)_2$	50	0.25 (b)	$C_6H_{13}Si(CH_3)_3$ (29)	981

Table 4.1

= $(CH_3)_3SiH$ = =2=

	$RhCl(PPh_3)_3$	20	96	$C_6H_{13}Si(CH_3)_3$ (2-77)	478
	$RhCl(PPh_3)_3$	50	0.25 (b)	$C_6H_{13}Si(CH_3)_3$ (82)	981
	$RhCl(Ph_2PMe)_3$	20	96	$C_6H_{13}Si(CH_3)_3$ (54-76)	478
	$RhCl_2(C_6H_{12})$ + phosphine	80	2 (b)	$C_6H_{13}Si(CH_3)_3$ (17-62)	981
	$RhH(CO)(PPh_3)_3$	50	0.25 (b)	$C_6H_{13}Si(CH_3)_3$ (21)	981
	$RhI(PPh_3)_3$	50	0.25 (b)	$C_6H_{13}Si(CH_3)_3$ (34)	981
C_8H_8	$C_6H_5CH=CH_2$				
	H_2PtCl_6	5-10	24	$C_6H_5CH_2CH_2Si(CH_3)_3$ (kin.)	966
	$[PtCl_2(PhCH=CH_2)]_2$	5-10	24	$C_6H_5C_2H_4Si(CH_3)_3$ (81-82)	965
	$RhCl(CO)(PPh_3)_2$	80	2 (b)	$C_6H_5CH_2CH_2Si(CH_3)_3$ + $C_6H_5CH(CH_3)Si(CH_3)_3$ (43)	981
	$RhCl(PPh_3)_3$	80	2 (b)	$C_6H_5CH_2CH_2Si(CH_3)_3$ + $C_6H_5CH(CH_3)Si(CH_3)_3$ (26)	981
	$RhH(CO)(PPh_3)_3$	80	2 (b)	$C_6H_5CH_2CH_2Si(CH_3)_3$ + $C_6H_5CH(CH_3)Si(CH_3)_3$ (30)	981
	$RhI(PPh_3)_3$	80	2 (b)	$C_6H_5CH_2CH_2Si(CH_3)_3$ + $C_6H_5CH(CH_3)Si(CH_3)_3$ (23)	981
C_8H_{16}	$CH_3(CH_2)_5CH=CH_2$				
	UV + $(t\text{-}BuO)_2$	40	12	$CH_3(CH_2)_7Si(CH_3)_3$ (9)	126
	UV + $Hg(SiMe_3)_2$	40	12	$CH_3(CH_2)_7Si(CH_3)_3$ (8)	126
	$Pd(PPh_3)_4$	-	-	$C_8H_{17}Si(CH_3)_3$ (15)	126
	$Pd(PPh_3)_4$	120	6	$C_8H_{17}Si(CH_3)_3$ (45)	1208
	$Rh_4(CO)_{12}$	45-50	1	$CH_3(CH_2)_7Si(CH_3)_3$ (95-99)	S -037
C_9H_{10}	$CH_3C(C_6H_5)=CH_2$				
	RhCl(diop)(S)	120	4	$CH_3CH(C_6H_5)CH_2Si(CH_3)_3$ (63)	1302
	$[Rh(BMPP)_2H_2L_2]ClO_4$	120	40	$CH_3CH(C_6H_5)CH_2Si(CH_3)_3$ (63)	1302

Table 4.2

= $(CH_3)_3SiH$ = =1=

C_4H_6 $CH_2=CHCH=CH_2$

Ni, Co + AlR_3	-	-	$CH_3CH=CHCH_2Si(CH_3)_3$ (90)	1356
UV + $Cr(CO)_6$	30	several days	$CH_3CH=CHCH_2Si(CH_3)_3$ (-)	1293
UV + $Fe(CO)_5$	-	165	$CH_3CH=CHCH_2Si(CH_3)_3$ + $CH_3CH_2CH=CHSi(CH_3)_3$ (28) $(CH_3)_3SiCH_2CH=CHCH_2Si(CH_3)_3$ (-) $C_4H_9Si(CH_3)_3$ (3)	366
$Co(acac)_2$ + $AlEt_3$	-	4	cis-$CH_3CH=CHCH_2Si(CH_3)_3$ + trans-$CH_3CH=CHCH_2Si(CH_3)_3$ (60-95)	1336
$Co(acac)_2$ + $AlEt_3$	60	4	$CH_3CH=CHCH_2Si(CH_3)_3$ (15) $CH_3CH=CHCH_2CH_2CH=CHCH_2Si(CH_3)_3$ (80)	1332
$Co(acac)_2$ + $AlEt_3$	85	4	cis-$CH_3CH=CHCH_2Si(CH_3)_3$ + trans-$CH_3CH=CHCH_2Si(CH_3)_3$ (95)	1330
$Ni(acac)_2$ + $AlEt_3$	-	4	cis-$CH_3CH=CHCH_2Si(CH_3)_3$ (90)	1336
$Ni(acac)_2$ + $AlEt_3$	60	4	cis-$CH_3CH=CHCH_2Si(CH_3)_3$ + trans-$CH_3CH=CHCH_2Si(CH_3)_3$ (90)	1332
$Ni(acac)_2$ + $AlEt_3$	85	4	cis-$CH_3CH=CHCH_2Si(CH_3)_3$ (until 90) $CH_3CH=CHCH_2CH_2CH=CHCH_2Si(CH_3)_3$ (until 3)	1330
$Ni(acac)_2$ + $AlEt_3$ + PPh_3	60	4	$CH_3CH=CHCH_2Si(CH_3)_3$ + $CH_3CH=CHCH_2CH_2CH=CHCH_2Si(CH_3)_3$ (50)	1332
$Ni(acac)_2$ + AlR_3 + PPh_3	60	5	cis-$CH_3CH=CHCH_2Si(CH_3)_3$ + cis,cis-$CH_3CH=CHCH_2CH_2CH=CHCH_2Si(CH_3)_3$ (until 60)	1336
$Ni(cod)_2$	80	1.5 (b)	$CH_3CH=CHCH_2Si(CH_3)_3$ (51) $CH_3CH=CHCH_2CH_2CH=CHCH_2Si(CH_3)_3$ (8)	205
Pd + PPh_3	-	-	$CH_3CH=CHCH_2CH_2CH=CHCH_2Si(CH_3)_3$ (90)	468
Pd + PPh_3	140	10	$CH_3CH=CHCH_2Si(CH_3)_3$ (77)	D -027
$Pd(\overline{OCOCH=CHCO})(PPh_3)_2$	65-100	4.5 (t)	$CH_3CH=CHCH_2CH_2CH=CHCH_2Si(CH_3)_3$ (74-98)	1176
$Pd(\overline{OCOCH=CHCO})(PPh_3)_2$	85	4.5 (THF)	$CH_3CH=CHCH_2CH_2CH=CHCH_2Si(CH_3)_3$ (96)	1176
$Pd(\overline{OCOCH=CHCO})(PPh_3)_2$	85	6	$CH_3CH=CHCH_2CH_2CH=CHCH_2Si(CH_3)_3$ (98)	J -004
$PdCl_2$	-	- (b)	$CH_3CH=CHCH_2CH_2CH=CHCH_2Si(CH_3)_3$ (-)	J -012
$PdCl_2(PPh_3)_2$	100	4 (b)	$CH_3CH=CHCH_2Si(CH_3)_3$ + $CH_3CH=CHCH_2CH_2CH=CHCH_2Si(CH_3)_3$ (75)	643
$PdCl_2(PPh_3)_2$	100	4	$CH_3CH=CHCH_2Si(CH_3)_3$ + $CH_3CH=CHCH_2CH_2CH=CHCH_2Si(CH_3)_3$ (79)	643

Table 4.2

= $(CH_3)_3SiH$ = =2=

$PdCl_2(PPh_3)_2$	100	4 (THF)	$CH_3CH{=}CHCH_2Si(CH_3)_3$ + $CH_3CH{=}CHCH_2CH_2CH{=}CHCH_2Si(CH_3)_3$ (68)	643
$PdCl_2(PhCN)_2$	100	4 (b)	$CH_3CH{=}CHCH_2Si(CH_3)_3$ + $CH_3CH{=}CHCH_2CH_2CH{=}CHCH_2Si(CH_3)_3$ (69)	643
$PdCl_2(PhCN)_2$	22	24 (b)	$CH_3CH{=}CHCH_2Si(CH_3)_3$ + $CH_3CH{=}CHCH_2CH_2CH{=}CHCH_2Si(CH_3)_3$ (77)	643
$PdCl_2$(c-HexNC)(PPh_3)	100	4 (b)	$CH_3CH{=}CHCH_2Si(CH_3)_3$ + $CH_3CH{=}CHCH_2CH_2CH{=}CHCH_2Si(CH_3)_3$ (52)	643
$PdCl_2$(c-HexNC)$_2$	100	4 (b)	$CH_3CH{=}CHCH_2Si(CH_3)_3$ (57)	643
$PdCl_2$(t-BuNC)(PPh_3)	100	4 (b)	$CH_3CH{=}CHCH_2Si(CH_3)_3$ + $CH_3CH{=}CHCH_2CH_2CH{=}CHCH_2Si(CH_3)_3$ (58)	643
$PdCl_2$(t-BuNC)$_2$	100	4 (b)	$CH_3CH{=}CHCH_2Si(CH_3)_3$ + $CH_3CH{=}CHCH_2CH_2CH{=}CHCH_2Si(CH_3)_3$ (55)	643
[PdCl(π-All)]$_2$	100	4 (b)	$CH_3CH{=}CHCH_2Si(CH_3)_3$ + $CH_3CH{=}CHCH_2CH_2CH{=}CHCH_2Si(CH_3)_3$ (74)	643
[PdCl(π-All)]$_2$	22	24	$CH_3CH{=}CHCH_2Si(CH_3)_3$ + $CH_3CH{=}CHCH_2CH_2CH{=}CHCH_2Si(CH_3)_3$ (89)	643
polymer/Pd complex	100	2	$CH_3CH{=}CHCH_2CH_2CH{=}CHCH_2Si(CH_3)_3$ (24) $(CH_3)_3Si(CH_2)_4Si(CH_3)_3$ (22)	1220
polymer/Pt complex	100	2	cis-$CH_3CH{=}CHCH_2Si(CH_3)_3$ (30) trans-$CH_3CH{=}CHCH_2Si(CH_3)_3$ (16)	1220
Rh(π-All)(PPh_3)]$_2$	80	0.25 (b)	$CH_2{=}CHCH_2CH_2Si(CH_3)_3$ + $CH_3CH{=}CHCH_2Si(CH_3)_3$ + $(CH_3)_3SiCH_2CH{=}CHCH_2Si(CH_3)_3$ (77)	982
$RhBr(PPh_3)_3$	80	0.25 (b)	$CH_2{=}CHCH_2CH_2Si(CH_3)_3$ + $CH_3CH{=}CHCH_2Si(CH_3)_3$ + $(CH_3)_3SiCH_2CH{=}CHCH_2Si(CH_3)_3$ (76)	982
$RhCl(CO)(PPh_3)_2$	80	0.25 (b)	$CH_2{=}CHCH_2CH_2Si(CH_3)_3$ + $CH_3CH{=}CHCH_2Si(CH_3)_3$ + $(CH_3)_3SiCH_2CH{=}CHCH_2Si(CH_3)_3$ (10)	982
$RhCl(PPh_3)_3$	80	0.25 (b)	$CH_2{=}CHCH_2CH_2Si(CH_3)_3$ + $CH_3CH{=}CHCH_2Si(CH_3)_3$ + $(CH_3)_3SiCH_2CH{=}CHCH_2Si(CH_3)_3$ (80)	982
$RhH(CO)(PPh_3)_3$	80	0.25 (b)	$CH_2{=}CHCH_2CH_2Si(CH_3)_3$ + $CH_3CH{=}CHCH_2Si(CH_3)_3$ + $(CH_3)_3SiCH_2CH{=}CHCH_2Si(CH_3)_3$ (6)	982
$RhI(PPh_3)_3$	80	0.25 (b)	$CH_2{=}CHCH_2CH_2Si(CH_3)_3$ + $CH_3CH{=}CHCH_2Si(CH_3)_3$ + $(CH_3)_3SiCH_2CH{=}CHCH_2Si(CH_3)_3$ (71)	982
$RhMe(PPh_3)_3$	80	0.25 (b)	$CH_2{=}CHCH_2CH_2Si(CH_3)_3$ + $CH_3CH{=}CHCH_2Si(CH_3)_3$ + $(CH_3)_3SiCH_2CH{=}CHCH_2Si(CH_3)_3$ (88)	982
polymer/Rh complex	100	2	$CH_2{=}CHCH_2CH_2Si(CH_3)_3$ (9) cis-$CH_3CH{=}CHCH_2Si(CH_3)_3$ (31) trans-$CH_3CH{=}CHCH_2Si(CH_3)_3$ (18) $(CH_3)_3Si(CH_2)_4Si(CH_3)_3$ (21)	1220

C_5H_8 $CH_2{=}C(CH_3)CH{=}CH_2$

	UV + $Cr(CO)_6$	30	several days	$(CH_3)_2C{=}CHCH_2Si(CH_3)_3$/$(CH_3)_3SiCH_2C(CH_3){=}CHCH_3$ (40/60)	1293
	$Ni(acac)_2$ + $AlEt_3$	-	4	$(CH_3)_2C{=}CHCH_2Si(CH_3)_3$ (25) $(CH_3)_2SiCH_2C(CH_3){=}CHCH_3$ (70)	1336
	$Ni(acac)_2$ + $AlEt_3$	85	4	$(CH_3)_2C{=}CHCH_2Si(CH_3)_3$ (-)	1332
	H_2PtCl_6 (i-PrOH)	165	17	$(CH_3)_3SiCH_2C(CH_3){=}CHCH_3$ (20) $CH_2{=}C(CH_3)CH_2CH_2Si(CH_3)_3$ (27) $(CH_3)_3SiCH_2CH(CH_3)CH_2CH_2Si(CH_3)_3$ (24)	122
C_5H_8	$CH_3CH{=}CHCH{=}CH_2$				
	Ni or Co salt + AlR_3	-	-	$CH_3CH_2CH{=}CHCH_2Si(CH_3)_3$ + $CH_3CH{=}CHCH_2CH_2Si(CH_3)_3$ (-)	1356
	Ni salt + AlR_3 + PPh_3	140	-	trans-$CH_3CH_2CH{=}CHCH_2SiCH_3$ + trans-$CH_3CH{=}CHCH_2CH_2Si(CH_3)_3$ + trans,trans-$CH_3CH{=}CHCH(CH_3)CH_2CH{=}CHCH(CH_3)$-$Si(CH_3)_3$ (30-50)	1324
	UV + $Cr(CO)_6$	30	several days	$CH_3CH_2CH{=}CHCH_2Si(CH_3)_3$/$CH_3CH{=}CHCH(CH_3)Si(CH_3)_3$ (90/10)	1293
	$Ni(acac)_2$ + $AlEt_3$	-	4	cis-$CH_3CH_2CH{=}CHCH_2Si(CH_3)$ + trans-$CH_3CH_2CH{=}CHCH_2Si(CH_3)_3$ (60-95)	1336
	$Ni(acac)_2$ + $AlEt_3$	85	4	$CH_3CH_2CH{=}CHCH_2Si(CH_3)_3$ + $CH_3CH{=}CHCH_2CH_2Si(CH_3)_3$ (94)	1332
C_6H_{10}	$CH_2{=}C(CH_3)C(CH_3){=}CH_2$				
	UV + $Cr(CO)_6$	30	several days	$(CH_3)_2C{=}C(CH_3)CH_2Si(CH_3)_3$ (-)	1293
C_8H_{11}	$CH_2{=}CHCH(CH_3)CH{=}CHCH{=}CH_2$				
	Ni or Co salt + AlR_3	-	-	$CH_2{=}CHCH(CH_3)CH_2CH{=}CHCH_2Si(CH_3)_3$ (-)	1356
	$Ni(acac)_2$ + $AlEt_3$	-	4	cis-$CH_2{=}CHCH(CH_3)CH_2CH{=}CHCH_2Si(CH_3)_3$ (60-95)	1336
	$Ni(acac)_2$ + $AlEt_3$	60	4	$CH_2{=}CHCH(CH_3)CH_2CH{=}CHCH_2Si(CH_3)_3$ (40)	1332
C_8H_{12}	Fig.34.				
	H_2PtCl_6 (i-PrOH)	45	36	Fig.201. (38)	1297
C_8H_{12}	Fig.35.				
	H_2PtCl_6 (i-PrOH)	45	36	Fig.202. (35)	1297
	Pt/C	200	16	Fig.202. (32)	1297
	$PtCl_2(C_2H_4)(py)$	45	16 (b)	Fig.202. (53)	1297
C_8H_{14}	$CH_2{=}CH(CH_2)_4CH{=}CH_2$				
	$[HPt(SiMe_2Bz)P(c\text{-}Hex)_3]_2$	20	2	$(CH_3)_3Si(CH_2)_8Si(CH_3)_3$ (86)	818

Table 4.3

= $(CH_3)_3SiH$ = =1=

C_2	$F_2C{=}CF_2$				
	UV+ Hg	-	-	$CHF_2CF_2Si(CH_3)_3$ + $CHF_2(CF_2)_3Si(CH_3)_3$ (kin.)	474
	gamma(^{60}Co)	20-30	-	$CHF_2CF_2Si(CH_3)_3$ (66) $CHF_2(CF_2)_3Si(CH_3)_3$ (34)	1360
C_2	$F_2C{=}CFCl$				
	gamma(^{60}Co)	20-30	-	$(CH_3)_3SiCF_2CHFCl$ (28) $(CH_3)_3SiCF_2CH_2F$ (24)	1360
C_2H_1	$HFC{=}CF_2$				
	gamma(^{60}Co)	20-30	-	$(CH_3)_3SiCHFCHF_2/CH_2FCF_2Si(CH_3)_3$ (50/50)	1360
C_2H_2	$Cl_3SiCH{=}CHSiCl_3$				
	H_2PtCl_6 (i-PrOH)	120-130	-	$Cl_3SiCH_2CH(SiCl_3)Si(CH_3)_3$ (54)	1053
C_2H_2	$HFC{=}CHCl$				
	UV	-	100	$(CH_3)_3SiCHFCH_2Cl$ (14) $(CH_3)_3SiCHFCH_3$ (27)	270
C_2H_3	$CH_2{=}CHF$				
	UV	-	100-300	$(CH_3)_3SiCH_2CH_2F$ (11-90)	270
C_2H_3	$Cl_3SiCH{=}CH_2$				
	H_2PtCl_6 (i-PrOH)	-	-	$Cl_3SiCH_2CH_2Si(CH_3)_3$ (52) $(CH_3)_3SiCH_2CH_2Si(CH_3)_3$ (4) $Cl_3SiCH_2CH_2SiCl_3$ (7)	1056
	H_2PtCl_6 (i-PrOH)	reflux	4	$Cl_3SiC_2H_4Si(CH_3)_3$ (-)	1059
C_3	$CF_3FC{=}CF_2$				
	gamma(^{60}Co)	20-30	-	$CF_3CHFCF_2Si(CH_3)_3/CF_3CF(CHF_2)Si(CH_3)_3$ (81/19)	1360
C_3H_1	$CF_3CH{=}CF_2$				
	gamma(^{60}Co)	40-60	-	$CF_3CH_2CF_2Si(CH_3)_3/CF_3CH(CHF_2)Si(CH_3)_3$ (56/44)	1360
C_3H_3	$CH_2{=}CHCN$				
	$Pd(acac)_2$ + PPh_3	120	6	$(CH_3)_3SiCH(CH_3)CN$ (46)	D -027
C_3H_6	$CH_2{=}CHSiCl_2CH_3$				
	H_2PtCl_6 (i-PrOH)	-	-	$(CH_3)_3SiCH_2CH_2SiCl_2CH_3$ (70) $CH_3Cl_2SiCH_2CH_2SiCl_2CH_3$ (3) $(CH_3)_3SiCH_2CH_2Si(CH_3)_3$ (1)	1056

Table 4.3

= $(CH_3)_3SiH$ = =2=

	H_2PtCl_6 (i-PrOH)	reflux	4	$(CH_3)_3SiC_2H_4SiCl_2CH_3$ (-)	1059
C_4	$\overline{CF_2CF_2CCl=C}Cl$				
	none	190	12	$\overline{CF_2CF_2CH(Cl)C}(Cl)Si(CH_3)_3$ + $\overline{CF_2CF_2C(Cl)[Si(CH_3)_3]C}HCl$ (60)	294
C_4	$\overline{CF_2CF_2CH=C}Cl$				
	none	250	4 days	$\overline{CF_2CF_2CH_2C}(Cl)Si(CH_3)_3$ + $\overline{CF_2CF_2CH[Si(CH_3)_3]C}HCl$ (60)	294
C_4H_3	$(CF_3)_2NCH=CH_2$				
	UV	-	48	$(CF_3)_2NCH_2CH_2Si(CH_3)_3$ (95)	11
C_4H_6	$CH_2=CHCOOCH_3$				
	$Co_2(CO)_8$	25	3 (b)	$(CH_3)_3SiCH=CHCOOCH_3$ (65) $(CH_3)_3SiCH_2CH_2COOCH_3$ (8)	1178
	H_2PtCl_6 (i-PrOH)	60	3-5	$(CH_3)_3SiCH_2CH_2COOCH_3$ (4) $(CH_3)_3SiCH(CH_3)COOCH_3$ (33)	1321
C_4H_6	$Cl_2Si(CH=CH_2)_2$				
	H_2PtCl_6 (i-PrOH)	reflux	4	$Cl_2Si(CH=CH_2)C_2H_4Si(CH_3)_3$ + $Cl_2Si[C_2H_4Si(CH_3)_3]_2$ (-)	1059
C_4H_8	$CH_3Cl_2SiCH=CHSiCl_2CH_3$				
	H_2PtCl_6 (i-PrOH)	120-130	-	$CH_3Cl_2SiCH_2CH(SiCl_2CH_3)Si(CH_3)_3$ (56)	1053
C_4H_9	$CH_2=CHSiCl(CH_3)_2$				
	H_2PtCl_6 (i-PrOH)	-	-	$(CH_3)_2ClSiCH_2CH_2Si(CH_3)_3$ (47) $(CH_3)_3SiCH_2CH_2Si(CH_3)_3$ (8) $(CH_3)_2ClSiCH_2CH_2SiCl(CH_3)_2$ (3)	1056
	H_2PtCl_6 (i-PrOH)	reflux	4	$(CH_3)_3SiC_2H_4SiCl(CH_3)_2$ (-)	1059
C_5H_5	$CH_2=CHCH_2N(CF_3)_2$				
	H_2PtCl_6	80	15	$(CH_3)_3Si(CH_2)_3N(CF_3)_2$ (98)	290
C_5H_8	$CH_2=C(CH_3)COOCH_3$				
	H_2PtCl_6 (i-PrOH)	60	3-5	$(CH_3)_3SiCH_2CH(CH_3)COOCH_3$ (69)	1321
	$RhCl(PPh_3)_3$	100	5	$(CH_3)_2C=C(OCH_3)OSi(CH_3)_3$ + $CH_2=C(CH_3)CH(OCH_3)OSi(CH_3)_3$ + $(CH_3)_3SiCH_2C(CH_3)COOCH_3$ (kin.)	984
	$RhCl(PPh_3)_3$	100	5	$(CH_3)_2C=C(OCH_3)OSi(CH_3)_3$ (50-60) $CH_2=C(CH_3)CH(OCH_3)OSi(CH_3)_3$ (-) $(CH_3)_3SiCH_2C(CH_3)COOCH_3$ (-)	984

Table 4.3

= $(CH_3)_3SiH$ =

	$RhCl(PPh_3)_3$	40-55	6-8	$(CH_3)_3SiCH_2CH(CH_3)COOCH_3$ (1-12) $(CH_3)_2C=C(OCH_3)[OSi(CH_3)_3]$ (41-80) $CH_2=C(CH_3)CH(OCH_3)[OSi(CH_3)_3]$ (2-10)	E -023
	$RhCl_3 \cdot 3\ H_2O$ (Me_2CO)	45-75	-	$(CH_3)_3SiCH_2CH(CH_3)COOCH_3$ (0-1) $(CH_3)_2C=C(OCH_3)[OSi(CH_3)_3]$ (45-86) $CH_2=C(CH_3)CH(OCH_3)[OSi(CH_3)_3]$ (5-11)	E -023
	$RhCl_3 \cdot 3\ H_2O$ (Me_2CO)	55	-	$(CH_3)_3SiCH_2CH(CH_3)COOCH_3$ (kin.) $(CH_3)_2C=C(OCH_3)[OSi(CH_3)_3]$ (kin.) $CH_2=C(CH_3)CH(OCH_3)[OSi(CH_3)_3]$ (kin.)	E -023
C_5H_8	$CH_3CH=CHCOOCH_3$				
	H_2PtCl_6 (i-PrOH)	60	3-5	$CH_3CH_2CH[Si(CH_3)_3]COOCH_3$ (17)	1321
	$RhCl(PPh_3)_3$	60	0.17	$CH_3CH_2CH[Si(CH_3)_3]COOCH_3$ (9) $CH_3CH_2CH=C(OCH_3)OSi(CH_3)_3$ (77)	1321
C_5H_9	$CH_2=C(CH_3)\overline{BOCH_2CH_2O}$				
	H_2PtCl_6	-	-	$(CH_3)_3SiCH_2CH(CH_3)\overline{BOCH_2CH_2O}$ (41)	162
C_5H_9	$ClSi(CH_3)(CH=CH_2)_2$				
	H_2PtCl_6 (i-PrOH)	reflux	4	$ClSi(CH_3)(CH=CH_2)C_2H_4Si(CH_3)_3$ (-)	1059
C_5H_{10}	$CH_2=CHCH(OCH_3)_2$				
	H_2PtCl_6 (i-PrOH)	115	72	$(CH_3)_3SiCH_2CH_2CH(OCH_3)_2$ (70)	135
C_5H_{12}	$CH_2=CHSi(CH_3)_3$				
	$Fe(CO)_4(CH_2=CHSiMe_3)$	70-80	3	$(CH_3)_3SiCH_2CH_2Si(CH_3)_3$ (4) $(CH_3)_3SiCH(CH_3)Si(CH_3)_3$ (2) $(CH_3)_3SiCH=CHSi(CH_3)_3$ (60)	399
	$Fe(CO)_4(CH_2=CHSiMe_3)$	70-80	3	$(CH_3)_3SiCH_2CH_2Si(CH_3)_3$ (4) $(CH_3)_3SiCH(CH_3)Si(CH_3)_3$ (2) $(CH_3)_3SiCH=CHSi(CH_3)_3$ (60)	398
	H_2PtCl_6	105-131	6	$(CH_3)_3SiCH_2CH_2Si(CH_3)_3$ (88-98)	1058
	H_2PtCl_6 (i-PrOH)	reflux	4	$(CH_3)_3SiCH_2CH_2Si(CH_3)_3$ + $(CH_3)_3SiCH(CH_3)Si(CH_3)_3$ (-)	1059
C_6H_9	$ClSi(CH=CH_2)_3$				
	H_2PtCl_6 (i-PrOH)	reflux	4	$ClSi(CH=CH_2)_2C_2H_4Si(CH_3)_3$ + $ClSi(CH=CH_2)[C_2H_4Si(CH_3)_3]_2$ + $ClSi[C_2H_4Si(CH_3)_3]_3$ (-)	1059
C_6H_{10}	$CH_3CH_2CH=CHCOOCH_3$				

Table 4.3

$= (CH_3)_3SiH =$

	H_2PtCl_6 (i-PrOH)	60	3-5	$CH_3CH_2CH_2CH[Si(CH_3)_3]COOCH_3$ (3) $(CH_3)_3SiOC(OCH_3)=C(CH_2CH_2CH_3)CH(CH_2CH_3)$- CH_2COOCH_3 (25)	1321
C_6H_{12}	$(CH_3)_2Si(CH=CH_2)_2$				
	H_2PtCl_6 (i-PrOH)	reflux	4	$(CH_3)_2Si(CH=CH_2)C_2H_4Si(CH_3)_3$ + $(CH_3)_2Si[C_2H_4Si(CH_3)_3]_2$ (-)	1059
C_6H_{12}	$CH_2=CH(CH_2)_3OCH_3$				
	H_2PtCl_6	100	4	$(CH_3)_3Si(CH_2)_5OCH_3$ (-)	1093
C_6H_{14}	$(CH_3)_2ClSiCH=CHSiCl(CH_3)_2$				
	H_2PtCl_6 (i-PrOH)	120-130	-	$(CH_3)_2ClSiCH_2CH[SiCl(CH_3)_2]Si(CH_3)_3$ (49)	1053
C_7H_{12}	$(CH_3)_2CHCH=CHCOOCH_3$				
	H_2PtCl_6 (i-PrOH)	60	3-5	$(CH_3)_2CHCH_2CH[Si(CH_3)_3]COOCH_3$ (3)	1321
C_7H_{12}	$CH_3Si(CH=CH_2)_3$				
	H_2PtCl_6 (i-PrOH)	reflux	4	$CH_3Si(CH=CH_2)_2C_2H_4Si(CH_3)_3$ + $CH_3Si(CH=CH_2)[C_2H_4Si(CH_3)_3]_2$ + $CH_3Si[C_2H_4Si(CH_3)_3]_3$ (-)	1059
C_7H_{14}	$CH_2=CHCH(OCH_2CH_3)_2$				
	H_2PtCl_6 (i-PrOH)	115	72	$(CH_3)_3SiCH_2CH_2CH(OCH_2CH_3)_2$ (86)	135
C_7H_{16}	$CH_2=CHCH_2CH_2Si(CH_3)_3$				
	$RhCl(PPh_3)_3$	80	2 (b)	$(CH_3)_3Si(CH_2)_4Si(CH_3)_3$ (42)	982
C_7H_{18}	$CH_2=CHSi(CH_3)_2Si(CH_3)_3$				
	$PtCl_2(C_2H_4)(py)$	45	20 (b)	$(CH_3)_3SiCH_2CH_2Si(CH_3)_2Si(CH_3)_3$ (88)	1305
C_8H_3	$C_6F_5CH=CH_2$				
	H_2PtCl_6 (i-PrOH)	20	70	$C_6F_5CH_2CH_2Si(CH_3)_3$ (88)	132
C_8H_{12}	$(CH_2=CH)_4Si$				
	H_2PtCl_6	163	5.5	$(CH_3)_3SiCH_2CH_2Si(CH=CH_2)_3$ (38) $(CH_3)_3SiCH(CH_3)Si(CH=CH_2)_3$ (9) $[(CH_3)_3SiCH_2CH_2]_2Si(CH=CH_2)_2$ (22) $[(CH_3)_3SiCH_2CH_2]_3SiCH=CH_2$ (7) $[(CH_3)_3SiCH_2CH_2]_4Si$ (5)	1058
C_8H_{12}	$(CH=CH_2)_4Si$				
	H_2PtCl_6 (i-PrOH)	reflux	4	$Si(CH=CH_2)_3C_2H_4Si(CH_3)_3$ + $Si(CH=CH_2)_2[C_2H_4Si(CH_3)_3]_2$ + $Si(CH=CH_2)[C_2H_4Si(CH_3)_3]_3$ + $Si[C_2H_4Si(CH_3)_3]_4$ (-)	1059
C_8H_{18}	$CH_2=CHSi(OCH_2CH_3)_3$				

	H_2PtCl_6 (i-PrOH)	50-200	0.58	$(CH_3)_3SiCH_2CH_2Si(OCH_2CH_3)_3$ (99)	575
C_8H_{20}	$(CH_3)_3SiCH=CHSi(CH_3)_3$				
	H_2PtCl_6 (i-PrOH)	120-130	-	$(CH_3)_3SiCH_2CH[Si(CH_3)_3]_2$ (32)	1053
$C_{10}H_{18}$	$CH_2=C(CH_3)COOCH_2CH_2OCH(CH_3)OCH_2CH_3$				
	$RhCl_3 \cdot 3\ H_2O$	45	10 (dcm)	$(CH_3)_2C=C[OSi(CH_3)_3]OCH_2CH_2OCH(CH_3)OCH_2CH_3$ (68) $CH_2=C(CH_3)CH[OSi(CH_3)_3]OCH_2CH_2OCH(CH_3)OCH_2CH_3$ (trace)	E -023

Table 4.4

= $(CH_3)_3SiH$ = =1=

C_2	$Cl_3SiC{\equiv}CSiCl_3$				
	H_2PtCl_6 (i-PrOH)	120-130	-	$Cl_3SiCH{=}C(SiCl_3)Si(CH_3)_3$ (56)	1053
C_2H_1	$CH{\equiv}CSiCl_3$				
	none	520	30s	$(CH_3)_3SiCH{=}CHSiCl_3$ (4)	1057
	H_2PtCl_6 (i-PrOH)	130	4	$(CH_3)_3SiCH{=}CHSiCl_3$ (70) $[(CH_3)_3Si]_2C_2H_3SiCl_3$ (10)	1057
C_3	$CF_3C{\equiv}CF$				
	UV	-	120	$CF_3CH{=}CFSi(CH_3)_3$ (82) $CF_3C[Si(CH_3)_3]{=}CHF$ (14)	481
C_3H_1	$CF_3C{\equiv}CH$				
	UV	-	120-130	$CF_3CH{=}CHSi(CH_3)_3$ (14-65) $CF_3C[Si(CH_3)_3]{=}CH_2$ (7-26)	481
C_4	$CF_3C{\equiv}CCF_3$				
	UV	-	11 days	$(CH_3)_3SiC(CF_3){=}CHCF_3$ (73) $(CH_3)_3SiCH[Si(CH_3)_3]CH[Si(CH_3)_3]CF_3$ (trace)	295
	none	235	7	$(CH_3)_3SiC(CF_3){=}CHCF_3$ + $(CH_3)_3SiCH[Si(CH_3)_3]CH[Si(CH_3)_3]CF_3$ (-)	295
C_4H_4	$CH{\equiv}CCH{=}CH_2$				
	H_2PtCl_6 (i-PrOH)	-	-	$(CH_3)_3SiCH{=}CHCH{=}CH_2$ (-)	1111
C_4H_6	$CH_3Cl_2SiC{\equiv}CSiCH_3Cl_2$				
	H_2PtCl_6 (i-PrOH)	120-130	-	$CH_3Cl_2SiCH{=}C(SiCl_2CH_3)Si(CH_3)_3$ (87)	1053
C_6H_{12}	$(CH_3)_2ClSiC{\equiv}CSiCl(CH_3)_2$				
	H_2PtCl_6 (i-PrOH)	120-130	-	$(CH_3)_2ClSiCH{=}C[SiCl(CH_3)_2]Si(CH_3)_3$ (82)	1053
C_7H_{14}	$BrCH_2Si(CH_3)_2C{\equiv}CSi(CH_3)_2Cl$				
	H_2PtCl_6 (i-PrOH)	80	5 days	$BrCH_2Si(CH_3)_2CH{=}C[Si(CH_3)_3]Si(CH_3)_2Cl$ (65)	387
C_8H_6	$C_6H_5C{\equiv}CH$				
	$(Bu_4N)_2PtCl_6$	80	1-3	$C_6H_5CH{=}CHSi(CH_3)_3$ + $C_6H_5C[Si(CH_3)_3]{=}CH_2$ (-)	673
	$[HPt(SiBzMe_2)P(c\text{-}Hex)_3]_2$	25	70	$C_6H_5CH{=}CHSi(CH_3)_3$ + $C_6H_5C[Si(CH_3)_3]{=}CH_2$ (80)	430
C_8H_{12}	$CH{\equiv}CCH{=}CHOC_4H_9$				
	H_2PtCl_6	80-90	6	$(CH_3)_3SiCH{=}CHCH{=}CHOC_4H_9$ (82)	1098
C_8H_{18}	$(CH_3)_3SiC{\equiv}CSi(CH_3)_3$				

Table 4.4

= $(CH_3)_3SiH$ = =2=

	H_2PtCl_6 (i-PrOH)	120-130	-	$(CH_3)_3SiCH=C[Si(CH_3)_3]_2$ (67)	1053
$C_{10}H_{18}$	$(CH_3)_3SiC\equiv CC\equiv CSi(CH_3)_3$				
	H_2PtCl_6 (i-PrOH)	100	2	$(CH_3)_3SiCH=C[Si(CH_3)_3]C\equiv CSi(CH_3)_3$ (40) $(CH_3)_3SiCH_2C[Si(CH_3)_3]=C=C[Si(CH_3)_3]_2$ (46)	633
	$Pt(PPh_3)_4$	90	12	$(CH_3)_3SiCH=C[Si(CH_3)_3]C\equiv CSi(CH_3)_3$ (1) $(CH_3)_3SiCH_2C[Si(CH_3)_3]=C=C[Si(CH_3)_3]_2$ (94)	633
	$Pt(PPh_3)_4$	90	1	$(CH_3)_3SiCH=C[Si(CH_3)_3]C\equiv CSi(CH_3)_3$ (69) $(CH_3)_3SiCH_2C[Si(CH_3)_3]=C=C[Si(CH_3)_3]_2$ (2)	633
	$RhCl(PPh_3)_3$	100	1-19	$(CH_3)_3SiCH_2C[Si(CH_3)_3]=C=C[Si(CH_3)_3]_2$ (90-95)	633
$C_{10}H_{22}$	$(CH_3)_3SiOCH_2C\equiv CCH_2OSi(CH_3)_3$				
	H_2PtCl_6 (THF)	(-20) - 100	2	$(CH_3)_3SiOCH_2C[Si(CH_3)_3]=CHCH_2OSi(CH_3)_3$ (88)	698
$C_{20}H_{24}$	Fig.203.				
	Pt/C	160	16	Fig.204. (-)	1050

Table 4.5

= $(CH_3)_3SiH$ = =1=

C_3	$(CF_2Cl)_2CO$				
	none	50	70	$(CF_2Cl)_2CHOSi(CH_3)_3$ (73)	304
C_3	$(CF_3)_2CO$				
	none	(-78)	33	$(CF_3)_2CHOSi(CH_3)_3$ (80) $(CF_3)_2CHOC(CF_3)_2OSi(CH_3)_3$ (9)	526
	none	145	5 days	$(CF_3)_2CHOSi(CH_3)_3$ (50)	296
	none	20	15	$(CF_3)_2CHOSi(CH_3)_3$ (68)	526
	none	50	24	$(CF_3)_2CHOSi(CH_3)_3$ (81)	304
C_4H_8	$CH_3CH_2COCH_3$				
	$[RhH_2(L)_2(BMPP)_2]ClO_4$	50	40	$CH_3CH_2(CH_3)CHOSi(CH_3)_3$ (79)	486
C_5H_8	$CH_2=C(CH_3)COOCH_3$				
	Rh/C (5%)	35-45	2	$(CH_3)_2C=C(OCH_3)OSi(CH_3)_3$ (45)	U -187
	Rh/C (5%) +hydroquinone	-	- (THF)	$(CH_3)_2C=C(OCH_3)OSi(CH_3)_3$ (70)	U -187
	$RhCl\cdot 3\ H_2O$ (Me_2CO)	45-75	-	$(CH_3)_2C=C(OCH_3)OSi(CH_3)_3$ (45-86) $CH_2=C(CH_3)CH(OCH_3)OSi(CH_3)_3$ (5-11) $(CH_3)_3SiCH_2C(CH_3)COOCH_3$ (0-1)	E -023
	$RhCl\cdot 3\ H_2O$ (Me_2CO)	55	-	$(CH_3)_2C=C(OCH_3)OSi(CH_3)_3$ + $CH_2=C(CH_3)CH(OCH_3)OSi(CH_3)_3$ + $(CH_3)_3SiCH_2C(CH_3)COOCH_3$ (kin.)	E -023
	$RhCl(PPh_3)_3$	100	5	$(CH_3)_2C=C(OCH_3)OSi(CH_3)_3$ (50-60) $CH_2=C(CH_3)CH(OCH_3)OSi(CH_3)_3$ (-) $(CH_3)_3SiCH_2C(CH_3)COOCH_3$ (-)	984
	$RhCl(PPh_3)_3$	100	5	$(CH_3)_2C=C(OCH_3)OSi(CH_3)_3$ + $CH_2=C(CH_3)CH(OCH_3)OSi(CH_3)_3$ + $(CH_3)_3SiCH_2C(CH_3)COOCH_3$ (kin.)	984
	$RhCl(PPh_3)_3$	40-55	6-8	$(CH_3)_2C=C(OCH_3)OSi(CH_3)_3$ (41-80) $CH_2=C(CH_3)CH(OCH_3)OSi(CH_3)_3$ (2-10) $(CH_3)_3SiCH_2C(CH_3)COOCH_3$ (1-12)	E -023
C_5H_8	$CH_3CH=CHCOOCH_3$				
	$RhCl(PPh_3)_3$	60	0.17	$CH_3CH_2CH[Si(CH_3)_3]COOCH_3$ (9) $CH_3CH_2CH=C(CH_3)OSi(CH_3)_3$ (77)	1321
C_6H_{12}	$(CH_3)_3CCOCH_3$				
	$[RhH_2(L)_2(BMPP)_2]ClO_4$	50	40	$(CH_3)_3CCH(CH_3)OSi(CH_3)_3$ (55)	486
C_6H_{12}	$CH_3(CH_2)_3COCH_3$				

Table 4.5

= $(CH_3)_3SiH$ = =2=

	$[RhH_2(L)_2(BMPP)_2]ClO_4$	50	40	$CH_3(CH_2)_3CH(CH_3)OSi(CH_3)_3$ (82)	486
C_7H_{10}	$CH_2=C(CH_3)COOCH_2CH=CH_2$				
	$RhCl(PPh_3)_3$	39-104	-	$(CH_3)_2C=C[OSi(CH_3)_3]_2$ (-) $CH_2=C(CH_3)COOSi(CH_3)_3$ (-)	U -183
	$RhCl_3 \cdot 3H_2O$ (THF)	41-65	-	$(CH_3)_2C=C[OSi(CH_3)_3]_2$ (2-56) $CH_2=C(CH_3)COOSi(CH_3)_3$ (5-61)	U -183
C_7H_{10}	$CH_2=C(CH_3)COOCH_2\overline{CHCH_2O}$				
	$RhCl_3 \cdot 6\,H_2O$	38-50	24	$(CH_3)_2C=C(OCH_2\overline{CHCH_2O})OSi(CH_3)_3$ (80)	984
C_7H_{14}	$CH_2=C(CH_3)COOSi(CH_3)_3$				
	$RhCl_3 \cdot 6\,H_2O$	-	- (THF)	$(CH_3)_2C=C[OSi(CH_3)_3]OSi(CH_3)_3$ (78)	984
C_8H_8	$C_6H_5COCH_3$				
	$[RhH_2(L)_2(BMPP)_2]ClO_4$	50	40	$C_6H_5CH(CH_3)OSi(CH_3)_3$ (100)	486
	$[RhH_2(L)_2(BMPP)_2]ClO_4$	50	40	$C_6H_5CH(CH_3)OSi(CH_3)_3$ (100)	1298
C_9H_{10}	$C_6H_5CH_2COCH_3$				
	$[RhH_2(L)_2(BMPP)_2]ClO_4$	50	40	$C_6H_5CH_2CH(CH_3)OSi(CH_3)_3$ (100)	486
C_9H_{10}	$C_6H_5COCH_2CH_3$				
	$RhCl(L)(diop)$	70	40	$C_6H_5CH(CH_2CH_3)OSi(CH_3)_3$ (62)	486
	$[RhCl(c\text{-}C_6H_{10})]_2$ + EPPF	50	-	$C_6H_5CH(CH_2CH_3)OSi(CH_3)_3$ (88)	488
	$[RhH_2(L)_2(BMPP)_2]ClO_4$	50	40	$C_6H_5CH(CH_2CH_3)OSi(CH_3)_3$ (92)	486
	$[RhH_2(L)_2(BMPP)_2]ClO_4$	50	40	$C_6H_5CH(CH_2CH_3)OSi(CH_3)_3$ (92)	1298
C_9H_{18}	$CH_2=C(CH_3)COOCH_2CH_2OSi(CH_3)_3$				
	$RhCl_3 \cdot 6\,H_2O$	-	- (THF)	$(CH_3)_2C=C[OCH_2CH_2OSi(CH_3)_3]OSi(CH_3)_3$ (77)	984
$C_{10}H_{12}$	$C_6H_5COCH(CH_3)_2$				
	$[RhH_2(L)_2(BMPP)_2]ClO_4$	50	40	$C_6H_5[(CH_3)_2CH]CHOSi(CH_3)_3$ (98)	486
$C_{10}H_{18}$	Fig.86.				
	$Rh(\pi\text{-}All)[P(OMe)_3]_3$	-	-	Fig.205. (-)	144
$C_{10}H_{20}$	$CH_2=C(CH_3)COO(CH_2)_3Si(OCH_3)_3$				
	$RhCl_3 \cdot 6\,H_2O$	-	- (THF)	$(CH_3)_2C=C[O(CH_2)_3Si(OCH_3)_3]OSi(CH_3)_3$ (66)	984
$C_{11}H_{12}$	$C_6H_5C(CH_3)=CHCOCH_3$				

Table 4.5
= $(CH_3)_3SiH$ = =3=

	$[RhH_2(L)_2(BMPP)_2]ClO_4$	20	-	$C_6H_5CH(CH_3)CH{=}C(CH_3)OSi(CH_3)_3$ (90)	487
$C_{11}H_{14}$	$C_6H_5COC(CH_3)_3$				
	RhCl(L)(diop)	90	40	$C_6H_5CH[C(CH_3)_3]OSi(CH_3)_3$ (83)	486
	$[RhH_2(L)_2(BMPP)_2]ClO_4$	50	40	$C_6H_5CH[C(CH_3)_3]OSi(CH_3)_3$ (81)	1298
	$[RhH_2(L)_2(BMPP)_2]ClO_4$	50	40	$C_6H_5CH[C(CH_3)_3]OSi(CH_3)_3$ (81)	486
$C_{14}H_{12}$	$C_6H_5CH_2COC_6H_5$				
	$[RhH_2(L)_2(BMPP)_2]ClO_4$	50	40	$C_6H_5CH_2CH(C_6H_5)OSi(CH_3)_3$ (70)	486
$C_{16}H_{14}$	$C_6H_5C(CH_3){=}CHCOC_6H_5$				
	$[RhCl(c\text{-}C_6H_{10})]_2$ + diop	0	-	$C_6H_5CH(CH_3)CH{=}C(C_6H_5)OSi(CH_3)_3$ + $C_6H_5C(CH_3){=}CHCH(C_6H_5)OSi(CH_3)_3$ (-)	487
	$[RhH_2(L)_2(BMPP)_2]ClO_4$	20	-	$C_6H_5CH(CH_3)CH{=}C(C_6H_5)OSi(CH_3)_3$ (94)	487

Table 4.6

= $(CH_3)_3SiH$ = =1=

C_1	CF_3NO				
	UV	20	18	$CF_3NHOSi(CH_3)_3$ (43) $(CF_3)_2NOSi(CH_3)_3$ (28)	305
	H_2PtCl_6	-	5 days	$CF_3NHOSi(CH_3)_3$ (74) $(CF_3)_2NOSi(CH_3)_3$ (16)	305
C_5H_5	Fig.206.				
	Ni-Raney	70-115	1-3 days	Fig.207 + Fig.208 + Fig.209 + Fig.210 + Fig.211. (1-5)	268
	Pd/C	0-80	4h-11 days	Fig.207 + Fig.208 + Fig.209 + Fig.210. (-)	268
	Pd/C	0-80	4h-11 days	Fig.207 + Fig.208 + Fig. 209 + Fig.210. (-)	F -014
	$PdCl_2$	20	1-6 days	Fig.207 + Fig.208 + Fig.209 + Fig.210 + Fig.211. (3-54)	268
	Pt/C	20	1-3 days	Fig.207 + Fig.208 + Fig.209 + Fig.210 + Fig.211. (trace)	268
	Rh/C	20	1-3 days	Fig.208. (23-72)	268
	Rh/C	27	0.20	Fig.208. (-)	F -014
C_6H_7	Fig.212.				
	Pd/C	50	48	Fig.213. (-)	F -014
C_6H_7	Fig.214.				
	Pd/C	24	18	Fig.215 + Fig.217. (-)	F -014
	Pd/C	24	18	Fig.215/Fig.216 + Fig.217. (70/30)	268
	Pd/C	32	0.67	Fig.215 + Fig.217. (-)	F -014
	Pd/C	40	20	Fig.215. (90)	F -014
C_6H_7	Fig.218.				
	Pd/C	24	5 days	Fig.219. (30) Fig.220. (18) Fig.221. (17)	268
	Pd/C	24	5 days	Fig.219. (30) Fig.220. (18) Fig.221.(17)	F -014
	Pd/C	35-50	5-40	Fig.219. (35) Fig.220.(5) Fig.221.(20)	268
C_7H_4	p-ClC_6H_4CN				

Table 4.6
= $(CH_3)_3SiH$ = =2=

	$Co_2(CO)_8$	60	48 (t)	p-$ClC_6H_4CH_2N[Si(CH_3)_3]_2$ (53)	802
C_7H_{11}	$(\overline{CH_2)_5C}HNC$				
	$Co(acac)_2$	100	5 (b)	$(\overline{CH_2)_5C}HN{=}CHSi(CH_3)_3$ (86)	1010
C_8H_4	p-NCC_6H_4CN				
	$Co_2(CO)_8$	60	20 (t)	p-$NCC_6H_4CH_2N[Si(CH_3)_3]_2$ (36)	802
C_8H_7	o-$CH_3C_6H_4CN$				
	$Co_2(CO)_8$	60	20 (t)	o-$CH_3C_6H_4CH_2N[Si(CH_3)_3]_2$ (22-68)	802
C_8H_7	p-$CH_3C_6H_4CN$				
	$Co_2(CO)_8$	60	20 (t)	p-$CH_3C_6H_4CH_2N[Si(CH_3)_3]_2$ (91)	802
C_8H_7	p-$CH_3OC_6H_4CN$				
	$Co_2(CO)_8$	60	20 (t)	p-$CH_3OC_6H_4CH_2N[Si(CH_3)_3]_2$ (88)	802
C_9H_7	p-$CH_3COC_6H_4CN$				
	$Co_2(CO)_8$	60	48 (t)	p-$CH_3COC_6H_4CH_2N[Si(CH_3)_3]_2$ (46)	802
C_9H_{10}	p-$(CH_3)_2NC_6H_4CN$				
	$Co_2(CO)_8$	60	20 (t)	p-$(CH_3)_2NC_6H_4CH_2N[Si(CH_3)_3]_2$ (73)	802
$C_{11}H_7$	$C_{10}H_7CN$				
	$Co_2(CO)_8$	60	20 (t)	p-$C_{10}H_7CH_2N[Si(CH_3)_3]_2$ (67)	802

Table 5.1

= $(CH_3O)_3SiH$ = =1=

C_2H_2	$CH\equiv CH$				
	-	40	2	$(CH_3O)_3SiCH=CH_2$ (98)	956
	H_2PtCl_6	120	1	$(CH_3O)_3SiCH_2CH_2Si(OCH_3)_3$ (76) $(CH_3O)_3SiCH=CH_2$ (22)	U -163
C_2H_4	$CH_2=CH_2$				
	$[RhCl(CO)_2]_2$	148-225	- (xy)	$CH_3CH_2Si(OCH_3)_3$ (78) $CH_2=CHSi(OCH_3)_3$ (12)	U -173
	$[RhCl(CO)_2]_2$	150	3 (xy)	$CH_3CH_2Si(OCH_3)_3$ (-) $CH_2=CHSi(OCH_3)_3$ (20)	U -173
C_4H_6	$CH_2=CHCH=CH_2$				
	$Ni(PPh_3)_4$	120	3	$C_4H_7Si(OCH_3)_3$ (38) $C_8H_{13}Si(OCH_3)_3$ (40)	D -070
	Pd + PPh_3	150	8	$CH_3CH=CHCH_2Si(OCH_3)_3$ (38)	D -027
C_5H_8	$CH_2=C(CH_3)CH=CH_2$				
	$Ni(acac)_2$ + $AlEt_3$	20	2	$CH_3CH=C(CH_3)CH_2Si(OCH_3)_3$ (92)	G -030
	$Ni(acac)_2$ + $AlEt_3$	20	2	$CH_3CH=C(CH_3)CH_2Si(OCH_3)_3$ (92)	663
	RhCl(CO)(1,4-diaza-1,3-diene)	r.t.	1 day(THF)	$CH_3CH=C(CH_3)CH_2Si(OCH_3)_3$ (87)	167
C_5H_8	$CH_3CH_2CH_2C\equiv CH$				
	$Ni(acac)_2$ + $AlEt_3$	20	3	$(CH_3O)_3SiCH=C(CH_2CH_2CH_3)C(CH_2CH_2CH_3)=CH_2$ (72)	663
C_6H_{10}	$(\overline{CH_2)_4CH=C}H$				
	H_2PtCl_6 ($2\text{-}EtC_6H_{10}OH$) + UV	70	7	$(\overline{CH_2)_5C}HSi(OCH_3)_3$ (91)	E -038
	polymer-Pt	-	-	$(\overline{CH_2)_4CH[Si(OCH_3)_3]C}H_2$ (kin.)	683
C_6H_{12}	$CH_3(CH_2)_3CH=CH_2$				
	$PdCl_2[P(c\text{-}Hex)_3]_2$	130	5	$CH_3(CH_2)_5Si(OCH_3)_3$ (30)	D -027
	$SiO_2\text{-}(CH_2)_nPPh_2$/Pt complex	-	-	$C_6H_{13}Si(OCH_3)_3$ (80-90)	1295
	polymer-PPh_2/Pt	40-60	-	$CH_3(CH_2)_5Si(OCH_3)_3$ (85)	240
	polymer-Pt	-	-	$C_6H_{13}Si(OCH_3)$ (kin.)	683
	$RuCl_2(PPh_3)_3$	80	2	$CH_3(CH_2)_5Si(OCH_3)_3$ (68)	Pl-001
	$RuCl_3(PPh_3)_3$	120	6	$CH_3(CH_2)_5Si(OCH_3)_3$ (80)	Pl-015
C_7H_8	Fig.222.				

= (CH₃O)₃SiH =   =2=

	H₂PtCl₆	90-110	5	Fig.223. (-)	U -110

C₈H₈	C₆H₅CH=CH₂				
	polymer-Pt	-	-	C₆H₅CH₂CH₂Si(OCH₃)₃ (kin.)	683
	Rh(acac)₃ + AlEt₃	60	-	C₆H₅CH₂CH₂Si(OCH₃)₃ (40)	271
				C₆H₅CH(CH₃)Si(OCH₃)₃ (10)	
				C₆H₅CH=CHSi(OCH₃)₃ (10)	

C₈H₁₂	$\overline{CH_2CH_2CH}$=CHCH₂CHCH=CH₂				
	H₂PtCl₆	-	-	$\overline{CH_2CH_2CH}$=CHCH₂CHCH₂CH₂Si(OCH₃)₃ (-)	D -044

C₈H₁₆	CH₃(CH₂)₅CH=CH₂				
	Co₂(CO)₈	0-60	until 24	CH₃(CH₂)₇Si(OCH₃)₃ (-)	470
	Co₂(CO)₈	20-130	17	CH₃(CH₂)₇Si(OCH₃)₃ (82)	D -009
	Pd(PPh₃)₄	120	6	CH₃(CH₂)₇Si(OCH₃)₃ (trace)	1208
	SiO₂-Pt(C₃H₅)	60	6	C₈H₁₇Si(OCH₃)₃ (25-46)	230

C₉H₁₂	Fig.107.				
	H₂PtCl₆	90-110	5	Fig.224. (84)	U -110

C₉H₁₂	Fig.226.				
	H₂PtCl₆	40-110	5	Fig.225. (-)	U -110

C₁₀H₁₂	Fig.42.				
	H₂PtCl₆	90-110	5	Fig.227. (-)	U -110

C₁₀H₁₆	(CH₃)₂C=CHCH₂CH=C(CH₃)CH=CH₂				
	RhCl(PPh₃)₃	110	22	(CH₃)₂C=CHCH₂CH₂C(CH₃)=CHCH₂Si(OCH₃)₃ +	851
				(CH₃)₂C=CHCH₂CH[Si(OCH₃)₃]C(CH₃)=CHCH₃ (92)	

C₁₀H₂₀	CH₃(CH₂)₇CH=CH₂				
	PtO₂(PPh₃)₂	35	-	CH₃(CH₂)₉Si(OCH₃)₃ (kin.)	1315
	polymer-PPh₂/Pt	40-60	-	CH₃(CH₂)₉Si(OCH₃)₃ (85)	240
	polymer-Pt	-	-	CH₃(CH₂)₉Si(OCH₃)₃ (kin.)	683

C₁₂H₂₄	CH₃(CH₂)₉CH=CH₂				
	polymer-PPh₂/Pt	40-60	-	CH₃(CH₂)₁₁Si(OCH₃)₃ (85)	240

Table 5.2

= $(CH_3O)_3SiH$ = =1=

C_3H_5	$CH_2{=}CHCH_2Cl$				
	$[IrCl(cod)]_2$	80	4 (xy)	$(CH_3O)_3Si(CH_2)_3Cl$ (75)	U -172
C_3H_7	$CH_2{=}CHCH_2NH_2$				
	H_2PtCl_6 (i-PrOH)	60	7	$(CH_3O)_3SiC_3H_6NH_2$ (74)	J -082
	H_2PtCl_6 (i-PrOH)	reflux	36	$(CH_3O)_3Si(CH_2)_3NH_2$ (60)	Dd-002
	$PtCl_2(acac)_2$	reflux	48	$(CH_3O)_3Si(CH_2)_3NH_2$ (80)	D -011
	$(Et_4N)_2Rh_6(CO)_{15}$ + CO	-	- (t)	$(CH_3O)_3Si(CH_2)_3NH_2$ (65) $(CH_3O)_3SiCH(CH_3)CH_2NH_2$ (5)	E -040
	$Rh_4(CO)_{12}$	-	-	$(CH_3O)_3Si(CH_2)_3NH_2$ (62) $(CH_3O)_3SiCH(CH_3)CH_2NH_2$ (4)	E -040
	$Rh_4(CO)_{12}$ + cod + CO	120	3 (xy)	$(CH_3O)_3Si(CH_2)_3NH_2$ (69) $(CH_3O)_3SiCH(CH_3)CH_2NH_2$ (7)	E -040
	$[Rh(\mu\text{-}PPh_2)(cod)]_2$	110	1 (t)	$(CH_3O)_3Si(CH_2)_3NH_2$ (70) $(CH_3O)_3SiCH(CH_3)CH_2NH_2$ (1)	E -048
C_4H_5	$CH_2{=}CHCH_2NCO$				
	H_2PtCl_6 ($MeOC_2H_4OMe$)	86-91	19	$(CH_3O)_3Si(CH_2)_3NCO$ (33)	U -039
	Rh catalyst	-	-	$(CH_3O)_3Si(CH_2)_3NCO$ (-)	J -024
C_5H_6	$N_2CHCOOCH_2CH{=}CH_2$				
	AIBN	50	16 (b)	$N_2CHCOO(CH_2)_3Si(OCH_3)_3$ (-)	G -016
C_5H_7	$ClCH_2COOCH_2CH{=}CH_2$				
	H_2PtCl_6 ($AllOCH_2COOAll$)	104-150	10.5	$ClCH_2COO(CH_2)_3Si(OCH_3)_3$ (65)	Pl-008
C_5H_8	$CH_2{=}CHOCH_2CH{=}CH_2$				
	H_2PtCl_6	20-40	8.5 (b)	$CH_2{=}CHO(CH_2)_3Si(OCH_3)_3$ (81)	D -047
	H_2PtCl_6	reflux	1 (b)	$CH_2{=}CHO(CH_2)_3Si(OCH_3)_3$ (65) $(CH_3O)_3SiCH_2CH_2O(CH_2)_3Si(OCH_3)_3$ (20)	D -047
C_5H_8	$CH_2{=}CHSCH_2CH{=}CH_2$				
	H_2PtCl_6 ($MeOC_2H_4OMe$)	reflux	16	$CH_2{=}CHS(CH_2)_3Si(OCH_3)_3$ (-)	B -001
C_5H_8	$CH_3COOCH_2CH{=}CH_2$				
	H_2PtCl_6 ($AllOCH_2COOAll$)	48-175	3.7	$CH_3COO(CH_2)_3Si(OCH_3)_3$ (79)	Pl-008
C_5H_{10}	$CH_2{=}CHCH(OCH_3)_2$				

Table 5.2

= $(CH_3O)_3SiH$ = =2=

	H_2PtCl_6 (i-PrOH)	-	2	$(CH_3O)_3SiCH_2CH_2CH(OCH_3)_2$ + $(CH_3O)_3SiCH(CH_3)CH(OCH_3)_3$ (-)	D -028
	H_2PtCl_6 (i-PrOH)	90	-	$(CH_3O)_3SiCH_2CH_2CH(OCH_3)_2$ (77)	308
C_5H_{12}	$(CH_3O)_3SiCH=CH_2$				
	H_2PtCl_6 (AcOH)	-	-	$(CH_3O)_3SiCH_2CH_2Si(OCH_3)_3$ (-)	740
	H_2PtCl_6 (AcOH)	-	-	$(CH_3O)_3SiCH_2CH_2Si(OCH_3)_3$ (-)	390
	H_2PtCl_6 (i-PrOH)	90	3	$(CH_3O)_3SiCH_2CH_2Si(OCH_3)_3$ (90)	E -039
	Rh complex	r.t.	3	$(CH_3O)_3SiCH_2CH_2Si(OCH_3)_3$ (47) $(CH_3O)_3SiCH(CH_3)Si(OCH_3)_3$ (47)	166
	$RhCl(PPh_3)_3$	120	6	$(CH_3O)_3SiCH_2CH_2Si(OCH_3)_3$ (47) $(CH_3O)_3SiCH(CH_3)Si(OCH_3)_3$ (17)	745
	$RhCl(PPh_3)_3$	120	6	$(CH_3O)_3SiCH_2CH_2Si(OCH_3)_3$ (47) $(CH_3O)_3SiCH(CH_3)Si(OCH_3)_3$ (17)	740
	$RhCl(PPh_3)_3$	reflux	2.5	$(CH_3O)_3SiCH_2CH_2Si(OCH_3)_3$ (71)	Pl-006
	$RhCl(PPh_3)_3$	reflux	2.5	$(CH_3O)_3SiCH_2CH_2Si(OCH_3)_3$ (71)	740
	$RhCl(PPh_3)_3$	reflux	2	$(CH_3O)_3SiCH_2CH_2Si(OCH_3)_3$ (71) $(CH_3O)_3SiCH(CH_3)Si(OCH_3)_3$ (5)	740
	$RhCl(PPh_3)_3$	reflux	2	$(CH_3O)_3SiCH_2CH_2Si(OCH_3)_3$ (71) $(CH_3O)_3SiCH(CH_3)Si(OCH_3)_3$ (5)	745
	$Ru(acac)_3$	120	6	$(CH_3O)_3SiCH_2CH_2Si(OCH_3)_3$ (62) $(CH_3O)_3SiCH(CH_3)Si(OCH_3)_3$ (5)	740
	$Ru(acac)_3$	120	6	$(CH_3O)_3SiCH_2CH_2Si(OCH_3)_3$ (62) $(CH_3O)_3SiCH(CH_3)Si(OCH_3)_3$ (5)	Pl-006
	$Ru(acac)_3$	120	6	$(CH_3O)_3SiCH_2CH_2Si(OCH_3)_3$ (62) $(CH_3O)_3SiCH(CH_3)Si(OCH_3)_3$ (5)	745
	$Ru(acac)_3$	reflux	2	$(CH_3O)_3SiCH_2CH_2Si(OCH_3)_3$ (72) $(CH_3O)_3SiCH(CH_3)Si(OCH_3)_3$ (4)	740
	$Ru(acac)_3$	reflux	2	$(CH_3O)_3SiCH_2CH_2Si(OCH_3)_3$ (72) $(CH_3O)_3SiCH(CH_3)Si(OCH_3)_3$ (4)	440
	$RuCl_2(PPh_3)_3$	120	6	$(CH_3O)_3SiCH_2CH_2Si(OCH_3)_3$ (66) $(CH_3O)_3SiCH(CH_3)Si(OCH_3)_3$ (1)	740
	$RuCl_2(PPh_3)_3$	120	6	$(CH_3O)_3SiCH_2CH_2Si(OCH_3)_3$ (66) $(CH_3O)_3SiCH(CH_3)Si(OCH_3)_3$ (1)	745
	$RuCl_2(PPh_3)_3$	reflux	2	$(CH_3O)_3SiCH_2CH_2Si(OCH_3)_3$ (66) $(CH_3O)_3SiCH(CH_3)Si(OCH_3)_3$ (2)	740

Table 5.2

= $(CH_3O)_3SiH$ = =3=

	$RuCl_2(PPh_3)_3$	reflux	2	$(CH_3O)_3SiCH_2CH_2Si(OCH_3)_3$ (66) $(CH_3O)_3SiCH(CH_3)Si(OCH_3)_3$ (2)	745
C_5H_{12}	$CH_2{=}CHCH_2NHCH_2CH_2NH_2$				
	$[PtCl_2(CHMe{=}CHCOMe)]_2$	reflux	2	$(CH_3O)_3Si(CH_2)_3NHCH_2CH_2NH_2$ + $(CH_3O)_3SiCH(CH_3)CH_2NHCH_2CH_2NH_2$ (20)	G -023
	$[PtCl_2(CHMe{=}CHCOMe)]_2$	reflux	2	$(CH_3O)_3Si(CH_2)_3NHCH_2CH_2NH_2$ + $(CH_3O)_3SiCH(CH_3)CH_2NHCH_2CH_2NH_2$ (20)	D -049
C_6H_8	$CH_2{=}C(CH_3)COOCH{=}CH_2$				
	H_2PtCl_6 (PhCOOMe)	105-112	2.5	$CH_2{=}C(CH_3)COOCH_2CH_2Si(OCH_3)_3$ (-)	U -016
C_6H_8	$CH_2{=}CHCH_2COCH{=}CH_2$				
	H_2PtCl_6 (i-PrOH)	75-80	0.67	$(CH_3O)_3Si(CH_2)_3COCH{=}CH_2$ (-)	D -081
C_6H_8	$CH_2{=}CHCOOCH_2CH{=}CH_2$				
	H_2PtCl_6 (PhCOOMe)	106-114	3 (t)	$CH_2{=}CHCOO(CH_2)_3Si(OCH_3)_3$ (-)	U -016
	H_2PtCl_6 (PhCOOMe)	106-114	3 (t)	$CH_2{=}CHCOO(CH_2)_3Si(OCH_3)_3$ (-)	G -001
C_6H_{10}	$CH_2{=}CHCH_2CH_2SCH{=}CH_2$				
	H_2PtCl_6 ($MeOC_2H_4OMe$)	reflux	16	$(CH_3O)_3Si(CH_2)_4SCH{=}CH_2$ (-)	B -001
C_6H_{10}	$CH_2{=}CHCH_2OCH_2CH{=}CH_2$				
	H_2PtCl_6	40	8	$CH_2{=}CHCH_2O(CH_2)_3Si(OCH_3)_3$ (50)	D -047
C_6H_{10}	$CH_2{=}CHCH_2OCH_2\overline{CHCH_2O}$				
	$PtCl_2(Me_2C{=}CHCOMe)$ (Me_2CO)	80	3	$(CH_3O)_3Si(CH_2)_3OCH_2\overline{CHCH_2O}$ (90)	E -042
C_6H_{10}	$CH_2{=}CHOCH_2CH_2SCH{=}CH_2$				
	H_2PtCl_6 ($MeOC_2H_4OMe$)	reflux	16	$(CH_3O)_3SiCH_2CH_2OCH_2CH_2SCH{=}CH_2$ (-)	B -001
C_6H_{10}	$CH_3OCH_2COOCH_2CH{=}CH_2$				
	H_2PtCl_6 ($AllOCH_2COOAll$)	150	7	$CH_3OCH_2COO(CH_2)_3Si(OCH_3)_3$ (79)	Pl-008
	SiO_2-$NCH_2CH_2OCH_2CH_2/H_2PtCl_6$ (ROH)	90-170	8.6	$CH_3OCH_2COO(CH_2)_3Si(OCH_3)_3$ (55)	Pl-007
C_6H_{10}	$\overline{OCH_2C}H(CH_3)OCHCH{=}CH_2$				
	H_2PtCl_6 (i-PrOH)	reflux	0.5	$\overline{OCH_2C}H(CH_3)OCHCH_2CH_2Si(OCH_3)_3$ (-)	D -028
C_6H_{10}	$\overline{OCH_2C}HCH_2OCH_2CH{=}CH_2$				
	H_2PtCl_6	110-150	-	$\overline{OCH_2C}HCH_2O(CH_2)_3Si(OCH_3)_3$ (70)	E -032

Table 5.2

= $(CH_3O)_3SiH$ = =4=

	H_2PtCl_6 (THF)	50	1.5	$\overline{OCH_2CH}CH_2O(CH_2)_3Si(OCH_3)_3$ (80)	E -037
	H_2PtCl_6 (i-PrOH)	reflux	2	$\overline{OCH_2CH}CH_2O(CH_2)_3Si(OCH_3)_3$ (66)	D -023
	Pt(IV)complex	-	-	$\overline{OCH_2CH}CH_2O(CH_2)_3Si(OCH_3)_3$ (-)	J -011
	$PtCl_2(acac)_2$	reflux	4	$\overline{OCH_2CH}CH_2O(CH_2)_3Si(OCH_3)_3$ (72)	D -023
	$[PtCl_2(Me_2C{=}CHCOMe)]_2$	130-140	0.17	$\overline{OCH_2CH}CH_2O(CH_2)_3Si(OCH_3)_3$ (-)	D -023
	polymer-$(CH_2)_n$-$N(COMe)_2/PtCl_x$	80	0.5	$\overline{OCH_2CH}CH_2O(CH_2)_3Si(OCH_3)_3$ (98)	505
	polymer-$(CH_2)_n$-$SCHC_6H_4N(CH_3)_2/PtCl_x$	80	0.5	$\overline{OCH_2CH}CH_2O(CH_2)_3Si(OCH_3)_3$ (96)	505
	polymer-$(CH_2)_nCOOH/PtCl_x$	80	0.5	$\overline{OCH_2CH}CH_2O(CH_2)_3Si(OCH_3)_3$ (72)	505
	siloxane-Pt	-	-	$\overline{OCH_2CH}CH_2O(CH_2)_3Si(OCH_3)_3$ (72-98)	679
	$Rh(CO)(acac)(PPh_3)$	80-100	-	$\overline{OCH_2CH}CH_2O(CH_2)_3Si(OCH_3)_3$ (60)	E -032
	$Rh(CO)_2(acac)$	110	-	$\overline{OCH_2CH}CH_2O(CH_2)_3Si(OCH_3)_3$ (70)	E -032
	Rh/C	110	-	$\overline{OCH_2CH}CH_2O(CH_2)_3Si(OCH_3)_3$ (75)	E -032
	Rh/C	110-150	-	$\overline{OCH_2CH}CH_2O(CH_2)_3Si(OCH_3)_3$ (70)	E -032
	Rh/C (5%)	80-100	-	$\overline{OCH_2CH}CH_2O(CH_2)_3Si(OCH_3)_3$ (70-80)	E -032
	$RhCl(PPh_3)_3$	110	-	$\overline{OCH_2CH}CH_2O(CH_2)_3Si(OCH_3)_3$ (65)	E -032
	$[RhCl(CO)_2]_2$	80-100	-	$\overline{OCH_2CH}CH_2O(CH_2)_3Si(OCH_3)_3$ (60)	E -032
	$[RhCl(cod)]_2$	80-100	-	$\overline{OCH_2CH}CH_2O(CH_2)_3Si(OCH_3)_3$ (60)	E -032
C_6H_{12}	$CH_3(CH_2)_3OCH{=}CH_2$				
	H_2PtCl_6	40	8	$CH_3(CH_2)_3OCH_2CH_2Si(OCH_3)_3$ (3-4)	D -047
C_6H_{12}	$CH_3CH_2CH_2OCH_2CH{=}CH_2$				
	H_2PtCl_6	40	8	$CH_3CH_2CH_2O(CH_2)_3Si(OCH_3)_3$ (55)	D -047
C_7H_8	$CH_2{=}C(CH_3)COOCH_2C{\equiv}CH$				
	H_2PtCl_6 (OctOH)	60-65	9	$(CH_3O)_3SiCH{=}CHCH_2OOCC(CH_3){=}CH_2$ (33) $(CH_3O)_3SiC({=}CH_2)CH_2OOCC(CH_3){=}CH_2$ (61)	U -057
C_7H_9	$CH_2{=}CHCH_2COOCH_2CH_2CN$				
	H_2PtCl_6 (OctOH)	90-130	14	$(CH_3O)_3Si(CH_2)_3COOCH_2CH_2CN$ (-)	U -047
C_7H_{10}	$CH_2{=}C(CH_3)COOCH_2CH{=}CH_2$				
	H_2PtCl_6	100-110	5	$CH_2{=}C(CH_3)COO(CH_2)_3Si(OCH_3)_3$ (71-85)	U -176

	H_2PtCl_6 (Me_2CO + p-HOC_6H_4OH)	68-72	0.5	$CH_2=C(CH_3)COO(CH_2)_3Si(OCH_3)_3$ (54)	D -078
	H_2PtCl_6 (PhCOOMe)	105-112	2.5	$CH_2=C(CH_3)COO(CH_2)_3Si(OCH_3)_3$ (-)	G -001
	H_2PtCl_6 (PhCOOMe)	105-112	2.5	$CH_2=C(CH_3)COO(CH_2)_3Si(OCH_3)_3$ (-)	U -016
	H_2PtCl_6 (THF)	reflux	1.5	$CH_2=C(CH_3)COO(CH_2)_3Si(OCH_3)_3$ (86)	E -037
	H_2PtCl_6 + carboxylic ester	90-100	-	$CH_2=C(CH_3)COO(CH_2)_3Si(OCH_3)_3$ (-)	U -176
	H_2PtCl_6 + pentanedione-2,4	3.1-3.3	-	$CH_2=C(CH_3)COO(CH_2)_3Si(OCH_3)_3$ (84-85)	U -176
	$Pt(PPh_3)_2(CH_2=CH_2)$	90-123	-	$CH_2=C(CH_3COO(CH_2)_3Si(OCH_3)_3$ (95)	Pl-034
	$PtCl_2(CH_3CN)_2$ + pentanedione-2,4	90-100	-	$CH_2=C(CH_3)COO(CH_2)_3Si(OCH_3)_3$ (-)	U -176
	$[PtCl_2(Me_2C=CHCOMe)]_2$	64	450s	$CH_2=C(CH_3)COO(CH_2)_3Si(OCH_3)_3$ (100)	D -023
C_7H_{10}	$CH_2=CHCH_2COC(CH_3)=CH_2$				
	H_2PtCl_6 (i-PrOH)	73-80	1	$(CH_3O)_3Si(CH_2)_3COC(CH_3)=CH_2$ (41)	D -081
C_7H_{10}	$CH_2=CHCH_2COOCH_2COCH_3$				
	-	-	-	$(CH_3O)_3Si(CH_2)_3COOCH_2COCH_3$ (-)	E -045
	H_2PtCl_6 ($PhCH_3$)	-	3	$(CH_3O)_3Si(CH_2)_3COOCH_2COCH_3$ (-)	E -041
C_7H_{10}	$CH_2=CHCH_2OCOOCH_2CH=CH_2$				
	H_2PtCl_6 (o-$C_6H_4(COOMe)_2$+p-$(PhNH)_2C_6H_4$	reflux	-	$CH_2=CHCH_2OCOO(CH_2)_3Si(OCH_3)_3$ (31)	U -023
C_7H_{10}	$CH_2=CHSCH_2COOCH_2CH=CH_2$				
	H_2PtCl_6 ($MeOC_2H_4OMe$)	reflux	16	$CH_2=CHSCH_2COO(CH_2)_3Si(OCH_3)_3$ (-)	B -001
C_7H_{10}	$\overline{CH_2OCOOC}HCH_2OCH_2CH=CH_2$				
	H_2PtCl_6 (MeCOMe)	70	1	$\overline{CH_2OCOOC}HCH_2O(CH_2)_3Si(OCH_3)_3$ (-)	D -048
C_7H_{11}	$CH_2=CHCH_2OCH(CH_3)CH_2CN$				
	H_2PtCl_6 (OctOH)	reflux	12	$(CH_3O)_3Si(CH_2)_3OCH(CH_3)CH_2CN$ (-)	G -017
C_7H_{11}	$CH_2=CHCH_2OCH_2CH(CH_3)CN$				
	H_2PtCl_6 (OctOH)	140-172	2.5	$(CH_3O)_3Si(CH_2)_3OCH_2CH(CH_3)CN$ (-)	G -017
	H_2PtCl_6 (OctOH)	140-172	2.5	$(CH_3O)_3Si(CH_2)_3OCH_2CH(CH_3)CN$ (-)	D -019
C_7H_{12}	$CH_2=CHCH_2OCH_2CH_2SCH=CH_2$				
	H_2PtCl_6 ($MeOC_2H_4OMe$)	reflux	16	$(CH_3O)_3Si(CH_2)_3OCH_2CH_2SCH=CH_2$ (-)	B -001
C_7H_{16}	$CH_2=CHCH_2NHPO(OCH_2CH_3)_2$				

	H_2PtCl_6	80	20.5 (b)	$(CH_3O)_3Si(CH_2)_3NHPO(OCH_2CH_3)_2$ (32)	D -058
C_8H_{10}	$CH_2=CHCHCH_2CHC(O)=CHCH_2$				
	H_2PtCl_6 (i-PrOH) + MeOH	50	3	$(CH_3O)_3SiCH_2CH_2CHCH_2CHC(O)=CHCH_2$ (99)	E -042
C_8H_{12}	$CH_2=CHCH_2CH_2COC(CH_3)=CH_2$				
	H_2PtCl_6 (i-PrOH)	75-78	0.78	$(CH_3O)_3Si(CH_2)_4COC(CH_3)=CH_2$ (-)	U -108
C_8H_{12}	$CH_2=CHCHCH_2CH_2CHCH_2CHO$				
	Rh/C (5%)	80-100	-	$(CH_3O)_3SiCH_2CH_2CHCH_2CH_2CHCH_2CHO$ (70)	E -032
	$RhCl(PPh_3)_3$	80-100	-	$(CH_3O)_3SiCH_2CH_2CHCH_2CH_2CHCH_2CHO$ (60)	E -032
	$[RhCl(CO)_2]_2$	80-100	-	$(CH_3O)_3SiCH_2CH_2CHCH_2CH_2CHCH_2CHO$ (70)	E -032
C_8H_{12}	$CH_3CH=CHCH_2COC(CH_3)=CH_2$				
	H_2PtCl_6 (i-PrOH)	85-90	0.75-0.83	$(CH_3O)_3SiCH(CH_3)CH_2CH_2COC(CH_3)=CH_2$ (35)	D -081
C_8H_{13}	$CH_2OCH_2C(CH_2Cl)CH_2OCH_2CH=CH_2$				
	H_2PtCl_6 (i-PrOH)	-	-	$CH_2OCH_2C(CH_2Cl)CH_2O(CH_2)_3Si(OCH_3)_3$ (-)	F -009
C_8H_{13}	$CH_3CH(CN)CH(CH_3)OCH_2CH=CH_2$				
	H_2PtCl_6 (OctOH)	reflux	12	$CH_3CH(CN)CH(CH_3)O(CH_2)_3Si(OCH_3)_3$ (-)	D -019
C_8H_{14}	$(CH_3)_3CCOOCH_2CH=CH_2$				
	H_2PtCl_6 (i-PrOH)	50	3	$(CH_3)_3CCOO(CH_2)_3Si(OCH_3)_3$ (30)	D -081
	H_2PtCl_6 (i-PrOH)	50	3	$(CH_3)_3CCOO(CH_2)_3Si(OCH_3)_3$ (30)	D -057
C_8H_{14}	$CH_2=CHCH_2OCH_2COOCH_2CH_2CH_3$				
	H_2PtCl_6 ($AllOCH_2COOAll$)	85-180	3.33	$(CH_3O)_3Si(CH_2)_3OCH_2COOCH_2CH_2CH_3$ (71)	Pl-008
C_8H_{14}	$O(CH_2)_3CHCH_2OCH_2CH=CH_2$				
	$[PtCl_2(Me_2C=CHCOMe)]_2$	130-140	0.33	$O(CH_2)_3CHCH_2O(CH_2)_3Si(OCH_3)_3$ (100)	D -023
C_8H_{16}	$CH_2=CHCH_2CH_2C(CH_3)=NN(CH_3)_2$				
	H_2PtCl_6 (OctOH)	80-130	1	$(CH_3O)_3Si(CH_2)_4C(CH_3)=NN(CH_3)_2$ (-)	U -050
	H_2PtCl_6 (OctOH)	80-130	1	$(CH_3O)_3Si(CH_2)_4C(CH_3)=NN(CH_3)_2$ (-)	U -078
C_8H_{18}	$(CH_3CH_2O)_3SiCH=CH_2$				
	$RuCl_2(PPh_3)_3$	120	6	$(CH_3CH_2O)_3SiCH_2CH_2Si(OCH_3)_3$ (-)	733
C_8H_{21}	$[N(CH_3)_2]_3SiCH=CH_2$				

Table 5.2

= $(CH_3O)_3SiH$ = =7=

	$RhCl(PPh_3)_3$	200	2 (xy)	$[N(CH_3)_2]_3SiCH_2CH_2Si(OCH_3)_3$ (-) $[N(CH_3)_2]_3SiCH{=}CHSi(OCH_3)_3$ (20)	U -173
C_9H_9	p,m-$ClCH_2C_6H_4CH{=}CH_2$				
	H_2PtCl_6	-	-	p,m-$ClCH_2C_6H_4CH_2CH_2Si(OCH_3)_3$ (-)	U -085
C_9H_{10}	$CH_2{=}CHCH_2OC_6H_4NO_2$-p				
	H_2PtCl_6	40	8	$(CH_3O)_3Si(CH_2)_3OC_6H_4NO_2$-p (-)	D -047
C_9H_{10}	$CH_2{=}CHCH_2OC_6H_5$				
	H_2PtCl_6	40	8	$(CH_3O)_3Si(CH_2)_3OC_6H_5$ (55)	D -047
	$Pt(PPh_3)_4$	-	-	$(CH_3O)_3Si(CH_2)_3OC_6H_5$ (-)	684
	polymer-PPh_2/Pt	40-60	-	$(CH_3O)_3Si(CH_2)_3OC_6H_5$ (85)	240
C_9H_{11}	$CH_2{=}CHCH_2NHSO_2C_6H_5$				
	H_2PtCl_6	40	8.5 (b)	$(CH_3O)_3Si(CH_2)_3NHSO_2C_6H_5$ (-)	D -058
C_9H_{12}	$CH_2{=}CHCH_2COOCH(COCH_3)_2$				
	H_2PtCl_6 ($PhCH_3$)	-	3	$(CH_3O)_3Si(CH_2)_3COOCH(COCH_3)_2$ (99)	E -041
C_9H_{14}	$CH_2{=}C(CH_3)COOCH_2CH_2OCH_2CH{=}CH_2$				
	H_2PtCl_6	65-75	2 (THF)	$CH_2{=}C(CH_3)COOCH_2CH_2O(CH_2)_3Si(OCH_3)_3$ (-)	U -016
C_9H_{16}	$\overline{CH_2OCH_2C}(CH_2CH_3)CH_2OCH_2CH{=}CH_2$				
	H_2PtCl_6 (i-PrOH)	100-130	6	$\overline{CH_2OCH_2C}(CH_2CH_3)CH_2O(CH_2)_3Si(OCH_3)_3$ (-)	F -009
C_9H_{18}	$CH_2{=}CHCH_2C(CH_3)_2CH{=}NN(CH_3)_2$				
	H_2PtCl_6 (OctOH)	reflux	42	$(CH_3O)_3Si(CH_2)_3C(CH_3)_2CH{=}NN(CH_3)_2$ (60)	U -050
	H_2PtCl_6 (OctOH)	reflux	42	$(CH_3O)_3Si(CH_2)_3C(CH_3)_2CH{=}NN(CH_3)_2$ (60)	U -078
C_9H_{18}	$CH_2{=}CHCH_2OCH_2CH_2OCH(CH_3)OCH_2CH_3$				
	H_2PtCl_6	-	-	$(CH_3O)_3Si(CH_2)_3OCH_2CH_2OCH(CH_3)OCH_2CH_3$ (-)	U -084
	H_2PtCl_6	-	-	$(CH_3O)_3Si(CH_2)_3OCH_2CH_2OCH(CH_3)OCH_2CH_3$ (-)	U -070
C_9H_{18}	$CH_3CH_2C(CH_2OH)_2CH_2OCH_2CH{=}CH_2$				
	H_2PtCl_6	100-130	-	$CH_3CH_2C(CH_2OH)_2CH_2O(CH_2)_3Si(OCH_3)_3$ (-)	N -004
$C_{10}H_{12}$	$CH_2{=}CHCH_2OCH_2C_6H_5$				
	H_2PtCl_6	80-107	1	$(CH_3O)_3Si(CH_2)_3OCH_2C_6H_5$ (63)	1023
$C_{10}H_{12}$	Fig.228.				

	H_2PtCl_6 (i-PrOH)	reflux	-	polymer (-)	F -006
$C_{10}H_{12}$	Fig.229.				
	Pt	reflux	- (b)	Fig.230. (-)	J -049
$C_{10}H_{16}$	$CH_3COOCH_2CH=CH(CH_2)_3CH=CH_2$				
	H_2PtCl_6 (i-PrOH)	50-60	2	$CH_3COOCH_2CH=CH(CH_2)_5Si(OCH_3)_3$ (85)	J -037
$C_{10}H_{18}$	$HN(CH_2)_4CH[N(CH_3)(CH_2CH=CH_2)]CO$				
	H_2PtCl_6	80	- (THF)	$HN(CH_2)_4CH\{N(CH_3)[(CH_2)_3Si(OCH_3)_3]\}CO$ (-)	J -105
$C_{10}H_{19}$	$(CH_3CH_2)_2NCH_2CH_2COOCH_2CH=CH_2$				
	H_2PtCl_6 (AllOCH$_2$COOAll)	104-142	36.3	$(CH_3CH_2)_2NCH_2CH_2COO(CH_2)_3Si(OCH_3)_3$ (-)	Pl-026
$C_{11}H_{12}$	$C_6H_5OCH_2COOCH_2CH=CH_2$				
	H_2PtCl_6 (AllOCH$_2$COOAll)	85-159	7	$C_6H_5OCH_2COO(CH_2)_3Si(OCH_3)_3$ (68)	Pl-008
$C_{11}H_{14}$	$CH_2=C(CH_2COOCH_2CH=CH_2)COOCH_2CH=CH_2$				
	H_2PtCl_6	110-120	3	$CH_2=C[CH_2COO(CH_2)_3Si(OCH_3)_3]COO(CH_2)_3Si(OCH_3)_3$ (-)	G -003
$C_{11}H_{14}$	p-$(CH_3)_2CHC_6H_4CH=CH_2$				
	H_2PtCl_6	-	-	p-$(CH_3)_2CHC_6H_4CH_2CH_2Si(OCH_3)_3$ (-)	D -044
$C_{11}H_{16}$	$CH_2OCH(CH=CH_2)OCH_2CCH_2OCH(CH=CH_2)OCH_2$				
	$PtCl_2(acac)_2$	55	0.5 (a)	$CH_2OCH(CH=CH_2)OCH_2CCH_2OCH[CH_2CH_2Si(OCH_3)_3]OCH_2$ (83)	D -011
$C_{11}H_{18}$	$CH_2=CHCH_2N(CH_3)COCH=CHN(CH_3)CH_2CH=CH_2$-trans				
	H_2PtCl_6	100-130	2 (xy)	$CH_2=CHCH_2N(CH_3)COCH=CHCON(CH_3)(CH_2)_3Si(OCH_3)_3$-trans + $(CH_3O)_3Si(CH_2)_3N(CH_3)COCH=CHCON(CH_3)$-$(CH_2)_3Si(OCH_3)_3$-trans (-)	67
$C_{11}H_{21}$	$CH_2=CH(CH_2)_8CH_2Cl$				
	polymer-PPh$_2$/Pt	40-60	-	$(CH_3O)_3Si(CH_2)_{11}Cl$ (85)	240
$C_{11}H_{21}$	$CH_2=CH(CH_2)_9Cl$				
	-	-	-	$Cl(CH_2)_{11}Si(OCH_3)_3$ (-)	1294
$C_{11}H_{22}$	$CH_2=CH(CH_2)_8SCH_3$				
	$Pt(PPh_3)_4$	-	-	$(CH_3O)_3Si(CH_2)_{10}SCH_3$ (-)	243
$C_{11}H_{22}$	$CH_2=CHCH_2C(CH_3)_2C(CH_2CH_3)=NN(CH_3)_2$				

	H_2PtCl_6 (OctOH)	-	-	$(CH_3O)_3Si(CH_2)_3C(CH_3)_2C(CH_2CH_3){=}NN(CH_3)_2$ (-)	U -078
$C_{12}H_{15}$	Fig.231.				
	H_2PtCl_6	85-100	-	Fig.232. (-) Fig.233. (-) Fig.234. (-)	D -017
$C_{12}H_{18}$	$CH_2{=}CHCH_2N(CH_3)COCH{=}CHCON(CH_3)CH_2CH{=}CH_2$-trans				
	H_2PtCl_6	100-130	2 (xy)	$CH_2{=}CHCH_2N(CH_3)COCH{=}CHCON(CH_3)(CH_2)_3Si(OCH_3)_3$-trans + $(CH_3O)_3Si(CH_2)_3N(CH_3)COCH{=}CHCON(CH_3)$-$(CH_2)_3Si(OCH_3)_3$-trans (-)	U -061
	H_2PtCl_6	100-130	2 (xy)	$CH_2{=}CHCH_2N(CH_3)COCH{=}CHCON(CH_3)(CH_2)_3Si(OCH_3)_3$-trans + $(CH_3O)_3Si(CH_2)_3N(CH_3)COCH{=}CHCON(CH_3)$-$(CH_2)_3Si(OCH_3)_3$-trans (-)	U -066
	H_2PtCl_6	100-130	2 (xy)	$CH_2{=}CHCH_2N(CH_3)COCH{=}CHCON(CH_3)(CH_2)_3Si(OCH_3)_3$-trans + $(CH_3O)_3Si(CH_2)_3N(CH_3)COCH{=}CHCON(CH_3)$-$(CH_2)_3Si(OCH_3)_3$-trans (-)	U -067
	H_2PtCl_6	100-130	2 (xy)	$CH_2{=}CHCH_2N(CH_3)COCH{=}CHCON(CH_3)(CH_2)_3Si(OCH_3)_3$-trans + $(CH_3O)_3Si(CH_2)_3N(CH_3)COCH{=}CHCON(CH_3)$-$(CH_2)_3Si(OCH_3)_3$-trans (-)	U -069
	H_2PtCl_6	100-130	2 (xy)	$CH_2{=}CHCH_2N(CH_3)COCH{=}CHCON(CH_3)(CH_2)_3Si(OCH_3)_3$-trans + $(CH_3O)_3Si(CH_2)_3N(CH_3)COCH{=}CHCON(CH_3)$-$(CH_2)_3Si(OCH_3)_3$-trans (-)	U -073
	H_2PtCl_6	100-130	2 (xy)	$CH_2{=}CHCH_2N(CH_3)COCH{=}CHCON(CH_3)(CH_2)_3Si(OCH_3)_3$-trans + $(CH_3O)_3Si(CH_2)_3N(CH_3)COCH{=}CHCON(CH_3)$-$(CH_2)_3Si(OCH_3)_3$-trans (-)	U -075
$C_{12}H_{22}$	$CH_2{=}CH(CH_2)_8COOCH_3$				
	H_2PtCl_6	-	-	$(CH_3O)_3Si(CH_2)_{10}COOCH_3$ (-)	F -028
$C_{12}H_{28}$	$(CH_3)_3SiOCH_2CH[OSi(CH_3)_3]CH_2OCH_2CH{=}CH_2$				
	H_2PtCl_6 (i-PrOH)	80-90	2	$(CH_3)_3SiOCH_2CH[OSi(CH_3)_3]CH_2O(CH_2)_3Si(OCH_3)_3$ (-)	G -013
$C_{13}H_{12}$	$CH_2{=}CHCH_2COCH{=}CHCOC_6H_5$				
	H_2PtCl_6	100-120	2 (xy)	$(CH_3O)_3Si(CH_2)_3COCH{=}CHCOC_6H_5$ (-)	U -060
	H_2PtCl_6	100-120	2 (xy)	$(CH_3O)_3Si(CH_2)_3COCH{=}CHCOC_6H_5$ (-)	U -061
	H_2PtCl_6	100-120	2 (xy)	$(CH_3O)_3Si(CH_2)_3COCH{=}CHCOC_6H_5$ (-)	U -066
	H_2PtCl_6	100-120	2 (xy)	$(CH_3O)_3Si(CH_2)_3COCH{=}CHCOC_6H_5$ (-)	U -067
	H_2PtCl_6	100-120	2 (xy)	$(CH_3O)_3Si(CH_2)_3COCH{=}CHCOC_6H_5$ (-)	U -069

	H_2PtCl_6	100-120	2 (xy)	$(CH_3O)_3Si(CH_2)_3COCH=CHCOC_6H_5$ (-)	U -073
	H_2PtCl_6	100-120	2 (xy)	$(CH_3O)_3Si(CH_2)_3COCH=CHCOC_6H_5$ (-)	U -075
$C_{13}H_{20}$	Fig.235.				
	H_2PtCl_6 (i-PrOH)	reflux	- (t)	Fig.236. (-)	F -006
$C_{15}H_{19}$	Fig.237.				
	H_2PtCl_6	-	-	Fig.238. (-)	U -139
$C_{15}H_{28}$	$CH_2=CH(CH_2)_8COOOC(CH_3)_3$				
	$PtCl_2(p\text{-}ClC_6H_4CN)_2$	25	40 (b)	$(CH_3O)_3Si(CH_2)_{10}COOOC(CH_3)_3$ (87)	D -045
$C_{16}H_{14}$	m-$C_6H_5COOC_6H_4OCH_2CH=CH_2$				
	H_2PtCl_6	90	17 (t)	m-$C_6H_5COOC_6H_4O(CH_2)_3Si(OCH_3)_3$ (-)	U -128
$C_{17}H_{26}$	Fig.72.				
	H_2PtCl_6 (i-PrOH)	180	16	Fig.239. (82)	F -027
	H_2PtCl_6 (i-PrOH)	180	16	Fig.239. (-)	U -136
$C_{19}H_9$	o-$C_{10}F_{19}OC_6H_4CH_2CH=CH_2$				
	$PtCl_2(SEt_2)_2$	reflux	6.5 (t)	o-$C_{10}F_{19}OC_6H_4(CH_2)_3Si(OCH_3)_3$ (-)	G -025
$C_{21}H_{24}$	Fig.240.				
	H_2PtCl_6 ($MeOC_2H_4OMe$)	65-260	8 (t)	Fig.241. (-)	U -087
$C_{22}H_{17}$	Fig.242.				
	Pt catalyst	90	17 (t)	Fig.243. (-)	U -128
$C_{22}H_{38}$	$[CH_2=CH(CH_2)_8COO]_2$				
	$PtCl_2(p\text{-}ClC_6H_4CN)_2$	25	- (b)	$[(CH_3O)_3Si(CH_2)_{10}COO]_2$ (73)	D -045
$C_{27}H_{30}$	Fig.244.				
	Pt	reflux	4 (t)	Fig.245. (96)	J -065

Table 5.3

= $(CH_3O)_3SiH$ = =1=

C_4H_6	$CH_3CH{=}CHCHO$				
	$Ni(cod)_2$	20	5	$CH_3CH{=}CHCH_2OSi(OCH_3)_3$ (84)	661
C_8H_8	$C_6H_5COCH_3$				
	$(LiOCHR')_2$	0	15-20(THF)	$C_6H_5CH(CH_3)OSi(OCH_3)_3$ (41-78)	596
	$LiOCH_2CH(NHLi)R$	r.t.	3 (THF)	$C_6H_5CH(CH_3)OSi(OCH_3)_3$ (62-100)	596
C_9H_{10}	$C_6H_5COCH_2CH_3$				
	$(LiOCHR')_2$	0	15-20(THF)	$C_6H_5CH(CH_3CH_2)OSi(OCH_3)_3$ (52-70)	596
	$LiOCH_2CH(NHLi)R$	r.t.	3 (THF)	$C_6H_5CH(CH_3CH_2)OSi(OCH_3)_3$ (44-88)	596
$C_{10}H_{12}$	$C_6H_5COCH(CH_3)_2$				
	$(LiOCHR')_2$	0	15-20(THF)	$C_6H_5CH[CH(CH_3)_2]OSi(OCH_3)_3$ (48-63)	596
	$LiOCH_2CH(NHLi)R$	r.t.	3 (THF)	$C_6H_5CH[CH(CH_3)_2]OSi(OCH_3)_3$ (55)	596
$C_{10}H_{12}$	$C_6H_5COCH_2CH_2CH_3$				
	$(LiOCHR')_2$	0	15-20(THF)	$C_6H_5CH(CH_3CH_2CH_2)OSi(OCH_3)_3$ (67-71)	596
	$LiOCH_2CH(NHLi)R$	r.t.	3 (THF)	$C_6H_5CH(CH_3CH_2CH_2)OSi(OCH_3)_3$ (27-52)	596
$C_{11}H_{14}$	$C_6H_5CO(CH_2)_3CH_3$				
	$(LiOCHR')_2$	0	15-20(THF)	$C_6H_5CH[CH_3(CH_2)_3]OSi(OCH_3)_3$ (51-64)	596
	$LiOCH_2CH(NHLi)R$	r.t.	3 (THF)	$C_6H_5CH[CH_3(CH_2)_3]OSi(OCH_3)_3$ (12-55)	596

Table 6.1

= $CH_3(C_2H_5O)_2SiH$ = =1=

C_4H_6	$CH_2=CHCH=CH_2$				
	$Pd(\overline{COCH=CHCOO})(PPh_3)_2$	-	- (b)	$CH_3CH=CHCH_2CH_2CH=CHCH_2Si(OC_2H_5)_2CH_3$ (-)	J -004
C_6H_{10}	$CH_2=CH(CH_3)C(CH_3)=CH_2$				
	$Cr(CO)_6$	100	24	$(CH_3)_2C=C(CH_3)CH_2Si(OC_2H_5)_2CH_3$ + $CH_2=C(CH_3)CH(CH_3)CH_2Si(OC_2H_5)_2CH_3$ (28)	202
	$Cr(CO)_6$ + t-BuOOH	100	24 (b)	$(CH_3)_2C=C(CH_3)CH_2Si(OC_2H_5)_2CH_3$ + $CH_2=C(CH_3)CH(CH_3)CH_2Si(OC_2H_5)_2CH_3$ (27)	202
	$Mo(CO)_6$	100	24	$(CH_3)_2C=C(CH_3)CH_2Si(OC_2H_5)_2CH_3$ + $CH_2=C(CH_3)CH(CH_3)CH_2Si(OC_2H_5)_2CH_3$ (85)	202
	$Mo(CO)_6$ + t-BuOOH	100	24 (b)	$(CH_3)_2C=C(CH_3)CH_2Si(OC_2H_5)_2CH_3$ + $CH_2=C(CH_3)CH(CH_3)CH_2Si(OC_2H_5)_2CH_3$ (89)	202
	$[RhCl(cod)]_2$ + DMPIC	100	6	$C_6H_{11}Si(OC_2H_5)_2CH_3$ (70)	884
C_6H_{12}	$CH_3(CH_2)_3CH=CH_2$				
	H_2PtCl_6 $[(\overline{CH_2)_5CO}]$	reflux	3	$CH_3(CH_2)_5Si(OC_2H_5)_2CH_3$ (82)	PI-011
	SiO_2-Pt/vinylsiloxane	80	0.16	$CH_3(CH_2)_5Si(OC_2H_5)_2CH_3$ (95)	1203
	$RhCl(PPh_3)_3$	reflux	3	$CH_3(CH_2)_5Si(OC_2H_5)_2CH_3$ (68)	PI-011
	$RuCl_3(PPh_3)_3$	120	6	$CH_3(CH_2)_5Si(OC_2H_5)_2CH_3$ (38)	PI-015
C_8H_{16}	$CH_3(CH_2)_5CH=CH_2$				
	$\overline{NMeCH_2CH_2NMeC}=Rh(PMe_2Ph)_2(CO)$	100	8	$CH_3(CH_2)_7Si(OC_2H_5)_2CH_3$ (52)	496
	$\overline{NMeCH_2CH_2NMeC}=RhCl(cod)$	100	8	$CH_3(CH_2)_7Si(OC_2H_5)_2CH_3$ (74)	496
	$\overline{NPhCH_2CH_2NPhC}=RhCl(cod)$	100	8	$CH_3(CH_2)_7Si(OC_2H_5)_2CH_3$ (40)	496
	$Rh_2(OAc)_4$	100	8	$CH_3(CH_2)_7Si(OC_2H_5)_2CH_3$ (53)	274
	$RhCl(PPh_3)_3$	80	2	$CH_3(CH_2)_7Si(OC_2H_5)_2CH_3$ (71)	PI-011
	$RhCl_2[P(c\text{-}Hex)_3]_2$	100	8	$CH_3(CH_2)_7Si(OC_2H_5)_2CH_3$ (98)	503
	$RhCl_2[P(o\text{-}C_6H_4Me)_3]_2$	100	8	$CH_3(CH_2)_7Si(OC_2H_5)_2CH_3$ (86)	503
	SiO_2-PPh_2/$RhCl_3$	70	3	$CH_3(CH_2)_7Si(OC_2H_5)_2CH_3$ (82)	C -007
$C_{10}H_{12}$	Fig.42.				
	$Co_2(CO)_8$	30-50	2	Fig.246. (83)	712
	$Rh_4(CO)_{12}$	30-50	1	Fig.246. (97)	712
$C_{10}H_{16}$	Fig.45.				
	Pt/C	90	16	Fig.250. (65)	U -164

Table 6.2

= $CH_3(C_2H_5O)_2SiH$ = =1=

C_3H_3	$CH_2=CHCN$				
	PPh_3	200	5	$CH_3Si(OC_2H_5)_2CH_2CH_2CN$ (-)	U -007
	PPh_3	200	5	$CH_3Si(OC_2H_5)_2CH_2CH_2CN$ (-)	U -015
C_3H_5	$CH_2=CHCH_2Cl$				
	$[IrCl(cod)]_2$	80	4 (xy)	$CH_3Si(OC_2H_5)_2(CH_2)_3Cl$ (70-75)	U -172
	$[RhCl(cod)]_2$	up to 135	4	$CH_3Si(OC_2H_5)_2(CH_2)_3Cl$ (-)	U -172
C_3H_7	$CH_2=CHCH_2NH_2$				
	H_2PtCl_6 (i-PrOH + AllOH)	until 120	23	$CH_3Si(OC_2H_5)_2C_3H_6NH_2$ (60)	S -019
	H_2PtCl_6 (i-PrOH + $CH_2=CHOBu$ + $(EtO)_2SiHMe$)	reflux	12.5	$CH_3Si(OC_2H_5)_2(CH_2)_3NH_2$ (40)	S -010
	H_2PtCl_6 (i-PrOH + Me_2CO)	until 120	14	$CH_3Si(OC_2H_5)_2C_3H_6NH_2$ (42)	S -011
	H_2PtCl_6 (i-PrOH)	146	40	$CH_3Si(OC_2H_5)_2(CH_2)_3NH_2$ (54)	Dd-002
	H_2PtCl_6 (i-PrOH)	reflux	40	$CH_3Si(OC_2H_5)_2(CH_2)_3NH_2$ (53)	91
	H_2PtCl_6 (i-PrOH)	reflux	12.5	$CH_3Si(OC_2H_5)_2(CH_2)_3NH_2$ (16)	S -010
	H_2PtCl_6 + Na_2CO_3	130	12	$(C_2H_5O)_2Si(CH_3)(CH_2)_3NH_2$ (54) $(C_2H_5O)_2Si(CH_3)CH_2CH(CH_3)NH_2$ (7)	E -008
	H_2PtCl_6 + Na_2CO_3	130	12	$(C_2H_5O)_2Si(CH_3)(CH_2)_3NH_2$ (54) $(C_2H_5O)_2Si(CH_3)CH_2CH(CH_3)NH_2$ (7)	U -144
	$RhH(CO)(PPh_3)_3$ + PPh_3	100	6	$CH_3Si(C_2H_5O)_2C_3H_6NH_2$ (70)	U -156
C_4H_5	$CH_2=CHCH_2NCO$				
	Rh catalyst	-	-	$CH_3Si(OC_2H_5)_2(CH_2)_3NCO$ (-)	J -024
C_5H_5	$CF_3COOCH_2CH=CH_2$				
	H_2PtCl_6 (i-PrOH)	50-150	10-12	$CF_3COO(CH_2)_3Si(OC_2H_5)_2CH_3$ (19)	S -005
	H_2PtCl_6 (i-PrOH)	reflux	10	$CF_3COO(CH_2)_3Si(OC_2H_5)_2CH_3$ (19)	931
C_5H_6	$CF_3CONHCH_2CH=CH_2$				
	$[PtCl_2(c\text{-}C_6H_{10})]_2$	reflux	7	$CF_3CONH(CH_2)_3Si(OC_2H_5)_2CH_3$ (-)	F -018
C_5H_8	$\overline{OCH_2C}HCH_2OCH=CH_2$				
	H_2PtCl_6	reflux	7-8	$\overline{OCH_2C}HCH_2OCH_2CH_2Si(OC_2H_5)_2CH_3$ (71)	1154
C_5H_{10}	$(CH_3O)_2CHCH=CH_2$				

Table 6.2

= $CH_3(C_2H_5O)_2SiH$ = =2=

	H_2PtCl_6 + $(EtO)_2SiHCl$	50	overnight	$(CH_3O)_2CHCH_2CH_2Si(OC_2H_5)_2CH_3$ (73)	D -021
C_5H_{11}	$CH_2=CHCH_2N(CH_3)_2$				
	H_2PtCl_6 (THF)	80-110	2-3	$CH_3Si(OC_2H_5)_2(CH_2)_3N(CH_3)_2$ (27)	553
	H_2PtCl_6 (THF)	80-110	2-3	$CH_3Si(OC_2H_5)_2(CH_2)_3N(CH_3)_2$ (27)	S -051
	$RhCl(PPh_3)_3$	until 140	2	$CH_3Si(OC_2H_5)_2(CH_2)_3N(CH_3)_2$ (65)	C -033
C_5H_{16}	$\overline{CHB_{10}H_{10}C}CH_2CH=CH_2$				
	Pt/C	reflux	14 days	$\overline{CHB_{10}H_{10}C}(CH_2)_3Si(OC_2H_5)_2CH_3$ (27)	1030
C_6H_9	$\overline{CO(CH_2)_3N}CH=CH_2$				
	H_2PtCl_6	100	8	$\overline{CO(CH_2)_3N}CH_2CH_2Si(OC_2H_5)_2CH_3$ + $\overline{CO(CH_2)_3N}CH=CHSi(OC_2H_5)_2CH_3$ (-)	J -022
	H_2PtCl_6	101-126	16	$\overline{CO(CH_2)_3N}CH_2CH_2Si(OC_2H_5)_2CH_3$ (17)	G -007
	Pt/C	100	8	$\overline{CO(CH_2)_3N}CH_2CH_2Si(OC_2H_5)_2CH_3$ + $\overline{CO(CH_2)_3N}CH=CHSi(OC_2H_5)_2CH_3$ (-)	J -022
	$PtCl_2(PPh_3)_2$	100	8	$\overline{CO(CH_2)_3N}CH_2CH_2Si(OC_2H_5)_2CH_3$ + $\overline{CO(CH_2)_3N}CH=CHSi(OC_2H_5)_2CH_3$ (-)	J -022
	$RhCl(PPh_3)_3$	100	8	$\overline{CO(CH_2)_3N}CH_2CH_2Si(OC_2H_5)_2CH_3$ + $\overline{CO(CH_2)_3N}CH=CHSi(OC_2H_5)_2CH_3$ (-)	J -022
	$RhCl_3$	100	8	$\overline{CO(CH_2)_3N}CH_2CH_2Si(OC_2H_5)_2CH_3$ + $\overline{CO(CH_2)_3N}CH=CHSi(OC_2H_5)_2CH_3$ (-)	J -022
	$RuCl_3$	100-140	11	$\overline{CO(CH_2)_3N}CH_2CH_2Si(OC_2H_5)_2CH_3$ + $\overline{CO(CH_2)_3N}CH=CHSi(OC_2H_5)_2CH_3$ (-)	J -023
C_6H_{10}	$\overline{OCH_2C}HCH_2OCH_2CH=CH_2$				
	H_2PtCl_6	-	-	$\overline{OCH_2C}HCH_2O(CH_2)_3Si(OC_2H_5)_2CH_3$ (62)	1153
	$Pt_2I_2Me_6(py)_6$	150	1	$\overline{OCH_2C}HCH_2O(CH_2)_3Si(OC_2H_5)_2CH_3$ (-)	D -010
C_6H_{12}	$CH_2=CHCOOSi(CH_3)_3$				
	H_2PtCl_6	80-120	3	$CH_3Si(OC_2H_5)_2CH_2CH_2COOSi(CH_3)_3$ (41)	1247
C_6H_{13}	$CH_3(C_2H_5O)P(O)OCH_2CH=CH_2$				
	H_2PtCl_6 (i-PrOH)	80-110	2	$CH_3(C_2H_5O)P(O)O(CH_2)_3Si(OC_2H_5)_2CH_3$ (25)	191
C_6H_{18}	$\overline{CHB_{10}H_{10}C}CH_2CH_2CH=CH_2$				
	Pt/C	95	68	$\overline{CHB_{10}H_{10}C}(CH_2)_4Si(OC_2H_5)_2CH_3$ (-)	U -026

Table 6.2

= $CH_3(C_2H_5O)_2SiH$ = =3=

	Pt/C	reflux	4 days	$\overline{CHB_{10}H_{10}C}(CH_2)_4Si(OC_2H_5)_2CH_3$ (53)	1030
	Pt/C	reflux	72	$\overline{CHB_{10}H_{10}C}(CH_2)_4Si(OC_2H_5)_2CH_3$ (-)	1030
	Pt/C	reflux	72	$\overline{CHB_{10}H_{10}C}(CH_2)_4Si(OC_2H_5)_2CH_3$ (-)	U -026
	Pt/C	reflux	-	$\overline{CHB_{10}H_{10}C}(CH_2)_4Si(OC_2H_5)_2CH_3$ (-)	U -032
C_7H_6	$C_3F_7CONHCH_2CH=CH_2$				
	$[PtCl_2(c\text{-}C_6H_{10})]_2$	reflux	3	$C_3F_7CONH(CH_2)_3Si(OC_2H_5)_2CH_3$ (-)	F -018
C_7H_{11}	$\overline{OCH(CH_2Cl)CH}CH_2C(CH_3)=CH_2$				
	H_2PtCl_6 (i-PrOH)	120-125	24	$\overline{OCH(CH_2Cl)CH}CH_2CH(CH_3)CH_2Si(OC_2H_5)_2CH_3$ (55)	S -013
C_7H_{14}	$(CH_3CH_2O)_2CHCH=CH_2$				
	H_2PtCl_6 (i-PrOH)	100-162	1	$(CH_3CH_2O)_2CHCH_2CH_2Si(OC_2H_5)_2CH_3$ (69)	1004
C_7H_{14}	$CH_2=C(CH_3)COOSi(CH_3)_3$				
	H_2PtCl_6	80-120	3	$CH_3Si(OC_2H_5)_2CH_2CH(CH_3)COOSi(CH_3)_3$ (16)	1247
	H_2PtCl_6 (i-PrOH)	100-180	7	$CH_3Si(OC_2H_5)_2CH_2CH(CH_3)COOSi(CH_3)_3$ (30)	355
C_7H_{14}	$CH_2=CHCH_2COOSi(CH_3)_3$				
	H_2PtCl_6 (i-PrOH)	100-180	7	$CH_3Si(OC_2H_5)_2(CH_2)_3COOSi(CH_3)_3$ (65)	355
C_7H_{15}	$(CH_3)_2N(CH_2)_3CH=CH_2$				
	H_2PtCl_6	-	-	$(CH_3)_2N(CH_2)_5Si(OC_2H_5)_2CH_3$ (43)	C -034
	$RhCl(PPh_3)_3$	until 150	3	$(CH_3)_2N(CH_2)_5Si(OC_2H_5)_2CH_3$ (51)	C -034
C_7H_{15}	$(CH_3CH_2)_2NCH_2CH=CH_2$				
	H_2PtCl_6 (THF)	90	3	$(CH_3CH_2)_2N(CH_2)_3Si(OC_2H_5)_2CH_3$ (50)	553
C_7H_{15}	$(CH_3CH_2O)_2P(O)OCH_2CH=CH_2$				
	H_2PtCl_6 (i-PrOH)	115-117	1	$(CH_3CH_2O)_2P(O)O(CH_2)_3Si(OC_2H_5)_2CH_3$ (22)	191
C_8H_9	$2\text{-}CH_2=CHCH_2OC_5H_4N$				
	H_2PtCl_6 $[o\text{-}C_6H_4(COOMe)_2]$	130-144	27 (xy)	$CH_3Si(OC_2H_5)_2(CH_2)_3OC_5H_4N\text{-}2$ (-)	F -010
C_8H_{11}	Fig.58.				
	H_2PtCl_6 (i-PrOH)	215	2	Fig.247. (59)	731
C_8H_{12}	$(CH_3CO)_2CHCH_2CH=CH_2$				

Table 6.2

= $CH_3(C_2H_5O)_2SiH$ = =4=

	trans-$PtCl_2(Et_2S)_2$	reflux	1.5 (t)	$(CH_3CO)_2CH(CH_2)_3Si(OC_2H_5)_2CH_3$ (-)	G -033
C_8H_{12}	Fig.248.				
	-	-	-	Fig.249. (-)	136
	H_2PtCl_6 (i-PrOH)	100-150	2	Fig.249. (78)	20
C_8H_{12}	$\overline{OCH_2C}(CH_2CH=CH_2)_2$				
	H_2PtCl_6	120-125	15	$\overline{OCH_2C}(CH_2CH=CH_2)(CH_2)_3Si(OC_2H_5)_2CH_3$ + $\overline{OCH_2C}[(CH_2)_3Si(OC_2H_5)_2CH_3]_2$ (64)	1156
C_8H_{18}	$(CH_3CH_2O)_3SiCH=CH_2$				
	$RhCl(PPh_3)_3$	120	6	$(CH_3CH_2O)_3SiCH_2CH_2Si(OC_2H_5)_2CH_3$ (80)	Pl-006
	$Ru(acac)_3$	120	6	$(CH_3CH_2O)_3SiCH_2CH_2Si(OC_2H_5)_2CH_3$ (52)	Pl-006
C_9H_{14}	$CH_2=CHCH_2CH(COCH_3)COOCH_2CH_3$				
	trans-$PtCl_2(Et_2S)_2$	reflux	1.5 (t)	$CH_3Si(OC_2H_5)_2(CH_2)_3CH(COCH_3)COOCH_2CH_3$ (-)	G -033
C_9H_{14}	$\overline{OCH_2C}HCH_2N(CH_2CH_2CN)CH_2CH=CH_2$				
	H_2PtCl_6	reflux	13	$\overline{OCH_2C}HCH_2N(CH_2CH_2CN)(CH_2)_3Si(OC_2H_5)_2CH_3$ (59)	1151
C_9H_{18}	$C_4H_9OCH(CH_3)OCH_2CH=CH_2$				
	H_2PtCl_6 (i-PrOH)	100-105	4	$C_4H_9OCH(CH_3)O(CH_2)_3Si(C_2H_5O)_2CH_3$ (50)	1005
$C_{10}H_{12}$	$CH_2=CHCH_2OCO\overline{CH(O)C}HCOOCH_2CH=CH_2$				
	Pt	reflux	- (b)	$CH_3Si(OC_2H_5)_2(CH_2)_3OCO\overline{CH(O)C}HCOO(CH_2)_3$-$Si(C_2H_5O)_2CH_3$ (-)	J -049
$C_{10}H_{16}$	$CH_2=CHCH_2CH(COOCH_2CH_3)_2$				
	H_2PtCl_6 (i-PrOH)	reflux	-	$CH_3Si(OC_2H_5)_2(CH_2)_3CH(COOCH_2CH_3)_2$ (60)	1008
$C_{11}H_{12}$	$C_6H_5COCH_2CH_2CH=CH_2$				
	H_2PtCl_6	20	0.5	$C_6H_5CO(CH_2)_4Si(OC_2H_5)_2CH_3$ (-)	1180
$C_{11}H_{17}$	Fig.251.				
	H_2PtCl_6 (i-PrOH)	80-138	1	Fig.252. (60)	1000
$C_{11}H_{21}$	$Br(CH_2)_9CH=CH_2$				
	$RhCl(PPh_3)_3$	80	3.5-9	$Br(CH_2)_{11}Si(OC_2H_5)_2CH_3$ (-)	1180
$C_{12}H_{20}$	$\overline{OCH_2C}HCH_2OCH(CH_2OCH_2CH=CH_2)_2$				

Table 6.2

= $CH_3(C_2H_5O)_2SiH$ = =5=

	H_2PtCl_6	120-125	1-6	$\overline{OCH_2C}HCH_2OCH(CH_2OCH_2CH=CH_2)CH_2O(CH_2)_3Si(OC_2H_5)_2$ $-CH_3$ (40)	1158
$C_{12}H_{20}$	$[-N=C(CH_3)CH_2CH_2CH=CH_2]_2$				
	H_2PtCl_6 (i-PrOH)	80-110	-	$[-N=C(CH_3)(CH_2)_4Si(OC_2H_5)_2CH_3]_2$ (63)	U -065
$C_{12}H_{22}$	$C_{10}H_{19}OSiCH=CH_2$				
	H_2PtCl_6 (i-PrOH)	150	1.5	$C_{10}H_{19}OSiCH_2CH_2Si(OC_2H_5)_2CH_3$ (66)	33
$C_{12}H_{22}$	$CH_2=CH(CH_2)_8COOCH_3$				
	H_2PtCl_6 (i-PrOH)	reflux	5	$CH_3Si(OC_2H_5)_2(CH_2)_{10}COOCH_3$ (85)	1011
	$RhCl(PPh_3)_3$	80	3.5-9	$CH_3Si(OC_2H_5)_2(CH_2)_{10}COOCH_3$ (-)	1180
$C_{12}H_{24}$	$[CH_3(CH_2)_3O]_2CHCH_2CH=CH_2$				
	H_2PtCl_6 (i-PrOH)	78-155	1	$[CH_3(CH_2)_3O]_2CH(CH_2)_3Si(OC_2H_5)_2CH_3$ (58)	1004
$C_{12}H_{24}$	$[OSi(CH_3)CH=CH_2]_4$				
	H_2PtCl_6	60-100	4	$[OSi(CH_3)CH_2CH_2Si(OC_2H_5)_2CH_3]_4$ (94)	S -050
$C_{13}H_{19}$	o-$(CH_3)_2NCH_2CH_2C_6H_4CH_2CH=CH_2$				
	H_2PtCl_6 (i-PrOH)	reflux	4	o-$(CH_3)_2NCH_2CH_2C_6H_4(CH_2)_3Si(OC_2H_5)_2CH_3$ (45)	C -034
$C_{13}H_{20}$	Fig.253.				
	H_2PtCl_6 ($EtOC_2H_4OEt$)	100-130	2-3	Fig.254. (79)	664
$C_{13}H_{24}$	$C_{10}H_{19}OCH_2CH=CH_2$				
	H_2PtCl_6 (i-PrOH)	150	1.5	$C_{10}H_{19}O(CH_2)_3Si(OC_2H_5)_2CH_3$ (85)	33
$C_{14}H_{18}$	Fig.255.				
	H_2PtCl_6 [$(MeOC_2H_4)_2O)$]	115-125	4.75	Fig.256. ( )	D -053
$C_{16}H_{24}$	Fig.257.				
	H_2PtCl_6 ($EtOC_2H_4OEt$)	100-130	2-3	Fig.258. (82)	664
$C_{16}H_{30}$	Crown-$CH_2OCH_2CH=CH_2$				
	H_2PtCl_6	reflux	16 (b)	Crown-$CH_2O(CH_2)_3SiCH_3(OC_2H_5)_2$ (-)	157
$C_{16}H_{32}$	$CH_2=C(CH_3)COO(CH_2)_3OTi[OCH(CH_3)_2]_3$				
	trans-$PtCl_2(Et_2S)_2$	reflux	1.5 (t)	$CH_3Si(OC_2H_5)_2CH_2CH(CH_3)COO(CH_2)_3OTi[OCH(CH_3)_2]_3$ (-)	D -069
$C_{20}H_{30}$	Fig.191.				

	H_2PtCl_6 (i-PrOH)	20-80	4 (t)	Fig.259. (73)	E -013
$C_{24}H_{33}$	Fig.260.				
	trans-$PtCl_2(Et_2S)_2$	95-110	0.5 (t)	$CH_3COC(CH_2CH{=}CH_2){=}C(CH_3)OAl\{OC(CH_3){=}C[(CH_2)_3$-$Si(OC_2H_5)_2CH_3]COCH_3\}_2$ (-)	G -033
$C_{30}H_{44}$	Crown-$CH_2OCH_2CH{=}CH_2$				
	H_2PtCl_6	reflux	16 (b)	Crown-$CH_2O(CH_2)_3SiCH_3(OC_2H_5)_2$ (-)	157

Table 6.3

= $CH_3(C_2H_5O)_2SiH$ = =1=

C_2H_2	$CH\equiv CH$				
	$Pd(PPh_3)_4$	80-100	7	$CH_2=CHSi(OC_2H_5)_2CH_3$ (19) $(C_2H_5O)_2Si(CH_3)CH_2CH_2Si(OC_2H_5)_2CH_3$ (21)	1290
	$Pd(PPh_3)_4$	80-100	7	$CH_2=CHSi(OC_2H_5)_2CH_3$ (19) $(C_2H_5O)_2Si(CH_3)CH_2CH_2Si(OC_2H_5)_2CH_3$ (21)	1288
	$PdCl_2(PPh_3)_2$	80-100	9	$CH_2=CHSi(OC_2H_5)_2CH_3$ (24) $(C_2H_5O)_2Si(CH_3)CH_2CH_2Si(OC_2H_5)_2CH_3$ (23)	1290
	$PdCl_2(PPh_3)_2$	80-100	9	$CH_2=CHSi(OC_2H_5)_2CH_3$ (24) $(C_2H_5O)_2Si(CH_3)CH_2CH_2Si(OC_2H_5)_2CH_3$ (23)	1288
	H_2PtCl_6 (i-PrOH)	20	1	$CH_2=CHSi(OC_2H_5)_2CH_3$ (5) $(C_2H_5O)_2Si(CH_3)CH_2CH_2Si(OC_2H_5)_2CH_3$ (78)	1290
	H_2PtCl_6 (i-PrOH)	20	1	$CH_2=CHSi(OC_2H_5)_2CH_3$ (5) $(C_2H_5O)_2Si(CH_3)CH_2CH_2Si(OC_2H_5)_2CH_3$ (78)	1288
	$Pt(PPh_3)_4$	60	6	$CH_2=CHSi(OC_2H_5)_2CH_3$ (38)	1290
	$Pt(PPh_3)_4$	60	6	$CH_2=CHSi(OC_2H_5)_2CH_3$ (38)	1288
	$PtCl_2(PPh_3)_2$	70	8	$CH_2=CHSi(OC_2H_5)_2CH_3$ (23)	1288
	$PtCl_2(PPh_3)_2$	70	8	$CH_2=CHSi(OC_2H_5)_2CH_3$ (23)	1290
	$RhCl(PPh_3)_3$	50	2	$CH_2=CHSi(OC_2H_5)_2CH_3$ (47) $(C_2H_5O)_2Si(CH_3)CH_2CH_2Si(OC_2H_5)_2CH_3$ (19)	1290
	$RhCl(PPh_3)_3$	50	2	$CH_2=CHSi(OC_2H_5)_2CH_3$ (47) $(C_2H_5O)_2Si(CH_3)CH_2CH_2Si(OC_2H_5)_2CH_3$ (19)	1288
	$RhH(PPh_3)_4$	20	1	$CH_2=CHSi(OC_2H_5)_2CH_3$ (80) $(C_2H_5O)_2Si(CH_3)CH_2CH_2Si(OC_2H_5)_2CH_3$ (18)	1290
	$RhH(PPh_3)_4$	20	1	$CH_2=CHSi(OC_2H_5)_2CH_3$ (80) $(C_2H_5O)_2Si(CH_3)CH_2CH_2Si(OC_2H_5)_2CH_3$ (18)	1288
	$RuCl_2(PPh_3)_3$	80	7	$CH_2=CHSi(OC_2H_5)_2CH_3$ (8) $(C_2H_5O)_2Si(CH_3)CH_2CH_2Si(OC_2H_5)_2CH_3$ (trace)	1288
	$RuCl_2(PPh_3)_3$	80	7	$CH_2=CHSi(OC_2H_5)_2CH_3$ (8) $(C_2H_5O)_2Si(CH_3)CH_2CH_2Si(OC_2H_5)_2CH_3$ (trace)	1290
C_6H_8	$\overline{OCH_2C}HCH_2OCH_2C\equiv CH$				
	H_2PtCl_6 (i-PrOH)	85	-	$\overline{OCH_2C}HCH_2OCH_2CH=CHSi(OC_2H_5)_2CH_3$ (57)	997
C_6H_{10}	$CH_3(CH_2)_3C\equiv CH$				
	H_2PtCl_6 (THF)	100	2	$CH_3(CH_2)_3CH=CHSi(OC_2H_5)_2CH_3$ + $CH_3(CH_2)_3C(=CH_2)Si(OC_2H_5)_2CH_3$ (-)	266

Table 6.3

= $CH_3(C_2H_5O)_2SiH$ = =2=

	H_2PtCl_6 (THF)	20	3	$C_6H_{11}Si(OC_2H_5)_2CH_3$ (kin.)	1252
C_6H_{12}	$\overline{OCH_2CH}CH_2OC(CH_3)_2C\equiv CH$				
	H_2PtCl_6 (i-PrOH)	70-150	4	$\overline{OCH_2CH}CH_2OC(CH_3)_2CH=CHSi(OC_2H_5)_2CH_3$ (-)	998
C_8H_6	$C_6H_5C\equiv CH$				
	SiO_2- $^+PPh_3/PtBr_4^-$	100	8	$C_6H_5CH=CHSi(OC_2H_5)_2CH_3$ + $C_6H_5C(=CH_2)Si(OC_2H_5)_2CH_3$ (75)	S -036
C_8H_{10}	$CH\equiv C(CH_2)_4C\equiv CH$				
	$Ni(acac)_2$	50	6 (b)	Fig.261. (68)	1182
C_8H_{12}	$\overline{OCH_2CH}CH_2OCH_2CH_2OCH_2C\equiv CH$				
	H_2PtCl_6 (i-PrOH)	reflux	15 (b)	$\overline{OCH_2CH}CH_2OCH_2CH_2OCH_2CH=CHSi(OC_2H_5)_2CH_3$ (90)	1162
C_9H_{13}	$\overline{OCH_2CH}CH_2OCH(CH_2Cl)CH_2OCH_2C\equiv CH$				
	H_2PtCl_6 (i-PrOH)	reflux	18 (b)	$\overline{OCH_2CH}CH_2OCH(CH_2Cl)CH_2OCH_2CH=CHSi(OC_2H_5)_2CH_3$(85)	1161
C_9H_{16}	$CH_3(CH_2)_3OCH(CH_3)OCH_2C\equiv CH$				
	H_2PtCl_6 (i-PrOH)	-	-	$CH_3(CH_2)_3OCH(CH_3)OCH_2CH=CHSi(OC_2H_5)_2CH_3$ (-)	1006
$C_{10}H_{14}$	$CH\equiv CCH_2CH(COOCH_2CH_3)_2$				
	H_2PtCl_6	reflux	-	$CH_3Si(OC_2H_5)_2CH=CHCH_2CH(COOCH_2CH_3)_2$ (65)	1009
$C_{10}H_{16}$	$\overline{OCH_2CH}CH_2OCH(CH_2OCH_3)CH_2OCH_2C\equiv CH$				
	H_2PtCl_6 (i-PrOH)	reflux	18 (b)	$\overline{OCH_2CH}CH_2OCH(CH_2OCH_3)CH_2OCH_2CH=CHSi(OC_2H_5)_2CH_3$ (83)	1161
$C_{10}H_{18}$	$CH_3(CH_2)_3C\equiv C(CH_2)_3CH_3$				
	H_2PtCl_6	20	0.5	$CH_3(CH_2)_3CH=C[(CH_2)_3CH_3]Si(OC_2H_5)_2CH_3$ (-)	1180
$C_{11}H_{10}$	$C_6H_5C\equiv CCH_2\overline{CHCH_2O}$				
	H_2PtCl_6	60-70	-	$C_6H_5CH=C[Si(OC_2H_5)_2CH_3]CH_2\overline{CHCH_2O}$ (82)	1195
$C_{11}H_{19}$	$\overline{OCH_2CH}CH_2OCH_2C\equiv CCH_2N(CH_2CH_3)_2$				
	H_2PtCl_6	reflux	18 (b)	$\overline{OCH_2CH}CH_2OCH_2C[Si(OC_2H_5)_2CH_3]=CHCH_2N(CH_2CH_3)_2$ (61)	1196
$C_{13}H_{14}$	$C_6H_5C\equiv CCH_2CH_2OCH_2\overline{CHCH_2O}$				
	H_2PtCl_6 (i-PrOH)	reflux	18 (b)	$C_6H_5CH=C[Si(OC_2H_5)_2CH_3]CH_2CH_2OCH_2\overline{CHCH_2O}$ (78)	1196
$C_{16}H_{22}$	Fig.262.				

Table 6.3

= $CH_3(C_2H_5O)_2SiH$ = =3=

	H_2PtCl_6 (EtOC$_2$H$_4$OEt)	140	2-3	Fig.263. (70)	664
$C_{16}H_{23}$	Fig.264.				
	H_2PtCl_6 (EtOC$_2$H$_4$OEt)	100-130	2-3	Fig.265.(80)	664

Table 6.4

= $CH_3(C_2H_5O)_2SiH$ =   =1=

C_5H_4	Fig.266.				
	KF	25	18	Fig.267. (-)	151
C_6H_{10}	$\overline{(CH_2)_5C}O$				
	HCOOK	80	5 (DMF)	$\overline{(CH_2)_5C}HOSi(OC_2H_5)_2CH_3$ (75)	262
	KF	60-80	1-1.5(DMF)	$\overline{(CH_2)_5C}HOSi(OC_2H_5)_2CH_3$ (70)	262
C_6H_{10}	$CH_2{=}CHCH_2CH_2COCH_3$				
	KF	60	5 (DMF)	$CH_2{=}CHCH_2CH_2CH(CH_3)OSi(OC_2H_5)_2CH_3$ (85)	262
	KF	60	5 (DMF)	$CH_2{=}CHCH_2CH_2CH(CH_3)OSi(OC_2H_5)_2CH_3$ (85)	286
C_7H_6	C_6H_5CHO				
	CsF	25	0.17	$C_6H_5CH_2OSi(OC_2H_5)_2CH_3$ (-)	151
	HCOOK	20	0.25(DMF)	$C_6H_5CH_2OSi(OC_2H_5)_2CH_3$ (90)	262
	KF	20	0.25 (DMF)	$C_6H_5CH_2OSi(OC_2H_5)_2CH_3$ (90)	286
	KF	20	0.25(DMF)	$C_6H_5CH_2OSi(OC_2H_5)_2CH_3$ (90)	262
C_7H_{14}	$CH_3(CH_2)_3COOCH_2CH_3$				
	CsF	120	3	$CH_3(CH_2)_3CH(OCH_2CH_3)OSi(OC_2H_5)_2CH_3$ (-)	154
C_7H_{14}	$CH_3(CH_2)_5CHO$				
	KF	10	1.75(DMF)	$CH_3(CH_2)_5CH_2OSi(OC_2H_5)_2CH_3$ (85)	262
	KF	10	1.75 (DMF)	$CH_3(CH_2)_5CH_2OSi(OC_2H_5)_2CH_3$ (85)	286
C_8H_8	$C_6H_5COCH_3$				
	CsF	100	2.5	$C_6H_5CH(CH_3)OSi(OC_2H_5)_2CH_3$ (-)	286
	CsF	100	2.5	$C_6H_5CH(CH_3)OSi(OC_2H_5)_2CH_3$ (-)	151
	KF	100	1-24	$C_6H_5CH(CH_3)OSi(OC_2H_5)_2CH_3$ (-)	151
	KF	80	1.5(DMF)	$C_6H_5CH(CH_3)OSi(OC_2H_5)_2CH_3$ (78)	262
	$LiOCH_2CH(NHLi)R$	r.t.	3 (THF)	$C_6H_5CH(CH_3)OSi(OC_2H_5)_2CH_3$ (42)	596
C_8H_{14}	$(CH_3)_2C{=}CHCH_2CH_2COCH_3$				
	KF	80	3 (DMF)	$(CH_3)_2C{=}CHCH_2CH_2CH(CH_3)OSi(OC_2H_5)_2CH_3$ (80)	262
C_9H_8	$C_6H_5CH{=}CHCHO$				

Table 6.4
= $CH_3(C_2H_5O)_2SiH$ = =2=

	CsF	25	2	$C_6H_5CH=CHCH_2OSi(OC_2H_5)_2CH_3$ (-)	286
	CsF	25	2	$C_6H_5CH=CHCH_2OSi(OC_2H_5)_2CH_3$ (-)	151
C_9H_{18}	$[(CH_3)_2CHCH_2]_2CO$				
	KF	60	5.5(DMF)	$[(CH_3)_2CHCH_2]_2CHOSi(OC_2H_5)_2CH_3$ (75)	262
$C_{10}H_{18}$	$(CH_3)_2C=CHCH_2CH_2CH(CH_3)CH_2CHO$				
	KF	0	0.75(DMF)	$(CH_3)_2C=CHCH_2CH_2CH(CH_3)CH_2CH_2OSi(OC_2H_5)_2CH_3$ (68)	262
	KF	25	18	$(CH_3)_2C=CHCH_2CH_2CH(CH_3)CH_2CH_2OSi(OC_2H_5)_2CH_3$ (-)	151
$C_{11}H_{12}$	$C_6H_5COCH_2CH_2COOCH_3$				
	CsF	25	2.5	$C_6H_5CH[OSi(OC_2H_5)_2CH_3]CH_2CH_2COOCH_3$ (-)	151
	CsF	25	2.5	$C_6H_5CH[OSi(OC_2H_5)_2CH_3]CH_2CH_2COOCH_3$ (-)	286
$C_{11}H_{20}$	$CH_3OCO(CH_2)_7COOCH_3$				
	CsF	120	2	$CH_3OCH[OSi(OC_2H_5)_2CH_3](CH_2)_7CH(OCH_3)O$-$Si(OC_2H_5)_2CH_3$ (-)	154
$C_{12}H_{22}$	$CH_2=CH(CH_2)_8COOCH_3$				
	CsF	80	48	$CH_2=CH(CH_2)_8CH(OCH_3)OSi(OC_2H_5)_2CH_3$ (-)	154
	KF	100	48 (DMF)	$CH_2=CH(CH_2)_8CH(OCH_3)OSi(OC_2H_5)_2CH_3$ (65)	262
$C_{13}H_{10}$	$C_6H_5COC_6H_5$				
	CsF	25	1.5	$(C_6H_5)_2CHOSi(OC_2H_5)_2CH_3$ (-)	151
	KF	20	0.5(DMF)	$(C_6H_5)_2CHOSi(OC_2H_5)_2CH_3$ (87)	262
	KF	20	0.5 (DMF)	$(C_6H_5)_2CHOSi(OC_2H_5)_2CH_3$ (87)	286
$C_{14}H_{28}$	$CH_3(CH_2)_{10}COOC_2H_5$				
	CsF	100	20	$CH_3(CH_2)_{10}CH(OCH_2CH_3)OSi(OC_2H_5)_2CH_3$ (-)	154
$C_{21}H_{23}$	Fig.268.				
	H_2PtCl_6	60	0.5	Fig.269. (100)	1320
	H_2PtCl_6 (i-PrOH)	r.t.	0.4 (t)	Fig.269. (-)	1322
$C_{28}H_{52}$	$CH_3(CH_2)_7CH=CH(CH_2)_7COOC_{10}H_{19}$				
	CsF	120	3	$CH_3(CH_2)_7CH=CH(CH_2)_7CH(OC_{10}H_{19})OSi(OC_2H_5)_2CH_3$(-)	154

Table 7

= $C_6H_5SiHCl_2$ = =1=

C_2H_3 $Cl_3SiCH=CH_2$				
H_2PtCl_6 (i-PrOH)	-	0.1	$Cl_3SiC_2H_4SiCl_2C_6H_5$ (83)	96
H_2PtCl_6 (i-PrOH)	150	4	$Cl_3SiCH_2CH_2SiCl_2C_6H_5$ (66)	893
H_2PtCl_6 (i-PrOH) + borane	20	170	$Cl_3SiC_2H_4SiCl_2C_6H_5$ (84)	96
C_3H_3 $CF_3CH=CH_2$				
H_2PtCl_6 (i-PrOH)	50-80	3	$CF_3CH_2CH_2SiCl_2C_6H_5$ (83)	D -079
C_3H_3 $CH_2=CHCN$				
NEt_3	150	5	$C_6H_5SiCl_2CH_2CH_2CN$ (23)	U -007
NEt_3	150	5	$C_6H_5SiCl_2CH_2CH_2CN$ (23)	U -015
Cu_2O + $(Me_2NCH_2)_2$	30	21	$C_6H_5Cl_2SiCH_2CH_2CN$ (32)	959
Cu_2O + $(Me_2NCH_2)_2$	reflux	0.5	$C_6H_5Cl_2SiCH_2CH_2CN$ (98-100)	959
Cu_2O + $(Me_2NCH_2)_2$; ultrasonic	30	4	$C_6H_5Cl_2SiCH_2CH_2CN$ (70-75)	959
C_3H_6 $CH_3SiCl_2CH=CH_2$				
H_2PtCl_6 (i-PrOH)	150	4	$CH_3SiCl_2CH_2CH_2SiCl_2C_6H_5$ (61)	893
C_4H_1 $CF_2CF_2CH=CCl$ [ring]				
gamma	-	-	$CF_2CF_2CH_2C(Cl)SiCl_2C_6H_5$ [ring] + $CHClCF_2CF_2CHSiCl_2C_6H_5$ [ring] (-)	1214
C_4H_5 $CH_2=CHCH_2CN$				
H_2PtCl_6	120	16	$C_6H_5SiCl_2(CH_2)_3CN$ (-)	U -070
Pt/Al_2O_3	200	2	$C_6H_5SiCl_2(CH_2)_3CN$ (19)	U -007
C_4H_6 $CH_2=CHCH_2CHO$				
H_2PtCl_6 (i-PrOH)	60-122	1	$C_6H_5SiCl_2(CH_2)_3CHO$ (36)	1052
C_4H_6 $CH_2=CHCH=CH_2$				
$Ni(acac)_2$ + $Na[AlH_2(OC_2H_4OMe)_2]$	82	5	$C_4H_7SiCl_2C_6H_5$ (30)	C -010
C_4H_6 $HCOOCH_2CH=CH_2$				
H_2PtCl_6 (i-PrOH)	60-112	1	$HCOO(CH_2)_3SiCl_2C_6H_5$ (56)	S -026
C_4H_8 $(CH_3)_2C=CH_2$				
H_2PtCl_6	60	-	$(CH_3)_2CHCH_2SiCl_2C_6H_5$ + $(CH_3)_2C(CH_3)SiCl_2C_6H_5$ (-)	1193
C_4H_8 $CH_3CH_2CH=CH_2$				

Table 7

= $C_6H_5SiHCl_2$ = =2=

	H_2PtCl_6	60	-	$CH_3(CH_2)_3SiCl_2C_6H_5$ + $CH_3CH_2CH(CH_3)SiCl_2C_6H_5$ (-)	1193
C_4H_8	$CH_3CH=CHCH_3$				
	H_2PtCl_6	60	-	$CH_3(CH_2)_3SiCl_2C_6H_5$ + $CH_3CH_2CH(CH_3)SiCl_2C_6H_5$ (-)	1193
C_5H_1	$(CF_2)_3CH=CCl$				
	gamma	-	-	$(CF_2)_3CH_2C(Cl)SiCl_2C_6H_5$ + $CHCl(CF_2)_3CHSiCl_2C_6H_5$ (-)	1214
C_5H_8	$CH_2=C(CH_3)COOCH_3$				
	H_2PtCl_6 (OctOH)	reflux	1 (t)	$C_6H_5SiCl_2CH_2CH(CH_3)COOCH_3$ (74)	U -014
C_5H_8	$CH_3OCOOCH_2CH=CH_2$				
	H_2PtCl_6 (i-PrOH)	50-170	0.3	$CH_3OCOO(CH_2)_3SiCl_2C_6H_5$ (67)	S -026
	H_2PtCl_6 (i-PrOH)	50-170	0.3	$CH_3OCOO(CH_2)_3SiCl_2C_6H_5$ (67)	1051
C_5H_{11}	$(CH_3)_2SiClCH_2CH=CH_2$				
	H_2PtCl_6	-	-	$(CH_3)_2SiCl(CH_2)_3SiCl_2C_6H_5$ (-)	499
C_5H_{16}	$H\overline{CB_{10}H_{10}C}C(CH_3)=CH_2$				
	none	260	-	$H\overline{CB_{10}H_{10}C}C_3H_6SiCl_2C_6H_5$ (-)	1045
C_6H_{10}	$CH_3CH_2OCH_2C\equiv CCH_3$				
	-	-	-	$CH_3CH_2OCH_2C(=CHCH_3)SiCl_2C_6H_5$ (-)	805
C_6H_{12}	$CH_3(CH_2)_3CH=CH_2$				
	gamma(^{60}Co)	41	-	$CH_3(CH_2)_5SiCl_2C_6H_5$ (kin.)	958
	$Co_2(CO)_8$	40	-	$CH_3(CH_2)_5SiCl_2C_6H_5$ (kin.)	225
	H_2PtCl_6 $[\overline{(CH_2)_5C}O]$	reflux	3	$CH_3(CH_2)_5SiCl_2C_6H_5$ (95)	Pl-011
	$[PtCl_2(C_2H_4)]_2$	reflux	2	$CH_3(CH_2)_5SiCl_2C_6H_5$ (76)	226
	$Rh(OCOHex)_3$	60	-	$CH_3(CH_2)_5SiCl_2C_6H_5$ (100)	F -004
	$RhCl_3$ + $(EtO)_3SiH$	100	2	$CH_3(CH_2)_5SiCl_2C_6H_5$ (-)	F -004
C_7H_5	$CH_2=CHCOOCH_2CF_2CF_2CF_3$				
	CuCl + $(Me_2CNH_2)_2$ + NBu_3	reflux	2	$C_6H_5SiCl_2CH_2CH_2COOCH_2CF_2CF_2CF_3$ (79)	489
C_7H_{14}	$CH_2=CHCH_2COOSi(CH_3)_3$				
	H_2PtCl_6 (i-PrOH)	reflux	2	$C_6H_5\overline{SiCl(CH_2)_3COO}$ (81)	774
C_8H_8	$C_6H_5CH=CH_2$				

Table 7

= $C_6H_5SiHCl_2$ = =3=

	H_2PtCl_6	80	2	$C_6H_5CH_2CH_2SiCl_2C_6H_5$ + $C_6H_5CH(CH_3)SiCl_2C_6H_5$ (-)	834
	H_2PtCl_6 (i-PrOH)	100	5	$C_6H_5CH_2CH_2SiCl_2C_6H_5$ (43)	257
C_8H_{14}	$\overline{(CH_2)_5C}HCH=CH_2$				
	$Co_2(CO)_8$	0-60	until 24	$\overline{(CH_2)_5C}HCH_2CH_2SiCl_2C_6H_5$ (-)	470
C_8H_{16}	$CH_3(CH_2)_5CH=CH_2$				
	$Co_2(CO)_8$	18-35	15.5 (t)	$CH_3(CH_2)_7SiCl_2C_6H_5$ (98)	U -019
	$Co_2(CO)_8$	18-35	15.5 (t)	$CH_3(CH_2)_7SiCl_2C_6H_5$ (98)	D -009
C_9H_{10}	$C_6H_5OCH_2CH=CH_2$				
	H_2PtCl_6 (i-PrOH)	110-130	2	$C_6H_5O(CH_2)_3SiCl_2C_6H_5$ (15)	53
C_9H_{12}	Fig.107.				
	H_2PtCl_6	140	20	Fig.270 + Fig.271 (83)	E -019
$C_{12}H_{22}$	$CH_2=CH(CH_2)_8COOCH_3$				
	H_2PtCl_6 (i-PrOH)	reflux	5	$C_6H_5SiCl_2(CH_2)_{10}COOCH_3$ (82)	1011
$C_{13}H_{24}$	$CH_2=CHCH_2OCH(CH_2)_9$				
	H_2PtCl_6 (i-PrOH)	reflux	3.5	$C_6H_5SiCl_2(CH_2)_3OCH(CH_2)_9$ (25)	33
$C_{14}H_{10}$	$C_6H_5C\equiv CC_6H_5$				
	H_2PtCl_6 (i-PrOH)	100	2	$C_6H_5CH=C(C_6H_5)SiCl_2C_6H_5$ (91)	837
$C_{14}H_{12}$	$(C_6H_5)_2C=CH_2$				
	H_2PtCl_6	200	5	$(C_6H_5)_2CHCH_2SiCl_2C_6H_5$ (30)	833
$C_{19}H_{36}$	$CH_2=CH(CH_2)_{15}COOCH_3$				
	$(t\text{-}BuO)_2$	reflux	5	$C_6H_5SiCl_2(CH_2)_{17}COOCH_3$ (38)	1011
	H_2PtCl_6 (i-PrOH)	reflux	5	$C_6H_5SiCl_2(CH_2)_{17}COOCH_3$ (70)	1011
	Pt/C	reflux	5	$C_6H_5SiCl_2(CH_2)_{17}COOCH_3$ (69)	1011

Table 8.1

= $(C_2H_5)_3SiH$ = =1=

C_2H_4	$CH_2=CH_2$				
	$(t\text{-}BuO)_2$	140	1	$CH_3CH_2Si(C_2H_5)_3$ (58) $CH_3(CH_2)_3Si(C_2H_5)_3$ (26) $CH_3(CH_2)_5Si(C_2H_5)_3$ (6) $CH_3(CH_2)_nSi(C_2H_5)_3$ (10)	1197
	UV + $(t\text{-}BuO)_2$	-	-	$CH_3CH_2Si(C_2H_5)_3$ (-)	623
	UV + $(t\text{-}BuO)_2$	-	-	$CH_3CH_2Si(C_2H_5)_3$ (-)	624
	t-BuOOH	140	-	$(C_2H_5)_4Si$ (-)	382
	$Fe(CO)_5$	140	1	$CH_3CH_2Si(C_2H_5)_3$ (90) $CH_3(CH_2)_nSi(C_2H_5)_3$ (10)	1197
	$[Ir(OMe)(cod)]_2$ + $AsPh_3$	25	24 (dc)	$C_2H_5Si(C_2H_5)_3$ / $CH_2=CHSi(C_2H_5)_3$ (62/38)	880
	$Mn_2(CO)_{10}$	140	1	$CH_3CH_2Si(C_2H_5)_3$ (90) $CH_3(CH_2)_nSi(C_2H_5)_3$ (10)	1197
	$Mn_2(CO)_{10}$	140	-	$(C_2H_5)_4Si$ (-)	382
	$Re_2(CO)_{10}$	140	-	$(C_2H_5)_4Si$ (-)	382
	$(C_5Me_5Rh)_2Cl_4$	40	- (DCE)	$CH_3CH_2Si(C_2H_5)_3$ (40-74) $CH_2=CHSi(C_2H_5)_3$ (26-60)	771
	$[HRu_3(CO)_{10}(SiEt_3)_2][N(PPh_3)_2]$	100	15	$CH_3CH_2Si(C_2H_5)_3$ (22) $CH_2=CHSi(C_2H_5)_3$ (52)	1144
C_3H_6	$CH_3CH=CH_2$				
	$[Ir(OMe)(cod)]_2$ + $AsPh_3$	25	24 (dc)	$C_3H_7Si(C_2H_5)_3$ / $C_3H_5Si(C_2H_5)_3$ (22/78)	880
	$RhCl(PPh_3)_3$	40	2-4 (b)	$CH_3CH_2CH_2Si(C_2H_5)_3$ (1-49)	314
	$[RhCl(C_8H_{14})_2]_2$	40	1 (b)	$CH_3CH_2CH_2Si(C_2H_5)_3$ (100)	314
	$[RhCl(C_8H_{14})_2]_2$ + PPh_3	40	1.5-6 (b)	$CH_3CH_2CH_2Si(C_2H_5)_3$ (0-74)	314
	$[HRu_3(CO)_{10}(SiEt_3)_2][N(PPh_3)_2]$	100	15	$CH_3CH_2CH_2Si(C_2H_5)_3$ (8) $CH_3CH=CHSi(C_2H_5)_3$ (38)	1144
C_5H_{10}	$CH_3CH_2CH_2CH=CH_2$				
	$Al_2O_3/\equiv SiCo(CO)_3P(OPh)_3$ + UV	25	0.5-4.63	$CH_3(CH_2)_4Si(C_2H_5)_3$ + $C_5H_9Si(C_2H_5)_3$ ( )	963
	$Al_2O_3/\equiv SiCo(CO)_4$ + UV	25	0.5-64	$CH_3(CH_2)_4Si(C_2H_5)_3$ (1-26)	963
	$Co(CO)_3SiEt_3$ + UV	-	39	$CH_3(CH_2)_4Si(C_2H_5)_3$ (60)	964
	$Co(CO)_3SiPh_3$ + UV	-	39	$CH_3(CH_2)_4Si(C_2H_5)_3$ (30-52)	964

Table 8.1

= $(C_2H_5)_3SiH$ = =2=

	$Co_2(CO)_8$	25	1.6-8.25	$CH_3(CH_2)_4Si(C_2H_5)_3$ (12-16)	963
	$Co_2(CO)_8$	30	-	$CH_3(CH_2)_4Si(C_2H_5)_3$ (kin.)	225
	$Co_2(CO)_8$ + UV	-	-	$CH_3(CH_2)_4Si(C_2H_5)_3$ (15-35)	964
	$SiO_2/{\equiv}SiCo(CO)_4$ + UV	25	0.5-64	$CH_3(CH_2)_4Si(C_2H_5)_3$ (1-26)	963
	$Fe(CO)_3(PPh_3)_2$ + UV	25	-	$CH_3(CH_2)_4Si(C_2H_5)_3$ (4) $CH_3CH_2CH_2CH{=}CHSi(C_2H_5)_3$ (30) $CH_3CH_2CH{=}CHCH_2Si(C_2H_5)_3$ (6)	914
	$Fe(CO)_4PPh_3$ + UV	25	-	$CH_3(CH_2)_4Si(C_2H_5)_3$ (3) $CH_3CH_2CH_2CH{=}CHSi(C_2H_5)_3$ (22) $CH_3CH_2CH{=}CHCH_2Si(C_2H_5)_3$ (5)	914
	$Fe(CO)_5$ + UV	25	-	$CH_3(CH_2)_4Si(C_2H_5)_3$ (14) $CH_3CH_2CH_2CH{=}CHSi(C_2H_5)_3$ (54) $CH_3CH_2CH{=}CHCH_2Si(C_2H_5)_3$ (12)	914
	$Fe(CO)_5$ + UV (laser)	25	-	$CH_3(CH_2)_4Si(C_2H_5)_3$ + $CH_3CH_2CH_2CH{=}CHSi(C_2H_5)_3$ + $CH_3CH_2CH{=}CHCH_2Si(C_2H_5)_3$ (-)	787
	$Fe_3(CO)_{12}$	-	-	$CH_3(CH_2)_4Si(C_2H_5)_3$ (-)	71
	$Fe_3(CO)_{12}$ + UV (laser)	25	-	$CH_3(CH_2)_4Si(C_2H_5)_3$ + $CH_3CH_2CH_2CH{=}CHSi(C_2H_5)_3$ + $CH_3CH_2CH{=}CHCH_2Si(C_2H_5)_3$ (-)	787
	polymer-$P(Ph)CH_2CH_2PPh_2/Fe(CO)_n$ + UV	25	- (xy)	$CH_3(CH_2)_4Si(C_2H_5)_3$ (3) $CH_3CH_2CH_2CH{=}CHSi(C_2H_5)_3$ (19) $CH_3CH_2CH{=}CHCH_2Si(C_2H_5)_3$ (5)	914
	polymer-$PPh_2/Fe(CO)_n$ + UV	25	-	$CH_3(CH_2)_4Si(C_2H_5)_3$ (8) $CH_3CH_2CH_2CH{=}CHSi(C_2H_5)_3$ (35) $CH_3CH_2CH{=}CHCH_2Si(C_2H_5)_3$ (7)	914
	$Os_3(CO)_{12}$	-	-	$CH_3(CH_2)_4Si(C_2H_5)_3$ (-)	71
	$Pt_2Mo_2(\mu\text{-}cp)_2(\mu_3CO)_2(\mu_2CO)_4(PEt_3)_2$ + $h\nu$	r.t.	48	$C_5H_{11}Si(C_2H_5)_3$ (4)	913
	$Pt_2Mo_2(\mu\text{-}cp)_2(\mu_3CO)_2(\mu_2CO)_4(PEt_3)_2$ + $h\nu$	r.t.	48	$C_5H_{11}Si(C_2H_5)_3$ (3)	913
	$[HPt(SiCl_3)P(c\text{-}Hex)_3]_2$	20	6 (h)	$CH_3(CH_2)_4Si(C_2H_5)_3$ (51)	429
	$Rh(acac)_3$	60	20	$CH_3(CH_2)_4Si(C_2H_5)_3$ (31)	274
	$Ru_3(CO)_{12}$	-	-	$CH_3(CH_2)_4Si(C_2H_5)_3$ (-)	71
C_5H_{10}	$CH_3CH_2CH{=}CHCH_3$				
	$Rh(acac)_3$ + $AlEt_3$	60	12	$CH_3(CH_2)_4Si(C_2H_5)_3$ (57)	271
C_6H_{10}	$(\overline{CH_2)_4CH{=}C}H$				

Table 8.1

= $(C_2H_5)_3SiH$ = =3=

	H_2PtCl_6 (THF) + $AlCl_3$	100	6	$\overline{(CH_2)_5CH}Si(C_2H_5)_3$ (30)	1262
	H_2PtCl_6 (THF) + $AlCl_3$	100	6	$\overline{(CH_2)_5CH}Si(C_2H_5)_3$ (30)	S -041
	H_2PtCl_6 (THF) + $AlCl_3$	100	1	$\overline{(CH_2)_5CH}Si(C_2H_5)_3$ (30)	1234
C_6H_{12}	$(CH_3)_2CHCH_2CH=CH_2$				
	H_2PtCl_6	58	2	$(CH_3)_2CH(CH_2)_3Si(C_2H_5)_3$ (11)	U -168
	H_2PtCl_6 + $HSiCl_3$	23-70	0.12	$(CH_3)_2CH(CH_2)_3Si(C_2H_5)_3$ (30)	U -168
	H_2PtCl_6 + $HSiMeCl_2$	55	1.5	$(CH_3)_2CH(CH_2)_3SiCl_2CH_3$ (-) $(CH_3)_2CH(CH_2)_3Si(C_2H_5)_3$ (-)	U -168
C_6H_{12}	$(CH_3)_3CCH=CH_2$				
	H_2PtCl_6 + $HSiCl_3$	57	0.06	$(CH_3)_3CCH_2CH_2SiCl_3$ (8) $(CH_3)_3CCH_2CH_2Si(C_2H_5)_3$ (60)	U -168
	H_2PtCl_6 + $HSiCl_3$	86.5	2	$(CH_3)_3CCH_2CH_2Si(C_2H_5)_3$ (38)	U -168
	polymer-$NMe_2/RhCl_3$	60	3	$(CH_3)_3CCH_2CH_2Si(C_2H_5)_3$ (79)	C -006
	SDVB-$NMe_2/[RhCl(CO)_2]_2$	50	2	$C_6H_{13}Si(C_2H_5)_3$ (-)	316
C_6H_{12}	$C_4H_9CH=CH_2$				
	$Mn_2(CO)_{10}$	145	5	$C_6H_{13}Si(C_2H_5)_3$ + $C_4H_9CH=CHSi(C_2H_5)_3$ + $C_3H_7CH=CHCH_2Si(C_2H_5)_3$ (22-31)	635
	$Re_2(CO)_{10}$	145	5	$C_6H_{13}Si(C_2H_5)_3$ (60-81)	635
C_6H_{12}	$CH_3(CH_2)_3CH=CH_2$				
	gamma(^{60}Co)	52	-	$CH_3(CH_2)_5Si(C_2H_5)_3$ (kin.)	958
	$Co_2(CO)_8$	40	1 (t)	$CH_3(CH_2)_5Si(C_2H_5)_3$ (kin.)	495
	$Co_2(CO)_8$	40	1 (THF)	$CH_3(CH_2)_5Si(C_2H_5)_3$ (kin.)	495
	$Co_2(CO)_8$	40	1 (d)	$CH_3(CH_2)_5Si(C_2H_5)_3$ (kin.)	495
	$Co_2(CO)_8PtCl_2$	reflux	2 (THF)	$CH_3(CH_2)_5Si(C_2H_5)_3$ (58)	U -053
	$Co_2(CO)_8PtCl_2$	reflux	2 (THF)	$CH_3(CH_2)_5Si(C_2H_5)_3$ (58)	U -102
	$Fe(CO)_5$	140	1	$CH_3(CH_2)_5Si(C_2H_5)_3$ (21-22)	1197
	$IrH_4(SiEt_3)(cod)(AsPh_3)$	60	1 (dcm)	$C_6H_{11}Si(C_2H_5)_3$ (46) $C_6H_{13}Si(C_2H_5)_3$ (8)	361
	$IrH_4(SiEt_3)(cod)(PPh_3)$	60	1 (dc)	$C_6H_{11}Si(C_2H_5)_3$ (41) $C_6H_{13}Si(C_2H_5)_3$ (18)	361

Table 8.1
= $(C_2H_5)_3SiH$ = =4=

$[Ir(OMe)(cod)]_2$+ N,P,As,Pb donors	60	1 (dc)	$C_6H_{11}Si(C_2H_5)_3$ (12-81) $C_6H_{13}Si(C_2H_5)_3$ (3-14)	880
$[IrCl(cod)]_2$ + P,As donors	60	1 (dc)	$C_6H_{11}Si(C_2H_5)_3$ (15-17) $C_6H_{13}Si(C_2H_5)_3$ (48-85)	880
$Mn_2(CO)_{10}$	140	1	$CH_3(CH_2)_5Si(C_2H_5)_3$ (32-39)	1197
$Mn_2(CO)_{10}$	40	1 (t)	$CH_3(CH_2)_5Si(C_2H_5)_3$ (kin.)	495
$Mn_2(CO)_8$	40	1 (THF)	$CH_3(CH_2)_5Si(C_2H_5)_3$ (kin.)	495
Al_2O_3-$PPh_2/PdCl_2(PhCN)_2$	150	4	$CH_3(CH_2)_5Si(C_2H_5)_3$ (21)	C -007
DVB-PPh_2/H_2PtCl_6 + AllCl	80	2	$CH_3(CH_2)_5Si(C_2H_5)_3$ (13)	217
$H_2PtCl_6 \cdot 6H_2O$ + magnetic field	20	1-72	$CH_3(CH_2)_5Si(C_2H_5)_3$ (-)	1265
PMM-CN/H_2PtCl_6	80	2	$CH_3(CH_2)_5Si(C_2H_5)_3$ (47)	217
PMM-NMe_2/H_2PtCl_6	80	2	$CH_3(CH_2)_5Si(C_2H_5)_3$ (27)	217
PMM-NMe_2/H_2PtCl_6	r.t.	24	$CH_3(CH_2)_5Si(C_2H_5)_3$ (82)	C -002
PMM-NMe_2/H_2PtCl_6	r.t.	24	$CH_3(CH_2)_5Si(C_2H_5)_3$ (82)	217
PMM-PPh_2/H_2PtCl_6	80	2	$CH_3(CH_2)_5Si(C_2H_5)_3$ (7-15)	217
Pt + sonic waves	-	-	$CH_3(CH_2)_5Si(C_2H_5)_3$ (-)	U -142
Pt/C + sonic waves	30	2	$CH_3(CH_2)_5Si(C_2H_5)_3$ (74)	467
SDVB-CN/H_2PtCl_6	80	2	$CH_3(CH_2)_5Si(C_2H_5)_3$ (30)	217
SDVB-NMe/H_2PtCl_6	80	2	$CH_3(CH_2)_5Si(C_2H_5)_3$ (49)	217
SDVB-NMe/H_2PtCl_6 (EtOH)	75	-	$CH_3(CH_2)_5Si(C_2H_5)_3$ (85)	C -002
SDVB-PPh_2/H_2PtCl_6	80	2	$CH_3(CH_2)_5Si(C_2H_5)_3$ (84)	217
SiO_2-$OSi(CH_2)_3SH$/Pt	64	1.5	$CH_3(CH_2)_5Si(C_2H_5)_3$ (95)	U -146
SiO_2-$OSi(CH_2)_3SH$/Pt	75	1	$CH_3(CH_2)_5Si(C_2H_5)_3$ (65)	U -146
SiO_2-$OSi(CH_2)_3SH$/Pt + $(MeO)_3Si(CH_2)_3SH$	75	1	$CH_3(CH_2)_5Si(C_2H_5)_3$ (31)	U -146
SiO_2-SH/H_2PtCl_6 (EtOH)	64	flow	$CH_3(CH_2)_5Si(C_2H_5)_3$ (-)	F -033
$[HPt(SiCl_3)P(c\text{-}Hex)_3]_2$	20	6 (h)	$CH_3(CH_2)_5Si(C_2H_5)_3$ (60)	429
$[PtCl_2(C_2H_4)]_2$	65-70	-	$CH_3(CH_2)_5Si(C_2H_5)_3$ (kin.)	226
$[PtCl_2(py)]_2$	-	3	$CH_3(CH_2)_5Si(C_2H_5)_3$ (40)	987

amberlyt-NMe_2/H_2PtCl_6	80	2	$CH_3(CH_2)_5Si(C_2H_5)_3$ (50)	217
polymer/Pt complex	100	2	$CH_3(CH_2)_5Si(C_2H_5)_3$ (12)	1220
$(C_5Me_5Rh)_2(OH)_3X$ X=Cl, PF_6	40	1 (DCE)	$CH_3(CH_2)_5Si(C_2H_5)_3/CH_3(CH_2)_3CH{=}CHSi(C_2H_5)_3/$ $CH_3CH_2CH_2CH{=}CHCH_2Si(C_2H_5)_3$ (28-30/54-55/16-17)	771
$(C_5Me_5Rh)_2Cl_4$	0-80	1 (DCE)	$CH_3(CH_2)_5Si(C_2H_5)_3/CH_3(CH_2)_3CH{=}CHSi(C_2H_5)_3/$ $CH_3CH_2CH_2CH{=}CHCH_2Si(C_2H_5)_3$ (23-25/24-63/13-24)	771
$(C_5Me_5Rh)_2Cl_4$	40	1 (b)	$CH_3(CH_2)_5Si(C_2H_5)_3/CH_3(CH_2)_3CH{=}CHSi(C_2H_5)_3/$ $CH_3CH_2CH_2CH{=}CHCH_2Si(C_2H_5)_3$ (47/31/22)	771
$(C_5Me_5Rh)_2Cl_4$	60	1(MeCOMe)	$CH_3(CH_2)_5Si(C_2H_5)_3/CH_3(CH_2)_3CH{=}CHSi(C_2H_5)_3/$ $CH_3CH_2CH_2CH{=}CHCH_2Si(C_2H_5)_3$ (67/22/11)	771
$(C_5Me_5Rh)_2Cl_4$ + $(t\text{-}BuO)_2$	40-42	0.25(DCE)	$CH_3(CH_2)_5Si(C_2H_5)_3/CH_3(CH_2)_3CH{=}CHSi(C_2H_5)_3/$ $CH_3CH_2CH_2CH{=}CHCH_2Si(C_2H_5)_3$ (18-19/58/23-24)	771
$(C_5Me_5Rh)_2Cl_4$ + AIBN	40-42	2.5 (DCE)	$CH_3(CH_2)_5Si(C_2H_5)_3/CH_3(CH_2)_3CH{=}CHSi(C_2H_5)_3$(10)/ $CH_3CH_2CH_2CH{=}CHCH_2Si(C_2H_5)_3$ (84/10/2)	771
$(C_5Me_5Rh)_2Cl_4$ + HBF_4	40	1 (DCE)	$CH_3(CH_2)_5Si(C_2H_5)_3/CH_3(CH_2)_3CH{=}CHSi(C_2H_5)_3$(41)/ $CH_3CH_2CH_2CH{=}CHCH_2Si(C_2H_5)_3$ (45/41/14)	771
$(C_5Me_5Rh)_2Cl_4$ + MeCN	40-42	0.33 (DCE)	$CH_3(CH_2)_5Si(C_2H_5)_3/CH_3(CH_2)_3CH{=}CHSi(C_2H_5)_3/$ $CH_3CH_2CH_2CH{=}CHCH_2Si(C_2H_5)_3$ (39/10/2)	771
$(C_5Me_5Rh)_2Cl_4$ + NEt_3	40	1 (DCE)	$CH_3(CH_2)_5Si(C_2H_5)_3/CH_3(CH_2)_3CH{=}CHSi(C_2H_5)_3/$ $CH_3CH_2CH_2CH{=}CHCH_2Si(C_2H_5)_3$ (35/50/15)	771
$(C_5Me_5Rh)_2Cl_4$ + ionol	40-42	0.33 (DCE)	$CH_3(CH_2)_5Si(C_2H_5)_3/CH_3(CH_2)_3CH{=}CHSi(C_2H_5)_3/$ $CH_3CH_2CH_2CH{=}CHCH_2Si(C_2H_5)_3$ (16-22/55-61/23)	771
$(C_5Me_5Rh)_2I_4$	40	1 (DCE)	$CH_3(CH_2)_5Si(C_2H_5)_3$ (99)	771
ACDVB-$PPh_2/RhCl_3$	80	2	$CH_3(CH_2)_5Si(C_2H_5)_3$ (68)	217
Al_2O_3-CN/$[RhCl(CO)_2]_2$	80	2.5	$CH_3(CH_2)_5Si(C_2H_5)_3$ (77)	Dd-003
Al_2O_3-CN/$[RhCl(CO)_2]_2$	80	2.5	$CH_3(CH_2)_5Si(C_2H_5)_3$ (77)	U -020
C + $C_6H_8(OCH_2CH_2CN)_6/[RhCl(CO)_2]_2$	80	2.5	$CH_3(CH_2)_5Si(C_2H_5)_3$ (78)	U -020
C + $C_6H_8(OCH_2CH_2CN)_6/[RhCl(CO)_2]_2$	80	2.5	$CH_3(CH_2)_5Si(C_2H_5)_3$ (78)	Dd-003
C/$RhCl_3$ + c-$C_{12}H_{18}$ + EtOH	80	2.5	$CH_3(CH_2)_5Si(C_2H_5)_3$ (65)	D -005
C/$RhCl_3$ + c-$C_{12}H_{18}$ + EtOH	80	2.5	$CH_3(CH_2)_5Si(C_2H_5)_3$ (65)	C -030
$C_5Me_5Rh(C_2H_4)_2$	40	1 (DCE)	$CH_3(CH_2)_5Si(C_2H_5)_3/CH_3(CH_2)_3CH{=}CHSi(C_2H_5)_3/$ $CH_3CH_2CH_2CH{=}CHCH_2Si(C_2H_5)_3$ (23/61/16)	771
$C_5Me_5RhH_2(SiEt_3)_2$	40	0.17 (DCE)	$CH_3(CH_2)_5Si(C_2H_5)_3/CH_3(CH_2)_3CH{=}CHSi(C_2H_5)_3/$ $CH_3CH_2CH_2CH{=}CHCH_2Si(C_2H_5)_3$ (59-97/3-34/trace-7)	771

Table 8.1

= $(C_2H_5)_3SiH$ = =6=

$C_5Me_5RhH_2(SiEt_3)_2$ + CF_3COOH	40	1 (DCE)	$CH_3(CH_2)_5Si(C_2H_5)_3$/$CH_3(CH_2)_3CH=CHSi(C_2H_5)_3$/ $CH_3CH_2CH_2CH=CHCH_2Si(C_2H_5)_3$ (53/38/9)	771
C-CN/$[RhCl(CO)_2]_2$	80	2.5	$CH_3(CH_2)_5Si(C_2H_5)_3$ (77)	Dd-003
C-CN/$[RhCl(CO)_2]_2$ + c-$C_{12}H_{18}$	80	2.5	$CH_3(CH_2)_5Si(C_2H_5)_3$ (55)	Dd-003
C-CN/$[RhCl(CO)_2]_2$ + c-$C_{12}H_{18}$	80	2.5	$CH_3(CH_2)_5Si(C_2H_5)_3$ (55)	U -020
$CaCO_3$-CN/$[RhCl(CO)_2]_2$	80	2.5	$CH_3(CH_2)_5Si(C_2H_5)_3$ (85)	U -020
$CaCO_3$-CN/$[RhCl(CO)_2]_2$	80	2.5	$CH_3(CH_2)_5Si(C_2H_5)_3$ (85)	Dd-003
PMM-CN/$RhCl_3$	80	2	$CH_3(CH_2)_5Si(C_2H_5)_3$ (99)	217
PMM-NMe_2/Rh complex	24	24	$CH_3(CH_2)_5Si(C_2H_5)_3$ (97)	217
PMM-NMe_2/Rh complex	80	2	$CH_3(CH_2)_5Si(C_2H_5)_3$ (93)	217
Rh(π-All)$[P(OMe)_3]_3$	-	-	$CH_3(CH_2)_5Si(C_2H_5)_3$ (-)	144
Rh(acac)$(CO)_2$	80	1.5	$CH_3(CH_2)_5Si(C_2H_5)_3$ (84)	C -009
Rh(acac)$(CO)_2$+ magnetic field	20	1-72	$CH_3(CH_2)_5Si(C_2H_5)_3$ (-)	1265
$Rh(acac)_3$	40-80	1-6 (DCE)	$CH_3(CH_2)_5Si(C_2H_5)_3$/$CH_3(CH_2)_3CH=CHSi(C_2H_5)_3$/ $CH_3CH_2CH_2CH=CHCH_2Si(C_2H_5)_3$ (53-90/trace-24/6-18)	771
$Rh(acac)_3$ + $AlEt_3$	60	18	$CH_3(CH_2)_5Si(C_2H_5)_3$ + $CH_3(CH_2)_3CH(CH_3)Si(C_2H_5)_3$ + $CH_3CH_2CH_2CH(CH_2CH_3)Si(C_2H_5)_3$ (64)	271
$RhCl(PPh_3)_3$	40	1 (DCE)	$CH_3(CH_2)_5Si(C_2H_5)_3$/$CH_3(CH_2)_3CH=CHSi(C_2H_5)_3$/ $CH_3CH_2CH_2CH=CHCH_2Si(C_2H_5)_3$ (20-75/trace-61/20-45)	771
$RhCl(PPh_3)_3$	60	6 days	$CH_3(CH_2)_5Si(C_2H_5)_3$ (60)	476
$RhCl(PPh_3)_3$	60	6 days	$CH_3(CH_2)_5Si(C_2H_5)_3$ (60)	496
$RhCl(PPh_3)_3$	60-80	1 (DCE)	$CH_3(CH_2)_5Si(C_2H_5)_3$/$CH_3(CH_2)_3CH=CHSi(C_2H_5)_3$/ $CH_3CH_2CH_2CH=CHCH_2Si(C_2H_5)_3$ (64-66/trace-17/15-25)	771
$RhCl(PPh_3)_3$	64-65	-	$CH_3(CH_2)_5Si(C_2H_5)_3$ (-)	U -160
$RhCl(PPh_3)_3$	80	2	$CH_3(CH_2)_5Si(C_2H_5)_3$ (49)	204
$RhCl(PPh_3)_3$	80-95	- (solv.)	$CH_3(CH_2)_5Si(C_2H_5)_3$ (-)	U -160
$RhCl(PPh_3)_3$	reflux	-	$CH_3(CH_2)_5Si(C_2H_5)_3$/$CH_3(CH_2)_3CH=CHSi(C_2H_5)_3$/ $CH_3CH_2CH_2CH=CHCH_2Si(C_2H_5)_3$ (31/55/14)	879
$RhCl(PPh_3)_3$	reflux	2	$CH_3(CH_2)_5Si(C_2H_5)_3$ (80)	227

Table 8.1

= $(C_2H_5)_3SiH$ = =7=

$RhCl(PPh_3)_3$ + $(t\text{-}BuO)_2$	40	2-3	$CH_3(CH_2)_5Si(C_2H_5)_3/CH_3(CH_2)_3CH{=}CHSi(C_2H_5)_3/$ $CH_3CH_2CH_2CH{=}CHCH_2Si(C_2H_5)_3$ (80/8/12)	771
$RhCl(PPh_3)_3$ + t-BuOH	50-55	-	$CH_3(CH_2)_5Si(C_2H_5)_3$ (kin.)	314
$RhCl(PR_3)_3$ $PR_3 = PPh_3$, PPh_2Me, PPh_2Et, PPh_2(c-Hex), PPh(c-Hex)$_2$, P(c-Hex)$_3$	50-60	untill 2.5	$CH_3(CH_2)_5Si(C_2H_5)_3$ (kin.)	478
$RhCl_3$ (EtOH)	80	2	$CH_3(CH_2)_5Si(C_2H_5)_3$ (62)	204
$RhH(CO)(PPh_3)_3$ + magnetic field	20	1-72	$CH_3(CH_2)_5Si(C_2H_5)_3$ (-)	1265
$RhH(PPh_3)_4$	100	-	$CH_3(CH_2)_5Si(C_2H_5)_3$ (66)	605
$RhHCl(SiCl_2Et)(PPh_3)_2$	60	6 days	$CH_3(CH_2)_5Si(C_2H_5)_3$ (60)	476
SDVB-CH_2Cl/$RhCl(CO)(PPh_3)_2$	80	2.5	$CH_3(CH_2)_5Si(C_2H_5)_3$ (71)	C -022
SDVB-CH_2Cl/RhCl(diphos)$_2$	80	2.5	$CH_3(CH_2)_5Si(C_2H_5)_3$ (26)	C -022
SDVB-CH_2Cl/RhCl(diphos)$_2$ + $NaBH_4$	80	2.5	$CH_3(CH_2)_5Si(C_2H_5)_3$ (25)	C -022
SDVB-CH_2Cl/$RhCl_3$ (THF)	80	2.5	$CH_3(CH_2)_5Si(C_2H_5)_3$ (83)	C -022
SDVB-NMe_2/$RhCl_3$	80	2	$CH_3(CH_2)_5Si(C_2H_5)_3$ (88)	217
SDVB-PPh_2/Rh complex	80	2	$CH_3(CH_2)_5Si(C_2H_5)_3$ (55)	217
SiO_2-PPh_2/$[RhCl(C_2H_4)_2]_2$	20	-	$CH_3(CH_2)_5Si(C_2H_5)_3$ (kin.)	769
$[RhCl(C_8H_{14})]_2$ + magnetic field	20	1-72	$CH_3(CH_2)_5Si(C_2H_5)_3$ (-)	1265
$[RhCl(CO)_2]_2$	40	1 (DCE)	$CH_3(CH_2)_5Si(C_2H_5)_3/CH_3(CH_2)_3CH{=}CHSi(C_2H_5)_3/$ $CH_3CH_2CH_2CH{=}CHCH_2Si(C_2H_5)_3$ (40-93/7-45/trace-15)	771
$[RhCl(CO)_2]_2$ + NEt_3	80	2.5	$CH_3(CH_2)_5Si(C_2H_5)_3$ (62)	C -022
$[RhCl(CO)_2]_2$ + PPh_3	40	1 (DCE)	$CH_3(CH_2)_5Si(C_2H_5)_3/CH_3(CH_2)_3CH{=}CHSi(C_2H_5)_3/$ $CH_3CH_2CH_2CH{=}CHCH_2Si(C_2H_5)_3$ (97/2/trace)	771
polymer/$RhCl_3$	100	2	$CH_3(CH_2)_5Si(C_2H_5)_3$ (15)	1220
porolith-CN/$[RhCl(CO)_2]_2$	80	2.5	$CH_3(CH_2)_5Si(C_2H_5)_3$ (84)	Dd-003
porolith-CN/$[RhCl(CO)_2]_2$	80	2.5	$CH_3(CH_2)_5Si(C_2H_5)_3$ (84)	U -020
support/Rh complex	80	2	$CH_3(CH_2)_5Si(C_2H_5)_3$ (0-44)	204
zeolite-CN/$RhCl(CO)(PPh_3)_2$	80	2.5	$CH_3(CH_2)_5Si(C_2H_5)_3$ (63)	Dd-003
zeolite-CN/$RhCl(CO)(PPh_3)_2$	80	2.5	$CH_3(CH_2)_5Si(C_2H_5)_3$ (63)	U -020

C_6H_{12} $CH_3CH_2CH_2CH{=}CHCH_3$

Table 8.1

= $(C_2H_5)_3SiH$ = 　　　　=8=

	C-CN/$RhCl_3$ + c-$C_{12}H_{18}$	80	2	$CH_3(CH_2)_5Si(C_2H_5)_3$ (31)	C -030
	$Rh(acac)_3$	60	18	$CH_3(CH_2)_5Si(C_2H_5)_3$ (29)	274
	SDVB-CH_2Cl/$RhCl(CO)(PPh_3)_2$	80	2	$CH_3(CH_2)_5Si(C_2H_5)_3$ (12)	C -022
C_7H_{10}	Fig.1.				
	$[HPt(SiBzMe_2)P(c\text{-}Hex)_3]_2$	20	80	Fig.272. (-)	429
C_7H_{14}	$(CH_3)_3CCH_2CH{=}CH_2$ + $CH_3(CH_2)_4CH{=}CH_2$				
	SDVB-NMe_2/$[RhCl(CO)_2]_2$	50	2	$C_7H_{15}Si(C_2H_5)_3$ (-)	316
C_7H_{14}	$CH_3(CH_2)_4CH{=}CH_2$				
	$Co_2(CO)_8$	120	-	$CH_3(CH_2)_6Si(C_2H_5)_3$ (kin.)	1169
	$NiCl_2$ + $(Me_3SiCH_2)_3PO$	120	20	$CH_3(CH_2)_6Si(C_2H_5)_3$ (40)	S -048
	H_2PtCl_6 (THF)	100	4	$CH_3(CH_2)_6Si(C_2H_5)_3$ (-)	214
	SiO_2-$L_2PtOCOCOO$	30	24 (h)	$CH_3(CH_2)_6Si(C_2H_5)_3$ (15)	938
	$(C_5Me_5Rh)_2Cl_4$	40	4	$CH_3(CH_2)_6Si(C_2H_5)_3$/$CH_3(CH_2)_4CH{=}CHSi(C_2H_5)_3$ (34-94/0-52)	771
	$RhCl(PPh_3)_3$	94-95	-	$CH_3(CH_2)_6Si(C_2H_5)_3$ (-)	U -160
	$RhCl(PPh_3)_3$	reflux	-	$CH_3(CH_2)_6Si(C_2H_5)_3$/$CH_3(CH_2)_4CH{=}CHSi(C_2H_5)_3$/ $CH_3(CH_2)_3CH{=}CHCH_2Si(C_2H_5)_3$ (75/19/6)	879
	SDVB-NMe_2/$RhCl_3$	80	-	$CH_3(CH_2)_6Si(C_2H_5)_3$ (81)	C -006
	SDVB-PPh_2/$RhCl_3$ (EtOH)	90	-	$CH_3(CH_2)_6Si(C_2H_5)_3$ (82)	D -056
	$[RhCl(C_2H_4)_2]_2$	85	2 (b)	$CH_3(CH_2)_6Si(C_2H_5)_3$ (60)	496
C_7H_{14}	$CH_3(CH_2)_4CH{=}CH_2$ + $(CH_3)_3CCH_2CH{=}CH_2$				
	SDVB-NMe_2/$[RhCl(CO)_2]_2$	50	2	$C_7H_{15}Si(C_2H_5)_3$ (-)	316
C_7H_{14}	$CH_3(CH_2)_4CH{=}CH_2$ + $(CH_3)_3CCH{=}CH_2$				
	SDVB-NMe_2/$[RhCl(CO)_2]_2$	50	2	$C_6H_{13}Si(C_2H_5)_3$ +$C_7H_{15}Si(C_2H_5)_3$ (-)	316
C_8H_8	$C_6H_5CH{=}CH_2$				
	$(Et_4N)[Pt(SnCl_3)_3(cod)]$	r.t.	22h-10 days	$C_6H_5CH_2CH_2Si(C_2H_5)_3$ + $C_6H_5CH(CH_3)Si(C_2H_5)_3$ + $C_6H_5CH{=}CHSi(C_2H_5)_3$ (-)	10
	$(Me_4N)_3[Pt(SnCl_3)_5]$	r.t.	2h-5.5 days	$C_6H_5CH_2CH_2Si(C_2H_5)_3$ + $C_6H_5CH(CH_3)Si(C_2H_5)_3$ + $C_6H_5CH{=}CHSi(C_2H_5)_3$ (-)	10

Table 8.1
= $(C_2H_5)_3SiH$ = =9=

$(Ph_4P)_2PtCl_4$ + $Ag(CF_3SO_3)$	r.t.	0.6-3	$C_6H_5CH_2CH_2Si(C_2H_5)_3$ + $C_6H_5CH(CH_3)Si(C_2H_5)_3$ + $C_6H_5CH{=}CHSi(C_2H_5)_3$ (-)	10
$(n\text{-}Bu_4N)_2Pt_2Cl_6$	r.t.	1h-9 days	$C_6H_5CH_2CH_2Si(C_2H_5)_3$ + $C_6H_5CH(CH_3)Si(C_2H_5)_3$ + $C_6H_5CH{=}CHSi(C_2H_5)_3$ (-)	10
$(n\text{-}Bu_4N)_2[PtCl_2(SnCl_3)_2]$	r.t.	2h-5.5 days	$C_6H_5CH_2CH_2Si(C_2H_5)_3$ + $C_6H_5CH(CH_3)Si(C_2H_5)_3$ + $C_6H_5CH{=}CHSi(C_2H_5)_3$ (-)	10
$(n\text{-}Bu_4N)[PtCl_3(PhCH{=}CH_2)]$	r.t.	3.5	$C_6H_5CH_2CH_2Si(C_2H_5)_3$ + $C_6H_5CH(CH_3)Si(C_2H_5)_3$ + $C_6H_5CH{=}CHSi(C_2H_5)_3$ (-)	10
H_2PtCl_6	-	-	$C_6H_5CH_2CH_2Si(C_2H_5)_3$ + $C_6H_5CH(CH_3)Si(C_2H_5)_3$ (-)	514
H_2PtCl_6	20	-	$C_6H_5CH_2CH_2Si(C_2H_5)_3$ (57) $C_6H_5CH(CH_3)Si(C_2H_5)_3$ (3)	172
H_2PtCl_6	22-29	0.5	$C_6H_5CH_2CH_2Si(C_2H_5)_3$ (2)	U -168
H_2PtCl_6 + $HSiCl_3$	22-39	0.6	$C_6H_5CH_2CH_2Si(C_2H_5)_3$ + $C_6H_5CH(CH_3)Si(C_2H_5)_3$ (24)	U -168
K_2PtCl_4	r.t.	3 days	$C_6H_5CH_2CH_2Si(C_2H_5)_3$ + $C_6H_5CH(CH_3)Si(C_2H_5)_3$ + $C_6H_5CH{=}CHSi(C_2H_5)_3$ (-)	10
$K[PtCl_3(PhCH{=}CH_2)]$	r.t.	25-4	$C_6H_5CH_2CH_2Si(C_2H_5)_3$ + $C_6H_5CH(CH_3)Si(C_2H_5)_3$ + $C_6H_5CH{=}CHSi(C_2H_5)_3$ (-)	10
$Pt(PPh_3)_4$	15	- (solv.)	$C_6H_5CH_2CH_2Si(C_2H_5)_3$ (kin.)	977
$Pt(PhCH{=}CH_2)_3$	-	-	$C_6H_5CH_2CH_2Si(C_2H_5)_3$ (73-75) $C_6H_5CH(CH_3)Si(C_2H_5)_3$ (1-2) $C_6H_5CH{=}CHSi(C_2H_5)_3$ (11-13)	221
$PtBr_2$	r.t.	1.25-16	$C_6H_5CH_2CH_2Si(C_2H_5)_3$ + $C_6H_5CH(CH_3)Si(C_2H_5)_3$ + $C_6H_5CH{=}CHSi(C_2H_5)_3$ (-)	10
$PtCl_2$	r.t.	0.6-4	$C_6H_5CH_2CH_2Si(C_2H_5)_3$ + $C_6H_5CH(CH_3)Si(C_2H_5)_3$ + $C_6H_5CH{=}CHSi(C_2H_5)_3$ (-)	10
$PtCl_2$	r.t.	4	$C_6H_5CH_2CH_2Si(C_2H_5)_3$ (74) $C_6H_5CH(CH_3)Si(C_2H_5)_3$ (12) $C_6H_5CH{=}CHSi(C_2H_5)_3$ (11) $C_6H_5CH_2CH_3$ (11)	1105
$PtCl_2(CH_3CN)_2$	r.t.	16	$C_6H_5CH_2CH_2Si(C_2H_5)_3$ + $C_6H_5CH(CH_3)Si(C_2H_5)_3$ + $C_6H_5CH{=}CHSi(C_2H_5)_3$ (-)	10
$PtCl_2(PhCH{=}CH_2)$	15-20	24	$C_6H_5C_2H_4Si(C_2H_5)_3$ (kin.)	965
$PtCl_2(PhCH{=}CH_2)_2$	-	-	$C_6H_5CH_2CH_2Si(C_2H_5)_3$ (73-75) $C_6H_5CH(CH_3)Si(C_2H_5)_3$ (1-2) $C_6H_5CH{=}CHSi(C_2H_5)_3$ (11-13)	221
$PtCl_2(cod)$	r.t.	0.6-4	$C_6H_5CH_2CH_2Si(C_2H_5)_3$ + $C_6H_5CH(CH_3)Si(C_2H_5)_3$ + $C_6H_5CH{=}CHSi(C_2H_5)_3$ (-)	10

Table 8.1
= $(C_2H_5)_3SiH$ = =10=

$PtCl_3(CH_2{=}CHCH_2PPh_3)$	20	-	$C_6H_5CH_2CH_2Si(C_2H_5)_3$ (5)	172
PtI_2	r.t.	21	$C_6H_5CH_2CH_2Si(C_2H_5)_3$ + $C_6H_5CH(CH_3)Si(C_2H_5)_3$ + $C_6H_5CH{=}CHSi(C_2H_5)_3$ (-)	10
$[HPt(SiEt_3)P(c\text{-}Hex)_3]_2$	20	3 days	$C_6H_5CH_2CH_2Si(C_2H_5)_3$ (75)	429
$[PtCl_2(py)_2]$	-	6	$C_6H_5CH_2CH_2Si(C_2H_5)_3$ (30)	987
$[SiO_2\text{-}(CH_2)_3PPh_3]_2PtCl_6$	50	-	$C_6H_5CH_2CH_2Si(C_2H_5)_3$ (43) $C_6H_5CH(CH_3)Si(C_2H_5)_3$ (2)	172
cis-$PtCl_2(PhCH{=}CH_2)_2$	r.t.	0.6-4	$C_6H_5CH_2CH_2Si(C_2H_5)_3$ + $C_6H_5CH(CH_3)Si(C_2H_5)_3$ + $C_6H_5CH{=}CHSi(C_2H_5)_3$ (-)	10
cis-$PtCl_2(p\text{-}ClC_6H_4CH{=}CH_2)_2$	r.t.	2.5	$C_6H_5CH_2CH_2Si(C_2H_5)_3$ + $C_6H_5CH(CH_3)Si(C_2H_5)_3$ + $C_6H_5CH{=}CHSi(C_2H_5)_3$ (-)	10
$\{Pt(\mu\text{-}Cl)Cl[P(C_6H_4CH_3)_3]\}_2$	r.t.	22h-28 days	$C_6H_5CH_2CH_2Si(C_2H_5)_3$ + $C_6H_5CH(CH_3)Si(C_2H_5)_3$ + $C_6H_5CH{=}CHSi(C_2H_5)_3$ (-)	10
$(C_5Me_5Rh)_2Cl_4$	40	36	$C_6H_5CH_2CH_2Si(C_2H_5)_3$ / $C_6H_5CH{=}CHSi(C_2H_5)_3$ (25-64/33-50)	771
$Rh(acac)_3$	60	16	$C_6H_5CH_2CH_2Si(C_2H_5)_3$ (24) $C_6H_5CH{=}CHSi(C_2H_5)_3$ (75)	274
$Rh(acac)_3$	60	16	$C_6H_5CH_2CH_2Si(C_2H_5)_3$ (24) $C_6H_5CH{=}CHSi(C_2H_5)_3$ (75)	271
$Rh(acac)_3$ + $AlEt_2(OEt)$ + PPh_3	60	5	$C_6H_5CH_2CH_2Si(C_2H_5)_3$ (44)	271
$Rh(acac)_3$ + $AlEt_2(OEt)$ + bipy	60	5	$C_6H_5CH_2CH_2Si(C_2H_5)_3$ (2) $C_6H_5CH{=}CHSi(C_2H_5)_3$ (5)	271
$Rh(acac)_3$ + $AlEt_2(OEt)$ + diphos	60	5	$C_6H_5CH_2CH_2Si(C_2H_5)_3$ (70) $C_6H_5CH{=}CHSi(C_2H_5)_3$ (16)	271
$Rh(acac)_3$ + $AlEt_3$	60	-	$C_6H_5CH_2CH_2Si(C_2H_5)_3$ (22) $C_6H_5CH(CH_3)Si(C_2H_5)_3$ (trace) $C_6H_5CH{=}CHSi(C_2H_5)_3$ (50)	271
$Rh(acac)_3$ + $AlEt_3$	60-160	5	$C_6H_5CH_2CH_2Si(C_2H_5)_3$ (12-26) $C_6H_5CH{=}CHSi(C_2H_5)_3$ (72-86)	271
$RhCl(PPh_3)_3$	50	3-17	$C_6H_5CH_2CH_2Si(C_2H_5)_3$ (25-34) $C_6H_5CH(CH_3)Si(C_2H_5)_3$ (2-47) $C_6H_5CH{=}CHSi(C_2H_5)_3$ (2-43)	878
$RhH(CO)(PPh_3)_3$	50	4.5-16 (b)	$C_6H_5CH_2CH_2Si(C_2H_5)_3$ (4-27) $C_6H_5CH(CH_3)Si(C_2H_5)_3$ (0-51) $C_6H_5CH{=}CHSi(C_2H_5)_3$ (1-44)	878
$[Rh(dmg)_2PPh_3]_2$	90	3	$C_6H_5CH_2CH_2Si(C_2H_5)_3$ + $C_6H_5CH{=}CHSi(C_2H_5)_3$ (50)	1048

Table 8.1

= $(C_2H_5)_3SiH$ = =11=

	$[Rh(dmg)_2PPh_3]_2$	90	3	$C_6H_5C_2H_4Si(C_2H_5)_3$ (45-55)	886
	$[Rh(dmg)_2PPh_3]_2$	90	3	$C_6H_5C_2H_4Si(C_2H_5)_3$ (45-55)	S -003
C_8H_{16}	$CH_3(CH_2)_4CH{=}CHCH_3$-cis				
	$Rh(acac)_3$	60	20	$CH_3(CH_2)_7Si(C_2H_5)_3$ (74)	274
	$Rh(acac)_3$ + $AlEt_3$	60	12	$CH_3(CH_2)_7Si(C_2H_5)_3$ (24)	271
C_8H_{16}	$CH_3(CH_2)_5CH{=}CH_2$				
	$(t\text{-}BuO)_2$	110-170	0.5-360	$CH_3(CH_2)_7Si(C_2H_5)_3$ (10-65)	347
	$(t\text{-}BuO)_2$	reflux	24	$CH_3(CH_2)_7Si(C_2H_5)_3$ (91)	347
	$Co_2(CO)_8$	0-60	until 24	$CH_3(CH_2)_7Si(C_2H_5)_3$ (-)	470
	$Co_2(CO)_8$	25	65	$CH_3(CH_2)_7Si(C_2H_5)_3$ (-)	225
	$Co_2(CO)_8$	r.t.	60 (t)	$CH_3(CH_2)_7Si(C_2H_5)_3$ (100)	U -019
	$Co_2(CO)_8$	r.t.	60 (t)	$CH_3(CH_2)_7Si(C_2H_5)_3$ (100)	D -009
	$(CH_2)_3SiMe_2Fe(CO)_4$	reflux	5-7 (b)	$CH_3(CH_2)_7Si(C_2H_5)_3$ (-)	299
	$Fe(CO)_5$	reflux	5-7 (b)	$CH_3(CH_2)_7Si(C_2H_5)_3$ (-)	299
	$[IrCl(c\text{-}C_8H_{14})_2]_2$ + $P(C_6H_4OMe\text{-}o)_3$	100	6	$CH_3(CH_2)_7Si(C_2H_5)_3$ (25)	59
	$[IrCl(c\text{-}C_8H_{14})_2]_2$ + PPh_3	100	6	$CH_3(CH_2)_7Si(C_2H_5)_3$ (23)	59
	$[IrCl(c\text{-}C_8H_{14})_2]_2$ + PPh_3	60	6	$CH_3(CH_2)_7Si(C_2H_5)_3$ (30)	59
	Pd + PPh_3	120	5	$CH_3(CH_2)_7Si(C_2H_5)_3$ (27)	D -027
	$Pd(PPh_3)_4$	120	5	$CH_3(CH_2)_7Si(C_2H_5)_3$ (30)	D -027
	H_2PtCl_6 (EtOH + EtONa)	130	-	$CH_3(CH_2)_7Si(C_2H_5)_3$ (48)	D -009
	H_2PtCl_6 (EtOH)	r.t.-130	16	$CH_3(CH_2)_7Si(C_2H_5)_3$ (48)	U -019
	$H_2PtCl_6 \cdot 6H_2O$	85	0.25 (t)	$CH_3(CH_2)_7Si(C_2H_5)_3$ (-)	876
	SDVB-NMe_2/H_2PtCl_6 + PPh_3	90	4	$CH_3(CH_2)_7Si(C_2H_5)_3$ (59)	C -002
	$Me_2NCH{=}RhHCl_2(PPh_3)_2$	100	8	$CH_3(CH_2)_7Si(C_2H_5)_3$ (61)	496
	$N(C_6H_4Me\text{-}p)CH_2CH_2N(C_6H_4Me\text{-}p)C{=}$ $RhCl(CO)(PPh_3)$	100	8	$CH_3(CH_2)_7Si(C_2H_5)_3$ (96-98)	496
	$NPhCH_2CH_2NPhC{=}RhCl(cod)$	100	8	$CH_3(CH_2)_7Si(C_2H_5)_3$ (73)	496
	$Rh(acac)_3$	60	17	$CH_3(CH_2)_7Si(C_2H_5)_3$ (93)	274

Table 8.1
= $(C_2H_5)_3SiH$ = =12=

$Rh(acac)_3$ + $AlEt_3$ or $Al(i\text{-}Bu)_3$	60	18	$CH_3(CH_2)_7Si(C_2H_5)_3$ + $CH_3(CH_2)_5CH(CH_3)Si(C_2H_5)_3$ + $CH_3(CH_2)_4CH(CH_2CH_3)Si(C_2H_5)_3$ + $CH_3(CH_2)_3CH(CH_2CH_2CH_3)Si(C_2H_5)_3$ (82-94)	271
$Rh_2(OAc)_4$	100	8	$CH_3(CH_2)_7Si(C_2H_5)_3$ (71)	274
$RhCl(CO)(PPh_3)_2$	20	6 (b)	$CH_3(CH_2)_7Si(C_2H_5)_3$ (15)	202
$RhCl(CO)(PPh_3)_2$ + m-ClC_6H_4COOH	20	8 (b)	$CH_3(CH_2)_7Si(C_2H_5)_3$ (44)	202
$RhCl(CO)(PPh_3)_2$ + t-BuOOH	20	6 (b)	$CH_3(CH_2)_7Si(C_2H_5)_3$ (52)	202
$RhCl(PPh_3)_3$	100	0.33 (t)	$CH_3(CH_2)_7Si(C_2H_5)_3$ (-)	876
$RhCl(PPh_3)_3$	61-62	-	$CH_3(CH_2)_7Si(C_2H_5)_3$ (-)	U -160
$RhCl(PPh_3)_3$	reflux	-	$CH_3(CH_2)_7Si(C_2H_5)_3$/$CH_3(CH_2)_5CH{=}CHSi(C_2H_5)_3$/ $CH_3(CH_2)_4CH{=}CHCH_2Si(C_2H_5)_3$ (79/14/7)	879
$RhCl(PPh_3)_3$ + Cum-OOH	20	8 (b)	$CH_3(CH_2)_7Si(C_2H_5)_3$ (44)	202
$RhCl(PPh_3)_3$ + H_2O_2	20	8 (b)	$CH_3(CH_2)_7Si(C_2H_5)_3$ (52)	202
$RhCl(PPh_3)_3$ + PhCOOOBu-t	20	8 (b)	$CH_3(CH_2)_7Si(C_2H_5)_3$ (30)	202
$RhCl(PPh_3)_3$ + m-ClC_6H_4COOH	20	8 (b)	$CH_3(CH_2)_7Si(C_2H_5)_3$ (32)	202
$RhCl_2[P(C_6H_4Me\text{-}o)_3]_2$	100	8	$CH_3(CH_2)_7Si(C_2H_5)_3$ (100)	503
$RhCl_2[P(c\text{-}Hex)_3]_2$	100	8	$CH_3(CH_2)_7Si(C_2H_5)_3$ (100)	503
$RhH(PPh_3)_4$	-	-	$CH_3(CH_2)_7Si(C_2H_5)_3$ (-)	J -019
SiO_2-$Rh(C_3H_5)_2$	100	2	$C_8H_{17}Si(C_2H_5)_3$ (50)	230
$[RhCl(cod)_2]_2$	20-60	-	$CH_3(CH_2)_7Si(C_2H_5)_3$ (63-73)	884
$[RhCl(cod)_2]_2$ + XNC	20-100	-	$CH_3(CH_2)_7Si(C_2H_5)_3$ (61-82)	884
$[RhCl(cod)]_2$ + XNC	20-100	24	$CH_3(CH_2)_7Si(C_2H_5)_3$ (0-82)	1
$\{RhCl[P(OC_6H_4Me)_3]_2\}_n$	100	2	$C_8H_7Si(C_2H_5)_3$ (53)	229
$\{RhCl[P(OC_6H_4Me)_3]_2\}_n$	25	18	$C_8H_7Si(C_2H_5)_3$ (34)	229
$\{RhCl[P(OMe)_3]_2\}_n$	100	2	$C_8H_7Si(C_2H_5)_3$ (38)	229
$\{RhCl[P(OMe)_3]_2\}_n$	25	18	$C_8H_7Si(C_2H_5)_3$ (48)	229
$\{RhH[P(OC_6H_4Me)_3]_2\}_n$	100	2	$C_8H_7Si(C_2H_5)_3$ (44)	229
$\{RhH[P(OC_6H_4Me)_3]_2\}_n$	25	18	$C_8H_7Si(C_2H_5)_3$ (48)	229
$\{RhH[P(OMe)_3]_2\}_3$	100	2	$C_8H_7Si(C_2H_5)_3$ (27)	229

	$\{RhH[P(OMe)_3]_2\}_3$	25	18	$C_8H_7Si(C_2H_5)_3$ (53)	229
C_9H_{10}	$C_6H_5C(CH_3)=CH_2$				
	$PtCl_2$	r.t.	96	$C_6H_5CH(CH_3)CH_2Si(C_2H_5)_3$ (4)	1105
C_9H_{10}	$C_6H_5CH_2CH=CH_2$				
	H_2PtCl_6	-	-	$C_6H_5(CH_2)_3Si(C_2H_5)_3$ + $C_6H_5CH_2CH(CH_3)Si(C_2H_5)_3$(-)	516
	$Rh(acac)_3$	60	18	$C_6H_5(CH_2)_3Si(C_2H_5)_3$ (46) $C_6H_5CH_2CH(CH_3)Si(C_2H_5)_3$ (2) $C_6H_5CH_2CH=CHSi(C_2H_5)_3$ (12)	274
	$Rh(acac)_3$	60	-	$C_6H_5(CH_2)_3Si(C_2H_5)_3$ (42) $C_6H_5CH_2CH(CH_3)Si(C_2H_5)_3$ (3) $C_6H_5CH_2CH=CHSi(C_2H_5)_3$ (15)	271
	$Rh(acac)_3$ + $AlEt_3$	60	-	$C_6H_5(CH_2)_3Si(C_2H_5)_3$ (46) $C_6H_5CH_2CH(CH_3)Si(C_2H_5)_3$ (2) $C_6H_5CH_2CH=CHSi(C_2H_5)_3$ (12)	271
C_9H_{18}	$CH_3(CH_2)_6CH=CH_2$				
	H_2PtCl_6	-	-	$CH_3(CH_2)_8Si(C_2H_5)_3$ (-)	515
	$(C_5Me_5Rh)_2Cl_4$	40	4	$CH_3(CH_2)_8Si(C_2H_5)_3$/$CH_3(CH_2)_6CH=CHSi(C_2H_5)_3$/ $CH_3(CH_2)_5CH=CHCH_2Si(C_2H_5)_3$ (34/46/20)	771
$C_{10}H_{12}$	$C_6H_5CH_2CH_2CH=CH_2$				
	H_2PtCl_6 (THF)	100	1	$C_6H_5(CH_2)_4Si(C_2H_5)_3$ (26) $C_6H_5CH_2CH_2CH(CH_3)Si(C_2H_5)_3$ (13)	1234
	H_2PtCl_6 (THF)	100	1	$C_6H_5(CH_2)_4Si(C_2H_5)_3$+$C_6H_5CH_2CH_2CH(CH_3)Si(C_2H_5)_3$ (39)	S -041
	H_2PtCl_6 (THF)	100	1	$C_6H_5(CH_2)_4Si(C_2H_5)_3$+$C_6H_5CH_2CH_2CH(CH_3)Si(C_2H_5)_3$ (39)	1262
	H_2PtCl_6 (THF) + $AlCl_3$	100	1	$C_6H_5(CH_2)_4Si(C_2H_5)_3$ (42) $C_6H_5CH_2CH_2CH(CH_3)Si(C_2H_5)_3$ (33)	1234
	H_2PtCl_6 (THF) + $AlCl_3$	100	1	$C_6H_5(CH_2)_4Si(C_2H_5)_3$+$C_6H_5CH_2CH_2CH(CH_3)Si(C_2H_5)_3$ (75)	S -041
	H_2PtCl_6 (THF) + $AlCl_3$	100	1	$C_6H_5(CH_2)_4Si(C_2H_5)_3$+$C_6H_5CH_2CH_2CH(CH_3)Si(C_2H_5)_3$ (75)	1262
$C_{10}H_{20}$	$CH_3(CH_2)_7CH=CH_2$				
	$RhCl(PPh_3)_3$	80-90	0.33-1 (t)	$CH_3(CH_2)_9Si(C_2H_5)_3$ + cis-,trans-$CH_3(CH_2)_7CH=CHSi(C_2H_5)_3$ + cis-,trans-$CH_3(CH_2)_6CH=CHCH_2Si(C_2H_5)_3$ (55-70)	879

	$RhCl(PPh_3)_3$	84-89	0.5	$CH_3(CH_2)_9Si(C_2H_5)_3$ + cis-,trans-$CH_3(CH_2)_7CH{=}CHSi(C_2H_5)_3$ + cis-,trans-$CH_3(CH_2)_6CH{=}CHCH_2Si(C_2H_5)_3$ (74)	879
	$RhCl(PPh_3)_3$	86-110	0.17-22	$CH_3(CH_2)_9Si(C_2H_5)_3$ + cis-,trans-$CH_3(CH_2)_7CH{=}CHSi(C_2H_5)_3$ + cis-,trans-$CH_3(CH_2)_6CH{=}CHCH_2Si(C_2H_5)_3$ (71-95)	879
$C_{12}H_{18}$	Fig.156.				
	H_2PtCl_6 (i-PrOH)	150	6	Fig.273. (62)	708
$C_{14}H_{28}$	$CH_3(CH_2)_{11}CH{=}CH_2$				
	$RhCl(PPh_3)_3$	90	6	$C_{14}H_{27}Si(C_2H_5)_3$ + $CH_3(CH_2)_{13}Si(C_2H_5)_3$ (-)	U -160

Table 8.2
= $(C_2H_5)_3SiH$ =

C_3H_4	$CH_2{=}C{=}CH_2$				
	$Pd(\overline{COCH{=}CHCOO})(PPh_3)_2$	120	6 (THF)	$CH_2{=}CHCH_2Si(C_2H_5)_3$ + $CH_2{=}C(CH_3)Si(C_2H_5)_3$ (48)	289
C_4H_6	$CH_2{=}CHCH{=}CH_2$				
	$Ni(cod)_2$	120	1.5	$CH_3CH{=}CHCH_2Si(C_2H_5)_3$ (70) $CH_3CH{=}CHCH_2CH_2CH{=}CHCH_2Si(C_2H_5)_3$ (26)	205
	Pd + PPh_3	150	6	$CH_3CH{=}CHCH_2Si(C_2H_5)_3$ (80)	D -027
	$Pd(\overline{COCH{=}CHCOO})(PPh_3)_2$	-	-	$CH_3CH{=}CHCH_2CH_2CH{=}CHCH_2Si(C_2H_5)_3$ (84)	1176
	$Pd(\overline{COCH{=}CHCOO})(PPh_3)_2$	-	-	$CH_3CH{=}CHCH_2CH_2CH{=}CHCH_2Si(C_2H_5)_3$ (-)	J -004
	$Pd(PPh_3)_4$	120	6	$CH_3CH{=}CHCH_2Si(C_2H_5)_3$ (1) $CH_3CH{=}CHCH_2CH_2CH{=}CHCH_2Si(C_2H_5)_3$ (95)	1208
	$RhCl[PH(C_6H_4Me\text{-}p)_2]_3$	60	2	$CH_3CH{=}CHCH_2Si(C_2H_5)_3$ (-)	C -004
C_5H_8	$CH_2{=}C(CH_3)CH{=}CH_2$				
	$Fe(acac)_3$ + $AlEt_3$	-	2	$(C_2H_5)_3SiCH_2C(CH_3){=}CHCH_3$ (41)	662
	Ni complex	110	8 (t)	$(C_2H_5)_3SiCH_2C(CH_3){=}CHCH_3$ (trace)	275
	$Ni(CO)_2(PPh_3)_2$	150	3	$(CH_3)_2C{=}CHCH_2Si(C_2H_5)_3$ (27) $(C_2H_5)_3SiCH_2C(CH_3){=}CHCH_3$ (73)	1222
	$Ni(PPh_3)_4$	150	3	$(CH_3)_2C{=}CHCH_2Si(C_2H_5)_3$ (29) $(C_2H_5)_3SiCH_2C(CH_3){=}CHCH_3$ (71)	1222
	$Ni(acac)_2$ + $AlEt_3$	-	2	$(C_2H_5)_3SiCH_2C(CH_3){=}CHCH_2$ (97)	663
	$Ni(acac)_2$ + $AlEt_3$	20	6 (THF)	$(C_2H_5)_3SiCH_2C(CH_3){=}CHCH_3$ (95)	G -030
	$Ni(acac)_2$ + $AlEt_3$	20	2	$(C_2H_5)_3SiCH_2C(CH_3){=}CHCH_3$ (97)	G -030
	$Ni(acac)_2$ + $AlEt_3$ or $LiAlH_4$	20	4(solv.)	$(C_2H_5)_3SiCH_2C(CH_3){=}CHCH_3$ (87-97)	663
	$Ni(cod)_2$	150	3	$(CH_3)_2C{=}CHCH_2Si(C_2H_5)_3$ (18) $(C_2H_5)_3SiCH_2C(CH_3){=}CHCH_3$ (66)	1222
	$Ni(cod)_2$ + PR_3	150	3	$(CH_3)_2C{=}CHCH_2Si(C_2H_5)_3$ (11-36) $(C_2H_5)_3SiCH_2C(CH_3){=}CHCH_3$ (47-69)	1222
	$NiCl_2$ + $AlEt_3$	reflux	6 (THF)	$(C_2H_5)_3SiCH_2CH(CH_3)CH{=}CH_2$ (95)	663
	$NiCl_2(PPh_3)_2$	150	3	$(CH_3)_2C{=}CHCH_2Si(C_2H_5)_3$ (14) $(C_2H_5)_3SiCH_2C(CH_3){=}CHCH_3$ (43)	1222
	$PdCl_2(PPh_3)_2$	100	6	$(CH_3)_2C{=}CHCH_2Si(C_2H_5)_3$ (trace) $(C_2H_5)_3SiCH_2C(CH_3){=}CHCH_3$ (trace)	1222

Table 8.2

= $(C_2H_5)_3SiH$ = =2=

	$PdCl_2(PhCN)_2$	100	6	$(CH_3)_2C=CHCH_2Si(C_2H_5)_3$ (11) $(C_2H_5)_3SiCH_2C(CH_3)=CHCH_3$ (6)	1222
	H_2PtCl_6 (i-PrOH)	150	3	$(CH_3)_2C=CHCH_2Si(C_2H_5)_3$ (20) $(C_2H_5)_3SiCH_2C(CH_3)=CHCH_3$ (80)	1222
	$Rh(acac)_3$	60	15	$CH_2=C(CH_3)CH_2CH_2Si(C_2H_5)_3$ (5) $(C_2H_5)_3SiCH_2C(CH_3)=CHCH_2Si(C_2H_5)_3$ (94)	274
	$Rh(acac)_3 + AlEt_3$	60	12	$(CH_3)_2C=CHCH_2Si(C_2H_5)_3$ (44) $CH_2=C(CH_3)CH_2CH_2Si(C_2H_5)_3$ + $(C_2H_5)_3SiCH_2CH(CH_3)CH=CH_2$ (41)	271
	$Rh(acac)_3 + AlEt_3$	60	6 (solv.)	$(C_2H_5)_3SiC_5H_9$ (83-99)	271
	RhCl(CO)(1,4-diaza-1,3-diene)	r.t.	1 day(THF)	$CH_3CH=C(CH_3)CH_2Si(C_2H_5)_3$ (85)	167
	$RhCl(CO)(1\text{-aza-}1,3\text{-diene})_2$	r.t.	1 day(THF)	$CH_3CH=C(CH_3)CH_2Si(C_2H_5)_3$ (86)	167
	RhCl(cod)(1,4-diaza-1,3-diene)	r.t.	1 day(THF)	$CH_3CH=C(CH_3)CH_2Si(C_2H_5)_3$ (75)	167
	RuHCl(cod)(1,4-diaza-1,3-diene)	r.t.	150 (dcm)	$CH_3CH=C(CH_3)CH_2Si(C_2H_5)_3$ (27-40) $(CH_3)_2C=CHCH_2Si(C_2H_5)_3$ (51-82) $CH_2=C(CH_3)CH_2CH_2Si(C_2H_5)_3$ (7-18) $CH_2=CHCH(CH_3)CH_2Si(C_2H_5)_3$ (9-13)	167
C_5H_8	$CH_3CH=CHCH=CH_2$				
	$Fe(acac)_3 + AlEt_3$	20	2	$CH_3CH_2CH=CHCH_2Si(C_2H_5)_3$ (41)	663
	$Ni(acac)_3 + AlEt_3$	-	3	$CH_3CH_2CH=CHCH_2Si(C_2H_5)_3$ (99)	662
	$Ni(acac)_3 + AlEt_3$	20	3	$CH_3CH_2CH=CHCH_2Si(C_2H_5)_3$ (99)	663
	$Ni(acac)_3 + AlEt_3$	20	3	$CH_3CH_2CH=CHCH_2Si(C_2H_5)_3$ (99)	G -030
C_6H_8	$\overline{CH_2CH_2CH=CHCH=CH}$				
	$Rh(acac)_3 + AlEt_3$	60	15	$(\overline{CH_2)_3CH=CHC}HSi(C_2H_5)_3$ (64)	271
C_6H_{10}	$CH_2=C(CH_3)C(CH_3)=CH_2$				
	$Cr(CO)_5(DMPIC)$	100	24	$(CH_3)_2C=C(CH_3)CH_2Si(C_2H_5)_3$ + $CH_2=C(CH_3)CH(CH_3)CH_2Si(C_2H_5)_3$ (3)	884
	$Cr(CO)_5CHNMe_2$	-	-	$(CH_3)_2C=C(CH_3)CH_2Si(C_2H_5)_3$ (-)	497
	$Cr(CO)_6$	100	24	$(CH_3)_2C=C(CH_3)CH_2Si(C_2H_5)_3$ + $CH_2=C(CH_3)CH(CH_3)CH_2Si(C_2H_5)_3$ (41)	202
	$Cr(CO)_6$	100	24	$(CH_3)_2C=C(CH_3)CH_2Si(C_2H_5)_3$ + $CH_2=C(CH_3)CH(CH_3)CH_2Si(C_2H_5))_3$ (41)	884
	$Cr(CO)_6$	100	24	$CH_2=C(CH_3)CH(CH_3)CH_2Si(C_2H_5)_3$ (6) $(CH_3)_2C=C(CH_3)CH_2Si(C_2H_5)_3$ (35)	1

Table 8.2
= $(C_2H_5)_3SiH$ = =3=

$Cr(CO)_6$ + $(PhCOO)_2$	100	24	$(CH_3)_2C=C(CH_3)CH_2Si(C_2H_5)_3$ + $CH_2=C(CH_3)CH(CH_3)CH_2Si(C_2H_5)_3$ (7)	202
$Cr(CO)_6$+ (t-BuOOH)	100	24	$(CH_3)_2C=C(CH_3)CH_2Si(C_2H_5)_3$ + $CH_2=C(CH_3)CH(CH_3)CH_2Si(C_2H_5)_3$ (100)	202
$Cr(XNC)(CO)_5$	100	24	$CH_2=C(CH_3)CH(CH_3)CH_2Si(C_2H_5)_3$ + $(CH_3)_2C=C(CH_3)CH_2Si(C_2H_5)_3$ (3)	1
$h\nu$ + $Fe(CO)_5$	-	241 (p)	$(CH_3)_2CHC(CH_3)=CHSi(C_2H_5)_3$ + $C_6H_{11}Si(C_2H_5)_3$ (47)	366
$h\nu$ + $Fe(CO)_5$	-	241 (p)	$(CH_3)_2CHC(CH_3)=CHSi(C_2H_5)_3$ + $C_6H_{11}Si(C_2H_5)_3$ (67)	366
$[IrCl(c\text{-}C_8H_{14})_2]_2$	60	6	$(CH_3)_2C=C(CH_3)CH_2Si(C_2H_5)_3$ (88) $CH_2=C(CH_3)CH(CH_3)CH_2Si(C_2H_5)_3$ (7)	59
$[IrCl(c\text{-}C_8H_{14})_2]_2$ + PPh_3	60	6	$(CH_3)_2C=C(CH_3)CH_2Si(C_2H_5)_3$ (60) $CH_2=C(CH_3)CH(CH_3)CH_2Si(C_2H_5)_3$ (5)	59
$Mo(CO)_4(DMPIC)_2$	100	24	$(CH_3)_2C=C(CH_3)CH_2Si(C_2H_5)_3$ + $CH_2=C(CH_3)CH(CH_3)CH_2Si(C_2H_5)_3$ (29)	884
$Mo(CO)_5(DMPIC)$	100	24	$(CH_3)_2C=C(CH_3)CH_2Si(C_2H_5)_3$ + $CH_2=C(CH_3)CH(CH_3)CH_2Si(C_2H_5)_3$ (63)	884
$Mo(CO)_6$	100	24	$(CH_3)_2C=C(CH_3)CH_2Si(C_2H_5)_3$ + $CH_2=C(CH_3)CH(CH_3)CH_2Si(C_2H_5)_3$ (96)	202
$Mo(CO)_6$	100	24	$(CH_3)_2C=C(CH_3)CH_2Si(C_2H_5)_3$ + $CH_2=C(CH_3)CH(CH_3)CH_2Si(C_2H_5)_3$ (96)	884
$Mo(CO)_6$	100	24	$CH_2=C(CH_3)CH(CH_3)CH_2Si(C_2H_5)_3$ (9) $(CH_3)_2C=C(CH_3)CH_2Si(C_2H_5)_3$ (87)	1
$Mo(CO)_6$+ (t-BuOOH)	100	24	$(CH_3)_2C=C(CH_3)CH_2Si(C_2H_5)_3$ + $CH_2=C(CH_3)CH(CH_3)CH_2Si(C_2H_5)_3$ (98)	202
$Mo(XNC)_2(CO)_4$	100	24	$CH_2=C(CH_3)CH(CH_3)CH_2Si(C_2H_5)_3$ (2) $(CH_3)_2C=C(CH_3)CH_2Si(C_2H_5)_3$ (27)	1
cis-$Mo(XNC)(CO)_5$	100	24	$CH_2=C(CH_3)CH(CH_3)CH_2Si(C_2H_5)_3$ (1) $(CH_3)_2C=C(CH_3)CH_2Si(C_2H_5)_3$ (62)	1
$Ni(1,5\text{-}C_8H_{12})_2$	150	3	$(CH_3)_2C=C(CH_3)CH_2Si(C_2H_5)_3$ (96) $CH_2=C(CH_3)CH(CH_3)CH_2Si(C_2H_5)_3$ (1)	1222
$Ni(1,5\text{-}C_8H_{12})_2$ + PR_3	150	3	$(CH_3)_2C=C(CH_3)CH_2Si(C_2H_5)_3$ (42-99) $CH_2=C(CH_3)CH(CH_3)CH_2Si(C_2H_5)_3$ (1-2)	1222
$Ni(CO)_2(PPh_3)_2$	150	2	$(CH_3)_2C=C(CH_3)CH_2Si(C_2H_5)_3$ (70) $CH_2=C(CH_3)CH(CH_3)CH_2Si(C_2H_5)_3$ (0)	1222

Table 8.2

= $(C_2H_5)_3SiH$ = =4=

$NiCl_2(PPh_3)_2$	150	3	$(CH_3)_2C{=}C(CH_3)CH_2Si(C_2H_5)_3$ (67) $CH_2{=}C(CH_3)CH(CH_3)CH_2Si(C_2H_5)_3$ (2)	1222
SiO_2-$Ni(C_3H_5)$	100	6	$(CH_3)_2C{=}C(CH_3)CH_2Si(C_2H_5)_3$ (60)	230
$PdCl_2(PPh_3)_2$	100	6	$(CH_3)_2C{=}C(CH_3)CH_2Si(C_2H_5)_3$ + $CH_2{=}C(CH_3)CH(CH_3)CH_2Si(C_2H_5)_3$ (-)	1222
$PdCl_2(PhCN)_2$	100	6	$(CH_3)_2C{=}C(CH_3)CH_2Si(C_2H_5)_3$ (14) $CH_2{=}C(CH_3)CH(CH_3)CH_2Si(C_2H_5)_3$ (4)	1222
H_2PtCl_6 (i-PrOH)	150	3	$(CH_3)_2C{=}C(CH_3)CH_2Si(C_2H_5)_3$ (87) $CH_2{=}C(CH_3)CH(CH_3)CH_2Si(C_2H_5)_3$ (13)	1222
SiO_2-$Pt(C_3H_5)$	60	2	$(CH_3)_2C{=}C(CH_3)CH_2Si(C_2H_5)_3$ (100)	230
$\overline{N(Ph)CH_2CH_2N(Ph)C}{=}RhCl(cod)$	100	8	$(CH_3)_2C{=}C(CH_3)CH_2Si(C_2H_5)_3$ (86) $CH_2{=}C(CH_3)CH(CH_3)CH_2Si(C_2H_5)_3$ (12)	496
$Rh(acac)_3$	60	12	$(CH_3)_2C{=}C(CH_3)CH_2Si(C_2H_5)_3$ (85)	274
$Rh(acac)_3$ + $AlClEt_2$	60	5	$C_6H_{11}Si(C_2H_5)_3$ (59)	271
$Rh(acac)_3$ + $AlEt_2(OEt)$	60	5	$C_6H_{11}Si(C_2H_5)_3$ (59)	271
$Rh(acac)_3$ + $AlEt_3$	60	6 (solv.)	$CH_2{=}C(CH_3)CH(CH_3)CH_2Si(C_2H_5)_3$ (52-56)	271
$Rh(acac)_3$ + $AlEt_3$	60	12	$(CH_3)_2C{=}C(CH_3)CH_2Si(C_2H_5)_3$ (54)	271
$Rh(acac)_3$ + $AlEt_3$	60	5	$C_6H_{11}Si(C_2H_5)_3$ (62)	271
$Rh(acac)_3$ + $LiAlH_4$	60	5	$C_6H_{11}Si(C_2H_5)_3$ (16)	271
$Rh(acac)_3$ + $LiAlH_4$	60	5 (e)	$C_6H_{11}Si(C_2H_5)_3$ (38)	271
$Rh(acac)_3$ + $NaBH_4$	60	5 (e)	$C_6H_{11}Si(C_2H_5)_3$ (29)	271
$Rh(acac)_3$ + $Na[AlH_2(OCH_2CH_2OMe)_2]$	60	5	$C_6H_{11}Si(C_2H_5)_3$ (64)	271
$Rh_2(OAc)_4$	100	8	$(CH_3)_2C{=}C(CH_3)CH_2Si(C_2H_5)_3$ (63) $CH_2{=}C(CH_3)CH(CH_3)CH_2Si(C_2H_5)_3$ (9)	274
$RhCl(CO)(PPh_3)_2$	20	6 (b)	$(CH_3)_2C{=}C(CH_3)CH_2Si(C_2H_5)_3$ + $CH_2{=}C(CH_3)CH(CH_3)CH_2Si(C_2H_5)_3$ (18)	202
$RhCl(CO)(PPh_3)_2$ + (t-BuOOH)	20	6 (b)	$(CH_3)_2C{=}C(CH_3)CH_2Si(C_2H_5)_3$ + $CH_2{=}C(CH_3)CH(CH_3)CH_2Si(C_2H_5)_3$ (77)	202
$RhCl(cod)_2$ + DIPPC	20	6	$C_6H_{11}Si(C_2H_5)_3$ (94)	884
$RhCl(cod)_2$ + DMPIC	60	6	$C_6H_{11}Si(C_2H_5)_3$ (100)	884
$RhCl_2[P(c\text{-}Hex)_3]_2$	100	8	$(CH_3)_2C{=}C(CH_3)CH_2Si(C_2H_5)_3$ (28) $CH_2{=}C(CH_3)CH(CH_3)CH_2Si(C_2H_5)_3$ (64)	503

	$RhCl_2[P(o\text{-}C_6H_4Me)_3]_2$	100	8	$(CH_3)_2C=C(CH_3)CH_2Si(C_2H_5)_3$ (17) $CH_2=C(CH_3)CH(CH_3)CH_2Si(C_2H_5)_3$ (48)	503
	$[RhCl(cod)]_2$ + XNC	20-60	6	$C_6H_{11}Si(C_2H_5)_3$ (94-100)	1
	$\{RhCl[P(OC_6H_4Me)_3]_2\}_n$	100	2	$CH_2=C(CH_3)CH(CH_3)CH_2Si(C_2H_5)_3$ / $CH_3C(CH_3)=C(CH_3)CH_2Si(C_2H_5)_3$ (17/83)	229
	$\{RhCl[P(OC_6H_4Me)_3]_2\}_n$	25	18	$CH_2=C(CH_3)CH(CH_3)CH_2Si(C_2H_5)_3$ / $CH_3C(CH_3)=C(CH_3)CH_2Si(C_2H_5)_3$ (10/90)	229
	$\{RhCl[P(OMe)_3]_2\}_n$	100	2	$CH_2=C(CH_3)CH(CH_3)CH_2Si(C_2H_5)_3$ / $CH_3C(CH_3)=C(CH_3)CH_2Si(C_2H_5)_3$ (40/60)	229
	$\{RhCl[P(OMe)_3]_2\}_n$	25	18	$CH_2=C(CH_3)CH(CH_3)CH_2Si(C_2H_5)_3$ / $CH_3C(CH_3)=C(CH_3)CH_2Si(C_2H_5)_3$ (30/70)	229
	$\{RhH[P(OC_6H_4Me)_3]_2\}_n$	25	18	$CH_2=C(CH_3)CH(CH_3)CH_2Si(C_2H_5)_3$ / $CH_3C(CH_3)=C(CH_3)CH_2Si(C_2H_5)_3$ (17/83)	229
	$\{RhH[P(OC_6H_4Me)_3]_2\}_n$	25	18	$CH_2=C(CH_3)CH(CH_3)CH_2Si(C_2H_5)_3$ / $CH_3C(CH_3)=C(CH_3)CH_2Si(C_2H_5)_3$ (10/90)	229
	$\{RhH[P(OMe)_3]_2\}_3$	100	2	$CH_2=C(CH_3)CH(CH_3)CH_2Si(C_2H_5)_3$ / $CH_3C(CH_3)=C(CH_3)CH_2Si(C_2H_5)_3$ (10/90)	229
	$\{RhH[P(OMe)_3]_2\}_3$	25	18	$CH_2=C(CH_3)CH(CH_3)CH_2Si(C_2H_5)_3$ / $CH_3C(CH_3)=C(CH_3)CH_2Si(C_2H_5)_3$ (10/90)	229
	$W(CO)_5$(DMPIC)	100	24	$(CH_3)_2C=C(CH_3)CH_2Si(C_2H_5)_3$ + $CH_2=C(CH_3)CH(CH_3)CH_2Si(C_2H_5)_3$ (3)	884
	$W(CO)_6$	100	24	$(CH_3)_2C=C(CH_3)CH_2Si(C_2H_5)_3$ + $CH_2=C(CH_3)CH(CH_3)CH_2Si(C_2H_5)_3$ (9)	884
	$W(CO)_6$	100	24	$(CH_3)_2C=C(CH_3)CH_2Si(C_2H_5)_3$ + $CH_2=C(CH_3)CH(CH_3)CH_2Si(C_2H_5)_3$ (9)	202
	$W(CO)_6$	100	24	$CH_2=C(CH_3)CH(CH_3)CH_2Si(C_2H_5)_3$ (2) $(CH_3)_2C=C(CH_3)CH_2Si(C_2H_5)_3$ (7)	1
	$W(CO)_6$ + (t-BuOOH)	100	24	$(CH_3)_2C=C(CH_3)CH_2Si(C_2H_5)_3$ + $CH_2=C(CH_3)CH(CH_3)CH_2Si(C_2H_5)_3$ (51)	202
	$W(XNC)(CO)_5$	100	24	$CH_2=C(CH_3)CH(CH_3)CH_2Si(C_2H_5)_3$ + $(CH_3)_2C=C(CH_3)CH_2Si(C_2H_5)_3$ (3)	1
C_6H_{10}	$CH_3CH_2CH=CHCH=CH_2$				
	H_2PtCl_6 (i-PrOH)	20	5	$CH_3CH_2CH_2CH=CHCH_2Si(C_2H_5)_3$ (13)	1110
C_7H_7	Fig.274.				
	H_2PtCl_6 (i-PrOH)	reflux	4	Fig.275.(-)	725
C_8H_{12}	Fig.34.				

Table 8.2

= $(C_2H_5)_3SiH$ = =6=

	$Co_2(CO)_8$	20	12	Fig.276. (65)	272
C_8H_{14}	$(CH_3)_3CCH=CHCH=CH_2$				
	H_2PtCl_6 (i-PrOH)	110-120	6	$(CH_3)_3CCH_2CH=CHCH_2Si(C_2H_5)_3$ (22)	1110
C_8H_{14}	$CH_2=CH(CH_2)_4CH=CH_2$				
	$[HPt(SiMe_2Bz)P(c\text{-}Hex)_3]_2$	20	2 days	$(C_2H_5)_3Si(CH_2)_8Si(C_2H_5)_3$ (60)	429

Table 8.3

= $(C_2H_5)_3SiH$ = =1=

C_2H_3	$CH_2=CHSiCl_3$				
	$Pt[(ViSiMe_2)_2O]_2$	-	-	$(C_2H_5)_3SiCH_2CH_2SiCl_3$ (kin.)	669
C_3H_3	$CH_2=CHCCl_3$				
	H_2PtCl_6 (i-PrOH)	20-25	48	$CH_3CH=CCl_2$ (55) $(C_2H_5)_3SiCCl=CHCH_3$ (9) $(C_2H_5)_3SiCH_2CH_2CCl_3$ (7)	400
C_3H_3	$CH_2=CHCF_3$				
	$RhCl(PPh_3)_3$	70	6	$(C_2H_5)_3SiCH_2CH_2CF_3$ (10) $(C_2H_5)_3SiCH=CHCF_3$ (67)	858
C_3H_3	$CH_2=CHCN$				
	SDVB-PPh_2/$RhCl(PPh_3)_3$	90	3	$(C_2H_5)_3SiCH(CH_3)CN$ (20)	C -008
C_3H_5	$CH_2=CHCH_2Cl$				
	H_2PtCl_6	27-70	18	$(C_2H_5)_3Si(CH_2)_3Cl$ (15)	U -168
	H_2PtCl_6 (i-PrOH)	50	6	$(C_2H_5)_3Si(CH_2)_3Cl$ (23)	705
	H_2PtCl_6 (i-PrOH)	reflux	55-290	$(C_2H_5)_3Si(CH_2)_3Cl$ (5)	98
	H_2PtCl_6 + H-siloxane	22-66	3	$(C_2H_5)_3Si(CH_2)_3Cl$ (traces)	U -168
	H_2PtCl_6 + $HSiCl_3$	19-66	2	$(C_2H_5)_3Si(CH_2)_3Cl$ (4) $Cl_3Si(CH_2)_3Cl$ (54)	U -168
C_3H_5	$CH_2=CHCH_2F$				
	H_2PtCl_6 (THF)	106-130	20	$(C_2H_5)_3Si(CH_2)_3F$ (23)	955
C_3H_6	$CH_2=CHCH_2OH$				
	H_2PtCl_6 (i-PrOH)	60	1	$(C_2H_5)_3Si(CH_2)_3OSi(C_2H_5)_3$ (11)	694
	anionite/H_2PtCl_6	12-25	-	$(C_2H_5)_3SiC_3H_6OH$ (kin.)	1347
C_3H_7	$CH_2=CHCH_2NH_2$				
	SDVB-PPh_2/$RhCl(PPh_3)_3$	120	2 (dcm)	$(C_2H_5)_3Si(CH_2)_3NH_2$ (24)	C -008
C_4H_1	$CF_2CF_2CH=CCl$				
	gamma	-	-	$CHClCF_2CF_2CHSi(C_2H_5)_3$ + $CH_2CF_2CF_2C(Cl)Si(C_2H_5)_3$ (-)	1214
C_4H_6	$CH_2=CHCOOCH_3$				
	$Mn_2(CO)_{10}$	140	-	$(C_2H_5)_3SiCH_2CH_2COOCH_3$ (-)	382

Table 8.3

= $(C_2H_5)_3SiH$ = =2=

	$RhCl(PPh_3)_3$	-	-	$(C_2H_5)_3SiCH_2CH_2COOCH_3$ (-)	J -021
	$RhCl(PPh_3)_3$	60	1 (b)	$(C_2H_5)_3SiCH_2CH_2COOCH_3$ (54)	868
C_4H_6	$CH_2=CHOCH=CH_2$				
	H_2PtCl_6 (i-PrOH)	70-80	3	$CH_2=CHOCH_2CH_2Si(C_2H_5)_3$ (77)	1212
C_4H_6	$CH_2=CHSCH=CH_2$				
	-	-	-	$CH_2=CHSCH_2CH_2Si(C_2H_5)_3+ CH_2=CHSCH(CH_3)Si(C_2H_5)_3$ (-)	1244
	H_2PtCl_6 (i-PrOH)	150	24	$CH_2=CHSCH_2CH_2Si(C_2H_5)_3$ (12) $CH_2=CHSCH(CH_3)Si(C_2H_5)_3$ (11)	1272
	H_2PtCl_6 (i-PrOH)	150	24	$CH_2=CHSCH_2CH_2Si(C_2H_5)_3$ (12) $CH_2=CHSCH(CH_3)Si(C_2H_5)_3$ (11)	1241
	H_2PtCl_6 (i-PrOH)	150	24	$CH_2=CHSCH_2CH_2Si(C_2H_5)_3$ (12) $CH_2=CHSCH(CH_3)Si(C_2H_5)_3$ (11)	1275
	$RhCl(PPh_3)_3$	100	8	$CH_2=CHSCH_2CH_2Si(C_2H_5)_3$ (2) $CH_2=CHSCH(CH_3)Si(C_2H_5)_3$ (2)	1241
	$RhCl(PPh_3)_3$	120	8	$CH_2=CHSCH_2CH_2Si(C_2H_5)_3$ (3) $CH_2=CHSCH(CH_3)Si(C_2H_5)_3$ (3)	1275
C_4H_6	$CH_3COOCH=CH_2$				
	$Rh_4(CO)_{12}$	100	-	$CH_3COOCH_2CH_2Si(C_2H_5)_3$ (-) $CH_3COOCH=CHSi(C_2H_5)_3$ (60)	U -173
	polymer/$RhCl_3$ (THF)	60-140	4	$CH_3COOCH_2CH_2Si(C_2H_5)_3$ (80)	S -029
C_4H_6	$HCOOCH_2CH=CH_2$				
	H_2PtCl_6 (i-PrOH)	60-127	25	$HCOO(CH_2)_3Si(C_2H_5)_3$ (48)	1052
	H_2PtCl_6 (i-PrOH)	60-127	25	$HCOO(CH_2)_3Si(C_2H_5)_3$ (48)	S -026
C_4H_7	$CH_2=C(CH_3)CH_2Cl$				
	H_2PtCl_6	reflux	50	$(C_2H_5)_3SiCH_2CH(CH_3)CH_2Cl$ (19)	U -168
	H_2PtCl_6 + $HSiCl_3$	1-6	1.6	$(C_2H_5)_3SiCH_2CH(CH_3)CH_2Cl$ (48)	U -168
	H_2PtCl_6 + $HSiCl_3$	reflux	4	$(C_2H_5)_3SiCH_2CH(CH_3)CH_2Cl$ (82)	U -168
	H_2PtCl_6+ $HSiMeCl_2$	84	0.07	$CH_3Cl_2SiCH_2CH(CH_3)CH_2Cl$ (67) $(C_2H_5)_3SiCH_2CH(CH_3)CH_2Cl$ (21)	U -168
	H_2PtCl_6 + $MeSiCl_3$	96	8	$(C_2H_5)_3SiCH_2CH(CH_3)CH_2Cl$ (49)	U -168

Table 8.3

= $(C_2H_5)_3SiH$ = =3=

	H_2PtCl_6 +halogenosilane	80	0.05-6	$(C_2H_5)_3SiCH_2CH(CH_3)CH_2Cl$ (4-88)	U -168
C_4H_7	$CH_2=CHCH(Cl)CH_3$				
	H_2PtCl_6	77	1.1	$(C_2H_5)_3SiCH_2CH_2CH(CH_3)Cl$ (8)	U -168
	H_2PtCl_6 + $HSiCl_3$	55	0.9	$(C_2H_5)_3SiCH_2CH_2CH(CH_3)Cl$ (74)	U -168
C_4H_7	$CH_2=CHCH_2CH_2Br$				
	$Rh(acac)_3$	60	12	$(C_2H_5)_3Si(CH_2)_4Br$ (22)	274
C_4H_7	$CH_2=CHCH_2CONH_2$				
	SDVB-CH_2Cl/$RhCl(CO)(PPh_3)_2$	80	6	$(C_2H_5)_3Si(CH_2)_3CONH_2$ (21)	C -022
C_4H_8	$CH_2=CHOCH_2CH_3$				
	Al_2O_3-NMe_2/$RhCl_3$	80	2	$(C_2H_5)_3SiCH_2CH_2OCH_2CH_3$ (64)	C -007
	$RhCl(PPh_3)_3$	80	2	$(C_2H_5)_3SiCH_2CH_2OCH_2CH_3$ (19)	204
	$RhCl_3$ (EtOH)	80	2	$(C_2H_5)_3SiCH_2CH_2OCH_2CH_3$ (31)	204
	support/Rh complex	80	2	$(C_2H_5)_3SiCH_2CH_2OCH_2CH_3$ (0-21)	204
C_4H_8	$CH_2=CHSCH_2CH_3$				
	gamma	50-120	until 1	$(C_2H_5)_3SiCH_2CH_2SCH_2CH_3$ + $(C_2H_5)_3SiCH(CH_3)SCH_2CH_3$ (19-31)	1216
	H_2PtCl_6	100-160	3-48	$(C_2H_5)_3SiCH_2CH_2SCH_2CH_3$ + $(C_2H_5)_3SiCH(CH_3)SCH_2CH_3$ (37-45)	1274
	H_2PtCl_6 (i-PrOH)	150-160	46	$(C_2H_5)_3SiCH_2CH_2SCH_2CH_3$ + $(C_2H_5)_3SiCH(CH_3)SCH_2CH_3$ (-)	1275
	$RhCl(PPh_3)_3$	100-160	3-48	$(C_2H_5)_3SiCH_2CH_2SCH_2CH_3$ + $(C_2H_5)_3SiCH(CH_3)SCH_2CH_3$ (37-45)	1274
C_4H_9	$(CH_3)_2SiClCH=CH_2$				
	H_2PtCl_6 (i-PrOH)	reflux	-	$(CH_3)_2SiClCH_2CH_2Si(C_2H_5)_3$ (66) $(CH_3)_2SiClCH_2CH_2SiCl(CH_3)_2$ (17) $(C_2H_5)_3SiCH_2CH_2Si(C_2H_5)_3$ (10)	912
C_5H_1	$\overline{C(F_2)_3CH=C}Cl$				
	gamma	-	-	$\overline{CHCl(CF_2)_3C}HSi(C_2H_5)_3$ (-)	1214
C_5H_7	$CH_2=CH(CH_2)_2CCl_3$				
	H_2PtCl_6 (i-PrOH)	20-25	48	$(C_2H_5)_3Si(CH_2)_4CCl_3$ (67)	400
C_5H_8	$CH_2=C(CH_3)COOCH_3$				

Table 8.3

= $(C_2H_5)_3SiH$ = =4=

	Ir/C (5%)	60-65	22	$(C_2H_5)_3SiCH_2CH(CH_3)COOCH_3$ (18) $(CH_3)_2C=C(OCH_3)OSi(C_2H_5)_3$ (traces)	U -187
	$Mn_2(CO)_{10}$	140	-	$(C_2H_5)_3SiCH_2CH(CH_3)COOCH_3$ (-)	382
	Pd/C (5%)	60-65	23	$(C_2H_5)_3SiCH_2CH(CH_3)COOCH_3$ (8) $(CH_3)_2C=C(OCH_3)OSi(C_2H_5)_3$ (traces)	U -187
	Pt/C (5%)	60-65	25	$(C_2H_5)_3SiCH_2CH(CH_3)COOCH_3$ (41-62) $(CH_3)_2C=C(OCH_3)OSi(C_2H_5)_3$ (0-1)	U -187
	$PtCl_2$	r.t.	24	$(C_2H_5)_3SiCH_2CH(CH_3)COOCH_3$ (84) $(CH_3)_2C=C(OCH_3)OSi(C_2H_5)_3$ (4) $(C_2H_5)_3SiCH=C(CH_3)COOCH_3$ (2)	1105
	$PtCl_2(PhCN)_2$	r.t.	24	$(C_2H_5)_3SiCH_2CH(CH_3)COOCH_3$ (83) $(CH_3)_2C=C(OCH_3)OSi(C_2H_5)_3$ (-) $(C_2H_5)_3SiCH=C(CH_3)COOCH_3$ (1)	1105
	$PtCl_2(PhCN)_2$ + phosphine	r.t.	24	$(C_2H_5)_3SiCH_2CH(CH_3)COOCH_3$ (20-72) $(CH_3)_2C=C(OCH_3)OSi(C_2H_5)_3$ (-) $(C_2H_5)_3SiCH=C(CH_3)COOCH_3$ (-)	1105
	$RhCl(PPh_3)_3$	55-60	25	$(C_2H_5)_3SiCH_2CH(CH_3)COOCH_3$ (2-16) $(CH_3)_2C=C(OCH_3)OSi(C_2H_5)_3$ (40-59)	U -187
	Ru/C (5%)	60-65	18	$(C_2H_5)_3SiCH_2CH(CH_3)COOCH_3$ (11) $(CH_3)_2C=C(OCH_3)OSi(C_2H_5)_3$ (traces)	U -187
C_5H_8	$CH_2=C=CHCH_2CH_2OH$				
	H_2PtCl_6	-	-	$(C_2H_5)_3SiCH_2CH=CHCH_2CH_2OH$ (-)	253
C_5H_8	$CH_2=CHCOOCH_2CH_3$				
	anionite/$Pt(NH_3)_4$	60	35	$(C_2H_5)_3SiCH_2CH_2COOCH_2CH_3$ + $(C_2H_5)_3SiCH(CH_3)COOCH_2CH_3$ (-)	S -016
	$RhCl(PPh_3)_3$	-	-	$(C_2H_5)_3SiCH_2CH_2COOCH_2CH_3$ (-)	J -021
	$RhCl(PPh_3)_3$	70	3 (b)	$(C_2H_5)_3SiCH_2CH_2COOCH_2CH_3$ (52)	868
C_5H_8	$CH_2=CHOCH_2\overline{CHCH_2O}$				
	H_2PtCl_6	reflux	7-8	$(C_2H_5)_3SiCH_2CH_2OCH_2\overline{CHCH_2O}$ (65)	1154
C_5H_8	$CH_3COOCH_2CH=CH_2$				
	H_2PtCl_6 (i-PrOH)	60-100	6	$CH_3COO(CH_2)_3Si(C_2H_5)_3$ (68)	1121
C_5H_8	$CH_3OCOOCH_2CH=CH_2$				
	H_2PtCl_6 (i-PrOH)	70-160	5	$CH_3OCOO(CH_2)_3Si(C_2H_5)_3$ (41)	S -026

	H_2PtCl_6 (i-PrOH)	70-160	5	$CH_3OCOO(CH_2)_3Si(C_2H_5)_3$ (41)	1051
C_5H_9	$CH_3CH{=}NOCH_2CH{=}CH_2$				
	H_2PtCl_6 (THF)	150	35	$CH_3CH{=}NO(CH_2)_3Si(C_2H_5)_3$ (34)	691
C_5H_{11}	$CH_2{=}CHCH_2NHCH_2CH_2OH$				
	H_2PtCl_6 (i-PrOH)	100-110	12	$(C_2H_5)_3Si(CH_2)_3NHCH_2CH_2OH$ (85)	690
C_5H_{11}	$CH_2{=}CHCH_2NHCH_2CH_3$				
	H_2PtCl_6 (i-PrOH)	120	56	$(C_2H_5)_3Si(CH_2)_3NHCH_2CH_3$ + $(C_2H_5)_3SiCH(CH_3)CH_2NHCH_2CH_3$ (84)	255
C_5H_{12}	$CH_2{=}CHSi(CH_3)_3$				
	Pt colloid	70	0.11	$(C_2H_5)_3SiCH_2CH_2Si(CH_3)_3$ (85)	667
	$Pt(cod)_2$	70	0.11	$(C_2H_5)_3SiCH_2CH_2Si(CH_3)_3$ (26)	667
	$Pt[(ViSiMe_2)_2O]_2$	-	-	$(C_2H_5)_3SiCH_2CH_2Si(CH_3)_3$ (kin.)	669
	$Pt[(ViSiMe_2)_2O]_2$	r.t.	20	$(CH_3)_3SiCH_2CH_2Si(C_2H_5)_3$ + $(CH_3)_3SiCH(CH_3)Si(C_2H_5)_3$ (74)	668
	$RuCl_2(PPh_3)_3$	120	6	$(C_2H_5)_3SiC_2H_4Si(CH_3)_3$ (-)	733
C_5H_{12}	$CH_2{=}CHSi(OCH_3)_3$				
	$Pt[(ViSiMe_2)_2O]_2$	-	-	$(C_2H_5)_3SiCH_2CH_2Si(OCH_3)_3$ (kin.)	669
C_6H_6	Fig.119.				
	Pd/C	100-110	100	Fig.277. (10)	693
	H_2PtCl_6	100-110	14-105	Fig.277. (29-41)	693
	Pt/C	100-160	15-100	Fig.277. (28-40)	693
C_6H_9	$CH_2{=}C(CH_2Cl)OCH_2\overline{CHCH_2O}$				
	H_2PtCl_6 (i-PrOH)	reflux	6 (b)	$(C_2H_5)_3SiCH_2CH(CH_2Cl)OCH_2\overline{CHCH_2O}$ (40)	1155
C_6H_{10}	$(CH_2{=}CHCH_2)_2S$				
	H_2PtCl_6 (i-PrOH)	100	2	$(C_2H_5)_3Si(CH_2)_3SCH_2CH{=}CH_2$ (58)	1241
	H_2PtCl_6 (i-PrOH)	100	2	$(C_2H_5)_3Si(CH_2)_3SCH_2CH{=}CH_2$ (58)	1268
	$RhCl(PPh_3)_3$	120	8	$(C_2H_5)_3Si(CH_2)_3SCH_2CH{=}CH_2$ (6)	1268
	$RhCl(PPh_3)_3$	120	8	$(C_2H_5)_3Si(CH_2)_3SCH_2CH{=}CH_2$ (6)	1241
C_6H_{10}	$CH_2{=}CHCH_2CH_2COCH_3$				

	$Fe(CO)_5$	112-189	-	$(C_2H_5)_3Si(CH_2)_4COCH_3$ (90)	15
	Ni	110-210	-	$(C_2H_5)_3Si(CH_2)_4COCH_3$ (86-98)	15
	Pd	110-176	-	$(C_2H_5)_3Si(CH_2)_4COCH_3$ (78)	15
	H_2PtCl_6	130-140	1.5-2	$(C_2H_5)_3Si(CH_2)_4COCH_3$ (34) $CH_2=CHCH_2CH_2CH(CH_3)OSi(C_2H_5)_3$ (40)	602
	$SnCl_2$	108-118	-	$(C_2H_5)_3Si(CH_2)_4COCH_3$ (9)	15
	$ZnCl_2$	110-193	-	$(C_2H_5)_3Si(CH_2)_4COCH_3$ (96)	15
C_6H_{10}	$CH_2=CHCH_2OCH_2\overline{CHCH_2O}$				
	H_2PtCl_6	-	-	$(C_2H_5)_3Si(CH_2)_3OCH_2\overline{CHCH_2O}$ (60)	1153
C_6H_{11}	$(CH_3)_2C=NOCH_2CH=CH_2$				
	H_2PtCl_6 (THF)	100-165	27-30	$(CH_3)_2C=NO(CH_2)_3Si(C_2H_5)_3$ (19-50)	691
	H_2PtCl_6 (THF)	140-150	30	$(CH_3)_2C=NO(CH_2)_3Si(C_2H_5)_3$ (44)	S -015
C_6H_{11}	$CH_2=CHCH_2NHCOOCH_2CH_3$				
	H_2PtCl_6 (i-PrOH)	130-170	10	$(C_2H_5)_3Si(CH_2)_3NHCOOCH_2CH_3$ (54)	783
C_6H_{11}	$NH(CH_2=CHCH_2)_2$				
	H_2PtCl_6 (i-PrOH)	125	5	$CH_2=CHCH_2NH(CH_2)_3Si(C_2H_5)_3$ (37) $NH[(CH_2)_3Si(C_2H_5)_3]_2$ (39)	255
C_6H_{12}	$CH_2=CHCOOSi(CH_3)_3$				
	H_2PtCl_6	80-120	3	$(C_2H_5)_3SiCH_2CH_2COOSi(CH_3)_3$ (31)	1247
C_6H_{12}	$CH_3CH_2CH_2SCH_2CH=CH_2$				
	H_2PtCl_6 (i-PrOH)	150	26	$CH_3CH_2CH_2S(CH_2)_3Si(C_2H_5)_3$ (82)	1268
	$RhCl(PPh_3)_3$	150	46	$CH_3CH_2CH_2S(CH_2)_3Si(C_2H_5)_3$ (24) $CH_3CH_2CH_2SCH(CH_3)CH_2Si(C_2H_5)_3$ (15)	1268
C_6H_{13}	$CH_2=CHCH_2N(CH_3)CH_2CH_2OH$				
	H_2PtCl_6 (i-PrOH)	25-206	1-7	$(C_2H_5)_3Si(CH_2)_3N(CH_3)CH_2CH_2OH$ (2-28) $(C_2H_5)_3Si(CH_2)_3N(CH_3)CH_2CH_2OSi(C_2H_5)_3$ (10-48)	689
C_6H_{14}	$(CH_3)_3SiCH_2SCH=CH_2$				
	H_2PtCl_6	-	-	$(CH_3)_3SiCH_2SCH_2CH_2Si(C_2H_5)_3$ + $(CH_3)_3SiCH_2SCH(CH_3)Si(C_2H_5)_3$ + $(C_2H_5)_2SiCH=CHSi(C_2H_5)_3$ (-)	1078
C_6H_{14}	$CH_2=CHSi(CH_3)_2OCH_2CH_3$				

	$RuCl_2(PPh_3)_3$	120	6	$(C_2H_5)_3SiC_2H_4Si(CH_3)_2OCH_2CH_3$ (-)	733
C_7H_7	2-$CH_2{=}CHC_5H_4N$				
	H_2PtCl_6 (i-PrOH)	160-170	7	2-$(C_2H_5)_3SiCH_2CH_2C_5H_4N$ + $(C_2H_5)_3SiCH(CH_3)C_5H_4N$ (46)	819
	cis-$PtCl_2(PhCH{=}CH_2)_2$	r.t.	7	2-$(C_2H_5)_3SiCH_2CH_2C_5H_4N$ (14) 2-$CH_3CH[Si(C_2H_5)_3]C_5H_4N$ (54)	220
C_7H_7	4-$CH_2{=}CHC_5H_4N$				
	H_2PtCl_6 (i-PrOH)	160-170	7	4-$(C_2H_5)_3SiCH_2CH_2C_5H_4N$ (46)	819
C_7H_{10}	$(CH_2{=}CHCH_2O)_2CO$				
	H_2PtCl_6	50-170	16	$[(C_2H_5)_3Si(CH_2)_3O]_2CO$ (14) $(C_2H_5)_3Si(CH_2)_3OSi(C_2H_5)_3$ (30)	1051
C_7H_{10}	$CH_2{=}C(CH_3)COOCH_2CH{=}CH_2$				
	H_2PtCl_6	-	-	$CH_2{=}C(CH_3)COO(CH_2)_3Si(C_2H_5)_3$ (-)	J -064
C_7H_{11}	$\overline{OCH(CH_2Cl)C}HCH_2C(CH_3){=}CH_2$				
	H_2PtCl_6 (i-PrOH)	120-125	12-24	$\overline{OCH(CH_2Cl)C}HCH_2CH(CH_3)CH_2Si(C_2H_5)_3$ (30)	92
C_7H_{12}	$(CH_3)_2C(OH)CH{=}CHCH{=}CH_2$				
	H_2PtCl_6	-	-	$(CH_3)_2C(OH)CH_2CH{=}CHCH_2Si(C_2H_5)_3$ + $(CH_3)_2C(OH)CH_2CH[Si(C_2H_5)_3]CH{=}CH_2$ + $(CH_3)_2C(OH)CH[Si(C_2H_5)_3]CH_2CH{=}CH_2$ + $(CH_3)_2C(OH)CH[Si(C_2H_5)_3]CH{=}CHCH_3$ (-)	246
C_7H_{13}	$(\overline{CH_2)_4N}CH_2CH{=}CH_2$				
	H_2PtCl_6 (i-PrOH)	100	12	$(\overline{CH_2)_4N}(CH_2)_3Si(C_2H_5)_3$ (64)	702
C_7H_{14}	$CH_2{=}C(CH_3)COOSi(CH_3)_3$				
	H_2PtCl_6	80-120	3	$(C_2H_5)_3SiCH_2CH(CH_3)COOSi(CH_3)_3$ (56)	1247
C_7H_{14}	$CH_2{=}CHCH(OCH_2CH_3)_2$				
	H_2PtCl_6 (i-PrOH)	100-162	1	$(C_2H_5)_3SiCH_2CH_2CH(OCH_2CH_3)_2$ (76)	1004
C_7H_{14}	$CH_2{=}CHSi(CH_3)_2CH_2CH{=}CH_2$				
	H_2PtCl_6 (i-PrOH)	until 140	0.5	$(C_2H_5)_3SiCH_2CH_2Si(CH_3)_2CH_2CH{=}CH_2$ + $CH_2{=}CHSi(CH_3)_2(CH_2)_3Si(C_2H_5)_3$ (65)	402
C_7H_{15}	$CH_2{=}CHCH_2N(CH_2CH_3)_2$				
	H_2PtCl_6 (i-PrOH)	110-115	60	$(C_2H_5)_3Si(CH_2)_3N(CH_2CH_3)_2$ + $(C_2H_5)_3SiCH(CH_3)CH_2N(CH_2CH_3)_2$ (66)	255
C_7H_{16}	$CH_2{=}CHCH_2OCH_2OSi(CH_3)_3$				

	H_2PtCl_6 (i-PrOH)	105	4	$(C_2H_5)_3Si(CH_2)_3OCH_2OSi(CH_3)_3$ (-)	809
C_7H_{16}	$CH_2=CHSi(CH_3)(OCH_2CH_3)_2$				
	$RuCl_2(PPh_3)_3$	120	6	$(C_2H_5)_3SiC_2H_4Si(CH_3)(OCH_2CH_3)_2$ (20)	733
C_8H_3	$C_6F_5CH=CH_2$				
	$RhCl(PPh_3)_3$	70	18	$C_6F_5CH_2CH_2Si(C_2H_5)_3$ (10) $C_6F_5CH=CHSi(C_2H_5)_3$ (81)	858
C_8H_7	p-$ClC_6H_4CH=CH_2$				
	$Pt(PPh_3)_4$	15	- (solv.)	p-$ClC_6H_4CH_2CH_2Si(C_2H_5)_3$ (55)	886
C_8H_9	Fig.52.				
	H_2PtCl_6 (i-PrOH)	160-170	7	Fig.278. (36)	819
C_8H_{10}	Fig.279.				
	H_2PtCl_6 (i-PrOH)	60	1	Fig.280.(-)	694
C_8H_{11}	Fig.58.				
	H_2PtCl_6 (i-PrOH)	215	2	Fig.281. (25)	730
C_8H_{12}	Fig.282.				
	H_2PtCl_6 (i-PrOH)	60-130	2	Fig.283. (58)	21
C_8H_{14}	$\overline{OCH_2CH_2OCH_2C}HCH_2OCH_2CH=CH_2$				
	H_2PtCl_6 (i-PrOH)	reflux	52	$\overline{OCH_2CH_2OCH_2C}HCH_2O(CH_2)_3Si(C_2H_5)_3$ (47)	554
C_8H_{15}	$(\overline{CH_2)_4N}CH_2CH_2CH=CH_2$				
	H_2PtCl_6	-	-	$(\overline{CH_2)_4N}(CH_2)_4Si(C_2H_5)_3$ (22)	703
C_8H_{15}	$(\overline{CH_2)_5N}CH_2CH=CH_2$				
	H_2PtCl_6 (i-PrOH)	100	0.5	$(\overline{CH_2)_5N}(CH_2)_3Si(C_2H_5)_3$ (80)	702
	H_2PtCl_6 (i-PrOH)	100-125	1	$(\overline{CH_2)_5N}(CH_2)_3Si(C_2H_5)_3$ (80)	S -006
C_8H_{15}	$CH_3COOCH_2CH_2N(CH_3)CH_2CH=CH_2$				
	H_2PtCl_6 (i-PrOH)	110-120	7	$CH_3COOCH_2CH_2N(CH_3)(CH_2)_3Si(C_2H_5)_3$ (44)	689
C_8H_{15}	$N(CH_2CH_2O)_3SiCH=CH_2$				
	Rh(acac)$(CO)_2$	120	8 (b)	$N(CH_2CH_2O)_3SiCH_2CH_2Si(C_2H_5)_3$ (69)	1237
C_8H_{16}	$CH_3CH_2S(CH_2)_4CH=CH_2$				

	H_2PtCl_6	150-160	48	$CH_3CH_2S(CH_2)_6Si(C_2H_5)_3$ (79)	1271
	H_2PtCl_6 (i-PrOH)	150	46	$CH_3CH_2S(CH_2)_6Si(C_2H_5)_3$ (26)	1271
	$RhCl(PPh_3)_3$	150	46	$CH_3CH_2S(CH_2)_6Si(C_2H_5)_3$ (9)	1272
	$RhCl(PPh_3)_3$	150-160	48	$CH_3CH_2S(CH_2)_6Si(C_2H_5)_3$ (23)	1271
C_8H_{17}	$CH_3N(OCH_2CH_2)_2Si(CH_3)CH=CH_2$				
	$Rh(acac)(CO)_2$	100	8	$CH_3N(OCH_2CH_2)_2Si(CH_3)CH_2CH_2Si(C_2H_5)_3$ + $CH_3N(OCH_2CH_2)_2Si(CH_3)_2CHSi(C_2H_5)_3$ (72)	1236
C_8H_{18}	$CH_2=CHSi(OC_2H_5)_3$				
	H_2PtCl_6	80-200	0.55	$(C_2H_5)_3SiCH_2CH_2Si(OC_2H_5)_3$ (92)	266
	$RuCl_2(PPh_3)_3$	120	6	$(C_2H_5)_3SiCH_2CH_2Si(OC_2H_5)_3$ (-)	733
C_9H_{10}	$C_6H_5OCH_2CH=CH_2$				
	H_2PtCl_6	22-55	3	$C_6H_5OCH_2CH_2CH_2Si(C_2H_5)_3$ (79)	U -168
	H_2PtCl_6	40-137	0.3	$C_6H_5OCH_2CH_2CH_2Si(C_2H_5)_3$ (79)	U -168
C_9H_{10}	$C_6H_5SeCH_2CH=CH_2$				
	H_2PtCl_6	150	15	$C_6H_5Se(CH_2)_3Si(C_2H_5)_3$ (65)	6
C_9H_{10}	p-$CH_3OC_6H_4CH=CH_2$				
	$Pt(PPh_3)_4$	15	- (solv.)	p-$CH_3OC_6H_4CH_2CH_2Si(C_2H_5)_3$ (kin.)	977
C_9H_{12}	Fig.284.				
	H_2PtCl_6 (i-PrOH)	reflux	-	Fig.285. (37)	1003
C_9H_{15}	$CH_2=CHCH_2OCH_2CH(CH_2Cl)OCH_2\overline{CHCH_2O}$				
	H_2PtCl_6	reflux	24 (b)	$(C_2H_5)_3Si(CH_2)_3OCH_2CH(CH_2Cl)OCH_2\overline{CHCH_2O}$ (73)	66
C_9H_{15}	$N(CH_2CH=CH_2)_3$				
	H_2PtCl_6 (i-PrOH)	110-115	60	$(CH_2=CHCH_2)_2N(CH_2)_3Si(C_2H_5)_3$ + $(CH_2=CHCH_2)_2NCH_2CH(CH_3)Si(C_2H_5)_3$ (70)	255
C_9H_{16}	$CH_2=CHCH_2OCH_2CH(CH_2OH)OCH_2\overline{CHCH_2O}$				
	H_2PtCl_6	reflux	40 (b)	$(C_2H_5)_3Si(CH_2)_3OCH_2CH(CH_2OH)OCH_2\overline{CHCH_2O}$ (48)	1150
C_9H_{16}	$CH_3CH_2CH_2CH=C=CHC(OH)(CH_3)_2$				
	H_2PtCl_6	80	-	$CH_3(CH_2)_3C[Si(C_2H_5)_3]=CHC(OH)(CH_3)_2$ + $CH_3CH_2CH_2CH=C[Si(C_2H_5)_3]CH_2C(OH)(CH_3)_2$ (40)	248
C_9H_{17}	$CH_3CH_2CH_2CH=C=C(CH_2CH_3)CH_2OH$				

Table 8.3
= $(C_2H_5)_3SiH$ =   =10=

	H_2PtCl_6	-	-	$(C_2H_5)_3SiCH(CH_2CH_2CH_3)CH=C(CH_2CH_3)CH_2OH$ (-)	253
C_9H_{19}	$(CH_3CH_2CH_2)_2NCH_2CH=CH_2$				
	H_2PtCl_6 (i-PrOH)	112-117	18	$(CH_3CH_2CH_2)_2N(CH_2)_3Si(C_2H_5)_3$ + $(CH_3CH_2CH_2)_2NCH_2CH(CH_3)Si(C_2H_5)_3$ (79)	255
C_9H_{20}	$CH_2=CHCH_2Si(C_2H_5)_3$				
	$Pt[(ViSiMe_2)_2O]_2$	-	-	$(C_2H_5)_3Si(CH_2)_3Si(C_2H_5)_3$ (kin.)	669
$C_{10}H_9$	Fig.288.				
	H_2PtCl_6 (THF)	140	5	Fig.289. (40)	607
$C_{10}H_9$	Fig.290.				
	H_2PtCl_6 (THF)	140	5	Fig.291. (70)	607
$C_{10}H_{12}$	$C_6H_5OCH_2C(CH_3)=CH_2$				
	H_2PtCl_6	21-86	2	$C_6H_5OCH_2CH(CH_3)CH_2Si(C_2H_5)_3$ (22)	U -168
	H_2PtCl_6 + $HSiCl_3$	23-131	0.025	$C_6H_5OCH_2CH(CH_3)CH_2Si(C_2H_5)_3$ (88)	U -168
$C_{10}H_{12}$	$C_6H_5SeCH_2C(CH_3)=CH_2$				
	H_2PtCl_6	150	15	$C_6H_5SeCH_2CH(CH_3)CH_2Si(C_2H_5)_3$ (40)	6
$C_{10}H_{12}$	m-$CH_3OC_6H_4OCH_2CH=CH_2$				
	H_2PtCl_6	-	-	m-$CH_3OC_6H_4O(CH_2)_3Si(C_2H_5)_3$ (-)	513
$C_{10}H_{13}$	Fig.292.				
	H_2PtCl_6 (THF)	100	3	Fig.293. (82)	943
$C_{10}H_{15}$	Fig.294.				
	H_2PtCl_6 (i-PrOH)	80-138	1	Fig.295. (53)	1000
$C_{10}H_{16}$	$CH_2=CHCH_2CH(COOCH_2CH_3)_2$				
	H_2PtCl_6 (i-PrOH)	reflux	-	$(C_2H_5)_3Si(CH_2)_3CH(COOCH_2CH_3)_2$ (42)	1008
$C_{10}H_{16}$	Fig.296.				
	H_2PtCl_6 (i-PrOH)	120-130	5	Fig.297. (30)	552
$C_{10}H_{18}$	$CH_2=CHCH_2OCH_2CH(CH_2OCH_3)OCH_2\overline{CHCH_2O}$				
	H_2PtCl_6	reflux	40 (b)	$(C_2H_5)_3Si(CH_2)_3OCH_2CH(CH_2OCH_3)OCH_2\overline{CHCH_2O}$ (43)	1150
$C_{10}H_{18}$	$CH_3CH_2CH_2CH=C=C(CH_2CH_3)CH(OH)CH_3$				

	H_2PtCl_6 (i-PrOH)	60	3	$CH_3CH_2CH_2CH=C[Si(C_2H_5)_3]CH(CH_2CH_3)CH(OH)CH_3$ + $CH_3(CH_2)_3C[Si(C_2H_5)_3]=C(CH_2CH_3)CH(OH)CH_3$ (75)	249
$C_{11}H_9$	Fig.298.				
	H_2PtCl_6 (i-PrOH)	reflux	10	Fig.299. (55)	699
$C_{11}H_{11}$	Fig.300.				
	H_2PtCl_6 (THF)	140	5	Fig.301. (65)	607
$C_{11}H_{18}$	$CH_2=CHCH_2OCH_2CH(CH_2OCOCH_3)OCH_2\overline{CHCH_2O}$				
	H_2PtCl_6	reflux	24-36 (b)	$(C_2H_5)_3Si(CH_2)_3OCH_2CH(CH_2OCOCH_3)OCH_2\overline{CHCH_2O}$ (60)	1149
	H_2PtCl_6	reflux	24-36 (b)	$(C_2H_5)_3Si(CH_2)_3OCH_2CH(CH_2OCOCH_3)OCH_2\overline{CHCH_2O}$ (60)	1150
$C_{11}H_{23}$	$[CH_3(CH_2)_3]_2NCH_2CH=CH_2$				
	H_2PtCl_6 (i-PrOH)	105-120	26	$[CH_3(CH_2)_3]_2N(CH_2)_3SiC_2H_5)_3$ + $[CH_3(CH_2)_3]_2NCH_2CH(CH_3)Si(C_2H_5)_3$ (43)	255
$C_{11}H_{25}$	$CH_2=CHCH_2NHCH_2CH_2OSi(C_2H_5)_3$				
	H_2PtCl_6 (i-PrOH)	100-110	24	$(C_2H_5)_3Si(CH_2)_3NHCH_2CH_2OSi(C_2H_5)_3$ (69)	690
$C_{12}H_{11}$	Fig.302.				
	H_2PtCl_6	100	10 (THF)	Fig.303. (76)	1202
	H_2PtCl_6 (THF)	100	3	Fig.303. (49)	943
	H_2PtCl_6 (THF)	140	5	Fig.303. (43)	607
	$RhCl(PPh_3)_3$	140	5	Fig.303. (50)	607
$C_{12}H_{13}$	Fig.304.				
	H_2PtCl_6 (THF)	140	5	Fig.305. (44)	607
$C_{12}H_{20}$	$\overline{OCH_2C}HCH_2OCH(CH_2OCH_2CH=CH_2)_2$				
	H_2PtCl_6	120-125	16	$\overline{OCH_2C}HCH_2OCH(CH_2OCH_2CH=CH_2)CH_2O(CH_2)_3Si(C_2H_5)_3$ (45)	1158
$C_{12}H_{20}$	$Si(OCH_2CH=CH_2)_4$				
	H_2PtCl_6 (i-PrOH)	reflux	1	$Si[O(CH_2)_3Si(C_2H_5)_3]_4$ (-)	432
$C_{12}H_{21}$	$(i-C_3H_7)_2CHCN=CHCH=CHCH_3$				
	$RhCl(CO)(MAD)_2$	r.t.	5	trans-$(i-C_3H_7)_2CHCN[Si(C_2H_5)_3]CH=CHCH_2CH_3$ (30) $(i-C_3H_7)_2CHN=CHCH_2CH(CH_3)Si(C_2H_5)_3$ (12)	165
$C_{12}H_{24}$	$CH_2=CHCH_2CH[O(CH_2)_3CH_3]_2$				

Table 8.3

= $(C_2H_5)_3SiH$ =

=12=

H_2PtCl_6 (i-PrOH)	78-155	1	$(C_2H_5)_3Si(CH_2)_3CH[O(CH_2)_3CH_3]_2$ (41)	1004
$C_{12}H_{27}$ $CH_2=CHCH_2N(CH_3)CH_2CH_2OSi(C_2H_5)_3$				
H_2PtCl_6 (i-PrOH)	110-120	7	$(C_2H_5)_3Si(CH_2)_3N(CH_3)CH_2CH_2OSi(C_2H_5)_3$ (40)	689
$C_{13}H_{13}$ Fig.286.				
H_2PtCl_6 (THF)	100	3	Fig.287. (79)	943
$C_{13}H_{13}$ Fig.306.				
H_2PtCl_6 (THF)	100	3	Fig.307. (77)	943
H_2PtCl_6 (THF)	140	5	Fig.307. (43)	607
$C_{13}H_{13}$ Fig.308.				
H_2PtCl_6 (THF)	140	5	Fig.309. (35)	607
$C_{13}H_{17}$ $(CH_3)_3CN=CHCH=CHC_6H_5$				
$RhCl(CO)(MAD)_2$	r.t.	5	cis + trans-$(CH_3)_3CN[Si(C_2H_5)_3]CH=CHCH_2C_6H_5$ (40) $(CH_3)_3CN=CHCH_2CH(C_6H_5)Si(C_2H_5)_3$ (10)	165
$C_{14}H_{24}$ $(CH_2)_5C(OH)C(CH_2CH_3)=C=CHCH_2CH_2CH_3$				
H_2PtCl_6	80	-	$(CH_2)_5C(OH)C(CH_2CH_3)=CHCH[Si(C_2H_5)_3]CH_2CH_2CH_3$ + $(CH_2)_5C(OH)C(CH_2CH_3)=C[Si(C_2H_5)_3](CH_2)_3CH_3$ (22)	248
$C_{16}H_{14}$ 2-C_6H_5CO-4-$HOC_6H_3OCH_2CH=CH_2$				
SiO_2-SH/H_2PtCl_6 (EtOH)	90	1	2-C_6H_5CO-4-$HOC_6H_3O(CH_2)_3Si(C_2H_5)_3$ (-)	F -033
$C_{16}H_{22}$ $CH_3CH_2CH_2CH=C=C(CH_2CH_3)C(OH)CH_3(C_6H_5)$				
H_2PtCl_6	80	-	$CH_3CH_2CH_2CH=C[Si(C_2H_5)_3]CH(CH_2CH_3)C(OH)CH_3$- (C_6H_5) (45)	248
$C_{16}H_{24}$ $CH_3CH_2CH_2CH(OH)CH=C(SC_6H_5)CH(OH)CH_2CH_2CH_3$				
H_2PtCl_6	-	-	$CH_3CH_2CH_2CH(OH)CH[Si(C_2H_5)_3]CH(SC_6H_5)CH(OH)$- $CH_2CH_2CH_3$ (-)	835
$C_{18}H_{26}$ $Fe[C_5H_4Si(CH_3)_2CH=CH_2]_2$				
H_2PtCl_6 (i-PrOH)	reflux	6	$Fe[C_5H_4Si(CH_3)_2CH_2CH_2Si(C_2H_5)_3]_2$ (70-95)	1123
$C_{22}H_{36}$ $(CH_3CH_2CH_2)_2C(OH)CH=C(SC_6H_5)C(OH)(CH_2CH_2CH_3)_2$				
H_2PtCl_6	-	-	$(CH_3CH_2CH_2)_2C(OH)CH[Si(C_2H_5)_3]CH(SC_6H_5)C(OH)$- $(CH_2CH_2CH_3)_2$ (-)	835
$C_{24}H_{24}$ $CH_3C(OH)(C_6H_5)CH=C(SC_6H_5)C(OH)(C_6H_5)CH_3$				

	H_2PtCl_6	-	-	$CH_3C(OH)(C_6H_5)CH[Si(C_2H_5)_3]CH(SC_6H_5)C(OH)$-$(C_6H_5)CH_3$ (-)	835
$C_{25}H_{26}$	$CH_3C(OH)(C_6H_5)CH=C(SCH_2C_6H_5)C(OH)(C_6H_5)CH_3$				
	H_2PtCl_6	-	-	$CH_3C(OH)(C_6H_5)CH[Si(C_2H_5)_3]CH(SCH_2C_6H_5)C(OH)$-$(C_6H_5)CH_3$ (-)	835
$C_{26}H_{28}$	$CH_3CH_2C(OH)(C_6H_5)CH=C(SC_6H_5)C(OH)(C_6H_5)CH_2CH_3$				
	H_2PtCl_6	-	-	$CH_3CH_2C(OH)(C_6H_5)CH[Si(C_2H_5)_3]CH(SC_6H_5)C(OH)$-$(C_6H_5)CH_2CH_3$ (-)	835

Table 8.4

= $(C_2H_5)_3SiH$ = =1=

C_2H_2 CH≡CH

-	20-70	0.5-1	CH_2=CHSi$(C_2H_5)_3$ (14-30) $(C_2H_5)_3SiCH_2CH_2Si(C_2H_5)_3$ (5-32)	956
$AlBr_3$	40	0.5	CH_2=CHSi$(C_2H_5)_3$ (90)	1238
$AlCl_3$	0	1-2	CH_2=CHSi$(C_2H_5)_3$ (50-70)	1238
H_2PtCl_6 (THF)	20	2	$(C_2H_5)_3SiCH_2CH_2Si(C_2H_5)_3$ (-) $(C_2H_5)_3$SiC≡CSi$(C_2H_5)_3$ (40)	1259
H_2PtCl_6 (THF) + $AlCl_3$	50	- (dc)	CH_2=CHSi$(C_2H_5)_3$ (93)	1234
H_2PtCl_6 + I_2 (or LiI or R_3SiI)	-	-	$(C_2H_5)_3SiCH_2CH_2Si(C_2H_5)_3$+$(C_2H_5)_3$SiC≡CSi$(C_2H_5)_3$ (-)	1256
H_2PtCl_6 + I_2 (or LiI or R_3SiI)	-	-	$(C_2H_5)_3SiCH_2CH_2Si(C_2H_5)_3$+$(C_2H_5)_3$SiC≡CSi$(C_2H_5)_3$ (-)	1255
Rh(OCCHMeCOMe)$(CO)_2$	20-30	3	$(C_2H_5)_3SiCH_2CH_2Si(C_2H_5)_3$ (69)	S -045
Rh(acac)$(CO)_2$	60	3	CH_2=CHSi$(C_2H_5)_3$ (12) $(C_2H_5)_3SiCH_2CH_2Si(C_2H_5)_3$ (56)	1257
RhHCO$(PPh_3)_3$	25	6	CH_2=CHSi$(C_2H_5)_3$ (27) $(C_2H_5)_3SiCH_2CH_2Si(C_2H_5)_3$ (14)	1257
$[RhCl(C_8H_{14})_2]_2$	50	6	CH_2=CHSi$(C_2H_5)_3$ (16) $(C_2H_5)_3SiCH_2CH_2Si(C_2H_5)_3$ (57)	1257
$[RhCl(C_8H_{14})_2]_2$	50	2	CH_2=CHSi$(C_2H_5)_3$ (35) $(C_2H_5)_3SiCH_2CH_2Si(C_2H_5)_3$ (50)	1257
$[RhCl(C_8H_{14})_2]_2$	70	1	CH_2=CHSi$(C_2H_5)_3$ (10) $(C_2H_5)_3SiCH_2CH_2Si(C_2H_5)_3$ (67)	941
$[RhCl(CO)(C_8H_{14})_2]_2$	r.t.	6 (THF)	$(C_2H_5)_3SiCH_2CH_2Si(C_2H_5)_3$ (93)	S -045
$[RhCl(CO)_2]_2$	20-30	2	$(C_2H_5)_3SiCH_2CH_2Si(C_2H_5)_3$ (83)	S -045
$[RhCl(CO)_2]_2$	30-40	2 (h)	CH_2=CHSi$(C_2H_5)_3$ (65)	1257
$[RhCl(CO)_2]_2$	50-60	2-4	CH_2=CHSi$(C_2H_5)_3$ (6-65) $(C_2H_5)_3SiCH_2CH_2Si(C_2H_5)_3$ (2-83)	1257
siloxane-$(CH_2)_2PPh_2/[RhCl(C_8H_{14})_2]_2$	40	1	CH_2=CHSi$(C_2H_5)_3$ (22)	941
siloxane-$(CH_2)_2PPh_2/[RhCl(C_8H_{14})_2]_2$	45	1	CH_2=CHSi$(C_2H_5)_3$ (46)	941
siloxane-$(CH_2)_2PPh_2/[RhCl(C_8H_{14})_2]_2$	70	1	CH_2=CHSi$(C_2H_5)_3$ (27) $(C_2H_5)_3SiCH_2CH_2Si(C_2H_5)_3$ (58)	941
siloxane-$(CH_2)_3PPh_2/[RhCl(C_8H_{14})_2]_2$	40	1	CH_2=CHSi$(C_2H_5)_3$ (87) $(C_2H_5)_3SiCH_2CH_2Si(C_2H_5)_3$ (7)	941

Table 8.4
= $(C_2H_5)_3SiH$ = =2=

	siloxane-$(CH_2)_3PPh_2$/$[RhCl(C_8H_{14})_2]_2$	45	1	$CH_2=CHSi(C_2H_5)_3$ (81)	941
	siloxane-$(CH_2)_3PPh_2$/$[RhCl(C_8H_{14})_2]_2$	70	1	$CH_2=CHSi(C_2H_5)_3$ (32) $(C_2H_5)_3SiCH_2CH_2Si(C_2H_5)_3$ (52)	941
	siloxane-CH_2PPh_2/$[RhCl(C_8H_{14})_2]_2$	45	1	$CH_2=CHSi(C_2H_5)_3$ (25)	941
	siloxane-CH_2PPh_2/$[RhCl(C_8H_{14})_2]_2$	70	1	$CH_2=CHSi(C_2H_5)_3$ (25) $(C_2H_5)_3SiCH_2CH_2Si(C_2H_5)_3$ (67)	941
C_4H_4	$CH_2=CHC\equiv CH$				
	H_2PtCl_6 (i-PrOH)	-	-	$CH_2=CHCH=CHSi(C_2H_5)_3$ (-)	1111
C_4H_6	$CH_3C\equiv CCH_3$				
	$[HPt(SiBzMe)_2P(c\text{-}Hex)_3]_2$	65	0.2	cis-$CH_3CH=C(CH_3)Si(C_2H_5)_3$ (92)	430
C_4H_6	$CH_3CH_2C\equiv CH$				
	$[HPt(SiBzMe)_2P(c\text{-}Hex)_3]_2$	50	0.5	trans-$CH_3CH_2CH=CHSi(C_2H_5)_3$ (88)	430
C_5H_6	$CH_2=C(CH_3)C\equiv CH$				
	H_2PtCl_6 (i-PrOH)	-	-	$CH_2=C(CH_3)CH=CHSi(C_2H_5)_3$ (-)	1111
C_5H_6	$CH_3CH=CHC\equiv CH$				
	H_2PtCl_6 (i-PrOH)	-	-	$CH_3CH=CHCH=CHSi(C_2H_5)_3$ (-)	1111
C_5H_8	$(CH_3)_2CHC\equiv CH$				
	H_2PtCl_6 (THF)	20	24 (THF)	$C_5H_9Si(C_2H_5)_3$ (kin.)	945
	H_2PtCl_6 (i-PrOH)	100	4	$(CH_3)_2CHCH=CHSi(C_2H_5)_3$/$(CH_3)_2CHC[Si(C_2H_5)_3]=CH_2$ (84/16)	951
C_5H_8	$CH_3CH_2C\equiv CCH_3$				
	$[HPt(SiBzMe_2)P(c\text{-}Hex)_3]_2$	65	0.75	$CH_3CH_2CH=C(CH_3)Si(C_2H_5)_3$ (63) $CH_3CH_2C[Si(C_2H_5)_3]=CHCH_3$ (34)	1205
C_5H_8	$CH_3CH_2CH_2C\equiv CH$				
	$[IrCl(c\text{-}C_8H_{14})_2]_2$	60	6	cis-$CH_3CH_2CH_2CH=CHSi(C_2H_5)_3$ (63) trans-$CH_3CH_2CH_2CH=CHSi(C_2H_5)_3$ (12)	59
	$[IrCl(c\text{-}C_8H_{14})_2]_2$ + $P(C_6H_4OMe)_3$	60	6	cis-$CH_3CH_2CH_2CH=CHSi(C_2H_5)_3$ (47) trans-$CH_3CH_2CH_2CH=CHSi(C_2H_5)_3$ (14)	59
	$[IrCl(c\text{-}C_8H_{14})_2]_2$ + PPh_3	60	6	cis-$CH_3CH_2CH_2CH=CHSi(C_2H_5)_3$ (70) trans-$CH_3CH_2CH_2CH=CHSi(C_2H_5)_3$ (30)	59
	$Ni(acac)_2$ + $AlEt_3$	20	6-12	$CH_2=C(CH_2CH_2CH_3)C(CH_2CH_2CH_3)=CHSi(C_2H_5)_3$ (20-90)	663

Table 8.4

= $(C_2H_5)_3SiH$ =

=3=

$Ni(acac)_2$ + $AlEt_3$	20	6	$CH_3CH_2CH_2CH{=}CHC(CH_2CH_2CH_3){=}CHSi(C_2H_5)_3$ + $CH_2{=}C(CH_2CH_2CH_3)C(CH_2CH_2CH_3){=}CHSi(C_2H_5)_3$ (90)	G -030
$Ni(acac)_2$ + $LiAlH_4$	20	6 (THF)	$CH_3CH_2CH_2CH{=}CHC(CH_2CH_2CH_3){=}CHSi(C_2H_5)_3$ + $CH_2{=}C(CH_2CH_2CH_3)C(CH_2CH_2CH_3){=}CHSi(C_2H_5)_3$ (87)	662
$Ni(acac)_2$ + $LiAlH_4$	20	6 (THF)	$CH_3CH_2CH_2CH{=}CHC(CH_2CH_2CH_3){=}CHSi(C_2H_5)_3$ + $CH_2{=}C(CH_2CH_2CH_3)C(CH_2CH_2CH_3){=}CHSi(C_2H_5)_3$ (87)	G -030
$Ni(acac)_2$ + MeLi	20	12	$CH_3CH_2CH_2CH{=}CHC(CH_2CH_2CH_3){=}CHSi(C_2H_5)_3$ + $CH_2{=}C(CH_2CH_2CH_3)C(CH_2CH_2CH_3){=}CHSi(C_2H_5)_3$ (20)	G -030
SiO_2-$Ni(C_3H_5)$	100	6	$CH_2{=}C(C_3H_7)C(C_3H_7)CH{=}CHSi(C_2H_5)_3$ + $C_3H_7CH{=}CHC(C_3H_7){=}CHSi(C_2H_5)_3$ (75)	230
$[HPt(SiBzMe_2)P(c\text{-}Hex)_3]_2$	65	5	trans-$CH_3CH_2CH_2CH{=}CHSi(C_2H_5)_3$ (88) $CH_3CH_2CH_2C[Si(C_2H_5)_3]{=}CH_2$ (4)	1205
anionite/H_2PtCl_6	85-100	-	trans-$CH_3CH_2CH_2CH{=}CHSi(C_2H_5)_3$ (kin.)	1346
$Rh(acac)_3$	60	12	$CH_3CH_2CH_2CH{=}CHSi(C_2H_5)_3$ (56) $CH_3CH_2CH_2C[Si(C_2H_5)_3]{=}CH_2$ (9)	274
$RhCl(PPh_3)_3$	40	2	$CH_3CH_2CH_2CH{=}CHSi(C_2H_5)_3$ (92)	870
$RhCl(cod)_2$ + DIPPC	100	6	$C_5H_9Si(C_2H_5)_3$ (60)	884
$RhCl(cod)_2$ + DIPPC	20	6	$C_5H_9Si(C_2H_5)_3$ (54)	884
$[RhCl(CO)PPh_3]_2$	20	6 (b)	$CH_3CH_2CH_2CH{=}CHSi(C_2H_5)_3$ (22)	202
$[RhCl(CO)PPh_3]_2$ + ClC_6H_4COOH	20	8 (b)	$CH_3CH_2CH_2CH{=}CHSi(C_2H_5)_3$ (77)	202
$[RhCl(CO)PPh_3]_2$ + t-BuOOH	20	6 (b)	$CH_3CH_2CH_2CH{=}CHSi(C_2H_5)_3$ (75)	202
$[RhCl(c\text{-}C_8H_{14})_2]_2$	60	6	cis-, trans-$CH_3CH_2CH_2CH{=}CHSi(C_2H_5)_3$ + $CH_3CH_2CH{=}CHCH_2Si(C_2H_5)_3$ (75)	160
$[RhCl(c\text{-}C_8H_{14})_2]_2$ + $P(C_6H_4Me)_3$	60	6	cis-$CH_3CH_2CH_2CH{=}CHSi(C_2H_5)_3$ (54) trans-$CH_3CH_2CH_2CH{=}CHSi(C_2H_5)_3$ (35) $CH_3CH_2CH{=}CHCH_2Si(C_2H_5)_3$ (11)	160
$[RhCl(c\text{-}C_8H_{14})_2]_2$ + $P(C_6H_4OMe)_3$	60	6	cis-$CH_3CH_2CH_2CH{=}CHSi(C_2H_5)_3$ (58-82) trans-$CH_3CH_2CH_2CH{=}CHSi(C_2H_5)_3$ (9-43) $CH_3CH_2CH{=}CHCH_2Si(C_2H_5)_3$ (9-10)	160
$[RhCl(c\text{-}C_8H_{14})_2]_2$ + $P(OR)_3$ (R=Me,Et,Ph)	60	6	cis-$CH_3CH_2CH_2CH{=}CHSi(C_2H_5)_3$ (30-39) trans-$CH_3CH_2CH_2CH{=}CHSi(C_2H_5)_3$ (56-63) $CH_3CH_2CH{=}CHCH_2Si(C_2H_5)_3$ (5-6)	160
$[RhCl(c\text{-}C_8H_{14})_2]_2$ + PPh_3	60	6	cis-$CH_3CH_2CH_2CH{=}CHSi(C_2H_5)_3$ (50-63) trans-$CH_3CH_2CH_2CH{=}CHSi(C_2H_5)_3$ (24-40) $CH_3CH_2CH{=}CHCH_2Si(C_2H_5)_3$ (10-14)	160

	$[RhCl(c\text{-}C_8H_{14})_2]_2$ + diphos	60	6	cis-$CH_3CH_2CH_2CH=CHSi(C_2H_5)_3$ (56-61) trans-$CH_3CH_2CH_2CH=CHSi(C_2H_5)_3$ (30-31) $CH_3CH_2CH=CHCH_2Si(C_2H_5)_3$ (12)	160
	$[RhCl(cod)]_2$ + XNC	20-100	6	$C_5H_9Si(C_2H_5)_3$ (54-60)	1
C_6H_{10}	$(CH_3)_2CHCH_2C\equiv CH$				
	$Rh(acac)_3$	60	15	$(CH_3)_2CHCH_2CH=CHSi(C_2H_5)_3$ (62) $(CH_3)_2CHCH_2C[Si(C_2H_5)_3]=CH_2$ (9)	274
C_6H_{10}	$(CH_3)_3CC\equiv CH$				
	none	200	2 (THF)	$(CH_3)_3CCH=CHSi(C_2H_5)_3$ + $(CH_3)_3CC[Si(C_2H_5)_3]=CH_2$ (-)	1263
	H_2PtCl_6	100	2 (THF)	$(CH_3)_3CCH=CHSi(C_2H_5)_3$ (78) $(CH_3)_3CC[Si(C_2H_5)_3]=CH_2$ (9)	954
	H_2PtCl_6	25	- (THF)	$(CH_3)_3CCH=CHSi(C_2H_5)_3$ + $(CH_3)_3CC[Si(C_2H_5)_3]=CH_2$ (kin.)	944
	H_2PtCl_6 (THF)	100	4 (THF)	$(CH_3)_3CCH=CHSi(C_2H_5)_3$ +$(CH_3)_3CC[Si(C_2H_5)_3]=CH_2$ (85)	1263
	H_2PtCl_6 (THF)	20	24	$(CH_3)_3CCH=CHSi(C_2H_5)_3$ +$(CH_3)_3CC[Si(C_2H_5)_3]=CH_2$ (kin.)	945
	H_2PtCl_6 (i-PrOH)	100	4	$(CH_3)_3CCH=CHSi(C_2H_5)_3$ +$(CH_3)_3CC[Si(C_2H_5)_3]=CH_2$ (85)	1253
	H_2PtCl_6 (i-PrOH)	100	4	$(CH_3)_3CCH=CHSi(C_2H_5)_3$ +$(CH_3)_3CC[Si(C_2H_5)_3]=CH_2$ (85)	951
	H_2PtCl_6 (i-PrOH)	100	- (THF)	$(CH_3)_3CCH=CHSi(C_2H_5)_3$ +$(CH_3)_3CC[Si(C_2H_5)_3]=CH_2$ (kin.)	1253
	H_2PtCl_6 (i-PrOH)	150	6	$(CH_3)_3CCH=CHSi(C_2H_5)_3$ (91) $(CH_3)_3CC[Si(C_2H_5)_3]=CH_2$ (5)	708
	$H_2[PtCl_2(OPr)(C\equiv CBu\text{-}t)]$	100	- (THF)	$(CH_3)_3CCH=CHSi(C_2H_5)_3$ +$(CH_3)_3CC[Si(C_2H_5)_3]=CH_2$ (kin.)	1253
	$H_2[PtCl_2(OPr)(C\equiv CBu\text{-}t)]$	100	4 (THF)	$(CH_3)_3CCH=CHSi(C_2H_5)_3$ +$(CH_3)_3CC[Si(C_2H_5)_3]=CH_2$ (95)	1253
	$H_2[PtCl_2(i\text{-}OPr)(C\equiv CBu\text{-}t)]$	-	-	$(CH_3)_3CCH=CHSi(C_2H_5)_3$ (-)	949
	$Pt_2Cl_4C[CH_2C(CH_3)_2OCH(CH_3)_2]_2$	100	2 (THF)	$(CH_3)_3CCH=CHSi(C_2H_5)_3$ (32) $(CH_3)_3CC[Si(C_2H_5)_3]=CH_2$ (48)	954
	$PtCl_2(i\text{-}OPr)_2(C\equiv CBu\text{-}t)_2$	100	4 (THF)	$(CH_3)_3CCH=CHSi(C_2H_5)_3$ +$(CH_3)_3CC[Si(C_2H_5)_3]=CH_2$ (-)	1253
C_6H_{10}	$C_4H_9C\equiv CH$				

Table 8.4

= $(C_2H_5)_3SiH$ = =5=

	$IrH_2(SiEt_3)(cod)(AsPh_3)$	60	1 (dc)	cis,trans-$C_4H_9CH{=}CHSi(C_2H_5)_3$ + $C_4H_9C{\equiv}CSi(C_2H_5)_3$ (-)	360
	$[Ir(OMe)(TFB)]_2$ +P or As donor ligand	60	1 (dc)	cis,trans-$C_4H_9CH{=}CHSi(C_2H_5)_3$ + $C_4H_9C{\equiv}CSi(C_2H_5)_3$ (-)	360
	$[Ir(OMe)(cod)]_2$ +P or As donor ligand	60	1 (dc)	cis,trans-$C_4H_9CH{=}CHSi(C_2H_5)_3$ + $C_4H_9C{\equiv}CSi(C_2H_5)_3$ (-)	360
C_6H_{10}	$CH_3(CH_2)_3C{\equiv}CH$				
	none	200	2 (THF)	$CH_3(CH_2)_3CH{=}CHSi(C_2H_5)_3$ + $CH_3(CH_2)_3C[Si(C_2H_5)_3]{=}CH_2$ (-)	1263
	$AlBr_3$	0	1 (ht)	trans-$CH_3(CH_2)_3CH{=}CHSi(C_2H_5)_3$ (30)	1238
	$AlBr_3$	20	1	cis-$CH_3(CH_2)_3CH{=}CHSi(C_2H_5)_3$ (80)	1238
	$[Pd_2Cl_2(PPh_3)_4]B_{10}Cl_{10}$	100	2	$CH_3(CH_2)_3CH{=}CHSi(C_2H_5)_3$ (90)	1268
	$[Pd_2Cl_2(PPh_3)_4]B_{10}Cl_{10}$	100	2	$CH_3(CH_2)_3CH{=}CHSi(C_2H_5)_3$ (90)	1258
	$[Pd_2Cl_2(PPh_3)_4]B_{12}Cl_{12}$	100	2	$CH_3(CH_2)_3CH{=}CHSi(C_2H_5)_3$ (65)	1258
	$[Pd_2Cl_2(PPh_3)_4]B_{12}Cl_{12}$	100	2	$CH_3(CH_2)_3CH{=}CHSi(C_2H_5)_3$ (65)	1268
	$[Pd_2Cl_2(PPh_3)_6]B_{10}Cl_{10}$	100	2	$CH_3(CH_2)_3CH{=}CHSi(C_2H_5)_3$ (89)	1258
	$[Pd_2Cl_2(PPh_3)_6]B_{10}Cl_{10}$	100	2	$CH_3(CH_2)_3CH{=}CHSi(C_2H_5)_3$ (89)	1268
	$[Pd_2Cl_2(PPh_3)_6]B_{12}Cl_{12}$	100	2	$CH_3(CH_2)_3CH{=}CHSi(C_2H_5)_3$ (86)	1258
	$[Pd_2Cl_2(PPh_3)_6]B_{12}Cl_{12}$	100	2	$CH_3(CH_2)_3CH{=}CHSi(C_2H_5)_3$ (86)	1268
	H_2PtCl_6	100	2 (THF)	$CH_3(CH_2)_3CH{=}CHSi(C_2H_5)_3$ (66) $CH_3(CH_2)_3C[Si(C_2H_5)_3]{=}CH_2$ (14)	954
	H_2PtCl_6	100	2	$CH_3(CH_2)_3CH{=}CHSi(C_2H_5)_3$ (78)	1258
	H_2PtCl_6 (THF)	100	4 (THF)	$CH_3(CH_2)_3CH{=}CHSi(C_2H_5)_3$ + $CH_3(CH_2)_3C[Si(C_2H_5)_3]{=}CH_2$ (-)	1263
	H_2PtCl_6 (THF)	17-45	- (THF)	$CH_3(CH_2)_3CH{=}CHSi(C_2H_5)_3$ + $CH_3(CH_2)_3C[Si(C_2H_5)_3]{=}CH_2$ (kin.)	944
	H_2PtCl_6 (THF)	20	-	trans-$CH_3(CH_2)_3CH{=}CHSi(C_2H_5)_3$ + $CH_3(CH_2)_3C[Si(C_2H_5)_3]{=}CH_2$ (kin.)	946
	H_2PtCl_6 (THF)	20	3 (THF)	$C_6H_{11}Si(C_2H_5)_3$ (kin.)	1252
	H_2PtCl_6 (THF)	r.t.	24 (THF)	$C_6H_{11}Si(C_2H_5)_3$ (kin.)	952
	H_2PtCl_6 (THF)	r.t.	24 (THF)	$C_6H_{11}Si(C_2H_5)_3$ (kin.)	945

Table 8.4
= $(C_2H_5)_3SiH$ = =6=

H_2PtCl_6 (THF) + LiI	70	1	$CH_3(CH_2)_3CH{=}CHSi(C_2H_5)_3$ (20)	1242
H_2PtCl_6 (i-PrOH)	-	- (THF)	$CH_3(CH_2)_3CH{=}CHSi(C_2H_5)_3$ + $CH_3(CH_2)_3C[Si(C_2H_5)_3]{=}CH_2$ (88)	1253
H_2PtCl_6 (i-PrOH)	-	- (THF)	$CH_3(CH_2)_3C[Si(C_2H_5)_3]{=}CH_2$ (kin)	1253
H_2PtCl_6 (i-PrOH)	100	4	$CH_3(CH_2)_3CH{=}CHSi(C_2H_5)_3$ + $CH_3(CH_2)_3C[Si(C_2H_5)_3]{=}CH_2$ (-)	951
H_2PtCl_6 (i-PrOH)	100	10	trans-$CH_3(CH_2)_3CH{=}CHSi(C_2H_5)_3$ (84)	947
H_2PtCl_6 (i-PrOH)	20	5	$CH_3(CH_2)_3CH{=}CHSi(C_2H_5)_3$ (45)	1110
H_2PtCl_6 (i-PrOH)	20	8	$CH_3(CH_2)_3CH{=}CHSi(C_2H_5)_3$ (40-60) $CH_3CH_2CH{=}CHCH_2Si(C_2H_5)_3$ (8)	1112
H_2PtCl_6 (i-PrOH)	24	-	$C_6H_{11}Si(C_2H_5)_3$ (kin.)	952
H_2PtCl_6 (i-PrOH)	50-60	1	$CH_3(CH_2)_3CH{=}CHSi(C_2H_5)_3$ (44) $CH_3CH_2CH{=}CHCH_2Si(C_2H_5)_3$ (24)	1112
H_2PtCl_6 (i-PrOH)	60	24	$CH_3(CH_2)_3CH{=}CHSi(C_2H_5)_3$ + $CH_3(CH_2)_3C{\equiv}CSi(C_2H_5)_3$ (-)	315
H_2PtCl_6 + I_2	20	-	$C_6H_{11}Si(C_2H_5)_3$ (kin.)	1259
$H_2[PtCl_2(OPr)(C{\equiv}CBu\text{-}t)]$	-	- (THF)	$C_6H_{11}Si(C_2H_5)_3$ (kin.)	1253
$H_2[PtCl_2(OPr)(C{\equiv}CBu\text{-}t)]$	-	- (THF)	$CH_3(CH_2)_3CH{=}CHSi(C_2H_5)_3$ + $CH_3(CH_2)_3C[Si(C_2H_5)_3]{=}CH_2$ (kin.)	1253
$Pt_2Cl_4C[CH_2C(CH_3)_2OCH(CH_3)_2]_2$	100	2 (THF)	$CH_3(CH_2)_3CH{=}CHSi(C_2H_5)_3$ (11) $CH_3(CH_2)_3C[Si(C_2H_5)_3]{=}CH_2$ (76)	954
$PtCl_2$(cod)	100	2 (THF)	$CH_3(CH_2)_3CH{=}CHSi(C_2H_5)_3$ (31) $CH_3(CH_2)_3C[Si(C_2H_5)_3]{=}CH_2$ (63)	954
$PtHCl(PPh_3)_2$	100	2 (THF)	$CH_3(CH_2)_3CH{=}CHSi(C_2H_5)_3$ (41) $CH_3(CH_2)_3C[Si(C_2H_5)_3]{=}CH_2$ (53)	954
SiO_2-$Pt(C_3H_5)$	60	6	$C_6H_{11}Si(C_2H_5)_3$ (81)	230
$[HPt(SiBzMe_2)P(c\text{-}Hex)_3]_2$	65	3.75	trans-$CH_3(CH_2)_3CH{=}CHSi(C_2H_5)_3$ (89) $CH_3(CH_2)_3C[Si(C_2H_5)_3]{=}CH_2$ (4)	1205
$[Pt_2Cl_2(PPh_3)_6]B_{10}Cl_{10}$	100	2	$CH_3(CH_2)_3CH{=}CHSi(C_2H_5)_3$ (85)	1268
$[Pt_2Cl_2(PPh_3)_6]B_{10}Cl_{10}$	100	2	$CH_3(CH_2)_3CH{=}CHSi(C_2H_5)_3$ (85)	1258
$[Pt_2Cl_2(PPh_3)_6]B_{12}Cl_{12}$	100	2	$CH_3(CH_2)_3CH{=}CHSi(C_2H_5)_3$ (90)	1268
$[Pt_2Cl_2(PPh_3)_6]B_{12}Cl_{12}$	100	2	$CH_3(CH_2)_3CH{=}CHSi(C_2H_5)_3$ (90)	1258

Table 8.4
= $(C_2H_5)_3SiH$ = =7=

$[PtCl(NH_3)_2(C_8H_8)]NO_3$	100	2 (THF)	$CH_3(CH_2)_3CH{=}CHSi(C_2H_5)_3$ (29) $CH_3(CH_2)_3C[Si(C_2H_5)_3]{=}CH_2$ (52)	954
$[PtCl(NH_3)_2(O_2)(C_8H_{14})]NO_3$	100	2 (THF)	$CH_3(CH_2)_3CH{=}CHSi(C_2H_5)_3$ (31) $CH_3(CH_2)_3C[Si(C_2H_5)_3]{=}CH_2$ (63)	954
$[PtCl(NH_3)_2(py)]NO_3$	100	2 (THF)	$CH_3(CH_2)_3CH{=}CHSi(C_2H_5)_3$ (19) $CH_3(CH_2)_3C[Si(C_2H_5)_3]{=}CH_2$ (67)	954
$Rh(acac)_3$	60	15	$CH_3(CH_2)_3CH{=}CHSi(C_2H_5)_3$ (80) $CH_3(CH_2)_3C[Si(C_2H_5)_3]{=}CH_2$ (7)	274
$Rh_2(OAc)_4$	100	8	$CH_3(CH_2)_3CH{=}CHSi(C_2H_5)_3$ (47)	274
$RhCl(CO)(PPh_3)_2$	100	10	$CH_3(CH_2)_3CH{=}CHSi(C_2H_5)_3$ + $CH_3CH_2CH{=}CHCH_2Si(C_2H_5)_3$ (85)	947
$RhCl(PPh_3)_3$	40	2	$CH_3(CH_2)_3CH{=}CHSi(C_2H_5)_3$ (95)	870
$RhCl(PPh_3)_3$	60	24	$CH_3(CH_2)_3CH{=}CHSi(C_2H_5)_3$ (-)	315
$RhCl(cod)_2$ + DIPPC	60	6	$C_6H_{11}Si(C_2H_5)_3$ (70)	884
$RhCl(cod)_2$ + DMPIC	100	6	$C_6H_{11}Si(C_2H_5)_3$ (75)	884
$RhCl_2[P(c\text{-}Hex)_3]_2$	100	8	$CH_3(CH_2)_3CH{=}CHSi(C_2H_5)_3$ (56)	503
$RhCl_2[P(o\text{-}MeC_6H_4)_3]_2$	100	8	$CH_3(CH_2)_3CH{=}CHSi(C_2H_5)_3$ (44)	503
$RhCl_3 \cdot 4H_2O$	-	-	cis-$CH_3(CH_2)_5CH{=}CH(C_2H_5)_3$/ trans-$CH_3(CH_2)_5CH{=}CHSi(C_2H_5)_3$ (80/20)	956
$RhCl_3 \cdot 4H_2O$	100	2 (MeCN)	$CH_3(CH_2)_5Si(C_2H_5)_3$ (91)	956
$RhCl[\overline{CNPh(CH_2)_2NPh}](cod)$	100	8	$CH_3(CH_2)_3CH{=}CHSi(C_2H_5)_3$ (49)	496
$RhH(CO)(PPh_3)_3$	100	10	$CH_3(CH_2)_3CH{=}CHSi(C_2H_5)_3$ + $CH_3CH_2CH{=}CHCH_2$ $Si(C_2H_5)_3$ (75)	947
$RhI(PPh_3)_3$	100	10	$CH_3(CH_2)_3CH{=}CHSi(C_2H_5)_3$ (79)	947
$SiO_2\text{-}Rh(C_3H_5)_2$	100	2	$CH_3(CH_2)_3CH{=}CHSi(C_2H_5)_3$ (5)	230
$[Rh(CO)_2Cl]_2$	-	-	cis-$CH_3(CH_2)_5CH{=}CH(C_2H_5)_3$ / trans-$CH_3(CH_2)_5CH{=}CHSi(C_2H_5)_3$ (79/21)	956
$[Rh(CO)_2Cl]_2$	100	2 (MeCN)	$CH_3(CH_2)_5Si(C_2H_5)_3$ (84)	956
$[RhCl(C_8H_{14})_2]_2$ + PPh_3	60	- (b)	$C_6H_{11}Si(C_2H_5)_3$ (kin.)	319
$[RhCl(cod)]_2$ + XNC	60-100	6	$C_6H_{11}Si(C_2H_5)_3$ (70-75)	1
$\{RhCl[P(OC_6H_3Me_2)_3]_2\}_2$	100	2	cis-$CH_3(CH_2)_3CH{=}CHSi(C_2H_5)_3$ (41) trans-$CH_3(CH_2)_3CH{=}CHSi(C_2H_5)_3$ (28)	229

	$\{RhCl[P(OC_6H_3Me_2)_3]_2\}_2$	25	18	cis-$CH_3(CH_2)_3CH{=}CHSi(C_2H_5)_3$ (-) trans-$CH_3(CH_2)_3CH{=}CHSi(C_2H_5)_3$ (38)	229
	$\{RhCl[P(OC_6H_4Bu\text{-}t)_3]_2\}_2$	100	2	trans-$CH_3(CH_2)_3CH{=}CHSi(C_2H_5)_3$ (82)	229
	$\{RhCl[P(OC_6H_4Bu\text{-}t)_3]_2\}_2$	25	18	trans-$CH_3(CH_2)_3CH{=}CHSi(C_2H_5)_3$ (29)	229
	$\{RhCl[P(OC_6H_4Me)_3]_2\}_n$	100	2	$CH_3(CH_2)_3CH{=}CHSi(C_2H_5)_3$ (68)	229
	$\{RhCl[P(OC_6H_4Me)_3]_2\}_n$	25	18	$CH_3(CH_2)_3CH{=}CHSi(C_2H_5)_3$ (41)	229
	$\{RhCl[P(OMe)_3]_2\}_2$	100	2	$CH_3(CH_2)_3CH{=}CHSi(C_2H_5)_3$ (65)	229
	$\{RhCl[P(OMe)_3]_2\}_2$	25	6	cis-$(C_2H_5)_3SiCH{=}CH(CH_2)_3CH_3$ (45) trans-$(C_2H_5)_3SiCH{=}CH(CH_2)_3(CH_3)_3$ (55)	229
	$\{RhCl[P(OMe)_3]_2\}_2$	25	6	cis-$(C_2H_5)_3SiCH{=}CH(CH_2)_3CH_3$ (9) trans-$(C_2H_5)_3SiCH{=}CH(CH_2)_3(CH_3)_3$ (14)	229
	$\{RhCl[P(OMe)_3]_2\}_2$	25	6	cis-$(C_2H_5)_3SiCH{=}CH(CH_2)_3CH_3$ (45) trans-$(C_2H_5)_3SiCH{=}CH(CH_2)_3(CH_3)_3$ (55)	229
	$\{RhCl[P(OMe)_3]_2\}_2$	25	18	$CH_3(CH_2)_3CH{=}CHSi(C_2H_5)_3$ (100)	229
	$\{RhCl[P(OMe)_3]_2\}_2$	25	6	cis-$(C_2H_5)_3SiCH{=}CH(CH_2)_3CH_3$ (22) trans-$(C_2H_5)_3SiCH{=}CH(CH_2)_3(CH_3)_3$ (32)	229
	$\{RhH[P(OC_6H_4Me)_3]_2\}_n$	100	2	$CH_3(CH_2)_3CH{=}CHSi(C_2H_5)_3$ (57)	229
	$\{RhH[P(OC_6H_4Me)_3]_2\}_n$	25	18	$CH_3(CH_2)_3CH{=}CHSi(C_2H_5)_3$ (27)	229
	$\{RhH[P(OMe)_3]_2\}_3$	100	2	$CH_3(CH_2)_3CH{=}CHSi(C_2H_5)_3$ (65)	229
	$\{RhH[P(OMe)_3]_2\}_3$	25	18	$CH_3(CH_2)_3CH{=}CHSi(C_2H_5)_3$ (96)	229
	$\{[Rh(CO)Cl]_2C_4H_6\}_n$	-	-	cis-$CH_3(CH_2)_5CH{=}CH(C_2H_5)_3$/ trans-$CH_3(CH_2)_5CH{=}CHSi(C_2H_5)_3$ (78/22)	956
	$\{[Rh(CO)Cl]_2C_4H_6\}_n$	100	2 (MeCN)	$CH_3(CH_2)_5Si(C_2H_5)_3$ (80)	956
C_6H_{10}	$CH_3CH_2C{\equiv}CCH_2CH_3$				
	$[HPt(SiBzMe_2)P(c\text{-}Hex)_3]_2$	65	0.5	cis-$CH_3CH_2CH{=}C[Si(C_2H_5)_3]CH_2CH_3$ (98)	1205
	$N(R){=}C(R)C(R){=}N(R)RhCl(CO)$	r.t.	-	$CH_3CH_2CH[Si(C_2H_5)_3]{=}CHCH_2CH_3$ (95)	166
	$Rh(acac)_3$	60	18	$CH_3CH_2CH{=}C[Si(C_2H_5)_3]CH_2CH_3$ (74)	274
	$Rh(acac)_3 + AlEt_3$	60	8	$CH_3CH_2CH{=}C[Si(C_2H_5)_3]CH_2CH_3$ (31)	271
C_6H_{10}	$CH_3CH_2CH_2C{\equiv}CCH_3$				
	$[HPt(SiBzMe_2)P(c\text{-}Hex)_3]_2$	65	1	$CH_3CH_2CH_2CH{=}C(CH_3)Si(C_2H_5)_3$ (58) $CH_3CH_2CH_2C[Si(C_2H_5)_3]{=}CHCH_3$ (38)	1205

Table 8.4
= $(C_2H_5)_3SiH$ = =9=

	$Rh(acac)_3$	60	18	$CH_3CH_2CH_2CH{=}C(CH_3)Si(C_2H_5)_3$ (38) $CH_3CH_2CH_2C[Si(C_2H_5)_3]{=}CHCH_3$ (44)	274
C_7H_{10}	$(\overline{CH_2)_4C}HC{\equiv}CH$				
	$[HPt(SiBzMe_2)P(c\text{-}Hex)_3]_2$	65	1.5	trans-$(\overline{CH_2)_4C}HCH{=}CHSi(C_2H_5)_3$ (97)	1205
C_7H_{12}	$CH_3(CH_2)_3C{\equiv}CCH_3$				
	$Rh(acac)_3$ + $AlEt_3$	60	8	$CH_3(CH_2)_3CH{=}C(CH_3)Si(C_2H_5)_3$ (35) $CH_3(CH_2)_3C[Si(C_2H_5)_3]{=}CHCH_3$ (25)	271
C_7H_{12}	$CH_3(CH_2)_4C{\equiv}CH$				
	$[HPt(SiBzMe_2)P(c\text{-}Hex)_3]_2$	65	2.5	$CH_3(CH_2)_4CH{=}CHSi(C_2H_5)_3$ (92) $CH_3(CH_2)_4C[Si(C_2H_5)_3]{=}CH_2$ (3)	1205
	C-CN/$Rh_2Cl_2(CO)_4$	80	2	$C_7H_{13}Si(C_2H_5)_3$ (20)	C -030
	SDVB-CH_2Cl/$RhCl(CO)(PPh_3)_2$	80	2	$C_7H_{13}Si(C_2H_5)_3$ (40)	C -022
C_8H_6	$C_6H_5C{\equiv}CH$				
	-	-	-	$C_6H_5C_2H_2Si(C_2H_5)_3$ (kin.)	512
	-	100	2 (solv.)	$C_6H_5CH{=}CHSi(C_2H_5)_3$ + $C_6H_5C[Si(C_2H_5)_3]{=}CH_2$ (90)	954
	none	200	2 (THF)	$C_6H_5CH{=}CHSi(C_2H_5)_3$ (31) $C_6H_5C[Si(C_2H_5)_3]{=}CH_2$ (64)	1263
	H_2OsCl_6 (THF)	110	4	$C_6H_5CH{=}CHSi(C_2H_5)_3$ (59) $C_6H_5C[Si(C_2H_5)_3]{=}CH_2$ (17)	608
	$[Pd_2Cl_2(PPh_3)_4]B_{10}Cl_{10}$	100	2	$C_6H_5CH{=}CHSi(C_2H_5)_3$ (17)	1258
	$[Pd_2Cl_2(PPh_3)_4]B_{10}Cl_{10}$	100	2	$C_6H_5CH{=}CHSi(C_2H_5)_3$ (17)	1268
	$[Pd_2Cl_2(PPh_3)_4]B_{12}Cl_{12}$	100	2	$C_6H_5CH{=}CHSi(C_2H_5)_3$ (29)	1258
	$[Pd_2Cl_2(PPh_3)_4]B_{12}Cl_{12}$	100	2	$C_6H_5CH{=}CHSi(C_2H_5)_3$ (29)	1268
	$[Pd_2Cl_2(PPh_3)_6]B_{10}Cl_{10}$	100	2	$C_6H_5CH{=}CHSi(C_2H_5)_3$ (24)	1258
	$[Pd_2Cl_2(PPh_3)_6]B_{10}Cl_{10}$	100	2	$C_6H_5CH{=}CHSi(C_2H_5)_3$ (24)	1268
	$[Pd_2Cl_2(PPh_3)_6]B_{12}Cl_{12}$	100	2	$C_6H_5CH{=}CHSi(C_2H_5)_3$ (32)	1268
	$[Pd_2Cl_2(PPh_3)_6]B_{12}Cl_{12}$	100	2	$C_6H_5CH{=}CHSi(C_2H_5)_3$ (32)	1258
	H_2PtCl_6	100	2	$C_6H_5CH{=}CHSi(C_2H_5)_3$ (91)	1258
	H_2PtCl_6	100	5	$C_6H_5CH{=}CHSi(C_2H_5)_3$ + $C_6H_5C[Si(C_2H_5)_3]{=}CH_2$ (93)	942
	H_2PtCl_6	60	-	$C_6H_5C_2H_2Si(C_2H_5)_3$ (kin.)	954

H_2PtCl_6	65	6.5	$C_6H_5CH{=}CHSi(C_2H_5)_3$ (36) $C_6H_5C[Si(C_2H_5)_3]{=}CH_2$ (10)	878
H_2PtCl_6(THF) + BF_3	100	1	$C_6H_5CH{=}CHSi(C_2H_5)_3$ + $C_6H_5C[Si(C_2H_5)_3]{=}CH_2$ (70)	S -041
H_2PtCl_6 (THF) + I_2	20	- (h)	$C_6H_5CH{=}CHSi(C_2H_5)_3$ (7) $C_6H_5C[Si(C_2H_5)_3]{=}CH_2$ (3) $C_6H_5C{\equiv}CSi(C_2H_5)_3$ (79)	1259
H_2PtCl_6 ($CH_2{=}CH$-N donor)	100	5	$C_6H_5CH{=}CHSi(C_2H_5)_3$ + $C_6H_5C[Si(C_2H_5)_3]{=}CH_2$ (66)	612
H_2PtCl_6 (THF)	100	1	$C_6H_5CH{=}CHSi(C_2H_5)_3$ (24)	S -041
H_2PtCl_6 (THF)	100	4	$C_6H_5CH{=}CHSi(C_2H_5)_3$ + $C_6H_5C[Si(C_2H_5)_3]{=}CH_2$ (-)	1263
H_2PtCl_6 (THF)	100	1	$C_6H_5CH{=}CHSi(C_2H_5)_3$ + $C_6H_5C[Si(C_2H_5)_3]{=}CH_2$ (24)	1234
H_2PtCl_6 (THF)	20	- (h)	$C_6H_5CH{=}CHSi(C_2H_5)_3$ (44) $C_6H_5C[Si(C_2H_5)_3]{=}CH_2$ (9) $C_6H_5C{\equiv}CSi(C_2H_5)_3$ (3)	1259
H_2PtCl_6 (THF)	20	24 (THF)	$C_6H_5C_2H_2Si(C_2H_5)_3$ (kin.)	945
H_2PtCl_6 (THF) + $AlBr_3$	100	1	$C_6H_5CH{=}CHSi(C_2H_5)_3$ + $C_6H_5C[Si(C_2H_5)_3]{=}CH_2$ (85)	S -041
H_2PtCl_6 (THF) + $AlBr_3$	20	- (h)	$C_6H_5CH{=}CHSi(C_2H_5)_3$ (48) $C_6H_5C[Si(C_2H_5)_3]{=}CH_2$ (14) $C_6H_5C{\equiv}CSi(C_2H_5)_3$ (2)	1259
H_2PtCl_6 (THF) + $AlCl_3$	100	1	$C_6H_5CH{=}CHSi(C_2H_5)_3$ + $C_6H_5C[Si(C_2H_5)_3]{=}CH_2$ (93)	1234
H_2PtCl_6 (THF) + $AlCl_3$	100	1	$C_6H_5CH{=}CHSi(C_2H_5)_3$ + $C_6H_5C[Si(C_2H_5)_3]{=}CH_2$ (85-98)	S -041
H_2PtCl_6 (THF) + $AlCl_3$	120	1	$C_6H_5C_2H_2Si(C_2H_5)_3$ (95)	S -041
H_2PtCl_6 (THF) + $AlCl_3$	20	- (h)	$C_6H_5CH{=}CHSi(C_2H_5)_3$ (69) $C_6H_5C[Si(C_2H_5)_3]{=}CH_2$ (11)	1259
H_2PtCl_6 (THF) + $AlCl_3$	50	2	$C_6H_5C_2H_2Si(C_2H_5)_3$ (54)	S -041
H_2PtCl_6 (THF) + $AlCl_3$ or $GeCl_4$	100	1	$C_6H_5CH{=}CHSi(C_2H_5)_3$ (2) $C_6H_5C[Si(C_2H_5)_3]{=}CH_2$ (21)	1262
H_2PtCl_6 (THF) + AlI_3	20	- (h)	$C_6H_5CH{=}CHSi(C_2H_5)_3$ (15) $C_6H_5C[Si(C_2H_5)_3]{=}CH_2$ (14) $C_6H_5C{\equiv}CSi(C_2H_5)_3$ (50)	1259
H_2PtCl_6 (THF) + CCl_4	100	1	$C_6H_5CH{=}CHSi(C_2H_5)_3$ + $C_6H_5C[Si(C_2H_5)_3]{=}CH_2$ (60)	S -041
H_2PtCl_6 (THF) + $CeCl_3$ or $NdCl_3$	100	1	$C_6H_5CH{=}CHSi(C_2H_5)_3$ (55) $C_6H_5C[Si(C_2H_5)_3]{=}CH_2$ (15)	1262
H_2PtCl_6 (THF) + CuI	20	- (h)	$C_6H_5CH{=}CHSi(C_2H_5)_3$ (47) $C_6H_5C[Si(C_2H_5)_3]{=}CH_2$ (10) $C_6H_5C{\equiv}CSi(C_2H_5)_3$ (4)	1259

Table 8.4
= $(C_2H_5)_3SiH$ =

=11=

H_2PtCl_6 (THF) + $GaCl_3$	100	1	$C_6H_5CH=CHSi(C_2H_5)_3$ + $C_6H_5C[Si(C_2H_5)_3]=CH_2$ (57)	S -041
H_2PtCl_6 (THF) + $GeCl_4$	100	1	$C_6H_5CH=CHSi(C_2H_5)_3$ + $C_6H_5C[Si(C_2H_5)_3]=CH_2$ (87)	S -041
H_2PtCl_6 (THF) + $HgCl_2$	100	1 (THF)	$C_6H_5CH=CHSi(C_2H_5)_3$ + $C_6H_5C[Si(C_2H_5)_3]=CH_2$ (2)	1264
H_2PtCl_6 (THF) + $InCl_3$	100	1	$C_6H_5CH=CHSi(C_2H_5)_3$ + $C_6H_5C[Si(C_2H_5)_3]=CH_2$ (58)	S -041
H_2PtCl_6 (THF) + KI	20	- (h)	$C_6H_5CH=CHSi(C_2H_5)_3$ (53) $C_6H_5C[Si(C_2H_5)_3]=CH_2$ (11) $C_6H_5C\equiv CSi(C_2H_5)_3$ (16)	1259
H_2PtCl_6(THF) + LiBr	20	- (h)	$C_6H_5CH=CHSi(C_2H_5)_3$ (72) $C_6H_5C[Si(C_2H_5)_3]=CH_2$ (14) $C_6H_5C\equiv CSi(C_2H_5)_3$ (4)	1259
H_2PtCl_6(THF) + LiCl	100	1	$C_6H_5CH=CHSi(C_2H_5)_3$ + $C_6H_5C[Si(C_2H_5)_3]=CH_2$ (42)	321
H_2PtCl_6(THF) + LiCl	20	- (h)	$C_6H_5CH=CHSi(C_2H_5)_3$ (76) $C_6H_5C[Si(C_2H_5)_3]=CH_2$ (14) $C_6H_5C\equiv CSi(C_2H_5)_3$ (1)	1259
H_2PtCl_6 (THF) + LiI	20	- (h)	$C_6H_5CH=CHSi(C_2H_5)_3$ (13) $C_6H_5C[Si(C_2H_5)_3]=CH_2$ (10) $C_6H_5C\equiv CSi(C_2H_5)_3$ (63)	1259
H_2PtCl_6 (THF) + LiI	60	1	$C_6H_5CH=CHSi(C_2H_5)_3$ (26)	1242
H_2PtCl_6 (THF) + LiI_3	20	- (h)	$C_6H_5CH=CHSi(C_2H_5)_3$ (6) $C_6H_5C[Si(C_2H_5)_3]=CH_2$ (6) $C_6H_5C\equiv CSi(C_2H_5)_3$ (74)	1259
H_2PtCl_6 (THF) + NaI	20	- (h)	$C_6H_5CH=CHSi(C_2H_5)_3$ (38) $C_6H_5C[Si(C_2H_5)_3]=CH_2$ (9) $C_6H_5C\equiv CSi(C_2H_5)_3$ (33)	1259
H_2PtCl_6 (THF) + Nb or Ta chlorides	100	1	$C_6H_5CH=CHSi(C_2H_5)_3$ (54) $C_6H_5C[Si(C_2H_5)_3]=CH_2$ (21)	1262
H_2PtCl_6 (THF) + NiI_2	20	- (h)	$C_6H_5CH=CHSi(C_2H_5)_3$ (47) $C_6H_5C[Si(C_2H_5)_3]=CH_2$ (15) $C_6H_5C\equiv CSi(C_2H_5)_3$ (24)	1259
H_2PtCl_6 (THF) + R_3SiI	20	- (h)	$C_6H_5CH=CHSi(C_2H_5)_3$ (5-11) $C_6H_5C[Si(C_2H_5)_3]=CH_2$ (trace) $C_6H_5C\equiv CSi(C_2H_5)_3$ (80-86)	1259
H_2PtCl_6 (THF) + $SiCl_4$	100	1	$C_6H_5CH=CHSi(C_2H_5)_3$ + $C_6H_5C[Si(C_2H_5)_3]=CH_2$ (83)	S -041
H_2PtCl_6 (THF) + chlorides	100	1	$C_6H_5CH=CHSi(C_2H_5)_3$ + $C_6H_5C[Si(C_2H_5)_3]=CH_2$ (1-94)	1234
H_2PtCl_6 (i-PrOH)	100	4	trans-$C_6H_5CH=CHSi(C_2H_5)_3$ / $C_6H_5C[Si(C_2H_5)_3]=CH_2$ (72/28)	951

Table 8.4

= $(C_2H_5)_3SiH$ = =12=

$H_2PtCl_6 + I_2$	20	- (solv.)	$C_6H_5CH=CHSi(C_2H_5)_3 + C_6H_5C[Si(C_2H_5)_3]=CH_2 +$ $C_6H_5C_2H_2Si(C_2H_5)_3$ (kin.)	1259
$H_2PtCl_6 + I_2$	20	-	$C_6H_5CH=CHSi(C_2H_5)_3$ (7) $C_6H_5C[Si(C_2H_5)_3]=CH_2$ (10) $C_6H_5C_2H_2Si(C_2H_5)_3$ (63)	1259
$H_2PtCl_6 + I_2$	40	-	$C_6H_5CH=CHSi(C_2H_5)_3$ (36) $C_6H_5C[Si(C_2H_5)_3]=CH_2$ (17) $C_6H_5C_2H_2Si(C_2H_5)_3$ (37)	1259
$H_2PtCl_6 + MX_n$ M = Li, Al, Na, K; X = Cl, Br, I	20	- (h)	$C_6H_5CH=CHSi(C_2H_5)_3 + C_6H_5C[Si(C_2H_5)_3]=CH_2 +$ $C_6H_5C\equiv CSi(C_2H_5)_3$ (-)	1269
$H_2PtCl_6 + R_3SiI$	20	- (h)	$C_6H_5CH=CHSi(C_2H_5)_3 + C_6H_5C[Si(C_2H_5)_3]=CH_2 +$ $C_6H_5C\equiv CSi(C_2H_5)_3$ (-)	1269
$PtCl(NH_3)_2(cod)NO_3$	100	2 (THF)	$C_6H_5CH=CHSi(C_2H_5)_3$ (57) $C_6H_5C[Si(C_2H_5)_3]=CH_2$ (13)	954
$PtCl_2$ (CH_2=CH-N donor)	100	5	$C_6H_5CH=CHSi(C_2H_5)_3 + C_6H_5C[Si(C_2H_5)_3]=CH_2$ (56-83)	612
$PtCl_2(PhCH=CH_2)_2$	r.t.	0.5	trans-$C_6H_5CH=CHSi(C_2H_5)_3$ (72) $C_6H_5[Si(C_2H_5)_3]=CH_2$ (14)	220
$PtCl_2(cod)$	100	2 (THF)	$C_6H_5CH=CHSi(C_2H_5)_3$ / $C_6H_5C[Si(C_2H_5)_3]=CH_2$(82/18)	954
$PtCl_2(cod)$	60	-	$C_6H_5C_2H_2Si(C_2H_5)_3$ (kin.)	954
$PtCl_4$ (CH_2=CH-N donor)	100	5	$C_6H_5CH=CHSi(C_2H_5)_3 + C_6H_5C[Si(C_2H_5)_3]=CH_2$ (51-71)	612
$PtHCl(PPh_3)_2$	100	2 (THF)	$C_6H_5CH=CHSi(C_2H_5)_3$ (55) $C_6H_5C[Si(C_2H_5)_3]=CH_2$ (10)	954
$[HPt(SiEtMe_2)P(c\text{-}Hex)_3]_2$	25	45	$C_6H_5CH=CHSi(C_2H_5)_3 + C_6H_5C[Si(C_2H_5)_3]=CH_2$ (90)	430
$[Pt_2Cl_2(PPh_3)_6]B_{10}Cl_{10}$	100	2	$C_6H_5CH=CHSi(C_2H_5)_3$ (78)	1268
$[Pt_2Cl_2(PPh_3)_6]B_{10}Cl_{10}$	100	2	$C_6H_5CH=CHSi(C_2H_5)_3$ (78)	1258
$[Pt_2Cl_2(PPh_3)_6]B_{12}Cl_{12}$	100	2	$C_6H_5CH=CHSi(C_2H_5)_3$ (88)	1258
$[Pt_2Cl_2(PPh_3)_6]B_{12}Cl_{12}$	100	2	$C_6H_5CH=CHSi(C_2H_5)_3$ (88)	1268
$[PtCl(NH_3)_2O_2(C_8H_{14})]NO_3$	100	2 (THF)	$C_6H_5CH=CHSi(C_2H_5)_3$ (69) $C_6H_5C[Si(C_2H_5)_3]=CH_2$ (13)	954
$[PtCl(NH_3)_2O_2(C_8H_{14})]NO_3$	60	-	$C_6H_5C_2H_2Si(C_2H_5)_3$ (kin.)	954
anionite/H_2PtCl_6	85-100	-	trans-$C_6H_5CH=CHSi(C_2H_5)_3$ (kin.)	1346
glass-NEt_3/H_2PtCl_6	100	8	$C_6H_5CH=CHSi(C_2H_5)_3$ (63) $C_6H_5C[Si(C_2H_5)_3]=CH_2$ (22)	S -036

Table 8.4

= $(C_2H_5)_3SiH$ =   =13=

$N(R)=C(R)C(R)=N(R)RhCl(CO)$	r.t.	-	$C_6H_5CH=CHSi(C_2H_5)_3$ (59) $C_6H_5C[Si(C_2H_5)_3]=CH_2$ (37)	166
$Rh(acac)(CO)_2$	100	5	$C_6H_5CH=CHSi(C_2H_5)_3$ + $C_6H_5C[Si(C_2H_5)_3]=CH_2$ (31)	942
$RhCl(CO)(PPh_3)_2$	100	5	$C_6H_5CH=CHSi(C_2H_5)_3$ + $C_6H_5C[Si(C_2H_5)_3]=CH_2$(40-80)	610
$RhCl(CO)(PPh_3)_2$	100	5	$C_6H_5CH=CHSi(C_2H_5)_3$ + $C_6H_5C[Si(C_2H_5)_3]=CH_2$ (12)	942
$RhCl(PPh_3)_3$	100	5	$C_6H_5CH=CHSi(C_2H_5)_3$ + $C_6H_5C[Si(C_2H_5)_3]=CH_2$ (35)	942
$RhCl(PPh_3)_3$	65	22 (b)	$C_6H_5CH=CHSi(C_2H_5)_3$ (15) $C_6H_5C\equiv CSi(C_2H_5)_3$ (3)	878
$RhCl(PPh_3)_3$	80	65 (b)	$C_6H_5CH=CHSi(C_2H_5)_3$ (44) $C_6H_5C[Si(C_2H_5)_3]=CH_2$ (6)	878
$RhCl(SbPh_3)_3$	100	5	$C_6H_5CH=CHSi(C_2H_5)_3$ + $C_6H_5C[Si(C_2H_5)_3]=CH_2$ (30)	942
$RhCl_3 \cdot 4H_2O$	-	-	cis-$C_6H_5CH=CH(C_2H_5)_3$/trans-$C_6H_5CH[Si(C_2H_5)_3]=CH_2$ (24/76)	956
$RhCl_3 \cdot 4H_2O$	100	2 (THF)	$C_6H_5CH[Si(C_2H_5)_3]=CH_2$ (16) $C_6H_5CH=CHSi(C_2H_5)_3$ (51)	956
$RhCl_3 \cdot 4H_2O$	100	2 (MeCN)	cis,trans-$C_6H_5C(=CH_2)Si(C_2H_5)_3$ + $C_6H_5CH=CHSi(C_2H_5)_3$ (85)	956
$RhCl[P(Ph_2CH)_3]_3$	100	5	$C_6H_5CH=CHSi(C_2H_5)_3$ + $C_6H_5C[Si(C_2H_5)_3]=CH_2$ (67)	942
$RhH(CO)(PPh_3)_3$	100	5	$C_6H_5CH=CHSi(C_2H_5)_3$ + $C_6H_5C[Si(C_2H_5)_3]=CH_2$ (31)	942
$RhH(CO)(PPh_3)_3$	20	24	$C_6H_5CH=CHSi(C_2H_5)_3$ (60)	1265
$RhH(CO)(PPh_3)_3$ + magnetic field	20	24	$C_6H_5CH=CHSi(C_2H_5)_3$ (99)	1265
$RhI(PPh_3)_3$	100	5	$C_6H_5CH=CHSi(C_2H_5)_3$ + $C_6H_5C[Si(C_2H_5)_3]=CH_2$ (31)	942
$[Rh(CO)_2Cl]_2$	-	-	cis-$C_6H_5CH=CH(C_2H_5)_3$ / trans-$C_6H_5CH[Si(C_2H_5)_3]=CH_2$ (6/94)	956
$[Rh(CO)_2Cl]_2$	100	2 (MeCN)	cis,trans-$C_6H_5C(=CH_2)Si(C_2H_5)_3$ + $C_6H_5CH=CHSi(C_2H_5)_3$ (87)	956
$[Rh(CO)_2Cl]_2$	100	2 (THF)	$C_6H_5CH[Si(C_2H_5)_3]=CH_2$ (16) $C_6H_5CH=CHSi(C_2H_5)_3$ (51)	956
cis-$RhCl[=\overline{CNR(CH_2)_2N}R](cod)$ + UV:R=Me,Ph	20	1	$C_6H_5CH=CHSi(C_2H_5)_3$ (60)	660
cis-$RhCl[=\overline{CNR(CH_2)_2N}R](cod)$; R = Me,Ph	20-100	0.5-2	$C_6H_5CH=CHSi(C_2H_5)_3$ + $C_6H_5C[Si(C_2H_5)_3]=CH_2$ (0-98)	660
trans-$RhCl[=\overline{CNR(CH_2)_2N}CH_3](PPh_3)_2$; R = Me, Ph	20-40	1-18 (dcm)	$C_6H_5CH=CHSi(C_2H_5)_3$ (2-100)	660

= $(C_2H_5)_3SiH$ = =14=

	trans-$RhCl[=CNR(CH_2)_2NR](PPh_3)_2$ R=Me,Ph	100	1	$C_6H_5CH=CHSi(C_2H_5)_3$ + $C_6H_5C[Si(C_2H_5)_3]=CH_2$ (99)	660
	$\{[Rh(CO)Cl]_2C_4H_6\}_n$	-	-	cis-$C_6H_5CH=CH(C_2H_5)_3$/trans-$C_6H_5CH[Si(C_2H_5)_3]=CH_2$ (27/73)	956
	$\{[Rh(CO)Cl]_2C_4H_6\}_n$	100	2 (MeCN)	cis,trans-$C_6H_5C(=CH_2)Si(C_2H_5)_3$ + $C_6H_5CH=CHSi(C_2H_5)_3$ (83)	956
	$\{[Rh(CO)Cl]_2C_4H_6\}_n$	100	2 (THF)	$C_6H_5CH[Si(C_2H_5)_3]=CH_2$ (16) $C_6H_5CH=CHSi(C_2H_5)_3$ (51)	956
C_8H_{14}	$CH_3(CH_2)_3C\equiv CCH_2CH_3$				
	$Rh(acac)_3$ + $AlEt_3$	60	10	$CH_3(CH_2)_3CH=C(CH_2CH_3)Si(C_2H_5)_3$ (27) $CH_3(CH_2)_3C[Si(C_2H_5)_3]=CHCH_2CH_3$ (25)	271
C_8H_{14}	$CH_3(CH_2)_4C\equiv CCH_3$				
	$Rh(acac)_3$ + $AlEt_3$	60	10	$CH_3(CH_2)_4CH=C(CH_3)Si(C_2H_5)_3$ (32) $CH_3(CH_2)_4C[Si(C_2H_5)_3]=CHCH_3$ (25)	271
C_8H_{14}	$CH_3(CH_2)_5C\equiv CH$				
	$AlBr_3$	0	1	cis-$CH_3(CH_2)_5CH=CHSi(C_2H_5)_3$ (90)	1238
	Rh complex	80	5 (b)	$CH_3(CH_2)_5CH=CHSi(C_2H_5)_3$ (85)	1342
C_8H_{14}	$CH_3CH_2CH_2C\equiv CCH_2CH_2CH_3$				
	$Rh(acac)_3$ + $AlEt_3$	60-120	5-11	$CH_3CH_2CH_2CH=C(CH_2CH_2CH_3)Si(C_2H_5)_3$ (94-98)	271
$C_{10}H_{10}$	p-$CH_3C_6H_4CH_2C\equiv CH$				
	H_2PtCl_6 (i-PrOH)	90-100	4	p-$CH_3C_6H_4CH_2CH=CHSi(C_2H_5)_3$ (-)	1077
$C_{12}H_{16}$	Fig.198.				
	H_2PtCl_6 (i-PrOH)	150	6	Fig.310. (84)	708
$C_{14}H_{10}$	$C_6H_5C\equiv CC_6H_5$				
	H_2PtCl_6 (i-PrOH)	100	2	$C_6H_5CH=C(C_6H_5)Si(C_2H_5)_3$ (80)	837
	$[HPt(SiBzMe_2)P(c\text{-}Hex)_3]_2$	50	0.75	$C_6H_5CH=C(C_6H_5)Si(C_2H_5)_3$ (95)	430

Table 8.5

= $(C_2H_5)_3SiH$ = =1=

C_3H_3	$CH \equiv CCH_2Cl$				
	$AlBr_3$	0	1 (ht)	cis-$(C_2H_5)_3SiCH=CHCH_3$ (25)	1238
	H_2PtCl_6 (THF)	20	24 (THF)	$(C_2H_5)_3SiC_2H_2CH_2Cl$ (kin.)	945
	H_2PtCl_6(THF) + I_2	10	-	$(C_2H_5)_3SiC \equiv CCH_2Cl$ + $(C_2H_5)_3SiCH=C=CH_2$ + $(C_2H_5)_3SiC \equiv CCH_3$ + $(C_2H_5)_3SiCH=CHCH_2Cl$ (61)	957
	H_2PtCl_6 (i-PrOH)	115	-	$(C_2H_5)_3SiCH=CHCH_2Cl+(C_2H_5)_3SiC(=CH_2)CH_2Cl$ (62)	1245
	H_2PtCl_6 (i-PrOH)	50	6	$(C_2H_5)_3SiCH=CHCH_2Cl$ (40) $(C_2H_5)_3SiC(=CH_2)CH_2Cl$ (37) $(C_2H_5)_3SiCH=CHCH_3$ (5)	705
	anionite/H_2PtCl_6	85-100	-	$(C_2H_5)_3SiCH=CHCH_2Cl$ (kin.)	1346
C_3H_4	$CH \equiv CCH_2OH$				
	H_2PtCl_6	-	-	$(C_2H_5)_3SiCH=CHCH_2OH$ + $CH_2=C(CH_2OH)Si(C_2H_5)_3$ (-)	1049
	H_2PtCl_6	115	-	$(C_2H_5)_3SiCH=CHCH_2OH+CH_2=C(CH_2OH)Si(C_2H_5)_3$ (67)	1245
	H_2PtCl_6 (THF)	100	4 (a)	$(C_2H_5)_3SiCH=CHCH_2OH$ (40) $CH_2=C(CH_2OH)Si(C_2H_5)_3$ (19) $CH_2=C[Si(C_2H_5)_3]CH_2OSi(C_2H_5)_3$ (40)	1263
	H_2PtCl_6 (THF)	100	4	$(C_2H_5)_3SiCH=CHCH_2OH$ (26) $CH_2=C(CH_2OH)Si(C_2H_5)_3$ (29) $CH_2=C[Si(C_2H_5)_3]CH_2OSi(C_2H_5)_3$ (34)	1263
	H_2PtCl_6 (THF)	100	4 (THF)	$(C_2H_5)_3SiCH=CHCH_2OH$ (51) $CH_2=C(CH_2OH)Si(C_2H_5)_3$ (41) $CH_2=C[Si(C_2H_5)_3]CH_2OSi(C_2H_5)_3$ (8)	1263
	H_2PtCl_6 (i-PrOH)	100	4	$(C_2H_5)_3SiCH=CHCH_2OH$ /$CH_2=C(CH_2OH)Si(C_2H_5)_3$ (48/52)	951
	H_2PtCl_6 (i-PrOH)	70-120	3	$(C_2H_5)_3SiCH=CHCH_2OH$ (80) $(C_2H_5)_3SiCH=CHCH_2OSi(C_2H_5)_3$ (13)	706
C_4H_4	$CH \equiv CCOCH_3$				
	H_2PtCl_6	70-80	1	$(C_2H_5)_3SiCH=CHCOCH_3$ + $CH_2=C(COCH_3)Si(C_2H_5)_3$ (62)	604
	H_2PtCl_6 (i-PrOH)	100	4	$(C_2H_5)_3SiCH=CHCOCH_3$ /$CH_2=C(COCH_3)Si(C_2H_5)_3$ (45/55)	951
C_4H_4	$CH \equiv CCOOCH_3$				
	N(R)=C(R)C(R)=N(R)RhCl(CO)	r.t.	3-6	$CH_2=C(COOCH_3)[Si(C_2H_5)_3]$ (16) $(C_2H_5)_3SiCH=CHCOOCH_3$ (14)	166
C_4H_4	$ClCH_2C \equiv CCH_2Cl$				

Table 8.5
= $(C_2H_5)_3SiH$ = =2=

	H_2PtCl_6	100	-	$ClCH_2CH=C(CH_2Cl)Si(C_2H_5)_3$ (~100)	83
	H_2PtCl_6 (i-PrOH)	90	1	$ClCH_2CH=C(CH_2Cl)Si(C_2H_5)_3$ (58)	730
C_4H_5	$ClCH_2OCH_2C\equiv CH$				
	H_2PtCl_6	60-100	0.5	$ClCH_2OCH_2CH=CHSi(C_2H_5)_3$ (60)	1086
C_4H_6	$CH_3CH(OH)C\equiv CH$				
	-	reflux	-	$CH_3CH(OH)CH=CHSi(C_2H_5)_3$ + $CH_3CH(OH)C(=CH_2)Si(C_2H_5)_3$ (-)	1119
	H_2PtCl_6	70-80	1	$CH_3CH(OH)CH=CHSi(C_2H_5)_3$ + $CH_3CH[OSi(C_2H_5)_3]CH=CHSi(C_2H_5)_3$ + $CH_3CH[OSi(C_2H_5)_3]C(=CH_2)Si(C_2H_5)_3$ (70)	604
	H_2PtCl_6 (THF)	100	4	$CH_3CH(OH)CH=CHSi(C_2H_5)_3$ + $CH_3CH(OH)C(=CH_2)Si(C_2H_5)_3$ (-)	1263
	H_2PtCl_6 (i-PrOH)	100	4	$CH_3CH(OH)CH=CHSi(C_2H_5)_3$ / $CH_3CH(OH)C(=CH_2)Si(C_2H_5)_3$ (55/45)	951
C_4H_7	$(CH_3)_2SiClC\equiv CH$				
	H_2PtCl_6 (i-PrOH)	70-80	1.5-2	$(CH_3)_2SiClCH=CHSi(C_2H_5)_3$ (45)	603
C_5H_5	$CCl_3CH(OH)CH_2C\equiv CH$				
	H_2PtCl_6	75-80	6	$CCl_3CH(OH)CH_2CH=CHSi(C_2H_5)_3$ (85)	1081
C_5H_6	$CH\equiv CCOOC_2H_5$				
	H_2PtCl_6 (THF)	20	24 (THF)	$(C_2H_5)_3SiC_2H_2COOC_2H_5$ (kin.)	945
C_5H_6	$CH_3C(OH)(CF_3)C\equiv CH$				
	H_2PtCl_6	reflux	-	$CH_3C(OH)(CF_3)CH=CHSi(C_2H_5)_3$ + $CH_3C(OH)(CF_3)C(=CH_2)Si(C_2H_5)_3$ (-)	1050
C_5H_6	$CH_3COOCH_2C\equiv CH$				
	H_2PtCl_6	115	-	$CH_3COOCH_2CH=CHSi(C_2H_5)_3$ + $CH_3COOCH_2C(=CH_2)Si(C_2H_5)_3$ (72)	1245
C_5H_6	$CH_3OCH=CHC\equiv CH$				
	H_2PtCl_6 (i-PrOH)	100	1.5	$CH_3OCH=CHCH=CHSi(C_2H_5)_3$ (57)	617
C_5H_7	$CH_2=CHSiCl(CH_3)C\equiv CH$				
	H_2PtCl_6 (i-PrOH)	78-80	1.5-2	$CH_2=CHSiCl(CH_3)CH=CHSi(C_2H_5)_3$ (32)	603
C_5H_7	$ClCH_2CH_2OCH_2C\equiv CH$				

	H_2PtCl_6 (i-PrOH)	70-80	3	$ClCH_2CH_2OCH_2CH=CHSi(C_2H_5)_3$ + $ClCH_2CH_2OCH_2C(=CH_2)Si(C_2H_5)_3$ (73)	549
C_5H_8	$(CH_3)_2C(OH)C\equiv CH$				
	H_2PtCl_6	reflux	-	$(CH_3)_2C(OH)CH=CHSi(C_2H_5)_3$ + $(CH_3)_2C(OH)C(=CH_2)Si(C_2H_5)_3$ (-)	1050
	H_2PtCl_6 (i-PrOH)	100	4	$(CH_3)_2C(OH)CH=CHSi(C_2H_5)_3$ / $(CH_3)_2C(OH)C(=CH_2)Si(C_2H_5)_3$ (61/39)	951
	$[HPt(SiMe_2Bz)P(c\text{-}Hex)_3]_2$	65	20	trans-$(CH_3)_2C(OH)CH=CHSi(C_2H_5)_3$ (82) · $(CH_3)_2C(OH)C(=CH_2)Si(C_2H_5)_3$ (6)	1205
C_5H_8	$C_2H_5OCH_2C\equiv CH$				
	H_2PtCl_6	115	-	$C_2H_5OCH_2CH=CHSi(C_2H_5)_3$ + $C_2H_5OCH_2C(=CH_2)Si(C_2H_5)_3$ (78)	1245
C_5H_8	$CH\equiv CCH_2OCH_2CH_2OH$				
	H_2PtCl_6	90	4	$(C_2H_5)_3SiCH=CHCH_2OCH_2CH_2OH$ (35) $(C_2H_5)_3SiCH=CHCH_2OCH_2CH_2OSi(C_2H_5)_3$ (18)	994
C_5H_9	$CH\equiv CCH_2N(CH_3)_2$				
	H_2PtCl_6 (THF)	150-160	-	$(C_2H_5)_3SiCH=CHCH_2N(CH_3)_2$ + $(C_2H_5)_3SiC(=CH_2)CH_2N(CH_3)_2$ + $(C_2H_5)_3SiCH_2CH=CHN(CH_3)_2$ (-)	1204
C_5H_{10}	$(CH_3)_3SiC\equiv CH$				
	H_2PtCl_6 (THF)	20	24 (THF)	$(CH_3)_3SiC_2H_2Si(C_2H_5)_3$ (kin.)	945
C_5H_{10}	$CH_3OSi(CH_3)_2C\equiv CH$				
	H_2PtCl_6 (i-PrOH)	80	0.5	trans-$CH_3OSi(CH_3)_2CH=CHSi(C_2H_5)_3$ + $CH_3OSi(CH_3)_2C(=CH_2)Si(C_2H_5)_3$ (77)	1309
C_6H_8	$C_2H_5OCH=CHC\equiv CH$				
	H_2PtCl_6	90	6-8	$C_2H_5OCH=CHCH=CHSi(C_2H_5)_3$ (72)	1205
C_6H_8	$CH\equiv CCH_2COOC_2H_5$				
	H_2PtCl_6 (THF)	20	24 (THF)	$(C_2H_5)_3SiC_2H_2CH_2COOC_2H_5$ (kin.)	945
C_6H_8	$CH\equiv CCOO\text{-}i\text{-}C_3H_7$				
	$N(R)=C(R)C(R)=N(R)RhCl(CO)$	r.t.	3-6	$CH_2=C(COO\text{-}i\text{-}C_3H_7)[Si(C_2H_5)_3]$ (14) $(C_2H_5)_3SiCH=CHCOO\text{-}i\text{-}C_3H_7$ (11)	166
C_6H_8	$CH_2=CHC\equiv CCH_2CH_2OH$				

Table 8.5

= $(C_2H_5)_3SiH$ =

=4=

	H_2PtCl_6	-	-	$CH_2{=}CHCH{=}C[Si(C_2H_5)_3]CH_2CH_2OH$ + $CH_2{=}CHCH{=}C[Si(C_2H_5)_3]CH_2CH_2OSi(C_2H_5)_3$ (-)	453
C_6H_8	$CH_3OCH_2C{\equiv}CCH{=}CH_2$				
	H_2PtCl_6	100	-	$CH_3OCH_2C[Si(C_2H_5)_3]{=}CHCH{=}CH_2$ (58-60) $CH_3OCH_2CH{=}C[Si(C_2H_5)_3]CH{=}CH_2$ (3-30) $CH_3OCH_2C[Si(C_2H_5)_3]{=}C{=}CHCH_3$ (9-10)	535
	H_2PtCl_6	100	-	$CH_3OCH_2CH{=}C[Si(C_2H_5)_3]CH{=}CH_2$ (-)	534
C_6H_8	$\overline{OCH_2C}HCH_2OCH_2C{\equiv}CH$				
	H_2PtCl_6 (i-PrOH)	85	-	$\overline{OCH_2C}HCH_2OCH_2CH{=}CHSi(C_2H_5)_3$ (84)	997
C_6H_9	$BrCH_2CH(OC_2H_5)C{\equiv}CH$				
	H_2PtCl_6 (i-PrOH)	100	1	$BrCH_2CH(OC_2H_5)CH{=}CHSi(C_2H_5)_3$ (60)	616
C_6H_9	$ClCH_2CH_2OCH_2OCH_2C{\equiv}CH$				
	H_2PtCl_6 (THF)	60-70	5	$ClCH_2CH_2OCH_2OCH_2CH{=}CHSi(C_2H_5)_3$ (78)	1082
	H_2PtCl_6 (i-PrOH)	70	4	$ClCH_2CH_2OCH_2OCH_2CH{=}CHSi(C_2H_5)_3$ (52)	12
C_6H_{10}	$(CH_3)_2CHCH(OH)C{\equiv}CH$				
	H_2PtCl_6 (i-PrOH)	2	5	$(CH_3)_2CHCH(OH)CH{=}CHSi(C_2H_5)_3$ (-)	1072
C_6H_{10}	$(CH_3)_3SiC{\equiv}CCHO$				
	H_2PtCl_6 (i-PrOH)	100	4	$(CH_3)_3SiCH{=}C(CHO)Si(C_2H_5)_3$ + $(CH_3)_3SiC[Si(C_2H_5)_3]{=}CHCHO$ (-)	953
C_6H_{10}	$C_2H_5OCH(CH_3)C{\equiv}CH$				
	H_2PtCl_6 (i-PrOH)	100	2	$C_2H_5OCH(CH_3)CH{=}CHSi(C_2H_5)_3$ (64)	616
C_6H_{10}	$C_2H_5OCH_2C{\equiv}CCH_3$				
	-	-	-	$C_2H_5OCH_2C[Si(C_2H_5)_3]{=}CHCH_3$ (-)	805
C_6H_{10}	$CH_3CH(OH)CH_2OCH_2C{\equiv}CH$				
	H_2PtCl_6	80-90	-	$CH_3CH(OH)CH_2OCH_2CH{=}CHSi(C_2H_5)_3$ (72)	546
C_6H_{10}	$CH_3CH_2C(OH)(CH_3)C{\equiv}CH$				
	$[HPt(SiMe_2Bz)P(c\text{-}Hex)_3]_2$	65	18	$CH_3CH_2C(OH)(CH_3)CH{=}CHSi(C_2H_5)_3$ (84) $CH_3CH_2C(OH)(CH_3)C({=}CH_2)Si(C_2H_5)_3$ (6)	1205
C_6H_{10}	$CH_3OC(CH_3)_2C{\equiv}CH$				
	H_2PtCl_6 (i-PrOH)	100	3	$CH_3OC(CH_3)_2CH{=}CHSi(C_2H_5)_3$ (64)	616
C_6H_{10}	$CH_3OCH_2C{\equiv}CCH_2OCH_3$				

	H_2PtCl_6 (i-PrOH)	70-90	2	$CH_3OCH_2CH{=}C[Si(C_2H_5)_3]CH_2OCH_3$ (51)	569
	$N(R){=}C(R)C(R){=}N(R)RhCl(CO)$	r.t.	-	$CH_3OCH_2CH{=}C[Si(C_2H_5)_3]CH_2OCH_3$ (97)	166
C_6H_{11}	$CH{\equiv}CCH_2CH_2N(CH_3)_2$				
	H_2PtCl_6 (i-PrOH)	130-140	6	$(C_2H_5)_3SiCH{=}CHCH_2CH_2N(CH_3)_2$ (27)	1137
C_6H_{12}	$CH_3CH_2OSi(CH_3)_2C{\equiv}CH$				
	H_2PtCl_6 (i-PrOH)	80	0.5	trans-$CH_3CH_2OSi(CH_3)_2CH{=}CHSi(C_2H_5)_3$ + $CH_3CH_2OSi(CH_3)_2C({=}CH_2)Si(C_2H_5)_3$ (79)	1309
C_7H_4	Fig.311.				
	H_2PtCl_6 (i-PrOH)	60-70	1	Fig.312. (47)	312
C_7H_9	$CCl_3CH_2OCH(CH_3)OCH_2C{\equiv}CH$				
	H_2PtCl_6 (i-PrOH)	76-80	6	$CCl_3CH_2OCH(CH_3)OCH_2CH{=}CHSi(C_2H_5)_3$ (32)	1076
C_7H_{10}	$CH{\equiv}CSi(CH_3)_2OCH_2C{\equiv}CH$				
	H_2PtCl_6 (i-PrOH)	-	-	$CH{\equiv}CSi(CH_3)_2OCH_2CH{=}CHSi(C_2H_5)_3$ (-)	542
C_7H_{10}	$CH_2{=}CHCH_2OCH_2OCH_2C{\equiv}CH$				
	H_2PtCl_6 (i-PrOH)	70	4	$CH_2{=}CHCH_2OCH_2OCH_2CH{=}CHSi(C_2H_5)_3$ (29) $(C_2H_5)_3Si(CH_2)_3OCH_2OCH_2CH{=}CHSi(C_2H_5)_3$ (11)	732
C_7H_{10}	$CH_3CH(OCH_3)C{\equiv}CCH{=}CH_2$				
	H_2PtCl_6	100	-	$CH_3CH(OCH_3)CH{=}C[Si(C_2H_5)_3]CH{=}CH_2$ (-)	534
C_7H_{10}	$CH_3OCH_2C{\equiv}CC(CH_3){=}CH_2$				
	H_2PtCl_6 (i-PrOH)	60-70	4	$CH_3OCH_2CH{=}C[C(CH_3){=}CH_2]Si(C_2H_5)_3$ (29)	1068
C_7H_{11}	$(CH_2)_4NCH_2C{\equiv}CH$				
	H_2PtCl_6 (THF)	150-240	3	$(CH_2)_4NCH_2CH{=}CHSi(C_2H_5)_3$ + $(CH_2)_4NC({=}CH_2)Si(C_2H_5)_3$ (60-70)	611
	H_2PtCl_6 (i-PrOH)	reflux	10	trans-$(CH_2)_4NCH_2CH{=}CHSi(C_2H_5)_3$ (28-53)	700
	$RhCl(CO)(AsPh_3)_2$	200	3	$(CH_2)_4NCH_2CH{=}CHSi(C_2H_5)_3$ + $(CH_2)_4NC({=}CH_2)Si(C_2H_5)_3$ (60-70)	611
	$RhCl(CO)(PPh_3)_2$	200	3	$(CH_2)_4NCH_2CH{=}CHSi(C_2H_5)_3$ + $(CH_2)_4NC({=}CH_2)Si(C_2H_5)_3$ (60-70)	611
	$RhCl(SbPh_3)_3$	200	3	$(CH_2)_4NCH_2CH{=}CHSi(C_2H_5)_3$ + $(CH_2)_4NC({=}CH_2)Si(C_2H_5)_3$ (60-70)	611

Table 8.5

= $(C_2H_5)_3SiH$ = =6=

	$RhI(PPh_3)_3$	200	3	$(\overline{CH_2)_4N}CH_2CH=CHSi(C_2H_5)_3$ + $(\overline{CH_2)_4N}C(=CH_2)Si(C_2H_5)_3$ (60-70)	611
C_7H_{11}	$(CH_3)_3SiOCH_2OCH_2C\equiv CH$				
	H_2PtCl_6 (i-PrOH)	105	4	$(CH_3)_3SiOCH_2OCH_2CH=CHSi(C_2H_5)_3$ (-)	809
C_7H_{11}	$O(CH_2CH_2)_2NCH_2C\equiv CH$				
	H_2PtCl_6 (i-PrOH)	reflux	10	trans-$O(CH_2CH_2)_2NCH_2CH=CHSi(C_2H_5)_3$ (12)	700
C_7H_{12}	$(CH_3)_2C(OH)C(OH)(CH_3)C\equiv CH$				
	H_2PtCl_6 (i-PrOH)	20-100	0.5-24(solv.)	$(CH_3)_2C(OH)C(OH)(CH_3)CH=CHSi(C_2H_5)_3$ (5-20) $(CH_3)_2C(OH)C(OH)(CH_3)C(=CH_2)Si(C_2H_5)_3$ (2-6)	247
	H_2PtCl_6 (i-PrOH)	90	4	$(CH_3)_2C(OH)C(OH)(CH_3)CH=CHSi(C_2H_5)_3$ (20) $(CH_3)_2C(OH)C(OH)(CH_3)C(=CH_2)Si(C_2H_5)_3$ (6)	252
C_7H_{12}	$(CH_3)_2CHCH_2CH(OH)C\equiv CH$				
	H_2PtCl_6 (i-PrOH)	80-100	6	$(CH_3)_2CHCH_2CH(OH)CH=CHSi(C_2H_5)_3$ (-)	1083
C_7H_{12}	$(CH_3)_3SiC\equiv CCOCH_3$				
	H_2PtCl_6 (i-PrOH)	100	4	$(CH_3)_3SiCH=C[Si(C_2H_5)_3]COCH_3$ + $(CH_3)_3SiC[Si(C_2H_5)_3=CHCOCH_3$ (-)	953
C_7H_{12}	$CH\equiv CCH_2O(CH_2)_4OH$				
	-	-	-	$(C_2H_5)_3SiCH=CHCH_2O(CH_2)_4OH$ (-)	545
C_7H_{12}	$CH_2=CHSi(CH_3)_2OCH_2C\equiv CH$				
	H_2PtCl_6 (i-PrOH)	-	-	$CH_2=CHSi(CH_3)_2OCH_2CH=CHSi(C_2H_5)_3$ (-)	542
C_7H_{12}	$CH_3OCH_2CH(OH)CH_2OCH_2C\equiv CH$				
	H_2PtCl_6 6 H_2O	-		$CH_3OCH_2CH[OSi(C_2H_5)_3]CH_2OCH_2CH=CHSi(C_2H_5)_3$ (-)	807
C_7H_{12}	$CH_3OCH_2CH_2C(OH)(CH_3)C\equiv CH$				
	H_2PtCl_6 (i-PrOH)	60-125	5	cis-$CH_3OCH_2CH_2C(OH)(CH_3)CH=CHSi(C_2H_5)_3$ (15) trans-$CH_3OCH_2CH_2C(OH)(CH_3)CH=CHSi(C_2H_5)_3$ (18) $CH_3OCH_2CH_2C(OH)(CH_3)C(=CH_2)Si(C_2H_5)_3$ (43)	445
C_7H_{12}	$CH_3OCH_2OCH_2CH_2OCH_2C\equiv CH$				
	H_2PtCl_6	-	-	$CH_3OCH_2OCH_2CH_2OCH_2CH=CHSi(C_2H_5)_3$ (-)	994
C_7H_{13}	$CH\equiv CCH_2N(C_2H_5)_2$				
	H_2PtCl_6 (THF)	20	24 (THF)	$(C_2H_5)_3SiC_2H_2CH_2N(C_2H_5)_2$ (kin.)	945

	$RhCl(CO)(AsPh_3)_2$	200	3	$(C_2H_5)_3SiCH=CHCH_2N(C_2H_5)_2$ + $(C_2H_5)_3SiC(=CH_2)CH_2N(C_2H_5)_2$ (60-70)	611
	$RhCl(CO)(PPh_3)_2$	200	3	$(C_2H_5)_3SiCH=CHCH_2N(C_2H_5)_2$ + $(C_2H_5)_3SiC(=CH_2)CH_2N(C_2H_5)_2$ (60-70)	611
	$RhCl(SbPh_3)_3$	200	3	$(C_2H_5)_3SiCH=CHCH_2N(C_2H_5)_2$ + $(C_2H_5)_3SiC(=CH_2)CH_2N(C_2H_5)_2$ (60-70)	611
	$RhI(PPh_3)_3$	200	3	$(C_2H_5)_3SiCH=CHCH_2N(C_2H_5)_2$ + $(C_2H_5)_3SiC(=CH_2)CH_2N(C_2H_5)_2$ (60-70)	611
C_8H_6	$C_6H_5OC\equiv CH$				
	H_2PtCl_6 (THF)	100	4-9	$C_6H_5OCH=CHSi(C_2H_5)_3$ + $C_6H_5OC(=CH_2)Si(C_2H_5)_3$ (-)	1263
	H_2PtCl_6 (THF)	100	4-9	$C_6H_5OCH=CHSi(C_2H_5)_3$ + $C_6H_5OC(=CH_2)Si(C_2H_5)_3$ (-)	1102
	H_2PtCl_6 (THF)	20	24 (THF)	$C_6H_5OC_2H_2Si(C_2H_5)_3$ (kin.)	945
	H_2PtCl_6 (i-PrOH)	100	4	$C_6H_5OCH=CHSi(C_2H_5)_3$ /$C_6H_5OC(=CH_2)Si(C_2H_5)_3$ (87/13)	951
C_8H_{10}	$(CH_2OCH_2C\equiv CH)_2$				
	H_2PtCl_6 (i-PrOH)	70-80	8	$[CH_2OCH_2CH=CHSi(C_2H_5)_3]_2$ (70)	493
C_8H_{10}	$CH_3CH(OCH_2C\equiv CH)_2$				
	H_2PtCl_6	-	-	$CH_3CH(OCH_2C\equiv CH)OCH_2CH=CHSi(C_2H_5)_3$ (53)	1087
C_8H_{11}	$CCl_3CH(OH)OCH(CH_2CH_2CH_3)C\equiv CH$				
	H_2PtCl_6 (i-PrOH)	50-55	6	$CCl_3CH(OH)OCH(CH_2CH_2CH_3)CH=CHSi(C_2H_5)_3$ (-)	67
C_8H_{12}	$CH\equiv CCH_2OCH_2CH(OH)CH_2OCOCH_3$				
	H_2PtCl_6 (i-PrOH)	reflux	40 (b)	$(C_2H_5)_3SiCH=CHCH_2OCH_2CH(OH)CH_2OCOCH_3$ (73)	1250
C_8H_{12}	$CH_2=CHCH_2OCH(CH_3)OCH_2C\equiv CH$				
	H_2PtCl_6 (i-PrOH)	70	4	$CH_2=CHCH_2OCH(CH_3)OCH_2CH=CHSi(C_2H_5)_3$ (39) $(C_2H_5)_3Si(CH_2)_3OCH(CH_3)OCH_2CH=CHSi(C_2H_5)_3$ (11)	732
C_8H_{12}	$CH_3CH_2CH(OCH_3)C\equiv CCH=CH_2$				
	H_2PtCl_6	100	-	$CH_3CH_2CH(OCH_3)CH=C[Si(C_2H_5)_3]CH=CH_2$ (-)	534
C_8H_{12}	$CH_3OC(CH_3)_2C\equiv CCH=CH_2$				
	H_2PtCl_6	25-150	-	$CH_3OC(CH_3)_2CH=C[Si(C_2H_5)_3]CH=CH_2$ + $CH_3OC(CH_3)_2C[Si(C_2H_5)_3]=CHCH=CH_2$ + $CH_3OC(CH_3)_2C[Si(C_2H_5)_3]=C=CHCH_3$ (78)	530

	H_2PtCl_6	25-150		$CH_3OC(CH_3)_2CH=C[Si(C_2H_5)_3]CH=CH_2$ + $CH_3OC(CH_3)_2C[Si(C_2H_5)_3]=CHCH=CH_2$ + $CH_3OC(CH_3)_2C[Si(C_2H_5)_3]=C=CHCH_3$ (78)	538
C_8H_{12}	$\overline{OCH_2C}HCH_2OCH_2CH_2OCH_2C\equiv CH$				
	H_2PtCl_6	reflux	15 (b)	$\overline{OCH_2C}HCH_2OCH_2CH_2OCH_2CH=CHSi(C_2H_5)_3$ (74)	1162
C_8H_{13}	$(C_2H_5)_2NCH=CHC\equiv CH$				
	H_2PtCl_6	100-110	12-14	$(C_2H_5)_2NCH=CHCH=CHSi(C_2H_5)_3$ (28)	1098
C_8H_{13}	$(\overline{CH_2)_5N}CH_2C\equiv CH$				
	H_2PtCl_6	reflux	0.5-10	$(\overline{CH_2)_5N}CH_2CH=CHSi(C_2H_5)_3$ (25-53)	701
	H_2PtCl_6 (THF)	150-240	3	$(\overline{CH_2)_5N}CH_2CH=CHSi(C_2H_5)_3$ + $(\overline{CH_2)_5N}CH_2C(=CH_2)Si(C_2H_5)_3$ (60-70)	611
	H_2PtCl_6 (i-PrOH)	reflux	10	trans-$(\overline{CH_2)_5N}CH_2CH=CHSi(C_2H_5)_3$ (27)	700
	$RhCl(CO)(AsPh_3)_2$	200	3	$(\overline{CH_2)_5N}CH_2CH=CHSi(C_2H_5)_3$ + $(\overline{CH_2)_5N}CH_2C(=CH_2)Si(C_2H_5)_3$ (60-70)	611
	$RhCl(CO)(PPh_3)_2$	200	3	$(\overline{CH_2)_5N}CH_2CH=CHSi(C_2H_5)_3$ + $(\overline{CH_2)_5N}CH_2C(=CH_2)Si(C_2H_5)_3$ (60-70)	611
	$RhCl(SbPh_3)_3$	200	3	$(\overline{CH_2)_5N}CH_2CH=CHSi(C_2H_5)_3$ + $(\overline{CH_2)_5N}CH_2C(=CH_2)Si(C_2H_5)_3$ (60-70)	611
	$RhI(PPh_3)_3$	200	3	$(\overline{CH_2)_5N}CH_2CH=CHSi(C_2H_5)_3$ + $(\overline{CH_2)_5N}CH_2C(=CH_2)Si(C_2H_5)_3$ (60-70)	611
C_8H_{13}	$ClCH_2CH_2OCH_2OC(CH_3)_2C\equiv CH$				
	H_2PtCl_6	60-70	6	$ClCH_2CH_2OCH_2OC(CH_3)_2CH=CHSi(C_2H_5)_3$ (56)	1066
	H_2PtCl_6 (THF)	60-70	5	$ClCH_2CH_2OCH_2OC(CH_3)_2CH=CHSi(C_2H_5)_3$ (75)	1082
C_8H_{14}	$(CH_3)_2C(OH)C\equiv CC(OH)(CH_3)_2$				
	$[HPt(SiMe_2Bz)P(c\text{-}Hex)_3]_2$	65	17	$(CH_3)_2C(OH)CH=C[Si(C_2H_5)_3]C(OH)(CH_3)_2$ (85)	1205
C_8H_{14}	$(CH_3)_2C(OH)C\equiv CCH_2CH_2CH_3$				
	-	-	-	$(CH_3)_2C(OH)CH=C[Si(C_2H_5)_3]CH_2CH_2CH_3$ + $(CH_3)_2C(OH)C[Si(C_2H_5)_3]=CHCH_2CH_2CH_3$ (-)	254
C_8H_{14}	$(CH_3)_2CHCH(OH)C(CH_3)(OH)C\equiv CH$				
	H_2PtCl_6 (i-PrOH)	-	-	$(CH_3)_2CHCH(OH)C(CH_3)(OH)CH=CHSi(C_2H_5)_3$ (61) $(CH_3)_2CHCH(OH)C(CH_3)(OH)C(=CH_2)Si(C_2H_5)_3$ (7)	251
C_8H_{14}	$C_2H_5OCH_2C\equiv CCH_2OC_2H_5$				

	H_2PtCl_6 (i-PrOH)	70-90	2	$C_2H_5OCH_2CH=C[Si(C_2H_5)_3]CH_2OC_2H_5$ (48)	569
C_8H_{14}	$C_2H_5OCH_2OCH_2CH_2OCH_2C\equiv CH$				
	H_2PtCl_6	-	-	$C_2H_5OCH_2OCH_2CH_2OCH_2CH=CHSi(C_2H_5)_3$ (-)	994
C_8H_{14}	$CH_3(CH_2)_3OCH_2OCH_2C\equiv CH$				
	H_2PtCl_6	80	6	$CH_3(CH_2)_3OCH_2OCH_2CH=CHSi(C_2H_5)_3$ (35)	1090
C_8H_{14}	$CH_3CH_2C(OH)(CH_3)C(OH)(CH_3)C\equiv CH$				
	H_2PtCl_6 (i-PrOH)	90	4	$CH_3CH_2C(OH)(CH_3)C(OH)(CH_3)CH=CHSi(C_2H_5)_3$ (36) $CH_3CH_2C(OH)(CH_3)C(OH)(CH_3)C(=CH_2)Si(C_2H_5)_3$ (9)	252
C_8H_{14}	$CH_3CH_2CH_2CH(OH)C(OH)(CH_3)C\equiv CH$				
	H_2PtCl_6	-	-	$CH_3CH_2CH_2CH(OH)C(OH)(CH_3)CH=CHSi(C_2H_5)_3$ + $CH_3CH_2CH_2CH(OH)C(OH)(CH_3)C(=CH_2)Si(C_2H_5)_3$ (-)	250
C_8H_{14}	$CH_3OCH_2OCH[CH(CH_3)_2]C\equiv CH$				
	H_2PtCl_6 (i-PrOH)	100	5	$CH_3OCH_2OCH[CH(CH_3)_2]CH=CHSi(C_2H_5)_3$ (36)	1092
C_8H_{15}	$(CH_3)_2NCH_2CH(OH)CH_2OCH_2C\equiv CH$				
	H_2PtCl_6 (i-PrOH)	150	8	$(CH_3)_2NCH_2CH(OH)CH_2OCH_2CH=CHSi(C_2H_5)_3$ (57)	806
C_8H_{15}	$ClCH_2CH_2OCH_2OC(CH_3)_2CH_2C\equiv CH$				
	H_2PtCl_6	60-70	6	$ClCH_2CH_2OCH_2OC(CH_3)_2CH_2CH=CHSi(C_2H_5)_3$ (-)	1066
C_8H_{15}	$ClCH_2Si(CH_3)_2OC(CH_3)_2C\equiv CH$				
	H_2PtCl_6	80-100	6	$ClCH_2Si(CH_3)_2OC(CH_3)_2CH=CHSi(C_2H_5)_3$ (64)	1064
C_8H_{15}	$ClCH_2Si(CH_3)_2OCH_2CH_2OCH_2C\equiv CH$				
	H_2PtCl_6	80-100	6	$ClCH_2Si(CH_3)_2OCH_2CH_2OCH_2CH=CHSi(C_2H_5)_3$ (65)	1064
C_8H_{16}	$(C_2H_5)_3SnC\equiv CH$				
	H_2PtCl_6 (THF)	20	24 (THF)	$(C_2H_5)_3SnC_2H_2Si(C_2H_5)_3$ (kin.)	945
C_8H_{16}	$(CH_3)_2Si(HO)C(CH_3)(C_2H_5)C\equiv CH$				
	H_2PtCl_6 (i-PrOH)	80-85	5	$(CH_3)_2Si(HO)C(CH_3)(C_2H_5)CH=CHSi(C_2H_5)_3$ (55)	436
C_9H_8	$C_6H_5CH(OH)C\equiv CH$				
	$[HPt(SiMe_2Bz)P(c\text{-}Hex)_3]_2$	65	40 (t)	trans-$C_6H_5CH(OH)CH=CHSi(C_2H_5)_3$ (68) $C_6H_5CH(OH)C(=CH_2)Si(C_2H_5)_3$ (13)	1205
C_9H_8	$C_6H_5OC\equiv CCH_3$				

= $(C_2H_5)_3SiH$ =   =10=

	H_2PtCl_6 (i-PrOH)	80-90	5	$C_6H_5OCH=C(CH_3)Si(C_2H_5)_3$ + $C_6H_5OC[Si(C_2H_5)_3]=CHCH_3$ (80)	709
C_9H_8	$C_6H_5OCH_2C\equiv CH$				
	H_2PtCl_6 (i-PrOH)	80-90	5	$C_6H_5OCH_2CH=CHSi(C_2H_5)_3$ + $C_6H_5OCH_2C(=CH_2)Si(C_2H_5)_3$ (80)	709
C_9H_8	$C_6H_5SCH_2C\equiv CH$				
	H_2PtCl_6 (i-PrOH)	90	5	$C_6H_5SCH_2CH=CHSi(C_2H_5)_3$ + $C_6H_5SCH_2C(=CH_2)Si(C_2H_5)_3$ (85)	709
C_9H_8	Fig.313.				
	H_2PtCl_6	-	1	Fig.314. (48)	1001
C_9H_{14}	$(\overline{CH_2)_4}C(OCH_3)CH_2C\equiv CH$				
	H_2PtCl_6 (i-PrOH)	60-70	3	$(\overline{CH_2)_4}C(OCH_3)CH_2CH=CHSi(C_2H_5)_3$ (69)	1080
C_9H_{14}	$(\overline{CH_2)_4}CHC(OH)(CH_3)C\equiv CH$				
	H_2PtCl_6 (i-PrOH)	90	4	$(\overline{CH_2)_4}CHC(OH)(CH_3)CH=CHSi(C_2H_5)_3$ (25) $(\overline{CH_2)_4}CHC(OH)(CH_3)C(=CH_2)Si(C_2H_5)_3$ (8)	252
C_9H_{14}	$(CH_3)_2C(OH)C(OH)(CH_3)C\equiv CCH=CH_2$				
	H_2PtCl_6 (i-PrOH)	50-60	24	$(CH_3)_2C(OH)C(OH)(CH_3)C[Si(C_2H_5)_3]=CHCH=CH_2$ (-)	830
C_9H_{14}	$CH_2=CHCH_2OCH_2CH(OH)CH_2OCH_2C\equiv CH$				
	H_2PtCl_6 (i-PrOH)	80	20	$CH_2=CHCH_2OCH_2CH(OH)CH_2OCH_2CH=CHSi(C_2H_5)_3$ (60-70)	1248
C_9H_{14}	$CH_2=CHCH_2OCH_2OCH_2CH_2OCH_2C\equiv CH$				
	H_2PtCl_6	-	-	$CH_2=CHCH_2OCH_2OCH_2CH_2OCH_2CH=CHSi(C_2H_5)_3$ (-)	994
C_9H_{14}	$\overline{CH_2CH_2OC(CH_3)_2CH_2}C(OH)C\equiv CH$				
	H_2PtCl_6	-	- (b)	$\overline{CH_2CH_2OC(CH_3)_2CH_2}C(OH)CH=CHSi(C_2H_5)_3$ + $\overline{CH_2CH_2OC(CH_3)_2CH_2}C(OH)C(=CH_2)Si(C_2H_5)_3$ (-)	73
C_9H_{14}	$CH_3CH_2CH_2CH(OCH_3)C\equiv CCH=CH_2$				
	H_2PtCl_6	100	-	$CH_3CH_2CH_2CH(OCH_3)CH-C[Si(C_2H_5)_3]CH=CH_2$ (-)	534
C_9H_{15}	$(\overline{CH_2)_6}NCH_2C\equiv CH$				
	H_2PtCl_6 (THF)	150-240	3	$(\overline{CH_2)_6}NCH_2CH=CHSi(C_2H_5)_3$ + $(\overline{CH_2)_6}NCH_2C(=CH_2)Si(C_2H_5)_3$ (60-70)	611
	$RhCl(CO)(AsPh_3)_3$	200	3	$(\overline{CH_2)_6}NCH_2CH=CHSi(C_2H_5)_3$ + $(\overline{CH_2)_6}NCH_2C(=CH_2)Si(C_2H_5)_3$ (60-70)	611

= (C₂H₅)₃SiH =
=11=

$= (C_2H_5)_3SiH =$

RhCl(CO)(PPh₃)₃	200	3	$(\overline{CH_2)_6}NCH_2CH=CHSi(C_2H_5)_3$ + $(\overline{CH_2)_6}NCH_2C(=CH_2)Si(C_2H_5)_3$ (60-70)	611
RhCl(SbPh₃)₃	200	3	$(\overline{CH_2)_6}NCH_2CH=CHSi(C_2H_5)_3$ + $(\overline{CH_2)_6}NCH_2C(=CH_2)Si(C_2H_5)_3$ (60-70)	611
RhI(PPh₃)₃	200	3	$(\overline{CH_2)_6}NCH_2CH=CHSi(C_2H_5)_3$ + $(\overline{CH_2)_6}NCH_2C(=CH_2)Si(C_2H_5)_3$ (60-70)	611

C₉H₁₆ $(CH_3)_2C(OH)C\equiv C(CH_2)_3CH_3$

-	-	-	$(CH_3)_2C(OH)CH=C[Si(C_2H_5)_3](CH_2)_3CH_3$ + $(CH_3)_2C(OH)C[Si(C_2H_5)_3]=CH(CH_2)_3CH_3$ (-)	254

C₉H₁₆ $CH_3CH_2CH_2OCH_2OCH_2CH_2OCH_2C\equiv CH$

H₂PtCl₆	-	-	$CH_3CH_2CH_2OCH_2OCH_2CH_2OCH_2CH=CHSi(C_2H_5)_3$ (-)	994

C₉H₁₆ $CH_3CH_2OCH_2OCH[CH(CH_3)_2]C\equiv CH$

H₂PtCl₆ (i-PrOH)	100	5	$CH_3CH_2OCH_2[OCH(CH_3)_2]CH=CHSi(C_2H_5)_3$ (-)	1092

C₉H₁₇ $ClCH_2Si(CH_3)_2OCH(CH_2CH_2CH_3)C\equiv CH$

H₂PtCl₆	80-100	6	$ClCH_2Si(CH_3)_2OCH(CH_2CH_2CH_3)CH=CHSi(C_2H_5)_3$ (58)	1064

C₉H₁₈ $(CH_3)_3SiOCH(CH_3)CH_2OCH_2C\equiv CH$

H₂PtCl₆	80-90	-	$(CH_3)_3SiOCH(CH_3)CH_2OCH_2CH=CHSi(C_2H_5)_3$ (53)	546

C₁₀H₁₁ $C_6H_5NHCH(CH_3)C\equiv CH$

H₂PtCl₆ (THF)	140	5(THF or d)	$C_6H_5NHCH(CH_3)CH=CHSi(C_2H_5)_3$ (96)	609

C₁₀H₁₄ $(\overline{CH_2)_4}CH=\overline{C}C\equiv CCH(OH)CH_3$

H₂PtCl₆	-	-	$(\overline{CH_2)_4}CH=\overline{C}C[Si(C_2H_5)_3]=CHCH(OH)CH_3$ + $(\overline{CH_2)_4}CH=\overline{C}CH=C[Si(C_2H_5)_3]CH(OH)CH_3$ (-)	449

C₁₀H₁₄ $(CH_2OCH_2OCH_2C\equiv CH)_2$

H₂PtCl₆	95-100	1	$[CH_2OCH_2OCH_2CH=CHSi(C_2H_5)_3]_2$ (48)	1070

C₁₀H₁₄ $\overline{CH_2CH_2}CH=CHCH_2\overline{C}HCH(OH)CH_2C\equiv CH$

H₂PtCl₆	-	-	$\overline{CH_2CH_2}CH=CHCH_2\overline{C}HCH(OH)CH_2CH=CHSi(C_2H_5)_3$ (76)	548

C₁₀H₁₅ $CCl_3CH(OH)OC(CH_3)[C(CH_3)_3]C\equiv CH$

H₂PtCl₆ (i-PrOH)	90-100	18	$CCl_3CH(OH)OC(CH_3)[C(CH_3)_3]CH=CHSi(C_2H_5)_3$ (28)	1063

C₁₀H₁₆ $(\overline{CH_2)_5}\overline{C}(OH)C\equiv CCH_2OCH_3$

H₂PtCl₆ (i-PrOH)	100	7	$(\overline{CH_2)_5}\overline{C}(OH)CH=C[Si(C_2H_5)_3]CH_2OCH_3$ (42)	1062

C₁₀H₁₆ $(\overline{CH_2)_5}\overline{C}HOC(OH)(CH_3)C\equiv CH$

	H_2PtCl_6 (i-PrOH)	90	4	$\overline{(CH_2)_5C}HOC(OH)(CH_3)CH=CHSi(C_2H_5)_3$ (21) $\overline{(CH_2)_5C}HOC(OH)(CH_3)C(=CH_2)Si(C_2H_5)_3$ (8)	252
$C_{10}H_{16}$	$CH_2=C(CH_3)C\equiv CCH_2O(CH_2)_3CH_3$				
	H_2PtCl_6 (i-PrOH)	60-70	4	$CH_2=C(CH_3)C[Si(C_2H_5)_3]=CHCH_2O(CH_2)_3CH_3$ (27)	463
$C_{10}H_{18}$	$(CH_3)_2CHOCH_2OCH[CH(CH_3)_2]C\equiv CH$				
	H_2PtCl_6 (i-PrOH)	100	5	$(CH_3)_2CHOCH_2OCH[CH(CH_3)_2]CH=CHSi(C_2H_5)_3$ (-)	1092
$C_{10}H_{18}$	$(CH_3)_3SiC\equiv CC\equiv CSi(CH_3)_3$				
	$Pd(PPh_3)_4$	100	14	$(CH_3)_3SiCH=C[Si(C_2H_5)_3]C\equiv CSi(CH_3)_3$ (8)	633
	$PdCl_2(PPh_3)_2$	90	35	$(CH_3)_3SiCH=C[Si(C_2H_5)_3]C\equiv CSi(CH_3)_3$ (18)	633
	H_2PtCl_6 (i-PrOH)	80	0.5	$(CH_3)_3SiCH_2C[Si(C_2H_5)_3]=C=C[Si(C_2H_5)_3]Si(CH_3)_3$ (100)	633
	$Pt(PPh_3)_4$	90	18	$(CH_3)_3SiCH=C[Si(C_2H_5)_3]C\equiv CSi(CH_3)_3$ (81) $(CH_3)_3SiCH_2C[Si(C_2H_5)_3]=C=C[Si(C_2H_5)_3]Si(CH_3)_3$ (18)	633
	$Pt(PPh_3)_4$	90	2	$(CH_3)_3SiCH=C[Si(C_2H_5)_3]C\equiv CSi(CH_3)_3$ (83) $(CH_3)_3SiCH_2C[Si(C_2H_5)_3]=C=C[Si(C_2H_5)_3]Si(CH_3)_3$ (6)	633
	$RhCl(PPh_3)_3$	90	5	$(CH_3)_3SiCH_2C[Si(C_2H_5)_3]=C=C[Si(C_2H_5)_3]Si(CH_3)_3$ (83)	633
	$RhCl(PPh_3)_3$	90	0.9	$(CH_3)_3SiCH=C[Si(C_2H_5)_3]C\equiv CSi(CH_3)_3$ (45) $(CH_3)_3SiCH_2C[Si(C_2H_5)_3]=C=C[Si(C_2H_5)_3]Si(CH_3)_3$ (28)	633
$C_{10}H_{18}$	$CH_3(CH_2)_3OCH_2OCH_2CH_2OCH_2C\equiv CH$				
	H_2PtCl_6	-	-	$CH_3(CH_2)_3OCH_2OCH_2CH_2OCH_2CH=CHSi(C_2H_5)_3$ (-)	994
$C_{10}H_{18}$	$CH_3(CH_2)_5OCH_2OCH_2C\equiv CH$				
	H_2PtCl_6 (i-PrOH)	70	4	$CH_3(CH_2)_5OCH_2OCH_2CH=CHSi(C_2H_5)_3$ (52)	732
$C_{10}H_{18}$	$CH_3CH(OC_2H_5)C\equiv CCH(OCH_2CH_3)CH_3$				
	H_2PtCl_6	100	0.5	$CH_3CH(OCH_2CH_3)CH=C[Si(C_2H_5)_3]CH(OCH_2CH_3)CH_3$ (60)	439
$C_{10}H_{18}$	$CH_3CH_2CH_2OCH_2C\equiv CCH_2OCH_2CH_2CH_3$				
	H_2PtCl_6 (i-PrOH)	70-90	2	$CH_3CH_2CH_2OCH_2CH=C[Si(C_2H_5)_3]CH_2OCH_2CH_2CH_3$ (50)	569
$C_{10}H_{18}$	$CH_3OCH_2CH_2C(OH)(CH_3)C\equiv CC(OH)(CH_3)_2$				

H_2PtCl_6	-	-	$CH_3OCH_2CH_2C(OH)(CH_3)CH=C[Si(C_2H_5)_3]C(OH)(CH_3)_2$ (35)	446
$C_{11}H_{12}$ p-$CH_3C_6H_4CH_2OCH_2C\equiv CH$				
H_2PtCl_6 (i-PrOH)	80-85	4	p-$CH_3C_6H_4CH_2OCH_2CH=CHSi(C_2H_5)_3$ (70)	1073
$C_{11}H_{13}$ o-$CH_3C_6H_4NHCH(CH_3)C\equiv CH$				
H_2PtCl_6 (THF)	140	5(THF or d)	o-$CH_3C_6H_4NHCH(CH_3)CH=CHSi(C_2H_5)_3$ (98)	609
$C_{11}H_{13}$ p-$CH_3C_6H_4NHCH(CH_3)C\equiv CH$				
H_2PtCl_6 (THF)	140	5(THF or d)	p-$CH_3C_6H_4NHCH(CH_3)CH=CHSi(C_2H_5)_3$ (81)	609
$C_{11}H_{14}$ $(CH_3O)_3SiC\equiv CC_6H_5$				
$Rh(acac)(CO)_2$	100	8	$C_6H_5CH=C[Si(C_2H_5)_3]Si(OCH_3)_3$ (21) $C_6H_5C[Si(C_2H_5)_3]=CHSi(OCH_3)_3$ (63)	1236
$C_{11}H_{14}$ p-$CH_3C_6H_4SiH(CH_3)OCH_2C\equiv CH$				
H_2PtCl_6 (i-PrOH)	90-100	4	p-$CH_3C_6H_4SiH(CH_3)OCH_2CH=CHSi(C_2H_5)_3$ (60)	1088
$C_{11}H_{16}$ $(\overline{CH_2)_4CH=C}C\equiv CC(CH_3)_2OH$				
H_2PtCl_6 (i-PrOH)	80-90	5	$(\overline{CH_2)_4CH=C}CH=C[Si(C_2H_5)_3]C(CH_3)_2OH$ + $(\overline{CH_2)_4CH=C}C[Si(C_2H_5)_3]=CHC(CH_3)_2OH$ (60)	447
$C_{11}H_{16}$ $(\overline{CH_2)_5C}(OCH_3)C\equiv CCH=CH_2$				
H_2PtCl_6 (i-PrOH)	-	-	$(\overline{CH_2)_5C}(OCH_3)CH=C[Si(C_2H_5)_3]CH=CH_2$ + $(\overline{CH_2)_5C}(OCH_3)C[Si(C_2H_5)_3]=CHCH=CH_2$ (62)	536
$C_{11}H_{16}$ $CH\equiv CCH_2OCH(CH_3)OCH_2C\equiv CC(CH_3)_2OH$				
H_2PtCl_6	-	-	$(C_2H_5)_3SiCH=CHCH_2OCH(CH_3)OCH_2C\equiv CC(CH_3)_2OH$ (46) $(C_2H_5)_3SiCH=CHCH_2OCH(CH_3)OCH_2CH=C[Si(C_2H_5)_3]$- $C(CH_3)_2OH$ (48)	1087
$C_{11}H_{17}$ $(\overline{CH_2)_4CH(OH)C}HC\equiv CCH_2OCH_2CH_2Cl$				
-	-	-	$(\overline{CH_2)_4CH(OH)C}HCH=C[Si(C_2H_5)_3]CH_2OCH_2CH_2Cl$ (35)	824
$C_{11}H_{17}$ $(\overline{CH_2)_5C}(C\equiv CH)OCH_2OCH_2CH_2Cl$				
H_2PtCl_6 (THF)	60-70	5	$(\overline{CH_2)_5C}[CH=CHSi(C_2H_5)_3]OCH_2OCH_2CH_2Cl$ (59)	1082
$C_{11}H_{18}$ $(\overline{CH_2)_4CH(OH)C}HOC(CH_3)_2C\equiv CH$				
H_2PtCl_6 (i-PrOH)	reflux	5	$(\overline{CH_2)_4CH(OH)C}HOC(CH_3)_2CH=CHSi(C_2H_5)_3$ (72)	547
$C_{11}H_{18}$ $(CH_3)_2CHC(CH_3)(OH)C(CH_3)(OH)C\equiv CCH=CH_2$				

Table 8.5

= $(C_2H_5)_3SiH$ = =14=

	H_2PtCl_6		24	$(CH_3)_2CHC(CH_3)(OH)C(CH_3)(OH)C[Si(C_2H_5)_3]=CH$- $CH=CH_2$ (-)	832
$C_{11}H_{18}$	$CH_2=C(CH_3)C\equiv CCH_2O(CH_2)_4CH_3$				
	H_2PtCl_6 (i-PrOH)	60-70	4	$CH_2=C(CH_3)C[Si(C_2H_5)_3]=CHCH_2O(CH_2)_4CH_3$ (30)	1068
$C_{11}H_{18}$	$CH_3SiH[OC(CH_3)_2C\equiv CH]_2$				
	H_2PtCl_6 (i-PrOH)	until 80	8	$HC\equiv CC(CH_3)_2OSiH(CH_3)OC(CH_3)_2CH=CHSi(C_2H_5)_3$ (31)	1075
$C_{11}H_{19}$	$(CH_3CH_2)_2NCH_2C\equiv CCH_2OCH_2\overline{CHCH_2O}$				
	H_2PtCl_6	reflux	18 (b)	$(CH_3CH_2)_2NCH_2CH=C[Si(C_2H_5)_3]CH_2OCH_2\overline{CHCH_2O}$ (53)	1196
$C_{11}H_{19}$	$CH\equiv CC(CH_3)[(CH_2)_3CH_3]OCH_2OCH_2CH_2Cl$				
	H_2PtCl_6 (THF)	60-70	5	$(C_2H_5)_3SiCH=CHC(CH_3)[(CH_2)_3CH_3]OCH_2OCH_2CH_2Cl$(62)	1082
$C_{11}H_{19}$	$CH\equiv CC(CH_3)[C(CH_3)_3]OCH_2OCH_2CH_2Cl$				
	H_2PtCl_6 (i-PrOH)	60-70	8	$(C_2H_5)_3SiCH=CHC(CH_3)[C(CH_3)_3]OCH_2OCH_2CH_2Cl$ (50)	1065
$C_{11}H_{20}$	$CH_3(CH_2)_3OCH_2OC(CH_3)(CH_2CH_3)C\equiv CH$				
	H_2PtCl_6	80	6	$CH_3(CH_2)_3OCH_2OC(CH_3)(CH_2CH_3)CH=CHSi(C_2H_5)_3$ (-)	1204
$C_{11}H_{20}$	$CH_3(CH_2)_3OCH_2OCH(CH_2CH_2CH_3)C\equiv CH$				
	H_2PtCl_6	80	6	$CH_3(CH_2)_3OCH_2OCH(CH_2CH_2CH_3)CH=CHSi(C_2H_5)_3$ (-)	1090
$C_{11}H_{20}$	$CH_3(CH_2)_3OCH_2OCH[CH(CH_3)_2]C\equiv CH$				
	H_2PtCl_6 (i-PrOH)	100	5	$CH_3(CH_2)_3OCH_2OCH[CH(CH_3)_2]CH=CHSi(C_2H_5)_3$ (-)	1092
$C_{11}H_{20}$	$CH_3(CH_2)_5OCH(CH_3)OCH_2C\equiv CH$				
	H_2PtCl_6 (i-PrOH)	70	5	$CH_3(CH_2)_5OCH(CH_3)OCH_2CH=CHSi(C_2H_5)_3$ (60)	732
$C_{11}H_{20}$	$CH_3CH_2C(CH_3)(OH)C\equiv CC(CH_3)(OH)CH_2CH_2OCH_3$				
	H_2PtCl_6 (i-PrOH)	90	6	$CH_3CH_2C(CH_3)(OH)C[Si(C_2H_5)_3]=CHC(CH_3)(OH)CH_2$- CH_2OCH_3 (24)	446
$C_{11}H_{20}$	$CH_3CH_2CH_2CH(OH)C\equiv CC(CH_3)(OH)CH_2CH_2OCH_3$				
	H_2PtCl_6	60-65	6	$CH_3CH_2CH_2CH(OH)C[Si(C_2H_5)_3]=CHC(CH_3)(OH)CH_2$- CH_2OCH_3 (35)	460
	H_2PtCl_6	85-90	7	$CH_3CH_2CH_2CH(OH)C[Si(C_2H_5)_3]=CHC(CH_3)(OH)CH_2$- CH_2OCH_3 (8) $CH_3CH_2CH_2CH[OSi(C_2H_5)_3]CH=C[Si(C_2H_5]_3C(CH_3)$- $(OH)CH_2CH_2OCH_3$ (4)	460

	Pt/C	40	19	$CH_3CH_2CH_2CH(OH)C[Si(C_2H_5)_3]{=}CHC(CH_3)(OH)CH_2$-$CH_2OCH_3$ (19)	460
$C_{11}H_{20}$	iso-$C_4H_9OCH_2OCH[CH(CH_3)_2]C{\equiv}CH$				
	H_2PtCl_6 (i-PrOH)	100	5	iso-$C_4H_9OCH_2OCH[CH(CH_3)_2]CH{=}CHSi(C_2H_5)_3$ (-)	1092
$C_{11}H_{22}$	$CH{\equiv}CCH_2OCH_2CH_2OSi(C_2H_5)_3$				
	H_2PtCl_6 (i-PrOH)	95-100	2	$(C_2H_5)_3SiCH{=}CHCH_2OCH_2CH_2OSi(C_2H_5)_3$ (41)	994
$C_{11}H_{23}$	$CH{\equiv}CC(CH_3)(CH_2CH_3)OSiH(CH_3)N(CH_2CH_3)_2$				
	Pt/C	120-130	30	$(C_2H_5)_3SiCH{=}CHC(CH_3)(CH_2CH_3)OSiH(CH_3)N(CH_2CH_3)_2$ (48)	436
$C_{11}H_{24}$	$CH{\equiv}CC(CH_3)_2OSi(CH_3)(CH_2CH_3)OSiH(CH_3)CH_2CH_3$				
	H_2PtCl_6 (i-PrOH)	reflux	2	$(C_2H_5)_3SiCH{=}CHC(CH_3)OSi(CH_3)(CH_2CH_3)OSiH(CH_3)$-$CH_2CH_3$ (20)	1069
$C_{11}H_{26}$	$\overline{(CH_2)_4C}(OCH_2CH{=}CH_2)OCH_2C{\equiv}CH$				
	H_2PtCl_6 (i-PrOH)	60-70	3	$\overline{(CH_2)_4C}(OCH_2CH{=}CH_2)OCH_2CH{=}CHSi(C_2H_5)_3$ (71)	1080
$C_{11}H_{26}$	$\overline{OCH_2C}HOCH_2OC(CH_3)_2C{\equiv}CCH_2CH{=}CH_2$				
	H_2PtCl_6 (i-PrOH)	reflux	18 (b)	$\overline{OCH_2C}HCH_2OC(CH_3)_2CH{=}C[Si(C_2H_5)_3]CH_2CH{=}CH_2$ (63)	1231
$C_{12}H_9$	$C_5H_5FeC_5H_4C{\equiv}CD$				
	H_2PtCl_6 (i-PrOH)	75-80	-	$C_5H_5FeC_5H_4CH{=}CDSi(C_2H_5)_3$ + $C_5H_5FeC_5H_4C({=}CHD)$-$Si(C_2H_5)_3$ (85)	63
$C_{12}H_{10}$	$C_5H_5FeC_5H_4C{\equiv}CH$				
	H_2PtCl_6	-	-	$C_5H_5FeC_5H_4CH{=}CHSi(C_2H_5)_3$ + $C_5H_5FeC_5H_4C[Si(C_2H_5)_3]{=}CH_2$ (86)	450
	H_2PtCl_6 (i-PrOH)	75-80	1.5	$C_5H_5FeC_5H_4CH{=}CHSi(C_2H_5)_3$ + $C_5H_5FeC_5H_4C[Si(C_2H_5)_3]{=}CH_2$ (92)	63
$C_{12}H_{10}$	$C_6H_5N(COCF_3)CH(CH_3)C{\equiv}CH$				
	H_2PtCl_6 (THF)	140	5(THF or d)	$C_6H_5N(COCF_3)CH(CH_3)CH{=}CHSi(C_2H_5)_3$ (15)	609
$C_{12}H_{13}$	$C_6H_5C(CH_3)(OH)C{\equiv}CCH{=}CH_2$				
	H_2PtCl_6	-	-	$C_6H_5C(CH_3)(OH)C[Si(C_2H_5)_3]{=}CHCH{=}CH_2$ (38)	452
	Pt/C	-	-	$C_6H_5C(CH_3)(OH)C[Si(C_2H_5)_3]{=}CHCH{=}CH_2$ (50)	452
$C_{12}H_{13}$	$C_6H_5N(COCH_3)CH(CH_3)C{\equiv}CH$				

	H_2PtCl_6 (THF)	140	5(THF or d)	$C_6H_5N(COCH_3)CH(CH_3)CH=CHSi(C_2H_5)_3$ (60)	609
$C_{12}H_{14}$	$C_6H_5C\equiv CC(CH_3)(OH)CH_2CH_3$				
	H_2PtCl_6 (i-PrOH)	60-70	1	$C_6H_5CH=C[Si(C_2H_5)_3]C(CH_3)(OH)CH_2CH_3$ (47)	437
$C_{12}H_{14}$	$C_6H_5CH_2Si(CH_3)_2OCH_2C\equiv CH$				
	H_2PtCl_6 (i-PrOH)	20	-	$C_6H_5CH_2Si(CH_3)_2OCH_2CH=CHSi(C_2H_5)_3$ (60)	39
$C_{12}H_{14}$	$C_6H_5COC\equiv CSi(CH_3)_3$				
	H_2PtCl_6 (i-PrOH)	100	4	$C_6H_5COCH=C[Si(CH_3)_3]Si(C_2H_5)_3$ + $C_6H_5COC[Si(C_2H_5)_3]=CHSi(CH_3)_3$ (-)	953
$C_{12}H_{14}$	$CH_3C_6H_4Si(CH_3)_2OCH_2C\equiv CH$				
	H_2PtCl_6 (i-PrOH)	100-120	3-4	$CH_3C_6H_4Si(CH_3)_2OCH_2CH=CHSi(C_2H_5)_3$ (61)	1078
$C_{12}H_{16}$	$CH\equiv CCH_2OCH_2CH(OCH_2\overline{CHCH_2O})CH_2OCH_2C\equiv CH$				
	H_2PtCl_6 (i-PrOH)	80	16 (b)	$(C_2H_5)_3SiCH=CHCH_2CH(OCH_2\overline{CHCH_2O})CH_2OCH_2$-$CH=CHSi(C_2H_5)_3$ (47)	626
$C_{12}H_{18}$	$(\overline{CH_2)_4CH=C}C\equiv CC(CH_3)(OH)CH_2CH_3$				
	H_2PtCl_6 (i-PrOH)	80-90	5	$(\overline{CH_2)_4CH=C}CH=C[Si(C_2H_5)_3]C(CH_3)(OH)CH_2CH_3$ + $(\overline{CH_2)_4CH=C}C[Si(C_2H_5)_3]=CHC(CH_3)(OH)CH_2CH_3$ (-)	447
$C_{12}H_{18}$	$(\overline{CH_2)_4CH=C}C\equiv CCH(OH)CH(CH_3)_2$				
	H_2PtCl_6	-	-	$(\overline{CH_2)_4CH=C}C[Si(C_2H_5)_3]=CHCH(OH)CH(CH_3)_2$ + $(\overline{CH_2)_4CH=C}CH=C[Si(C_2H_5)_3]CH(OH)CH(CH_3)_2$ (-)	449
$C_{12}H_{20}$	$(\overline{CH_2)_5C}(OH)C\equiv CCH_2OCH_2CH_2CH_3$				
	H_2PtCl_6 (i-PrOH)	100	7	$(\overline{CH_2)_5C}(OH)CH=C[Si(C_2H_5)_3]CH_2OCH_2CH_2CH_3$ (42)	1062
$C_{12}H_{22}$	$(C_2H_5)_2Si(CH_3)CH_2CH_2Si(CH_3)(C\equiv CH)_2$				
	H_2PtCl_6	150-160	3	$(C_2H_5)_2Si(CH_3)CH_2CH_2Si(CH_3)[CH=CHSi(C_2H_5)_3]_2$ (25)	636
$C_{12}H_{22}$	$[CH\equiv CCH_2OSi(CH_3)CH_2CH_3]_2O$				
	H_2PtCl_6 (i-PrOH)	reflux	2	$[(C_2H_5)_3SiCH=CHCH_2OSi(CH_3)CH_2CH_3]_2O$ (41)	1069
$C_{12}H_{26}$	$CH_3CH_2CH_2CH(OH)C\equiv CH$				
	H_2PtCl_6 (i-PrOH)	80-100	6	$CH_3CH_2CH_2CH(OH)CH=CHSi(C_2H_5)_3$ (50)	1083
$C_{13}H_{12}$	o-$CH_3C_6H_4N(COF_3)CH(CH_3)C\equiv CH$				
	H_2PtCl_6 (THF)	140	5(THF or d)	o-$CH_3C_6H_4N(COCF_3)CH(CH_3)CH=CHSi(C_2H_5)_3$ (15)	609
$C_{13}H_{14}$	$(\overline{CH_2)_4C}(OH)C\equiv CC_6H_5$				

	H_2PtCl_6 (i-PrOH)	60-70	10	$(CH_2)_4C(OH)C[Si(C_2H_5)_3]=CHC_6H_5$ + $(CH_2)_3CH=CC[Si(C_2H_5)_3]=CHC_6H_5$ (-)	452
$C_{13}H_{14}$	$p\text{-}HOC_6H_4C(CH_3)_2C\equiv CCH=CH_2$				
	H_2PtCl_6	40-80	3	$p\text{-}HOC_6H_4C(CH_3)_3C[Si(C_2H_5)_3]=CHCH=CH_2$ (65)	13
	H_2PtCl_6	60-70	-	$p\text{-}HOC_6H_4C(CH_3)_3C[Si(C_2H_5)_3]=CHCH=CH_2$ + $p\text{-}(C_2H_5)_3SiOC_6H_4C(CH_3)_2C[Si(C_2H_5)_3]$ (-)	531
$C_{13}H_{16}$	$CH_3C_6H_4Si(CH_3)(CH=CH_2)OCH_2C\equiv CH$				
	H_2PtCl_6	90-95	4	$CH_3C_6H_4Si(CH_3)(CH=CH_2)OCH_2CH=CHSi(C_2H_5)_3$ (72)	1084
$C_{13}H_{18}$	$(CH_2)_4CH=CC\equiv CC(OH)(CH_2)_4$				
	-	-	-	$(CH_2)_4CH=CCH=C[Si(C_2H_5)_3]C(OH)(CH_2)_4$ + $(CH_2)_4CH=CC[Si(C_2H_5)_3=CHC(OH)(CH_2)_4$ (-)	448
$C_{13}H_{18}$	$p\text{-}CH_3C_6H_4Si(C_2H_5)_2C\equiv CH$				
	H_2PtCl_6	60-70	3	$p\text{-}CH_3C_6H_4Si(C_2H_5)_2CH=CHSi(C_2H_5)_3$ (78)	1074
$C_{13}H_{18}$	$p\text{-}CH_3C_6H_4SiH(CH_3)OC(CH_3)_2C\equiv CH$				
	H_2PtCl_6 (i-PrOH)	90-100	4	$p\text{-}CH_3C_6H_4SiH(CH_3)OC(CH_3)_2CH=CHSi(C_2H_5)_3$ (62)	1088
$C_{13}H_{20}$	$(CH_2)_4CH=CC\equiv CC(CH_2CH_3)_2OH$				
	-	-	-	$(CH_2)_4CH=CCH=C[Si(C_2H_5)_3]=CHC(CH_2CH_3)_2OH$ (-)	448
$C_{13}H_{22}$	$CH_3(CH_2)_3C(OH)(CH_2CH_3)C(CH_3)(OH)C\equiv CCH=CH_2$				
	H_2PtCl_6	reflux	24	$CH_3(CH_2)_3C(OH)(CH_2CH_3)C(CH_3)(OH)C[Si(C_2H_5)_3]=$ $=CHCH=CH_2$ (-)	831
$C_{13}H_{24}$	$(CH_3CH_2)_3SiCH_2CH_2Si(CH_3)(C\equiv CH)_2$				
	H_2PtCl_6	150-160	3	$(C_2H_5)_3SiCH_2CH_2Si(CH_3)[CH=CHSi(C_2H_5)_3]_2$ (30)	636
$C_{13}H_{24}$	$CH\equiv CSi(CH_2CH_3)_2CH_2CH_2Si(CH_3)(CH_2CH_3)C\equiv CH$				
	H_2PtCl_6	150-160	3	$(C_2H_5)_3SiCH=CHSi(CH_3CH_2)_2CH_2CH_2Si(CH_3)(CH_2CH_3)$- $CH=CHSi(C_2H_5)_3$ (50)	636
$C_{13}H_{24}$	$CH_3(CH_2)_3OCH(CH_3)C(CH_3)(CH_2CH_2CH_3)C\equiv CH$				
	H_2PtCl_6 (i-PrOH)	75-80	4	$CH_3(CH_2)_3OCH(CH_3)C(CH_3)(CH_2CH_2CH_3)$- $CH=CHSi(C_2H_5)_3$ (-)	1061
$C_{14}H_{12}$	$1\text{-}C_{10}H_7C(CH_3)(OH)C\equiv CH$				
	-	-	-	$1\text{-}C_{10}H_7C(CH_3)(OH)CH=CHSi(C_2H_5)_3$ (-)	443
$C_{14}H_{16}$	$(CH_2)_5C(OH)C\equiv CC_6H_5$				

	H_2PtCl_6 (i-PrOH)	60-70	10	$(CH_2)_5C(OH)C[Si(C_2H_5)_3]=CHC_6H_5$ (-)	438
$C_{14}H_{16}$	Fig.315.				
	H_2PtCl_6	60-70	-	Fig.316. (-)	531
$C_{14}H_{16}$	Fig.317.				
	H_2PtCl_6	40-80	3	Fig.318. (62)	13
	H_2PtCl_6	40-80	3	Fig.319. (-)	517
	H_2PtCl_6	60-70	-	Fig.318 + Fig.319. (-)	531
$C_{14}H_{16}$	Fig.320.				
	H_2PtCl_6 (i-PrOH)	100	5	Fig.321. (72)	533
$C_{14}H_{16}$	Fig.322.				
	H_2PtCl_6	40-80	3	Fig.323. (69)	13
	H_2PtCl_6	60-70	-	Fig.323 + Fig.324. (-)	531
$C_{14}H_{16}$	$p\text{-}CH_3OC_6H_4C(CH_3)_2C\equiv CCH=CH_2$				
	H_2PtCl_6	-	-	$p\text{-}CH_3OC_6H_4C(CH_3)_2CH=C[Si(C_2H_5)_3]CH=CH_2$ / $p\text{-}CH_3OC_6H_4C(CH_3)_2C[Si(C_2H_5)_3]=CHCH=CH_2$ / $p\text{-}CH_3OC_6H_4C(CH_3)_2C[Si(C_2H_5)_3]=C=C=CHCH_3$ (65/10/25)	539
	H_2PtCl_6	100	6	$p\text{-}CH_3OC_6H_4C(CH_3)_2CH=C[Si(C_2H_5)_3]CH=CH_2$ (80)	532
	H_2PtCl_6	40-80	3	$p\text{-}CH_3OC_6H_4C(CH_3)_2C[Si(C_2H_5)_3]=CHCH=CH_2$ (70)	13
	H_2PtCl_6	75	10	$p\text{-}CH_3OC_6H_4C(CH_3)_2CH=C[Si(C_2H_5)_3]CH=CH_2$ (70)	540
$C_{14}H_{17}$	$C_6H_5C\equiv CSi(OCH_2CH_2)_3N$				
	$Rh(acac)(CO)_2$	100	8	$(C_2H_5)_3Si(C_6H_5)C=CHSi(OCH_2CH_2)_3N$ (25)	1236
$C_{14}H_{18}$	$CH\equiv CSi(CH_3)_2C_6H_4Si(CH_3)_2C\equiv CH\text{-}p$				
	H_2PtCl_6	-	-	$p\text{-}(C_2H_5)_3SiCH=CHSi(CH_3)_2C_6H_4Si(CH_3)_2CH=CH\text{-}Si(C_2H_5)_3$ (-)	463
	H_2PtCl_6	150-160	3	$p\text{-}(C_2H_5)_3SiCH=CHSi(CH_3)_2C_6H_4Si(CH_3)_2CH=CH\text{-}Si(C_2H_5)_3$ (50)	636
$C_{14}H_{20}$	$(CH_2)_4CH=CC\equiv CC(OH)(CH_2)_5$				
	-	-	-	$(CH_2)_4CH=CC[Si(C_2H_5)_3]=CHC(OH)(CH_2)_5$ (-)	448
$C_{14}H_{20}$	$C_6H_5Si(C_2H_5)_2OCH_2OCH_2C\equiv CH$				

	H_2PtCl_6 (i-PrOH)	105	4	$C_6H_5Si(C_2H_5)_2OCH_2OCH_2CH=CHSi(C_2H_5)_3$ (45)	809
$C_{14}H_{20}$	$CH_3C_6H_4Si(CH_3)_2OC(CH_3)_2C\equiv CH$				
	H_2PtCl_6 (i-PrOH)	100-120	3-4	$CH_3C_6H_4Si(CH_3)_2OC(CH_3)_2CH=CHSi(C_2H_5)_3$ (61)	1079
$C_{14}H_{20}$	$CH_3C_6H_4SiH(CH_3)OC(CH_3)(CH_2CH_3)C\equiv CH$				
	H_2PtCl_6 (i-PrOH)	90-100	4	$CH_3C_6H_4SiH(CH_3)OC(CH_3)(CH_2CH_3)CH=CHSi(C_2H_5)_3$(61)	1088
$C_{14}H_{20}$	$\overline{S(CH_2)_3C}HC(CH_3)(OH)C\equiv CC(CH_3)(OH)\overline{CH(CH_2)_3S}$				
	H_2PtCl_6	90	-	$\overline{S(CH_2)_3C}HC(CH_3)(OH)CH=C[Si(C_2H_5)_3]C(CH_3)$- $(OH)\overline{CH(CH_2)_3S}$ (66)	440
$C_{14}H_{20}$	p-$CH_3C_6H_4Si(C_2H_5)_2OCH_2C\equiv CH$				
	H_2PtCl_6 (i-PrOH)	90-100	4	p-$CH_3C_6H_4Si(C_2H_5)_2OCH_2CH=CHSi(C_2H_5)_3$ (76)	797
$C_{14}H_{22}$	$(\overline{CH_2)_5C}(OH)C\equiv C\overline{C(OH)(C}H_2)_5$				
	Pt/C or H_2PtCl_6	reflux	10	$(\overline{CH_2)_5C}(OH)C[Si(C_2H_5)_3]=CH\overline{C(OH)(C}H_2)_5$ (33)	451
$C_{14}H_{23}$	$CH_3(CH_2)_3OCH(CH_3)OCH(CCl_3)OCH(CH_2CH_2CH_3)C\equiv CH$				
	H_2PtCl_6 (i-PrOH)	45-50	7	$CH_3(CH_2)_3OCH(CH_3)OCH(CCl_3)OCH(CH_2CH_2CH_3)CH=CH$- $Si(C_2H_5)_3$ (59)	1067
$C_{14}H_{24}$	$CH_3(CH_2)_3OCH(CH_3)\overline{C(C}H_2)_5C\equiv CH$				
	H_2PtCl_6 (i-PrOH)	75-80	4	$CH_3(CH_2)_3OCH(CH_3)\overline{C(C}H_2)_5CH=CHSi(C_2H_5)_3$ (-)	1061
$C_{14}H_{26}$	$[CH\equiv CCH_2OSi(CH_3)CH_2CH_2CH_3]_2O$				
	H_2PtCl_6 (i-PrOH)	reflux	2	$[(C_2H_5)_3SiCH=CHCH_2OSi(CH_3)CH_2CH_2CH_3]_2O$ (39)	1069
$C_{15}H_{18}$	$C_5H_5FeC_5H_4C\equiv CSi(CH_3)_3$				
	H_2PtCl_6	90-95	-	$C_5H_5FeC_5H_4CH=C[Si(CH_3)_3]Si(C_2H_5)_3$ (13) $C_5H_5FeC_5H_4C[Si(C_2H_5)_3]=CHSi(CH_3)_3$ (31) $C_5H_5FeC_5H_4C(=CH_2)Si(C_2H_5)_3$ (17)	62
$C_{15}H_{18}$	Fig.325.				
	H_2PtCl_6	-	-	Fig.326./Fig.327./Fig.328. (78/10/12)	539
	H_2PtCl_6	100	-	Fig.329. (-)	517
	H_2PtCl_6	100	6	Fig.326. (78)	532
$C_{15}H_{20}$	$CH_3C_6H_4Si(CH_3)(CH=CH_2)OC(CH_3)_2C\equiv CH$				
	H_2PtCl_6	90-95	4	$CH_3C_6H_4Si(CH_3)(CH=CH_2)OC(CH_3)_2CH=CHSi(C_2H_5)_3$(65)	1084
$C_{15}H_{22}$	p-$CH_3C_6H_4Si(CH_3)_2OC(CH_3)(CH_2CH_3)C\equiv CH$				

	H_2PtCl_6 (i-PrOH)	100-120	3-4	p-$CH_3C_6H_4Si(CH_3)_2OC(CH_3)(CH_2CH_3)CH=CH$-$Si(C_2H_5)_3$ (50)	1078
$C_{16}H_{10}$	Fig.330.				
	H_2PtCl_6	60	6	Fig.331. (65)	459
$C_{16}H_{16}$	2 $C_{10}H_7C(CH_3)(OH)C\equiv CCH(OH)CH_3$				
	H_2PtCl_6	90	-	2-$C_{10}H_7C(CH_3)(OH)CH=C[Si(C_2H_5)_3]CH(OH)CH_3$ + 2-$C_{10}H_7C(CH_3)(OH)C[Si(C_2H_5)_3]=CHCH(OH)CH_3$ (38) 2-$C_{10}H_7C(CH_3)=C[Si(C_2H_5)_3]COCH=CH_2$ + 2-$C_{10}H_7C[Si(C_2H_5)_3]=CHCOCH=CH_2$ (41)	441
$C_{16}H_{30}$	$[CH\equiv CC(CH_3)_2OSi(CH_3)(CH_2CH_2CH_3)]_2O$				
	H_2PtCl_6 (i-PrOH)	reflux	2	$[(C_2H_5)_3CH=CHC(CH_3)_2OSi(CH_3)(CH_2CH_2CH_3)]_2O$ (42)	1069
$C_{17}H_{16}$	p-$HOC_6H_4C(CH_3)_2C\equiv CC_6H_5$				
	H_2PtCl_6 (THF)	100	-	p-$HOC_6H_4C(CH_3)_2C[Si(C_2H_5)_3]=CHC_6H_5$ (73-94)	537
$C_{17}H_{18}$	1-$C_{10}H_7C(CH_3)(OH)C\equiv CC(CH_3)_2OH$				
	H_2PtCl_6 (i-PrOH)	90	4	1-$C_{10}H_7C(CH_3)(OH)C[Si(C_2H_5)_3]=CH(CH_3)C=CH_2$ (53) Fig.332. (42)	444
$C_{17}H_{18}$	2-$C_{10}H_7C(CH_3)(OH)C\equiv CC(CH_3)_2OH$				
	H_2PtCl_6	100	2	Fig.333. (62)	442
	H_2PtCl_6	90	8	2-$C_{10}H_7C(CH_3)(OH)CH=C[Si(C_2H_5)_3]=C(CH_3)=CH_2$ (73)	442
$C_{18}H_{16}$	$(C_6H_5)_2Si(C\equiv CCH_2OH)_2$				
	-	-	-	$(C_6H_5)_2Si\{CH=C[Si(C_2H_5)_3]CH_2OH\}_2$ (-)	340
$C_{18}H_{18}$	p-$CH_3OC_6H_4C(CH_3)_2C\equiv CC_6H_5$				
	H_2PtCl_6 (THF)	100	-	p-$CH_3OC_6H_4C(CH_3)_2C[Si(C_2H_5)_3]=CHC_6H_5$ (73-94)	537
$C_{18}H_{20}$	1-$C_{10}H_7C(CH_3)(OH)C\equiv CCCH_3(OH)CH_2CH_3$				
	H_2PtCl_6 (i-PrOH)	90	4	1-$C_{10}H_7C(CH_3)(OH)C[Si(C_2H_5)_3]=CHCCH_3(OH)CH_2CH_3$ (31) Fig.334. (54)	444
$C_{18}H_{20}$	2-$C_{10}H_7C(CH_3)(OH)C\equiv CCH(OH)CH_2CH_2CH_3$				
	H_2PtCl_6 (i-PrOH)	90	-	2-$C_{10}H_7C(CH_3)[Si(C_2H_5)_3]CH_2COCH=CHCH_2CH_3$ + 2-$C_{10}H_7CH(CH_3)CH[Si(C_2H_5)_3]COCH=CHCH_2CH_3$ (42)	441
$C_{18}H_{24}$	$C_5H_5FeC_5H_4C\equiv CSi(C_2H_5)_3$				

Table 8.5

= $(C_2H_5)_3SiH$ =

H_2PtCl_6	90-95	-	$C_5H_5FeC_5H_4C[Si(C_2H_5)_3]=CHSi(C_2H_5)_3$ (60) $C_5H_5FeC_5H_4CH=C[Si(C_2H_5)_3]_2$ (18) $C_5H_5FeC_5H_4C(=CH_2)Si(C_2H_5)_3$ (2)	62
$C_{18}H_{26}$ $CH\equiv CSi(C_2H_5)_2C_6H_4Si(C_2H_5)_2C\equiv CH$				
H_2PtCl_6	150-160	3	$(C_2H_5)_3SiCH=CHSi(C_2H_5)_2C_6H_4Si(C_2H_5)_2CH=CH$- $Si(C_2H_5)_3$-p (40)	636
$C_{18}H_{34}$ $[CH\equiv CCH(CH_2CH_2CH_3)OSi(CH_3)CH_2CH_3]_2O$				
H_2PtCl_6 (i-PrOH)	reflux	2	$[(C_2H_5)_3SiCH=CHCH(CH_2CH_2CH_3)OSi(CH_3)CH_2CH_3]_2O$ (39)	1069
$C_{20}H_{24}$ Fig.335.				
Pt/C	160	16	Fig.336. (-)	F -024
$C_{22}H_{22}$ Fig.337.				
-	-	-	Fig.338. (20-26)	455
$C_{22}H_{22}$ Fig.339.				
H_2PtCl_6 (i-PrOH)	-	-	Fig.340. (-)	454
$C_{24}H_{26}$ Fig.341.				
-	-	-	Fig.342. (20-26)	455
$C_{26}H_{22}$ 2-$C_{10}H_7C(CH_3)(OH)C\equiv CC(CH_3)(OH)C_{10}H_7$-2				
H_2PtCl_6	90	5	Fig.343. (35)	444
$C_{26}H_{24}$ $(C_5H_5FeC_5H_4C\equiv C)_2Ge(CH_3)_2$				
-	90-95	-	$\{C_5H_5FeC_5H_4C[Si(C_2H_5)_3]=CH\}_2Ge(CH_3)_2$ + $C_5H_5FeC_5H_4C[Si(C_2H_5)_3]=CHGe(CH_3)_2C[Si(C_2H_5)_3]=$ $=CHC_5H_4FeC_5H_5$ (38) + $C_5H_5FeC_5H_4C(=CH_2)Si(C_2H_5)_3$ + $C_5H_5FeC_5H_4CH=CHSi(C_2H_5)_3$ (7)	64
$C_{26}H_{24}$ $(C_5H_5FeC_5H_4C\equiv C)_2Si(CH_3)_2$				
H_2PtCl_6	90-95	-	$\{C_5H_5FeC_5H_4C[Si(C_2H_5)_3]=CH\}_2Si(CH_3)_2$ (35) $C_5H_5FeC_5H_4C[Si(C_2H_5)_3]=CHSi(CH_3)_2C[Si(C_2H_5)_3]=$ $=CHC_5H_4FeC_5H_5$ (53)	62
$C_{28}H_{28}$ $(C_5H_5FeC_5H_4C\equiv C)_2Si(C_2H_5)_2$				
H_2PtCl_6	90-95	-	$\{C_5H_5FeC_5H_4C[Si(C_2H_5)_3]=CH\}_2Si(C_2H_5)_2$ (60) $C_5H_5FeC_5H_4C[Si(C_2H_5)_3]=CHSi(C_2H_5)_2C[Si(C_2H_5)_3]=$ $=CHC_5H_4FeC_5H_5$ (32)	62
$C_{28}H_{30}$ Fig.344.				

	-	-	-	Fig.345. (20-26)	455
$C_{30}H_{24}$	$C_5H_5FeC_5H_4C{\equiv}CGe(C_6H_5)_3$				
	H_2PtCl_6 (i-PrOH)	90-95	-	$C_5H_5FeC_5H_4C[Si(C_2H_5)_3]{=}CHGe(C_6H_5)_3$ + $C_5H_5FeC_5H_4CH{=}C[Si(C_2H_5)_3]Ge(C_6H_5)_3$ (68)	64
$C_{30}H_{24}$	$C_5H_5FeC_5H_4C{\equiv}CSi(C_6H_5)_3$				
	H_2PtCl_6	90-95	-	$C_5H_5FeC_5H_4C[Si(C_2H_5)_3]{=}CHSi(C_6H_5)_3$ (43) $C_5H_5FeC_5H_4CH{=}C[Si(C_2H_5)_3]Si(C_6H_5)_3$ (53)	62
$C_{36}H_{28}$	$(C_5H_5FeC_5H_4C{\equiv}C)_2Ge(C_6H_5)_2$				
	H_2PtCl_6 (i-PrOH)	90-95	-	$C_5H_5FeC_5H_4C{\equiv}CGe(C_6H_5)_2C[Si(C_2H_5)_3]{=}CHC_5H_4FeC_5H_5$ (9) $\{C_5H_5FeC_5H_4CH{=}C[Si(C_2H_5)_3]\}_2Ge(C_6H_5)_2$ (11) $C_5H_5FeC_5H_4CH{=}C[Si(C_2H_5)_3]Ge(C_2H_5)_2CH{=}C$-$[Si(C_2H_5)_3]C_5H_4FeC_5H_5$ (16)	64
$C_{36}H_{28}$	$(C_5H_5FeC_5H_4C{\equiv}C)_2Si(C_6H_5)_2$				
	H_2PtCl_6	90-95	-	$\{C_5H_5FeC_5H_4C[Si(C_2H_5)_3]{=}CH\}_2Si(C_6H_5)_2$ (28) $C_5H_5FeC_5H_4C[Si(C_2H_5)_3]{=}CHSi(C_6H_5)_2C[Si(C_2H_5)_3]{=}$ $=CHC_5H_4Fe(C_5H_5$ (52)	62
$C_{42}H_{56}$	$(C_6H_5)_2Si\{C{\equiv}CC[\overline{CH(CH_2)_5}]_2OH\}_2$				
	-	-	-	$(C_6H_5)_2Si\{CH{=}C[Si(C_2H_5)_3]C[\overline{CH(CH_2)_5}]_2OH\}_2$ (-)	340

C_1	CO_2				
	$[HRu_3(CO)_{10}(SiEt_3)]^-$	60	24	$HCOOSi(C_2H_5)_3$ (-)	1145
	$[HRu_3(CO)_{11}]^-$	60	24	$HCOOSi(C_2H_5)_3$ (-)	1145
C_2H_4	CH_3CHO				
	Ni	100-140	2-3 (b)	$CH_3CH_2OSi(C_2H_5)_3$ (-)	146
	Ni +Et_2S	110	-	$CH_3CH_2OSi(C_2H_5)_3$ / $CH_2=CHOSi(C_2H_5)_3$ (85/15)	F -008
C_3	CF_3COCF_3				
	none	150	6	$(CF_3)_2CHOSi(C_2H_5)_3$ (-)	163
C_3H_3	CH_3COCN				
	$RhCl(PPh_3)_3$	80	2	$CH_3CH[OSi(C_2H_5)_3]CN$ (88)	873
C_3H_4	$CH_2=CHCHO$				
	Ni	110-120	3 (b)	$CH_2=CHCH_2OSi(C_2H_5)_3$ + $CH_3CH=CHOSi(C_2H_5)_3$ + $CH_3CH_2CH_2OSi(C_2H_5)_3$ (79)	148
	Pd/C	reflux	2 (b)	$CH_2=CHCH_2OSi(C_2H_5)_3$ + $CH_3CH=CHOSi(C_2H_5)_3$ + $CH_3CH_2CH_2OSi(C_2H_5)_3$ (86)	148
	H_2PtCl_6	reflux	2 (b)	$CH_2=CHCH_2OSi(C_2H_5)_3$ + $CH_3CH=CHOSi(C_2H_5)_3$ + $CH_3CH_2CH_2OSi(C_2H_5)_3$ (60)	148
	$Rh(acac)_3$	60	15	$CH_3CH=CHOSi(C_2H_5)_3$ (42)	274
C_3H_6	CH_3CH_2CHO				
	Ni	-	-	$CH_3CH_2CH_2OSi(C_2H_5)_3$ + $CH_3CH=CHOSi(C_2H_5)_3$ (85)	416
	Ni	-	-	$CH_3CH_2CH_2OSi(C_2H_5)_3$ + $CH_3CH=CHOSi(C_2H_5)_3$ (-)	374
	Ni	100-110	2-3 (b)	$CH_3CH_2CH_2OSi(C_2H_5)_3$ (84)	372
	Ni	180	-	$CH_3CH_2CH_2OSi(C_2H_5)_3$ (83)	146
	Ni +Et_2S	100-110	2-3	$CH_3CH_2CH_2OSi(C_2H_5)_3$ + $CH_3CH=CHOSi(C_2H_5)_3$ (78)	372
C_3H_6	CH_3COCH_3				
	UV	-	20	$(CH_3)_2CHOSi(C_2H_5)_3$ (-)	U -129
	$Co_2(CO)_6(PPh_3)_2$ + UV	-	20	$(CH_3)_2CHOSi(C_2H_5)_3$ (-)	S -049
	$Co_2(CO)_8$ + UV	29	20	$(CH_3)_2CHOSi(C_2H_5)_3$ (25)	1313
	$Co_2(CO)_8$ + UV	29	20	$(CH_3)_2CHOSi(C_2H_5)_3$ (25)	U -129

Table 8.6
= $(C_2H_5)_3SiH$ = =2=

$Co_4(CO)_{12}$	29	20	$(CH_3)_2CHOSi(C_2H_5)_3$ (-)	U -140
$Co_4(CO)_{12}$	80	20	$(CH_3)_2CHOSi(C_2H_5)_3$ (3)	U -140
$Co_4(CO)_{12}$	80	20	$(CH_3)_2CHOSi(C_2H_5)_3$ (3)	1313
$Co_4(CO)_{12}$ + UV	29	20	$(CH_3)_2CHOSi(C_2H_5)_3$ (30)	U -129
$Co_4(CO)_{12}$ + UV	29	20	$(CH_3)_2CHOSi(C_2H_5)_3$ (30)	1313
$Cr(CO)_6$ + UV	29	20	$(CH_3)_2CHOSi(C_2H_5)_3$ (25)	U -017
$Cr(CO)_6$ + UV	29	20	$(CH_3)_2CHOSi(C_2H_5)_3$ (54)	1313
$Fe_2(CO)_9$ + UV	-	20	$(CH_3)_2CHOSi(C_2H_5)_3$ (-)	U -129
$Fe_3(CO)_{12}$	29	20	$(CH_3)_2CHOSi(C_2H_5)_3$ (-)	U -140
$Fe_3(CO)_{12}$	80	20	$(CH_3)_2CHOSi(C_2H_5)_3$ (34)	U -140
$Fe_3(CO)_{12}$	80	20	$(CH_3)_2CHOSi(C_2H_5)_3$ (34)	1313
$Fe_3(CO)_{12}$ + UV	29	20	$(CH_3)_2CHOSi(C_2H_5)_3$ (5)	U -017
$Fe_3(CO)_{12}$ + UV	29	20	$(CH_3)_2CHOSi(C_2H_5)_3$ (5)	1313
$GaCl_3$	-	-	$(CH_3)_2CHOSi(C_2H_5)_3$ (77)	195
$InCl_3$	-	-	$(CH_3)_2CHOSi(C_2H_5)_3$ (trace)	195
$Ir(CO)_2$(acac) + UV	-	20	$(CH_3)_2CHOSi(C_2H_5)_3$ (-)	U -129
$Ir_4(CO)_{12}$	29	20	$(CH_3)_2CHOSi(C_2H_5)_3$ (-)	U -140
$Ir_4(CO)_{12}$	80	20	$(CH_3)_2CHOSi(C_2H_5)_3$ (6)	1313
$Ir_4(CO)_{12}$	80	20	$(CH_3)_2CHOSi(C_2H_5)_3$ (6)	U -140
$Ir_4(CO)_{12}$ + UV	29	20	$(CH_3)_2CHOSi(C_2H_5)_3$ (86)	U -129
$Ir_4(CO)_{12}$ + UV	29	20	$(CH_3)_2CHOSi(C_2H_5)_3$ (86)	1313
$Mn_2(CO)_{10}$ + UV	29	20	$(CH_3)_2CHOSi(C_2H_5)_3$ (5)	1313
$Mo(CO)_6$ + UV	29	20	$(CH_3)_2CHOSi(C_2H_5)_3$ (1)	1313
Ni	-	-	$(CH_3)_2CHOSi(C_2H_5)_3 + CH_2{=}C(CH_3)OSi(C_2H_5)_3$ (87)	416
Ni + Et_2S	reflux	6.5 (b)	$(CH_3)_2CHOSi(C_2H_5)_3$ / $CH_2{=}C(CH_3)OSi(C_2H_5)_3$(10/90)	F -008
$Os_3(CO)_{12}$	29	20	$(CH_3)_2CHOSi(C_2H_5)_3$ (-)	U -140
$Os_3(CO)_{12}$	80	20	$(CH_3)_2CHOSi(C_2H_5)_3$ (76)	U -140

Table 8.6
= $(C_2H_5)_3SiH$ = =3=

	$Os_3(CO)_{12}$	80	20	$(CH_3)_2CHOSi(C_2H_5)_3$ (76)	1313
	$Os_3(CO)_{12}$ + UV	29	20	$(CH_3)_2CHOSi(C_2H_5)_3$ (83)	1313
	$Os_3(CO)_{12}$ + UV	29	20	$(CH_3)_2CHOSi(C_2H_5)_3$ (83)	U -129
	$PdCl_2$	20	6	$(CH_3)_2CHOSi(C_2H_5)_3$ (48) $CH_2=C(CH_3)OSi(C_2H_5)_3$ (38)	811
	$Re_2(CO)_{10}$	29	20	$(CH_3)_2CHOSi(C_2H_5)_3$ (-)	U -140
	$Re_2(CO)_{10}$	80	20	$(CH_3)_2CHOSi(C_2H_5)_3$ (10)	U -140
	$Re_2(CO)_{10}$	80	20	$(CH_3)_2CHOSi(C_2H_5)_3$ (10)	1313
	$Re_2(CO)_{10}$ + UV	-	20	$(CH_3)_2CHOSi(C_2H_5)_3$ (100)	U -129
	$Re_2(CO)_{10}$ + UV	20	20	$(CH_3)_2CHOSi(C_2H_5)_3$ (87-89)	U -129
	$Re_2(CO)_{10}$ + UV	20	20	$(CH_3)_2CHOSi(C_2H_5)_3$ (87-89)	1313
	$Rh(\pi\text{-All})[P(OMe)_3]_3$	-	-	$(CH_3)_2CHOSi(C_2H_5)_3$ (-)	144
	$Rh_6(CO)_{16}$ + UV	-	20	$(CH_3)_2CHOSi(C_2H_5)_3$ (-)	U -129
	$RhCl(PPh_3)_3$	r.t.	0.17	$(CH_3)_2CHOSi(C_2H_5)_3$ (85-95)	F -023
	$RhCl(PPh_3)_3$	r.t.	0.17	$(CH_3)_2CHOSi(C_2H_5)_3$ (85-95)	873
	$Ru_3(CO)_{12}$	29	20	$(CH_3)_2CHOSi(C_2H_5)_3$ (-)	U -140
	$Ru_3(CO)_{12}$	80	20	$(CH_3)_2CHOSi(C_2H_5)_3$ (100)	1313
	$Ru_3(CO)_{12}$	80	20	$(CH_3)_2CHOSi(C_2H_5)_3$ (100)	U -140
	$Ru_3(CO)_{12}$ + UV	20	20	$(CH_3)_2CHOSi(C_2H_5)_3$ (87)	1313
	$Ru_3(CO)_{12}$ + UV	29	20	$(CH_3)_2CHOSi(C_2H_5)_3$ (87)	U -135
	$W(CO)_6$ + UV	29	20	$(CH_3)_2CHOSi(C_2H_5)_3$ (5)	1313
	$ZnCl_2$	-	-	$(CH_3)_2CHOSi(C_2H_5)_3$ (68)	195
C_4	$(CF_3)_2C=C=O$				
	H_2PtCl_6	-	-	$(CF_3)_2C=CHOSi(C_2H_5)_3$ (75)	1054
C_4H_6	$CH_2=C(CH_3)CHO$				
	Ni	reflux	2 (b)	$CH_2=C(CH_3)CH_2OSi(C_2H_5)_3$ + $(CH_3)_2C=CHOSi(C_2H_5)_3$ (85)	148
	Pd/C	reflux	2 (b)	$CH_2=C(CH_3)CH_2OSi(C_2H_5)_3$ + $(CH_3)_2C=CHOSi(C_2H_5)_3$ (89)	148

Table 8.6

= $(C_2H_5)_3SiH$ = =4=

	H_2PtCl_6	reflux	2 (b)	$CH_2{=}C(CH_3)CH_2OSi(C_2H_5)_3$ + $(CH_3)_2C{=}CHOSi(C_2H_5)_3$ (66)	148
C_4H_6	$CH_2{=}CHCOOCH_3$				
	$RhCl(PPh_3)_3$	100	60s	$CH_2{=}CHCH[OSi(C_2H_5)_3]OCH_3$ (39) $CH_3CH{=}C(OCH_3)OSi(C_2H_5)_3$ (32)	1321
C_4H_6	$CH_3CH{=}CHCHO$				
	Ni	-	-	$CH_3CH{=}CHCH_2OSi(C_2H_5)_3$ + $CH_3CH_2CH{=}CHOSi(C_2H_5)_3$ (-)	374
	Ni	-	-	$[(CH_3)_2C{=}COSi(C_2H_5)_3]_2$ (80)	416
	Ni	reflux	2 (b)	$CH_3CH{=}CHCH_2OSi(C_2H_5)_3$ + $CH_3CH_2CH{=}CHOSi(C_2H_5)_3$ (88)	148
	Ni + Et_2S	reflux	2 (b)	$CH_3CH{=}CHCH_2OSi(C_2H_5)_3$ + $CH_3CH_2CH{=}CHOSi(C_2H_5)_3$ (57)	148
	Pd/C	reflux	2 (b)	$CH_3CH{=}CHCH_2OSi(C_2H_5)_3$ + $CH_3CH_2CH{=}CHOSi(C_2H_5)_3$ (92)	148
	H_2PtCl_6	reflux	2 (b)	$CH_3CH{=}CHCH_2OSi(C_2H_5)_3$ + $CH_3CH_2CH{=}CHOSi(C_2H_5)_3$ (77)	148
	$Rh(acac)_3$	60	15	$CH_3CH_2CH{=}CHOSi(C_2H_5)_3$ (31)	274
	$RhCl(PPh_3)_3$	60	0.25	$CH_3CH_2CH{=}CHOSi(C_2H_5)_3$ (95)	F -023
C_4H_6	$CH_3COCOCH_3$				
	$RhCl(PPh_3)_3$	70	2	$CH_3CH[OSi(C_2H_5)_3]COCH_3$ + $(C_2H_5)_3SiOCH(CH_3)CH(CH_3)OSi(C_2H_5)_3$ (-)	873
C_4H_8	$(CH_3)_2CHCHO$				
	Ni	100-110	2-3 (b)	$(CH_3)_2CHCH_2OSi(C_2H_5)_3$ + $(CH_3)_2C{=}CHOSi(C_2H_5)_3$ (87)	372
	Ni	180	-	$(CH_3)_2CHCH_2OSi(C_2H_5)_3$ (84)	146
	Ni + Et_2S	100-110	2-3 (b)	$(CH_3)_2CHCH_2OSi(C_2H_5)_3$ + $(CH_3)_2C{=}CHOSi(C_2H_5)_3$ (79)	372
C_4H_8	$CH_3CH_2CH_2CHO$				
	$Co_2(CO)_6(PPh_3)_2$	29	20	$CH_3(CH_2)_3OSi(C_2H_5)_3$ (87)	1313
	$Co_2(CO)_6(PPh_3)_2$ + UV	29	20	$CH_3(CH_2)_3OSi(C_2H_5)_3$ + $CH_3CH_2CH{=}CHOSi(C_2H_5)_3$ (100)	U -129
	$Co_4(CO)_{12}$ + UV	29	20	$CH_3(CH_2)_3OSi(C_2H_5)_3$ + $CH_3CH_2CH{=}CHOSi(C_2H_5)_3$ (100)	U -129

Table 8.6

= $(C_2H_5)_3SiH$ = =5=

	$Co_4(CO)_{12}$ + UV	29	20	$CH_3(CH_2)_3OSi(C_2H_5)_3$ (90)	1313
	$Cr(CO)_6$	29	20	$CH_3(CH_2)_3OSi(C_2H_5)_3$ (9)	1313
	$Fe_3(CO)_{12}$	29	20	$CH_3(CH_2)_3OSi(C_2H_5)_3$ (8)	1313
	$Ir_4(CO)_{12}$ + UV	29	20	$CH_3(CH_2)_3OSi(C_2H_5)_3$ (7)	1313
	Ni	-	-	$CH_3(CH_2)_3OSi(C_2H_5)_3$ + $CH_3CH_2CH{=}CHOSi(C_2H_5)_3$ (79)	416
	Ni	100-110	2-3 (b)	$CH_3(CH_2)_3OSi(C_2H_5)_3$ + $CH_3CH_2CH{=}CHOSi(C_2H_5)_3$ (81)	372
	Ni	180	-	$CH_3(CH_2)_3OSi(C_2H_5)_3$ (80)	146
	Ni + Et_2S	100-110	2-3	$CH_3(CH_2)_3OSi(C_2H_5)_3$ + $CH_3CH_2CH{=}CHOSi(C_2H_5)_3$ (84)	372
	$Re_2(CO)_{10}$ + UV	29	20	$CH_3(CH_2)_3OSi(C_2H_5)_3$ (53)	U -017
	$Re_2(CO)_{10}$ + UV	29	20	$CH_3(CH_2)_3OSi(C_2H_5)_3$ (53)	1313
C_4H_8	$CH_3CH_2COCH_3$				
	Ni + CS_2	reflux	8	$CH_3CH_2CH[OSi(C_2H_5)_3]CH_3$ + $CH_3CH{=}C(CH_3)OSi(C_2H_5)_3$ (-)	370
	Ni + Et_2S	120	64	$CH_3CH_2CH[OSi(C_2H_5)_3]CH_3$ + $CH_3CH{=}C(CH_3)OSi(C_2H_5)_3$ (74)	370
	Ni + NiS	130	17	$CH_3CH_2CH[OSi(C_2H_5)_3]CH_3$ + $CH_3CH{=}C(CH_3)OSi(C_2H_5)_3$ (80)	370
	$PdCl_2$	20	6	$CH_3CH_2CH[OSi(C_2H_5)_3]CH_3$ (19) $CH_3CH{=}C(CH_3)OSi(C_2H_5)_3$ (66)	811
	cis-RhCl(=$\overline{CNMeCH_2CH_2NMe}$)(cod)	60-120	1.5-4	$CH_3CH_2CH[OSi(C_2H_5)_3]CH_3$ (5-99)	660
	trans-RhCl(=$\overline{CNMeCH_2CH_2NMe}$)$(PPh_3)_2$	40-100	5	$CH_3CH_2CH[OSi(C_2H_5)_3]CH_3$ (18-50)	660
C_4H_8	$CH_3COOCH_2CH_3$				
	Ni	reflux	14	$CH_3CH[OSi(C_2H_5)_3]OCH_2CH_3$ (30)	377
C_5H_4	Fig.266.				
	Ni	-	-	Fig.346. +Fig.347. (-)	147
	Ni	-	-	Fig.346 + Fig.347. (-)	275
	Ni	100-110	- (b)	Fig.346. (34) Fig.347. (29)	380

	Ni	100-140	17 (THF)	Fig.346. (32)	411
	Ni + Et_2S	120-130	-	Fig.346. (23) Fig.347. (57)	380
C_5H_8	$(CH_2)_4CO$				
	Ni + CS_2	reflux	3	$(CH_2)_4CHOSi(C_2H_5)_3$ + $(CH_2)_3CH{=}COSi(C_2H_5)_3$ (-)	370
	Ni + Et_2S	110	7	$(CH_2)_4CHOSi(C_2H_5)_3$ + $(CH_2)_3CH{=}COSi(C_2H_5)_3$ (98)	370
	Ni + NiS	60	1	$(CH_2)_4CHOSi(C_2H_5)_3$ + $(CH_2)_3CH{=}COSi(C_2H_5)_3$ (87)	370
	Ni + PhSH	reflux	3	$(CH_2)_4CHOSi(C_2H_5)_3$ + $(CH_2)_3CH{=}COSi(C_2H_5)_3$ (-)	370
	$PdCl_2$	20	6	$(CH_2)_4CHOSi(C_2H_5)_3$ (42) $(CH_2)_3CH{=}COSi(C_2H_5)_3$ (34)	811
C_5H_8	$CH_2{=}C(CH_3)COOCH_3$				
	RhCl($CNPhCH_2CH_2NPh$)(cod)	100	8	$(CH_3)_2C{=}C[OSi(C_2H_5)_3]OCH_3$ (98)	496
	$RhCl(PPh_3)_3$	100	60s	$(CH_3)_2C{=}C[OSi(C_2H_5)_3]OCH_3$ (70)	1321
	$RhCl(PPh_3)_3$	70	2 (b)	$(CH_3)_2C{=}C[OSi(C_2H_5)_3]OCH_3$ + $CH_2{=}C(CH_3)CH[OSi(C_2H_5)_3]OCH_3$ (80)	868
	$RhCl(PPh_3)_3$	reflux	2 (b)	$(CH_3)_2C{=}C[OSi(C_2H_5)_3]OCH_3$ (94) $CH_2{=}C(CH_3)CH[OSi(C_2H_5)_3]OCH_3$ (trace)	1319
	$RhCl_2[P(c\text{-}Hex)_3]_2$	100	8	$(CH_3)_2C{=}C[OSi(C_2H_5)_3]OCH_3$ (75)	503
C_5H_8	$CH_3CH{=}CHCOCH_3$				
	$PtCl_2(py)_2$	-	1	$CH_3CH_2CH{=}C[OSi(C_2H_5)_3]CH_3$ (100)	987
C_5H_8	$CH_3CH{=}CHCOOCH_3$				
	$RhCl(PPh_3)_3$	100	60s	$CH_3CH_2CH{=}C[OSi(C_2H_5)_3]OCH_3$ (86)	1319
	$RhCl(PPh_3)_3$	60	1	$CH_3CH{=}CHCH[OSi(C_2H_5)_3]OCH_3$ + $CH_3CH_2CH{=}C[OSi(C_2H_5)_3]OCH_3$ (95)	868
C_5H_8	$CH_3COCH_2COCH_3$				
	$RhCl(PPh_3)_3$	75	5	$CH_3COCH{=}C(CH_3)OSi(C_2H_5)_3$ (92)	873
C_5H_8	$CH_3COCH_2COOCH_3$				
	$RhCl(PPh_3)_3$	60	12	$CH_3C[OSi(C_2H_5)_3]{=}CHCOOCH_3$ + $CH_3CH[OSi(C_2H_5)_3]CH_2COOCH_3$ (-)	873
	$RhCl(PPh_3)_3$ + Et_3SiSPh	reflux	12-24 (b)	$CH_3C[OSi(C_2H_5)_3]{=}CHCOOCH_3$ (-)	853
C_5H_9	$CH_3(CH_2)_3NCO$				

	$PdCl_2$	-	-	$CH_3(CH_2)_3NCHOSi(C_2H_5)_3$ (-)	J -014
C_5H_{10}	$(CH_3)_2CHCOCH_3$				
	$PdCl_2$	20	6	$(CH_3)_2CHCH[OSi(C_2H_5)_3]CH_3$ (37) $(CH_3)_2C{=}C[OSi(C_2H_5)_3]CH_3$ (52)	811
C_5H_{10}	$(CH_3)_3CCHO$				
	Ni	100-110	2-3 (b)	$(CH_3)_3CCH_2OSi(C_2H_5)_3$ (82)	372
	Ni +Et_2S	100-110	2-3 (b)	$(CH_3)_3CCH_2OSi(C_2H_5)_3$ (84)	372
C_5H_{10}	$CH_3(CH_2)_3CHO$				
	Ni	-	-	$CH_3(CH_2)_4OSi(C_2H_5)_3$ + $CH_3CH_2CH_2CH{=}CHOSi(C_2H_5)_3$ (71)	416
	Ni	100-110	2-3 (b)	$CH_3(CH_2)_4OSi(C_2H_5)_3$ + $CH_3CH_2CH_2CH{=}CHOSi(C_2H_5)_3$ (84)	372
	Ni	180	-	$CH_3(CH_2)_4OSi(C_2H_5)_3$ (76)	146
C_5H_{10}	$CH_3CH_2CH_2COCH_3$				
	Ni +Et_2S	120	23	$CH_3CH_2CH_2CH[OSi(C_2H_5)_3]CH_3$ + $CH_3CH_2CH{=}C(CH_3)[OSi(C_2H_5)_3]$ (93)	370
	Ni + NiS	120	3	$CH_3CH_2CH_2CH[OSi(C_2H_5)_3]CH_3$ + $CH_3CH_2CH{=}C(CH_3)[OSi(C_2H_5)_3]$ (71)	370
C_5H_{10}	$CH_3CH_2COCH_2CH_3$				
	$GaCl_3$	-	-	$CH_3CH_2CH[OSi(C_2H_5)_3]CH_2CH_3$ (98)	195
	Ni	-	-	$CH_3CH_2CH[OSi(C_2H_5)_3]CH_2CH_3$ + $CH_3CH_2C[OSi(C_2H_5)_3]{=}CHCH_3$ (92)	416
	Ni +CS_2	reflux	7	$CH_3CH_2CH[OSi(C_2H_5)_3]CH_2CH_3$ + $CH_3CH_2C[OSi(C_2H_5)_3]{=}CHCH_3$ (-)	370
	Ni +Et_2S	130	26	$CH_3CH_2CH[OSi(C_2H_5)_3]CH_2CH_3$ + $CH_3CH_2C[OSi(C_2H_5)_3]{=}CHCH_3$ (70)	370
	Ni + NiS	140	90	$CH_3CH_2CH[OSi(C_2H_5)_3]CH_2CH_3$ + $CH_3CH_2C[OSi(C_2H_5)_3]{=}CHCH_3$ (82)	370
	Ni + PhSH	reflux	7	$CH_3CH_2CH[OSi(C_2H_5)_3]CH_2CH_3$ + $CH_3CH_2C[OSi(C_2H_5)_3]{=}CHCH_3$ (-)	370
	$PdCl_2$	20	6	$CH_3CH_2CH[OSi(C_2H_5)_3]CH_2CH_3$ (12) $CH_3CH_2C[OSi(C_2H_5)_3]{=}CHCH_3$ (42)	811
	Pt complex	20	5	$CH_3CH_2CH[OSi(C_2H_5)_3]CH_2CH_3$ (-)	353
C_6H_4	Fig.348.				

	H_2PtCl_6 (i-PrOH)	-	2-1	1,4-$C_6H_4[OSi(C_2H_5)_3]_2$ (64)	654
C_6H_5	2-C_5H_4NCHO				
	Ni	100-110	- (b)	2-$C_5H_4NCH_2OSi(C_2H_5)_3$ (37) {2-$C_5H_4NCH[OSi(C_2H_5)_3]\}_2$ (48)	380
C_6H_8	$(CH_2)_3CH=CHCO$				
	$[IrCl(c\text{-}C_8H_{14})_2]_2$	25	2	$(CH_2)_3CH=CHCHOSi(C_2H_5)_3$ (76) $(CH_2)_4CH=COSi(C_2H_5)_3$ (24)	59
	$[IrCl(c\text{-}C_8H_{14})_2]_2$	60	2	$(CH_2)_3CH=CHCHOSi(C_2H_5)_3$ (29) $(CH_2)_4CH=COSi(C_2H_5)_3$ (13)	59
	$[IrCl(c\text{-}C_8H_{14})_2]_2 + PPh_3$	60	2	$(CH_2)_3CH=CHCHOSi(C_2H_5)_3$ (30) $(CH_2)_4CH=COSi(C_2H_5)_3$ (47)	59
	$[IrCl(cod)]_2$	25	2	$(CH_2)_3CH=CHCHOSi(C_2H_5)_3$ (29) $(CH_2)_4CH=COSi(C_2H_5)_3$ (11)	59
	$Rh(\pi\text{-All})[P(OMe)_3]_3$	-	-	$(CH_2)_3CH=CHCHOSi(C_2H_5)_3$ (-)	144
C_6H_{10}	$(CH_2)_5CO$				
	$GaCl_3$	-	-	$(CH_2)_5CHOSi(C_2H_5)_3$ (98)	195
	$[IrCl(c\text{-}C_8H_{14})_2]_2$	60	6	$(CH_2)_5CHOSi(C_2H_5)_3$ (64)	59
	$[IrCl(c\text{-}C_8H_{14})_2]_2 + PPh_3$	60	6	$(CH_2)_5CHOSi(C_2H_5)_3$ (2-76)	59
	Ni	-	-	$(CH_2)_4CH=COSi(C_2H_5)_3$ (68) $[(CH_2)_4CHCHOSi(C_2H_5)_3]_2$ (24)	416
	Ni	reflux	6	$(CH_2)_5CHOSi(C_2H_5)_3 + (CH_2)_4CH=COSi(C_2H_5)_3$ (-)	370
	Ni + CS_2	reflux	6	$(CH_2)_5CHOSi(C_2H_5)_3 + (CH_2)_4CH=COSi(C_2H_5)_3$ (-)	370
	Ni + Et_2S	120	5	$(CH_2)_5CHOSi(C_2H_5)_3 + (CH_2)_4CH=COSi(C_2H_5)_3$ (83)	370
	Ni + Et_2S	reflux	-	$(CH_2)_4CH=COSi(C_2H_5)_3$ (96)	223
	Ni + NiS	60-80	1	$(CH_2)_5CHOSi(C_2H_5)_3 + (CH_2)_4CH=COSi(C_2H_5)_3$ (97)	370
	Ni + PhSH	90	1	$(CH_2)_5CHOSi(C_2H_5)_3$ (11) $(CH_2)_4CH=COSi(C_2H_5)_3$ (84)	F -008
	Ni + PhSH	reflux	6	$(CH_2)_5CHOSi(C_2H_5)_3 + (CH_2)_4CH=COSi(C_2H_5)_3$ (-)	370
	Ni + $SiEt_2Me_2$	reflux	6	$(CH_2)_5CHOSi(C_2H_5)_3 + (CH_2)_4CH=COSi(C_2H_5)_3$ (-)	370
	$PdCl_2$	20	6	$(CH_2)_5CHOSi(C_2H_5)_3$ (40) $(CH_2)_4CH=COSi(C_2H_5)_3$ (35)	811

Table 8.6
= $(C_2H_5)_3SiH$ =

=9=

	$Rh(\pi\text{-}All)[P(OMe)_3]_3$	-	-	$\overline{(CH_2)_5C}HOSi(C_2H_5)_3$ (-)	144
	$Rh_2(OAc)_4$	100	8	$\overline{(CH_2)_5C}HOSi(C_2H_5)_3$ (51)	274
	$RhCl(CO)(PPh_3)_2$	20	6 (b)	$\overline{(CH_2)_5C}HOSi(C_2H_5)_3$ (4)	202
	$RhCl(CO)(PPh_3)_2$ + t-BuOOH	20	6 (b)	$\overline{(CH_2)_5C}HOSi(C_2H_5)_3$ (21)	202
	$RhCl(PPh_3)_3$	r.t.	0.08	$\overline{(CH_2)_5C}HOSi(C_2H_5)_3$ (98)	873
	$RhCl(PPh_3)_3$	r.t.	0.08	$\overline{(CH_2)_5C}HOSi(C_2H_5)_3$ (98)	872
	$RhCl_2[P(c\text{-}Hex)_3]_2$	100	8	$\overline{(CH_2)_5C}HOSi(C_2H_5)_3$ (82)	503
	$RhCl_2[P(o\text{-}MeC_6H_4)_3]_2$	100	8	$\overline{(CH_2)_5C}HOSi(C_2H_5)_3$ (74)	503
	$[RhCl(cod)]_2$ + DMPIC	100	6	$\overline{(CH_2)_5C}HOSi(C_2H_5)_3$ (58)	884
	$[RhCl(cod)]_2$ + DMPIC	20	6	$\overline{(CH_2)_5C}HOSi(C_2H_5)_3$ (79)	884
	$[RhCl(cod)]_2$ + XNC	20-100	6	$\overline{(CH_2)_5C}HOSi(C_2H_5)_3$ (58-79)	1
	$\{RhCl[P(OC_6H_4Me)_3]_2\}_n$	100	2	$\overline{(CH_2)_5C}HOSi(C_2H_5)_3$ (86)	229
	$\{RhCl[P(OC_6H_4Me)_3]_2\}_n$	25	6	$\overline{(CH_2)_5C}HOSi(C_2H_5)_3$ (5)	229
	$\{RhCl[P(OMe)_3]_2\}_n$	100	2	$\overline{(CH_2)_5C}HOSi(C_2H_5)_3$ (51)	229
	$\{RhCl[P(OMe)_3]_2\}_n$	25	6	$\overline{(CH_2)_5C}HOSi(C_2H_5)_3$ (41)	229
	$\{RhH[P(OC_6H_4Me)_3]_2\}_n$	100	2	$\overline{(CH_2)_5C}HOSi(C_2H_5)_3$ (100)	229
	$\{RhH[P(OC_6H_4Me)_3]_2\}_n$	25	6	$\overline{(CH_2)_5C}HOSi(C_2H_5)_3$ (78)	229
	$\{RhH[P(OMe)_3]_2\}_3$	100	2	$\overline{(CH_2)_5C}HOSi(C_2H_5)_3$ (84)	229
	$\{RhH[P(OMe)_3]_2\}_3$	25	6	$\overline{(CH_2)_5C}HOSi(C_2H_5)_3$ (10)	229
	$RuCl_2(PPh_3)_3$	80	3	$\overline{(CH_2)_5C}HOSi(C_2H_5)_3$ (74)	339
	$ZnCl_2$	reflux	2 (e)	$[\overline{(CH_2)_4C}HCHOSi(C_2H_5)_3]_2$ (26)	656
C_6H_{10}	$(CH_3)_2C=CHCOCH_3$				
	Cosalt	110-190	4.6	$(CH_3)_2CHCH=C[OSi(C_2H_5)_3]CH_3$ (72)	15
	$[IrCl(c\text{-}C_8H_{14})_2]_2$	25	2	$(CH_3)_2CHCH=C[OSi(C_2H_5)_3]CH_3$ (25) $(CH_3)_2C=CCH(CH_3)OSi(C_2H_5)_3$ (75)	59
	$RhCl(PPh_3)_3$	50	1	$(CH_3)_2CH=C[OSi(C_2H_5)_3]CH_3$ (95)	850
	$RhCl(PPh_3)_3$	50	0.25	$(CH_3)_2CH=C[OSi(C_2H_5)_3]CH_3$ (94)	F -023
	$RhCl(PPh_3)_3$	50	0.25	$(CH_3)_2CH=C[OSi(C_2H_5)_3]CH_3$ (94)	873

	$\{RhCl[P(OC_6H_4Me)_3]_2\}_n$	100	2	$(CH_3)_2CHCH{=}C[OSi(C_2H_5)_3]CH_3$ (95)	229
	$\{RhCl[P(OMe)_3]_2\}_n$	100	2	$(CH_3)_2CHCH{=}C[OSi(C_2H_5)_3]CH_3$ (100)	229
	$\{RhCl[P(OMe)_3]_2\}_n$	20	6	$(CH_3)_2CHCH{=}C[OSi(C_2H_5)_3]CH_3$ (90)	229
	$\{RhH[P(OC_6H_4Me)_3]_2\}_n$	100	2	$(CH_3)_2CHCH{=}C[OSi(C_2H_5)_3]CH_3$ (61)	229
	$\{RhH[P(OMe)_3]_2\}_3$	100	2	$(CH_3)_2CHCH{=}C[OSi(C_2H_5)_3]CH_3$ (95)	229
	$\{RhH[P(OMe)_3]_2\}_3$	20	6	$(CH_3)_2CHCH{=}C[OSi(C_2H_5)_3]CH_3$ (67)	229
C_6H_{10}	$(CH_3)_3CCOCH_3$				
	Ni	58-112	5	$(CH_3)_3CCH(CH_3)OSi(C_2H_5)_3$ (46)	414
	Ni + Et_2S	120	23	$C_3H_7CH{=}C(CH_3)OSi(C_2H_5)_3$ (91)	370
	Ni + NiS	100-120	2	$C_3H_7CH{=}C(CH_3)OSi(C_2H_5)_3$ (89)	370
	$PdCl_2$	20	6	$(CH_3)_3CCH(CH_3)OSi(C_2H_5)_3$ (27) $(CH_3)_3CC({=}CH_2)OSi(C_2H_5)_3$ (29)	811
C_6H_{10}	$CH_2{=}C(CH_3)COOCH_2CH_3$				
	$RhCl(PPh_3)_3$	80	2 (b)	$CH_3CH{=}CHCH[OSi(C_2H_5)_3]OCH_2CH_3$ + . $CH_3CH_2CH{=}C[OSi(C_2H_5)_3]OCH_2CH_3$ (80)	868
C_6H_{10}	$CH_2{=}CHCH_2CH_2COCH_3$				
	H_2PtCl_6	130-140	1.5-2	$CH_2{=}CHCH_2CH_2CH[OSi(C_2H_5)_3]CH_3$ (40) $(C_2H_5)_3Si(CH_2)_4COCH_3$ (34)	602
C_6H_{10}	$CH_3CH_2CH{=}CHCOOCH_3$				
	$RhCl(PPh_3)_3$	100	60s	$CH_3CH_2CH_2CH{=}C[OSi(C_2H_5)_3]OCH_3$ (75)	1321
C_6H_{10}	$CH_3CH{=}CHCOOCH_2CH_3$				
	$RhCl(PPh_3)_3$	50	1	$CH_3CH_2CH{=}C[OSi(C_2H_5)_3]OCH_2CH_3$ (95)	868
C_7H_4	Fig.349.				
	Ni	reflux	3-4	Fig.350 Fig.351. (-)	645
C_7H_5	C_6H_5NCO				
	$PdCl_2$	-	-	$C_6H_5NCHOSi(C_2H_5)_3$ (-)	J -014
C_7H_5	Fig.352.				
	Ni	reflux	3-4.5	Fig.353. (61) Fig.354. (32)	645
C_7H_5	o-, m-, p-ClC_6H_4CHO				

	Ni	100-110	- (b)	m-$ClC_6H_4CH_2OSi(C_2H_5)_3$ (77)	380
	Ni	100-110	- (b)	o-$ClC_6H_4CH_2OSi(C_2H_5)_3$ (79)	380
	Ni + Et_2S	120-130	-	o-$ClC_6H_4CH_2OSi(C_2H_5)_3$ (trace) o-$ClC_6H_4CH[OSi(C_2H_5)_3]_2$ (88)	380
	Ni + Et_2S	120-130	-	p-$ClC_6H_4CH_2OSi(C_2H_5)_3$ (trace) p-$ClC_6H_4CH[OSi(C_2H_5)_3]_2$ (90)	380
	$RuCl_2(PPh_3)_3$	90	1.25	m-$ClC_6H_4CH_2OSi(C_2H_5)_3$ (64)	339
	$RuCl_2(PPh_3)_3$	90	1.25	o-$ClC_6H_4CH_2OSi(C_2H_5)_3$ (61)	339
C_7H_6	C_6H_5CHO				
	F^-	20	10-12	$C_6H_5CH_2OSi(C_2H_5)_3$ (-)	307
	$LiClO_4$, Bu_4NClO_4 (CH_2Cl_2)+electrolysis	r.t.	-	$C_6H_5CH_2OSi(C_2H_5)_3$ (91)	1201
	CsF	r.t.	10 (MeCN)	$C_6H_5CH_2OSi(C_2H_5)_3$ (96)	8
	Ni	-	-	$C_6H_5CH_2OSi(C_2H_5)_3$ + $C_6H_5CH[OSi(C_2H_5)_3]CH(C_6H_5)OSi(C_2H_5)_3$ (-)	374
	Ni	-	-	$C_6H_5CH_2OSi(C_2H_5)_3$ (23-38) $C_6H_5CH[OSi(C_2H_5)_3]CH(C_6H_5)OSi(C_2H_5)_3$ (11-59)	412
	Ni	-	-	$C_6H_5CH_2OSi(C_2H_5)_3$ (75)	373
	Ni	-	-	$C_6H_5CH[OSi(C_2H_5)_3]CH(C_6H_5)OSi(C_2H_5)_3$ (-)	413
	Ni	100-137	0.5 (b)	$C_6H_5CH_2OSi(C_2H_5)_3$ (24)	411
	Ni	20	24	$C_6H_5CH_2OSi(C_2H_5)_3$ (27)	380
	Ni	85-90	3.2 (b)	$C_6H_5CH_2OSi(C_2H_5)_3$ (36)	411
	Ni + Et_2S	120-130	-	$C_6H_5CH_2OSi(C_2H_5)_3$ (5) $C_6H_5CH[OSi(C_2H_5)_3]CH(C_6H_5)OSi(C_2H_5)_3$ (84)	380
	Ni Raney	-	-	$C_6H_5CH[OSi(C_2H_5)_3]CH(C_6H_5)OSi(C_2H_5)_3$ (59)	417
	$Rh(\pi\text{-All})[P(OMe)_3]_3$	-	-	$C_6H_5CH_2OSi(C_2H_5)_3$ (-)	144
	$RhCl(PPh_3)_3$	-	-	$C_6H_5CH_2OSi(C_2H_5)_3$ (90)	872
	$RhCl(PPh_3)_3$	50	0.5	$C_6H_5CH_2OSi(C_2H_5)_3$ (85)	F -023
	$RhCl(PPh_3)_3$	r.t.	0.08	$C_6H_5CH_2OSi(C_2H_5)_3$ (95)	873
	$Ru(CO)_3(PPh_3)_2$ + $h\nu$ (C_6D_6)	-	-	$C_6H_5CH_2OSi(C_2H_5)_3$ (-)	424
	$Ru(CO)_4(PPh_3)$ + $h\nu$ (C_6D_6)	-	-	$C_6H_5CH_2OSi(C_2H_5)_3$ (-)	424

Table 8.6
= $(C_2H_5)_3SiH$ = =12=

	$RuCl_2(PPh_3)_3$	90	1.25	$C_6H_5CH_2OSi(C_2H_5)_3$ (74)	339
	$SnCl_2$	-	-	$C_6H_5CH_2OSi(C_2H_5)_3$ (12)	412
	$ZnCl_2$	until 144	1.5	$C_6H_5CH_2OSi(C_2H_5)_3$ (72)	412
C_7H_6	m-HOC_6H_4CHO				
	Ni	85-90	2.5-3	m-$(C_2H_5)_3SiOC_6H_4CH_2OSi(C_2H_5)_3$ (59)	645
	Ni	reflux	4-4.5	m-$(C_2H_5)_3SiOC_6H_4CH_2OSi(C_2H_5)_3$ (61)	645
C_7H_6	o-HOC_6H_4CHO				
	Ni	85-90	2.5-3	o-$HOC_6H_4CH_2OSi(C_2H_5)_3$ (42)	645
	Ni	reflux	4-4.5	o-$HOC_6H_4CH_2OSi(C_2H_5)_3$ (47)	645
C_7H_{10}	$(CH_2)_2CH{=}CHCH_2CHCHO$				
	Ni	reflux	12 (b)	$(CH_2)_2CH{=}CHCH_2CHCH_2OSi(C_2H_5)_3$ (26) $(CH_2)_5CHCH_2OSi(C_2H_5)_3$ (35) $(CH_2)_5C{=}CHOSi(C_2H_5)_3$ (12) $C_6H_5CH_2OSi(C_2H_5)_3$ (1)	145
	Ni + Et_2S	reflux	12	$(CH_2)_2CH{=}CHCH_2CHCH_2OSi(C_2H_5)_3$ (28) $(CH_2)_5C{=}CHOSi(C_2H_5)_3$ (65)	145
C_7H_{10}	$(CH_2)_3C(CH_3){=}CHCO$				
	$RhCl(PPh_3)_3$	50	30	$C_7H_{11}OSi(C_2H_5)_3$ (-)	850
C_7H_{10}	$(CH_2)_4CH{=}CCHO$				
	Ni	reflux	2 (b)	$(CH_2)_5CHCH_2OSi(C_2H_5)_3$ (30) $(CH_2)_5C{=}CHOSi(C_2H_5)_3$ (39) $C_6H_5CH_2OSi(C_2H_5)_3$ (14)	145
	Ni + Et_2S	reflux	2	$(CH_2)_4CH{=}CCH_2OSi(C_2H_5)_3$ (28) $(CH_2)_5C{=}CHOSi(C_2H_5)_3$ (66)	145
C_7H_{11}	$(CH_2)_5CHN{=}C{=}O$				
	$PtCl_2$	-	-	$(CH_2)_5CHN{=}CHOSi(C_2H_5)_3$ (-)	J -011
C_7H_{12}	$(CH_2)_3CH(CH_3)CH_2CO$				
	$RhCl(PPh_3)_3$	20	3	$(CH_2)_3CH(CH_3)CH_2CHOSi(C_2H_5)_3$ (42)	519
	SiO_2-PPh_2/Rh complex	20	3	$(CH_2)_3CH(CH_3)CH_2CHOSi(C_2H_5)_3$ (27-29)	519
C_7H_{12}	$(CH_2)_4CH(CH_3)CO$				
	Ni + CS_2	reflux	5	$(CH_2)_4CH(CH_3)CHOSi(C_2H_5)_3$ + $(CH_2)_4C(CH_3){=}COSi(C_2H_5)_3$ (-)	370

	Ni + Et_2S	100-120	8	$(CH_2)_4CH(CH_3)CHOSi(C_2H_5)_3$ + $(CH_2)_4C(CH_3)=COSi(C_2H_5)_3$ (81)	370
	Ni + NiS	60	14	$(CH_2)_4CH(CH_3)CHOSi(C_2H_5)_3$ + $(CH_2)_4C(CH_3)=COSi(C_2H_5)_3$ (88)	370
	Ni + PhSH	reflux	5	$(CH_2)_4CH(CH_3)CHOSi(C_2H_5)_3$ + $(CH_2)_4C(CH_3)=COSi(C_2H_5)_3$ (-)	370
	$RhCl(PPh_3)_3$	20	3	$(CH_2)_4CH(CH_3)CHOSi(C_2H_5)_3$ (35)	519
	SiO_2-PPh_2/Rh complex	20	3	$(CH_2)_4CH(CH_3)CHOSi(C_2H_5)_3$ (37)	519
C_7H_{12}	$(CH_2)_5CHCHO$				
	$RhCl(PPh_3)_3$	r.t.	-	$(CH_2)_5CHCH_2OSi(C_2H_5)_3$ (98)	F -023
C_7H_{12}	$(CH_2)_6CO$				
	$Ir_4(CO)_{12}$ + UV	-	20	$(CH_2)_6CHOSi(C_2H_5)_3$ (18)	U -129
C_7H_{12}	$(CH_3)_2CHCH=CHCOOCH_3$				
	$RhCl(PPh_3)_3$	100	0.02	$(CH_3)_2CHCH_2CH=C(OCH_3)OSi(C_2H_5)_3$ (76)	1321
C_7H_{12}	$CH_2CH_2CH(CH_3)CH_2CH_2CO$				
	Ni + CS_2	reflux	4	$CH_2CH_2CH(CH_3)CH_2CH_2CHOSi(C_2H_5)_3$ + $CH_2CH_2CH(CH_3)CH_2CH=COSi(C_2H_5)_3$ (-)	370
	Ni + Et_2S	100-120	12	$CH_2CH_2CH(CH_3)CH_2CH_2CHOSi(C_2H_5)_3$ + $CH_2CH_2CH(CH_3)CH_2CH=COSi(C_2H_5)_3$ (83)	370
	Ni + NiS	80-90	1.5	$CH_2CH_2CH(CH_3)CH_2CH_2CHOSi(C_2H_5)_3$ + $CH_2CH_2CH(CH_3)CH_2CH=COSi(C_2H_5)_3$ (87)	370
	Ni + PhSH	reflux	4	$CH_2CH_2CH(CH_3)CH_2CH_2CHOSi(C_2H_5)_3$ + $CH_2CH_2CH(CH_3)CH_2CH=COSi(C_2H_5)_3$ (-)	370
	$RhCl(PPh_3)_3$	20	3	$CH_2CH_2CH(CH_3)CH_2CH_2CHOSi(C_2H_5)_3$ (40)	519
	SiO_2-PPh_2/Rh complex	20	3	$CH_2CH_2CH(CH_3)CH_2CH_2CHOSi(C_2H_5)_3$ (32-35)	519
C_7H_{12}	$CH_3CO(CH_3)_2COOCH_3$				
	$ZnCl_2$	105-110	2-3	$(C_2H_5)_3SiOCH(CH_3)C(CH_3)_2COOCH_3$ (71)	646
	$ZnCl_2$	105-110	2-3	$(C_2H_5)_3SiOCH(CH_3)C(CH_3)_2COOCH_3$ (71)	S -030
C_7H_{12}	$CH_3COCH_2CH_2COOC_2H_5$				
	Ni	108-120	3-4	$(C_2H_5)_3SiOCH(CH_3)CH_2CH_2COOC_2H_5$ (81)	646
C_7H_{14}	$(CH_3)_3CCOCH_2CH_3$				

	Ni + Et_2S	120-130	24	$(CH_3)_3CC(=CHCH_3)OSi(C_2H_5)_3$ (82)	370
	Ni + NiS	160-180	6	$(CH_3)_3CC(=CHCH_3)OSi(C_2H_5)_3$ (86)	370
C_7H_{14}	$(CH_3CH_2CH_2)_2CO$				
	$ZnCl_2$	reflux	2 (c)	$(CH_3CH_2CH_2)_2CHOSi(C_2H_5)_3$ (79)	656
C_7H_{14}	$CH_3(CH_2)_4COCH_3$				
	$Ir_4(CO)_{12}$ + UV	-	20	$CH_3(CH_2)_4CH(CH_3)OSi(C_2H_5)_3$ (-)	U -129
	$Ru_2(CO)_{10}$ + UV	-	20	$CH_3(CH_2)_4CH(CH_3)OSi(C_2H_5)_3$ (-)	U -129
	$RuCl_2(PPh_3)_3$	90	3	$CH_3(CH_2)_4CH(CH_3)OSi(C_2H_5)_3$ (71)	346
C_7H_{14}	$CH_3(CH_2)_5CHO$				
	Ni	-	-	$CH_3(CH_2)_6OSi(C_2H_5)_3$ + $CH_3(CH_2)_4CH=CHOSi(C_2H_5)_3$ (-)	374
	Ni	100-110	2-3	$CH_3(CH_2)_6OSi(C_2H_5)_3$ + $CH_3(CH_2)_4CH=CHOSi(C_2H_5)_3$ (87)	372
	Ni	180	-	$CH_3(CH_2)_6OSi(C_2H_5)_3$ (93)	146
C_8H_5	C_6H_5COCN				
	$PdCl_2$	110	18	$C_6H_5CH(CN)OSi(C_2H_5)_3$ (45)	873
	$RhCl(PPh_3)_3$	110	18	$(C_2H_5)_3SiOC(C_6H_5)(CN)C(C_6H_5)(CN)OSi(C_2H_5)_3$ (50)	873
C_8H_7	p-$CH_3OC_6H_4COCl$				
	$PtCl_2(PPh_3)_2$	120	5	p-$CH_3OC_6H_4CH_2OSi(C_2H_5)_3$ (27)	310
C_8H_8	$C_6H_5CH_2CHO$				
	Ni	100-110	2-3 (b)	$C_6H_5CH_2CH_2OSi(C_2H_5)_3$ + $C_6H_5CH=CHOSi(C_2H_5)_3$ (85)	372
	Ni	180	-	$C_6H_5CH_2CH_2OSi(C_2H_5)_3$ (78)	146
	Ni + Et_2S	100-110	2-3	$C_6H_5CH_2CH_2OSi(C_2H_5)_3$ + $C_6H_5CH=CHOSi(C_2H_5)_3$ (81)	372
C_8H_8	$C_6H_5COCH_3$				
	Et_4NF	20	10 (MeCN)	$C_6H_5CH(CH_3)OSi(C_2H_5)_3$ + $(C_2H_5)_3SiOC(CH_3)(C_6H_5)C(CH_3)(C_6H_5)OSi(C_2H_5)_3$ (-)	307
	Fig.355.	25	-	$C_6H_5CH(CH_3)OSi(C_2H_5)_3$ + $C_6H_5C(=CH_2)OSi(C_2H_5)_3$(-)	915
	Fig.356.	25	-	$C_6H_5CH(CH_3)OSi(C_2H_5)_3$ + $C_6H_5C(=CH_2)OSi(C_2H_5)_3$(-)	915

	Fig.357.	25	-	$C_6H_5C(=CH_2)OSi(C_2H_5)_3$ (-)	915
	$GaCl_3$	-	-	$C_6H_5CH(CH_3)OSi(C_2H_5)_3$ (78)	195
	Ni	110-160	1	$C_6H_5CH(CH_3)OSi(C_2H_5)_3$ (50) $(C_2H_5)_3SiOC(CH_3)(C_6H_5)C(CH_3)(C_6H_5)OSi(C_2H_5)_3$ (43)	414
	Ni	reflux	-	$C_6H_5CH(CH_3)OSi(C_2H_5)_3$ (40) $(C_2H_5)_3SiOC(CH_3)(C_6H_5)C(CH_3)(C_6H_5)OSi(C_2H_5)_3$ (25)	1104
	Ni + Et_2S	reflux	-	$(C_2H_5)_3SiCO(CH_3)(C_6H_5)C(CH_3)(C_6H_5)OSi(C_2H_5)_3$ (84)	1104
	Ni + NiS	reflux	-	$C_6H_5CH(CH_3)OSi(C_2H_5)_3$ (16) $(C_2H_5)_3SiOC(CH_3)(C_6H_5)C(CH_3)(C_6H_5)OSi(C_2H_5)_3$(42) $C_6H_5C(=CH_2)OSi(C_2H_5)_3$ (22)	1104
	$PdCl_2$	20	6	$C_6H_5CH(CH_3)OSi(C_2H_5)_3$ (12) $C_6H_5C(=CH_2)OSi(C_2H_5)_3$ (15)	811
	$Rh(\pi\text{-All})[P(OMe)_3]_3$	-	-	$C_6H_5CH(CH_3)OSi(C_2H_5)_3$ (-)	144
	$RhCl(PPh_3)_3$	60	0.25	$C_6H_5CH(CH_3)OSi(C_2H_5)_3$ + $(C_2H_5)_3SiOC(CH_3)(C_6H_5)C(CH_3)(C_6H_5)OSi(C_2H_5)_3$ (97)	872
	$RhCl(PPh_3)_3$	80	0.5	$C_6H_5CH(CH_3)OSi(C_2H_5)_3$ (97)	F -023
	$RhCl(cod)[:\overline{CN(R)CH_2CH_2NR}]$; R=Me,Ph,Bz	40-120	4	$C_6H_5CH(CH_3)OSi(C_2H_5)_3$ (10-98)	660
	$RhCl(cod)[:\overline{CN(R)CH_2CH_2NR}]$; R=Me,Ph,Bz	40-120	1-4	$C_6H_5CH(CH_3)OSi(C_2H_5)_3$ (0-97)	660
	$[RhCl(C_2H_4)_2]_2$ + diop	50	26 (MeCN)	$C_6H_5CH(CH_3)OSi(C_2H_5)_3$ (-)	336
	$RuCl_2(PPh_3)_3$	100	6	$C_6H_5CH(CH_3)OSi(C_2H_5)_3$ (62)	339
	$RuCl_2[:\overline{CN(R)CH_2CH_2NR}]$; R = Me,Ph,Bz	100	1.5-4	$C_6H_5CH(CH_3)OSi(C_2H_5)_3$ (58-70)	660
	$ZnCl_2$	108-135	3	$C_6H_5CH(CH_3)OSi(C_2H_5)_3$ (70)	414
C_8H_8	o-,m-,p-$CH_3OC_6H_4CHO$				
	Ni	-	-	$(C_2H_5)_3SiOCH(C_6H_4OCH_3\text{-}p)CH(C_6H_4OCH_3\text{-}p)$-$OSi(C_2H_5)_3$ (-)	374
	Ni	100-110	- (b)	m-$CH_3OC_6H_4CH_2OSi(C_2H_5)_3$ (58)	380
	Ni	100-110	- (b)	p-$CH_3OC_6H_4CH_2OSi(C_2H_5)_3$ (78)	380
	Ni	100-110	- (b)	o-$CH_3OC_6H_4CH_2OSi(C_2H_5)_3$ (77)	380

	Ni	85-126	0.5-1	p-$CH_3OC_6H_4CH_2OSi(C_2H_5)_3$ (42)	411
	Ni +Et_2S	120-130	-	p-$CH_3OC_6H_4CH_2OSi(C_2H_5)_3$ (17) $(C_2H_5)_3SiOCH(C_6H_4OCH_3$-p)$CH(C_6H_4OCH_3$-p)- $OSi(C_2H_5)_3$ (74)	380
	Ni +Et_2S	120-130	-	o-$CH_3OC_6H_4CH_2OSi(C_2H_5)_3$ (17) $(C_2H_5)_3SiOCH(C_6H_4OCH_3$-o)$CH(C_6H_4OCH_3$-o)- $OSi(C_2H_5)_3$ (76)	380
	Ni +Et_2S	120-130	-	m-$CH_3OC_6H_4CH_2OSi(C_2H_5)_3$ (trace) $(C_2H_5)_3SiOCH(C_6H_4OCH_3$-m)$CH(C_6H_4OCH_3$-m)- $OSi(C_2H_5)_3$ (90)	380
	$RuCl_2(PPh_3)_3$	120	40	p-$CH_3OC_6H_4CH_2OSi(C_2H_5)_3$ (88)	374
	$RuCl_2(PPh_3)_3$	90	1.25	p-$CH_3OC_6H_4CH_2OSi(C_2H_5)_3$ (74)	339
	$RuCl_2(PPh_3)_3$	90	1.25	o-$CH_3OC_6H_4CH_2OSi(C_2H_5)_3$ (69)	339
C_8H_8	o-$CH_3C_6H_4CHO$				
	Ni	100-110	- (b)	o-$CH_3C_6H_4CH_2OSi(C_2H_5)_3$ (85)	380
	Ni +Et_2S	120-130	-	o-$CH_3C_6H_4CH_2OSi(C_2H_5)_3$ (trace) $(C_2H_5)_3SiOCH(C_6H_4CH_3$-o)$CH(C_6H_4CH_3$-o)- $Si(C_2H_5)_3$ (87)	380
C_8H_8	p-$CH_3C_6H_4CHO$				
	Ni	-	-	p-$CH_3C_6H_4CH_2OSi(C_2H_5)_3$ + $(C_2H_5)_3SiOCH(C_6H_4CH_3$-p)$CH(C_6H_4CH_3$-p)- $Si(C_2H_5)_3$ (-)	374
	Ni	100-110	- (b)	p-$CH_3C_6H_4CH_2OSi(C_2H_5)_3$ (80)	380
	Ni	85-137	0.5-3 (b)	p-$CH_3C_6H_4CH_2OSi(C_2H_5)_3$ (43-50)	411
	Ni +Et_2S	120-130	-	p-$CH_3C_6H_4CH_2OSi(C_2H_5)_3$ (trace) $(C_2H_5)_3SiOCH(C_6H_4CH_3$-p)$CH(C_6H_4CH_3$-p)- $Si(C_2H_5)_3$ (79)	380
C_8H_{12}	$\overline{(CH_2)_4CH=C}COCH_3$				
	$RhCl(PPh_3)_3$	50	5 (b)	$C_8H_{13}OSi(C_2H_5)_3$ (-)	850
C_8H_{14}	$\overline{CH_2CH_2CH(OCH_3)CH_2CH_2C}HCHO$				
	Et_4NF	20	10 (MeCN)	$\overline{CH_2CH_2CH(OCH_3)CH_2CH_2C}HCH_2OSi(C_2H_5)_3$ (-)	307
C_8H_{14}	$CH_3CH(CH_3)CH=CHCOOCH_2CH_3$				
	$RhCl(PPh_3)_3$	r.t.	8 (b)	$CH_3CH(CH_3)CH_2CH=C(OCH_2CH_3)OSi(C_2H_5)_3$ (67)	678
C_8H_{14}	$CH_3CH_2CH_2CH=C(CH_2CH_3)CHO$				

	Ni	reflux	2 (b)	$CH_3CH_2CH_2C(CH_2CH_3)=CHOSi(C_2H_5)_3$ + $CH_3CH_2CH_2CH=C(CH_2CH_3)CH_2OSi(C_2H_5)_3$ (89)	148
	Ni +Et_2S	reflux	2 (b)	$CH_3CH_2CH_2C(CH_2CH_3)=CHOSi(C_2H_5)_3$ + $CH_3CH_2CH_2CH=C(CH_2CH_3)CH_2OSi(C_2H_5)_3$ (76)	148
	Pd/C	reflux	2 (b)	$CH_3CH_2CH_2C(CH_2CH_3)=CHOSi(C_2H_5)_3$ + $CH_3CH_2CH_2CH=C(CH_2CH_3)CH_2OSi(C_2H_5)_3$ (94)	148
	H_2PtCl_6	reflux	2 (b)	$CH_3CH_2CH_2C(CH_2CH_3)=CHOSi(C_2H_5)_3$ + $CH_3CH_2CH_2CH=C(CH_2CH_3)CH_2OSi(C_2H_5)_3$ (67)	148
C_8H_{14}	$CH_3COC(CH_3)_2COOCH_2CH_3$				
	$ZnCl_2$	105-110	2-3	$(C_2H_5)_3SiOCH(CH_3)C(CH_3)_2COOCH_2CH_3$ (69)	646
C_8H_{16}	$CH_3(CH_2)_6CHO$				
	Ni	100-110	2-3	$CH_3(CH_2)_7OSi(C_2H_5)_3$ + $CH_3(CH_2)_5CH=CHOSi(C_2H_5)_3$ (84)	372
	Ni	180	-	$CH_3(CH_2)_7OSi(C_2H_5)_3$ (72)	146
	Ni +Et_2S	100-110	2-3	$CH_3(CH_2)_7OSi(C_2H_5)_3$ + $CH_3(CH_2)_5CH=CHOSi(C_2H_5)_3$ (77)	372
C_9H_6	Fig.358.				
	$RhCl(PPh_3)_3$	r.t.	16 (b)	Fig.359. (87)	678
C_9H_7	$C_6H_5COCH_2CN$				
	$RhCl(PPh_3)_3$ + $PhSSiEt_3$	reflux	12-24 (b)	$C_6H_5CH=C(CN)OSi(C_2H_5)_3$ (98)	853
C_9H_8	$C_6H_5CH=CHCHO$				
	-	45	1 (b)	$C_6H_5CH=CHCH_2OSi(C_2H_5)_3$ (-)	850
	CsF	20	10 (MeCN)	$C_6H_5CH=CHCH_2OSi(C_2H_5)_3$ (82)	307
	Ni	100-132	0.25-2(THF)	$C_6H_5CH=CHCH_2OSi(C_2H_5)_3$ (28-40)	411
	Ni	reflux	2 (b)	$C_6H_5CH=CHCH_2OSi(C_2H_5)_3$ + $C_6H_5CH_2CH=CHOSi(C_2H_5)_3$ + $C_6H_5(CH_2)_3OSi(C_2H_5)_3$ (27)	148
	Ni +Et_2S	reflux	2 (b)	$C_6H_5CH=CHCH_2OSi(C_2H_5)_3$ + $C_6H_5CH_2CH=CHOSi(C_2H_5)_3$ + $C_6H_5(CH_2)_3OSi(C_2H_5)_3$ (8)	148
	Pd/C	reflux	2 (b)	$C_6H_5CH=CHCH_2OSi(C_2H_5)_3$ + $C_6H_5CH_2CH=CHOSi(C_2H_5)_3$ + $C_6H_5(CH_2)_3OSi(C_2H_5)_3$ (49)	148
	H_2PtCl_6	reflux	2 (b)	$C_6H_5CH=CHCH_2OSi(C_2H_5)_3$ + $C_6H_5CH_2CH=CHOSi(C_2H_5)_3$ + $C_6H_5(CH_2)_3OSi(C_2H_5)_3$ (72)	148

	H_2PtCl_6 (i-PrOH)	20	10 (MeCN)	$C_6H_5CH_2CH{=}CHOSi(C_2H_5)_3$ (60)	307
	$[HPt(SiMe_2Ph)(PPh_3)_3]_2$	60-100	1	$C_6H_5CH_2CH{=}CHOSi(C_2H_5)_3$ (-)	79
	$[PtCl_2(py)]_2$	-	2	$C_6H_5CH_2CH{=}CHOSi(C_2H_5)_3$ (85)	987
	$RhCl(PPh_3)_3$	60	0.25	$C_6H_5CH_2CH{=}CHOSi(C_2H_5)_3$ (96)	873
	$RhCl(PPh_3)_3$	60	0.25	$C_6H_5CH_2CH{=}CHOSi(C_2H_5)_3$ (96)	F -023
C_9H_{10}	$C_6H_5CH(CH_3)CHO$				
	Ni	100-110	2-3	$C_6H_5CH(CH_3)CH_2OSi(C_2H_5)_3$ + $C_6H_5C(CH_3){=}CHOSi(C_2H_5)_3$ (78)	372
	Ni +Et_2S	100-110	2-3	$C_6H_5CH(CH_3)CH_2OSi(C_2H_5)_3$ + $C_6H_5C(CH_3){=}CHOSi(C_2H_5)_3$ (93)	372
C_9H_{10}	$C_6H_5CH_2CH_2CHO$				
	Ni	100-110	2-3 (b)	$C_6H_5(CH_2)_3OSi(C_2H_5)_3$ + $C_6H_5CH_2CH{=}CHOSi(C_2H_5)_3$ (85)	372
	Ni	180	-	$C_6H_5(CH_2)_3OSi(C_2H_5)_3$ (81)	146
	Ni +Et_2S	100-110	2-3	$C_6H_5(CH_2)_3OSi(C_2H_5)_3$ + $C_6H_5CH_2CH{=}CHOSi(C_2H_5)_3$ (84)	372
C_9H_{10}	$C_6H_5CH_2COCH_3$				
	Ni +Et_2S	120-140	5	$C_6H_5CH_2CH(CH_3)OSi(C_2H_5)_3$ + $C_6H_5CH{=}C(CH_3)OSi(C_2H_5)_3$ (91)	370
	Ni +Et_2S	130	5	$C_6H_5CH_2CH(CH_3)OSi(C_2H_5)_3$ + $C_6H_5CH{=}C(CH_3)OSi(C_2H_5)_3$ (80)	370
	Ni + NiS	100-120	2	$C_6H_5CH_2CH(CH_3)OSi(C_2H_5)_3$ + $C_6H_5CH{=}C(CH_3)OSi(C_2H_5)_3$ (80)	370
	Ni + NiS	60-80	2	$C_6H_5CH_2CH(CH_3)OSi(C_2H_5)_3$ + $C_6H_5CH{=}C(CH_3)OSi(C_2H_5)_3$ (70)	370
	$[RhCl(C_2H_4)_2]_2$ + diop	50	20	$C_6H_5CH_2CH(CH_3)OSi(C_2H_5)_3$ (-)	336
C_9H_{10}	$C_6H_5COCH_2CH_3$				
	$RhCl(PPh_3)_3$	80	0.5	$C_6H_5CH(CH_2CH_3)OSi(C_2H_5)_3$ (87)	F -023
C_9H_{10}	Fig.360.				
	$RhCl(PPh_3)_3$	r.t.	8 (b)	Fig.361. (87)	678
C_9H_{11}	p-$(CH_3)_2NC_6H_4CHO$				

	Ni	80-160	1,5-3	p-$(CH_3)_2NC_6H_4CH_2OSi(C_2H_5)_3$ (38-41)	411
C_9H_{14}	Fig.362.				
	$RhCl(PPh_3)_3$	20	3	Fig.363. (-)	519
	SiO_2-PPh_2/Rh complex	20	3	Fig.363. (-)	519
C_9H_{14}	Fig.364.				
	$RhCl(PPh_3)_3$	-	-	Fig.365. (-)	863
C_9H_{15}	$\overline{CH_2C(CH_3)_2CH_2CH(CH_3)CH_2C}O$				
	$RhCl(PPh_3)_3$	20	3	$\overline{CH_2C(CH_3)_2CH_2CH(CH_3)CH_2C}HOSi(C_2H_5)_3$ (-)	519
	SiO_2-PPh_2/Rh complex	20	3	$\overline{CH_2C(CH_3)_2CH_2CH(CH_3)CH_2C}HOSi(C_2H_5)_3$ (-)	519
C_9H_{18}	$[CH_3(CH_2)_3]_2CO$				
	$ZnCl_2$	reflux	3 (THF)	$[CH_3(CH_2)_3]_2CHOSi(C_2H_5)_3$ (57)	656
$C_{10}H_7$	$C_6H_5COCH_2COOCN$				
	$RhCl(PPh_3)_3$	60	12	$C_6H_5CH[OSi(C_2H_5)_3]CH_2COOCN$ + $C_6H_5C[OSi(C_2H_5)_3]=CHCOOCN$ (90)	873
$C_{10}H_9$	o-$ClC_6H_4COCOOCH_2CH_3$				
	Ni	160-170	6-7 (xy)	o-$ClC_6H_4CH[OSi(C_2H_5)_3]COOCH_2CH_3$ (68)	650
	Ni	160-170	6-7 (xy)	o-$ClC_6H_4CH[OSi(C_2H_5)_3]COOCH_2CH_3$ (68)	649
$C_{10}H_{10}$	$C_6H_5C(CH_2CH_3)=C=O$				
	Ni	reflux	5	E-$C_6H_5C(CH_2CH_3)=CHOSi(C_2H_5)_3$ + Z-$C_5H_5C(CH_2CH_3)=CHOSi(C_2H_5)_3$ (80)	376
$C_{10}H_{10}$	$C_6H_5CH_2CH=CHCHO$				
	$RhCl(PPh_3)_3$	60	0.25(MeCN)	$C_6H_5CH_2CH_2CH=CHOSi(C_2H_5)_3$ (96)	866
$C_{10}H_{10}$	$C_6H_5CH=CHCOCH_3$				
	-	45	15	$C_6H_5CH_2CH=C(CH_3)OSi(C_2H_5)_3$ (-)	850
	$[PtCl_2(py)]_2$	-	1	$C_6H_5CH_2CH=C(CH_3)OSi(C_2H_5)_3$ (75)	347
	$RhCl(PPh_3)_3$	60	0.5	$C_6H_5CH_2CH=C(CH_3)OSi(C_2H_5)_3$ (80)	F -023
$C_{10}H_{10}$	$C_6H_5COCOOCH_2CH_3$				
	Ni	reflux	3-4	$(C_2H_5)_3SiOC(C_6H_5)(COOCH_2CH_3)C(C_6H_5)(COOCH_2CH_3)$- $OSi(C_2H_5)_3$ (44)	422
$C_{10}H_{10}$	$CH_3CH=CHCOC_6H_5$				

	-	45	15	$CH_3CH_2CH{=}C(C_6H_5)OSi(C_2H_5)_3$ (-)	850
	$RhCl(PPh_3)_3$	60	0.5	$CH_3CH_2CH{=}C(C_6H_5)OSi(C_2H_5)_3$ (90)	873
	$RhCl(PPh_3)_3$	60	0.5	$CH_3CH_2CH{=}C(C_6H_5)OSi(C_2H_5)_3$ (82)	F -023
$C_{10}H_{12}$	$C_6H_5CH_2COCH_2CH_3$				
	$ZnCl_2$	reflux	3 (THF)	$C_6H_5CH_2CH(CH_2CH_3)OSi(C_2H_5)_3$ (29)	656
$C_{10}H_{12}$	$C_6H_5CH_2COOCH_2CH_3$				
	Ni + Et_2S	reflux	72	$C_6H_5CH{=}C[OSi(C_2H_5)_3]_2$ (32) $C_6H_5CH_2COOSi(C_2H_5)_3$ (51)	890
$C_{10}H_{12}$	p-$(CH_3)_2CHC_6H_4CHO$				
	Ni	100-110	- (b)	p-$(CH_3)_2CHC_6H_4CH_2OSi(C_2H_5)_3$ (78)	380
	Ni + Et_2S	120-130	-	p-$(CH_3)_2CHC_6H_4CH_2OSi(C_2H_5)_3$ (16) $(C_2H_5)_3SiOCH[C_6H_4CH(CH_3)_2$-p]$CH[C_6H_4CH(CH_3)_2$-p]- $OSi(C_2H_5)_3$ (73)	380
$C_{10}H_{14}$	$C_6H_5COSi(CH_3)_3$				
	$InCl_3$	reflux	40	$C_6H_5CH[Si(CH_3)_3]OSi(C_2H_5)_3$ (57)	201
$C_{10}H_{14}$	Fig.366.				
	$RhCl(PPh_3)_3$	50	1	Fig.367. (92)	873
$C_{10}H_{14}$	Fig.368.				
	$RhCl(PPh_3)_3$	80	25	$C_{10}H_{15}OSi(C_2H_5)_3$ (-)	850
$C_{10}H_{14}$	p-$(CH_3)_3SiC_6H_4CHO$				
	Ni	-	-	p-$(CH_3)_3SiC_6H_4CH_2OSi(C_2H_5)_3$ (10-22) $(C_2H_5)_3SiOCH[(C_6H_4Si(CH_3)_3$-p]$CH[C_6H_4Si(CH_3)_3$-p]- $OSi(C_2H_5)_3$ (47-80)	412
	$SnCl_2$	-	3	p-$(CH_3)_3SiC_6H_4CH_2OSi(C_2H_5)_3$ (12)	412
	$ZnCl_2$	145	4	p-$(CH_3)_3SiC_6H_4CH_2OSi(C_2H_5)_3$ (10-22)	412
$C_{10}H_{16}$	$(CH_3)_2C{=}CHCH_2CH_2C(CH_3){=}CHCHO$				
	Ni	reflux	2 (b)	$(CH_3)_2C{=}CHCH_2CH_2CH(CH_3)CH{=}CHOSi(C_2H_5)_3$ + $(CH_3)_2C{=}CHCH_2CH_2C(CH_3){=}CHCH_2OSi(C_2H_5)_3$ (77)	148
	Ni + Et_2S	reflux	2 (b)	$(CH_3)_2C{=}CHCH_2CH_2CH(CH_3)CH{=}CHOSi(C_2H_5)_3$ + $(CH_3)_2C{=}CHCH_2CH_2C(CH_3){=}CHCH_2OSi(C_2H_5)_3$ (69)	148

Table 8.6

= $(C_2H_5)_3SiH$ = =21=

$RhCl(PPh_3)_3$	20	1	$(CH_3)_2C{=}CHCH_2CH_2CH(CH_3)CH{=}CHOSi(C_2H_5)_3$ + $(CH_3)_2C{=}CHCH_2CH_2C(CH_3){=}CHCH_2OSi(C_2H_5)_3$ (-)	148
$RhCl(PPh_3)_3$	50	2	$(CH_3)_2C{=}CHCH_2CH_2CH(CH_3)CH{=}CHOSi(C_2H_5)_3$ (97)	863
$RhCl(PPh_3)_3$	50	2	$(CH_3)_2C{=}CHCH_2CH_2CH(CH_3)CH{=}CHOSi(C_2H_5)_3$ (97)	F -023
$C_{10}H_{16}$ $CH_3CH_2C({=}CH_2)CH{=}C(CH_3)COOCH_2CH_3$				
$RhCl(PPh_3)_3$	80	10 (b)	$CH_3CH_2C(CH_3){=}CHC(CH_3){=}C(OCH_2CH_3)OSi(C_2H_5)_3$ (54)	678
$C_{10}H_{16}$ Fig.369.				
$RhCl(PPh_3)_3$	0-80	-	Fig.370. (-)	871
$C_{10}H_{18}$ $CH_2CH(CH_3)CH_2CH_2CH[CH(CH_3)_2]CO$				
$RhCl(PPh_3)_3$	0-80	-	$CH_2CH(CH_3)CH_2CH_2CH[CH(CH_3)_2]CHOSi(C_2H_5)_3$ (-)	871
$RhCl(PPh_3)_3$	20	3	$CH_2CH(CH_3)CH_2CH_2CH[CH(CH_3)_2]CHOSi(C_2H_5)_3$ (-)	519
$RhCl(PPh_3)_3$	70	0.5	$CH_2CH(CH_3)CH_2CH_2CH[CH(CH_3)_2]CHOSi(C_2H_5)_3$ (92)	F -023
$RhCl(PPh_3)_3$	80	1	$CH_2CH(CH_3)CH_2CH_2CH[CH(CH_3)_2]CHOSi(C_2H_5)_3$ (76)	F -041
SiO_2-PPh_2/Rh complex	20	3	$CH_2CH(CH_3)CH_2CH_2CH[CH(CH_3)_2]CHOSi(C_2H_5)_3$ (-)	519
$C_{10}H_{18}$ $CH_2CH_2CH[C(CH_3)_3]CH_2CH_2CO$				
$Rh(C_3H_5)[P(OMe)_3]_3$	-	-	$CH_2CH_2CH[C(CH_3)_3]CH_2CH_2CHOSi(C_2H_5)_3$ (-)	144
$RhCl(PPh_3)_3$	20	3	$CH_2CH_2CH[C(CH_3)_3]CH_2CH_2CHOSi(C_2H_5)_3$ (-)	519
$RhCl(PPh_3)_3$	25	3	$CH_2CH_2CH[C(CH_3)_3]CH_2CH_2CHOSi(C_2H_5)_3$ (until 100)	1034
SiO_2-PPh_2/Rh complex	20	3	$CH_2CH_2CH[C(CH_3)_3]CH_2CH_2CHOSi(C_2H_5)_3$ (-)	519
$RuCl_2(PPh_3)_3$	110	20 (b)	$CH_2CH_2CH[C(CH_3)_3]CH_2CH_2CHOSi(C_2H_5)_3$ (-)	1034
$RuCl_2(PPh_3)_3$	80	8 (t)	$CH_2CH_2CH[C(CH_3)_3]CH_2CH_2CHOSi(C_2H_5)_3$ (-)	1034
$C_{10}H_{18}$ $CH_2CH[C(CH_3)_3](CH_2)_3CO$				
$RhCl(PPh_3)_3$	20	3	$CH_2CH[C(CH_3)_3](CH_2)_3CHOSi(C_2H_5)_3$ (-)	519
SiO_2-PPh_2/Rh complex	20	3	$CH_2CH[C(CH_3)_3](CH_2)_3CHOSi(C_2H_5)_3$ (-)	519
$C_{10}H_{18}$ $CH_3(CH_2)_3COCOO(CH_2)_3CH_3$				
Ni	108-120	3-4	$CH_3(CH_2)_3CH[OSi(C_2H_5)_3]COO(CH_2)_3CH_3$ (66)	646
Ni	110-130	5-7	$CH_3(CH_2)_3CH[OSi(C_2H_5)_3]COO(CH_2)_3CH_3$ (66)	S -031
$ZnCl_2$	105-110	2-3	$CH_3(CH_2)_3CH[OSi(C_2H_5)_3]COO(CH_2)_3CH_3$ (56)	646
$C_{10}H_{18}$ $CH_3CH_2CH_2COC(CH_3)_2COOCH_2CH_3$				

Table 8.6

= $(C_2H_5)_3SiH$ = =22=

	$ZnCl_2$	105-110	2-3	$CH_3CH_2CH_2CH[OSi(C_2H_5)_3]C(CH_3)_2COOCH_2CH_3$ (73)	646
	$ZnCl_2$	105-110	2-3	$CH_3CH_2CH_2CH[OSi(C_2H_5)_3]C(CH_3)_2COOCH_2CH_3$ (73)	S -030
$C_{10}H_{18}$	$\overline{CH[C(CH_3)_3](CH_2)_4CO}$				
	$RhCl(PPh_3)_3$	20	3	$\overline{CH[C(CH_3)_3](CH_2)_4C}HOSi(C_2H_5)_3$ (-)	519
	SiO_2-PPh_2/Rh complex	20	3	$\overline{CH[C(CH_3)_3](CH_2)_4C}HOSi(C_2H_5)_3$ (-)	519
$C_{11}H_7$	1-$C_{10}H_7N{=}C{=}O$				
	$PdCl_2$	80	6	1-$C_{10}H_7N(CHO)Si(C_2H_5)_3$ (-)	J -014
$C_{11}H_{12}$	$C_6H_5C(CH_2CH_2CH_3){=}C{=}O$				
	Ni	reflux	5	cis-, trans-$C_6H_5C(CH_2CH_2CH_3){=}CHOSi(C_2H_5)_3$ (85)	376
$C_{11}H_{12}$	$C_6H_5CH{=}CHCOOCH_2CH_3$				
	$RhCl(PPh_3)_3$	r.t.	8 (b)	$C_6H_5CH_2CH{=}C(OCH_2CH_3)OSi(C_2H_5)_3$ (98)	678
$C_{11}H_{12}$	o-$CH_3C_6H_4COCOOCH_2CH_3$				
	Ni	160-170	6-7	o-$CH_3C_6H_4CH[OSi(C_2H_5)_3]COOCH_2CH_3$ (48-66)	649
$C_{11}H_{12}$	p-$CH_3C_6H_4COCOOCH_2CH_3$				
	Ni	160-170	6-7 (xy)	$\{$p-$CH_3C_6H_4[(C_2H_5)_3SiO]C(COOCH_2CH_3)\}_2$ (48)	650
	Ni	reflux	3-4	$\{$p-$CH_3C_6H_4[(C_2H_5)_3SiO]C(COOCH_2CH_3)\}_2$ (27)	645
$C_{11}H_{12}$	p-$CH_3OC_6H_4COCOOCH_2CH_3$				
	Ni	160-170	6-7 (xy)	p-$CH_3OC_6H_4CH[OSi(C_2H_5)_3]COOCH_2CH_3$ (66)	650
$C_{11}H_{14}$	$(CH_3)_3CCOC_6H_5$				
	$InCl_3$	reflux	14	$(CH_3)_3CCH(C_6H_5)OSi(C_2H_5)_3$ (61)	201
$C_{11}H_{14}$	p-$(CH_3)_3CC_6H_4CHO$				
	Ni	-	-	p-$(CH_3)_3CC_6H_4CH_2OSi(C_2H_5)_3$ (53) $\{$p-$(CH_3)_3CC_6H_4CH[OSi(C_2H_5)_3]\}_2$ (36)	412
	$SnCl_2$	-	3	p-$(CH_3)_3CC_6H_4CH_2OSi(C_2H_5)_3$ (15)	412
	$ZnCl_2$	152	1.5	p-$(CH_3)_3CC_6H_4CH_2OSi(C_2H_5)_3$ (72)	412
$C_{11}H_{20}$	$CH_2{=}CH(CH_2)_8CHO$				
	Ni	reflux	4 (b)	$CH_3CH{=}CH(CH_2)_7CH_2OSi(C_2H_5)_3$ (92)	148
	Ni + Et_2S	reflux	4 (b)	$CH_3CH{=}CH(CH_2)_7CH_2OSi(C_2H_5)_3$ + $CH_3CH{=}CH(CH_2)_6CH{=}CHOSi(C_2H_5)_3$ (86)	148
$C_{11}H_{20}$	$CH_3(CH_2)_6COCOOCH_2CH_3$				

	$ZnCl_2$	105-110	2-3	$CH_3(CH_2)_6CH[OSi(C_2H_5)_3]COOCH_2CH_3$ (43)	646
$C_{11}H_{20}$	$CH_3CH_2CH_2COC(CH_3)_2COOCH(CH_3)_2$				
	$ZnCl_2$	105-110	2-3	$CH_3CH_2CH_2CH[OSi(C_2H_5)_3]C(CH_3)_2COOCH(CH_3)_2$ (46)	646
$C_{11}H_{20}$	$CH_3CH=CH(CH_2)_7CHO$				
	Ni	reflux	4 (b)	$CH_3CH=CH(CH_2)_7CH_2OSi(C_2H_5)_3$ + $CH_3CH=CH(CH_2)_6CH=CHOSi(C_2H_5)_3$ (91)	148
	Ni +Et_2S	reflux	4 (b)	$CH_3CH=CH(CH_2)_7CH_2OSi(C_2H_5)_3$ + $CH_3CH=CH(CH_2)_6CH=CHOSi(C_2H_5)_3$ (88)	148
$C_{11}H_{22}$	$(CH_3CH_2)_3SiOCOCH_2CH_2COCH_3$				
	Ni	110	4	$(CH_3CH_2)_3SiOCOCH_2CH_2CH(CH_3)OSi(C_2H_5)_3$ (86)	S -040
$C_{11}H_{22}$	$[CH_3(CH_2)_4]_2CO$				
	$ZnCl_2$	reflux	5 (e)	$[CH_3(CH_2)_4]_2CHOSi(C_2H_5)_3$ (59)	656
$C_{12}H_3$	$CH_3COCCo_3(CO)_9$				
	none	reflux	5 (THF)	$C_{12}H_4Co_3O_{10}Si(C_2H_5)_3$ (-)	1042
$C_{12}H_{14}$	$CH_3(CH_2)_3C(C_6H_5)=C=O$				
	Ni	reflux	5	cis-,trans-$CH_3(CH_2)_3C(C_6H_5)=CHOSi(C_2H_5)_3$ (88)	376
$C_{12}H_{20}$	$CH_3(CH_2)_4C(=CH_2)CH=CHCOOCH_2CH_3$				
	$RhCl(PPh_3)_3$	80	8 (b)	$CH_3(CH_2)_4C(CH_3)=CHCH=C(OCH_2CH_3)OSi(C_2H_5)_3$ (40)	678
$C_{12}H_{22}$	$CH_3(CH_2)_5COCOO(CH_2)_3CH_3$				
	$ZnCl_2$	105-110	2-3	$CH_3(CH_2)_5CH[OSi(C_2H_5)_3]COO(CH_2)_3CH_3$ (37)	646
$C_{12}H_{26}$	$(C_2H_5)_3SiOC(CH_3)(CH_2CH_3)COCH_3$				
	Ni	reflux	8	$(C_2H_5)_3SiOC(CH_3)(CH_2CH_3)CH(CH_3)OSi(C_2H_5)_3$ (53)	652
$C_{13}H_5$	$CH_3CH_2COCCo_3(CO)_9$				
	none	reflux	5 (THF)	$C_{13}H_6Co_3O_{10}Si(C_2H_5)_3$ (-)	1042
$C_{13}H_{10}$	$(C_6H_5)_2CO$				
	Ni	110-145	1	$(C_6H_5)_2CHOSi(C_2H_5)_3$ (40) $(C_2H_5)_3SiOC(C_6H_5)_2C(C_6H_5)_2OSi(C_2H_5)_3$ (57)	414
$C_{13}H_{14}$	$C_6H_5CH=CHCH=COOCH_2CH_3$				
	$RhCl(PPh_3)_3$	80	16 (b)	$C_6H_5CH_2CH=CHCH=C(OCH_2CH_3)OSi(C_2H_5)_3$ (68)	678
$C_{13}H_{16}$	Fig.371.				

	Ni	160-170	6-7 (xy)	Fig.372. (42)	649
	Ni	160-170	6-7 (xy)	Fig.372. (42)	650
$C_{13}H_{20}$	Fig.373.				
	$RhCl(PPh_3)_3$	50	2	Fig.374. (74)	850
	$RhCl(PPh_3)_3$	50	2	Fig.374 + Fig.375. (-)	863
$C_{13}H_{20}$	Fig.376.				
	$RhCl(PPh_3)_3$	50	2	Fig.377. (95-96)	863
	$RhCl(PPh_3)_3$	50	2	Fig.377. (95-96)	F -023
$C_{14}H_7$	$(CH_3)_2CHCOCCo_3(CO)_9$				
	none	reflux	5 (THF)	$C_{14}H_8Co_3O_{10}Si(C_2H_5)_3$ (-)	1042
$C_{14}H_7$	$CH_3CH_2CH_2COCCo_3(CO)_9$				
	none	reflux	5 (THF)	$C_{14}H_8Co_3O_{10}Si(C_2H_5)_3$ (-)	1042
$C_{14}H_{10}$	$C_6H_5COCOC_6H_5$				
	$RhCl(PPh_3)_3$	reflux	4 (h)	$(C_2H_5)_3SiOCH(C_6H_5)CH(C_6H_5)OSi(C_2H_5)_3$ (83)	873
$C_{14}H_{12}$	$(C_6H_5)_2CHCHO$				
	Ni	180	-	$(C_6H_5)_2CHCH_2OSi(C_2H_5)_3 + (C_6H_5)_2C{=}CHOSi(C_2H_5)_3$ (91)	372
	Ni + Et_2S	100-110	2-3	$(C_6H_5)_2CHCH_2OSi(C_2H_5)_3 + (C_6H_5)_2C{=}CHOSi(C_2H_5)_3$ (92)	372
$C_{14}H_{20}$	Fig.378.				
	$RhCl(PPh_3)_3$	80	10 (b)	Fig.379. (80)	678
$C_{14}H_{22}$	Fig.380.				
	$RhCl(PPh_3)_3$	r.t.	5 (b)	Fig.381. (89)	678
$C_{14}H_{24}$	Fig.382.				
	-	reflux	- (solv.)	Fig.383. (92)	1142
$C_{14}H_{30}$	$(C_2H_5)_3SiOCH(CH_2CH_2CH_3)CH_2COCH_2CH_3$				
	Ni	50-60	2	$(C_2H_5)_3SiOCH(CH_2CH_2CH_3)CH_2CH(CH_2CH_3)OSi(C_2H_5)_3$ (-)	651
$C_{15}H_9$	$CH_3(CH_2)_3COCCo_3(CO)_9$				

Table 8.6

= $(C_2H_5)_3SiH$ =

	none	reflux	5 (THF)	$C_{15}H_{10}Co_3O_{10}Si(C_2H_5)_3$ (-)	1042
$C_{15}H_{10}$	$(C_6H_5)_2C=C=CO$				
	Ni	reflux	5	$(C_6H_5)_2C=C=CHOSi(C_2H_5)_3$ (90)	376
$C_{15}H_{12}$	$C_6H_5CH=CHCOC_6H_5$				
	$RhCl(PPh_3)_3$	25	15 (b)	$C_6H_5CH_2CH=C(C_6H_5)OSi(C_2H_5)_3$ (-)	850
$C_{15}H_{22}$	Fig.384.				
	Ni	85-90	2.5-3	Fig.385. (70)	645
	Ni	reflux	3-4	Fig.386. (30)	645
$C_{17}H_4$	p-$BrC_6H_4COCCo_3(CO)_9$				
	none	reflux	5 (THF)	$C_{17}H_5BrCo_3O_{10}Si(C_2H_5)_3$ (-)	1042
$C_{17}H_5$	$C_6H_5COCCo_3(CO)_9$				
	none	reflux	5 (THF)	$C_{17}H_6Co_3O_{10}Si(C_2H_5)_3$ (-)	1042
$C_{17}H_{11}$	$\overline{(CH_2)_5CH}COCCo_3(CO)_9$				
	none	reflux	5 (THF)	$C_{17}H_{12}Co_3O_{10}Si(C_2H_5)_3$ (-)	1042
$C_{17}H_{13}$	$CH_3(CH_2)_5COCCo_3(CO)_9$				
	none	reflux	5 (THF)	$C_{17}H_{14}Co_3O_{10}Si(C_2H_5)_3$ (-)	1042
$C_{17}H_{14}$	$C_6H_5CH=CHCOCH=CHC_6H_5$				
	$[HPt(SiMe_2Ph)(PPh_3)]_2$	60-100	1	$C_6H_5CH_2CH=C[OSi(C_2H_5)_3]CH=CHC_6H_5$ (-)	79
$C_{18}H_7$	p-$CH_3C_6H_4COCCo_3(CO)_9$				
	none	reflux	5 (THF)	$C_{18}H_8Co_3O_{10}Si(C_2H_5)_3$ (-)	1042
$C_{20}H_{36}$	Fig.387.				
	$RhCl(PPh_3)_3$	50	2 (t)	Fig.388. (64)	1189
$C_{22}H_{32}$	Fig.389.				
	H_2PtCl_6 (i-PrOH)	150	5 (t)	Fig.391. (-)	1320
	H_2PtCl_6 (i-PrOH)	150-160	- (t)	FIg.390. (80)	1320
$C_{29}H_{31}$	Fig.392.				
	$BF_3{\cdot}Et_2O$, CF_3COOH	-25	1.5 (anisole)	Fig.393. (71)	592
	$BF_3{\cdot}Et_2O$, CF_3COOH	0	12 (anisole)	Fig.394. (64)	592

Table 8.7

= $(C_2H_5)_3SiH$ = =1=

C_4H_6	$CH_3OCON{=}NCOOCH_3$				
	$(t\text{-}BuO)_2$	reflux	11 days (b)	$CH_3OCON[Si(C_2H_5)_3]NHCOOCH_3$ (54)	675
C_4H_7	$(CH_3)_2CHCN$				
	$ZnCl_2$	-	-	$(CH_3)_2CHCH{=}NSi(C_2H_5)_3$ (90)	381
	$ZnCl_2$	reflux	90	$(CH_3)_2C{=}CHNHSi(C_2H_5)_3$ (20)	681
C_5H_9	$CH_3(CH_2)_3NCO$				
	Pd/C	-	-	$CH_3(CH_2)_3N(CHO)Si(C_2H_5)_3$ (46)	859
	$PdCl_2$	130	12	$CH_3(CH_2)_3N(CHO)Si(C_2H_5)_3$ (46)	849
	$PdCl_2$	130	12	$CH_3(CH_2)_3N(CHO)Si(C_2H_5)_3$ (46)	859
C_5H_9	$CH_3CH_2CH(CH_3)CN$				
	$ZnCl_2$	-	-	$CH_3CH_2CH(CH_3)CH{=}NSi(C_2H_5)_3$ (90)	381
C_6H_{10}	$CH_3CH_2OCON{=}NCOOCH_2CH_3$				
	$(t\text{-}BuO)_2$	reflux	11 days (b)	$CH_3CH_2OCON[Si(C_2H_5)_3]NHCOOCH_2CH_3$ (47)	675
	$(t\text{-}BuO)_2$ + UV	20	3.5 (b)	$CH_3CH_2OCON[Si(C_2H_5)_3]NHCOOCH_2CH_3$ (26)	675
	UV	10-15	-	$CH_3CH_2OCON[Si(C_2H_5)_3]NHCOOCH_2CH_3$ (4)	675
	none	reflux	11 days (b)	$CH_3CH_2OCON[Si(C_2H_5)_3]NHCOOCH_2CH_3$ (28)	675
C_6H_{11}	$(C_2H_5)_2CHCN$				
	$ZnCl_2$	-	-	$(C_2H_5)_2CHCH{=}NSi(C_2H_5)_3$ (90)	381
C_7H_4	$ClCH_2C_6H_4NCO$				
	$PdCl_2$	80	12	$ClCH_2C_6H_4N(CHO)Si(C_2H_5)_3$ (85)	849
C_7H_5	C_6H_5NCO				
	Pd/C	130	48	$C_6H_5N(CHO)Si(C_2H_5)_3$ (88)	859
	Pd/C	130	48	$C_6H_5N(CHO)Si(C_2H_5)_3$ (88)	849
	$PdCl_2$	130	12	$C_6H_5N(CHO)Si(C_2H_5)_3$ (95)	849
	$PdCl_2$	130	12	$C_6H_5N(CHO)Si(C_2H_5)_3$ (95)	859
C_7H_5	$C_6H_5SO_2NCO$				
	none	120-130	3 (t)	$C_6H_5SO_2N(CHO)Si(C_2H_5)_3$ (89)	311
C_7H_{11}	$\overline{(CH_2)_5C}HNCO$				

Table 8.7

= $(C_2H_5)_3SiH$ = =2=

	Pd/C	180	48	$(CH_2)_5CHN(CHO)Si(C_2H_5)_3$ (10)	849
	$PdCl_2$	80	12	$(CH_2)_5CHN(CHO)Si(C_2H_5)_3$ (90)	849
C_7H_{14}	$(CH_3)_2CHN{=}C{=}NCH(CH_3)_2$				
	$PdCl_2$	140	15	$(CH_3)_2CHN[Si(C_2H_5)_3]CH{=}NCH(CH_3)_2$ (85-98)	J -015
	$PdCl_2$	140	15	$(CH_3)_2CHN[Si(C_2H_5)_3]CH{=}NCH(CH_3)_2$ (85-98)	860
	$PdCl_2$	140	15	$(CH_3)_2CHN[Si(C_2H_5)_3]CH{=}NCH(CH_3)_2$ (85-98)	849
C_8H_9	$C_6H_5CH{=}NCH_3$				
	$PdCl_2$	55	24	$C_6H_5CH_2N(CH_3)Si(C_2H_5)_3$ (90)	862
	$RhCl(PPh_3)_3$	100	20	$C_6H_5CH_2N(CH_3)Si(C_2H_5)_3$ (65)	862
	$RhCl(PPh_3)_3$	r.t.	0.5	$C_6H_5CH_2N(CH_3)Si(C_2H_5)_3$ (-)	J -010
C_8H_{15}	$(CH_3CH_2CH_2)_2CHCN$				
	$ZnCl_2$	170	36	$(CH_3CH_2CH_2)_2C{=}CHNHSi(C_2H_5)_3$ (54)	681
C_9H_9	$C_6H_5CH(CH_3)CN$				
	$ZnCl_2$	-	-	$C_6H_5CH(CH_3)CH{=}NSi(C_2H_5)_3$ (90)	381
	$ZnCl_2$	reflux	90	$C_6H_5C(CH_3){=}CHNHSi(C_2H_5)_3$ (28)	681
$C_{10}H_{11}$	$CH_3CH_2CH(C_6H_5)CN$				
	$ZnCl_2$	-	-	$CH_3CH_2CH(C_6H_5)CH{=}NSi(C_2H_5)_3$ (90)	381
	$ZnCl_2$	reflux	90	$CH_3CH_2C(C_6H_5){=}CHNHSi(C_2H_5)_3$ (12)	681
$C_{10}H_{19}$	$[CH_3(CH_2)_3]_2CHCN$				
	$ZnCl_2$	-	-	$[CH_3(CH_2)_3]_2C{=}CHNHSi(C_2H_5)_3$ (90)	381
	$ZnCl_2$	170	36	$[CH_3(CH_2)_3]_2C{=}CHNHSi(C_2H_5)_3$ (53)	681
$C_{11}H_7$	$1\text{-}C_{10}H_7NCO$				
	Pd/C	80	2-3	$1\text{-}C_{10}H_7N(CHO)Si(C_2H_5)_3$ (-)	859
	$PdCl_2$	80	2-3	$1\text{-}C_{10}H_7N(CHO)Si(C_2H_5)_3$ (-)	859
	$PdCl_2$	80	4-6	$1\text{-}C_{10}H_7N(CHO)Si(C_2H_5)_3$ (97)	849
$C_{12}H_{15}$	$CH_3(CH_2)_3CH(C_6H_5)CN$				
	$ZnCl_2$	170	36	$CH_3(CH_2)_3C(C_6H_5){=}CHNHSi(C_2H_5)_3$ (42)	681
$C_{12}H_{21}$	$(i\text{-}C_3H_7)_2CHCN{=}CHCH{=}CHCH_3$				

	$RhCl(CO)(MAD)_2$	r.t.	5	trans-$(i\text{-}C_3H_7)_2CHCN[Si(C_2H_5)_3]CH{=}CHCH_2CH_3$ (30) $(i\text{-}C_3H_7)_2CHN{=}CHCH_2CH(CH_3)Si(C_2H_5)_3$ (12)	165
$C_{13}H_{10}$	$C_6H_5CH{=}NC_6H_4Br$-m				
	$RhCl(PPh_3)_3$	75	- (b)	$C_6H_5CH_2N(C_6H_4Br\text{-}m)Si(C_2H_5)_3$ (kin.)	42
$C_{13}H_{10}$	$C_6H_5CH{=}NC_6H_4Br$-p				
	$RhCl(PPh_3)_3$	75	- (b)	$C_6H_5CH_2N(C_6H_4Br\text{-}p)Si(C_2H_5)_3$ (kin.)	42
$C_{13}H_{11}$	$C_6H_5CH{=}NC_6H_5$				
	$IrCl_6[NH(Oct)_3]_2$	130	16	$C_6H_5CH_2N(C_6H_5)Si(C_2H_5)_3$ (4-8)	17
	$PtCl_6[NH(Oct)_3]_2$	130	16	$C_6H_5CH_2N(C_6H_5)Si(C_2H_5)_3$ (4-8)	17
	$RhCl(PPh_3)_3$	100	15	$C_6H_5CH_2N(C_6H_5)Si(C_2H_5)_3$ (91)	862
	$RhCl(PPh_3)_3$	130	16(b or none)	$C_6H_5CH_2N(C_6H_5)Si(C_2H_5)_3$ (60)	17
	$RhCl(PPh_3)_3$	75	- (b)	$C_6H_5CH_2N(C_6H_5)Si(C_2H_5)_3$ (kin.)	42
	$RhCl(PPh_3)_3$	r.t.	0.5	$C_6H_5CH_2N(C_6H_5)Si(C_2H_5)_3$ (-)	J -010
	$RhHCl(PPh_3)_2(SiEt_3)$	60	5 days (b)	$C_6H_5CH_2N(C_6H_5)Si(C_2H_5)_3$ (13-67)	23
	$RuCl_6[NH(Oct)_3]_2$	130	16	$C_6H_5CH_2N(C_6H_5)Si(C_2H_5)_3$ (4-8)	17
	$ZnCl_2$	120-125	16	$C_6H_5CH_2N(C_6H_5)Si(C_2H_5)_3$ (8-49)	18
$C_{13}H_{22}$	$(CH_2)_5CHN{=}C{=}NCH(CH_2)_5$				
	$PdCl_2$	200	48	$(CH_2)_5CHN[Si(C_2H_5)_3]CH{=}NCH(CH_2)_5$ (96)	849
	$PdCl_2$	200	48	$(CH_2)_5CHN[Si(C_2H_5)_3]CH{=}NCH(CH_2)_5$ (96)	J -015
	$PdCl_2$	200	48	$(CH_2)_5CHN[Si(C_2H_5)_3]CH{=}NCH(CH_2)_5$ (96)	860
	$RhCl(PPh_3)_3$	140	15	$(CH_2)_5CHN[Si(C_2H_5)_3]CH{=}NCH(CH_2)_5$ (75)	860
	$RhCl(PPh_3)_3$	140	15	$(CH_2)_5CHN[Si(C_2H_5)_3]CH{=}NCH(CH_2)_5$ (75)	849
$C_{14}H_{11}$	$(C_6H_5)_2CHCN$				
	$ZnCl_2$	-	-	$(C_6H_5)_2CHCH{=}NSi(C_2H_5)_3$ (90)	381
$C_{14}H_{13}$	$C_6H_5CH{=}NC_6H_4CH_3$-m				
	$RhCl(PPh_3)_3$	75	- (b)	$C_6H_5CH_2N(C_6H_4CH_3\text{-}m)Si(C_2H_5)_3$ (kin.)	42
$C_{14}H_{13}$	$C_6H_5CH{=}NC_6H_4CH_3$-p				
	$RhCl(PPh_3)_3$	75	- (b)	$C_6H_5CH_2N(C_6H_4CH_3\text{-}p)Si(C_2H_5)_3$ (kin.)	42
$C_{17}H_{19}$	$(C_6H_5)_2C{=}C{=}NSi(CH_3)_3$				

Table 8.7

= $(C_2H_5)_3SiH$ =   =4=

Ni	reflux	72	$(C_6H_5)_2C{=}CHN[Si(CH_3)_3]Si(C_2H_5)_3$ (75-80)	680
$PtCl_4$	reflux	8	$(C_6H_5)_2C{=}CHN[Si(CH_3)_3]Si(C_2H_5)_3$ (75-80)	680

Table 9.1

= $(C_2H_5O)_3SiH$ =

=1=

C_2H_4	$CH_2=CH_2$				
	$[RhCl(CO)_2]_2$	148-225	- (xy)	$CH_3CH_2Si(OC_2H_5)_3$ (28) $CH_2=CHSi(OC_2H_5)_3$ (62)	U -173
	$[RhCl(CO)_2]_2$	20	65	$CH_3CH_2Si(OC_2H_5)_3$ (95)	227
	$[RhCl(CO)_2]_2$	200	3 (xy)	$CH_3CH_2Si(OC_2H_5)_3$ (30) $CH_2=CHSi(OC_2H_5)_3$ (-)	U -173
	$Ru_3(CO)_{12}$	-	-	$CH_3CH_2Si(OC_2H_5)_3$ (19) $CH_2=CHSi(OC_2H_5)_3$ (41)	1032
	$Ru_3(CO)_{12}$	80	5 (b)	$CH_3CH_2Si(OC_2H_5)_3$ (19) $CH_2=CHSi(OC_2H_5)_3$ (41)	J -043
C_3H_6	$CH_3CH=CH_2$				
	$Fe(CO)_5$	150	2-3	$CH_3CH_2CH_2Si(OC_2H_5)_3$ (-)	B -003
	H_2PtCl_6	150	2-3	$CH_3CH_2CH_2Si(OC_2H_5)_3$ (-)	B -003
	$[RhCl(CO)_2]_2$	225	0.33	$CH_3CH=CHSi(OC_2H_5)_3$ (60) $CH_3CH_2CH_2Si(OC_2H_5)_3$ (-)	U -173
C_4H_8	$(CH_3)_2C=CH_2$				
	H_2PtCl_6 $[(\overline{CH_2)_5CO}]$	42	12	$CH_3(CH_2)_3Si(OC_2H_5)_3/(CH_3)_2CHCH_2Si(OC_2H_5)_3$ (23/77)	Pl-029
	$[SiO_2\text{-}]_3N/H_2PtCl_6$	56	16	$CH_3(CH_2)_3Si(OC_2H_5)_3/(CH_3)_2CHCH_2Si(OC_2H_5)_3$ (36/64)	Pl-029
C_4H_8	$CH_3CH_2CH=CH_2$				
	H_2PtCl_6 $[(\overline{CH_2)_5CO}]$	46	10	$CH_3(CH_2)_3Si(OC_2H_5)_3/(CH_3)_2CHCH_2Si(OC_2H_5)_3$ (95/5)	Pl-029
	$[SiO_2\text{-}]_3N/H_2PtCl_6$	40	16 (flow)	$CH_3(CH_2)_3Si(OC_2H_5)_3/(CH_3)_2CHCH_2Si(OC_2H_5)_3$ (95/5)	Pl-029
C_5H_{10}	$(CH_3)_2CHCH=CH_2$				
	$RhCl(PHPh_2)_3$	85	2	$(CH_3)_2CHCH_2CH_2Si(OC_2H_5)_3$ (90)	1166
C_5H_{10}	$CH_3CH_2CH_2CH=CH_2$				
	$Co_2(CO)_8$	30	-	$CH_3(CH_2)_4Si(OC_2H_5)_3$ (kin.)	225
	$\{HPt[Si(OEt)_3]P(c\text{-}Hex)_3\}_2$	20	17	$CH_3(CH_2)_4Si(OC_2H_5)_3$ (92)	429
	$RhCl(PPh_3)_3$	25	7 days	$CH_3(CH_2)_4Si(OC_2H_5)_3$ (84)	227
	$RuCl_2(PPh_3)_3$	120	6	$CH_3(CH_2)_4Si(OC_2H_5)_3$ (48)	733
C_5H_{10}	$CH_3CH_2CH=CHCH_3$				

$RuCl_2(PPh_3)_3$	120	6	$CH_3(CH_2)_4Si(OC_2H_5)_3$ (26)	1194
C_6H_{10} $(CH_2)_4CH{=}CH$				
C-CN/$RhCl_3$ + c-$C_{12}H_{18}$	80	8	$(CH_2)_5CHSi(OC_2H_5)_3$ (9)	C -030
$RhCl(PHPh_2)_3$	100	2	$(CH_2)_5CHSi(OC_2H_5)_3$ (-)	C -004
SDVB-CH_2Cl/$RhCl(CO)(PPh_3)_2$	80	2	$(CH_2)_5CHSi(OC_2H_5)_3$ (9)	C -022
C_6H_{12} $(CH_3)_2CHCH_2CH{=}CH_2$				
Pt/C + sonic waves	30	1	$(CH_3)_2CH(CH_2)_3Si(OC_2H_5)_3$ (93)	467
C_6H_{12} $(CH_3)_3CCH{=}CH_2$				
SDVB-PPh_2/$RhCl(PPh_3)_3$	85	2	$(CH_3)_3CCH_2CH_2Si(OC_2H_5)_3$ (65)	1168
polymer-$PPh_2RhCl(PPh_3)_2$	80	3	$(CH_3)_3CCH_2CH_2Si(OC_2H_5)_3$ (58)	C -008
C_6H_{12} $CH_3(CH_2)_3CH{=}CH_2$				
Fig.395.	90	2	$CH_3(CH_2)_5Si(OC_2H_5)_3$ (53)	S -017
$Co_2(CO)_8$	120	2	$CH_3(CH_2)_5Si(OC_2H_5)_3$ (kin.)	1169
$Co_2(CO)_8$	20-68	0.3	$CH_3(CH_2)_5Si(OC_2H_5)_3$ (-)	61
$Co_2(CO)_8$	35-40	1	$CH_3(CH_2)_5Si(OC_2H_5)_3$ (89)	715
$Co_2(CO)_8$	40	1	$CH_3(CH_2)_5Si(OC_2H_5)_3$ (kin.)	495
$Co_2(CO)_8$ + Et_2O	20-30	0.25 - 0.33	$CH_3(CH_2)_5Si(OC_2H_5)_3$ (95-98)	S -044
$Co_2(CO)_8$ + KI	20-30	0.25 - 0.33	$CH_3(CH_2)_5Si(OC_2H_5)_3$ (94-96)	S -044
$Co_2(CO)_8$ + PhC≡CPh	25-35	0.33 - 0.5	$CH_3(CH_2)_5Si(OC_2H_5)_3$ (93-98)	S -038
$Co_2(CO)_8$ + RNH_2	30	0.5 - 0.66	$CH_3(CH_2)_5Si(OC_2H_5)_3$ (90)	716
$Co_2(CO)_8$ + THF	20-30	0.33 - 0.41	$CH_3(CH_2)_5Si(OC_2H_5)_3$ (92-95)	S -044
$CoH(N_2)L_3$	20-55	0.3	$CH_3(CH_2)_5Si(OC_2H_5)_3$ (-)	61
$CoH_2[Si(OEt)_3](PPh_3)_3$	25-65	0.17	$CH_3(CH_2)_5Si(OC_2H_5)_3$ (40-75)	60
$CoH_2[Si(OEt)_3]L_3$	20-60	0.3	$CH_3(CH_2)_5Si(OC_2H_5)_3$ (-)	61
CoH_3L_3	20-50	0.3	$CH_3(CH_2)_5Si(OC_2H_5)_3$ (-)	61
$Mn_2(CO)_{10}$	140	-	$CH_3(CH_2)_5Si(OC_2H_5)_3$ (-)	382
$Mn_2(CO)_{10}$	40	1	$CH_3(CH_2)_5Si(OC_2H_5)_3$ (kin.)	495

Table 9.1
= $(C_2H_5O)_3SiH$ = =3=

$PdCl_2$ + $P(c\text{-}Hex)_3$	130	5	$CH_3(CH_2)_5Si(OC_2H_5)_3$ (28)	D -027
siloxane-SH/Pd complex	80	2	$CH_3(CH_2)_5Si(OC_2H_5)_3$ (41)	1286
DVBAC-PPh_2/H_2PtCl_6	80	2	$CH_3(CH_2)_5Si(OC_2H_5)_3$ (90)	217
PMM-CN/H_2PtCl_6	80	2	$CH_3(CH_2)_5Si(OC_2H_5)_3$ (99)	217
PMM-NMe_2/H_2PtCl_6	24-80	2-24	$CH_3(CH_2)_5Si(OC_2H_5)_3$ (85-97)	217
PMM-PPh_2/H_2PtCl_6	80	2	$CH_3(CH_2)_5Si(OC_2H_5)_3$ (90-93)	217
$Pt(O_2)(PPh_3)_2$	30-90	-	$CH_3(CH_2)_5Si(OC_2H_5)_3$ (kin.)	1349
$Pt(PPh_3)_4$	30-90	-	$CH_3(CH_2)_5Si(OC_2H_5)_3$ (kin.)	1349
SDVB-CN/H_2PtCl_6	80	2	$CH_3(CH_2)_5Si(OC_2H_5)_3$ (95)	217
SDVB-NMe_2/H_2PtCl_6	80	2	$CH_3(CH_2)_5Si(OC_2H_5)_3$ (99)	217
SDVB-PPh_2/H_2PtCl_6	25	26	$CH_3(CH_2)_5Si(OC_2H_5)_3$ (96-98)	217
SDVB-PPh_2/H_2PtCl_6 (EtOH)	r.t.	4	$CH_3(CH_2)_5Si(OC_2H_5)_3$ (97)	C -002
SiO_2-NH_2/H_2PtCl_6	25	16	$CH_3(CH_2)_5Si(OC_2H_5)_3$ (kin.)	1285
SiO_2-NH_2/H_2PtCl_6	25	16	$CH_3(CH_2)_5Si(OC_2H_5)_3$ (kin.)	509
SiO_2-NH_2/H_2PtCl_6	80	2	$CH_3(CH_2)_5Si(OC_2H_5)_3$ (100)	1285
SiO_2-NH_2/H_2PtCl_6	80	2	$CH_3(CH_2)_5Si(OC_2H_5)_3$ (100)	509
$[HPt(SiBzMe_2)P(c\text{-}Hex)_3]_2$	20	24	$CH_3(CH_2)_5Si(OC_2H_5)_3$ (55)	429
amberlyt/H_2PtCl_6	80	2	$CH_3(CH_2)_5Si(OC_2H_5)_3$ (90)	217
polymer/Pt complex	100	2	$CH_3(CH_2)_5Si(OC_2H_5)_3$ (85-91)	1220
siloxane-SH/Pt complex	20	8	$CH_3(CH_2)_5Si(OC_2H_5)_3$ (90)	1286
siloxane-SH/Pt complex	20	8	$CH_3(CH_2)_5Si(OC_2H_5)_3$ (90)	509
siloxane-SH/Pt complex	20-80	-	$CH_3(CH_2)_5Si(OC_2H_5)_3$ (-)	1284
siloxane-SH/Pt complex	80	2	$CH_3(CH_2)_5Si(OC_2H_5)_3$ (96)	1286
siloxane-SH/Pt complex	80	2	$CH_3(CH_2)_5Si(OC_2H_5)_3$ (96)	509
support/Pt complex	80	2	$CH_3(CH_2)_5Si(OC_2H_5)_3$ (3-99)	204
$\{HPt[Si(OEt)_3]P(c\text{-}Hex)_3\}_2$	20	24	$CH_3(CH_2)_5Si(OC_2H_5)_3$ (60)	429
Al_2O_3-$CN/[RhCl(CO)_2]_2$	80	2.5	$CH_3(CH_2)_5Si(OC_2H_5)_3$ (82)	C -030

Al_2O_3-CN/$[RhCl(CO)_2]_2$	80	2.5	$CH_3(CH_2)_5Si(OC_2H_5)_3$ (82)	Dd-003
C + $C_6H_8(OCH_2CH_2CN)_6[RhCl(CO)_2]_2$	80	2.5	$CH_3(CH_2)_5Si(OC_2H_5)_3$ (81)	Dd-003
C + $C_6H_8(OCH_2CH_2CN)_6[RhCl(CO)_2]_2$	80	2.5	$CH_3(CH_2)_5Si(OC_2H_5)_3$ (81)	C -030
C/$RhCl_3$ + c-$C_{12}H_{18}$ (EtOH)	25	24	$CH_3(CH_2)_5Si(OC_2H_5)_3$ (83)	C -030
C/$RhCl_3$ + c-$C_{12}H_{18}$ (EtOH)	25	24	$CH_3(CH_2)_5Si(OC_2H_5)_3$ (83)	Dd-003
C/$RhCl_3$ + c-$C_{12}H_{18}$ (EtOH)	80	3 (solv.)	$CH_3(CH_2)_5Si(OC_2H_5)_3$ (82-87)	Dd-003
C/$RhCl_3$ + c-$C_{12}H_{18}$ (EtOH)	80	3 (solv.)	$CH_3(CH_2)_5Si(OC_2H_5)_3$ (82-87)	C -030
C/$RhCl_3$ + c-$C_{12}H_{18}$ (EtOH)	80	2.5	$CH_3(CH_2)_5Si(OC_2H_5)_3$ (76)	C -030
C/$RhCl_3$ + c-$C_{12}H_{18}$ (EtOH)	80	2.5	$CH_3(CH_2)_5Si(OC_2H_5)_3$ (76)	Dd-003
C/$[RhCl(CO)_2]_2$ + c-$C_{12}H_{18}$ (EtOH)	80	2.5	$CH_3(CH_2)_5Si(OC_2H_5)_3$ (78)	Dd-003
C-CN/$[RhCl(CO)_2]_2$	25	24	$CH_3(CH_2)_5Si(OC_2H_5)_3$ (47)	C -030
C-CN/$[RhCl(CO)_2]_2$	25	24	$CH_3(CH_2)_5Si(OC_2H_5)_3$ (47)	Dd-003
C-CN/$[RhCl(CO)_2]_2$	80	2.5	$CH_3(CH_2)_5Si(OC_2H_5)_3$ (84)	Dd-003
C-CN/$[RhCl(CO)_2]_2$	80	2.5	$CH_3(CH_2)_5Si(OC_2H_5)_3$ (84)	C -030
C-CN/$[RhCl(CO)_2]_2$	80	2	$CH_3(CH_2)_5Si(OC_2H_5)_3$ (81)	Dd-003
C-CN/$[RhCl(CO)_2]_2$ + c-$C_{12}H_{18}$	80	2.5	$CH_3(CH_2)_5Si(OC_2H_5)_3$ (78)	C -030
$CaCO_3$-CN/$[RhCl(CO)_2]_2$	80	2.5	$CH_3(CH_2)_5Si(OC_2H_5)_3$ (85)	Dd-00
$CaCO_3$-CN/$[RhCl(CO)_2]_2$	80	2.5	$CH_3(CH_2)_5Si(OC_2H_5)_3$ (85)	C -030
DVBAC-PPh_2/$RhCl_3$	80	2	$CH_3(CH_2)_5Si(OC_2H_5)_3$ (95)	217
PMM-CN/$RhCl_3$	80	2	$CH_3(CH_2)_5Si(OC_2H_5)_3$ (99)	217
PMM-NMe_2/$RhCl_3$	80	2	$CH_3(CH_2)_5Si(OC_2H_5)_3$ (97)	217
Rh complex	r.t.	-	$CH_3(CH_2)_5Si(OC_2H_5)_3$ (97)	166
Rh(acac)$(CO)_2$	-	-	$CH_3(CH_2)_5Si(OC_2H_5)_3$ (-)	236
Rh(acac)$(CO)_2$	70-80	-	$CH_3(CH_2)_5Si(OC_2H_5)_3$ (100)	S -017
Rh(acac)$(CO)_2$ + PPh_3	-	-	$CH_3(CH_2)_5Si(OC_2H_5)_3$ (-)	236
$Rh_4(CO)_{12}$	35-45	1	$CH_3(CH_2)_5Si(OC_2H_5)_3$ (97)	S -037
$RhBr(PPh_3)_3$	100	2 (b)	$CH_3(CH_2)_5Si(OC_2H_5)_3$ (52)	981

Table 9.1

= $(C_2H_5O)_3SiH$ = =5=

$RhCl(CO)(PPh_3)_2$	100	2 (b)	$CH_3(CH_2)_5Si(OC_2H_5)_3$ (55)	981
$RhCl(PHPh_2)_3$	20	20	$CH_3(CH_2)_5Si(OC_2H_5)_3$ (87)	1166
$RhCl(PHPh_2)_3$	85	2	$CH_3(CH_2)_5Si(OC_2H_5)_3$ (95)	1166
$RhCl(PHPh_2)_3$	r.t.	20	$CH_3(CH_2)_5Si(OC_2H_5)_3$ (-)	C -004
$RhCl(PPh_3)_2$	80	1.5	$CH_3(CH_2)_5Si(OC_2H_5)_3$ (68)	333
$RhCl(PPh_3)_3$	120	2	$CH_3(CH_2)_5Si(OC_2H_5)_3$ (76)	739
$RhCl(PPh_3)_3$	50	1.5	$CH_3(CH_2)_5Si(OC_2H_5)_3$ (61)	333
$RhCl(PPh_3)_3$	80	2	$CH_3(CH_2)_5Si(OC_2H_5)_3$ (96)	204
$RhCl(PPh_3)_3$	80	2	$CH_3(CH_2)_5Si(OC_2H_5)_3$ (74)	739
$RhCl(PPh_3)_3$	80	2 (b)	$CH_3(CH_2)_5Si(OC_2H_5)_3$ (64)	981
$RhCl(PPh_3)_3$	80	2	$CH_3(CH_2)_5Si(OC_2H_5)_3$ (kin.)	1169
$RhCl(PPh_3)_3$	80-100	2	$CH_3(CH_2)_5Si(OC_2H_5)_3$ (72-74)	750
$RhCl(cod)(PPh_3)$	20	-	$CH_3(CH_2)_5Si(OC_2H_5)_3$ (kin.)	741
$RhCl(cod)(PPh_3)$	50	1.5	$CH_3(CH_2)_5Si(OC_2H_5)_3$ (31)	333
$RhCl(cod)(PPh_3)$	80	1.5	$CH_3(CH_2)_5Si(OC_2H_5)_3$ (48)	333
$RhCl(cod)_2$	50	1.5	$CH_3(CH_2)_5Si(OC_2H_5)_3$ (17)	333
$RhCl(cod)_2$	80	1.5	$CH_3(CH_2)_5Si(OC_2H_5)_3$ (19)	333
$RhCl_3$ (EtOH)	80	2	$CH_3(CH_2)_5Si(OC_2H_5)_3$ (86)	204
$RhCl[PPh_2C_2H_4Si(OEt)_3]_3$	80-100	2	$CH_3(CH_2)_5Si(OC_2H_5)_3$ (73-77)	750
$RhCl[PPh_2C_3H_6Si(OEt)_3]_3$	80-100	2	$CH_3(CH_2)_5Si(OC_2H_5)_3$ (73-76)	750
$RhH(CO)(PPh_3)_3$	100	2 (b)	$CH_3(CH_2)_5Si(OC_2H_5)_3$ (54)	981
$RhH(PPh_3)_4$	100	-	$CH_3(CH_2)_5Si(OC_2H_5)_3$ (80)	458
$RhI(PPh_3)_3$	100	2 (b)	$CH_3(CH_2)_5Si(OC_2H_5)_3$ (36)	981
SDVB-$CH_2Cl/RhCl(CO)(PPh_3)_2$	25	24	$CH_3(CH_2)_5Si(OC_2H_5)_3$ (85)	C -022
SDVB-$CH_2Cl/RhCl(CO)(PPh_3)_2$	80	3 (solv.)	$CH_3(CH_2)_5Si(OC_2H_5)_3$ (29-82)	C -022
SDVB-$CH_2Cl/RhCl(CO)(PPh_3)_2$	80	2-2,5	$CH_3(CH_2)_5Si(OC_2H_5)_3$ (81-82)	C -022
SDVB-CH_2Cl/RhCl(diphos)$_2$	80	2,5	$CH_3(CH_2)_5Si(OC_2H_5)_3$ (61)	C -022

Table 9.1

= $(C_2H_5O)_3SiH$ = =6=

SDVB-CH_2Cl/$RhCl(diphos)_2$	80	2,5	$CH_3(CH_2)_5Si(OC_2H_5)_3$ (63)	C -022
SDVB-CH_2Cl/$RhCl_3$ (THF)	80	2,5	$CH_3(CH_2)_5Si(OC_2H_5)_3$ (31)	C -022
SDVB-NMe_2/$RhCl_3$	80	2	$CH_3(CH_2)_5Si(OC_2H_5)_3$ (97)	217
SDVB-PPh_2/$RhCl(PPh_3)_3$	80	2	$CH_3(CH_2)_5Si(OC_2H_5)_3$ (19)	216
SDVB-PPh_2/$RhCl_3$	24	24	$CH_3(CH_2)_5Si(OC_2H_5)_3$ (33)	217
SDVB-PPh_2/$RhCl_3$	80	2	$CH_3(CH_2)_5Si(OC_2H_5)_3$ (67)	217
SDVB-PPh_2/$RhCl_3$ (EtOH)	r.t.	24	$CH_3(CH_2)_5Si(OC_2H_5)_3$ (-)	D -056
$[RhCl(1,5\text{-}C_8H_{12})_2]$ + phosphine	100	1	$CH_3(CH_2)_5Si(OC_2H_5)_3$ (31-67)	981
$[RhCl(C_2H_4)_2]_2$	24-85	2-24 (b)	$CH_3(CH_2)_5Si(OC_2H_5)_3$ (100)	1167
$[RhCl(CO)_2]_2$+(4-OH-3-$HCOC_6H_4)_2CH_2$+NEt_3	80	2.5	$CH_3(CH_2)_5Si(OC_2H_5)_3$ (84)	C -022
$[RhCl(cod)Ph_2P(CH_2)_nSiMe_2]_2O$	50	1.5	$CH_3(CH_2)_5Si(OC_2H_5)_3$ (38-52)	333
$[RhCl(cod)Ph_2P(CH_2)_nSiMe_2]_2O$	80	1.5	$CH_3(CH_2)_5Si(OC_2H_5)_3$ (48-62)	333
$[RhCl(cod)Ph_2PCH_2CH_2SiMe_2O]_2SiMeCH_2CH_2$ $\cdot PPh_2RhCl(cod)$	50	1.5	$CH_3(CH_2)_5Si(OC_2H_5)_3$ (51)	333
$[RhCl(cod)Ph_2PCH_2CH_2SiMe_2O]_2SiMeCH_2CH_2$ $\text{-}PPh_2RhCl(cod)$	80	1.5	$CH_3(CH_2)_5Si(OC_2H_5)_3$ (64)	333
$[RhCl(cod)Ph_2PCH_2CH_2SiMe_2]_2O$	20	-	$CH_3(CH_2)_5Si(OC_2H_5)_3$ (kin.)	332
$[RhCl(cod)Ph_2PCH_2CH_2SiMe_2]_2O$	20-30	3-14.5	$CH_3(CH_2)_5Si(OC_2H_5)_3$ (kin.)	741
$[RhCl(cod)Ph_2PCH_2CH_2SiMe_2]_2O$ + $[Ph_2PCH_2CH_2SiMe_2]_2O$	20	-	$CH_3(CH_2)_5Si(OC_2H_5)_3$ (kin.)	332
$[RhCl(cod)]_2$	80	1.5	$CH_3(CH_2)_5Si(OC_2H_5)_3$ (19)	Pl-025
$[RhCl(cod)]_2$ + $[Ph_2PCH_2SiMe_2]_2O$	80	1.5	$CH_3(CH_2)_5Si(OC_2H_5)_3$ (62)	Pl-025
asbestos-PPh_2/$RhCl(PPh_3)_3$	80	2	$CH_3(CH_2)_5Si(OC_2H_5)_3$ (-)	Pl-020
chrysotile-PPh_2/$RhCl(PPh_3)_3$	120	2	$CH_3(CH_2)_5Si(OC_2H_5)_3$ (54-78)	739
glass-CN/$[RhCl(CO)_2]_2$	80	2.5	$CH_3(CH_2)_5Si(OC_2H_5)_3$ (85)	C -030
polymer/Rh complex	100	2	$CH_3(CH_2)_5Si(OC_2H_5)_3$ (92)	1220
polymer-NMe_2/$RhCl_3$	-	20	$CH_3(CH_2)_5Si(OC_2H_5)_3$ (95)	C -006
polymer-PPh_2/$RhCl(PPh_3)_3$	80	2	$CH_3(CH_2)_5Si(OC_2H_5)_3$ (66)	C -008
polymer-PPh_2/$RhCl(PPh_3)_3$	80	2	$CH_3(CH_2)_5Si(OC_2H_5)_3$ (22)	216

Table 9.1
= $(C_2H_5O)_3SiH$ = =7=

polymer-PPh_2/$RhCl(PPh_3)_3$	85	2	$CH_3(CH_2)_5Si(OC_2H_5)_3$ (85)	1168
polymer-PPh_2/$RhCl(PPh_3)_3$	reflux	23 (THF)	$CH_3(CH_2)_5Si(OC_2H_5)_3$ (93)	1168
polymer-PPh_2/$RhCl_3$	80	2	$CH_3(CH_2)_5Si(OC_2H_5)_3$ (91)	216
polymer-PPh_2/$[RhCl(C_2H_4)_2]_2$	80	2	$CH_3(CH_2)_5Si(OC_2H_5)_3$ (99)	216
porolith-CN/$[RhCl(CO)_2]_2$	80	2.5	$CH_3(CH_2)_5Si(OC_2H_5)_3$ (86)	Dd-003
siloxane-SH/Rh complex	80	2	$CH_3(CH_2)_5Si(OC_2H_5)_3$ (61)	1286
siloxane-SH/Rh complex	80	2	$CH_3(CH_2)_5Si(OC_2H_5)_3$ (61)	509
silsesquioxane/$RhCl(PPh_3)_3$	80-100	2	$CH_3(CH_2)_5Si(OC_2H_5)_3$ (58-89)	750
zeolite-CN/$RhCl(CO)(PPh_3)_2$	85	2.5	$CH_3(CH_2)_5Si(OC_2H_5)_3$ (85)	Dd-003
$RuCl_2(CO)_2(PPh_3)_2$	120	6	$CH_3(CH_2)_5Si(OC_2H_5)_3$ (15)	733
$RuCl_2(PPh_3)_3$	60-120	6	$CH_3(CH_2)_5Si(OC_2H_5)_3$ (48-68)	733
$RuCl_2(PPh_3)_3$	80	2	$CH_3(CH_2)_5Si(OC_2H_5)_3$ (66)	Pl-001
$RuCl_3$	120	6	$CH_3(CH_2)_5Si(OC_2H_5)_3$ (77)	733
$RuCl_3 + PPh_3$	120	6	$CH_3(CH_2)_5Si(OC_2H_5)_3$ (86-96)	733
$RuCl_3 + Ph_3PO$	120	6	$CH_3(CH_2)_5Si(OC_2H_5)_3$ (81)	733
$RuCl_3(NO)(PPh_3)_2$	120	6	$CH_3(CH_2)_5Si(OC_2H_5)_3$ (28)	733
$RuCl_3(PBu_3)_3$	120	6	$CH_3(CH_2)_5Si(OC_2H_5)_3$ (69)	733
$RuCl_3(PBu_3)_3$	120	6	$CH_3(CH_2)_5Si(OC_2H_5)_3$ (69)	Pl-013
$RuCl_3(PBu_3)_3$	120	6	$CH_3(CH_2)_5Si(OC_2H_5)_3$ (69)	Pl-015
$RuCl_3(PEt_3)_3$	120	6	$CH_3(CH_2)_5Si(OC_2H_5)_3$ (24)	733
$RuCl_3(PPh_3)_2(MeOH)$	120	6	$CH_3(CH_2)_5Si(OC_2H_5)_3$ (87)	Pl-013
$RuCl_3(PPh_3)_2(MeOH)$	120	6	$CH_3(CH_2)_5Si(OC_2H_5)_3$ (85-87)	733
$RuCl_3(PPh_3)_2(MeOH)$	120	6	$CH_3(CH_2)_5Si(OC_2H_5)_3$ (85-87)	Pl-015
$RuCl_3(PPh_3)_3$	120	6	$CH_3(CH_2)_5Si(OC_2H_5)_3$ (86-88)	733
$RuCl_3(PPh_3)_3$	120	6	$CH_3(CH_2)_5Si(OC_2H_5)_3$ (86-88)	Pl-015
$RuCl_3(PPh_3)_3$	120	6	$CH_3(CH_2)_5Si(OC_2H_5)_3$ (62-88)	Pl-013
$RuCl_3(PPh_3)_3$	reflux	6	$CH_3(CH_2)_5Si(OC_2H_5)_3$ (68)	Pl-013

	$RuCl_3(PPh_3)_3$	reflux	6	$CH_3(CH_2)_5Si(OC_2H_5)_3$ (68)	Pl-015
	RuH(acac)$(PPh_3)_3$	120	6	$CH_3(CH_2)_5Si(OC_2H_5)_3$ (24)	733
	$RuH_2(PPh_3)_4$	120	6	$CH_3(CH_2)_5Si(OC_2H_5)_3$ (16)	733
	$RuHCl(PPh_3)_3$	120	6	$CH_3(CH_2)_5Si(OC_2H_5)_3$ (16)	733
	$RuHCl(PPh_3)_3$	80	2	$CH_3(CH_2)_5Si(OC_2H_5)_3$ (15)	Pl-001
	RuHCl(cod)(DAD)	r.t.	- (dcm)	$CH_3(CH_2)_5Si(OC_2H_5)_3$ (-)	1363
	$RuH[Si(OEt)_3(PPh_3)_2$	120	6	$CH_3(CH_2)_5Si(OC_2H_5)_3$ (5)	733
C_6H_{12}	$CH_3CH_2CH_2CH{=}CHCH_3$				
	$C/RhCl_3$ + c-$C_{12}H_{18}$ (EtOH)	80	2	$CH_3(CH_2)_5Si(OC_2H_5)_3$ (15-31)	Dd-003
	C-CN/$RhCl_3$ + c-$C_{12}H_{18}$	80	2	$CH_3(CH_2)_5Si(OC_2H_5)_3$ (15)	C -030
	SDVB-CH_2Cl/RhCl(CO)$(PPh_3)_2$	80	2	$CH_3(CH_2)_5Si(OC_2H_5)_3$ (15)	C -022
	SDVB-PPh_2/$RhCl(PPh_3)_2$	80	2	$CH_3(CH_2)_3CH(CH_3)Si(OC_2H_5)_3$ (5)	216
	polymer-PPh_2/$RhCl_x$	80	2	$CH_3(CH_2)_3CH(CH_3)Si(OC_2H_5)_3$ (5)	216
C_7H_{14}	$CH_3(CH_2)_3CH{=}CHCH_3$				
	$RhCl(PHPh_2)_3$	85	2	trans-$CH_3(CH_2)_6Si(OC_2H_5)_3$ (5)	1166
	$RhCl(PHPh_2)_3$	85	2	cis-$CH_3(CH_2)_6Si(OC_2H_5)_3$ (60)	1166
C_7H_{14}	$CH_3(CH_2)_4CH{=}CH_2$				
	H_2PtCl_6 (THF)	100	4	$CH_3(CH_2)_6Si(OC_2H_5)_3$ (-)	214
	H_2PtCl_6 (i-PrOH)	-	0.1	$CH_3(CH_2)_6Si(OC_2H_5)_3$ (75)	96
	H_2PtCl_6 (i-PrOH) + borate	20	24	$CH_3(CH_2)_6Si(OC_2H_5)_3$ (78)	96
	SiO_2-$L_2Pt(C_2O_4)$	30	24 (h)	$CH_3(CH_2)_6Si(OC_2H_5)_3$ (50)	165
	C-CN/$[RhCl(CO)_2]_2$	80	2	$CH_3(CH_2)_6Si(OC_2H_5)_3$ (81)	C -030
	$RhCl(PHPh_2)_3$	85	2	$CH_3(CH_2)_6Si(OC_2H_5)_3$ (-)	1166
	SiO_2-$\overline{N(CH_2)_4}$/$RhCl(PPh_3)_3$	100	4	$CH_3(CH_2)_6Si(OC_2H_5)_3$ (24)	746
	SiO_2-$\overline{N(CH_2)_5}$/$RhCl(PPh_3)_3$	100	4	$CH_3(CH_2)_6Si(OC_2H_5)_3$ (10)	746
	SiO_2-$\overline{NCH_2CH_2OCH_2CH_2}$/$RhCl(PPh_3)_3$	100	4	$CH_3(CH_2)_6Si(OC_2H_5)_3$ (10)	746
	SiO_2-NEt_2/$RhCl(PPh_3)_3$	100	4	$CH_3(CH_2)_6Si(OC_2H_5)_3$ (38)	746

	SiO_2-NPh_2/$RhCl(PPh_3)_3$	100	4	$CH_3(CH_2)_6Si(OC_2H_5)_3$ (40)	746
	$RuCl_2(PPh_3)_3$	120	6	$CH_3(CH_2)_6Si(OC_2H_5)_3$ (53)	733
C_8H_8	$C_6H_5CH{=}CH_2$				
	$Co_2(CO)_8$	30-40	1	$C_6H_5CH_2CH_2Si(OC_2H_5)_3$ (91)	715
	$Co_2(CO)_8$	30-40	1	$C_6H_5CH_2CH_2Si(OC_2H_5)_3$ (91)	724
	$Co_2(CO)_8$ + Et_2O	20	0.25	$C_6H_5CH_2CH_2Si(OC_2H_5)_3$ (98)	S -044
	$Co_2(CO)_8$ + KI	20	0.25	$C_6H_5CH_2CH_2Si(OC_2H_5)_3$ (96)	S -044
	$Co_2(CO)_8$ + $PhC{\equiv}CPh$	25-35	0.25-0.41	$C_6H_5CH_2CH_2Si(OC_2H_5)_3$ (92-98)	S -038
	$Co_2(CO)_8$ + THF	20	0.33	$C_6H_5CH_2CH_2Si(OC_2H_5)_3$ (93)	S -044
	$Pt[(ViMe_2Si)_2O]_2$	-	-	$C_6H_5CH_2CH_2Si(OC_2H_5)_3$ + $C_6H_5CH(CH_3)Si(OC_2H_5)_3$ (kin.)	669
	support/Pt complex	80	2	$C_6H_5CH_2CH_2Si(OC_2H_5)_3$ (0-42) $C_6H_5CH(CH_3)Si(OC_2H_5)_3$ (0-6)	204
	$Rh(PPh_3)_4$	100	-	$C_6H_5C_2H_4Si(OC_2H_5)_3$ (-)	605
	$Rh(acac)(CO)_2$	r.t.	24	$C_6H_5CH_2CH_2Si(OC_2H_5)_3$ (18) $C_6H_5CH(CH_3)Si(OC_2H_5)_3$ (72)	C -009
	$Rh(acac)_3$ + $AlEt_3$	60	-	$C_6H_5CH_2CH_2Si(OC_2H_5)_3$ (39) $C_6H_5CH(CH_3)Si(OC_2H_5)_3$ (7) $C_6H_5CH{=}CHSi(OC_2H_5)_3$ (15)	271
	$Rh_2Cl_2(CO)_4$ + phenothiazine	120-150	-	$C_6H_5CH_2CH_2Si(OC_2H_5)_3$ (2) $C_6H_5CH{=}CHSi(OC_2H_5)_3$ (1)	U -173
	$Rh_4(CO)_{12}$	20-30	70	$C_6H_5CH_2CH_2Si(OC_2H_5)_3$ (91-94)	715
	$Rh_4(CO)_{12}$	30-40	1	$C_6H_5CH_2CH_2Si(OC_2H_5)_3$ (94)	719
	$RhCl(CO)(PPh_3)_2$	100	2 (b)	$C_6H_5CH(CH_3)Si(OC_2H_5)_3$ (62)	981
	$RhCl(PPh_3)_3$	100	4 (t)	$C_6H_5CH_2CH_2Si(OC_2H_5)_3$ (30) $C_6H_5CH(CH_3)Si(OC_2H_5)_3$ (11) $C_6H_5CH{=}CHSi(OC_2H_5)_3$ (7)	878
	$RhCl(PPh_3)_3$	100	2 (b)	$C_6H_5CH(CH_3)Si(OC_2H_5)_3$ (46)	981
	$RhH(CO)(PPh_3)_3$	100	2 (b)	$C_6H_5CH(CH_3)Si(OC_2H_5)_3$ (41)	981
	$RhI(PPh_3)_3$	100	2 (b)	$C_6H_5CH(CH_3)Si(OC_2H_5)_3$ (40)	981
	SDVB-NMe_2/$[RhCl(CO)_2]_2$	80	2	$C_6H_5CH_2CH_2Si(OC_2H_5)_3$ + $C_6H_5CH(CH_3)Si(OC_2H_5)_3$ (-)	316
C_8H_{16}	$CH_3(CH_2)_5CH{=}CH_2$				

Table 9.1

= $(C_2H_5O)_3SiH$ = =10=

$Co_2(CO)_8$	(-20) -20	17	$CH_3(CH_2)_7Si(OC_2H_5)_3$ (82)	U -019
$[IrCl(c\text{-}C_8H_{14})_2]_2 + PPh_3$	100	6	$CH_3(CH_2)_7Si(OC_2H_5)_3$ (4)	59
H_2PtCl_6 (i-PrOH)	25	30	$CH_3(CH_2)_7Si(OC_2H_5)_3$ (66)	748
H_2PtCl_6 (i-PrOH)	25	15	$CH_3(CH_2)_7Si(OC_2H_5)_3$ (58)	748
H_2PtCl_6 (i-PrOH)	25	6	$CH_3(CH_2)_7Si(OC_2H_5)_3$ (48)	748
H_2PtCl_6 (i-PrOH)	50	2	$CH_3(CH_2)_7Si(OC_2H_5)_3$ (26)	748
H_2PtCl_6 (i-PrOH)	50	6	$CH_3(CH_2)_7Si(OC_2H_5)_3$ (47)	748
H_2PtCl_6 (i-PrOH)	50	4	$CH_3(CH_2)_7Si(OC_2H_5)_3$ (44)	748
$Pt(PPh_3)_2(CH_2=CH_2)$	25	15	$CH_3(CH_2)_7Si(OC_2H_5)_3$ (13)	748
$Pt(PPh_3)_2(CH_2=CH_2)$	25	30	$CH_3(CH_2)_7Si(OC_2H_5)_3$ (21)	748
$Pt(PPh_3)_2(CH_2=CH_2)$	50	2	$CH_3(CH_2)_7Si(OC_2H_5)_3$ (89)	PI-030
$Pt(PPh_3)_2(CH_2=CH_2)$	50	4	$CH_3(CH_2)_7Si(OC_2H_5)_3$ (83)	748
$Pt(PPh_3)_2(CH_2=CH_2)$	50	2	$CH_3(CH_2)_7Si(OC_2H_5)_3$ (89)	748
$Pt(PPh_3)_2(CH_2=CH_2)$	50	6	$CH_3(CH_2)_7Si(OC_2H_5)_3$ (85)	748
$PtCl_2(Ph_2PCH_2PPh_2)$	50	10	$CH_3(CH_2)_7Si(OC_2H_5)_3$ (82)	PI-031
$PtCl_2(PhCN)_2$	50	2	$CH_3(CH_2)_7Si(OC_2H_5)_3$ (76)	PI-031
$SiO_2\text{-}Pt(C_3H_5)$	60	6	$CH_3(CH_2)_7Si(OC_2H_5)_3$ (65)	230
$Rh(acac)_3$	60	20	$CH_3(CH_2)_7Si(OC_2H_5)_3$ (82)	274
$Rh(acac)_3 + AlEt_3$	60	20	$CH_3(CH_2)_7Si(OC_2H_5)_3$ (60)	271
$Rh_2(OAc)_4$	100	8	$CH_3(CH_2)_7Si(OC_2H_5)_3$ (98)	274
$RhCl(\overline{CNMeCH_2CH_2NMe})(cod)$	100	8	$CH_3(CH_2)_7Si(OC_2H_5)_3$ (62)	496
$RhCl(\overline{CNPhCH_2CH_2NPh})(cod)$	100	8	$CH_3(CH_2)_7Si(OC_2H_5)_3$ (92)	496
$RhCl(CO)(PPh_3)_2$	20	6 (b)	$CH_3(CH_2)_7Si(OC_2H_5)_3$ (4)	202
$RhCl(CO)(PPh_3)_2$ + t-BuOOH	20	6 (b)	$CH_3(CH_2)_7Si(OC_2H_5)_3$ (50-75)	202
$RhCl_2[P(MeOC_6H_4)_3]_2$	100	8	$CH_3(CH_2)_7Si(OC_2H_5)_3$ (81)	503
$RhCl_2[P(c\text{-}Hex)_3]_2$	100	8	$CH_3(CH_2)_7Si(OC_2H_5)_3$ (86)	503
$RhH(PPh_3)_4$	80	1	$CH_3(CH_2)_7Si(OC_2H_5)_3$ (70)	J -019

	SiO_2-$\overline{N(CH_2)_4}$/$RhCl(PPh_3)_3$	100	4	$CH_3(CH_2)_7Si(OC_2H_5)_3$ (20)	746
	SiO_2-$\overline{N(CH_2)_5}$/$RhCl(PPh_3)_3$	100	4	$CH_3(CH_2)_7Si(OC_2H_5)_3$ (8)	746
	SiO_2-$\overline{NCH_2CH_2OCH_2CH_2}$/$RhCl(PPh_3)_3$	100	4	$CH_3(CH_2)_7Si(OC_2H_5)_3$ (7)	746
	SiO_2-NEt_2/$RhCl(PPh_3)_3$	100	4	$CH_3(CH_2)_7Si(OC_2H_5)_3$ (30)	746
	SiO_2-NPh_2/$RhCl(PPh_3)_3$	100	4	$CH_3(CH_2)_7Si(OC_2H_5)_3$ (35)	746
	$[RhCl(c\text{-}C_8H_{14})_2]_2$	80	2	$CH_3(CH_2)_7Si(OC_2H_5)_3$ (78)	743
	[RhCl(cod)] + diphos	80	0.67	$CH_3(CH_2)_7Si(OC_2H_5)_3$ (-)	J -020
	$[RhCl(cod)]_2$ + DIPPC	20	-	$CH_3(CH_2)_7Si(OC_2H_5)_3$ (40)	884
	$[RhCl(cod)]_2$ + XNC	20	24	$CH_3(CH_2)_7Si(OC_2H_5)_3$ (40)	1
	$[RhCl\{P(OC_6H_4Me)_3\}_2]_n$	25	6	$CH_3(CH_2)_7Si(OC_2H_5)_3$ (67)	229
	$[RhCl\{P(OMe)_3\}_2]_n$	100	2	$CH_3(CH_2)_7Si(OC_2H_5)_3$ (37)	229
	$[RhCl\{P(OMe)_3\}_2]_n$	25	6	$CH_3(CH_2)_7Si(OC_2H_5)_3$ (36)	229
	$[RhH\{P(OMe)_3\}_2]_3$	100	2	$CH_3(CH_2)_7Si(OC_2H_5)_3$ (6)	229
	siloxane-$N(All)_2$/$[RhCl(c\text{-}C_8H_{14})_2]_2$	80	2	$CH_3(CH_2)_7Si(OC_2H_5)_3$ (64-72)	743
	$RuCl_2(PPh_3)_3$	120	6	$CH_3(CH_2)_7Si(OC_2H_5)_3$ (72)	733
	$RuCl_2(PPh_3)_3$	reflux	5	$CH_3(CH_2)_7Si(OC_2H_5)_3$ (67)	Pl-001
	$RuCl_3(PPh_3)_3$	120	6	$CH_3(CH_2)_7Si(OC_2H_5)_3$ (42)	Pl-015
	$RuCl_3(PPh_3)_3$	120	6	$CH_3(CH_2)_7Si(OC_2H_5)_3$ (42)	Pl-013
C_9H_{10}	$C_6H_5CH_2CH=CH_2$				
	$Rh(acac)_3$ + $AlEt_3$	60	-	$C_6H_5(CH_2)_3Si(OC_2H_5)_3$ (38) $C_6H_5CH_2CH(CH_3)Si(OC_2H_5)_3$ (6) $C_6H_5CH_2CH=CHSi(OC_2H_5)_3$ (7)	271
C_9H_{10}	p-$CH_3C_6H_4CH=CH_2$				
	$Pt[(ViMe_2Si)_2O]_2$	-	-	p-$CH_3C_6H_4CH_2CH_2Si(OC_2H_5)_3$ + p-$CH_3C_6H_4CH(CH_3)Si(OC_2H_5)_3$ (kin.)	669
C_9H_{18}	$CH_3(CH_2)_6CH=CH_2$				
	$Co_2(CO)_8$ + PhC≡CPh	25-35	0.33	$C_9H_{19}Si(OC_2H_5)_3$ (98)	S -038
$C_{10}H_{20}$	$CH_3(CH_2)_7CH=CH_2$				
	support/Pd complex	80	2	$C_{10}H_{21}Si(OC_2H_5)_3$ (26)	204

H_2PtCl_6 (i-PrOH)	80	2	$C_{10}H_{21}Si(OC_2H_5)_3$ (40)	204
$Pt(PPh_3)_4$	-	-	$C_{10}H_{21}Si(OC_2H_5)_3$ (85)	243
polymer-CN/$Pt(PPh_3)_4$	100	3	$C_{10}H_{21}Si(OC_2H_5)_3$ (52)	C -013
support/Pt complex	80	2	$C_{10}H_{21}Si(OC_2H_5)_3$ (0-20)	204
$(RhL_2PPh_3)_2$	70-80	1	$C_{10}H_{21}Si(OC_2H_5)_3$ (81)	886
$(RhL_2PPh_3)_2$	70-80	1	$C_{10}H_{21}Si(OC_2H_5)_3$ (81)	S -003
$[RhCl(c\text{-}C_8H_{14})_2]_2$	80	2	$CH_3(CH_2)_9Si(OC_2H_5)_3$ (77)	743
siloxane-$N(All)_2$/$[RhCl(c\text{-}C_8H_{14})_2]_2$	80	2	$CH_3(CH_2)_9Si(OC_2H_5)_3$ (68-75)	743
$C_{12}H_{24}$ $CH_3(CH_2)_9CH=CH_2$				
$RuCl_2(PPh_3)_3$	120	6	$CH_3(CH_2)_{11}Si(OC_2H_5)_3$ (15)	733
$C_{14}H_{28}$ $CH_3(CH_2)_{11}CH=CH_2$				
$RhCl(PPh_3)_3$	reflux	2 (b)	$CH_3(CH_2)_{13}Si(OC_2H_5)_3$ (63)	G -011
$RhCl(PPh_3)_3$	reflux	2 (b)	$CH_3(CH_2)_{13}Si(OC_2H_5)_3$ (63)	N -005
$C_{18}H_{36}$ $CH_3(CH_2)_{15}CH=CH_2$				
H_2PtCl_6 (i-PrOH)	80	6	$CH_3(CH_2)_{17}Si(OC_2H_5)_3$ (30)	748
$Pt(PPh_3)_2(CH_2=CH_2)$	50	2	$CH_3(CH_2)_{17}Si(OC_2H_5)_3$ (83)	Pl-030
$Pt(PPh_3)_2(CH_2=CH_2)$	80	6	$CH_3(CH_2)_{17}Si(OC_2H_5)_3$ (83)	748
$Pt(PPh_3)_4$	80	6	$CH_3(CH_2)_{17}Si(OC_2H_5)_3$ (60)	748
$PtCl_2(PPh_3)_2$	80	6	$CH_3(CH_2)_{17}Si(OC_2H_5)_3$ (87)	Pl-031
$PtCl_2(PhCN)_2$	80	6	$CH_3(CH_2)_{17}Si(OC_2H_5)_3$ (65)	Pl-031
$C_{20}H_{40}$ $CH_3(CH_2)_{17}CH=CH_2$				
$[RhCl(c\text{-}C_8H_{14})_2]_2$	80	2	$CH_3(CH_2)_{19}Si(OC_2H_5)_3$ (72)	743
siloxane-$N(All)_2$/$[RhCl(c\text{-}C_8H_{14})_2]_2$	80	2	$CH_3(CH_2)_{19}Si(OC_2H_5)_3$ (64-69)	743

Table 9.2

= $(C_2H_5O)_3SiH$ = =1=

C_3H_4	$CH_2=C=CH_2$				
	$Rh(acac)(CO)_2$	-	-	$CH_2=CHCH_2Si(OC_2H_5)_3$ + $(C_2H_5O)_3Si(CH_2)_3Si(OC_2H_5)_3$ + $(C_2H_5O)_3SiCH_2CH(CH_3)Si(OC_2H_5)_3$ + $(CH_3)_2C[Si(OC_2H_5)_3]_2$ (-)	236
	$Rh(acac)(CO)_2$	60	2	$CH_2=CHCH_2Si(OC_2H_5)_3$ (98)	S -014
	$Rh(acac)(CO)_2$	70	2	$(C_2H_5O)_3Si(CH_2)_3Si(OC_2H_5)_3$ + $(C_2H_5O)_3SiCH_2CH(CH_3)Si(OC_2H_5)_3$ + $(CH_3)_2C[Si(OC_2H_5)_3]_2$ (79-94)	S -034
	$Rh(acac)(CO)_2$	r.t.	1.5-2	$CH_2=CHCH_2Si(OC_2H_5)_3$ (100)	S -014
C_4H_6	C_4 fraction				
	$Ni(cod)_2$	120	3	$C_4H_7Si(OC_2H_5)_3$ (58-61)	C -018
	$Ni[CH_3CH(OH)COO]_2 + PPh_3$	125	4	$C_4H_7Si(OC_2H_5)_3$ (52)	C -018
	$PdCl_2(\pi\text{-All})_2$	120	3	$C_4H_7Si(OC_2H_5)_3$ (18)	C -017
	$PdCl_2(\pi\text{-All})_2$	80	2	$C_4H_7Si(OC_2H_5)_3$ (10)	C -017
	$PdCl_2(PPh_3)_2$	125	4	$C_4H_7Si(OC_2H_5)_3$ (8)	C -017
	$PdCl_2(PhCN)_2$	80	2	$C_4H_7Si(OC_2H_5)_3$ (12)	C -017
C_4H_6	$CH_2=CHCH=CH_2$				
	$Co(acac)_2$	120	3	$CH_2=CHCH_2CH_2Si(OC_2H_5)_3$ (10) cis-$CH_3CH=CHCH_2Si(OC_2H_5)_3$ (19) trans-$CH_3CH=CHCH_2Si(OC_2H_5)_3$ (13)	C -016
	$Co(acac)_2$	150	2	$CH_2=CHCH_2CH_2Si(OC_2H_5)_3$ (16) cis-$CH_3CH=CHCH_2Si(OC_2H_5)_3$ (28) trans-$CH_3CH=CHCH_2Si(OC_2H_5)_3$ (24)	C -016
	$Co(acac)_2$	80	8	$CH_2=CHCH_2CH_2Si(OC_2H_5)_3$ (8) cis-$CH_3CH=CHCH_2Si(OC_2H_5)_3$ (16) trans-$CH_3CH=CHCH_2Si(OC_2H_5)_3$ (13)	C -016
	$Co(acac)_2 + NaAlH_2(OC_2H_4OEt)_2$	120	3 (b)	$CH_2=CHCH_2CH_2Si(OC_2H_5)_3$ (2) cis-$CH_3CH=CHCH_2Si(OC_2H_5)_3$ (23) trans-$CH_3CH=CHCH_2Si(OC_2H_5)_3$ (27)	C -016
	$Co(acac)_2 + NaAlH_2(OC_2H_4OEt)_2$	120	3 (THF)	$CH_2=CHCH_2CH_2Si(OC_2H_5)_3$ (16) cis-$CH_3CH=CHCH_2Si(OC_2H_5)_3$ (22) trans-$CH_3CH=CHCH_2Si(OC_2H_5)_3$ (8)	C -016
	$Co(acac)_2 + NaAlH_2(OC_2H_4OEt)_2$	120	3 (ht)	$CH_2=CHCH_2CH_2Si(OC_2H_5)_3$ (10) cis-$CH_3CH=CHCH_2Si(OC_2H_5)_3$ (19) trans-$CH_3CH=CHCH_2Si(OC_2H_5)_3$ (27)	C -016

Table 9.2
= $(C_2H_5O)_3SiH$ =

=2=

$Co(acac)_2$ + $NaAlH_2(OC_2H_4OEt)_2$	120	3	$CH_2=CHCH_2CH_2Si(OC_2H_5)_3$ (17) cis-$CH_3CH=CHCH_2Si(OC_2H_5)_3$ (41) trans-$CH_3CH=CHCH_2Si(OC_2H_5)_3$ (14)	C -016
$Co(acac)_2$ + $NaAlH_2(OC_2H_4OEt)_2$	120	3	$CH_2=CHCH_2CH_2Si(OC_2H_5)_3$ (8) cis-$CH_3CH=CHCH_2Si(OC_2H_5)_3$ (33) trans-$CH_3CH=CHCH_2Si(OC_2H_5)_3$ (17)	C -016
$Co(acac)_2$ + $NaAlH_2(OC_2H_4OEt)_2$ + $AsPh_3$	120	3	$CH_2=CHCH_2CH_2Si(OC_2H_5)_3$ (3) cis-$CH_3CH=CHCH_2Si(OC_2H_5)_3$ (33) trans-$CH_3CH=CHCH_2Si(OC_2H_5)_3$ (22)	C -016
$Co(acac)_2$ + $NaAlH_2(OC_2H_4OEt)_2$ + PBu_3	120	3	$CH_2=CHCH_2CH_2Si(OC_2H_5)_3$ (2) cis-$CH_3CH=CHCH_2Si(OC_2H_5)_3$ (9) trans-$CH_3CH=CHCH_2Si(OC_2H_5)_3$ (7)	C -016
$Co(acac)_2$ + $NaAlH_2(OC_2H_4OEt)_2$ + PPh_3	120	3	$CH_2=CHCH_2CH_2Si(OC_2H_5)_3$ (2) cis-$CH_3CH=CHCH_2Si(OC_2H_5)_3$ (28) trans-$CH_3CH=CHCH_2Si(OC_2H_5)_3$ (27)	C -016
$Co(acac)_2$ + $NaAlH_2(OC_2H_4OMe)_2$	120	3	$CH_2=CHCH_2CH_2Si(OC_2H_5)_3$ (5) $CH_3CH=CHCH_2Si(OC_2H_5)_3$ (50) $CH_3CH=CHCH_2CH_2CH=CHCH_2Si(OC_2H_5)_3$ (5)	208
$Co(acac)_2$ + $NaAlH_2(OC_2H_4OMe)_2$ + PR_3	120	3	$CH_2=CHCH_2CH_2Si(OC_2H_5)_3$ (1-33) $CH_3CH=CHCH_2Si(OC_2H_5)_3$ (0-50) $CH_3CH=CHCH_2CH_2CH=CHCH_2Si(OC_2H_5)_3$ (0-6)	208
$Co(acac)_2$+$NaAlH_2(OC_2H_4OEt)_2$+P(c-Hex)$_3$	120	3	$CH_2=CHCH_2CH_2Si(OC_2H_5)_3$ (3) cis-$CH_3CH=CHCH_2Si(OC_2H_5)_3$ (33) trans-$CH_3CH=CHCH_2Si(OC_2H_5)_3$ (22)	C -016
$CoBr_2$ + $NaAlH_2(OC_2H_4OEt)_2$	120	3	$CH_2=CHCH_2CH_2Si(OC_2H_5)_3$ (23) cis-$CH_3CH=CHCH_2Si(OC_2H_5)_3$ (28) trans-$CH_3CH=CHCH_2Si(OC_2H_5)_3$ (17)	C -016
$CoCl_2$ + $NaAlH_2(OC_2H_4OEt)_2$	120	3	$CH_2=CHCH_2CH_2Si(OC_2H_5)_3$ (23) cis-$CH_3CH=CHCH_2Si(OC_2H_5)_3$ (24) trans-$CH_3CH=CHCH_2Si(OC_2H_5)_3$ (22)	C -016
$CoCl_2(PPh_3)_2$ + $NaAlH_2(OC_2H_4OEt)_2$	120	3	$CH_2=CHCH_2CH_2Si(OC_2H_5)_3$ (12) cis-$CH_3CH=CHCH_2Si(OC_2H_5)_3$ (19) trans-$CH_3CH=CHCH_2Si(OC_2H_5)_3$ (15)	C -016
CoX_2 + $NaAlH_2(OC_2H_4OMe)_2$	120	3	$CH_2=CHCH_2CH_2Si(OC_2H_5)_3$ (1-11) $CH_3CH=CHCH_2Si(OC_2H_5)_3$ (6-76) $CH_3CH=CHCH_2CH_2CH=CHCH_2Si(OC_2H_5)_3$ (0-5)	208
$Co[CH_3CH(OH)COO]_2$ +$NaAlH_2(OC_2H_4OEt)_2$	120	3	$CH_2=CHCH_2CH_2Si(OC_2H_5)_3$ (3) cis-$CH_3CH=CHCH_2Si(OC_2H_5)_3$ (12) trans-$CH_3CH=CHCH_2Si(OC_2H_5)_3$ (6)	C -016
$Cr(CO)_6$ + UV	30	several days	$CH_3CH=CHCH_2Si(OC_2H_5)_3$ (-)	1293

Table 9.2

= $(C_2H_5O)_3SiH$ = =3=

$Ni(PPh_3)_4$ + $NaAlH_2(OC_2H_4OMe)_2$	70	3	$CH_3CH{=}CHCH_2Si(OC_2H_5)_3$ (91)	207
$Ni(PPh_3)_4$ + $NaAlH_2(OC_2H_4OMe)_2$ + PPh_3	70	3	$CH_3CH{=}CHCH_2Si(OC_2H_5)_3$ (79)	207
$Ni(acac)_2$	120	3	$CH_3CH{=}CHCH_2Si(OC_2H_5)_3$ (92) $CH_3CH{=}CHCH_2CH_2CH{=}CHCH_2Si(OC_2H_5)_3$ (5)	205
$Ni(acac)_2$ + $NaAlH_2(OC_2H_4OMe)_2$	40-70	3	$CH_3CH{=}CHCH_2Si(OC_2H_5)_3$ (87)	207
$Ni(acac)_2$ + $NaAlH_2(OC_2H_4OMe)_2$ + PPh_3	40-70	3	$CH_3CH{=}CHCH_2Si(OC_2H_5)_3$ (6-70)	207
$Ni(acac)_2$ + PPh_3	120	3	$CH_3CH{=}CHCH_2Si(OC_2H_5)_3$ (61) $CH_3CH{=}CHCH_2CH_2CH{=}CHCH_2Si(OC_2H_5)_3$ (5)	205
$Ni(acac)_2$ + PPh_3	125-150	2-4	$CH_3CH{=}CHCH_2Si(OC_2H_5)_3$ (34-86)	210
$Ni(cod)_2$	40-70	3	$CH_3CH{=}CHCH_2Si(OC_2H_5)_3$ (68-78)	207
$Ni(cod)_2$	80	1.5 (b)	$CH_3CH{=}CHCH_2Si(OC_2H_5)_3$ (96) $CH_3CH{=}CHCH_2CH_2CH{=}CHCH_2Si(OC_2H_5)_3$ (6)	205
$Ni(cod)_2$ + PPh_3	40-70	3	$CH_3CH{=}CHCH_2Si(OC_2H_5)_3$ (20-80)	207
$Ni(cpd)_2(CO)_2$	80	2	$C_4H_7Si(OC_2H_5)_3$ (80)	D -070
$NiBr_2$	120	3	$CH_3CH{=}CHCH_2Si(OC_2H_5)_3$ (28)	205
$NiBr_2$ + $Na[AlH_2(OC_2H_4OMe)_2]$	70	3	$CH_3CH{=}CHCH_2Si(OC_2H_5)_3$ (85)	207
$NiBr_2$ + $Na[AlH_2(OC_2H_4OMe)_2]$ + PPh_3	70	3	$CH_3CH{=}CHCH_2Si(OC_2H_5)_3$ (61)	207
$NiBr_2$+ PPh_3	120	3	$CH_3CH{=}CHCH_2Si(OC_2H_5)_3$ (54) $CH_3CH{=}CHCH_2CH_2CH{=}CHCH_2Si(OC_2H_5)_3$ (4)	205
$NiBr_2(PPh_3)_2$	120	3	$C_4H_7Si(OC_2H_5)_3$ (52) $C_8H_{13}Si(OC_2H_5)_3$ (18)	D -070
$NiCl_2$	120	3	$CH_3CH{=}CHCH_2Si(OC_2H_5)_3$ (61) $CH_3CH{=}CHCH_2CH_2CH{=}CHCH_2Si(OC_2H_5)_3$ (4)	205
$NiCl_2$+ Et_3N	120	3	$C_4H_7Si(OC_2H_5)_3$ (28) $C_8H_{13}Si(OC_2H_5)_3$ (8)	D -070
$NiCl_2$+ PPh_3	120	3	$CH_3CH{=}CHCH_2Si(OC_2H_5)_3$ (70) $CH_3CH{=}CHCH_2CH_2CH{=}CHCH_2Si(OC_2H_5)_3$ (6)	205
$NiCl_2(PPh_3)_2$	120	3	$C_4H_7Si(OC_2H_5)_3$ (46) $C_8H_{13}Si(OC_2H_5)_3$ (17)	D -070
$NiCl_2(PPh_3)_2$ (PhMe)	100	6	$C_4H_7Si(OC_2H_5)_3$ (57) $C_8H_{13}Si(OC_2H_5)_3$ (2)	C -010
$NiCl_2(PPh_3)_2$ + C_5H_5N	120	3	$C_4H_7Si(OC_2H_5)_3$ (54) $C_8H_{13}Si(OC_2H_5)_3$ (14)	D -070

Table 9.2
= $(C_2H_5O)_3SiH$ =

=4=

$NiCl_2(PPh_3)_2$ + Et_3N	120	3	$C_4H_7Si(OC_2H_5)_3$ (62) $C_8H_{13}Si(OC_2H_5)_3$ (11)	D -070
$NiCl_2(PPh_3)_2$ + $Na[AlH_2(OC_2H_4OMe)_2]$	100	6	$C_4H_7Si(OC_2H_5)_3$ (57) $C_8H_{13}Si(OC_2H_5)_3$ (2)	C -010
NiF_2	120	3	$CH_3CH{=}CHCH_2Si(OC_2H_5)_3$ (28)	205
NiF_2+ PPh_3	120	3	$CH_3CH{=}CHCH_2Si(OC_2H_5)_3$ (77) $CH_3CH{=}CHCH_2CH_2CH{=}CHCH_2Si(OC_2H_5)_3$ (1)	205
NiI_2+ PPh_3	120	3	$CH_3CH{=}CHCH_2Si(OC_2H_5)_3$ (60) $CH_3CH{=}CHCH_2CH_2CH{=}CHCH_2Si(OC_2H_5)_3$ (5)	205
$Ni[CH_3CH(OH)COO]_2$	120	3	$CH_3CH{=}CHCH_2Si(OC_2H_5)_3$ (69) $CH_3CH{=}CHCH_2CH_2CH{=}CHCH_2Si(OC_2H_5)_3$ (6)	205
$Ni[CH_3CH(OH)COO]_2$ + C_5H_5N	120	3	$C_4H_7Si(OC_2H_5)_3$ (68) $C_8H_{13}Si(OC_2H_5)_3$ (4)	D -070
$Ni[CH_3CH(OH)COO]_2$ + C_5H_5N	90	3	$C_4H_7Si(OC_2H_5)_3$ (52) $C_8H_{13}Si(OC_2H_5)_3$ (13)	D -070
$Ni[CH_3CH(OH)COO]_2$ + PBu_3	120	3	$C_4H_7Si(OC_2H_5)_3$ (68) $C_8H_{13}Si(OC_2H_5)_3$ (3)	D -070
$Ni[CH_3CH(OH)COO]_2$ + PBu_3	120	3	$CH_3CH{=}CHCH_2Si(OC_2H_5)_3$ (90) $CH_3CH{=}CHCH_2CH_2CH{=}CHCH_2Si(OC_2H_5)_3$ (2)	205
$Ni[CH_3CH(OH)COO]_2$ + PPh_3	120	3	$C_4H_7Si(OC_2H_5)_3$ (28) $C_8H_{13}Si(OC_2H_5)_3$ (2)	D -070
$Ni[CH_3CH(OH)COO]_2$ + PPh_3	120	3	$CH_3CH{=}CHCH_2Si(OC_2H_5)_3$ (76) $CH_3CH{=}CHCH_2CH_2CH{=}CHCH_2Si(OC_2H_5)_3$ (2)	207
$Ni[CH_3CH(OH)COO]_2$ + PPh_3 ($CH_2{=}CHBu$)	150	2	$C_4H_7Si(OC_2H_5)_3$ (84)	C -018
$Ni[CH_3CH(OH)COO]_2$ + PPh_3 ($CH_2{=}CHBu$)	150	2	cis-$CH_3CH{=}CHCH_2Si(OC_2H_5)_3$ (72)	C -018
SiO_2-CN/$NiCl_2$(EtOH) + PPh_3	120	3	$C_4H_7Si(OC_2H_5)_3$ (79) $C_8H_{13}Si(OC_2H_5)_3$ (4)	D -070
Pd($\overline{COCH{=}CHCOO}$)($PPh_3)_2$	-	-	$CH_3CH{=}CHCH_2CH_2CH{=}CHCH_2Si(OC_2H_5)_3$ (-)	J -004
$PdCl_2(\pi\text{-All})_2$	120	3 (t)	$C_4H_7Si(OC_2H_5)_3$ (17)	C -017
$PdCl_2(\pi\text{-}C_3H_4Me)_2$	120	3 (t)	$C_4H_7Si(OC_2H_5)_3$ (18)	C -017
$PdCl_2(PPh_3)_2$	100	4 (b)	$CH_3CH{=}CHCH_2Si(OC_2H_5)_3$ (63)	644
$PdCl_2(PPh_3)_2$	70	3	$CH_3CH{=}CHCH_2Si(OC_2H_5)_3$ (12)	210
$PdCl_2(PhCN)_2$	100	3	$CH_3CH{=}CHCH_2Si(OC_2H_5)_3$ (21) $CH_3CH{=}CHCH_2CH_2CH{=}CHCH_2Si(OC_2H_5)_3$ (56)	U -080

Table 9.2

= $(C_2H_5O)_3SiH$ = =5=

$PdCl_2(PhCN)_2$	100	3	$CH_3CH=CHCH_2Si(OC_2H_5)_3$ (21) $CH_3CH=CHCH_2CH_2CH=CHCH_2Si(OC_2H_5)_3$ (56)	C -001
$PdCl_2(PhCN)_2$	100	3	$CH_3CH=CHCH_2Si(OC_2H_5)_3$ (21) $CH_3CH=CHCH_2CH_2CH=CHCH_2Si(OC_2H_5)_3$ (56)	D -060
$PdCl_2(PhCN)_2$	130	3	$CH_3CH=CHCH_2Si(OC_2H_5)_3$ (29) $CH_3CH=CHCH_2CH_2CH=CHCH_2Si(OC_2H_5)_3$ (51)	C -001
$PdCl_2(PhCN)_2$	130	3	$CH_3CH=CHCH_2Si(OC_2H_5)_3$ (29) $CH_3CH=CHCH_2CH_2CH=CHCH_2Si(OC_2H_5)_3$ (51)	U -080
$PdCl_2(PhCN)_2$	130	3	$CH_3CH=CHCH_2Si(OC_2H_5)_3$ (29) $CH_3CH=CHCH_2CH_2CH=CHCH_2Si(OC_2H_5)_3$ (51)	D -060
$PdCl_2(PhCN)_2$	20	12	$CH_3CH=CHCH_2Si(OC_2H_5)_3$ (10) $CH_3CH=CHCH_2CH_2CH=CHCH_2Si(OC_2H_5)_3$ (32)	U -080
$PdCl_2(PhCN)_2$	20	12	$CH_3CH=CHCH_2Si(OC_2H_5)_3$ (10) $CH_3CH=CHCH_2CH_2CH=CHCH_2Si(OC_2H_5)_3$ (32)	D -060
$PdCl_2(PhCN)_2$	20	12	$CH_3CH=CHCH_2Si(OC_2H_5)_3$ (10) $CH_3CH=CHCH_2CH_2CH=CHCH_2Si(OC_2H_5)_3$ (32)	C -001
$PdCl_2(PhCN)_2$	22	24	$CH_3CH=CHCH_2CH_2CH=CHCH_2Si(OC_2H_5)_3$ (84)	644
$PdCl_2(PhCN)_2$	25-100	2-24 (THF)	$CH_3CH=CHCH_2Si(OC_2H_5)_3$ (4-13) $CH_3CH=CHCH_2CH_2CH=CHCH_2Si(OC_2H_5)_3$ (37-84)	1221
$PdCl_2(PhCN)_2$	50	5	$CH_3CH=CHCH_2Si(OC_2H_5)_3$ (28) $CH_3CH=CHCH_2CH_2CH=CHCH_2Si(OC_2H_5)_3$ (52)	C -001
$PdCl_2(PhCN)_2$	50	5	$CH_3CH=CHCH_2Si(OC_2H_5)_3$ (28) $CH_3CH=CHCH_2CH_2CH=CHCH_2Si(OC_2H_5)_3$ (52)	U -080
$PdCl_2(PhCN)_2$	50	5	$CH_3CH=CHCH_2Si(OC_2H_5)_3$ (28) $CH_3CH=CHCH_2CH_2CH=CHCH_2Si(OC_2H_5)_3$ (52)	D -060
$PdCl_2(PhCN)_2$	80	3	$CH_3CH=CHCH_2Si(OC_2H_5)_3$ (45) $CH_3CH=CHCH_2CH_2CH=CHCH_2Si(OC_2H_5)_3$ (19)	D -060
$PdCl_2(PhCN)_2$	80	3	$CH_3CH=CHCH_2Si(OC_2H_5)_3$ (45) $CH_3CH=CHCH_2CH_2CH=CHCH_2Si(OC_2H_5)_3$ (19)	C -001
$PdCl_2(PhCN)_2$	80	2	$CH_3CH=CHCH_2Si(OC_2H_5)_3$ (7)	210
$PdCl_2(PhCN)_2$	80	3	$CH_3CH=CHCH_2Si(OC_2H_5)_3$ (45) $CH_3CH=CHCH_2CH_2CH=CHCH_2Si(OC_2H_5)_3$ (19)	U -080
SDVB-CN/$PdCl_2$ (H_2O)	85	3	$CH_3CH=CHCH_2CH_2CH=CHCH_2Si(OC_2H_5)_3$ (32)	C -001
SDVB-NMe_2/$PdCl_2$ (H_2O)	until 80	2	$C_4H_7Si(OC_2H_5)_3$ (52)	U -080
SDVB-NMe_2/$PdCl_2$ (H_2O)	until 80	2	$C_4H_7Si(OC_2H_5)_3$ (52)	D -060

Table 9.2
= $(C_2H_5O)_3SiH$ =

=6=

SDVB-NMe_2/$PdCl_2$ (H_2O)	until 80	2	$C_4H_7Si(OC_2H_5)_3$ (52)	C -001
$[PdCl(\pi\text{-All})_2]_2$	22	24 (b)	$CH_3CH{=}CHCH_2Si(OC_2H_5)_3$ (-)	644
$[PdCl(\pi\text{-All})_2]_2$	80	2	$CH_3CH{=}CHCH_2Si(OC_2H_5)_3$ (45) $CH_3CH{=}CHCH_2CH_2CH{=}CHCH_2Si(OC_2H_5)_3$ (23)	D -060
$[PdCl(\pi\text{-All})_2]_2$	80	2	$CH_3CH{=}CHCH_2Si(OC_2H_5)_3$ (45) $CH_3CH{=}CHCH_2CH_2CH{=}CHCH_2Si(OC_2H_5)_3$ (23)	C -001
$[PdCl(\pi\text{-All})_2]_2$	80	2	$CH_3CH{=}CHCH_2Si(OC_2H_5)_3$ (45) $CH_3CH{=}CHCH_2CH_2CH{=}CHCH_2Si(OC_2H_5)_3$ (23)	U -080
$[PdCl(\pi\text{-All})_2]_2$	80-150	2-3	$CH_3CH{=}CHCH_2Si(OC_2H_5)_3$ (13-22)	210
polymer/Pd complex	100	2	cis-$CH_3CH{=}CHCH_2Si(OC_2H_5)_3$ (30) trans-$CH_3CH{=}CHCH_2Si(OC_2H_5)_3$ (61)	1220
H_2PtCl_6 (i-PrOH)	150	8	$C_4H_7Si(OC_2H_5)_3$ (52)	1267
polymer/Pt complex	100	2	cis-$CH_2{=}CHCH_2CH_2Si(OC_2H_5)_3$ (28) trans-$CH_3CH{=}CHCH_2Si(OC_2H_5)_3$ (16) $(C_2H_5O)_3SiC_4H_8Si(OC_2H_5)_3$ (8)	1220
C/$RhCl_3$ $3H_2O$ (EtOH + c-$C_{12}H_{18}$)	80	3	$C_4H_7Si(OC_2H_5)_3$ (82)	Dd-003
C-N/$RhCl_3$ + c-$C_{12}H_{18}$	80	3	$C_4H_7Si(OC_2H_5)_3$ (82)	C -030
$Rh(\pi\text{-All})(PPh_3)_2$	100	0.5	$CH_2{=}CHCH_2CH_2Si(OC_2H_5)_3$ + $CH_3CH{=}CHCH_2Si(OC_2H_5)_3$ + $CH_3CH{=}CHCH_2CH_2CH{=}CHCH_2Si(OC_2H_5)_3$ (55)	982
$Rh(acac)(CO)_2$	80	2	$CH_3CH{=}CHCH_2Si(OC_2H_5)_3$ (67)	C -009
$RhBr(PPh_3)_3$	100	0.5	$CH_2{=}CHCH_2CH_2Si(OC_2H_5)_3$ + $CH_3CH{=}CHCH_2Si(OC_2H_5)_3$ + $CH_3CH{=}CHCH_2CH_2CH{=}CHCH_2Si(OC_2H_5)_3$ (17)	982
$RhCl(CO)(PPh_3)_2$	100	0.5	$CH_2{=}CHCH_2CH_2Si(OC_2H_5)_3$ + $CH_3CH{=}CHCH_2Si(OC_2H_5)_3$ + $CH_3CH{=}CHCH_2CH_2CH{=}CHCH_2Si(OC_2H_5)_3$ (92)	982
$RhCl(PPh_3)_3$	100	0.5	$CH_2{=}CHCH_2CH_2Si(OC_2H_5)_3$ + $CH_3CH{=}CHCH_2Si(OC_2H_5)_3$ + $CH_3CH{=}CHCH_2CH_2CH{=}CHCH_2Si(OC_2H_5)_3$ (36)	982
$RhH(CO)(PPh_3)_3$	100	0.5	$CH_2{=}CHCH_2CH_2Si(OC_2H_5)_3$ + $CH_3CH{=}CHCH_2Si(OC_2H_5)_3$ + $CH_3CH{=}CHCH_2CH_2CH{=}CHCH_2Si(OC_2H_5)_3$ (96)	982
$RhMe(PPh_3)_3$	100	0.5	$CH_2{=}CHCH_2CH_2Si(OC_2H_5)_3$ + $CH_3CH{=}CHCH_2Si(OC_2H_5)_3$ + $CH_3CH{=}CHCH_2CH_2CH{=}CHCH_2Si(OC_2H_5)_3$ (52)	982
SDVB-CH_2Cl/$RhCl(CO)(PPh_3)_2$	80	3	$C_4H_7Si(OC_2H_5)_3$ (72)	C -022
SDVB-NMe/$[RhCl(CO)_2]_2$	100	4	cis-$CH_3CH{=}CHCH_2Si(OC_2H_5)_3$ + trans-$CH_3CH{=}CHCH_2Si(OC_2H_5)_3$ (-)	277
SDVB-PPh_2/$RhCl_3$	60	2 (dcm)	$CH_3CH{=}CHCH_2Si(OC_2H_5)_3$ (-)	D -056

Table 9.2

= $(C_2H_5O)_3SiH$ = =7=

	$[CH_2CH\text{-}p\text{-}C_6H_4PPh_2RhCl(PPh_3)_2]_n$	60	2	$CH_3CH{=}CHCH_2Si(OC_2H_5)_3$ (94)	C -008
	polymer/Rh complex	100	2	cis-$CH_3CH{=}CHCH_2Si(OC_2H_5)_3$ (68) trans-$CH_3CH{=}CHCH_2Si(OC_2H_5)_3$ (12)	1220
	polyolefine/$RhCl_3$ (THF)	100	3	$CH_3CH{=}CHCH_2Si(OC_2H_5)_3$ (95)	S -029
C_5H_6	Fig.25.				
	$Co_2(CO)_8$	30-35	1	Fig.396. (93-95)	715
	$Rh(acac)(CO)_2$	20	20	Fig.397. (40)	878
	$Rh(acac)(CO)_2 + PPh_3$	-	-	Fig.398. (90)	878
	$Rh_4(CO)_{12}$	30-35	1	Fig.396. (93-95)	715
C_5H_8	$CH_2{=}C(CH_3)CH{=}CH_2$				
	Co complex	0-110	- (t)	$(C_2H_5O)_3SiCH_2C(CH_3){=}CHCH_3$ (13-28)	275
	$Co(acac)_3 + AlEt_3$	20	6	$(C_2H_5O)_3SiCH_2CH(CH_3)CH{=}CH_2$ (44)	662
	Fe	0-110	- (t)	$(C_2H_5O)_3SiCH_2C(CH_3){=}CHCH_3$ (trace-23)	275
	Ni	0	- (t)	$(C_2H_5O)_3SiCH_2C(CH_3){=}CHCH_3$ (79-100)	275
	$Ni(CO)_2(PPh_3)_2$	150	3	$(C_2H_5O)_3SiCH_2C(CH_3){=}CHCH_3$ (62) $(CH_3)_2C{=}CHCH_2Si(OC_2H_5)_3$ (38)	1222
	$Ni(PPh_3)_4$	150	3	$(C_2H_5O)_3SiCH_2C(CH_3){=}CHCH_3$ (64) $(CH_3)_2C{=}CHCH_2Si(OC_2H_5)_3$ (36)	1222
	$Ni(acac)_2 + AlEt_3$	20	2	$(C_2H_5O)_3SiCH_2C(CH_3){=}CHCH_3$ (94)	G -030
	$Ni(acac)_2 + AlEt_3$	20	2	$(C_2H_5O)_3SiCH_2C(CH_3){=}CHCH_3$ (94)	662
	$Ni(cod)_2$	150	3	$(C_2H_5O)_3SiCH_2C(CH_3){=}CHCH_3$ (65) $(CH_3)_2C{=}CHCH_2Si(OC_2H_5)_3$ (35)	1222
	$NiCl_2(PPh_3)_2$	150	3	$(C_2H_5O)_3SiCH_2C(CH_3){=}CHCH_3$ (21) $(CH_3)_2C{=}CHCH_2Si(OC_2H_5)_3$ (14)	1222
	$PdCl_2(PPh_3)_2$	100	6	$(C_2H_5O)_3SiCH_2C(CH_3){=}CHCH_3$ (4) $(CH_3)_2C{=}CHCH_2Si(OC_2H_5)_3$ (22)	1222
	$PdCl_2(PhCN)_2$	100	6	$(C_2H_5O)_3SiCH_2C(CH_3){=}CHCH_3$ (5) $(CH_3)_2C{=}CHCH_2Si(OC_2H_5)_3$ (35)	1222
	H_2PtCl_6	150	8	$CH_2{=}C(CH_3)CH_2CH_2Si(OC_2H_5)_3$ + $(C_2H_5O)_3SiCH_2C(CH_3){=}CHCH_3$ + $(CH_3)_2C{=}CHCH_2Si(OC_2H_5)_3$ (80)	1267
	H_2PtCl_6 (i-PrOH)	150	3	$(C_2H_5O)_3SiCH_2C(CH_3){=}CHCH_3$ (20) $(CH_3)_2C{=}CHCH_2Si(OC_2H_5)_3$ (68)	1222

Rh(acac)$(CO)_2$	r.t.	12	$C_5H_9Si(OC_2H_5)_3$ (82)	618
Rh(acac)$(CO)_2$	r.t.	12	$C_5H_9Si(OC_2H_5)_3$ (82)	S -017
$Rh(acac)_3$	60	12	$(C_2H_5O)_3SiCH_2C(CH_3)=CHCH_2Si(OC_2H_5)_3$ (80)	274
$Rh(acac)_3$ + $AlEt_3$	60	15	$CH_2=C(CH_3)CH_2CH_2Si(OC_2H_5)_3$ + $(C_2H_5O)_3SiCH_2CH(CH_3)CH=CH_2$ (6) $(CH_3)_2C=CHCH_2Si(OC_2H_5)_3$ (43)	271
Rh(cod)+ diphos	-	-	$C_5H_9Si(OC_2H_5)_3$ (-)	J -020
RhCl(CO)(1,4-diaza-1,3-diene)	r.t.	1 day (THF)	$(C_2H_5O)_3SiCH_2C(CH_3)=CHCH_3$ (87)	167
$RhCl(PPh_3)_3$	110	2	$(C_2H_5O)_3SiCH_2C(CH_3)=CHCH_3$ (24) $(CH_3)_2C=CHCH_2Si(OC_2H_5)_3$ (58)	851
$RhCl(PPh_3)_3$	110	2	$(C_2H_5O)_3SiCH_2C(CH_3)=CHCH_3$ (24) $(CH_3)_2C=CHCH_2Si(OC_2H_5)_3$ (58)	852
$RhH(PPh_3)_4$	100	-	$C_5H_9Si(OC_2H_5)_3$ (-)	605
$[RhCl(CO)(1\text{-aza-}1,3\text{-diene})]_2$	r.t.	1 day (THF)	$(C_2H_5O)_3SiCH_2C(CH_3)=CHCH_3$ (87)	167
RuHCl(cod)(1,4-diaza-1,3-diene)	50	150	$CH_2=C(CH_3)CH_2CH_2Si(OC_2H_5)_3$ (31-33) $(C_2H_5O)_3SiCH_2C(CH_3)=CHCH_3$ (15-22) $(CH_3)_2C=CHCH_2Si(OC_2H_5)_3$ (46-47) $(C_2H_5O)_3SiCH_2CH(CH_3)CH=CH_2$ (0-23)	167
RuHCl(cod)(1,4-diaza-1,3-diene)	r.t.	150 (dcm)	$CH_2=C(CH_3)CH_2CH_2Si(OC_2H_5)_3$ (6-15) $(C_2H_5O)_3SiCH_2C(CH_3)=CHCH_3$ (0-18) $(CH_3)_2C=CHCH_2Si(OC_2H_5)_3$ (35-82) $(C_2H_5O)_3SiCH_2CH(CH_3)CH=CH_2$ (14-54)	167
RuHCl(cod)(DAD)	r.t.	- (dcm)	$(CH_3)_2C=CHCH_2Si(OC_2H_5)_3$ (80)	1363
$[RuCl_2L]_2$	25	150 (dcm)	$CH_2=C(CH_3)CH_2CH_2Si(OC_2H_5)_3$ (50)	167
Ti + Et_2AlCl	0	- (t)	$(C_2H_5O)_3SiCH_2C(CH_3)=CHCH_3$ (trace)	275
Ti + Et_2AlCl	110	6 (t)	$(C_2H_5O)_3SiCH_2C(CH_3)=CHCH_3$ (trace)	275
C_5H_8 $CH_3CH=CHCH=CH_2$				
$Co(acac)_n$ + $AlEt_3$	r.t.	4	$CH_3CH_2CH=CHCH_2Si(OC_2H_5)_3$ (48)	663
$Ni(acac)_2$ + $AlEt_3$	20	3	$CH_3CH_2CH=CHCH_2Si(OC_2H_5)_3$ + $CH_3CH=CHCH(CH_3)Si(OC_2H_5)_3$ (96)	662
$Ni(acac)_2$ + $AlEt_3$	20	3	$CH_3CH_2CH=CHCH_2Si(OC_2H_5)_3$ + $CH_3CH=CHCH(CH_3)Si(OC_2H_5)_3$ (96)	G -030
$Ni(acac)_2$ + $AlEt_3$	20	-	$CH_3CH_2CH=CHCH_2Si(OC_2H_5)_3$ + $CH_3CH=CHCH(CH_3)Si(OC_2H_5)_3$ (85)	273

	$Ni(acac)_2 + AlEt_3$	r.t.	2-3	$CH_3CH_2CH=CHCH_2Si(OC_2H_5)_3$ + $CH_3CH=CHCH(CH_3)Si(OC_2H_5)_3$ (91-96)	663
	$Ni(acac)_2$ + BuMgBr	20	3	$CH_3CH_2CH=CHCH_2Si(OC_2H_5)_3$ + $CH_3CH=CHCH(CH_3)Si(OC_2H_5)_3$ (96)	G -030
	$Ni(acac)_2$ + BuMgBr	20	3	$CH_3CH_2CH=CHCH_2Si(OC_2H_5)_3$ + $CH_3CH=CHCH(CH_3)Si(OC_2H_5)_3$ (96)	662
	$Ni(acac)_2 + Et_2AlOEt$	20	3	$CH_3CH_2CH=CHCH_2Si(OC_2H_5)_3$ + $CH_3CH=CHCH(CH_3)Si(OC_2H_5)_3$ (91)	662
	$Ni(acac)_2 + Et_2AlOEt$	20	3	$CH_3CH_2CH=CHCH_2Si(OC_2H_5)_3$ + $CH_3CH=CHCH(CH_3)Si(OC_2H_5)_3$ (91)	G -030
	$NiCl_2(PR_3)_2$	20	4 (t)	$CH_3CH_2CH=CHCH_2Si(OC_2H_5)_3$ + $CH_3CH=CHCH(CH_3)Si(OC_2H_5)_3$ (45-77)	273
	H_2PtCl_6 (i-PrOH)	200	8	$CH_3CH_2CH=CHCH_2Si(OC_2H_5)_3$ (17)	1267
C_6H_8	Fig.30.				
	-	60	12	Fig.399. (69)	274
	$Co_2(CO)_8$	20	12	Fig.399. (88)	272
	$Ni(acac)_2 + Et_2AlOEt$	20	8	Fig.399. (97)	273
C_6H_8	Fig.33.				
	$Co_2(CO)_8$	20	12	Fig.400. (48)	272
	$Ni(acac)_2 + Et_2AlOEt$	20	8	Fig.400. (52)	273
C_6H_{10}	$CH_2=C(CH_3)C(CH_3)=CH_2$				
	$Cr(CO)_5$(DMPIC)	100	24	$(CH_3)_2C=C(CH_3)CH_2Si(OC_2H_5)_3$ (51) $CH_2=C(CH_3)CH(CH_3)CH_2Si(OC_2H_5)_3$ (2)	884
	$Cr(CO)_6$	100	24	$(CH_3)_2C=C(CH_3)CH_2Si(OC_2H_5)_3$ (1)	884
	$Cr(CO)_6$	100	24	$(CH_3)_2C=C(CH_3)CH_2Si(OC_2H_5)_3$ (1) $CH_2=C(CH_3)CH(CH_3)CH_2Si(OC_2H_5)_3$ (-)	1
	$Cr(CO)_6$ + t-BuOOH	100	24 (b)	$(CH_3)_2C=C(CH_3)CH_2Si(OC_2H_5)_3$ (3)	202
	$Cr(XNC)(CO)_5$	100	24	$(CH_3)_2C=C(CH_3)CH_2Si(OC_2H_5)_3$ (51) $CH_2=C(CH_3)CH(CH_3)CH_2Si(OC_2H_5)_3$ (2)	1
	$[IrCl(c\text{-}C_8H_{14})_2]_2$	60	6	$(CH_3)_2C=C(CH_3)CH_2Si(OC_2H_5)_3$ (10) $CH_2=C(CH_3)CH(CH_3)CH_2Si(OC_2H_5)_3$ (6)	59
	$[IrCl(c\text{-}C_8H_{14})_2]_2 + PPh_3$	60	6	$(CH_3)_2C=C(CH_3)CH_2Si(OC_2H_5)_3$ (7) $CH_2=C(CH_3)CH(CH_3)CH_2Si(OC_2H_5)_3$ (5)	59

Table 9.2

= $(C_2H_5O)_3SiH$ = =10=

$Mo(CO)_4(DMPIC)_2$	100	24	$(CH_3)_2C=C(CH_3)CH_2Si(OC_2H_5)_3$ (57)	884
$Mo(CO)_5(DMPIC)$	100	24	$(CH_3)_2C=C(CH_3)CH_2Si(OC_2H_5)_3$ (73)	884
$Mo(CO)_6$	100	24	$(CH_3)_2C=C(CH_3)CH_2Si(OC_2H_5)_3$ + $CH_2=C(CH_3)CH(CH_3)CH_2Si(OC_2H_5)_3$ (71)	202
$Mo(CO)_6$	100	24	$(CH_3)_2C=C(CH_3)CH_2Si(OC_2H_5)_3$ (71)	1
$Mo(CO)_6$ + t-BuOOH	100	24	$(CH_3)_2C=C(CH_3)CH_2Si(OC_2H_5)_3$ + $CH_2=C(CH_3)CH(CH_3)CH_2Si(OC_2H_5)_3$ (72)	202
$Mo(XNC)_2(CO)_4$	100	24	$(CH_3)_2C=C(CH_3)CH_2Si(OC_2H_5)_3$ (57)	1
cis-$Mo(XNC)(CO)_5$	100	24	$(CH_3)_2C=C(CH_3)CH_2Si(OC_2H_5)_3$ (73)	1
$Ni(CO)_2(PPh_3)_2$	150	3	$(CH_3)_2C=C(CH_3)CH_2Si(OC_2H_5)_3$ (33)	1222
$Ni(acac)_2$ + $AlEt_3$	20	4	$(CH_3)_2C=C(CH_3)CH_2Si(OC_2H_5)_3$ (94)	273
$Ni(cod)_2$	150	3	$(CH_3)_2C=C(CH_3)CH_2Si(OC_2H_5)_3$ (55)	1222
$Ni(cod)_2$	150	3	$(CH_3)_2C=C(CH_3)CH_2Si(OC_2H_5)_3$ (67)	1222
$NiCl_2(PPh_3)_2$	150	3	$(CH_3)_2C=C(CH_3)CH_2Si(OC_2H_5)_3$ (12)	1222
$PdCl_2(PPh_3)_2$	100	6	$(CH_3)_2C=C(CH_3)CH_2Si(OC_2H_5)_3$ (-) $CH_2=C(CH_3)CH(CH_3)CH_2Si(OC_2H_5)_3$ (-)	1222
$PdCl_2(PhCN)_2$	100	6	$(CH_3)_2C=C(CH_3)CH_2Si(OC_2H_5)_3$ (11) $CH_2=C(CH_3)CH(CH_3)CH_2Si(OC_2H_5)_3$ (2)	1222
H_2PtCl_6 (i-PrOH)	150	3	$(CH_3)_2C=C(CH_3)CH_2Si(OC_2H_5)_3$ (24) $CH_2=C(CH_3)CH(CH_3)CH_2Si(OC_2H_5)_3$ (21)	1222
$MeNCH_2CH_2N(Me)C-RhBr(CO)(PMe_2Ph)_2$	100	8	$(CH_3)_2C=C(CH_3)CH_2Si(OC_2H_5)_3$ (77) $CH_2=C(CH_3)CH(CH_3)CH_2Si(OC_2H_5)_3$ (4)	496
$MeNCH_2CH_2N(Me)C-RhCl(cod)$	100	8	$(CH_3)_2C=C(CH_3)CH_2Si(OC_2H_5)_3$ (35) $CH_2=C(CH_3)CH(CH_3)CH_2Si(OC_2H_5)_3$ (35)	496
$PhNCH_2CH_2N(Ph)C-RhCl(cod)$	100	8	$CH_2=C(CH_3)CH(CH_3)CH_2Si(OC_2H_5)_3$ (24)	496
$Rh(acac)_3$ + $AlEt_3$	60	15	$(CH_3)_2C=C(CH_3)CH_2Si(OC_2H_5)_3$ (40) $CH_2=C(CH_3)CH(CH_3)CH_2Si(OC_2H_5)_3$ (13)	271
$Rh_2(CH_3COO)_4$	100	8	$(CH_3)_2C=C(CH_3)CH_2Si(OC_2H_5)_3$ (24) $CH_2=C(CH_3)CH(CH_3)CH_2Si(OC_2H_5)_3$ (8)	274
$RhCl(cod)_2(DMPIC)$	100	6	$(CH_3)_2C=C(CH_3)CH_2Si(OC_2H_5)_3$ + $CH_2=C(CH_3)CH(CH_3)CH_2Si(OC_2H_5)_3$ (16)	884
$RhCl_2[P(c\text{-}Hex)_3]_2$	100	8	$(CH_3)_2C=C(CH_3)CH_2Si(OC_2H_5)_3$ (22) $CH_2=C(CH_3)CH(CH_3)CH_2Si(OC_2H_5)_3$ (24)	503

Table 9.2
= $(C_2H_5O)_3SiH$ = =11=

	$RhCl_2[P(o\text{-}C_6H_4CH_3)_3]_2$	100	8	$(CH_3)_2C=C(CH_3)CH_2Si(OC_2H_5)_3$ (15) $CH_2=C(CH_3)CH(CH_3)CH_2Si(OC_2H_5)_3$ (19)	503
	$[RhCl(cod)]_2$ + XNC	100	6	$C_6H_{11}Si(OC_2H_5)_3$ (16)	1
	$[RhCl\{P(OC_6H_4Me)_3\}_2]_n$	100	2	$(CH_3)_2C=C(CH_3)CH_2Si(OC_2H_5)_3$ + $CH_2=C(CH_3)CH(CH_3)CH_2Si(OC_2H_5)_3$ (66)	229
	$[RhCl\{P(OMe)_3\}_2]_n$	100	2	$(CH_3)_2C=C(CH_3)CH_2Si(OC_2H_5)_3$ + $CH_2=C(CH_3)CH(CH_3)CH_2Si(OC_2H_5)_3$ (70)	229
	$[RhH\{P(OC_6H_4Me)_3\}_2]_n$	100	2	$(CH_3)_2C=C(CH_3)CH_2Si(OC_2H_5)_3$ + $CH_2=C(CH_3)CH(CH_3)CH_2Si(OC_2H_5)_3$ (8)	229
	$[RhH\{P(OMe)_3\}_2]_3$	100	2	$(CH_3)_2C=C(CH_3)CH_2Si(OC_2H_5)_3$ + $CH_2=C(CH_3)CH(CH_3)CH_2Si(OC_2H_5)_3$ (64)	229
	$W(CO)_5$(DMPIC)	100	24	$(CH_3)_2C=C(CH_3)CH_2Si(OC_2H_5)_3$ (11) $CH_2=C(CH_3)CH(CH_3)CH_2Si(OC_2H_5)_3$ (10)	884
	$W(CO)_6$	100	24	$(CH_3)_2C=C(CH_3)CH_2Si(OC_2H_5)_3$ + $CH_2=C(CH_3)CH(CH_3)CH_2Si(OC_2H_5)_3$ (1)	202
	$W(CO)_6$	100	24	$(CH_3)_2C=C(CH_3)CH_2Si(OC_2H_5)_3$ (1)	1
	$W(CO)_6$ + t-BuOOH	100	24	$(CH_3)_2C=C(CH_3)CH_2Si(OC_2H_5)_3$ + $CH_2=C(CH_3)CH(CH_3)CH_2Si(OC_2H_5)_3$ (2)	202
	$W(XNC)(CO)_5$	100	24	$(CH_3)_2C=C(CH_3)CH_2Si(OC_2H_5)_3$ (11) $CH_2=C(CH_3)CH(CH_3)CH_2Si(OC_2H_5)_3$ (10)	1
C_6H_{10}	$CH_2=CHCH_2CH_2CH=CH_2$				
	H_2PtCl_6 (EtOH)	50	24	$CH_2=CH(CH_2)_4Si(OC_2H_5)_3$ (44) $(C_2H_5O)_3Si(CH_2)_6Si(OC_2H_5)_3$ (39)	U -118
	H_2PtCl_6 (EtOH)	50	24	$CH_2=CH(CH_2)_4Si(OC_2H_5)_3$ (44) $(C_2H_5O)_3Si(CH_2)_6Si(OC_2H_5)_3$ (39)	D -068
	H_2PtCl_6 (EtOH)	50	24	$CH_2=CH(CH_2)_4Si(OC_2H_5)_3$ (44) $(C_2H_5O)_3Si(CH_2)_6Si(OC_2H_5)_3$ (39)	U -109
	H_2PtCl_6 (i-PrOH)	120-150	3	$CH_2=CH(CH_2)_4Si(OC_2H_5)_3$ (57) $(C_2H_5O)_3Si(CH_2)_6Si(OC_2H_5)_3$ (30)	1266
	$PtCl_2(Et_2S)_2$	reflux	24 (b)	$(C_2H_5O)_3Si(CH_2)_6Si(OC_2H_5)_3$ (-)	133
C_6H_{10}	$CH_3CH=CHCH_2CH=CH_2$				
	H_2PtCl_6 (EtOH)	50	24	$CH_3CH=CH(CH_2)_3Si(OC_2H_5)_3$ (88) $(C_2H_5O)_3Si(CH_2)_6Si(OC_2H_5)_3$ (8)	D -068
	H_2PtCl_6 (EtOH)	70	1	$CH_3CH=CH(CH_2)_3Si(OC_2H_5)_3$ (88)	U -109

	H_2PtCl_6 (EtOH)	70	1	$CH_2{=}CH(CH_2)_4Si(OC_2H_5)_3$ (88)	U -118
C_8H_{12}	$\overline{CH{=}CH(CH_2)_2CH{=}CH(CH_2)_2}$				
	RuHCl(cod)(DAD)	r.t.	- (dcm)	c-$C_8H_{13}Si(OC_2H_5)_3$ (-)	1363
C_8H_{12}	Fig.34.				
	$Co_2(CO)_8$	20	8	Fig.401. (82)	272
C_8H_{12}	Fig.36.				
	$NiV_2(Mes)_6$	77	4	Fig.402. (85)	Dd-008
	H_2PtCl_6 (THF) + $OPPh_3$(PhH)	80-150	-	Fig.402. (92)	C -039
	H_2PtCl_6 (i-PrOH)	80-160	1	Fig.402. (89)	D -085
	H_2PtCl_6 (i-PrOH)	until 130	5	Fig.402. (81)	Dd-007
	support/Pt + Re	-	-	Fig.402. (73)	Dd-007
C_8H_{14}	$CH_2{=}CH(CH_2)_4CH{=}CH_2$				
	-	-	-	$CH_2{=}CH(CH_2)_6Si(OC_2H_5)_3$ + $(C_2H_5)_3Si(CH_2)_8Si(OC_2H_5)_3$ (-)	881
	H_2PtCl_6 (EtOH)	50	26	$CH_2{=}CH(CH_2)_6Si(OC_2H_5)_3$ (79) $(C_2H_5)_3Si(CH_2)_8Si(OC_2H_5)_3$ (10)	U -118
	H_2PtCl_6 (EtOH)	50	26	$CH_2{=}CH(CH_2)_6Si(OC_2H_5)_3$ (79) $(C_2H_5)_3Si(CH_2)_8Si(OC_2H_5)_3$ (10)	U -109
	H_2PtCl_6 (EtOH)	50	26	$CH_2{=}CH(CH_2)_6Si(OC_2H_5)_3$ (79) $(C_2H_5)_3Si(CH_2)_8Si(OC_2H_5)_3$ (10)	D -068
	$[HPt\{Si(OEt)_3\}P(c\text{-}Hex)_3]_2$	20	24	$(C_2H_5)_3Si(CH_2)_8Si(OC_2H_5)_3$ (65)	429
C_9H_{12}	Fig.107.				
	H_2PtCl_6	60	20	Fig.403. +Fig.404. (-)	J -069
$C_{10}H_{12}$	Fig.42.				
	$Co_2(CO)_8$	30-50	2	Fig.405. (80)	712
	$Rh_4(CO)_{12}$	30-50	1	Fig.405. (96)	712
$C_{10}H_{16}$	$(CH_3)_2C{=}CHCH_2CH_2C({=}CH_2)CH{=}CH_2$				
	$PdCl_2(PhCN)_2$) + PPh_3	100	24	$(CH_3)_2C{=}CHCH_2CH_2C[CH_2Si(OC_2H_5)_3]{=}CHCH_3$ (29)	851
	$RhCl(PPh_3)_3$	100	24	$(CH_3)_2C{=}CHCH_2CH_2C(CH_3){=}CHCH_2Si(OC_2H_5)_3$ (48) $(CH_3)_2C{=}CHCH_2CH_2C[CH_2Si(OC_2H_5)_3]{=}CHCH_3$ (38)	851
$C_{10}H_{16}$	$(CH_3)_2C{=}CHCH_2CH{=}C(CH_3)CH{=}CH_2$				

Table 9.2

= $(C_2H_5O)_3SiH$ = =13=

	$RhCl(PPh_3)_3$	100	24	$(CH_3)_2C=CHCH_2CH_2C(CH_3)=CHCH_2Si(OC_2H_5)_3$ (96) $(CH_3)_2C=CHCH_2CH[Si(OC_2H_5)_3]C(CH_3)=CHCH_3$ (2)	851
	$RhCl(PPh_3)_3$	100-110	12	$(CH_3)_2C=CHCH_2CH_2C(CH_3)=CHCH_2Si(OC_2H_5)_3$ (90)	J -036
$C_{10}H_{16}$	Fig.45.				
	Pt/C	80	16	Fig.406. (65)	U -164

C_2H_3	$CH_2=CHSiCl_3$				
	$Pt[(ViMe_2Si)_2O]_2$	-	-	$(C_2H_5O)_3SiCH_2CH_2SiCl_3$ (kin.)	669
C_3H_3	$CF_3CH=CH_2$				
	$RhCl(PPh_3)_3$	120	24	$CF_3CH_2CH_2Si(OC_2H_5)_3$ (85)	858
	$Ru_3(CO)_{12}$	150	24	$CF_3CH_2CH_2Si(OC_2H_5)_3$ (52)	858
C_3H_3	$CH_2=CHCN$				
	SDVB-NMe_2/H_2PtCl_6 (EtOH)	140	2	$(C_2H_5O)_3SiCH(CH_3)CN$ (27)	C -002
	Rh(acac)$(CO)_2$	95-150	0.5-1	$(C_2H_5O)_3SiCH(CH_3)CN$ (56)	S -018
	$RhCl(PPh_3)_3$	120	96	$(C_2H_5O)_3SiCH(CH_3)CN$ (66)	868
	RhCl(cod) + diphos	-	-	$(C_2H_5O)_3SiC_2H_4CN$ (-)	J -020
	$RhH(PPh_3)_4$	100	-	$(C_2H_5O)_3SiC_2H_4CN$ (-)	605
	SDVB-NMe_2/$RhCl_3$	80	3	$(C_2H_5O)_3SiCH(CH_3)CN$ (19)	C -006
	SDVB-PPh_2/$RhCl(PPh_3)_3$	80	2	$(C_2H_5O)_3SiCH_2CH_2CN$ (14)	216
	SDVB-PPh_2/$RhCl(PPh_3)_3$	85	2	$(C_2H_5O)_3SiCH_2CH_2CN$ (8)	1168
	SDVB-PPh_2/$RhCl_3$	80	2	$(C_2H_5O)_3SiCH_2CH_2CN$ (18)	216
	$[RhCl(c\text{-}C_8H_{13})_2]_2$	130	24	$(C_2H_5O)_3SiCH_2CH_2CN$ (90)	Pl-035
	$[RhCl(cod)]_2$	130	24	$(C_2H_5O)_3SiCH_2CH_2CN$ (85)	Pl-035
C_3H_5	$CH_2=CHCH_2Cl$				
	$[IrCl(cod)]_2$	80	4 (xy)	$(C_2H_5O)_3Si(CH_2)_3Cl$ (75)	U -172
	$[IrCl(cod)]_2$	80-135	4 (xy)	$(C_2H_5O)_3Si(CH_2)_3Cl$ (70-75)	U -172
	H_2PtCl_6	100	4	$(C_2H_5O)_3Si(CH_2)_3Cl$ (10)	U -172
	H_2PtCl_6 (i-PrOH)	175	3 (solv.)	$(C_2H_5O)_3Si(CH_2)_3Cl$ (13)	98
	$PtCl_2(PPh_3)_2$	150	4	$(C_2H_5O)_3Si(CH_2)_3Cl$ (-)	U -172
	polysiloxane-$NHCH_2CH_2NH_2$/H_2PtCl_6	60-85	3	$(C_2H_5O)_3Si(CH_2)_3Cl$ (70)	D -071
	$RhH(PPh_3)_4$	100	-	$(C_2H_5O)_3Si(CH_2)_3Cl$ (-)	605
	$RuCl_2(PPh_3)_3$	80	2	$(C_2H_5O)_3Si(CH_2)_3Cl$ (30)	Pl-001
C_3H_7	$CH_2=CHCH_2NH_2$				

Table 9.3
= $(C_2H_5O)_3SiH$ = =2=

$Co_2(CO)_8$	r.t.	5	$(C_2H_5O)_3Si(CH_2)_3N[Si(OC_2H_5)_3]_2$ (53)	715
$Co_2(CO)_8$	r.t.	5	$(C_2H_5O)_3Si(CH_2)_3N[Si(OC_2H_5)_3]_2$ (53)	259
$Cu_2Cl_2+NEt_3+Me_2NC_2H_4NMe_2+Si(OEt)_4$	150	38	$(C_2H_5O)_3Si(CH_2)_3NH_2$ (-)	Dd-002
CuCl + $Me_2NC_2H_4NMe_2$	74-130	42	$(C_2H_5O)_3Si(CH_2)_3NH_2$ + $(C_2H_5O)_3SiCH(CH_3)CH_2NH_2$ (8)	93
$AllNH_3{}^+PtCl_3{}^-$	85	7	$(C_2H_5O)_3Si(CH_2)_3NH_2$ (30) $(C_2H_5O)_3SiCH(CH_3)CH_2NH_2$ (5)	D -119
$AllNH_3{}^+PtCl_3{}^-$	until 120	24	$(C_2H_5O)_3Si(CH_2)_3NH_2$ (54) $(C_2H_5O)_3SiCH(CH_3)CH_2NH_2$ (11)	1282
$AllNH_3{}^+PtCl_3{}^-$	until 130	7	$(C_2H_5O)_3Si(CH_2)_3NH_2$ (51) $(C_2H_5O)_3SiCH(CH_3)CH_2NH_2$ (9)	C -012
$AllNH_3{}^+PtCl_3{}^-$	until 150	7	$(C_2H_5O)_3Si(CH_2)_3NH_2$ (46) $(C_2H_5O)_3SiCH(CH_3)CH_2NH_2$ (10)	C -012
H_2PtCl_6	120-130	-	$(C_2H_5O)_3SiC_3H_6NH_2$ (80-85)	Dd-009
H_2PtCl_6	130	6.5	$(C_2H_5O)_3Si(CH_2)_3NH_2$ (71)	U -144
H_2PtCl_6	80-125	0.17-2	$(C_2H_5O)_3Si(CH_2)_3NH_2$ (kin.)	1094
H_2PtCl_6	reflux	7	$(C_2H_5O)_3SiCH(CH_3)CH_2NH_2$ (9)	C -012
H_2PtCl_6 (i-PrOH)	until 130	5	$(C_2H_5O)_3Si(CH_2)_3NH_2$ + $(C_2H_5O)_3SiCH(CH_3)CH_2NH_2$ (58-63)	Dd-007
$H_2PtCl_6[\overline{(CH_2)_5CO}]$	reflux	6	$(C_2H_5O)_3Si(CH_2)_3NH_2$ (17)	748
H_2PtCl_6 ($AllOCH_2COOAll$)	71-136	51.5	$(C_2H_5O)_3Si(CH_2)_3NH_2$ (64)	Pl-002
H_2PtCl_6 ($AllOCH_2COOAll$)	71-136	51.5	$(C_2H_5O)_3Si(CH_2)_3NH_2$ (64)	Pl-008
H_2PtCl_6 (EtOH + All_2O)	115	2	$(C_2H_5O)_3SiC_3H_6NH_2$ (55)	97
H_2PtCl_6 (EtOH)	115	2	$(C_2H_5O)_3SiC_3H_6NH_2$ (35)	97
H_2PtCl_6 (EtOH)	reflux	12-15	$(C_2H_5O)_3SiC_3H_6NH_2$ (51-57)	S -021
H_2PtCl_6 (OctOH)	until 120	12	$(C_2H_5O)_3SiC_3H_6NH_2$ (57)	S -021
H_2PtCl_6 (THF)	90-100	5	$(C_2H_5O)_3Si(CH_2)_3NH_2$ (62)	C -026
H_2PtCl_6 (i-PentOH)	until 120	13	$(C_2H_5O)_3SiC_3H_6NH_2$ (55)	S -021
H_2PtCl_6 (i-PrOH + AllOH)	until 120	12-27.7	$(C_2H_5O)_3Si(CH_2)_3NH_2$ + $(C_2H_5O)_3SiCH(CH_3)CH_2NH_2$ (66-67)	S -019
H_2PtCl_6 (i-PrOH)	120	9	$(C_2H_5O)_3Si(CH_2)_3NH_2$ (44) $(C_2H_5O)_3SiCH(CH_3)CH_2NH_2$ (10)	E -040

Table 9.3
= (C$_2$H$_5$O)$_3$SiH =   =3=

H$_2$PtCl$_6$ (i-PrOH)	120	9	(C$_2$H$_5$O)$_3$Si(CH$_2$)$_3$NH$_2$ (40) (C$_2$H$_5$O)$_3$SiCH(CH$_3$)CH$_2$NH$_2$ (10)	E -048
H$_2$PtCl$_6$ (i-PrOH)	86-146	7.5	(C$_2$H$_5$O)$_3$Si(CH$_2$)$_3$NH$_2$ (64)	C -026
H$_2$PtCl$_6$ (i-PrOH)	reflux	24	(C$_2$H$_5$O)$_3$Si(CH$_2$)$_3$NH$_2$ (51) (C$_2$H$_5$O)$_3$SiCH(CH$_3$)CH$_2$NH$_2$ (11)	93
H$_2$PtCl$_6$ (i-PrOH)	reflux	6	(C$_2$H$_5$O)$_3$Si(CH$_2$)$_3$NH$_2$ (3)	748
H$_2$PtCl$_6$ (i-PrOH)	reflux	26.5	(C$_2$H$_5$O)$_3$Si(CH$_2$)$_3$NH$_2$ (48)	C -025
H$_2$PtCl$_6$ (i-PrOH)	reflux	56	(C$_2$H$_5$O)$_3$Si(CH$_2$)$_3$NH$_2$ (79)	Dd-002
H$_2$PtCl$_6$ (i-PrOH)	reflux	24	(C$_2$H$_5$O)$_3$SiC$_3$H$_6$NH$_2$ (65)	S -035
H$_2$PtCl$_6$ (i-PrOH)	reflux	16	(C$_2$H$_5$O)$_3$SiC$_3$H$_6$NH$_2$ (32)	S -010
H$_2$PtCl$_6$ (i-PrOH)	reflux	24	(C$_2$H$_5$O)$_3$SiC$_3$H$_6$NH$_2$ (57)	S -010
H$_2$PtCl$_6$ (i-PrOH)	reflux	24	(C$_2$H$_5$O)$_3$SiC$_3$H$_6$NH$_2$ (53)	S -011
H$_2$PtCl$_6$ (i-PrOH)	reflux	30	(C$_2$H$_5$O)$_3$SiC$_3$H$_6$NH$_2$ (70)	S -026
H$_2$PtCl$_6$ (i-PrOH)	until 120	9.7	(C$_2$H$_5$O)$_3$Si(CH$_2$)$_3$NH$_2$ (58) (C$_2$H$_5$O)$_3$SiCH(CH$_3$)CH$_2$NH$_2$ (11)	1282
H$_2$PtCl$_6$ (i-PrOH)	until 120	24	(C$_2$H$_5$O)$_3$Si(CH$_2$)$_3$NH$_2$ (41) (C$_2$H$_5$O)$_3$SiCH(CH$_3$)CH$_2$NH$_2$ (8)	1282
H$_2$PtCl$_6$ (i-PrOH)	until 120	11.5-13.5	(C$_2$H$_5$O)$_3$SiC$_3$H$_6$NH$_2$ (69-79)	S -021
H$_2$PtCl$_6$ (i-PrOH) +7,8-C$_2$B$_9$H$_{12}$$^-K^+$	until 120	6-19	(C$_2$H$_5$O)$_3$SiC$_3$H$_6$NH$_2$ (39-70)	S -035
H$_2$PtCl$_6$ (i-PrOH) +7,8-C$_2$B$_9$H$_{12}$$^-K^+$	until 120	6-19	(C$_2$H$_5$O)$_3$SiC$_3$H$_6$NH$_2$ (39-70)	96
H$_2$PtCl$_6$ (i-PrOH)+7,8-Me$_2$C$_2$B$_9$H$_{10}$$^-K^+$	until 120	6.5	(C$_2$H$_5$O)$_3$SiC$_3$H$_6$NH$_2$ (63)	S -035
H$_2$PtCl$_6$ (i-PrOH) + AllOAc	until 125	16	(C$_2$H$_5$O)$_3$Si(CH$_2$)$_3$NH$_2$ (57)	S -010
H$_2$PtCl$_6$ (i-PrOH) + AllOCH$_2$$\overline{\text{CHCH}_2\text{O}}$	until 120	14	(C$_2$H$_5$O)$_3$Si(CH$_2$)$_3$NH$_2$ (56)	S -010
H$_2$PtCl$_6$ (i-PrOH) + AsPh$_3$	until 120	9	(C$_2$H$_5$O)$_3$Si(CH$_2$)$_3$NH$_2$ (62) (C$_2$H$_5$O)$_3$SiCH(CH$_3$)CH$_2$NH$_2$ (12)	1282
H$_2$PtCl$_6$ (i-PrOH) + BuOCH=CH$_2$	reflux	9-18	(C$_2$H$_5$O)$_3$Si(CH$_2$)$_3$NH$_2$ (48-67)	S -010
H$_2$PtCl$_6$ (i-PrOH) + BuOCH=CH$_2$	reflux	12	(C$_2$H$_5$O)$_3$SiC$_3$H$_6$NH$_2$ (58)	S -012
H$_2$PtCl$_6$ (i-PrOH) + CF$_3$COOAll	until 125	17	(C$_2$H$_5$O)$_3$Si(CH$_2$)$_3$NH$_2$ (50)	S -010
H$_2$PtCl$_6$ (i-PrOH) + EtOCH=CH$_2$	reflux	16	(C$_2$H$_5$O)$_3$Si(CH$_2$)$_3$NH$_2$ (56)	S -010
H$_2$PtCl$_6$ (i-PrOH) + $\overline{\text{OCH}_2\text{CHCH}_2}$Cl	reflux	11-14	(C$_2$H$_5$O)$_3$SiC$_3$H$_6$NH$_2$ (65-70)	S -020

Table 9.3

= $(C_2H_5O)_3SiH$ = =4=

H_2PtCl_6 (i-PrOH) + $\overline{OCH_2CH}CH_2Cl$	reflux	11-14	$(C_2H_5O)_3SiC_3H_6NH_2$ (65-70)	S -035
H_2PtCl_6 (i-PrOH) + RCOOH	until 120	8-16	$(C_2H_5O)_3SiC_3H_6NH_2$ (33-61)	S -011
H_2PtCl_6 (i-PrOH) + RR^1CO	until 120	12-14.5	$(C_2H_5O)_3SiC_3H_6NH_2$ (40-60)	S -011
H_2PtCl_6 (i-PrOH) + $SbPh_3$	until 120	9	$(C_2H_5O)_3Si(CH_2)_3NH_2$ (63) $(C_2H_5O)_3SiCH(CH_3)CH_2NH_2$ (12)	1282
H_2PtCl_6 (i-PrOH) + $Si(OEt)_4$	160	19	$(C_2H_5O)_3Si(CH_2)_3NH_2$ (88)	Dd-002
H_2PtCl_6 (i-PrOH) + $Si(OEt)_4$	180	32	$(C_2H_5O)_3Si(CH_2)_3NH_2$ (85)	Dd-002
H_2PtCl_6 (i-PrOH) + carbonyl deriv.	until 120	13-24	$(C_2H_5O)_3Si(CH_2)_3NH_2$ (19-56) $(C_2H_5O)_3SiCH(CH_3)CH_2NH_2$ (5-13)	1282
H_2PtCl_6 (i-PrOH) + $o\text{-}HOC_6H_4COOH$	until 120	6.5-10.5	$(C_2H_5O)_3SiC_3H_6NH_2$ (50-57)	S -011
H_2PtCl_6(i-PrOH)+8-Me-7,9-$C_2B_9H_{11}^-K^+$	until 120	6	$(C_2H_5O)_3SiC_3H_6NH_2$ (62)	S -035
H_2PtCl_6 (i-PrOH)+PR_3 R=Ph,OPh,Bu,c-Hex	until 120	6.5-9	$(C_2H_5O)_3Si(CH_2)_3NH_2$ (10-64) $(C_2H_5O)_3SiCH(CH_3)CH_2NH_2$ (0-12)	1282
H_2PtCl_6 (t-BuOH)	until 120	12	$(C_2H_5O)_3Si(CH_2)_3NH_2$ (61)	S -021
H_2PtCl_6 + NR_3 (R=c-Hex, Ph)	-	-	$(C_2H_5O)_3Si(CH_2)_3NH_2$ (53-64)	C -031
H_2PtCl_6+ Na_2CO_3	130	6.5	$(C_2H_5O)_3Si(CH_2)_3NH_2$ (49) $(C_2H_5O)_3SiCH(CH_3)CH_2NH_2$ (13)	E -008
H_2PtCl_6+ Na_2CO_3	130	6.5	$(C_2H_5O)_3Si(CH_2)_3NH_2$ (49) $(C_2H_5O)_3SiCH(CH_3)CH_2NH_2$ (13)	U -144
H_2PtCl_6+ Na_2CO_3	130	12	$(C_2H_5O)_3SiC_3H_6NH_2$ (60-75)	U -156
H_2PtCl_6+ $P(c\text{-}Hex)_3$	reflux	9.8	$(C_2H_5O)_3Si(CH_2)_3NH_2$ (56)	C -031
H_2PtCl_6 + PPh_3	reflux	6.5-7.3	$(C_2H_5O)_3Si(CH_2)_3NH_2$ (54-64)	C -031
H_2PtCl_6 [$\overline{(CH_2)_5CO}$]	reflux	10	$(C_2H_5O)_3SiC_3H_6NH_2$ (66-70)	Pl-022
H_2PtCl_6 [$\overline{(CH_2)_5CO}$] + $CH_2=CHSi(OR)_3$] R=Me,Et,Pr	reflux	10	$(C_2H_5O)_3SiC_3H_6NH_2$ (70-76)	Pl-022
H_2PtCl_6 [$\overline{(CH_2)_5CO}$] + $CH_2=CHSi(OR)_3$] R=Me,Et,Pr	reflux	10	$(C_2H_5O)_3SiC_3H_6NH_2$ (70-76)	Pl-021
Pt(1,2-$C_2B_9H_{11})_2$	reflux	5	$(C_2H_5O)_3SiC_3H_6NH_2$ (63)	96
$Pt(CH_2=CH_2)(PPh_3)_2$	reflux	6	$(C_2H_5O)_3Si(CH_2)_3NH_2$ (47)	748
$Pt(CH_2=CH_2)(PPh_3)_2$	reflux	8	$(C_2H_5O)_3SiC_3H_6NH_2$ (66-68)	Pl-033

Table 9.3
= $(C_2H_5O)_3SiH$ = =5=

$Pt(PPh_3)_4$	130	2	$(C_2H_5O)_3Si(CH_2)_3NH_2$ (71)	C -025
$Pt(PPh_3)_4$	130	12	$(C_2H_5O)_3SiC_3H_6NH_2$ (60)	U -156
$Pt(PPh_3)_4$	reflux	17	$(C_2H_5O)_3SiC_3H_6NH_2$ (70)	96
$Pt(PPh_3)_4$	reflux	6	$(C_2H_5O)_3Si(CH_2)_3NH_2$ (18)	748
$Pt(PPh_3)_4$	until 120	15	$(C_2H_5O)_3Si(CH_2)_3NH_2$ (60)	C -025
$Pt(PPh_3)_4$	until 120	6	$(C_2H_5O)_3Si(CH_2)_3NH_2$ (72)	C -025
$Pt(PPh_3)_4$	until 120	15.5	$(C_2H_5O)_3Si(CH_2)_3NH_2$ (60) $(C_2H_5O)_3SiCH(CH_3)CH_2NH_2$ (11)	1282
$Pt(PPh_3)_4 + Na_2CO_3$	-	-	$(C_2H_5O)_3Si(CH_2)_3NH_2$ (42) $(C_2H_5O)_3SiCH(CH_3)CH_2NH_2$ (7)	E -008
$Pt(PPh_3)_4 + Na_2CO_3$	-	-	$(C_2H_5O)_3Si(CH_2)_3NH_2$ (42) $(C_2H_5O)_3SiCH(CH_3)CH_2NH_2$ (7)	U -144
$Pt(acac)_2$	until 120	12	$(C_2H_5O)_3SiC_3H_6NH_2$ (70)	96
Pt/C	reflux	21	$(C_2H_5O)_3Si(CH_2)_3NH_2$ + $(C_2H_5O)_3SiCH(CH_3)CH_2NH_2$ (60)	93
$PtCl_2(AllNH_2)_4$	115	2	$(C_2H_5O)_3SiC_3H_6NH_2$ (35)	97
$PtCl_2(C_2H_4)_2$	until 120	18	$(C_2H_5O)_3Si(CH_2)_3NH_2$ (60) $(C_2H_5O)_3SiCH(CH_3)CH_2NH_2$ (11)	1282
$PtCl_2(\overline{CH_2CH=CHCH_2CH_2C}HCH=CH_2)$	until 120	7	$(C_2H_5O)_3Si(CH_2)_3NH_2$ (68) $(C_2H_5O)_3SiCH(CH_3)CH_2NH_2$ (12)	1282
$PtCl_2(\overline{CH_2CH=CHCH_2CH_2C}HCH=CH_2)$	until 120	7	$(C_2H_5O)_3Si(CH_2)_3NH_2$ (68) $(C_2H_5O)_3SiCH(CH_3)CH_2NH_2$ (12)	C -027
$PtCl_2(PPh_3)_2$	-	-	$(C_2H_5O)_3Si(CH_2)_3NH_2$ (44) $(C_2H_5O)_3SiCH(CH_3)CH_2NH_2$ (7)	E -008
$PtCl_2(PPh_3)_2$	-	-	$(C_2H_5O)_3Si(CH_2)_3NH_2$ (44) $(C_2H_5O)_3SiCH(CH_3)CH_2NH_2$ (7)	U -144
$PtCl_2(PPh_3)_2$	reflux	6	$(C_2H_5O)_3Si(CH_2)_3NH_2$ (21)	748
$PtCl_2(PPh_3)_2$	until 120	18 (b)	$(C_2H_5O)_3Si(CH_2)_3NH_2$ (57)	C -025
$PtCl_2(PPh_3)_2$	until 120	18.5	$(C_2H_5O)_3Si(CH_2)_3NH_2$ (58) $(C_2H_5O)_3SiCH(CH_3)CH_2NH_2$ (12)	747
$PtCl_2(PPh_3)_2$ (MeCOMe)	120	6	$(C_2H_5O)_3Si(CH_2)_3NH_2$ (64)	C -025
$PtCl_2(PhCH=CH_2)_2$	reflux	6.5	$(C_2H_5O)_3Si(CH_2)_3NH_2$ (69)	C -027

Table 9.3

= $(C_2H_5O)_3SiH$ = =6=

$PtCl_2(PhCH{=}CH_2)_2$	until 120	7.3	$(C_2H_5O)_3Si(CH_2)_3NH_2$ (60) $(C_2H_5O)_3SiCH(CH_3)CH_2NH_2$ (12)	747
$PtCl_2(PhCN)_2$	reflux	20	$(C_2H_5O)_3SiC_3H_6NH_2$ (70)	96
$PtCl_2(acac)_2$ (MeCOMe)	-	-	$(C_2H_5O)_3Si(CH_2)_3NH_2$ (-)	D -011
$PtCl_2(c\text{-}C_6H_{10})$	115	2	$(C_2H_5O)_3SiC_3H_6NH_2$ (55)	97
$PtCl_6(AllNH_3)_2$	115	2	$(C_2H_5O)_3SiC_3H_6NH_2$ (35)	97
$PtI(PPh_3)_2$	until 120	18	$(C_2H_5O)_3Si(CH_2)_3NH_2$ (58) $(C_2H_5O)_3SiCH(CH_3)CH_2NH_2$ (12)	1282
$Pt[(CH_3)_2C{=}CHCOCH_3]$ + phenothiazine	110	-	$(C_2H_5O)_3Si(CH_2)_3NH_2$ (75)	J -072
$[PtCl_2(C_2H_4)]_2$	reflux	7	$(C_2H_5O)_3Si(CH_2)_3NH_2$ (70)	C -027
$[PtCl_2(C_2H_4)]_2 + C_5H_5N$	reflux	24	$(C_2H_5O)_3Si(CH_2)_3NH_2$ + $(C_2H_5O)_3SiCH(CH_3)CH_2NH_2$ (63)	93
$[PtCl_2(C_2H_4)]_2 + PPh_3$	until 120	7	$(C_2H_5O)_3Si(CH_2)_3NH_2$ (70) $(C_2H_5O)_3SiCH(CH_3)CH_2NH_2$ (11)	1282
$[PtCl_2(Me_2C{=}CHCOMe)]_2$	reflux	2	$(C_2H_5O)_3Si(CH_2)_3NH_2$ + $(C_2H_5O)_3SiCH(CH_3)CH_2NH_2$ (20)	U -144
$[PtCl_2(Me_2C{=}CHCOMe)]_2$	reflux	2	$(C_2H_5O)_3Si(CH_2)_3NH_2$ + $(C_2H_5O)_3SiCH(CH_3)CH_2NH_2$ (20)	E -008
cis-$PtCl_2(NH_3)_2$	-	-	$(C_2H_5O)_3Si(CH_2)_3NH_2$ (47) $(C_2H_5O)_3SiCH(CH_3)CH_2NH_2$ (9)	U -144
cis-$PtCl_2(NH_3)_2$	-	-	$(C_2H_5O)_3Si(CH_2)_3NH_2$ (47) $(C_2H_5O)_3SiCH(CH_3)CH_2NH_2$ (9)	E -008
cis-$PtCl_2(NH_3)_2$	130	12	$(C_2H_5O)_3SiC_3H_6NH_2$ (60)	U -156
$(Et_4N)_2[Rh_4(CO)_{11}]$	-	(xy)	$(C_2H_5O)_3Si(CH_2)_3NH_2$ (68) $(C_2H_5O)_3SiCH(CH_3)CH_2NH_2$ (4)	E -040
$(Et_4N)_2[Rh_6(CO)_{15}]$	85-110	0.5 (xy)	$(C_2H_5O)_3Si(CH_2)_3NH_2$ (65) $(C_2H_5O)_3SiCH(CH_3)CH_2NH_2$ (4)	E -040
$Rh(CO)_2(acac) + PPh_3$	130	0.25	$(C_2H_5O)_3SiC_3H_6NH_2$ (70)	U -156
$Rh_4(CO)_{12}$	20	5	$(C_2H_5O)_3Si(CH_2)_3N[Si(OC_2H_5)_3]_2$ (54-72)	715
$Rh_4(CO)_{12}$	20	5	$(C_2H_5O)_3Si(CH_2)_3N[Si(OC_2H_5)_3]_2$ (54-72)	259
$Rh_4(CO)_{12}$	85-110	0.5 (t)	$(C_2H_5O)_3Si(CH_2)_3NH_2$ (68) $(C_2H_5O)_3SiCH(CH_3)CH_2NH_2$ (5)	E -040
$Rh_4(CO)_{12}$ + CO	110	1	$(C_2H_5O)_3Si(CH_2)_3NH_2$ (62) $(C_2H_5O)_3SiCH(CH_3)CH_2NH_2$ (4)	E -040

Table 9.3
= $(C_2H_5O)_3SiH$ = =7=

	$Rh_4(CO)_{12}$ + cod	110	1.5	$(C_2H_5O)_3Si(CH_2)_3NH_2$ (78) $(C_2H_5O)_3SiCH(CH_3)CH_2NH_2$ (6)	E -040
	$Rh_4(CO)_{12}$ + cod + CO	110	2.5	$(C_2H_5O)_3Si(CH_2)_3NH_2$ (58) $(C_2H_5O)_3SiCH(CH_3)CH_2NH_2$ (5)	E -040
	$Rh_4(CO)_{12}$ + cod + CO	120	3 (xy)	$(C_2H_5O)_3Si(CH_2)_3NH_2$ (75) $(C_2H_5O)_3SiCH(CH_3)CH_2NH_2$ (8)	E -040
	$Rh_6(CO)_{16}$	120	3 (xy)	$(C_2H_5O)_3Si(CH_2)_3NH_2$ (76) $(C_2H_5O)_3SiCH(CH_3)CH_2NH_2$ (7)	E -040
	$Rh_6(CO)_{16}$	85-110	0.5 (xy)	$(C_2H_5O)_3Si(CH_2)_3NH_2$ (71) $(C_2H_5O)_3SiCH(CH_3)CH_2NH_2$ (4)	E -040
	$Rh_6(CO)_{16}$ + CO	-	-	$(C_2H_5O)_3Si(CH_2)_3NH_2$ (63) $(C_2H_5O)_3SiCH(CH_3)CH_2NH_2$ (4)	E -040
	$RhCl(CO)(PPh_3)_3$	110	6	$(C_2H_5O)_3Si(CH_2)_3NH_2$ (71) $(C_2H_5O)_3SiCH(CH_3)CH_2NH_2$ (7)	E -040
	$RhH(CO)(PPh_3)_3$ + PPh_3	110	3	$(C_2H_5O)_3Si(CH_2)_3NH_2$ (71) $(C_2H_5O)_3SiCH(CH_3)CH_2NH_2$ (7)	E -048
	$RhH(CO)(PPh_3)_3$ + PPh_3	125	5	$(C_2H_5O)_3SiC_3H_6NH_2$ (70)	U -156
	$RhH(PPh_3)_4$	130	12	$(C_2H_5O)_3SiC_3H_6NH_2$ (70)	U -156
	$[Rh(\mu\text{-}PPh_2)(cod)]_2$	110	1 (xy,THF)	$(C_2H_5O)_3Si(CH_2)_3NH_2$ (92) $(C_2H_5O)_3SiCH(CH_3)CH_2NH_2$ (1)	E -048
	$[Rh(\mu\text{-}PPh_2)(cod)]_2$	110	- (xy)	$(C_2H_5O)_3Si(CH_2)_3NH_2$ (81) $(C_2H_5O)_3SiCH(CH_3)CH_2NH_2$ (1)	E -048
	$Ru(acac)_3$	175	- (xy)	$(C_2H_5O)_3SiC_3H_6NH_2$ (-) $(C_2H_5O)_3SiCH{=}CHCH_2NH_2$ (50)	U -173
	$RuCl_2(CO)_2(PPh_3)_2$	175	4	$(C_2H_5O)_3SiC_3H_6NH_2$ (-) $(C_2H_5O)_3SiCH{=}CHCH_2NH_2$ (40)	U -173
	$RuCl_2(PPh_3)_3$	175	4 (xy)	$(C_2H_5O)_3SiC_3H_6NH_2$ (-) $(C_2H_5O)_3SiCH{=}CHCH_2NH_2$ (50)	U -173
C_4H_5	$CH_2{=}CHCH_2CN$				
	C-CN/$[RhCl(CO)_2]_2$	80	6	$(C_2H_5O)_3Si(CH_2)_3CN$ (11)	Dd-003
C_4H_5	$CH_2{=}CHCH_2NCO$				
	Rh catalyst	-	-	$(C_2H_5O)_3Si(CH_2)_3NCO$ (-)	J -024
C_4H_6	$CH_2{=}CHCOOCH_3$				
	$Co_2(CO)_8$	25	3	$(C_2H_5O)_3SiCH{=}CHCOOCH_3$ (43) $(C_2H_5O)_3SiCH_2CH_2COOCH_3$ (4)	1178

Table 9.3

= $(C_2H_5O)_3SiH$ = =8=

	$RhCl(PHPh_2)_3$	85	2	$(C_2H_5O)_3SiCH_2CH_2COOCH_3$ (-)	C -004
	RhCl(cod) + diphos	-	-	$(C_2H_5O)_3SiCH_2CH_2COOCH_3$ (-)	J -020
	$RhH(PPh_3)_4$	100	-	$(C_2H_5O)_3SiCH_2CH_2COOCH_3$ (-)	605
	SDVB-NMe_2/$RhCl_3$	100	5	$(C_2H_5O)_3SiCH_2CH_2COOCH_3$ (-)	C -006
	SDVB-PPh_2/$RhCl(PPh_3)_3$	85	2	$(C_2H_5O)_3SiCH_2CH_2COOCH_3$ (3)	1168
	$RuCl_2(PPh_3)_3$	100	2	$(C_2H_5O)_3SiCH_2CH_2COOCH_3$ (42)	PI-001
	$RuCl_3(PPh_3)_3$	reflux	6	$(C_2H_5O)_3SiCH_2CH_2COOCH_3$ (74)	PI-013
C_4H_6	$CH_2=CHOCH=CH_2$				
	H_2PtCl_6 (i-PrOH)	120-150	3	$CH_2=CHOCH_2CH_2Si(OC_2H_5)_3$ (47) $[(C_2H_5O)_3SiCH_2CH_2]_2O$ (30)	1266
C_4H_6	$CH_2=CHSCH=CH_2$				
	H_2PtCl_6 (i-PrOH)	150	46	$CH_2=CHSCH_2CH_2Si(OC_2H_5)_3$ (16) $CH_2=CHSCH(CH_3)Si(OC_2H_5)_3$ (5)	1272
	H_2PtCl_6 (i-PrOH)	150-160	46	$CH_2=CHSCH_2CH_2Si(OC_2H_5)_3$ + $CH_2=CHSCH(CH_3Si(OC_2H_5)_3$ (15)	1241
	H_2PtCl_6 (i-PrOH)	160	46	polymer (-)	1275
	$RhCl(PPh_3)_3$	100-120	46	$CH_2=CHSCH_2CH_2Si(OC_2H_5)_3$ + $CH_2=CHSCH(CH_3)Si(OC_2H_5)_3$ (15)	1241
	$RhCl(PPh_3)_3$	150	46	$CH_2=CHSCH_2CH_2Si(OC_2H_5)_3$ (14) $CH_2=CHSCH(CH_3)Si(OC_2H_5)_3$ (1)	1272
C_4H_6	$CH_3COOCH=CH_2$				
	$RuCl_3(PPh_3)_3$	reflux	6	$CH_3COOCH_2CH_2Si(OC_2H_5)_3$ (74)	PI-013
C_4H_6	$HCOOCH_2CH=CH_2$				
	H_2PtCl_6 (i-PrOH)	55-142	0.5	$HCOO(CH_2)_3Si(OC_2H_5)_3$ (49)	1052
	H_2PtCl_6 (i-PrOH)	55-142	0.5	$HCOO(CH_2)_3Si(OC_2H_5)_3$ (49)	S -026
C_4H_7	$CH_2=CHCH_2CONH_2$				
	C-CN/$[RhCl(CO)_2]_2$	80	6	$(C_2H_5O)_3Si(CH_2)_3CN$ (11)	C -030
C_4H_8	$CH_3CH_2OCH=CH_2$				
	Al_2O_3-NMe_2/H_2PtCl_6	80	2.5 (b)	$CH_3CH_2OCH_2CH_2Si(OC_2H_5)_3$ (80)	C -007
	PMM-CN/H_2PtCl_6 (EtOH)	80	2	$CH_3CH_2OCH_2CH_2Si(OC_2H_5)_3$ (80)	C -002

	Al_2O_3-NMe_2/$RhCl_3$	80	2	$CH_3CH_2OCH_2CH_2Si(OC_2H_5)_3$ (51)	C -007
	C-CN/$[RhCl(CO)_2]_2$	80	3	$CH_3CH_2OCH_2CH_2Si(OC_2H_5)_3$ (11)	Dd-00
	C-CN/$[RhCl(CO)_2]_2$ + c-$C_{12}H_{18}$	80	3	$CH_3CH_2OCH_2CH_2Si(OC_2H_5)_3$ (69)	C -030
	PMM-CN/$RhCl_3$	80	2	$CH_3CH_2OCH_2CH_2Si(OC_2H_5)_3$ (90)	C -006
	$RhCl(PPh_3)_3$	80	2	$CH_3CH_2OCH_2CH_2Si(OC_2H_5)_3$ (92)	201
	$RhCl_3$ (EtOH)	80	2	$CH_3CH_2OCH_2CH_2Si(OC_2H_5)_3$ (85)	201
	SDVB-CH_2Cl/$RhCl(CO)(PPh_3)_2$	80	3	$CH_3CH_2OCH_2CH_2Si(OC_2H_5)_3$ (29)	C -022
	SDVB-PPh_2/$RhCl(PPh_3)_3$	80	2	$CH_3CH_2OCH_2CH_2Si(OC_2H_5)_3$ (50)	216
	SDVB-PPh_2/$RhCl(PPh_3)_3$	85	2	$CH_3CH_2OCH_2CH_2Si(OC_2H_5)_3$ (15)	1168
	SDVB-PPh_2/$RhCl_3$ (EtOH)	80	2	$CH_3CH_2OCH_2CH_2Si(OC_2H_5)_3$ (-)	D -056
	polymer/$RhCl_3$ (THF)	reflux	1.5	$CH_3CH_2OCH_2CH_2Si(OC_2H_5)_3$ (90)	S -029
	polymer-$PPh_2RhCl(PPh_3)_2$	80	3	$CH_3CH_2OCH_2CH_2Si(OC_2H_5)_3$ (34)	C -008
	support/Rh complex	80	2	$CH_3CH_2OCH_2CH_2Si(OC_2H_5)_3$ (0-67)	201
C_4H_8	$CH_3CH_2SCH{=}CH_2$				
	H_2PtCl_6	100-160	3-48	$CH_3CH_2SCH_2CH_2Si(OC_2H_5)_3$ + $CH_3CH_2SCH(CH_3)Si(OC_2H_5)_3$ (37-45)	1274
	H_2PtCl_6 (i-PrOH)	150	44-46	$CH_3CH_2SCH_2CH_2Si(OC_2H_5)_3$ (33) $CH_3CH_2SCH(CH_3)Si(OC_2H_5)_3$ (5) $CH_3CH_2SSi(OC_2H_5)_3$ (22)	1275
	H_2PtCl_6 (i-PrOH)	150	44-46	$CH_3CH_2SCH_2CH_2Si(OC_2H_5)_3$ (33) $CH_3CH_2SCH(CH_3)Si(OC_2H_5)_3$ (5) $CH_3CH_2SSi(OC_2H_5)_3$ (22)	1272
	$RhCl(PPh_3)_3$	100-160	3-48	$CH_3CH_2SCH_2CH_2Si(OC_2H_5)_3$ + $CH_3CH_2SCH(CH_3)Si(OC_2H_5)_3$ (37-45)	1274
	$RhCl(PPh_3)_3$	150	44-46	$CH_3CH_2SCH_2CH_2Si(OC_2H_5)_3$ (44) $CH_3CH_2SCH(CH_3)Si(OC_2H_5)_3$ (4) $CH_3CH_2SSi(OC_2H_5)_3$ (15)	1272
	$RhCl(PPh_3)_3$	150	44-46	$CH_3CH_2SCH_2CH_2Si(OC_2H_5)_3$ (44) $CH_3CH_2SCH(CH_3)Si(OC_2H_5)_3$ (4) $CH_3CH_2SSi(OC_2H_5)_3$ (15)	1275
C_4H_{14}	$\overline{CHB_{10}H_{10}C}CH{=}CH_2$				
	$Co_2(CO)_8$	20-30	60	$\overline{CHB_{10}H_{10}C}CH_2CH_2Si(OC_2H_5)_3$ (35-50)	715

Table 9.3

= $(C_2H_5O)_3SiH$ = =10=

	$Rh_4(CO)_{12}$	20-30	60	$\overline{CHB_{10}H_{10}C}CH_2CH_2Si(OC_2H_5)_3$ (35-50)	715
C_5H_5	$CF_3COOCH_2CH=CH_2$				
	H_2PtCl_6 (i-PrOH)	50-150	10-12	$CF_3COO(CH_2)_3Si(OC_2H_5)_3$ (43)	S -005
	H_2PtCl_6 (i-PrOH)	reflux	10	$CF_3COO(CH_2)_3Si(OC_2H_5)_3$ (42)	931
C_5H_6	$CH_2=CHCOOCH_2CH_3$				
	$Ru_3(CO)_{12}$	150	4 (xy)	$(C_2H_5O)_3SiCH_2CH_2COOCH_2CH_3$ (-) $(C_2H_5O)_3SiCH=CHCOOCH_2CH_3$ (62)	U -173
C_5H_7	$ClCH_2COOCH_2CH=CH_2$				
	H_2PtCl_6 (AllOCH$_2$COOAll)	183-205	6.67	$ClCH_2COO(CH_2)_3Si(OC_2H_5)_3$ (-)	Pl-008
	$Pt(PPh_3)_4$	20-100	6.5	$ClCH_2COO(CH_2)_3Si(OC_2H_5)_3$ (53)	Pl-005
	SiO_2-$\overline{N(CH_2)_3C}H_2/H_2PtCl_6$	20-100	6.5	$ClCH_2COO(CH_2)_3Si(OC_2H_5)_3$ (54)	Pl-007
	SiO_2-$\overline{N(CH_2)_4C}H_2/H_2PtCl_6$	20-100	6.5	$ClCH_2COO(CH_2)_3Si(OC_2H_5)_3$ (53-57)	Pl-007
	SiO_2-$\overline{N(CH_2)_4C}H_2/Pt(PPh_3)_4$	20-100	6.5	$ClCH_2COO(CH_2)_3Si(OC_2H_5)_3$ (25)	Pl-007
	SiO_2-$\overline{NCH_2CH_2OCH_2C}H_2/H_2PtCl_6$	20-100	6.5	$ClCH_2COO(CH_2)_3Si(OC_2H_5)_3$ (55-58)	Pl-007
	SiO_2-$\overline{NCH_2CH_2OCH_2C}H_2/Pt(PPh_3)_4$	20-100	6.5	$ClCH_2COO(CH_2)_3Si(OC_2H_5)_3$ (38-44)	Pl-007
C_5H_8	$CH_2=C(CH_3)COOCH_3$				
	$RhCl(PHPh_2)_3$	120	1	$(C_2H_5O)_3SiCH_2CH(CH_3)COOCH_3$ (61)	C -004
	$[SDVB\text{-}PPh_2RhCl(PPh_3)_2]_n$	80	2 (b)	$(C_2H_5O)_3SiCH_2CH(CH_3)COOCH_3$ (58)	C -008
C_5H_8	$CH_2=CHOCH_2CH=CH_2$				
	H_2PtCl_6	18-40	8.5 (b)	$CH_2=CHO(CH_2)_3Si(OC_2H_5)_3$ (65)	D -047
C_5H_8	$CH_3COOCH_2CH=CH_2$				
	H_2PtCl_6 (AllOCH$_2$COOAll)	160-176	6.74	$CH_3COO(CH_2)_3Si(OC_2H_5)_3$ (56)	Pl-007
	H_2PtCl_6 (i-PrOH)	reflux	10	$CH_3COO(CH_2)_3Si(OC_2H_5)_3$ (53)	931
C_5H_8	$CH_3OCOOCH_2CH=CH_2$				
	H_2PtCl_6 (i-PrOH)	55-180	0.2	$CH_3OCOO(CH_2)_3Si(OC_2H_5)_3$ (41)	1051
	H_2PtCl_6 (i-PrOH)	55-180	0.2	$CH_3OCOO(CH_2)_3Si(OC_2H_5)_3$ (41)	S -026
C_5H_9	$CH_2=CH(CH_2)_3Br$				
	$RhCl(PPh_3)_3$	reflux	2.5	$(C_2H_5O)_3Si(CH_2)_5Br$ (75)	224
C_5H_9	$CH_3CH=NOCH_2CH=CH_2$				

= (C₂H₅O)₃SiH =

	H₂PtCl₆ (THF)	120-160	3-27	CH₃CH=NO(CH₂)₃Si(OC₂H₅)₃ (38)	691

C_5H_{11} $CH_2=CHCH_2N(CH_3)_2$

	Co₂(CO)₈	20	5	(C₂H₅O)₃Si(CH₂)₃N(CH₃)₂ (86)	259
	Co₂(CO)₈	20	5	(C₂H₅O)₃Si(CH₂)₃N(CH₃)₂ (86)	715
	H₂PtCl₆ (THF)	until 140	2	(C₂H₅O)₃Si(CH₂)₃N(CH₃)₂ (70) (C₂H₅O)₃SiCH₂CH(CH₃)N(CH₃)₂ (1)	C -033
	Rh₄(CO)₁₂	20	5	(C₂H₅O)₃Si(CH₂)₃N(CH₃)₂ (86)	715
	Rh₄(CO)₁₂	20	5	(C₂H₅O)₃Si(CH₂)₃N(CH₃)₂ (86)	259
	RhCl(PPh₃)₃	until 140	2	(C₂H₅O)₃Si(CH₂)₃N(CH₃)₂ (80) (C₂H₅O)₃SiCH₂CH(CH₃)N(CH₃)₂ (1)	C -033
	[RhCl(CO)₂]₂	until 140	2	(C₂H₅O)₃Si(CH₂)₃N(CH₃)₂ (36) (C₂H₅O)₃SiCH₂CH(CH₃)N(CH₃)₂ (42)	C -033

C_5H_{11} $CH_2=CHCH_2NHC_2H_5$

	H₂PtCl₆ (i-PrOH)	125	12	(C₂H₅O)₃Si(CH₂)₃NH(C₂H₅) + (C₂H₅O)₃SiCH(CH₃)CH₂NH(C₂H₅) (87)	255

C_5H_{12} $(CH_3)_3SiCH=CH_2$

	Co₂(CO)₈	20-40	-	(CH₃)₃SiCH₂CH₂Si(OC₂H₅)₃ (kin.)	713
	Co₂(CO)₈	20-40	-	(CH₃)₃SiCH₂CH₂Si(OC₂H₅)₃ (kin.)	717
	Co₂(CO)₈	30	2	(CH₃)₃SiCH₂CH₂Si(OC₂H₅)₃ (80)	S -044
	Co₂(CO)₈	30	-	(CH₃)₃SiCH₂CH₂Si(OC₂H₅)₃ (kin.)	714
	Co₂(CO)₈	30-40	1	(CH₃)₃SiCH₂CH₂Si(OC₂H₅)₃ (95-96)	715
	Co₂(CO)₈	65-70	1	(CH₃)₃SiCH₂CH₂Si(OC₂H₅)₃ (77)	31
	Co₂(CO)₈ + RNH₂ R=Et,Pr	30	0.5-0.67	(CH₃)₃SiCH₂CH₂Si(OC₂H₅)₃ (95)	716
	Co₂(CO)₈ + Et₂O	20-30	0.25-0.5	(CH₃)₃SiCH₂CH₂Si(OC₂H₅)₃ (90-96)	S -044
	Co₂(CO)₈ + KI	20-30	0.25-0.5	(CH₃)₃SiCH₂CH₂Si(OC₂H₅)₃ (90-98)	S -044
	Co₂(CO)₈ + PhC≡CPh	5-35	0.25-0.5	(CH₃)₃SiCH₂CH₂Si(OC₂H₅)₃ (90-98)	S -038
	Co₂(CO)₈ + THF	20-30	0.25-0.5	(CH₃)₃SiCH₂CH₂Si(OC₂H₅)₃ (93-98)	S -044
	Co₂(CO)₈ + cocatalyst	30	-	(CH₃)₃SiCH₂CH₂Si(OC₂H₅)₃ (kin.)	719
	Co₂(CO)₈ + cocatalyst	30	-	(CH₃)₃SiCH₂CH₂Si(OC₂H₅)₃ (kin.)	720

Table 9.3

= $(C_2H_5O)_3SiH$ = =12=

	$Co_2(CO)_8$ + cocatalyst	30	-	$(CH_3)_3SiCH_2CH_2Si(OC_2H_5)_3$ (kin.)	714
	$Co_4(CO)_{12}$	20-40	-	$(CH_3)_3SiCH_2CH_2Si(OC_2H_5)_3$ (kin.)	717
	$Co_6(CO)_{16}$	20-40	-	$(CH_3)_3SiCH_2CH_2Si(OC_2H_5)_3$ (kin.)	717
	$[Co(CO)_4]_2SnCl_2$	20-40	-	$(CH_3)_3SiCH_2CH_2Si(OC_2H_5)_3$ (kin.)	717
	$[Co(CO)_4]_4Sn$	20-40	-	$(CH_3)_3SiCH_2CH_2Si(OC_2H_5)_3$ (kin.)	717
	Fe colloid	r.t.	45	$(CH_3)_3SiCH_2CH_2Si(OC_2H_5)_3$ (52)	668
	Os colloid	r.t.	60	$(CH_3)_3SiCH_2CH_2Si(OC_2H_5)_3$ (3)	668
	H_2PtCl_6 (i-PrOH)	50-193	0.58	$(CH_3)_3SiCH_2CH_2Si(OC_2H_5)_3$ (99)	575
	Pt colloid	r.t.	10	$(CH_3)_3SiCH_2CH_2Si(OC_2H_5)_3$ (100)	668
	$Pt[(ViMe_2Si)_2O]_2$	-	-	$(CH_3)_3SiCH_2CH_2Si(OC_2H_5)_3$ (kin.)	669
	Rh colloid	r.t.	40	$(CH_3)_3SiCH_2CH_2Si(OC_2H_5)_3$ (60)	668
	$Rh_4(CO)_{12}$	20-40	-	$(CH_3)_3SiCH_2CH_2Si(OC_2H_5)_3$ (kin.)	717
	$Rh_4(CO)_{12}$	20-40	-	$(CH_3)_3SiCH_2CH_2Si(OC_2H_5)_3$ (kin.)	713
	$Rh_4(CO)_{12}$	30	1	$(CH_3)_3SiCH_2CH_2Si(OC_2H_5)_3$ (85)	S -044
	$Rh_4(CO)_{12}$	30-40	1	$(CH_3)_3SiCH_2CH_2Si(OC_2H_5)_3$ (95-96)	715
	$Rh_4(CO)_{12}$	65-70	1	$(CH_3)_3SiCH_2CH_2Si(OC_2H_5)_3$ (92)	31
	$Rh_6(CO)_{16}$	20-40	-	$(CH_3)_3SiCH_2CH_2Si(OC_2H_5)_3$ (kin.)	717
	Ru colloid	r.t.	30	$(CH_3)_3SiCH_2CH_2Si(OC_2H_5)_3$ (74)	668
	$Ru(acac)_3$	120	6	$(CH_3)_3SiCH_2CH_2Si(OC_2H_5)_3$ (19)	745
	$RuCl_2(PPh_3)_3$	120	6	$(CH_3)_3SiCH_2CH_2Si(OC_2H_5)_3$ (12)	733
C_5H_{12}	$(CH_3O)_3SiCH{=}CH_2$				
	$RuCl_2(PPh_3)_3$	120	6	$(CH_3O)_3SiCH_2CH_2Si(OC_2H_5)_3$ (-)	733
C_5H_{12}	$CH_2{=}CHCH_2NHCH_2CH_2NH_2$				
	$Rh_4(CO)_{12}$	150	-	$(C_2H_5O)_3Si(CH_2)_3NHCH_2CH_2NH_2$ (63)	E -040
	$Rh_4(CO)_{12}$ + CO	150	-	$(C_2H_5O)_3Si(CH_2)_3NHCH_2CH_2NH_2$ (60) $(C_2H_5O)_3SiCH(CH_3)CH_2NHCH_2CH_2NH_2$ (5)	E -040
	$[Rh(\mu\text{-}PPh_2)(cod)]_2$	110	-	$(C_2H_5O)_3Si(CH_2)_3NHCH_2CH_2NH_2$ (67) $(C_2H_5O)_3SiCH(CH_3)CH_2NHCH_2CH_2NH_2$ (1)	E -048
C_5H_{12}	$CH_2{=}CHOSi(CH_3)_3$				

Table 9.3

= $(C_2H_5O)_3SiH$ = =13=

	H_2PtCl_6 (i-PrOH)	until 152	-	$(C_2H_5O)_3SiCH_2CH_2OSi(CH_3)_3$ (60)	D -014
C_5H_{12}	$CH_2=CHSi(CH_3)_3$				
	H_2PtCl_6 (i-PrOH)	-	8	$(CH_3)_3SiCH_2CH_2Si(C_2H_5)_3$ (100)	D -142
	H_2PtCl_6 (i-PrOH)	r.t.	1	$(CH_3)_3SiCH_2CH_2Si(OC_2H_5)_3$ (34)	667
	Pt colloid	-	0.08	$(CH_3)_3SiCH_2CH_2Si(OC_2H_5)_3$ (100)	D -142
	Pt colloid	25	5.1	$(CH_3)_3SiCH_2CH_2Si(OC_2H_5)_3$ (100)	667
	Pt colloid	r.t.	1	$(CH_3)_3SiCH_2CH_2Si(OC_2H_5)_3$ (69)	667
	$Pt(cod)_2$	25	0.25	$(CH_3)_3SiCH_2CH_2Si(OC_2H_5)_3$ (100)	667
	$PtCl_2(cod)$	25	19	$(CH_3)_3SiCH_2CH_2Si(OC_2H_5)_3$ (97)	667
	$PtCl_2(cod)$ (CH_2Cl_2)	r.t.	1	$(CH_3)_3SiCH_2CH_2Si(OC_2H_5)_3$ (7)	667
	$Pt[(CH_2=CHSiMe_2)_2O]$	-	8	$(CH_3)_3SiCH_2CH_2Si(OC_2H_5)_3$ (-)	D -142
C_5H_{16}	$CHB_{10}H_{10}CC(CH_3)=CH_2$				
	$Co_2(CO)_8$	20-30	60	$CHB_{10}H_{10}CCH(CH_3)CH_2Si(OC_2H_5)_3$ (33-45)	715
	$Rh_4(CO)_{12}$	20-30	60	$CHB_{10}H_{10}CCH(CH_3)CH_2Si(OC_2H_5)_3$ (33-45)	715
C_6H_5	$CF_3CF_2COOCH_2CH=CH_2$				
	H_2PtCl_6 (i-PrOH)	50-150	10-12	$CF_3CF_2COO(CH_2)_3Si(OC_2H_5)_3$ (35)	931
	H_2PtCl_6 (i-PrOH)	50-150	10-12	$CF_3CF_2COO(CH_2)_3Si(OC_2H_5)_3$ (35)	S -005
C_6H_8	$CH_2=CHCOOCH_2CH=CH_2$				
	H_2PtCl_6 (PhCOOMe)	106-114	3 (THF)	$CH_2=CHCOO(CH_2)_3Si(OC_2H_5)_3$ (-)	U -016
	H_2PtCl_6 $[(CH_2)_5CO$, $CH_2=CHSi(OEt)_3]$	reflux	-	$CH_2=CHCOO(CH_2)_3Si(OC_2H_5)_3$ (97)	Pl-034
	$Pt(CH_2=CH_2)(PPh_3)_2$	90-147	-	$CH_2=CHCOO(CH_2)_3Si(OC_2H_5)_3$ (98)	Pl-034
C_6H_9	$CH_2=CHCH_2CHCH(CH_2Cl)O$				
	H_2PtCl_6 (i-PrOH)	120-125	20	$(C_2H_5O)_3Si(CH_2)_3CHCH(CH_2Cl)O$ (50)	92
C_6H_{10}	$(CH_2=CHCH_2)_2S$				
	H_2PtCl_6 (i-PrOH)	120	6	$CH_2=CHCH_2S(CH_2)_3Si(OC_2H_5)_3$ (33) $CH_2=CHCH_2SCH_2CH(CH_3)Si(OC_2H_5)_3$ (15)	1241
	H_2PtCl_6 (i-PrOH)	120	6	$CH_2=CHCH_2S(CH_2)_3Si(OC_2H_5)_3$ (33) $CH_2=CHCH_2SCH_2CH(CH_3)Si(OC_2H_5)_3$ (15)	1272

Table 9.3

= $(C_2H_5O)_3SiH$ = =14=

	$RhCl(PPh_3)_3$	120	6	$CH_2=CHCH_2S(CH_2)_3Si(OC_2H_5)_3$ (32) $CH_2=CHCH_2SCH_2CH(CH_3)Si(OC_2H_5)_3$ (3-8)	1241
	$RhCl(PPh_3)_3$	120	6	$CH_2=CHCH_2S(CH_2)_3Si(OC_2H_5)_3$ (32) $CH_2=CHCH_2SCH_2CH(CH_3)Si(OC_2H_5)_3$ (3-8)	1272
C_6H_{10}	$CH_2=CHCH_2OCH_2CHCH_2O$				
	H_2PtCl_6	120-165	1.4	$(C_2H_5O)_3Si(CH_2)_3OCH_2CHCH_2O$ (50-65)	Dd-007
	H_2PtCl_6 (i-PrOH) + EtOH	80	3	$(C_2H_5O)_3Si(CH_2)_3OCH_2CHCH_2O$ (88)	E -042
C_6H_{10}	$CH_2=CHCH_2OCH_2COOCH_3$				
	$PtCl_2(SEt_2)_2$	100-150	5	$(C_2H_5O)_3Si(CH_2)_3OCH_2COOCH_3$ (-)	D -022
C_6H_{10}	$CH_3OCH_2COOCH_2CH=CH_2$				
	$Pt(PPh_3)_4$	138-204	7	$CH_3OCH_2COO(CH_2)_3Si(OC_2H_5)_3$ (59)	Pl-005
	SiO_2-$N(CH_2)_3CH_2/H_2PtCl_6$ (MeOH)	20-100	6.5	$CH_3OCH_2COO(CH_2)_3Si(OC_2H_5)_3$ (64)	Pl-007
	SiO_2-$N(CH_2)_4CH_2/H_2PtCl_6$ (MeOH)	20-100	6.5	$CH_3OCH_2COO(CH_2)_3Si(OC_2H_5)_3$ (57)	Pl-007
	SiO_2-$NCH_2CH_2OCH_2CH_2/H_2PtCl_6$	20-100	6.5	$CH_3OCH_2COO(CH_2)_3Si(OC_2H_5)_3$ (61)	Pl-007
C_6H_{11}	$(CH_2=CHCH_2)_2NH$				
	H_2PtCl_6 (i-PrOH)	reflux	24	$CH_2=CHCH_2NH(CH_2)_3Si(OC_2H_5)_3$ (39)	91
C_6H_{11}	$(CH_3)_2C=NOCH_2CH=CH_2$				
	H_2PtCl_6 (THF)	130-160	3-27	$(CH_3)_2C=NO(CH_2)_3Si(OC_2H_5)_3$ (40)	691
	H_2PtCl_6 (THF)	140-160	3-27	$(CH_3)_2C=NO(CH_2)_3Si(OC_2H_5)_3$ (41)	691
C_6H_{11}	$(CH_3)_2NCOOCH_2CH=CH_2$				
	H_2PtCl_6	135-184	15	$(CH_3)_2NCOO(CH_2)_3Si(OC_2H_5)_3$ (46)	783
C_6H_{11}	$CH_3CH_2OCONHCH_2CH=CH_2$				
	H_2PtCl_6	165-189	6	$CH_3CH_2OCONH(CH_2)_3Si(OC_2H_5)_3$ (30)	783
C_6H_{12}	$CH_2=CHCH_2OCH_2CH(OH)CH_2OH$				
	H_2PtCl_6	110	18	$(C_2H_5O)_3Si(CH_2)_3OCH_2CH(OH)CH_2OH$ (-)	G -009
	H_2PtCl_6	r.t.	46	$(C_2H_5O)_3Si(CH_2)_3OCH_2CH(OH)CH_2OH$ (-)	N -003
C_6H_{12}	$CH_2=CHCH_2OCH_2CH_2OCH_3$				
	H_2PtCl_6	reflux	2	$(C_2H_5O)_3Si(CH_2)_3OCH_2CH_2OCH_3$ (90)	E -012
C_6H_{12}	$CH_2=CHSi(CH_3)_2CH=CH_2$				

	$PtCl_2(Et_2S)_2$	reflux	48 (b)	$(CH_3)_2Si[CH_2CH_2Si(OC_2H_5)_3]_2$ (-)	133
C_6H_{12}	$CH_3CH_2CH_2SCH_2CH=CH_2$				
	H_2PtCl_6 (i-PrOH)	150	46	$CH_3CH_2CH_2S(CH_2)_3Si(OC_2H_5)_3$ (23) $CH_3CH_2CH_2SCH_2CH(CH_3)Si(OC_2H_5)_3$ (22)	1272
	$RhCl(PPh_3)_3$	150	46	$CH_3CH_2CH_2S(CH_2)_3Si(OC_2H_5)_3$ (43) $CH_3CH_2CH_2SCH_2CH(CH_3)Si(OC_2H_5)_3$ (-)	1272
C_6H_{13}	$CH_2=CHCH_2OPO(CH_3)OCH_2CH_3$				
	H_2PtCl_6 (i-PrOH)	110	0.8	$(C_2H_5O)_3Si(CH_2)_3OPO(CH_3)OCH_2CH_3$ (47)	191
C_6H_{14}	$(CH_3)_3SiOCH_2CH=CH_2$				
	H_2PtCl_6	-	-	$(CH_3)_3SiO(CH_2)_3Si(OC_2H_5)_3$ (53)	927
	Pt/C	160	6	$(C_2H_5O)_2\overline{SiO(C}H_2)_3$ (75)	D -005
C_6H_{14}	$CH_2=CHCH_2COOSi(CH_3)_3$				
	H_2PtCl_6 (i-PrOH)	reflux	1.5	$(C_2H_5O)_3Si(CH_2)_3COOSi(CH_3)_3$ (85)	778
C_6H_{14}	$CH_3CH_2OSi(CH_3)_2CH=CH_2$				
	$RuCl_2(PPh_3)_3$	120	6	$CH_3CH_2OSi(CH_3)_2CH_2CH_2Si(OC_2H_5)_3$ (60)	733
C_7H_5	$CF_3CF_2CF_2COOCH_2CH=CH_2$				
	H_2PtCl_6 (i-PrOH)	reflux	10	$CF_3CF_2CF_2COO(CH_2)_3Si(OC_2H_5)_3$ (34)	931
C_7H_8	$\overline{CH_2COOCOC}HCH_2CH=CH_2$				
	$PtCl_4(CH_2=CHHex)$	90	2	$\overline{CH_2COOCOC}H(CH_2)_3Si(OC_2H_5)_3$ (89)	D -119
	$PtCl_4(CH_2=CHHex)$	90	2	$\overline{CH_2COOCOC}H(CH_2)_3Si(OC_2H_5)_3$ (89)	E -005
C_7H_9	Fig.121.				
	H_2PtCl_6	140-160	-	Fig.407. (71)	729
C_7H_{10}	$CH_2=C(CH_3)COOCH_2CH=CH_2$				
	H_2PtCl_6	40-50	3	$CH_2=C(CH_3)COO(CH_2)_3Si(OC_2H_5)_3$ (87)	G -011
	H_2PtCl_6 (PhCOOMe)	105-112	2.5	$CH_2=C(CH_3)COO(CH_2)_3Si(OC_2H_5)_3$ (-)	U -016
	H_2PtCl_6 (i-PrOH) + $CH_2=CHOBu$	21	0.25	$CH_2=C(CH_3)COO(CH_2)_3Si(OC_2H_5)_3$ (73)	S -012
	H_2PtCl_6 (i-PrOH) + $CH_2=CHOBu$	60	0.5	$CH_2=C(CH_3)COO(CH_2)_3Si(OC_2H_5)_3$ (84)	S -012
	$PtBr_2[AcCH(CH_3)COOMe]$	40-50	3	$CH_2=C(CH_3)COO(CH_2)_3Si(OC_2H_5)_3$ (97)	G -011

= $(C_2H_5O)_3SiH$ =   =16=

	$PtCl_2(acac)_2$	40-50	3	$CH_2=C(CH_3)COO(CH_2)_3Si(OC_2H_5)_3$ (97)	D -011

C_7H_{11}	$CH_2=C(CH_3)CH_2\overline{CHCH(CH_2Cl)O}$				
	H_2PtCl_6 (i-PrOH)	120-125	20	$(C_2H_5O)_3SiCH_2CH(CH_3)CH_2\overline{CHCH(CH_2Cl)O}$ (50)	S -013
	H_2PtCl_6 (i-PrOH)	120-125	37	$(C_2H_5O)_3SiCH_2CH(CH_3)CH_2\overline{CHCH(CH_2Cl)O}$ (53)	S -013
	H_2PtCl_6 (i-PrOH)	120-125	5-37	$(C_2H_5O)_3SiCH_2CH(CH_3)CH_2\overline{CHCH(CH_2Cl)O}$ (53-65)	92
	H_2PtCl_6 (i-PrOH) + $CH_2=CHOBu$	90-100	4	$(C_2H_5O)_3SiCH_2CH(CH_3)CH_2\overline{CHCH(CH_2Cl)O}$ (66)	S -012

C_7H_{12}	$CH_3CH_2OCH_2COOCH_2CH=CH_2$				
	$SiO_2-\overline{N(CH_2)_5}/H_2PtCl_6$	20-100	0.5	$CH_3CH_2OCH_2COO(CH_2)_3Si(OC_2H_5)_3$ (25-39)	PI-007
	$SiO_2-\overline{N(CH_2)_5}/Pt(PPh_3)_4$	20-100	0.5	$CH_3CH_2OCH_2COO(CH_2)_3Si(OC_2H_5)_3$ (12)	PI-007
	$SiO_2-\overline{NCH_2CH_2OCH_2CH_2}/H_2PtCl_6$	20-100	0.5	$CH_3CH_2OCH_2COO(CH_2)_3Si(OC_2H_5)_3$ (27)	PI-007
	$SiO_2-\overline{NCH_2CH_2OCH_2CH_2}/Pt(PPh_3)_4$	20-100	0.5	$CH_3CH_2OCH_2COO(CH_2)_3Si(OC_2H_5)_3$ (3-6)	PI-007

C_7H_{14}	$(C_2H_5O)_2CHCH=CH_2$				
	H_2PtCl_6 (i-PrOH)	100-162	1	$(C_2H_5O)_2CHCH_2CH_2Si(OC_2H_5)_3$ (65)	1004

C_7H_{14}	$(CH_3)_3SiOCOC(CH_3)=CH_2$				
	H_2PtCl_6	80-120	3	$(CH_3)_3SiOCOCH(CH_3)CH_2Si(OC_2H_5)_3$ (11)	1247
	H_2PtCl_6 (i-PrOH)	100-180	7	$(CH_3)_3SiOCOCH(CH_3)CH_2Si(OC_2H_5)_3$ (29)	355

C_7H_{14}	$(CH_3)_3SiOCOCH_2CH=CH_2$				
	H_2PtCl_6 (i-PrOH)	100-180	7	$(CH_3)_3SiOCO(CH_2)_3Si(OC_2H_5)_3$ (76)	355

C_7H_{15}	$(CH_3)_2N(CH_2)_3CH=CH_2$				
	H_2PtCl_6	-	-	$(CH_3)_2N(CH_2)_5Si(OC_2H_5)_3$ (51)	C -034
	$RhCl(PPh_3)_3$	until 150	3	$(CH_3)_2N(CH_2)_5Si(OC_2H_5)_3$ (43)	C -034

C_7H_{15}	$(CH_3CH_2)_2NCH_2CH=CH_2$				
	H_2PtCl_6 (i-PrOH)	reflux	24	$(CH_3CH_2)_2N(CH_2)_3Si(OC_2H_5)_3$ (56)	91

C_7H_{15}	$CH_2=CHCH_2OPO(OC_2H_5)_2$				
	H_2PtCl_6 (i-PrOH)	105-115	2	$(C_2H_5O)_3Si(CH_2)_3OPO(OC_2H_5)_3$ (26)	191

C_7H_{15}	$CH_3(CH_2)_3NHCH_2CH=CH_2$				
	H_2PtCl_6 (i-PrOH)	125	12	$CH_3(CH_2)_3NH(CH_2)_3Si(OC_2H_5)_3$ + $CH_3(CH_2)_3NHCH_2CH(CH_3)$ (72)	255

C_7H_{16}	$(C_2H_5O)_2Si(CH_3)CH=CH_2$				

	$RuCl_2(PPh_3)_3$	120	6	$(C_2H_5O)_2Si(CH_3)CH_2CH_2Si(OC_2H_5)_3$ (62)	733
C_7H_{16}	$CH_2{=}C(CH_3)COOCH_2CH{=}CH_2$				
	$Pt(PPh_3)_2(CH_2{=}CH_2)$	-	2	$CH_2{=}C(CH_3)COO(CH_2)_3Si(OC_2H_5)_3$ (98)	748
C_8H_5	$CF_3(CF_2)_3COOCH_2CH{=}CH_2$				
	H_2PtCl_6 (i-PrOH)	50-150	10-12	$CF_3(CF_2)_3COO(CH_2)_3Si(OC_2H_5)_3$ (34)	S -005
C_8H_7	p-$ClC_6H_4CH{=}CH_2$				
	$Pt[(ViMe_2Si)_2O]_2$	-	-	p-$ClC_6H_4CH_2CH_2Si(OC_2H_5)_3$ + p-$ClC_6H_4CH(CH_3)Si(OC_2H_5)_3$ (kin.)	669
C_8H_9	p-$NH_2C_6H_4CH{=}CH_2$				
	H_2PtCl_6 (i-PrOH)	-	-	p-$NH_2C_6H_4CH_2CH_2Si(OC_2H_5)_3$ (80)	751
C_8H_{11}	Fig.58.				
	H_2PtCl_6 (i-PrOH)	215	2	Fig.408. (65)	731
C_8H_{12}	$(CH_3CO)_2CHCH_2CH{=}CH_2$				
	Pt/C	100	24	$(CH_3CO)_2CH(CH_2)_3Si(OC_2H_5)_3$ (60)	14
C_8H_{12}	$CH_2{=}C(CH_3)COOCH_2C(CH_3){=}CH_2$				
	Pt/Al_2O_3	-	-	$CH_2{=}C(CH_3)COOCH_2CH(CH_3)CH_2Si(OC_2H_5)_3$ (-)	U -016
C_8H_{12}	Fig.282.				
	H_2PtCl_6	r.t	1.4	Fig.409. (-)	J -041
C_8H_{14}	$\overline{OCH_2CH_2OCH_2C}HCH_2OCH_2CH{=}CH_2$				
	H_2PtCl_6 (i-PrOH)	reflux	52	$\overline{OCH_2CH_2OCH_2C}HCH_2O(CH_2)_3Si(OC_2H_5)_3$ (53)	554
C_8H_{15}	$(\overline{CH_2)_5N}CH_2CH{=}CH_2$				
	H_2PtCl_6 (i-PrOH)	100	4	$(\overline{CH_2)_5N}(CH_2)_3Si(OC_2H_5)_3$ (67)	702
C_8H_{15}	$CH_2{=}CHSi(OCH_2CH_2)_3N$				
	$Rh(CO)_2(acac)$	120	8 (b)	$(C_2H_5O)_3SiCH_2CH_2Si(OCH_2CH_2)_3N$ (85)	1237
C_8H_{16}	$(CH_3)_2CHN{=}CHCH{=}NCH(CH_3)_2$				
	[Rh(CO)Cl(DAD)]	r.t.	5	$(CH_3)_2CHNCH{=}CHN[CH(CH_3)_2]Si(OC_2H_5)_3$ (20) $NH{=}CHCH_2N[CH(CH_3)_2]Si(OC_2H_5)_3$ (15)	165
C_8H_{16}	$CH_2{=}CHCH_2O(CH_2CH_2O)_2CH_3$				

	H_2PtCl_6	reflux	2	$(C_2H_5O)_3Si(CH_2)_3O(CH_2CH_2O)_2CH_3$ (82)	E -012
C_8H_{16}	$CH_3CH_2S(CH_2)_4CH{=}CH_2$				
	H_2PtCl_6 (i-PrOH)	150	46	$CH_3CH_2S(CH_2)_4CH(CH_3)Si(OC_2H_5)_3$ (79)	1272
	H_2PtCl_6 (i-PrOH)	150-160	48	$CH_3CH_2S(CH_2)_6Si(OC_2H_5)_3$ (9)	1271
	$RhCl(PPh_3)_3$	150	46	$CH_3CH_2S(CH_2)_4CH(CH_3)Si(OC_2H_5)_3$ (23)	1272
	$RhCl(Ph_3)_3$	150-160	48	$CH_3CH_2S(CH_2)_6Si(OC_2H_5)_3$ (23)	1271
	polymer/Rh catalyst	110-120	7	$CH_3CH_2S(CH_2)_6Si(OC_2H_5)_3$ (77)	S -046
C_8H_{18}	$(C_2H_5O)_3SiCH{=}CH_2$				
	$Ni(acac)_2$	120	2	$(C_2H_5O)_3SiCH_2CH_2Si(OC_2H_5)_3$ (6) $(C_2H_5O)_3SiCH(CH_3)Si(OC_2H_5)_3$ (29) $(C_2H_5O)_3SiCH{=}CHSi(OC_2H_5)_3$ (10)	745
	$Ni(acac)_2$	60	2	$(C_2H_5O)_3SiCH_2CH_2Si(OC_2H_5)_3$ (6) $(C_2H_5O)_3SiCH(CH_3)Si(OC_2H_5)_3$ (21) $(C_2H_5O)_3SiCH{=}CHSi(OC_2H_5)_3$ (13)	745
	H_2PtCl_6	80	2	$(C_2H_5O)_3SiCH_2CH_2Si(OC_2H_5)_3$ (84) $(C_2H_5O)_3SiCH(CH_3)Si(OC_2H_5)_3$ (11)	748
	H_2PtCl_6 (i-PrOH)	120	6	$(C_2H_5O)_3SiCH_2CH_2Si(OC_2H_5)_3$ (82) $(C_2H_5O)_3SiCH(CH_3)Si(OC_2H_5)_3$ (8)	748
	H_2PtCl_6 (i-PrOH)	80	2	$(C_2H_5O)_3SiCH_2CH_2Si(OC_2H_5)_3$ (78) $(C_2H_5O)_3SiCH(CH_3)Si(OC_2H_5)_3$ (8)	748
	H_2PtCl_6 + $(NPCl_2)_3$	50-120	- (THF)	$(C_2H_5O)_3SiCH_2CH_2Si(OC_2H_5)_3$ (kin.)	615
	H_2PtCl_6 + $N_3P_3(NMe_2)_4Cl_2$	50-120	- (THF)	$(C_2H_5O)_3SiCH_2CH_2Si(OC_2H_5)_3$ (kin.)	615
	H_2PtCl_6 + $[NP(NHBu)_2]_3$	50-120	- (THF)	$(C_2H_5O)_3SiCH_2CH_2Si(OC_2H_5)_3$ (kin.)	615
	H_2PtCl_6 + $[NP(NHC_6H_{13})_2]_3$	50-120	- (THF)	$(C_2H_5O)_3SiCH_2CH_2Si(OC_2H_5)_3$ (kin.)	615
	H_2PtCl_6 + $[NP(NHEt)_2]_3$	50-120	- (THF)	$(C_2H_5O)_3SiCH_2CH_2Si(OC_2H_5)_3$ (kin.)	615
	H_2PtCl_6 + siloxyphosphazene	50-120	- (THF)	$(C_2H_5O)_3SiCH_2CH_2Si(OC_2H_5)_3$ (kin.)	615
	H_2PtCl_6 $[\overline{(CH_2)_5CO}]$	120	6	$(C_2H_5O)_3SiCH_2CH_2Si(OC_2H_5)_3$ (61) $(C_2H_5O)_3SiCH(CH_3)Si(OC_2H_5)_3$ (7)	745
	H_2PtCl_6 $[\overline{(CH_2)_5CO}]$	120	6	$(C_2H_5O)_3SiCH_2CH_2Si(OC_2H_5)_3$ (61) $(C_2H_5O)_3SiCH(CH_3)Si(OC_2H_5)_3$ (7)	740
	H_2PtCl_6 $[\overline{(CH_2)_5CO}]$	120	6	$(C_2H_5O)_3SiCH_2CH_2Si(OC_2H_5)_3$ (87) $(C_2H_5O)_3SiCH(CH_3)Si(OC_2H_5)_3$ (6)	748

Table 9.3

= $(C_2H_5O)_3SiH$ = =19=

H_2PtCl_6 [$(CH_2)_5CO$]	80	2	$(C_2H_5O)_3SiCH_2CH_2Si(OC_2H_5)_3$ (62) $(C_2H_5O)_3SiCH(CH_3)Si(OC_2H_5)_3$ (7)	PI-006
H_2PtCl_6 [$(CH_2)_5CO$]	80	2	$(C_2H_5O)_3SiCH_2CH_2Si(OC_2H_5)_3$ (62) $(C_2H_5O)_3SiCH(CH_3)Si(OC_2H_5)_3$ (7)	745
H_2PtCl_6 [$(CH_2)_5CO$]	80	2	$(C_2H_5O)_3SiCH_2CH_2Si(OC_2H_5)_3$ (62) $(C_2H_5O)_3SiCH(CH_3)Si(OC_2H_5)_3$ (7)	740
H_2PtCl_6 [$(CH_2)_5CO$]	reflux	2	$(C_2H_5O)_3SiCH_2CH_2Si(OC_2H_5)_3$ (75) $(C_2H_5O)_3SiCH(CH_3)Si(OC_2H_5)_3$ (2)	740
H_2PtCl_6 [$(CH_2)_5CO$]	reflux	2	$(C_2H_5O)_3SiCH_2CH_2Si(OC_2H_5)_3$ (75) $(C_2H_5O)_3SiCH(CH_3)Si(OC_2H_5)_3$ (2)	745
$Pt(PPh_3)_2(CH_2=CH_2)$	120	6	$(C_2H_5O)_3SiCH_2CH_2Si(OC_2H_5)_3$ (79) $(C_2H_5O)_3SiCH(CH_3)Si(OC_2H_5)_3$ (7)	748
$Pt(PPh_3)_2(CH_2=CH_2)$	80	2	$(C_2H_5O)_3SiCH_2CH_2Si(OC_2H_5)_3$ (89) $(C_2H_5O)_3SiCH(CH_3)Si(OC_2H_5)_3$ (3)	748
$Pt(PPh_3)_4$	120	6	$(C_2H_5O)_3SiCH_2CH_2Si(OC_2H_5)_3$ (69) $(C_2H_5O)_3SiCH(CH_3)Si(OC_2H_5)_3$ (6)	748
$PtCl_2(Et_2S)_2$	120	48 (b)	$(C_2H_5O)_3SiCH_2CH_2Si(OC_2H_5)_3$ (58)	133
$PtCl_2(PPh_3)_2$	120	6	$(C_2H_5O)_3SiCH_2CH_2Si(OC_2H_5)_3$ (78) $(C_2H_5O)_3SiCH(CH_3)Si(OC_2H_5)_3$ (8)	748
$PtCl_2(PPh_3)_2$	80	2	$(C_2H_5O)_3SiCH_2CH_2Si(OC_2H_5)_3$ (22) $(C_2H_5O)_3SiCH(CH_3)Si(OC_2H_5)_3$ (2)	748
siloxane-$NHCH_2CH_2NH_2/H_2PtCl_6$	60-85	3	$(C_2H_5O)_3SiCH_2CH_2Si(OC_2H_5)_3$ (85)	D -071
Rh complex	r.t.	3	$(C_2H_5O)_3SiCH_2CH_2Si(OC_2H_5)_3$ (49) $(C_2H_5O)_3SiCH(CH_3)Si(OC_2H_5)_3$ (49)	166
$RhCl(PPh_3)_3$	120	2-6	$(C_2H_5O)_3SiCH_2CH_2Si(OC_2H_5)_3$ (68-74) $(C_2H_5O)_3SiCH(CH_3)Si(OC_2H_5)_3$ (7)	PI-006
$RhCl(PPh_3)_3$	120	2-6	$(C_2H_5O)_3SiCH_2CH_2Si(OC_2H_5)_3$ (68-74) $(C_2H_5O)_3SiCH(CH_3)Si(OC_2H_5)_3$ (7)	745
$RhCl(PPh_3)_3$	120	2	$(C_2H_5O)_3SiCH_2CH_2Si(OC_2H_5)_3$ (90) $(C_2H_5O)_3SiCH(CH_3)Si(OC_2H_5)_3$ (6)	743
$RhCl(PPh_3)_3$	80	2	$(C_2H_5O)_3SiCH_2CH_2Si(OC_2H_5)_3$ (70) $(C_2H_5O)_3SiCH(CH_3)Si(OC_2H_5)_3$ (5)	745
$RhCl(PPh_3)_3$	80	2	$(C_2H_5O)_3SiCH_2CH_2Si(OC_2H_5)_3$ (70) $(C_2H_5O)_3SiCH(CH_3)Si(OC_2H_5)_3$ (5)	740
$RhCl(PPh_3)_3$	reflux	6	$(C_2H_5O)_3SiCH_2CH_2Si(OC_2H_5)_3$ (75) $(C_2H_5O)_3SiCH(CH_3)Si(OC_2H_5)_3$ (6)	745

Table 9.3

= $(C_2H_5O)_3SiH$ = =20=

$RhCl(PPh_3)_3$	reflux	6	$(C_2H_5O)_3SiCH_2CH_2Si(OC_2H_5)_3$ (75) $(C_2H_5O)_3SiCH(CH_3)Si(OC_2H_5)_3$ (6)	740
$[RhCl(c\text{-}C_8H_{14})_2]_2$	120	2	$(C_2H_5O)_3SiCH_2CH_2Si(OC_2H_5)_3$ (82) $(C_2H_5O)_3SiCH(CH_3)Si(OC_2H_5)_3$ (13)	743
asbestos-PPh_2/$RhCl(PPh_3)_3$	120	6	$(C_2H_5O)_3SiCH_2CH_2Si(OC_2H_5)_3$ (67) $(C_2H_5O)_3SiCH(CH_3)Si(OC_2H_5)_3$ (20)	PI-020
chrysotile-PPh_2/$RhCl(PPh_3)_3$	120	6	$(C_2H_5O)_3SiCH_2CH_2Si(OC_2H_5)_3$ (58) $(C_2H_5O)_3SiCH(CH_3)Si(OC_2H_5)_3$ (19)	PI-020
polymer/Rh complex	110-120	7	$(C_2H_5O)_3SiCH_2CH_2Si(OC_2H_5)_3$ (77)	932
polymer/Rh complex	110-120	7	$(C_2H_5O)_3SiCH_2CH_2Si(OC_2H_5)_3$ (77)	S -046
siloxane-$N(All)_2$/$[RhCl(c\text{-}C_8H_{14})_2]_2$	120	2	$(C_2H_5O)_3SiCH_2CH_2Si(OC_2H_5)_3$ (58-70) $(C_2H_5O)_3SiCH(CH_3)Si(OC_2H_5)_3$ (19-26)	743
siloxane-$N(All)_2$/$[RhCl(c\text{-}C_8H_{14})_2]_2$	80	2	$(C_2H_5O)_3SiCH_2CH_2Si(OC_2H_5)_3$ (67-73) $(C_2H_5O)_3SiCH(CH_3)Si(OC_2H_5)_3$ (20-25)	743
vermiculite-PPh_2/$RhCl(PPh_3)_3$	120	6	$(C_2H_5O)_3SiCH_2CH_2Si(OC_2H_5)_3$ (54) $(C_2H_5O)_3SiCH(CH_3)Si(OC_2H_5)_3$ (19)	PI-020
$Ru(acac)_3$	120	6	$(C_2H_5O)_3SiCH_2CH_2Si(OC_2H_5)_3$ (72)	745
$Ru(acac)_3$	120	6	$(C_2H_5O)_3SiCH_2CH_2Si(OC_2H_5)_3$ (72)	PI-015
$Ru(acac)_3$	80	2	$(C_2H_5O)_3SiCH_2CH_2Si(OC_2H_5)_3$ (62)	745
$Ru(acac)_3$	80	2	$(C_2H_5O)_3SiCH_2CH_2Si(OC_2H_5)_3$ (62)	740
$Ru(acac)_3$	reflux	2	$(C_2H_5O)_3SiCH_2CH_2Si(OC_2H_5)_3$ (90)	740
$Ru(acac)_3$	reflux	3	$(C_2H_5O)_3SiCH_2CH_2Si(OC_2H_5)_3$ (94)	740
$Ru(acac)_3$	reflux	3	$(C_2H_5O)_3SiCH_2CH_2Si(OC_2H_5)_3$ (94)	PI-006
$Ru(acac)_3$	reflux	2	$(C_2H_5O)_3SiCH_2CH_2Si(OC_2H_5)_3$ (90)	745
$RuCl_2(PPh_3)_3$	120	6	$(C_2H_5O)_3SiCH_2CH_2Si(OC_2H_5)_3$ (60)	PI-001
$RuCl_2(PPh_3)_3$	120	6	$(C_2H_5O)_3SiCH_2CH_2Si(OC_2H_5)_3$ (60)	740
$RuCl_2(PPh_3)_3$	120	6-24	$(C_2H_5O)_3SiCH_2CH_2Si(OC_2H_5)_3$ (66-73) $(C_2H_5O)_3SiCH{=}CHSi(OC_2H_5)_3$ (5-6)	745
$RuCl_2(PPh_3)_3$	120	6-24	$(C_2H_5O)_3SiCH_2CH_2Si(OC_2H_5)_3$ (66-73) $(C_2H_5O)_3SiCH{=}CHSi(OC_2H_5)_3$ (5-6)	733
$RuCl_2(PPh_3)_3$	80	2	$(C_2H_5O)_3SiCH_2CH_2Si(OC_2H_5)_3$ (57)	745
$RuCl_2(PPh_3)_3$	80	2	$(C_2H_5O)_3SiCH_2CH_2Si(OC_2H_5)_3$ (57)	740

Table 9.3

= $(C_2H_5O)_3SiH$ = =21=

	$RuCl_2(PPh_3)_3$	reflux	0.5	$(C_2H_5O)_3SiCH_2CH_2Si(OC_2H_5)_3$ (46) $(C_2H_5O)_3SiCH{=}CHSi(OC_2H_5)_3$ (14)	745
	$RuCl_2(PPh_3)_3$	reflux	2	$(C_2H_5O)_3SiCH_2CH_2Si(OC_2H_5)_3$ (78) $(C_2H_5O)_3SiCH{=}CHSi(OC_2H_5)_3$ (6)	745
	$RuCl_2(PPh_3)_3$	reflux	2	$(C_2H_5O)_3SiCH_2CH_2Si(OC_2H_5)_3$ (78) $(C_2H_5O)_3SiCH{=}CHSi(OC_2H_5)_3$ (6)	740
	$RuCl_2(PPh_3)_3$	reflux	0.5	$(C_2H_5O)_3SiCH_2CH_2Si(OC_2H_5)_3$ (46) $(C_2H_5O)_3SiCH{=}CHSi(OC_2H_5)_3$ (14)	740
	$RuCl_3$	120	6	$(C_2H_5O)_3SiCH_2CH_2Si(OC_2H_5)_3$ (77)	740
	$RuCl_3$	120	6	$(C_2H_5O)_3SiCH_2CH_2Si(OC_2H_5)_3$ (77)	PI-015
	$RuCl_3 + PPh_3$	120	6	$(C_2H_5O)_3SiCH_2CH_2Si(OC_2H_5)_3$ (86)	740
	$RuCl_3 + PPh_3$	120	6	$(C_2H_5O)_3SiCH_2CH_2Si(OC_2H_5)_3$ (86)	PI-015
	$RuCl_3{\cdot}H_2O$	120	6	$(C_2H_5O)_3SiCH_2CH_2Si(OC_2H_5)_3$ (77)	PI-013
	$RuCl_3{\cdot}H_2O + PPh_3$	120	6	$(C_2H_5O)_3SiCH_2CH_2Si(OC_2H_5)_3$ (86)	PI-013
	$RuCl_3(PPh_3)_3$	120	6	$(C_2H_5O)_3SiCH_2CH_2Si(OC_2H_5)_3$ (75)	740
	$RuCl_3(PPh_3)_3$	120	6	$(C_2H_5O)_3SiCH_2CH_2Si(OC_2H_5)_3$ (75-86)	PI-013
	$RuCl_3(PPh_3)_3$	120	6	$(C_2H_5O)_3SiCH_2CH_2Si(OC_2H_5)_3$ (75)	PI-015
	$RuCl_3(PPh_3)_3$	reflux	6	$(C_2H_5O)_3SiCH_2CH_2Si(OC_2H_5)_3$ (86)	PI-015
	$RuCl_3(PPh_3)_3$	reflux	6	$(C_2H_5O)_3SiCH_2CH_2Si(OC_2H_5)_3$ (86)	740
C_8H_{19}	$NH[Si(CH_3)_2CH{=}CH_2]_2$				
	H_2PtCl_6 $[CH_2{=}CHSiMe_2]_2O$	-	0.33	$NH[Si(CH_3)_2C_2H_4Si(OC_2H_5)_3]$ (87)	E -050
C_9H_5	$C_6Cl_5OCH_2CH{=}CH_2$				
	H_2PtCl_6 ($\overline{C_6H_4NHC_6H_4S}$)	80-100	0.25	$C_6Cl_5O(CH_2)_3Si(OC_2H_5)_3$ (-)	U -132
C_9H_5	$CF_3CF_2CF_2OCF(CF_3)COOCH_2CH{=}CH_2$				
	H_2PtCl_6	80	3	$CF_3CF_2CF_2OCF(CF_3)COO(CH_2)_3Si(OC_2H_5)_3$ (79)	J -049
C_9H_9	$o\text{-}BrC_6H_4OCH_2CH{=}CH_2$				
	H_2PtCl_6 (i-PrOH)	reflux	4	$o\text{-}BrC_6H_4O(CH_2)_3Si(OC_2H_5)_3$ (30)	102
	H_2PtCl_6 (i-PrOH)	reflux	4	$o\text{-}BrC_6H_4O(CH_2)_3Si(OC_2H_5)_3$ (30)	S -008
C_9H_9	$o\text{-}ClC_6H_4OCH_2CH{=}CH_2$				

	H_2PtCl_6 (i-PrOH)	reflux	4	o-$ClC_6H_4O(CH_2)_3Si(OC_2H_5)_3$ (46)	102
	H_2PtCl_6 (i-PrOH)	reflux	4	o-$ClC_6H_4O(CH_2)_3Si(OC_2H_5)_3$ (46)	S -008
C_9H_9	o-$O_2NC_6H_4OCH_2CH=CH_2$				
	H_2PtCl_6 (i-PrOH)	reflux	4	o-$O_2NC_6H_4O(CH_2)_3Si(OC_2H_5)_3$ (53)	102
C_9H_9	p-$O_2NC_6H_4OCH_2CH=CH_2$				
	H_2PtCl_6	80-110	2	p-$O_2NC_6H_4O(CH_2)_3Si(OC_2H_5)_3$ (81)	D -075
C_9H_{10}	$C_6H_5OCH_2CH=CH_2$				
	H_2PtCl_6 (i-PrOH + 7,8-$C_2B_9H_{12}^-K^+$)	70	3	$C_6H_5OC_3H_6Si(OC_2H_5)_3$ (78)	96
	H_2PtCl_6 (i-PrOH)	100	5	$C_6H_5OC_3H_6Si(OC_2H_5)_3$ (76)	96
	H_2PtCl_6 (i-PrOH)	reflux	4	$C_6H_5O(CH_2)_3Si(OC_2H_5)_3$ (70)	102
	H_2PtCl_6 (i-PrOH)	reflux	4	$C_6H_5O(CH_2)_3Si(OC_2H_5)_3$ (70)	S -008
	polymer/Rh (dg)$(PPh_3)_3$	100-105	12	$C_6H_5O(CH_2)_3Si(OC_2H_5)_3$ (70-74)	S -046
	polymer/Rh complex	100-115	12	$C_6H_5O(CH_2)_3Si(OC_2H_5)_3$ (67-74)	932
	polymer/Rh complex	100-115	12	$C_6H_5O(CH_2)_3Si(OC_2H_5)_3$ (67-74)	S -046
	polymer/Rh complex	100-115	12	$C_6H_5O(CH_2)_3Si(OC_2H_5)_3$ (78-83)	S -046
	polymer/Rh complex	100-115	12	$C_6H_5O(CH_2)_3Si(OC_2H_5)_3$ (78-83)	932
C_9H_{10}	o-$HOC_6H_4CH_2CH=CH_2$				
	H_2PtCl_6 ($MeOC_2H_4OMe$)	reflux	2 (b)	o-$HOC_6H_4(CH_2)_3Si(OC_2H_5)_3$ (-)	F -006
C_9H_{10}	p-$CH_3OC_6H_4CH=CH_2$				
	$Pt[(ViMe_2Si)_2O]_2$	-	-	p-$CH_3OC_6H_4CH_2CH_2Si(OC_2H_5)_3$ + p-$CH_3OC_6H_4CH(CH_3)Si(OC_2H_5)_3$ (kin.)	669
C_9H_{11}	$C_6H_5NHCH_2CH=CH_2$				
	H_2PtCl_6 (i-PrOH)	reflux	24	$C_6H_5NH(CH_2)_3Si(OC_2H_5)_3$ (72)	91
	H_2PtCl_6+ Na_2CO_3	10	7	$C_6H_5NH(CH_2)_3Si(OC_2H_5)_3$ (-) $C_6H_5NHCH_2CH(CH_3)Si(OC_2H_5)_3$ (-)	E -008
	H_2PtCl_6+ Na_2CO_3	10	7	$C_6H_5NH(CH_2)_3Si(OC_2H_5)_3$ (-) $C_6H_5NHCH_2CH(CH_3)Si(OC_2H_5)_3$ (-)	U -144
C_9H_{14}	$\overline{OCH_2C}HCH_2N(CH_2CH_2CN)CH_2CH=CH_2$				
	H_2PtCl_6	reflux	13	$\overline{OCH_2C}HCH_2N(CH_2CH_2CN)(CH_2)_3Si(OC_2H_5)_3$ (39)	1151
C_9H_{15}	$(CH_2=CHCH_2)_3N$				

	H_2PtCl_6	130	0.25	$(C_2H_5O)_3SiC_3H_6N(CH_2CH{=}CH_2)_2$ + $[(C_2H_5O)_3SiC_3H_6]_2NCH_2CH{=}CH_2$ + $[(C_2H_5O)_3SiC_3H_6]_3N$ (61)	96
	H_2PtCl_6 (i-PrOH)	reflux	24	$(C_2H_5O)_3SiC_3H_6N(CH_2CH{=}CH_2)_2$ + $[(C_2H_5)_3SiC_3H_6]_2NCH_2CH{=}CH_2$ + $[(C_2H_5O)_3SiC_3H_6]_3N$ (81)	91
	H_2PtCl_6 (i-PrOH) +7,8-$C_2B_9H_{12}^{-}K^{+}$	130	0.5	$(C_2H_5O)_3SiC_3H_6N(CH_2CH{=}CH_2)_2$ + $[(C_2H_5O)_3SiC_3H_6]_2NCH_2CH{=}CH_2$ + $[(C_2H_5O)_3SiC_3H_6]_3N$ (60)	96
C_9H_{15}	$\overline{CH_2CH_2O(CH_2CH_2)_2N}CH_2COOCH_2CH{=}CH_2$				
	H_2PtCl_6 ($AllOCH_2COOAll$)	143-236	2.6	$\overline{CH_2CH_2O(CH_2CH_2)_2N}CH_2COO(CH_2)_3Si(OC_2H_5)_3$ (-)	Pl-028
C_9H_{15}	$\overline{CH_2CH_2OCH_2CH_2N}CH_2COOCH_2CH{=}CH_2$				
	H_2PtCl_6 ($AllOCH_2COOAll$)	143-236	2.5	$\overline{CH_2CH_2OCH_2CH_2N}CH_2COO(CH_2)_3Si(OC_2H_5)_3$ (62)	Pl-027
	H_2PtCl_6 ($AllOCH_2COOAll$)	143-236	2.5	$\overline{CH_2CH_2OCH_2CH_2N}CH_2COO(CH_2)_3Si(OC_2H_5)_3$ (62)	Pl-028
C_9H_{18}	$CH_2{=}CHC(CH_3)_2CH_2CH{=}NN(CH_3)_2$				
	H_2PtCl_6 (OctOH)	until 185	12	$(C_2H_5O)_3SiCH_2CH_2C(CH_3)_2CH_2CH{=}NN(CH_3)_2$ (-)	U -078
C_9H_{18}	$CH_2{=}CHCH_2O(CH_2CH_2O)_2CH_2CH_3$				
	H_2PtCl_6	reflux	2	$(C_2H_5O)_3Si(CH_2)_3O(CH_2CH_2O)_2CH_2CH_3$ (91)	E -012
C_9H_{18}	$CH_3(CH_2)_3OCH(CH_3)OCH_2CH{=}CH_2$				
	H_2PtCl_6 (i-PrOH)	100-105	4	$CH_3(CH_2)_3OCH(CH_3)O(CH_2)_3Si(OC_2H_5)_3$ (60)	1005
C_9H_{18}	$[OSi(CH_3)CH{=}CH_2]_3$				
	H_2PtCl_6	90-120	3	$[OSi(CH_3)CH_2CH_2Si(OC_2H_5)_3]_3$ (99)	S -050
C_9H_{19}	$(CH_3CH_2CH_2)_2NCH_2CH{=}CH_2$				
	H_2PtCl_6 (i-PrOH)	110-115	38	$(CH_3CH_2CH_2)_2N(CH_2)_3Si(OC_2H_5)_3$ + $(CH_3CH_2H_2)_2NCH_2CH(CH_3)Si(OC_2H_5)_3$ (69)	255
C_9H_{19}	$CH_3(CH_2)_3N(CH_3CH_2)CH_2CH{=}CH_2$				
	H_2PtCl_6 (i-PrOH)	120	50	$CH_3(CH_2)_3N(CH_3CH_2)(CH_2)_3Si(OC_2H_5)_3$ + $CH_3(CH_2)_3N(CH_3CH_2)[CH_2CH(CH_3)]Si(OC_2H_5)_3$ (29)	255
C_9H_{20}	$(C_2H_5O)_3SiCH_2CH{=}CH_2$				
	polymer/Rh complex	120	4	$(C_2H_5O)_3Si(CH_2)_3Si(OC_2H_5)_3$ (50)	932
	polymer/Rh complex	120	4	$(C_2H_5O)_3Si(CH_2)_3Si(OC_2H_5)_3$ (50)	S -046
$C_{10}H_{10}$	$C_6H_4(CH{=}CH_2)_2$				

Table 9.3

= $(C_2H_5O)_3SiH$ = =24=

	H_2PtCl_6 (i-PrOH)	80	6	$C_6H_4[CH_2CH_2Si(OC_2H_5)_3]_2$ + $C_6H_4[CH_2CH_2Si(OC_2H_5)_3][CH(CH_3)Si(OC_2H_5)_3]$ + $C_6H_4[CH(CH_3)Si(OC_2H_5)_3]_2$ (93)	D -131
$C_{10}H_{12}$	$CH_2=CHCH_2OCOCHCH(COOCH_2CH=CH_2)O$				
	Pt	reflux	3 (b)	$(C_2H_5O)_3Si(CH_2)_3OCOCHCH[COO(CH_2)_3Si(OC_2H_5)_3]O$ (-)	J -047
$C_{10}H_{12}$	o-$CH_3C_6H_4OCH_2CH=CH_2$				
	H_2PtCl_6 (i-PrOH)	reflux	4	o-$CH_3C_6H_4O(CH_2)_3Si(OC_2H_5)_3$ (50)	102
	H_2PtCl_6 (i-PrOH)	reflux	4	o-$CH_3C_6H_4O(CH_2)_3Si(OC_2H_5)_3$ (50)	S -008
$C_{10}H_{13}$	$C_6H_5CH_2N(H)CH_2CH=CH_2$				
	H_2PtCl_6 (i-PrOH)	125	27	$C_6H_5CH_2N(H)(CH_2)_3Si(OC_2H_5)_3$ + $C_6H_5CH_2N(H)CH_2CH(CH_3)Si(OC_2H_5)_3$ (43)	255
$C_{10}H_{13}$	Fig.410.				
	H_2PtCl_6 ($CHCl_3$ - THF)	reflux	12	Fig.411. (-)	J -110
$C_{10}H_{13}$	Fig.412.				
	$PtCl_2(cpd)_2$ (CH_2Cl_2)	100	4 (t)	Fig.413. +Fig.414. (-)	D -138
$C_{10}H_{16}$	$CH_2=CHCH_2CH(COOCH_2CH_3)_2$				
	H_2PtCl_6 (i-PrOH)	reflux	-	$(C_2H_5O)_3Si(CH_2)_3CH(COOCH_2CH_3)_2$ (60)	1008
$C_{10}H_{18}$	$OCH_2CHCH_2OCH(CH_2OCH_3)CH_2OCH_2CH=CH_2$				
	H_2PtCl_6	reflux	40 (b)	$OCH_2CHCH_2OCH(CH_2OCH_3)CH_2O(CH_2)_3Si(OC_2H_5)_3$ (56)	1150
$C_{10}H_{20}$	$(CH_3)_3CN=CHCH=NC(CH_3)_3$				
	[Rh(CO)Cl(DAD)]	r.t.	5	$(CH_3)_3CHNCH=CHN[C(CH_3)_3]Si(OC_2H_5)_3$ (4)	165
$C_{10}H_{20}$	$CH_2=CHCH_2C(CH_3)_2CH_2CH=NN(CH_3)_2$				
	H_2PtCl_6 (OctOH)	until 185	12	$(C_2H_5O)_3Si(CH_2)_3C(CH_3)_2CH_2CH=NN(CH_3)_2$ (-)	U -050
	H_2PtCl_6 (OctOH)	until 185	12	$(C_2H_5O)_3Si(CH_2)_3C(CH_3)_2CH_2CH=NN(CH_3)_2$ (-)	U -078
$C_{10}H_{20}$	$CH_2=CHCH_2O(CH_2CH_2O)_3CH_3$				
	H_2PtCl_6	reflux	2	$(C_2H_5O)_3Si(CH_2)_3O(CH_2CH_2O)_3CH_3$ (92)	E -012
$C_{11}H_9$	Fig.298.				
	H_2PtCl_6 (i-PrOH)	reflux	10	Fig.419. + Fig.420. (27)	699
$C_{11}H_{12}$	$C_6H_5OCH_2COOCH_2CH=CH_2$				

$Pt(PPh_3)_4$	136-140	4	$C_6H_5OCH_2COO(CH_2)_3Si(OC_2H_5)_3$ (58)	Pl-005
$C_{11}H_{12}$ Fig.415.				
H_2PtCl_6	reflux	48 (b)	Fig.416. (66)	520
$C_{11}H_{17}$ Fig.417.				
H_2PtCl_6 (i-PrOH)	80-138	1	Fig.418. (50)	1000
$C_{11}H_{20}$ $(CH_2{=}CHCH_2)_2C(CH_3)CH{=}NN(CH_3)_2$				
H_2PtCl_6 (OctOH)	125-140	18	$(C_2H_5O)_3Si(CH_2)_3C(CH_3)(CH_2CH{=}CH_2)CH{=}NN(CH_3)_2$ (55)	U -050
H_2PtCl_6 (OctOH)	125-140	18	$(C_2H_5O)_3Si(CH_2)_3C(CH_3)(CH_2CH{=}CH_2)CH{=}NN(CH_3)_2$ (55)	U -078
$C_{11}H_{20}$ $CH_2{=}CH(CH_2)_8CHO$				
$PtCl_2(CH_2{=}CH_2)_2$ $(CHCl_3)$	-	1	$(C_2H_5O)_3Si(CH_2)_{10}CHO$ (-)	1060
$C_{11}H_{22}$ $CH_2{=}CHCH_2O(CH_2CH_2O)_2(CH_2)_3CH_3$				
H_2PtCl_6	reflux	2	$(C_2H_5O)_3Si(CH_2)_3O(CH_2CH_2O)_2(CH_2)_3CH_3$ (80)	E -012
$C_{11}H_{23}$ $[CH_3(CH_2)_3]_2NCH_2CH{=}CH_2$				
H_2PtCl_6 (i-PrOH + AllOH)	50-120	13	$[CH_3(CH_2)_3]_2NC_3H_6Si(OC_2H_5)_3$ (63)	S -019
H_2PtCl_6 (i-PrOH)	reflux	24	$[CH_3(CH_2)_3]_2N(CH_2)_3Si(OC_2H_5)_3$ (60)	91
$C_{11}H_{24}$ $(CH_3CH_2CH_2O)_3SiCH{=}CH_2$				
$Ru(acac)_3$	120	6	$(CH_3CH_2CH_2O)_3SiCH_2CH_2Si(OC_2H_5)_3$ (35)	745
$C_{11}H_{24}$ $(CH_3OCH_2CH_2O)_3SiCH{=}CH_2$				
$Ru(acac)_3$	120	6	$(CH_3OCH_2CH_2O)_3SiCH_2CH_2Si(OC_2H_5)_3$ (54)	745
$RuCl_2(PPh_3)_3$	120	6	$(CH_3OCH_2CH_2O)_3SiCH_2CH_2Si(OC_2H_5)_3$ (34)	733
$C_{11}H_{24}$ $[(CH_3)_2CHO]_3SiCH{=}CH_2$				
$Ru(acac)_3$	120	6	$[(CH_3)_2CHO]_3SiCH_2CH_2Si(OC_2H_5)_3$ (20)	745
$C_{12}H_{15}$ $(CH_3)_2CHN{=}CHCH{=}CHC_6H_5$				
$[Rh(CO)Cl(MAD)_2]$	r.t.	5	trans-$(CH_3)_2CHN[Si(OC_2H_5)_3]CH{=}CHCH_2C_6H_5$ (68) $(CH_3)_2CHN{=}CHCH_2CH(C_6H_5)Si(OC_2H_5)_3$ (32)	165
$C_{12}H_{16}$ $CH_2{=}CHCH_2OCH(CH_2CN)CH(CH_2CN)OCH_2CH{=}CH_2$				
$RhCl(PPh_3)_3$	reflux	4	$(C_2H_5O)_3Si(CH_2)_3OCH(CH_2CN)CH(CH_2CN)O(CH_2)_3$-$Si(OC_2H_5)_3$ (20)	224
$C_{12}H_{16}$ p-$CH_3C_6H_4SO_3(CH_2)_3CH{=}CH_2$				

	$RhCl(PPh_3)_3$	reflux	6	$p-CH_3C_6H_4SO_3(CH_2)_5Si(OC_2H_5)_3$ (-)	224
$C_{12}H_{24}$	$[CH_3(CH_2)_3O]_2CHCH_2CH=CH_2$				
	H_2PtCl_6 (i-PrOH)	78-155	1	$[CH_3(CH_2)_3O]_2CH(CH_2)_3Si(OC_2H_5)_3$ (60)	1004
$C_{12}H_{24}$	$[OSi(CH_3)CH=CH_2]_4$				
	$Co_2(CO)_8$	90-120	3	$[OSi(CH_3)CH_2CH_2Si(OC_2H_5)_3]_4$ (97)	S -050
	H_2PtCl_6	90	-	$[OSi(CH_3)CH_2CH_2Si(OC_2H_5)_3]_4$ (kin.)	606
	H_2PtCl_6	90-129	4	$[OSi(CH_3)CH_2CH_2Si(OC_2H_5)_3]_4$ (99)	S -050
	Pt complex	90	-	$[(CH_3)SiO(CH=CH_2)]_3[CH_3Si(O)CH_2CH_2Si(OC_2H_5)_3]$ (kin.)	426
	$Rh_4(CO)_{12}$	90-120	4	$[OSi(CH_3)CH_2CH_2Si(OC_2H_5)_3]_4$ (99)	S -050
$C_{13}H_{18}$	$CH_2=CHCH_2CH(CH_2COOCH_2CH=CH_2)COOCH_2CH=CH_2$				
	$PtCl_2(cpd)_2/(CH_2Cl_2)$	100	4 (t)	$(C_2H_5O)_3Si(CH_2)_3CH(CH_2COOCH_2CH=CH_2)COO-CH_2CH=CH_2 + CH_2=CHCH_2CH[CH_2COO(CH_2)_3-Si(OC_2H_5)_3]COOCH_2CH=CH_2$ (-)	D -138
$C_{13}H_{19}$	$o-(CH_3)_2NCH_2CH_2C_6H_4CH_2CH=CH_2$				
	H_2PtCl_6 (i-PrOH)	until 150	4	$o-(CH_3)_2NCH_2CH_2C_6H_4(CH_2)_3Si(OC_2H_5)_3$ (48)	C -034
$C_{13}H_{24}$	$C_{10}H_{19}OCH_2CH=CH_2$				
	H_2PtCl_6 (i-PrOH)	50	1.5	$C_{10}H_{19}O(CH_2)_3Si(OC_2H_5)_3$ (16)	33
$C_{13}H_{24}$	$CH_3CH_2OCO(CH_2)_8CH=CH_2$				
	$RhCl(PPh_3)_3$	reflux	12 (b)	$CH_3CH_2OCO(CH_2)_{10}Si(OC_2H_5)_3$ (58)	N -005
$C_{13}H_{27}$	$(CH_3)_2N(CH_2)_9CH=CH_2$				
	H_2PtCl_6 (i-PrOH)	120	8	$(CH_3)_2N(CH_2)_{11}Si(OC_2H_5)_3$ (60)	1138
$C_{14}H_{14}$	Fig.421.				
	H_2PtCl_6 $(MeOC_2H_4)_2O$	115-125	4.75	Fig.422. (83)	D -053
$C_{14}H_{14}$	$o-C_6H_4(COOCH_2CH=CH_2)_2$				
	$PtCl_2(acac)_2$	45	3.5 (a)	$o-CH_2=CHCH_2OCOC_6H_4COO(CH_2)_3Si(OC_2H_5)_3$ (97)	G -011
	$PtCl_2(acac)_2$	45	3.5 (a)	$o-CH_2=CHCH_2OCOC_6H_4COO(CH_2)_3Si(OC_2H_5)_3$ (97)	D -011
	$PtCl_2(acac)_2$ (MeCOMe)	-	-	$o-CH_2=CHCH_2OCOC_6H_4COO(CH_2)_3Si(OC_2H_5)_3$ (-)	D -011
$C_{14}H_{18}$	Fig.255.				

	H_2PtCl_6 $(MeOC_2H_4)_2O$	115-125	4.75	Fig.423. (-)	D -053
$C_{14}H_{22}$	$(CH_2{=}CHCH_2O)_2CHCH(OCH_2CH{=}CH_2)_2$				
	H_2PtCl_6 (i-PrOH)	90-115	-	$(CH_2{=}CHCH_2O)_2CHCH(OCH_2CH{=}CH_2)O(CH_2)_3Si(OC_2H_5)_3$ (100)	D -088
$C_{14}H_{22}$	$[(CH_2{=}CHCH_2O)_2CH]_2$				
	$PtCl_2(cpd)_2/(CH_2Cl_2)$	100	4 (t)	$(C_2H_5O)_3Si(CH_2)_3OCH(OCH_2CH{=}CH_2)CH(OCH_2CH{=}CH_2)_2$ (-)	D -138
$C_{14}H_{24}$	Fig.424.				
	H_2PtCl_6 (THF)	120	24	Fig.425. (45)	C -037
$C_{14}H_{26}$	$CH_3(CH_2)_6CH(CH_2CH_3)COOCH_2CH{=}CH_2$				
	$[PtCl_2(Me_2C{=}CH_2COMe)]_2$	reflux	0.25	$CH_3(CH_2)_6CH(CH_2CH_3)COO(CH_2)_3Si(OC_2H_5)_3$ (92)	D -023
$C_{14}H_{30}$	$(CH_3)_3SiO(CH_2)_9CH{=}CH_2$				
	H_2PtCl_6 (i-PrOH)	120	8	$(CH_3)_3SiO(CH_2)_{11}Si(OC_2H_5)_3$ (60)	S -047
$C_{15}H_{30}$	$[OSi(CH_3)(CH{=}CH_2)]_5$				
	H_2PtCl_6	90-120	3	$[OSi(CH_3)CH_2CH_2Si(OC_2H_5)_3]_5$ (97)	S -050
$C_{15}H_{35}$	$[(C_2H_5O)_3Si]_2NCH_2CH{=}CH_2$				
	$Co_2(CO)_8$	20	5	$[(C_2H_5O)_3Si]_2N(CH_2)_3Si(OC_2H_5)_3$ (83)	715
	$Rh_4(CO)_{12}$	20	5	$[(C_2H_5O)_3Si]_2N(CH_2)_3Si(OC_2H_5)_3$ (83)	715
$C_{16}H_{14}$	Fig.426.				
	H_2PtCl_6 $[(CH_2{=}CHSiMe_2)_2O]$	110	0.5 (t)	Fig.427. (96)	U -119
	Pt catalyst	-	- (t)	Fig.427. (-)	D -120
$C_{16}H_{14}$	Fig.428.				
	SiO_2-$OSi(CH_2)_3SH$/Pt	60-100	1-16	Fig.429. (50-95)	U -146
	SiO_2-$OSi(CH_2)_3SH$/Pt	90	1	Fig.429. (50)	U -146
$C_{16}H_{15}$	Fig.430.				
	Pt catalyst	60	1 (t)	Fig.431. (-)	U -125
	Pt catalyst	60	1 (t)	Fig.431. (-)	F -031
$C_{16}H_{16}$	$C_6H_5CH{=}C(COOCH_2CH{=}CH_2)_2$				

	Pt catalyst	60	1	$C_6H_5CH=C[COO(CH_2)_3Si(OC_2H_5)_3]_2$ (-)	U -124
$C_{17}H_{19}$	$(C_6H_5CH_2)_2NCH_2CH=CH_2$				
	H_2PtCl_6 (i-PrOH)	130	24	$(C_6H_5CH_2)_2N(CH_2)_3Si(OC_2H_5)_3$ + $(C_6H_5CH_2)_2NCH_2CH(CH_3)Si(OC_2H_5)_3$ (27)	255
$C_{19}H_{10}$	p-$C_{10}F_{19}OC_6H_4SO_2NHCH_2CH=CH_2$				
	$PtCl_2(SEt_2)_2$	-	-	p-$C_{10}F_{19}OC_6H_4SO_2NH(CH_2)_3Si(OC_2H_5)_3$ (-)	G -025
$C_{19}H_{15}$	$(C_6H_5)_2C=C(CN)COOCH_2CH=CH_2$				
	Pt catalyst	60	1	$(C_6H_5)_2C=C(CN)COO(CH_2)_3Si(OC_2H_5)_3$ (-)	U -124
$C_{19}H_{26}$	Fig.432.				
	H_2PtCl_6 (i-PrOH)	110	15.25 (t)	Fig.433. (60)	D -101
	H_2PtCl_6 (i-PrOH)	110	15.25 (t)	Fig.433. (60)	D -114
$C_{24}H_{30}$	Fig.434.				
	$RhCl(PPh_3)_3$	reflux	7.5	Fig.435. (-)	224
$C_{28}H_{42}$	$(CH_3)_2C_{10}H_5CH(C_3H_7)NHCONH(CH_2)_9CH=CH_2$				
	H_2PtCl_6	90	-	$(CH_3)_2C_{10}H_5CH(C_3H_7)NHCONH(CH_2)_{11}Si(OC_2H_5)_3$ (38)	909
$C_{34}H_{36}$	Fig.436.				
	$RhCl(PPh_3)_3$	reflux	-	Fig.437. (-)	C -023

Table 9.4

= $(C_2H_5O)_3SiH$ = =1=

C_2H_2 CH≡CH

-	40	2	CH_2=CHSi$(OC_2H_5)_3$ (86)	956
-	55	2	CH_2=CHSi$(OC_2H_5)_3$ (10) $(C_2H_5O)_3SiCH_2CH_2Si(OC_2H_5)_3$ (1)	956
$Co_2(CO)_8$	15-20	1-2	CH_2=CHSi$(OC_2H_5)_3$ (65-86)	S -043
$Co_2(CO)_8$ + Et_2O	15-20	1.16-1.33	CH_2=CHSi$(OC_2H_5)_3$ (95-98)	S -043
$Co_2(CO)_8$ + KI	15-20	1-1.5	CH_2=CHSi$(OC_2H_5)_3$ (90-96)	S -043
EDM-MCE/H_2PtCl_6	50	flow	CH_2=CHSi$(OC_2H_5)_3$ (90)	C -003
H_2PtCl_6	150	1	CH_2=CHSi$(OC_2H_5)_3$ (25) $(C_2H_5O)_3SiCH_2CH_2Si(OC_2H_5)_3$ (72)	U -163
H_2PtCl_6 (i-PrOH)	40	0.67 (b)	CH_2=CHSi$(OC_2H_5)_3$ (42) $(C_2H_5O)_3SiCH_2CH_2Si(OC_2H_5)_3$ (37)	1290
H_2PtCl_6 (i-PrOH)	40	0.67 (b)	CH_2=CHSi$(OC_2H_5)_3$ (42) $(C_2H_5O)_3SiCH_2CH_2Si(OC_2H_5)_3$ (37)	1288
MEE/H_2PtCl_6	40	4	CH_2=CHSi$(OC_2H_5)_3$ (-)	C -011
$Pt(PPh_3)_4$	120	1.16 (t)	CH_2=CHSi$(OC_2H_5)_3$ (96)	J -042
$Pt(PPh_3)_4$	80	5 (b)	CH_2=CHSi$(OC_2H_5)_3$ (74) $(C_2H_5O)_3SiCH_2CH_2Si(OC_2H_5)_3$ (5)	1290
$Pt(PPh_3)_4$	80	5 (b)	CH_2=CHSi$(OC_2H_5)_3$ (74) $(C_2H_5O)_3SiCH_2CH_2Si(OC_2H_5)_3$ (5)	1288
$PtCl_2(PPh_3)_2$	80	10 (b)	CH_2=CHSi$(OC_2H_5)_3$ (97) $(C_2H_5O)_3SiCH_2CH_2Si(OC_2H_5)_3$ (3)	1290
$PtCl_2(PPh_3)_2$	80	10 (b)	CH_2=CHSi$(OC_2H_5)_3$ (97) $(C_2H_5O)_3SiCH_2CH_2Si(OC_2H_5)_3$ (3)	1288
SDVB-CN/H_2PtCl_6	100	flow	CH_2=CHSi$(OC_2H_5)_3$ (100)	C -003
siloxane-SH/Pt complex	20	-	CH_2=CHSi$(OC_2H_5)_3$ (60) $(C_2H_5O)_3SiCH_2CH_2Si(OC_2H_5)_3$ (20)	1286
siloxane-SH/Pt complex	20	-	CH_2=CHSi$(OC_2H_5)_3$ + $(C_2H_5O)_3SiCH_2CH_2Si(OC_2H_5)_3$ (-)	1284
siloxane-SH/Pt complex	20	-	CH_2=CHSi$(OC_2H_5)_3$ (60) $(C_2H_5O)_3SiCH_2CH_2Si(OC_2H_5)_3$ (20)	509
siloxane-SH/Pt complex	80	-	CH_2=CHSi$(OC_2H_5)_3$ (50) $(C_2H_5O)_3SiCH_2CH_2Si(OC_2H_5)_3$ (50)	1286

Table 9.4

= $(C_2H_5O)_3SiH$ = | =2=

	siloxane-SH/Pt complex	80	-	$CH_2=CHSi(OC_2H_5)_3$ (50) $(C_2H_5O)_3SiCH_2CH_2Si(OC_2H_5)_3$ (50)	509
	$RhCl(PPh_3)_3$	75	3 (b)	$CH_2=CHSi(OC_2H_5)_3$ (54) $(C_2H_5O)_3SiCH_2CH_2Si(OC_2H_5)_3$ (10)	1288
	$RhCl(PPh_3)_3$	75	3 (b)	$CH_2=CHSi(OC_2H_5)_3$ (54) $(C_2H_5O)_3SiCH_2CH_2Si(OC_2H_5)_3$ (10)	1290
	$RhH(PPh_3)_4$	80	7 (b)	$CH_2=CHSi(OC_2H_5)_3$ (65) $(C_2H_5O)_3SiCH_2CH_2Si(OC_2H_5)_3$ (16)	1288
	$RhH(PPh_3)_4$	80	7 (b)	$CH_2=CHSi(OC_2H_5)_3$ (65) $(C_2H_5O)_3SiCH_2CH_2Si(OC_2H_5)_3$ (16)	1290
	$[Rh(CO)_2Cl]_2$	45	2	$(C_2H_5O)_3SiCH_2CH_2Si(OC_2H_5)_3$ (28)	1257
	$[Rh(CO)_2Cl]_2$	45	2	$(C_2H_5O)_3SiCH_2CH_2Si(OC_2H_5)_3$ (28)	1288
	$[Rh(CO)_2Cl]_2$	45	1	$CH_2=CHSi(OC_2H_5)_3$ (64)	1288
	siloxane-$(CH_2)_2PPh_2/[RhCl(C_8H_{14})_2]_2$	40	1	$CH_2=CHSi(OC_2H_5)_3$ (64)	941
	siloxane-$(CH_2)_2PPh_2/[RhCl(C_8H_{14})_2]_2$	45	1	$CH_2=CHSi(OC_2H_5)_3$ (42)	941
	siloxane-$(CH_2)_3PPh_2/[RhCl(C_8H_{14})_2]_2$	40	1	$CH_2=CHSi(OC_2H_5)_3$ (90)	941
	siloxane-$(CH_2)_3PPh_2/[RhCl(C_8H_{14})_2]_2$	45	1	$CH_2=CHSi(OC_2H_5)_3$ (90) $(C_2H_5O)_3SiCH_2CH_2Si(OC_2H_5)_3$ (3)	941
	siloxane-$CH_2PPh_2/[RhCl(C_8H_{14})_2]_2$	45	1	$CH_2=CHSi(OC_2H_5)_3$ (42)	941
	$RuCl_2(PPh_3)_3$	80-100	13 (b)	$CH_2=CHSi(OC_2H_5)_3$ (traces)	1290
	$RuCl_2(PPh_3)_3$	80-100	13 (b)	$CH_2=CHSi(OC_2H_5)_3$ (traces)	1288
C_4H_6	$CH_3C\equiv CCH_3$				
	$[HPt(SiMe_2Bz)P(c\text{-}Hex)_3]_2$	65	2	cis-$CH_3CH=C(CH_3)Si(OC_2H_5)_3$ (87)	430
C_4H_6	$CH_3CH_2C\equiv CH$				
	$[HPt(SiMe_2Bz)P(c\text{-}Hex)_3]_2$	65	1	trans-$CH_3CH_2CH=CHSi(OC_2H_5)_3$ + $CH_3CH_2C(=CH_2)Si(OC_2H_5)_3$ (83)	430
C_5H_8	$CH_3CH_2CH_2C\equiv CH$				
	$\overline{(CO)_4FeSiMe_2(CH_2)_3}$	reflux	5-7 (b)	$C_5H_9Si(OC_2H_5)_3$ (10)	299
	$Fe(CO)_5$	reflux	5-7 (b)	$C_5H_9Si(OC_2H_5)_3$ (10)	299
	$Ni(acac)_2$ + $AlEt_3$	r.t.	3	$CH_3CH_2CH_2CH=CHC(CH_2CH_2CH_3)=CHSi(OC_2H_5)_3$ (19) $CH_2=C(CH_2CH_2CH_3)C(CH_2CH_2CH_3)=CHSi(OC_2H_5)_3$ (72)	662

	$Ni(acac)_2 + AlEt_3$	r.t.	3	$CH_3CH_2CH_2CH{=}CHC(CH_2CH_2CH_3){=}CHSi(OC_2H_5)_3$ (19) $CH_2{=}C(CH_2CH_2CH_3)C(CH_2CH_2CH_3){=}CHSi(OC_2H_5)_3$ (72)	G -030
	$Ni(acac)_2 + AlEt_3$	r.t.	5	$CH_3CH_2CH_2CH{=}CHC(CH_2CH_2CH_3){=}CHSi(OC_2H_5)_3$ (10) $CH_2{=}C(CH_2CH_2CH_3)C(CH_2CH_2CH_3){=}CHSi(OC_2H_5)_3$ (51)	G -030
	$Ni(acac)_2 + AlEt_3$	r.t.	6 (b)	$CH_3CH_2CH_2CH{=}CHC(CH_2CH_2CH_3){=}CHSi(OC_2H_5)_3$ (17) $CH_2{=}C(CH_2CH_2CH_3)C(CH_2CH_2CH_3){=}CHSi(OC_2H_5)_3$ (64)	G -030
	$Ni(acac)_2 + AlEt_3$	r.t.	12	$CH_3CH_2CH_2CH{=}CHC(CH_2CH_2CH_3){=}CHSi(OC_2H_5)_3$ + $CH_2{=}C(CH_2CH_2CH_3)C(CH_2CH_2CH_3){=}CHSi(OC_2H_5)_3$ (21-89)	663
	$Ni(acac)_2$ + BuMgBr	r.t.	4 (e)	$CH_3CH_2CH_2CH{=}CHC(CH_2CH_2CH_3){=}CHSi(OC_2H_5)_3$ (11) $CH_2{=}C(CH_2CH_2CH_3)C(CH_2CH_2CH_3){=}CHSi(OC_2H_5)_3$ (40)	G -030
	$NiCl_2 + AlEt_3$	r.t.	6	$CH_3CH_2CH_2CH{=}CHC(CH_2CH_2CH_3){=}CHSi(OC_2H_5)_3$ (12) $CH_2{=}C(CH_2CH_2CH_3)C(CH_2CH_2CH_3){=}CHSi(OC_2H_5)_3$ (68)	G -030
	$NiCl_2 + AlEt_3$	r.t.	6	$CH_3CH_2CH_2CH{=}CHC(CH_2CH_2CH_3){=}CHSi(OC_2H_5)_3$ (12) $CH_2{=}C(CH_2CH_2CH_3)C(CH_2CH_2CH_3){=}CHSi(OC_2H_5)_3$ (68)	662
	$[RhCl(cod)]_2$ + DIPPC	20	6	$C_5H_9Si(OC_2H_5)_3$ (78)	884
	$[RhCl(cod)]_2$ + DMPIC	100	6	$C_5H_9Si(OC_2H_5)_3$ (30)	884
	$[RhCl(cod)]_2$ + XNC	100	6	$C_5H_9Si(OC_2H_5)_3$ (30)	1
	$[RhCl(cod)]_2$ + XNC	20	6	$C_5H_9Si(OC_2H_5)_3$ (78)	1
C_6H_{10}	$C_2H_5C{\equiv}CC_2H_5$				
	$N(R){=}C(R)C(R){=}N(R)RhCl(CO)$	-	-	$C_2H_5C[Si(OC_2H_5)_3]{=}CHC_2H_5$ (98)	166
C_6H_{10}	$CH_3(CH_2)_3C{\equiv}CH$				
	H_2PtCl_6 (THF)	100	2	$CH_3(CH_2)_3CH{=}CHSi(OC_2H_5)_3$ + $CH_3(CH_2)_3C({=}CH_2)Si(OC_2H_5)_3$ (-)	266
	H_2PtCl_6 (THF)	20	3 (THF)	$C_6H_{11}Si(OC_2H_5)_3$ (kin.)	1252
	$N(R){=}C(R)C(R){=}N(R)RhCl(CO)$	r.t.	-	$CH_3(CH_2)_3CH{=}CHSi(OC_2H_5)_3$ (63) $CH_3(CH_2)_3C({=}CH_2)Si(OC_2H_5)_3$ (29)	166
	$RhCl(CO)(PPh_3)_2$	100	10	cis-$CH_3(CH_2)_3CH{=}CHSi(OC_2H_5)_3$ + trans-$CH_3(CH_2)_3CH{=}CHSi(OC_2H_5)_3$ (51)	235
	$RhCl(PPh_3)_3$	100	10	cis-$CH_3(CH_2)_3CH{=}CHSi(OC_2H_5)_3$ + trans-$CH_3(CH_2)_3CH{=}CHSi(OC_2H_5)_3$ (57)	235
	$RhH(CO)(PPh_3)_3$	100	10	cis-$CH_3(CH_2)_3CH{=}CHSi(OC_2H_5)_3$ + trans-$CH_3(CH_2)_3CH{=}CHSi(OC_2H_5)_3$ + $CH_3(CH_2)_3C({=}CH_2)Si(OC_2H_5)_3$ (41)	235

	$RhI(PPh_3)_3$	100	10	cis-$CH_3(CH_2)_3CH=CHSi(OC_2H_5)_3$ + trans-$CH_3(CH_2)_3CH=CHSi(OC_2H_5)_3$ (61)	235
C_7H_{12}	$CH_3(CH_2)_4C\equiv CH$				
	C-CN/$[Rh(CO)_2Cl]_2$	80	2	$CH_3(CH_2)_4CH=CHSi(OC_2H_5)_3$ (20)	Dd-003
C_8H_6	$C_6H_5C\equiv CH$				
	zeolites	160	10	$C_6H_5CH=CHSi(OC_2H_5)_3$ + $C_6H_5C(=CH_2)Si(OC_2H_5)_3$ (-)	1251
	H_2OsCl_6 (THF)	110	4	$C_6H_5CH=CHSi(OC_2H_5)_3$ (48) $C_6H_5C(=CH_2)Si(OC_2H_5)_3$ (11)	608
	H_2PtCl_6 ($CH_2=CH$-N donor)	100	5	$C_6H_5CH=CHSi(OC_2H_5)_3$+$C_6H_5C(=CH_2)Si(OC_2H_5)_3$ (75)	612
	H_2PtCl_6 (THF)	110	4	$C_6H_5CH=CHSi(OC_2H_5)_3$ (37) $C_6H_5C(=CH_2)Si(OC_2H_5)_3$ (28)	608
	H_2PtCl_6 + I_2	20	- (h)	$C_6H_5CH=CHSi(OC_2H_5)_3$ + $C_6H_5C(=CH_2)Si(OC_2H_5)_3$ (82) $C_6H_5C\equiv CSi(OC_2H_5)_3$ (16)	1259
	H_2PtCl_6 + I_2	20	- (h)	$C_6H_5CH=CHSi(OC_2H_5)_3$ + $C_6H_5C(=CH_2)Si(OC_2H_5)_3$ (82) $C_6H_5C\equiv CSi(OC_2H_5)_3$ (16)	1269
	$PtCl_2$ ($CH_2=CH$-N donor)	100	5	$C_6H_5CH=CHSi(OC_2H_5)_3$+$C_6H_5C(=CH_2)Si(OC_2H_5)_3$ (71-84)	612
	$PtCl_4$ ($CH_2=CH$-N donor)	100	5	$C_6H_5CH=CHSi(OC_2H_5)_3$+$C_6H_5(=CH_2)Si(OC_2H_5)_3$ (75-79)	612
	$[HPt(SiMe_2Bz)P(c\text{-}Hex)_3]_2$	65	24	trans-$C_6H_5CH=CHSi(OC_2H_5)_3$ (6) $C_6H_5C(=CH_2)Si(OC_2H_5)_3$ (80)	430
	$N(R)=C(R)C(R)=N(R)RhCl(CO)$	r.t.	-	cis,trans-$C_6H_5CH=CHSi(OC_2H_5)_3$ + $C_6H_5C(=CH_2)Si(OC_2H_5)_3$ (-)	166
	$[Rh(CO)_2Cl]_2$	r.t.	-	$C_6H_5CH=CHSi(OC_2H_5)_3$ (69) $C_6H_5C(=CH_2)Si(OC_2H_5)_3$ (13)	166
C_8H_{10}	$CH\equiv C(CH_2)_4C\equiv CH$				
	$Ni(acac)_2$	50	6 (b)	Fig.438. (70)	1182
$C_{10}H_{14}$	$CH_3C\equiv C(CH_2)_4C\equiv CCH_3$				
	$Ni(acac)_2$ + PPh_3	100	12 (t)	Fig.439. (22)	1182
$C_{14}H_{10}$	$C_6H_5C\equiv CC_6H_5$				
	$[HPt(SiMe_2Bz)P(c\text{-}Hex)_3]_2$	65	2	cis-$C_6H_5CH=C(C_6H_5)Si(OC_2H_5)_3$ (82)	698

Table 9.5

= $(C_2H_5O)_3SiH$ = =1=

C_4H_4	$CH\equiv CCOOCH_3$				
	$N(R)=C(R)C(R)=N(R)RhCl(CO)$	r.t.	3-6	$CH_2=C(COOCH_3)[Si(OC_2H_5)_3]$ (6) $(C_2H_5O)_3SiCH=CHCOOCH_3$ (11)	166
C_5H_7	$CH\equiv CCH_2OCH_2CH_2Cl$				
	H_2PtCl_6 (i-PrOH)	70-80	3	$(C_2H_5O)_3SiCH=CHCH_2OCH_2CH_2Cl$ + $CH_2=C[Si(OC_2H_5)_3]CH_2OCH_2CH_2Cl$ (44)	549
C_6H_8	$CH\equiv C(COO\text{-}i\text{-}C_3H_7)$				
	$N(R)=C(R)C(R)=N(R)RhCl(CO)$	r.t.	3-6	$CH_2=C(COO\text{-}i\text{-}C_3H_7)[Si(OC_2H_5)_3]$ (11) $(C_2H_5O)_3SiCH=CH(COO\text{-}i\text{-}C_3H_7)$ (15)	166
C_6H_8	$CH\equiv CCH_2OCH_2\overline{CHCH_2O}$				
	H_2PtCl_6 (i-PrOH)	85	-	$(C_2H_5O)_3SiCH=CHCH_2OCH_2\overline{CHCH_2O}$ (78)	997
C_6H_{10}	$CH\equiv CCOOSi(CH_3)_3$				
	H_2PtCl_6 (i-PrOH)	50	1.5	$(C_2H_5O)_3SiCH=CHCOOSi(CH_3)_3$ (59)	355
C_6H_{10}	$CH_3OCH_2C\equiv CCH_2OCH_3$				
	$N(R)=C(R)C(R)=N(R)RhCl(CO)$	-	-	$CH_3OCH_2CH=C[Si(OC_2H_5)_3]CH_2OCH_3$ (98)	166
C_6H_{12}	$CH\equiv CCH_2OSi(CH_3)_3$				
	H_2PtCl_6	70	2	$(C_2H_5O)_3SiCH=CHCH_2OSi(CH_3)_3$ (-)	D -029
C_7H_{10}	$CH\equiv CCH_2OCH_2OCH(CH_2Cl)_2$				
	H_2PtCl_6	70-80	-	$(C_2H_5O)_3SiCH=CHCH_2OCH_2OCH(CH_2Cl)_2$ (68)	550
C_7H_{14}	$C_2H_5OCH_2C\equiv CCH_2OCH_3$				
	$N(R)=C(R)C(R)=N(R)RhCl(CO)$	-	-	$C_2H_5OCH_2C[Si(OC_2H_5)_3]=CHCH_2OCH_3$ (96)	166
C_8H_{12}	$\overline{OCH_2C}HCH_2OC(CH_3)_2C\equiv CH$				
	H_2PtCl_6 (i-PrOH)	70-150	4	$\overline{OCH_2C}HCH_2OC(CH_3)_2CH=CHSi(OC_2H_5)_3$ (-)	998
C_8H_{13}	$(\overline{CH_2)_5N}CH_2C\equiv CH$				
	H_2PtCl_6 (i-PrOH)	100-115	1.5	$(\overline{CH_2)_5N}CH_2CH=CHSi(OC_2H_5)_3$ (26) $(\overline{CH_2)_5N}CH_2C[Si(OC_2H_5)_3]=CH_2$ (24)	701
C_8H_{14}	$CH_3OCH(CH_3)C\equiv CCH(CH_3)OCH_3$				
	$N(R)=C(R)C(R)=N(R)RhCl(CO)$	-	- (dcm)	$CH_3OCH(CH_3)C[Si(OC_2H_5)_3]=CH(CHOCH_3)$ (48)	166
C_9H_{16}	$CH\equiv CCH_2OCH(CH_3)OC_4H_9$				

	H_2PtCl_6 (i-PrOH)	-	-	$(C_2H_5O)_3SiCH{=}CHCH_2OCH(CH_3)OC_4H_9$ (-)	1006
$C_{10}H_{11}$	$C_6H_5NHCH(CH_3)C{\equiv}CH$				
	H_2PtCl_6 (THF)	140	5 (THF)	$C_6H_5NHCH(CH_3)CH{=}CHSi(OC_2H_5)_3$ (60)	609
$C_{10}H_{14}$	$CH{\equiv}CCH_2CH(COOCH_2CH_3)_2$				
	H_2PtCl_6	reflux	-	$(C_2H_5O)_3SiCH{=}CHCH_2CH(COOCH_2CH_3)_2$ (66)	1009
$C_{10}H_{14}$	$CH_2{=}CHC{\equiv}CC(CH_3)_2OCH_2CHCH_2O$				
	Rh(acac)(CO)(PPh_3)	25-40	3-5	$CH_2{=}CHC[Si(OC_2H_5)_3]{=}CHC(CH_3)_2OCH_2CHCH_2O$ (65-81)	S -053
$C_{10}H_{16}$	$(CH_2)_5C(OCH_2OCH_3)C{\equiv}CH$				
	H_2PtCl_6	20	12	$(CH_2)_5C(OCH_2OCH_3)CH{=}CHSi(OC_2H_5)_3$ (34)	1091
$C_{11}H_{12}$	$CH_3CH(C_6H_5)OCH_2C{\equiv}CH$				
	H_2PtCl_6 (i-PrOH)	-	-	$CH_3CH(C_6H_5)OCH_2CH{=}CHSi(OC_2H_5)_3$ (56)	544
$C_{11}H_{12}$	Fig.440.				
	H_2PtCl_6 (i-PrOH)	reflux	18 (b)	Fig.441. (62)	1002
$C_{11}H_{12}$	p-$CH_3C_6H_4CH_2OCH_2C{\equiv}CH$				
	H_2PtCl_6 (i-PrOH)	80-85	4	p-$CH_3C_6H_4CH_2OCH_2CH{=}CHSi(OC_2H_5)_3$ (69)	1073
$C_{11}H_{15}$	$CH_2{=}CHCH_2C{\equiv}CC(CH_3)_2OCH_2CH_2CN$				
	H_2PtCl_6 (i-PrOH)	reflux	18	$CH_2{=}CHCH_2C[Si(OC_2H_5)_3]{=}CHC(CH_3)_2OCH_2CH_2CN$ (60)	1231
$C_{11}H_{16}$	$CH_2{=}CHCH_2C{\equiv}CC(CH_3)_2OCH_2CHCH_2O$				
	Rh(acac)(CO)(PPh_3)	25-40	3-5	$CH_2{=}CHCH_2C[Si(OC_2H_5)_3]{=}CHC(CH_3)_2OCH_2CHCH_2O$ (69-85)	S -053
$C_{11}H_{19}$	$(CH_3CH_2)_2NCH_2C{\equiv}CCH_2OCH_2CHCH_2O$				
	H_2PtCl_6	reflux	18 (b)	$(CH_3CH_2)_2NCH_2CH{=}C[Si(OC_2H_5)_3]CH_2OCH_2CHCH_2O$ (60)	1196
$C_{12}H_{16}$	$CH{\equiv}CCH_2OSi(CH_3)_2CH_2C_6H_5$				
	H_2PtCl_6 (i-PrOH)	20	-	$(C_2H_5O)_3SiCH{=}CHCH_2OSi(CH_3)_2CH_2C_6H_5$ (58)	39
$C_{12}H_{20}$	$(CH_2)_5C(C{\equiv}CH)OCH_2OCH(CH_3)_2$				
	H_2PtCl_6	20	12	$(CH_2)_5C[CH{=}CHSi(OC_2H_5)_3]OCH_2OCH(CH_3)_2$ (-)	1091
$C_{13}H_{14}$	$C_6H_5C{\equiv}CCH_2CH_2OCH_2CHCH_2O$				
	H_2PtCl_6	reflux	18 (b)	$C_6H_5CH{=}C[Si(OC_2H_5)_3]CH_2CH_2OCH_2CHCH_2O$ (76)	1196
$C_{13}H_{22}$	$(CH_2)_5C(C{\equiv}CH)OCH_2O(CH_2)_3CH_3$				

	H_2PtCl_6	20	12	$\overline{(CH_2)_5C}[CH=CHSi(OC_2H_5)_3]OCH_2O(CH_2)_3CH_3$ (-)	1091
$C_{14}H_{16}$	Fig.317.				
	H_2PtCl_6	25-100	-	Fig.442. (-)	517
$C_{14}H_{17}$	$C_6H_5C\equiv CSi(OCH_2CH_2)_3N$				
	$Rh(acac)(CO)_2$	100	8	$(C_2H_5O)_3Si(C_6H_5)C=CHSi(OCH_2H_2)_3N$ (49)	1236
$C_{14}H_{20}$	$CH\equiv CCH_2OSi(CH_2CH_3)_2CH_2C_6H_5$				
	H_2PtCl_6 (i-PrOH)	20	-	$(C_2H_5O)_3SiCH=CHCH_2OSi(CH_2CH_3)_2CH_2C_6H_5$ (59)	39

Table 9.6

= $(C_2H_5O)_3SiH$ = =1=

C_3	CF_3COCF_3				
	none	175	6	$C_2H_5OSiH[OC(CF_3)_2OC_2H_5]_2$ (73)	163
C_4H_6	trans-$CH_3CH{=}CHCHO$				
	Ni(PhCN)(bpy)	20-100	7	$CH_3CH{=}CHCH_2OSi(OC_2H_5)_3$ (75)	273
	$Ni(acac)_2 + AlEt_3$	20	16	trans-$CH_3CH{=}CHCH_2OSi(OC_2H_5)_3$ (5)	661
	$Ni(cod)_2$	20	5	trans-$CH_3CH{=}CHCH_2OSi(OC_2H_5)_3$ (80)	661
	$Ni(cod)_2 + PPh_3$	20	6	trans-$CH_3CH{=}CHCH_2OSi(OC_2H_5)_3$ (85)	661
	$Ni[P(OEt)_3]_4$	80-100	2-5	trans-$CH_3CH{=}CHCH_2OSi(OC_2H_5)_3$ (30-67)	661
	$Rh(acac)_3$	60	12	$CH_3CH_2CH{=}CHOSi(OC_2H_5)_3$ (28)	274
	$RhCl(PPh_3)_4$	90	24	$CH_3CH_2CH{=}CHOSi(OC_2H_5)_3$ (84)	661
C_5H_4	Fig.266.				
	KF	25	1	Fig.444. (-)	151
C_5H_8	$(CH_2)_4CO$				
	$RuCl_2(PPh_3)_3$	60	5	$(CH_2)_4CHOSi(OC_2H_5)_3$ (-)	760
C_5H_{10}	$CH_3CH_2COCH_2CH_3$				
	$RuCl_2(PPh_3)_3$	100	3	$(CH_3CH_2)_2CHOSi(OC_2H_5)_3$ (-)	760
C_6H_{10}	$(CH_2)_5CO$				
	CsF	25	60 s	$(CH_2)_5CHOSi(OC_2H_5)_3$ (-)	154
	$(CO)_4FeSiMe_2(CH_2)_3$	reflux	6 (b)	$(CH_2)_5CHOSi(OC_2H_5)_3$ (40)	299
	$Fe(CO)_5$	reflux	6 (b)	$(CH_2)_5CHOSi(OC_2H_5)_3$ (40)	299
	KF	100	12	$(CH_2)_5CHOSi(OC_2H_5)_3$ (-)	151
	$Rh_2(OAc)_4$	100	8	$(CH_2)_5CHOSi(OC_2H_5)_3$ (58)	274
	$RhBr(CNMeCH_2CH_2NMe)(CO)(PMe_2Ph)_2$	100	8	$(CH_2)_5CHOSi(OC_2H_5)_3$ (29)	496
	$RhCl(CNMeCH_2CH_2NMe)(cod)$	100	8	$(CH_2)_5CHOSi(OC_2H_5)_3$ (60)	496
	$RhCl(CNPhCH_2CH_2NPh)(cod)$	100	8	$(CH_2)_5CHOSi(OC_2H_5)_3$ (88)	496
	$RhCl_2[P(c\text{-}Hex)_3]_2$	100	8	$(CH_2)_5CHOSi(OC_2H_5)_3$ (78)	503
	$RhCl_2[P(o\text{-}MeC_6H_4)_3]_2$	100	8	$(CH_2)_5CHOSi(OC_2H_5)_3$ (54)	503

Table 9.6

= $(C_2H_5O)_3SiH$ = =2=

	$[RhCl(cod)]_2$ + DIPPC	100	6	$\overline{(CH_2)_5CH}OSi(OC_2H_5)_3$ (47)	884
	$[RhCl(cod)]_2$ + DMPIC	20	6	$\overline{(CH_2)_5CH}OSi(OC_2H_5)_3$ (20)	884
	$[RhCl(cod)]_2$ + XNC	100	6	$\overline{(CH_2)_5CH}OSi(OC_2H_5)_3$ (47)	1
	$[RhCl(cod)]_2$ + XNC	20	6	$\overline{(CH_2)_5CH}OSi(OC_2H_5)_3$ (20)	1
	$RuCl_2(PPh_3)_3$	100	10	$\overline{(CH_2)_5CH}OSi(OC_2H_5)_3$ (-)	760
C_6H_{10}	$CH_2=CHCH_2CH_2COCH_3$				
	CsF	0	0.25	$CH_2=CHCH_2CH_2CH(CH_3)OSi(OC_2H_5)_3$ (-)	286
C_6H_{12}	$(CH_3)_3CCOCH_3$				
	CsF	25	1	$(CH_3)_3CCH(CH_3)OSi(OC_2H_5)_3$ (-)	151
	KF	100	24	$(CH_3)_3CCH(CH_3)OSi(OC_2H_5)_3$ (-)	151
C_7H_5	m-$NO_2C_6H_4CHO$				
	CsF	25	1	m-$NO_2C_6H_4CH_2OSi(OC_2H_5)_3$ (-)	151
C_7H_5	p-BrC_6H_4CHO				
	KF	100	1	p-$BrC_6H_4CH_2OSi(OC_2H_5)_3$ (-)	151
C_7H_5	p-$NO_2C_6H_4CHO$				
	KF	100	2	p-$NO_2C_6H_4CH_2OSi(OC_2H_5)_3$ (-)	151
	KF	100	2	p-$NO_2C_6H_4CH_2OSi(OC_2H_5)_3$ (-)	286
C_7H_6	C_6H_5CHO				
	KF	25	6	$C_6H_5CH_2OSi(OC_2H_5)_3$ (-)	286
	KF	25	6	$C_6H_5CH_2OSi(OC_2H_5)_3$ (-)	151
	$RuCl_2(PPh_3)_3$	120	6	$C_6H_5CH_2OSi(OC_2H_5)_3$ (-)	760
C_7H_{14}	$CH_3(CH_2)_5CHO$				
	CsF	25	60 s	$CH_3(CH_2)_6OSi(OC_2H_5)_3$ (-)	154
	KF	25	4	$CH_3(CH_2)_6OSi(OC_2H_5)_3$ (-)	151
	KF	25	4	$CH_3(CH_2)_6OSi(OC_2H_5)_3$ (-)	286
C_8H_8	$C_6H_5COCH_3$				
	CsF	0	0.5	$C_6H_5CH(CH_3)OSi(OC_2H_5)_3$ (-)	286

Table 9.6
= $(C_2H_5O)_3SiH$ = =3=

	CsF	0	0.5	$C_6H_5CH(CH_3)OSi(OC_2H_5)_3$ (-)	151
	KF	100	1	$C_6H_5CH(CH_3)OSi(OC_2H_5)_3$ (-)	151
	$LiOCH_2CH(NHLi)R$	r.t.	3 (THF)	$C_6H_5CH(CH_3)OSi(OC_2H_5)_3$ (75)	166
	$[RhCl(C_2H_4)_2]_2$ + diop	50	24 (b)	$C_6H_5CH(CH_3)OSi(OC_2H_5)_3$ (-)	336
	$[RhCl(cod)]_2$	50	72	$C_6H_5CH(CH_3)OSi(OC_2H_5)_3$ (20)	177
	$[RhCl(cod)]_2$ + N,P donors	50	46-85	$C_6H_5CH(CH_3)OSi(OC_2H_5)_3$ (2-64)	177
	$RuCl_2(PPh_3)_3$	120	10	$C_6H_5CH(CH_3)OSi(OC_2H_5)_3$ (-)	760
C_8H_{10}	Fig.445.				
	KF	70	1	Fig.446. (-)	151
C_9H_8	$C_6H_5CH=CHCHO$				
	CsF	25	1	$C_6H_5CH=CHCH_2OSi(OC_2H_5)_3$ (-)	151
	KF	25	24	$C_6H_5CH=CHCH_2OSi(OC_2H_5)_3$ (-)	151
	KF	25	24	$C_6H_5CH=CHCH_2OSi(OC_2H_5)_3$ (-)	286
C_9H_9	$C_6H_5COCH(CH_3)Br$				
	CsF	25	0.5	$C_6H_5CH[OSi(OC_2H_5)_3]CH(CH_3)Br$ (-)	151
	CsF	25	0.5	$C_6H_5CH[OSi(OC_2H_5)_3]CH(CH_3)Br$ (-)	286
C_9H_9	$CH_3CONHC_6H_4CHO$				
	CsF	80	-	$CH_3CONHC_6H_4CH_2OSi(OC_2H_5)_3$ (-)	151
C_9H_{10}	$C_6H_5CH_2COCH_3$				
	CsF	25	1	$C_6H_5CH_2CH(CH_3)OSi(OC_2H_5)_3$ (-)	151
	$[RhCl(C_2H_4)_2]_2$	50	48 (b)	$C_6H_5CH_2CH(CH_3)OSi(OC_2H_5)_3$ (-)	336
C_9H_{10}	$C_6H_5COOCH_2CH_3$				
	CsF	100	3	$C_6H_5CH(OCH_2CH_3)OSi(OC_2H_5)_3$ (-)	154
C_9H_{18}	$[(CH_3)_2CHCH_2]_2CO$				
	CsF	25	1	$[(CH_3)_2CHCH_2]_2CHOSi(OC_2H_5)_3$ (-)	151
$C_{10}H_{11}$	$C_6H_5NHCOCH_2COCH_3$				
	CsF	25	0.17	$C_6H_5NHCOCH_2CH(CH_3)OSi(OC_2H_5)_3$ (-)	151

Table 9.6

= $(C_2H_5O)_3SiH$ = =4=

	CsF	25	0.17	$C_6H_5NHCOCH_2CH(CH_3)OSi(OC_2H_5)_3$ (-)	286
$C_{10}H_{16}$	Fig.369.				
	CsF	25	60 s	Fig.447. (-)	151
$C_{10}H_{18}$	$(CH_3)_2C=CHCH_2CH_2CH(CH_3)CH_2CHO$				
	KF	25	1	$(CH_3)_2C=CHCH_2CH_2CH(CH_3)CH_2CH_2OSi(OC_2H_5)_3$ (-)	151
$C_{10}H_{18}$	$\overline{CH_2CH_2CH[C(CH_3)_3]CH_2CH_2CO}$				
	$RhCl(PPh_3)_3$	80	20 (b)	$\overline{CH_2CH_2CH[C(CH_3)_3]CH_2CH_2C}HOSi(OC_2H_5)_3$ (-)	1037
$C_{11}H_{14}$	$(CH_3)_3CCOC_6H_5$				
	CsF	25	2	$(CH_3)_3CCH[OSi(OC_2H_5)_3]C_6H_5$ (-)	151
$C_{11}H_{18}$	$(CH_3)_2CHOCOCH_2CH_2COCOOCH(CH_3)_2$				
	CsF	25	0.5	$(CH_3)_2CHOCOCH_2CH_2CH[OSi(OC_2H_5)_3]COOCH(CH_3)_2$ (-)	151
$C_{11}H_{20}$	$CH_3OCO(CH_2)_7COOCH_3$				
	CsF	25	12	$CH_3OCH[OSi(OC_2H_5)_3](CH_2)_7CH(OCH_3)OSi(OC_2H_5)_3$ (-)	154
$C_{12}H_{22}$	$CH_2=CH(CH_2)_8COOCH_3$				
	CsF	25-60	0.5-20	$CH_2=CH(CH_2)_8CH(OCH_3)OSi(OC_2H_5)_3$ (-)	154
$C_{13}H_{10}$	$C_6H_5COC_6H_5$				
	CsF	25	0.08	$(C_6H_5)_2CHOSi(OC_2H_5)_3$ (-)	151
	CsF	25	0.08	$(C_6H_5)_2CHOSi(OC_2H_5)_3$ (-)	286
	KF	100	10	$(C_6H_5)_2CHOSi(OC_2H_5)_3$ (-)	151
$C_{14}H_{28}$	$CH_3(CH_2)_{10}COOCH_2CH_3$				
	CsF	60	0.5	$CH_3(CH_2)_{10}CH(OCH_2CH_3)OSi(OC_2H_5)_3$ (-)	154
$C_{19}H_{24}$	Fig.448.				
	RhCl(S) (diop)	22	168 (b)	Fig.449. (trace)	420
$C_{19}H_{36}$	$CH_3(CH_2)_7CH=CH(CH_2)_7COOCH_3$				
	CsF	25	4	$CH_3(CH_2)_7CH=CH(CH_2)_7CH(OCH_3)OSi(OC_2H_5)_3$ (-)	154
$C_{21}H_{38}$	$CH_2=CH(CH_2)_8COOC_{10}H_{19}$				
	CsF	25	72	$CH_2=CH(CH_2)_8CH(OC_{10}H_{19})OSi(OC_2H_5)_3$ (-)	154

Table 9.6

= $(C_2H_5O)_3SiH$ = =5=

	CsF	60	9	$CH_2=CH(CH_2)_8CH(OC_{10}H_{19})OSi(OC_2H_5)_3$ (-)	154
$C_{23}H_{29}$	Fig.149.				
	H_2PtCl_6 (i-PrOH)	60	0.5	Fig.450. (100)	1320
$C_{28}H_{52}$	$CH_3(CH_2)_7CH=CH(CH_2)_7COOC_{10}H_{19}$				
	CsF	25	72	$CH_3(CH_2)_7CH=CH(CH_2)_7CH(OC_{10}H_{19})OSi(OC_2H_5)_3$ (-)	154

Table 10.1

= $(CH_3)_2C_6H_5SiH$ = =1=

C_2H_4	$CH_2=CH_2$				
	$Co_2(CO)_8$	-	- (h)	$CH_3CH_2Si(CH_3)_2C_6H_5$ (kin.)	214
	H_2PtCl_6 (THF)	100	-	$CH_3CH_2Si(CH_3)_2C_6H_5$ (-)	214
	H_2PtCl_6 (THF)	15-30	-	$CH_3CH_2Si(CH_3)_2C_6H_5$ (kin.)	214
	$RhCl(PPh_3)_3$	60	-	$CH_3CH_2Si(CH_3)_2C_6H_5$ (kin.)	214
C_5H_{10}	$CH_3CH_2CH_2CH=CH_2$				
	$[HPt(SiEtMe_2)P(c\text{-}Hex)_3]_2$	20	1	$CH_3(CH_2)_4Si(CH_3)_2C_6H_5$ (91)	429
	glass-$^+PPh_3/PtCl_6^{-2}$	35	1.67	$CH_3(CH_2)_4Si(CH_3)_2C_6H_5$ (60)	S -036
	$RhCl(PPh_3)_3$	reflux	0.33	$CH_3(CH_2)_4Si(CH_3)_2C_6H_5$ (80)	227
C_5H_{10}	$CH_3CH_2CH=CHCH_3$				
	$RhCl(PPh_3)_3$	25	2 days	$CH_3(CH_2)_4Si(CH_3)_2C_6H_5$ (75)	227
C_6H_{12}	$(CH_3)_3CCH=CH_2$				
	polymer/H_2PtCl_6 (EtOH)	80	2	$(CH_3)_3CCH_2CH_2Si(CH_3)_2C_6H_5$ (80)	C -002
C_6H_{12}	$C_4H_9CH=CH_2$				
	$Co_2(CO)_8$	70-90	4	$C_4H_9CH_2CH_2Si(CH_3)_2C_6H_5$ (65)	718
C_6H_{12}	$CH_3(CH_2)_3CH=CH_2$				
	anionite/$[Ni(NH_3)_6]^{+2}$	until 110	20	$CH_3(CH_2)_5Si(CH_3)_2C_6H_5$ (40)	S -033
	SiO_2-$PPh_2/Pt(PPh_3)_4$	45	2 (b)	$CH_3(CH_2)_5Si(CH_3)_2C_6H_5$ (96)	767
	$[HPt(SiEtMe_2)(P\text{-}Hex)_3]_2$	20	1 (h)	$CH_3(CH_2)_5Si(CH_3)_2C_6H_5$ (90)	429
	$[PtCl_2(py)]_2$	-	1.5	$CH_3(CH_2)_5Si(CH_3)_2C_6H_5$ (95)	987
	anionite/$PtCl_6^{-2}$	15-40	-	$CH_3(CH_2)_5Si(CH_3)_2C_6H_5$ (kin.)	974
	anionite/$PtCl_6^{-2}$	15-40	-	$CH_3(CH_2)_5Si(CH_3)_2C_6H_5$ (kin.)	363
	anionite/$[Pt(NH_3)_4NH_2Cl]^+$	50	10	$CH_3(CH_2)_5Si(CH_3)_2C_6H_5$ (95-98)	S -016
	anionite/$[Pt(NH_3)_4]^{+2}$	40	20	$CH_3(CH_2)_5Si(CH_3)_2C_6H_5$ (65)	S -016
	anionite/$[Pt(NH_3)_4]^{+2}$	40	1-3	$CH_3(CH_2)_5Si(CH_3)_2C_6H_5$ (7-36)	S -016
	cis-$PtCl_2(PhCH=CH_2)_2$	r.t.	0.33	$CH_3(CH_2)_5Si(CH_3)_2C_6H_5$ (94)	220
	polyamide-$(CH_2)_nN/H_2PtCl_6$	10-20	-	$CH_3(CH_2)_5Si(CH_3)_2C_6H_5$ (-)	765

Table 10.1
= $(CH_3)_2C_6H_5SiH$ = =2=

	polyamide-$(CH_2)_nN/PtCl_2(CH_3CN)_2$	10-20	-	$CH_3(CH_2)_5Si(CH_3)_2C_6H_5$ (-)	765
	polyamide-$(CH_2)_nNPPh_2/H_2PtCl_6$	10-20	-	$CH_3(CH_2)_5Si(CH_3)_2C_6H_5$ (-)	765
	polyamide-$(CH_2)_nNPPh_2/PtCl_2(CH_2CN)_2$	10-20	-	$CH_3(CH_2)_5Si(CH_3)_2C_6H_5$ (-)	765
	Fig.451.	30	-	$CH_3(CH_2)_5Si(CH_3)_2C_6H_5$ (kin.)	768
	$RhCl(CO)(PPh_3)_3$	45	18 (b)	$CH_3(CH_2)_5Si(CH_3)_2C_6H_5$ (24)	767
	$RhCl(PPh_3)_3$	45	25 (b)	$CH_3(CH_2)_5Si(CH_3)_2C_6H_5$ (20)	767
	$RhH(CO)(PPh_3)_3$	45	0.75 (b)	$CH_3(CH_2)_5Si(CH_3)_2C_6H_5$ (73)	767
	$RhH(PPh_3)_4$	100	-	$CH_3(CH_2)_5Si(CH_3)_2C_6H_5$ (94)	605
	SiO_2-$(CH_2)_nPPh_2/[RhCl(C_2H_4)_2]_2$ n=1-6	10-40	-	$CH_3(CH_2)_5Si(CH_3)_2C_6H_5$ (kin.)	769
	SiO_2-$PPh_2/RhCl(CO)(PPh_3)_3$	45	2 (b)	$CH_3(CH_2)_5Si(CH_3)_2C_6H_5$ (79-85)	767
	SiO_2-$PPh_2/RhCl(PPh_3)_3$	45	2 (b)	$CH_3(CH_2)_5Si(CH_3)_2C_6H_5$ (82)	767
	SiO_2-$PPh_2/RhH(CO)(PPh_3)_3$	45	2 (b)	$CH_3(CH_2)_5Si(CH_3)_2C_6H_5$ (82)	767
	polyamide-$(CH_2)_nN/[RhCl(CO)_2]_2$	10-20	-	$CH_3(CH_2)_5Si(CH_3)_2C_6H_5$ (-)	765
	polyamide-$(CH_2)_nPPh_2/[RhCl(CO)_2]_2$	10-20	-	$CH_3(CH_2)_5Si(CH_3)_2C_6H_5$ (-)	765
C_7H_{10}	Fig.1.				
	$[HPt(SiEtMe_2)P(c\text{-}Hex)_3]_2$	20	0.5	Fig.452. (75)	429
C_7H_{14}	$CH_3(CH_2)_4CH=CH_2$				
	anionite/$PtCl_6^{-2}$	-	-	$CH_3(CH_2)_6Si(CH_3)_2C_6H_5$ (kin.)	1296
	anionite/$PtCl_6^{-2}$	-	-	$CH_3(CH_2)_6Si(CH_3)_2C_6H_5$ (kin.)	1344
	anionite/$PtCl_6^{-2}$	-	-	$CH_3(CH_2)_6Si(CH_3)_2C_6H_5$ (kin.)	1343
	anionite/$PtCl_6^{-2}$	15-45	-	$CH_3(CH_2)_6Si(CH_3)_2C_6H_5$ (kin.)	363
	anionite/$PtCl_6^{-2}$	15-45	-	$CH_3(CH_2)_6Si(CH_3)_2C_6H_5$ (kin.)	791
	anionite/$PtCl_6^{-2}$	15-45	-	$CH_3(CH_2)_6Si(CH_3)_2C_6H_5$ (kin.)	1345
	anionite/$PtCl_6^{-2}$	15 45	-	$CH_3(CH_2)_6Si(CH_3)_2C_6H_5$ (kin.)	974
	anionite/$PtCl_6^{-2}$	45-66	0-30	$CH_3(CH_2)_6Si(CH_3)_2C_6H_5$ (kin.)	1348
	anionite/$PtCl_6^{-2}$	45-66	0-30	$CH_3(CH_2)_6Si(CH_3)_2C_6H_5$ (kin.)	1232
	anionite/$PtCl_6^{-2}$	60	9 (THF)	$CH_3(CH_2)_6Si(CH_3)_2C_6H_5$ (96)	980

	(anionite-$NMe_3)_2Rh_2Cl_9$	65	-	$CH_3(CH_2)_6Si(CH_3)_2C_6H_5$ (kin.)	973
	(anionite-$NMe_3)_2RhCl_4(DESO)_2$	65	-	$CH_3(CH_2)_6Si(CH_3)_2C_6H_5$ (kin.)	973
	(anionite-$NMe_3)_2RhCl_4(DMSO)_2$	65	-	$CH_3(CH_2)_6Si(CH_3)_2C_6H_5$ (kin.)	973
	SDVB-NMe_2/H_2PtCl_6	45	4	$CH_3(CH_2)_6Si(CH_3)_2C_6H_5$ (79-98)	S -025
	$SiO_2[OSi(CH_2)_3ER_3]_2PtCl_6$; ER_3 = NMe_3, NEt_3, NBu_3, PBu_3, PPh_3	18-40	-	$CH_3(CH_2)_6Si(CH_3)_2C_6H_5$ (kin.)	1106
C_8H_8	$C_6H_5CH=CH_2$				
	$(Bu_4N)[PtCl_3(PhCH=CH_2)]$	r.t.	1	$C_6H_5CH_2CH_2Si(CH_3)_2C_6H_5$ (94) $C_6H_5CH(CH_3)Si(CH_3)_2C_6H_5$ (2)	220
	$K[PtCl_3(CH_2=CH_2)]$	r.t.	1	$C_6H_5CH_2CH_2Si(CH_3)_2C_6H_5$ (96) $C_6H_5CH(CH_3)Si(CH_3)_2C_6H_5$ (2)	220
	$PtCl_2$(1,5-cod)	r.t.	1	$C_6H_5CH_2CH_2Si(CH_3)_2C_6H_5$ (93) $C_6H_5CH(CH_3)Si(CH_3)_2C_6H_5$ (2)	220
	$[HPt(SiEtMe_2)P(c\text{-}Hex)_3]_2$	20	1.5 (h)	$C_6H_5CH_2CH_2Si(CH_3)_2C_6H_5$ (83)	429
	$[PtCl_2(py)]_2$	-	2	$C_6H_5CH_2CH_2Si(CH_3)_2C_6H_5$ (85)	987
	anionite/$PtCl_6^{-2}$	14-40	-	$C_6H_5CH_2CH_2Si(CH_3)_2C_6H_5$ (kin.)	974
	anionite/$PtCl_6^{-2}$	14-40	-	$C_6H_5CH_2CH_2Si(CH_3)_2C_6H_5$ (kin.)	363
	cis-$PtCl_2(PhCH=CH_2)_2$	r.t.	1	$C_6H_5CH_2CH_2Si(CH_3)_2C_6H_5$ (95) $C_6H_5CH(CH_3)Si(CH_3)_2C_6H_5$ (3)	220
	Rh complex	60	3 (THF)	$C_6H_5CH_2CH_2Si(CH_3)_2C_6H_5$ + $C_6H_5CH(CH_3)Si(CH_3)_2C_6H_5$ (-)	1339
	Rh complex	80	5 (b)	$C_6H_5CH_2CH_2Si(CH_3)_2C_6H_5$ + $C_6H_5CH(CH_3)Si(CH_3)_2C_6H_5$ (75)	1342
	$RhCl(PPh_3)_3$	60	3 (THF)	$C_6H_5CH_2CH_2Si(CH_3)_2C_6H_5$ + $C_6H_5CH(CH_3)Si(CH_3)_2C_6H_5$ (-)	1339
	$[RhH_2(BMPP)_2(S)_2]ClO_4$	110	168	$C_6H_5CH_2CH_2Si(CH_3)_2C_6H_5$ (70)	1302
C_8H_{16}	$CH_3(CH_2)_5CH=CH_2$				
	anionite/$[Co(NH_3)_6]^{+2}]$	120	10	$CH_3(CH_2)_7Si(CH_3)_2C_6H_5$ (75)	S -033
	$[IrCl(c\text{-}C_8H_{14})_2]_2$	60	6	$CH_3(CH_2)_7Si(CH_3)_2C_6H_5$ (18)	59
	$[IrCl(c\text{-}C_8H_{14})_2]_2$ + PPh_3	60	6	$CH_3(CH_2)_7Si(CH_3)_2C_6H_5$ (trace)	59

	$Rh_2(OAc)_4$	100	8	$CH_3(CH_2)_7Si(CH_3)_2C_6H_5$ (47)	274
	RhBr($\overline{CNMeCH_2CH_2NMe}$)(CO)$(PMe_2Ph)_2$	100	8	$CH_3(CH_2)_7Si(CH_3)_2C_6H_5$ (96)	496
	RhCl($\overline{CNMeCH_2CH_2NMe}$)(cod)	100	8	$CH_3(CH_2)_7Si(CH_3)_2C_6H_5$ (99)	496
	RhCl($\overline{CNPhCH_2CH_2NPh}$)(cod)	100	8	$CH_3(CH_2)_7Si(CH_3)_2C_6H_5$ (98)	496
	$RhCl_2[P(o\text{-}Hex)_3]_2$	100	8	$CH_3(CH_2)_7Si(CH_3)_2C_6H_5$ (100)	503
	$RhCl_2[P(o\text{-}MeC_6H_4)_3]_2$	100	8	$CH_3(CH_2)_7Si(CH_3)_2C_6H_5$ (100)	503
	$[RhCl(cod)_2]$ + DIPPC	20	-	$CH_3(CH_2)_7Si(CH_3)_2C_6H_5$ (81)	884
	$[RhCl(cod)]_2$ + XNC	20	24	$CH_3(CH_2)_7Si(CH_3)_2C_6H_5$ (81)	1
C_9H_8	Fig.453.				
	cis-$PtCl_2(PhCH{=}CH_2)_2$	r.t.	2	Fig.454. (85)	220
C_9H_{10}	$C_6H_5C(CH_3){=}CH_2$				
	cis-$PtCl_2(PhCH{=}CH_2)_2$	r.t.	0.25	$C_6H_5CH(CH_3)CH_2Si(CH_3)_2C_6H_5$ (13)	220
	RhCl(S)(diop)	120	40	$C_6H_5CH(CH_3)CH_2Si(CH_3)_2C_6H_5$ (19)	1302
	$[RhH_2(S)_2(BMPP)_2]ClO_4$	120	40	$C_6H_5CH(CH_3)CH_2Si(CH_3)_2C_6H_5$ (25)	1302
C_9H_{10}	$C_6H_5CH_2CH{=}CH_2$				
	anionite/$PtCl_6^{-2}$	30-50	-	$C_6H_5C_3H_6Si(CH_3)_2C_6H_5$ (kin.)	974
C_9H_{18}	$(CH_3)_2CH(CH_2)_3C(CH_3){=}CH_2$				
	anionite/$PtCl_6^{-2}$	90	-	$(CH_3)_2CH(CH_2)_3CH(CH_3)CH_2Si(CH_3)_2C_6H_5$ (kin.)	974
$C_{10}H_{12}$	$CH_3C_6H_4C(CH_3){=}CH_2$				
	anionite/$PtCl_6^{-2}$	90	-	$CH_3C_6H_4CH(CH_3)CH_2Si(CH_3)_2C_6H_5$ (kin.)	974
$C_{12}H_{18}$	Fig.156.				
	H_2PtCl_6 (i-PrOH)	150	6	Fig.458. (87)	708

Table 10.2

= $(CH_3)_2C_6H_5SiH$ = =1=

C_4H_6 $CH_2=CHCH=CH_2$					
	$Pd(PPh_3)_2(\overline{OCOCH=CHCO})$	-	-	$CH_2=CHCH_2CH_2Si(CH_3)_2C_6H_5$ (-)	1176
C_5H_8 $CH_2=C(CH_3)CH=CH_2$					
	$RhCl(PPh_3)_3$	80	2	$C_6H_5Si(CH_3)_2CH_2C(CH_3)=CHCH_3$ (27) $C_6H_5Si(CH_3)_2CH_2CH=C(CH_3)_2$ (71)	851
	$RhCl(PPh_3)_3$	80	2	$C_6H_5Si(CH_3)_2CH_2C(CH_3)=CHCH_3$ (27) $C_6H_5Si(CH_3)_2CH_2CH=C(CH_3)_2$ (71)	852
C_6H_{10} $CH_2=C(CH_3)C(CH_3)=CH_2$					
	$Cr(CO)_5(DMPIC)$	100	24	$CH_2=C(CH_3)CH(CH_3)CH_2Si(CH_3)_2C_6H_5$ (1) $(CH_3)_2C=C(CH_3)CH_2Si(CH_3)_2C_6H_5$ (48)	884
	$Cr(CO)_6$	100	24	$(CH_3)_2C=C(CH_3)CH_2Si(CH_3)_2C_6H_5$ (7)	884
	$Cr(CO)_6$	100	24	$CH_2=C(CH_3)CH(CH_3)Si(CH_3)_2C_6H_5$ (0) $(CH_3)_2C=C(CH_3)CH_2Si(CH_3)_2C_6H_5$ (7)	1
	$Cr(XNC)(CO)_5$	100	24	$CH_2=C(CH_3)CH(CH_3)Si(CH_3)_2C_6H_5$ (0) $(CH_3)_2C=C(CH_3)CH_2Si(CH_3)_2C_6H_5$ (49)	1
	$Mo(CO)_4(DMPIC)_2$	100	24	$(CH_3)_2C=C(CH_3)CH_2Si(CH_3)_2C_6H_5$ (23)	884
	$Mo(CO)_6$	100	24	$CH_2=C(CH_3)CH(CH_3)CH_2Si(CH_3)_2C_6H_5$ (3) $(CH_3)_2C=C(CH_3)CH_2Si(CH_3)_2C_6H_5$ (57)	884
	$Mo(CO)_6$	100	24	$CH_2=C(CH_3)CH(CH_3)Si(CH_3)_2C_6H_5$ (3) $(CH_3)_2C=C(CH_3)CH_2Si(CH_3)_2C_6H_5$ (57)	1
	$Mo(XNC)_2(CO)_4$	100	24	$CH_2=C(CH_3)CH(CH_3)Si(CH_3)_2C_6H_5$ (0) $(CH_3)_2C=C(CH_3)CH_2Si(CH_3)_2C_6H_5$ (23)	1
	$W(CO)_5(DMPIC)$	100	24	$CH_2=C(CH_3)CH(CH_3)CH_2Si(CH_3)_2C_6H_5$ (3) $(CH_3)_2C=C(CH_3)CH_2Si(CH_3)_2C_6H_5$ (10)	884
	$W(CO)_6$	100	24	$CH_2=C(CH_3)CH(CH_3)Si(CH_3)_2C_6H_5$ (3) $(CH_3)_2C=C(CH_3)CH_2Si(CH_3)_2C_6H_5$ (10)	1
	$W(CO)_6(DMPIC)$	100	24	$CH_2=C(CH_3)CH(CH_3)CH_2Si(CH_3)_2C_6H_5$ (2) $(CH_3)_2C=C(CH_3)CH_2Si(CH_3)_2C_6H_5$ (51)	884
	$W(XNC)(CO)_5$	100	24	$CH_2=C(CH_3)CH(CH_3)Si(CH_3)_2C_6H_5$ (2) $(CH_3)_2C=C(CH_3)CH_2Si(CH_3)_2C_6H_5$ (51)	1
C_6H_{10} $CH_2=CHCH_2CH_2CH=CH_2$					
	$[HPt(SiEtMe_2)P(c\text{-}Hex)_3]_2$	20	0.5	$C_6H_5Si(CH_3)_2(CH_2)_6Si(CH_3)_2C_6H_5$ (100)	429
	$[HPt(SiEtMe_2)P(c\text{-}Hex)_3]_2$	20	0.5	$C_6H_5Si(CH_3)_2(CH_2)_6Si(CH_3)_2C_6H_5$ (100)	428
C_6H_{10} $CH_3CH=CHCH_2CH=CH_2$					

	$Pt(PPh_3)_4$	45-52	4	$CH_3CH=CH(CH_2)_3Si(CH_3)_2C_6H_5$ (98)	364
C_8H_{12}	$\overline{CH_2CH_2CH=CHCH_2CH}CH=CH_2$				
	$[HPt(SiEtMe_2)P(c\text{-}Hex)_3]_2$	20	0.5	$\overline{CH_2CH_2CH=CHCH_2CH}CH_2CH_2Si(CH_3)_2C_6H_5$ (75)	428
	$[HPt(SiEtMe_2)P(c\text{-}Hex)_3]_2$	20	0.5	$\overline{CH_2CH_2CH=CHCH_2CH}CH_2CH_2Si(CH_3)_2C_6H_5$ (75)	429
C_8H_{14}	$CH_2=CH(CH_2)_4CH=CH_2$				
	$[HPt(SiEtMe_2)P(c\text{-}Hex)_3]_2$	20	0.5	$C_6H_5Si(CH_3)_2(CH_2)_8Si(CH_3)_2C_6H_5$ (88)	429
$C_{10}H_{16}$	$(CH_3)_2C=CHCH_2CH_2C(=CH_2)CH=CH_2$				
	$RhCl(PPh_3)_3$	100	36	$(CH_3)_2C=CHCH_2CH_2C(CH_3)=CHCH_2Si(CH_3)_2C_6H_5$ (50) $(CH_3)_2C=CHCH_2CH_2C(=CHCH_3)CH_2Si(CH_3)_2C_6H_5$ (15)	851
$C_{10}H_{16}$	$(CH_3)_2C=CHCH_2CH=C(CH_3)CH=CH_2$				
	$RhCl(PPh_3)_3$	100	24	$(CH_3)_2C=CHCH_2CH_2C(CH_3)=CHCH_2Si(CH_3)_2C_6H_5$ (51) $(CH_3)_2C=CHCH_2CH[Si(CH_3)_2C_6H_5]C(CH_3)=CHCH_3$ (16)	851

Table 10.3
= $(CH_3)_2C_6H_5SiH$ = =1=

C_2H_3	$Cl_3SiCH{=}CH_2$				
	H_2PtCl_6 (i-PrOH)	-	-	$Cl_3SiCH_2CH_2Si(CH_3)_2C_6H_5$ (92) $Cl_3SiCH_2CH_2SiCl_3$ (4)	1056
	H_2PtCl_6 (i-PrOH)	reflux	4	$Cl_3SiCH_2CH_2Si(CH_3)_2C_6H_5$ / $Cl_3SiCH(CH_3)Si(CH_3)_2C_6H_5$ (76/24)	1059
C_3H_3	$CF_3CH{=}CH_2$				
	$RhCl(PPh_3)_3$	70	3	$CF_3CH_2CH_2Si(CH_3)_2C_6H_5$ (59) $CF_3CH{=}CHSi(CH_3)_2C_6H_5$ (11)	858
	$Ru_3(CO)_{12}$	70	3	$CF_3CH_2CH_2Si(CH_3)_2C_6H_5$ (12) $CF_3CH{=}CHSi(CH_3)_2C_6H_5$ (68)	858
C_3H_3	$CH_2{=}CHCN$				
	$RhCl(PPh_3)_3$	100	3	$C_6H_5Si(CH_3)_2CH(CH_3)CN$ (72)	J -018
	$RhCl(PPh_3)_3$	100	3	$C_6H_5Si(CH_3)_2CH(CH_3)CN$ (87)	868
	$RhCl(PPh_3)_3$	100	3	$C_6H_5Si(CH_3)_2CH(CH_3)CN$ (87)	869
	$RhH(CO)(PPh_3)_3$	reflux	0.25	$C_6H_5Si(CH_3)_2CH(CH_3)CN$ (80)	227
C_3H_5	$CH_2{=}CHCH_2Cl$				
	H_2PtCl_6 (i-PrOH)	50	6	$C_6H_5Si(CH_3)_2(CH_2)_3Cl$ (22)	705
	H_2PtCl_6 (i-PrOH)	reflux	9	$C_6H_5Si(CH_3)_2(CH_2)_3Cl$ (19)	99
	$[HPt(SiBzMe_2)P(c\text{-}Hex)_3]_2$	20	43	$C_6H_5Si(CH_3)_2(CH_2)_3Cl$ (18)	429
C_3H_6	$CH_2{=}CHCH_2OH$				
	H_2PtCl_6	-	46 (t-BuOH)	$C_6H_5Si(CH_3)_2(CH_2)_3Cl$ (-)	N -003
	H_2PtCl_6	-	46 (t-BuOH)	$C_6H_5Si(CH_3)_2(CH_2)_3Cl$ (-)	G -009
	anionite/$PtCl_6^{-2}$	10-20	-	$C_6H_5Si(CH_3)_2(CH_2)_3Cl$ (kin.)	1347
C_3H_7	$CH_2{=}CHCH_2NH_2$				
	H_2PtCl_6	reflux	24	$C_6H_5Si(CH_3)_2(CH_2)_3NH_2$ (55)	91
C_4H_5	$CH_2{=}CHCH_2CN$				
	anionite/$PtCl_6^{-2}$	50-69	-	$C_6H_5Si(CH_3)_2CH(CH_2CH_3)CN$ (kin.)	974
C_4H_5	$CH_3CH{=}CHCN$				
	$RhCl(PPh_3)_3$	-	-	$CH_3CH_2CH[Si(CH_3)_2C_6H_5]CN$ (-)	J -018

	$RhCl(PPh_3)_3$	60	4	$CH_3CH_2CH[Si(CH_3)_2C_6H_5]CN$ (82)	869
	$RhCl(PPh_3)_3$	60	4	$CH_3CH_2CH[Si(CH_3)_2C_6H_5]CN$ (87)	868
C_4H_6	$CH_2=CHCOOCH_3$				
	$Co_2(CO)_8$	25	3 (b)	$C_6H_5Si(CH_3)_2CH=CHCOOCH_3$ (69) $C_6H_5Si(CH_3)_2CH_2CH_2COOCH_3$ (8)	1178
	$RhCl(PPh_3)_3$	-	-	$C_6H_5Si(CH_3)_2CH_2CH_2COOCH_3$ (-)	J -021
	$RhCl(PPh_3)_3$	60	12	$C_6H_5Si(CH_3)_2CH_2CH_2COOCH_3$ (80)	868
C_4H_8	$CH_2=C(CH_3)CH_2OH$				
	H_2PtCl_6	r.t.	3 days	$C_6H_5Si(CH_3)_2CH_2CH(CH_3)CH_2OH$ (91)	N -003
	H_2PtCl_6	r.t.	3 days	$C_6H_5Si(CH_3)_2CH_2CH(CH_3)CH_2OH$ (91)	G -009
C_4H_8	$CH_2=CHOCH_2CH_3$				
	cis-$PtCl_2(PhCH=CH_2)_2$	r.t.	0.33	$C_6H_5(CH_3)_2SiCH_2CH_2OCH_2CH_3$ (81)	220
C_5H_8	$CH_2=CHCOOCH_2CH_3$				
	anionite/$PtCl_6^{-2}$	60	10 (b)	$C_6H_5Si(CH_3)_2CH_2CH_2COOCH_2CH_3$ + $C_6H_5Si(CH_3)_2CH(CH_3)COOCH_2CH_3$ (74)	980
	$RhCl(PPh_3)_3$	80	2.5	$C_6H_5Si(CH_3)_2CH_2CH_2COOCH_2CH_3$ (-)	J -021
	$RhCl(PPh_3)_3$	80	2	$C_6H_5Si(CH_3)_2CH_2CH_2COOCH_2CH_3$ (85)	868
C_5H_{11}	$BrCH_2Si(CH_3)_2CH=CH_2$				
	H_2PtCl_6 (i-PrOH)	80-100	-	$BrCH_2Si(CH_3)_2CH_2CH_2Si(CH_3)_2C_6H_5$ (84)	387
C_5H_{11}	$ClCH_2Si(CH_3)_2CH=CH_2$				
	H_2PtCl_6 (i-PrOH)	80-100	-	$ClCH_2Si(CH_3)_2CH_2CH_2Si(CH_3)_2C_6H_5$ (87)	387
C_5H_{12}	$(CH_3)_3SiCH=CH_2$				
	$Co_2(CO)_8$	70-90	4	$(CH_3)_3SiCH_2CH_2Si(CH_3)_2C_6H_5$ (68)	718
	H_2PtCl_6	r.t.	42	$(CH_3)_3SiCH_2CH_2Si(CH_3)_2C_6H_5$ (80)	G -009
	H_2PtCl_6 (i-PrOH)	-	-	$(CH_3)_3SiCH_2CH_2Si(CH_3)_2C_6H_5$ (89) $(CH_3)_3SiCH_2CH_2Si(CH_3)_3$ (5)	1056
	H_2PtCl_6 (i-PrOH)	reflux	4	$(CH_3)_3SiCH_2CH_2Si(CH_3)_2C_6H_5$ / $(CH_3)_3SiCH(CH_3)Si(CH_3)_2C_6H_5$ (97/3)	1059
	cis-$PtCl_2(PhCH=CH_2)_2$	r.t.	0.33	$(CH_3)_3SiCH_2CH_2Si(CH_3)_2C_6H_5$ (100)	220
C_6H_9	$CH_2=C(CH_2Cl)OCH_2\overline{CHCH_2O}$				

	H_2PtCl_6 (i-PrOH)	reflux	6 (b)	$C_6H_5Si(CH_3)_2CH_2CH(CH_2Cl)OCH_2\overline{CHCH_2O}$ (43)	1155
C_7H_{11}	$CH_2=C(CH_3)CH_2\overline{CHCH(CH_2Cl)O}$				
	H_2PtCl_6 (i-PrOH)	120-125	12	$C_6H_5Si(CH_3)_2CH_2CH(CH_3)CH_2\overline{CHCH(CH_2Cl)O}$ (60)	92
C_8H_3	$C_6F_5CH=CH_2$				
	$RhCl(PPh_3)_3$	100	15	$C_6F_5CH_2CH_2Si(CH_3)_2C_6H_5$ (55) $C_6F_5CH=CHSi(CH_3)_2C_6H_5$ (24)	858
	$Ru_3(CO)_{12}$	70	18	$C_6F_5CH_2CH_2Si(CH_3)_2C_6H_5$ (5) $C_6F_5CH=CHSi(CH_3)_2C_6H_5$ (75)	858
C_8H_7	$ClC_6H_4CH=CH_2$				
	anionite-$NMe_2/PtCl_6^{-2}$	20-34	-	$ClC_6H_4CH_2CH_2Si(CH_3)_2C_6H_5$ (kin.)	974
C_8H_7	m-$NO_2C_6H_4CH=CH_2$				
	H_2PtCl_6 (i-PrOH)	80-150	1.5	m-$NO_2C_6H_4CH_2CH_2Si(CH_3)_2C_6H_5$ (59)	48
C_8H_8	$CH_2=CHSi(OCH_2CH_3)_3$				
	H_2PtCl_6 (i-PrOH)	50-200	0.55	$C_6H_5Si(CH_3)_2CH_2CH_2Si(OCH_2CH_3)_3$ (87)	575
C_8H_9	2-$C_5H_4NOCH_2CH=CH_2$				
	H_2PtCl_6	120-150	42	2-$C_5H_4NO(CH_2)_3Si(CH_3)_2C_6H_5$ (-)	F -010
C_8H_{12}	Fig.248.				
	H_2PtCl_6 (i-PrOH)	60-130	2	Fig.455. (55)	21
C_8H_{12}	$Si(CH=CH_2)_4$				
	H_2PtCl_6 (i-PrOH)	-	-	$(CH_2=CH)_3SiCH_2CH_2Si(CH_3)_2C_6H_5$ (78)	1056
	H_2PtCl_6 (i-PrOH)	reflux	4	$Si(CH=CH_2)_3C_2H_4Si(CH_3)_2C_6H_5$ + $Si(CH=CH_2)_2[C_2H_4Si(CH_3)_2C_6H_5]_2$ + $Si(CH=CH_2)[C_2H_4Si(CH_3)_2C_6H_5]_3$ + $Si[C_2H_4Si(CH_3)_2C_6H_5]_4$ (-)	1059
C_9H_7	$C_6H_5CH=CHCN$				
	$RhCl(PPh_3)_3$	100	12	$C_6H_5CH_2CH[Si(CH_3)_2C_6H_5]CN$ (72)	868
	$RhCl(PPh_3)_3$	100	12	$C_6H_5CH_2CH[Si(CH_3)_2C_6H_5]CN$ (72)	869
C_9H_{15}	$N(CH_2CH=CH_2)_3$				
	H_2PtCl_6 (i-PrOH)	120-125	8	$(CH_2=CHCH_2)_2N(CH_2)_3Si(CH_3)_2C_6H_5$ + $(CH_2=CHCH_2)_2N(CH_2)_3CH(CH_3)Si(CH_3)_2C_6H_5$ (67)	255
$C_{10}H_{24}$	$[\overline{OSi(CH_3)(CH=CH_2)]_2[OSi}(CH_3)_2]_2$				

	H_2PtCl_6 (THF)	-	-	$[\overline{OSi(CH_3)CH_2CH_2Si(CH_3)_2C_6H_5]_2[OSi}(CH_3)_2]_2$ (72)	52
$C_{10}H_{24}$	$[OSi(CH_3)_2][OSi(CH_3)(CH=CH_2)][OSi(CH_3)_2]OSi(CH_3)CH=CH_2$				
	H_2PtCl_6 (THF)	-	-	$[OSi(CH_3)_2\{OSi(CH_3)[CH_2CH_2Si(CH_3)_2C_6H_5]\}\{[O$-$Si(CH_3)_2][OSi(CH_3)CH_2CH_2Si(CH_3)_2C_6H_5]\}$ (72)	27
$C_{11}H_{14}$	Fig.456.				
	H_2PtCl_6	reflux	6 (b)	Fig.457. (80)	1152
$C_{12}H_{20}$	$(CH_2=CHCH_2OCH_2)_2CHOCH_2\overline{CHCH_2O}$				
	H_2PtCl_6	120-125	16	$C_6H_5Si(CH_3)_2(CH_2)_3OCH_2CH(CH_2OCH_2CH=CH_2)O$-$CH_2\overline{CHCH_2O}$ (43)	1158

Table 10.4

= $(CH_3)_2C_6H_5SiH$ = =1=

C_2H_2	$CH{\equiv}CH$				
	MCE/$RhCl_3$	20	24	$CH_2{=}CHSi(CH_3)_2C_6H_5$ (34)	C -006
C_4H_6	$CH_3C{\equiv}CCH_3$				
	$[HPt(SiBzMe_2)P(c\text{-}Hex)_3]_2$	40	0.25	cis-$CH_3CH{=}C(CH_3)Si(CH_3)_2C_6H_5$ (97)	430
C_4H_6	$CH_3CH_2C{\equiv}CH$				
	$[HPt(SiBzMe_2)P(c\text{-}Hex)_3]_2$	40	0.1	trans-$CH_3CH_2CH{=}CHSi(CH_3)_2C_6H_5$ (74)	430
C_5H_8	$CH_3CH_2CH_2C{\equiv}CH$				
	cis-$PtCl_2(PhCH{=}CH_2)_2$	r.t.	0.5	trans-$CH_3CH_2CH_2CH{=}CHSi(CH_3)_2C_6H_5$ (74) $CH_3CH_2CH_2C[Si(CH_3)_2C_6H_5]{=}CH_2$ (19)	220
	$RhCl(PPh_3)_3$	40	2	$CH_3CH_2CH_2CH{=}CHSi(CH_3)_2C_6H_5$ (94)	870
C_6H_{10}	$(CH_3)_3CC{\equiv}CH$				
	H_2PtCl_6 (i-PrOH)	150	6	$(CH_3)_3CCH{=}CHSi(CH_3)_2C_6H_5$ (87) $(CH_3)_3CC[Si(CH_3)_2C_6H_5]{=}CH_2$ (5)	708
C_6H_{10}	$CH_3(CH_2)_3C{\equiv}CH$				
	H_2PtCl_6 (THF)	100	2	$CH_3(CH_2)_3CH{=}CHSi(CH_3)_2C_6H_5$ + $CH_3(CH_2)_3C[SiCH_3)_2C_6H_5]{=}CH_2$ (-)	266
	H_2PtCl_6 (THF)	20	3 (THF)	$CH_3(CH_2)_3CH{=}CHSi(CH_3)_2C_6H_5$ + $CH_3(CH_2)_3C[Si(CH_3)_2C_6H_5]{=}CH_2$ (kin.)	1252
	H_2PtCl_6 (i-PrOH)	50	6	$CH_3(CH_2)_3CH{=}CHSi(CH_3)_2C_6H_5$ (24) $CH_3(CH_2)_3C[Si(CH_3)_2C_6H_5]{=}CH_2$ (55)	706
	$Rh_2(OAc)_4$	100	8	$CH_3(CH_2)_3CH{=}CHSi(CH_3)_2C_6H_5$ (71)	274
	$RhCl(\overline{CNPhCH_2CH_2N}Ph)(cod)$	100	8	$CH_3(CH_2)_3CH{=}CHSi(CH_3)_2C_6H_5$ (77)	496
	$RhCl(PPh_3)_3$	-	-	$CH_3(CH_2)_3CH{=}CHSi(CH_3)_2C_6H_5$ (98)	870
	$RhCl_2[P(c\text{-}Hex)_3]_2$	100	8	$CH_3(CH_2)_3CH{=}CHSi(CH_3)_2C_6H_5$ (63)	503
	$RhCl_2[P(o\text{-}MeC_6H_4)_3]_2$	100	8	$CH_3(CH_2)_3CH{=}CHSi(CH_3)_2C_6H_5$ (46)	503
C_8H_6	$C_6H_5C{\equiv}CH$				
	$(Bu_4N)_2PtCl_6$	120	1-3	$C_6H_5CH{=}CHSi(CH_3)_2C_6H_5$ +$C_6H_5C[Si(CH_3)_2C_6H_5]{=}CH_2$ (-)	673
	$[HPt(SiEtMe_2)P(c\text{-}Hex)_3]_2$	25	19 (h)	$C_6H_5CH{=}CHSi(CH_3)_2C_6H_5$ +$C_6H_5C[Si(CH_3)_2C_6H_5]{=}CH_2$ (84)	430
	cis-$PtCl_2(PhCH{=}CH_2)_2$	r.t.	0.75	trans-$C_6H_5CH{=}CHSi(CH_3)_2C_6H_5$ (72) $C_6H_5C[Si(CH_3)_2C_6H_5]{=}CH_2$ (28)	220

Table 10.4

= $(CH_3)_2C_6H_5SiH$ = =2=

	Rh complex	60	2 (THF)	$C_6H_5CH=CHSi(CH_3)_2C_6H_5$ (-)	1274
	$RhCl(CO)(PPh_3)_2$	40-80	1-24	$C_6H_5CH=CHSi(CH_3)_2C_6H_5$ (57-94)	1291
	$RhCl(PPh_3)_3$	40-80	2-52	$C_6H_5CH=CHSi(CH_3)_2C_6H_5$ (45-100)	1291
$C_{12}H_{16}$	Fig.198.				
	H_2PtCl_6 (i-PrOH)	150	6	Fig.459. (77)	708
$C_{14}H_{10}$	$C_6H_5C\equiv CC_6H_5$				
	$[HPt(SiBzMe_2)P(c\text{-}Hex)_3]_2$	40	0.5	Z-$C_6H_5CH=C(C_6H_5)[Si(CH_3)_2C_6H_5]$ (86)	430

Table 10.5

= $(CH_3)_2C_6H_5SiH$ = =1=

C_3H_3	$CH\equiv CCH_2Cl$				
	H_2PtCl_6 (i-PrOH)	50	6	$C_6H_5Si(CH_3)_2CH=CHCH_2Cl$ (45) $C_6H_5Si(CH_3)_2C(=CH_2)CH_2Cl$ (12) $C_6H_5Si(CH_3)_2CH=CHCH_3$ (10)	705
C_3H_4	$CH\equiv CCH_2OH$				
	H_2PtCl_6 (i-PrOH)	50	6	trans-$C_6H_5Si(CH_3)_2CH=CHCH_2OH$ (27) $C_6H_5Si(CH_3)_2C(=CH_2)CH_2OH$ (22) trans-$C_6H_5Si(CH_3)_2CH=CHCH_2OSi(CH_3)_2C_6H_5$ (18)	706
	$Pt(CH_2=CH_2)_2[P(c\text{-}Hex)_3]$	reflux	2 (THF)	$C_6H_5Si(CH_3)_2CH=CHCH_2OH$ + $CH_2=C(CH_2OH)Si(CH_3)_2C_6H_5$ (82)	804
	$Pt(Nb)_2P(Bu\text{-}t)_3$	reflux	- (THF)	$C_6H_5Si(CH_3)_2CH=CHCH_2OH$ (88)	803
	$Pt(Nb)_2P(Bu\text{-}t)_3$	reflux	1.5 (THF)	$C_6H_5Si(CH_3)_2CH=CHCH_2OH$ + $CH_2=C(CH_2OH)Si(CH_3)_2C_6H_5$ (88)	804
	$PtP(Bu\text{-}t)_3[(CH_2=CH)_2Si(OMe)_2]$	reflux	1.5 (THF)	$C_6H_5Si(CH_3)_2CH=CHCH_2OH$ + $CH_2=C(CH_2OH)Si(CH_3)_2C_6H_5$ (76)	804
C_4H_4	$CH\equiv CCOOCH_3$				
	$Pt(CH_2=CH_2)_2[P(c\text{-}Hex)_3]$	reflux	20 (THF)	$C_6H_5Si(CH_3)_2CH=CHCOOCH_3$ + $CH_2=C[Si(CH_3)_2C_6H_5]COOCH_3$ (71)	804
C_4H_6	$CH\equiv CCH_2CH_2OH$				
	$Pt(CH_2=CH_2)_2[P(c\text{-}Hex)_3]$	reflux	20 (THF)	$C_6H_5Si(CH_3)_2CH=CHCH_2CH_2OH$ + $CH_2=C(CH_2CH_2OH)Si(CH_3)_2C_6H_5$ (94)	804
C_4H_6	$CH\equiv CCH_2OCH_3$				
	H_2PtCl_6 (i-PrOH)	50	6	trans-$C_6H_5Si(CH_3)_2CH=CHCH_2OCH_3$ (49) $C_6H_5Si(CH_3)_2C(=CH_2)CH_2OCH_3$ (39)	706
C_4H_6	$CH\equiv CCH(CH_3)OH$				
	$Pt(CH_2=CH_2)_2[P(c\text{-}Hex)_3]$	reflux	20 (THF)	$C_6H_5Si(CH_3)_2CH=CHCH(CH_3)OH$ + $CH_2=C[Si(CH_3)_2C_6H_5]CH(CH_3)OH$ (96)	804
	$Pt(Nb)_2P(Bu\text{-}t)_3$	reflux	3 (THF)	$C_6H_5Si(CH_3)_2CH=CHCH(CH_3)OH$ + $CH_2=C[Si(CH_3)_2C_6H_5]CH(CH_3)OH$ (70)	804
	$PtP(Bu\text{-}t)_3[(CH_2=CH)_2Si(OMe)_2]$	reflux	2.75 (THF)	$C_6H_5Si(CH_3)_2CH=CHCH(CH_3)OH$ + $CH_2=C[Si(CH_3)_2C_6H_5]CH(CH_3)OH$ (94)	804
C_4H_6	$CH_3C\equiv CCH_2OH$				
	$Pt(CH_2=CH_2)_2[P(c\text{-}Hex)_3]$	reflux	4 (THF)	$CH_3C[Si(CH_3)_2C_6H_5]=CHCH_2OH$ + $CH_3CH=C[Si(CH_3)_2C_6H_5]CH_2OH$ (91)	804

Table 10.5

= $(CH_3)_2C_6H_5SiH$ = =2=

	$Pt(Nb)_2P(Bu\text{-}t)_3$	reflux	1 (THF)	$CH_3C[Si(CH_3)_2C_6H_5]=CHCH_2OH$ + $CH_3CH=C[Si(CH_3)_2C_6H_5]CH_2OH$ (91)	804
C_5H_6	$CH\equiv CCH_2COOCH_3$				
	H_2PtCl_6 (i-PrOH)	150	6	trans-$C_6H_5Si(CH_3)_2CH=CHCH_2COOCH_3$ (46) $C_6H_5Si(CH_3)_2C(=CH_2)CH_2COOCH_3$ (30)	706
C_5H_8	$CH\equiv CC(CH_3)_2OH$				
	$Pt(CH_2=CH_2)_2[P(c\text{-}Hex)_3]$	reflux	24 (THF)	$C_6H_5Si(CH_3)_2CH=CHC(CH_3)_2OH$ + $CH_2=C[Si(CH_3)_2C_6H_5]C(CH_3)_2OH$ (98)	804
C_5H_9	$BrCH_2Si(CH_3)_2C\equiv CH$				
	H_2PtCl_6 (i-PrOH)	80-100	-	$BrCH_2Si(CH_3)_2CH=CHSi(CH_3)_2C_6H_5$ + $BrCH_2Si(CH_3)_2C[Si(CH_3)_2C_6H_5]=CH_2$ (86)	387
C_6H_{10}	$CH\equiv CCOOSi(CH_3)_3$				
	H_2PtCl_6 (i-PrOH)	50	1.5	$C_6H_5Si(CH_3)_2CH=CHCOOSi(CH_3)_3$ (66)	355
C_7H_6	$CH\equiv CCH_2\overline{C=CHCH=CHS}$				
	H_2PtCl_6 (i-PrOH)	50	6	trans-$C_6H_5Si(CH_3)_2CH=CHCH_2\overline{C=CHCH=CHS}$ (55) $C_6H_5Si(CH_3)_2C(=CH_2)CH_2\overline{C=CHCH=CHS}$ (12)	706
C_7H_{11}	$CH\equiv CCH_2\overline{N(CH_2)_4}$				
	H_2PtCl_6 (i-PrOH)	150	6	trans-$C_6H_5Si(CH_3)_2CH=CHCH_2\overline{N(CH_2)_4}$ (64) $C_6H_5Si(CH_3)_2C(=CH_2)CH_2\overline{N(CH_2)_4}$ (14)	706
	H_2PtCl_6 (i-PrOH)	200	6	trans-$C_6H_5Si(CH_3)_2CH=CHCH_2\overline{N(CH_2)_4}$ (86) $C_6H_5Si(CH_3)_2C(=CH_2)CH_2\overline{N(CH_2)_4}$ (10)	706
C_7H_{13}	$CH\equiv CCH_2N(CH_2CH_3)_2$				
	H_2PtCl_6 (i-PrOH)	150	6	trans-$C_6H_5Si(CH_3)_2CH=CHCH_2N(CH_2CH_3)_2$ (58) $C_6H_5Si(CH_3)_2C(=CH_2)CH_2N(CH_2CH_3)_2$ (14)	706
	H_2PtCl_6 (i-PrOH)	200	6	trans-$C_6H_5Si(CH_3)_2CH=CHCH_2N(CH_2CH_3)_2$ (59) $C_6H_5Si(CH_3)_2C(=CH_2)CH_2N(CH_2CH_3)_2$ (17)	706
C_8H_6	$CH\equiv CC_6H_5$				
	H_2PtCl_6 (i-PrOH)	50	6	trans-$C_6H_5Si(CH_3)_2CH=CHC_6H_5$ (55) $C_6H_5Si(CH_3)_2C(=CH_2)C_6H_5$ (24)	706
C_8H_{12}	$CH\equiv CCH_2OCH_2CH_2OCH_2\overline{CHCH_2O}$				
	H_2PtCl_6 (i-PrOH)	reflux	15 (b)	$C_6H_5Si(CH_3)_2CH=CHCH_2OCH_2CH_2OCH_2\overline{CHCH_2O}$ (75)	1162
C_8H_{13}	$CH\equiv CCH_2\overline{N(CH_2)_5}$				

Table 10.5

= $(CH_3)_2C_6H_5SiH$ =   =3=

	H_2PtCl_6 (i-PrOH)	150	6	trans-$C_6H_5Si(CH_3)_2CH{=}CHCH_2N(CH_2)_5$ (62) $C_6H_5Si(CH_3)_2C({=}CH_2)CH_2N(CH_2)_5$ (13)	706
	H_2PtCl_6 (i-PrOH)	200	6	trans-$C_6H_5Si(CH_3)_2CH{=}CHCH_2N(CH_2)_5$ (92) $C_6H_5Si(CH_3)_2C({=}CH_2)CH_2N(CH_2)_5$ (6)	706
C_8H_{13}	$CH{\equiv}CSi(OCH_2CH_2)_3N$				
	H_2PtCl_6 $6H_2O$	50	6	$(CH_3)_2C_6H_5SiCH{=}CHSi(OCH_2CH_2)_3N$ (62)	704
C_8H_{14}	$C_5H_{11}C{\equiv}CCH_2OH$				
	$Pt(CH_2{=}CH_2)_2[P(c\text{-}Hex)_3]$	reflux	2 (THF)	$C_5H_{11}C[Si(CH_3)_2C_6H_5]{=}CHCH_2OH$ + $C_5H_{11}CH{=}C[Si(CH_3)_2C_6H_5]CH_2OH$ (99)	804
C_8H_{14}	$CH{\equiv}CCH(OH)C_5H_{11}$				
	$Pt(CH_2{=}CH_2)_2[P(c\text{-}Hex)_3]$	reflux	0.1 (THF)	$C_6H_5Si(CH_3)_2CH{=}CHC(OH)C_5H_{11}$ + $CH_2{=}C[Si(CH_3)_2C_6H_5]CH(OH)C_5H_{11}$ (99)	804
C_8H_{14}	$HOC(CH_3)_2C{\equiv}CC(CH_3)_2OH$				
	$[HPt(SiBzMe_2)P(c\text{-}Hex)_3]_2$	65	15	cis-$HOC(CH_3)_2C[Si(CH_3)_2C_6H_5]{=}CHC(CH_3)_2OH$ (83)	1205
C_9H_{13}	$CH{\equiv}CCH_2OCH_2CH(CH_2Cl)OCH_2CHCH_2O$				
	H_2PtCl_6 (i-PrOH)	reflux	18 (b)	$C_6H_5Si(CH_3)_2CH{=}CHCH_2OCH_2CH(CH_2Cl)OCH_2CHCH_2O$ (76)	1161
C_9H_{15}	$CH{\equiv}CCH_2N(CH_2)_6$				
	H_2PtCl_6 (i-PrOH)	150	6	trans-$C_6H_5Si(CH_3)_2CH{=}CHCH_2N(CH_2)_6$ (55) $C_6H_5Si(CH_3)_2C({=}CH_2)CH_2N(CH_2)_6$ (15)	706
	H_2PtCl_6 (i-PrOH)	200	6	trans-$C_6H_5Si(CH_3)_2CH{=}CHCH_2N(CH_2)_6$ (55) $C_6H_5Si(CH_3)_2C({=}CH_2)CH_2N(CH_2)_6$ (15)	706
$C_{10}H_{16}$	$CH{\equiv}CCH_2OCH_2CH(CH_2OCH_3)OCH_2CHCH_2O$				
	H_2PtCl_6 (i-PrOH)	reflux	18 (b)	$C_6H_5Si(CH_3)_2CH{=}CHCH_2OCH_2CH(CH_2OCH_3)OCH_2CHCH_2O$ (80)	1161
$C_{10}H_{18}$	$(CH_3)_3SiC{\equiv}CC{\equiv}CSi(CH_3)_3$				
	H_2PtCl_6 (i-PrOH)	r.t.	1.5	$(CH_3)_3SiCH{=}C[Si(CH_3)_2C_6H_5]C{\equiv}CSi(CH_3)_3$ (70) $(CH_3)_3SiCH_2C[Si(CH_3)_2C_6H_5]{=}C{=}C[Si(CH_3)_3]$- $Si(CH_3)_2C_6H_5$ (10)	633
	$RhCl(PPh_3)_3$	100	2	$(CH_3)_3SiCH_2C[Si(CH_3)_2C_6H_5]{=}C{=}C[Si(CH_3)_3]$- $Si(CH_3)_2C_6H_5$ (86)	633
	$RhCl(PPh_3)_3$	90	1	$(CH_3)_3SiCH{=}C[Si(CH_3)_2C_6H_5]C{\equiv}CSi(CH_3)_3$ (27) $(CH_3)_3SiCH_2C[Si(CH_3)_2C_6H_5]{=}C{=}C[Si(CH_3)_3]$- $Si(CH_3)_2C_6H_5$ (36)	633
$C_{10}H_{22}$	$(CH_3)_3SiOCH_2C{\equiv}CCH_2OSi(CH_3)_3$				

Table 10.5

= $(CH_3)_2C_6H_5SiH$ =

	H_2PtCl_6 (THF)	-	0.5	$(CH_3)_3SiOCH_2CH=C[Si(CH_3)_2C_6H_5]CH_2OSi(CH_3)_3$ (-)	698
$C_{11}H_{10}$	$C_6H_5C\equiv CCH_2\overline{CHCH_2O}$				
	H_2PtCl_6	60-70	-	$C_6H_5CH=C[Si(CH_3)_2C_6H_5]CH_2\overline{CHCH_2O}$ (85)	1195
$C_{11}H_{12}$	$CH\equiv CCH(OH)CH_2OCH_2C_6H_5$				
	$Pt(CH_2=CH_2)_2[P(c\text{-}Hex)_3]$	reflux	6 (THF)	$C_6H_5Si(CH_3)_2CH=CHC(OH)CH_2OCH_2C_6H_5$ + $CH_2=C[Si(CH_3)_2C_6H_5]CH(OH)CH_2OCH_2C_6H_5$ (75)	804
$C_{11}H_{19}$	$\overline{OCH_2CHCH_2OCH_2C}\equiv CCH_2N(CH_2CH_3)_2$				
	H_2PtCl_6	reflux	18 (b)	$\overline{OCH_2CH}CH_2OCH_2C[Si(CH_3)_2C_6H_5]=CHCH_2N(CH_2CH_3)_2$ (62)	1196
$C_{13}H_{14}$	$C_6H_5C\equiv CCH_2CH_2OCH_2\overline{CHCH_2O}$				
	H_2PtCl_6	reflux	18 (b)	$C_6H_5CH=C[Si(CH_3)_2C_6H_5]CH_2CH_2OCH_2\overline{CHCH_2O}$ (72)	1196

Table 10.6

= $(CH_3)_2C_6H_5SiH$ = =1=

C_3H_3	CH_3COCN				
	$RhCl(PPh_3)_3$	80	2	$CH_3CH[OSi(CH_3)_2C_6H_5]CN$ (87)	873
C_4H_8	$CH_3CH_2COCH_3$				
	$n\text{-}Bu_4NClO_4 + LiClO_4(CH_2Cl_2) + C_6H_5(CH_2)_3\text{-}OSi(CH_3)_3$ + electrolysis	r.t.	-	$CH_3CH_2CH(CH_3)OSi(CH_2)_3C_6H_5$ (95)	1201
C_4H_8	$CH_3COCH_2CH_3$				
	$[RhCl(C_2H_4)_2]_2$ + O-(t-Bu)$_2PC_6H_4CH(CH_3)N\text{-}(CH_3)_2$	r.t.	7 days (b)	$CH_3CH[OSi(CH_3)_2C_6H_5]CH_2CH_3$ (15-21)	762
C_5H_4	Fig.266.				
	-	r.t.	3 (dcm)	Fig.460. (57)	418
C_5H_4	Fig.461.				
	-	r.t.	3 (dcm)	Fig.462. (54)	418
C_5H_8	$CH_2=C(CH_3)COOCH_3$				
	Rh/C (5%)	20-25	24	$(CH_3)_2C=C(OCH_3)OSi(CH_3)_2C_6H_5$ (71)	U -187
	$RhCl(PPh_3)_3$	70	12 (b)	$(CH_3)_2C=C(OCH_3)OSi(CH_3)_2C_6H_5$ (82)	868
C_5H_8	$CH_3CH=CHCOCH_3$				
	$[PtCl_2(py)]_2$	-	5	$CH_3CH_2CH=C(CH_3)OSi(CH_3)_2C_6H_5$ (100)	987
C_5H_8	$CH_3CH=CHCOOCH_3$				
	$RhCl(PPh_3)_3$	70	2 (b)	$CH_3CH_2CH=C(OCH_3)OSi(CH_3)_2C_6H_5$ (70)	868
C_6H_6	Fig.463.				
	-	r.t.	0.5 (dcm)	Fig.464. (58)	418
C_6H_6	Fig.465.				
	-	r.t.	10 (dcm)	Fig.466. (61)	418
C_6H_6	Fig.467.				
	-	r.t.	11 (dcm)	Fig.468. (45)	418
C_6H_8	$\overline{(CH_2)_3CH=CHCO}$				
	$RhCl(PPh_3)_3$	81	3 (b)	$\overline{(CH_2)_3CH=CHC}HOSi(CH_3)_2C_6H_5$ + $\overline{(CH_2)_4CH=C}[OSi(CH_3)_2C_6H_5]$ (kin.)	1046
C_6H_{10}	$\overline{(CH_2)_5CO}$				

Table 10.6

= $(CH_3)_2C_6H_5SiH$ = =2=

	$Co_2(CO)_8$	50	4	$\overline{(CH_2)_5C}HOSi(CH_3)_2C_6H_5$ + $\overline{(CH_2)_4CH=C}OSi(CH_3)_2C_6H_5$ (-)	1013
	Rh complex	reflux	8(b or THF)	$\overline{(CH_2)_5C}HOSi(CH_3)_2C_6H_5$ (29-84) $\overline{(CH_2)_4CH=C}OSi(CH_3)_2C_6H_5$ (16-71)	1340
	Rh complex + Et_3N	reflux	8-10	$\overline{(CH_2)_5C}HOSi(CH_3)_2C_6H_5$ (6-71) $\overline{(CH_2)_4CH=C}OSi(CH_3)_2C_6H_5$ (29-94)	1340
C_6H_{10}	$(CH_3)_2C=CHCOCH_3$				
	$RhCl(PPh_3)_3$	-	-	$(CH_3)_2CHCH=C(CH_3)OSi(CH_3)_2C_6H_5$ (98)	850
	$RhCl(PPh_3)_3$	50	0.5	$(CH_3)_2CHCH=C(CH_3)OSi(CH_3)_2C_6H_5$ (98)	873
	$RhCl(PPh_3)_3$	50	0.5	$(CH_3)_2CHCH=C(CH_3)OSi(CH_3)_2C_6H_5$ (98)	F -023
C_6H_{10}	$CH_2=CHCH_2CH_2COCH_3$				
	Bu_4NF (HMPA)	0-r.t.	0.5-12	$CH_2=CHCH_2CH_2CH(CH_3)OSi(CH_3)_2C_6H_5$ (57)	393
C_6H_{10}	$CH_3COOCH(CH_3)COCH_3$				
	$Bu_4N^+Cl^-$	-	20	$CH_3COOCH(CH_3)CH(CH_3)OSi(CH_3)_2C_6H_5$ (-)	J -062
C_6H_{12}	$(CH_3)_3CCOCH_3$				
	$RhCl(C_6H_{10})_2$ + PPFA	50	-	$(CH_3)_3CCH(CH_3)OSi(CH_3)_2C_6H_5$ (71)	488
	$[RhH_2(L)_2(BMPP)_2]ClO_4$	50	40	$(CH_3)_3CCH(CH_3)OSi(CH_3)_2C_6H_5$ (46)	486
C_6H_{12}	$CH_3(CH_2)_3COCH_3$				
	$RhCl(1,5\text{-}C_6H_8)_2$ + BMPP	-	-	$CH_3(CH_2)_3CH(CH_3)OSi(CH_3)_2C_6H_5$ (-)	864
	$[RhH_2(L)_2(BMPP)_2]ClO_4$	50	40	$CH_3(CH_2)_3CH(CH_3)OSi(CH_3)_2C_6H_5$ (64)	486
C_6H_{12}	$CH_3COC(CH_3)_2CH_3$				
	$[RhCl(C_2H_4)_2]_2$ + O-(t-Bu)$_2PC_6H_4CH(CH_3)N$-$(CH_3)_2$	r.t.	7 days (b)	$CH_3CH[OSi(CH_3)_2C_6H_5]C(CH_3)_2CH_3$ (18-25)	762
C_7H_5	4-ClC_6H_4CHO				
	$LiClO_4$,Bu_4NClO_4 (CH_2Cl_2) + electrolysis	r.t.	-	4-$ClC_6H_4CH_2OSi(CH_3)_2C_6H_5$ (86)	1201
C_7H_6	C_6H_5CHO				
	-	r.t.	11 (dcm)	$C_6H_5CH_2OSi(CH_3)_2C_6H_5$ (57)	418
	Bu_4NF (HMPA)	0-r.t.	0.5-12	$C_6H_5CH_2OSi(CH_3)_2C_6H_5$ (91)	393
	$LiClO_4$,Bu_4NClO_4 (CH_2Cl_2) + electrolysis	r.t.	-	$C_6H_5CH_2OSi(CH_3)_2C_6H_5$ (96-99)	1201

Table 10.6
= (CH$_3$)$_2$C$_6$H$_5$SiH =   . =3=

	RhCl(PPh$_3$)$_3$	r.t.	0.08	C$_6$H$_5$CH$_2$OSi(CH$_3$)$_2$C$_6$H$_5$ (94)	866
	RhCl(PPh$_3$)$_3$	r.t.	0.08	C$_6$H$_5$CH$_2$OSi(CH$_3$)$_2$C$_6$H$_5$ (94)	873
	RhCl(PPh$_3$)$_3$	r.t.	0.08	C$_6$H$_5$CH$_2$OSi(CH$_3$)$_2$C$_6$H$_5$ (94)	F -023

C$_7$H$_{12}$ $\overline{(CH_2)_4CH(CH_3)}$CO

	Bu$_4$NF (HMPA)	0-r.t.	0.5-12	$\overline{(CH_2)_4CH(CH_3)}$CHOSi(CH$_3$)$_2C_6H_5$ (76)	393

C$_7$H$_{12}$ (CH$_3$)$_2$CHCH=CHCOCH$_3$

	RhCl(PPh$_3$)$_3$	50	0.5	(CH$_3$)$_2$CHCH=C(CH$_3$)CH$_2$OSi(CH$_3$)$_2$C$_6$H$_5$ (98)	866

C$_7$H$_{12}$ CH$_3$COCH(CH$_3$)COOCH$_2$CH$_3$

	CF$_3$COOH	0	10	CH$_3$CH[OSi(CH$_3$)$_2$C$_6$H$_5$]CH(CH$_3$)COOC$_2$H$_5$ (90)	392

C$_7$H$_{14}$ (CH$_3$)$_2$CHNCNCH(CH$_3$)$_2$

	PdCl$_2$	-	-	(CH$_3$)$_2$CHN[Si(CH$_3$)$_2$C$_6$H$_5$]CHNCH(CH$_3$)$_2$ (-)	J -015
	PdCl$_2$	200	48	(CH$_3$)$_2$CHN[Si(CH$_3$)$_2$C$_6$H$_5$]CHNCH(CH$_3$)$_2$ (83)	860
	RhCl(PPh$_3$)$_3$	150	36	(CH$_3$)$_2$CHN[Si(CH$_3$)$_2$C$_6$H$_5$]CHNCH(CH$_3$)$_2$ (80)	860

C$_7$H$_{14}$ CH$_3$(CH$_2$)$_5$CHO

	n-Bu$_4$NClO$_4$+LiClO$_4$(CH$_2$Cl$_2$) +C$_6$H$_5$(CH$_2$)$_3$- OSi(CH$_3$)$_3$+ electrolysis	r.t.	-	CH$_3$(CH$_2$)$_5$CH$_2$OSi(CH$_2$)$_3$C$_6$H$_5$ (82)	1201

C$_7$H$_{14}$ n-C$_6$H$_{13}$CHO

	LiClO$_4$,Bu$_4$NClO$_4$ (CH$_2$Cl$_2$) + electrolysis	r.t.	-	n-C$_6$H$_{13}$CH$_2$OSi(CH$_3$)$_2$C$_6$H$_5$ (92)	1201

C$_8$H$_5$ C$_6$H$_5$COCF$_3$

	Me$_2$SO	62	35	C$_6$H$_5$CH(CF$_3$)OSi(CH$_3$)$_2$C$_6$H$_5$ (22)	1307
	Me$_2$SO, Bu$_4$NF	62	35	C$_6$H$_5$CH(CF$_3$)OSi(CH$_3$)$_2$C$_6$H$_5$ (86)	1307

C$_8$H$_7$ C$_6$H$_5$COCH$_2$F

	Me$_2$SO	40	12	C$_6$H$_5$CH(CH$_2$F)OSi(CH$_3$)$_2$C$_6$H$_5$ (73-89)	1307

C$_8$H$_8$ 4-CH$_3$C$_6$H$_4$CHO

	LiClO$_4$,Bu$_4$NClO$_4$ (CH$_2$Cl$_2$) + electrolysis	r.t.	-	4-CH$_3$C$_6$H$_4$CH$_2$OSi(CH$_3$)$_2$C$_6$H$_5$ (94)	1201

C$_8$H$_8$ C$_6$H$_5$COCH$_3$

	(Et$_2$N)$_3$S$^+$Me$_3$SiF$_2$$^-$	-23	12	C$_6$H$_5$CH[OSi(CH$_3$)$_2$C$_6$H$_5$]CH$_3$ (-)	395

	-	r.t.	7 (dcm)	$C_6H_5CH(CH_3)OSi(CH_3)_2C_6H_5$ (54)	418
	$n\text{-}Bu_4NClO_4 + LiClO_4(CH_2Cl_2) + C_6H_5(CH_2)_3\text{-}OSi(CH_3)_3$ + electrolysis	r.t.	-	$C_6H_5CH(CH_3)OSi(CH_2)_3C_6H_5$ (80)	1201
	$Co_2(CO)_8$ + py	50	2	$C_6H_5CH(CH_3)OSi(CH_3)_2C_6H_5$ (56)	1013
	Rh complex	reflux	15 (THF)	$C_6H_5CH(CH_3)OSi(CH_3)_2C_6H_5$ (-)	1341
	Rh complex	reflux	8-10 (solv.)	$C_6H_5CH(CH_3)OSi(CH_3)_2C_6H_5$ (57-97) $C_6H_5C[OSi(CH_3)_2C_6H_5]{=}CH_2$ (3-43)	1340
	$RhCl(L)(BMPP)_2$	40-50	-	$C_6H_5CH(CH_3)OSi(CH_3)_2C_6H_5$ (-)	587
	$RhCl(L)(BMPP)_2$	40-50	-	$C_6H_5CH(CH_3)OSi(CH_3)_2C_6H_5$ (-)	865
	$RhCl(L)(BMPP)_2$	50	8 (b)	$C_6H_5CH(CH_3)OSi(CH_3)_2C_6H_5$ (92)	864
	$RhCl(L)(BMPP)_2$	50	8 (b)	$C_6H_5CH(CH_3)OSi(CH_3)_2C_6H_5$ (92)	854
	RhCl(L)(diop)	70	40	$C_6H_5CH(CH_3)OSi(CH_3)_2C_6H_5$ (35)	486
	$[RhCl(C_2H_4)_2]_2$ + O-(t-Bu)$_2PC_6H_4CH(CH_3)N\text{-}(CH_3)_2$	r.t.	7 days (b)	$C_6H_5CH(CH_3)OSi(CH_3)_2C_6H_5$ (57-61)	762
	$[RhH_2(L)_2(BMPP)_2]ClO_4$	50	40	$C_6H_5CH(CH_3)OSi(CH_3)_2C_6H_5$ (97)	486
	$[RhH_2(L)_2(BMPP)_2]ClO_4$	50	40	$C_6H_5CH(CH_3)OSi(CH_3)_2C_6H_5$ (97)	1298
C_8H_{16}	$CH_3(CH_2)_5COCH_3$				
	Bu_4NF (HMPA)	0-r.t.	0.5-12	$CH_3(CH_2)_5CH(CH_3)OSi(CH_3)_2C_6H_5$ (87)	393
C_9H_8	$C_6H_5CH{=}CHCHO$				
	Bu_4NF (HMPA)	0-r.t.	0.5-12	$C_6H_5CH{=}CHCH_2OSi(CH_3)_2C_6H_5$ (100)	393
	$[PtCl_2(py)]_2$	-	3.5	$C_6H_5CH_2CH{=}CHOSi(CH_3)_2C_6H_5$ (95)	1216
C_9H_{10}	$C_6H_5CH_2COCH_3$				
	$[RhH_2(BMPP)_2]ClO_4$	50	40	$C_6H_5CH_2CH(CH_3)OSi(CH_3)_2C_6H_5$ (69)	486
C_9H_{10}	$C_6H_5COCH_2CH_3$				
	Rh complex + NEt_3	reflux	8-10 (solv.)	$C_6H_5CH(CH_2CH_3)OSi(CH_3)_2C_6H_5$ (7-78) $C_6H_5C[OSi(CH_3)_2C_6H_5]{=}CHCH_3$ (22-93)	1340
	$RhCl(C_6H_{10})_2$ + PPFA	50	-	$C_6H_5CH(CH_2CH_3)OSi(CH_3)_2C_6H_5$ (40-61)	488
	$RhCl(L)(BMPP)_2$	-	-	$C_6H_5CH(CH_2CH_3)OSi(CH_3)_2C_6H_5$ (96)	854
	$RhCl(L)(BMPP)_2$	40-50	-	$C_6H_5CH(CH_2CH_3)OSi(CH_3)_2C_6H_5$ (-)	587

	CF_3COOH	0	4	$C_6H_5CH[OSi(CH_3)_2C_6H_5]CH(CH_3)NHCOOCH_2CH_3$ (87)	J -061
$C_{13}H_{11}$	$C_6H_5CHNC_6H_5$				
	$RhCl(PPh_3)_3$	20	77	$C_6H_5CH_2N[Si(CH_3)_2C_6H_5]C_6H_5$ (34)	43
	$ZnCl_2$	120	16	$C_6H_5CH_2N[Si(CH_3)_2C_6H_5]C_6H_5$ (8)	43
$C_{13}H_{13}$	$C_6H_5COCH(CH_3)CONC(O)O(CH_2)_2$				
	CF_3COOH	0	5	$C_6H_5CH[OSi(CH_3)_2C_6H_5]CH(CH_3)CONC(O)O(CH_2)_2$ (98)	392
$C_{13}H_{16}$	$(CH_2)_5CHCOC_6H_5$				
	$RhCl(L)(BMPP)_2$	40-50	-	$(CH_2)_5CHCH[OSi(CH_3)_2C_6H_5]C_6H_5$ (90)	587
	$RhCl(L)(BMPP)_2$	40-50	-	$(CH_2)_5CHCH[OSi(CH_3)_2C_6H_5]C_6H_5$ (90)	865
$C_{13}H_{17}$	Fig.471.				
	CF_3COOH	0	0.25	Fig.472. (-)	393
$C_{13}H_{20}$	Fig.473.				
	$RhCl(PPh_3)_3$	-	-	Fig.474. (81)	850
	$RhCl(PPh_3)_3$	70	3	Fig.474. (93)	F -023
$C_{14}H_{12}$	$C_6H_5CH_2COC_6H_5$				
	$[RhH_2(L)_2(BMPP)_2]ClO_4$	50	40	$C_6H_5CH_2CH[OSi(CH_3)_2C_6H_5]C_6H_5$ (44)	486
$C_{14}H_{16}$	$p\text{-}ClC_6H_4COCH(CH_3)CON(CH_2)_4$				
	CF_3COOH	0	6	$p\text{-}ClC_6H_4CH[OSi(CH_3)_2C_6H_5]CH(CH_3)CON(CH_2)_4$ (98)	392
	F^-	0	16	$p\text{-}ClC_6H_4CH[OSi(CH_3)_2C_6H_5]CH(CH_3)CON(CH_2)_4$ (86)	392
$C_{14}H_{17}$	$C_6H_5COCH(CH_3)CON(CH_2)_4$				
	CF_3COOH	0	3	$C_6H_5CH[OSi(CH_3)_2C_6H_5]CH(CH_3)CON(CH_2)_4$ (99)	392
	F^-	0	22	$C_6H_5CH[OSi(CH_3)_2C_6H_5]CH(CH_3)CON(CH_2)_4$ (91)	392
$C_{14}H_{19}$	$C_6H_5COCH(CH_3)CON(CH_2CH_3)_2$				
	CF_3COOH	0	4	$C_6H_5CH[OSi(CH_3)_2C_6H_5]CH(CH_3)CON(CH_2CH_3)_2$ (98)	392
	F^-	0	12	$C_6H_5CH[OSi(CH_3)_2C_6H_5]CH(CH_3)CON(CH_2CH_3)_2$ (98)	392
$C_{15}H_{15}$	$C_6H_5COCH(CH_3)NHSO_2C_6H_5$				
	CF_3COOH	0	20	$C_6H_5Si(CH_3)_2OCH(C_6H_5)CH(CH_3)NHSO_2C_6H_5$ (-)	393
$C_{15}H_{19}$	$C_6H_5CH{=}CHCOCH(CH_3)OCHNH(CH_2)_3$				

= (CH₃)₂C₆H₅SiH =

$= (CH_3)_2C_6H_5SiH =$ $= 8 =$

Bu₄NF (HMPA)	0	16	$C_6H_5CH=CHCH[OSi(CH_3)_2C_6H_5]CH(CH_3)O\overline{CHNH(CH_2)}_3$ (-)	393

$C_{15}H_{19}$ $p\text{-}CH_3OC_6H_4COCH(CH_3)CO\overline{N(CH_2)}_4$

F⁻	0	16	$p\text{-}CH_3OC_6H_4CH[OSi(CH_3)_2C_6H_5]CH(CH_3)CO\overline{N(CH_2)}_4$ (92)	392

$C_{15}H_{20}$ $C_6H_5CH=CHCOCH(CH_3)OC(CH_3)_3$

Bu₄NF (HMPA)	0	20	$C_6H_5CH=CHCH[OSi(CH_3)_2C_6H_5]CH(CH_3)OC(CH_3)_3$ (-)	393

$C_{16}H_{14}$ $C_6H_5C(CH_3)=CHCOC_6H_5$

RhCl(c-C₆H₁₀)₂ + diop	20	-	$C_6H_5CH(CH_3)CH=C[OSi(CH_3)_2C_6H_5]C_6H_5$ (83)	487
[RhH₂(L)₂(BMPP)₂]ClO₄	20	-	$C_6H_5CH(CH_3)CH=C[OSi(CH_3)_2C_6H_5]C_6H_5$ (72)	487

$C_{16}H_{14}$ $C_6H_5COCH(CH_3)OCOC_6H_5$

Bu₄NF (HMPA)	0	6	$C_6H_5Si(CH_3)_2OCH(C_6H_5)CH(CH_3)OCOC_6H_5$ (-)	393
CF₃COOH	0	6	$C_6H_5Si(CH_3)_2OCH(C_6H_5)CH(CH_3)OCOC_6H_5$ (-)	393

$C_{20}H_{19}$ $C_6H_5COCH(CH_3)CO\overline{NC(O)OCH(C_6H_5)}\dot{C}H(CH_3)$

CF₃COOH	0	4	$C_6H_5CH[OSi(CH_3)_2C_6H_5]CH(CH_3)CO\text{-}$ $\overline{NC(O)CH(C_6H_5)}\dot{C}H(CH_3)$ (98)	392

Table 11.1

= $(C_3H_7)_3SiH$ = =1=

C_3H_5	$CH_2=CHCH_2Cl$				
	SDVB-PPh_2/$RhCl(PPh_3)_3$	80	3	$(C_3H_7)_3Si(CH_2)_3Cl + (C_3H_7)_3SiCH(CH_3)CH_2Cl$ (-)	D -056
C_4H_6	$CH_2=CHCH=CH_2$				
	$PdCl(C_3H_5)_2$	100	4	$CH_3CH=CHCH_2CH_2CH=CHCH_2Si(C_3H_7)_3$ (12)	644
	$PdCl(C_3H_5)_2$	22	24	$CH_3CH=CHCH_2CH_2CH=CHCH_2Si(C_3H_7)_3$ (90)	644
	$PdCl_2(PPh_3)_2$	100	4 (b)	$CH_3CH=CHCH_2CH_2CH=CHCH_2Si(C_3H_7)_3$ (62)	644
	SDVB-NMe_2/$RhCl_3$	20	24	$CH_3CH=CHCH_2Si(C_3H_7)_3$ (81)	C -006
C_6H_{10}	$CH_2=C(CH_3)C(CH_3)=CH_2$				
	$Cr(CO)_6$ + UV	10-60	4-7	$CH_3C(CH_3)=C(CH_3)CH_2Si(C_3H_7)_3$ (0-1)	1293
C_6H_{10}	$CH_2=CHCH_2CH_2COCH_3$				
	Ni	141-221	1.5	$(C_3H_7)_3Si(CH_2)_4COCH_3$ (88)	15
	$ZnCl_2$	138-205	0.75	$(C_3H_7)_3Si(CH_2)_4COCH_3$ (74)	15
C_6H_{12}	$CH_3(CH_2)_3CH=CH_2$				
	$Co_2(CO)_8$	120	2	$CH_3(CH_2)_5Si(C_3H_7)_3$ (kin.)	1169
	H_2PtCl_6 (THF)	20	3 (THF)	$CH_3(CH_2)_5Si(C_3H_7)_3$ (kin.)	1252
	$RhCl(PHPh_2)_3$	85	2	$CH_3(CH_2)_5Si(C_3H_7)_3$ (78)	1166
	$RhCl(PPh_3)_3$	80	2	$CH_3(CH_2)_5Si(C_3H_7)_3$ (78)	1169
	SDVB-NMe_2/$RhCl_3$	reflux	4	$CH_3(CH_2)_5Si(C_3H_7)_3$ (83)	C -006
	polymer-PPh_2/$RhCl(PPh_3)_3$	85	2	$CH_3(CH_2)_5Si(C_3H_7)_3$ (22)	1168
	polymer-PPh_2/$RhCl(PPh_3)_3$	reflux	13	$CH_3(CH_2)_5Si(C_3H_7)_3$ (70)	1168
C_7H_{14}	$CH_2=CHCH(OCH_2CH_3)_2$				
	H_2PtCl_6 (i-PrOH)	110	72	$(C_3H_7)_3SiCH_2CH_2CH(OCH_2CH_3)_2$ (57)	135
C_7H_{14}	$CH_3(CH_2)_4CH=CH_2$				
	H_2PtCl_6 (THF)	100	4	$CH_3(CH_2)_6Si(C_3H_7)_3$ (-)	214
C_8H_8	$C_6H_5CH=CH_2$				
	$(t\text{-}BuO)_2$	140	-	$C_6H_5CH_2CH_2Si(C_3H_7)_3$ (1)	347
C_8H_{14}	$\overline{(CH_2)_6CH=CH}$				

Table 11.1

= $(C_3H_7)_3SiH$ = =2=

	$(t\text{-}BuO)_2$	140	-	$\overline{(CH_2)_7C}HSi(C_3H_7)_3$ (70)	347
C_8H_{15}	$\overline{(CH_2)_5N}CH_2CH=CH$				
	H_2PtCl_6 (i-PrOH)	100	17	$\overline{(CH_2)_5N}(CH_2)_3Si(C_3H_7)_3$ (49)	702
C_8H_{16}	$CH_3(CH_2)_4CH=CHCH_3$				
	$(t\text{-}BuO)_2$	140	-	$C_8H_{17}Si(C_3H_7)_3$ (60)	347
C_8H_{16}	$CH_3(CH_2)_5CH=CH_2$				
	$(t\text{-}BuO)_2$	140	-	$CH_3(CH_2)_7Si(C_3H_7)_3$ (90)	347
	SDVB-NMe_2/H_2PtCl_6	reflux	3 (b)	$CH_3(CH_2)_7Si(C_3H_7)_3$ (73)	C -002
C_8H_{18}	$(CH_3CH_2O)_3SiCH=CH_2$				
	H_2PtCl_6 (i-PrOH)	80-200	0.35	$(CH_3CH_2O)_3SiCH_2CH_2Si(C_3H_7)_3$ (85)	575
C_9H_{10}	$C_6H_5CH_2CH=CH_2$				
	H_2PtCl_6	-	-	$C_6H_5C_3H_6Si(C_3H_7)_3$ (-)	516
C_9H_{16}	$CH_3CH_2CH_2CH=C=C(CH_2CH_3)CH_2OH$				
	H_2PtCl_6	-	-	$(C_3H_7)_3SiCH(CH_2CH_2CH_3)CH=C(CH_2CH_3)CH_2OH$ + $CH_3CH_2CH_2CH=C=C(CH_2CH_3)CH_2OSi(C_3H_7)_3$ (-)	253
$C_{10}H_{12}$	$C_6H_5CH_2CH_2CH=CH_2$				
	H_2PtCl_6	-	-	$C_6H_5(CH_2)_4Si(C_3H_7)_3$ (-)	515
$C_{10}H_{12}$	m-$CH_3OC_6H_4OCH_2CH=CH_2$				
	H_2PtCl_6	-	-	m-$CH_3OC_6H_4O(CH_2)_3Si(C_3H_7)_3$ (-)	513
$C_{10}H_{13}$	Fig.475.				
	H_2PtCl_6 (THF)	100	3	Fig.476. (86)	943
$C_{10}H_{16}$	Fig.479.				
	$(t\text{-}BuO)_2$	140	-	Fig.480. (45)	347
$C_{10}H_{16}$	Fig.505.				
	$(t\text{-}BuO)_2$	140	-	Fig.478. (40)	347
$C_{10}H_{18}$	$CH_3CH(OH)C(CH_2CH_3)=C=CHCH_2CH_2CH_3$				
	H_2PtCl_6 (i-PrOH)	60	3	$CH_3CH(OH)C(CH_2CH_3)=C[Si(C_3H_7)_3](CH_2)_3CH_3$ + $CH_3CH(OH)C(CH_2CH_3)C[Si(C_3H_7)_3]=CHCH_2CH_2CH_3$ (88)	249
$C_{10}H_{18}$	$CH_3CH(OH)CH=C=CH(CH_2)_4CH_3$				

Table 11.1

= $(C_3H_7)_3SiH$ = =3=

	H_2PtCl_6 (i-PrOH)	60	3	$CH_3CH(OH)CH=C[Si(C_3H_7)_3](CH_2)_5CH_3$ + $CH_3CH(OH)CH_2C[Si(C_3H_7)_3]=CH(CH_2)_4CH_3$ (57)	249
$C_{12}H_{11}$	Fig.302.				
	H_2PtCl_6 (THF)	100	3	Fig.481. (68)	943
$C_{13}H_{13}$	Fig.306.				
	H_2PtCl_6 (THF)	100	3	Fig.482. (79)	943

Table 11.2

$= (C_3H_7)_3SiH =$ =1=

C_6H_8	$CH_2{=}CHC{\equiv}CCH_2OCH_3$				
	H_2PtCl_6	100	-	$CH_2{=}CHC[Si(C_3H_7)_3]{=}CHCH_2OCH_3$ + $CH_2{=}CHCH{=}C[Si(C_3H_7)_3]CH_2OCH_3$ (-)	534
	H_2PtCl_6	100	-	$CH_2{=}CHCH{=}C[Si(C_3H_7)_3]CH_2OCH_3$ (58-60) $CH_2{=}CHC[Si(C_3H_7)_3]{=}CHCH_2OCH_3$ (3-30) $CH_3CH{=}C{=}C[Si(C_3H_7)_3]CH_2OCH_3$ (-)	535
C_6H_{10}	$CH_3(CH_2)_3C{\equiv}CH$				
	H_2PtCl_6 (THF)	100	2	$CH_3(CH_2)_3CH{=}CHSi(C_3H_7)_3$ + $CH_2{=}C[Si(C_3H_7)_3](CH_2)_3CH_3$ (-)	266
C_6H_{10}	$CH_3CH_2CH_2CH(OH)C{\equiv}CH$				
	H_2PtCl_6 (i-PrOH)	80-100	2	$CH_3CH_2CH_2CH(OH)CH{=}CHSi(C_3H_7)_3$ (-)	1083
C_7H_{10}	$CH_2{=}CHC{\equiv}CCH(CH_3)OCH_3$				
	H_2PtCl_6	100	-	$CH_2{=}CHC[Si(C_3H_7)_3]{=}CHCH(CH_3)OCH_3$ + $CH_2{=}CHCH{=}C[Si(C_3H_7)_3]CH(CH_3)OCH_3$ (-)	534
C_7H_{12}	$(CH_3)_2C(OH)C{\equiv}CCH_2OCH_3$				
	H_2PtCl_6 (i-PrOH)	45-50	4	$(CH_3)_2C(OH)CH{=}C(CH_2OCH_3)Si(C_3H_7)_3$ (50)	1071
C_7H_{12}	$(CH_3)_2CHCH_2CH(OH)C{\equiv}CH$				
	H_2PtCl_6 (i-PrOH)	80-100	6	$(CH_3)_2CHCH_2CH(OH)CH{=}CHSi(C_3H_7)_3$ (-)	1083
C_7H_{12}	$CH_3CH(OH)C{\equiv}CCH_2CH_2CH_3$				
	H_2PtCl_6	-	-	$CH_3CH(OH)CH{=}C[Si(C_3H_7)_3]CH_2CH_2CH_3$ + $CH_3CH(OH)C[Si(C_3H_7)_3]{=}CHCH_2CH_2CH_3$ + $CH_3CH[OSi(C_3H_7)_3]CH{=}C[Si(C_3H_7)_3]CH_2CH_2CH_3$ (-)	254
C_8H_6	$C_6H_5C{\equiv}CH$				
	-	-	-	$C_6H_5C_2H_2Si(C_3H_7)_3$ (kin.)	512
C_8H_{12}	$CH_3CH_2CH(OCH_3)C{\equiv}CCH{=}CH_2$				
	H_2PtCl_6	100	-	$CH_3CH_2CH(OCH_3)CH{=}C[Si(C_3H_7)_3]CH{=}CH_2$ + $CH_3CH_2CH(OCH_3)C[Si(C_3H_7)_3]{=}CHCH{=}CH_2$ (-)	534
C_8H_{12}	$CH_3OC(CH_3)_2C{\equiv}CCH{=}CH_2$				
	H_2PtCl_6 (i-PrOH)	25-125	-	$CH_3OC(CH_3)_2C[Si(C_3H_7)_3]{=}CHCH{=}CH_2$ + $CH_3OC(CH_3)_2CH{=}C[Si(C_3H_7)_3]CH{=}CH_2$ + $CH_3OC(CH_3)_2CH{=}C[Si(C_3H_7)_3]CH_2CH_3$ (75)	530
	H_2PtCl_6 (i-PrOH)	25-125	-	$CH_3OC(CH_3)_2C[Si(C_3H_7)_3]{=}CHCH{=}CH_2$ + $CH_3OC(CH_3)_2CH{=}C[Si(C_3H_7)_3]CH{=}CH_2$ + $CH_3OC(CH_3)_2CH{=}C[Si(C_3H_7)_3]CH_2CH_3$ (75)	538
C_9H_{14}	$CH_3CH_2CH_2CH(OCH_3)C{\equiv}CCH{=}CH_2$				

	H_2PtCl_6	100	-	$CH_3CH_2CH_2CH(OCH_3)CH=C[Si(C_3H_7)_3]CH=CH_2$ + $CH_3CH_2CH_2CH(OCH_3)C[Si(C_3H_7)_3]=CHCH=CH_2$ (-)	534
$C_{11}H_{16}$	$\overline{(CH_2)_5C}(OCH_3)C\equiv CCH=CH_2$				
	H_2PtCl_6	100	4	$\overline{(CH_2)_5C}(OCH_3)C[Si(C_3H_7)_3]=CHCH=CH_2$ + $\overline{(CH_2)_5C}(OCH_3)CH=C[Si(C_3H_7)_3]CH=CH_2$ (63)	536
$C_{11}H_{16}$	$ClCH_2CH_2OCH_2OC(CH_3)[C(CH_3)_3]C\equiv CH$				
	H_2PtCl_6 (i-PrOH)	60-70	8	$ClCH_2CH_2OCH_2OC(CH_3)[C(CH_3)_3]CH=CHSi(C_3H_7)_3$ (-)	1065
$C_{14}H_{10}$	$C_6H_5C\equiv CC_6H_5$				
	H_2PtCl_6 (i-PrOH)	100	2	$C_6H_5CH=C(C_6H_5)Si(C_3H_7)_3$ (73)	837
$C_{14}H_{16}$	p-$CH_3OC_6H_4C(CH_3)_2C\equiv CCH=CH_2$				
	H_2PtCl_6	-	-	p-$CH_3OC_6H_4C(CH_3)_2CH=C[Si(C_3H_7)_3]CH=CH_2$ (-)	539
	H_2PtCl_6 (i-PrOH)	100	6	p-$CH_3OC_6H_4C(CH_3)_2CH=C[Si(C_3H_7)_3]CH=CH_2$ (75)	532
	H_2PtCl_6 (i-PrOH)	75	10	p-$CH_3OC_6H_4C(CH_3)_2CH=C[Si(C_3H_7)_3]CH=CH_2$ (67)	540
$C_{15}H_{18}$	Fig.325.				
	H_2PtCl_6	-	-	Fig.483/Fig.484/Fig.485. (82/10/8)	539
	H_2PtCl_6 (i-PrOH)	100	6	Fig.483. (76)	532
$C_{20}H_{24}$	Fig.335.				
	Pt/C	160	16 (dg)	Fig.486. (-)	F -024

Table 11.3

= $(C_3H_7)_3SiH$ = =1=

C_2H_4	CH_3CHO				
	Ni	190	-	$CH_3CH_2OSi(C_3H_7)_3$ (55)	146
C_3H_6	CH_3CH_2CHO				
	Ni	200	-	$CH_3CH_2CH_2OSi(C_3H_7)_3$ (86)	146
C_4H_6	$CH_3COCOCH_3$				
	Ni	50-60	3-4	$[CH(CH_3)OSi(C_3H_7)_3]_2$ (72)	936
C_4H_8	$(CH_3)_2CHCHO$				
	Ni	200	-	$(CH_3)_2CHCH_2OSi(C_3H_7)_3$ (80)	146
C_4H_8	$CH_3CH_2CH_2CHO$				
	Ni	200	-	$CH_3(CH_2)_3OSi(C_3H_7)_3$ (83)	146
C_5H_8	$CH_3CH=CHCOOCH_3$				
	$RhCl(PPh_3)_3$	100	70s	$CH_3CH_2CH=C(OCH_3)OSi(C_3H_7)_3$ (74)	1321
C_6H_{10}	$(CH_3)_2C=CHCOCH_3$				
	Ni	120-180	0.17	$(CH_3)_2CHCH=C(CH_3)OSi(C_3H_7)_3$ (85)	15
C_6H_{10}	$CH_3CH_2OCOCOOCH_2CH_3$				
	$ZnCl_2$	50-60	2	$CH_3CH_2OCOCH(OCH_2CH_3)OSi(C_3H_7)_3$ (51)	658
C_9H_{14}	Fig.364.				
	$RhCl(PPh_3)_3$	-	-	Fig.487. (-)	863
$C_{10}H_{12}$	$C_6H_5CH(OH)COOCH_2CH_3$				
	$ZnCl_2$	reflux	5-6	$C_6H_5CH[OSi(C_3H_7)_3]CH(OCH_2CH_3)OSi(C_3H_7)_3$ (52)	657
$C_{11}H_{14}$	$m-CH_3C_6H_4CH(OH)COOCH_2CH_3$				
	$ZnCl_2$	reflux	5-6	$m-CH_3C_6H_4CH[OSi(C_3H_7)_3]CH(OCH_2CH_3)OSi(C_3H_7)_3$ (56)	657
$C_{13}H_{20}$	Fig.473.				
	$RhCl(PPh_3)_3$	-	-	Fig.488. (-)	863
$C_{14}H_{10}$	$(C_6H_5)_2C=C=O$				
	Ni	reflux	5	$(C_6H_5)_2C=CHOSi(C_3H_7)_3$ (70)	477
$C_{16}H_{26}$	$C_6H_5CH[OSi(CH_2CH_3)_3]COOCH_2CH_3$				

= (C$_3$H$_7$)$_3$SiH =   =2=

	ZnCl$_2$	reflux	5-6	C$_6$H$_5$CH[OSi(CH$_2$CH$_3$)$_3$]CH(OCH$_2$CH$_3$)OSi(C$_3$H$_7$)$_3$ (64)	657
C$_{17}$H$_{28}$	m-CH$_3$C$_6$H$_4$CH[OSi(CH$_2$CH$_3$)$_3$]COOCH$_2$CH$_3$				
	ZnCl$_2$	reflux	5-6	m-CH$_3$C$_6$H$_4$CH[OSi(CH$_2$CH$_3$)$_3$]CH(OCH$_2$CH$_3$)OSi(C$_3$H$_7$)$_3$ (63)	657
C$_{18}$H$_{36}$	(iso-C$_5$H$_{11}$COOCH$_2$O)$_2$Si(CH$_2$CH$_3$)$_2$				
	ZnCl$_2$	50-60	2	iso-C$_5$H$_{11}$[OSi(C$_3$H$_7$)$_3$]OCH$_2$OSi(CH$_2$CH$_3$)$_2$OCH$_2$OCO-C$_5$H$_{11}$-iso (52)	658
C$_{23}$H$_{30}$	Fig.149.				
	H$_2$PtCl$_6$ (i-PrOH)	150-160	- (t)	Fig.489. (70)	1320
	H$_2$PtCl$_6$ (i-PrOH)	160	7 (t)	Fig.489. (-)	1322

Table 11.4

= $(iso-C_3H_7)_3SiH$ = =1=

C_6H_{10}	$CH_3CH_2CH_2CH(OH)C{\equiv}CH$				
	H_2PtCl_6 (i-PrOH)	80-100	6	$CH_3CH_2CH_2CH(OH)CH{=}CHSi(C_3H_7\text{-iso})_3$ (-)	1083
C_8H_8	$C_6H_5CH{=}CH_2$				
	$RhCl(PPh_3)_3$	65	6 (b)	$C_6H_5CH_2CH_2Si(C_3H_7\text{-iso})_3$ (1) $C_6H_5CH(CH_3)Si(C_3H_7\text{-iso})_3$ (1) $C_6H_5CH{=}CHSi(C_3H_7\text{-iso})_3$ (30)	878
C_9H_{10}	$C_6H_5CH_2CH{=}CH_2$				
	H_2PtCl_6	-	-	$C_6H_5C_3H_6Si(C_3H_7\text{-iso})_3$ (-)	516
$C_{10}H_{12}$	$C_6H_5CH_2CH_2CH{=}CH_2$				
	H_2PtCl_6	-	-	$C_6H_5(CH_2)_4Si(C_3H_7\text{-iso})_3$ (-)	515
$C_{10}H_{12}$	m-$CH_3OC_6H_4OCH_2CH{=}CH_2$				
	H_2PtCl_6	-	-	m-$CH_3OC_6H_4O(CH_2)_3Si(C_3H_7\text{-iso})_3$ (-)	513
$C_{10}H_{18}$	$(CH_3)_3SiC{\equiv}CC{\equiv}CSi(CH_3)_3$				
	H_2PtCl_6 (i-PrOH)	90	8	$(CH_3)_3SiCH{=}C[Si(C_3H_7\text{-iso})_3]C{\equiv}CSi(CH_3)_3$ (92)	633

Table 12.1

= $(C_4H_9)_3SiH$ = =1=

C_3H_4	$HC\equiv CCH_2OH$				
	$Pt(CH_2=CH_2)_2[P(c\text{-}Hex)_3]_2$	reflux	8 (THF)	$(C_4H_9)_3SiCH=CHCH_2OH$ + $(C_4H_9)_3SiC(CH_2OH)=CH_2$(84)	804
C_3H_5	$CH_2=CHCH_2Cl$				
	siloxane-$(NHCH_2CH_2NH_2)/H_2PtCl_6$	60-85	3	$(C_4H_9)_3Si(CH_2)_3Cl$ (72)	D -057
C_4H_5	$CH_2=CHCH_2NCO$				
	Rh complex	-	-	$(C_4H_9)_3Si(CH_2)_3NCO$ (-)	J -024
C_5H_{11}	$CH_2=CHCH_2N(CH_3)_2$				
	H_2PtCl_6	reflux	-	$(C_4H_9)_3Si(CH_2)_3N(CH_3)_2$ (-)	S -012
C_6H_8	$CH_3OCH_2C\equiv CCH=CH_2$				
	H_2PtCl_6	100	-	$CH_3OCH_2C[Si(C_4H_9)_3]=CHCH=CH_2$ (58-60) $CH_3OCH_2CH=C[Si(C_4H_9)_3]CH=CH_2$ (3-30) $CH_3OCH_2C[Si(C_4H_9)_3]=C=CHCH_3$ (3-10)	535
	H_2PtCl_6 (i-PrOH)	100	-	$CH_3OCH_2CH=C[Si(C_4H_9)_3]CH=CH_2$ (-)	534
C_6H_{10}	$(CH_3)_2CHCH(OH)C\equiv CH$				
	H_2PtCl_6 (i-PrOH)	reflux	2	$(CH_3)_2CHCH(OH)CH=CHSi(C_4H_9)_3$ (-)	1072
C_6H_{10}	$CH_3CH_2CH_2CH(OH)C\equiv CH$				
	H_2PtCl_6 (i-PrOH)	80-100	6	$CH_3CH_2CH_2CH(OH)CH=CHSi(C_4H_9)_3$ (-)	1083
C_6H_{10}	$CH_3COCH_2CH_2CH=CH_2$				
	Ni	157-236	0.17	$CH_3CO(CH_2)_4Si(C_4H_9)_3$ (90)	15
	$ZnCl_2$	156-230	1	$CH_3CO(CH_2)_4Si(C_4H_9)_3$ (75)	15
C_7H_{10}	$CH_3CH(OCH_3)C\equiv CCH=CH_2$				
	H_2PtCl_6 (i-PrOH)	100	-	$CH_3CH(OCH_3)CH=C[Si(C_4H_9)_3]CH=CH_2$ (-)	534
C_8H_6	$C_6H_5C\equiv CH$				
	-	-	-	$C_6H_5C_2H_2Si(C_4H_9)_3$ (kin.)	512
C_8H_{12}	$CH_3CH_2CH(OCH_3)C\equiv CCH=CH_2$				
	H_2PtCl_6 (i-PrOH)	100	-	$CH_3CH_2CH(OCH_3)CH=C[Si(C_4H_9)_3]CH=CH_2$ (-)	534
C_8H_{12}	$CH_3COOCH_2CH(OH)CH_2OCH_2C\equiv CH$				
	H_2PtCl_6 (i-PrOH)	reflux	40 (b)	$CH_3COOCH_2CH(OH)CH_2OCH_2CH=CHSi(C_4H_9)_3$ (72)	1250
C_8H_{12}	$CH_3OC(CH_3)_2C\equiv CCH=CH_2$				

Table 12.1

= $(C_4H_9)_3SiH$ =   =2=

	H_2PtCl_6 (i-PrOH)	25-125	-	$CH_3OC(CH_3)_2C[Si(C_4H_9)_3]=CHCH=CH_2$ + $CH_3OC(CH_3)_2CH=C[Si(C_4H_9)_3]CH=CH_2$ + $CH_3OC(CH_3)_2C[Si(C_4H_9)_3]=C=CHCH_3$ (72)	530
	H_2PtCl_6 (i-PrOH)	25-125	-	$CH_3OC(CH_3)_2C[Si(C_4H_9)_3]=CHCH=CH_2$ + $CH_3OC(CH_3)_2CH=C[Si(C_4H_9)_3]CH=CH_2$ + $CH_3OC(CH_3)_2C[Si(C_4H_9)_3]=C=CHCH_3$ (72)	538
C_8H_{15}	$(CH_2)_5NCH_2CH=CH_2$				
	H_2PtCl_6 (i-PrOH)	100	17	$(CH_2)_5N(CH_2)_3Si(C_4H_9)_3$ (-)	702
C_8H_{15}	$(CH_3)_2NCH_2CH(OH)CH_2OCH_2C\equiv CH$				
	H_2PtCl_6 (i-PrOH)	150	8	$(CH_3)_2NCH_2CH(OH)CH_2OCH_2CH=CHSi(C_4H_9)_3$ (61)	1249
	H_2PtCl_6 (i-PrOH)	reflux	40 (b)	$(CH_3)_2NCH_2CH(OH)CH_2OCH_2CH=CHSi(C_4H_9)_3$ (61)	806
C_8H_{18}	$(CH_3CH_2O)_3SiCH=CH_2$				
	H_2PtCl_6 (i-PrOH)	50-200	0.55	$(CH_3CH_2O)_3SiCH_2CH_2Si(C_4H_9)_3$ (86)	575
C_9H_{10}	$C_6H_5CH_2CH=CH_2$				
	H_2PtCl_6	-	-	$C_6H_5C_3H_6Si(C_4H_9)_3$ (-)	516
C_9H_{14}	$CH_2=CHCH_2OCH_2CH(OH)CH_2OCH_2C\equiv CH$				
	H_2PtCl_6 (i-PrOH)	80	20	$CH_2=CHCH_2OCH_2CH(OH)CH_2OCH_2CH=CHSi(C_4H_9)_3$ (60-70)	1248
C_9H_{14}	$CH_3CH_2CH_2CH(OCH_3)C\equiv CCH=CH_2$				
	H_2PtCl_6 (i-PrOH)	100	-	$CH_3CH_2CH_2CH(OCH_3)CH=C[Si(C_4H_9)_3]CH=CH_2$ (-)	534
$C_{10}H_{12}$	$C_6H_5CH_2CH_2CH=CH_2$				
	H_2PtCl_6	-	-	$C_6H_5(CH_2)_4Si(C_4H_9)_3$ (-)	515
$C_{10}H_{12}$	m-$CH_3OC_6H_4OCH_2CH=CH_2$				
	H_2PtCl_6	-	-	m-$CH_3OC_6H_4O(CH_2)_3Si(C_4H_9)_3$ (-)	516
$C_{11}H_{16}$	$(CH_2)_5C(OCH_3)C\equiv CCH=CH_2$				
	H_2PtCl_6	100	5	$(CH_2)_5C(OCH_3)C[Si(C_4H_9)_3]=CHCH=CH_2$ + $(CH_2)_5C(OCH_3)CH=C[Si(C_4H_9)_3]CH=CH_2$ (65)	536
$C_{11}H_{19}$	$ClCH_2CH_2OCH_2OC(CH_3)[C(CH_3)_3]C\equiv CH$				
	H_2PtCl_6 (i-PrOH)	60-70	8	$ClCH_2CH_2OCH_2OC(CH_3)[C(CH_3)_3]CH=CHSi(C_4H_9)_3$ (-)	1065
$C_{12}H_{18}$	$(CH_2)_4C(OH)C\equiv CC(OH)(CH_2)_4$				

	H_2PtCl_6 (i-PrOH)	100	7	$(CH_2)_4C(OH)CH{=}CH[Si(C_4H_9)_3]C(OH)(CH_2)_4$ (18)	190
$C_{14}H_{10}$	$C_6H_5C{\equiv}CC_6H_5$				
	H_2PtCl_6 (i-PrOH)	100	2	$C_6H_5CH{=}C(C_6H_5)Si(C_4H_9)_3$ (78)	837
$C_{14}H_{16}$	Fig.490.				
	H_2PtCl_6	-	-	Fig.491. (90/10)	539
$C_{14}H_{16}$	p-$CH_3OC_6H_4C(CH_3)_2C{\equiv}CCH{=}CH_2$				
	H_2PtCl_6	-	-	p-$CH_3OC_6H_4C(CH_3)_2CH{=}C[Si(C_4H_9)_3]CH{=}CH_2$ (-)	539
	H_2PtCl_6	100	6	p-$CH_3OC_6H_4C(CH_3)_2CH{=}C[Si(C_4H_9)_3]CH{=}CH_2$ (74)	532
	H_2PtCl_6	75	10	p-$CH_3OC_6H_4C(CH_3)_2CH{=}C[Si(C_4H_9)_3]CH{=}CH_2$ (65)	540
$C_{14}H_{22}$	$(CH_2)_5C(OH)C{\equiv}CC(OH)(CH_2)_5$				
	H_2PtCl_6	reflux	5	$(CH_2)_5C(OH)CH{=}C[Si(C_4H_9)_3]C(OH)(CH_2)_5$ (43)	451
	Pt/C	reflux	5	$(CH_2)_5C(OH)CH{=}C[Si(C_4H_9)_3]C(OH)(CH_2)_5$ (43)	451
$C_{15}H_{18}$	Fig.325.				
	H_2PtCl_6	-	-	Fig.494.(85/10/5)	539
$C_{15}H_{18}$	Fig.492.				
	H_2PtCl_6	100	6	Fig.493. (84)	532
$C_{20}H_{24}$	Fig.335.				
	Pt/C	160	16 (dg)	Fig.495. (-)	F -024

Table 12.2
= (iso-$C_4H_9)_3SiH$ = =1=

C_6H_{10}	$CH_3(CH_2)_3C\equiv CH$				
	H_2PtCl_6	100	2	$CH_3(CH_2)_3CH=CHSi(C_4H_9\text{-iso})_3$ + $CH_3(CH_2)_3C(=CH_2)Si(C_4H_9\text{-iso})_3$ (-)	266
C_6H_{10}	$CH_3COCH_2CH_2CH=CH_2$				
	Ni	145-235	1	$CH_3CO(CH_2)_4Si(C_4H_9\text{-iso})_3$ (75)	15
C_8H_{18}	$(CH_3CH_2O)_3SiCH=CH_2$				
	H_2PtCl_6 (i-PrOH)	50-200	0.55	$(CH_3CH_2O)_3SiCH_2CH_2Si(C_4H_9\text{-iso})_3$ (82)	575
C_9H_{10}	$C_6H_5CH_2CH=CH_2$				
	H_2PtCl_6	-	-	$C_6H_5C_3H_6Si(C_4H_9\text{-iso})_3$ (-)	516
$C_{10}H_{12}$	$C_6H_5CH_2CH_2CH=CH_2$				
	H_2PtCl_6	-	-	$C_6H_5(CH_2)_4Si(C_4H_9\text{-iso})_3$ (-)	515
$C_{10}H_{12}$	m-$CH_3OC_6H_4OCH_2CH=CH_2$				
	H_2PtCl_6	-	-	m-$CH_3OC_6H_4O(CH_2)_3Si(C_4H_9\text{-iso})_3$ (-)	513

Table 12.4
= $(C_4H_9)_3SiH$ = =1=

C_6H_{10}	$(CH_3)_2C{=}CHCOCH_3$				
	Ni	130-215	5	$(CH_3)_2CHCH{=}C(CH_3)OSi(C_4H_9\text{-iso})_3$ (-)	15
C_7H_{12}	$\overline{CH_2CH_2CH(CH_3)CH_2CH_2CO}$				
	CF_3COOH	0	60 days	$\overline{CH_2CH_2CH(CH_3)CH_2CH_2C}HOSi(C_4H_9\text{-tert})_3$ (-)	328
$C_{10}H_9$	o-$ClC_6H_4COCOOCH_2CH_3$				
	Ni	160-170	6-7 (x)	o-$ClC_6H_4CH(COOCH_2CH_3)OSi(C_4H_9\text{-iso})_3$ (65)	649
	Ni	160-170	6-7 (x)	o-$ClC_6H_4CH(COOCH_2CH_3)OSi(C_4H_9\text{-iso})_3$ (65)	650
$C_{10}H_{18}$	$\overline{CH_2CH_2CH[C(CH_3)_3]CH_2CH_2CO}$				
	CF_3COOH	20	-	$\overline{CH_2CH_2CH[C(CH_3)_3]CH_2CH_2C}HOSi(C_4H_9\text{-tert})_3$ (-)	328
$C_{11}H_{12}$	o-$CH_3C_6H_4COCOOCH_2CH_3$				
	Ni	160-170	6-7 (x)	o-$CH_3C_6H_4CH(COOCH_2CH_3)OSi(C_4H_9\text{-iso})_3$ (46)	650
	Ni	160-170	6-7 (x)	o-$CH_3C_6H_4CH(COOCH_2CH_3)OSi(C_4H_9\text{-iso})_3$ (46)	649
$C_{11}H_{12}$	o-$CH_3OC_6H_4COCOOCH_2CH_3$				
	Ni	160-170	6-7 (x)	o-$CH_3OC_6H_4CH(COOCH_2CH_3)OSi(C_4H_9\text{-iso})_3$ (55)	650
$C_{13}H_{16}$	2,4,6-$(CH_3)_3C_6H_2COCOOCH_2CH_3$				
	Ni	160-170	6-7 (x)	2,4,6-$(CH_3)_3C_6H_2CH(COOCH_2CH_3)OSi(C_4H_9\text{-iso})_3$ (40)	649
	Ni	160-170	6-7 (x)	2,4,6-$(CH_3)_3C_6H_2CH(COOCH_2CH_3)OSi(C_4H_9\text{-iso})_3$ (40)	650

C_3H_5	$CH_2{=}CHCH_2Cl$				
	H_2PtCl_6 (i-PrOH)	50	6	$(C_6H_5)_2Si(CH_3)(CH_2)_3Cl$ (15)	705
	H_2PtCl_6 (i-PrOH)	reflux	7	$(C_6H_5)_2Si(CH_3)(CH_2)_3Cl$ (10)	99
C_5H_8	$CH_2{=}CHCH_2OCOCH_3$				
	H_2PtCl_6 (i-PrOH)	90	4	$(C_6H_5)_2Si(CH_3)(CH_2)_3OCOCH_3$ (-)	U -113
C_5H_{11}	$CH_3CH_2NHCH_2CH{=}CH_2$				
	H_2PtCl_6 (i-PrOH)	115	30	$CH_3CH_2NH(CH_2)_3SiCH_3(C_6H_5)_2$ + $CH_3CH_2NHCH_2CH(CH_3)SiCH_3(C_6H_5)_2$ (90)	255
C_5H_{12}	$(CH_3)_3SiCH{=}CH_2$				
	$Co_2(CO)_8$	60-70	4	$(CH_3)_3SiCH_2CH_2SiCH_3(C_6H_5)_2$ (50)	718
	H_2PtCl_6 (i-PrOH)	reflux	4	$(CH_3)_3SiCH_2CH_2SiCH_3(C_6H_5)_2$ (-)	1059
	$H_2PtCl_6{\cdot}6H_2O$	50	6	$(CH_3)_3SiCH_2CH_2SiCH_3(C_6H_5)_2$ (92)	704
C_6H_{10}	$CH_2{=}C(CH_3)C(CH_3){=}CH_2$				
	$Cr(CO)_3$(disilaarene) + UV	80	4 (h)	$(CH_3)_2C{=}C(CH_3)CH_2CH_2SiCH_3(C_6H_5)_2$ (16)	3
	$Cr(CO)_3$(disilaarene) + UV	80	3 (h)	$(CH_3)_2C{=}C(CH_3)CH_2CH_2SiCH_3(C_6H_5)_2$ (75)	3
	$Cr(CO)_3$(silaarene)	80	3	$(CH_3)_2C{=}C(CH_3)CH_2CH_2SiCH_3(C_6H_5)_2$ (-)	3
	$Cr(CO)_3$(silaarene) + UV	80	4 (h)	$(CH_3)_2C{=}C(CH_3)CH_2CH_2SiCH_3(C_6H_5)_2$ (10-28)	3
	$Cr(CO)_3$(silane) + UV	80	4 (h)	$(CH_3)_2C{=}C(CH_3)CH_2CH_2SiCH_3(C_6H_5)_2$ (60)	3
C_6H_{10}	$CH_3COCH_2CH_2CH{=}CH_2$				
	Ni	160-225	0.3	$CH_3CO(CH_2)_4SiCH_3(C_6H_5)_2$ (88)	15
C_6H_{10}	$OCH_2CHCH_2OCH_2CH{=}CH_2$				
	H_2PtCl_6	-	-	$OCH_2CHCH_2O(CH_2)_3SiCH_3(C_6H_5)_2$ (55)	1153
C_6H_{12}	$C_4H_9CH{=}CH_2$				
	$Co_2(CO)_8$	70-90	4	$C_4H_9CH_2CH_2SiCH_3(C_6H_5)_2$ (52)	718
C_6H_{12}	$CH_3(CH_2)_3CH{=}CH_2$				
	$[PtCl_2(py)]_2$	-	3	$CH_3(CH_2)_5SiCH_3(C_6H_5)_2$ (90)	987
	SiO_2-PPh_2/$[RhCl(C_2H_4)_2]_2$	20	-	$CH_3(CH_2)_5SiCH_3(C_6H_5)_2$ (kin.)	769
C_7H_{13}	$CH_2CH_2OCH_2CH_2NCH_2CH{=}CH_2$				

Table 13.1

= $CH_3(C_6H_5)_2SiH$ = =2=

	H_2PtCl_6 (THF)	150-160	20	$\overline{CH_2CH_2OCH_2CH_2N}(CH_2)_3SiCH_3(C_6H_5)_2$ (63)	697
C_7H_{14}	$CH_3(CH_2)_4CH=CH_2$				
	anionite/$PtCl_6^{-2}$	-	-	$CH_3(CH_2)_6SiCH_3(C_6H_5)_2$ (kin.)	1344
	anionite/$PtCl_6^{-2}$	70-95	-	$CH_3(CH_2)_6SiCH_3(C_6H_5)_2$ (kin.)	975
	anionite/$PtCl_6^{-2}$	70-95	-	$CH_3(CH_2)_6SiCH_3(C_6H_5)_2$ (kin.)	1345
C_7H_{15}	$(CH_3CH_2)_2NCH_2CH=CH_2$				
	H_2PtCl_6 (i-PrOH)	until 140	10	$(CH_3CH_2)_2N(CH_2)_3SiCH_3(C_6H_5)_2$ (70)	386
C_8H_7	m-$O_2NC_6H_4CH=CH_2$				
	H_2PtCl_6 (i-PrOH)	80-150	1.5	m-$O_2NC_6H_4CH_2CH_2SiCH_3(C_6H_5)_2$ (46)	48
C_8H_8	$C_6H_5CH=CH_2$				
	$Co_2(CO)_8$	30-40	1	$C_6H_5CH_2CH_2SiCH_3(C_6H_5)_2$ (90)	719
	$[PtCl_2(py)]_2$	-	1-1.5	$C_6H_5CH_2CH_2SiCH_3(C_6H_5)_2$ (30-100)	987
	$Rh_4(CO)_{12}$	30-40	1	$C_6H_5CH_2CH_2SiCH_3(C_6H_5)_2$ (95)	719
C_8H_{12}	Fig.282.				
	H_2PtCl_6 (i-PrOH)	60-130	2	Fig.497. (-)	21
C_8H_{14}	$\overline{CH_2OCH_2CH_2OC}HCH_2OCH_2CH=CH_2$				
	H_2PtCl_6 (i-PrOH)	reflux	52	$\overline{CH_2OCH_2CH_2OC}HCH_2O(CH_2)_3SiCH_3(C_6H_5)_2$ (59)	554
C_8H_{16}	$CH_3N(CH_2CH_2)_2NCH_2CH=CH_2$				
	H_2PtCl_6 (THF)	150-160	20	$CH_3N(CH_2CH_2)_2N(CH_2)_3SiCH_3(C_6H_5)_2$ (55)	697
C_8H_{17}	$(CH_3CH_2)_2NCH_2CH_2OCH=CH_2$				
	H_2PtCl_6 (i-PrOH)	70	7	$(CH_3CH_2)_2NCH_2CH_2OCH_2CH_2SiCH_3(C_6H_5)_2$ (25)	723
	H_2PtCl_6 (i-PrOH)	70	7	$(CH_3CH_2)_2NCH_2CH_2OCH_2CH_2SiCH_3(C_6H_5)_2$ (25)	Dd-006
C_8H_{18}	$(CH_3CH_2O)_3SiCH=CH_2$				
	H_2PtCl_6	50-200	0.58	$(CH_3CH_2O)_3SiCH_2CH_2SiCH_3(C_6H_5)_2$ (90)	575
C_9H_{10}	$C_6H_5CH=CHCH_3$-trans				
	H_2PtCl_6	80	3.5	$C_6H_5C_2H_3(CH_3)SiCH_3(C_6H_5)_2$ (kin.)	919
C_9H_{15}	$\overline{OCH_2C}HCH_2OCH(CH_2Cl)CH_2OCH_2CH=CH_2$				

= $CH_3(C_6H_5)_2SiH$ = =3=

	H_2PtCl_6	reflux	24 (b)	$OCH_2CHCH_2OCH(CH_2Cl)CH_2O(CH_2)_3SiCH_3(C_6H_5)_2$ (53)	66
$C_{10}H_{16}$	Fig.498.				
	H_2PtCl_6 (i-PrOH)	120-130	5	Fig.499. (77)	552
$C_{10}H_{18}$	$OCH_2CHCH_2OCH(CH_2OCH_3)CH_2OCH_2CH{=}CH_2$				
	H_2PtCl_6	reflux	40 (b)	$OCH_2CHCH_2OCH(CH_2OCH_3)CH_2O(CH_2)_3SiCH_3(C_6H_5)_2$ (66)	1150
$C_{11}H_{14}$	Fig.500.				
	H_2PtCl_6	reflux	6 (b)	Fig.501. (86)	1152
$C_{11}H_{18}$	$OCH_2CHCH_2OCH(CH_2OCOCH_3)CH_2OCH_2CH{=}CH_2$				
	H_2PtCl_6	reflux	36 (b)	$OCH_2CHCH_2OCH(CH_2OCOCH_3)CH_2O(CH_2)_3SiCH_3(C_6H_5)_2$ (67)	1150
$C_{12}H_{10}$	1-$C_{10}H_7CH{=}CH_2$				
	H_2PtCl_6 (THF)	50	1-2	1-$C_{10}H_7CH_2CH_2SiCH_3(C_6H_5)_2$ (kin.)	923
	H_2PtCl_6 (THF)	80	-	1-$C_{10}H_7CH_2CH_2SiCH_3(C_6H_5)_2$ (kin.)	922
$C_{12}H_{12}$	$C_5H_5FeC_5H_4CH{=}CH_2$ + $C_5H_5FeC_5H_4C(CH_3){=}CH_2$				
	H_2PtCl_6 (i-PrOH)	100-150	7.5	$C_5H_5FeC_5H_4CH_2CH_2SiCH_3(C_6H_5)_2$ (24) $C_5H_5FeC_5H_4CH(CH_3)CH_2SiCH_3(C_6H_5)_2$ (26)	1123
$C_{12}H_{12}$	$C_5H_5FeC_5H_4CH{=}CH_2$ + $C_5H_5FeC_5H_4Si(CH_3)(C_6H_5)CH{=}CH_2$				
	H_2PtCl_6 (i-PrOH)	100-150	7.5	$C_5H_5FeC_5H_4CH_2CH_2SiCH_3(C_6H_5)_2$ / $C_5H_5FeC_5H_4Si(CH_3)(C_6H_5)CH_2CH_2SiCH_3(C_6H_5)_2$(19/81)	1123
$C_{12}H_{12}$	$C_5H_5FeC_5H_4CH{=}CH_2$ + $C_5H_5FeC_5H_4Si(CH_3)(CH_2CH_3)CH{=}CH_2$				
	H_2PtCl_6 (i-PrOH)	100-150	7.5	$C_5H_5FeC_5H_4CH_2CH_2SiCH_3(C_6H_5)_2$ / $C_5H_5FeC_5H_4Si(CH_3)(CH_2CH_3)CH_2CH_2SiCH_3(C_6H_5)_2$ (32/68)	1123
$C_{12}H_{12}$	$C_5H_5FeC_5H_4CH{=}CH_2$ + $C_5H_5FeC_5H_4Si(CH_3)_2CH{=}CH_2$				
	H_2PtCl_6 (i-PrOH)	100-150	7.5	$C_5H_5FeC_5H_4CH_2CH_2SiCH_3(C_6H_5)_2$ / $C_5H_5FeC_5H_4Si(CH_3)_2CH_2CH_2SiCH_3(C_6H_5)_2$ (34/66)	1123
$C_{13}H_{12}$	1-$C_{10}H_7CH_2CH{=}CH_2$				
	H_2PtCl_6 (THF)	80	-	1-$C_{10}H_7(CH_2)_3SiCH_3(C_6H_5)_2$ (kin.)	922
$C_{13}H_{12}$	2-$C_{10}H_7CH{=}CHCH_3$				

	H_2PtCl_6	150-225	-	$2\text{-}C_{10}H_7C_3H_6SiCH_3(C_6H_5)_2$ (kin.)	919
$C_{13}H_{14}$	$C_5H_5FeC_5H_4CH{=}CH_2$ + $C_5H_5FeC_5H_4C(CH_3){=}CH_2$				
	H_2PtCl_6 (i-PrOH)	100-150	7.5	$C_5H_5FeC_5H_4CH_2CH_2SiCH_3(C_6H_5)_2$ (24) $C_5H_5FeC_5H_4CH(CH_3)CH_2SiCH_3(C_6H_5)_2$ (26)	1123
$C_{13}H_{26}$	$CH_3(CH_2)_{10}CH{=}CH_2$				
	H_2PtCl_6 (THF)	20-80	-	$CH_3(CH_2)_{12}SiCH_3(C_6H_5)_2$ (kin.)	922
	H_2PtCl_6 (THF)	50	1-2	$CH_3(CH_2)_{12}SiCH_3(C_6H_5)_2$ (kin.)	923
$C_{13}H_{28}$	$[CH_3(CH_2)_4]_2Si(CH_3)CH{=}CH_2$				
	H_2PtCl_6 (THF)	50	1-2	$[CH_3(CH_2)_4]_2Si(CH_3)CH_2CH_2SiCH_3(C_6H_5)_2$ (kin.)	923
$C_{14}H_{18}$	$C_5H_5FeC_5H_4Si(CH_3)_2CH{=}CH_2$ + $C_5H_5FeC_5H_4CH{=}CH_2$				
	H_2PtCl_6 (i-PrOH)	100-150	7.5	$C_5H_5FeC_5H_4CH_2CH_2SiCH_3(C_6H_5)_2$/ $C_5H_5FeC_5H_4Si(CH_3)_2CH_2CH_2SiCH_3(C_6H_5)_2$ (34/66)	1123
$C_{15}H_{16}$	$(C_6H_5)_2Si(CH_3)CH{=}CH_2$				
	H_2PtCl_6	80-120	-	$(C_6H_5)_2Si(CH_3)CH_2CH_2SiCH_3(C_6H_5)_2$ (kin.)	921
	H_2PtCl_6	80-120	-	$(C_6H_5)_2Si(CH_3)CH_2CH_2SiCH_3(C_6H_5)_2$ (kin.)	1282
	H_2PtCl_6 + Bu_2O	80	-	$(C_6H_5)_2Si(CH_3)CH_2CH_2SiCH_3(C_6H_5)_2$ (kin.)	1282
	H_2PtCl_6 + $Ti(OBu)_2(acac)_2$	80	-	$(C_6H_5)_2Si(CH_3)CH_2CH_2SiCH_3(C_6H_5)_2$ (kin.)	1282
	H_2PtCl_6 + $Ti(OBu)_4$	80	-	$(C_6H_5)_2Si(CH_3)CH_2CH_2SiCH_3(C_6H_5)_2$ (kin.)	1282
	H_2PtCl_6 (THF)	50	1-2	$(C_6H_5)_2Si(CH_3)CH_2CH_2SiCH_3(C_6H_5)_2$ (kin.)	923
	H_2PtCl_6 (THF)	80	-	$(C_6H_5)_2Si(CH_3)CH_2CH_2SiCH_3(C_6H_5)_2$ (kin.)	922
	H_2PtCl_6 (THF)	80	1.5	$(C_6H_5)_2Si(CH_3)CH_2CH_2SiCH_3(C_6H_5)_2$ (55)	924
	H_2PtCl_6 (THF) + $Ti(OBu)_4$	60	1.5	$(C_6H_5)_2Si(CH_3)CH_2CH_2SiCH_3(C_6H_5)_2$ (76)	924
$C_{15}H_{20}$	$C_5H_5FeC_5H_4Si(CH_3)(CH_2CH_3)CH{=}CH_2$ + $C_5H_5FeC_5H_4CH{=}CH_2$				
	H_2PtCl_6 (i-PrOH)	100-150	7.5	$C_5H_5FeC_5H_4CH_2CH_2SiCH_3(C_6H_5)_2$ / $C_5H_5FeC_5H_4Si(CH_3)(CH_2CH_3)CH_2CH_2SiCH_3(C_6H_5)_2$ (32/68)	1123
$C_{17}H_{20}$	$(C_6H_5CH_2)_2Si(CH_3)CH{=}CH_2$				
	H_2PtCl_6	80-120	1-2	$(C_6H_5CH_2)_2Si(CH_3)CH_2CH_2SiCH_3(C_6H_5)_2$ (kin.)	921
$C_{19}H_{20}$	$C_5H_5FeC_5H_4Si(CH_3)(C_6H_5)CH{=}CH_2$				

Table 13.1

= $CH_3(C_6H_5)_2SiH$ = =5=

	H_2PtCl_6 (i-PrOH)	reflux	7	$C_5H_5FeC_5H_4Si(CH_3)(C_6H_5)CH_2CH_2SiCH_3(C_6H_5)_2$ (69)	1122
$C_{19}H_{20}$	$C_5H_5FeC_5H_4Si(CH_3)(C_6H_5)CH{=}CH_2$ + $C_5H_5FeC_5H_4CH{=}CH_2$				
	H_2PtCl_6 (i-PrOH)	100-150	7.5	$C_5H_5FeC_5H_4CH_2CH_2SiCH_3(C_6H_5)_2$ / $C_5H_5FeC_5H_4Si(CH_3)(C_6H_5)CH_2CH_2SiCH_3(C_6H_5)_2$(18/81)	1123
$C_{20}H_{18}$	1-$C_{10}H_7C(OH)(CH_3)CH{=}CHC_6H_5$				
	-	-	-	1-$C_{10}H_7C(OH)(CH_3)C_2H_3(C_6H_5)SiCH_3(C_6H_5)_2$ (-)	231

Table 13.2

= $CH_3(C_6H_5)_2SiH$ = 　　　　=1=

C_3H_3	$CH\equiv CCH_2Cl$				
	H_2PtCl_6 (i-PrOH)	50	6	$(C_6H_5)_2Si(CH_3)CH{=}CHCH_2Cl$ (31) $(C_6H_5)_2Si(CH_3)C({=}CH_2)CH_2Cl$ (18) $(C_6H_5)_2Si(CH_3)CH{=}CHCH_3$ (5)	705
C_3H_4	$CH\equiv CCH_2OH$				
	H_2PtCl_6 (i-PrOH)	50	6	trans-$CH_3Si(C_6H_5)_2CH{=}CHCH_2OH$ (43) $CH_3Si(C_6H_5)_2C({=}CH_2)CH_2OH$ (32)	706
C_4H_6	$CH\equiv CCH_2OCH_3$				
	H_2PtCl_6 (i-PrOH)	50	6	trans-$CH_3Si(C_6H_5)_2CH{=}CHCH_2OCH_3$ (59) $CH_3Si(C_6H_5)_2C({=}CH_2)CH_2OCH_3$ (28)	706
C_5H_6	$CH\equiv CCH_2COOCH_3$				
	H_2PtCl_6 (i-PrOH)	150	6	trans-$CH_3Si(C_6H_5)_2CH{=}CHCH_2COOCH_3$ (34) $CH_3Si(C_6H_5)_2C({=}CH_2)CH_2COOCH_3$ (26)	706
C_6H_{10}	$CH_3CH(OH)CH_2OCH_2C\equiv CH$				
	H_2PtCl_6	80-90	-	$CH_3CH(OH)CH_2OCH_2CH{=}CHSiCH_3(C_6H_5)_2$ (70)	546
C_7H_6	$CH\equiv CCH_2\overline{C{=}CHCH{=}CHS}$				
	H_2PtCl_6 (i-PrOH)	50	6	trans-$CH_3Si(C_6H_5)_2CH{=}CHCH_2\overline{C{=}CHCH{=}CHS}$ (54) $CH_3Si(C_6H_5)_2C({=}CH_2)CH_2\overline{C{=}CHCH{=}CHS}$ (18)	706
C_7H_{11}	$CH\equiv CCH_2\overline{N(CH_2)_4}$				
	H_2PtCl_6 (i-PrOH)	150	20	trans-$CH_3Si(C_6H_5)_2CH{=}CHCH_2\overline{N(CH_2)_4}$ (29) $CH_3Si(C_6H_5)_2C({=}CH_2)CH_2\overline{N(CH_2)_4}$ (3)	706
	H_2PtCl_6 (i-PrOH)	200	6	trans-$CH_3Si(C_6H_5)_2CH{=}CHCH_2\overline{N(CH_2)_4}$ (45) $CH_3Si(C_6H_5)_2C({=}CH_2)CH_2\overline{N(CH_2)_4}$ (15)	706
C_7H_{13}	$CH\equiv CCH_2N(CH_3CH_2)_2$				
	H_2PtCl_6 (i-PrOH)	150	6	trans-$CH_3Si(C_6H_5)_2CH{=}CHCH_2N(CH_2CH_3)_2$ (8) $CH_3Si(C_6H_5)_2C({=}CH_2)CH_2N(CH_2CH_3)_2$ (2)	706
	H_2PtCl_6 (i-PrOH)	200	6	trans-$CH_3Si(C_6H_5)_2CH{=}CHCH_2N(CH_2CH_3)_2$ (46) $CH_3Si(C_6H_5)_2C({=}CH_2)CH_2N(CH_2CH_3)_2$ (11)	706
C_8H_6	$C_6H_5C\equiv CH$				
	$(Bu_4N)_2PtCl_6$	120	1-3	$C_6H_5CH{=}CHSiCH_3(C_6H_5)_2$ + $C_6H_5C[SiCH_3(C_6H_5)_2]{=}CH_2$ (-)	673
	H_2PtCl_6 (i-PrOH)	50	6	$C_6H_5CH{=}CHSiCH_3(C_6H_5)_2$ (20) $C_6H_5C({=}CH_2)SiCH_3(C_6H_5)_2$ (71)	706
C_8H_{12}	$CH\equiv CCH_2OCH_2CH(OH)CH_2OCOCH_3$				

	H_2PtCl_6 (i-PrOH)	reflux	40 (b)	$CH_3(C_6H_5)_2SiCH{=}CHCH_2OCH_2CH(OH)CH_2OCOCH_3$ (79)	1250
C_8H_{13}	$CH{\equiv}CCH_2N(CH_2)_5$				
	H_2PtCl_6 (i-PrOH)	150	6	trans-$CH_3Si(C_6H_5)_2CH{=}CHCH_2N(CH_2)_5$ (48) $CH_3Si(C_6H_5)_2C({=}CH_2)CH_2N(CH_2)_5$ (7)	706
	H_2PtCl_6 (i-PrOH)	200	6	trans-$CH_3Si(C_6H_5)_2CH{=}CHCH_2N(CH_2)_5$ (69) $CH_3Si(C_6H_5)_2C({=}CH_2)CH_2N(CH_2)_5$ (16)	706
C_8H_{13}	$CH{\equiv}CSi(OCH_2CH_2)_3N$				
	$H_2PtCl_6{\cdot}6H_2O$	50	6	$(C_6H_5)_2Si(CH_3)CH{=}CHSi(OCH_2CH_2)_3N$ (39)	704
C_8H_{15}	$CH{\equiv}CCH_2OCH_2CH(OH)CH_2N(CH_3)_2$				
	H_2PtCl_6 (i-PrOH)	150	8	$CH_3(C_6H_5)_2SiCH{=}CHCH_2OCH_2CH(OH)CH_2N(CH_3)_2$ (67)	806
	H_2PtCl_6 (i-PrOH)	150	8	$CH_3(C_6H_5)_2SiCH{=}CHCH_2OCH_2CH(OH)CH_2N(CH_3)_2$ (67)	1249
C_9H_{13}	$CH{\equiv}CCH_2OCH_2CH(CH_2Cl)OCH_2CHCH_2O$				
	H_2PtCl_6 (i-PrOH)	reflux	18 (b)	$(C_6H_5)_2Si(CH_3)CH{=}CHCH_2OCH_2CH(CH_2Cl)OCH_2CHCH_2O$ (79)	1161
C_9H_{14}	$CH{\equiv}CCH_2OCH_2CH(OH)CH_2OCH_2CH{=}CH_2$				
	H_2PtCl_6 (i-PrOH)	80	20	$CH_3(C_6H_5)_2SiCH{=}CHCH_2OCH_2CH(OH)CH_2OCH_2CH{=}CH_2$ (60-70)	1248
C_9H_{15}	$CH{\equiv}CCH_2N(CH_2)_6$				
	H_2PtCl_6 (i-PrOH)	150	6	trans-$CH_3Si(C_6H_5)_2CH{=}CHCH_2N(CH_2)_6$ (43) $CH_3Si(C_6H_5)_2C({=}CH_2)CH_2N(CH_2)_6$ (5)	706
	H_2PtCl_6 (i-PrOH)	200	6	trans-$CH_3Si(C_6H_5)_2CH{=}CHCH_2N(CH_2)_6$ (54) $CH_3Si(C_6H_5)_2C({=}CH_2)CH_2N(CH_2)_6$ (10)	706
$C_{10}H_{16}$	$CH{\equiv}CCH_2OCH_2CH(CH_2OCH_3)OCH_2CHCH_2O$				
	H_2PtCl_6 (i-PrOH)	reflux	18 (b)	$(C_6H_5)_2Si(CH_3)CH{=}CHCH_2OCH_2CH(CH_2OCH_3)OCH_2CHCH_2O$ (86)	1161
$C_{11}H_{19}$	$(CH_3CH_2)_2NCH_2C{\equiv}CCH_2OCH_2CHCH_2O$				
	H_2PtCl_6	reflux	18 (b)	$(CH_3CH_2)_2NCH_2CH{=}C[SiCH_3(C_6H_5)_2]CH_2OCH_2CHCH_2O$ (64)	1196
$C_{13}H_{20}$	Fig.502.				
	H_2PtCl_6	reflux	5	Fig.503. (63)	547
$C_{14}H_{10}$	$C_6H_5C{\equiv}CC_6H_5$				

Table 13.2

= $CH_3(C_6H_5)_2SiH$ = =3=

	Pt/C	reflux	4	$C_6H_5CH=C(C_6H_5)SiCH_3(C_6H_5)_2$ (88)	169
$C_{20}H_{16}$	1-$C_{10}H_7C(OH)(CH_3)C\equiv CC_6H_5$				
	-	-	-	1-$C_{10}H_7C(OH)(CH_3)CH=C(C_6H_5)SiCH_3(C_6H_5)_2$ (-)	231

Table 13.3

= $CH_3(C_6H_5)_2SiH$ = =1=

C_3H_4	$CH_2{=}CHCHO$				
	$[HPt(SiMe_2Ph)(PPh_3)]_2$	60-100	1	$CH_3CH{=}CHOSiCH_3(C_6H_5)_2$ (-)	79
C_4H_6	$CH_3CH{=}CHCHO$				
	$[HPt(SiMe_2Ph)(PPh_3)_2]_2$	60-100	1	$CH_3CH_2CH{=}CHOSiCH_3(C_6H_5)_2$ (-)	79
C_5H_8	$CH_3CH{=}CHCOCH_3$				
	$[PtCl_2(py)]_2$	-	18	$CH_3CH_2CH{=}C(CH_3)OSiCH_3(C_6H_5)_2$ (30)	987
C_6H_8	$(CH_2)_3CH{=}CHCO$				
	$[HPt(SiMe_2Ph)(PPh_3)_2]_2$	60-100	1	$(CH_2)_4CH{=}COSiCH_3(C_6H_5)_2$ (-)	79
C_6H_{10}	$(CH_3)_2C{=}CHCOCH_3$				
	$[HPt(SiMe_2Ph)(PPh_3)_2]_2$	60-100	1	$(CH_3)_2CHCH{=}C(CH_3)OSiCH_3(C_6H_5)_2$ (-)	79
C_7H_{12}	$CH_3CH(CH_2)_4CO$				
	Bu_4NF (HMPA)	0-r.t.	0.5-12	$CH_3CH(CH_2)_4CHOSiCH_3(C_6H_5)_2$ (94)	394
	Bu_4NF (HMPA)	0-r.t.	0.5-12	$CH_3CH(CH_2)_4CHOSiCH_3(C_6H_5)_2$ (94)	393
C_8H_8	$C_6H_5COCH_3$				
	$[RhCl(cod)]_2$	100	23	$C_6H_5CH(CH_3)OSiCH_3(C_6H_5)_2$ (60)	177
C_9H_8	$C_6H_5CH{=}CHCHO$				
	$[HPt(SiMe_2Ph)(PPh_3)_2]_2$	60-100	1	$C_6H_5CH_2CH{=}CHOSiCH_3(C_6H_5)_2$ (-)	79
	$[PtCl_2(py)]_2$	-	23	$C_6H_5CH_2CH{=}CHOSiCH_3(C_6H_5)_2$ (100)	987
C_9H_{10}	$C_6H_5COCH_2CH_3$				
	$[RhH_2(L)_2(BMPP)]ClO_4$	8	120	$C_6H_5CH(CH_2CH_3)OSiCH_3(C_6H_5)_2$ (65)	486
$C_{10}H_{10}$	$C_6H_5CH{=}C(CH_3)CHO$				
	$[HPt(SiMe_2Ph)(PPh_3)_2]_2$	60-100	1	$C_6H_5CH_2C(CH_3){=}CHOSiCH_3(C_6H_5)_2$ (-)	79
$C_{10}H_{10}$	$C_6H_5CH{=}CHCOCH_3$				
	$[HPt(SiMe_2Ph)(PPh_3)_2]_2$	60-100	1	$C_6H_5CH_2CH{=}C(CH_3)OSiCH_3(C_6H_5)_2$ (-)	79
	$[PtCl_2(py)]_2$	-	90	$C_6H_5CH_2CH{=}C(CH_3)OSiCH_3(C_6H_5)_2$ (15)	987

Table 14.1

= $(C_6H_5)_3SiH$ = =1=

C_2H_4	$CH_2=CH_2$				
	UV + $(t\text{-}BuO)_2$	-60	-	$CH_3CH_2Si(C_6H_5)_3$ (-)	623
	$[RhCl(CO)_2]_2$	235	11 (xy)	$CH_3CH_2Si(C_6H_5)_3$ (-) $CH_2=CHSi(C_6H_5)_3$ (25)	U -173
C_3H_5	$CH_2=CHCH_2Cl$				
	H_2PtCl_6 (i-PrOH)	50	6	$(C_6H_5)_3SiC_3H_6Cl$ (-)	705
C_4H_5	$\overline{OCH_2C}ClCH=CH_2$				
	H_2PtCl_6 (i-PrOH)	100	3-15	$(C_6H_5)_3SiOCH_2CH_2CH(CH_3)Si(C_6H_5)_3$ (-)	1047
C_4H_6	$\overline{OCH_2C}HCH=CH_2$				
	H_2PtCl_6 (i-PrOH)	100	3-15	$(C_6H_5)_3SiOCH_2CH_2CH(CH_3)Si(C_6H_5)_3$ (-)	1047
C_4H_8	$(CH_3)_2C=CH_2$				
	UV + $(t\text{-}BuO)_2$	-90	-	$(CH_3)_2CHCH_2Si(C_6H_5)_3$ (-)	623
C_5H_8	$CH_3COOC(CH_3)=CH_2$				
	$PtCl_2$	r.t.	96	$CH_3COOCH(CH_3)CH_2Si(C_6H_5)_3$ (96)	1105
C_5H_8	$CH_3COOCH_2CH=CH_2$				
	H_2PtCl_6 (i-PrOH)	r.t.	3	$CH_3COO(CH_2)_3Si(C_6H_5)_3$ (92)	U -181
	H_2PtCl_6 (i-PrOH)	r.t.	-	$CH_3COO(CH_2)_3Si(C_6H_5)_3$ (92)	J -099
C_5H_8	$\overline{OCH(CH_3)C}HCH=CH_2$				
	H_2PtCl_6 (i-PrOH)	-	18-20	$(C_6H_5)_3SiOCH(CH_3)CH_2CH(CH_3)Si(C_6H_5)_3$ (-)	825
C_5H_8	$\overline{OCH_2C}(CH_3)CH=CH_2$				
	H_2PtCl_6 (i-PrOH)	100	3-15	$(C_6H_5)_3SiOCH_2CH(CH_3)CH(CH_3)Si(C_6H_5)_3$ (-)	1047
C_5H_{12}	$(CH_3)_3SiCH=CH_2$				
	H_2PtCl_6 (i-PrOH)	reflux	4	$(CH_3)_3SiCH_2CH_2Si(C_6H_5)_3$ / $(CH_3)_3SiCH(CH_3)Si(C_6H_5)_3$ (95/5)	1059
	$H_2PtCl_6\ 6H_2O$	50	6	$(CH_3)_3SiCH_2CH_2Si(C_6H_5)_3$ (92)	704
C_6H_{10}	$\overline{(CH_2)_4CH=C}H$				
	$(PhCOO)_2$	-	48	$\overline{(CH_2)_5C}HSi(C_6H_5)_3$ (3)	883
	AIBN	75	48-72	$\overline{(CH_2)_5C}HSi(C_6H_5)_3$ (12-34)	883

	gamma(^{60}Co)	-	-	$(CH_2)_5CHSi(C_6H_5)_3$ (-)	344
C_6H_{10}	$CH_2=C(CH_3)C(CH_3)=CH_2$				
	$Fe(CO)_5$ + UV	-	125 (pn)	$C_6H_{11}Si(C_6H_5)_3$ (74)	366
C_6H_{10}	$CH_2=CHCH_2OCH_2CH=CH_2$				
	H_2PtCl_6 (i-PrOH)	r.t.	3	$CH_2=CHCH_2O(CH_2)_3Si(C_6H_5)_3$ (80)	U -181
C_6H_{12}	$CH_3(CH_2)_3CH=CH_2$				
	gamma(^{60}Co)	52	-	$C_6H_{13}Si(C_6H_5)_3$ (kin.)	958
	$RhCl(CO)(PPh_2CH_2CH_2SiO_{3/2})_2(SiO_2)_x$-$(SiMe_3)_y$	60	-	$CH_3(CH_2)_5Si(C_6H_5)_3$ (kin.)	1029
	$RhCl(CO)(PPh_2CH_2CH_2SiO_{3/2})_2(SiO_2)_x$-$(SiMe_3)_y$	60	-	$CH_3(CH_2)_5Si(C_6H_5)_3$ (kin.)	1029
	$RhCl(CO)(PPh_3)_2$	60	6 days	$C_6H_{13}Si(C_6H_5)_3$ (75)	476
	$RhCl(CO)[PPh_2CH_2CH_2Si(OEt)_3)]_2$	60	-	$CH_3(CH_2)_5Si(C_6H_5)_3$ (kin.)	1029
	$RhCl(PPh_3)_3$	60	6 days	$C_6H_{13}Si(C_6H_5)_3$ (40-100)	478
	$RhCl(PPh_3)_3$	60	6 days	$C_6H_{13}Si(C_6H_5)_3$ (40-100)	476
	$RhHCl[Si(OEt)_3](PR_3)_2$	60	6 days	$C_6H_{13}Si(C_6H_5)_3$ (38-81)	478
	SiO_2-PPh_2/$[RhCl(C_2H_4)_2]_2$	20	-	$C_6H_{13}Si(C_6H_5)_3$ (kin.)	769
C_7H_{14}	$CH_3(CH_2)_4CH=CH_2$				
	$(PhCOO)_2$	75-78	14-48 (h)	$C_7H_{15}Si(C_6H_5)_3$ (0-33)	883
	AIBN	75-78	14 (h)	$C_7H_{15}Si(C_6H_5)_3$ (40)	883
C_8H_8	$C_6H_5CH=CH_2$				
	$PtCl_2$	r.t.	96	$C_6H_5CH_2CH_2Si(C_6H_5)_3$ (36) $C_6H_5CH=CHSi(C_6H_5)_3$ (30) $C_6H_5CH_2CH_3$ (31)	1105
	$Rh(acac)_3$ + $AlEt_3$	60	-	$C_6H_5CH_2CH_2Si(C_6H_5)_3$ (94)	271
	$RhCl(PPh_3)_3$	65	6.5 (b)	$C_6H_5CH_2CH_2Si(C_6H_5)_3$ (12) $C_6H_5CH(CH_3)Si(C_6H_5)_3$ (-) $C_6H_5CH=CHSi(C_6H_5)_3$ (-)	878
C_8H_{12}	Fig.36.				
	H_2PtCl_6 (i-PrOH) + PPh_3 (PhH)	105-115	-	Fig.504. (97)	C -039
C_8H_{16}	$CH_3(CH_2)_5CH=CH_2$				

Table 14.1

= $(C_6H_5)_3SiH$ = =3=

	gamma(^{60}Co)	65	63	$CH_3(CH_2)_7Si(C_6H_5)_3$ (-)	344
	$Rh(acac)_3 + AlEt_3$	60	10	$CH_3(CH_2)_7Si(C_6H_5)_3$ (45)	271
C_8H_{18}	$(CH_3CH_2O)_3SiCH{=}CH_2$				
	H_2PtCl_6 (i-PrOH)	50-200	0.58	$(CH_3CH_2O)_3SiCH_2CH_2Si(C_6H_5)_3$ (86)	575
$C_{10}H_{16}$	Fig.477.				
	$(t\text{-}BuO)_2$	140	-	Fig.506. (55)	347

Table 14.2

= $(C_6H_5)_3SiH$ = =1=

C_3H_3	$CH{\equiv}CCH_2Cl$				
	H_2PtCl_6 (i-PrOH)	50	6	$(C_6H_5)_3SiCH{=}CHCH_2Cl$ (7)	705
C_3H_4	$CH{\equiv}CCH_2OH$				
	H_2PtCl_6 (i-PrOH)	50	6	trans-$(C_6H_5)_3SiCH{=}CHCH_2OH$ (59) $(C_6H_5)_3SiC({=}CH_2)CH_2OH$ (32)	706
	$Pt(CH_2{=}CH_2)_2[P(c\text{-}Hex)_3]$	reflux	36 (THF)	$(C_6H_5)_3SiCH{=}CHCH_2OH$ + $CH_2{=}C(CH_2OH)Si(C_6H_5)_3$ (54)	804
	$Pt(Nb)_2[P(Bu\text{-}t)_3]$	reflux	1 (THF)	$(C_6H_5)_3SiCH{=}CHCH_2OH$ + $CH_2{=}C(CH_2OH)Si(C_6H_5)_3$ (67)	804
C_4H_6	$CH{\equiv}CCH(OH)CH_3$				
	$Pt(CH_2{=}CH_2)_2[P(c\text{-}Hex)_3]$	reflux	20 (THF)	$(C_6H_5)_3SiCH{=}CHCH(OH)CH_3$ + $CH_2{=}C[Si(C_6H_5)_3]CH(OH)CH_3$ (94)	804
C_4H_6	$CH{\equiv}CCH_2OCH_3$				
	H_2PtCl_6 (i-PrOH)	50	6	trans-$(C_6H_5)_3SiCH{=}CHCH_2OCH_3$ (59) $(C_6H_5)_3SiC({=}CH_2)CH_2OCH_3$ (21)	706
C_5H_6	$CH{\equiv}CCH_2COOCH_3$				
	H_2PtCl_6 (i-PrOH)	150	6	trans-$(C_6H_5)_3SiCH{=}CHCH_2COOCH_3$ (31) $(C_6H_5)_3SiC({=}CH_2)CH_2COOCH_3$ (31)	706
C_5H_{10}	$(CH_3)_3SiC{\equiv}CH$				
	H_2PtCl_6	35-40	0.5	$(CH_3)_3SiCH{=}CHSi(C_6H_5)_3$ (91)	620
C_6H_8	$(CH_3)_2Si(C{\equiv}CH)_2$				
	H_2PtCl_6	50	2.5	$(CH_3)_2Si[CH{=}CHSi(C_6H_5)_3]_2$ (96)	620
C_6H_{10}	$CH_3CH_2OCH_2C{\equiv}CCH_3$				
	-	-	-	$CH_3CH_2OCH_2C[Si(C_6H_5)_3]{=}CHCH_3$ + $CH_3CH_2OCH_2CH{=}C(CH_3)Si(C_6H_5)_3$ (-)	805
C_6H_{10}	$CH_3OCH_2C{\equiv}CCH_2OCH_3$				
	$N(R){=}C(R)C(R){=}N(R)RhCl(CO)$	r.t.	-	$CH_3OCH_2C[Si(C_6H_5)_3]{=}CHCH_2OCH_3$ (98)	166
C_7H_6	$CH{\equiv}CCH_2C{=}CHCH{=}CHS$				
	H_2PtCl_6 (i-PrOH)	50	6	trans-$(C_6H_5)_3SiCH{=}CHCH_2C{=}CHCH{=}CHS$ (37) $(C_6H_5)_3SiC({=}CH_2)CH_2C{=}CHCH{=}CHS$ (9)	706
C_7H_{11}	$CH{\equiv}CCH_2N(CH_2)_4$				

Table 14.2

= $(C_6H_5)_3SiH$ = =2=

	H_2PtCl_6 (i-PrOH)	150	6	trans-$(C_6H_5)_3SiCH{=}CHCH_2\overline{N(C}H_2)_4$ (76) $(C_6H_5)_3SiC({=}CH_2)CH_2\overline{N(C}H_2)_4$ (6)	706
C_7H_{13}	$CH{\equiv}CCH_2N(CH_3CH_2)_2$				
	H_2PtCl_6 (i-PrOH)	150	6	trans-$(C_6H_5)_3SiCH{=}CHCH_2N(CH_3CH_2)_2$ (31) $(C_6H_5)_3SiC({=}CH_2)CH_2N(CH_3CH_2)_2$ (6)	706
C_8H_6	$C_6H_5C{\equiv}CH$				
	$(Bu_4N)_2PtCl_6$	140	1-3	trans-$C_6H_5CH{=}CHSi(C_6H_5)_3$ (-)	673
	H_2PtCl_6 (i-PrOH)	50	6	$C_6H_5CH{=}CHSi(C_6H_5)_3$ (9) $C_6H_5C({=}CH_2)Si(C_6H_5)_3$ (67-78)	706
	$RhCl(PPh_3)_3$	80	23 (t)	cis-$C_6H_5CH{=}CHSi(C_6H_5)_3$ (7) trans-$(C_6H_5CH{=}CHSi(C_6H_5)_3$ (19) $C_6H_5C({=}CH_2)Si(C_6H_5)_3$ (2)	878
C_8H_{13}	$CH{\equiv}CCH_2\overline{N(C}H_2)_5$				
	H_2PtCl_6 (i-PrOH)	150	6	trans-$(C_6H_5)_3SiCH{=}CHCH_2\overline{N(C}H_2)_5$ (84) $(C_6H_5)_3SiC({=}CH_2)CH_2\overline{N(C}H_2)_5$ (1)	706
C_8H_{13}	$CH{\equiv}CSi(OCH_2CH_2)_3N$				
	$H_2PtCl_6{\cdot}6H_2O$	50	6	$(C_6H_5)_3Si(CH_3)CH{=}CHSi(OCH_2CH_2)_3N$ (47)	704
C_9H_{15}	$CH{\equiv}CCH_2\overline{N(C}H_2)_6$				
	H_2PtCl_6 (i-PrOH)	150	6	trans-$(C_6H_5)_3SiCH{=}CHCH_2\overline{N(C}H_2)_6$ (40) $(C_6H_5)_3SiC({=}CH_2)CH_2\overline{N(C}H_2)_6$ (2)	706
C_9H_{16}	$CH{\equiv}CC(CH_3)_2OOC(CH_3)_3$				
	H_2PtCl_6 (i-PrOH)	50	1.5 (h)	$CH_2{=}C[Si(C_6H_5)_3]C(CH_3)_2OOC(CH_3)_3$ (66)	788
$C_{10}H_{18}$	$CH{\equiv}CC(CH_3)_2OOC(CH_3)_2C_2H_5$				
	H_2PtCl_6 (i-PrOH)	40-50	5-11 (h)	$CH_2{=}C[Si(C_6H_5)_3]C(CH_3)_2OOC(CH_3)_2C_2H_5$ (64)	788
$C_{11}H_{10}$	$C_6H_5Si(CH_3)(C{\equiv}CH)_2$				
	H_2PtCl_6	90-100	20	$C_6H_5Si(CH_3)[CH{=}CHSi(C_6H_5)_3]_2$ (47)	620
$C_{11}H_{20}$	$CH{\equiv}CC(CH_3)_2OOC(CH_3)_2C_3H_7$				
	H_2PtCl_6 (i-PrOH)	40-50	5-11 (h)	$CH_2{=}C[Si(C_6H_5)_3]C(CH_3)_2OOC(CH_3)_2C_3H_7$ (51)	788
$C_{12}H_9$	$C_5H_5FeC_5H_4C{\equiv}CD$				
	H_2PtCl_6 (i-PrOH)	75-80	-	$C_5H_5FeC_5H_4CH{=}CDSi(C_6H_5)_3$ (88)	63
$C_{12}H_{10}$	$C_5H_5FeC_5H_4C{\equiv}CH$				

Table 14.2

= $(C_6H_5)_3SiH$ = =3=

	H_2PtCl_6	-	-	trans-$C_5H_5FeC_5H_4CH=CHSi(C_6H_5)_3$ (73-85)	450
	H_2PtCl_6 (i-PrOH)	75-80	2.33	$C_5H_5FeC_5H_4CH=CHSi(C_6H_5)_3$ (90)	63
$C_{13}H_{24}$	$CH\equiv CC(CH_3)_2OOC(CH_3)_2(CH_2)_4CH_3$				
	H_2PtCl_6 (i-PrOH)	40-50	5-11 (h)	$CH_2=C[Si(C_6H_5)_3]C(CH_3)_2OOC(CH_3)_2(CH_2)_4CH_3$ (57)	788
$C_{14}H_{10}$	$C_6H_5C\equiv CC_6H_5$				
	H_2PtCl_6	110	8	cis-$C_6H_5CH=C(C_6H_5)Si(C_6H_5)_3$ (61)	169
$C_{14}H_{16}$	p-$CH_3OC_6H_4C(CH_3)_2C\equiv CCH=CH_2$				
	H_2PtCl_6	75	10	p-$CH_3OC_6H_4C(CH_3)_2CH=C[Si(C_6H_5)_3]CH=CH_2$ (20)	540
$C_{15}H_{14}$	$CH_3Si(C_6H_5)_2C\equiv CH$				
	H_2PtCl_6	90-100	20	$CH_3Si(C_6H_5)_2CH=CHSi(C_6H_5)_3$ (48)	620
$C_{15}H_{18}$	$C_5H_5FeC_5H_4C\equiv CSi(CH_3)_3$				
	H_2PtCl_6	90-95	-	$C_5H_5FeC_5H_4CH=C[Si(CH_3)_3]Si(C_6H_5)_3$ (38) $C_5H_5FeC_5H_4C[Si(C_6H_5)_3]=CHSi(CH_3)_3$ (31)	62
$C_{16}H_{10}$	Fig.507.				
	H_2PtCl_6	90-100	8	Fig.508. (27)	459
$C_{16}H_{12}$	$(C_6H_5)_2Si(C\equiv CH)_2$				
	H_2PtCl_6	90-100	20	$(C_6H_5)_2Si[CH=CHSi(C_6H_5)_3]_2$ (42)	620
$C_{18}H_{24}$	$C_5H_5FeC_5H_4C\equiv CGe(CH_2CH_3)_3$				
	H_2PtCl_6 (i-PrOH)	90-95	-	$C_5H_5FeC_5H_4C[Si(C_6H_5)_3]=CHGe(CH_2CH_3)_3$ + $C_5H_5FeC_5H_4CH=C[Si(C_6H_5)_3]Ge(CH_2CH_3)_3$ (35)	64
$C_{18}H_{24}$	$C_5H_5FeC_5H_4C\equiv CSi(CH_2CH_3)_3$				
	H_2PtCl_6	90-95	-	$C_5H_5FeC_5H_4CH=C[Si(CH_3CH_2)_3]Si(C_6H_5)_3$ (38) $C_5H_5FeC_5H_4C[Si(C_6H_5)_3]=CHSi(CH_3CH_2)_3$ (31)	62
$C_{20}H_{24}$	Fig.335.				
	Pt/C	160	16 (dg)	Fig.509. (-)	F -024
$C_{26}H_{24}$	$(C_5H_5FeC_5H_4C\equiv C)_2Ge(CH_3)_2$				
	H_2PtCl_6 (i-PrOH)	90-95	-	α,α -$[C_5H_5FeC_5H_4C_2HSi(C_6H_5)_3]_2Ge(CH_3)_2$ (13) α,β -$[C_5H_5FeC_5H_4C_2HSi(C_6H_5)_3]_2Ge(CH_3)_2$ (13) β,β -$[C_5H_5FeC_5H_4C_2HSi(C_6H_5)_3]_2Ge(CH_3)_2$ (13) $C_5H_5FeC_5H_4CH=CHSi(C_6H_5)_3$ (26)	64
$C_{26}H_{24}$	$(C_5H_5FeC_5H_4C\equiv C)_2Si(CH_3)_2$				

H_2PtCl_6	90-95	-	$[C_5H_5FeC_5H_4CH{=}C[Si(C_6H_5)_3]Si(CH_3)_2$ (18) $\{C_5H_5FeC_5H_4C[Si(C_6H_5)_3]{=}CH\}_2Si(CH_3)_2$ (27) $C_5H_5FeC_5H_4C[Si(C_6H_5)_3]{=}CHSi(C{\equiv}CC_5H_4FeC_5H_5)$- $(CH_3)_2$ (45)	62
$C_{28}H_{28}$ $(C_5H_5FeC_5H_4C{\equiv}C)_2Si(CH_2CH_3)_2$				
H_2PtCl_6	90-95	-	$\{C_5H_5FeC_5H_4C[Si(C_6H_5)_3]{=}CH\}_2Si(CH_2CH_3)_2$ (51) $C_5H_5FeC_5H_4C[Si(C_6H_5)_3]{=}CHSi(CH_2CH_3)_2C[Si(C_6H_5)_3]{=}$ $={CHC_5H_4FeC_5H_5}$ (34)	62
$C_{30}H_{24}$ $C_5H_5FeC_5H_4C{\equiv}CGe(C_6H_5)_3$				
H_2PtCl_6 (i-PrOH)	90-95	-	$C_5H_5FeC_5H_4C[Si(C_6H_5)_3]{=}CHGe(C_6H_5)_3$ + $C_5H_5FeC_5H_4CH{=}CH[Si(C_6H_5)_3]Ge(C_6H_5)_3$ (-)	64
$C_{30}H_{24}$ $C_5H_5FeC_5H_4C{\equiv}CSi(C_6H_5)_3$				
H_2PtCl_6 (i-PrOH)	90-95	-	$C_5H_5FeC_5H_4C[Si(C_6H_5)_3]{=}CHSi(C_6H_5)_3$ (49) $C_5H_5FeC_5H_4CH{=}CH[Si(C_6H_5)_3]_2$ (33)	64
H_2PtCl_6 (i-PrOH)	90-95	-	$C_5H_5FeC_5H_4C[Si(C_6H_5)_3]{=}CHSi(C_6H_5)_3$ (49) $C_5H_5FeC_5H_4CH{=}CH[Si(C_6H_5)_3]_2$ (33)	62
$C_{36}H_{28}$ $(C_5H_5FeC_5H_4C{\equiv}C)_2Ge(C_6H_5)_2$				
H_2PtCl_6 (i-PrOH)	90-95	-	α,α -$[C_5H_5FeC_5H_4C_2HSi(C_6H_5)_3]_2Ge(C_6H_5)_2$ (8) α,β -$[C_5H_5FeC_5H_4C_2HSi(C_6H_5)_3]_2Ge(C_6H_5)_2$ (14) β,β -$[C_5H_5FeC_5H_4C_2HSi(C_6H_5)_3]_2Ge(C_6H_5)_2$ (8)	64
$C_{36}H_{28}$ $(C_5H_5FeC_5H_4C{\equiv}C)_2Si(C_6H_5)_2$				
H_2PtCl_6	90-95	-	$\{C_5H_5FeC_5H_4CH{=}C[Si(C_6H_5)_3]\}_2Si(C_6H_5)_2$ (81)	62

Table 14.3

= $(C_6H_5)_3SiH$ = =1=

C_3H_6	CH_3COCH_3				
	$RhCl(PPh_3)_3$	reflux	2	$(CH_3)_2CHOSi(C_6H_5)_3$ (98)	F -023
	$RhCl(PPh_3)_3$	reflux	2	$(CH_3)_2CHOSi(C_6H_5)_3$ (98)	873
	$RhCl(PPh_3)_3$	reflux	2	$(CH_3)_2CHOSi(C_6H_5)_3$ (98)	866
C_4H_6	$CH_3CH{=}CHCHO$				
	$RhCl(PPh_3)_3$	60	0.5	$CH_3CH{=}CHCH_2OSi(C_6H_5)_3$ (97)	F -023
	$RhCl(PPh_3)_3$	60	0.5	$CH_3CH{=}CHCH_2OSi(C_6H_5)_3$ (97)	866
	$RhCl(PPh_3)_3$	60	0.5	$CH_3CH{=}CHCH_2OSi(C_6H_5)_3$ (97)	873
C_7H_6	C_6H_5CHO				
	$NiCl_2$	-	-	$(C_6H_5)_3SiOCH(C_6H_5)CH(C_6H_5)OSi(C_6H_5)_3$ (-)	413
C_7H_{12}	$CH_3\overline{CH(CH_2)_4C}O$				
	Bu_4NF	0-r.t.	0.5-12	$CH_3\overline{CH(CH_2)_4C}HOSi(C_6H_5)_3$ (95)	393
C_8H_8	$C_6H_5COCH_3$				
	$K[PtCl_3(C_2H_4)]$	50	72	$C_6H_5CH(CH_3)OSi(C_6H_5)_3$ (15)	177
	$[RhCl(C_2H_4)_2]_2$ + diop	50	20 (h)	$C_6H_5CH(CH_3)OSi(C_6H_5)_3$ (-)	336
	$[RhCl(cod)]_2$	50	22	$C_6H_5CH(CH_3)OSi(C_6H_5)_3$ (5)	177
$C_{10}H_{18}$	$\overline{CH_2CH_2CH[C(CH_3)_3]CH_2CH_2C}O$				
	$RhCl(PPh_3)_3$	25-80	64-24 (b)	$\overline{CH_2CH_2CH[C(CH_3)_3]CH_2CH_2C}HOSi(C_6H_5)_3$ (to 100)	1037
	$RuCl_2(PPh_3)_3$	80	24 (b)	$\overline{CH_2CH_2CH[C(CH_3)_3]CH_2CH_2C}HOSi(C_6H_5)_3$ (-)	1037
$C_{13}H_{10}$	$C_6H_5COC_6H_5$				
	UV	16	3.5	$(C_6H_5)_2C(OH)C(C_6H_5)_2OSi(C_6H_5)_3$ (56)	84
$C_{19}H_{24}$	Fig.448.				
	RhCl(diop)(S)	22	168 (b)	Fig.510. (traces-12)	420
$C_{27}H_{20}$	Fig.511.				
	none	330-340	0.08	Fig.512. (45)	84
$C_{28}H_{40}$	Fig.513.				
	none	300-310	0.25	Fig.514. (69)	84

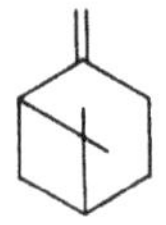

Figure 15

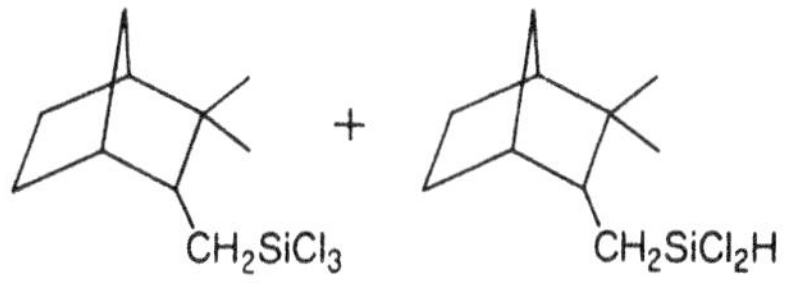

Figure 16

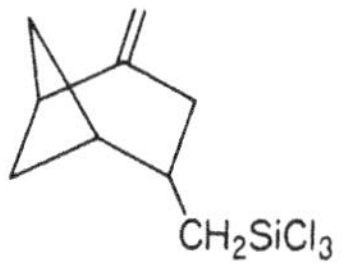

Figure 17

Figure 18

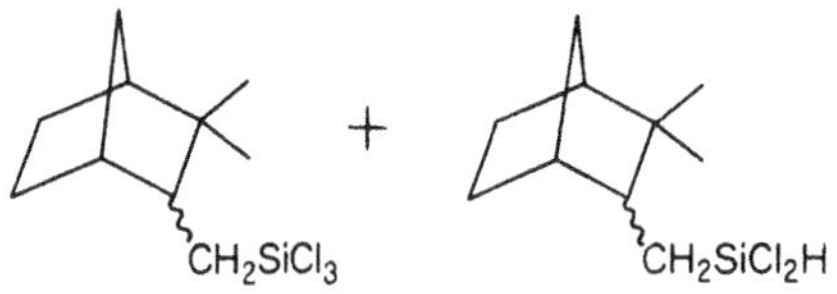

Figure 19

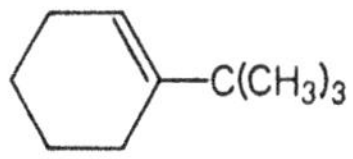

Figure 20

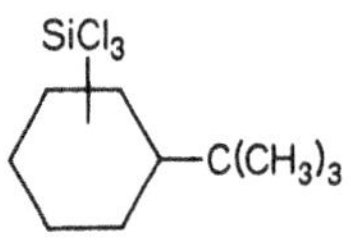

Figure 21

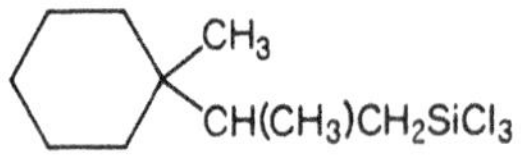

Figure 22

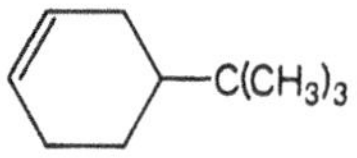

Figure 23

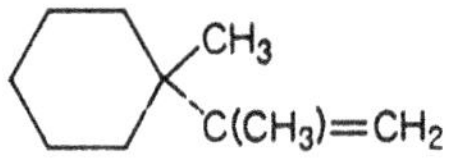

Figure 24

Figure 25

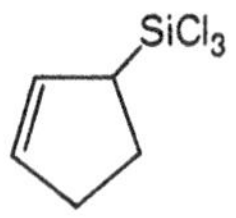

Figure 26

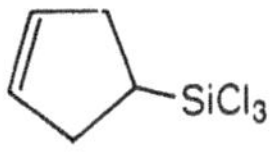

Figure 27

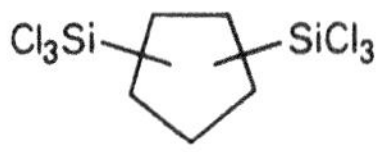

Figure 28

$C_{10}H_{13}SiCl_3$

Figure 29

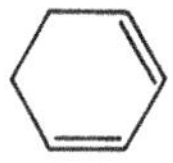

Figure 30

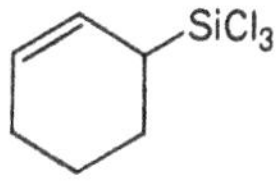

Figure 31

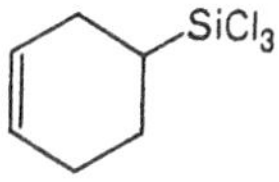

Figure 32

Figure 33

Figure 34

Figure 35

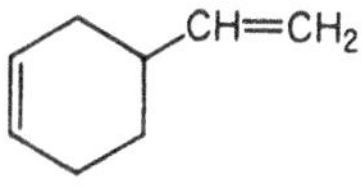

Figure 36

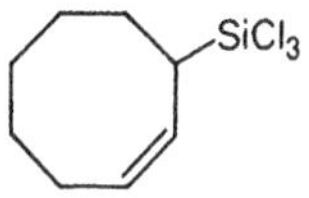

Figure 37

$c\text{-}C_8H_{13}SiCl_3$

Figure 38

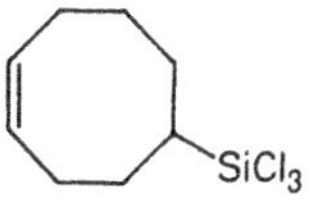

Figure 39

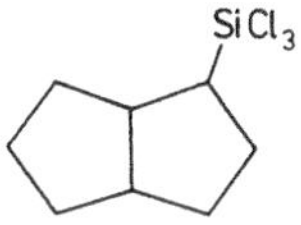

Figure 40

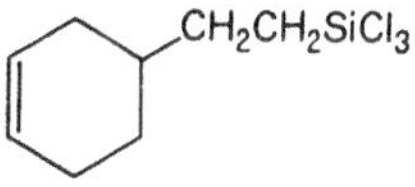

Figure 41

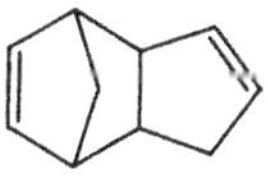

Figure 42

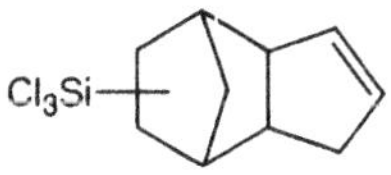

Figure 43

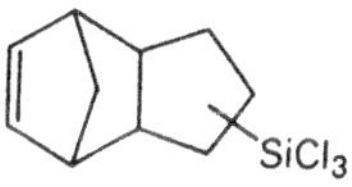

Figure 44

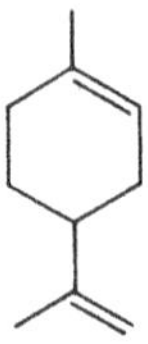

Figure 45

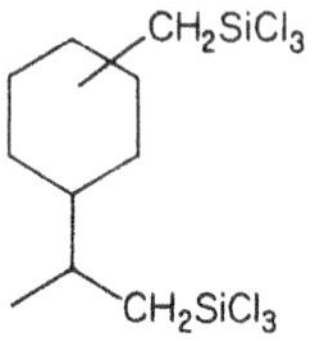

Figure 46

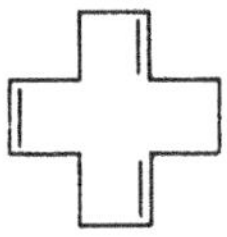

Figure 47

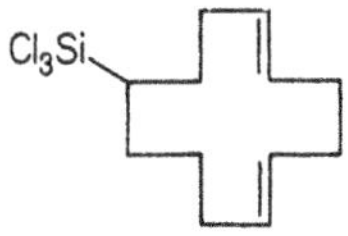

Figure 48

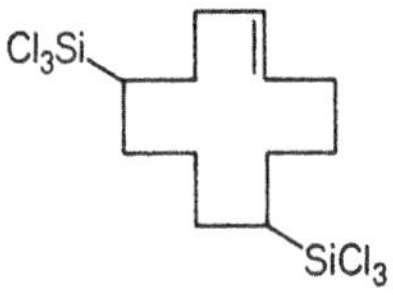

Figure 49

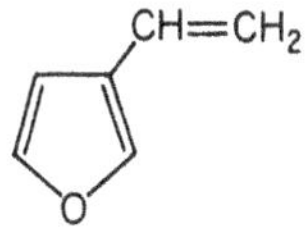

Figure 50

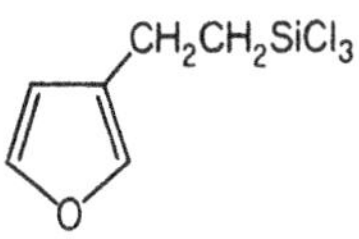

Figure 51

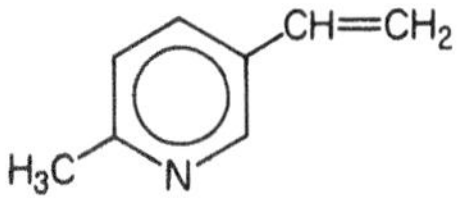

Figure 52

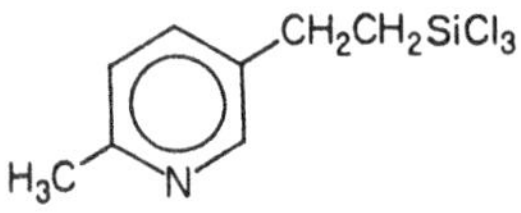

Figure 53

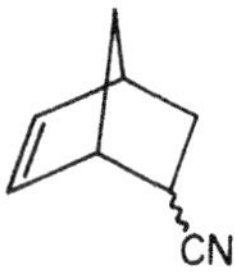

Figure 54

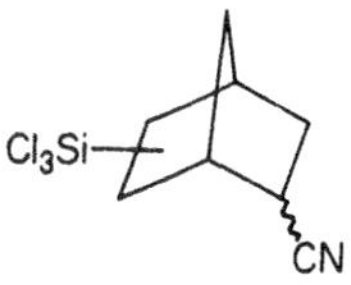

Figure 55

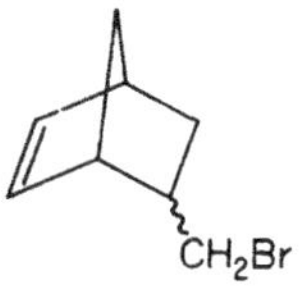

Figure 56

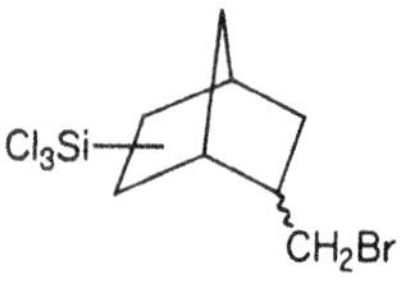

Figure 57

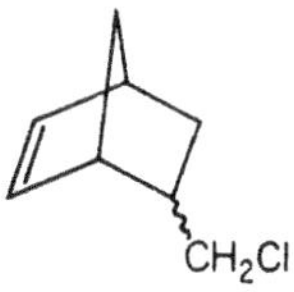

Figure 58

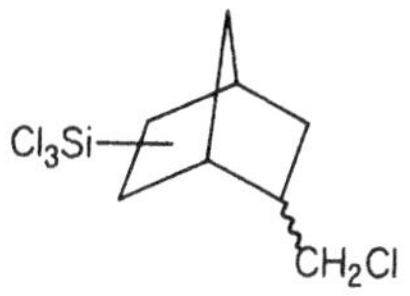

Figure 59

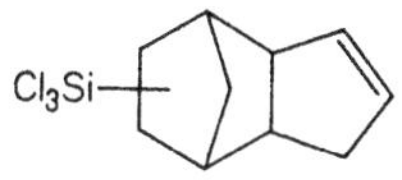

Figure 60

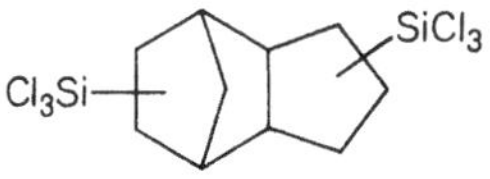

Figure 61

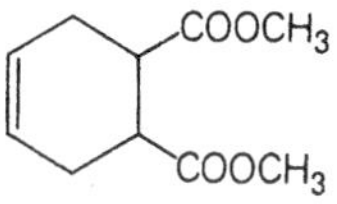

Figure 62

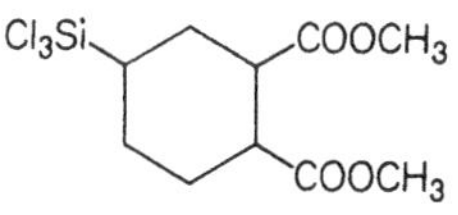

Figure 63

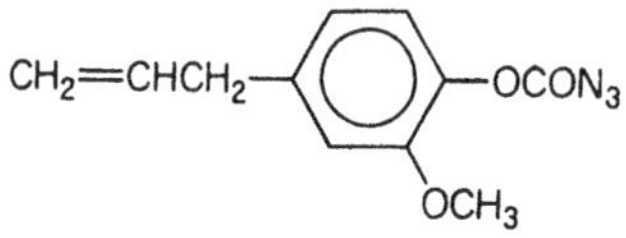

Figure 64

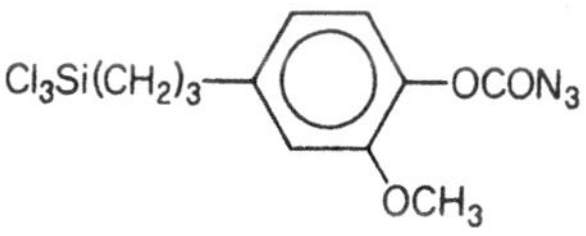

Figure 65

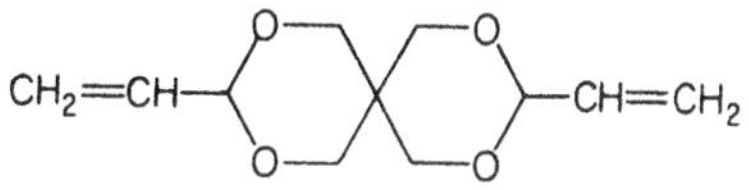

Figure 66

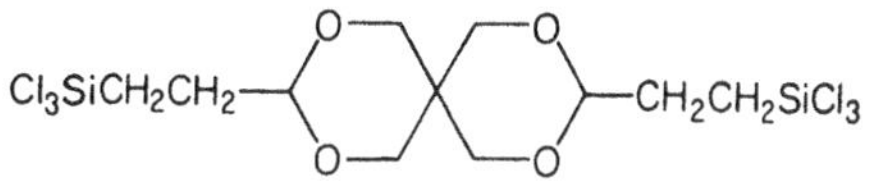

Figure 67

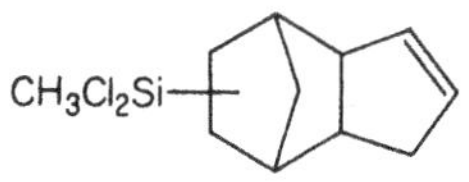

Figure 68

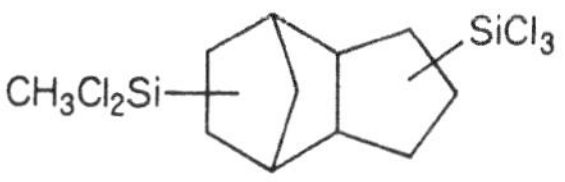

Figure 69

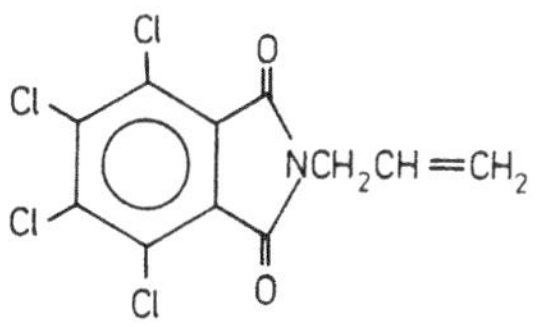

Figure 70

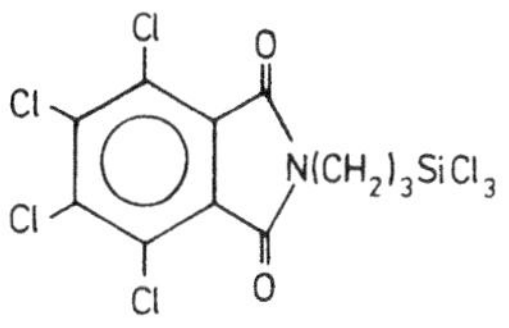

Figure 71

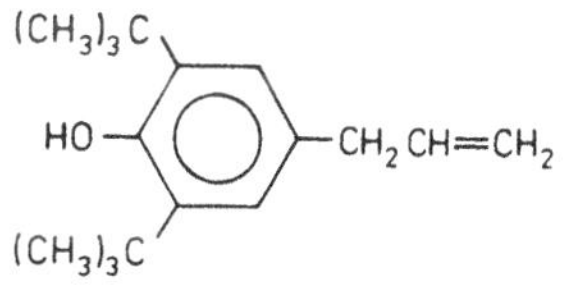

Figure 72

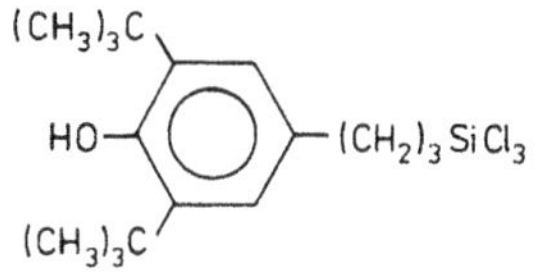

Figure 73

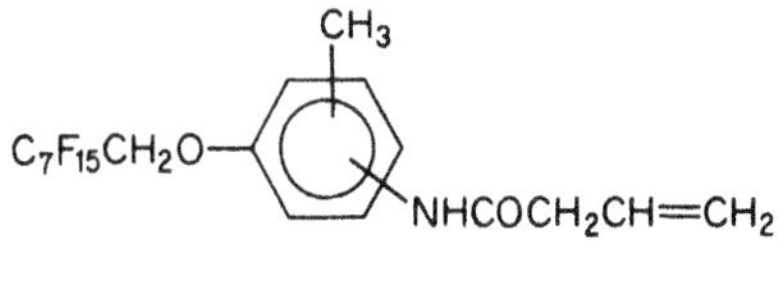

Figure 74

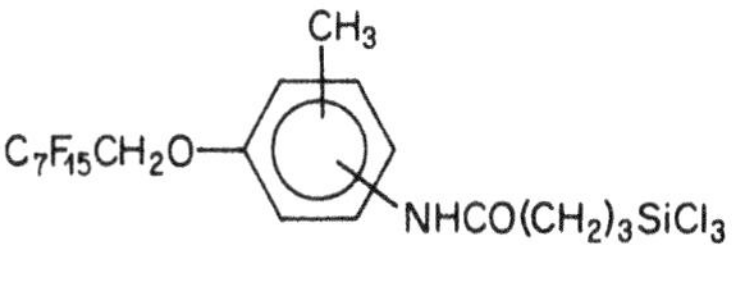

Figure 75

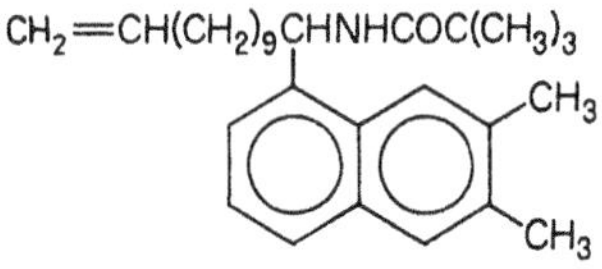

Figure 76

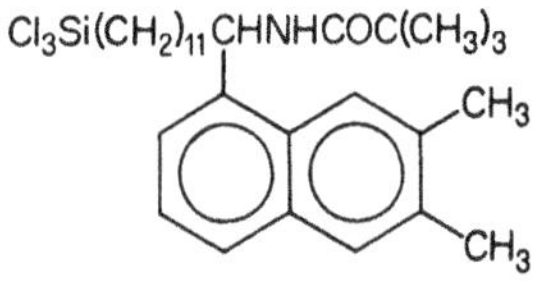

Figure 77

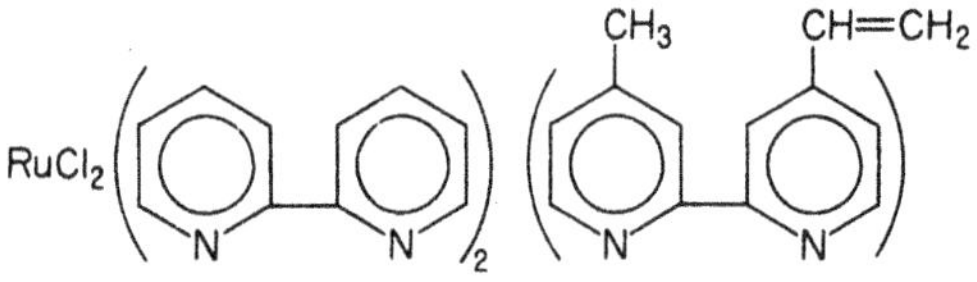

Figure 78

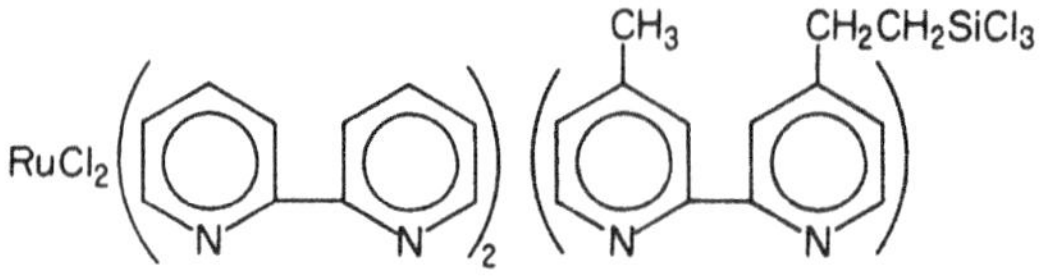

Figure 79

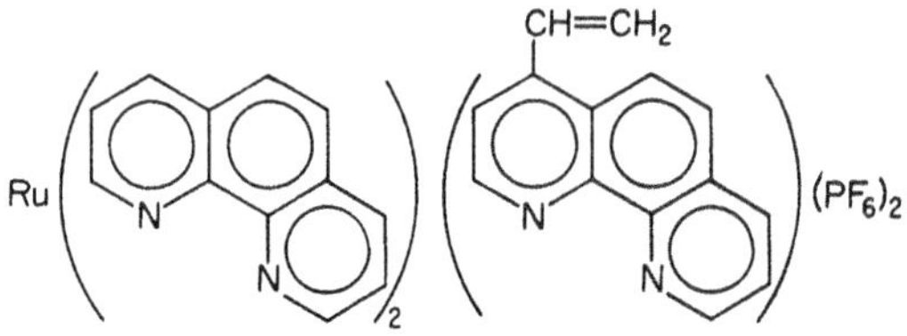

Figure 80

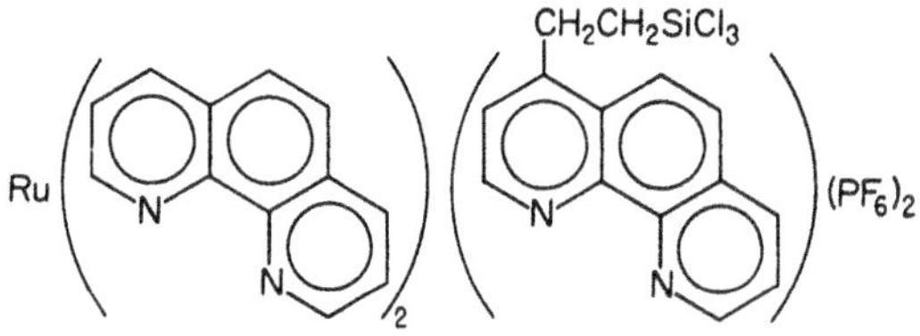

Figure 81

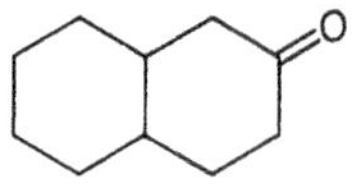

Figure 82

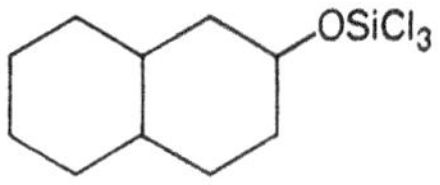

Figure 83

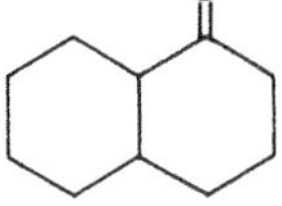

Figure 84

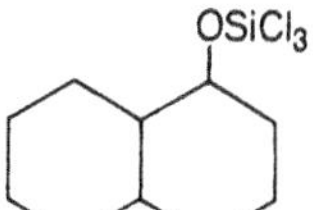

Figure 85

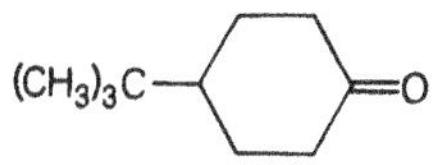

Figure 86

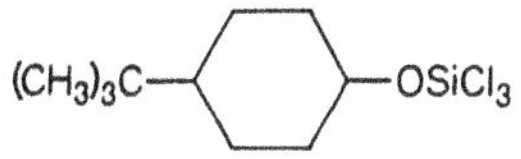

Figure 87

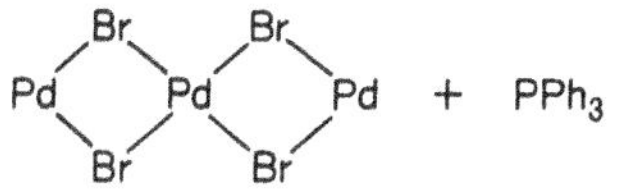

Figure 88

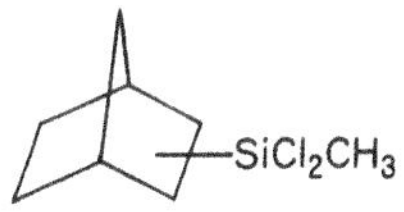

Figure 89

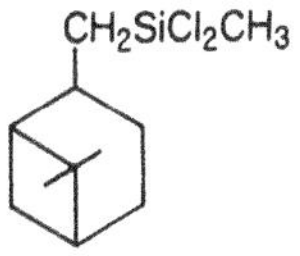

Figure 90

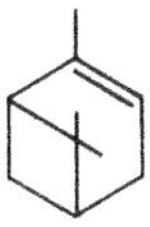

Figure 91

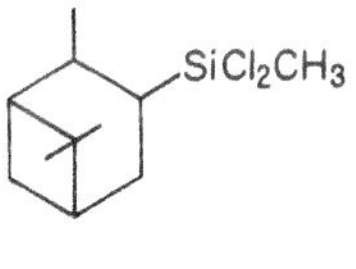

Figure 92

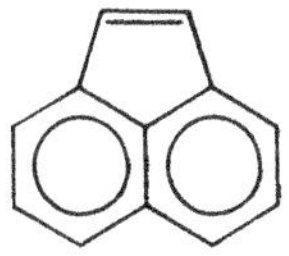

Figure 93

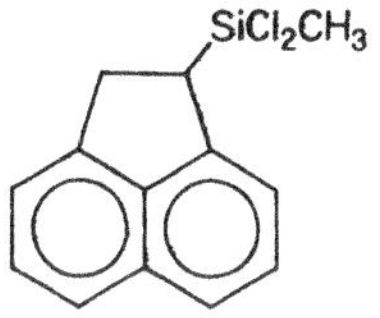

Figure 94

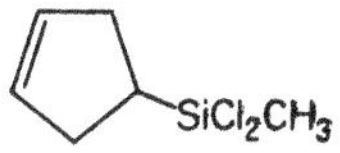

Figure 95

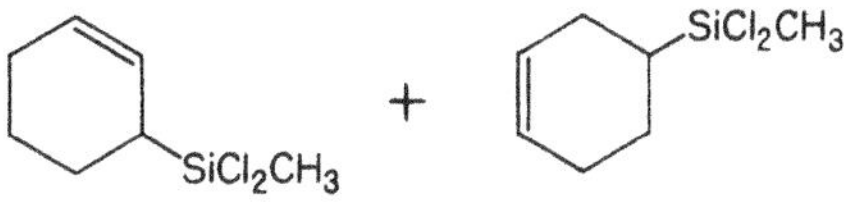

Figure 96

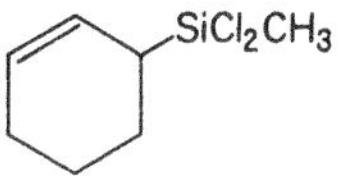

Figure 97

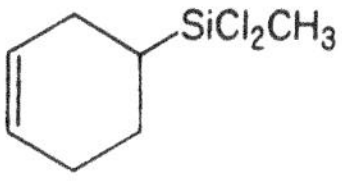

Figure 98

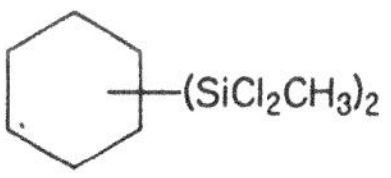

Figure 99

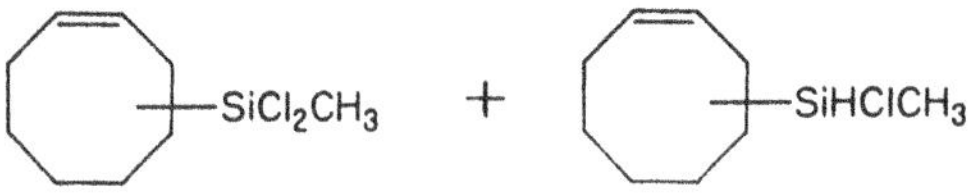

Figure 100

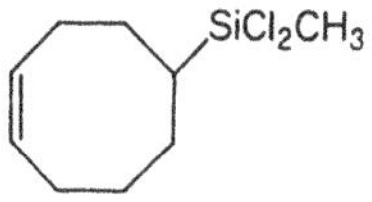

Figure 101

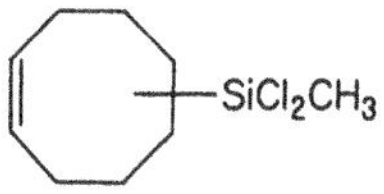

Figure 102

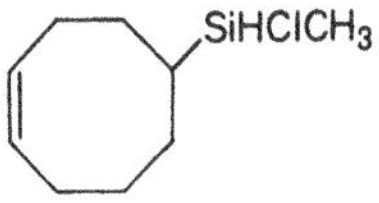

Figure 103

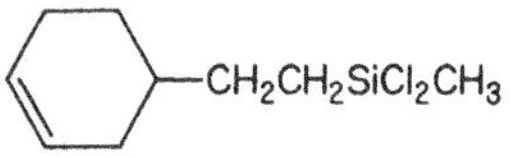

Figure 104

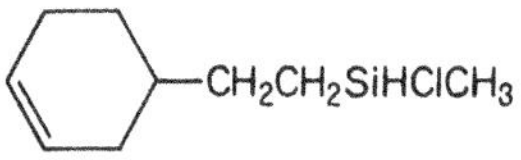

Figure 105

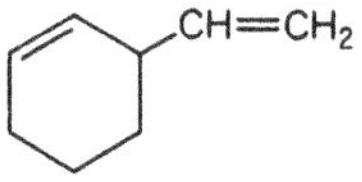

Figure 106

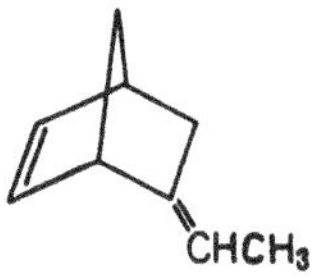

Figure 107

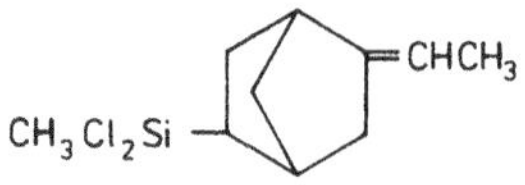

Figure 108

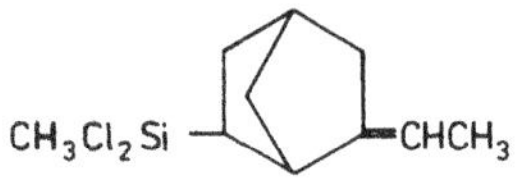

Figure 109

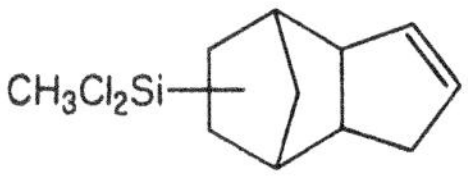

Figure 110

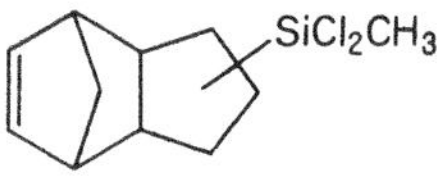

Figure 111

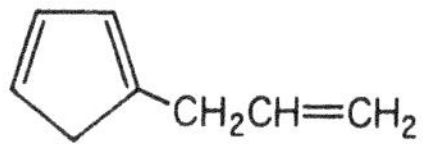

Figure 112

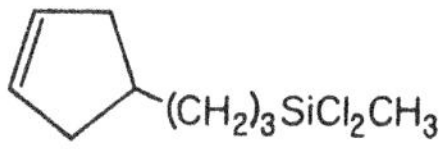

Figure 113

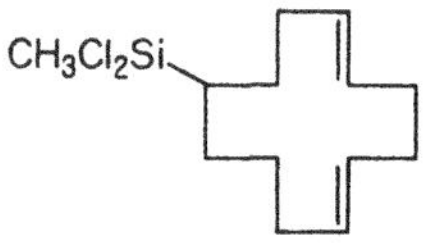

Figure 114

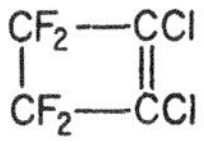

Figure 115

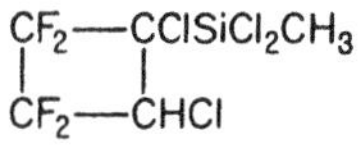

Figure 116

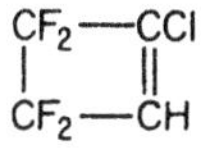

Figure 117

Figure 118

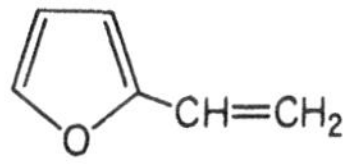

Figure 119

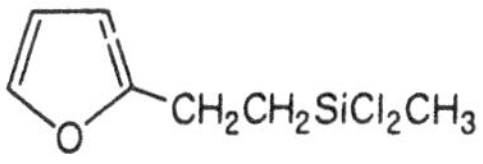

Figure 120

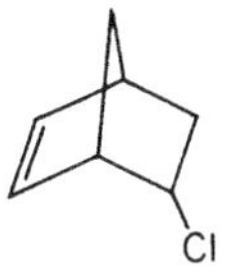

Figure 121

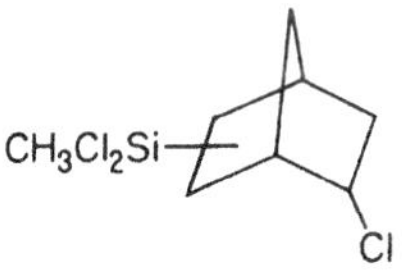

Figure 122

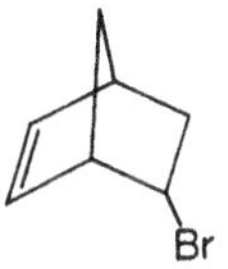

Figure 123

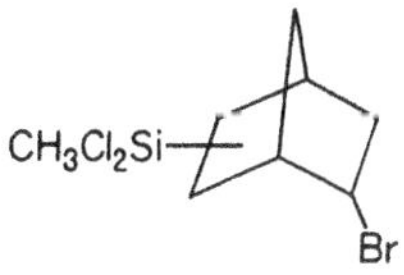

Figure 124

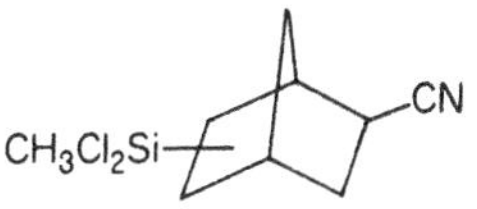

Figure 125

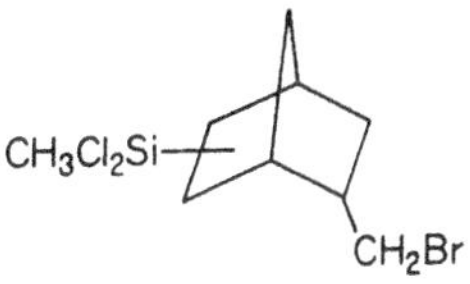

Figure 126

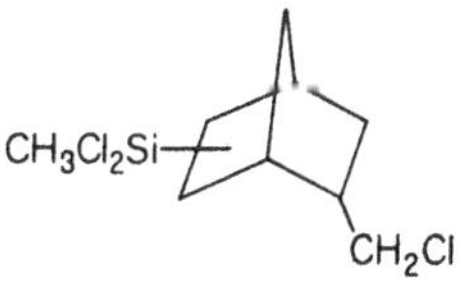

Figure 127

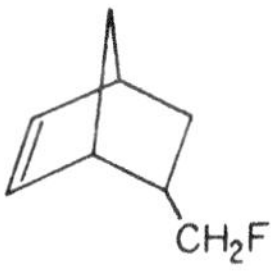

Figure 128

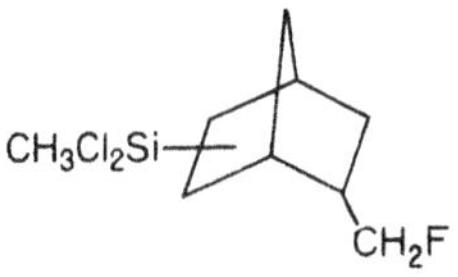

Figure 129

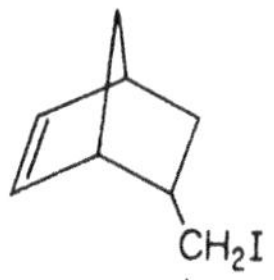

Figure 130

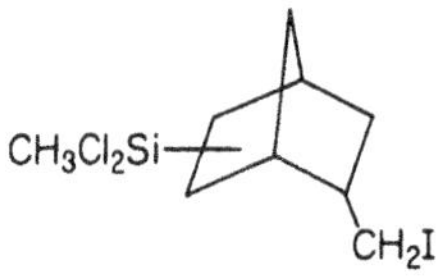

Figure 131

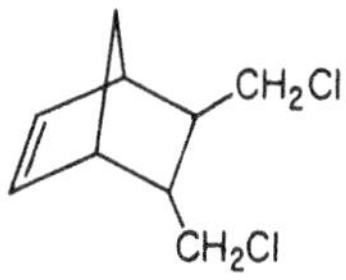

Figure 132

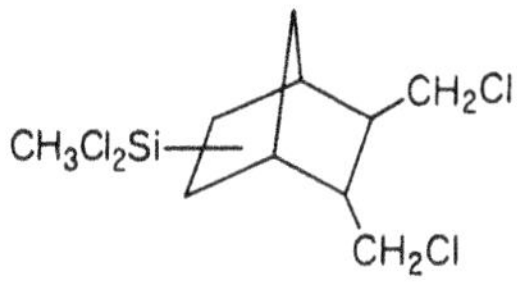

Figure 133

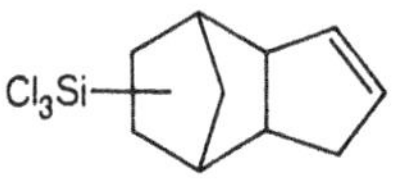

Figure 134

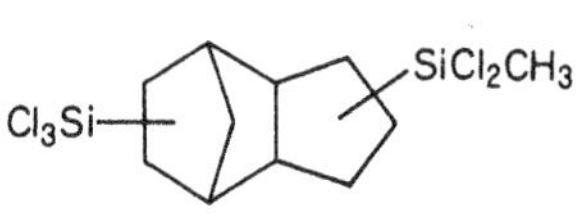

Figure 135

Figure 136

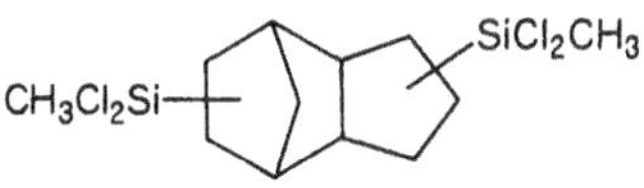

Figure 137

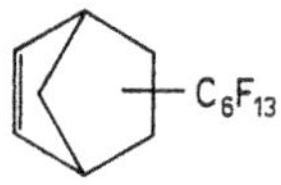

Figure 138

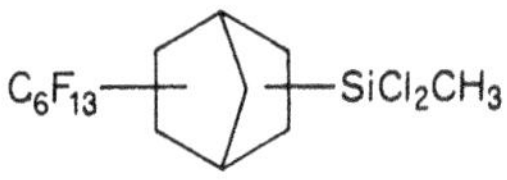

Figure 139

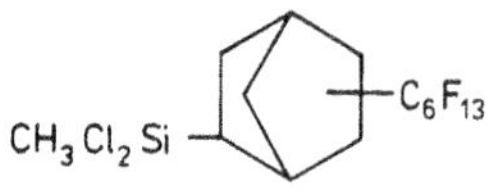

Figure 140

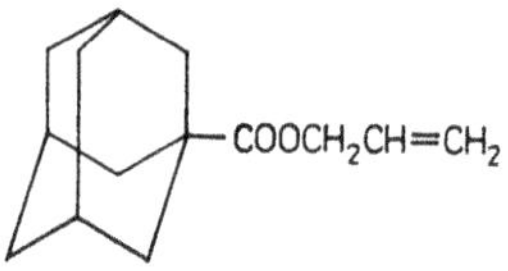

Figure 141

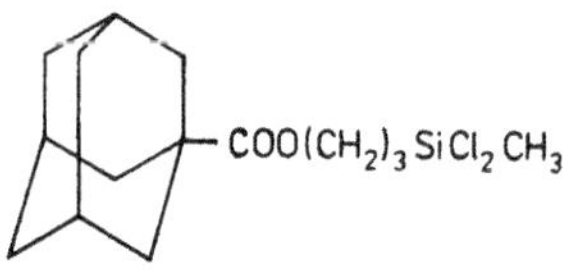

Figure 142

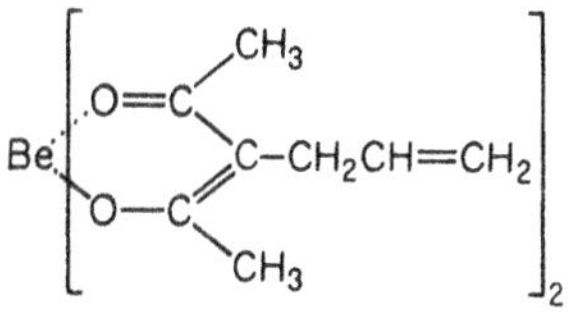

Figure 143

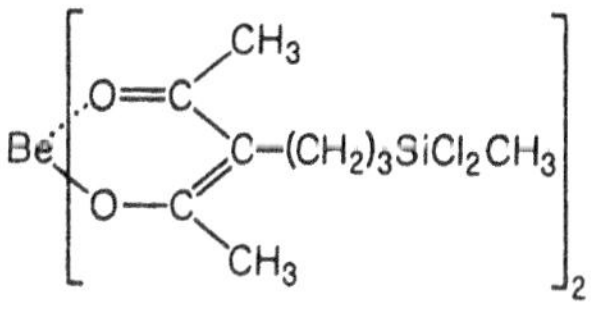

Figure 144

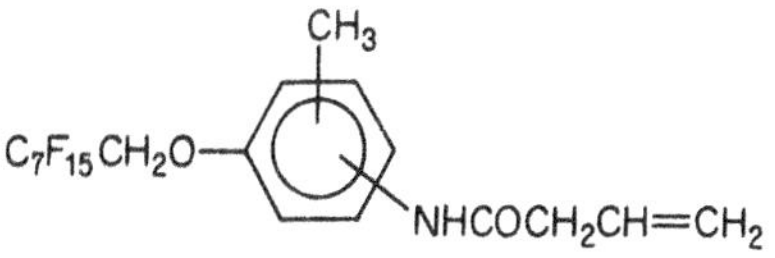

Figure 145

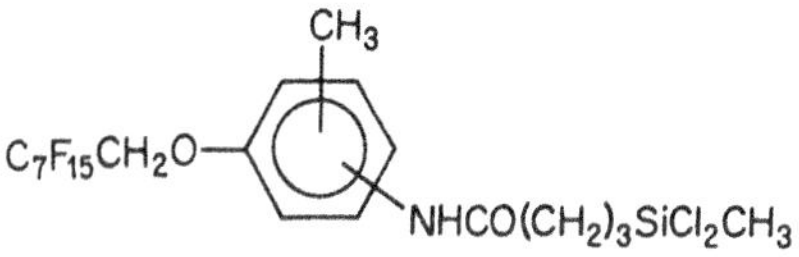

Figure 146

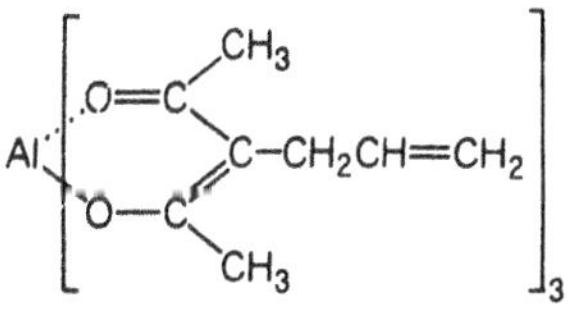

Figure 147

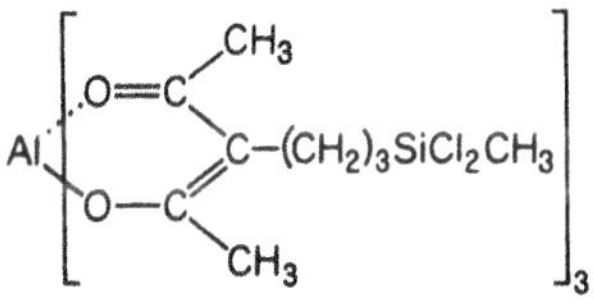

Figure 148

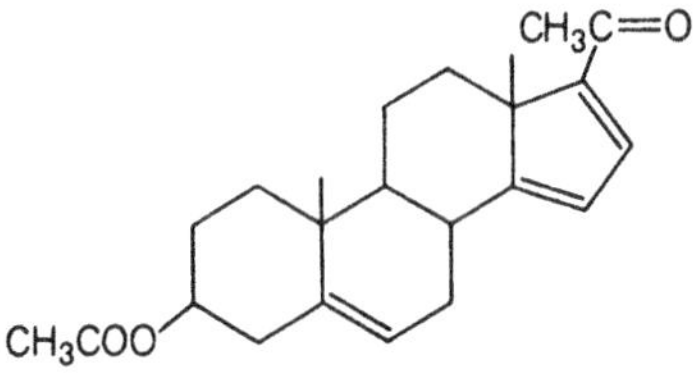

Figure 149

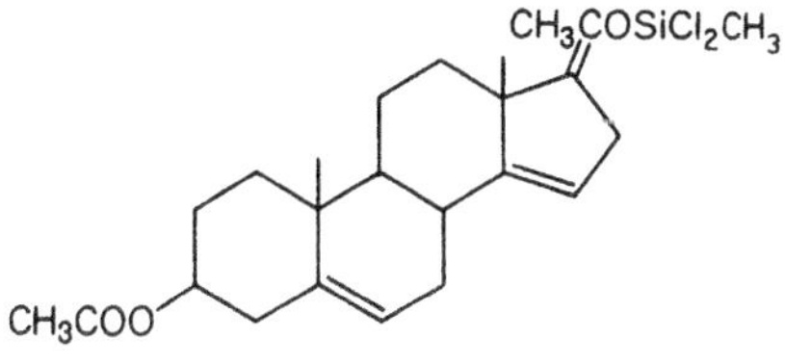

Figure 150

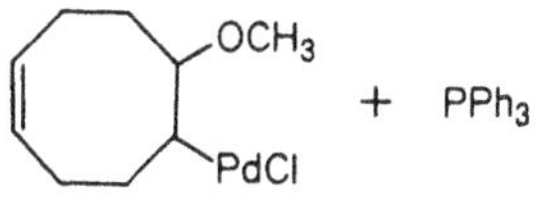

Figure 151

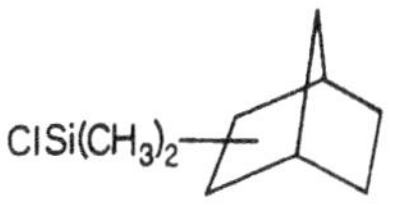

Figure 152

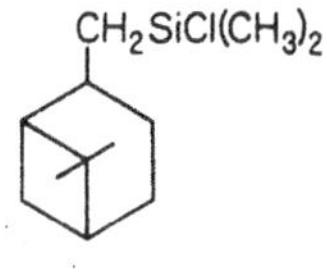

Figure 153

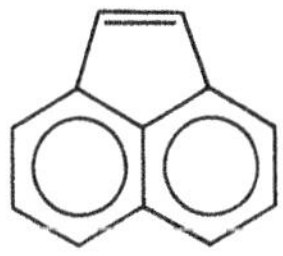

Figure 154

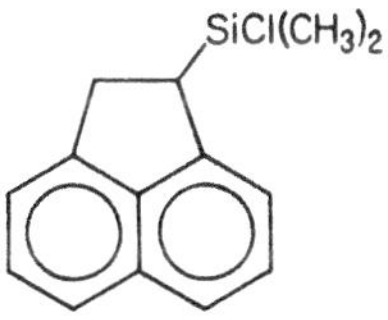

Figure 155

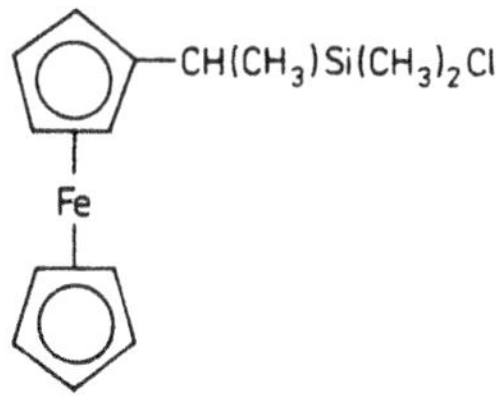

Figure 185

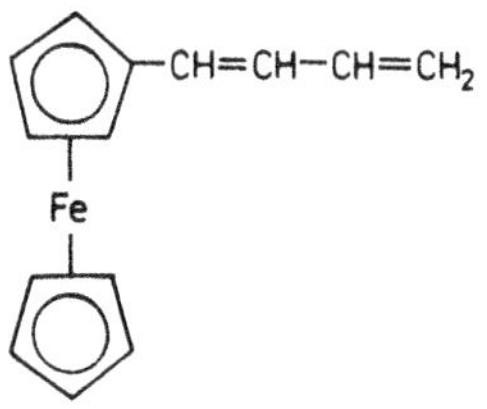

Figure 186

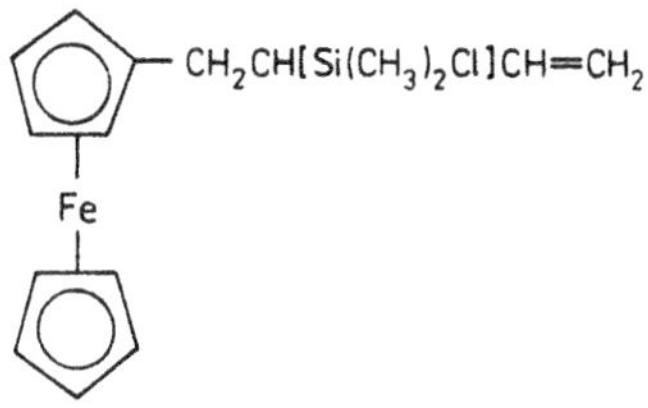

Figure 187

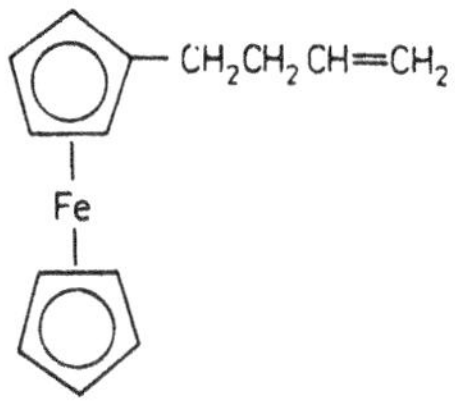

Figure 188

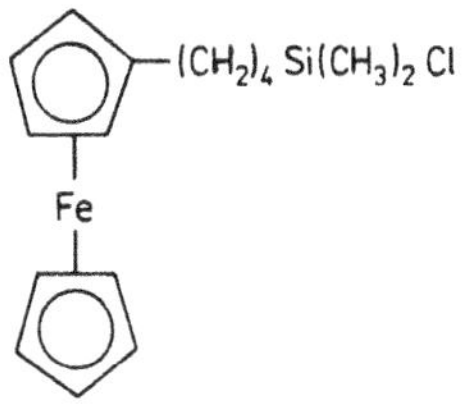

Figure 189

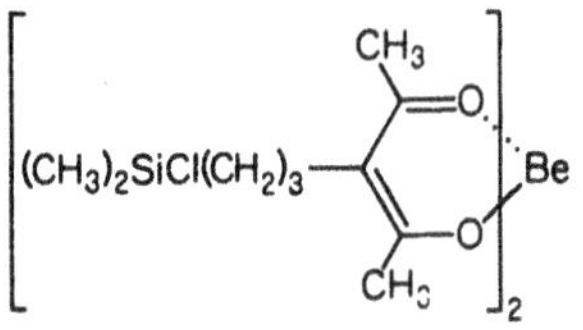

Figure 190

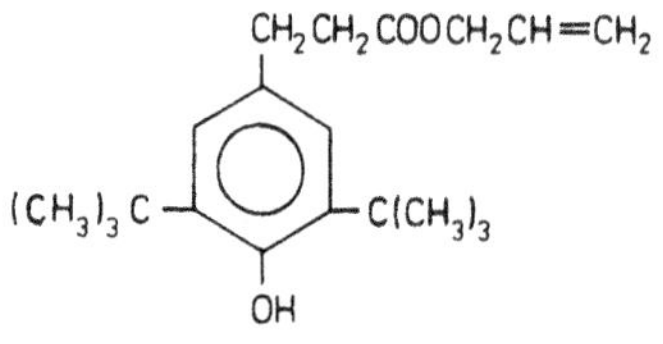

Figure 191

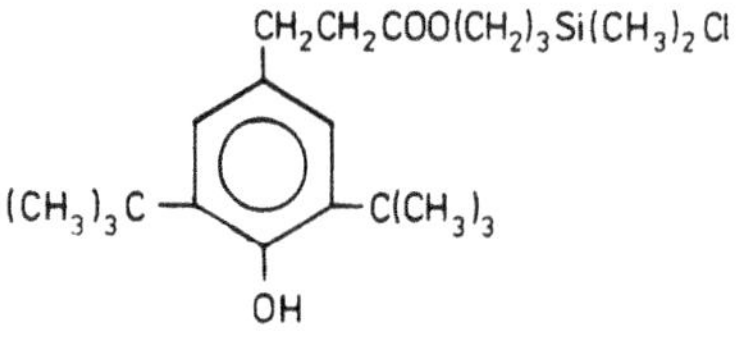

Figure 192

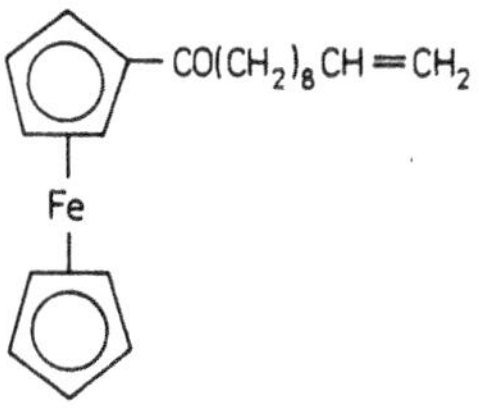

Figure 193

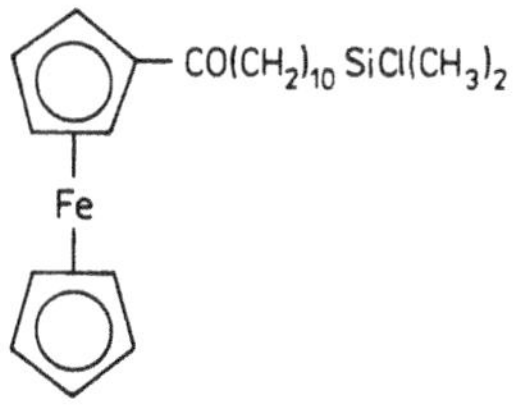

Figure 194

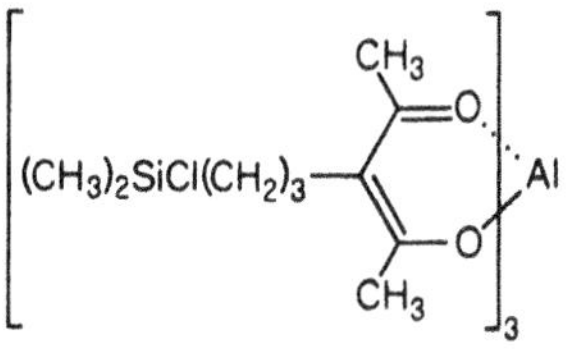

Figure 195

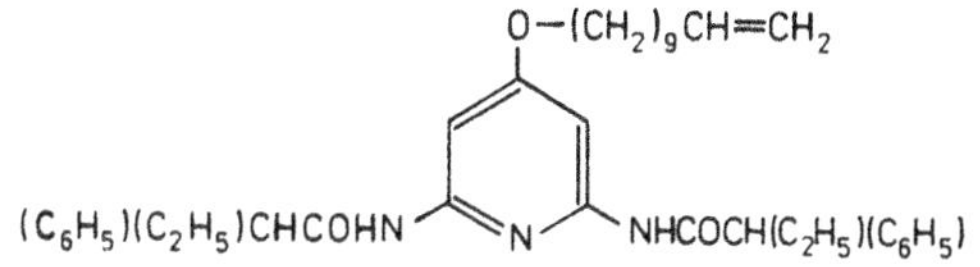

Figure 196

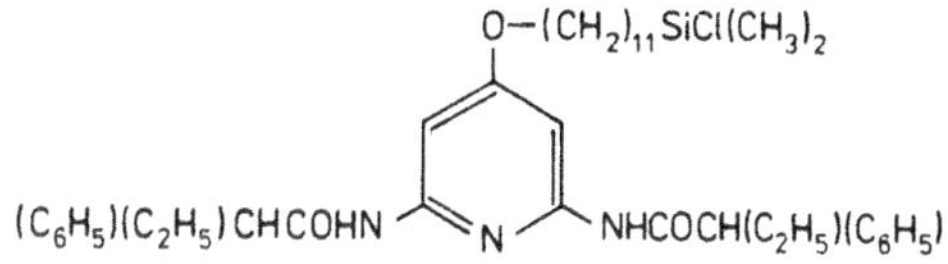

Figure 197

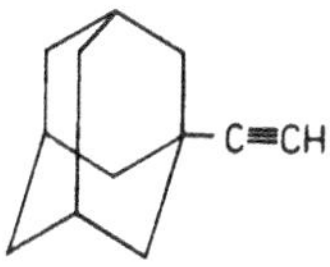

Figure 198

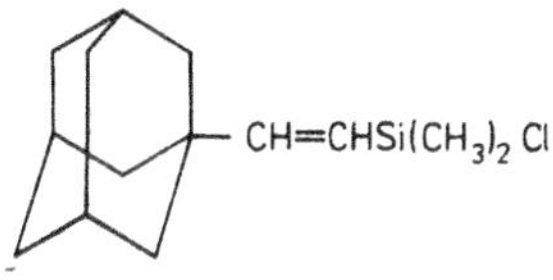

Figure 199

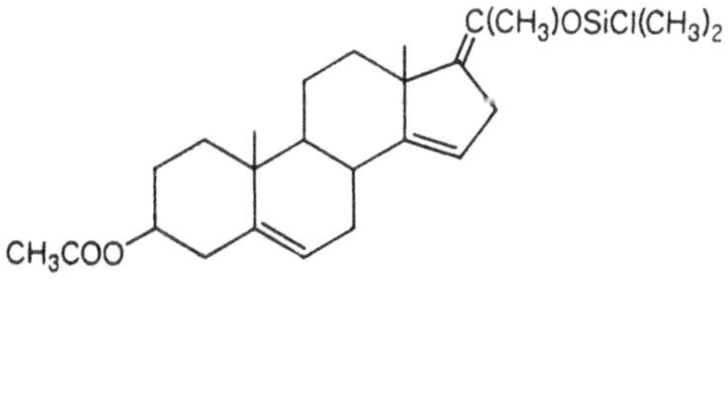

Figure 200

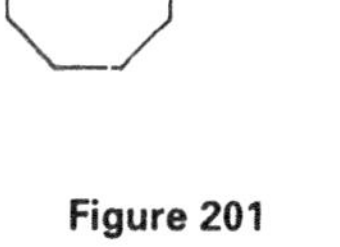

Figure 201

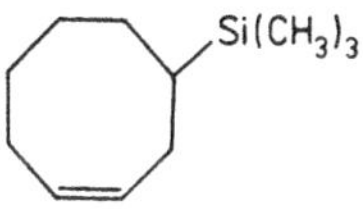

Figure 202

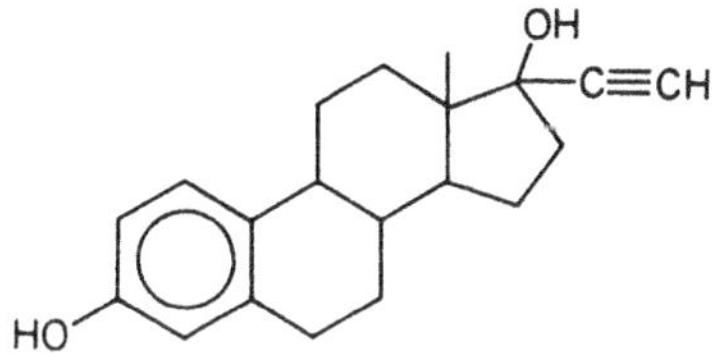

Figure 203

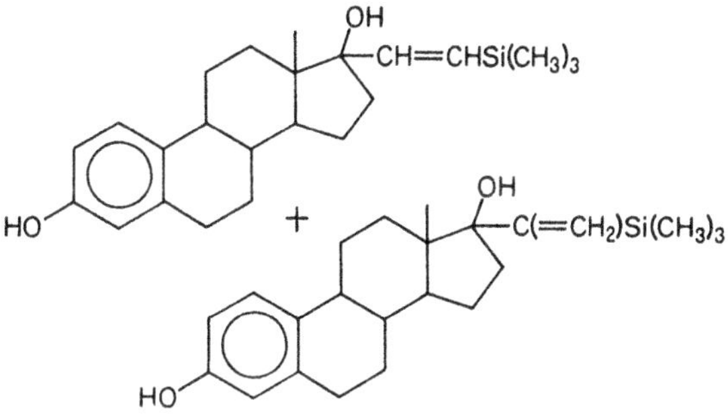

Figure 204

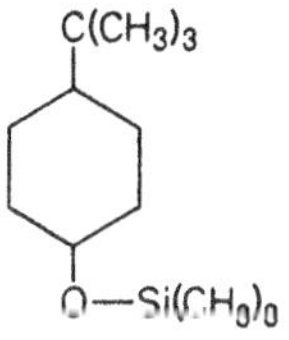

Figure 205

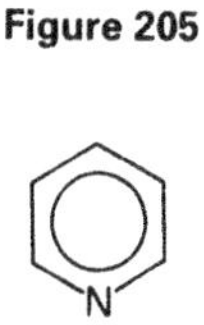

Figure 206

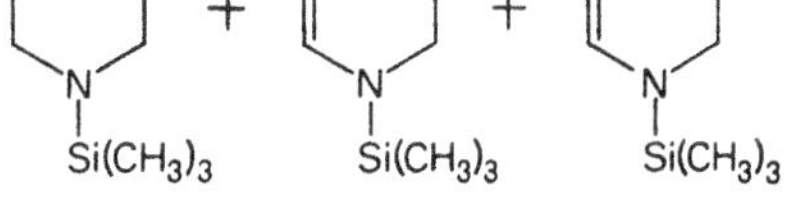

Figure 207

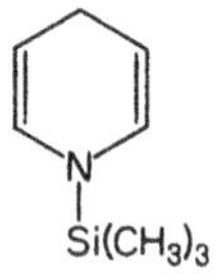

Figure 208

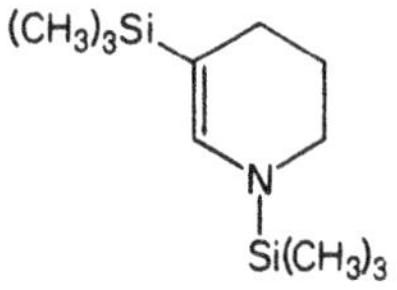

Figure 209

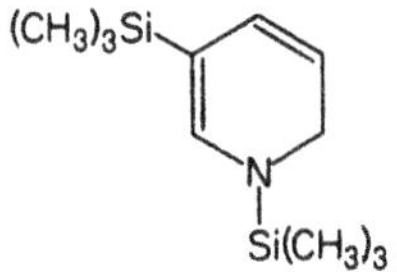

Figure 210

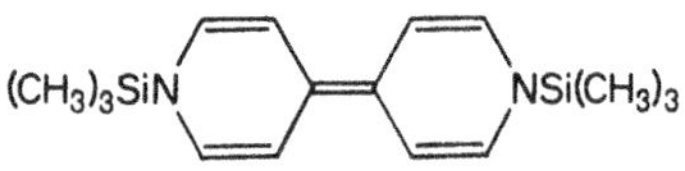

Figure 211

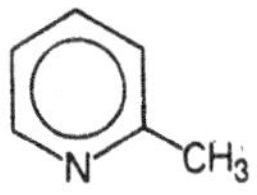

Figure 212

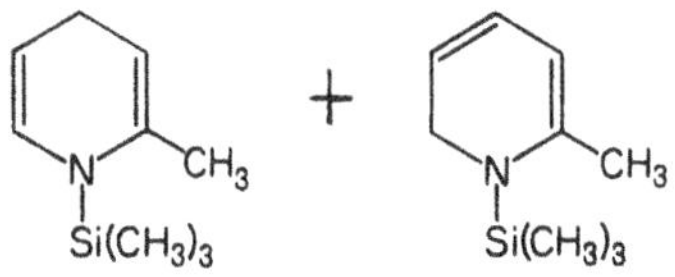

Figure 213

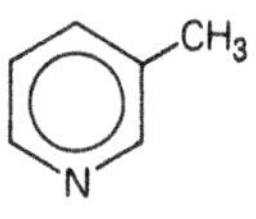

Figure 214

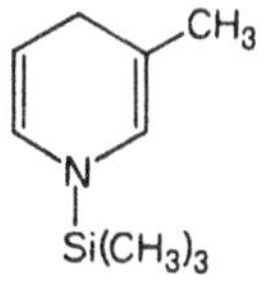

Figure 215

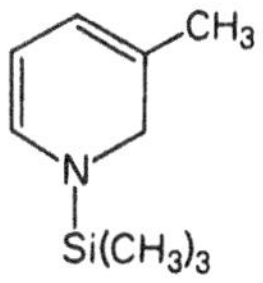

Figure 216

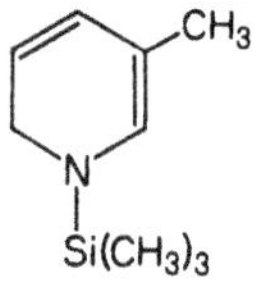

Figure 217

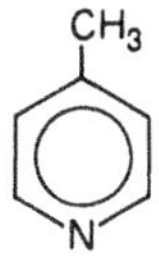

Figure 218

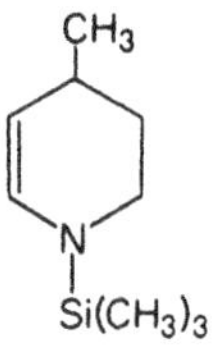

Figure 219

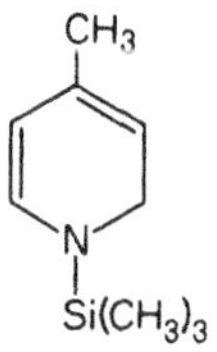

Figure 220

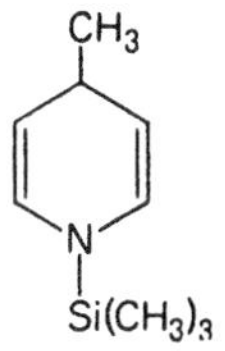

Figure 221

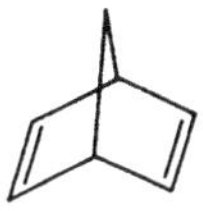

Figure 222

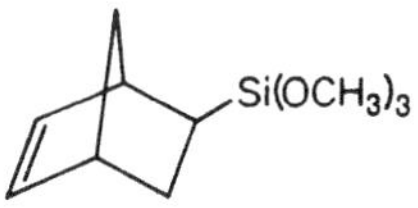

Figure 223

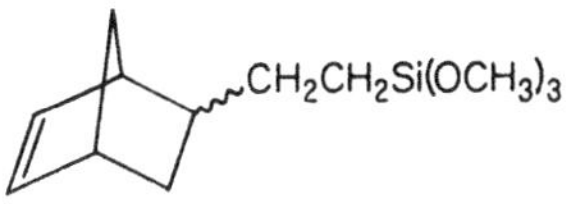

Figure 224

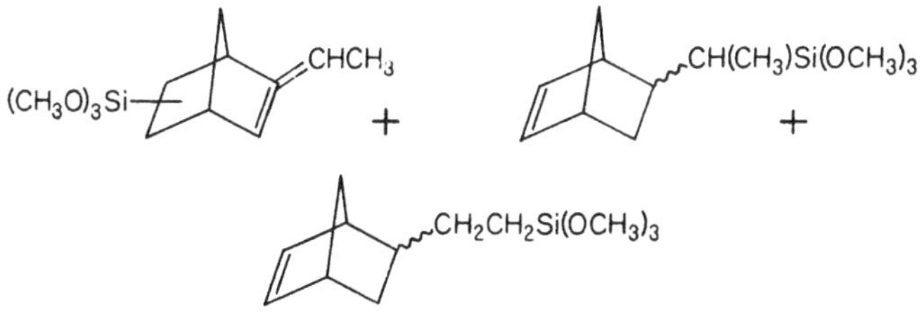

Figure 225

Figure 226

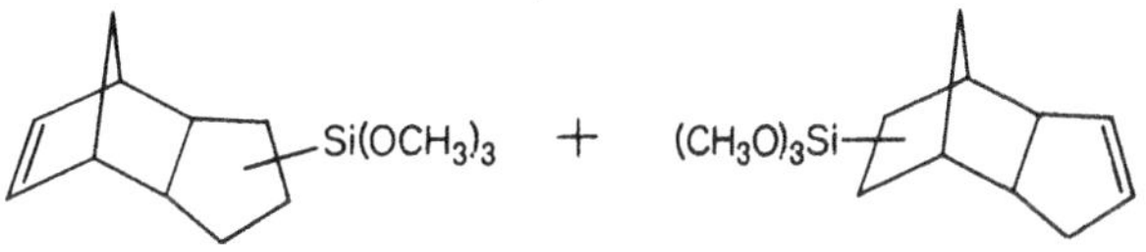

Figure 227

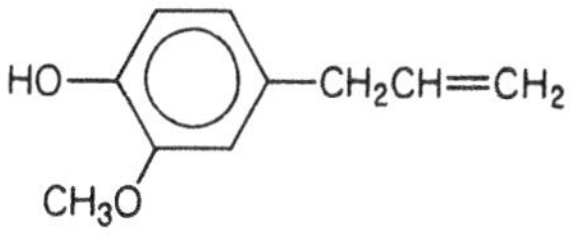

Figure 228

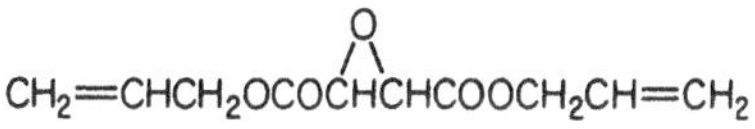

Figure 229

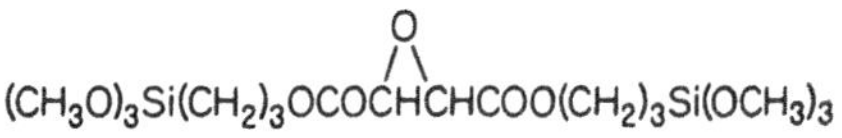

Figure 230

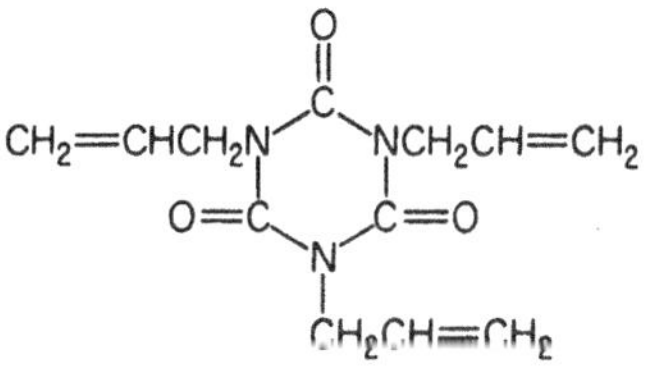

Figure 231

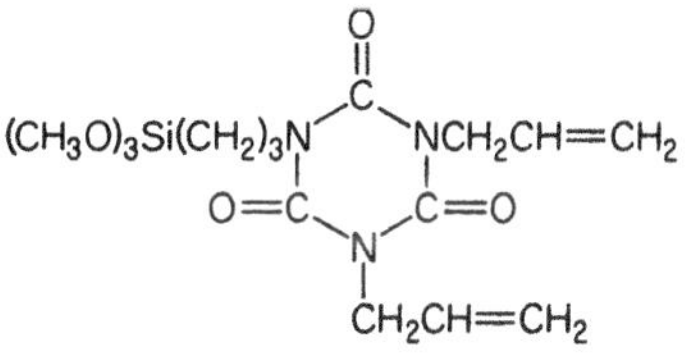

Figure 232

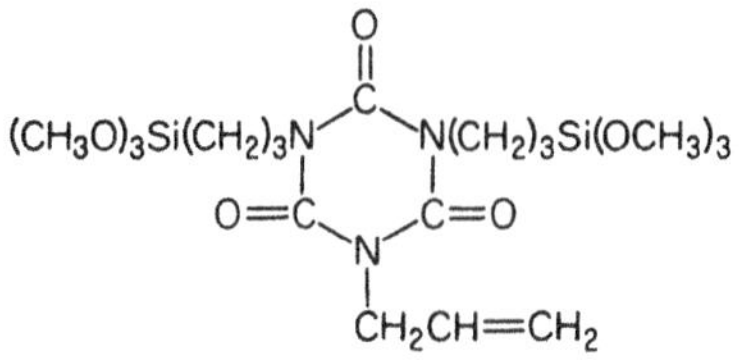

Figure 233

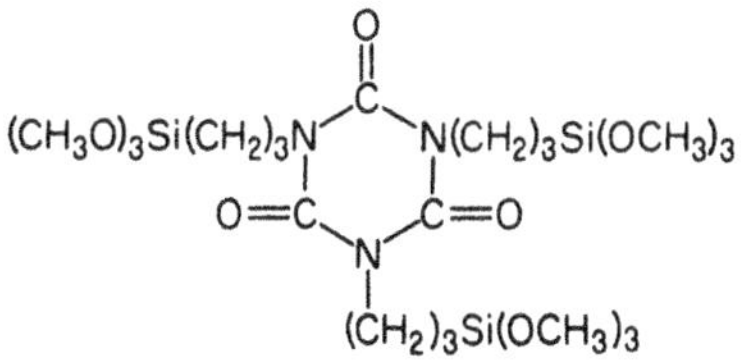

Figure 234

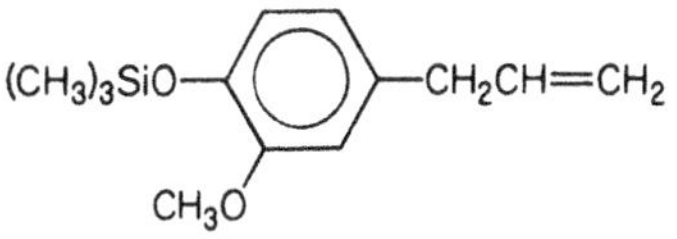

Figure 235

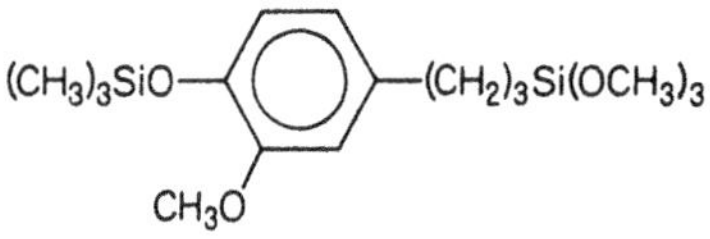

Figure 236

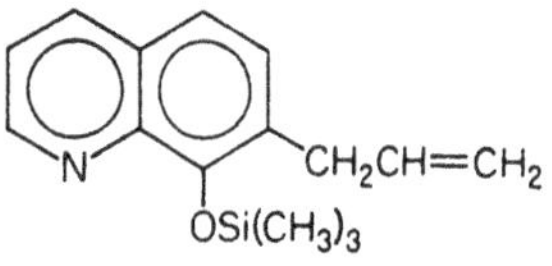

Figure 237

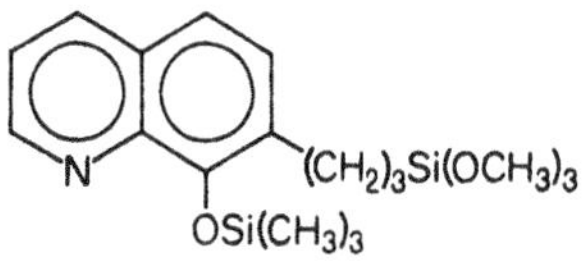

Figure 238

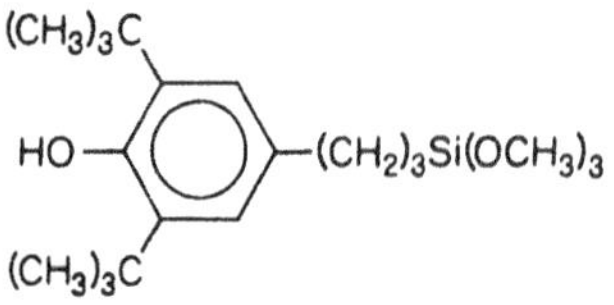

Figure 239

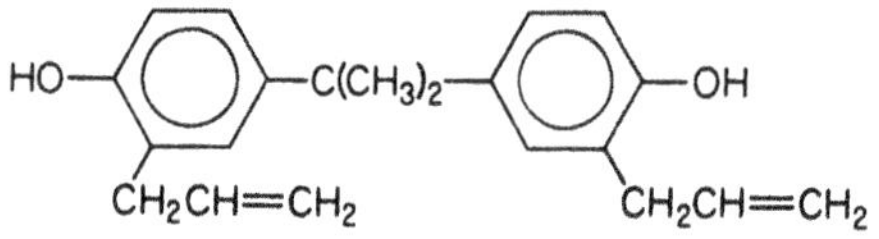

Figure 240

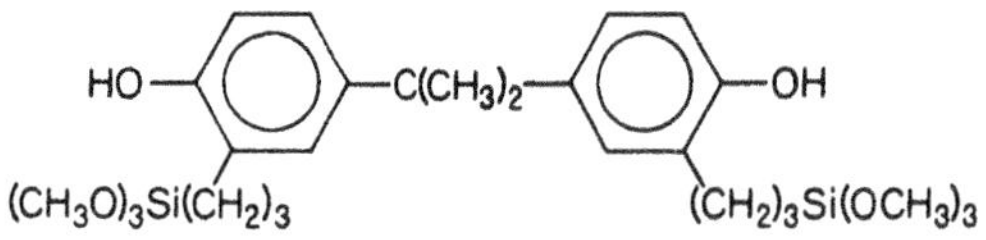

Figure 241

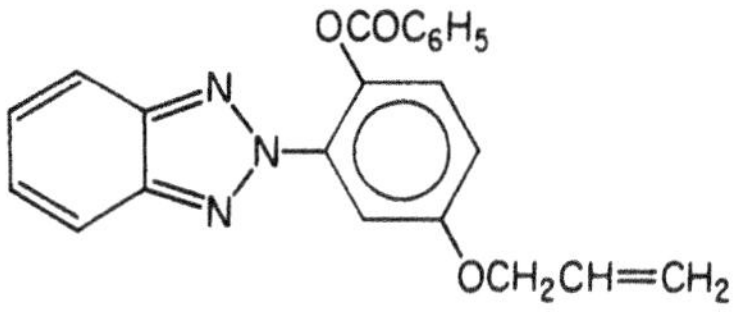

Figure 242

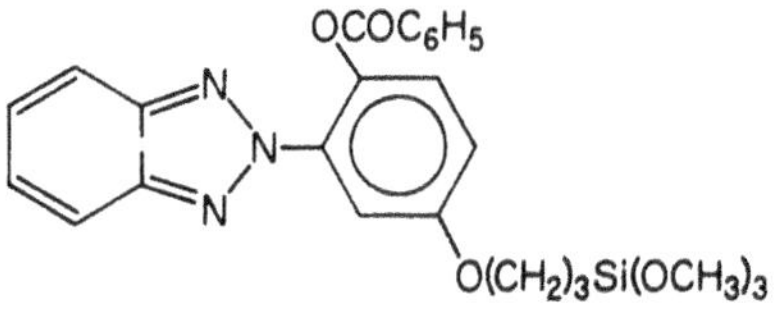

Figure 243

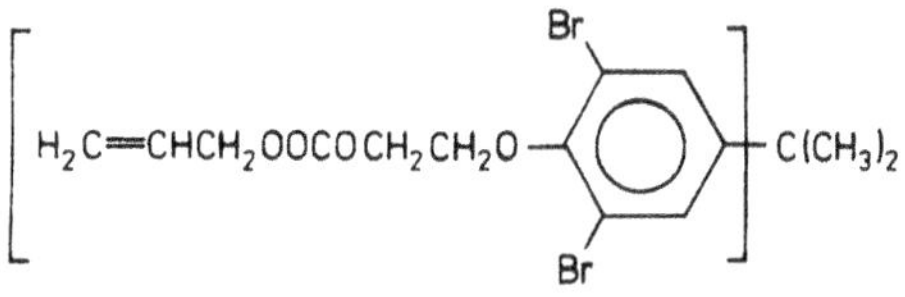

Figure 244

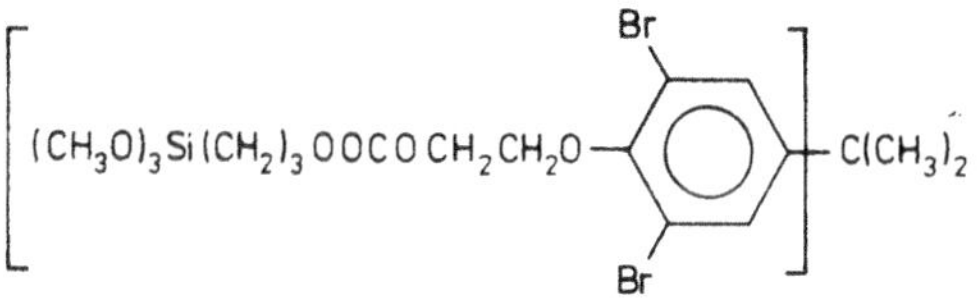

Figure 245

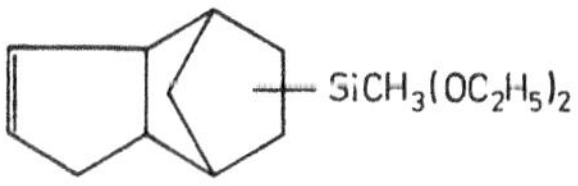

Figure 246

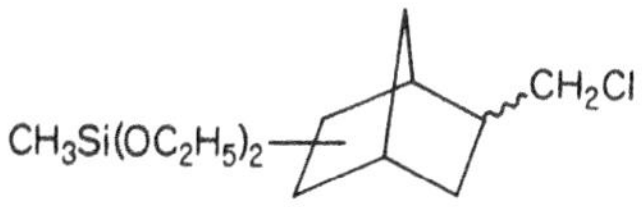

Figure 247

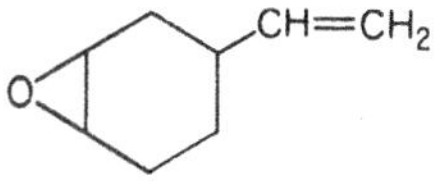

Figure 248

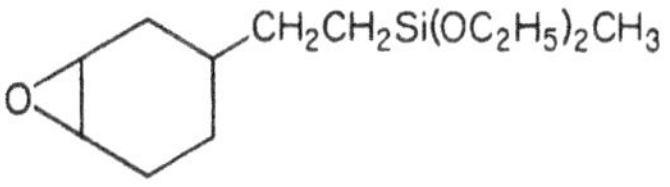

Figure 249

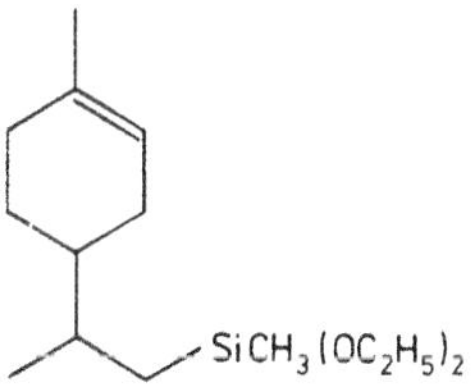

Figure 250

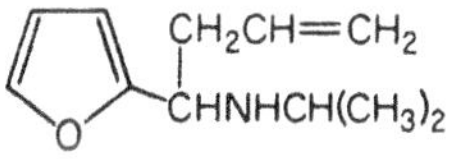

Figure 251

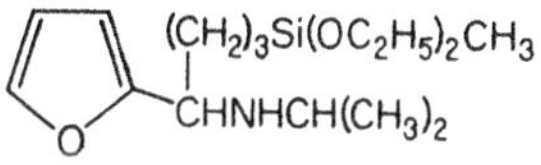

Figure 252

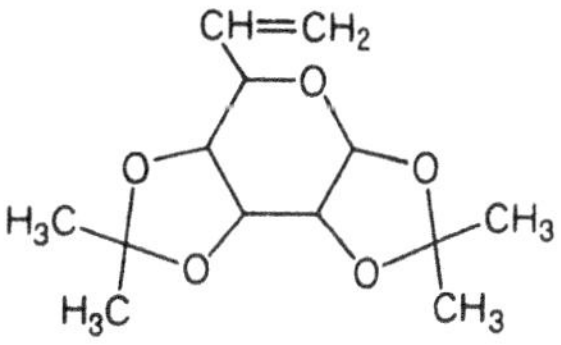

Figure 253

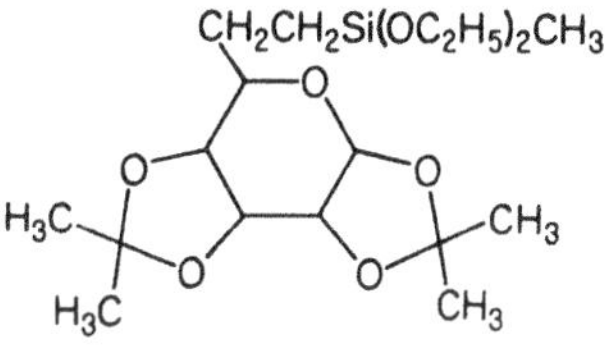

Figure 254

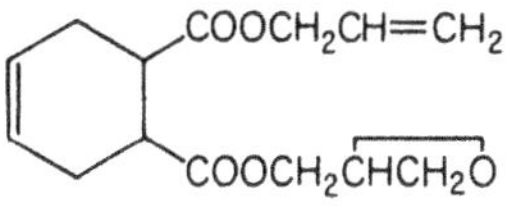

Figure 255

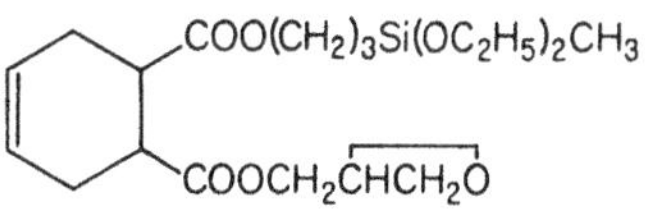

Figure 256

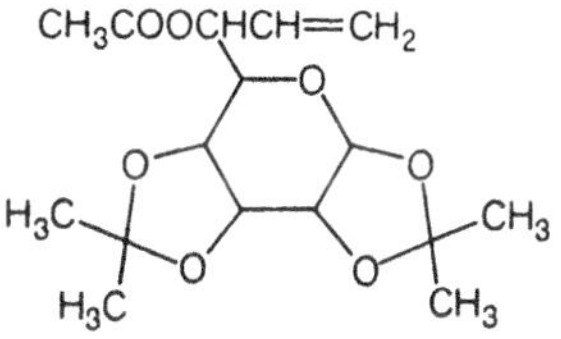

Figure 257

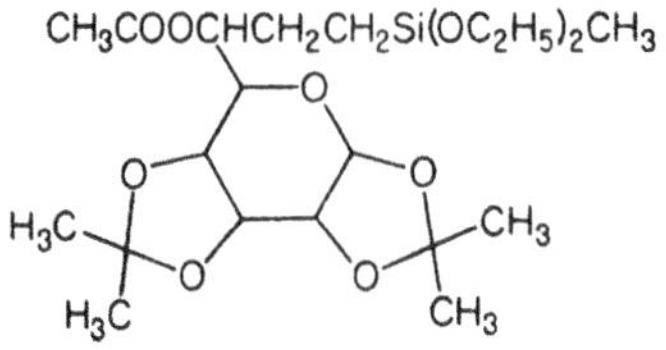

Figure 258

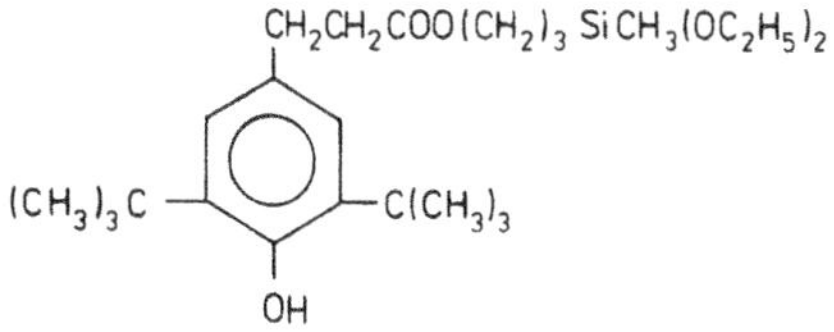

Figure 259

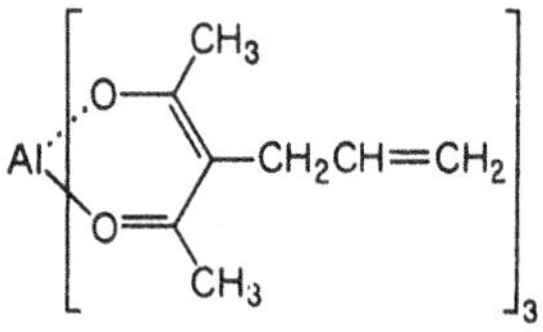

Figure 260

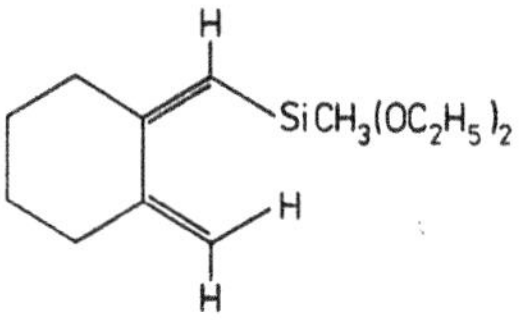

Figure 261

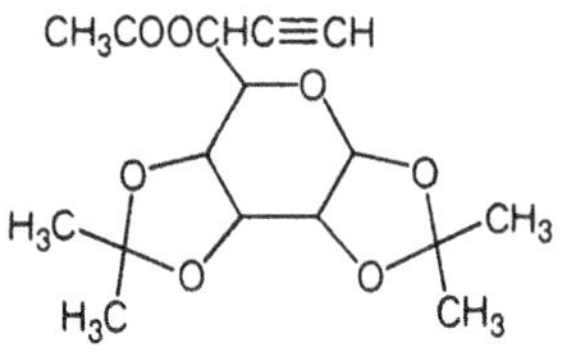

Figure 262

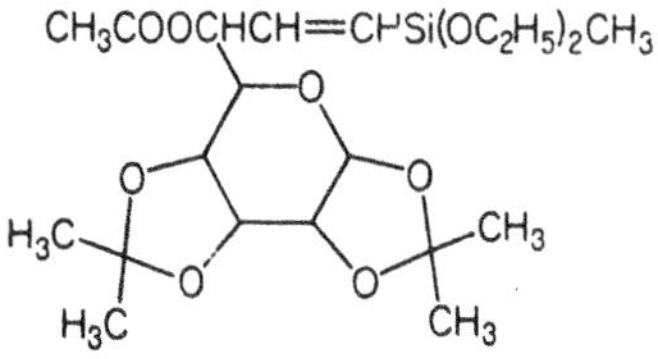

Figure 263

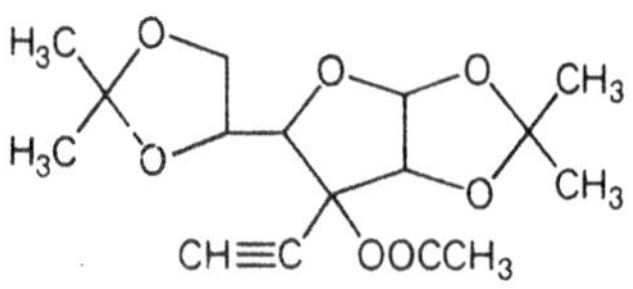

Figure 264

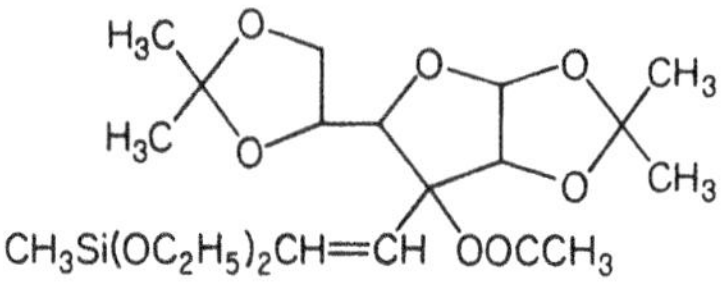

Figure 265

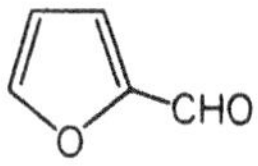

Figure 266

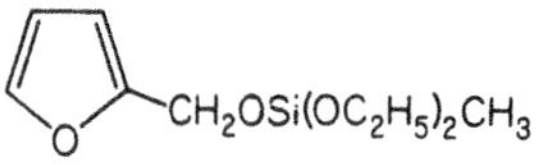

Figure 267

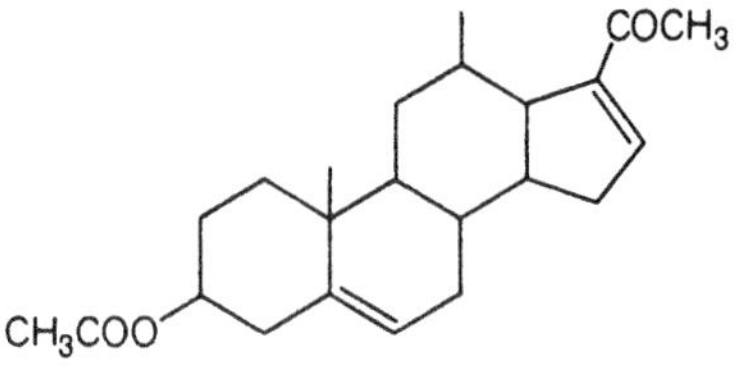

Figure 268

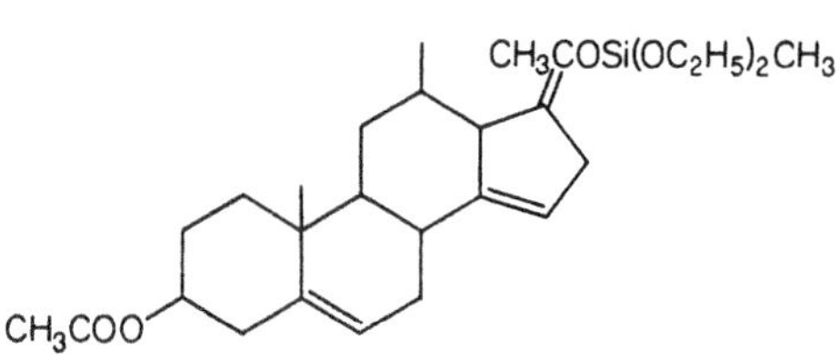

Figure 269

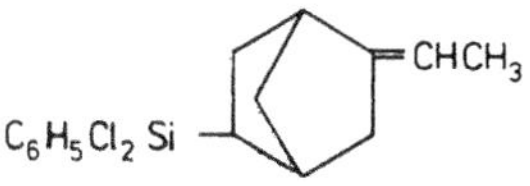

Figure 270

Figure 271

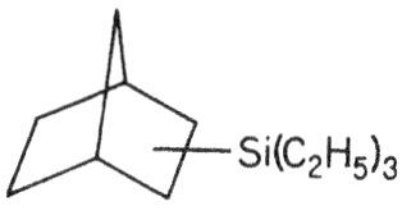

Figure 272

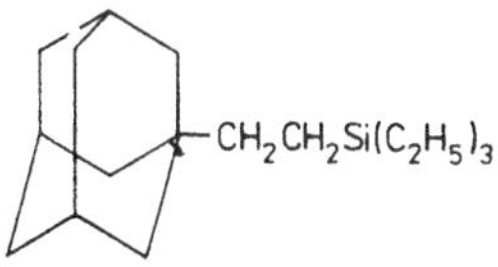

Figure 273

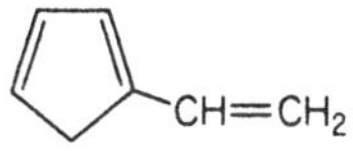

Figure 274

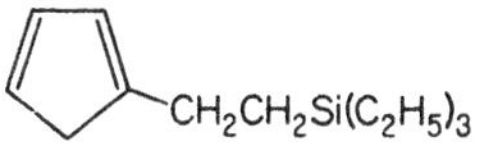

Figure 275

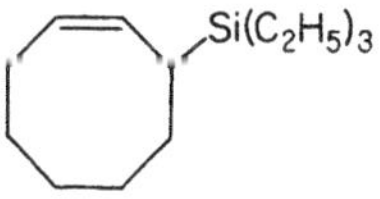

Figure 276

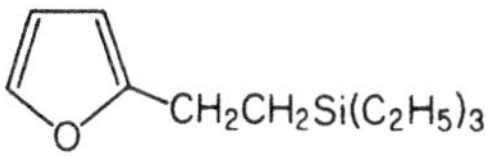

Figure 277

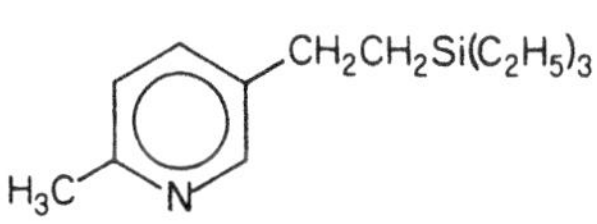

Figure 278

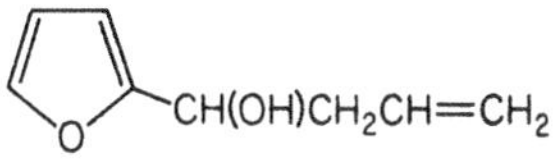

Figure 279

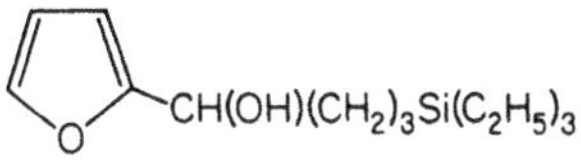

Figure 280

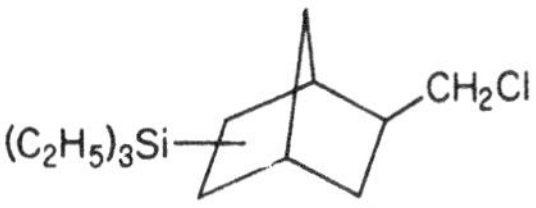

Figure 281

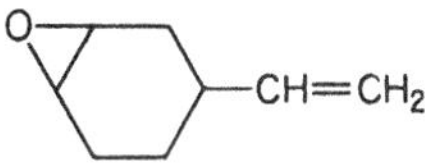

Figure 282

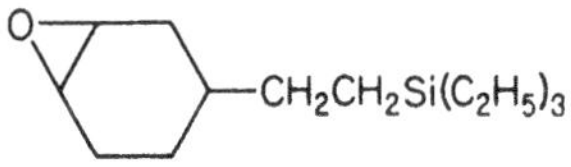

Figure 283

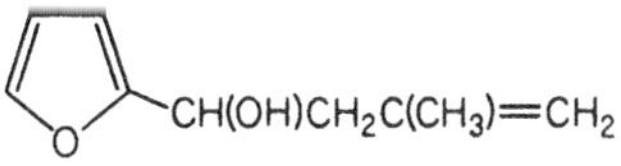

Figure 284

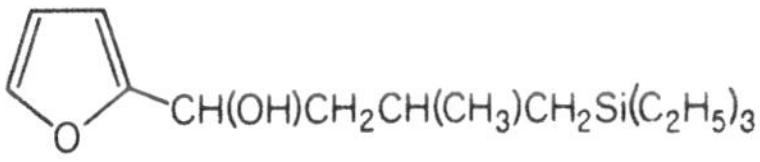

Figure 285

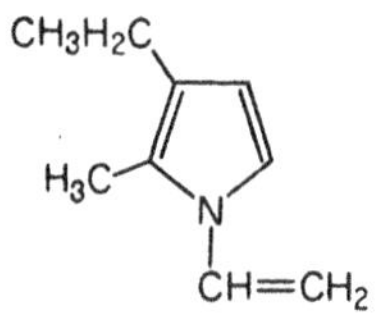

Figure 286

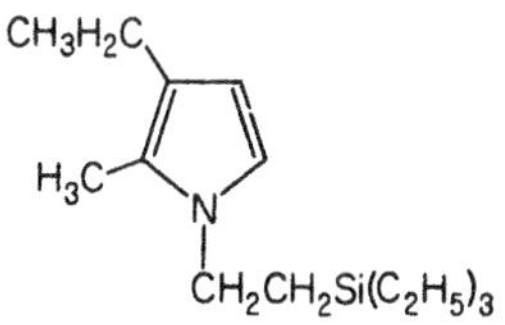

Figure 287

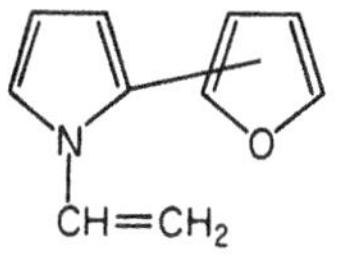

Figure 288

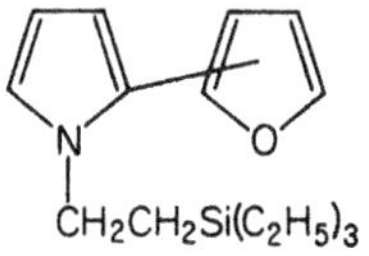

Figure 289

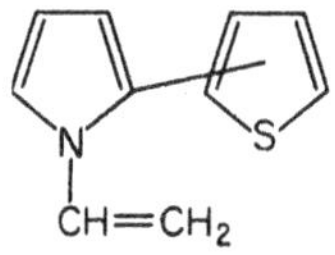

Figure 290

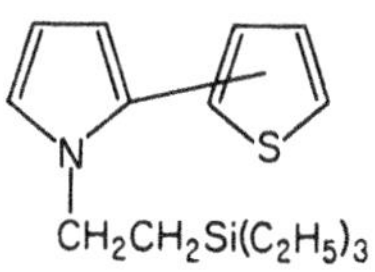

Figure 291

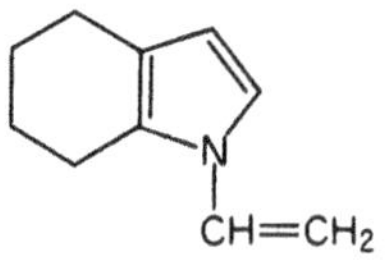

Figure 292

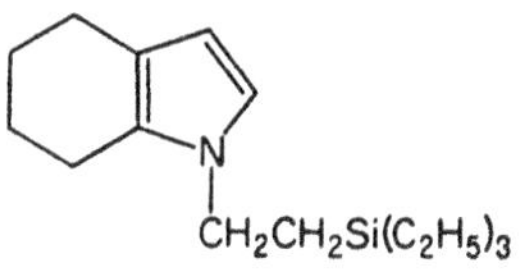

Figure 293

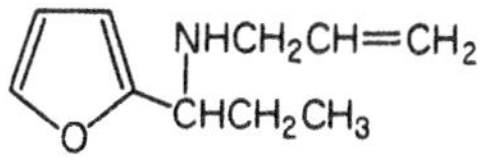

Figure 294

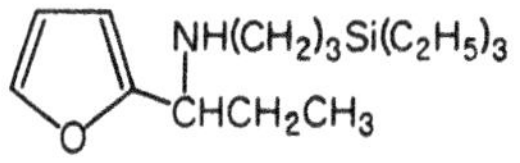

Figure 295

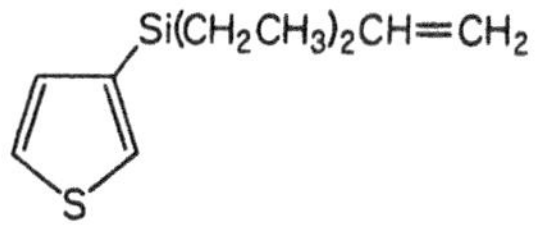

Figure 296

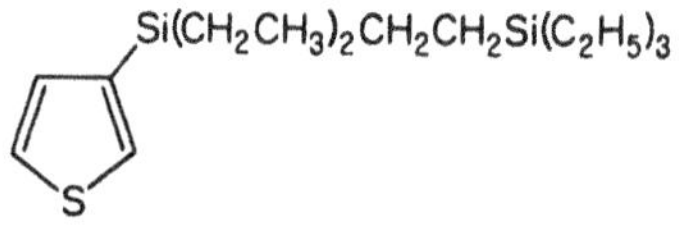

Figure 297

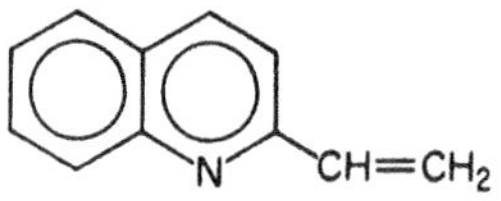

Figure 298

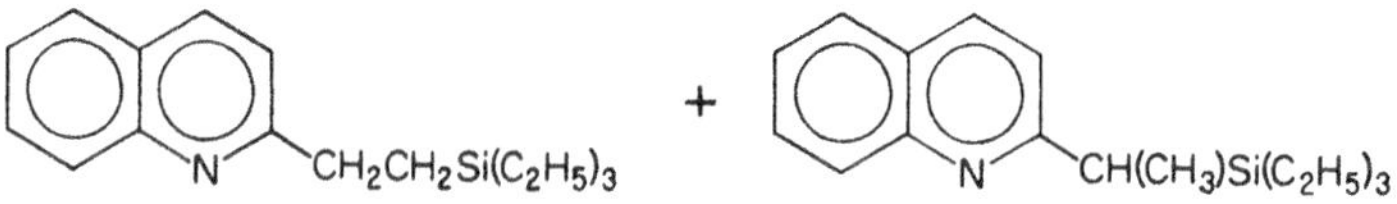

Figure 299

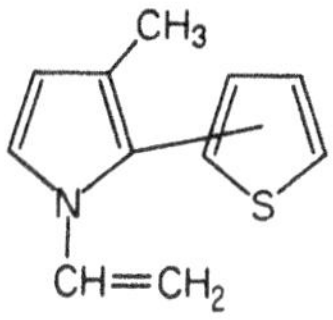

Figure 300

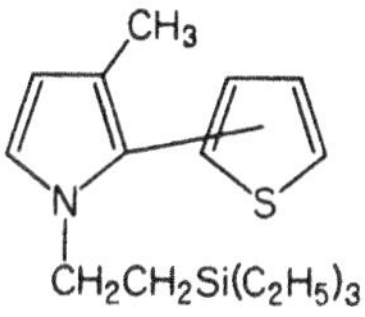

Figure 301

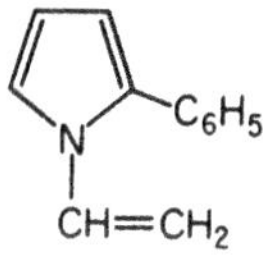

Figure 302

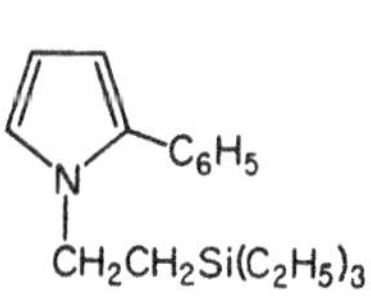

Figure 303

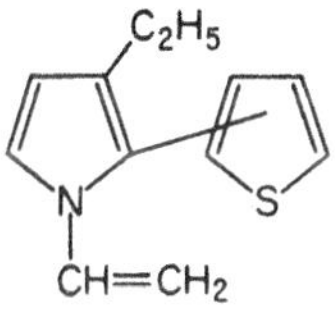

Figure 304

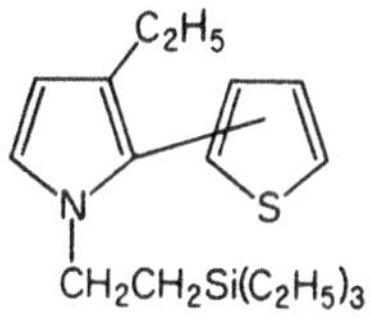

Figure 305

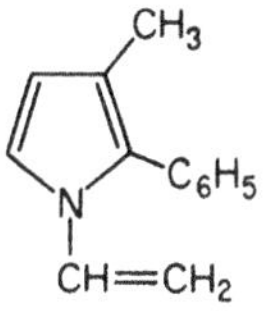

Figure 306

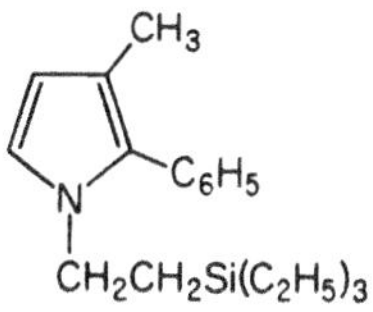

Figure 307

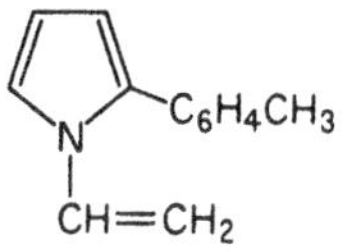

Figure 308

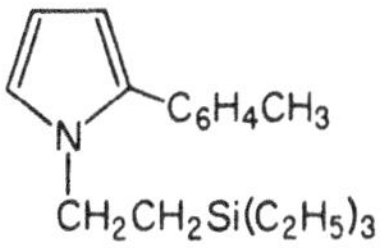

Figure 309

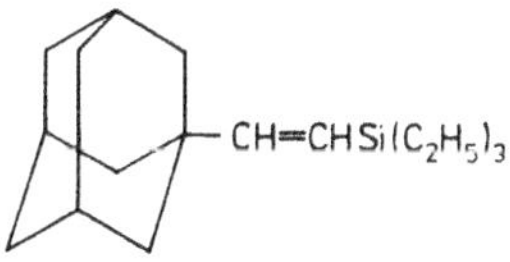

Figure 310

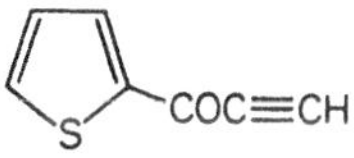

Figure 311

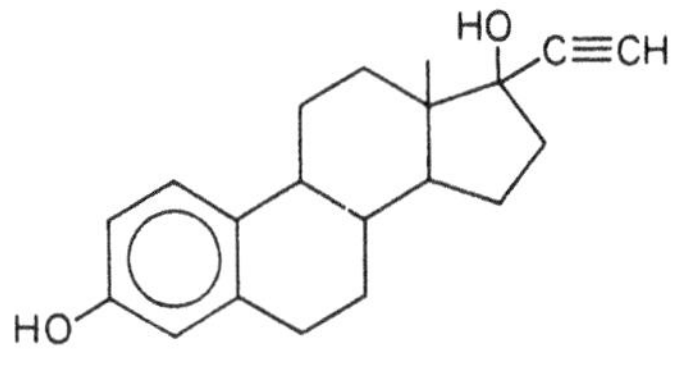

Figure 335

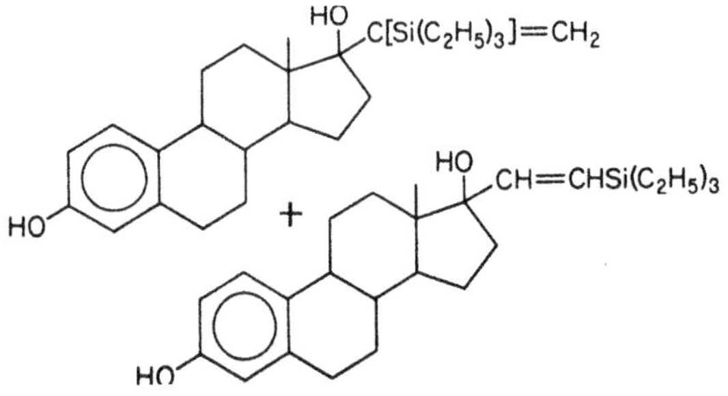

Figure 336

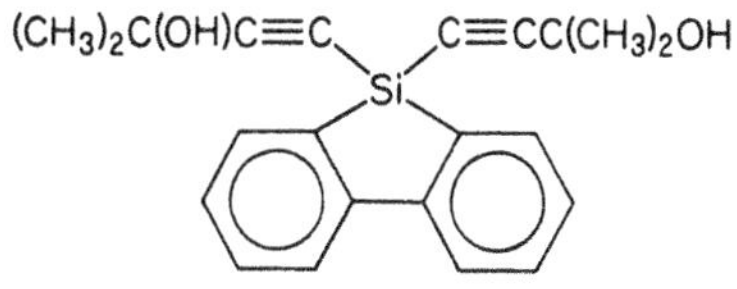

Figure 337

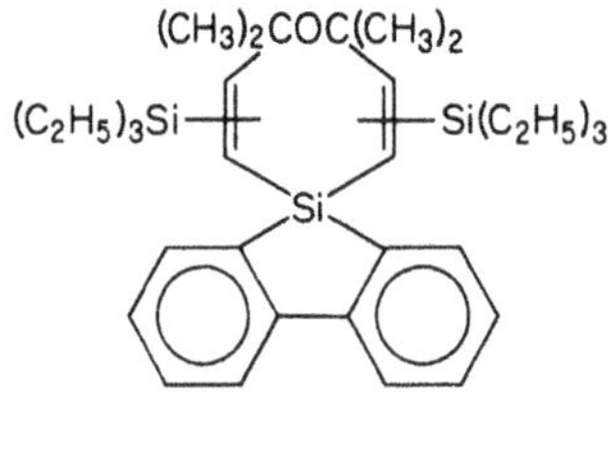

Figure 338

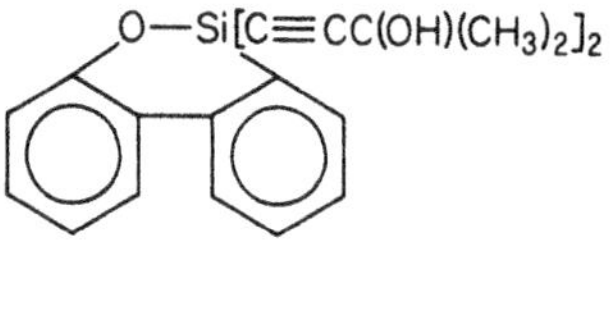

Figure 339

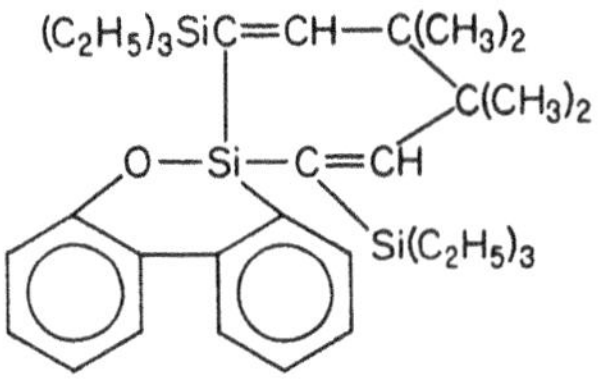

Figure 340

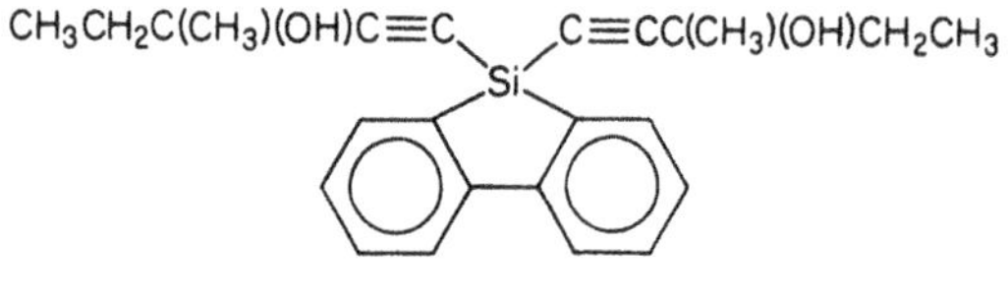

Figure 341

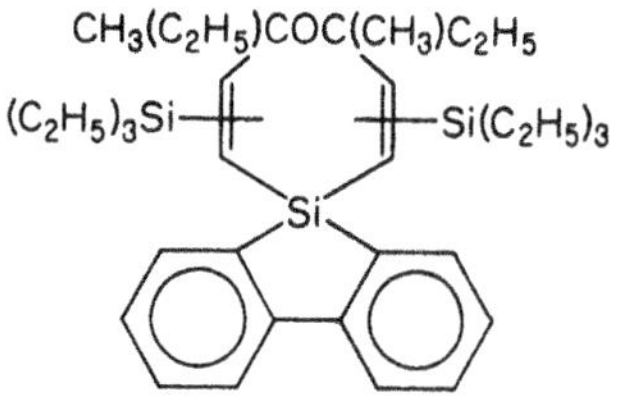

Figure 342

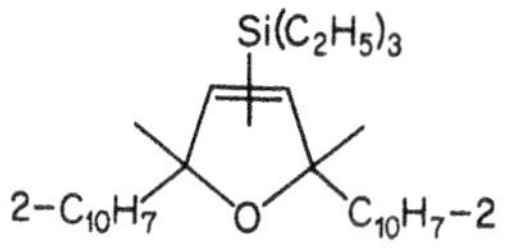

Figure 343

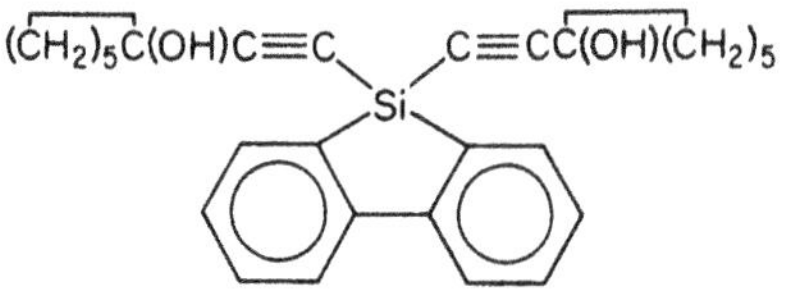

Figure 344

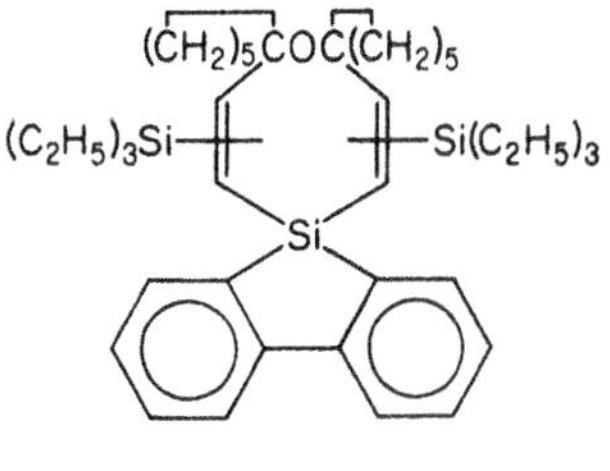

Figure 345

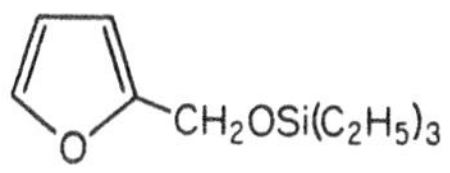

Figure 346

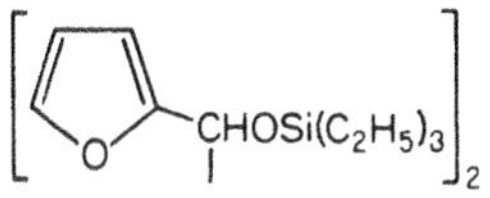

Figure 347

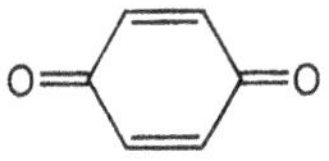

Figure 348

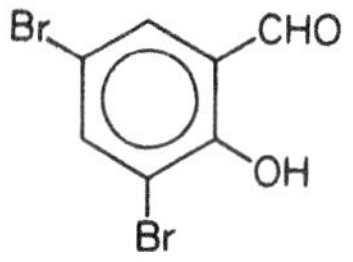

Figure 349

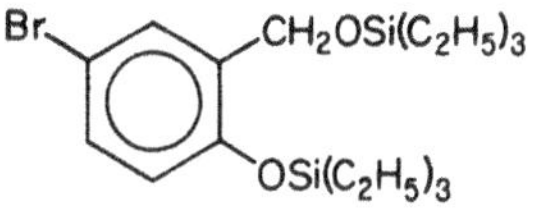

Figure 350

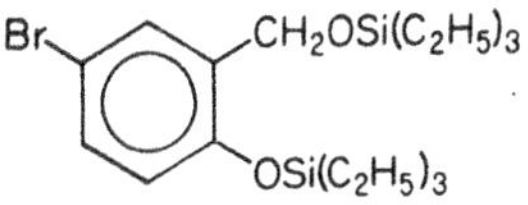

Figure 351

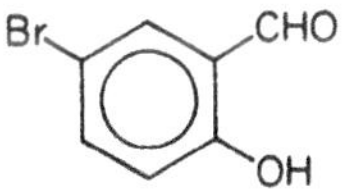

Figure 352

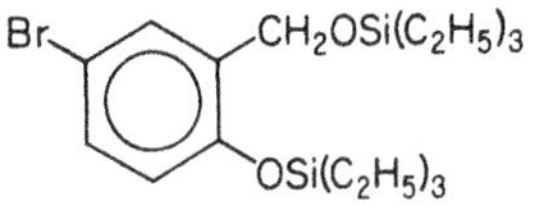

Figure 353

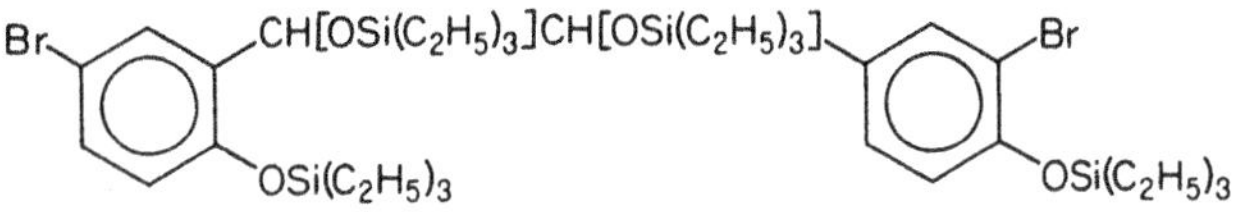

Figure 354

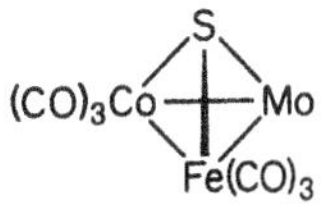

Figure 355

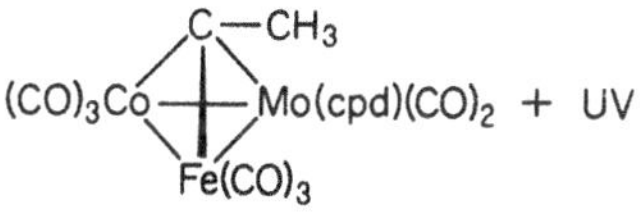

Figure 356

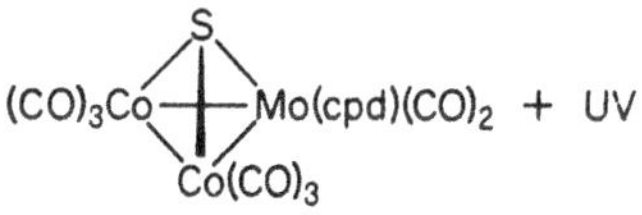

Figure 357

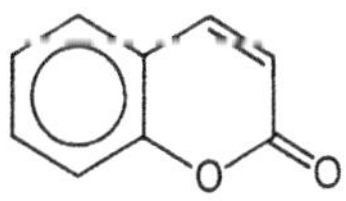

Figure 358

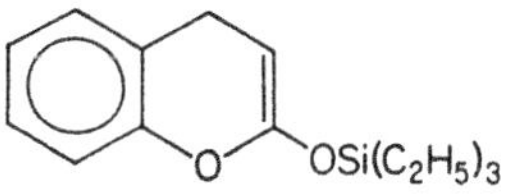

Figure 359

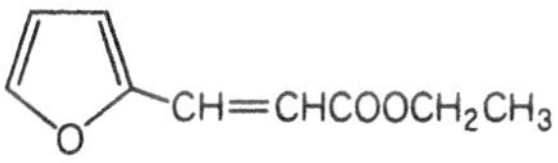

Figure 360

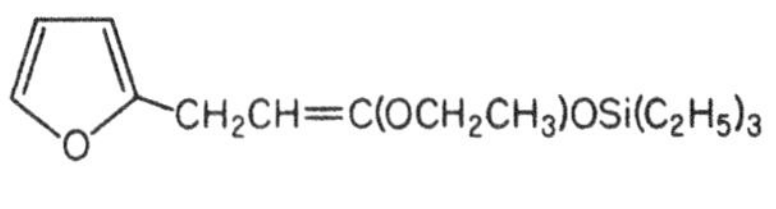

Figure 361

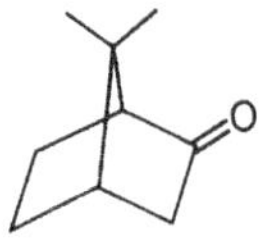

Figure 362

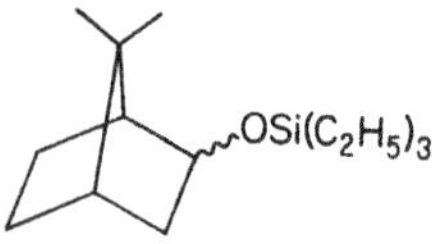

Figure 363

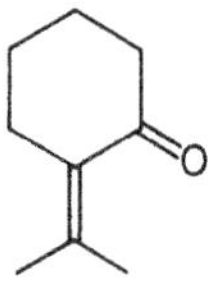

Figure 364

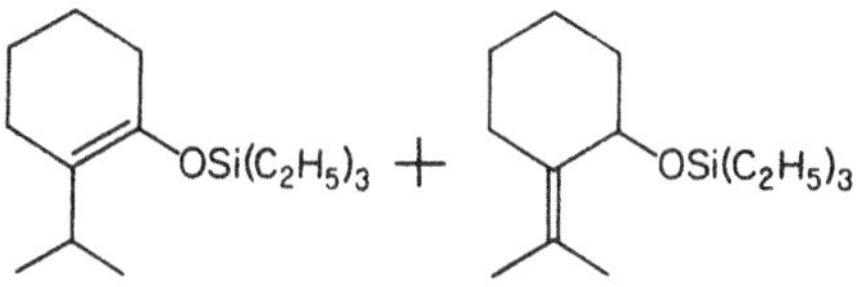

Figure 365

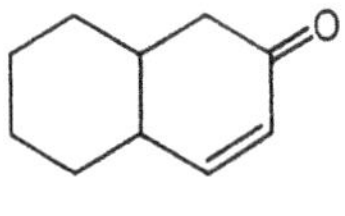

Figure 366

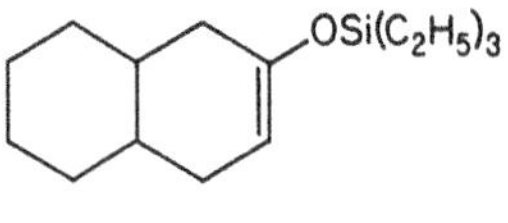

Figure 367

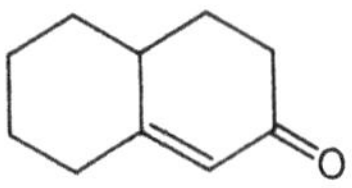

Figure 368

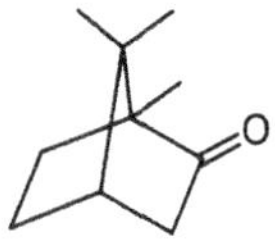

Figure 369

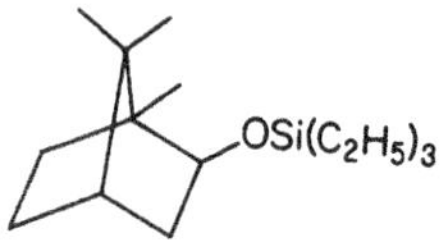

Figure 370

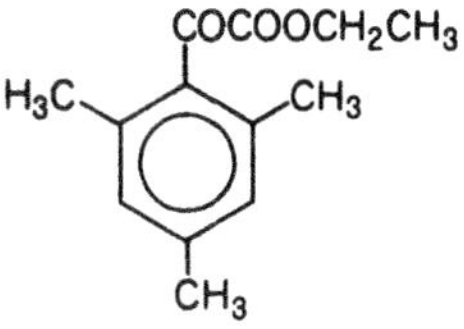

Figure 371

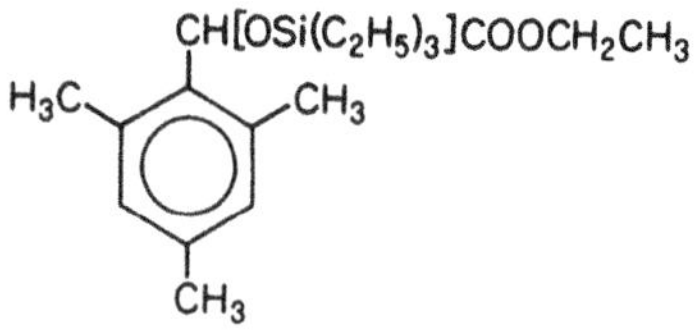

Figure 372

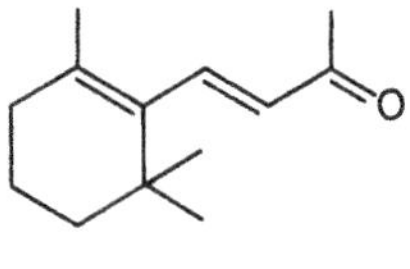

Figure 373

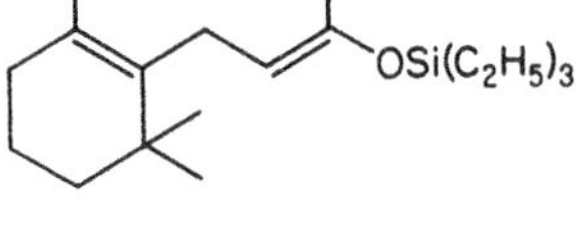

Figure 374

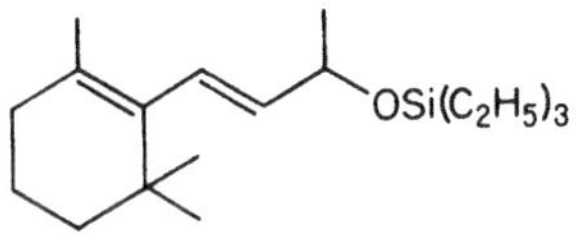

Figure 375

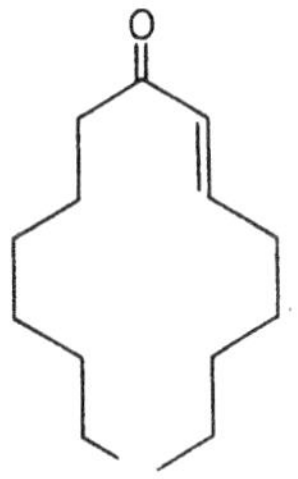

Figure 382

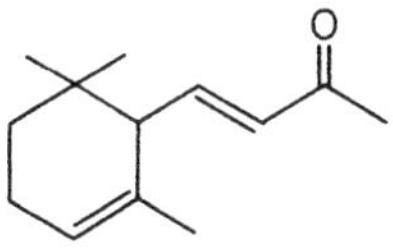

Figure 376

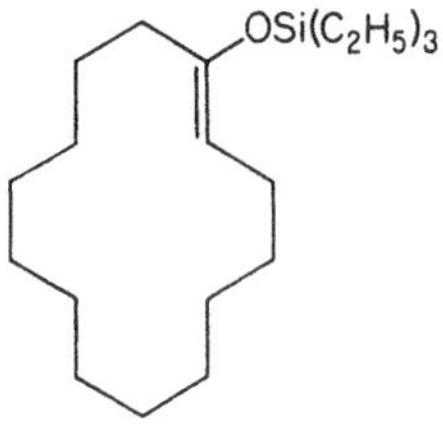

Figure 383

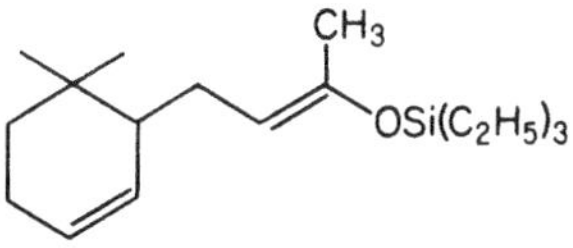

Figure 377

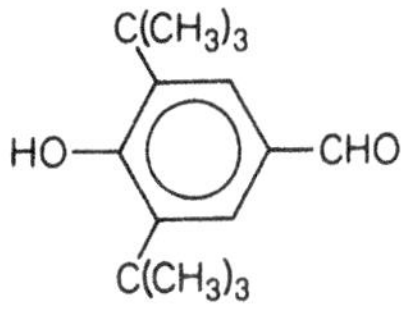

Figure 384

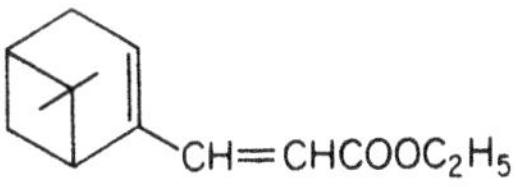

Figure 378

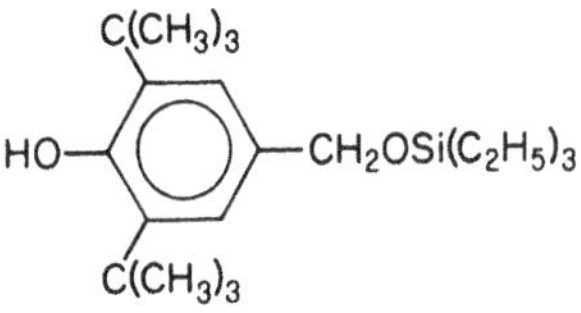

Figure 385

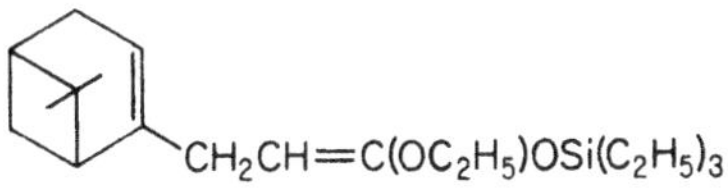

Figure 379

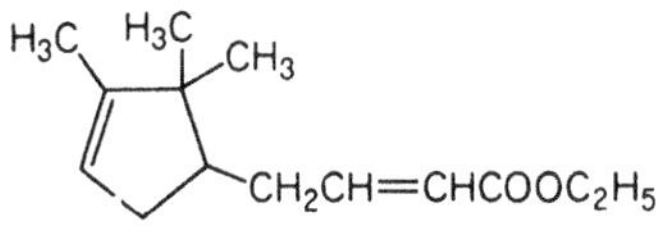

Figure 380

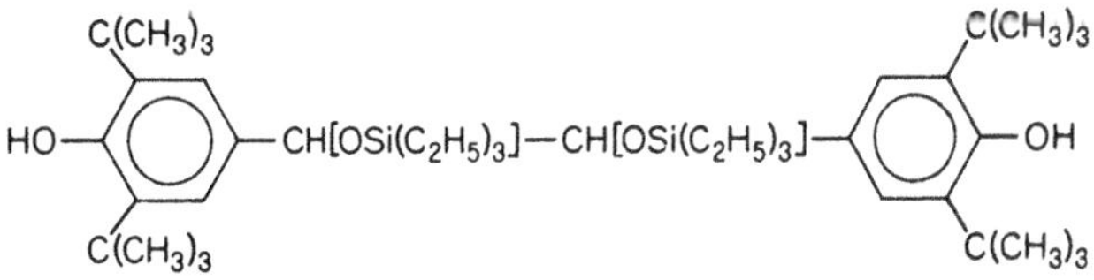

Figure 386

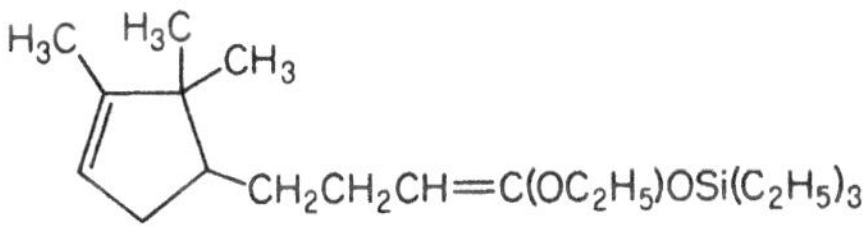

Figure 381

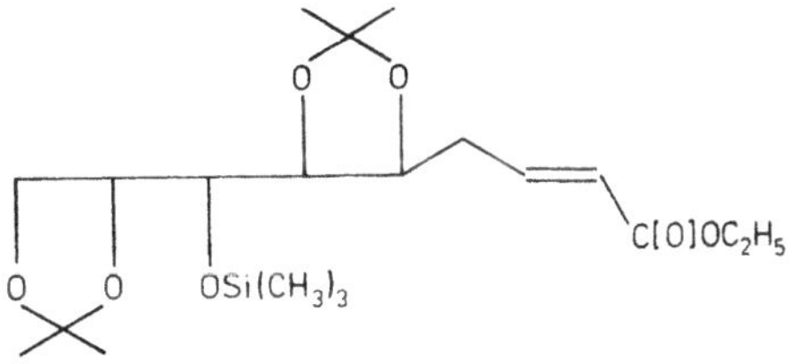

Figure 387

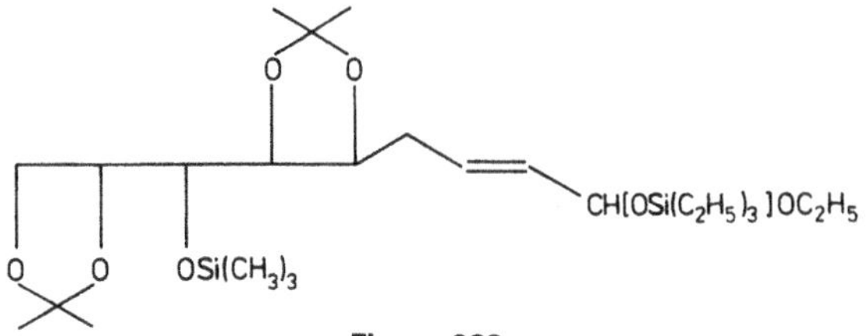

Figure 388

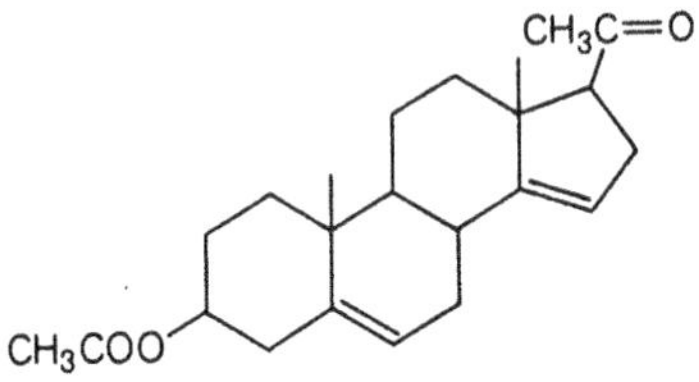

Figure 389

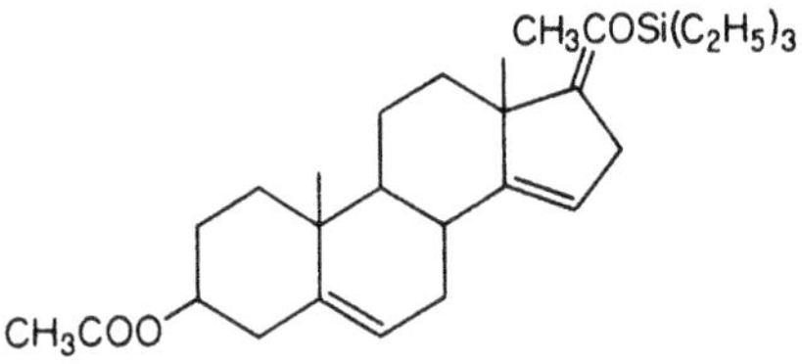

Figure 390

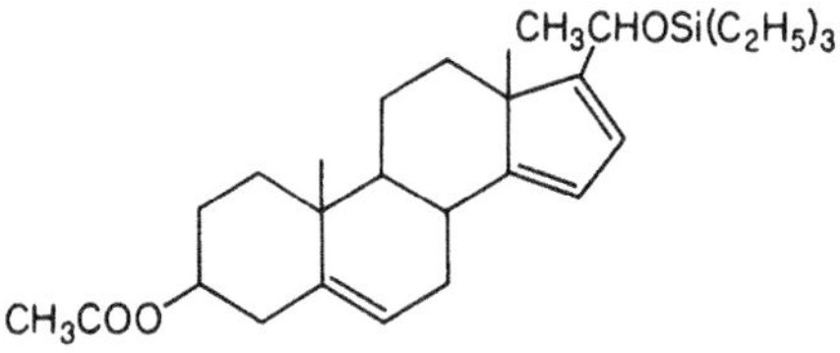

Figure 391

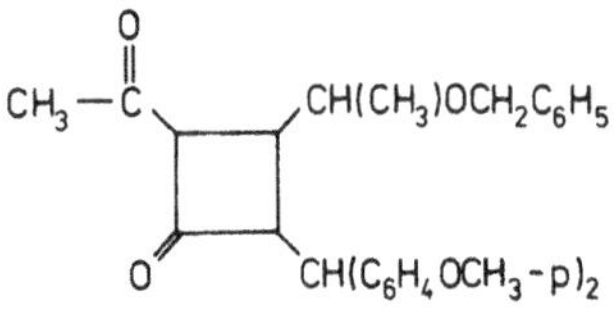

Figure 392

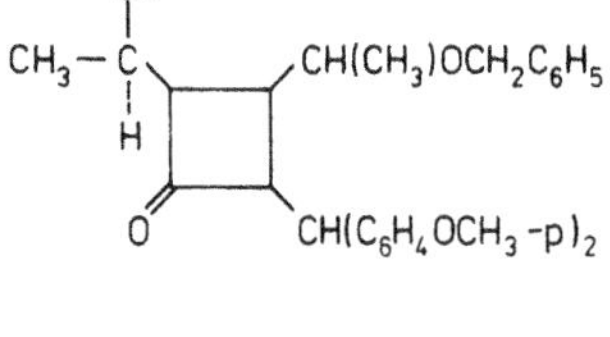

Figure 393

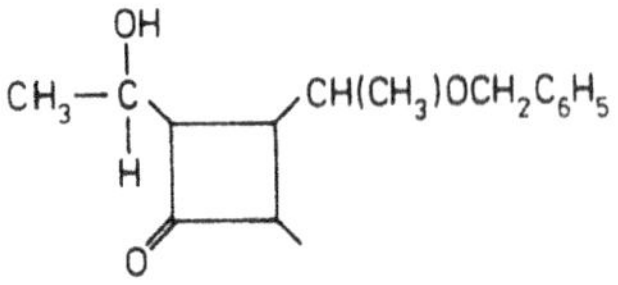

Figure 394

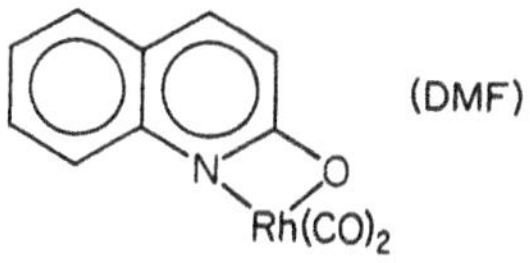

Figure 395

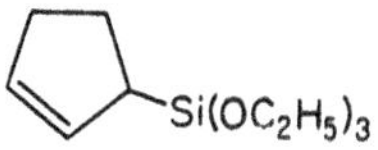

Figure 396

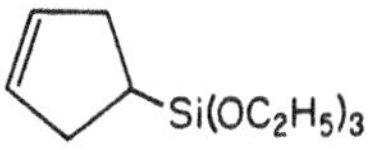

Figure 397

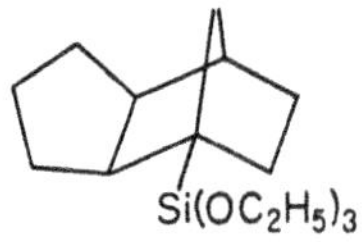

Figure 398

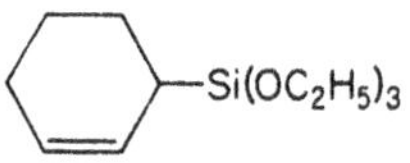

Figure 399

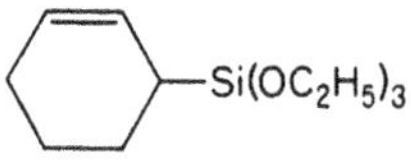

Figure 400

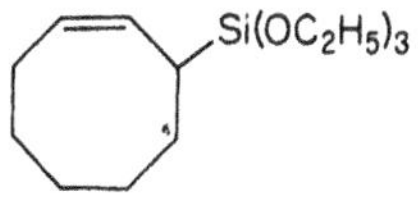

Figure 401

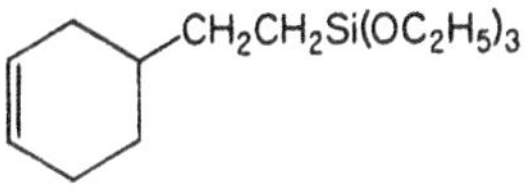

Figure 402

Figure 403

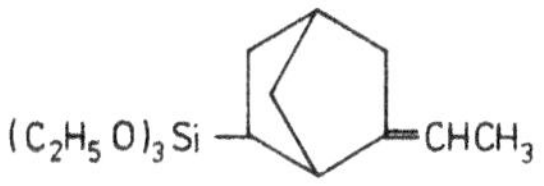

Figure 404

Figure 405

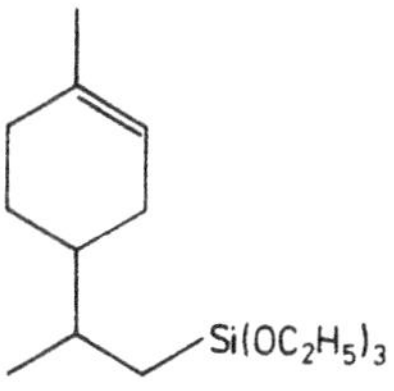

Figure 406

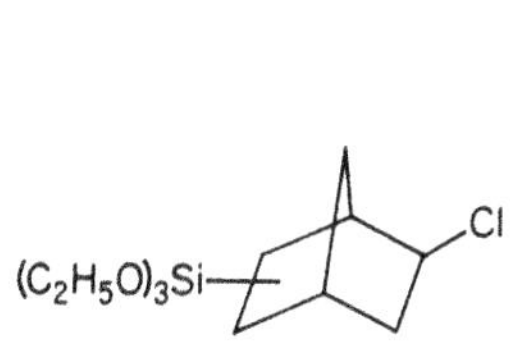

Figure 407

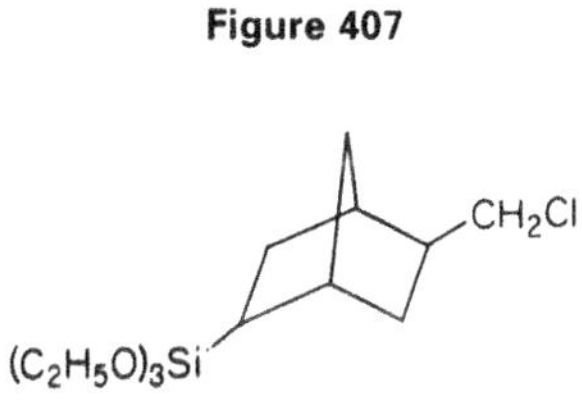

Figure 408

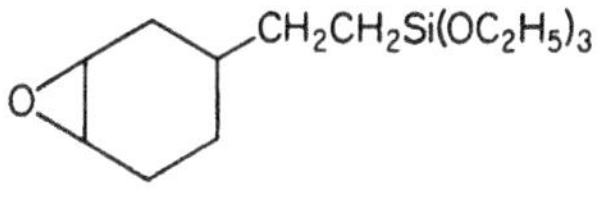

Figure 409

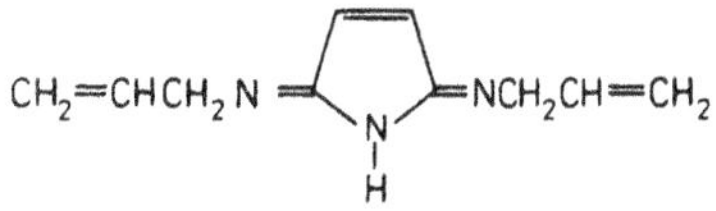

Figure 410

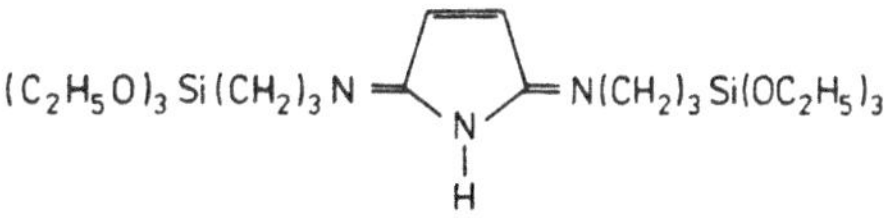

Figure 411

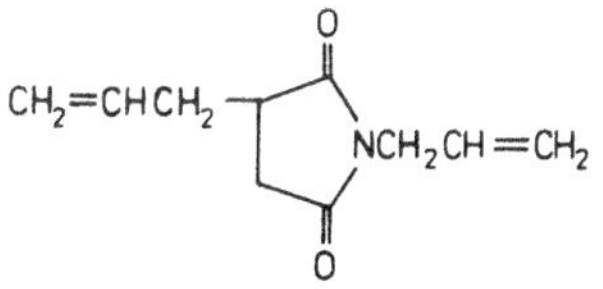

Figure 412

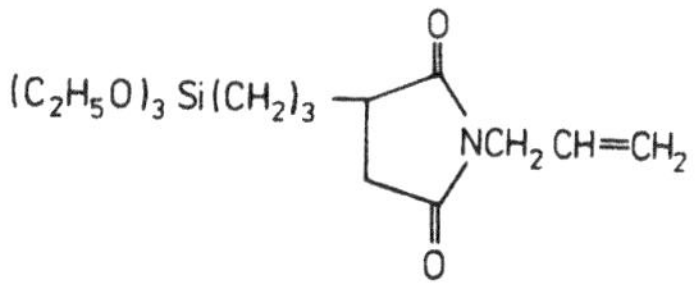

Figure 413

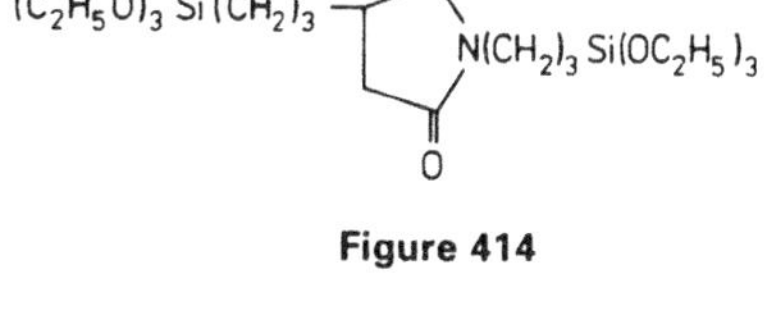

Figure 414

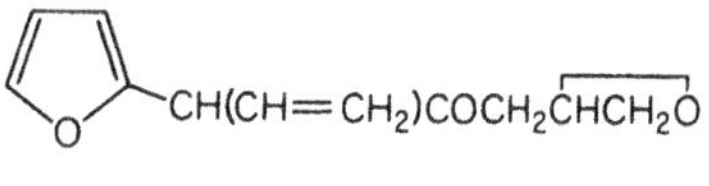

Figure 415

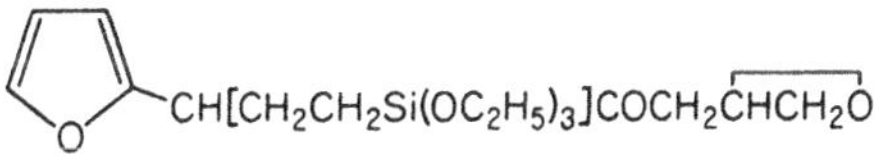

Figure 416

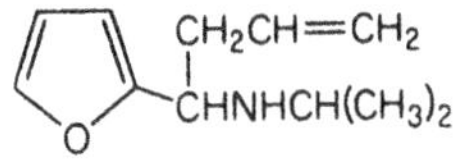

Figure 417

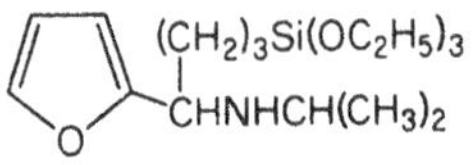

Figure 418

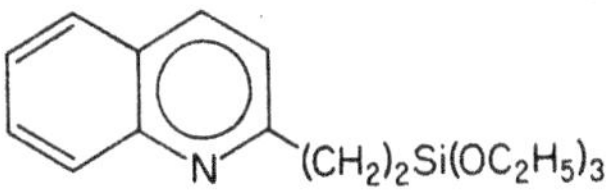

Figure 419

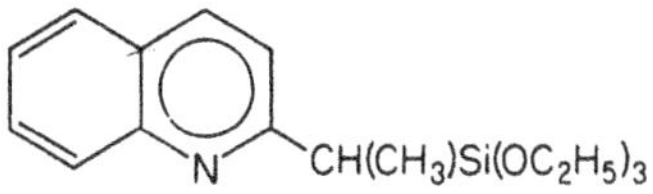

Figure 420

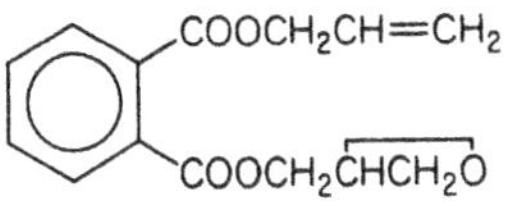

Figure 421

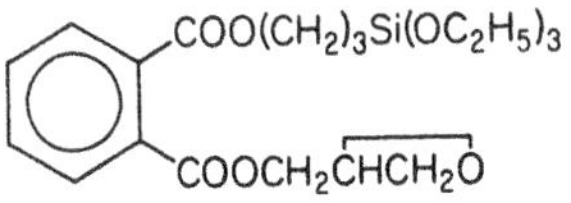

Figure 422

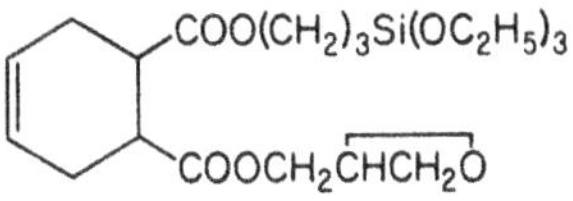

Figure 423

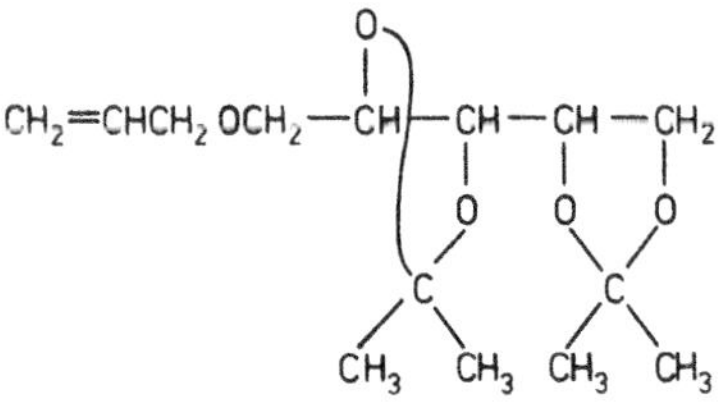

Figure 424

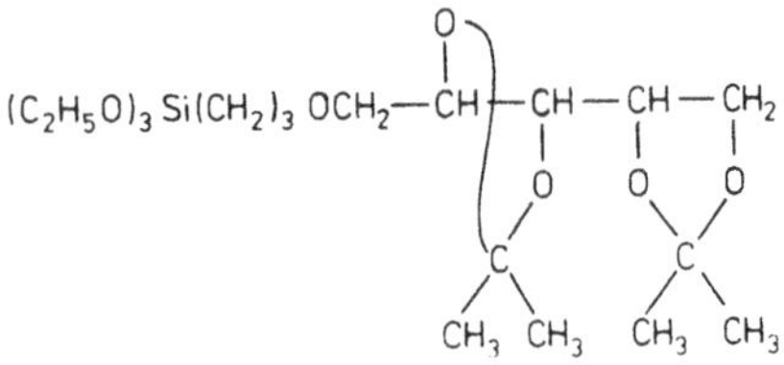

Figure 425

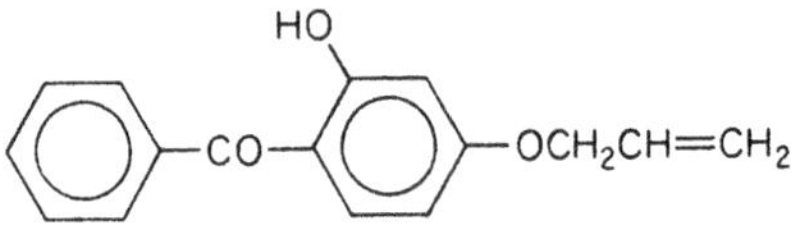

Figure 426

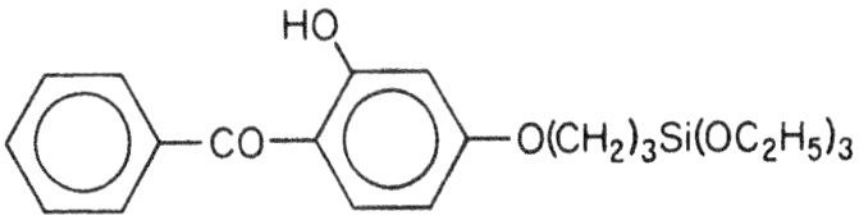

Figure 427

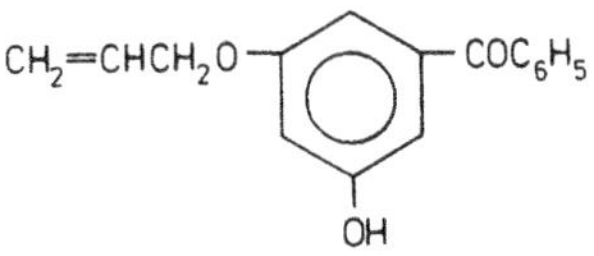

Figure 428

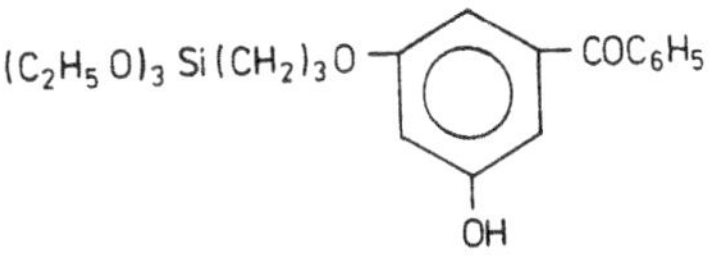

Figure 429

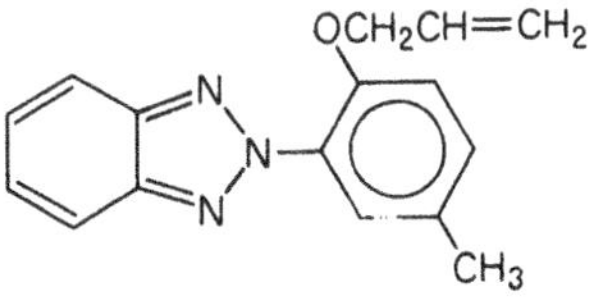

Figure 430

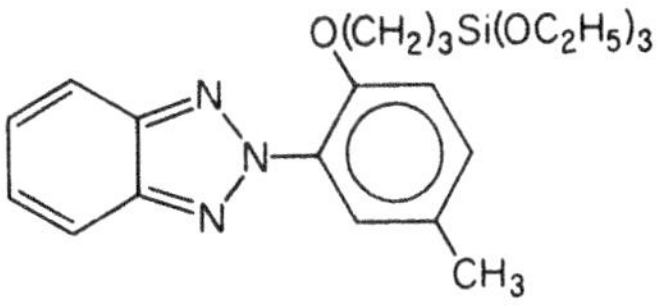

Figure 431

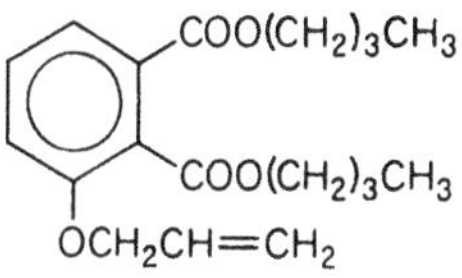

Figure 432

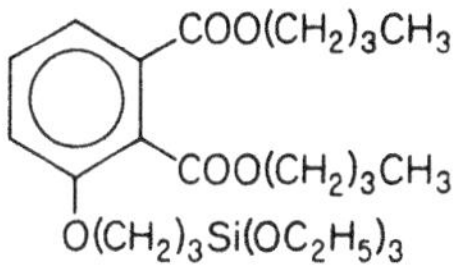

Figure 433

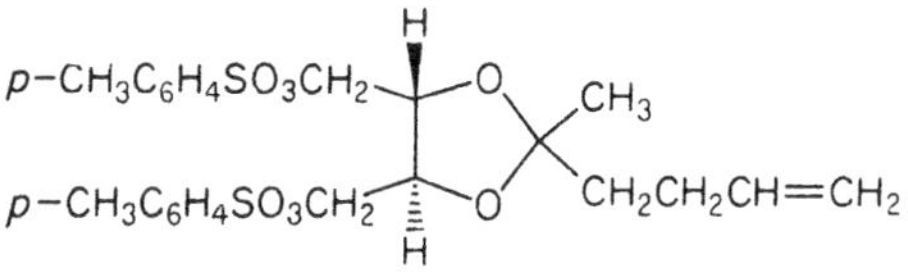

Figure 434

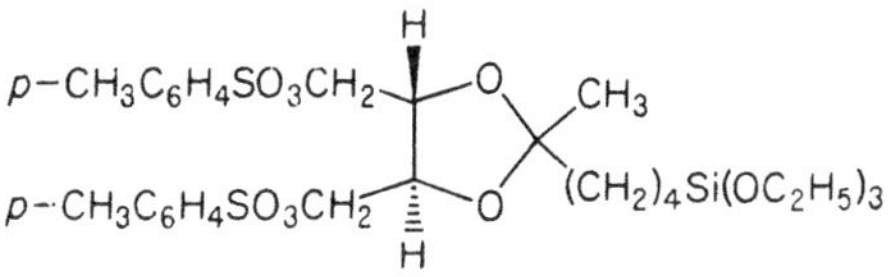

Figure 435

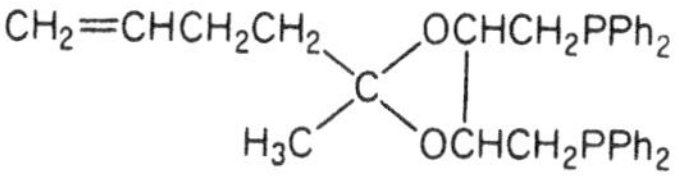

Figure 436

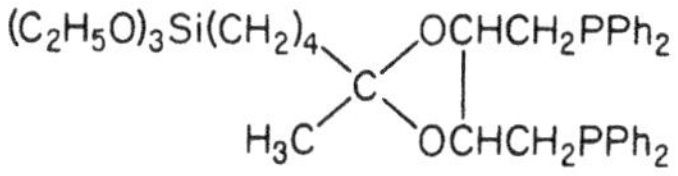

Figure 437

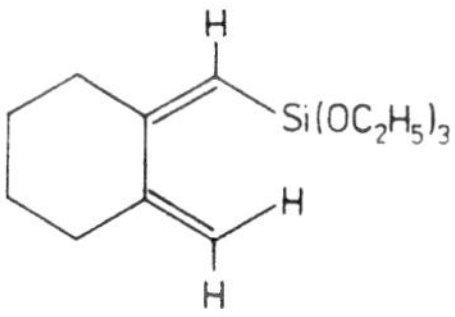

Figure 438

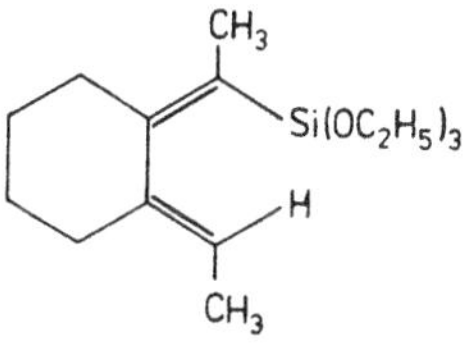

Figure 439

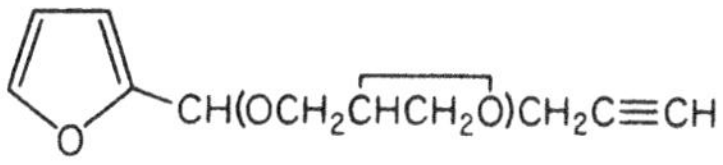

Figure 440

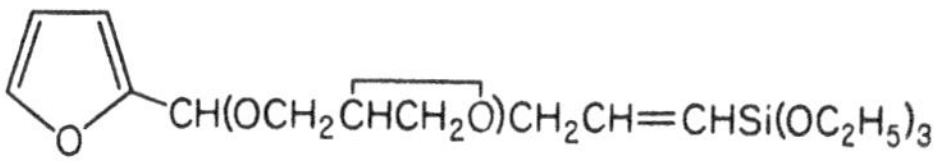

Figure 441

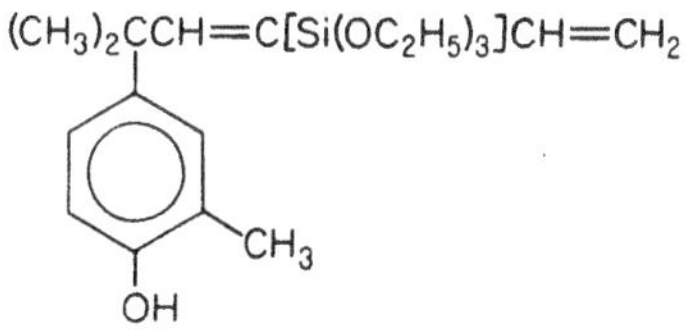

Figure 442

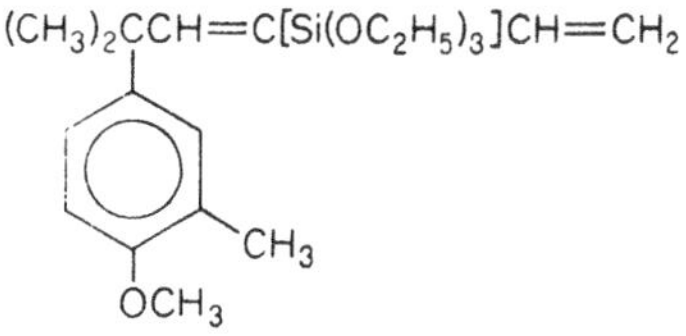

Figure 443

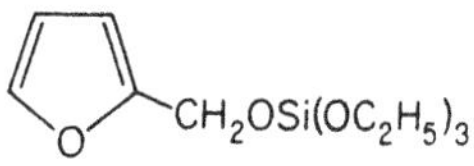

Figure 444

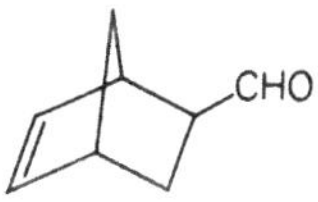

Figure 445

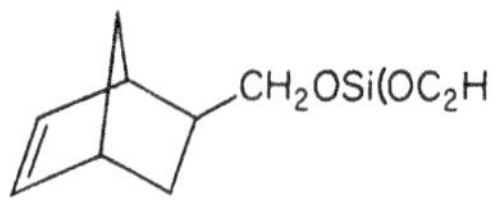

Figure 446

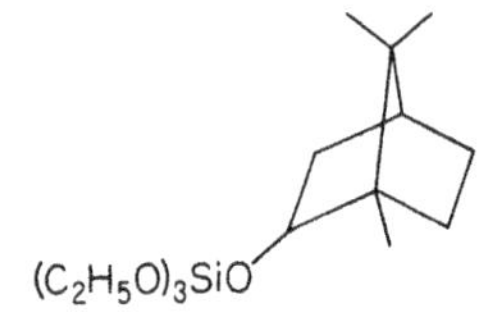

Figure 447

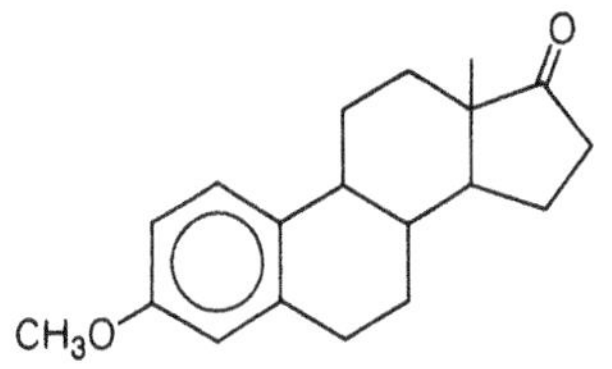

Figure 448

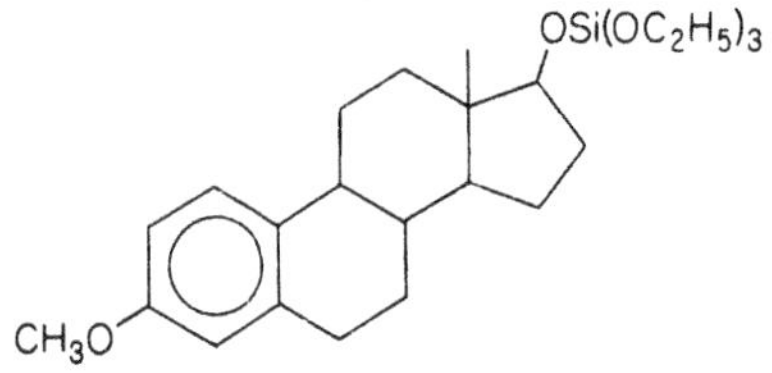

Figure 449

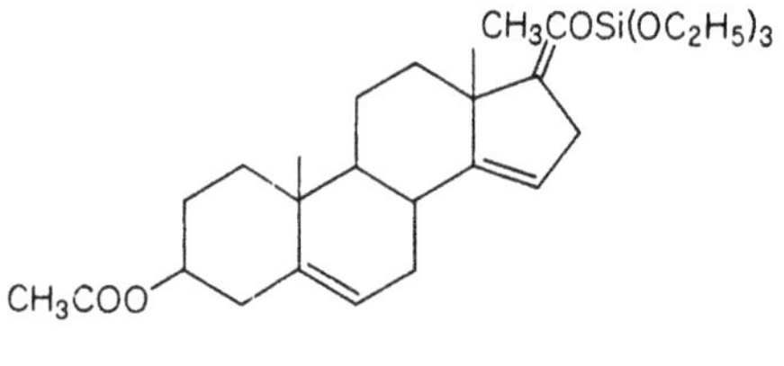

Figure 450

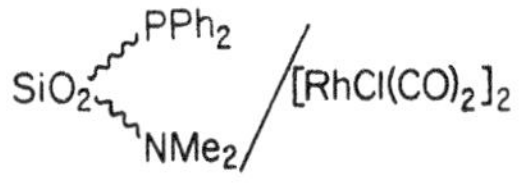

Figure 451

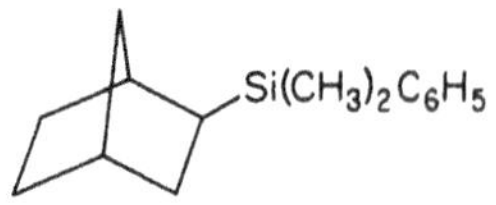

Figure 452

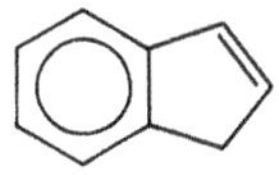

Figure 453

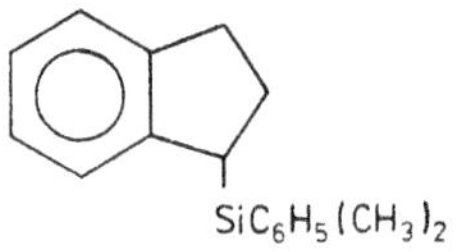

Figure 454

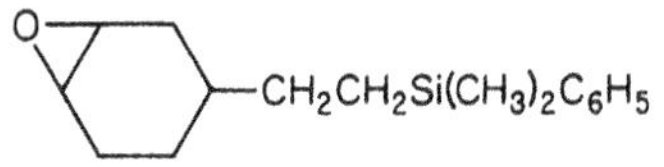

Figure 455

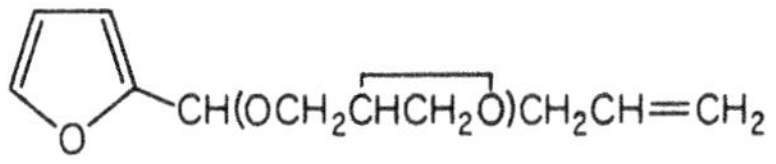

Figure 456

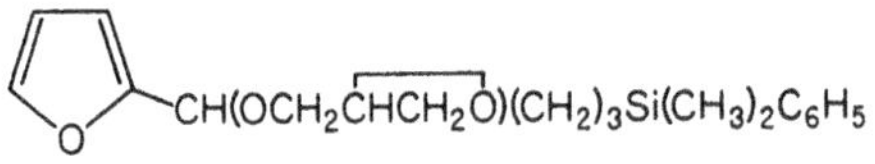

Figure 457

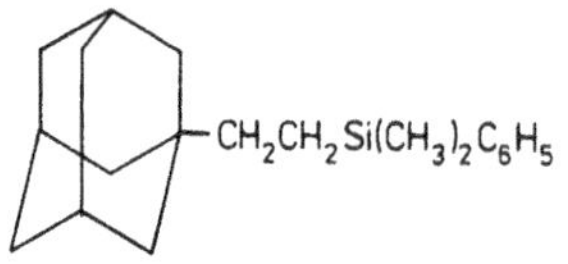

Figure 458

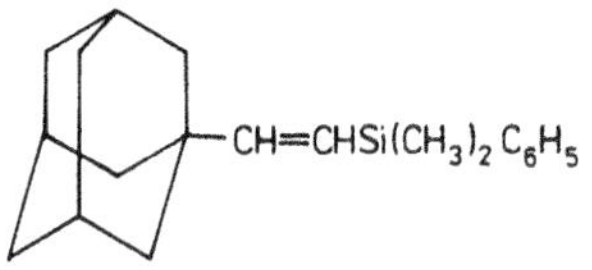

Figure 459

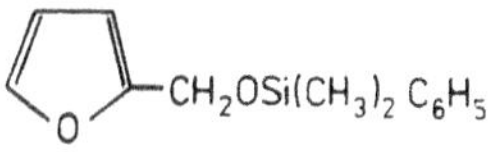

Figure 460

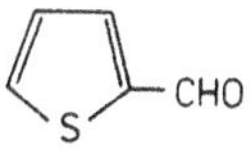

Figure 461

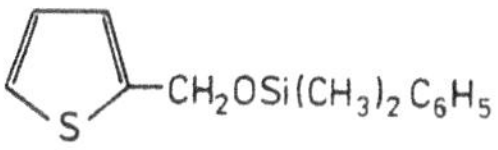

Figure 462

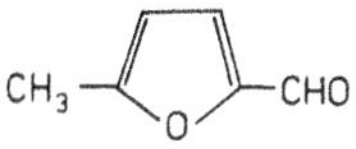

Figure 463

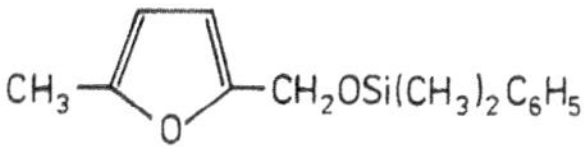

Figure 464

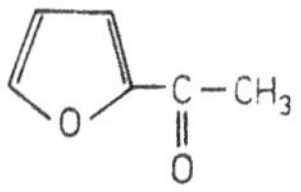

Figure 465

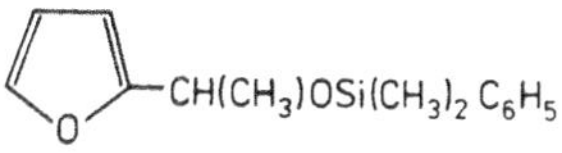

Figure 466

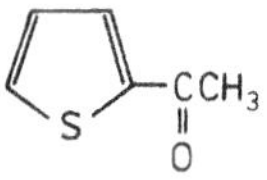

Figure 467

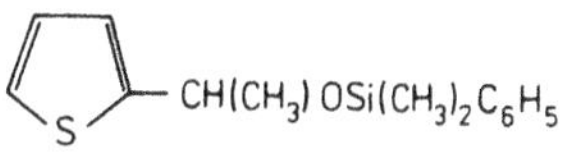

Figure 468

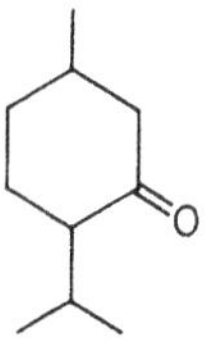

Figure 469

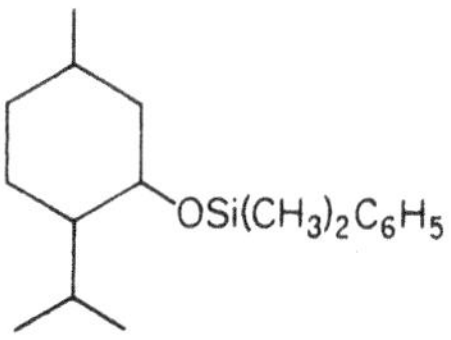

Figure 470

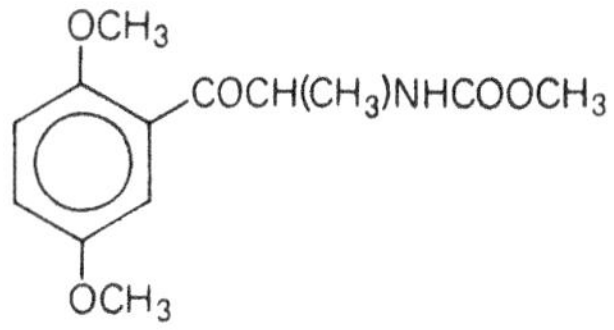

Figure 471

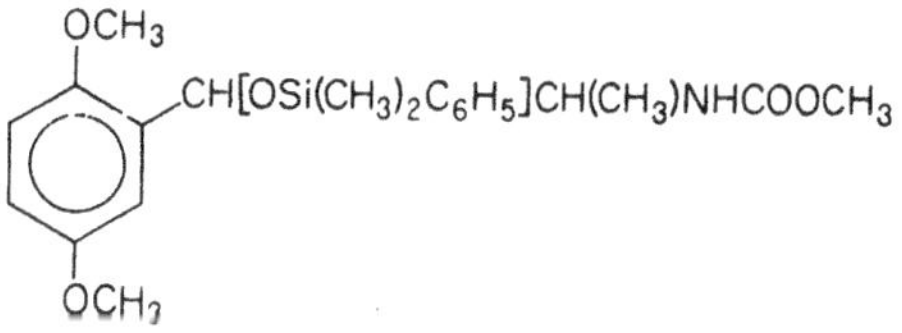

Figure 472

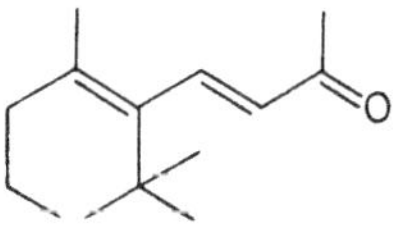

Figure 473

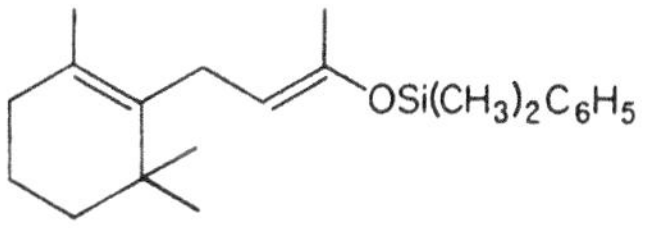

Figure 474

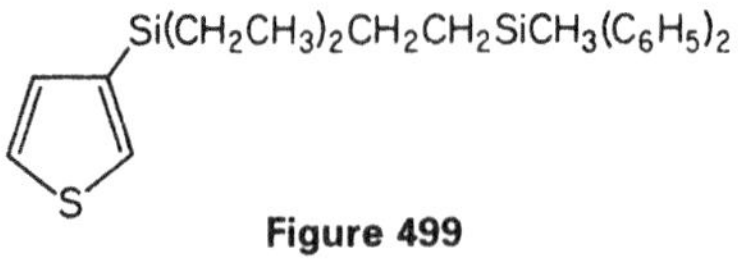

Figure 499

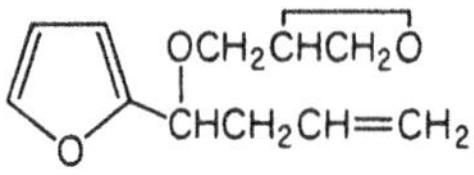

Figure 500

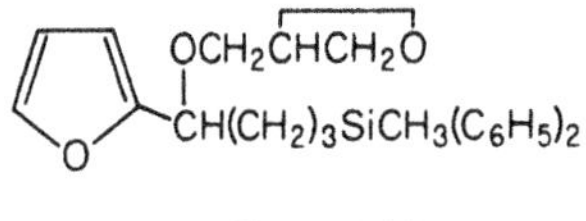

Figure 501

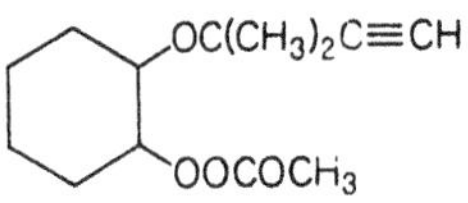

Figure 502

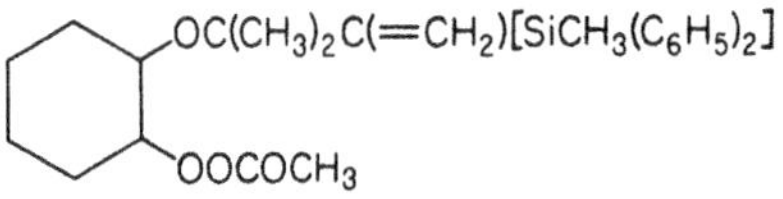

Figure 503

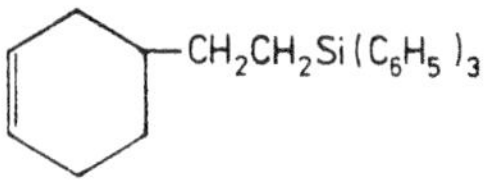

Figure 504

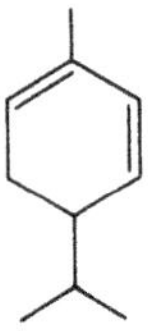

Figure 505

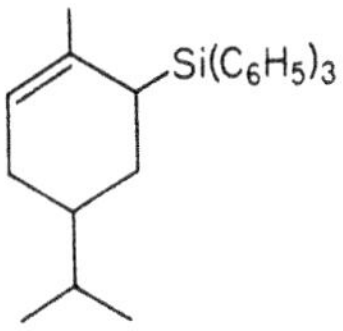

Figure 506

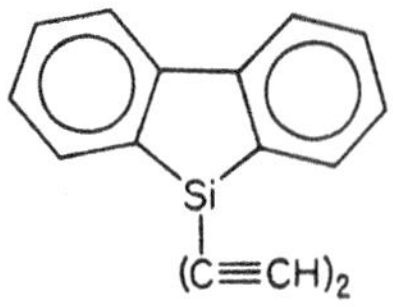

Figure 507

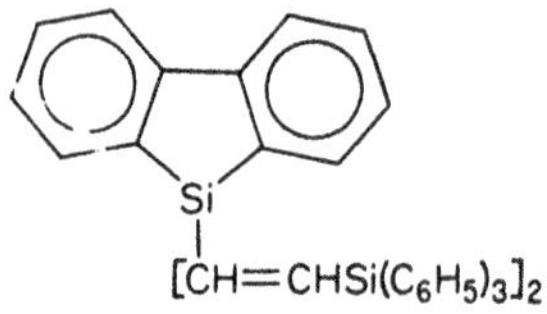

Figure 508

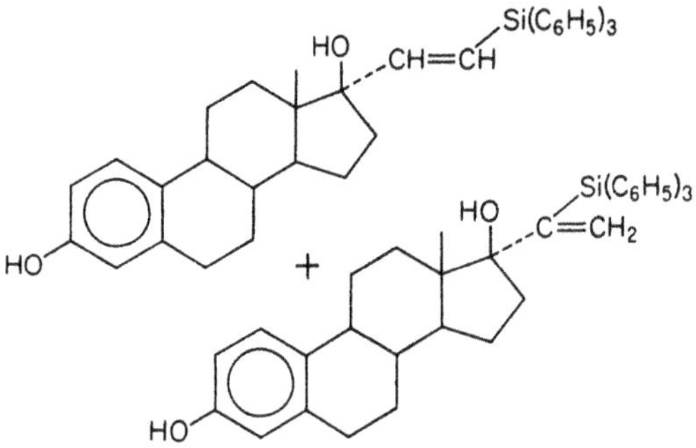

Figure 509

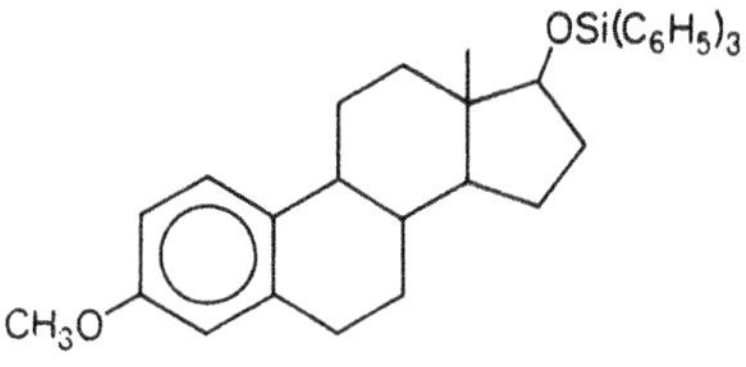

Figure 510

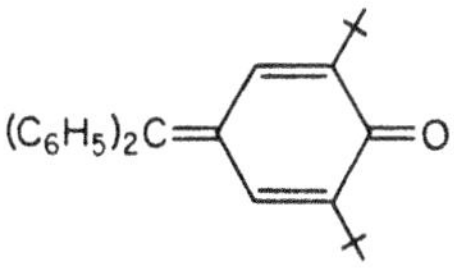

Figure 511

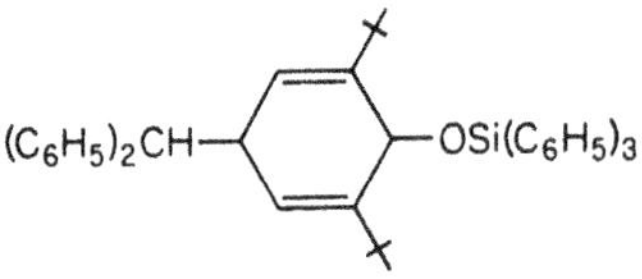

Figure 512

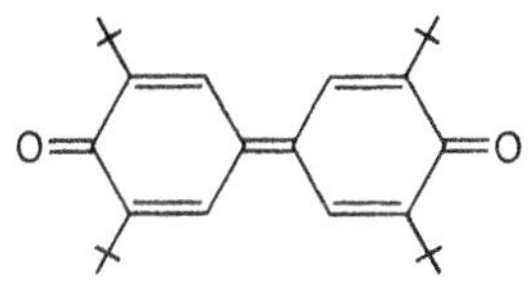

Figure 513

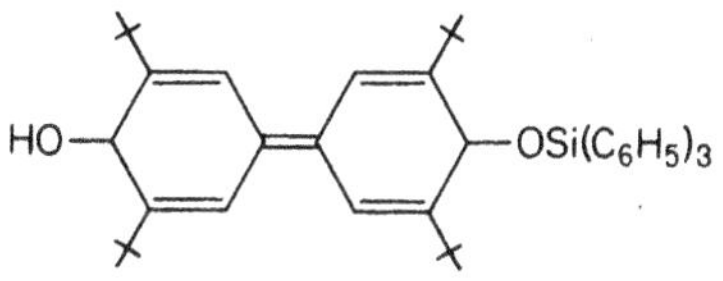

Figure 514

Bibliography

Reviews

Rev. 1. Alberti A. and Pedulli G.F., *Revs. Chem. Intern.*, **8**, 207 (1987).
Rev. 2. Andrianov K.A., Soucek J. and Khananashvili L.M., *Usp. Khim.*, **48**, 1233 (1979).
Rev. 3. Arkles B. and Crosby J., in *Silicon-Based Polymer Science, Comprehensive Resource* (J.M. Zeigler and F.W. Gordon Fearon, Eds), Adv. Chem., Ser. 224, 1990, Chapt. 10.
Rev. 4. Bazant V., Chvalovsky V. and Rathousky J., *Chemistry of Organosilicon Compounds*, Acad. Press, New York, 1965, Vol. 1, p. 139.
Rev. 5. Benkeser R.A., *Pure Appl. Chem.*, **13**, 133 (1966).
Rev. 6. Belyakova Z.V., Pomerantseva M.G. and Chernyshev E.A., *Katalizatory Gidrosililirovaniya* (Russ.), NIITEKHIM, Moskva, 1982.
Rev. 7. Brunner H., *Angew. Chem., Int. Ed. Engl.*, **22**, 897 (1983).
Rev. 8. Calas R., *Pure Appl. Chem.*, **13**, 61 (1966).
Rev. 9. Chalk A.J., *Trans. N.Y. Acad. Sci.*, Ser. II, **32**, 481 (1970).
Rev.10. Collman J.P. and Hegedus L.S., *Principles and Applications of Organotransition Metal Chemistry*, Univ. Sci. Books, Mill Valley, 1980, Chapt. 6.8.
Rev.11. Cundy C.S., Kingston B.M. and Lappert M.F., in *Adv. Organometal. Chem.* (F.A.A. Stone and R. West, Eds), Acad. Press, New York, 1973, Vol. 2, p. 297.
Rev.12. Eaborn C., *Organosilicon Compounds*, Butterworth, London, 1960, p. 45.
Rev.13. Eaborn C. and Bott R.W., in *Organometallic Compounds of the Group IV Elements*, (M.G. MacDiarmid, Ed.), Marcel Dekker, New York, 1968, Vol. 1, p. [illegible]
Rev.14. Frainnet E., *Pure Appl. Chem.*, **19**, 489 (1969).
Rev.15. Gorshkov A.V. and Dontsov A.A., *Kauch. Rezina*, **37** (1983).
Rev.16. Hajos A., *Complex Hydrides and Related Reducing Agents in Organic Synthesis*, Akademia Kiado, Budapest, 1979, Chapt. 7, p. 177.
Rev.17. Harrod J.F. and Chalk A.J., *Org. Synth. Met. Carbonyls* (I. Wender and P. Pino, Eds), J. Wiley and Sons Inc., New York, 1977, Vol. 2, p. 673.
Rev.18. Itoh K., Sakai S. and Ishii Y., *Yuki Gosei Kyokai Shi*, (Jap.) **24**, 729 (1966); C.A., **65**, 169988 (1966).
Rev.19. Kagan H.B., *Pure Appl. Chem.*, **43**, 401 (1975).
Rev.20. Lukevics E.Y. and Voronkov M.F. *Gidrosililirovanie, Gidrogermilirovanie, Gidrostannilirovanie* (Russ.), Izd. Akad. Nauk Latv. SSR, Riga, 1964, Part I.
Rev.21. Lukevics E.Y., *Usp. Khim.*, **46**, 507 (1977).
Rev.22. Lukevics E.Y., Belyakova Z.V., Pomerantseva M.G. and Voronkov M.G., in *J. Organometal. Rev.*, Vol. V, (D. Seyferth *et al.*, Eds) Elsevier, Amsterdam, 1977.
Rev.23. Marciniec B., in *Coordination Chemistry and Catalysis* (J.J. Ziółkowski, Ed.), World Sci., Singapore, 1988, p. 226.
Rev.24. Marciniec B., in *Wybrane Zagadnienia Chemii Krzemu* (Pol.) (B. Marciniec Ed.) A. Mickiewicz University Press, Poznan, 1985, Ser. Chem. 47, p. 93.
Rev.25. Marciniec B., in *Osiagniecia chemii i technologii organicznej* (Pol.) (B. Marciniec Ed.), A. Mickiewicz University Press, Poznan, 1989, Ser. Chem. 57, p. 327.
Rev.26. Marciniec B., *Hydrosililowanie* (Pol.), PWN Warszawa-Poznan, 1981.
Rev.27. Meals R.N., *Pure Appl. Chem.*, **13**, 141 (1966).

Rev.28. Nagai Y., *Org. Prep. Proc. Int.*, **12**, 13 (1980).
Rev.29. Nogradi M., *Stereoselective Synthesis*, Verlag Chemie, Weinheim, 1987, Chapt. 2.2, p. 90.
Rev.30. Ojima I. and Hirai K., in *Asymmetric Synthesis* (J.D. Morrison Ed.), Acad. Press Inc., New York, 1985, Vol. 5, p. 103.
Rev.31. Ojima I. and Kogure T., *Revs Silicon, Germanium, Tin, Lead Compounds*, **5**, 7 (1981).
Rev.32. Ojima I., in *The Chemistry of Organic Silicon Compounds* (S. Patai and Z. Rappoport, Eds), J. Wiley & Sons, Chichester, 1989, Part 2, Chapt. 25, p. 1479.
Rev.33. Ojima I., in *Fundamental Research in Homogeneous Catalysis* (I. Ishi and M. Tsutsui, Eds), Plenum Press, New York, 1978, Vol. 2, p. 181.
Rev.34. Ojima I., *Pure Appl. Chem.*, **56**, 99 (1984).
Rev.35. Ojima I., Yamamoto K. and Kumada M., in *Aspects of Homogeneous Catalysis* (R. Ugo, Ed.), Reidel, Dordrecht, 1977, Vol. 3, p. 185.
Rev.36. Petrov A.D., Mironov V.F., Ponomarenko V.A. and Chernyshev E.A., *Synthesis of Organosilicon Monomers*, Consultants Bureau New York, 1964, p. 386.
Rev.37. Pomerantseva M.G., Belyakova Z.V., Golubtsov S.A. and Shvarts N.S., *Poluchenie Karbofunktsional'nykh Organosilanov po Reaktsii Prisoedineniya* (Russ.), NIITEKHIM, Moskva, 1971.
Rev.38. Speier J.L., *Adv. Organometal. Chem.*, **17**, 407 (1979).
Rev.39. Voorhoeve R.J.H., *Organohalosilanes – Precursors to Silicones*, Elsevier, Amsterdam, 1967, pp. 23, 81.
Rev.40. Voronkov M.G. and Pukhnarevich V.B., *Izv. Akad. Nauk SSSR, Ser. Khim.*, 1056 (1982).
Rev.41. Voronkov M.G. and Pukhnarevich V.B., in *Kompleksy Metallov Platinovoy Gruppy v Sintezie i Katalizie* (Russ.). Sov. Acad. Sci. Press, Chernogolovka, 1983, p. 16.
Rev.42. Wiberg E. and Amberger E., *Hydrides of the Elements of Main Groups I–IV*, Elsevier, Amsterdam, 1971, p. 532.
Rev.43. Yur'ev V.P. and Salimgareeva I.M., *Reaktsia Gidrosililirovaniya Olefinov* (Russ.), Nauka, Moskva, 1982.
Rev.44. Wright A.P., in *Comprehensive Organometallic Chemistry* (F.G.A. Stone et al, Eds), Pergamon Press, Oxford, 1982, Chapt. 9.3, p. 310.

Papers

1. Adams K.P., Joyce J.A., Nile T.A., Patel A.I., Reid C.D. and Walters J.M., *J. Mol. Catal.*, **29**, 201 (1985).
2. Adams N.W., Bradshaw J.S., Beyona J.M., Markides K.E. and Lee M.E., *Mol. Cryst. Lig. Cryst.*, **147**, 4360 (1987).
3. Afanasova O.B., Zubarev Y.E., Sokolova T.M. and Chernyshev E.A., *Zh. Obshch. Khim.*, **57**, 1909 (1987); *C.A.*, **109**, 93108y (1988).
4. Aggarwal S.K., Bradshaw J.S., Eguchi M., Parry S., Rossiter B.E., Markides K.E. and Lee M.L., *Tetrahedron*, **43**, 462 (1987).
5. Aitken C., Harrod J.F. and Samuel E., *J. Organometal. Chem.*, **279**, C11 (1985).
6. Akhmedov I.M., Radzhabov D.T., Guseinov M.M., Sultanov R.A. and Shakhtakhtinskii T.N., *Zh. Obshch. Khim.*, **46**, 442 (1976).
7. Akhmetov R.R., *Issled. Obl. Khim. Vysokomol. Soedon. Neftekhim.*, **9** (1977).
8. Akhrem I.S., Deneux M. and Vol'pin M.E., *Izv. Akad. Nauk SSSR, Ser. Khim.*, 932 (1973).
9. Akita M., Mitani O. and Moro-oka Y., *J. Chem. Soc., Chem. Commun.*, 527 (1989).
10. Albinati A., Caseri W.R. and Pregosin P.S., *Organometallics*, **6**, 788 (1987).
11. Alexander E.S., Haszeldine R.N., Newlands M.J. and Tipping A.E., *J. Chem. Soc. (A)*, 2285 (1970).
12. Ali-Zade I.G., Shikhieva M.I., Salimov M.A., Abdullaev N.D. and Shikhiev I.A., *Dokl. Akad. Nauk Azerb. SSR*, **21**, 28 (1965).
13. Aliev M.I., Shikhiev I.A., Salimov M.A. and Kurbanalieva T.K., *Dokl. Akad. Nauk Azerb. SSR*, **21**, 26 (1965); *C.A.*, **63**, 9976h (1965).

14. Allum K.G., Hancock R.D., Howell I.V., McKenzie S., Pitkethly R.C. and Robinson P.J., *J. Organometal. Chem.*, **87**, 203 (1975).
15. Andreeva N.A., Khudobin Yu.I. and Kharitonov N.P., *Kremniiorg. Mater.*, **53** (1971); *Ref. Zh. Khim.*, Abstr. 7Zh518 (1972).
16. Andrianov K.A. and Movsum-Zade Z.M., *Dokl. Akad. Nauk SSSR*, **229**, 863 (1976).
17. Andrianov K.A. and Sidorov V.I., *Izv. Akad. Nauk SSSR, Ser. Khim.*, **24**, 2128 (1975).
18. Andrianov K.A. and Sidorov V.I., *Dokl. Akad. Nauk SSSR*, **220**, 349 (1975).
19. Andrianov K.A., Abkhazova I.I. and Khananashvili L.M., *Zh. Obshch. Khim.*, **44**, 1919 (1974).
20. Andrianov K.A., Bochkareva G.P. and Golubenko M.A., *Zh. Obshch. Khim.*, **37**, 398 (1967).
21. Andrianov K.A., Bochkareva G.P. and Golubenko M.A., *Zh. Obshch. Khim.*, **38**, 2312 (1968).
22. Andrianov K.A., Delazari N.V., Vol'kova L.M. and Chumayevskii N.A., *Dokl. Akad. Nauk SSSR*, **160**, 1307 (1965).
23. Andrianov K.A., Filimonova M.I. and Sidorov V.I., *J. Organometal. Chem.*, **144**, 27 (1978).
24. Andrianov K.A., Gavrikova L.A. and Rodinova E.F., *Vysokomol. Soedin., Ser. A*, **13**, 937 (1971).
25. Andrianov K.A., Gorshkov A.V., Kruglikov A.M., Dontsov A.A., Kopylov V.M. and Shkol'nik O.V., *Vysokomol. Soedin., Ser. A*, **21**, 1348 (1979).
26. Andrianov K.A., Kotov V.N. and Pryakhina T.A., *Izv. Akad. Nauk SSSR, Ser. Khim.*, **410** (1977).
27. Andrianov K.A., Kotov V.M. and Pryakhina T.A., *Vysokomol. Soedin, Ser. B*, **18**, 254 (1976).
28. Andrianov K.A., Kotov V.M. and Pryakhina T.A., *Izv. Akad. Nauk SSSR, Ser. Khim.*, 129 (1975).
29. Andrianov K.A., Kotov V.M. and Pryakhina T.A., *Izv. Akad. Nauk SSSR, Ser. Khim.*, 2055 (1975).
30. Andrianov K.A., Magomedov G.I., Shkol'nik, O.V., Izmalov B.A., Morozova L.V. and Kalinin V.N., *Dokl. Akad. Nauk SSSR*, **228**, 1094 (1976).
31. Andrianov K.A., Magomedov G.I., Shkol'nik O.V., Syrkin V.G. and Kamaritskii B.A., *Zh. Obshch. Khim.*, **46**, 2048 (1976).
32. Andrianov K.A., Makarova L.I. and Zima T.A., *Izv. Akad. Nauk SSSR, Ser. Khim.*, 832 (1968).
33. Andrianov K.A., Mamedov A.A., Vol'kova L.M. and Klabunovskii E.I., *Izv. Akad. Nauk SSSR, Ser. Khim.*, 356 (1968).
34. Andrianov K.A., Mamedov A.A., Vol'kova L.M. and Klabunovskii E.I., *Izv. Akad. Nauk SSSR, Ser. Khim.*, 1481 (1967).
35. Andrianov K.A., Movsum-Zade Z.M., Shabanova D.A. and Shikhiev I.A., *Dokl. Akad. Nauk SSSR*, **234**, 1082 (1977).
36. Andrianov K.A., Petrashko A.I., Asnovich L.Z. and Gashnikova N.P., *Izv. Akad. Nauk, SSSR, Ser. Khim.*, 1267 (1967).
37. Andrianov K.A., Prudnik I.M., Kotrelev G.V. and Konov A.M., *Zh. Obshch. Khim.*, **44**, 2147 (1974).
38. Andrianov K.A., Rad'kova O.M., Golubtsov S.A. and Morozov N.G., *Zh. Obshch. Khim.*, **41**, 1569 (1971).
39. Andrianov K.A., Shikhiev I.A., Alieva S.Z. and Karaev S.F., *Dokl. Akad. Nauk SSSR*, **226**, 331 (1976).
40. Andrianov K.A., Shkol'nik M.I., Kopylov V.M., Khananashvili L.M. and Prikhodko P.L., *Vysokomol. Soedin., Ser. A*, **21**, 907 (1979).
41. Andrianov K.A., Shkol'nik, M.I., Kopylov V.M., Khananashvili L.M., Minakov V.T. and Zaitseva M.G., *Dokl. Akad, Nauk SSSR*, **249**, 363 (1979).
42. Andrianov K.A., Sidorov V.I. and Filimonova M.I., *Izv. Akad. Nauk SSSR, Ser. Khim.*, 460 (1978).

104. Belyakova Z.V., Yakusheva T.M. and Golubtsov S.A., *Zh. Obshch. Khim.*, **34**, 1480 (1964).
105. Benes J. and Hetflejs J., *Collect. Czech. Chem. Commun.*, **41**, 2264 (1976).
106. Benkeser R.A. and Cunico R.F., *J. Organometal. Chem.*, **6**, 441 (1966).
107. Benkeser R.A. and Cunico R.F., *J. Organometal. Chem.*, **4**, 284 (1965).
108. Benkeser R.A. and Ehler D.F., *J. Org. Chem.*, **38**, 3660 (1973).
109. Benkeser R.A. and Ehler D.F., *Org. Synth.*, **53**, 159 (1973).
110. Benkeser R.A. and Ehler D.F., *J. Organometal. Chem.*, **69**, 193 (1974).
111. Benkeser R.A. and Hickner R.A., *J. Am. Chem. Soc.*, **80**, 5298 (1958).
112. Benkeser R.A. and Kang J., *J. Organometal. Chem.*, **185**, C9 (1980).
113. Benkeser R.A. and Muench W.C., *J. Organometal. Chem.*, **184**, C3 (1980).
114. Benkeser R.A. and Muench W.C., *J. Am. Chem. Soc.*, **95**, 285 (1973).
115. Benkeser R.A. and Snyder D.C., *J. Organometal. Chem.*, **225**, 107 (1982).
116. Benkeser R.A., *Pure Appl. Chem.*, **24**, 31 (1973).
117. Benkeser R.A., Cunico R.F., Dunny S., Jones P.R. and Nerlekar P.G., *J. Org. Chem.*, **32**, 2634 (1967).
118. Benkeser R.A., Dunny S. and Jones P.R., *J. Organometal. Chem.*, **4**, 338 (1965).
119. Benkeser R.A., Dunny S., Li G.S., Nerlekar P.G. and Work S.D., *J. Am. Chem. Soc.*, **90**, 1871 (1968).
120. Benkeser R.A., Foley K.M., Gaul J.W. and Li G.S., *J. Am. Chem. Soc.*, **92**, 3232 (1970).
121. Benkeser R.A, Foley K.M., Grutzner J.B. and Smith W.E., *J. Am. Chem. Soc.*, **92**, 697 (1970).
122. Benkeser R.A., Merritt II F.M. and Roche R.T., *J. Organometal. Chem.*, **156**, 235 (1978).
123. Benkeser R.A., Mozdzen E.C., Muench W.C., Roche R.T. and Siklosi M.P., *J. Org. Chem.*, **44**, 1370 (1979).
124. Benkeser R.A., Nelson L.E. and Swisher J.V., *J. Am. Chem. Soc.*, **83**, 4385 (1961).
125. Bennett E.W. and Orenski P.J., *J. Organometal. Chem.*, **28**, 137 (1971).
126. Bennett E.W., Eaborn C. and Jackson R.A., *J. Organometal. Chem.*, **21**, 79 (1970).
127. Berendsen G.E., Pikaart K.A. and De Galan L., *J. Liq. Chromat.*, **3**, 1437 (1980).
128. Bertrand M.J., Stefenidis S. and Sarrasin B., *J. Chromat.*, **351**, 47 (1986).
129. Bevan W.I., Haszeldine R.N., Middleton J. and Tipping A.E., *J. Chem. Soc., Dalton Trans.*, 2305 (1974).
130. Bevan W.I., Haszeldine R.N., Middleton J. and Tipping A.E., *J. Chem. Soc., Dalton Trans.*, 620 (1975).
131. Billingham N.C., Jackson R.A. and Malek F., *J. Chem. Soc., Chem. Commun.*, 344 (1977).
132. Birchall J.M., Haszeldine R.N., Newlands M.J., Rolfe P.H., Scott D.L., Tipping A.E. and Ward D., *J. Chem. Soc. (A)*, 3760 (1971).
133. Birkofer L. and Grafen K., *J. Organometal. Chem.*, **299**, 143 (1986).
134. Birkofer L. and Kuehn T., *Chem. Ber.*, **111**, 3119 (1978).
135. Birkofer L. and Quittmann W., *Chem. Ber.*, **118**, 2874 (1985).
135a. Bluestein B.A., *J. Am. Chem. Soc.*, **83**, 1000 (1961).
136. Bochkareva G.P., Andrianova G.P. and Golubenko M.A., *Kremniiorg. Soedin., Tr. Sovesch.*, **82** (1967); *Ref. Zh. Khim.*, Abstr. 24S467 (1967).
137. Bock H. and Seidl H., *J. Organometal. Chem.*, **13**, 87 (1968).
138. Bock H. and Seidl H., *Angew. Chem.*, **79**, 1106 (1967).
139. Bock H. and Seidl H., *J. Am. Chem. Soc.*, **90**, 5694 (1968).
140. Bocksanyi L., Liardon O. and Kovats E., *Helv. Chim. Acta*, **59**, 717 (1976).
141. Borisov S.N., Timofeeva N.P., Yuzhelevskii Yu.A., Kagan E.G. and Kozlova N.V., *Zh. Obshch. Khim.*, **42**, 382 (1972).
142. Boroni P.J., Corriu R.J.P. and Guerin C., *J. Organometal. Chem.*, **104**, C17 (1976).
143. Botteghi C., Schionato A., Chelucci G., Brunner H., Kurzinger A. and Obermann U., *J. Organometal. Chem.*, **370**, 17 (1989).
144. Bottrill M. and Green M., *J. Organometal. Chem.*, **111**, C6 (1976).
145. Bourhis R. and Frainnet E., *C.R. Acad. Sci., Ser. C*, **272**, 1153 (1971).

146. Bourhis R. and Frainnet E., *J. Organometal. Chem.*, **86**, 205 (1975).
147. Bourhis R., Frainnet E. and Barsacq S., *Bull. Soc. Chim. France*, 2698 (1965).
148. Bourhis R., Frainnet E. and Moulines F., *J. Organometal. Chem.*, **141**, 157 (1977).
149. Bouterin B., Pietrasanta Y. and Youssef B., *J. Fluorine Chem.*, **31**, 57 (1986); *C.A.*, **106**, 68479a (1987).
150. Boyer J., Breliere C., Corriu R.J.P., Kpoton A., Poirier M. and Royo G., *J. Organometal. Chem.*, **311**, C39 (1986).
151. Boyer J, Corriu R.J.P., Perz R. and Reye C., *Tetrahedron*, **37**, 2165 (1981).
152. Boyer J., Corriu R.J.P., Perz R. and Reye C., *J. Organometal. Chem.*, **148**, C1 (1978).
153. Boyer J., Corriu R.J.P., Perz R. and Reye C., *J. Organometal. Chem.*, **172**, 143 (1979).
154. Boyer J., Corriu R.J.P., Perz R., Poirier M. and Reye C., *Synthesis*, 558 (1981).
155. Bradshaw J.S., Adams N.W., Johnson R.S., Tarbet B.J., Schregenberger C.M., Pulsipher M.A., Andrus M.B., Markides K.E. and Lee M.L., *J. High Resol. Chromat., Chromat. Commun.*, **8**, 678 (1985).
156. Bradshaw J.S., Aggarwal S.K., Rouse Ch.A., Tarbet B.J., Markides K.E., Lee M.L., *J. Chromatography*, **405**, 169 (1987).
157. Bradshaw J.S., Bruening R.L., Krakowiak K.E., Tarbet B.J., Bruening M.L., Izatt R.M. and Christensen J.J., *J. Chem. Soc., Chem. Commun.*, 812 (1988).
158. Bradshaw J.S., Krakowiak K.E., Bruening R.L., Tarbet B.J., Savage P.B. and Izatt R.M., *J. Org. Chem.*, **53**, 3190 (1988).
159. Bradshaw J.S., Schregenberger C.M., Chang K.H.C., Markides K.E. and Lee M.L., *J. Chromat.*, **358**, 95 (1986).
160. Brady K.A. and Nile T.A., *J. Organometal. Chem.*, **206**, 299 (1981)
161 Braun F., Willner L., Hess M. and Kosfeld R., *J. Organometal Chem.*, **332,** 63 (1987).
162. Braun J., *C.R. Acad. Sci., Ser. C*, **260**, 218 (1965).
163. Braun R.A., *Inorg. Chem.*, **5**, 1831 (1966).
164. Brennan T. and Gilman H., *J. Organometal. Chem.*, **16**, 63 (1969).
165. Brockmann M. and tom Dieck H., *J. Organometal. Chem.*, **314**, 75 (1986).
166. Brockmann M., tom Dieck H. and Klaus J., *J. Organometal. Chem.*, **301,** 209 (1986).
167. Brockmann M., tom Dieck H. and Kleinwachter I., *J. Organometal. Chem.*, **309**, 345 (1986).
168. Brook A.G. and Pierce J.B., *J. Org. Chem.*, **30**, 2566 (1965).
169. Brook A.G., Pannell K.H. and Anderson D.G., *J. Am. Chem. Soc.*, **90**, 4374 (1968).
170. Brovko V.S., Skvortsov N.K. and Makarov A.L., *Khim. i Prakt. Primenenie Kremnii- i Fosfor. Soedin., L.*, 58 (1980); *Ref. Zh. Khim.*, Abstr. 11B1047 (1981).
171. Brovko V.S., Skvortsov N.K. and Reikhsfel'd V.O., *Zh. Obshch. Khim.*, **51**, 487 (1981).
172. Brovko V.S., Skvortsov N.K. and Reikhsfel'd V.O., *Zh. Obshch. Khim.*, **51**, 411 (1981).
173. Brovko V.S., Skvortsov N.K., Ivanov A.Yu. and Reikhsfel'd V.O., *Zh. Obshch. Khim.*, **53**, 1831 (1983).
174. Brown-Wensley K.A., *Organometallics*, **6**, 1590 (1987).
175. Brunner H. and Fisch H., *J. Organometal. Chem.*, **335**, 15 (1987).
176. Brunner H. and Fisch H., *J. Organometal. Chem.*, **335**, 1 (1987).
177. Brunner H. and Fisch H., *Monatsch. Chem.*, **119**, 525 (1988).
178. Brunner H. and Knott A., *Z. Naturforsch.*, **40B**, 1243 (1985).
179. Brunner H. and Kurzinger A., *J. Organometal. Chem.*, **346**, 413 (1988).
180. Brunner H. and Leyerer H., *J. Organometal. Chem.*, **334**, 369 (1987).
181. Brunner H. and Obermann U., *Chem. Ber.*, **122**, 499 (1989).
182. Brunner H. and Weber H., *Chem. Ber.*, **118**, 3380 (1985).
183. Brunner H., *J. Organometal. Chem.*, **275**, C17 (1984).
184. Brunner H., *Angew. Chem., Int. Ed. Engl.*, **22**, 331 (1983).
185. Brunner H., Becker R. and Riepl G., *Orgnaometallics*, **3**, 1354 (1984).
186. Brunner H., Becker R. and Sonja G., *Organometallics*, **5**, 739 (1986).
187. Brunner J., Kuerzinger A., Mahboobi S. and Wiegrebe W., *Arch. Pharm.*, **321**, 736 (1988).

188. Brunner H., Reiter B. and Riepl G., *Chem. Ber.*, **117**, 1330 (1984).
189. Bryantseva N.V., Lugovi Y.M., Quinanes Garcia I. and Shostenko A.G., *Khim. Vys. Energ.*, **21**, 422 (1987); *C.A.*, **109**, 54829q (1988).
190. Buachidze M.A. and Guntsadze T.P., *Izv. Akad. Nauk Gruz. SSR, Ser. Khim.*, **14**, 105 (1988); *C.A.*, **111**, 57909t (1989).
191. Bugarenko E.F., Petukhova A.S. and Chernyshev E.A., *Zh. Obshch. Khim.*, **40**, 606 (1970).
192. Burkhard C.A. and Kriebie R.H., *J. Am. Chem. Soc.*, **69**, 2687 (1947).
193. Bush R.P., Lloyd N.C. and Pearce C.A., *J. Chem. Soc.* (*A*), 253 (1969).
194. Calas R. and Duffaut N., *Bull. Soc. Chim. France*, 896 (1954).
195. Calas R., *C.R. Acad. Sci., Ser. C,* **264,** 1402 (1967).
196. Calas R., *J. Organometal. Chem.*, **200**, 11 (1980).
197. Calas R., Duffaut N. and Bardot C., *C.R. Acad. Sci., Ser. C,* **249,** 1682 (1959).
198. Calas R., Duffaut N. and Valade J., *Bull. Soc. Chim. France*, 439 (1954).
199. Calas R., Duffaut N. and Valade J., *Bull. Soc. Chim. France*, 790 (1955).
200. Calas R., Frainnet E. and Bonastre J., *C.R. Acad. Sci., Ser. C,* **251,** 2987 (1960).
201. Calas R., Picard J.P. and Bernadou F., *J. Organometal. Chem.*, **60**, 49 (1973).
202. Calhoun A.D., Lung K.R., Nile T.A., Stokes L.L. and Smith S.C., *Trans. Met. Chem.*, **8**, 365 (1983).
203. Cameron G.G. and Qureshi M.Y., *Makromol. Chem., Rapid Commun.*, **2**, 287 (1981).
204. Capka M. and Hetflejs J., *Collect. Czech. Chem. Commun.*, **39**, 154 (1974).
205. Capka M. and Hetflejs J., *Collect. Czech. Chem. Commun.*, **40**, 3020 (1975).
206. Capka M. and Hetflejs J., *Collect. Czech. Chem. Commun.*, **40**, 2073 (1975).
207. Capka M. and Hetflejs J., *Collect. Czech. Chem. Commun.*, **40**, 3186 (1975).
208. Capka M. and Hetflejs J., *Collect. Czech. Chem. Commun.*, **41**, 1024 (1976).
209. Capka M., *Collect. Czech. Chem. Commun.*, **42**, 3410 (1977).
210. Capka M., Hetflejs J. and Macho V., *Chem. Prumysl*, **26**, 522 (1976).
211. Capka M., Hetflejs J. and Selke R., *React. Kinet. Catal. Lett.*, **10**, 225 (1979).
212. Capka M., Schraml J. and Jancke H., *Collect. Czech. Chem. Commun.*, **43**, 3347 (1978).
213. Capka M., Svoboda P. and Hetflejs J., *Collect. Czech. Chem. Commun.*, **38**, 3830 (1973).
214. Capka M., Svoboda P., Bazant V. and Chvalovsky V., *Collect. Czech. Chem. Commun.*, **36**, 2785 (1971).
215. Capka M., Svoboda P., Bazant V. and Chvalovsky V., *Chem. Prumysl*, **21**, 324 (1971).
216. Capka M., Svoboda P., Cerny M. and Hetflejs J., *Tetrahedron Lett.*, 4787 (1971).
217. Capka M., Svoboda P., Kraus M. and Hetflejs J., *Chem. Ind.* (*London*), 650 (1972).
218. Cardin D.J. and Dixon K.R., *Ann. Rep. Prog. Chem., Sect. A, Phys. Inorg. Chem.*, 211 (1976).
219. Carlini C. and Sbrana G., *J. Macromol. Sci., Chem.*, **A16**, 323 (1981).
220. Caseri W. and Pregosin P.S., *Organometallics*, **7**, 1373 (1988).
221. Caseri W. and Pregosin P.S., *J. Organometal. Chem.*, **356**, 259 (1988).
222. Cavezzan J. and Soula G., *Proc. VII Int. Symp. Organosilicon Chem., Kyoto, Japan*, 1984, p. 193.
223. Cazeau P., Moulines F., Frainnet E. and Calas R., *Bull. Soc. Chim. France*, 3365 (1966).
224. Cerny M., *Collect. Czech. Chem. Commun.*, **42**, 3069 (1977).
225. Chalk A.J. and Harrod J.F., *J. Am. Chem. Soc.*, **89**, 1640 (1967).
226. Chalk A.J. and Harrod J.F., *J. Am. Chem. Soc.*, **87**, 16 (1965).
227. Chalk A.J., *J. Organometal. Chem.*, **21**. 207 (1970).
228. Chalk A.J., *J. Chem. Soc., Chem. Commun.*, 1207 (1969).
229. Chance J.M. and Nile T.A., *J. Mol. Catal.*, **42**, 91 (1987).
230. Chance J.M., Linebarrier D.L. and Nile T.A., *Trans. Met. Chem.*, **12**, 276 (1987).
230a. Chandra G., Lo P.Y., Hitchock P.B. and Lappert M.F., *Organometallics,* **6**, 191 (1987).
231. Chanturiya M.D., *Soobshch. Akad. Nauk Gruz. SSR*, **108**, 565 (1982).
232. Charentenay F., Osborn J.A. and Wilkinson G., *J. Chem. Soc.* (*A*), 787 (1968).

233. Chaumont P., Beinert G., Herz J. and Rempp P., *Eur. Polym. J.*, **15**, 459 (1979).
234. Chaumont P., Beinert G., Herz J. and Rempp P., *Polymer*, **22**, 663 (1981).
235. Chekrii P.S., Kotova N.N. and Kochetkova S.A., *Proc. IV Int. Symp. Homogeneous Catal., Leningrad, USSR*, 1984, Book I, p. 302.
236. Chekrii P.S., Kotova N.N., Karasev Yu.Z. and Kalechits I.V., *Nov. Metody Sint. Org. Soedin. Osn. Neftekhim. Syr'ya*, 103 (1982).
237. Chen I., Zhang Ch. and Du Z., *Gaodeng Xuexiao Xuaxue Xuebao*, **8**, 283 (1987); *C.A.*, **108**, 186817y (1988).
238. Chen L., Wang J. and Huang W., *Huaxue Xuebao*, **41**, 860 (1983); *C.A.*, **100**, 103435h (1984).
239. Chen S. and Lu F., *Boruxue Zazhi*, **47**, 7 (1987); *C.A.*, **108**, 131900q (1988).
240. Chen Y. and Lu X., *Gaofenzi Xuebao*, **6**, 464 (1988); *C.A.*, **111**, 235518z (1989).
241. Chen Y., *Gaodeng Xuexiao Huaxue Xuebao*, **4**, 739 (1983); *C.A.*, **100**, 121709c (1984).
242. Chen Y., Liu J., Lin Y., Ni J., You J., Xiao Ch. and Wang Y., *Wuhan Daxue Xuebao Ziran Kexueban*, **61** (1981); *C.A.*, **97**, 7147q (1982).
243. Chen Y., Lu X., Mei I. and Yang Y., *Youji Huaxue*, **8**, 502 (1988); *C.A.*, **110**, 127543z (1989).
244. Chen Y., Lu X., Wang Z. and Yang Y., *Yingyong Huaxue*, **6**, 1 (1989); *C.A.*, **112**, 98793q (1990).
245. Chen Y., Mei I. and Feng H., *Cuihua Xuebao*, **9**, 444 (1988); *C.A.*, **110**, 241557g (1989).
246. Cherkezishvili K.I., Gelashvili K.Sh. and Dugashvili N.O., *Soobshch. Akad. Nauk. Gruz. SSR*, **93**, 609 (1979).
247. Cherkezishvili K.I., Gverdtsiteli I.M. and Kublashvili R.I., *Zh. Obshch. Khim.*, **44**, 1049 (1974).
248. Cherkezishvili K.I., Gverdtsiteli I.M. and Taktakishvili M.O., *Zh. Obshch. Khim.*, **45**, 1802 (1975).
249. Cherkezishvili K.I., Gverdtsiteli I.M. and Taktakishvili M.O., *Zh. Obshch. Khim.*, **46**, 1297 (1976).
250. Cherkezishvili K.I., Gverdtsiteli I.M., Kublashvili R.I. and Evdokimov A.M., *Soobshch. Akad. Nauk Gruz. SSR*, **63**, 906 (1971).
251. Cherkezishvili K.I., Kublashvili R.I. and Gverdtsiteli I.M., *Soobshch. Akad. Nauk Gruz. SSR*, **61**, 321 (1971).
252. Cherkezishvili K.I., Kublashvili R.I. and Gverdtsiteli I.M., *Zh. Obshch. Khim.*, **41**, 2051 (1971).
253. Cherkezishvili K.I., Taktakishvili M.O. and Gverdtsiteli I.M., *Soobshch. Akad. Nauk Gruz. SSR*, **77**, 609 (1975).
254. Cherkezishvili K.I., Taktakishvili M.O. and Gverdtsiteli I.M., *Soobshch. Akad. Nauk Gruz. SSR*, **78**, 101 (1975).
255. Chernyshev E.A., Belyakova Z.V., Shevchenko V.M., Yagodina L.A., Kisin A.V., Bernadskii A.A., Sheludyakov V.D. and Bochkarev V.N., *Zh. Obshch. Khim.*, **54**, 2031 (1984).
256. Chernyshev E.A. and Shchepinov S.A., *Zh. Obshch. Khim.*, **40**, 1747 (1970).
257. Chernyshev E.A., Dolgaya M.E. and Lubuzh E.D., *Izv. Akad. Nauk SSR., Ser Khim.*, 650 (1965).
258. Chernyshev E.A., Kohalenkova N.G., Bashkirova S.A. and Sokolov V.V., *Zh. Obshch. Khim.*, **48**, 830 (1978).
259. Chernyshev E.A., Magomedov G.I., Shkol'nik O.V., Syrkin V.G. and Belyakova Z.V., *Zh. Obshch. Khim.*, **48**, 1742 (1978).
260. Chogovadze T.N., Nogaideli A.I., Khananashvili L.M., Nakaidze L.I., Tskhovrebasvili V.S., Gusev A.I. and Nestorov D.Yu., *Dokl. Akad. Nauk SSSR*, **246**, 891 (1979).
261. Choo K.K. and Gaspar P.P., *J. Am. Chem. Soc.*, **96**, 1284 (1974).
262. Chuit C., Corriu R.J.P., Perz R. and Reye C., *Synthesis*, 981 (1982).
263. Chujo Y., Ihara E., Ihara H. and Saegusa T., *Macromolecules*, **22**, 2040 (1989).
264. Chujo Y., Ihara E., Ihara H. and Saegusa T., *Polym. Bull.*, **19**, 435 (1988).
265. Chukovskaya E.Ts., Kuz'mina N.A. and Rozhkova M.I., *Zh. Obshch. Khim.*, **36**, 2170 (1966).

266. Chvalovsky V., Pola J., Pukhnarevich V.B., Kopylova L.I., Tsetlina E.O., Pestunovich V.A., Trofimov B.A. and Voronkov M.G., *Collect. Czech. Chem. Commun.*, **41**, 391 (1976).
267. Citron J.D., *J. Appl. Polym. Sci.*, **18**, 3259 (1974).
268. Cook N.C. and Lyons J.E., *J. Am. Chem. Soc.*, **88**, 3396 (1966).
269. Cook N.C. and Lyons J.E., *J. Am. Chem. Soc.*, **87**, 3283 (1965).
270. Cooper D., Haszeldine R.N. and Newlands M.J., *J. Chem. Soc.* (*A*), 2098 (1967).
271. Cornish A.J., and Lappert M.F., *J. Organometal. Chem.*, **271**, 153 (1984).
272. Cornish A.J., Lappert M.F. and Nile T.A., *J. Organometal. Chem.*, **136**, 73 (1977).
273. Cornish A.J., Lappert M.F. and Nile T.A., *J. Organometal. Chem.*, **132**, 133 (1977).
274. Cornish A.J., Lappert M.F., Filatovs G.L. and Nile T.A., *J. Organometal. Chem.*, **172**, 153 (1979).
275. Cornish A.J., Lappert M.F., Macquitty J.J. and Maskell R.K., *J. Organometal. Chem.*, **177**, 153 (1979).
276. Corriu R.J.P. and Masse J.J.E., *Bull. Soc. Chim. France*, 3045 (1974).
277. Corriu R.J.P. and Moreau J.J.E., *J. Organometal. Chem.*, **64**, C51 (1974).
278. Corriu R.J.P. and Moreau J.J.E., *J. Organometal. Chem.*, **21**, 207 (1970).
279. Corriu R.J.P. and Moreau J.J.E., *J. Chem. Soc., Chem. Commun.*, 38 (1973).
280. Corriu R.J.P. and Moreau J.J.E., *J. Organometal. Chem.*, **85**, 19 (1975).
281. Corriu R.J.P. and Moreau J.J.E., *J. Organometal. Chem.*, **91**, C27 (1975).
282. Corriu R.J.P. and Moreau J.J.E., *Nouv. J. Chim.*, **1**, 71 (1977).
283. Corriu R.J.P., Moreau J.J.E. and Pataud-Sat M., *J. Organometal. Chem.*, **228**, 301 (1982).
284. Corriu R.J.P., Moreau J.J.E. and Praet H., *Organometallics*, **9**, 2086 (1990).
285. Corriu R.J.P., Moreau J.J.E. and Praet H., *J. Organometal. Chem.*, **376**, C39 (1989).
286. Corriu R.J.P., Perz R. and Reye C., *Tetrahedron*, **39**, 999 (1983).
287. Cotton F.A. and Wilkinson G., *Advanced Inorganic Chemistry*, J. Wiley and Sons, Inc., London, 1980.
288. Cotton F.A. and Wilkinson G., *Basic Inorganic Chemistry*, J. Wiley and Sons, Inc., London, 1984, Chapts. 28–30.
289. Coulson D.R., *J. Org. Chem.*, **38**, 1483 (1973).
290. Coy D.H., Fitton F., Haszeldine R.N., Newlands M.J. and Tipping A.E., *J. Chem. Soc., Dalton Trans.*, 1852 (1974).
291. Crawford E.J., Hanna P.K. and Cutler A.R., *J. Am. Chem. Soc.*, **111**, 6891 (1989).
292. Crivello J.V., Lee J.L. and Conlou D.A., *Adv. Elastom. Rubber Elasticity*, 57 (1986).
293. Cullen W.R. and Han N.F., *J. Organometal. Chem.*, **333**, 269 (1987).
294. Cullen W.R. and Styan G.E., *J. Organometal. Chem.*, **6**, 633 (1966).
295. Cullen W.R. and Styan G.E., *J. Organometal. Chem.*, **6**, 117 (1966).
296. Cullen W.R. and Styan G.E., *Inorg. Chem.*, **4**, 1437 (1965).
297. Cullen W.R. and Wickenheiser E.B., *J. Organometal. Chem.*, **370**, 141 (1989).
298. Cullen W.R., Evans S.V., Han N.F. and Trotter J., *Inorg. Chem.*, **26**, 514 (1987).
299. Cundy C.S. and Lappert M.F., *J. Chem. Soc., Dalton Trans.*, 665 (1978).
300. Curry J.W. and Harrison G.W., *J. Org. Chem.*, **23**, 627(1958).
301. Curtis M.D. and Epstein P.S., *Adv. Organometal. Chem.*, **19**, 213 (1981).
302. Damrauer R., Davis R.A., Burke M.T., Karn R.A. and Goodman G.T., *J. Organometal. Chem.*, **43**, 121 (1972).
303. Danilov K.E., Lugovoi Yu.M., Tarasova N.P., Churkin A.A., Shostenko A.G., Uranov L.G., Filliatre C. and Bourgeois G., *React. Kinet. Catal. Lett.*, **41**, 205 (1990).
304. Dear R.E.A., *J. Org. Chem.*, **33**, 3959 (1968).
305. Delany A.C., Haszeldine R.N. and Tipping A.E., *J. Chem. Soc.* (*A*), 2537 (1968).
306. Den T.G. and Lee C.H., *Fresenius Z. Anal. Chem.*, **330**, 683 (1988); *C.A.*, **109**, 243273q (1988).
307. Deneux M., Akhrem I., Avetisyan D.V., Mysov E.I. and Vol'pin M.E., *Bull. Soc. Chem. France*, 2638 (1973).
308. Dennis W.E. and Ryan J.W., *J. Org. Chem.*, **35**, 4180 (1970).
309. Dennis W.E. and Speier J.L., *J. Org. Chem.*, **35**, 3879 (1970).
310. Dent S.P., Eaborn C. and Pidcock A., *J. Chem. Soc., Dalton Trans.*, 2646 (1975).

311. Dergunov Yu.I., Mysin N.I. and Mushkin Yu.I., *Zh. Obshch. Khim.*, **45**, 1280 (1975).
312. Deryagina E.N., Nakhmanovich A.S., Elokhina V.N. and Yarosh O.G., *Khim. Geterotsikl. Soedin.*, 891 (1970).
313. Deschler U., Kleinschmit P. and Panster P., *Proc. VII Int. Symp. Organosilicon Chemistry*, Gunma Symp., Organosilicon Chem. Directed to Practical Use, Kiryu-Gunma, Japan, 1984, p. 12.
313a. Deschler U., Kleinschmit P. and Panster P., *Angew. Chem. Int. Ed.*, **25**, 236 (1986).
314. Dickers H.M., Haszeldine R.N., Malkin L.S., Mather A.P. and Parish R.V., *J. Chem. Soc., Dalton Trans.*, 308 (1980).
315. Dickers H.M., Haszeldine R.N., Mather A.P. and Parish R.V., *J. Organometal. Chem.*, **161**, 91 (1978).
316. Dietzmann I., Tomanova D. and Hetflejs J., *Collect. Czech. Chem. Commun.*, **39**, 123 (1974).
317. Dimitrieva G.V. and Sokolov B.A., *Gidrodinam. i Yavleniya Perenosa v Dvukhfaz. Dispers. Sistemakh*, Irkutsk, USSR, 77 (1982); *Ref. Zh. Khim.*, *Abstr.* 24Zh464 (1983).
318. Dohmaru T. and Nagata Y., *J. Phys. Chem.*, **86**, 4522 (1982).
319. Dohmaru T. and Nagata Y., *J. Chem. Soc., Faraday Trans.* 1, **78**, 1141 (1982).
320. Dohmaru T. and Nagata Y., *Bull. Chem. Soc. Japan*, **55**, 323 (1982).
321. Dohmaru T. and Nagata Y., *Bull. Chem. Soc. Japan*, **56**, 1847 (1983).
322. Dohmaru T. and Nagata Y., *J. Chem. Soc., Faraday Trans.* 1, **75**, 2617 (1979).
323. Dohmaru T. and Nagata Y., *Ann. Rep. Radiat. Cent. Osaka Prefect.*, **23**, 51 (1982).
324. Dohmaru T. and Nagata Y., *Bull. Chem. Soc. Japan*, **56**, 2387 (1983).
325. Dohmaru T. and Nagata Y. and Tsurugi J., *Chem. Lett.*, 1031 (1973).
326. Dohmaru T. and Nagata Y. and Tsurugi J., *Bull. Chem. Soc. Japan*, **45**, 2660 (1972).
327. Dontsov A.A., Kuz'mina E.F. and Shapatin A.S., *Sb. Tr. Leningr. Tekhnol. In-ta im. Lensoveta*, 99 (1975); *Ref. Zh. Khim.*, Abstr. 12T 340 (1976).
328. Doyle M.P. and West Ch.T., *J. Org. Chem.*, **40**, 3829 (1975).
329. Doyle M.P. and West Ch.T., *J. Org. Chem.*, **40**, 3821 (1975).
330. Drake A.F., Fung M.A. and Simpson C.F., *J. Chromatogr.*, **476**, 159 (1989).
331. Duczmal W., Marciniec B. and Sliwinska E., *Transition Met. Chem.*, **14**, 105 (1989).
331a. Duczmal W., Marciniec B., Sliwinska E. and Urbaniak W., *Transition Met. Chem.*, **14**, 407 (1989).
332. Duczmal W., Marciniec B. and Urabniak W., *J. Organometal. Chem.*, **327**, 295 (1987).
333. Duczmal W., Urbaniak W. and Marciniec B., *J. Organometal. Chem.*, **317**, 85 (1986).
334. Duffant N. and Calas R., *Bull. Soc. Chim. France*, 283 (1957).
335. Duffant N. and Calas R., *Bull. Soc. Chim. France*, 241 (1962).
336. Dumont W., Paulin J.G., Dang T.P. and Kagan H.B., *J. Am. Chem. Soc.*, **95** 8295 (1973).
337. Dzurinskaya N.T., Mironov V.F. and Petrov A.D., *Dokl. Akad. Nauk SSR*, **138**, 1107 (1961).
338. Eaborn C., Harrison M.R. and Walton D.R.M., *J. Organometal. Chem.*, **31**, 43 (1971).
339. Eaborn C., Odell K. and Pidcock A., *J. Organometal. Chem.*, **63**, 93 (1973).
340. Ediberidze D.A. and Chernyshev E.A., *Izv. Akad. Nauk Gruz. SSR, Ser. Khim.*, **8**, 116 (1982).
341. Efremova L.A., Belyakova Z.V., Popkov K.K. and Golubtsov S.A., *Zh. Prikl. Spektr.*, **19**, 367 (1973).
342. Eisch J.J. and Galle J.E., *J. Organometal. Chem.*, **341**, 293 (1988).
343. Ejike E.N., and Parish R.V., *J. Organometal. Chem.*, **321**, 135 (1987).
344. El-Abbady A.M., *J. Chem. U.A.R.*, **9**, 281 (1966).
345. El-Abbady A.M. and Anderson L.C., *J. Am. Chem. Soc.*, **80**, 1737 (1958).
346. El-Attar A.A.A. and Cerny M., *Collect. Czech. Chem. Commun.*, **40**, 2806 (1975).
347. El-Durini N.M.K. and Jackson R.A., *J. Organometal. Chem.*, **232**, 117 (1982).
348. Engelbrecht L., Sonnek G., Zhdanov A.A. and Makarova L.I., *J. Prakt. Chem.*, **329**, 901 (1987).
348a. Erwin J.W., Ring M.A. and O'Neal H.E., *Int. J. Chem. Kin.*, **17**, 1067 (1985).

349. Evans G.O., Pittman Jr. C.U., McMillan R., Beach R.T. and Jones R., *J. Organometal. Chem.*, **67**, 295 (1974).
350. Faltynek R.A., *Inorg. Chem.*, **20**, 1357 (1981).
351. Fang T.R. and Kennedy J.P., *Polym. Bull. (Berlin)*, **10**, 82 (1983).
352. Farmer E.H., *J. Soc. Chem. Ind.*, **66**, 86 (1947).
353. Fedorova G.T., Kharitonov N.P. and Nechaev B.P., *Zh. Obshch. Khim.*, **44**, 121 (1974).
354. Fedotov N.S., Koziukov V.P., Goler G.E. and Mironov V.F., *Zh. Obshch. Khim.*, **42**, 358 (1972).
355. Fedotov N.S., Luk'yanova I.A., Rybalka I.G. and Mironov V.F., *Zh. Obshch. Khim.*, **49**, 817 (1969).
356. Fedotov N.S., Rusakova N.I., Ryasin G.V. and Mironov V.F., *Zh. Obshch. Khim.*, **46**, 1041 (1976).
357. Fedotov N.S., Rybalka I.G., Korolev V.A. and Mironov V.F., *Zh. Obshch. Khim.*, **47**, 1350 (1977).
358. Felfoldi K., Kapocsi I. and Bartok M., *J. Organometal. Chem.*, **362**, 411 (1989).
359. Feoktistov A.E. and Mironov B.F., *Zh. Obshch. Khim.*, **56**, 2585 (1986).
360. Fernandez M.J. and Oro L.A., *J. Mol. Catal.*, **45**, 7 (1988).
361. Fernandez M.J., Esteruelas M.A., Jimenez M.S. and Oro L.A., *Organometallics*, **5**, 1519 (1986).
362. Fessenden R.J. and Kray W.D., *J. Org. Chem.*, **38**, 87 (1973).
363. Filippov N.A., Reikhsfel'd V.O., Zaslavskaya T.N. and Kuz'mina G.A., *Zh. Obshch. Khim.*, **47**, 1374 (1977).
364. Fink W., *Helv. Chim. Acta*, **54**, 1304 (1971).
365. Fischer E. and Palm C., *Chem. Ber.*, **91**, 1725 (1958).
366. Fischler I. and Grevels F.W., *J. Organometal. Chem.*, **204**, 181 (1980).
367. Fitton P. and McKeon J.E., *J. Chem. Soc., Chem. Commun.*, 4 (1968).
368. Fitton P., McKeon J.E. and Osborn J.A., *J. Chem. Soc., Chem. Commun.*, 1231 (1968).
369. Fountain R. and Kardos J.L., *J. Adhes.*, **8**, 21 (1976).
370. Frainnet E., Martel-Siegfried V., Brousse E. and Dedier J., *J. Organometal. Chem.*, **85**, 297 (1975).
371. Frainnet E. and Arretz E., *Bull. Soc. Chim. France*, 2414 (1970).
372. Frainnet E. and Bourhis R., *J. Organometal. Chem.*, **93**, 309 (1975).
373. Frainnet E. and Bourhis R., *Bull. Soc. Chim. France*, 574 (1965).
374. Frainnet E. and Bourhis R., *Bull. Soc. Chim. France*, 2134 (1966).
375. Frainnet E. and Calas R., *C.R. Acad. Sci., Ser. C*, 240, 203 (1955).
376. Frainnet E. and Causse J., *Bull. Soc. Chim. France*, 3034 (1968).
377. Frainnet E. and Paul M., *Bull. Soc. Chim. France*, 1172 (1966).
378. Frainnet E., *Bull. Soc. Chim. France*, 792 (1963).
379. Frainnet E., Bazouin A. and Calas R., *C.R. Acad. Sci., Ser. C*, **257**, 1304 (1963).
380. Frainnet E., Bourhis R., Simonin F. and Moulines F., *J. Organometal. Chem.*, **105**, 17 (1976).
381. Frainnet E., Llonch J.P., Dubourdin F. and Calas R., *Bull. Soc. Chim. France*, 1172 (1966).
382. Friedlina R.X., Terent'ev A.B., Retrova R.G., Churkina T.D. and Moskulenko M.A., *Dokl. Akad. Nauk SSSR*, **288**, 1436 (1986).
383. Friedlina R.K., Chukovskaya E.C., Tsao J. and Nesmeyanov A.N., *Dokl. Akad. Nauk SSSR*, **132**, 374 (1960).
384. Friedmann G. and Brossas J., *Polym. Bull. (Berlin)*, **10**, 28 (1983).
385. Friedmann G., Herz J. and Brossas J., *Polym. Bull. (Berlin)*, **6**, 252 (1982).
386. Friedrich G., Bartsch R. and Ruehlmann K., *Pharmazie*, **32**, 394 (1977).
387. Fritz G. and Schober P., *Z. anorg. allg. Chemie*, **372**, 21 (1970).
388. Fritz G., *Z. Naturforsch.*, **76**, 329 (1952).
389. Fritz G., Wilhelm H. and Oleownik A., *Z. anorg. allg. Chemie*, **478**, 97 (1981).
390. Frye J.L. and Collins W.T., *J. Org. Chem.*, 35, 2964 (1970).

391. Frye J.L.. *Chem. Commun.*, 45 (1974).
392. Fujita M. and Hiyama T., *J. Am. Chem. Soc.*, **107**, 8294 (1985).
393. Fujita M. and Hiyama T., *J. Am. Chem. Soc.*, **106**, 4629 (1984).
394. Fujita M. and Hiyama T., *Proc. VII Int. Symp. Organosilicon Chem., Kyoto, Japan*, 1984, p. 178.
395. Fujita M. and Hiyama T., *Tetrahedron Lett.*, **28**, 2263 (1987).
396. Fujita M. and Hiyama T., *J. Org. Chem.*, **53**, 5405 (1988).
397. Furuya N. and Sukawa T., *J. Organometal. Chem.*, **96**, C1 (1975).
398. Gailiunas G., Nurtdinova G.V., Yur'ev V.P., Tolstikov G.A. and Rafikov S.R., *Proc. IV Int. Symp. Homogeneous Catal., Leningrad, USSR*, 1984, Book I, p. 255.
399. Gailiunas G., Nurtdinova G.V., Yur'ev V.P., Tolstikov G.A. and Rafikov S.R., *Izv. Akad. Nauk SSSR, Ser. Khim.*, 914 (1982).
400. Galkina M.A., Tshukovskaya E.C. and Freidlina R.H., *Izv. Akad. Nauk SSSR, Ser. Khim.*, 1095 (1987).
401. Gandoda M.E. and Gilpin R.K., *Labelled Comp. Radiopharm.*, **19**, 1051 (1982).
402. Gar T.K., Buyakov A.A., Kisin A.V. and Mironov V.F., *Zh. Obshch. Khim.*, **41**, 1589 (1971).
403. Gar T.K., Nosova V.M., Kisin A.V. and Mironov V.F., *Zh. Obshch. Khim.*, **48**, 838 (1978).
404. Garrido L., Mark J.F., Clarson S.I. and Semlyen J.A., *Polym. Commun.*, **26**, 53 (1985).
405. Garrido L., Mark J.F., Clarson S.I. and Semlyen J.A., *Polym. Commun.*, **26**, 55, (1985).
406. Gentle T.M., Grassian V.H., Klarup D.G. and Muetterties E.L., *J. Am. Chem. Soc.*, **105**, 6766 (1983).
407. Geyer A.E., and Haszeldine R.N., *J. Chem. Soc.*, 3925 (1957).
408. Geyer A.E., Haszeldine R.N., Leedham K. and Marklow R.J., *J. Chem. Soc.*, 4472 (1957).
409. Ghose B.N., *J. Organometal. Chem.*, **164**, 11 (1979).
410. Ghosh P.K. and Spiro T.G., *J. Electrochem. Sci.*, **128**, 1281 (1981).
411. Glushkova N.E. and Kharitonov N.P., *Zh. Obshch. Khim.*, **45**, 2018 (1975).
412. Glushkova N.E. and Kharitonov N.P., *Izv. Akad. Nauk SSSR, Ser. Khim.*, 88 (1967).
413. Glushkova N.E. and Kharitonov N.P., Khim. Prakt. Prim. Kremnii- org. Soedin., *Tr. Sovesch.*, 33 (1966); *C.A.*, **72**, 31909u (1970).
414. Glushkova N.E. and Kharitonov N.P., *Khim. Prakt. Prim. Kremnii- org. Soedin.*, *Tr.* 1075 (1973).
415. Glushkova N.E., Kharitonov N.P. and Zakharova E.I., *Kremniiorg. Mater.* 79 (1971); *Ref. Zh. Khim,.*, Abstr. 8Zh543 (1972).
416. Glushkova N.E., Kharitonov N.P. and Zakharova E.I., *Khim. i Prakt. Primen. Kremnii- i Fosforoorg. Soedin. L.*, 103 (1978); *Ref. Zh. Khim.*, Abstr. 4Zh294 (1980).
417. Glushkova N.E., Kharitonov N.P., Volchinskaya N.I., Dintses A.I., Tregubova V.N. and Stolyar Z.M., *Kremniiorg. Mater.*, 76 (1971); *Ref. Zh. Khim.*, Abstr. 10Zh440 (1972).
418. Goldberg Y., Abele E., Shymanska M. and Lukevics E., *J. Organometal. Chem.*, **372**, C9 (1989).
419. Golubtsov S.A., Pomerantseva M.G., Belyakova Z.V., Schraml J. and Chvalovsky V., *Chem. Prum.*, **24**, 291 (1974).
420. Gondos G. and Orr J.C., *J. Chem. Soc., Chem. Commun.*, 1238 (1982).
421. Gondos G., Gera L., Bartok M. and Orr J.C., *J. Organometal. Chem.*, **373**, 365 (1989).
422. Goodman L., Silverstein R.N. and Benitez A., *J. Am. Chem. Soc.*, **79**, 3073 (1957).
423. Goodman L., Silverstein R.N. and Shodery J.N., *J. Am. Chem. Soc.*, **78**, 26 (1956).
424. Gordon E.M. and Eisenberg R., *J. Mol. Catal.*, **45**, 57 (1988).
426. Gorshkov A.B., Kopylov V.M., Khazen L.Z. and Dontsov A.A., *Kauch. Rez.*, **1**, 25 (1989).
427. Greber G. and Hallensleben M.L., *Angew. Chem.*, **77**, 511 (1965).

428. Green M., Howard J.A.K., Proud J., Spencer J.L., Stone F.G.A. and Tsipis C.A., *J. Chem. Soc., Chem. Commun.*, 671 (1976).
429. Green M., Spencer J.L., Stone F.G.A. and Tsipis C.A., *J. Chem. Soc., Dalton Trans.*, 1519 (1977).
430. Green M., Spencer J.L., Stone F.G.A. and Tsipis C.A., *J. Chem. Soc., Dalton Trans.*, 1525 (1977).
431. Grishko A.N. and Lenskaya O.T., *Gidrodinam. i Yavleniya Perenosa v Dvukhfaz. Dispers. Sistemakh., Irkutsk*, 81 (1982); *Ref. Zh. Khim.*, Abstr. 24Zh465 (1983).
432. Grishko A.N. and Samokhalov N.M., *Zh. Obshch. Khim.*, **47**, 377 (1977).
433. Grueniger H.R. and Calzaferri G., *Helv. Chim. Acta*, **62**, 2547 (1979).
434. Guanpu Y., Caiyun L., Shurong Y. and Zhengxiao X., *Cuihua Xuebao*, **9**, 52 (1988); *C.A.*, **110**, 135314r (1989).
435. Gustafson W.R. and Kapner R.S., *Chem. Eng. Progr., Symp. Ser.*, **64**, 30 (1968).
436. Gverdtsiteli I.M. and Baramidze L.V., *Zh. Obshch. Khim.*, **42**, 2019 (1972).
437. Gverdtsiteli I.M. and Baramidze L.V., *Soobshch. Akad. Nauk Gruz. SSR*, **50**, 83 (1968).
438. Gverdtsiteli I.M. and Baramidze L.V., *Zh. Obshch. Khim.*, **38**, 1598 (1968).
439. Gverdtsiteli I.M. and Buachidze M.A., *Soobshch. Akad. Nauk Gruz. SSR*, **48**, 571 (1967).
440. Gverdtsiteli I.M. and Chanturiya M.D., *Zh. Obshch. Khim.*, **45**, 2349 (1975).
441. Gverdtsiteli I.M. and Chanturiya M.D., *Zh. Obshch. Khim.*, **46**, 865 (1976).
442. Gverdtsiteli I.M. and Chanturiya M.D., *Zh. Obshch. Khim.*, **40**, 2719 (1970).
443. Gverdtsiteli I.M. and Chanturiya M.D., *Soobshch. Akad. Nauk Gruz. SSR*, **88**, 593 (1977).
444. Gverdtsiteli I.M. and Chanturiya M.D., *Zh. Obshch. Khim.*, **42**, 1767 (1972).
445. Gverdtsiteli I.M. and Gelashvili E.S., *Zh. Obshch. Khim.*, **41**, 2061 (1971).
446. Gverdtsiteli I.M. and Gelashvili E.S., *Zh. Obshch. Khim.*, **37**, 2297 (1967).
447. Gverdtsiteli I.M. and Talakvadze T.G., *Tr. Tbilis. Univ.*, **A8**, 153 (1974); *Ref. Zh. Khim.*, Abstr. 24Zh438 (1974).
448. Gverdtsiteli I.M. and Talakvadze T.G., *Soobshch. Akad. Nauk Gruz. SSR*. **79**, 601 (1975).
449. Gverdtsiteli I.M. and Talakvadze T.G., *Soobshch. Akad. Nauk Gruz. SSR*. **79**, 89 (1975).
450. Gverdtsiteli I.M., Asatiani L.P. and Zurabishvili D.S. *Soobshch. Akad. Nauk Gruz. SSR*, **76**, 93 (1974).
451. Gverdtsiteli I.M., Buachidze M.A. and Guntsadze T.P., *Tr. Tbilis. Univ.*, 57 (1976); *Ref. Zh. Khim.*, Abstr. 1Zh303 (1978).
452. Gverdtsiteli I.M., Cherkezishvili K.I. and Asatiani L.P., *Tr. Tbilis. Univ.*, **A1**, 175 (1971); *Ref. Zh. Khim.*, Abstr. 18Zh419 (1971).
453. Gverdtsiteli I.M., Cherkezishvili K.I. and Melua M.S., *Soobshch. Akad. Nauk Gruz. SSR*, **59**, 73 (1970).
454. Gverdtsiteli I.M., Chernyshev E.A. and Dzotsenidze L.A., *Soobshch. Akad. Nauk Gruz. SSR*, **74**, 93 (1974).
455. Gverdtsiteli I.M., Chernyshev E.A. and Ediberidze D.A., *Soobshch. Akad. Nauk Gruz. SSR*, **71**, 361 (1973).
456. Gverdtsiteli I.M., Doksopulo T.P. and Chikovani E.I., *Zh. Obshch. Khim.*, **47**, 352 (1977).
457. Gverdtsiteli I.M., Doksopulo T.P., Gorelashvili N.P. and Andranikashvili G.G., *Zh. Obshch. Khim.*, **46**, 2531 (1976).
458. Gverdtsiteli I.M., Dzheliya M.I. and Baramidze L.V., *Soobshch. Akad. Nauk Gruz. SSR*, **84**, 381 (1976).
459. Gverdtsiteli I.M., Ediberidze A.A. and Chernyshev E.A., *Zh. Obshch. Khim.*, **44**,. 2449 (1974).
460. Gverdtsiteli I.M., Gelashvili E.S., Khmiadishvili T.I. and Dondua K.V., *Tr. Tbilis. Univ.*, **A9**, 125 (1975); *Ref. Zh. Khim.*, Abstr. 7Zh401 (1976).

461. Gverdtsiteli I.M., Melua M.S. and Doksopulo T.P., *Zh. Obshch. Khim.*, **42**, 1777 (1972).
462. Gverdtsiteli I.M., Melua M.S. and Doksopulo T.P., *Zh. Obshch. Khim.*, **42**, 2022 (1972).
463. Gverdtsiteli I.M., Melua M.S. and Doksopulo T.P. and Chegelishvili V.A., *Soobshch. Akad. Nauk Gruz. SSR*, **62**, 317 (1971).
464. Gverdtsiteli I.M., Nguyen Dang Quang and Doksopulo T.P., *Soobshch. Akad. Nauk Gruz. SSR*, **77**, 85 (1975).
465. Haas A. and Koehler J., *J. Fluorine Chem.*, **17**, 531 (1981).
466. Hadjichristidis N., Guyot A. and Fetters L.J., *Macromolecules*, **11**, 668 (1978).
467. Han B.H. and Boudjouk P., *Organometallics*, **2**, 769 (1983).
468. Hara M., Ohno K. and Tsuji J., *J. Chem. Soc. (D)*, 247 (1971).
469. Hardin S. and Turney T.W., *J. Mol. Catal.*, **39**, 237 (1987).
470. Harrod J.F. and Chalk A.J., *J. Am. Chem. Soc.*, **87**, 1133 (1965).
470a. Harrod J.F. and Smith C.A., *Can. J. Chem.*, **48**, 870 (1970).
471. Harrod J.F., Gilson D.F. and Charles R., *Can. J. Chem.*, **47**, 2205 (1969).
472. Harrod J.F. and Yun S.S., *Organometallics*, **6**, 1381 (1987).
473. Hart D.W. and Schwartz J., *J. Organometal. Chem.*, **87**, C11 (1975).
474. Haszeldine R.N., Lythgoe S. and Robinson P.J., *J. Chem. Soc. (B)*, 1634 (1970).
474a. Haszeldine R.N., Malkin L.S. and Parish R.V., *J. Organometal. Chem.*, **182**, 323 (1979).
475. Haszeldine R.N., Newlands M.J. and Plumb J.B., *J. Chem. Soc.*, 2101 (1965).
476. Haszeldine R.N., Parish R.V. and Parry D.J., *J. Chem. Soc. (A)*, 683 (1969).
477. Haszeldine R.N., Parish R.V. and Parry D.J., *J. Organometal. Chem.*, **9**, P13 (1967).
478. Haszeldine R.N., Parish R.V. and Taylor R.J., *J. Chem. Soc., Dalton Trans.*, 2311 (1974).
479. Haszeldine R.N., Pool C.R. and Tipping A.E., *J. Chem. Soc., Dalton Trans.*, 2292 (1975).
480. Haszeldine R.N., Pool C.R. and Tipping A.E., *J. Chem. Soc., Dalton Trans.*, 2177 (1975).
481. Haszeldine R.N., Pool C.R. and Tipping A.E., *J. Chem. Soc., Perkin Trans.* 1, 2293 (1974).
482. Hayashi T. and Kabata K., *Proc. VII Int. Symp. Organosilicon Chem., Kyoto, Japan*, 1984, p. 186.
483. Hayashi T. and Kabata K., *Tetrahedron Lett.*, **26**, 3023 (1985).
484. Hayashi T., Kabata K., Yamamoto T., Tamao K. and Kumada M., *Tetrahedron Lett.*, **24**, 5661 (1983).
485. Hayashi T., Tamao K., Katsuro Y., Nakae I. and Kumada M., *Tetrahedron Lett.*, **21**, 1871 (1980).
486. Hayashi T., Yamamoto K., Kasuga K., Omizu H. and Kumada M., *J. Organometal. Chem.*, **113**, 127 (1976).
487. Hayashi T., Yamamoto K. and Kumada M., *Tetrahedron Lett.*, 3 (1975).
488. Hayashi T., Yamamoto K. and Kumada M., *Tetrahedron Lett.*, 4405 (1974).
489. Hayashi T., Yamamoto K. and Kumada M., *Tetrahedron Lett.*, 331 (1974).
490. Hayashi T., Yamamoto K. and Kumada M., *J. Organometal. Chem.*, **112**, 253 (1976).
491. Heidingsfeldova M. and Capka M., *J. Appl. Polym. Science*, **30**, 1837 (1985).
492. Heidingsfeldova M. and Schatz M., *Sci. Papers Prague Inst. Chem. Technol.*, (5), 65 (1981), Prague (Czechoslovakia).
493. Heimbach P., Jolly P.W. and Wilke G., *Adv. Organometal. Chem.*, **8**, 29 (1970).
494. Henrici-Olive G. and Olive S., *Coordination and Catalysis*, Verlag Chemie GmbH, Weinheim, 1977.
495. Hilal H.S., Abu-Eid M., Al-Subu M. and Khalaf S., *J. Mol. Catal.*, **39**, 1 (1987).
496. Hill J.E. and Nile T.A., *J. Organometal. Chem.*, **137**, 293 (1977).
497. Hill J.E. and Nile T.A., *Trans. Met. Chem.*, **3**, 315 (1978).
498. Hofer A., Kuckertz H. and Sander M., *Macromol. Chem.*, **90**, 49 (1966).
499. Hong M., and Zhou X., *Huaxue Xuebao*, 190 (1981); *C.A.*, **98**, 179469h (1983).

500. Hong M., Zhao Y. and Zhou X., *Gaodeng Xuexiao, Huaxue Xuebao*, **4**, 735 (1983); *C.A.*, **100**, 209948t (1984).
501. Hosomi A., Hayashida H., Kohra S. and Tominaga Y., *J. Chem. Soc., Chem. Commun.*, 1411 (1986).
502. Howard E.G., Sargeant P.B. and Krespan C.G., *J. Am. Chem. Soc.*, **89**, 1422 (1967).
503. Howe J.P., Lung K. and Nile T., *J. Organometal. Chem.*, **208**, 401 (1981).
504. Hu C., Han X., and Jiang Y., *J. Molec. Catal.*, **35**, 329 (1986).
505. Hu C., Han X., Jiang Y., Liu J.G. and Shi T.Y., *J. Macromol. Sci. Chem.*, **A26**, 349 (1989).
506. Hu C., Han X., Jiang Y., *Kexue Tongbao*, **33**, 843 (1988); *C.A.*, **111**, 97331e (1989).
507. Hu C., Yang R. and Jiang Y., *Fenzi Cuihua*, **2**, 38 (1988); *C.A.*, **111**, 78085m (1989).
508. Hu C., Zhao D. and Jiang Y., *Cuihua Xuebao*, **10**, 213 (1989); *C.A.*, **111**, 235512t (1989).
509. Huang M.Y., Ren C.Y., Zhon Y.Z., Zhao L., Cao X.P., Dong D.L., Liu T. and Jiang Y.Y., in *Organosilicon and Bioorganosilicon Chemistry*, (H. Sakurai, Ed.), Ellis Horwood, New York, 1985, p. 275.
510. Hudrlik P.F., Schwartz R.H. and Hogan J.C., *J. Org. Chem.*, **44**, 155 (1979).
511. Hudrlik P.F. and Wan Chung-Nan, *Synth. Commun.*, **9**, 333 (1979).
512. Ioramashvili D.Sh., Machavariani D.N. and Sepiashvili L.M. *Soobshch. Akad. Nauk Gruz. SSR*, **111**, 309 (1983).
513. Ioramashvili D.Sh. and Kakhniashvili A.I., *Izv. Akad. Nauk Gruz. SSR*, **6**, 129 (1980).
514. Ioramashvili D.Sh. and Shudra O.S., *Soobshch. Akad. Nauk Gruz. SSR*, **94**, 81 (1979).
515. Ioramashvili D.Sh. *Tr. Tbilis. Univ.*, 52 (1983); *Ref. Zh. Khim.*, Abstr. 14Zh55 (1984).
516. Ioramashvili D.Sh., *Soobshch. Akad. Nauk Gruz. SSR*, **97**, 357 (1980).
517. Ioramashvili D.Sh. and Kakhniashvili A.I., *Soobshch. Akad. Nauk Gruz. SSR*, **94**, 349 (1979).
517a. Iovel I.G., Goldberg Y.S., Shymanska M.V. and Lukevics E., *J. Chem. Soc., Chem. Commun.*, 31 (1987).
518. Ishiyama I., Seuda Y. and Imaizumi S., *Nippon Kagaku Kaishi*, **6**, 834 (1986); *C.A.*, **106**, 119060s (1987).
519. Ishiyama J., Senda Y., Shinoda I. and Imaizumi S., *Bull. Chem. Soc. Jpn.*, **52**, 2353 (1979).
520. Israfilov Ya.M., Sultanov R.A. and Sadykh-Zade S.I., *Zh. Obshch. Khim.*, **46**, 2747 (1976).
521. Jackson R.A., *Adv. Free Radical Chem.*, **3**, 231 (1969).
522. Jadaud M., Caude M., Rosset R., *J. Chromatography*, **393**, 39 (1987).
523. Jaeger, A., Kochs H., Rosenberg A. and Heusinger H., *Farbe Lack*, **88**, 632 (1982).
524. James B.R., *Homogeneous Hydrogenation*, J. Wiley and Sons, New York (1973).
525. Jancke H., Engelhardt G., Kriegsmann H., Volkova L.M., Delazari N.V. and Andrianov K.A., *Z. anorg. allgem. Chemie*, **402**, 97 (1973).
526. Janzen A.F. and Willis C.J., *Can. J. Chem.*, **43**, 3063 (1965).
527. Johnson T.H., Klein K.C. and Thomen S., *J. Mol. Catal.*, **12**, 37 (1981).
528. Kagan E.G., *Zh. Obshch. Khim.*, **37**, 1692 (1967).
529. Kagan H.B., Langlois N. and Dang T.P., *J. Organometal. Chem.*, **90**, 353 (1975).
530. Kakhniashvili A.I. and Ioramashvili D.Sh. *Zh. Obshch. Khim.*, **40**, 1552 (1970).
531. Kakhniashvili A.I. and Ioramashvili D.Sh. *Zh. Obshch. Khim.*, **37**, 1411 (1967).
532. Kakhniashvili A.I. and Ioramashvili D.Sh. *Zh. Obshch. Khim.*, **40**, 1556 (1970).
533. Kakhniashvili A.I. and Ioramashvili D.Sh. *Tr. Tbilis. Univ.*, 195 (1973); *Ref. Zh. Khim.*, Abstr. 2Zh405 (1974).
534. Kakhniashvili A.I., Gvaliya T.Sh. and Ioramashvili D.Sh., *Zh. Obshch. Khim.*, **40**, 2518 (1970).
535. Kakhniashvili A.I., Gvaliya T.Sh. and Ioramashvili D.Sh., *Soobshch. Akad. Nauk Gruz. SSR*, **68**, 609 (1972).
536. Kakhniashvili A.I., Ioramashvili D.Sh. and Gvaliya T.Sh., *Zh. Obshch. Khim.*, **42**, 356 (1972).

537. Kakhniashvili A.I., Ioramashvili D.Sh. and Nadirashvili M.D. *Soobshch. Akad. Nauk Gruz. SSR*, **87**, 81 (1977).
538. Kakhniashvili A.I., Ioramashvili D.Sh.. Fedin E.I., Petrovskii P.V. and Rubin I.D., *Soobshch. Akad. Nauk Gruz. SSR*, **53**, 573 (1969).
539. Kakhniashvili A.I., Ioramashvili D.Sh.. Fedin E.I., Petrovskii P.V. and Rubin I.D., *Soobshch. Akad. Nauk Gruz. SSR*, **54**, 337 (1969).
540. Kakhniashvili A.I., Ioramashvili D.Sh.. Fedin E.I., Petrovskii P.V., Rubin I.D. and Gvaliya T.Sh., *Soobshch. Akad. Nauk Gruz. SSR*, **51**, 79 (1968).
541. Kanazashi M., *Bull. Chem. Soc. Jpn.*, **26**, 493 (1953).
542. Karaev S.F., *Proc. VII Int. Symp. Organosilicon Chem., Kyoto, Japan*, 1984, p. 158.
543. Karaev S.F., Babaev B.G. and Aliev A.K., *Azerb. Khim. Zh.*, 72 (1987); *C.A.*, **110**, 57144n (1989).
544. Karaev S.F., Dzhafarov D.S. and Askerov M.E., *Zh. Org. Khim.*, **16**, 928 (1980).
545. Karaev S.F., Mamedov E.A., Ivanov E.S. and Thalikova Z.M.,, *Dokl. Akad. Nauk Azerb. SSR*, **39**, 43 (1983); *C.A.*, **100**, 5784b (1984).
546. Karaev S.F., Movsum-Zade M.M., Askerov M.E. and Agimirzoev N.A., *Izv. Vyssh. Ucheb. Zaved., Khim. Khim. Technol.*, **17**, 565 (1974).
547. Karaev S.F., Movsum-Zade M.M., and Agamirzoev N.A., *Zh. Obshch. Khim.*, **44**, 131 (1974).
548. Karaev S.F., Shikhiev I.A. and Khabibova A.K., *Izv. Vyssh. Ucheb. Zaved., Khim. Khim. Tekhnol.*, **19**, 208 (1976).
549. Karaev S.F., Shikhiev I.A. and Tsalikova Z.M., *Zh. Obshch. Khim.*, **43**, 109 (1973).
550. Karaev S.F., Shikhiev I.A., Tsalikova Z.M., Khabibova A.K. and Mukhtarov F.G., *Izv. Vyssch. Ucheb. Zaved., Khim. Khim. Tekhnol.*, **17**, 1035 (1974).
551. Karim A., Mortreux A. and Petit F., *Tetrahedron Lett.*, **27**, 345 (1986).
552. Kartasheva L.I., Nametkin N.S. and Chernysheva T.I., *Zh. Obshch. Khim.*, **40**, 1262 (1970).
553. Kashutina E.A., Andrianov A.K., Zhdanov A.A., Olenin A.V. and Zubov V.P., *Zh. Obshch. Khim.*, **54**, 657 (1984).
554. Kasumov F.Ya, Askerov O.V., Aronova L.L., Sadykh-Zade S.I. and Sultanov T.A., *Azerb. Khim. Zh.*, 35 (1974).
555. Kavan V. and Capka M., *Collect. Czech. Chem. Commun.*, **45**, 2100 (1980).
556. Kaverin V.V., *Khim. Vysokomol. Soedin, Neftekhim.*, 36 (1975); *C.A.*, **85**, 94439e (1976).
557. Kaverin V.V., Salimgareeva I.M., Brui I.N., Kayumov F.F., Kuchin A.V., Spivak A.Yu. and Yur'ev V.P., *Izv. Akad. Nauk SSSR, Ser. Khim.*, 2657 (1980).
558. Kaverin V.V., Salimgareeva I.M., Kovaleva I.V. and Yur'ev V.P., *Izv. Akad. Nauk SSSR, Ser. Khim.*, 2119 (1981).
559. Kaverin V.V., Salimgareeva I.M., Kovaleva I.V. and Yur'ev V.P., *Zh. Obshch. Khim.*, **49**, 1793 (1979).
560. Kaverin V.V., Salimgareeva I.M. and Yur'ev V.P., *Zh. Obshch. Khim.*, **48**, 122 (1978).
561. Kaverin V.V. and Brui I.N., *Issled. Obl. Khim. Vysokomol. Soedin. Neftekhim.*, 7 (1977); *C.A.*, **93**, 26490e (1980).
562. Kawakami Y., Aswatha Murthy R.A. and Yamashita Y., *Makromol. Chem.*, **185**, 9 (1984).
563. Keinan E. and Greenspoon N., *J. Am. Chem. Soc.*, **108**, 7314 (1986).
564. Keinan E. and Perez D., *J. Org. Chem.*, **52**, 2576 (1987).
565. Kennedy J.P. and Carlson G.M., *J. Polym. Sci., Polym. Chem. Ed.*, **21**, 3551 (1983).
566. Kennedy J.P. and Chang V.S.C., *Polym. Prepr. Am. Chem. Soc., Div. Polym. Chem.*, **21**, 146 (1980).
567. Kennedy J.P. and Chang V.S.C., *Adv. Polym. Sci.*, **43**, 1 (1982).
568. Khabibova A.K., Timurova R.A., Kazieva S.T. and Karaev S.F., *Azerb. Khim. Zh.*, 47 (1987); *C.A.*, **110**, 172782n (1989).
569. Khalilova E.M. and Sadykh-Zade S., *Azerb. Khim. Zh.*, 131 (1966).
570. Khananashvili L.M., Akhobadze D.Sh., Dzhanishvili L.K. and Andronikashvili G.G., *Zh. Obshch. Khim.*, **60**, 390 (1990).

571. Khananashvili L.M., Buziashvili V.I., Tskhovrebashvili V.S. and Kisin A.V., *Izv. Akad. Nauk Gruz. SSR, Ser. Khim.*, **9**, 305 (1982).
572. Khankhodzhaeva D.A. and Reikhsfel'd V.O., *Uzb. Khim. Zh.*, **11**, 38 (1967); *C.A.*, **68**, 502770v (1967).
573. Khapicheva A.A., Berberova N.T., Klimov E.S. and Okhlobystin O.YU., *Zh. Obshch. Chim.*, **55**, 1533 (1985).
574. Khatuntsev G.D., Sheludyakov V.D. and Mironov V.F., *Zh. Obshch. Khim.*, **44**, 2150 (1974).
575. Khudobin Yu.I., Makarskaya V.M., Makarskii V.V., Kharitonov N.P. and Voronkov M.G., *Izv. Akad. Nauk SSSR, Ser. Khim.*, 1622 (1976).
576. Khvatova T.P., Reikhsfel'd V.O. and Astrakhanov M.J., *Zh. Obshch. Khim.*, **49**, 1271 (1979).
577. Khvatova T.P., Reikhsfel'd V.O. and Romanyuk T.G., *Zh. Obshch. Khim.*, **49**, 1275 (1979).
578. Kiladze T.K., Galtseva T.D., Melnitski I.A., Karakhanov R.A. and Kantor E.A., *Zh. Org. Khim.*, **21**, 1584 (1985).
579. Kiladze T.K., Mel'nitskii I.M., Mironov V.F., Kantor E.A., Rakhmankulov D.L. and Kavakhanov R.A., *Arm. Khim. Zh.*, **38**, 248 (1985); *C.A.*, **104**, 34127 (1986).
580. Kim Y.K., Pierce O.R., Bajzer W.X. and Smith A.G., *Polym. Prepr. Am. Chem. Soc., Div. Polym. Chem.*, **12**, 482 (1971).
581. Kim Y.K., Smith A.G. and Pierce O.R., *J. Org. Chem.*, **38**, 1615 (1973).
582. King R.B., *Adv. Organometal. Chem.*, **2**, 157 (1964).
583. Kinting A., *Z. Chem.*, **26**, 180 (1986).
584. Kinting A., Kreuzfeld H.J., *J. Organometal. Chem.*, **370**, 343 (1989).
585. Kireev V.V., Kobyazin V.A., Kopylov V.M., Zaitseva M.G. and Kostylev I.M., *Zh. Obshch. Khim.*, **54**, 367 (1984).
586. Kirpichenko S.V., Keiko V.V. and Voronkov M.G., *Proc. IV Int. Symp. Homogeneous Catal., Leningrad, USSR*, 1984, vol. 1, p. 271.
587. Kirpichnikova A.A., Noskov V.G., Sokol'skii M.A. and Englin M.A., *Zh. Obshch. Khim.*, **43**, 1862 (1973).
588. Kiso Y., Kumada M., Maeda K., Sumitani K. and Tamao K., *J. Organometal. Chem.*, **50**, 311 (1973).
589. Kiso Y., Kumada M., Tamao K. and Umeno M., *J. Organometal. Chem.*, **50**, 297 (1973).
589a. Kiso Y., Tamao K. and Kumada M., *J. Chem. Soc., Chem. Commun.*, 105 (1972).
590. Kiso Y., Yamamoto K., Tamao K. and Kumada M., *J. Am. Chem. Soc.*, **94**, 4373 (1972).
591. Kitazume T., Kobayashi K., Yamamoto T. and Yamazaki T., *J. Org. Chem.*, **52** (1987) 3218.
592. Kobayashi Y., Ito Y. and Terashima S., *Bull. Chem. Soc. Jpn.*, **62**, 3041 (1989).
593. Kobrakov K.I., Chernysheva T.I., Nametkin N.S. and Fedorov L.A., *Dokl. Akad. Nauk SSSR*, **193**, 1072 (1970).
594. Kobrakov K.I., Chernysheva T.I., and Nametkin N.S. *Dokl. Akad. Nauk SSSR*, **198**, 1340 (1971).
595. Kogure T. and Ojima I., *J. Organometal. Chem.*, **234**, 249 (1982).
596. Kohra S., Kayashida H., Tominaga Y. and Hosomi A., *Tetrahedron Lett.*, **29**, 89 (1988).
597. Kojima K., Iwabuchi S., Sangane T., Miyoshi K. and Shimomura K., *Kogakubu Kenkyu Hokoku*, **38**, 33 (1987); *C.A.*, **107**, 219187c (1987).
598. Kolb I. and Hetflejs J., *Collect. Czech. Chem. Commun.*, **45**, 2224 (1980).
599. Kolb I., Cerny M. and Hetflejs J., *React. Kinet. Catal. Lett.*, **7**, 199 (1977).
600. Komarov N.V. and Igonina I.I., *Zh. Obshch. Khim.*, **37**, 2108 (1967).
601. Komarov N.V. and Kirpichenko S.V., *Izv. Akad. Nauk SSSR, Ser. Khim.*, 2288 (1970).
602. Komarov N.V. and Roman V.K., *Zh. Obshch. Khim.*, **35**, 2017 (1965).
603. Komarov N.V., Ol'khnovskaya L.I. and Baranova O.I., *Zh. Obshch. Khim.*, **50**, 412 (1980).

604. Komarov N.V., Pukhnarevich V.B., Sushchinskaya S.P., Kalabin G.A. and Sakharovskii V.G., *Izv. Akad. Nauk SSSR, Ser. Khim.*, 839 (1968).
605. Kono H., Wakao N., Ojima I. and Nagai Y., *Chem. Lett.*, 189 (1975).
605a. Kono H., Wakao N., Ito K. and Nagai Y., *J. Organometal. Chem.*, **132**, 53 (1977).
606. Kopylov V.M., Koryazina T.G., Buskeva T.M., Sinitsyn N.M., Kireev V.V. and Gorshkov A.V., *Zh. Obshch. Khim.*, **57**, 117 (1987).
607. Kopylova L.I., Korostova S.E., Sobenina L.N., Nesterenko R.N., Sigalov M.V., Mikhaleva A.I., Trofimov B.A. and Voronkov M.G., *Zh. Obshch. Khim.*, **51**, 1778 (1981).
608. Kopylova L.I., Lukevics E. and Voronkov M.G., *Zh. Obshch. Khim.*, **54**, 115 (1984).
609. Kopylova L.I., Malysheva S.F., Vyalykh E.P. and Voronkov M.G. *Zh. Obshch. Khim.*, **52**, 2259 (1982).
610. Kopylova L.I., Pukhnarevich V.B., Tsykhanskaya I.I., Satsuk E.N., Timokhin B.V., Dmitriev V.I., Chvalovsky V., Capka M., Kalabina A.V. and Voronkov M.G., *Zh. Obshch. Khim.*, **51**, 1851 (1981).
611. Kopylova L.I., Sigalov M.V., Satsuk E.N., Capka M., Chvalovsky V., Pukhnarevich V.B., Lukevics E. and Voronkov M.G., *Zh. Obshch. Khim.*, **51**, 385 (1981).
612. Kopylova L.I., Ivanova D., Voropayev V.N., Domnina E.S., Skvortsova G.G. and Voronkov M.G., *Zh. Obshch. Chim.*, **55**, 1036 (1985).
613. Kornetka Z.W., unpublished results.
614. Koroleva G.N. and Reikhsfel'd V.O., *Zh. Obshch. Khim.*, **37**, 2768 (1967).
615. Koryazin W.A., Kopylov W.M., Kireev W.W., Kisin A.B., Alekseev B. and Gorshkov A.W., *Zh. Obshch. Khim.*, **57**, 1735 (1987).
616. Koshutin V.I., Stadnichuk M.D. and Petrov A.A., *Zh. Obshch. Khim.*, **40**, 1787 (1970).
617. Koshutin V.I., Stadnichuk M.D. and Petrov A.A., *Zh. Obshch. Khim.*, **41**, 848 (1971).
618. Kotova N.N., Chekrin P.S., Kalechits I.V. and Vostrikova V.N., *Porysh. Effekt i Soversh. Tekchnol. Pr-va Monom. i Rastv. na Baze Olefin. Syr'ya, M.*, 53 (1983); *Ref. Zh. Khim.*, Abstr. 23Zh363 (1983).
619. Kozlikov V.L., Fedotov N.S. and Mironov V.F., *Zh. Obshch. Khim.*, **39**, 2284 (1969).
620. Kraihanzel C.S. and Losee L., *J. Organometal. Chem.*, **10**, 427 (1967).
621. Kraus M., *Collect. Czech. Chem. Commun.*, **39**, 1318 (1974).
622. Krueger C.R., *Inorg. Nucl. Chem. Lett.*, **1**, 85 (1965).
623. Krusic P.J. and Kochi J.K., *J. Am. Chem. Soc.*, **93**, 846 (1971).
624. Krusic P.J. and Kochi J.K., *J. Am. Chem. Soc.*, **91**, 6161 (1969).
625. Krusic P.J. and Kochi J.K., *J. Am. Chem. Soc.*, **91**, 3938 (1969).
626. Kuli-Zade F.A., Ragimov A.A., Khudayarov I.A. and Askerov O.V., *Azerb. Khim. Zh.*, 73 (1986); *C.A.*, **109**, 54824j (1988).
627. Kuliev A.M., Dzhafarov A.A. and Mamedov F.N., *Azerb. Khim. Zh.*, 84 (1968).
628. Kumada M., Kiso Y. and Umeno M., *J. Chem. Soc. (D)*, 611 (1970).
629. Kumada M., Naka K. and Yamamoto Y., *Bull. Chem. Soc. Jpn.*, **37**, 871 (1964).
629a. Kumada M., Sumitami K., Kiso Y. and Tamao K., *J. Organometal. Chem.*, **50**, 319 (1973).
630. Kuncova G. and Chvalovsky V., *Collect. Czech. Chem. Commun.*, **45**, 2085 (1980).
631. Kuropka R., Muller B., Hocker H. and Berndt H., *J. Chromatogr.*, **481**, 380 (1989).
632. Kursanov D.N., Parnes Z.N. and Loim N.M., *Synthesis*, 633 (1974).
633. Kusumoto T. and Hiyama T., *Chem. Lett.*, 1405 (1985).
634. Kusumoto T. and Hiyama T., *Tetrahedron. Lett.*, **28**, 1811 (1987).
635. Kuzmina P.A., Ilinskaya L.V., Gasanov R.G., Tshukovskaya E.C., Freidlina R.H., *Izv. Akad. Nauk. SSSR*, 212 (1986).
636. Kuznetsova V.P., Smetankina N.P., Belogolivina G.N., Oprya V.Ya. and Kudinova M.A., *Zh. Obshch. Khim.*, **35**, 1636 (1965).
637. Laane, J., *J. Am. Chem. Soc.*, **89**, 1144 (1967).
638. Laguerre M., Felix G., Dunogues J. and Calas R., *J. Org. Chem.*, **44**, 4275 (1979).
639. Lahaye J. and Lagarde R., *Bull. Soc. Chim. France*, 2999 (1974).
640. Lampe F.W., Snyderman J.S. and Johnston W.H., *J. Phys. Chem.*, **70**, 3934 (1966).

641. Lane T.H. and Frye C.L., *J. Organometal. Chem.*, **172**, 213 (1979).
642. Langlois N., Dang T.P. and Kagan H.B., *Tetrahedron Lett.*, 4865 (1973).
643. Langova J. and Hetflejs J., *Collect. Czech. Chem. Commun.*, **40**, 432 (1975).
644. Langova J. and Hetflejs J., *Collect. Czech. Chem. Commun.*, **40**, 420 (1975).
645. Lapkin I.I. and Dvinskikh V.V., *Zh. Obshch. Khim.*, **51**, 1354 (1981).
646. Lapkin I.I. and Dvinskikh V.V., *Zh. Obshch. Khim.*, **48**, 2509 (1978).
647. Lapkin I.I. and Dvinskikh V.V., *Sintezy na Osnove Magnii i Tsinkoorgan. Soedin., Mezhvuz Sb. Nauchnykh Trudov, Permskii Univ.*, 43 (1980); *Ref. Zh. Khim.*, Abstr. 10Zh317 (1981).
648. Lapkin I.I. and Povarnitsyna T.N., *Zh. Obshch. Khim.*, **38**, 643 (1968).
649. Lapkin I.I., Dvinskikh V.V. and Povarnitsyna T.N., *Zh. Obshch. Khim.*, **53**, 116 (1983).
650. Lapkin I.I., Pidemskii E.L., Povarnitsyna T.N., Goleneva A.F., Mardanova L.G. and Dvinskikh V.V., *Khim. Pharm. Zh.*, **17**, 817 (1983).
651. Lapkin I.I., Povarnitsyna T.N. and Denisova T.Yu., *Zh. Obshch. Khim.*, **44**, 123 (1974).
652. Lapkin I.I., Povarnitsyna T.N. and Dvinskikh V.V., *Zh. Obschch. Khim.*, **47**, 2284 (1977).
653. Lapkin I.I., Povarnitsyna T.N. and Dvinskikh V.V., *Zh. Obshch. Khim.*, **48**, 607 (1978).
654. Lapkin I.I., Povarnitsyna T.N., Dvinskikh V.V. and Rogozina E.N., *Zh. Obshch. Khim.*, **54**, 1114 (1984).
655. Lapkin I.I., Povarnitsyna T.N. and Fomin V.V., *Zh. Obshch. Khim.*, **51**, 1797 (1981).
656. Lapkin I.I., Povarnitsyna T.N. and Kostareva L.A., *Zh. Obshch. Khim.*, **38**, 1578 (1968).
657. Lapkin I.I., Povarnitsyna T.N. and Kozlova N.A., *Zh. Obshch. Khim.*, **46**, 2522 (1976).
658. Lapkin I.I., Povarnitsyna T.N. and Kozlova N.A., *Zh. Obshch. Khim.*, **46**, 1567 (1976).
659. Lapkin I.I., Povarnitsyna T.N. and Kozlova N.A., *Zh. Obshch. Khim.*, **51**, 1091 (1981).
660. Lappert M.F. and Maskell R.K., *J. Organometal. Chem.*, **264**, 217 (1984).
661. Lappert M.F. and Nile T.A., *J. Organometal. Chem.*, **102**, 543 (1975).
662. Lappert M.F. and Takahashi S., *J. Chem. Soc., Chem. Commun.*, 1272 (1972).
663. Lappert M.F., Nile T.A. and Takahashi S., *J. Organometal. Chem.*, **72**, 425 (1974).
664. Lehmann J. and Schafer H., *Chem. Ber.*, **105**, 969 (1972).
665. Lester L., Boileau S. and Charadame H., *Polym. Prepr. Am. Chem. Soc. Div. Polym. Chem.*, **30**, 133 (1989).
666. Levin V.Yu, Andrianov K.A., Slonimskii G.L., Zhdanov A.A., Lubavskaya E.A. and Malykhin A.P., *Vysokomol. Soedin., Ser. A*, **16**, 1951 (1974).
667. Lewis L.N. and Lewis N., *J. Am. Chem. Soc.*, **108**, 7228 (1986).
668. Lewis L.N. and Uriarte R.J., *Organometallics*, **9**, 621 (1990).
669. Lewis L.N., *J. Am. Chem. Soc.*, **112**, 5998 (1990).
670. Li C., Yuan G., Yan S. and Xie Z., *Yingyong Huaxue*, **5**, 96 (1988); *C.A.*, **109**, 47333r (1988).
671. Li D.C., Sun H. and Wu G.L., *Kao Fen Tzu T'ung Hsun*, 190 (1980); *C.A.*, **94,** 66823t (1981).
672. Licchelli M. and Greco A., *Tetrahedron Lett.*, **28**, 3719 (1987).
673. Liepins E., Goldberg Y., Iovel I. and Lukevics E., *J. Organometal. Chem.*, **335**, 301 (1987).
674. Lin X., Hong M. and Cai Q., *Xiamen Daxue Xuebao, Ziran Kexueban*, **27**, 558 (1988); *C.A.*, **111**, 78097s (1989).
675. Linke K.H. and Goehausen H.J., *Chem. Ber.*, **106**, 3438 (1973).
676. Linke K.H. and Goehausen H.J., *Angew. Chem., Int. Ed. Engl.*, **10**, 408 (1971).
677. Liu H., Wang X., Du Z., Ma I. and Wang X., *Shandong Daxuo Xuobao, Ziran Kexueban*, **21**, 100 (1986).
678. Liu H.J. and Ramani B., *Synth. Commun.*, **15**, 965 (1985).
679. Liu S., Shi T., Hu Ch. and Iang Y., *Gaofenzi Xuebao*, 290 (1988); *C.A.*, **110**, 115859u (1988).
680. Llonch J.P. and Frainnet E., *C.R. Acad. Sci., Ser. C*, **276**, 1803 (1973).
681. Llonch J.P. and Frainnet E., *C.R. Acad. Sci., Ser. C*, **274**, 70 (1972).
682. Lou X., Liu Z. and Zhou R., *Gaofenzi Xuebao*, 226 (1987); *C.A.*, **107**, 237324z (1987).

683. Lu F., Wang Y., Wang D. and Yu Y., *Gaofenzi Tongxun*, 154 (1985); *C.A.*, **104**, 19914p (1986).
684. Lu X., Mei J. and Meng L., *Wullan Daxue Xuebao, Ziran Kexueban*, 70 (1988); *C.A.*, **110**, 144667k (1989).
685. Lugovoi Yu.M., Shostenko A.G. and Myshkin V.E., *React. Kinet. Catal. Lett.*, **12**, 441 (1979).
686. Lugovoi Yu.M., Shostenko A.G., Myshkin V.E. and Krapchatov V.P., *Kinet. Katal.*, **22**, 1055 (1981).
687. Lugovoi Yu.M., Shostenko A.G., Tarasova N.P., Zaboluev I.A., Danilov K.E., Filliatre C. and Bourgeois G., *React. Kinet. Catal. Lett.*, **39**, 379 (1989).
688. Lugovoi Yu.M., Shostenko A.G., Zagorets P.A. and Krapchatov V.P., *React. Kinet. Catal. Lett.*, **24**, 329 (1984).
689. Lukevics E. and Khokhlova L.N., *Zh. Obshch. Khim.*, **48**, 826 (1978).
690. Lukevics E. and Liberts L., *Zh. Obshch. Khim.*, **43**, 778 (1973).
691. Lukevics E. and Popova E.P., *Zh. Obshch. Khim.*, **47**, 112 (1977).
692. Lukevics E. and Pudova O.A., Sturkovich R.Ya. and Gaukhman A.P., *Proc. VII Int. Symp. Organosilicon Chem., Kyoto, Japan*, 1984, p. 97.
693. Lukevics E. and Voronkov M.G., *Khim Geterotsikl. Soedin., Akad. Nauk Latv. SSR*, 490 (1965).
694. Lukevics E. and Voronkov M.G., *Khim Geterotsikl. Soedin., Akad. Nauk Latv. SSR*, 179 (1965).
695. Lukevics E., in *Organosilicon and Bioorganosilicon Chemistry* (H. Sakurai, Ed), Ellis Horwood Ltd, New York, 1985, p. 243.
696. Lukevics E., *Proc. VII Int. Symp. Organosilicon Chem., Kyoto, Japan*, 1984, p. 94.
697. Lukevics E., Germane S., Erchak N.P. and Popova E.P., *Khim. Pharm.Zh.*, **12**, 67 (1978).
698. Lukevics E., Gevorgyan V.N., Goldberg Y.S. and Shymanska M.V. *J. Organometal. Chem.*, **263**, 283 (1984).
699. Lukevics E., Lapina T., Liepins E. and Segals J., *Khim. Geterotsikl. Soedin.*, 962 (1977).
700. Lukevics E., Pestunovich A.E., Pestunovich V.A. and Voronkov M.G., *Zh. Obshch. Khim.*, **40**, 624 (1970).
701. Lukevics E., Pestunovich A.E., Pestunovich V.A., Liepins E. and Voronkov M.G., *Zh. Obshch. Khim.*, **41**, 1585 (1971).
702. Lukevics E., Pestunovich A.E., Voronkov M.G., and Kupcse T., *Khim. Geterotsikl. Soedin.*, 485 (1972).
703. Lukevics E., Pestunovich A.E. and Voronkov M.G., *Izv. Akad. Nauk Latv. SSR*, 447 (1972).
704. Lukevics E., Pudova O., Sturkovich R. and Gaukhman A., *J. Org. Chem.*, **346,** 297 (1988).
705. Lukevics E., Pudova O.A., Sturkovich R.Ya. and Gaukhman A.P., *Zh. Obshch. Khim.*, **55**, 1520 (1985).
706. Lukevics E., Sturkovich R.Ya. and Pudova O.A., *J. Organometal. Chem.*, **292**, 151 (1985).
707. Lukevics E., Sturkovich R.Ya., Pudova O.A. and Gaukhman A.P., *Zh. Obshch. Khim.*, **56**, 140 (1986).
708. Lukevics E., Sturkovich R.Ya., Pudova O.A. and Gaukhman A.P., *Zh. Obshch. Khim.*, **55**, 624 (1985).
709. Lyashenko G.S., Filippova A.Kh., Kalikhman I.D., Keiko V.V., Kruglaya O.A. and Vyazankin N.S., *Izv. Akad. Nauk SSSR, Ser. Khim.*, 874 (1981).
710. Lyashenko G.S., Nametkin N.S. and Chernysheva T.I., *Izv. Akad. Nauk SSSR, Ser. Khim.*, 2628 (1969).
711. Macosko C.W. and Saam I.C., *Polym. Prepr., Am. Chem. Soc., Dir. Polym. Chem.*, **26**, 48 (1985).
712. Magomedov G.K.I., Shkol'nik O.B., Khananashvili L.M., Nakaidze L.I. and Katsitadze M.G., *Zh. Obshch. Khim.*, **58**, 104 (1988).
713. Magomedov G.K.I. and Shkol'nik O.V., *Zh. Obshch. Khim.*, **50**, 1103 (1980).

714. Magomedov G.K.I. and Shkol'nik O.V., *Zh. Obshch. Khim.*, **51**, 841 (1981).
715. Magomedov G.K.I., Andrianov K.A., Shkol'nik O.V., Izmailov B.A. and Kalinin V.N., *J. Organometal. Chem.*, **149**, 29 (1978).
716. Magomedov G.K.I., Druzhkova G.V., Shkol'nik O.V., Shestakova T.A. and Bizyukova N.M., *Zh. Obshch. Khim.*, **53**, 1823 (1983).
717. Magomedov G.K.I., Druzhkova G.V., Syrkin V.G. and Shkol'nik O.V., *Koord. Khim.*, **6**, 767 (1980).
718. Magomedov G.K.I., Frenkel A.S., Shkol'nik O.V., Bulyheva E.V. and Sheludyakov V.D., *Metaloorg. Khim.*, **1**, 938 (1988); *C.A.*, **111**, 134393z (1989).
719. Magomedov G.K.I., Shkol'nik O.V. and Druzhkova G.V., *Zh. Obshch. Khim.*, **53**, 392 (1983).
720. Magomedov G.K.I., Shkol'nik O.V., Izmailov B.A. and Sigachev S.A., *Koord. Khim.*, **6**, 761 (1980).
721. Magomedov G.K.I., Shkol'nik O.V., Khananashvili L.M., Nakaidze L.I. and Ketsitadze M.G., *Metaloorg. Khim.*, **1**, 232 (1988); *C.A.*, **110**, 212899q (1989).
722. Maitlis P.M., *Chem. Soc. Rev.*, **10**, 1 (1981).
723. Makarenko I.A., Ivanov V.I., Khoroshilova E.V., Shapatin A.S., Severnyi V.V. and Zhinkin D.Ya., *Prom-st' Sintetich. Kauchuka Nauch.-Tekhn. Sb.*, 12 (1975); *Ref. Zh. Khim.*, Abstr. 15T463 (1975).
724. Makarenko I.A., Khoroshilova E.V. and Severnyi V.V., *Kauch. Rezina*, 20 (1976).
725. Maksimova L.N. and Koshutin V.I., *Zh. Obshch. Khim.*, **44**, 2180 (1974).
726. Mamedalev J.G., Mamedov M.A., Sadykh-Zade S.I. and Akhmedov I.M., *Azerb. Khim. Zh.*, 9 (1962).
727. Mamedov M.A., and Akhmedov I.M., *Azerb. Khim. Zh.*, 105 (1968).
728. Mamedov M.A., Sadykh-Zade S.I. and Akhmedov I.M., *Zh. Obshch. Khim.*, **36**, 2018 (1966).
729. Mamedov M.A., Akhmedov I.M., and Guseinov M.M., *Azerb. Khim. Zh.*, 93 (1966).
730. Mamedov M.A., Akhmedov I.M., Guseinov M.M. and Sadykh-Zade S.I., *Zh. Obshch. Khim.*, **35**, 461 (1965).
731. Mamedov M.A., Akhmedov I.M., and Sadykh-Zade S.I., *Azerb. Khim. Zh.*, 46 (1965).
732. Mamedov M.A., Shikhieva M.I. and Shikhiev I.A., *Azerb. Khim. Zh.*, 85 (1968).
733. Marciniec B. and Gulinski J., *J. Organometal. Chem.*, **253**, 349 (1983).
734. Marciniec B. and Gulinski J., in *Niektore aspekty fizycznej chemii nieorganicznej*, Pol. Akad. Nauk, Poznan, Poland, 1976, p. 141.
735. Marciniec B. and Gulinski J., *J, Mol. Catal.*, **10**, 123 (1981).
736. Marciniec B. and Gulinski J., in *Fundamental Res. Homogeneous Catal.* (A.E. Shilov, Ed), Gordon and Breach, New York, 1986, Vol. 5, p. 1079.
737. Marciniec B. and Gulinski J., *J. Organometal. Chem.*, **266**, C19 (1984).
738. Marciniec B. and Mackowska E., *J. Mol. Catal.*, **51**, 41 (1989).
739. Marciniec B. and Urbaniak W., *J. Mol. Catal.*, **18**, 49 (1983).
740. Marciniec B., in *Organosilicon and Bioorganosilicon Chemistry* (H. Sakurai, Ed), Ellis Horwood Ltd, New York, 1985, p. 183.
741. Marciniec B., Duczmal W., Urbaniak W. and Sliwinska E., *J. Organometal. Chem.*, **385**, 319 (1990).
742. Marciniec B., Foltynowicz Z. and Urbaniak W., *Appl. Organometal. Chem.*, **1**, 459 (1987).
743. Marciniec B., Foltynowicz Z. and Urbaniak W., *J. Mol. Catal.*, **42**, 195 (1987).
744. Marciniec B., Foltynowicz Z. and Urbaniak W. and Perkowski J., *Appl. Organometal. Chem.*,**1**, 267 (1987).
745. Marciniec B., Gulinski J. and Urbaniak W., *Pol. J. Chem.*, **56**, 287 (1982).
746. Marciniec B., Kornetka Z.W. and Urbaniak W., *J. Mol. Catal.*, **12**, 221 (1981).
747. Marciniec B., Gulinski J. and Urbaniak W., *Synth. React. Inorg. Metal.-Org. Chem.*, **12**, 139 (1982).
747a. Marciniec B., Gulinski J., Mirecki J. and Foltynowicz Z., *Polimery (Pol.)*, 213 (1990).

748. Marciniec B., Gulinski J., Urbaniak W., Nowicka T. and Mirecki J., *Appl. Organometal. Chem.*, **4**, 27 (1990).
749. Marciniec B., Mackowska E., Gulinski J. and Urbaniak W., *Z. anorg. allg. Chemie*, **529**, 222 (1985).
750. Marciniec B., Urbaniak W. and Pawlak P., *J. Mol. Catal.*, **14**, 323 (1982).
751. Markin S.V., Kudryavtsev G.V., Yudin A.V., Vertinskaya T.E. and Lisichkin G.V., *Zh. Obshch. Khim.*, **55**, 2761 (1985).
752. Maroshin Yu.V., Komarov N.V. and Kalabin G.A., *Khim. Atsetilena*, 155 (1968); *C.A.*, **70**, 115208m (1969).
753. Mashlyakovskii L.N. and Chelpanova L.F., *Zh. Obshch. Khim.*, **35**, 2009 (1965).
754. Massol M., Barrov J., Satge J. and Bovyssieres B., *J. Organometal. Chem.*, **80**, 47 (1974).
755. Masters C., *Transition-Metal Catalysis*, Chapman and Hall, New York, 1981.
756. Masuda T. and Ibuki H., *Polym. J.*, **12**, 143 (1980).
757. Masuda T. and Stille J.K., *J. Am. Chem. Soc.*, **100**, 268 (1978).
758. Matsuda I., Ogiso A., Sato S. and Izumi Y., *J. Am. Chem. Soc.*, **111**, 2332 (1989).
759. Matsumoto H., Hoshino Y. and Nagai Y., *Chem. Lett.*, 1663 (1982).
760. Matsumoto H., Hoshino Y. and Nagai Y., *Bull. Chem. Soc. Jpn.*, **54**, 1279 (1981).
761. Mayes N., Green J. and Cohen M.S., *J. Polym. Sci.*, A-1, 365 (1967).
762. McKay I.D. and Payne N.C., *Can. J. Chem.*, **64**, 1930 (1986).
763. Meen R.H. and Gilman H., *J. Org. Chem.*, **22**, 684 (1957).
764. Mejstrikova M., Rericha R. and Kraus M., *Collect. Czech. Chem. Commun.*, **39**, 135 (1974).
765. Michalska Z.M. and Ostaszewski B., *J. Organometal. Chem.*, **299**, 259 (1986).
766. Michalska Z.M. and Ostaszewski B., *Proc. IV Int. Symp. Homogeneous Catal., Leningrad, USSR*, 1984, Vol. 3, p. 158.
767. Michalska Z.M., *J. Mol. Catal.*, **3**, 125 (1977).
768. Michalska Z.M., *J. Mol. Catal.*, **19**, 345 (1983).
769. Michalska Z.M., Capka M. and Stoch J., *J. Mol. Catal.*, **11**, 323 (1981).
770. Mikautidze A.S., Chernomordlik Yu.A., Zawin B.G., Zhdanov A.A., Sergeev V.A., Khananashvili L.M. and Lekishvili N.G., *Vysokomol. Soedin., Ser. (A)*, **26**, 196 (1984).
771. Millan A., Fernandez M.J., Bentz P. and Maitlis P.M., *J. Mol. Catal.*, **26**, 89 (1984).
772. Millan A., Towns E. and Maitlis P.M., *J. Chem. Soc., Chem. Commun.*, 673 (1981).
773. Miller R.B. and McGarvey G., *J. Org. Chem.*, **43**, 4424 (1978).
774. Mironov V.F. and Fedotov N.S., *Khim. Geterotsikl. Soedin.*, 179 (1967).
775. Mironov V.F., *Dokl. Akad. Nauk SSSR*, **153**, 848 (1963).
776. Mironov V.F., Buyanov A.A. and Gar T.K., *Zh. Obshch. Khim.*, **41**, 2223 (1971).
777. Mironov V.F., Fedotov N.S. and Kozlikov V.L., *Khim. Geterotsikl. Soedin.*, 354 (1968).
778. Mironov V.F., Fedotov N.S. and Rybalka I.G., *Khim. Geterotsikl. Soedin.*, 440 (1969).
779. Mironov V.F., Kozlikov V.L. and Fedotov N.S., *Zh. Obshch. Khim.*, **41**, 1077 (1971).
780. Mironov V.F., Kozlikov V.L. and Fedotov N.S., *Zh. Obshch. Khim.*, **39**, 966 (1969).
781. Mironov V.F., Kozlikov V.L., Fedotov N.S., Khatuntsev G.D. and Sheludyakov V.D., *Zh. Obshch. Khim.*, **42**, 1365 (1972).
782. Mironov V.F., Kozlikov V.L., Kozyukov V.P., Fedotov N.S., Khatuntsev G.D. and Sheludyakov V.D., *Zh. Obshch. Khim.*, **41**, 2470 (1971).
783. Mironov V.F., Kozyukov V.P. and Sheludyakov V.D., *Dokl. Akad. Nauk SSSR*, **178**, 358 (1968).
784. Mironov V.F., Maksimova N.G. and Nepomnina V.V., *Izv. Akad. Nauk SSSR, Ser. Khim.*, 329 (1967).
785. Mironov V.F., Sheludyakov V.D. and Khatuntsev G.D., *Zh. Obshch. Khim.*, **42**, 2118 (1972).
786. Mironov V.F., Sheludyakov V.D., Khatuntsev G.D. and Kozyukov V.P., *Zh. Obshch. Khim.*, **42**, 2710 (1972).
787. Mitchener J.C. and Wrighton M.S., *J. Am. Chem. Soc.*, **103**, 975 (1981).

788. Moiseichuk K.L., Livshits F.Z., Yuvchenko A.P., Dikusar E.A. and Ol'dekop Y.A., *Zh. Obshch. Khim.*, **56**, 967 (1986).
789. Mokhlesur Rahman A.F.M. and Wild S.B., *J. Mol. Catal.*, **39**, 155 (1987).
790. Moldavskaya N.A., Chebrakov Yu.V., Skvortsov N.K., Krysenko L.V., Avetikyan G.B., Reikhsfel'd V.O. and Kukushkin Yu.N., *Zh. Obshch. Khim.*, **51**, 2279 (1981).
791. Moldavskaya N.A., Khvatova T.P., Skvortsov N.K. and Reikhsfel'd V.O., *Zh. Obshch. Khim.*, **50**, 851 (1980).
792. Moldavskaya N.A., Peshkova G.G., Sedova G.N., Skvortsov N.K. and Reikhsfel'd V.O., *Zh. Obshch. Khim.*, **51**, 2654 (1981).
793. Moldavskaya N.A., Skvortsov N.K., Voloshina N.F. and Rekhsfel'd V.O., *Zh. Obshch. Khim.*, **51**, 1621 (1981).
794. Moldavskaya N.A., Skvortsov N.K. and Reikhsfel'd V.O., *Proc. IV Int. Symp. Homogeneous Catal., Leningrad, USSR*, 1984, Vol. 1, p. 142.
795. Movsum-Zade E.M. and Narimanbekov O.A., *Azerb. Khim.Zh.*, **52**, 39 (1972).
796. Movsum-Zade E.M. and Shabanova D.A., *Uch. Zap.*, Ser. 9, 68 (1978) *Ref. Zh. Khim.*, Abstr. 1Zh364 (1979).
797. Movsum-Zade E.M., *Azerb. Khim. Zh.*, 135 (1977).
798. Movsum-Zade E.M., *Azerb. Khim. Zh.*, 90 (1981).
799. Movsum-Zade E.M., Narimanbekov O.A. and Shikhiev I.A., *Azerb. Neft. Khim.*, 38 (1970); *C.A.*, **74**, 3701t (1971).
800. Munstedt R. and Wannagat U., *Monatsh. Chem.*, **116**, 693 (1985).
801. Murai S. and Sonoda N., *Angew. Chem. Int. Ed. Engl.*, **18**, 837 (1979).
802. Murai T., Sakane T. and Kato S., *Tetrahedron Lett.*, **26**, 5145 (1985).
803. Murphy P.J., and Procter G., *Tetrahedron Lett.*, **31**, 1059 (1990).
804. Murphy P.J., Spencer J.L. and Procter G. *Tetrahedron Lett.*, **31**, 1051 (1990).
805. Mushegyan N.G., Vartanyan A.G. and Melikyan M.O., *Arm. Khim. Zh.*, **32**, 120 (1979); *C.A.* **91**, 193354d (1979).
806. Mustafaev R.M., Kulieva L.G. and Sadykh-Zade S.I., *Zh. Veses, Khim. O-va im D.I. Mendeleeva*, **31**, 474 (1986); *C.A.*, **107**, 39313c (1987).
807. Mustafaev R.M., Kulieva L.G., Sadykh-Zade S.I. and Mustafaev S.N., *Izv. Vyssh. Uchebn. Zaved., Khim. Khim. Tekhnol.*, **28**, 24 (1985).
808. Mustafaev R.M., Kulieva L.G., Sadykh-Zade S.I. and Mustafaev S.N., *Izv. Vyssh. Ucheb. Zaved., Khim. Khim. Tekhnol.*, **29**, 120 (1986); *C.A.*, **106**, 196489r (1987).
809. Mustafaev R.M., Mustafaev S.N., Shikhieva M.I. and Shikhiev I.A., *Zh. Obshch. Khim.*, **42**, 2705 (1972).
810. Naaktegeboren A.J., Nolte R.J.M. and Drenth W., *J. Am. Chem. Soc.*, **102**, 3350 (1980).
811. Nagai Y., Uetake K., Yoshikava T. and Matsumoto H., *J. Synth. Org., Chem. J.*, **31**, 759 (1973); *C.A.*, **80**, 48091j (1974).
812. Nagashima H., Tatebe K., Ishibashi T., Sakakibara J. and Itoh K., *Organometallics*, **8**, 2495 (1989).
813. Nagata Y., Dohmaru T. and Tsurugi J., *J. Org. Chem.*, **38**, 795 (1973).
814. Nakamura A. and Tsutsui M., *Principy i Primenienie Gomogennogo Kataliza*, Mir, Moskva, 1983.
815. Nakano T. and Nagai Y., *Chem. Lett.*, 481 (1988).
816. Nakao R., Fukumoto T. and Tsurugi J., *Bull. Chem. Soc. Jpn.*, **47**, 932 (1974).
817. Nametkin N.S., Gusel'nikov L.E., Ushakova R.L., Startseva O.M. and Vdovin V.M., *Izv. Akad. Nauk SSSR, Ser. Khim.*, 494 (1970).
818. Nametkin, N.S., Kobrakov K.I., Kuz'min O.V. and Sorokin G.V., *Vysokomol. Soedin., Ser. B*, **12,** 425 (1970).
819. Nametkin N.S., Lyashenko L.N., Chernysheva T.I., Borisov S.N. and Pestunovich V.A., *Dokl. Nauk SSSR*, **174**, 1105 (1967).
820. Nametkin N.S., Topchiev A.V., Chernysheva T.I. and Lyashenko I.N., *Dokl. Akad. Nauk SSSR*, **140**, 384 (1961).
821. Nametkin N.S., Vdovin V.M., Grinberg P.L. and Babich E.D., *Dokl. Akad. Nauk SSSR*, **161**, 358 (1965).

822. Nametkin N.S., Vdovin V.M. and Archipov T.N.. *Dokl. Akad. Nauk SSSR*, **159**, 146 (1964).
823. Nasiak L.D. and Post H.W., *J. Organometal. Chem.*, **23**, 91 (1970).
824. Nasirova M.M., Shikhiev I.A. and Rzaeva Sh.M., *Uch. Zap., Azerb. Univ., Ser. Khim.*, 64 (1971); *Ref. Zh. Khim.*, Abstr. 12Zh151 (1972).
825. Nekhorosheva E.V. and Al'bitskaya V.M., *Zh. Obshch. Khim.*, **38**, 1511 (1968).
826. Nesmeyanov A.N., Friedlina R.K., and Chukovskaya E.C., *Dokl. Akad. Nauk SSSR*, **113**, 120 (1957).
827. Nesmeyanov A.N., Friedlina R.K., Chukovskaya E.C., Petrov R.G. and Belyavsky A.B., *Tetrahedron*, **17**, 61 (1962).
828. Nikitin A.W. and Reikhsfel'd V.O., *Zh. Obshch. Khim.*, **55**, 2079 (1985).
829. Nishiyama H., Sakaguchi H., Nakamura T., Horihata M., Kondo M. and Itoh. K., *Organometallics*, **8**, 846 (1989).
830. Nogaideli A.I. and Rtveliashili N.A., *Tr. Tbilis. Univ.*, 101 (1972); *Ref. Zh. Khim.*, Abstr. 6Zh461 (1973).
831. Nogaideli A.I. and Rtveliashili N.A., *Tr. Tbilis. Univ.*, 141 (1974); *Ref. Zh. Khim.*, Abstr. 24Zh440 (1974).
832. Nogaideli A.I. and Rtveliashili N.A., *Tr. Tbilis. Univ.*, 177 (1973); *Ref. Zh. Khim.*, Abstr. 2Zh406 (1974).
833. Nogaideli A.I., Nakaidze L.I. and Tskhovrebashvili V.S., *Zh. Obshch. Khim.*, **44**, 1763 (1974).
834. Nogaideli A.I., Nakaidze L.I. and Tskhovrebashvili V.S., *Zh. Obshch. Khim.*, **45**, 1309 (1975).
835. Nogaideli A.I., Tkeshelashvili R.Sh. and Makharashvili N.P. *Soobshch. Akad. Nauk Gruz. SSR*, **76**, 621 (1974).
836. Nogaideli A.I., Tkeshelashvili R.Sh., Nogaideli G.A. and Chogovadze T.V., *Soobshch. Akad. Nauk Gruz. SSR*, **66**, 601 (1972).
837. Nogaideli A.I., Zhinkin D.Ya., Nakaidze L.I., Shapatin A.S. and Tskhovrebashvili V.S., *Zh. Obshch. Khim.*, **46**, 1048 (1976).
838. Novikov S.S. and Sevost'yanova V.V., *Izv. Akad. Nauk SSSR, Ser. Khim.*, 1485 (1962).
839. Novikova L.M., Ospanova K.M., Nazarova O.V., Dmitrenko A.V., Zubko N.V. and Rostovskii E.N., *Izv. Akad. Nauk Kaz. SSR*, **28**, 36 (1978).
840. Nozakura S. and Konotsune S., *Bull. Chem. Soc. Jpn*, **29**, 322 (1956).
841. Nozakura S. and Konotsune S., *Bull. Chem. Soc. Jpn*, **29**, 326 (1956).
842. Nozakura S., *Bull. Chem. Soc. Jpn*, **29**, 660 (1956).
843. Nozakura S., *Bull. Chem. Soc. Jpn*, **29**, 789 (1956).
844. Nurtdinova G.V., Gailunas G. and Yur'ev V., *Izv. Akad. Nauk SSSR, Ser. Khim.*, 2652 (1981).
845. Odabashyan G.V. and Tarashenko V.N., *Tr. Mosk. Khim.-Tekhnol. Inst.*, 60 (1969); *C.A.*, **73**, 13777r (1970).
846. Odabashyan G.V., Romashkin I.V., Rogachevskii V.L. and Kirichenko L.Ya., *Zh. Obshch. Khim.*, **38**, 2331 (1968).
847. Odabashyan G.V., Zhuravleva T.A., Golovin B.A. and Petrov A.D., *Tr. Mosk. Khim.-Tekhnol. Inst.*, 151 (1965); *C.A.*, **65**, 15413 (1966).
848. Oertle K. and Wetter H., *Tetrahedron Lett.*, **26**, 5511 (1985).
849. Ojima I. and Inaba S., *J. Organometal. Chem.*, **140**, 97 (1977).
850. Ojima I. and Kogure T., *Organometallics*, **1**, 1390 (1982).
851. Ojima I. and Kumagai M., *J. Organometal. Chem.*, **157**, 359 (1978).
852. Ojima I. and Kumagai M., *J. Organometal. Chem.*, **134**, C6 (1977).
853. Ojima I. and Nagai Y., *J. Organometal. Chem.*, **57**, C42 (1973).
854. Ojima I. and Nagai Y., *Chem. Lett.*, 233 (1974).
855. Ojima I., *J. Organometal. Chem.*, **134**, C1 (1977).
856. Ojima I., *Organotransition Met. Chem.*, 255 (1974).
857. Ojima I., *Strem Chemiker*, **8**, 1 (1980).
858. Ojima I., Fuchikami T. and Yatabe M., *J. Organometal. Chem.*, **260**, 335 (1984).

859. Ojima I., Inaba S. and Nagai Y., *Tetrahedron Lett.*, 4363 (1973).
860. Ojima I., Inaba S. and Nagai Y., *J. Organometal. Chem.*, **72**, C11 (1974).
861. Ojima I., Kogure T. and Kumagai M., *J. Org. Chem.*, **42**, 1671 (1977).
862. Ojima I., Kogure T. and Nagai Y., *Tetrahedron Lett.*, 2475 (1973).
863. Ojima I., Kogure T. and Nagai Y., *Tetrahedron Lett.*, 5025 (1972).
864. Ojima I., Kogure T. and Nagai Y., *Chem. Lett.*, 541 (1973).
865. Ojima I., Kogure T., Kumagai M., Horiuchi S. and Sato T., *J. Organometal. Chem.*, **122**, 83 (1976).
866. Ojima I., Kogure T., Nihonyanagi M. and Nagai Y., *Bull. Chem. Soc. Jpn*, **45**, 3506 (1972).
867. Ojima I., Kumagai M. and Miyazawa Y., *Tetrahedron Lett.*, 1385 (1977).
868. Ojima I., Kumagai M. and Nagai Y., *J. Organometal. Chem.*, **111**, 43 (1976).
869. Ojima I., Kumagai M. and Nagai Y., *Tetrahedron Lett.*, 4005 (1974).
870. Ojima I., Kumagai M. and Nagai Y., *J. Organometal. Chem.*, **66**, C14 (1974).
871. Ojima I., Nihonyanagi M. and Nagai Y., *Bull. Chem. Soc. Jpn*, **45**, 3722 (1971).
872. Ojima I., Nihonyanagi M. and Nagai Y., *J. Chem. Soc., Chem. Commun.*, 938 (1972).
873. Ojima I., Nihonyanagi M., Kogure T., Kumagai M., Horiuchi S. and Nakatsugawa K., *J. Organometal. Chem.*, **94**, 449 (1975).
874. Oka K., Nakao R., Abe Y. and Dohmaru T., *J. Organometal. Chem.*, **381**, 155 (1990).
875. Okinoshima H., Yamamoto K. and Kumada M., *J. Organometal. Chem.*, **86**, C27 (1975).
876. Onopchenko A. and Sabourin E.T., *J. Org. Chem.*, **52**, 4118 (1987).
877. Onopchenko A. and Sabourin E.T., *J. Chem. Eng. Data*, **32**, 64 (1988).
878. Onopchenko A., Dnopchenko A., Sabourin E.T. and Beach D.L., *J. Org. Chem.*, **48**, 5101 (1983).
879. Onopchenko A., Sabourin E.T. and Beach D.L., *J. Org. Chem.*, **49**, 3389 (1984).
880. Oro L.A., Fernandez M.J., Esteruelas M.A. and Jimenez M.S., *J. Mol. Catal.*, **37**, 151 (1986).
881. Oswald A.A., Murrel L.L. and Boucher L.J., *Prepr., Div. Pet. Chem., Am. Chem. Soc.*, **19**, 155 (1974); *C.A.*, **84**, 105685q (1976).
882. Ovchinnikova T.A. and Yakubchik A.I., *Vestn. Leningrad. Univ. Fiz. Khim.*, 134 (1971); *C.A.*, **76**, 142050j (1972).
883. Ovchinnikova T.A., Yakubchik A.I. and Ummurti G., *Vestn. Leningrad. Univ. Fiz. Khim.*, 114 (1971); *C.A.*, **76**, 46237e (1972).
884. Paige A.K., Joyce J.A., Nile T.A., Patel A.I., Reid C.D. and Walters J.M., *J. Mol. Catal.*, **29**, 201 (1985).
885. Pannell K.H., Rozell J.M., Lii J. and Tien-Mayr S.Y., *Organometallics*, **7**, 2524 (1988).
886. Panov V.B., Khidekel M.L. and Shchepinov S.A., *Izv. Akad. Nauk SSSR, Ser. Khim.*, 2397 (1968).
887. Panster P. and Kleinschmit P., *Proc. VII Int. Symp. Organosilicon Chem., Kyoto, Japan*, 1984, p. 150.
888. Park I.H., *Pollimo*, **4**, 358 (1989); *C.A.*, **111**, 154471b (1989).
889. Parshall G.W., *Inorg. Chem.*, **4**, 52 (1965).
890. Paul M. and Frainnet E., *J. Organometal. Chem.*, **30**, C64 (1971).
891. Payne N.C. and Stephan D.W., *Inorg. Chem.*, **21**, 182 (1982).
892. Perevalova V.I., Shimova N.B., Al't L.Ya. and Duplyakin Zh., *Obshch. Khim.*, **58**, 1694 (1988).
893. Petrashko A.I., Andrianov K.A., Korneeva G.P., Kuptsova Z.M., Moiseenko A.P. and Asnovich L.Z., *Vysokomol. Soedin., Ser. A.*, **9**, 2034 (1967).
894. Petrov A.D. and Sadykh-Zade S.I., *Zh. Obshch. Khim.*, **29**, 319 (1959).
895. Petrov A.D. and Sadykh-Zade S.I., *Dokl. Akad. Nauk SSSR*, **121**, 119 (1958).
896. Petrov A.D., Minachev K.M., Ponomarenko V.A., Sokolov B.A. and Odabashyan G.V., *Izv. Akad. Nauk SSSR, Ser. Khim.*, 1150 (1956).
897. Petrov A.D., Minachev K.M., Ponomarenko V.A., Sokolov B.A. and Odabashyan G.V., *Dokl. Akad. Nauk SSSR*, **112**, 273 (1957).

898. Petrov A.D., Mironov C.F., Vol'pin V.M. and Sadykh-Zade S.I., *Izv. Akad. Nauk SSSR, Ser. Khim.*, 250 (1956).
899. Petrov A.D., Ponomarenko V.A., Sokolov B.A. and Odabashyan G.V. and Krochmalev S.I., *Dokl. Akad. Nauk SSSR*, **124**, 838 (1959).
900. Petrov A.D., Ponomarenko V.A., Sokolov B.A. and Odabashyan G.V., *Izv. Akad. Nauk SSSR, Ser. Khim.*, 1206 (1957).
901. Petrov A.D., Sadykh-Zade S.I. and Filarova E.I., *Zh. Obshch. Khim.*, **29**, 2936 (1959).
902. Peyronel J.F., Fiaud J.C. and Kagan H.B., *J. Chem. Res. Synop.*, 320 (1980).
903. Pietrusza E.V., Sommer L.H. and Whitemore F.C., *J. Am. Chem. Soc.*, **70**, 484 (1948).
904. Pike R.A., *J. Org. Chem.*, **27**, 2186 (1962).
905. Pike R.A., McMahon G.E., and Jex V.B., *J. Org. Chem.*, **24**, 1939 (1959).
906. Pillot J.P., Dunogues J., Gerval J., The M.D. and Thanh M.V., *Eur. Polym. J.*, **25**, 285 (1989).
907. Pinazzi C.P., Soutif J.C. and Brosse J.C., *Bull. Soc. Chim. France*, 2166 (1974).
908. Pinazzi C.P., Soutif J.C. and Brosse J.C., *Eur. Polym. J.*, 523 (1975).
909. Pirkle W.H. and Hyun M.H., *J. Chromatogr.*, **322**, 295 (1985).
910. Pirkle W.H. and Pochapsky T.C., *J. Am. Chem. Soc.*, **108**, 352 (1986).
911. Pirkle W.H., Hyun M.H. and Bank B., *J. Chromatogr.*, **316**, 585 (1984).
912. Pitt C.G. and Skillern K.R., *J. Organometal. Chem.*, **7**, 525 (1967).
913. Pittman Jr. C.U., Honnick W., Absi-Halabi M., Richmond, M.G., Bender R. and Braunstein, P., *J. Mol. Catal.*, **32**, 177 (1985).
914. Pittman Jr. C.U., Honnick W.D., Wrighton M.S., Sanner R.D. and Austin R.G., in *Fundam. Res. Homogeneous Catalysis*, (M. Tsutsui, Ed.), Plenum Press, New York, 1979, Vol. 3, p. 603.
915. Pittman Jr. C.U., Richmond M.G., Absi-Halabi M., Beurich H., Richter F. and Wahrenkamp H., *Angew. Chem.*, **94**, 805 (1982).
916. Plueddemann E.D., Chemistry of silane coupling agents, in *Silylated Surfaces*, Gordon and Breach Sci., Publ. Inc., New York, 1980, p. 31.
917. Plueddemann E.D., *Silane Coupling Agents*, Plenum Press, New York, 1982.
918. Podol'skii A.V., Bulatov M.A. and Ezhova N.Yu., *Vysokomol. Soedin.*, **20**, 119 (1978).
919. Podol'skii A.V., Cherezova T.G., Kachalkov V.P. and Kodess M.I., *Izv. Akad. Nauk SSSR, Ser. Khim.*, 215 (1979).
920. Podol'skii A.V., Cherezova T.G., Podgornaya I.V., Popova V.A. and Kachalkov V.P., *Zh. Obshch. Khim.*, **50**, 1570 (1980).
921. Podol'skii A.V., Cherezova T.G. and Bulatov M.A., *Zh. Obshch. Khim.*, **47**, 1527 (1977).
922. Podol'skii A.V., Cherezova T.G. and Kachalkov V.P., *Izv. Akad. Nauk SSSR, Ser. Khim.*, 401 (1981).
923. Podol'skii A.V., Sukin A.V., Cherezova T.G., Kachalkov V.P. and Siniteyn V.V., *Izv. Akad. Nauk SSSR, Ser. Khim.*, 1945 (1978).
924. Podol'skii A.V., Suvorov A.L., Cherezova T.G. and Fridman L.I., *Zh. Obshch. Khim.*, **47**, 1532 (1977).
925. Podol'skii A.V., Suvorov A.L. and Cherezova T.G., *Izv. Akad. Nauk SSSR, Ser. Khim.*, 2778 (1979).
926. Pola J., Bazant V. and Chvalovsky V., *Collect. Czech. Chem. Commun.*, **37**, 3885 (1972).
927. Pola J., Bazant V. and Chvalovsky V., *Collect. Czech. Chem. Commun.*, **38**, 1528 (1973).
928. Polmanteer K.E., *Rub. Chem. Technol.*, **61**, 470 (1988).
929. Polyakova A.M., Suchkova M.D., Korshak V.V. and Vdovin V.M., *Izv. Akad. Nauk SSSR, Ser. Khim.*, 1267 (1965).
930. Pomerantseva M.G., Belyakova Z.V., Golubtsov S.A. and Efremova L.A., *Zh. Obshch. Khim.*, **40**, 1089 (1970).

931. Pomerantseva M.G., Belyakova Z.V., Golubtsov S.A., Zubkov V.I., Ainshtein A.A. and Baranova G.G., *Zh. Obshch. Khim.*, **42**, 862 (1972).
932. Pomerantseva M.G., Belyakova Z.V., Shchepinov S.A., Efremova L.A. and Chernyshev E.A., *Zh. Obshch. Khim.*, **54**, 354 (1984).
933. Ponec R., Chvalovsky V., Cerkayev V.G. and Komarenkova N.C., *Collect. Czech. Chem. Commun.*, **39**, 1177 (1974).
934. Ponomarenko V.A., Cherkayev V.G. and Zadrozhnii N.A., *Izv. Akad. Nauk SSSR, Ser. Khim.*, 1610 (1960).
935. Ponomarev A.I. and Klebanskii A.L., *Zh. Prikl. Khim.*, **46**, 960 (1973).
936. Povarnitsyna T.N., Dvinskikh V.V. and Lapkin I.I., *Zh. Obshch. Khim.*, **47**, 1538 (1977).
937. Pratt S.L. and Faltynek R.A., *J. Mol. Catal.*, **24**, 47 (1984).
938. Prignano A.L. and Trogler W.C., *J. Am. Chem. Soc.*, **109**, 3586 (1987).
939. Prignano A.L. and Trogler W.C., *Monatsh. Chem.*, **117**, 617 (1986).
940. Prud'homme C., Bargain M. and Lefort M., *Proc. VII Inst. Symp. Organosilicon Chem., Kyoto, Japan*, 1984, p. 195.
941. Pukhnarevich V.B., Burnashova T.D., Omel'chenko G.P., Tsykhanskaya I.I., Capka M. and Voronkov M.G., *Zh. Obshch. Khim.*, **56**, 2092 (1986).
942. Pukhnarevich V.B., Kopylova L.I., Capka M., Hetflejs J., Satsuk E.N., Sigalov M.V., Chvalovsky V. and Voronkov M.G., *Zh. Obshch. Khim.*, **50**, 1554 (1980).
943. Pukhnarevich V.B., Kopylova L.I., Korostova S.E., Mikhaleva A.I., Balabanova L.N., Vasil'ev A.N., Sigalov M.V., Trofimov B.A. and Voronkov M.G., *Zh. Obshch. Khim.*, **49**, 116 (1979).
944. Pukhnarevich V.B., Kopylova L.I., Trofimov B.A. and Voronkov M.G., *Zh. Obshch. Khim.*, **43**, 593 (1973).
945. Pukhnarevich V.B., Kopylova L.I., Trofimov B.A. and Voronkov M.G., *Zh. Obshch. Khim.*, **45**, 89 (1975).
946. Pukhnarevich V.B., Kopylova L.I., Trofimov B.A. and Voronkov M.G., *Zh. Obshch. Khim.*, **45**, 2638 (1975).
947. Pukhnarevich V.B., Kopylova L.I., Tsetlina E.O., Pestunovich V.A., Chvalovsky V., Hetflejs J. and Voronkov M.G., *Dokl. Akad. Nauk SSSR*, **231**, 1366 (1976).
948. Pukhnarevich V.B., Struchkov Yu.T., Aleksandrov G.G., Sushchinskaya S.P., Tsetlina E.O. and Voronkov M.G., *Koord. Khim.*, **5**, 1535 (1979).
949. Pukhnarevich V.B., Sushchinskaya S.P. and Voronkov M.G., *Zh. Obshch. Khim.*, **41**, 1879 (1971).
950. Pukhnarevich V.B., Sushchinskaya S.P., Druzhdzh P.V., Kopylova, L.I., Vainshtein B.A. and Voronkov M.G., *Khim. Prakt. Prim. Kremnii- i Fosfororg. Soedin., L.*, 37 (1978); *Ref. Zh. Khim.*, Abstr. 17B1033 (1979).
951. Pukhnarevich V.B., Sushchinskaya S.P., Pestunovich V.A. and Voronkov M.G., *Zh. Obshch. Khim.*, **43**, 1283 (1973).
952. Pukhnarevich V.B., Trofimov B.A., Kopylova L.I. and Voronkov M.G., *Zh. Obshch. Khim.*, **43**, 2691 (1973).
953. Pukhnarevich V.B., Tsykhanskaya I.I. and Voronkov M.G., *Izv. Akad. Nauk SSSR, Ser. Khim.*, 427 (1984).
954. Pukhnarevich V.B., Tsykhanskaya I.I., Ushakova N.I., Gelfman M.I. and Voronkov M.G., *Izv. Akad. Nauk SSSR, Ser. Khim.*, 2774 (1984).
955. Pukhnarevich V.B., Vcelak J., Voronkov M.G. and Chvalovsky V., *Collect. Czech. Chem. Commun.*, **39**, 2616 (1974).
956. Pukhnarevich V.B., Ushakova N.I., Adamovich S.N., Tsykhanskaya I.I., Varshavskii Y.S. and Voronkov M.G., *Izv. Akad. Nauk SSSR, Ser. Khim.*, **11**, 2589 (1985).
957. Pukhnarevich V.B., Ushakova N.I., Tsykhanskaya I.I., Albanov A.I. and Voronkov M.G., *Izv. Akad. Nauk SSSR, Ser. Khim.*, 2137 (1986).
958. Rabilloud G., *Bull. Soc. Chim. France*, 2152 (1965).
959. Rajkumar A.B. and Boudjouk P., *Organometallics*, **8**, 549 (1989).
960. Randolph C. and Wrighton M.S., *J. Am. Chem. Soc.*, **108**, 3366 (1986).

961. Reed J., Eisenberger P., Teo B.K. and Kincaid B.M., *J. Am. Chem. Soc.*, **99**, 5217 (1977).
962. Reed J., Eisenberger P., Teo B.K. and Kincaid B.M., *J. Am. Chem. Soc.*, **100**, 2375 (1978).
963. Reichel C.L. and Wrighton, M.S., *J. Am Chem. Soc.*, **103**, 7180 (1981).
964. Reichel C.L. and Wrighton, M.S., *Inorg. Chem.*, **19**, 3858 (1980).
965. Reikhsfel'd V.O. and Astrakhanov M.I., *Zh. Obshch. Khim.*, **47**, 1493 (1977).
966. Reikhsfel'd V.O. and Astrakhanov M.I., *Zh. Obshch. Khim.*, **47**, 1497 (1977).
967. Reikhsfel'd V.O. and Astrakhanov M.I., *Kremniiorg. Mat.*, 50 (1971); *Ref. Zh. Khim.*, Abstr. 7B1086 (1972).
968. Reikhsfel'd V.O. and Astrakhanov M.I., *Zh. Obshch. Khim.*, **43**, 2431 (1973).
969. Reikhsfel'd V.O. and Koroleva G.N., *Zh. Obshch. Khim.*, **37**, 2774 (1967).
970. Reikhsfel'd V.O. and Koroleva G.N., *Zh. Obshch. Khim.*, **36**, 1474 (1966).
971. Reikhsfel'd V.O. and Lebedev E.P., *Zh. Obshch. Khim.*, **40**, 615 (1970).
972. Reikhsfel'd V.O., Astrakhanov M.I. and Kagan E.G., *Zh. Obshch. Khim.*, **40**, 699 (1970).
973. Reikhsfel'd V.O., Filippov N.A. and Nikitin A.V., *Zh. Obshch. Khim.*, **47**, 959 (1977).
974. Reikhsfel'd V.O., Filippov N.A. and Zaslavskaya T.N. and Kuzmina G.A., *Zh. Obshch. Khim.*, **47**, 1488 (1977).
975. Reikhsfel'd V.O., Flerova N.I., Filippov N.A. and Zaslavskaya T.N., *Zh. Obshch. Khim.*, **50**, 2017 (1980).
976. Reikhsfel'd V.O., Gel'fman M.I., Khvatova T.P., Astrakhanov M.I. and Gavrilova I.V., *Zh. Obshch. Khim.*, **47**, 2093 (1977).
977. Reikhsfel'd V.O., Khvatova T.P., and Astrakhanov M.I., *Zh. Obshch. Khim.*, **47**, 2625 (1977).
978. Reikhsfel'd V.O., Khvatova T.P., Astrakhanov M.I. and Saenko G.I., *Zh. Obshch. Khim.*, **47**, 726 (1977).
979. Reikhsfel'd V.O., Saratov I.E. and Shpak I.V., *Zh. Obshch. Khim.*, **48**, 2383 (1978).
980. Reikhsfel'd V.O., Vinogradov V.N. and Filippov N.A., *Zh. Obshch. Khim.*, **43**, 2216 (1973).
981. Rejhon J. and Hetflejs J., *Collect. Czech. Chem. Commun.*, **40**, 3680 (1975).
982. Rejhon J. and Hetflejs J., *Collect. Czech. Chem. Commun.*, **40**, 3190 (1975).
983. Rericha R. and Capka M., *Collect. Czech. Chem. Commun.*, **39**, 144 (1974).
984. Revis A. and Hilty T.K., *J. Org. Chem.*, **55**, 2972 (1990).
985. Rijken F., Janssen M.J., Drenth W. and van der Kerk G.J.M., *J. Organometal. Chem.*, **2**, 347 (1964).
986. Ruiz J, Bentz P.O., Mann B.E., Spencer C.M., Taylor B.F., Maitlis P.M., *J. Chem. Soc., Dalton Trans.*, 2709 (1987).
987. Rumin R., *J. Organometal. Chem.*, **247**, 351 (1983).
988. Rustamov K.M. and Sultanov R.A., *Zh. Vses. Khim., O-va im D.I. Mendeleeva*, **31**, 473 (1986).
989. Rustamov K.M., Dzhalilov R.A., Sultanov R.A. and Saryev G.A., *Dokl. Akad. Nauk Azerb. SSR*, **44**, 42 (1988).
990. Rustamov K.M., Sultanov R.A., Saryer G.A., Mamedova R.I., Sultanova M.Sh., *Dokl. Akad. Nauk Azerb. SSR*, **43**, 39 (1987); *C.A.*, **110**, 135303m (1989).
991. Ryan J.W. and Speier J.L., *J. Org. Chem.*, **31**, 2698 (1966).
992. Ryan J.W. and Speier J.L., *J. Am. Chem. Soc.*, **86**, 895 (1964).
993. Ryu I., Kusumoto N., Ogawa A., Kambe N. and Sonoda N., *Organometallics*, **8**, 2279 (1989).
994. Rzaeva S.A., Shikhiev I.A., Askerov M.E. and Askerov G.F., *Azerb. Khim. Zh.*, 89 (1971).
995. Saam J.C. and Speier J.L., *J. Am. Chem. Soc.*, **80**, 4104 (1958).
996. Saam J.C. and Speier J.L., *J. Am. Chem. Soc.*, **83**, 1351 (1961).
997. Sadykh-Zade, S.I. and Babaeva R.B. and Salimov A., *Zh. Obshch. Khim.*, **36**, 695 (1966).

998. Sadykh-Zade, S.I. and Babaeva R.B., *Azerb. Khim. Zh.*, 38 (1966).
999. Sadykh-Zade, S.I. and Sultanova M.Sh., *Zh. Obshch. Khim.*, **44**, 1784 (1974).
1000. Sadykh-Zade, S.I., Sultanov R.A. and Mamedova B.A., *Azerb. Khim. Zh.*, 45 (1968).
1001. Sadykh-Zade, S.I., Sultanov R.A. and Mamedova B.A., *Dokl. Akad. Nauk Azerb. SSR*, **21**, 23 (1965).
1002. Sadykh-Zade, S.I., Sultanov R.A., Israfilov Ya.M. and Khudayarov I.A., *Zh. Obshch. Khim.*, **43**, 2248 (1973).
1003. Sadykh-Zade, S.I., Mamedova B.A. and Sultanov R.A., *Sin Prevrashch. Monomer. Soedin.*, 110 (1967); *C.A.*, **71**, 37876 (1969).
1004. Sadykh-Zade, S.I., Mardanov M.A., Sultanova Z.B. and Sultanov R.A., *Azerb. Khim. Zh.*, 29 (1966).
1005. Sadykh-Zade, S.I., Mardanov M.A., and Sultanova Z.B., *Vop. Neftekhim.*, 267 (1971); *C.A.*, 76, 99754c (1972).
1006. Sadykh-Zade, S.I., Mardanov M.A., and Sultanova Z.B., *Dokl. Akad. Nauk Azerb. SSR*, **26**, 23 (1970).
1007. Sadykh-Zade, S.I., Mustafaev R.M. and Kulieva L.G., *Dokl. Akad. Nauk Azerb. SSR*, **37**, 51 (1981).
1008. Sadykh-Zade, S.I., Sultanova Z.B. and Mardanov M.A., *Sint. Prevrashch. Monomer. Soedin.*, 91 (1967); *C.A.*, **70**, 37871f (1969).
1009. Sadykh-Zade, S.I., Sultanova Z.B. and Mardanov M.A., *Azerb. Khim. Zh.*, 83 (1967).
1010. Saegusa T., Ito Y., Kobayashi S. and Hirota K., *J. Am. Chem. Soc.*, **89**, 2240 (1967).
1011. Sagain N. and Gertner D., *J. Am. Oil Chem. Soc.*, **51**, 363 (1974).
1012. Sakurai H., Hirose T. and Hosomi A., *J. Organometal. Chem.*, **86**, 197 (1975).
1013. Sakurai H., Miyoshi K. and Nakadaira Y., *Tetrahedron Lett.*, 2671 (1977).
1014. Salimgareeva I.M., Kaverin V.V. and Yur'ev V.P., *J. Organometal. Chem.*, 148, 23 (1978).
1015. Salimgareeva I.M., Kaverin V.V., Kozhevnikova G.A. and Yur'ev V.P., *Zh. Obshch. Khim.*, **47**, 2626 (1977).
1016. Salimgareeva I.M., Kaverin V.V., Panasenko A.A., Khalilov L.M. and Yur'ev V.P., *Zh. Obshch. Khim.*, **48**, 930 (1978).
1017. Salimgareeva I.M., Kaverin V.V., Panasenko A.A., Rafikov S.R., Fedotov N.S. and Yur'ev V.P., *Dokl. Akad. Nauk SSSR*, **236**, 131 (1977).
1018. Salimgareeva I.M., Zhebarov O.Zh., Akhmetov R.R., Khalilov L.M. and Yur'ev V.P., *Izv. Akad. Nauk SSSR, Ser. Khim.*, 418 (1979).
1019. Salimgareeva I.M., Zhebarov O.Zh., Bogatova N.G., Lakhtin V.G. and Yur'ev V.P., *Izv. Akad. Nauk SSSR, Ser. Khim.*, 407 (1979).
1020. Salomon R.G., *Tetrahedron*, **39**, 485 (1983).
1021. Sassaman M.B., Kotion K.D., Prakasch G.K.S. and Olah G.A., *J. Org. Chem.*, **52**, 4314 (1987).
1022. Sato F., Vchiyama H. and Samaddar A.K., *Chem. Ind.*, 743 (1984).
1023. Schilling Jr C.L., *J. Organometal. Chem.*, **29**, 93 (1971).
1024. Schmidbaur H., Ebenhoch J. and Muller G., *Z. Naturforsch.*, **42B**, 142 (1987).
1025. Schmidt J.F. and Lampe F.W., *J. Phys. Chem.*, **73**, 2706 (1969).
1026. Schott G. and Berge H., *Chem. Techn.*, **6**, 503 (1954).
1027. Schott G. and Fischer E., *Chem. Ber.*, **93**, 2525 (1960).
1028. Schroeder M.A. and Wrighton M.S., *J. Organometal. Chem.*, **128**, 345 (1977).
1029. Schubert U., Egger Ch., Rose K. and Alt Ch., *J. Mol. Catal.*, **55**, 330 (1989).
1030. Schwartz N.S., O'Brien E., Karlan S. and Fein M.M., *Inorg. Chem.*, **4**, 661 (1965).
1031. Seitz F. and Wrighton M.S., *Angew. Chem.*, **27**, 289 (1988).
1032. Seki Y., Takeshita K., Kawamoto K., Murai S. and Sonoda N., *J. Org. Chem.*, **51**, 3890 (1986).
1033. Seki Y., Takeshita K., Kawamoto K., Murai S. and Sonoda N., *Angew. Chem., Int. Ed. Engl.*, **19**, 928 (1980).
1034. Seliga M., *Chem. Prumysl*, **16**, 481 (1966).
1035. Selin T.G. and West R., *J. Am. Chem. Soc.*, **84**, 1860 (1962).
1036. Selin T.G. and West R., *J. Am. Chem. Soc.*, **84**, 1863 (1962).

1037. Semmelhack M.F. and Misra R.N., *J. Org. Chem.*, **47**, 2469 (1982).
1038. Sergeev V.A., Shitikov V.K., Abbasov G.U., Bairamov M.R., Zhdanov A.A., Astapova T.V. and Aliev S.M., *Zh. Obshch. Khim.*, **52**, 1846 (1982).
1039. Sergeev V.A., Zhdanov A.A., Chernomordik Yu.A., Zawin B.G., Lekishvili N.G., Mikautidze A.S. and Korshak V.V., *Vysokomol. Soedin., Ser. A*, **23**, 1581 (1981).
1040. Severnyj V.V., Flaks E.Yu., Zhdanov A.A., Vlasova V.A., Andrianov K.A., and Vishnevskii F.N., *Vysokomol. Soedin.*, *Ser. A*, **16**, 419 (1974).
1041. Seyferth D. and Rochov E.G., *J. Am. Chem. Soc.*, **77**, 907 (1955).
1042. Seyferth D., Hung, P.L.K. and Hallgren J.E., *J. Organometal. Chem.*, **44**, C55 (1972).
1043. Seyferth D., Jula T.F., and Dertouzos H., *J. Organometal. Chem.*, **17**, 485 (1969).
1044. Seyferth D., Jula T.F., Dertouzos H. and Pereyre M., *J. Organometal. Chem.*, **11**, 63 (1968).
1045. Shapatin A.S., Golubtsov S.A., Solov'ev A.A., Zhigach A.F. and Siryatskaya V.N., *Plast. Massy*, **12**, 19 (1965).
1046. Sharf V.A., Shekoyan I.S. and Krutij V.V., *Izv. Akad. Nauk SSSR, Ser. Khim.*, 1087 (1977).
1047. Sharikova I.E. and Al'bitskaya V.M., *Izv. Vyssh. Ucheb. Zaved., Khim. Khim. Tekhnol.*, **9**, 595 (1966).
1048. Shchepinov S.A., Khidekel M.L. and Lagodzinskaya G.V., *Izv. Akad. Nauk SSSR, Ser. Khim.*, 2165 (1968).
1049. Shchukovskaya L.L. and Pal'chik R.I., *Zh. Obshch. Khim.*, **35**, 1122 (1965).
1050. Shchukovskaya L.L., Pal'chik R.I. and Petrov A.D., *Dokl. Akad. Nauk SSSR*, **160**, 621 (1965).
1051. Sheludyakov V.D., Khatuntsev G.D. and Mironov V.F., *Zh. Obshch. Khim.*, **43**. 314 (1973).
1052. Sheludyakov V.D., Khatuntsev G.D. and Mironov V.F., *Zh. Obshch. Khim.*, **42**, 2209 (1972).
1053. Sheludyakov V.D., Korshunov A.J., Lakhtin V.G., Timofeev V.S., Shynsarenko T.F., Nosova V.M. and Gradova E.V., *Zh. Obshch. Khim.*, **56**, 2743 (1986).
1054. Sheludyakov V.D., Kozyukov V.P., Rybakov E.A. and Mironov V.F., *Zh. Obshch. Khim.*, **37**, 2141 (1967).
1055. Sheludyakov V.D., Lachshin V.G., Nosova V.M., Kisin A.V., Stolyarova O.V., Korshunov A.J., Alekseev N.V., *Zh. Obshch. Khim.*, **59**, 1280 (1989).
1056. Sheludyakov V.D., Zhun V.I. and Vlasenko S.D., *Zh. Obshch. Khim.*, **55**, 1795 (1985); *C.A.*, **105**, 42907t (1986).
1057. Sheludyakov V.D., Zhun V.I., Lakhtin V.G., Bochkarev V.N., Slusarenko T.F., Nosova V.M. and Kisin A.V., *Zh. Obshch. Khim.*, **54**, 640 (1984).
1058. Sheludyakov V.D., Zhun V.I., Vlasenko S.D., Bochkarev V.N., Slusarenko T.F., Kisin A.V., Nosova V.M., Turkel'taub G.N. and Chernyshev E.A., *Zh. Obshch. Khim.*, **51**, 2022 (1981).
1059. Sheludyakov V.D., Zhun V.I., Vlasenko S.D., Bochkarev V.N., Slusarenko T.F., Turkel'taub G.N., Kisin A.V. and Nosova V.M., *Zh. Obshch. Khim.*, **55**, 1544 (1985)
1060. Shieh C.H., Karger B.L., Gelber L.R., Feibush B., *J. Chromatogr.*, **406**, 343 (1987).
1061. Shikhiev I.A. and Nasirova M.M., *Uch. Zap. Azerb. Univ., Ser. Khim. Nauk*, 61 (1968); *Ref. Zh. Khim.*, Abstr. 12Zh476 (1969).
1062. Shikhiev I.A. and Nasirova M.M., *Azerb. Khim. Zh.*, 23 (1968).
1063. Shikhiev I.A. and Rzaeva S.A., *Uch. Zap. Azerb. Inst. Neft. Khim.*, **9**, 116 (1971); *Ref. Zh. Khim.*, Abstr. 13Zh389 (1971).
1064. Shikhiev I.A., Abbasova G.A. and Gasanova R.Yu., *Azerb. Khim. Zh.*, 33 (1974).
1065. Shikhiev I.A., Abdulaev N.D. and Akhundova G.Yu., *Azerb. Khim. Zh.*, 87 (1968).
1066. Shikhiev I.A., Abdulaev N.D., Mustafaev R.M. and Shikhieva M.I., *Zh. Obshch. Khim.*, **36**, 1843 (1966).
1067. Shikhiev I.A., Aliev M.I. and Yusufov B.G., *Zh. Obshch. Khim.*, **35**, 1654 (1965).
1068. Shikhiev I.A., Askerov F.V. and Garaeva Sh.V. *Azerb. Khim. Zh.*, 33 (1966).
1069. Shikhiev I.A., Askerov F.V. and Garaeva Sh.V., *Zh. Obshch. Khim.*, **38**, 639 (1968).
1070. Shikhiev I.A., Askerov F.V. and Rzaeva S.A., *Zh. Obshch. Khim.*, **41**, 872 (1971).

1071. Shikhiev I.A., Askerov F.V., Garaeva Sh.V. and Aliev M.I., *Zh. Obshch. Khim.*, **37**, 317 (1967).
1072. Shikhiev I.A., Aslanov I.A. and Verdieva S.Sh., *Zh. Obshch. Khim.*, **36**, 355 (1966).
1073. Shikhiev I.A., Dzhafarov D.S. and Karaev S.F., *Zh. Obshch. Khim.*, **45**, 1340 (1975).
1074. Shikhiev I.A., Dzhafarov D.S. and Karaev S.F., *Zh. Obshch. Khim.*, **43**, 1994 (1973).
1075. Shikhiev I.A., Garaeva Sh.V. and Shikhieva M.I., *Zh. Obshch. Khim.*, **37**, 2778 (1967).
1076. Shikhiev I.A., Garaeva Sh.V., Mamedova L.D. and Karaev S.F., *Zh. Org. Khim.*, **13**, 512 (1977).
1077. Shikhiev I.A., Gasanova R.Yu., Askerov G.F. and Mustafaev R.M., *Azerb. Khim. Zh.*, 117 (1970).
1078. Shikhiev I.A., Gasanova R.Yu., Askerov G.F. and Rzaeva S.A., *Zh. Obshch. Khim.*, **40**, 817 (1970).
1079. Shikhiev I.A., Guseinova M.A., Rzaeva S.A., Gasanova R.Yu. and Shakhverdieva F.M., *Uch. Zap. Azerb. Inst. Neft. Khim.*, **9**, 66 (1972); *Ref. Zh. Khim.*, Abstr. 3Zh385 (1973).
1080. Shikhiev I.A., Ibragimov V.A. and Karaev S.F., *Uch. Zap. Azerb. Univ., Ser. Khim. Nauk*, 36 (1972); *Ref. Zh. Khim.*, Abstr. 12Zh415 (1973).
1081. Shikhiev I.A., Karaev S.F. and Garaeva Sh.V., *Zh. Obshch. Khim.*, **38**, 2570 (1968).
1082. Shikhiev I.A., Karaev S.F., Kazieva S.T. and Tsalikova Z.M., *Azerb. Khim. Zh.*, 43 (1975).
1083. Shikhiev I.A., Mekhmandarova N.T. and Aslanova I.A., *Azerb. Khim. Zh.*, 59 (1965).
1084. Shikhiev I.A., Movsum-Zade E.M., Shabanova D.A., Nasirov Ya.F. and Rasulbekova T.I., *Zh. Obshch. Khim.*, **43**, 614 (1973).
1085. Shikhiev I.A., Mukharamova Kh.F., Aliev M.I. and Alieva A.R., *Azerb. Neft. Khoz.*, **45**, 41 (1966); *C.A.*, **65**, 16992g (1966).
1086. Shikhiev I.A., Nasirov Ya.F., Gasanova R.Yu., Shikhieva M.I. and Movsum-Zade E.M., *Zh. Obshch. Khim.*, **45**, 100 (1975).
1087. Shikhiev I.A., Rzaeva S.A., Dadasheva Ya.A. and Guseinova M.A., in *Khim. Elementoorg. Soedin.*, Nauka, Leningrad, 1976, p. 71.
1088. Shikhiev I.A., Rzaeva S.A. and Guseinova M.A., *Zh. Obshch. Khim.*, **42**, 1525 (1972).
1089. Shikhiev I.A., Shiraliev V.M. and Askerov G.F., *Zh. Obshch. Khim.*, **39**, 1299 (1969).
1090. Shikhiev I.A., Vatankha A.A. and Gusein-Zade B.M., *Zh. Obshch. Khim.*, **35**, 812 (1965).
1091. Shikhiev I.A., Vatankha A.A. and Gusein-Zade B.M., *Zh. Obshch. Khim.*, **36**, 1293 (1966).
1092. Shikhiev I.A., Verdieva S.Sh., Aslanov I.A. and Akhmedova F.M., *Azerb. Khim. Zh.*, 41 (1966).
1093. Shikhmamedbekova A.Z. and Khudiev A.K., *Zh. Org. Khim.*, **12**, 335 (1976).
1094. Shlapa J., Loetzsch P. and Fischer B., *Chem. Tech.*, **35**, 404 (1983).
1095. Shostakovskii M.F. and Kochkin D.A., *Dokl. Akad. Nauk SSSR*, **109**, 113 (1956).
1096. Shostakovskii M.F. and Kochkin D.A., *Izv. Akad. Nauk SSSR, Ser. Khim.*, 1150 (1956).
1097. Shostakovskii M.F., Atavin A.S., Vyalykh E.P., Gavrilova G.M., Kalabin G.A. and Trofimov B.A., *Zh. Obshch. Khim.*, **41**, 613 (1971).
1098. Shostakovskii M.F., Komarov N.V. and Burnashova T.D., *Izv. Akad. Nauk SSSR, Ser. Khim.*, 629 (1968).
1099. Shostakovskii M.F., Komarov N.V. and Roman V.K., *Zh. Obshch. Khim.*, **38**, 382 (1968).
1100. Shostakovskii M.F., Komarov N.V., Yarosh O.G. and Balashenko L.V., *Izv. Akad. Nauk SSSR, Ser. Khim.*, 1478 (1971).
1101. Shostakovskii M.F., Sokolov B.A., Dimitreev T.V. and Alekseeva T.M., *Zh. Obshch. Khim.*, **33**, 3778 (1963).
1102. Shostakovskii M.F., Sushchinskaya S.P., Pukhnarevich V.B., Borysova A.I., Filippova A.N. and Sakharovskii V.G., *Izv. Akad. Nauk SSSR, Ser. Khim.*, 2599 (1969).
1103. Shvekhgeimer G.A. and Kruchkova A.P., *Zh. Obshch. Khim.*, **36**, 1852 (1966).

1104. Siegfried V., Frainnet E., Barthelemy J.C. and Calas R., *Bull. Soc. Chim. France*, 2698 (1965).
1105. Skoda-Foldes R., Kollar L. and Heil B., *J. Organometal. Chem.*, **366**, 275 (1989).
1106. Skvortsov N.K., Brovko V.S. and Reikhsfel'd V.O., *Zh. Obshch. Khim.*, **54**, 2310 (1984).
1107. Skvortsov N.K., Patrushina N.L. and Moldavskaya N.A., *Proc. IV Int. Symp. Homogeneous Catal., Leningrad, USSR*, 1984, Vol. 1 p. 173.
1108. Skvortsov N.K., Voloshina N.F., Moldavskaya N.A. and Reikhsfel'd V.O., *Zh. Obshch. Khim.*, **58**, 1831 (1988).
1109. Skvortsov N.K., Voloshina N.F., Moldavskaya N.A. and Reikhsfel'd V.O., *Zh. Obshch. Khim.*, **55**, 2323 (1985).
1110. Sleta T.M. and Stadnichuk M.D., *Zh. Obshch. Khim.*, **39**, 2031 (1969).
1111. Sleta T.M., Stadnichuk M.D. and Petrov A.A., *Zh. Obshch. Khim.*, **38**, 374 (1968).
1112. Sleta T.M., Stadnichuk M.D. and Petrov A.A., *Zh. Obshch. Khim.*, **39**, 1173 (1969).
1113. Smith A.G., Ryan J.W. and Speier J.L., *J. Org. Chem.*, **27**, 2183 (1962).
1114. Snyder D.C., *J. Organometal. Chem.*, **301**, 137 (1986).
1115. Sobolev E.S., Antipin L.M. and Mironov V.F., *Zh. Obshch. Khim.*, **39**, 2691 (1969).
1116. Sobolevskii M.V., Malyrovskaya O.M., Muzovskaya O.A. and Popeleva G.S., *Svoistva i Oblasti Primenenya Kremnoorganicheskikh Productov*, Khimiya, Moskva, 1975.
1117. Sobolevskii M.V., Zhigach A.F., Grinevich K.P., Sarishvili I.G., Siryatskaya V.N. and Kozyreva E.M., *Plast. Massy*, 21 (1966).
1118. Sokolov B.A. and Grishko A.N., *Zh. Obshch. Khim.*, **37**, 2117 (1967).
1119. Sokolov B.A., Alekseeva G.M. and Dimitrieva G.V., *Zh. Obshch. Khim.*, **35**, 1839 (1965).
1120. Sokolov B.A., Grishko A.N., Kuznetsova T.A. and Sultangareeva R.G., *Zh. Obshch. Khim.*, **37**, 260 (1967).
1121. Sokolov B.A., Kail'ko O.N., Zhivotova M.M. and Kositsyna E.I., *Zh. Obshch. Khim.*, **36**, 108 (1966).
1122. Sokolova E.B., Massarskaya S.M. and Varfolomeeva N.A., *Zh. Obshch. Khim.*, **40**, 1762 (1970).
1123. Sokolova E.B., Massarskaya S.M., Dyatlova V.G. and Prokopova L.K., *Zh. Obshch. Khim.*, **43**, 2706 (1973).
1124. Sommer L.H. and Lyons J.E., *J. Am. Chem. Soc.*, **90**, 4197 (1968).
1125. Sommer L.H., Lyons J.E. and Fujimoto H., *J. Am. Chem. Soc.*, **91**, 7051 (1969).
1126. Sommer L.H., Lyons J.E., Michael K.W. and Fujimoto H., *J. Am. Chem. Soc.*, **89**, 5483 (1967).
1127. Sommer L.H., McKay F.P., Steward O.W. and Campbell P.G., *J. Am. Chem. Soc.*, **79**, 2764 (1957).
1128. Sommer L.H., Michael K.W. and Fujimoto H., *J. Am. Chem. Soc.*, **89**, 1519 (1967).
1129. Sommer L.H., Pietrusza E.W. and Whitmore F.C., *J. Am. Chem. Soc.*, **69**, 188 (1947).
1130. Souchek I., Andrianov K.A., Khananashvili L.M. and Myasina V.M., *Dokl. Akad. Nauk SSSR*, **222**, 128 (1975).
1131. Souchek I., Andrianov K.A., Khananashvili L.M. and Myasina V.M., *Dokl. Akad. Nauk SSSR*, **227**, 98 (1976).
1132. Speier J.L. and Webster J.A., *J. Org. Chem.*, **21**, 1044 (1956).
1133. Speier J.L., Webster J.A. and Barnes G.H., *J. Am. Chem. Soc.*, **79**, 974 (1957).
1134. Speier J.L., Zimmerman R.E. and Webster J.A., *J. Am. Chem. Soc.*, **78**, 2278 (1956).
1135. Spevak V.N., Lopez X., Area O., Voloshina N.F., Skvortsov N.K., Reikhsfel'd V.O., *Zh. Obshch. Khim.*, **56**, 479 (1986).
1136. Spialter L. and O'Brien D.H., *J. Org. Chem.*, **32**, 222 (1967).
1137. Stadnichuk M.D., Sleta T.M. and Petrov A.A., *Zh. Obshch. Khim.*, **38**, 2573 (1968).
1138. Staroverov S.M., Nesterenko P.N. and Lisichkin G.V., *Zh. Obshch. Khim.*, **49**, 2487 (1979).
1139. Steinmetz M.G. and Udayakumar B.S., *J. Organometal. Chem.*, **378**, 1 (1989).
1140. Steinmetz M.G., Mayes R.T. and Udayakumar B.S., *J. Chem. Soc. Chem. Commun.*, 759 (1987).

1141. Steward O.W., Lutkus A.G. and Greenshields J.B., *J. Organometal. Chem.*, **144**, 147 (1978).
1142. Stork G. and MacDonald T.L., *J. Am. Chem. Soc.*, **97**, 1264 (1975).
1143. Struchkov Yu.T., Aleksandrov G.G., Pukhnarevich V.B., Sushchinskaya S.P. and Voronkov M.G., *J. Organometal. Chem.*, **172**, 269 (1979).
1144. Suess-Fink G. and Reiner J., *J. Mol. Catal.*, **16**, 231 (1982).
1145. Suess-Fink G. and Reiner J., *J. Organometal. Chem.*, **221**, C36 (1981).
1146. Sultanov R.A. and Rustamov K.M., *Dokl. Akad. Nauk Azerb. SSR*, **43,** 44 (1987); *C.A.*, **110,** 114911t (1989).
1147. Sultanov R.A., Abasova G.G. and Dzhalilov R.A., *Dokl. Akad. Nauk Azerb. SSR*, **44,** 35 (1988); *C.A.*, **111**, 174201s (1989).
1148. Sultanov R.A., Abasova G.G. and Dzhalilov R.A., *Dokl. Akad. Nauk Azerb. SSR*, **44,** 33 (1988); *C.A.*, **112**, 98625m (1990).
1149. Sultanov R.A., Askerov O.V., Aronova L.A., Khudayarov I.A. and Sadykh-Zade S.I., *Zh. Obshch. Khim.*, **45**, 2102 (1975).
1150. Sultanov R.A., Askerov O.V., Bairamov G.K. and Tarverdiev Sh.A., *Zh. Obshch. Khim.*, **46**, 1806 (1976).
1151. Sultanov R.A., Gasanova F.A. and Sadykh-Zade S.I., *Zh. Obshch. Khim.*, **37**, 739 (1967).
1152. Sultanov R.A., Israfilov Ya.M. and Sadykh-Zade S.I., *Zh. Obshch. Khim.*, **42**, 160 (1972).
1153. Sultanov R.A., Kadyrova M.B., Khudayarov I.A. and Sadykh-Zade S.I., *Uch. Zap. Azerb. Univ., Ser. Khim. Nauk*, 37 (1971); *Ref. Zh. Khim.*, 24Zh730 (1971).
1154. Sultanov R.A., Khudayarov I.A. and Sadykh-Zade S.I., *Zh. Obshch. Khim.*, **39**, 396 (1969).
1155. Sultanov R.A., Khudayarov I.A., Khalilova E.M. and Sadykh-Zade S.I., *Azerb. Khim. Zh.*, 97 (1969).
1156. Sultanov R.A., Kulizade F.A. and Sadykh-Zade S.I., *Azerb. Khim Zh.*, 73 (1970).
1157. Sultanov R.A., Kulizade F.A., Sadykh-Zade S.I. and Bairamov G.K., *Zh. Obshch. Khim.*, **42**, 2038 (1972).
1158. Sultanov R.A., Kulizade F.A., Sadykh-Zade S.I., Rzaeva A.S. and Shukyurova M.B., *Dokl. Akad. Nauk Azerb. SSR*, **29**, 21 (1973).
1159. Sultanov R.A., Radzhabov M.I., Tarverdiev Sh.A. and Bairamov G.K., *Zh. Obshch. Khim.*, **45**, 2669 (1975).
1160. Sultanov R.A., Rustamov K.M., Pashaer Z.M. and Mamedova R.I., *Azerb. Khim. Zh.*, 78 (1987); *C.A.*, **110**, 95336r (1989).
1161. Sultanov R.A., Tarverdiev Sh.A., Askerov O.V. and Sadykh-Zade S.I., *Zh. Obshch. Khim.*, **42**, 2707 (1972).
1162. Sultanov R.A., Tarverdiev Sh.A., Novruzov Sh.M. and Sadykh-Zade S.I., *Zh. Obshch. Khim.*, **42**, 1058 (1972).
1163. Suryanarayanan B., Peace B.W. and Mayhon K.G., *J. Organometal. Chem.*, **55**, 65 (1973).
1164. Suyma Y., Marra R.M., Haggerty J.S. and Bowen K., *Am. Ceram. Soc. Bull.*, **64**, 1356 (1985).
1165. Svoboda P. and Hetflejs J., *Collect. Czech. Chem. Commun.*, **38**, 3834 (1973).
1166. Svoboda P., Capka M. and Hetflejs J., *Collect. Czech. Chem. Commun.*, **38**, 1235 (1973).
1167. Svoboda P., Capka M. and Hetflejs J., *Collect. Czech. Chem. Commun.*, **37**, 3059 (1972).
1168. Svoboda P., Capka M., Chvalovsky V., Bazant V., Hetflejs J., Jahr H. and Pracejus H., *Z. Chem..*, **12**, 153 (1972).
1169. Svoboda P., Capka M., Hetflejs J. and Chvalovsky V., *Collect. Czech. Chem. Commun.*, **37**, 1585 (1972).
1170. Svoboda P., Rericha R. and Hetflejs J., *Collect. Czech. Chem. Commun.*, **39**, 1324 (1974).

1171. Svoboda P., Sedlmayer P. and Hetflejs J., *Collect. Czech. Chem. Commun.*, **38**, 1783 (1973).
1172. Swisher J.V. and Chen H.H., *J. Organometal. Chem.*, **69**, 83 (1974).
1173. Swisher J.V. and Zullig Jr Ch., *J. Org. Chem.*, **38**, 3353 (1973).
1174. Takahashi H. and Hagihara N., *Kogyo Kagaku Zasshi*, **72**, 1637 (1969); *C.A.*, **72**, 11726d (1970).
1175. Takahashi H., *Osaka Kogyo Gijutsu Shikensho Hokoku*, 1 (1965); *C.A.*, **67**, 53755q (1967).
1176. Takahashi H., Shibano T. and Hagihara N., *J. Chem. Soc., Chem. Commun.*, 161 (1969).
1177. Takahashi H., Shibano T., Kojima H. and Hagihara N., *Organometal. Chem. Synth.*, **1**, 193 (1971).
1178. Takeshita K., Seki Y., Kawamoto K., Murai S. and Naboru S., *J. Org. Chem.*, **52**, 4864 (1987).
1179. Tamao K., Akita M. and Kumada M., *J. Organometal. Chem.*, **254**, 13 (1983).
1180. Tamao K., Ishida N., Tanaka T. and Kumada M., *Organometallics*, **2**, 1694 (1983).
1181. Tamao K., Iyoda J. and Shiihara I., *Kogyo Kagaku Zasshi*, **71**, 923 (1968).
1182. Tamao K., Kobayashi K. and Ito Y., *J. Am. Chem. Soc.*, **111**, 6478 (1989).
1183. Tamao K., Maeda K., Tanaka T. and Ito Y., *Tetrahedron Lett.*, **29**, 6955 (1988).
1184. Tamao K., Miyake N., Kiso Y. and Kumada M., *J. Am. Chem. Soc.*, **97**, 5603 (1975).
1185. Tamao K., Nakagawa Y., Arai H., Higuchi N. and Ito Y., *J. Am. Chem. Soc.*, **110**, 3712 (1988).
1186. Tamao K., Nakajima T. and Tanaka T., *Proc. VII Int. Symp. Organosilicon Chem., Kyoto, Japan*, 1984, p. 234.
1187. Tamao K., Nakajima T., Sumiya R., Arai H., Higuchi N. and Ito Y., *J. Am. Chem. Soc.*, **108**, 6090 (1986).
1188. Tamao K., Nakajo E. and Ito Y., *J. Org. Chem.*, **53**, 414 (1988).
1189. Tamao K., Nakajo E. and Ito Y., *J. Org. Chem.*, **52**, 4414 (1987).
1190. Tamao K., Tanaka T., Nakajima T., Sumiya R., Arai H. and Ito Y., *Tetrahedron Lett.*, **27**, 3377 (1986).
1191. Tamao K., Yamauchi T. and Ito Y., *Chem. Lett.*, 171 (1987).
1192. Tamao K., Yoshida J., Yamamoto H., Kakui T., Matsumoto H., Takahashi M., Kurita A., Murata M. and Kumada M., *Organometallics*, **1**, 355 (1982).
1193. Tarasenko V. and Odabashyan G.V., *Tr. Mosk. Khim. Tekhnol. Inst.*, **70**, 138 (1972).
1194. Tarrant P. and Oliver W.H., *J. Org. Chem.*, **31**, 1143 (1966).
1195. Tarverdiev Sh.A. and Sadykh-Zade R.A., *Dokl. Akad. Nauk Azerb. SSR*, **29**, 34 (1973).
1196. Tarverdiev Sh.A., Sultanov R.A., Bairamov G.K. and Khudayarov I.A., *Zh. Obshch. Khim.*, **44**, 1511 (1974).
1197. Terentev A.B., Moskalenko M.A. and Freidlina R.Kh., *Izv. Akad. Nauk SSSR*, 2825 (1984).
1198. Thames S.F., Bufkin B.G., Jen S.J., Evans J.M. and Long J.S., *J. Coat. Technol.*, **48**, 46 (1976).
1199. Tolchinskaya R.E., Nanush'yan S.R., Gorbatkina Yu.A. and Alekseeva E.A., *Mekh. Kompoz. Mater.*, 25 (1987). *C.A.*, **106,** 177911f (1987).
1200. Topchiev A.V., Nametkin N.S. and Chernysheva T.I., *Dokl. Akad. Nauk SSSR*, **118**, 517 (1958).
1201. Torii S., Takagishi S., Inokuchi T. and Okumoto H., *Bull. Chem. Soc. Jpn*, **60**, 775 (1987).
1202. Trofimov B.A., Mikhaleva A.I., Belyaevskii A.I., Volyanskaya N.A., Korostova S.E., Balabanov L.N., Vasil'ev A.N. and Volyanskii Yu.L., *Khim. Pharm. Zh.*, **15**, 25 (1981).
1203. Trogler W.C., *J. Chem. Educ.*, **65**, 294 (1988).
1204. Tsetlina E.O., Pukhnarevich V.B., Lukevics E. and Voronkov M.G., *Zh. Obshch. Khim.*, **46**, 2155 (1976).

1205. Tsipis C.A., *J. Organometal. Chem.*, **187**, 427 (1980).
1206. Tsipis C.A., *J. Organometal. Chem.*, **188**, 53 (1980).
1207. Tsuji J., *Kagaku (Kyoto)*, **27**, 292 (1972); *C.A.*, **78**, 143245e (1973).
1208. Tsuji J., Hara M. and Ohno K., *Tetrahedron*, **30**, 2143 (1974).
1209. Tsunin E.D., Volin Yu.M., Lazarov S.Y. and Kalitsyn I.V., *Zh. Prikl. Khim.*, 1650 (1988).
1210. Tsurugi J., Nakao R. and Fukumoto T., *J. Org. Chem.*, **35**, 833 (1970).
1211. Tsurugi J., Nakao R. and Fukumoto T., *J. Am. Chem. Soc.*, **91**, 4587 (1969).
1212. Tsykhanskaya I.I., Vlasova N.N., Keiko V.V., Sakharovskii V.G. and Balashenko L.V., *Izv. Akad. Nauk SSSR, Ser. Khim.*, 2761 (1970).
1213. Ueda T., Inukai K. and Muramatsu H., *Kogyo Kagaku Zasshi*, **74**, 1676 (1971); *C.A.*, **75**, 98614z (1971).
1214. Ueda T., Muramatsu H. and Inukai K., *Nippon Kagaku Kaishi*, 602 (1972); *C.A.*, **77**, 5566p (1972).
1215. Urenovitch J.V. and West R., *J. Organometal. Chem.*, **3**, 138 (1965).
1216. Vainshtein B.I., Bol'shakova S.A., Vlasova N.N. and Voronkov M.G., *Zh. Obshch. Khim.*, **49**, 719 (1979).
1217. Vaisarova V. and Hetflejs J., *Collect. Czech. Chem. Commun.*, **41**, 1906 (1976).
1218. Vaisarova V. and Hetflejs J., *Synth. Inorg. Metal.-Org. Chem.*, **13**, 977 (1983).
1219. Vaisarova V., Capka M. and Hetflejs J., *Synth. Inorg. Metal.-Org. Chem.*, **2**, 289 (1972).
1220. Vaisarova V., Hetflejs J., Krause H.W. and Pracejus H., *Z. Chem.*, **14**, 105 (1974).
1221. Vaisarova V., Langova J., Hetflejs J., Oehme G. and Pracejus H., *Z. Chem.*, **14**, 64 (1974).
1222. Vaisarova V., Schraml J. and Hetflejs J., *Collect. Czech. Chem. Commun.*, **43**, 265 (1978).
1223. Valade J. and Calas R., *Bull. Soc. Chim. France*, 473 (1958).
1224. Valier M.G., Chekrii P.S., Guseinov M.M., Kalechits I.V., Taber A.M. and Garmanov A.M., *Zh. Vses. Khim. O-va im D.I. Mendeleeva*, **31**, 591 (1986).
1225. Valles E.M. and Macosko C.W., *Am. Chem. Soc., Div. Org. Coat. Plast. Chem.*, **35**, 44 (1975); *C.A.*, **87**, 6502h (1977).
1226. Valles E.M. and Macosko C.W., *Macromolecules*, **12**, 521 (1979).
1227. Vannoorenberghe Y. and Buono G., *Tetrahedron Lett.*, **29**, 3235 (1988).
1228. Vcelak J., Chvalovsky V., Voronkov M.G., Pukhnarevich V.B. and Pestunovich V.A., *Collect. Czech. Chem. Commun.*, **41**, 386 (1976).
1229. Vdovin V.M. and Petrov A.D., *Zh. Obshch. Khim.*, **30**, 838 (1959).
1230. Vdovin V.M., Fedorov V.E., Pritula N.A. and Fedorova G.K., *Izv. Akad. Nauk SSSR*, 2663 (1981).
1231. Veliev M.G., Garamanov A.M., Kuli-Zade F.A. and Guseinov M.M., *Azerb. Khim. Zh.*, 60 (1987).
1232. Vinogradov V.N., Reikhsfel'd V.O., Filippov N.A., Zaslavskaya T.N. and Korichev G.L., *Zh. Obshch. Khim.*, **45**, 2031 (1975).
1233. Voronkov M.G. and Romanova N.G., *Zh. Obshch. Khim.*, **28**, 2122 (1958).
1234. Voronkov M.G. and Sushchinskaya S.P., *Zh. Obshch. Khim.*, **56**, 627 (1986).
1235. Voronkov M.G., Abramovitsch S.N. and Kubiakov N.I., *Izv. Akad. Nauk SSSR, Ser. Khim.*, 488 (1986).
1236. Voronkov M.G., Adamovich S.N., Khramtsova S.Yu., Shterenberg B.Z., Rakhlin V.I. and Mirskov R.G., *Izv. Akad. Nauk SSSR, Ser. Khim.*, **6**, 1424 (1987); *C.A.*, **109**, 23058h (1988).
1237. Voronkov M.G., Adamovich S.N., Koblyakov I.M., Khramtsova S.Yu., Rakhlin V.I. and Mirskov R.G., *Izv. Akad. Nauk*, **2**, 488 (1986).
1238. Voronkov M.G., Adamovich S.N., Sherstyannikova L.V. and Pukhnarevich V.B., *Zh. Obshch. Khim.*, **53**, 806 (1983).
1239. Voronkov M.G., Barton T.D., Kirpichenko S.V., Keiko V.V. and Pestunovich V.A., *Izv. Akad. Nauk SSSR, Ser. Khim.*, 710 (1976).
1240. Voronkov M.G., Burnashova T.D. and Yarosh O.G., *Zh. Obshch. Khim.*, **52**, 614 (1982).
1241. Voronkov M.G., Chvalovsky V., Kirpichenko S.V., Vlasova N.A., Bol'shakova S.T.,

Kuncova G., Keiko V.V. and Tsetlina E.O., *Collect. Czech. Chem. Commun.*, **44**, 742 (1979).
1242. Voronkov M.G., Druzhdzh P.V. and Pukhnarevich V.B., *Zh. Obshch. Khim.*, **50**, 1662 (1980).
1243. Voronkov M.G., Keiko N.A., Kuznetsova T.A., Svishcheva I.N., Pestunovich V.A., Tsetlina E.O. and Balashenko L.V., *Zh. Obshch. Khim.*, **45**, 813 (1975).
1244. Voronkov M.G., Kirpichenko S.V., Barton T.D., Keiko V.V., Pestunovich V.A. and Trofimov B.A., *Tezisy Dokl. Nauchn. Sess. Khim. Tekhnol. Org. Soedin. Sery Sernitykh Nefti* (I.G. Bakhtalze Ed), Zinatne, Riga, USSR, 1976, p. 147.
1245. Voronkov M.G., Kirpichenko S.V., Keiko V.V., Sherstyannikova L.V., Pestunovich V.A. and Tsetlina E.O., *Izv. Akad. Nauk SSSR, Ser. Khim.*, 390 (1975).
1246. Voronkov M.G., Kletsko F.P., Vlasova N.N., Tsetlina E.O., Keiko V.V., Kaigorodova V.I. and Pestunovich V.A., *Zh. Obshch. Khim.*, **35**, 1191 (1975).
1247. Voronkov M.G., Mirskov R.G., Ishchenko O.S. and Sitnikova S.P., *Zh. Obshch. Khim.*, **45**, 2634 (1975).
1248. Voronkov M.G., Mustafaev R.M., Kuleva L.G. and Sadykh-Zade S.I., *Zh. Obshch. Khim.*, **55**, 2091 (1985); *C.A.*, **105**, 97545d (1986).
1249. Voronkov M.G., Mustafaev R.M., Kuleva L.G. and Sadykh-Zade S.I., *Zh. Obshch. Khim.*, **56**, 1520 (1986).
1250. Voronkov M.G., Mustafaev R.M., Kuleva L.G. and Sadykh-Zade S.I., *Zh. Obshch. Khim.*, **56**, 1827 (1986).
1251. Voronkov M.G., Pukhnarevich V.B., Kopylova L.I., Burnashova T.D. and Vysotskii A.V., *Zh. Obshch. Khim.*, **59**, 2150 (1989).
1252. Voronkov M.G., Pukhnarevich V.B., Kopylova L.I., Pestunovich V.A., Tsetlina E.O., Trofimov B.A., Pola I. and Chvalovsky V., *Dokl. Akad. Nauk SSSR*, **227**, 91 (1976).
1253. Voronkov M.G., Pukhnarevich V.B., Sushchinskaya S.P., Kopylova L.I. and Trofimov B.A., *Zh. Obshch. Khim.*, **42**, 2687 (1972).
1254. Voronkov M.G., Pukhnarevich V.B., Sushchinskaya S.P., Kopylova L.I. and Trofimov B.A., *Zh. Obshch. Khim.*, **41**, 2102 (1971).
1255. Voronkov M.G., Pukhnarevich V.B., Tsykhanskaya I.I. and Ushakova N.I., *Proc. VII Int. Symp. Organosilicon Chem., Kyoto, Japan*, 1984, p. 193.
1256. Voronkov M.G., Pukhnarevich V.B., Tsykhanskaya I.I. and Ushakova N.I., *Proc. IV Int. Symp. Homogeneous Catal., Leningrad, USSR*, 1984, Vol. 1, p. 73.
1257. Voronkov M.G., Pukhnarevich V.B., Tsykhanskaya I.I. and Varshavskii Yu.S., *Dokl. Akad. Nauk SSSR*, **254**, 887 (1980).
1258. Voronkov M.G., Pukhnarevich V.B., Tsykhanskaya I.I., Ushakova N.I., Gaft Yu.L. and Zakharova I.A., *Inorg. Chim. Acta*, **68**, 103 (1983).
1259. Voronkov M.G., Pukhnarevich V.B., Ushakova N.I., Tsykhanskaya I.I., Albinov A.I. and Vitkovskii V.Yu., *Zh. Obshch. Khim.*, **55**, 94 (1985).
1260. Voronkov M.G., Rakhlin V.I., Adamovich S.N., Petukhov L.P., Kirskov R.G., Yarosh N.K., Kuznetsov A.L. and Sigalov M.V., *Izv. Akad. Nauk. SSSR, Ser. Khim.*, 899 (1986).
1261. Voronkov M.G., Shchukina L.V., Yarosh O.G. and Tsetlina E.O., *Izv. Akad. Nauk SSSR, Ser. Khim.*, 1874 (1976).
1262. Voronkov M.G., Sushchinskaya S.P. and Pukhnarevich V.B., *Zh. Obshch. Khim.*, **51**, 242 (1981).
1263. Voronkov M.G., Sushchinskaya S.P., Pukhnarevich V.B. and Randin O.I., *Zh. Obshch. Khim.*, **49**, 1281 (1979).
1264. Voronkov M.G., Sushchinskaya S.P. and Pukhnarevich V.B., *Zh. Obshch. Khim.*, **49**, 1171 (1979).
1265. Voronkov M.G., Tsykhanskaya I.I., Mishim V.M. and Pukhnarevich V.B., *Zh. Obshch. Khim.*, **57**, 2642 (1987).
1266. Voronkov M.G., Tsykhanskaya I.I., Vlasova N.N., Kaliberado L.M., Gont L.N., Korotaeva I.M. and Satsuk S.I., *Izv. Akad. Nauk SSSR, Ser. Khim.*, 1368 (1976).
1267. Voronkov M.G., Tsykhanskaya I.I., Vlasova N.N., Tsetlina E.O., Keiko V.V. and Kaigorodova V.I., *Izv. Akad. Nauk SSSR, Ser. Khim.*, 1883 (1977).

1268. Voronkov M.G., Ushakova N.I., Pukhnarevich V.B. and Tsykhanskaya I.I., *Gidrodinam. i Yavl. Perenosa v Dvukh. Dispers. Sist., Irkutsk, USSR*, 1982, p. 72; *Ref. Zh. Khim.*, Abstr. 1B, 4139 (1984).
1269. Voronkov M.G., Ushakova N.I., Tsykhanskaya I.I. and Pukhnarevich V.B., *J. Organometal. Chem.*, **264**, 39 (1984).
1270. Voronkov M.G., Vainshtein B.I., Pukhnarevich V.B., Druzhdzh P.V. and Kopylova L.I., *Dokl. Akad. Nauk SSSR*, **237**, 1101 (1977).
1271. Voronkov M.G., Vlasova N.N. and Bol'shakova S.A., *Izv. Akad. Nauk SSSR, Ser. Khim.*, 170 (1981).
1272. Voronkov M.G., Vlasova N.N., Bol'shakova S.A. and Kirpichenko S.V., *J. Organometal. Chem.*, **190**, 335 (1980).
1273. Voronkov M.G., Vlasova N.N., Bol'shakova S.A., Gosarova N.K. and Efremova G.G., *Zh. Obshch. Khim.*, **55**, 1034 (1985).
1274. Voronkov M.G., Vlasova N.N., Kirpichenko S.V., Bol'shakova S.A. and Pestunovich A.E., *Katal. Sint. Org. Soedin. Sery*, 7 (1979); *C.A.*, **93**, 239514c (1980).
1275. Voronkov M.G., Vlasova N.N., Kirpichenko S.V., Bol'shakova S.A., Keiko V.V., Tsetlina E.O., Amosova S.V., Trofimov B.A. and Chvalovsky V., *Izv. Akad. Nauk SSSR, Ser. Khim.*, 422 (1979).
1276. Voronkov M.G., Yarosh O.G., Ivanova Z.G., Roman V.K. and Albanov A.I., *Izv. Akad. Nauk.*, **6**, 1403 (1987).
1277. Voronkov M.G., Yarosh O.G., Orlova T.M. and Albanov A.I., *Metaloorg. Khim.*, **2**, 466 (1989).
1278. Voronkov M.G., Yarosh O.G., Pestunovich V.A., Sigalov M.V. and Tsvetaeva L.V., *Izv. Akad. Nauk SSSR, Ser. Khim.*, 2066 (1974).
1279. Voronkov M.G., Yarosh O.G., Shchukina L.V. and Kuznetsova E.E., *Izv. Akad. Nauk SSSR, Ser. Khim.*, 2611 (1984).
1280. Voronkov M.G., Yarosh O.G., Shchukina L.V., Tsetlina E.O., Tandura S.N. and Korotaeva I.M., *Zh. Obshch. Khim.*, **49**, 614 (1979).
1281. Voronkov M.G., Yarosh O.G., Tsvetaeva L.V., Sigalov M.V. and Gromkova R.A., *Zh. Obshch. Khim.*, **44**, 1747 (1974).
1282. Vybiral V., Svoboda P. and Hetflejs J., *Collect. Czech. Chem. Commun.*, **44**, 866 (1979).
1283. Wang D. and Chan T.H., *Tetrahedron Lett.*, **24**, 1573 (1983).
1284. Wang D. and Jiang Y., *Gaofenzi Tongxun*, 78 (1983); *C.A.*, **99**, 195191 (1983).
1285. Wang D. and Jiang Y., *J. Catal. China*, **2**, 236 (1981); *C.A.*, **96**, 68095 (1982).
1286. Wang D. and Jiang Y., *J. Organometal. Chem.*, **251**, 39 (1983).
1287. Warrick E.L., Pierce O.R., Polmanteer K.E. and Saam J.C., *Rubber Chem. Technol.*, **52**, 449 (1979).
1288. Watanabe H., *Proc. VII Int. Symp. Organosilicon Chem., Gunma Symp.-Organosilicon Chem. Directed to Pract. Use, Kiryu-Gunma, Japan*, 1984, p. 25.
1289. Watanabe H., Aoki M., Saurai N., Watanabe K. and Nagai Y., *J. Organometal. Chem.*, **160**, C1 (1978).
1290. Watanabe H., Asami M. and Nagai Y., *J. Organometal. Chem.*, **195**, 363 (1980).
1291. Watanabe H., Kitahara T., Motegi T. and Nagai Y., *J. Organometal. Chem.*, **139**, 215 (1977).
1292. White D.G. and Rochov E.G., *J. Am. Chem. Soc.*, **76**, 3897 (1954).
1293. Wrighton M.S. and Schroeder M.A., *J. Am. Chem. Soc.*, **96**, 6235 (1974).
1294. Xiao Ch., Jing Z., Lu H. and Lin Y., *Huaxue Shiji*, **10**, 188 (1988); *C.A.*, **111**, 57845 (1989).
1295. Xiao Ch., Lin Y., Luo G., Tian Z. and Chen Y., *Wuhan Daxue Xuebao, Ziran Xuebao*, 81 (1988); *C.A.*, **110**, 231731r (1989).
1296. Yakukhina O.M. and Reikhsfel'd V.O., *Khim. Prakt. Prim. Kremnii. i Fosfor. Soedin.*, *L*, 67 (1980); *C.A.*, **95**, 62307a (1981).
1297. Yamamoto K. and Kumada M., *J. Organometal. Chem.*, **13**, 131 (1968).
1298. Yamamoto K., Hayashi T. and Kumada M., *J. Organometal. Chem.*, **54**, C45 (1973).
1299. Yamamoto K., Hayashi T. and Kumada M., *J. Organometal. Chem.*, **46**, C65 (1972).

1300. Yamamoto K., Hayashi T. and Kumada M., *J. Organometal. Chem.*, **28**, C37 (1971).
1301. Yamamoto K., Hayashi T. and Kumada M., *J. Am. Chem. Soc.*, **93**, 5301 (1971).
1302. Yamamoto K., Hayashi T., Uramoto Y., Ito R. and Kumada M., *J. Organometal. Chem.*, **118**, 331 (1976).
1303. Yamamoto K., Hayashi T., Zembayashi M. and Kumada M., *J. Organometal. Chem.*, **118**, 161 (1976).
1304. Yamamoto K., Kiso Y., Ito R., Tamao K. and Kumada M., *J. Organometal. Chem.*, **210**, 9 (1981).
1305. Yamamoto K., Kumada M., Nakajima I., Maeda K. and Imaki N., *J. Organometal. Chem.*, **13**, 329 (1968).
1306. Yamamoto K., Uramoto Y. and Kumada M., *J. Organometal. Chem.*, **31**, C9 (1971).
1307. Yang D. and Tanner D.D., *J. Org. Chem.*, **51**, 2267 (1986).
1308. Yarosh O.G., Burnashova T.D., Albanov A.I. and Voronkov M.G., *Metaloorg. Khim.*, **1**, 773 (1988); *C.A.*, **111**, 194839j (1989).
1309. Yarosh O.G., Komarov N.V. and Ivanova Z.G., *Izv. Akad. Nauk SSSR, Ser. Khim.*, 2751 (1972).
1310. Yarosh O.G., Modanov V.B. and Voronkov M.G., *Metaloorg. Khim.*, **1**, 712 (1988); *C.A.*, **111**, 115293a (1989).
1311. Yarosh O.G., Shchukina L.V., Tsetlina E.O. and Voronkov M.G., *Zh. Obshch. Khim.*, **48**, 2059 (1978).
1312. Yarosh O.G., Turkina G.U., Albanov A.I. and Voronkov M.G., *Metaloorg. Khim.*, **1**, 708 (1988); *C.A.*, **110**, 231701k (1989).
1313. Yates R.L., *J. Catal.*, **78**, 111 (1982).
1314. Yeh H.Ch., Eichinger B.E. and Anderson N.H., *J. Polym. Sci., Polym. Chem. Ed.*, **20**, 2575 (1982).
1315. Yi X., Xu Q., Zhang X., Wang Ch., *Wuhan Daxue Xuebao, Ziran Kexueban*, **4**, 108 (1985); *C.A.*, **105**, 191209 (1986).
1316. Yokoyama T. and Kinjo N., *J. Appl. Polym. Sci.*, **29**, 2665 (1984).
1317. Yokoyama T., Kinjo N. and Wakashima Y., *Proc. VII Int. Symp. Organosilicon Chem., Kyoto, Japan*, 1984, p. 194.
1318. Yoshida J., Tamao K., Takahashi M. and Kumada M., *Tetrahedron Lett.*, 2161 (1978).
1319. Yoshii E. and Takeda K., *Chem. Pharm. Bull.*, **31**, 4586 (1983).
1320. Yoshii E., Ikeshima H. and Ozaki K., *Chem. Pharm. Bull.*, **20**, 1827 (1972).
1321. Yoshii E., Kobayashi Y., Koizumi T. and Oribe T., *Chem. Pharm. Bull.*, **22**, 2767 (1974).
1322. Yoshii E., Koizumi T., Ikeshima H., Ozaki K. and Hayashi J., *Chem. Pharm. Bull.*, **23**, 2496 (1975).
1323. Yu X., Ding L., Liu X. and Hong M., *Bopuxue Zazhi*, **3**, 349 (1986); *C.A.*, **107**, 198443 (1987).
1324. Yur'ev V.P. and Salimgareeva I.M., *Izv. Akad. Nauk SSSR, Ser. Khim.*, 2135 (1975).
1325. Yur'ev V.P., Salimgareeva I.M. and Kaverin V.V., *Izv. Akad. Nauk SSSR, Ser. Khim.*, 1673 (1976).
1326. Yur'ev V.P., Salimgareeva I.M. and Kaverin V.V., *Zh. Obshch. Khim.*, **47**, 592 (1977).
1327. Yur'ev V.P., Salimgareeva I.M., Kaverin V.V. and Tolstikov G.A., *Zh. Obshch. Khim.*, **47**, 355 (1977).
1328. Yur'ev V.P., Salimgareeva I.M., Kaverin V.V. and Zhebarov O.Zh., *Zh. Org. Khim.*, **12**, 2470 (1976).
1329. Yur'ev V.P., Salimgareeva I.M., Kaverin V.V., Khalilov K.M. and Panasenko A.A., *J. Organometal. Chem.*, **171**, 167 (1979).
1330. Yur'ev V.P., Salimgareeva I.M., Tolstikov G.A. and Zhebarov O.Zh., *Zh. Obshch. Khim.*, **45**, 955 (1975).
1331. Yur'ev V.P., Salimgareeva I.M., Zhebarov O.Zh. and Khalilov L.M., *Zh. Obshch. Khim.*, **47**, 1541 (1977).
1332. Yur'ev V.P., Salimgareeva I.M., Zhebarov O.Zh. and Tolstikov G.A., *Zh. Obshch. Khim.*, **46**, 372 (1976).

1333. Yur'ev V.P., Salimgareeva I.M., Zhebarov O.Zh. and Tolstikov G.A., *Zh. Obshch. Khim.*, **45**, 2568 (1975).
1334. Yur'ev V.P., Salimgareeva I.M., Zhebarov O.Zh. and Tolstikov G.A., *Izv. Akad. Nauk SSSR, Ser. Khim.*, 1888 (1975).
1335. Yur'ev V.P., Salimgareeva I.M., Zhebarov O.Zh., Kaverin V.V. and Rafikov S.R., *Dokl. Akad. Nauk SSSR*, **229**, 892 (1976).
1336. Yur'ev V.P., Salimgareeva I.M., Zhebarov O.Zh., Tolstikov G.A. and Rafikov S.R., *Dokl. Akad. Nauk SSSR*, **224**, 1092 (1975).
1337. Yusupova F.G., Gailunas G.A., Isaeva L.S., Peganova T.A., Kajumov F.F. and Yur'ev V.P., *Proc. IV Int. Symp. Homogeneous Catal., Leningrad, USSR*, 1984, Vol. 3, p. 226.
1338. Yuzhelevskii Yu.A., Timofeeva N.P. and Novikova G.A., *Vysokomol. Soedin., Ser. B*, **21**, 746 (1979).
1339. Zakharkin L.I. and Agakhanova T.B., *Izv. Akad. Nauk SSSR, Ser. Khim.*, 1208 (1980).
1340. Zakharkin L.I. and Agakhanova T.B., *Izv. Akad. Nauk SSSR, Ser. Khim.*, 2151 (1978).
1341. Zakharkin L.I. and Agakhanova T.B., *Izv. Akad. Nauk SSSR, Ser. Khim.*, 2632 (1977).
1342. Zakharkin L.I., Pisareva I.V. and Agakhanova T.B., *Izv. Akad. Nauk SSSR, Ser. Khim.*, 2389 (1977).
1343. Zaslavskaya T.N., Filippov N.A. and Reikhsfel'd V.O., *Khim. Prakt. Prim. Kremnii i Fosfor. Soedin.*, *L*, 101 (1977); *Ref. Zh. Khim., Abstr.*, 14B,1161 (1978).
1344. Zaslavskaya T.N., Filippov N.A. and Reikhsfel'd V.O., *Zh. Obshch. Khim.*, **52**, 80 (1982).
1345. Zaslavskaya T.N., Filippov N.A. and Reikhsfel'd V.O., *Zh. Obshch. Khim.*, **51**, 107 (1981).
1346. Zaslavskaya T.N., Filippov N.A., Skvortsov N.K., Maretina E.Yu. and Reikhsfel'd V.O., *Khim. Prakt. Prim. Kremnii i Fosfor Soedin.*, *L*, 64 (1978); *Ref. Zh. Khim. Abstr.*, 16B 1072(1979).
1347. Zaslavskaya T.N., Reikhsfel'd V.O., Filippov N.A. and Shornik N.A., *Zh. Obshch. Khim.*, **50**, 2478 (1980).
1348. Zaslavskaya T.N., Reikhsfel'd V.O. and Filippov N.A., *Zh. Obshch. Khim.*, **50**, 2286 (1980).
1349. Zhang X., Lu X., Zhany R. and Duan H., *Cuihua Xuebao*, **7**, 378 (1986); *C.A.*, **107**, 198441p (1987).
1350. Zhang Z., Liu H. and Wang I., *Cuihua Xuebao*, **9**, 325 (1988); *C.A.*, **110**, 153517 (1989).
1351. Zhang Z., Liu H. and Wang I., *Youji Xuaxue*, **8**, 528 (1988); *C.A.*, **111**, 97341h (1988).
1352. Zhdanov A.A. and Andrianov K.A., *Dokl. Akad. Nauk SSSR*, **211**, 1104 (1973).
1353. Zhdanov A.A., *Makromol. Chem. Suppl.*, **6**, 227 (1984).
1354. Zhdanov A.A., Andrianov K.A. and Malykhin A.P., *Vysokomol. Soedin., Ser. A*, **16**, 1765 (1974).
1355. Zhdanov A.A., Kotov V.M., Lavrykhin B.D. and Pryakhina T.A., *Izv. Akad. Nauk SSSR, Ser. Khim.*, 1851 (1984).
1356. Zhebarov O.Zh., *Khim. Vysokomol. Soedin. Neftekhim.*, 34 (1975); *C.A.*, **85**, 94438d (1976).
1357. Zhebarov O.Zh., Salimgareeva I.M. and Yur'ev V.P., *Zh. Obshch. Khim.*, **48**, 233 (1978).
1358. Zhou G.B., Khan I.M. and Smid J., *Polym. Prepr.*, **30**, 416 (1989); *C.A.*, **111**, 7900n (1989).
1359. Zhun V.I., Tsvetkov A.L., Bochkarev V.N., Slusarenko F., Turkel'taub G.N., and Sheludyakov V.D., *Zh. Obshch. Khim.*, **59**, 390 (1989).
1360. Zimin A.V., Matyuk V.M., Yankelevich A.Z. and Shapet'ko N.N., *Dokl. Akad. Nauk SSSR, Ser. Khim.*, **231**, 870 (1976).
1361. Zisman W.A., *Ind. Eng. Chem.*, **57**, 26 (1965).

1362. Feibush B., Figueroa A., Charles R., Oman K.D., Feibush P. and Karger B.L., *J. Am. Chem. Soc.*, **108**, 3310 (1986).
1363. Dieck tom H., Kleinwachter I. and Haupt E.T.K., *J. Organometal. Chem.*, **321**, 237 (1987).

Patents

A: Austrian Patents

A-001. 285 628 (1970).
A-002. 297 033 (1972).
A-003. 299 975 (1972).
A-004. 300 839 (1972).

B: Belgian Patents

B-001. 628 951 (1963); *C.A.*, **60**, 16103f (1964).
B-002. 648 690 (1964); *C.A.*, **63**, 13504d (1965).
B-003. 668 907 (1966); *C.A.*, **65**, 3906h (1966).

C: Czechoslovakian Patents

C-001. 156 898 (1975); *C.A.*, **83**, 114632b (1975).
C-002. 157 917 (1975).
C-003. 157 948 (1975).
C-004. 158 944 (1975); *C.A.*, **84**, 165016g (1976).
C-005. 159 140 (1975); *C.A.*, **84**, 165017h (1976).
C-006. 159 412 (1975).
C-007. 159 689 (1975); *C.A.*, **84**, 165015f (1976).
C-008. 159 908 (1975); *C.A.*, **85**, 5868d (1976).
C-009. 161 177 (1975); *C.A.*, **85**, 33182u (1976).
C-010. 162 588 (1976); *C.A.*, **85**, 177609m (1976).
C-011. 164 584 (1975); *C.A.*, **86**, 141262z (1977).
C-012. 165 746 (1976); *C.A.*, **87**, 39650h (1977).
C-013. 170 439 (1977); *C.A.*, **89**, 43754d (1978).
C-014. 171 581 (1978); *C.A.*, **89**, 24529t (1978).
C-015. 171 582 (1978); *C.A.*, **89**, 43761d (1978).
C-016. 171 597 (1978); *C.A.*, **89**, 215545e (1978).
C-017. 174 583 (1978); *C.A.*, **90**, 87657z (1979).
C-018. 174 585 (1978); *C.A.*, **90**, 104114q (1979).
C-019. 176 909 (1979).
C-020. 176 910 (1979); *C.A.*, **91**, 20703p (1979).
C-021. 179 188 (1979); *C.A.*, **92**, 59426w (1980).
C-022. 182 315 (1980); *C.A.*, **94**, 30911f (1981).
C-023. 185 139 (1980).
C-024. 187 167 (1981); *C.A.*, **96**, 6863n (1982).
C-025. 193 448 (1981); *C.A.*, **96**, 218019s (1982).
C-026. 193 623 (1981); *C.A.*, **96**, 162935t (1982).
C-027. 194 149 (1982); *C.A.*, **99**, 22680t (1983).
C-028. 195 549 (1982).
C-029. 196 917 (1981); *C.A.*, **97**, 23996w (1982).
C-030. 199 329 (1982); *C.A.*, **98**, 107543e (1982).
C-031. 200 379 (1983); *C.A.*, **99**, 158621g (1983).
C-032. 206 394 (1983).

C-033. 213 696 (1984).
C-034. 215 743 (1984); *C.A.*, **103**, 71492t (1985).
C-035. 224 790 (1984); *C.A.*, **103**, 196230n (1985).
C-036. 228 499 (1986); *C.A.*, **106**, 50441x (1987).
C-037. 229 088 (1986); *C.A.*, **105**, 209334t (1986).
C-038. 240 639 (1987); *C.A.*, **111**, 241176c (1989).
C-039. 240 704 (1987); *C.A.*, **109**, 149782h (1988).
C-040. 240 948 (1987).
C-041. 241 605 (1988); *C.A.*, **109**, 231277f (1988).
C-042. 254 484 (1988); *C.A.*, **111**, 97495m (1989).
C-043. 257 650 (1989); *C.A.*, **111**, 116268h (1989).

Ch: Swiss Patents

Ch-001. 497 464 (1970).
Ch-002. 507 988 (1971).

D: Federal Republic of Germany Patents

D-001. 1 069 148.
D-002. 1 158 071 (1964); *C.A.*, **61**, 8340d (1964).
D-003. 1 161 270 (1964); *C.A.*, **60**, 10717f (1964).
D-004. 1 163 818
D-005. 1 165 027 (1964); *C.A.*, **60**, 14540d (1964).
D-006. 1 165 028 (1964); *C.A.*, **60**, 14540g (1964).
D-007. 1 203 776 (1964); *C.A.*, **64**, 2126b (1966).
D-008. 1 210 844 (1966); *C.A.*, **64**, 15922b (1966).
D-009. 1 245 956 (1967); *C.A.*, **67**, 82543z (1967).
D-010. 1 259 887 (1968); *C.A.*, **68**, 95946d (1968).
D-011. 1 271 712 (1968); *C.A.*, **70**, 57988c (1969).
D-012. 1 281 445 (1968); *C.A.*, **70**, 47586f (1969).
D-013. 1 643 926 (1971); *C.A.*, **76**, 46294w (1972).
D-014. 1 668 605 (1971).
D-015. 1 668 637 (1971).
D-016. 1 792 046.
D-017. 1 795 251.
D-018. 1 795 262.
D-019. 1 816 392 (1969).
D-020. 1 900 969 (1969); *C.A.*, **71**, 82127z (1969).
D-021. 1 936 068 (1970); *C.A.*, **72**, 100873s (1970).
D-022. 1 936 745 (1970); *C.A.*, **72**, 91149f (1970).
D-023. 1 937 904 (1970); *C.A.*, **74**, 100217x (1971).
D-024. 1 938 743 (1970); *C.A.*, **75**, 20588z (1971).
D-025. 1 939 300 (1970); *C.A.*, **72**, 101642c (1970).
D-026. 1 941 411 (1970); *C.A.*, **74**, 100519x (1971).
D-027. 1 942 798 (1970); *C.A.*, **73**, 15497y (1970).
D-028. 1 950 960 (1970); *C.A.*, **72**, 133590a (1970).
D-029. 1 954 961 (1970); *C.A.*, **73**, 14978u (1970).
D-030. 1 964 609 (1970); *C.A.*, **73**, 121246a (1970).
D-031. 2 001 303 (1971).
D-032. 2 008 427 (1971); *C.A.*, **73**, 110629t (1970).
D-033. 2 012 229 (1971); *C.A.*, **76**, 14700j (1972).
D-034. 2 030 653 (1971); *C.A.*, **74**, 77232n (1971).
D-035. 2 033 666 (1971); *C.A.*, **74**, 125833h (1971).

D-036. 2 034 909 (1971); *C.A.*, **74**, 91744n (1971).
D-037. 2 037 146 (1971).
D-038. 2 055 787 (1971); *C.A.*, **75**, 77517e (1971).
D-039. 2 062 816 (1974); *C.A.*, **82**, 73162g (1975).
D-040. 2 121 578 (1972).
D-041. 2 131 741 (1972); *C.A.*, **78**, 72355c (1973).
D-042. 2 131 742 (1972); *C.A.*, **78**, 72359g (1973).
D-043. 2 133 397 (1972).
D-044. 2 152 275 (1972).
D-045. 2 152 286 (1972); *C.A.*, **77**, 49648w (1972).
D-046. 2 157 405 (1973); *C.A.*, **79**, 43066z (1973).
D-047. 2 159 723 (1973); *C.A.*, **79**, 67022h (1973).
D-048. 2 159 991 (1973); *C.A.*, **79**, 78952h (1973).
D-049. 2 161 716 (1973); *C.A.*, **79**, 66562x (1973).
D-050. 2 210 380 (1972); *C.A.*, **78**, 5649p (1973).
D-051. 2 212 239.
D-052. 2 215 629 (1973); *C.A.*, **80**, 3622s (1974).
D-053. 2 231 261 (1973); *C.A.*, **78**, 112245a (1973).
D-054. 2 235 889 (1974).
D-055. 2 244 278 (1974).
D-056. 2 245 187 (1973); *C.A.*, **79**, 67025m (1973).
D-057. 2 247 885 (1973); *C.A.*, **79**, 19634n (1973).
D-058. 2 250 921 (1974); *C.A.*, **81**, 13646n (1974).
D-059. 2 251 297 (1974); *C.A.*, **81**, 106571z (1974).
D-060. 2 260 260 (1973); *C.A.*, **79**, 78953p (1973).
D-061. 2 302 231 (1973); *C.A.*, **79**, 105400q (1973).
D-062. 2 305 097 (1973).
D-063. 2 307 085 (1973); *C.A.*, **80**, 16134j (1974).
D-064. 2 308 238 (1974); *C.A.*, **82**, 58848s (1975).
D-065. 2 311 879 (1973); *C.A.*, **80**, 121681h (1974).
D-066. 2 317 985 (1973); *C.A.*, **80**, 84243u (1974).
D-067. 2 330 308 (1974); *C.A.*, **80**, 121737f (1974).
D-068. 2 332 167 (1974); *C.A.*, **80**, 83252j (1974).
D-069. 2 337 641 (1974).
D-070. 2 355 181 (1974); *C.A.*, **81**, 78082v (1974).
D-071. 2 400 039 (1974).
D-072. 2 405 274 (1974); *C.A.*, **81**, 152406x (1974).
D-073. 2 408 480 (1975); *C.A.*, **84**, 75014a (1976).
D-074. 2 411 326 (1975); *C.A.*, **84**, 19124z (1976).
D-075. 2 418 387 (1974); *C.A.*, **82**, 73165k (1975).
D-076. 2 429 772 (1975); *C.A.*, **82**, 172750d (1975).
D-077. 2 459 901 (1975), *C.A.*, **83**, 131115c (1975).
D-078. 2 508 931 (1976).
D-079. 2 511 187 (1975); *C.A.*, **84**, 31237w (1976).
D-080. 2 517 601 (1975).
D-081. 2 519 720 (1976); *C.A.*, **86**, 90848t (1977).
D-082. 2 550 659 (1976); *C.A.*, **85**, 143291m (1976).
D-083. 2 550 660 (1976); *C.A.*, **85**, 25895p (1976).
D-084. 2 558 191 (1976); *C.A.*, **85**, 178754s (1976).
D-085. 2 601 913 (1976); *C.A.*, **86**, 43809p (1977).
D-086. 2 607 714 (1976); *C.A.*, **85**, 161591p (1976).
D-087. 2 635 601 (1978); *C.A.*, **91**, 22206c (1979).
D-088. 2 655 877 (1978); *C.A.*, **89**, 111929s (1978).
D-089. 2 731 870 (1978); *C.A.*, **88**, 111218p (1978).
D-090. 2 752 973 (1979); *C.A.*, **91**, 108086a (1979).
D-091. 2 753 124 (1979); *C.A.*, **91**, 140986c (1979).

D-092. 2 804 204 (1979); *C.A.*, **91**, 193413x (1979).
D-093. 2 809 874 (1979); *C.A.*, **91**, 212163j (1979).
D-094. 2 809 875 (1979); *C.A.*, **92**, 24026p (1980).
D-095. 2 810 032 (1978); *C.A.*, **90**, 23240a (1979).
D-096. 2 815 316 (1979); *C.A.*, **91**, 193414y (1979).
D-097. 2 815 978 (1979); *C.A.*, **92**, 164081c (1980).
D-098. 2 834 691 (1980).
D-099. 2 836 054 (1980); *C.A.*, **93**, 114693d (1980).
D-100. 2 837 830 (1979); *C.A.*, **91**, 21899u (1979).
D-101. 2 838 842 (1979); *C.A.*, **90**, 205249t (1979).
D-102. 2 846 621 (1980).
D-103. 2 851 456 (1980); *C.A.*, **93**, 220931q (1980).
D-104. 2 856 906 (1979); *C.A.*, **91**, 93136r (1979).
D-105. 2 918 254 (1980); *C.A.*, **94**, 67419c (1981).
D-106. 2 934 550 (1980); *C.A.*, **93**, 114694e (1980).
D-107. 2 934 578 (1980); *C.A.*, **93**, 47817g (1980).
D-108 2 948 760 (1980).
D-109. 2 950 402 (1981).
D-110. 3 000 768 (1981); *C.A.*, **95**, 115744z (1981).
D-111. 3 002 238 (1981); *C.A.*, **95**, 187411m (1981).
D-112. 3 018 674 (1980); *C.A.*, **94**, 48950c (1981).
D-113. 3 029 599 (1982).
D-114. 3 120 847 (1982).
D-115. 3 123 676 (1982).
D-116. 3 138 235 (1983); *C.A.*, **99**, 70978t (1983).
D-117. 3 138 236 (1983); *C.A.*, **99**, 53953p (1983).
D-118. 3 202 493 (1982).
D-119. 3 301 807 (1984); *C.A.*, **102**, 6795j (1985).
D-120. 3 315 281 (1983); *C.A.*, **100**, 69987c (1984).
D-121. 3 327 795 (1985).
D-122. 3 331 515 (1985); *C.A.*, **103**, 88066e (1985).
D-123. 3 331 682 (1985); *C.A.*, **103**, 71497g (1985).
D-124. 3 404 703 (1985); *C.A.*, **103**, 196232q (1985).
D-126. 3 423 608 (1986); *C.A.*, **105**, 79866e (1986).
D-127. 3 431 075 (1986); *C.A.*, **105**, 98202b (1986).
D-128. 3 507 424 (1986); *C.A.*, **106**, 67481h (1987).
D-129. 3 517 615 (1986); *C.A.*, **106**, 139632h (1987).
D-130. 3 518 605 (1986); *C.A.*, **106**, 138941w (1987).
D-131. 3 518 878 (1986); *C.A.*, **106**, 102807t (1987).
D-132. 3 524 235.
D-133. 3 532 686 (1987); *C.A.*, **107**, 8319j (1987).
D-134. 3 539 714 (1986); *C.A.*, **106**, 138937z (1987).
D-135. 3 602 490 (1987); *C.A.*, **108**, 7024t (1988).
D-136. 3 616 575 (1986); *C.A.*, **106**, 68883c (1987).
D-137. 3 628 319 (1988); *C.A.*, **109**, 23808c (1988).
D-138. 3 631 125 (1988); *C.A.*, **109**, 74985r (1988).
D-139. 3 632 869 (1988); *C.A.*, **109**, 150298m (1988).
D-140. 3 637 273 (1988).
D-141. 3 637 837 (1988).
D-142. 3 642 058 (1987); *C.A.*, **107**, 206008v (1987).
D-143. 3 718 588 (1988); *C.A.*, **110**, 194720t (1989).
D-144. 3 742 000 (1988).
D-145. E 824 130 (1989); *C.A.*, **111**, 79994u (1989).

Dd: German Democratic Republic Patents

Dd-001. 32938 (1964); *C.A.*, **63**, 9987g (1965).
Dd-002. 72788 (1970); *C.A.*, **74**, 42481r (1971).
Dd-003. 103902 (1974); *C.A.*, **81**, 49816g (1974).
Dd-004. 103903 (1974); *C.A.*, **81**, 49815f (1974).
Dd-005. 117021 (1975); *C.A.*, **85**, 21606x (1976).
Dd-006. 130353 (1978); *C.A.*, **91**, 74699k (1979).
Dd-007. 144413 (1980); *C.A.*, **95**, 62399g (1981).
Dd-008. 147628 (1981); *C.A.*, **96**, 149889m (1982).
Dd-009. 151944 (1981); *C.A.*, **97**, 39138d (1982).
Dd-010. 239596 (1986).
Dd-011. 255737 (1988); *C.A.*, **110**, 193118k (1989).

E: European Patents

E-001. 28665 (1981); *C.A.*, **95**, 204142u (1981).
E-002. 51384 (1982); *C.A.*, **97**, 74244g (1982).
E-003. 54725 (1982); *C.A.*, **97**, 129496y (1982).
E-004. 61241 (1982); *C.A.*, **98**, 180051x (1983).
E-005. 114636, see also D-119.
E-006. 122008 (1984); *C.A.*, **102**, 47489f (1985).
E-007. 130731 (1985); *C.A.*, **102**, 150314u (1985).
E-008. 135813, see also U-144.
E-009. 151991 (1985).
E-010. 153700 (1985).
E-011. 154867 (1985); *C.A.*, **103**, 196234s (1985).
E-012. 160988 (1985).
E-013. 162523 (1985).
E-014. 173512 (1986); *C.A.*, **105**, 134159d (1986).
E-015. 176085.
E-016. 179355 (1986); *C.A.*, **105**, 79156y (1986).
E-017. 182611 (1986); *C.A.*, **105**, 153341y (1986).
E-018. 184829 (1986); *C.A.*, **107**, 8241e (1987).
E-019. 195997 (1986); *C.A.*, **106**, 33300h (1987).
E-020. 202542 (1986); *C.A.*, **106**, 196975w (1987).
E-021. 204171 (1986).
E-022. 217539 (1987); *C.A.*, **107**, 176756t (1987).
E-023. 219322 (1987); *C.A.*, **107**, 40071h (1987).
E-024. 219720 (1987); *C.A.*, **107**. 98346t (1987).
E-025. 231420 (1987); *C.A.*, **107**, 242673s (1987).
E-026. 249416 (1987); *C.A.*, **108**, 222560v (1988).
E-027. 249944 (1987); *C.A.*, **108**, 211167w (1988).
E-028. 251678 (1988); *C.A.*, **108**, 80721x (1988).
E-029. 255170 (1988); *C.A.*, **109**, 55459f (1988).
E-030. 257970 (1988); *C.A.*, **109**, 55958t (1988).
E-031. 260103 (1988); *C.A.*, **109**, 56748m (1988).
E-032. 262642 (1988); *C.A.*, **109**, 6732f (1988).
E-033. 265929 (1988); *C.A.*, **109**, 94235z (1988).
E-035. 266895 (1988); *C.A.*, **109**, 130928p (1988).
E-036. 276986 (1988); *C.A.*, **110**, 58775n (1989).
E-037. 277023 (1988); *C.A.*, **109**, 198322s (1988).
E-038. 278863 (1988); *C.A.*, **109**, 211217h (1988).
E-039. 279706 (1988); *C.A.*, **110**, 25261p (1989).
E-040. 284447 (1988); *C.A.*, **110**, 57848b (1989).

E-041. 286387 (1988); *C.A.*, **110**, 75793j (1989).
E-042. 288286 (1988); *C.A.*, **110**, 57853z (1989).
E-043. 291427 (1988); *C.A.*, **110**, 155079w (1989).
E-044. 292760 (1988); *C.A.*, **110**, 232327e (1989).
E-045. 295657 (1988); *C.A.*, **111**, 79993t (1989).
E-046. 300146 (1989); *C.A.*, **111**, 8992f (1989).
E-047. 300645 (1989); *C.A.*, **112**, 37309q (1990).
E-048. 302672 (1989); *C.A.*, **111**, 154099m (1989).
E-049. 306177 (1989); *C.A.*, **111**, 58017n (1989).
E-050. 307098 (1989); *C.A.*, **111**, 116975m (1989).
E-051. 308835 (1989); *C.A.*, **111**, 196054y (1989).
E-052. 308836 (1989); *C.A.*, **111**, 79325b (1989).
E-053. 310129 (1989); *C.A.*, **111**, 234880z (1989).
E-054. 314903 (1989); *C.A.*, **111**, 135012m (1989).
E-055. 315836 (1989); *C.A.*, **111**, 234850q (1989).
E-056. 318899 (1989); *C.A.*, **111**, 156059x (1989).
E-057. 342809 (1989).
E-058. 345965 (1989).
E-059. 350951 (1990).

F: French Patents

F-001. 1 352 325 (1964); *C.A.*, **62**, 2890g (1965).
F-002. 1 361 690 (1964); *C.A.*, **61**, 12153g (1964).
F-003. 1 361 705 (1964); *C.A.*, **61**, 13451b (1964).
F-004. 1 366 279 (1964); *C.A.*, **62**, 2795b (1965).
F-005. 1 371 405 (1964); *C.A.*, **62**, 664c (1965).
F-006. 1 387 338 (1965); *C.A.*, **64**, 2127a (1966).
F-007. 1 390 999 (1965).
F-008. 1 391 833 (1965); *C.A.*, **62**, 16297f (1965).
F-009. 1 397 043 (1965); *C.A.*, **63**, 8563b (1965).
F-010. 1 437 281 (1966); *C.A.*, **66**, 28884c (1967).
F-011. 1 437 798 (1966); *C.A.*, **66**, 19033x (1967).
F-012. 1 438 344 (1966); *C.A.*, **66**, 3525a (1967).
F-013. 1 466 547 (1967); *C.A.*, **67**, 108748x (1967).
F-014. 1 481 448 (1967); *C.A.*, **68**, 12860x (1968).
F-015. 1 509 761 (1968); *C.A.*, **70**, 29048c (1969).
F-016. 1 520 444 (1968); *C.A.*, **70**, 107091c (1969).
F-017. 1 549 776 (1969); *C.A.*, **72**, 21780f (1970).
F-018. 1 558 735 (1969); *C.A.*, **72**, 43851g (1970).
F-019. 1 579 903 (1970); *C.A.*, **72**, 132943f (1970).
F-020. 1 582 748 (1969); *C.A.*, **73**, 46424b (1970).
F-021. 2 162 114 (1973).
F-022. 2 178 218 (1973); *C.A.*, **80**, 108668q (1974).
F-023. 2 187 798 (1974); *C.A.*, **81**, 25799c (1974).
F-024. 2 241 320 (1975); *C.A.*, **83**, 179410x (1975).
F-025. 2 243 022 (1975); *C.A.*, **83**, 137539r (1975).
F-026. 2 268 020 (1975).
F-027. 2 433 533 (1980).
F-028. 2 465 741 (1981).
F-029. 2 474 040 (1981); *C.A.*, **96**, 7967m (1982).
F-030. 2 474 890 (1981); *C.A.*, **96**, 124298e (1982).
F-031. 2 483 435 (1981); *C.A.*, **96**, 182925t (1982).
F-032. 2 550 201 (1985).
F-033. 2 551 066 (1985).

F-034. 2 554 117 (1985); *C.A.*, **104**, 6607r (1986).
F-035. 2 567 890 (1986); *C.A.*, **105**, 153340x (1986).
F-036. 2 575 086 (1986); *C.A.*, **107**, 8245g (1987).
F-037. 2 579 982 (1986); *C.A.*, **107**, 7857w (1987).
F-038. 2 595 363 (1987); *C.A.*, **108**, 151209z (1988).
F-039. 2 595 364 (1987); *C.A.*, **108**, 151210t (1988).
F-040. 2 609 993 (1988); *C.A.*, **110**, 115906g (1989).
F-041. 2 611 728 (1988); *C.A.*, **110**, 136338p (1989).

G: British Patents

G-001. 949 126 (1970); *C.A.*, **75**, 140972s (1971).
G-002. 968 470 (1964); *C.A.*, **61**, 12035f (1964).
G-003. 1 031 914 (1966).
G-004. 1 042 677 (1966); *C.A.*, **65**, 18407a (1966).
G-005. 1 054 658 (1967).
G-006. 1 066 346 (1966).
G-007. 1 097 453 (1968).
G-008. 1 104 206 (1968); *C.A.*, **68**, 105349c (1968).
G-009. 1 107 192 (1968).
G-010. 1 141 868 (1968); *C.A.*, **70**, 78790x (1969).
G-011. 1 158 510 (1969).
G-012. 1 160 644 (1969); *C.A.*, **71**, 102003m (1969).
G-013. 1 205 929 (1970).
G-014. 1 212 410 (1970).
G-015. 1 221 156 (1971); *C.A.*, **74**, 115909y (1971).
G-016. 1 275 120 (1972).
G-017. 1 281 352 (1972).
G-018. 1 292 077 (1972).
G-019. 1 377 214 (1974); *C.A.*, **82**, 171182h (1975).
G-020. 1 384 923 (1975).
G-021. 1 400 713 (1975).
G-022. 1 402 637 (1975).
G-023. 1 409 483 (1975); *C.A.*, **84**, 123587p (1976).
G-024. 1 414 662 (1975); *C.A.*, **84**, 90300j (1976).
G-025. 1 418 465 (1975).
G-026. 1 419 769 (1975); *C.A.*, **84**, 135818c (1976).
G-027. 1 420 928 (1976); *C.A.*, **84**, 165013d (1976).
G-028. 1 428 552 (1976).
G-029. 1 431 506 (1976).
G-030. 1 438 949 (1976); *C.A.*, **85**, 161580j (1976).
G-031. 1 448 826 (1976); *C.A.*, **86**, 72861e (1977).
G-032. 1 460 699 (1977).
G-033. 1 473 335 (1977); *C.A.*, **87**, 135899s (1977).
G-034. 1 476 314 (1977).
G-035. 1 581 850 (1980).
G-036. 2 013 207 (1979); *C.A.*, **93**, 150364f (1980).
G-037. 2 065 152 (1981).
G-038. 2 209 169 (1989); *C.A.*, **111**, 196315j (1989).

J: Japanese Patents

J-001. 7110846 (1971); *C.A.*, **75**, 49311q (1971).
J-002. 7111645 (1971); *C.A.*, **75**, 20599d (1971).
J-003. 7111646 (1971); *C.A.*, **75**, 20598c (1971).

J-004. 73 07416 (1973); *C.A.*, **79,** 53532n (1973).
J-005. 73 92453 (1973); *C.A.*, **80,** 109608 (1974).
J-006. 74 25649 (1974); *C.A.*, **82**, 156489v (1975).
J-007. 74 28189 (1974); *C.A.*, **82**, 125804g (1975).
J-008. 74 35336 (1974); *C.A.*, **81**, 120782a (1974).
J-009. 74 47325 (1974); *C.A.*, **81**, 120780y (1974).
J-010. 74110632 (1974); *C.A.*, **82**, 156487t (1975).
J-011. 75 24947 (1975); *C.A.*, **84**, 165010a (1976).
J-012. 75 24948 (1975); *C.A.*, **84**, 105749p (1976).
J-013. 75 35062 (1975); *C.A.*, **85**, 110398y (1976).
J-014. 75 70314 (1975); *C.A.*, **83**, 177508m (1975).
J-015. 75116428 (1975); *C.A.*, **84**, 74424x (1976).
J-016. 75124712 (1975); *C.A.*, **84**, 61483r (1976).
J-017. 76 9738 (1976);
J-018. 76 32528 (1976); *C.A.*, **85**, 192872k (1976).
J-019. 76 36425 (1976); *C.A.*, **85**, 124132u (1976).
J-020. 76 88923 (1976); *C.A.*, **86**, 106766v (1977).
J-021. 76 98225 (1976); *C.A.*, **86**, 106768x (1977).
J-022. 77 42873 (1977); *C.A.*, **87**, 117773s (1977).
J-023. 77 46065 (1977); *C.A.*, **87**, 68490m (1977).
J-024. 77 65225 (1977); *C.A.*, **87**, 135906s (1977).
J-025. 77 79884 (1977); *C.A.*, **88**, 38925k (1978).
J-026. 77 93468 (1977);
J-027. 77 93718 (1977); *C.A.*, **88**, 23134c (1978).
J-028. 77105464 (1977); *C.A.*, **90**, 6532e (1979).
J-029. 78 12814 (1978); *C.A.*, **88**, 191045t (1978).
J-030. 78144396 (1978);
J-031. 78146993 (1978); *C.A.*, **90**, 153308q (1979).
J-032. 78147029 (1978); *C.A.*, **90**, 204241x (1979).
J-033. 79 37184 (1979); *C.A.*, **91**, 41040k (1979).
J-034. 79 76530 (1979); *C.A.*, **91**, 211572e (1979).
J-035. 79 92902 (1979); *C.A.*, **92**, 22025p (1980).
J-036. 79100355 (1979); *C.A.*, **92**, 6684x (1980).
J-037. 80 49387 (1980); *C.A.*, **94**, 30910e (1981).
J-038. 80145693 (1980); *C.A.*, **94**, 175250z (1981).
J-039. 81 46888 (1981); *C.A.*, **95**, 187410k (1981).
J-040. 81 87587 (1981); *C.A.*, **95**, 204146y (1981).
J-041. 81 90092 (1981); *C.A.*, **96**, 6859r (1982).
J-042. 82 4995 (1982); *C.A.*, **96**, 218022n (1982).
J-043. 82109795 (1982); *C.A.*, **98**, 4661n (1983).
J-044. 82128693 (1982); *C.A.*, **98**, 4664r (1983).
J-045. 83 14929 (1983); *C.A.*, **99**, 6656j (1983).
J-046. 83124792 (1983); *C.A.*, **100**, 6834e (1984).
J-047. 83132022 (1983); *C.A.*, **99**, 213934p (1983).
J-048. 83216195 (1983); *C.A.*, **100**, 210126e (1984).
J-049. 83219188 (1983); *C.A.*, **100**, 210128g (1984).
J-050. 84 12760 (1984); *C.A.*, **101**,24479t (1984).
J-051. 84118794 (1984); *C.A.*, **101**, 211455u (1984).
J-052. 84144789 (1984); *C.A.*, **102**, 46124w (1985).
J-053. 84144790 (1984); *C.A.*, **102**, 149518g (1985).
J-054. 84144791 (1984); *C.A.*, **102**, 46123v (1985).
J-055. 84181286 (1984); *C.A.*, **102**, 113731g (1985).
J-056. 84184190 (1984); *C.A.*, **102**, 113732h (1985).
J-057. 85 6747 (1985); *C.A.*, **102**, 205582n (1985).
J-058. 85 88068 (1985); *C.A.*, **103**, 88854k (1985).
J-059. 85101146 (1985); *C.A.*, **103**, 142861a (1985).

J-060. 85115661 (1985); *C.A.*, **103**, 161961z (1985).
J-061. 85172953 (1985); *C.A.*, **104**, 68442s (1986).
J-062. 85184028 (1985); *C.A.*, **104**, 168106d (1986).
J-063. 85208332 (1985); *C.A.*, **104**, 169703h (1986).
J-064. 85214793 (1985); *C.A.*, **104**, 111158t (1986).
J-065. 86 27992 (1986); *C.A.*, **105**, 115198z (1986).
J-066. 86 51058 (1986); *C.A.*, **105**, 115929p (1986).
J-067. 86 55178 (1986); *C.A.*, **105**, 80246j (1986).
J-068. 86100587 (1986); *C.A.*, **105**, 226993z (1986).
J-069. 86129288 (1986); *C.A.*, **106**, 18810f (1987).
J-070. 86171754 (1986); *C.A.*, **106**, 6568v (1987).
J-071. 86185527 (1986); *C.A.*, **105**, 288226a (1986).
J-072. 86205287 (1986); *C.A.*, **106**, 33304 (1987).
J-073. 86229886 (1986); *C.A.*, **106**, 156996x (1987).
J-074. 86236786 (1986); *C.A.*, **106**, 138626x (1987).
J-075. 86238851 (1986); *C.A.*, **106**, 197383p (1987).
J-076. 86247723 (1986); *C.A.*, **107**, 24255r (1987).
J-077. 86254625 (1986); *C.A.*, **106**, 177083m (1987).
J-078. 86277901 (1986); *C.A.*, **106**, 157681 (1987).
J-079. 86286393 (1986); *C.A.*, **107**, 31341u (1987).
J-079a. 87 15929 (1987); *C.A.*, **110**, 173964 (1988).
J-080. 87 39658 (1987); *C.A.*, **107**, 8542b (1987).
J-081. 87 39659 (1987); *C.A.*, **107**, 8541a (1987).
J-082. 87111991 (1987); *C.A.*, **107**, 176225x (1987).
J-083. 87129375 (1987); *C.A.*, **108**, 23007v (1988).
J-084. 87149685 (1987); *C.A.*, **107**, 237000r (1987).
J-085. 87179506 (1987); *C.A.*, **108**, 205255g (1988).
J-086. 87181357 (1987); *C.A.*, **108**, 113559f (1988).
J-087. 87195389 (1987); *C.A.*, **108**, 113206p (1988).
J-088. 87225533 (1987); *C.A.*, **108**, 76116m (1988).
J-089. 87227424 (1987); *C.A.*, **108**, 113861 (1988).
J-090. 87240357 (1987); *C.A.*, **108**, 113643 (1988).
J-091. 87252489 (1987); *C.A.*, **108**, 95839r (1988).
J-092. 87257919 (1987); *C.A.*, **108**, 113207q (1988).
J-093. 88 10738 (1988); *C.A.*, **109**, 128580a (1988).
J-094. 88 22822 (1988); *C.A.*, **109**, 111727f (1988).
J-095. 88 33384 (1988); *C.A.*, **109**, 211215f (1988).
J-096. 88 86787 (1988); *C.A.*, **109**, 74878h (1988).
J-097. 88121004 (1988); *C.A.*, **109**, 151055s (1988).
J-098. 88132942 (1988); *C.A.*, **109**, 231765p (1988).
J-099. 88145287 (1988); *C.A.*, **110**, 39513j (1988).
J-100. 88145386 (1988), *C.A.*, **109**, 151098h (1988).
J-101. 88171678 (1988); *C.A.*, **109**, 232404 (1988).
J-102. 88189437 (1988); *C.A.*, **110**, 9136w. (1988).
J-103. 88192791 (1988); *C.A.*, **110**, 31179e. (1988).
J-104. 88255288 (1988); *C.A.*, **110**, 135489b. (1988).
J-105. 88258887 (1988); *C.A.*, **110**, 213589a. (1988).
J-106. 88270784 (1988); *C.A.*, **111**, 40836. (1989).
J-107. 89 11160 (1989); *C.A.*, **111**, 98441c. (1989).
J-108. 89 14273 (1989); *C.A.*, **111**, 155406w. (1989).
J-109. 89 29385 (1989); *C.A.*, **111**, 98796x. (1989).
J-110. 89 31792 (1989).
J-111. 89 36674 (1989); *C.A.*, **111**, 116781. (1989).
J-112. 89 50887 (1989); *C.A.*, **111**, 134918z. (1989).
J-113. 89 75524 (1989); *C.A.*, **111**, 175724. (1989).
J-114. 89125390 (1989); *C.A.*, **111**, 233944m. (1989).

N: Netherlands Patents

N-001. 6 510 184; *C.A.*, **65**, 749a (1966).
N-002. 6 511 871; *C.A.*, **65**, 3907c (1966).
N-003. 6 601 308; *C.A.*, **66**, 28883t (1967).
N-004. 6 609 970; *C.A.*, **68**, 13167p (1968).
N-005. 6 612 227; *C.A.*, **67**, 74522n (1967).
N-006. 8 001 994; *C.A.*, **94**, 67052c (1981).

Pl: Polish Patents

Pl-001. 111 276 (1978).
Pl-002. 115 644 (1978).
Pl-003. 115 645 (1978); *C.A.*, **99**, 44149d (1983).
Pl-004. 115 678 (1978); *C.A.*, **99**, 195213w (1983).
Pl-005. 116 320 (1978); *C.A.*, **99**, 176046c (1983).
Pl-006. 116 960 (1979); *C.A.*, **99**, 105510p (1983).
Pl-007. 117 627 (1978); *C.A.*, **99**, 88364z (1983).
Pl-008. 119 230 (1978); *C.A.*, **99**, 140147x (1983).
Pl-009. 119 366 (1978); *C.A.*, **99**, 122642a (1983).
Pl-010. 124 589 (1979).
Pl-011. 127 132 (1981).
Pl-012. 133 261 (1982).
Pl-013. 133 265 (1982).
Pl-014. 133 266 (1982).
Pl-015. 138 308 (1983).
Pl-016. 140 988 (1984).
Pl-017. 140 989 (1984).
Pl-018. 140 990 (1984).
Pl-019. 140 991 (1984).
Pl-020. 142 385 (1984).
Pl-021. 143 118 (1984).
Pl-022. 144 753 (1984).
Pl-023. 145 670 (1985).
Pl-024. 145 671 (1985).
Pl-025. 148 273 (1985).
Pl-026. 133 108 (1981).
Pl-027. 133 109 (1981).
Pl-028. 131 983 (1981).
Pl-029. 155 837 (1988).
Pl-030. 155 839 (1988).
Pl-031. 156 240 (1988).
Pl-032. 156 241 (1988).
Pl-033. 155 840 (1988).
Pl-034. P 279 596.
Pl-035. P 284 990.
Pl-036. P 285 570.

S: USSR Patents

S-001. 115 992.
S-002. 170 495 (1964); *C.A.*, **63**, 9987h (1965).
S-003. 234 364 (1969); *C.A.*, **70**, 118593u (1969).
S-004. 319 600 (1972).

S-005. 319 602 (1972); *C.A.*, **76**, 46297z (1972).
S-006. 320 500 (1972).
S-007. 353 547 (1972).
S-008. 368 270 (1973); *C.A.*, **79**, 53538u (1973).
S-009. 371 791 (1974); *C.A.*, **80**, 15061w (1974).
S-010. 372 228 (1973).
S-011. 415 268 (1974); *C.A.*, **80**, 133618f (1974).
S-012. 426 483 (1975); *C.A.*, **84**, 22705v (1976).
S-013. 449 057 (1975).
S-014. 461 931 (1975); *C.A.*, **83**, 10380b (1975).
S-015. 467 598 (1975).
S-016. 469 705 (1975); *C.A.*, **83**, 59025d (1975).
S-017. 499 891 (1976); *C.A.*, **84**, 141310x (1976).
S-018. 503 879 (1976); *C.A.*, **84**, 135815z (1976).
S-019. 505 647 (1976); *C.A.*, **85**, 124130s (1976).
S-020. 534 459 (1977).
S-021. 555 998 (1977).
S-022. 585 168 (1977); *C.A.*, **88**, 105555e (1978).
S-023. 598 902 (1978); *C.A.*, **89**, 43756f (1978).
S-024. 598 903 (1978); *C.A.*, **89**, 6411u (1978).
S-025. 602 218 (1978); *C.A.*, **89**, 12776h (1978).
S-026. 630 256 (1978); *C.A.*, **90**, 39033w (1979).
S-027. 632 699 (1978); *C.A.*, **90**, 87654w (1979).
S-028. 632 700 (1978); *C.A.*, **90**, 87655x (1979).
S-029. 657 839 (1979); *C.A.*, **91**, 19872m (1979).
S-030. 694 020 (1979).
S-031. 694 021 (1979).
S-032. 702 037 (1979); *C.A.*, **92**, 94905y (1980).
S-033. 709 628 (1980); *C.A.*, **92**, 146901r (1980).
S-034. 715 582 (1980); *C.A.*, **93**, 8286p (1980).
S-035. 724 515 (1980); *C.A.*, **93**, 105371f (1980).
S-036. 727 653 (1980); *C.A.*, **93**, 168403c (1980).
S-037. 732 268 (1980); *C.A.*, **93**, 204840d (1980).
S-038. 734 210 (1980); *C.A.*, **93**, 204841e (1980).
S-039. 743 717 (1980); *C.A.*, **94**, 8195d (1981).
S-040. 756 810 (1980).
S-041. 765 272 (1980); *C.A.*, **94**, 47472m (1981).
S-042. 767 112 (1980); *C.A.*, **94**, 65828m (1981).
S-043. 810 704 (1981); *C.A.*, **95**, 43337u (1981).
S-044. 810 705 (1981); *C.A.*, **95**, 25262m (1981.
S-045. 810 706 (1981); *C.A.*, **95**, 25261k (1981).
S-046. 845 444 (1983); *C.A.*, **99**, 195214x (1983).
S-047. 846 546 (1981).
S-048. 891 676 (1981); *C.A.*, **96**, 199871v (1982).
S-049. 910 641 (1982); *C.A.*, **97**, 92553m (1982).
S-050. 914 580 (1982); *C.A.*, **97**, 93046k (1982).
S-051. 924 049 (1982).
S-052. 935 508 (1982).
S-053. 1 006 439 (1983); *C.A.*, **99**, 88363y (1983).
S-054. 1 128 978 (1984); *C.A.*, **102**, 138490r (1985).

U: USA Patents

U-001. 2 839 557 (1958); *C.A.*, **52**, 13777 (1958).
U-002. 3 020 260.

U-003. 3 071 561 (1963).
U-004. 3 099 670 (1963); *C.A.*, **60**, 4183b (1964).
U-005. 3 159 662 (1964); *C.A.*, **61**, 14862a (1964).
U-006. 3 162 663 (1964); *C.A.*, **63**, 1904g (1965).
U-007. 3 168 544 (1965); *C.A.*, **62**, 11853a (1965).
U-008. 3 170 891 (1965).
U-009. 3 177 236 (1965).
U-010. 3 180 882 (1966); *C.A.*, **65**, 8959c (1966).
U-011. 3 185 719 (1965).
U-012. 3 188 299 (1965).
U-013. 3 188 300 (1965); *C.A.*, **63**, 7043d (1965).
U-014. 3 220 972 (1965); *C.A.*, **64**, 8237d (1966).
U-015. 3 257 440 (1966); *C.A.*, **65**, 8960a (1966).
U-016. 3 258 477 (1966); *C.A.*, **65**, 12237g (1966).
U-017. 3 267 135 (1966); *C.A.*, **66**, 11039x (1967).
U-018. 3 269 983 (1966); *C.A.*, **65**, 20298b (1966).
U-019. 3 271 362 (1966); *C.A.*, **65**, 18617h (1966).
U-020. 3 278 465 (1966).
U-021. 3 313 773 (1967); *C.A.*, **67**, 3316b (1967).
U-022. 3 331 813 (1967).
U-023. 3 338 869 (1967); *C.A.*, **67**, 91323e (1967).
U-024. 3 344 111 (1967); *C.A.*, **67**, 109343y (1967).
U-025. 3 354 193 (1967); *C.A.*, **68**, 22053a (1968).
U-026. 3 355 478 (1967); *C.A.*, **68**, 22504y (1968).
U-027. 3 410 886 (1968); *C.A.*, **70**, 47584d (1969).
U-028. 3 419 593 (1968).
U-029. 3 422 131 (1969); *C.A.*, **70**, 68492n (1969).
U-030. 3 423 234 (1969).
U-031. 3 427 271 (1969); *C.A.*, **70**, 78566d (1969).
U-032. 3 431 234 (1969); *C.A.*, **70**, 88455f (1969).
U-033. 3 441 431 (1969); *C.A.*, **71**, 14148g (1969).
U-034. 3 445 420 (1969).
U-035. 3 453 233 (1969).
U-036. 3 453 234 (1969); *C.A.*, **71**, 71368c (1969).
U-037. 3 483 241 (1969); *C.A.*, **72**, 32010f (1970).
U-038. 3 488 319 (1970); *C.A.*, **72**, 79828u (1970).
U-039. 3 511 866 (1970); *C.A.*, **73**, 25649p (1970).
U-040. 3 565 937 (1971); *C.A.*, **75**, 20603a (1971).
U-041. 3 598 852 (1971).
U-042. 3 609 174 (1971); *C.A.*, **76**, 35053r (1972).
U-043. 3 627 801 (1971); *C.A.*, **77**, 6915p (1972).
U-044. 3 631 086 (1971).
U-045. 3 639 156 (1972).
U-046. 3 644 315 (1972); *C.A.*, **76**, 155306x (1972).
U-047. 3 668 229 (1972).
U-048. 3 681 418 (1972); *C.A.*, **77**, 115363c (1972).
U-049. 3 699 142 (1972); *C.A.*, **78**, 16992w (1973).
U-050. 3 700 711 (1972); *C.A.*, **78**, 43688h (1973).
U-051. 3 700 714 (1972).
U-052. 3 700 716 (1972).
U-053. 3 714 212 (1973); *C.A.*, **78**, 115763d (1973).
U-054. 3 715 387 (1973); *C.A.*, **79**, 79454p (1973).
U-055. 3 723 497 (1973); *C.A.*, **79**, 6549p (1973).
U-056. 3 729 445 (1973); *C.A.*, **79**, 54630e (1973).
U-057. 3 746 734 (1973).
U-058. 3 755 354 (1973).

U-059. 3 759 869 (1973); *C.A.*, **80**, 27849x (1974).
U-060. 3 759 968 (1973); *C.A.*, **80**, 27942x (1974).
U-061. 3 772 066 (1973).
U-062. 3 773 817 (1973); *C.A.*, **80**, 37272t (1974).
U-063. 3 775 386 (1973); *C.A.*, **80**, 122130q (1974).
U-064. 3 775 452 (1973).
U-065. 3 778 430 (1973); *C.A.*, **80**, 60026z (1974).
U-066. 3 778 459 (1973); *C.A.*, **80**, 70952n (1974).
U-067. 3 780 080 (1973); *C.A.*, **80**, 70953p (1974).
U-068. 3 780 127 (1973); *C.A.*, **80**, 82065p (1974).
U-069. 3 793 361 (1974); *C.A.*, **80**, 108667p (1974).
U-070. 3 794 673 (1974); *C.A.*, **81**, 92452x (1974).
U-071. 3 795 656 (1974); *C.A.*, **81**, 92240b (1974).
U-072. 3 798 252 (1974).
U-073. 3 806 532 (1974); *C.A.*, **81**, 13644k (1974).
U-074. 3 806 549 (1974).
U-075. 3 808 248 (1974); *C.A.*, **81**, 13642h (1974).
U-076. 3 809 783 (1974).
U-077. 3 828 087 (1974).
U-078. 3 845 127 (1974); *C.A.*, **82**, 86401j (1975).
U-079. 3 846 462 (1974).
U-080. 3 864 372 (1975).
U-081. 3 876 677 (1975); *C.A.*, **83**, 43492t (1975).
U-082. 3 881 536 (1975).
U-083. 3 903 123 (1975).
U-084. 3 907 851 (1975).
U-085. 3 925 434 (1975); *C.A.*, **84**, 74427a (1976).
U-086. 3 929 851 (1975).
U-087. 3 931 267 (1976); *C.A.*, **84**, 105755n (1976).
U-088. 3 933 880 (1976); *C.A.*, **84**, 150753w (1976).
U-089. 3 933 882 (1976); *C.A.*, **84**, 165011b (1976).
U-090. 3 944 519 (1976).
U-091. 3 956 565 (1976); *C.A.*, **86**, 30944d (1977).
U-092. 3 979 420 (1976).
U-093. 3 989 666 (1976).
U-094. 3 989 667 (1976).
U-095. 3 992 355 (1976).
U-096. 3 996 165 (1976).
U-097. 3 997 580 (1976).
U-098. 4 006 176 (1977).
U-099. 4 011 194 (1977).
U-100. 4 028 391 (1977).
U-102. 4 033 902 (1977); *C.A.*, **87**, 91405w (1977).
U-103. 4 035 453 (1977).
U-104. 4 044 037 (1977).
U-105. 4 057 566 (1977).
U-106. 4 061 609 (1977).
U-107. 4 063 911 (1977); *C.A.*, **89**, 36219e (1978).
U-108. 4 064 154 (1977).
U-109. 4 083 803 (1978).
U-110. 4 100 172 (1978).
U-111. 4 108 833 (1978).
U-112. 4 130 574 (1980).
U-113. 4 139 403 (1979).
U-114. 4 150 048 (1979); *C.A.*, **91**, 92243m (1979).
U-115. 4 156 689 (1979); *C.A.*, **91**, 91761s (1979).

U-116. 4 234 713 (1980).
U-117. 4 256 616 (1981); *C.A.*, **95**, 8596f (1981).
U-118. 4 268 682 (1981); *C.A.*, **95**, 150878t (1981).
U-119. 4 278 804 (1981); *C.A.*, **96**, 35526k (1982).
U-120. 4 282 336 (1981).
U-121. 4 292 433 (1981); *C.A.*, **95**, 204149b (1981).
U-122. 4 304 920 (1981).
U-123. 4 306 073 (1981).
U-124. 4 307 240 (1981).
U-125. 4 316 033 (1982).
U-126. 4 319 034 (1982).
U-127. 4 322 518 (1982).
U-128. 4 328 346 (1982).
U-129. 4 332 654 (1982); *C.A.*, **97**, 1632243y (1982).
U-130. 4 336 364 (1982).
U-131. 4 336 395 (1982).
U-132. 4 338 454 (1982); *C.A.*, **97**, 127805t (1982).
U-133. 4 347 346 (1982).
U-134. 4 360 687 (1982); *C.A.*, **98**, 89650f (1983).
U-135. 4 362 820 (1982).
U-136. 4 374 742 (1983).
U-137. 4 374 967 (1983).
U-138. 4 377 706 (1983); *C.A.*, **99**, 5814x (1983).
U-139. 4 379 931 (1983); *C.A.*, **99**, 5816z (1983).
U-140. 4 383 120 (1983); *C.A.*, **99**, 22685y (1983).
U-141. 4 417 069 (1983); *C.A.*, **100**, 51828s (1984).
U-142. 4 447 633 (1984); *C.A.*, **101**, 152064r (1984).
U-143. 4 469 881 (1984); *C.A.*, **102**, 6797m (1985).
U-144. 4 481 364 (1984); *C.A.*, **102**, 95815 (1985); see also E-008.
U-145. 4 489 881 (1984).
U-146. 4 503 160 (1985); *C.A.*, **102**, 173414r (1985).
U-147. 4 503 208 (1985); *C.A.*, **103**, 6900k (1985).
U-148. 4 504 645 (1985); *C.A.*, **102**, 186724s (1985).
U-149. 4 510 094 (1985); *C.A.*, **103**, 38798z (1985).
U-150. 4 520 160 (1985); *C.A.*, **103**, 161286h (1985).
U-151. 4 526 996 (1986).
U-152. 4 529 553 (1985); *C.A.*, **103**, 161668j (1985).
U-153. 4 536 265 (1985); *C.A.*, **104**, 6677p (1986).
U-154. 4 542 226 (1985); *C.A.*, **103**, 196235t (1985).
U-155. 4 549 003 (1985).
U-156. 4 556 722 (1985); *C.A.*, **104**, 225041e (1986).
U-157. 4 558 111 (1985); *C.A.*, **104**, 207444 (1986).
U-158. 4 558 146 (1985); *C.A.*, **104**, 207443h (1986).
U-159. 4 568 492 (1986); *C.A.*, **105**, 60820p (1986).
U-160. 4 572 791 (1986); *C.A.*, **105**, 9078a (1986).
U-161. 4 574 149 (1986).
U-162. 4 578 497 (1986); *C.A.*, **104**, 227629b (1986).
U-163. 4 579 965 (1986); *C.A.*, **105**, 6638x (1986).
U-164. 4 579 966 (1986); *C.A.*, **106**, 5274j (1987).
U-165. 4 584 393 (1986); *C.A.*, **105**, 60756a (1986).
U-166. 4 587 276 (1986); *C.A.*, **105**, 115549q (1986).
U-167. 4 590 222 (1986); *C.A.*, **105**, 80410h (1986).
U-168. 4 614 812 (1986); *C.A.*, **106**, 18808m (1987).
U-169. 4 623 700 (1986); *C.A.*, **106**, 103947a (1987).
U-170. 4 639 501 (1987); *C.A.*, **106**, 120443a (1987).
U-171. 4 640 956 (1987); *C.A.*, **106**, 121188h (1987).

U-172. 4 658 050 (1987).
U-173. 4 668 812 (1987).
U-174. 4 670 574 (1987).
U-175. 4 677 161 (1987); *C.A.*, **107**, 200155n (1987).
U-176. 4 709 067 (1987); *C.A.*, **108**, 132463m (1988).
U-177. 4 720 533 (1988); *C.A.*, **108**, 187475x (1988).
U-178. 4 736 049 (1988); *C.A.*, **109**, 54954b (1988).
U-179. 4 737 562 (1988); *C.A.*, **109**, 94885m (1988).
U-180. 4 740 607 (1988).
U-181. 4 742 136 (1988); *C.A.*, **109**, 94421g (1988).
U-182. 4 743 667 (1988); *C.A.*, **110**, 82535e (1989).
U-183. 4 746 750 (1988); *C.A.*, **109**, 110646s (1988).
U-184. 4 746 751 (1988); *C.A.*, **109**, 75214a (1988).
U-185. 4 753 978 (1988); *C.A.*, **109**, 130608j (1988).
U-186. 4 784 879 (1988); *C.A.*, **110**, 116044n (1989).
U-187. 4 785 126 (1988); *C.A.*, **110**, 154566r (1989).
U-188. 4 788 312 (1988); *C.A.*, **110**, 157490x (1989).
U-189. 4 801 659 (1989); *C.A.*, **110**, 232580g (1989).
U-190. 4 803 244 (1989); *C.A.*, **110**, 174945s (1989).
U-191. 4 808 664 (1989); *C.A.*, **111**, 8773k (1989).
U-192. 4 814 473 (1989); *C.A.*, **111**, 97496n (1989).
U-193. 4 814 475 (1989); *C.A.*, **111**, 195566e (1989).
U-194. 4 826 710 (1989); *C.A.*, **111**, 196902y (1989).
U-194a. 4 837 339 (1989); *C.A.*, **111**, 233199r (1989).
U-195. 4 840 974 (1989); *C.A.*, **111**, 155571w (1989).
U-196. 4 847 228 (1989); *C.A.*, **112**, 8360u (1990).
U-197. 4 851 452 (1989); *C.A.*, **112**, 21947k (1990).
U-198. 4 857 564 (1989).
U-199. 4 898 910 (1988).
U-200. 4 918 200 (1990).

Index

Printed and bound by CPI Group (UK) Ltd, Croydon, CR0 4YY
03/10/2024
01040323-0013